KT-461-727

213 027

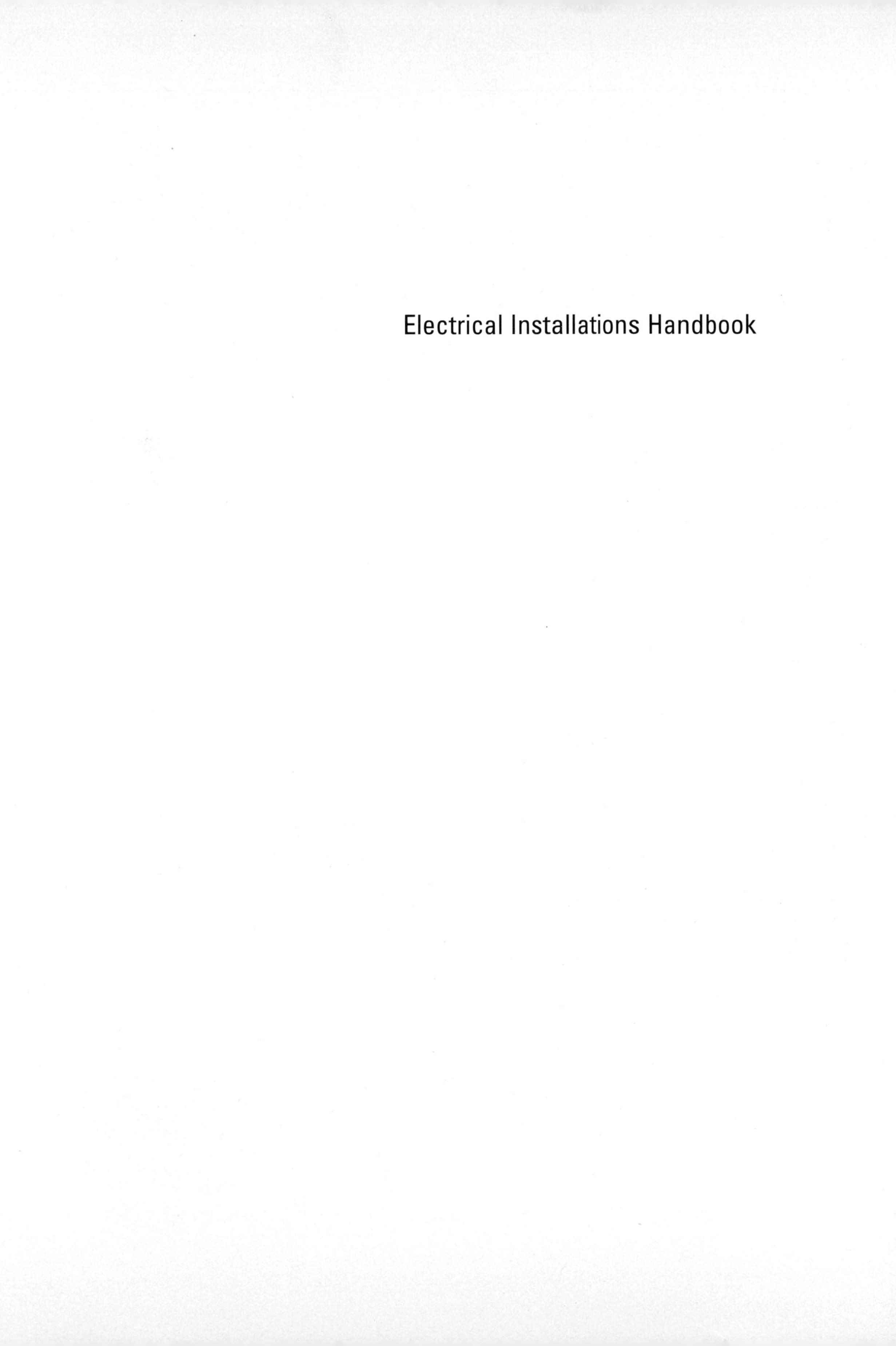

Electrical Installations Handbook

Subeditor: Werner Sturm

Authors	Section
Stefan Bingel	4.3.2.3
Walter Fuhrmann	1.4.1
Dr. Horst Gerlach	4.4.1–4.4.1.4
Siegbert Gern	4.1.1, 4.1.4, 4.3.2.4–4.3.2.4.4
Harald Heil	4.1.2
Werner Hörmann	2–2.5, 2.7
Franz Kammerl	4.4
Olaf Koppe	1.6–1.6.4
Helmut Kotulla	1.4.2–1.4.2.2
Dr. Winfried Kristen	4.4.5–4.4.5.2, 6–6.6
Georg Luber	3–3.4
Christian Mauracher	1.4.3–1.4.3.2
Erich Möller	1.5–1.5.7, 1.7.3, 1.8–1.8.3, 5.3, 5.4, 5.5, 5.7
Norbert Pantenburg	1.1.2–1.1.4, 5.1–5.1.2, 5.6
Wolfgang Pilsl	2.6, 4.1.3–4.1.3.12
Hans-Jörg Schab	4.3.1.1, 4.3.1.2, 4.3.2–4.3.2.2.3, 5.2–5.2.5, 5.8–5.8.2, 5.8.4
Bernd Schade	4.4.4.5–4.4.4.6.1, 5.1.4.10–5.1.4.10.3
Ralf Schmidt	1.4.4–1.4.4.2, 1.4.5–1.4.5.2, 5.1.4–5.1.4.9.2, 5.8.3
Horst Scholz	5.9–5.9.13, 7.1
Hans-Ulrich Schweer	4.2–4.2.14
August Sensing	5.1.3–5.1.3.2
Reinhard Solleder	1.3.2.1, 4.1.3
Herbert Stich	7.3
Klaus-Jürgen Stöber	1.2–1.2.4, 1.3–1.3.5.2
Werner Sturm	7.2
Thomas-Michael Stutzer	1.4.6
Walter Symontschyk	4.3.1
Dieter Teich	1.1.1
Arno Valerius	4.4.2–4.4.4.4
Jörg Westerholt	1.4.2.3
Dr. Johannes Wolf	1.3.1, 1.3.2
Jörn Zimmermann	4.3.2.3

NORWICH CITY COLLEGE LIBRARY
Stock No. 213027
Class 621.31924 SEIP
Cat. M Proc. 3m

Electrical Installations Handbook

Power Supply and Distribution
Protective Measures
Electromagnetic Compatibility
Electrical Installation Equipment and Systems
Application Examples for Electrical Installation Systems
Building Management

Executive Editor: Günter G. Seip

Third Edition, 2000

Publicis MCD Verlag John Wiley & Sons

Die Deutsche Bibliothek – CIP-Cataloguing-in-Publication-Data
A catalogue record for this publication is available from Die Deutsche Bibliothek

Library of Congress Cataloging-in-Publication Data
A catalog record for this book has been applied for

This book was carefully produced. Nevertheless, author and publisher do not warrant the information contained therein to be free of errors. Neither the author nor the publisher can assume any liability or legal responsibility for omissions or errors. Terms reproduced in this book may be registered trademarks, the use of which by third parties for their own purposes may violate the rights of the owners of those trademarks.

ISBN 3-89578-061-8 (Publicis MCD Verlag)

ISBN 0-471-49435-6 (John Wiley & Sons)

Third edition, 2000

Editor: Siemens Aktiengesellschaft, Berlin and Munich
Publisher: Publicis MCD Verlag, Erlangen and Munich; John Wiley & Sons Chichester
© 2000 by Publicis MCD Werbeagentur GmbH, Munich
This publication and all parts thereof are protected by copyright. All rights reserved.
Any use of it outside the strict provisions of the copyright law without the consent of the publisher is forbidden and will incur penalties. This applies particularly to reproduction, translation, microfilming or other processing, and to storage or processing in electronic systems. It also applies to the use of extracts from the text.

Printed in Germany

Preface

The content and structure of the "Electrical Installations Handbook" have been completely revised for the third edition. The manual offers a basic introduction to constructing and dimensioning electrical distribution systems, applying protective measures, as well as using conventional and state-of-the-art installation equipment, with particular reference to building services automation and building system engineering for residential and functional buildings. One of the focal points of interest here is communication-capable installation equipment, particularly in networks with the *instabus®* *EIB* bus system, which has won awards in a number of countries including the United States. This clearly shows that the manual is not primarily concerned with presenting individual devices and systems, but rather with standardized technology. Many of the applications discussed in the manual are illustrated by means of practical examples.

All the topics are presented on the basis of international, European, and German standards, which makes the manual suitable for use throughout the world.

The information presented in the manual – which is both extensive and fully up to date – provides a solid foundation for engineers and technicians involved in planning/designing, erecting, or operating electrical installation systems. As a comprehensive reference work, it is sure to be a useful source of secondary material for students all over the world. It will also be of use to readers who wish to acquire a solid grounding in the principles of particular areas.

Franz-Josef Wissing
General Executive Manager

Zentralverband Elektrotechnik- und
Elektroindustrie (ZVEI) e.V.

Editor's Preface

Recent innovations and advances in technology and the sustained interest in this book prompted us to restructure and revise it extensively, and define a fresh new approach.

The first three chapters give readers, be they planners, installation engineers, owners, students or trainees, all the information they need for planning and setting up electrical installations.

The next three chapters provide a comprehensive overview of the installation equipment and systems to be implemented, and give details of latest developments in systems engineering. All points are illustrated using a variety of application examples. Special emphasis is placed on installation equipment with communication capability and, in particular, on the way in which this equipment is networked with the *instabus*® *EIB* bus system for a wide range of applications in residential and functional buildings.

In view of the globalization of the market place and the standardization efforts being pursued at both an international and European level, reference has been made, when necessary, to current international, European and German norms, regulations and standards.

This edition, now published as a single volume, gives readers faster and easier access to the information they require. At the same time, the scope of the information provided in the key areas has been retained and, in certain cases, consolidated. References to other Siemens publications from related areas have been provided to facilitate further research.

Many new authors have contributed to this book and brought with them many fresh ideas. Mr Sturm was once again responsible for editing, a task which he carried out with inimitable competence. To those mentioned above, and to all the others who co-operated in producing this book, as well as to the publisher, I would like to express my sincere gratitude.

Günter G. Seip
Executive Editor

Content

1 Power supply and distribution

1.1 Planning and design of electrical distribution systems in buildings

1.1.1 Structural design

The purpose of electrical installations in buildings is to supply and distribute power. Buildings that are used for different purposes have different requirements in this respect. For this reason, the structure of the buildings must be known before the power supply and distribution systems can be planned and designed.

Figures 1.1/1 to 1.1/4 illustrate the basic forms of these systems in buildings that are used for different purposes.

Fig. 1.1/1 shows the basic structure of the power supply system for *industrial*, *functional* and *residential* buildings.

Industrial and functional buildings are usually supplied with electrical energy from a electricity generation (e.g.) system; the supply voltage for the distribution system is generated via transformers in this case. The transformer station required for this purpose is usually part of the owner's system. A utilities substation with switching, protection and measuring equipment must also be provided for the public utility.

In the case of *residential buildings*, the public utility provides the supply voltage for the distribution systems via separate transformer stations. Each residential building is connected to the power supply system via a low voltage cable.

Industrial/ functional buildings

Fig. 1.1/2 shows the most important components that are required for distributing power in an industrial or functional building.

Distribution levels

Fig. 1.1/3 shows these components again at the three distribution levels:

▷ main distribution level
▷ sub-distribution level
▷ load level

Power is thus supplied to the electrical loads via the low-voltage switching station, main distribution board, sub-distribution board and small distribution board (see Sections 1.4.2.2 and 1.4.5.1) by means of a low voltage, e.g. 230/400 V. The electrical connections between these components are established via different cables and electric lines (see Section 1.7) as well as busbar trunking and installation systems (see Sections 1.4.4 and 5.1.4).

The power distribution components are classified on the basis of the relevant current intensities at each power distribution level (see Fig. 1.1/3). At the *main distribution level*, SIVACON low-voltage switchboards are used for up to 6300 A (see Sections 1.4.2.2 and 1.4.2.3) and SIKUS or STRATUM modular distribution boards for up to 3200 A (see Section 1.4.5.2).

Main distribution boards

Main distribution boards are used first and foremost for:

▷ Safety disconnection
▷ Coupling busbar sections
▷ Protecting busbars
▷ Selectivity vis-à-vis upstream protection equipment

They are primarily equipped with:

- circuit-breakers and non-automatic circuit-breakers
- tie circuit-breakers and
- fuses

Sub-distribution boards

At the sub-distribution level, permanently-installed STAB, SIKUS and SIPRO floor-mounting distribution boards are used for up to 2500 A (see Section 1.4.5.1).

These distribution boards are used for:

▷ Safety disconnection
▷ Switching electrical loads, e.g. lighting systems and motors

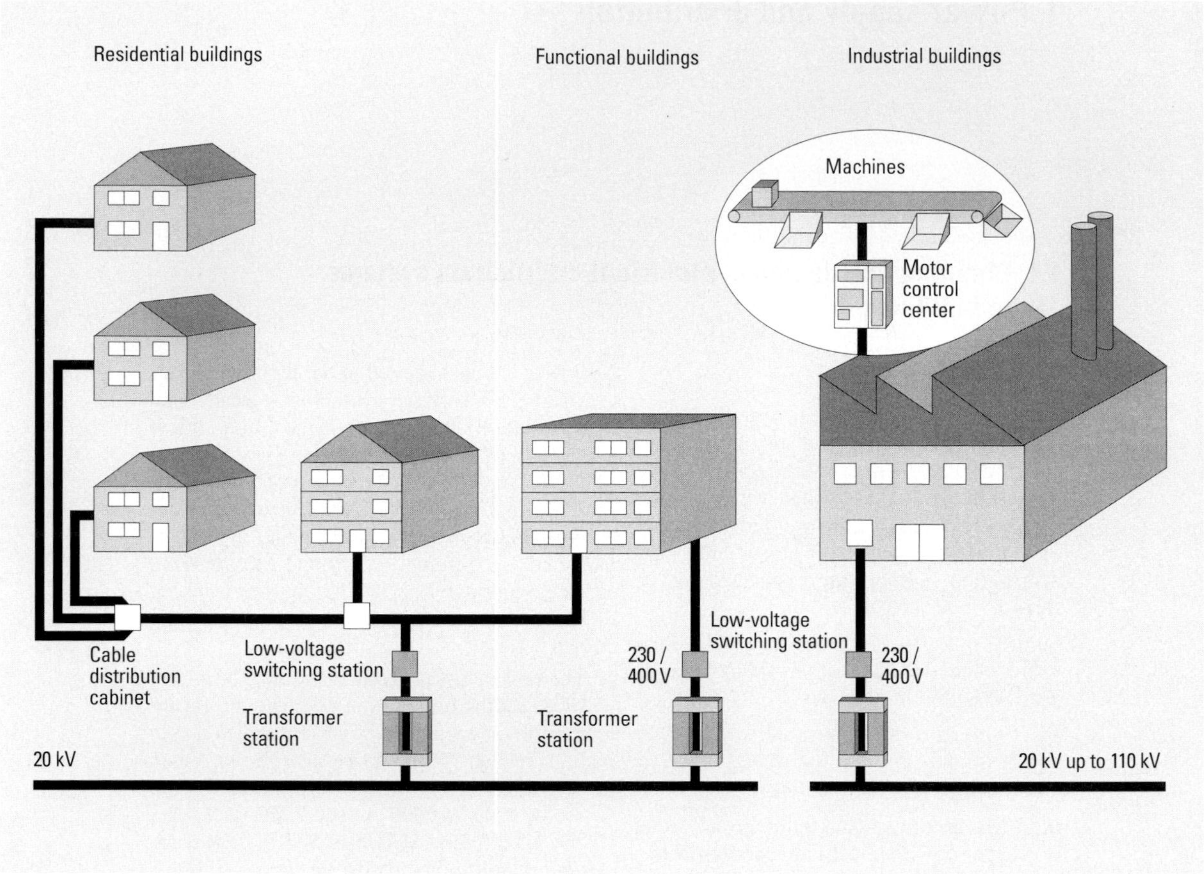

Fig. 1.1/1
Structure of the power supply system for industrial, functional and residential buildings

- ▷ Protecting cables, electric lines and loads
- ▷ Back-up protection and selectivity vis-à-vis upstream and downstream protection equipment
- ▷ Overvoltage protection
- ▷ Control, metering and measuring purposes

The following devices are integrated in the distribution boards in order to carry out these functions:

- Circuit-breakers, switch-disconnectors and fuse switch-disconnectors.
- Miniature circuit-breakers (see Section 4.1.2)
- Fuses (see Section 4.1.1)
- Modular built-in equipment for control, metering and measuring purposes (see Section 4.2)

Distribution boards at the load level

STAB wall-mounting distribution boards and SIMBOX 63 small distribution boards (see Section 1.4.5.1) are mainly used at the load level.

These distribution boards are used for:

- ▷ Protecting persons and property
- ▷ Protecting electrical loads
- ▷ Protecting cables and electric lines
- ▷ Overvoltage protection
- ▷ Safety disconnection
- ▷ Monitoring and signaling
- ▷ Open and closed-loop control
- ▷ Metering, measuring and display purposes

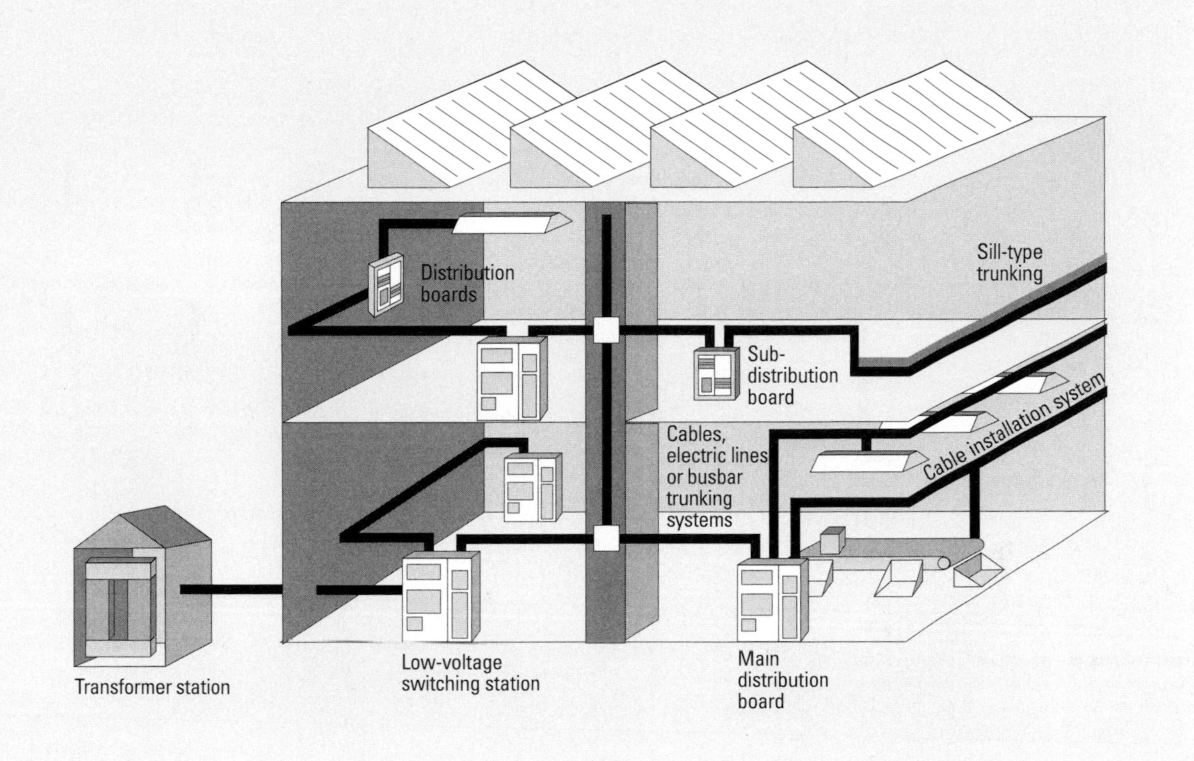

Fig. 1.1/2
Most important components for power distribution in industrial and functional buildings

The following devices are integrated in the distribution boards in order to carry out these functions:

- Circuit-breakers
- Switch-disconnectors and fuse switch-disconnectors
- Residual-current devices (see Section 4.1.3)
- Miniature circuit-breakers (see Section 4.1.2)
- Fuse systems (see Section 4.1.1)
- Earth-leakage monitors (see Section 4.1.4)
- Mechanical, electromechanical and electronic built-in devices (see Section 4.2)
- Time switches (see Section 4.2)
- <u>*instabus*</u> *EIB* components (see Section 4.4)

Residential buildings

Fig. 1.1/4 shows the basic structure of the distribution levels in *residential buildings.* Once the power has been supplied by the public utility, only two levels are relevant here, namely the main distribution and load level. This is due to the fact that the main distribution boards and sub-distribution boards are usually combined to form one *power distribution level* in residential buildings which is referred to as a *meter distribution board.*

<u>*instabus*</u> *EIB*

Figures 1.1/3 and 1.1/4 illustrate how the <u>*instabus*</u> *EIB* (European Installation Bus System) is implemented at the load level. Unlike conventional installation systems, *instabus* offers key benefits in that it simplifies cable installation while, at the same time, supporting a wide range of functions and applications (see Section 4.4).

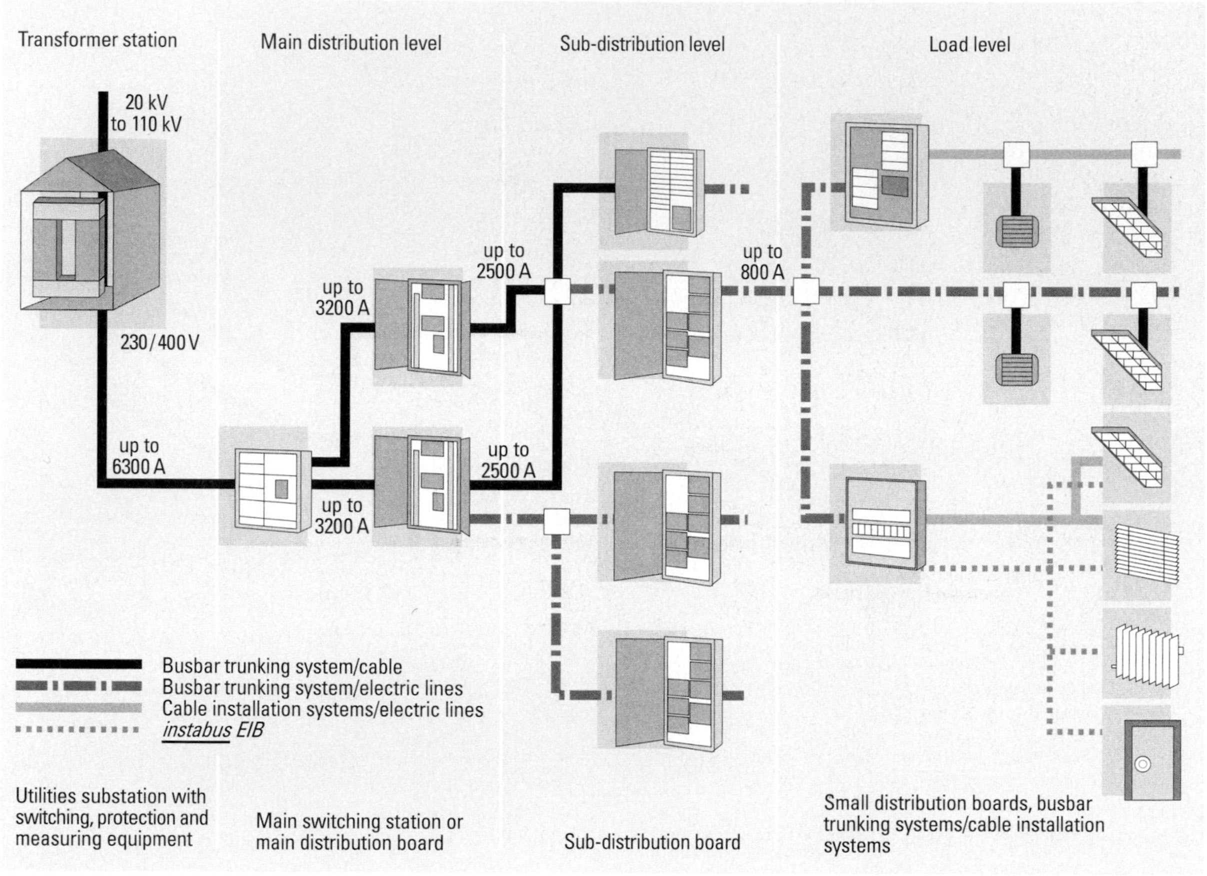

Transformer station
Main distribution level
Sub-distribution level
Load level
20 kV to 110 kV
230/400 V
up to 6300 A
up to 3200 A
up to 3200 A
up to 2500 A
up to 2500 A
up to 800 A
Busbar trunking system/cable
Busbar trunking system/electric lines
Cable installation systems/electric lines
instabus EIB
Utilities substation with switching, protection and measuring equipment
Main switching station or main distribution board
Sub-distribution board
Small distribution boards, busbar trunking systems/cable installation systems

Fig. 1.1/3
Distribution levels in industrial and functional buildings

1.1.2 Planning fundamentals

Objectives

The purpose of the planning stage is to determine the most suitable electrotechnical infrastructure for the type of building concerned. This infrastructure must satisfy the technical requirements for the respective power supply and the applicable standards. It must, however, also be possible to erect and operate the infrastructure economically. The electrical equipment, such as transformers, switchgear, cables and conductors, as well as control and monitoring devices, must be dimensioned and selected such that it represents a cost-effective and state-of-the-art solution both when considered in terms of individual components and as a complete system. The electrotechnical infrastructure should, therefore, be selected and dimensioned in line with the principles of system engineering, i.e. where preference is given to compatible products from one manufacturer. This electrical equipment must be sufficiently dimensioned to cope with the loads encountered both under normal operating conditions and in the event of a fault.

Standards

Electrical power-supply systems must always be planned, erected, and operated in compliance with recognized codes of practice and technical regulations which are defined in national, European, and international standards. These form the basis for safe operation of the electrical installations and specify minimum requirements for the technical data of the equipment and its fields of application. Compliance with these standards ensures the safety of personnel and in-

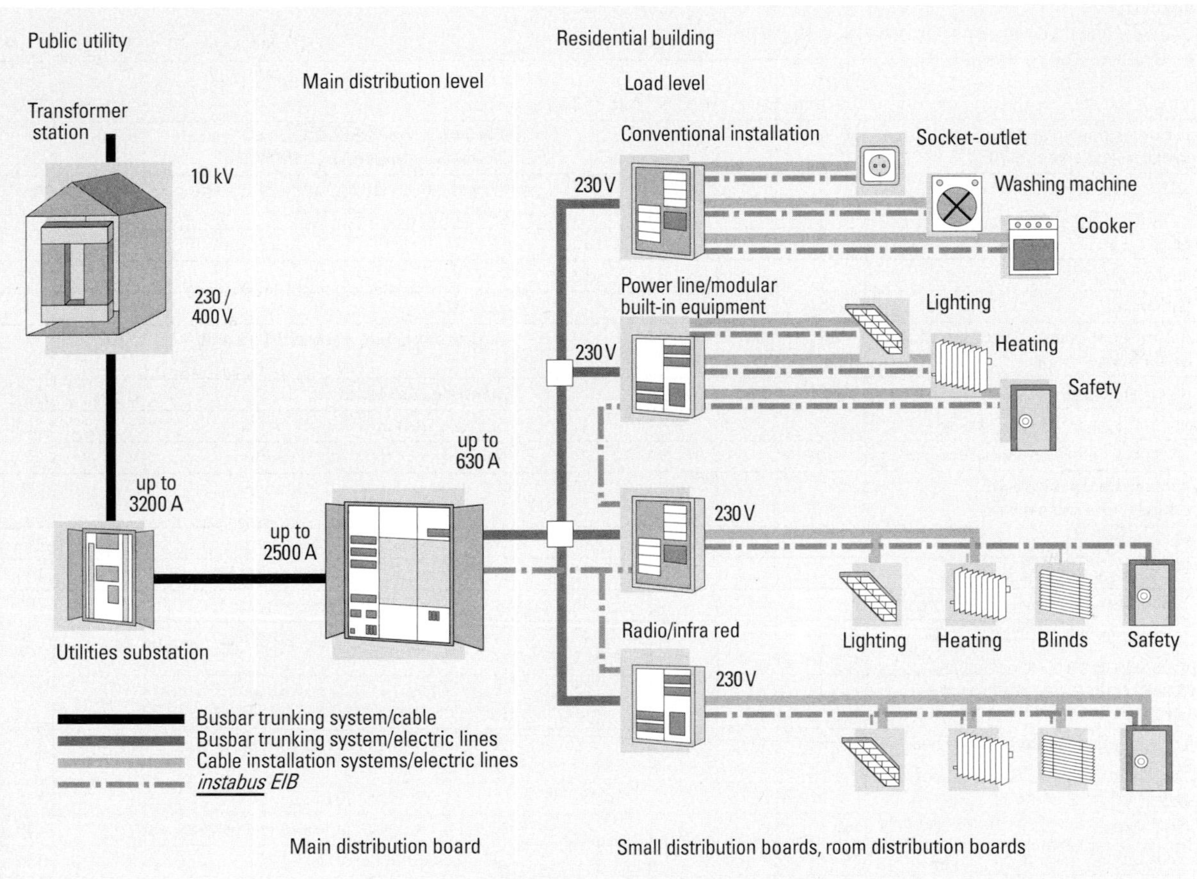

Fig. 1.1/4
Distribution levels in residential buildings

stallations. In Germany, these are the standards published by DIN VDE which, in the CENELEC member states, correspond to the European Standards (EN). As a result of harmonization, these documents are identical to international standards (IEC publications), with the exception of agreed amendments. The most important standards (CENELEC harmonization documents HD, EN, IEC, and DIN VDE) are listed in Table 1.1/1.

Deviations from these standards are permissible in certain cases if the same degree of safety can be achieved by different means. It is, however, often difficult to verify this. Additional provisions beyond the requirements specified in these standards may be necessary if the safety of personnel and of the installations requires this or if these are prescribed by legal requirements, directives, or contracts.

Additional standards

These include:

- ▷ National legal requirements
- ▷ Legal requirements specified by the European Union (EU)
- ▷ Accident prevention regulations of trade associations
- ▷ Requirements laid down by building and property insurance companies
- ▷ Technical supply conditions of the local public utility (TAB)
- ▷ Conditions imposed by authorities responsible for certain operating plants
- ▷ Special requirements of subsequent users (described in the contract or specifications)

Table 1.1/1
List of most important standards for erecting electrical installations

HD, EN, IEC						DIN VDE		
Titel	IEC-Pub.	HD	EN	Part	Chapter	Titel	Gruppe	Teil
Erection of power installations with nominal voltages up to 1000 V	60364-	384-		1		**Errichten von Starkstromanlagen mit Nennspannungen bis 1000 V**	**0100-**	
Scope				1	11	Anwendungsbereich, Allgemeine Anforderungen		100
Object					12			
Fundamental Principles					13			
Definitions	60364-	384-		2		**Begriffe**	0100-	200
Assessment of general characteristics of installations	60364-mod	384-		3		**Bestimmung allgemeiner Merkmale**	0100-	300
Protection for safety	60364-	384-		4		**Schutzmaßnahmen**	0100-	400
Protection against electric shock					41	Schutz gegen elektrischen Schlag		410
Protection against thermal effects					42	Schutz gegen thermische Einflüsse		420
Protection against overcurrent (for conductors and cables)					43	Schutz von Kabel und Leitungen		430
Protection against overvoltage					44	Schutz gegen Überspannungen		440
Protection against undervoltage					45	Schutz gegen Überspannungen		450
Isolation and switching					46	Schutz durch Trennen und Schalten		460
Application of protective measures for safety					47	Anwendung von Schutzmaßnahmen		470
Choice of protective measures as a function of external influences					48	Anwendung von Schutzmaßnahmen unter Berücksichtigung der äußeren Einflüsse		480
Selection and erection of electrical equipment	60364-	384-		5		**Auswahl und Errichtung elektrischer Betriebsmittel**	0100-	500
Common rules					51	Allgemeine Bestimmungen		510
Wiring systems					52	Kabel, Leitungen, Stromschienen		520
Switchgear and control gear					53	Schalt- und Steuergeräte		530
Earthing arrangements and protective conductors					54	Erdung, Schutzleiter, Potentialausgleichsleiter		540
Other equipment					55	Sonstige elektrische Betriebsmittel		550
Supplies for safety services					56	Elektrische Anlagen für Sicherheitszwecke		560
Verifications	60364-	384-		6		**Prüfungen**	0100-	600
Initial verification (prior to commissioning for the installation)					61	Prüfungen, Erstprüfung		610
Requirements for special installations or locations	60364-	384-		7		**Bestimmungen für Betriebsstätten, Räume und Anlagen besonderer Art**	0100-	700
Location containing a bath tube or shower basin					701	Räume für Badewanne oder Dusche		701
Swimming pools and other basins					702	Schwimmbäder		702
Locations containing sauna heaters					703	Sauna-Anlagen		703
Construction and demolition site installations					704	Baustellen		704
Agricultural and horticultural premises					705	Landwirtschaftliche und gartenbauliche Anwesen		705
Restrictive conducting locations					706	Leitfähige Bereiche mit begrenzter Bewegungsfreiheit		706
Earthing requirements for the installation of processing equipment					707	Erdungsanforderungen für das Errichten von Betriebsmitteln der Informationstechnik		707

Table 1.1/1 (continued)
List of most important standards for erecting electrical installations

HD, EN, IEC						DIN VDE		
Titel	IEC-Pub.	HD	EN	Part	Chapter	Titel	Gruppe	Teil
Electrical installations in caravan parks and caravans					708	Elektrische Anlagen auf Campingplätzen und in Caravans		708
	60909-					**Berechnung von Kurzschlußströmen**	0102-	BB1
	60781-					Anwendungsleitfaden für die Berechnung in Niederspannungsstrahlennetzen		BB2
	60781-					Beispiele		
Short-circuit currents – calculations in three-phase a.c. systems	60909-					Kurzschlußströme – Berechnung der Ströme in Drehstromanlagen	0102-	
Currents during two separate simultaneous single phase line-to-earth short-circuits and partial short-circuit currents flowing through earth	60909-			3		Doppelerdkurzschlußströme und Teilkurzschlußströme über Erde		3
						Berechnung der Wirkung		1
Short-circuit currents – Calculation of effects						**Kurzschlußströme – Berechnung der Wirkung**	0103-	
Definitions and calculation methods	60865-		60865-	1		Begriffe und Berechnungsverfahren		1
Operation of electrical installations			50110-	1		**Betrieb von elektrischen Anlagen**	0105-	
Insulation coordination for electrical equipment within low-voltage systems						**Isolationskoordinierung für elektrische Betriebsmittel in Niederspannungsanlagen**	0110-	1
	60664-	625-		1		Grundsätze, Anforderungen und Prüfungen		1
Low-voltage switchgear and controlgear assemblies						**Niederspannung-Schaltgerätekombinationen**	0660-	
Type-tested and partially type-tested assemblies	60439-		60439-	1		Typgeprüfte und partiell typgeprüfte Kombinationen		500
Particular requirements for assemblies for construction sites (ACS)	60439-		60439-	4		Besondere Anforderungen an Baustromverteiler		501
	60439-		60439-	2		Besondere Anforderungen an Schienenverteiler		502
	60439-		60439-	5		Besondere Anforderungen an Niederspannung-Schaltgerätekombinationen, die im Freien an öffentlich zugänglichen Plätzen aufgestellt werden – Kabelverteilerschränke (KVS)		503
	60439-		60439-	3		Besondere Anforderungen an Niederspannung-Schaltgerätekombinationen, zu deren Bedienung Laien Zutritt haben – Installationsverteiler		504
						Bestimmungen für Hausanschlußkästen und Sicherungskästen		505
						Schaltgerätekombinationen, Schaltanlagen; Verdrahtungskanäle; Anforderungen und Prüfung		506
A method of temperature-rise assessment by extrapolation for partially type-tested assemblies (PTTA) of low-voltage switchgear and controlgear	60890	528-S2				Verfahren zur Ermittlung der Erwärmung von partiell typgeprüften Niederspannung-Schaltgerätekombinationen (PTSK) durch Extrapolation		507
						Verfahren zur Ermittlung der Kurzschlußfestigkeit von partiell typgeprüften Niederspannung-Schaltgerätekombinationen (PTSK) durch Extrapolation		
	60898		60898			**Leitungsschutzschalter bis 63 A Bemessungsstrom, bis 415 V AC und 440 V DC**	0641	
Low-voltage fuses	60269-	630-		2		Niederspannungssicherungen	0636	

Important directives for EU member states are:

- ▷ Machine directive
- ▷ EMC directive
- ▷ Low-voltage directive
- ▷ Telecommunications terminal equipment directive

Integrated planning

Early cooperation between the electrical system planner and the architects, specialist planners, different trades, the responsible public utility, and the purchasing authorities is essential during the project definition and integrated planning stages.

The following points must be clarified here:

- ▷ Type, use, and form of the buildings (residential, functional, industrial, high-rise, low-profile, number of storeys, etc.) as well as their location to allow the demarcation of individual supply zones,
- ▷ Standards and conditions imposed by the building authorities with regard to the erection of the installation and permit-granting procedures,
- ▷ Connection queries and conditions of the responsible public utility. These determine the supply voltage (high or low-voltage), the design of the public utility's substation, as well as the measuring and metering equipment. They also influence dimensioning of the installation, e.g. power system protection (starting currents, grading times); please refer to Section 1.3.
- ▷ Declaration of power requirements, rates, and connection charges,
- ▷ Determination of the building-related connected loads based on specific surface-area loading in accordance with the use of the building for different project stages,
- ▷ Determination of load centers, possible supply routes, and locations for transformers, main switching stations, and distribution systems,
- ▷ Requirements for necessary emergency power-supply systems and/or UPS systems with regard to the permissible interruption times, and for additional safety supply systems to DIN VDE 0107 and DIN VDE 0108. An international standard is currently in preparation.

Variants

Each power-supply problem can always be solved in a number of different ways. Each of these solutions must be evaluated with regard to economic and technical implications as well as the requirements of the user. In order to avoid expensive interim solutions, plans should always allow for maximum configuration of the installation but also permit intermediate expansion levels which fulfill current requirements.

Requirements

The requirements of the building users with regard to electrical installations have changed considerably in recent years with respect to convenience, flexible room and building utilization, and economic efficiency. Nowadays, the electrical infrastructure of a building must be designed in such a way that it satisfies the planned utilization requirements and can be adapted quickly and with minimum outlay to take account of changing requirements with regard to room and building utilization throughout the service life of the building.

This involves aspects such as:

- ▷ Load management to reduce energy costs,
- ▷ Environmental compatibility of the building during its utilization and upon demolition,
- ▷ Reduction of operating costs by using standardized, low-maintenance equipment (no special solutions)

Building system engineering

The requirements regarding a modern, upgradeable electrotechnical building infrastructure can only be fulfilled with systems which intelligently link the different trades in the building.

Building system engineering, however, changes the design and layout of the electrical power-supply systems in the building and affects planning and project design procedures.

Functional, low-cost solutions should, therefore, take account of the following general requirements vis-à-vis the electrical power supply bearing in mind the needs of the user:

- ▷ Transparent, uncomplicated system design to facilitate system management,
- ▷ Simple adaptation to changing utilization and operational conditions,
- ▷ Installations with high degree of supply and operational reliability, even if individual components fail (back-up systems, emergency power-supply systems, selectivity of the power-system protection, high degree of availability),

▷ High level of supply quality, i.e. low voltage fluctuations caused by load changes with sufficient voltage symmetry and low harmonic loads

Determining the connected load (power demand)

Power demand P_{max}

Estimating the future power demand P_{max} accurately for a building or a building complex influences expenditure on electrical equipment. The reference power to be requested from the public utility also depends on the power demand. It only corresponds to the sum of the installed loads in exceptional cases, e.g. with individual machines. It is usually derived from the sum of individual rated outputs (P_i) for all installed loads, multiplied by their utilization factors (a_i) and the coincidence factor (g):

$$P_{max} = \sum_{i=1}^{n} (P_i \, a_i) \, g.$$

Reference power S_{max}

Bearing in mind the reference power factor ($\cos \varphi_B$), the apparent reference power for which the transmission equipment is to be dimensioned is calculated with:

$$S_{max} = P_{max} \cos \varphi_B.$$

Utilization factor a_i with motors

The utilization factor (a_i) of electrical equipment is the difference between the rated output and actual reference power. Motor drives are designed according to the mechanical power to be output at their shafts. Based on the power ratings offered by the manufacturers, the motor with the next highest output is then installed. Since a motor only draws the effective power output plus internal losses from the power system, the installed rated output of the motor is always greater than the actual power consumption. A deterioration of the power factor ($\cos \varphi$) must be taken into consideration for partial motor loads.

Coincidence factor g

The coincidence factor (g), also frequently referred to as the demand factor, allows for the fact that not all of the loads in an installation are switched on at the same time. It is, therefore, always < 1 and decreases with a growing number of loads in the installation. Consequently, it drops increasingly in the direction from the load, sub-distribution boards, main distribution board, main substation to the incoming supply of the public utility (see Fig. 1.1/5).

Table 1.1/2 contains empirical coincidence factors (g) for a number of load groups in three common types of building.

The ratio of the connected load to be requested from the public utility to the installed load decreases as the size of a building or building complex increases (referred to the total power consumption measured). This has a positive effect on the tariff condi-

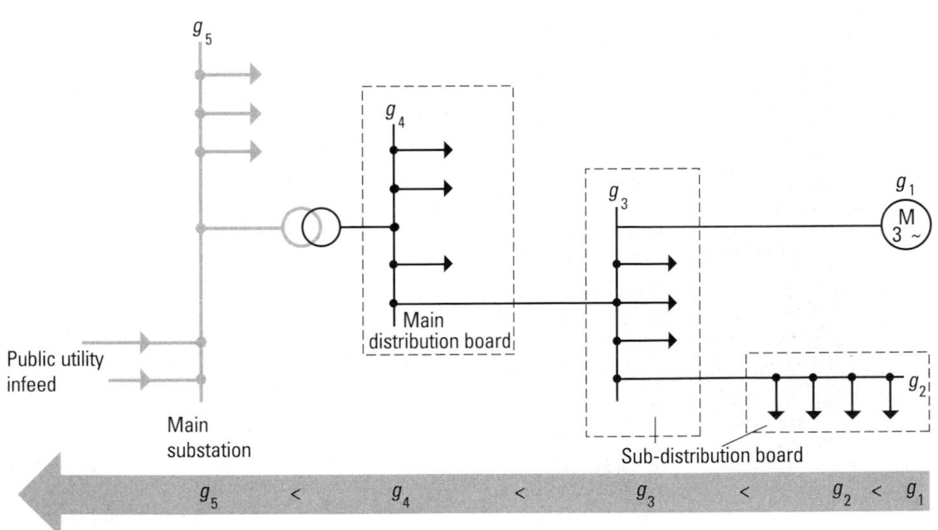

Fig. 1.1/5 Behavior of the coincidence factor (g) in a distribution system

Table 1.1/2
Coincidence factors (g) for individual load groups in three common types of building

Load groups	Office building	Hospitals	Department stores
Lighting	0.85 . . 0.95	0.7 . . . 0.9	0.85 . . 0.95
Air conditioning	1	0.9 . . . 1	0.9 . . . 1
Kitchens	0.5 . . . 0.85	0.6 . . . 0.8	0.6 . . . 0.8
Elevators/escalators	0.7 . . . 1	0.5 . . . 1	0.7 . . . 1
Sockets	0.1 . . . 0.15	0.1 . . . 0.2	0.2

tions and connection charges of the public utility.

A guideline for calculating the coincidence factor in accordance with HD 384-3 S2/ 60364-3/DIN VDE 0100-300 is currently in preparation.

Specific surface-area loading P_m (W/m²)

At the project definition and integration planning stages, the specific surface-area loading P_m (W/m²) with an estimated average power factor (cos φ) is frequently used as the basis for determining the anticipated system load. Approximate planning values for different building types with proportionate power demand for the various load groups are given in Section 5.3.

It should, however, be pointed out that the values for both the coincidence factor (g) and the power demand of residential and functional buildings (hotels, administrative buildings, department stores, schools, hospitals, theaters, airport terminals, etc.) are largely determined by local climatic conditions (locations in the tropics or polar regions) and the level of comfort in the buildings themselves. No globally applicable information has yet been published.

Effect of short-circuit current loads

Maximum short-circuit current load

The magnitude of the short-circuit current plays a decisive role in determining the required short-circuit current strength of the installations and equipment and selecting the rated currents and settings for the short-circuit protective devices used. It is also a measure of the supply quality at a given distribution point.

The short-circuit currents in three-phase systems are calculated in accordance with IEC 60909 and DIN VDE 0102. Both the highest and lowest short-circuit currents, which can occur as a result of possible system switching operations (parallel operation or stand-alone operation of incoming feeders, transformers, generators, and cables), must always be calculated (see Section 1.2).

Motors increase short-circuit current load

High-voltage and low-voltage motors contribute to the initial symmetrical short-circuit current (I_k''), the peak short-circuit current (i_p) and, with asymmetrical faults, the sustained short-circuit current (I_k).

Loads on installations and equipment

In both h. v. (> 1 kV) and l. v. (≤ 1 kV) systems, the maximum short-circuit currents generated determine the required making and breaking capacity of the switchgear as well as the thermal and dynamic strength for which the installations and equipment are to be dimensioned. The following conditions must be fulfilled:

$I_{b\ max} \leq I_{cn}$

$I_{p\ max} \leq I_{cm}$ and I_{pk}

$I_{th\ max} \leq I_{cw}$ and $I_{th\ per}$

$I_{b\ max}$ Maximum short-circuit breaking current (root-mean-square value)
$I_{p\ max}$ Maximum peak short-circuit current (max. possible instantaneous value)
$I_{th\ max}$ Maximum thermal short-circuit current (root-mean-square value)
I_{cn} Short-circuit breaking capacity
I_{cm} Short-circuit making capacity
I_{cw} Short-time current
I_{pk} Impulse withstand current
$I_{th\ per}$ Permissible thermal short-circuit current carrying capacity

The highest short-circuit current load should normally be expected in the event of a three-phase short circuit in l. v. systems. In the immediate vicinity of transformer infeeds, the single-phase fault current may be greater than the three-phase short-circuit current with very short l. v. connections between the infeed and main distribution board, with high system impedance in the upstream e. g. system, and with TN systems on the low-voltage side. In such cases, the maximum single-phase short-circuit current as well as the maximum three-phase initial symmetrical short-circuit current (I_k'') and the three-phase peak short-circuit current (i_p) must be taken into consideration when dimensioning the installations and equipment. A conductor temperature of 20° Celsius must be allowed for when calculating the maximum fault currents.

The protective devices used in the building power-supply system must detect short-circuit faults at the end of an electric circuit in l.v. systems reliably and trigger at times which would not cause thermal overloading of the transmission equipment (short-circuit protection). The lowest short-circuit current (I_{kerf}) is calculated as follows:

Lowest short-circuit current

$$I_{kerf} = \sqrt{\frac{K^2\, S^2}{t}}$$

K Material coefficient
S Cross section of conductor
t Short-circuit clearance time (≤ 5 s)

The lowest value should normally be expected for a single-phase short-circuit current in networks with TN systems. It, therefore, determines the maximum permissible rated currents and settings for the protective devices used in the system. The cross section must often be increased, particularly for long connections, to ensure that the protective devices of the system respond and trigger correctly in the event of faults at the end of the electric circuit.

Due to the expense involved, the return conductor (PEN, PE, and N) should, in such cases, have the same cross section as the external conductor.

With very short l.v. connections, the lowest short-circuit current must be expected in the immediate vicinity of transformer and generator infeeds in the event of a double-phase-to-earth fault. In such cases, this short-circuit current plays a decisive role in determining the maximum rated currents and settings for the protective devices of the system.

Rated system voltage

Voltage values to IEC 38

The voltage values were adjusted worldwide to IEC 38 as part of the harmonization process. The values and notation used in IEC 38 have also been adopted by the majority of CENELEC countries (HD 472.1 amended). Table 1.1/3 specifies a number of voltage values from IEC 38 for three-phase, three-wire, and four-wire systems ≤ 35 kV. As of 2003, the current voltage levels of 220/380 V and 240/415 V will be replaced worldwide by the standard value 240/400 V. A permissible tolerance of $\pm 10\%$ is recommended for the new standard value. During the transition period, a permissible tolerance of $+6\%$ and -10% will apply in countries which currently have a rated system voltage of 220/380 V $\pm 5\%$ (this includes Germany). A tolerance of $+10\%$ and -6% will apply in countries which currently have 240/415 V.

Highest and lowest values for 230/400 V voltage level

The permissible tolerances describe the highest and lowest voltages under normal operating conditions at a given time and at a given point in the system. Approximate values for the highest and lowest permissible voltages for systems, which currently have a rated voltage of 220/380 V, are specified in Table 1.1/4.

The same considerations apply for the new standard voltage level 400/690 V. Discussions regarding a possible reduction of the permissible tolerances after 2003 are still taking place.

Effect on filament lamps and asynchronous motors

Continuous overshooting of the rated system voltage usually reduces the service life of the equipment, whereas continuous undershooting is often associated with a reduction in performance. With filament lamps (I proportional U), for example, a 5% higher voltage

Table 1.1/3
Rated system voltages ≤ 35 kV to IEC 38

Systems ≤ 1000 V	230/400, 277/480, 400/690, 1000 V
Systems ≤ 35 kV	3, 6, 10, (15), 20, 35 kV

Table 1.1/4
Highest and lowest rated system voltages in low-voltage three-phase four-wire systems

Countries currently with 220/380 V		U_{min} (V)	U_{max} (V)
Now	220/380 +/ − 5%	209/360	231/400
Transition	230/400 +6/ − 10%	209/360	245/424
2003	230/400 +/ − 10%	209/360	254/440

will reduce the service life by half and a 5% lower voltage will reduce luminous efficiency by approximately 20%. With asynchronous motors, however, lower voltages result in higher power consumption. Violation of the tolerances (undershooting) specified by the manufacturer (usually P_n with U_n – 5%) frequently causes overloading of the asynchronous motors.

The electrical power supply must always be planned such that the permissible voltage tolerances of the equipment (manufacturer specifications) are observed at the point at which the load is connected.

Permissible voltage tolerances

Using a 230/400 V system as an example, Fig. 1.1/6 illustrates the possible distribution of the permissible drop in voltage starting at a 110 kV infeed and ending at the load for a complex building installation for the transition period up until 2003.

Voltage variation by means of 110/20 kV variable-voltage transformer

The voltage of the public utility's e. g. system is corrected under load at the infeed point of the 110/20 kV variable-voltage transformer. The setpoint voltage of the controller is set to a rated system voltage of 105%. With a step voltage of 1.65% and a controller hysteresis of 20%, this results in a maximum voltage variation of 2% or ± 1% from the setpoint value.

Voltage drop H.v. system

Voltage drops in e. g. transmission systems with identical transmission performance depend to a large extent on the selected voltage level (6 kV, 10 kV, 20 kV, 35 kV) and expansion. (A value ≤ 2% has been assumed in the example shown in Fig. 1.1/6.)

Voltage drop Distribution transformer

An approximate value for the voltage drop via a transformer can be determined using the following equation:

$$\Delta u = n\,u' + \frac{1}{2}\,\frac{(n\,u'')^2}{100}$$

$$u' = u_{\mathrm{Rr}} \cos\varphi + u_{\mathrm{Xr}} \sin\varphi$$

$$u'' = u_{\mathrm{Rr}} \sin\varphi - u_{\mathrm{Xr}} \cos\varphi$$

Δu Relative voltage variation (%)
u_{Rr} Rated value of resistance voltage (%)
u_{Xr} Rated value of reactance drop (%)
n Ratio of load to rated transformer output (S/S_{rT})
φ Load angle

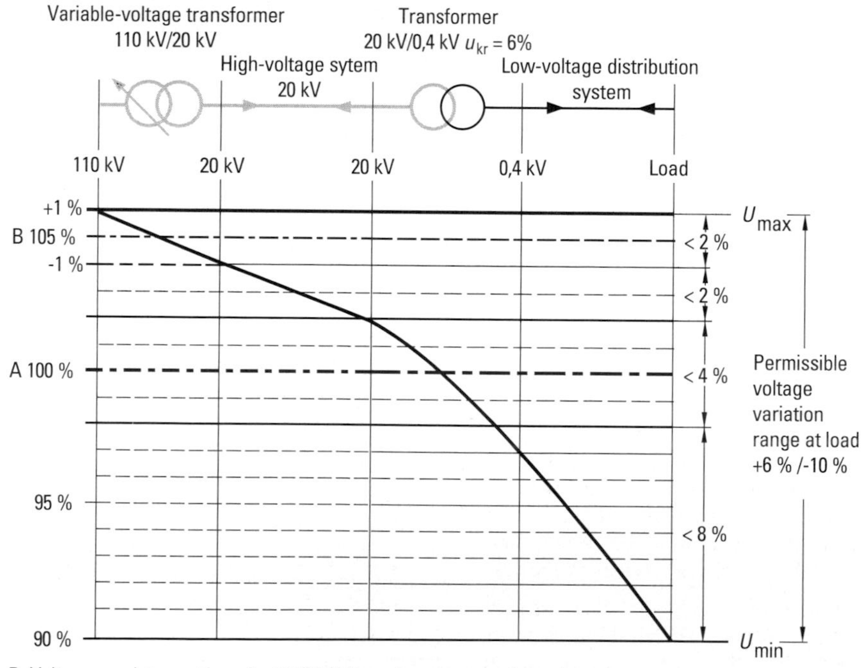

Fig. 1.1/6 Voltage ratios in a 110 kV/20 kV/0.4 kV system

If distribution transformers with a rated short-circuit voltage of $u_{kr} = 6\%$ are loaded with the rated output (S/S_{rT}), the maximum voltage drop with a power factor $\cos \varphi \geq 0.8$ will be less than 4%.

Voltage drop L.v. system

When a l.v. distribution system is dimensioned, a permissible voltage drop of approximately 8% (up to approximately 10% with vector addition of the individual voltage drops) is possible without the permissible limits specified in IEC 38 being overshot or undershot. If lower voltage tolerances are required by the loads used (e.g. ± 5% for asynchronous motors), they must be allowed for when the distribution system is being dimensioned.

At present, the standards do not contain generally applicable international reference values for the partial voltage drops of a supply concept (such as the one shown in Fig. 1.1/6). Reference to this is only made in national standards (e.g. DIN VDE 0100-520 "Selection and erection of electrical equipment; wiring systems" recommends a voltage drop of 4% between the distribution system and electrical equipment). The technical supply conditions of the public utilities define certain limit values for the permissible voltage drop in the case of supply lines upstream of the metering devices.

The connection of large loads (e.g. when starting large motors, peak currents in welding equipment, arc furnaces, etc.) causes considerable momentary variations in the power-frequency voltage.

Rhythmic voltage dips

Frequent, rhythmic variations will cause lighting to flicker and result in faults at voltage-sensitive loads (such as communications equipment, television sets, automation systems, etc.).

Fig. 1.1/7 shows the permissible voltage variations relative to the frequency at which the rhythmic voltage dips occur over time.

Voltage variations caused by commutation functions

Periodic momentary voltage variations in the a.c. supply voltage may also be caused by commutation functions at a power converter.

Permissible values are specified in EN 60146-1/IEC 60146-1/VDE 0558 Part 11 (Short-time dips for commutation) and in EN 61136-1 (Electrical operating conditions). These requirements result in a minimum short-circuit power at the connecting point in accordance with the rated output and commutation impedance of the power converter. The amplitudes of these voltage variations are inversely proportionate to the short-circuit power at the point at which the load is connected to the power system (corresponds to lower resistance of the upstream power system).

Reduction measures

Measures aimed at reducing system perturbation on voltage-sensitive loads include:

- ▷ Separate l.v. systems for voltage-sensitive and voltage-insensitive loads (point of connection to the e.g. system)
- ▷ Use of distribution transformers with lower rated short-circuit voltages, e.g. $u_{kr} = 4\%$ instead of 6% and parallel operation of several transformers (increase in short-circuit power)
- ▷ Connection of voltage-sensitive devices to an uninterruptible power system (UPS) or fast-response compensation devices

Estimating the voltage dip

The approximate magnitude of anticipated rhythmic voltage dips can be calculated for infeeds with transformers using the following equation:

$$\Delta u \approx \frac{\Delta S_L}{S_{rT}} u_{kr} \sin \varphi_L$$

Δu Permissible rhythmic voltage dip
S_{rT} Rated transformer output (kVA)
ΔS_L Load surge (kVA)
u_{kr} Rated short-circuit voltage
φ_L Angle of load surge

Example

Example

Given:
Administrative building with three 630 kVA transformers, $u_{kr} = 6\%$ connected in parallel.

Total rated transformer output
$S_{rT} = 3 \cdot 630 \text{ kVA} = 1890 \text{ kVA}$
Load surge $\Delta S_L = 500 \text{ kVA}$
$\cos \varphi_L = 0.85$; $\sin \varphi_L = 0.52$

Required:
Δu at 6 load surges per hour

Result:

$$\Delta u \approx \frac{500 \text{ kVA}}{1890 \text{ kVA}} \cdot 6\% \cdot 0.52 = 0.83\%$$

According to Fig. 1.1/7, curve 1 (e.g. administrative buildings), the value lies below the perception limit for filament lamps and is, therefore, acceptable for lighting. In addition to these voltage variations caused by rhyth-

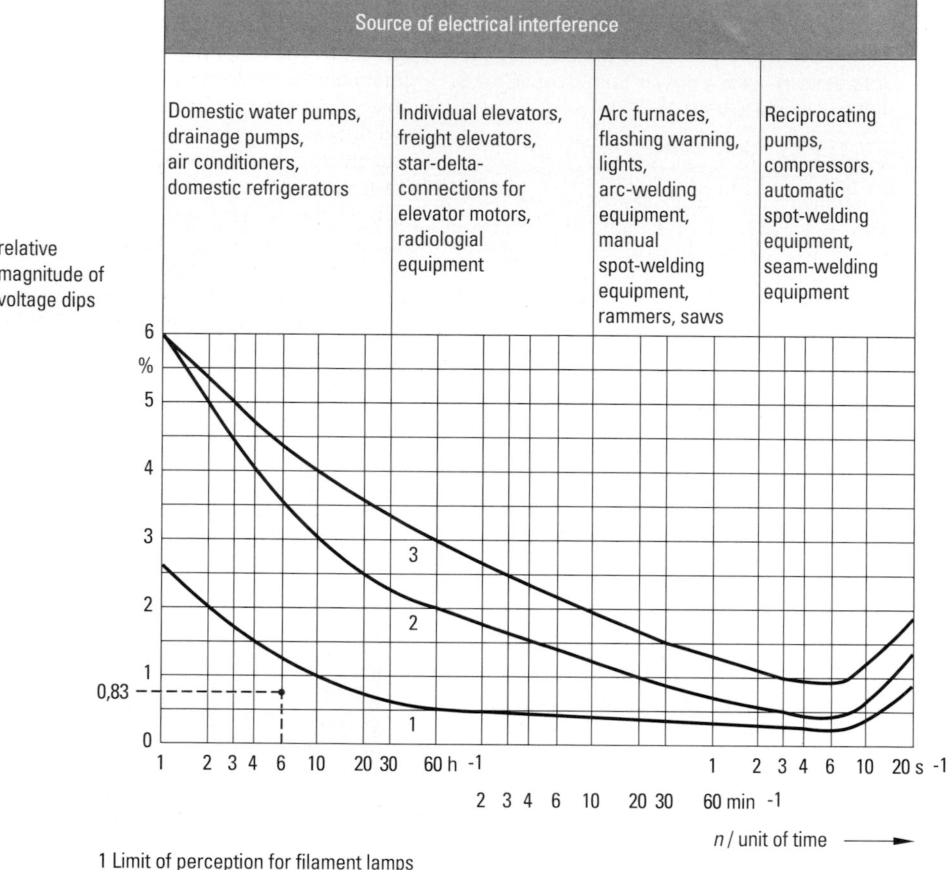

Fig. 1.1/7
Acceptable voltage dips (Δu), expressed as percentage of rated system voltage, relative to the number of voltage fluctuations (n) over time

mic loads, a voltage dip must also be expected for the period leading up to disconnection by the power system protection device whenever a short-circuit-type system fault occurs. If important loads are affected by this, they must be supplied via uninterruptible power systems.

Effects of electromagnetic compatibility (EMC)

Owing to the increasing use of modern communications equipment in buildings, a high level of electromagnetic compatibility (EMC) is becoming an increasingly important factor in the planning and design of electricity supply systems (see Chapter 3).

Limit values for the emitted interference and interference susceptibility of devices are defined by the law governing the electromagnetic compatibility of devices (regulation for EU member states) and the relevant associated standards. Sensitive information technology installations in buildings can, nevertheless, be influenced by electrical power installations. Early coordination between the different trades is essential to ensure a high level of electromagnetic compatibility in buildings. Improvements and corrective measures are usually very costly.

Types of l. v. distribution system

Special importance is attached to the type of l. v. distribution system used.

According to HD 384 Part 3 and IEC 60364, l. v. systems are classified as *IT*, *TT*, and *TN systems* depending on the grounding connection system used. General power-supply system networks are usually designed as TT or TN systems. The TT system is used predominantly in rural areas since long supply lines do not require any equipment grounding conductors (PE) and are thus less expensive (see Chapter 2). The TN system is preferred in densely built-up supply areas. A distinction is made here between the *TN-C system* with combined equipment grounding and neutral conductor (PEN) and the *TN-S system* with separate equipment grounding conductor (PE) and neutral conductor (N) (see Chapter 2).

Building currents in TN-C system

As Fig. 1.1/8 clearly illustrates, the TN-S system is regarded as EMC "friendly" and the TN-C system as EMC "unfriendly". In TN-C systems, a potential difference is caused by the return conductor current along the common PEN conductor. This potential difference results in operating currents along communications line shielding grounded at both ends, and also building currents along metallic loops. Due to the building current, the infeed and return conductor currents differ from each other along the power transmission routes. This results in an increased electromagnetic field around the power transmission cable. These fields can easily cause interference in visual display units. Although installing equipotential bonding conductors (a method frequently used) does indeed reduce the operating current content along the shielding of communications cables, it also increases the difference between the infeed (sum of L1 + L2 + L3) and return conductor current (PEN) and, in turn, the magnetic fields around the power transmission routes.

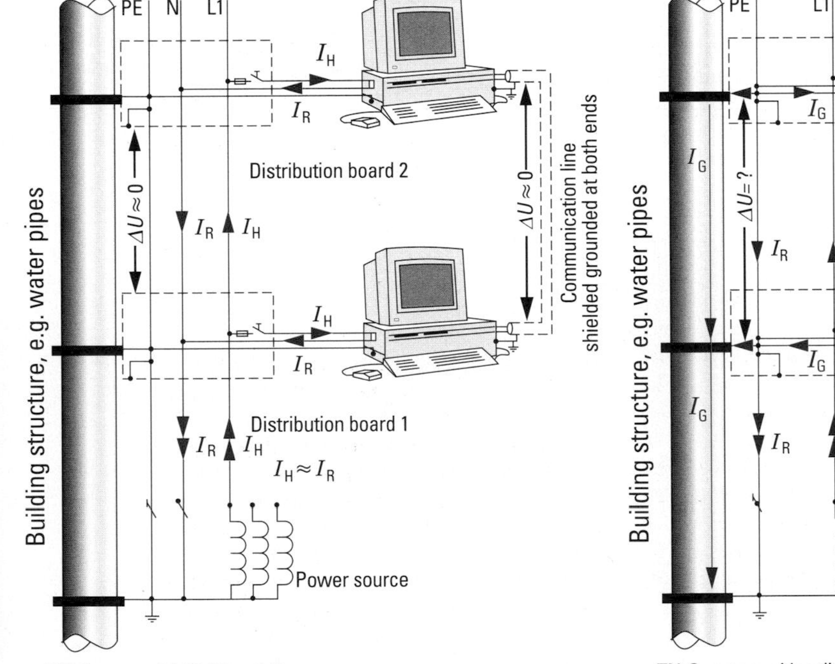

TN-S system (EMC "friendly")

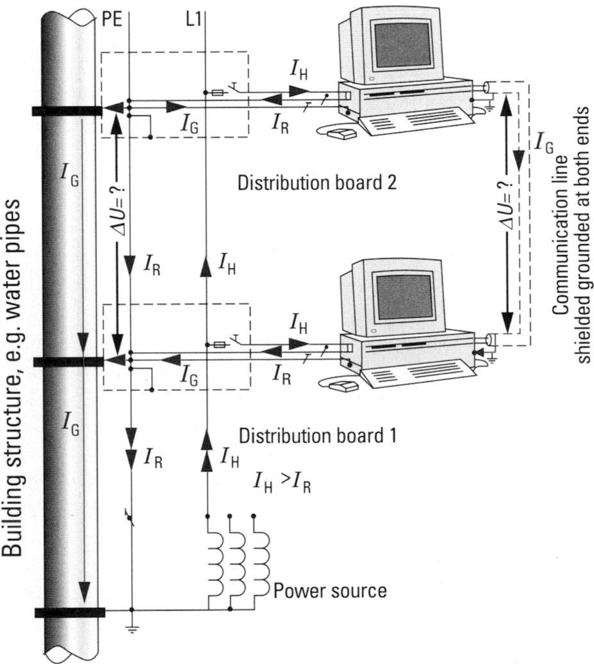

TN-C system with split return conductor currents

I_G = Building and shield currents
I_R = Infeed conductor currents (sum of L1 + L2 + L3)
I_H = Return conductor currents in PEN bzw. N

Fig. 1.1/8
Building currents in a TN-C system and TN-S system

TN-S system only permissible with single infeed or central arrangement

The PE conductor is potential free in TN-S systems. Building currents are thus prevented. The magnetic fields around the power transmission cables depend solely on their respective geometrical arrangements. According to HD 384.5.54 and IEC 60364-5-54, a TN-S system is only permitted for systems in which the infeeds (transformers and generators) are arranged centrally (see also Chapter 2). If the infeeds are distributed, four-pole infeeds with switchover devices (no parallel operation) are required for L1, L2, L3, and N conductors in TN-S systems (see Fig. 1.1/13).

Effect of building system engineering

Nowadays, a large number of different automation, monitoring, and control tasks have to be performed in practically every type of building, be this residential, functional, or industrial.

These tasks include:

▷ Heating, air-conditioning, and ventilation control
▷ Building security with intruder detection, fire alarm, and access control systems
▷ Lighting, roller blind shutters, and blind control
▷ Load monitoring and load management

In the past, several transmission systems, in addition to the power-supply and distribution systems, were installed in order to deal with the individual tasks. This, coupled with the increasing complexity of modern communications technology, resulted in complicated, costly solutions which increased expenditure even at the project planning stage. The power transmission routes and the devices for controlling them are usually linked in these building concepts which means that the intended use of the area in question must be known at the project definition stage.

Service life of the building

With conventional installation techniques, the costs involved in adapting the building's infrastructure to changing user requirements with regard to the building as a whole, sections of the building, or individual rooms can be quite considerable throughout the service life of the building (from erection and utilization to demolition). The costs incurred while the conversion work is being carried out, during which the building or parts thereof cannot be used, must also be taken into consideration.

instabus EIB

Nowadays, building system engineering with *instabus EIB* (see Section 4.4) is used for buildings with a high utility value. The actuators and sensors in the *instabus* system use the same transmission route. Leading manufacturers have opted for building system engineering equipment based on the **E**uropean **I**nstallation **B**us.

Lower costs for cables and conductors

This technology relieves the electrical power transmission systems of control and monitoring tasks. They can, therefore, be dimensioned according to the technical and cost-related requirements of power transmission alone. The cable routes between the distribution boards and loads are shorter. It is often possible to install cables and conductors with smaller cross sections or to use the existing circuits more efficiently with regard to the ratio of maximum load current to rated current.

Lower fire load

Both reduce the number and complexity of cables and conductors and, therefore, the fire load for the useful floor areas.

Busbar trunking systems

Busbar trunking systems (see Section 1.4.4) are being used to an increasing extent both for the primary power supply between l. v. main switchgear and storey or sub-distribution boards, and for supplying power to the load circuit. The power distribution systems for a supply section are dimensioned according to the requirements defined for the intended use (planning phase) and for subsequent changes in the usage of the building. It is, therefore, no longer necessary to know the precise location of the individual useful floor areas during the project definition or integration planning stage. The percentage of overall utilization is often all that is required to dimension the power demand and, in turn, the power transmission equipment based on surface-area loads. Frequent replanning, which costs both time and money, is therefore avoided.

***EIB* tool software**

In the case of monitoring and control systems, the sensors are not assigned to the actuators and to the loads to be monitored and controlled until the commissioning stage. The appropriate *EIB* tool software is used for this purpose.

Cross-installation system integration

Information from the individual installations is exchanged via the same transmission route. Information from the building monitoring system, e.g. window "open" or "closed", can be used to control heating and air-conditioning systems in individual rooms, and blind "up" or "down" to save energy by switching off lighting rows at windows. An access control system can monitor personnel leaving and entering buildings or sections of buildings. At the same time, this information can be used to save energy, e. g. lights/computers/heating "on" or "off", or to activate security equipment such as room monitoring devices, intruder detection systems, etc.

Planning and cost optimization for users and planners

Planning measures and costs are thus optimized for subsequent users and planners with reduced planning outlay during initial project phases (project definition, system integration planning, and implementation planning).

1.1.3 Design of building networks

Effect on power system configuration

The configuration of l. v. distribution systems in a complex building and for individual supply areas depends on the following factors and requirements:

▷ Level of power demand (determines the voltage level for the power supplied by the public utility)
▷ Structure of supply area, e. g.
  – density (low and/or high)
  – building type (low-profile, high-rise, etc.)
  – purpose
▷ Size, number, and physical location of the load centers in individual supply areas
▷ Possible arrangements of the transformers and associated l. v. main distribution boards
▷ Possible routing for the main distribution system
▷ General and special requirements of the investor and subsequent user with regard to
  – supply reliability and supply quality
  – low investment costs and high cost-effectiveness during operation

When configuring building power systems, it is important not just to choose the most cost-effective equipment to be used in the building but to find the best possible overall solution – both from a technical and cost-related point of view – not only for the point in time at which the building is erected but also for the total service life of the building.

General power supply

Small building power systems with a power requirement of up to 200 kW–300 kW are usually supplied from the l. v. distribution system of the public utility.

Supply for special-tariff customers

Special-tariff customers, such as hospitals, large hotels, administrative buildings, banks, large computer centers, sports stadiums, research institutes, universities, airports, industrial plants, etc., with a high power demand are supplied from the e. g. system of the public utility via transformer substations.

A ring feeder cable, to which other customers are connected, is frequently used for power demands of up to 1.5 MW. With power demands in excess of 1.5 MW, the power is usually supplied directly from the nearest public utility transformer substation via special cables.

Supply from h. v. system

If power is supplied from the h. v. system, the buildings are provided with a transformer substation with the public utility infeed, main circuit-breaker, metering equipment (sequence is determined by the public utility), transformer feeders and, if the power system of the special-tariff customer is particularly large, additional cable feeders to satellite substations (see Fig. 1.1/9).

Low-cost switchpanels with switch disconnectors and high-voltage high-breaking-capacity (h. v. h. b. c.) fuses are used for connecting the transformers in systems with a rated output of up to 800 kVA. Circuit-breakers with secondary relays must be provided for higher rated outputs and short-circuit current loads of 16 kA to 20 kA.

H. v. h. b. c. fuses and switch-disconnectors

In addition to the lower cost for the h. v. switchgear, the use of switch-disconnectors with h. v. h. b. c. fuses instead of circuit-breakers also has the following benefits:

▷ The voltage dip caused by short-circuit-type faults downstream of fuses is very small.
With circuit-breakers, however, a voltage dip of almost 100% and load disturbance must be expected within the short-circuit clearance time.
▷ The energy released in the event of a short-circuit (proportional to $I^2 \cdot t$) is dampened considerably by the current-limiting effect of the fuse and the associated short clearance times. Connecting cables with the smallest commercially

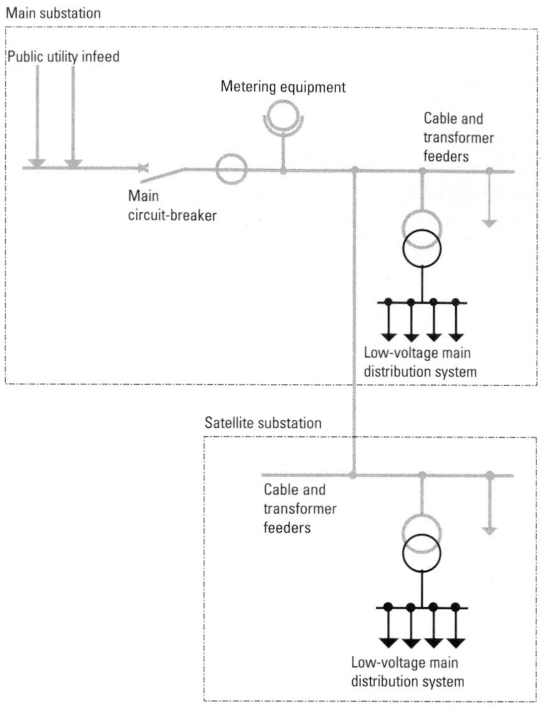

Fig. 1.1/9
Simplified diagram showing main substation and satellite substation

available cross section (approximately 25 mm² with 10 kV and 20 kV) can, therefore, usually be installed.
With circuit-breakers, the cross sections must normally be dimensioned according to the required short-circuit strength (e.g. minimum cross section 95 mm², XLPE cable, with $I_k'' = 20$ kA and $t_k = 0.5$ s).

Circuit-breakers and secondary relays

If circuit-breakers are used, vent outlets must be provided for the transformer and switchgear rooms. The advantages of using circuit-breakers include less complicated system management (no fuses need to be changed), simpler configuration, and the high number of switching cycles between maintenance inspections.

Dimensioning h.v. h.c.b. fuses

Both the *inrush current* of the transformers and the *starting frequency* are crucial factors in determining the smallest permissible h.v. h.c.b. fuse. The lowest short-circuit current that occurs on the low-voltage side of the transformer which must be disconnected in the event of a fault is used to determine the largest permissible h.v. h.c.b. fuse. The area to be supplied with power by a transformer (expansion and surface-area loading) and the reserve power to be provided if a transformer fails determine the rated output to be selected for the transformer.

Rated output of transformers

Transformers with a rated output of between 400 kVA and 1000 kVA are usually selected for the surface-area loads encountered nowadays. More powerful transformers of up to approximately 3.5 MVA are used especially in industrial buildings with a particularly high load density or large individual loads. Depending on the required level of supply reliability (partial or full reserve, instantaneous reserve), the necessary transformer output is divided into several smaller units which can be operated either in parallel or individually. The required reserve power increases with the rated output of the transformers. This results in a lower capacity utilization of the transformers under normal operating conditions.

The total output of transformers operating parallel to each other and feeding power into an l.v. system is restricted with respect to the short-circuit current load. In the case of parallel infeeds (as shown in Fig. 1.1/10 or Fig. 1.1/11), the short-circuit currents in the outgoing feeder panels are always higher than those in the individual incoming feeder panels or coupler units between the infeeds. The equipment in the infeeds and coupler units can be configured for a short-circuit current load from a specific number (n) of parallel transformers.

Table 1.1/5 specifies the maximum number of transformers, with a rated short-circuit voltage of $u_{kr} = 6\%$, which can be connected in parallel if the switchgear and the combinations of switching devices used in this are to be configured in accordance with EN 60439-1/IEC 60439-1/DIN VDE 0660-500 or with a rated conditional short-circuit current (I_{cc}) of 80 kA or 50 kA and a rated impulse withstand current (I_{pk}) of 176 kA or 110 kA. Asynchronous motors are not taken into consideration here and an initial symmetrical short-circuit power of approximately 250 MVA is assumed for the h.v. system.

Type	Busbar with tie breaker	Split busbar with transfer / auxiliary busbar
Tie breakers	in busbar for I_{rT}	to auxiliary busbar for I_{rT}
Load balancing	in busbar	via auxiliary busbar
Failure of one external transformer	Main load on busbar of middle transformer	Equal load on all busbars
Transformer reserve with $^2/_3$ capacity utilization $\cong 66\,^2/_3\,\%$ $= 1$ transformer unit	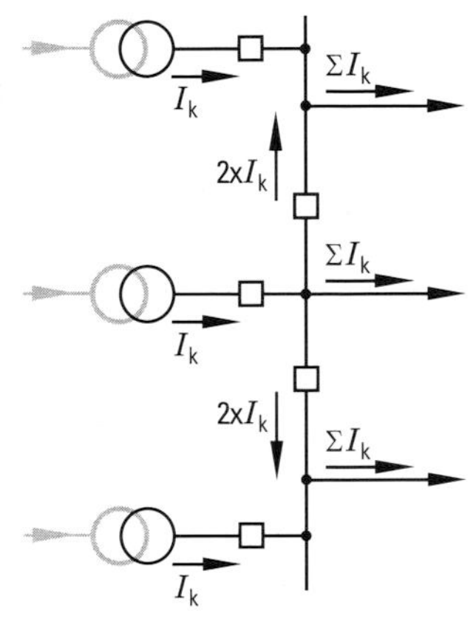	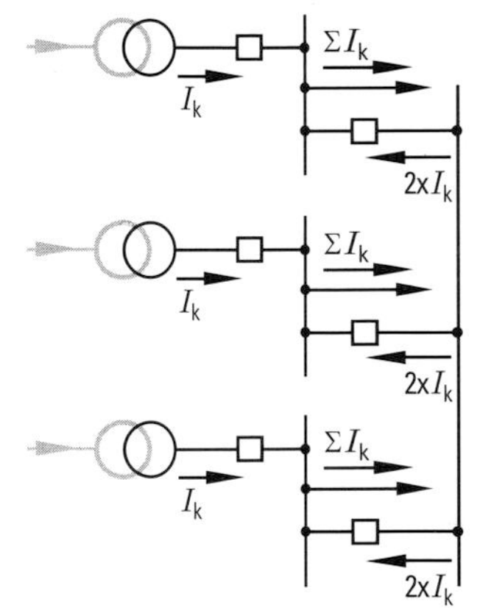
Assessment	Simple design of main distribution board	Greater flexibility at higher cost

-□- Circuit-breaker, closed
I_k Short-circuit current of one transformer
$2 \times I_k$ Short-circuit current of 2 transformers without asynchronous motor component
ΣI_k Total short-circuit current

Fig. 1.1/10 Central arrangement of transformers at load center with three units

Table 1.1/5
Number and maximum permissible system load with parallel transformers ($u_{kr} = 6\%$) taking permissible short-circuit current load I_{cc}/I_{pk} into consideration

Rated transformer output (kVA)	$I_{cc}/I_{pk} \leq 80$ kA/176 kA					$I_{cc}/I_{pk} \leq 50$ kA/110 kA				
	n	ΣS_{nT} kVA	$\Sigma S_{nT\ n-1}$ kVA	I''_k kA	a %	n	ΣS_{nT} kVA	$\Sigma S_{nT\ n-1}$ kVA	I''_k kA	a %
500	8	4000	3500	<76	87.5	4	2000	1500	<43	75
630	6	3780	3150	<73	83.3	3	1890	1260	<41	66.7
800	5	4000	3200	<76	80	3	2400	1600	<50	66.7
1000	4	4000	3000	<76	75	2	2000	1000	<43	50
1250	3	3750	2500	<73	66.7	1	1250	*U*	<28	*U*
1600	2	3200	1600	<64	50	1	1600	*U*	<35	*U*

n Number of parallel transformers
ΣS_{nT} Sum of rated transformer outputs
$\Sigma S_{nT\ n-1}$ Permissible load if one transformer fails
a Capacity utilization (%) of transformers under normal operating conditions
U Only transfer reserve possible (capacity utilization under normal operating conditions $\leq 50\%$)
I''_k Maximum initial symmetrical short-circuit current
I_{cc} Rated conditional short-circuit current
I_{pk} Rated impulse withstand current

System configuration	*Open* system with full transfer reserve (ring-cable switch "open")				*Closed* system with instantaneous reserve (ring-cable switch "closed")			
	Max. sub-system output S_v	Max. system output $g \cdot \Sigma S_v$	Trans-former capacity utiliza-tion normal	Ring cable rating	Max. sub-system output S_v	Max. system output $g \cdot \Sigma S_v$	Transformer capacity utilization normal	Ring cable rating
2 areas supplied 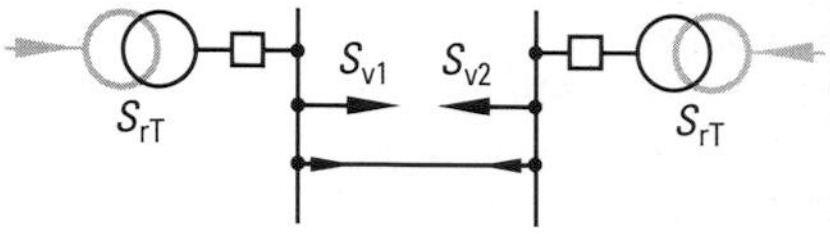	$< S_{rT}$	$< \frac{n}{2} \cdot S_{rT}$ $< S_{rT}$	Irregular $\widehat{=}$ S_{v1}, S_{v2}	$> S_{v\,max}$ $\geq S_{rT}$	$< S_{rT}$	$< (n-1) \cdot S_{rT}$ $< S_{rT}$	approx. 50% S_{rT}	1 cable $> S_{v\,max}$ $\leq S_{rT}$ 2 cables $> 0.5 \cdot S_{v\,max}$ $\leq 0.5 \cdot S_{rT}$
3 areas supplied 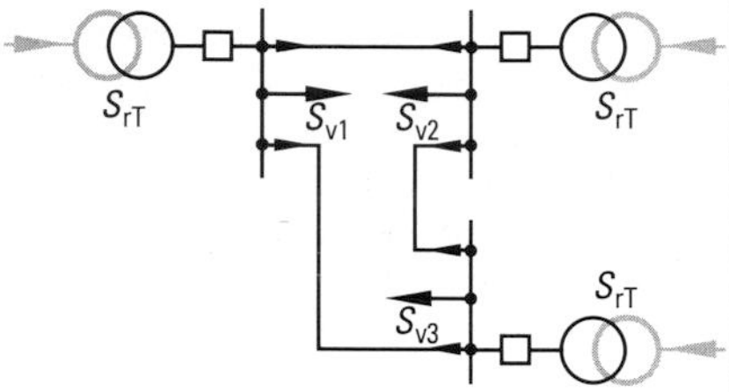	$< S_{rT}$	$< \frac{n}{2} \cdot S_{rT}$ $< 1.5 \cdot S_{rT}$	Irregular $\widehat{=}$ S_{v1}, S_{v2} S_{v3}	$> S_{v\,max}$ $\geq S_{rT}$	$< S_{rT}$	$< (n-1) \cdot S_{rT}$ $< 2 \cdot S_{rT}$	approx. 66% S_{rT}	Open ring with 2 cables $> S_{v\,max}$ $\leq S_{rT}$ Closed ring with 3 cables $> 0.5 \cdot S_{v\,max}$ $\leq 0.5 \cdot S_{rT}$
4 areas supplied 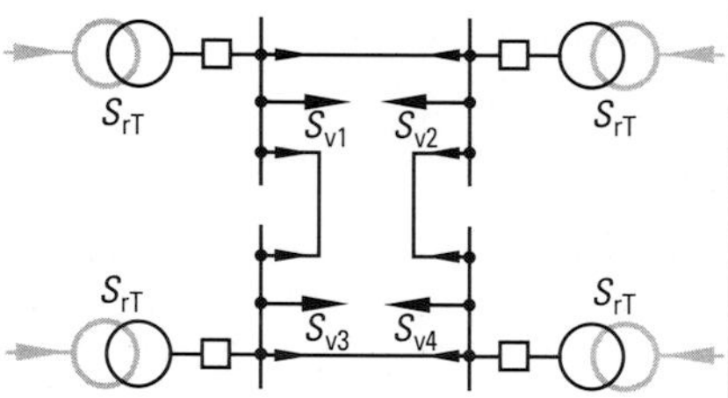	$< S_{rT}$	$< \frac{n}{2} \cdot S_{rT}$ $< S_{rT}$	Irregular $\widehat{=}$ S_{v1}, S_{v2} S_{v3}, S_{v4}	$> S_{v\,max}$ $\geq S_{rT}$	$< S_{rT}$	$< (n-1) \cdot S_{rT}$ $< 3 \cdot S_{rT}$	approx. 75% S_{rT}	Open ring with 3 cables $> S_{v\,max}$ $\leq S_{rT}$ Closed ring with 4 cables $> 0.5 \cdot S_{v\,max}$ $\leq 0.5 \cdot S_{rT}$

-□- Circuit-breaker, closed
S_{rT} Transformer output
S_v Peak load of subsystem
n Number of transformers
g Coincidence factor (corresponds to g_5 in Fig. 1.1/5)

Fig. 1.1/11
Distributed arrangements of transformers at load centers of subsystems with open or closed ring feeders (interconnecting cables) under normal operating conditions

Transformers with u_{kr} = 4%

If transformers with $u_{kr} = 4\%$ instead of $u_{kr} = 6\%$ are used, either the short-circuit current load is increased by a maximum of 50% or, if the short-circuit current load remains unchanged, only 2/3 (max.) of the number of transformers specified in the table can be operated in parallel.

Segmentation into several subsystems

Segmentation into several subsystems to prevent high short-circuit currents and the associated high costs for the switchgear is recommended if open circuit-breakers with low rated currents are used in the outgoing feeders.

Arrangement of transformers

Central arrangement of transformers

The physical location of the load centers, the size, number, and arrangement of the transformers (centralized or distributed) as well as their operating modes (individual or parallel) play a decisive role in determining the configuration of the power system, the subsequent operating costs, as well as the cost and complexity of the measures for reducing electromagnetic interference caused by operating currents. The following points must be observed for central transformer arrangements located at the load center of a supply area with power fed to one busbar section (Fig. 1.1/10):

- ▷ Tie breakers in the busbar run or for transfer busbars should permit individual operation of transformers or busbar sections both under normal operating conditions and when maintenance or expansion work is being carried out.
- ▷ To reduce the short-circuit current load in the busbars, the transformer infeeds should always be in the end sections of a busbar.

Advantages of distributed arrangement

Distributed arrangement of the transformers (see Fig. 1.1/11) has both technical and cost-related benefits if there are several load centers in one supply area.

These benefits include:

- ▷ lower voltage drops and lower transmission losses along the supply lines to the loads owing to the shorter distances involved,
- ▷ constant short-circuit currents at the infeed points of the transformers and, in turn, lower system perturbation with load variations and harmonic currents,
- ▷ weaker electromagnetic fields around the power routes to the transformers owing to the lower transmission currents at the high-voltage level (I proportional $1/U$)

Transformer load-center substations

Particularly economical solutions are possible by using transformer load-center substations with encapsulated-winding dry-type transformers which can be installed in "general operating areas" (see Section 5.9.11), e.g. in the production areas of industrial plants. No special provisions (e.g. fire walls) are required if encapsulated-winding dry-type transformers are used.

Transmission lines to satellite substations

The transmission lines to the satellite substations and the loads can be configured as an "EMC-friendly" TN-S system with five conductors (L1, L2, L3, N, and PE) for both central and distributed transformer arrangements (see Fig. 1.1/8). Distributed transformers must, however, only be interconnected using a TN-C system (HD 384.5.54/IEC 60364-5-54) with a common PEN conductor.

Connections in distributed arrangement

The busbars in the transformer stations must, therefore, also be configured as a TN-C system with L1, L2, L3, and PEN conductors. In each transformer station, an N conductor may be branched off from the PEN conductor separately for each busbar section or individual panel. In this case, however, the N conductors must no longer be connected to each other (see Fig. 1.1/12).

Four-phase switching devices

If the distributed transformers are not operated in parallel and the connections between the substations are only used as transfer connections, the connections may be configured as a five-phase TN-S system with L1, L2, L3, N, and PE. Four-phase switching devices (L1, L2, L3, and N) are, however, required in this case (see Fig. 1.1/13). A four-phase transfer device is also required if satellite substations are supplied by two system infeeds in a TN-S system (see Fig. 1.1/13).

1.1.4 Stand-by supply

Necessity of stand-by supply systems

An stand-by supply for providing buildings with electricity if the general power supply fails or is faulty can be necessary for two reasons:

- ▷ official regulations for electrical installations which supply power for important safety equipment, for example:
  - safety lighting systems
  - elevators for fire-fighting personnel
  - passenger elevators used for evacuation, e.g. high-rise buildings
  - fire-fighting installations, e.g. booster pumps, sprinkler systems, etc.

  (The relevant national regulations and/or standards must be observed in these cases)
- ▷ more stringent cost-related and technical requirements of the user, for example:
  - continued operation even after the public utility supply has failed
  - continued operation even with brief interruptions or transfers ($t \leq 15$ s)
  - greater demands with regard to voltage and frequency tolerances, e.g. with EDP and communications equipment

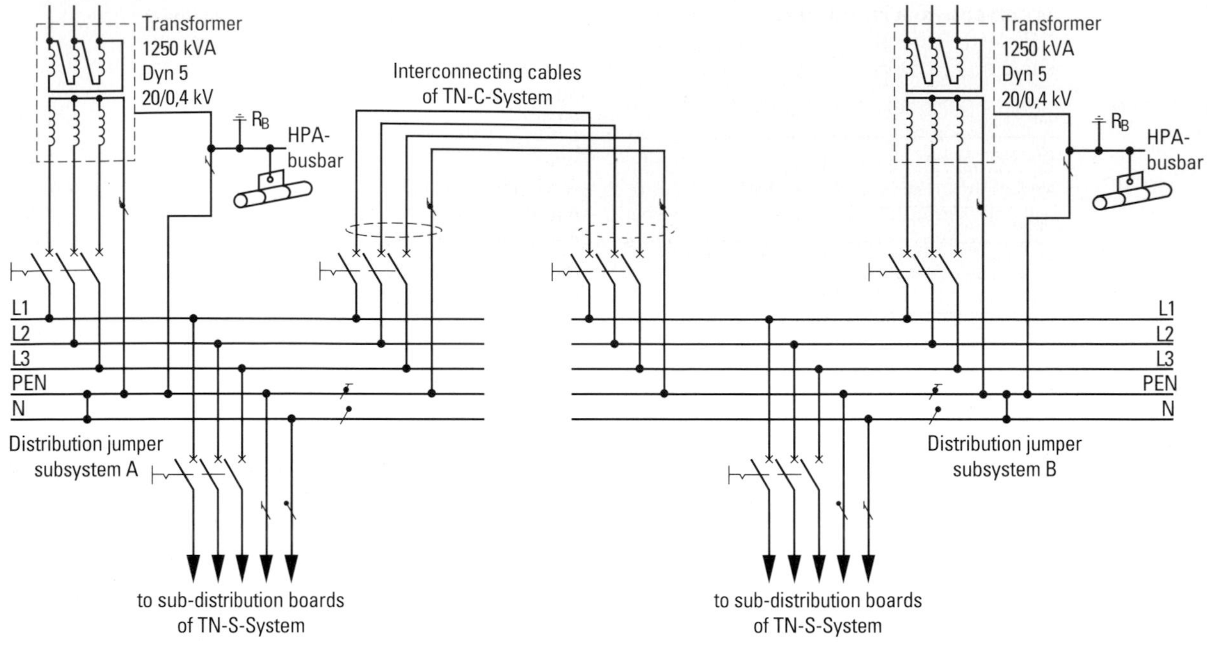

Fig. 1.1/12 TN-C system with several infeeds (distributed arrangement)

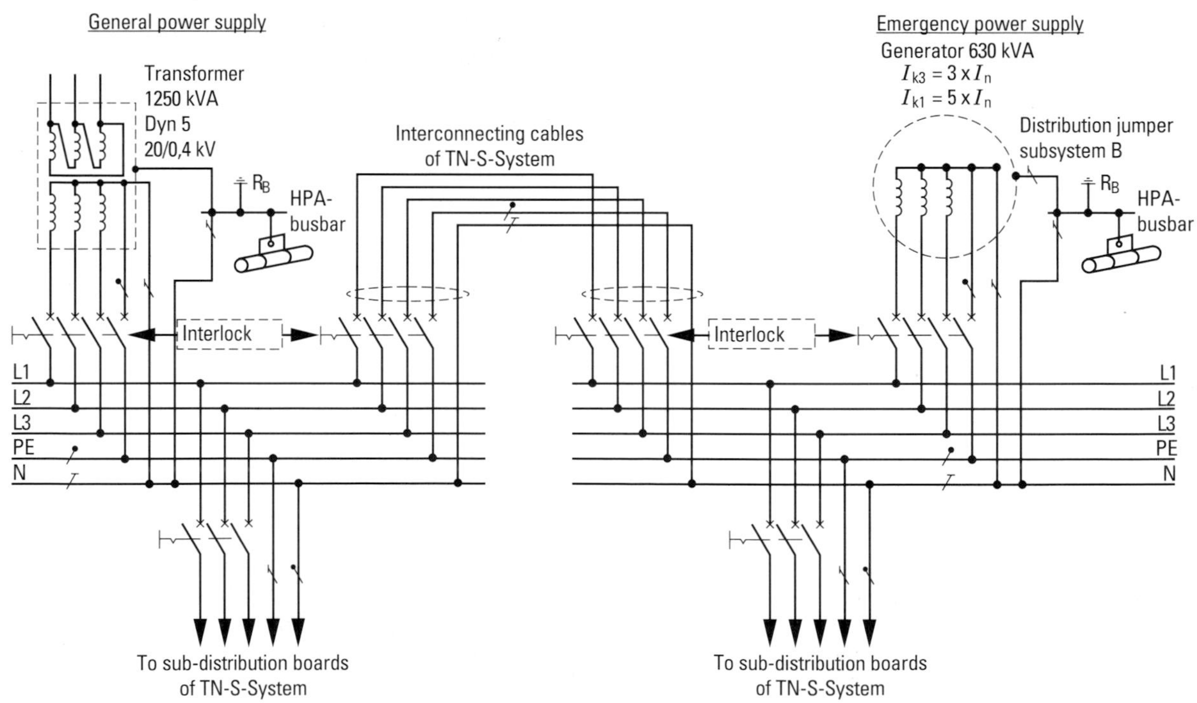

Fig. 1.1/13 TN-S system with several infeeds (distributed arrangement) with 4-phase transfer

Interruption time

In compliance with HD 384.3/IEC 60364-3/ DIN VDE 0100-300, power sources are classified according to interruption times:

Interruption	Interruption time
No interruption	0
Very short	$t \leq 0.15$ s
Short	0.15 s $\leq t \leq$ 0.5 s
Medium	0.5 s $\leq t \leq$ 15 s
Long	$t \geq 15$ s

Load transfer times of power sources

Experience has shown that the various power sources listed below have the following approximate load transfer times:

Power source	Load transfer time
UPS installation	0
Standard emergency generating set	$t \leq 15$ s
Uninterruptible power set (out of service/stand-by)	$t < 15$ s/$t < 1$ s
Second, independent system infeed with automatic transfer	$t < 1$ s

System disturbances of less than 0.5 s can be caused by short-circuit clearing, brief interruptions, or transfers in the public power supply system. They are, therefore, not regarded as system failures and are not included in fault statistics.

Generator infeed

Generating sets

Generating sets are categorized according to four operating modes:

1. Base-load generation
2. Peak-load generation
3. Supply of power for loads connected to general power supply according to user demand
4. Supply of power for important safety equipment

Unit-type district-heating power stations

The base load is predominantly supplied by unit-type district-heating power stations. Cogeneration of heat and power allows them to achieve a particularly high level of efficiency. In unit-type district-heating power stations, the generation of electrical power depends on the generation of heat. If unit-type district-heating power stations are also used as peak-load or emergency power-supply units (electricity generation independent of heat requirements), additional heat-exchangers with cooling towers or heat accumulators are required for cooling.

Unit-type district-heating power stations are operated in parallel to the general system infeed. They must, therefore, always be taken into consideration when calculating the maximum short-circuit current load (initial symmetrical short-circuit current and peak short-circuit current).

Operating modes

Standard emergency generating sets have three operating modes based on parallel operation with the general system infeed:

1. No parallel operation
2. Short-time parallel operation
3. Parallel operation over long periods of time

As with unit-type district-heating power stations, the emergency generating set component must be taken into consideration when calculating the maximum short-circuit current load if parallel operation with the general power supply is intended.

Equipment for parallel operation

The following equipment must be provided in addition to the standard generator protection devices:

▷ Automatic synchronizer for synchronization with the general power supply

▷ Disconnecting element for parallel operation faults, comprising:
- vector snap-acting relay
- overvoltage/undervoltage relay
- overfrequency/underfrequency relay

In large generating sets, sudden-power-change relays are used instead of vector snap-acting relays.

▷ Devices for active-power load and voltage balancing in parallel operation

▷ Power-factor correction often has to be provided for generators for parallel operation with low-capacity power systems.

Trial run under load

A trial run under load for this operating mode can be carried out at any time for the connected loads without first having to interrupt the voltage supply.

Short-time parallel operation

With short-time parallel operation, the generating set is only operated in parallel with the general power supply during the transition from the general power supply to the emergency power supply during a trial run, when the system switches back to the general power supply (after this has been resumed) or at the end of a trial run. The switchover takes place without the power supply to the connected loads being interrupted.

Dimensioning the generating sets

The generating set must be configured for 100% load transfer with short-time parallel operation, otherwise the load must be gradually shed or connected via load shedding or load connection devices. This will, however, mean an interruption in the voltage supply for at least some of the loads.

Since parallel operation with the general power-supply system only occurs momentarily, the permissible short-circuit current carrying capacity of the equipment may, according to the current standards, be overshot during the transfer phase. Additional provisions must, however, be made, e. g. remote operation, which ensure that personnel are not injured during the transfer phase in the event of a fault.

Equipment without parallel operation

If the generating set is not intended for parallel operation with the general power supply, the power supply will always be interrupted when the general power supply is switched over to the emergency power supply.

The automatic synchronizer and the additional protective devices required for disconnection are not necessary.

The general power-supply system, the emergency power supply, and the connection between the busbar sections may only be configured as a TN-S system with separate PE and N conductors (see Fig. 1.1/14) if the generator is not intended for parallel operation with the general power supply.

Four-phase switching devices (L1, L2, L3, and N) must be provided as transfer devices.

Short-circuit values

In all three operating modes, the lowest short-circuit current must be expected at the emergency power-supply busbar for generating sets in isolated operation.

If current-dependent, time-delay short-circuit protective devices such as fuses or circuit-breakers with short-time-delay releases are used, this is usually the lowest single-pole, two-pole, or three-pole sustained short-circuit current of the generator.

It determines the maximum usable rated currents and settings for the protective devices at the outgoing feeders of the emergency power-supply busbar. The three-pole sustained short-circuit current is generally three times the rated current, whereas the single-pole sustained short-circuit current in the most unfavorable phase can be expected to be approximately 4.5 to 5 times the rated current. The exact values must be obtained from the manufacturer of the emergency

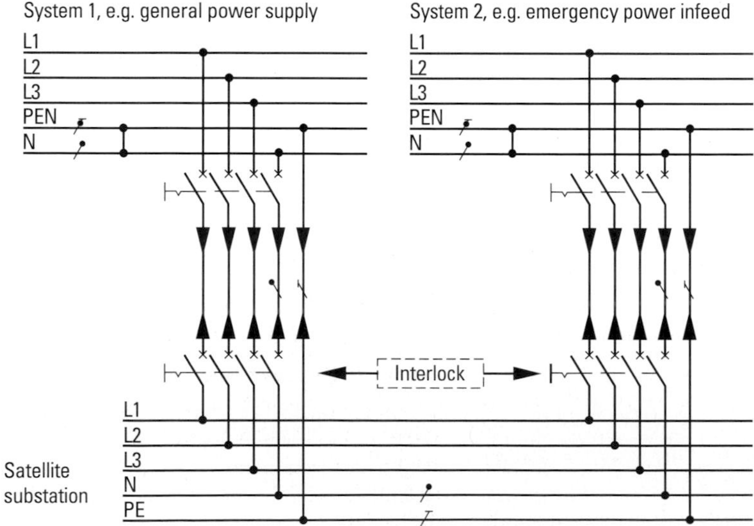

Fig. 1.1/14
Connection of satellite substations for TN-S system with several infeeds

power generating set prior to planning. According to information supplied by the manufacturer, values approximately 5 times greater than the rated current are also possible for the three-pole sustained short-circuit current. This results either in greater permissible rated currents and setting values for the protective devices at the outgoing feeders of the emergency power-supply busbar or frequently in a reduction of the required cable cross sections.

It is essential that:

- the entire generating set (generator and drive motor) supplies the required sustained short-circuit current for the necessary short-circuit clearance time (usually between 3 s and 5 s),
- the connection for parallel operation is only configured as a TN-C system with PEN conductors (see Fig. 1.1/12).

Parallel system operation

If generators are operated in parallel with transformers, 150 Hz circulating currents are produced between the generator and transformer neutral point. If the voltage of the generator is not sinusoidal (high third harmonic content), this can cause overloading, especially of the PEN conductor.

Neutral reactor

Neutral reactors must be installed along the PEN conductor connection of the generator to reduce the circulating currents.

Owing to the increased use of switched-mode, single-phase power-supply units for many loads, higher loading caused by *150 Hz harmonic currents* must be expected in the building power supply. With emergency power-supply operation, the neutral reactor must, therefore, be frequently jumpered to prevent supply voltage asymmetry and neutral reactor overloading.

Uninterruptible power system (UPS)

The *load transfer time* of standard emergency power generating sets ranges between 10 s and 15 s.

Uninterruptible power set

If uninterruptible power sets with energy storage mechanisms, e. g. flywheels, are used, the load transfer time in active standby operation drops to below 1 s.

Uninterruptible power system (UPS)

If the associated interruptions in power supplied to important loads cause injury or damage, or if considerable financial losses are to be expected for the user, e. g. loss of production, these loads must be connected to an uninterruptible power system (UPS). Loads should also be connected to a UPS if the permissible voltage and frequency tolerances are lower than the standard tolerances of the general power-supply system, e. g. the power supply for EDP and communications systems.

A distinction is made between static and rotary (or dynamic) UPS installations.

Rotary UPS installations

Rotary UPS installations generally provide a higher short-circuit current in the time range >0 ms to approximately 20 ms.

Short-circuit current values 7 to 10 times greater than the rated current are possible depending on the manufacturer and design of the installation.

Static UPS installations

A static UPS installation is represented in the form of a block diagram in Fig. 1.1/15.

General requirements

The general requirements for uninterruptible power systems for use outside electrical operating areas are specified in EN 50091-1/VDE 0558 Part 511. According to this standard, protection against overcurrent, short circuits, and ground faults must be provided either in the UPS installation or at the load. The root-mean-square values of the fault current possible under the least favorable conditions must be specified by the manufacturer so that protection and transmission equipment (cables or busbars) can be correctly dimensioned.

Inverter operation

With static UPS installations, the following short-circuit currents must be expected for inverter operation without a connected static bypass switch:

$2.1 \cdot I_n$	three-pole	for $t \leq 0.02$ s
$3.0 \cdot I_n$	single-pole	for $t \leq 0.02$ s

Fault currents

These fault currents are limited to

approximately $1.5 \cdot I_n$ for $t \leq 5$ s

and approximately $1.35 \cdot I_n$ for $t \leq 30$ s

to protect the power electronics (overrange).

Dimensioning cables and conductors

The minimum fault currents at the end of an electric circuit are crucial for dimensioning cables and conductors in compliance with HD 384.4.41/IEC 60364-4-41/DIN VDE 0100-410 “Protection against electric shock” and HD 384.4.42/DIN VDE 0100-420 “Protection against thermal effects”. With system-commutated inverter operation where

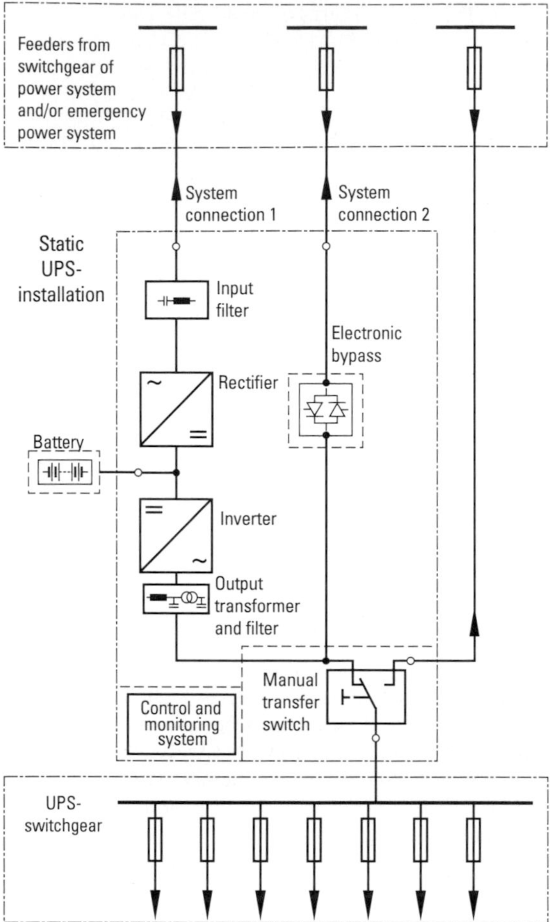

Fig. 1.1/15
Block diagram of a static UPS installation

the static bypass switch (SBS) can be connected, the largest usable fuse to be selected depends on:

- selective characteristics for protection in the SBS (factor 1.6 fuses),
- the maximum permissible clearance time for faults without causing interference to other loads.

Protective devices for EDP systems

If the UPS installation is used for supplying EDP systems, protective devices must be selected which clear faults in less than 10 ms. Currents which are more than 25 times the rated fuse current are required for fuses with a gL characteristic (upper tolerance band).

If clearance times of less than 0.02 s and 5 s are also required for self-commutated inverter operation, the limited UPS installation fault currents for 0.02 s and 5 s must be taken as a basis when dimensioning cables and conductors. This, however, presupposes the failure of the power-supply system or that the voltage and frequency tolerances exceed the permissible limits (SBS disabled). Table 1.1/6 specifies the minimum required rated UPS currents for clearance times of less than 0.02 s and 5 s for three fuse sizes with gL characteristic (upper tolerance band).

Selection of system configuration for distribution system

The following two configurations for the distribution system are possible with both central (see Fig. 1.1/10) and distributed (see Fig. 1.1/11) arrangement of infeeds and l. v. main distribution boards for the general and emergency power supply:

▷ Radial feeder system with/without transfer connection
▷ Closed meshed system

Closed meshed system

Closed meshed systems with TN-C system (L1, L2, L3, and PEN) are nowadays only used for industrial applications. The associated building currents and the more intense electromagnetic fields around the power routes are usually acceptable because the devices used in the production area have a greater interference immunity. The closed meshed system provides a *higher level of supply reliability and supply quality* for the loads right down to the sub-distribution board level. It also permits greater flexibility with regard to changing operating and load conditions. The increasing use of automation and communications equipment in production areas means that low-frequency operating currents must be reduced via the building structure and bilaterally-shielded communications lines in industrial systems as well. This problem can only be solved by using

Table 1.1/6
Minimum required rated currents and fault currents of UPS installations with clearance via fuses of utilization category "gL"

Rated current of outgoing fuse	Required fault currents for t_a		Rated UPS current where $1.5 \cdot I_n\, t < 5$ s / $3 \cdot I_n\, t < 0.02$ s	
	0.02 s	5 s	for $t_a < 0.02$ s	for $t_a < 5$ s
25 A	> 500 A	> 120 A	166 A	80 A
63 A	> 1300 A	> 340 A	430 A	230 A
100 A	> 2500 A	> 560 A	830 A	370 A

optical fibers for communication cables or by changing over to radial feeder configurations with TN-S systems.

Radial feeder system

As already mentioned in Section 1.1.2, radial feeder systems are nowadays preferred for the l. v. distribution system, supply conductors to the sub-distribution boards, and branch circuits in building systems due to the low-frequency electromagnetic interference.

If reserve infeeds or infeeds from a central emergency power-supply system are required for the sub-distribution boards in order to ensure supply reliability, they should be configured as TN-S systems with separate PE and N conductors. HD 384.5.54 and IEC 60364-5-54 specify four-phase switching devices for the transfer equipment (Fig. 1.1/14).

Dimensioning criteria

Cable runs between l. v. main distribution and sub-distribution boards and the protective device must be dimensioned in accordance with the following criteria:

▷ Overload protection
▷ Short-circuit protection
▷ Voltage drop (static/dynamic)
▷ Protection against electric shock in the event of indirect contact
▷ Selective behavior of upstream and downstream protective devices

If safety devices are supplied with power, verification of the selectivity of the protective devices must be provided. Building users are also expressing a need for selective protective devices for the general power supply as well. Selective protective devices considerably reduce the total number of loads affected by the fault and also allow the fault to be located very quickly.

Fuseless technology

Both fuses and circuit-breakers are used as protective devices. A distinction is made between the following two circuit-breaker types for fuseless technology with circuit-breakers:
▷ Circuit-breakers with current-zero cut-off
▷ Current-limiting circuit-breakers

Circuit-breakers with current-zero cut-off

Time-delay short-circuit releases are used to ensure the selective behavior of circuit-breakers with current-zero cut-off.

The following types of release are used:

▷ Overload releases (current-dependent characteristic)
▷ Time-delay short-circuit releases
▷ Instantaneous releases

Two types are currently available for time-delay short-circuit releases:

- Releases with defined release time (UMZ)
- Releases with current-dependent time delay (AMZ), e. g. $I^2 \cdot t$ releases

The selective behavior of the current-dependent time-delay release characteristics (e. g. for $I^2 \cdot t$ release) can be adapted more effectively to the release characteristics of fuses.

The delay times required from the load to the infeed are becoming longer for circuit-breakers with defined release time delay. This has a negative effect on the short-circuit energy released at the fault location since the short-circuit currents in the vicinity of the infeed will be higher than at the sub-distribution boards. If circuit-breakers with time-reduced selectivity control are used, it is also possible to clear faults in the vicinity of the infeeds with short delay times.

Current-limiting circuit-breakers

If current-limiting circuit-breakers with current limiting are used (rated short-circuit breaking capacity ≤ 100 kA for circuit-breakers with a low rated current I_n), selective behavior with respect to downstream circuit-breakers is only possible if the selectivity limits (current selectivity) are utilized. Manufacturers only specify the selectivity limits between an upstream and downstream circuit-breaker for their own products. If the selectivity limits are used, it is, therefore, essential that all the circuit-breakers used in a power-supply concept are from the same manufacturer.

1.2 Calculating short-circuit currents in three-phase systems

1.2.1 Introduction

The relevant standards (EN/IEC/DIN VDE) apply when dimensioning and selecting the electrical equipment to be used in switchgear and supply systems. These standards specify that the loads and effects resulting from short circuits must be taken into consideration in addition to measuring the continuous stress caused by the operating current and operating voltage, for example, under normal operating conditions. Since the short-circuit currents are usually many times greater than the rated currents, high dynamic and thermal stress and, in certain cases, unacceptable voltages which are liable to cause danger must be expected. Since these could destroy equipment and endanger personnel, it is essential, for safety reasons, to evaluate the loads to be expected in the event of a short circuit. The short-circuit current values must also be known. For this reason, the "Verband Deutscher Elektrotechniker" (VDE) (German association of electrical engineers) has published the following specification for calculating significant short-circuit currents:

Standards

DIN VDE 0102 – Short-circuit current calculation in three-phase a.c. systems.

The international standard IEC 60909 "Short-circuit calculation in three-phase a.c. systems" has been incorporated in the German DIN VDE specifications.

The terms listed in the following section are, in certain cases, used and explained in these standards.

It is particularly important to note that the person performing the calculations must always have access to these standards since they can only be explained here by means of examples and not reproduced in full. A number of important diagrams, which are directly associated with the numerical calculations, are shown on pages 71 to 75.

The values for the d.c. resistance per unit length (r') and inductance per unit length (l') for cables can be found in the Siemens publication "Power Cables and their Application".

Explanation of terms [1)]

Short-circuit current is the current that flows through an electrical system when a short circuit occurs. A distinction must be made between the short-circuit current at the fault location and the transferred short-circuit currents in the system branches.

Prospective short-circuit current is the current which would occur if the short circuit was replaced by an ideal connection with negligible impedance without the infeed being changed.

Initial symmetrical short-circuit current (I''_k) is the r.m.s. value of the a.c. component of a prospective short-circuit current at the instant at which the short circuit occurs if the impedance remains unchanged at "zero".

Maximum asymmetrical short-circuit current (i_p) is the maximum possible instantaneous value of the prospective short-circuit current.

Symmetrical breaking current (I_b) is the r.m.s. value, i.e. the average value for a cycle of the a.c. component of the prospective short-circuit current which occurs at the instant of contact separation at the extinguishing pole of a switching device.

Sustained short-circuit current (I_k) is the r.m.s. value of the short-circuit current after all transient reactions have decayed.

Rated system voltage (U_n) is the voltage between the conductors which is specified for a system and to which certain operating characteristics refer.

Initial symmetrical short-circuit power (S''_k) is a fictitious value which is the product of the initial symmetrical short-circuit current (I''_k), the rated system voltage (U_n), and the factor $\sqrt{3}$: $S''_k = \sqrt{3} \cdot U_n \cdot I''_k$.

Minimum switching delay (t_{min}) is the shortest time between the start of the short-circuit current and contact separation at the opening pole of a switching device.

Short circuit close to generator terminals occurs if at least one synchronous machine supplies a prospective initial symmetrical short-circuit current which is more than twice the rated current of the machine, or if a short circuit occurs and asynchronous mo-

[1)] See also the explanation in Section 7.2.
The explanation of terms is included in the draft of the forthcoming standard DIN VDE 0102-100.

tors account for more than 5% of the initial symmetrical short-circuit current (I_k'') not including motors.

Short circuit remote from generator terminals occurs if the a. c. component of the prospective short-circuit current remains more or less constant.

Short-circuit positive-sequence impedance ($\underline{Z}_{(1)}$) of a three-phase system is the impedance in the positive-sequence system relative to the short-circuit location.

Short-circuit negative-sequence impedance ($\underline{Z}_{(2)}$) of a three-phase system is the impedance of the negative-sequence system relative to the short-circuit location.

Short-circuit zero-sequence impedance ($\underline{Z}_{(0)}$) of a three-phase system is the impedance of the zero-sequence system relative to the short-circuit location. It is three times the impedance $\underline{Z}_N$ between the neutral point and ground.

Dynamic stress – Effect of electromechanical forces.

Thermal stress – Effect of Joulean heat.

Hazardous voltages – Pace and touch voltages, interference voltages.

Ground-fault current extinction – Systems with ground-fault neutralizers (Petersen coils).

Direct grounding – Transformer neutrals with direct grounding (also "effective grounding").

Ground-fault current – Short-circuit current or short-circuit current component which flows back into the system via ground.

Emergency voltage source ($c \cdot U_n/\sqrt{3}$) is the voltage from an ideal source which is fed in at the short-circuit location in the positive-sequence system and is used to calculate the short-circuit current. This is the only effective voltage in the system.

Symbols used

A	Initial value of aperiodic component $i_{d.c.}$
I_b	Symmetrical breaking current
I_k''	Initial symmetrical short-circuit current
I_k	Sustained short-circuit current
I_r (I_n)	Rated current
i_p	Maximum asymmetrical short-circuit current
$i_{d.c.}$	Decaying aperiodic component of short-circuit current
K_G	Correction factor for generator impedance
R and r	Resistance, absolute and specific value
S_k''	Initial symmetrical short-circuit power (apparent power)
S_r	Rated apparent power of electrical device
U_n	Rated system voltage, conductor–conductor (r.m.s. value)
U_r	Rated voltage, conductor–conductor (r.m.s. value)
U_{rHV}	Rated voltage of h. v. side of transformer
U_{rLV}	Rated voltage of l. v. side of transformer
X and x	Reactance, absolute and specific value
X_{rD}	Reactance of reactor
X_d'' and x_g''	Direct-axis and quadrature-axis subtransient reactance of a synchronous machine (saturated value)
Z and z	Impedance, absolute and specific value
Z_k	Short-circuit impedance of a three-phase system
c	Voltage factor to IEC 909/DIN VDE 0102
i_0	No-load current (instantaneous value)
l	Length of conductor
q	Factor for calculating symmetrical breaking current of asynchronous motors
r'	Linear resistance (resistance per unit length)
t_{min}	Minimum switching delay
u_{kr}	Rated short-circuit voltage of transformer (%)
u_{Rr}	Rated effective component of short-circuit voltage of transformer (%)
u_{Xr}	Rated inductive reactive component of short-circuit voltage of transformer (%)
x'	Linear reactance (inductive reactance per unit length)

x''_d	Subtransient reactance of synchronous machine, specific value
κ	Factor for calculating maximum asymmetrical short-circuit current
λ	Factor for calculating sustained short-circuit current
μ	Factor for calculating symmetrical breaking current

Other indices

G	Generator
M	Motor
Ka	Cable (Ka1=Cable 1 etc.)
Q	System infeed
T	Transformer (T1=Transformer 1 etc.)
$\alpha, \beta, \gamma \ldots$	Equivalent impedances for series and parallel connections
GK	Impedance correction for generators
(1)	Positive-sequence system
(2)	Negative-sequence system
(0)	Zero-sequence system
max	Maximum value
min	Minimum value
k, k3	Three-phase short circuit
k1	Phase-to-ground fault
k2	Phase-to-phase ungrounded fault

Examples

$\underline{Z}_{(1)\alpha}$	Equivalent positive-sequence impedance
$I''_{kG3\,max}$	Generator contribution to maximum initial symmetrical short-circuit current with three-phase short circuit
$\underline{Z}_{(0)T}$	Zero-sequence impedance of transformer T
$\mu_{0.05}$	Factor μ to IEC 909/DIN VDE 0102 with minimum switching delay $t_{min} = 0.05$ s

Types of fault

According to the standards IEC 60909/DIN VDE 0102, four types of fault can occur in three-phase installations. These faults and their associated short-circuit currents are illustrated in Fig. 1.2/1. In addition to these types of fault, other double faults can occur which only need to be taken into consideration in special cases, however, when dimensioning and selecting the electrical equipment. These faults are not dealt with here.

Symmetrical faults
Three-phase short circuit

The types of fault illustrated in Fig. 1.2/1 can be divided into symmetrical and asymmetrical faults. The three-phase short circuit is the least complex and easiest of these fault types to calculate. The three voltages are zero at the fault location and all three conductors are loaded symmetrically by the symmetrical short-circuit currents. Neither ground nor a grounding electrode conductor are involved in current transfer. No other short-circuit currents would be produced even if the neutral of the system in which the three-phase short circuit occurs were connected to ground. The short-circuit currents can, therefore, only be calculated for one conductor (as is the case when calculating symmetrical load relationships). The equations and the calculation to be applied are illustrated using examples as of page 49. Statistically speaking, three-phase short circuits represent a relatively small proportion of the faults that occur. It should, however, be noted that the highest short-circuit currents usually occur at a given fault location in a three-phase short circuit and that these values are, therefore, decisive when dimensioning the electrical equipment.

Asymmetrical faults

The electrical relationships of asymmetrical faults are, at first sight, much more complicated, particularly if the ground connection has to be taken into consideration. In such cases, not all voltages at the fault locations are zero. Furthermore, owing to the asymmetry, greater or lesser degrees of coupling occur between the phase conductors and between the phase conductors and ground, together with the grounding electrode conductors which are often present. The equations required for determining the asymmetrical short-circuit currents can be formulated by means of special mathematical methods; the symmetrical component method is normally used on account of its clarity. A detailed knowledge of this method is not, however, necessary if the equations and explanations in the standards IEC 60909/DIN VDE 0102 are used. In exceptional cases, anyone not familiar with the method is advised to seek the advice of an expert since it is not possible to

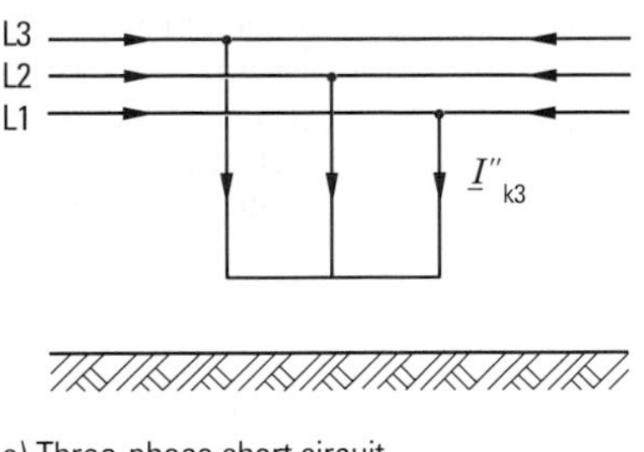

a) Three-phase short circuit
(symmetrical fault)

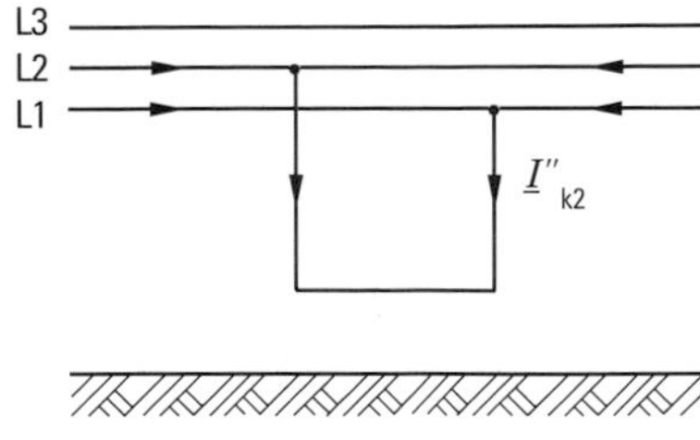

b) Phase-to-phase
ungrounded fault

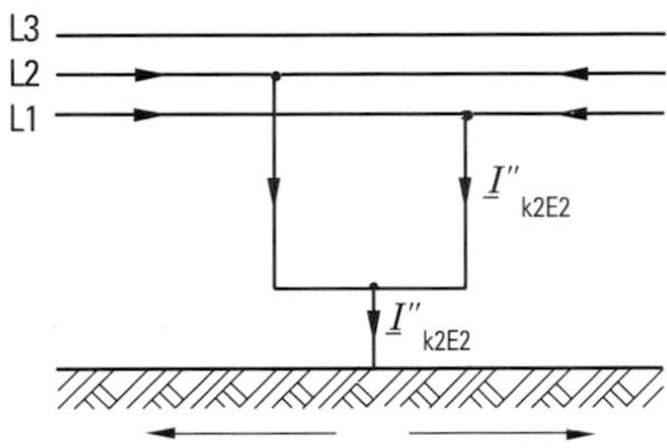

c) Phase-to-phase
grounded fault

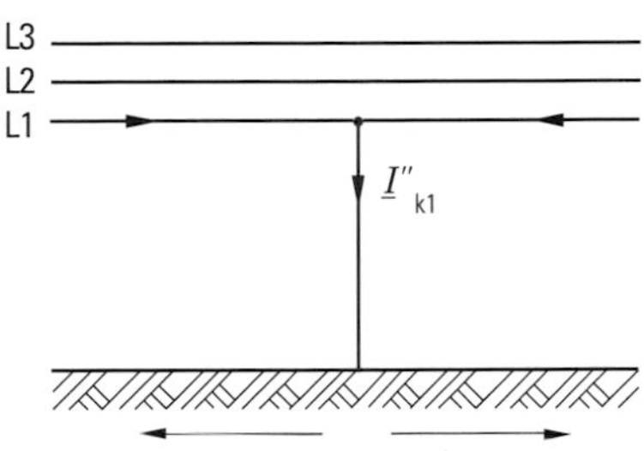

d) Phase-to-ground fault

Transferred short-circuit currents in conductors and ground return

Short-circuit current

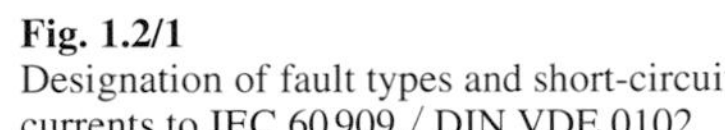

Fig. 1.2/1
Designation of fault types and short-circuit currents to IEC 60909 / DIN VDE 0102

Transferred short-circuit currents can also flow through conductors which are free of defects. The directions of the current arrows have been randomly selected.

include all solutions in the standards. The significance of the asymmetrical faults with regard to dimensioning and selecting equipment is only dealt with briefly here.

Phase-to-phase ungrounded fault

In the case of a phase-to-phase ungrounded fault, initial symmetrical short-circuit currents, which are lower than those resulting from a three-phase short circuit, occur at the fault location. If, however, the fault occurs in the vicinity of synchronous machines and/or asynchronous machines with comparable power ratings, the phase-to-phase fault current may, in certain cases, be higher than the three-phase short-circuit current in the latter stages of the fault. Under these conditions, the phase-to-phase fault current may be a determining factor when dimensioning the switching devices with regard to the necessary symmetrical breaking current or when selecting the protective devices.

Phase-to-phase grounded fault

Double ground fault

Similar relationships may result when phase-to-phase grounded faults occur. With a phase-to-phase grounded fault, the initial symmetrical short-circuit current lies between the values associated with three-phase short circuits and phase-to-ground faults. The double ground fault is only important in systems with ground-fault current extinction, or with isolated neutral, for checking protection and interference-related aspects since the resulting short-circuit current cannot be greater than that caused by a phase-to-phase ungrounded or grounded fault. Double ground faults are not usually possible in l. v. systems (solid grounding).

Phase-to-ground fault

The phase-to-ground fault is the most important of the asymmetrical faults. This type of fault is not only the most frequently encountered fault in h. v. systems with effective neutral grounding and in l. v. systems with directly grounded neutrals, but also exhibits the largest spread of current values. In exceptional cases, the phase-to-ground fault current may even exceed the three-phase short-circuit current. Example 2 from page 57 onwards illustrates such a case. Whereas the three-phase short-circuit current plays, in most cases, a decisive role when dimensioning the equipment in the normal operating circuit, the phase-to-ground fault current is of particular importance when determining the pace and touch voltages as well as in matters concerning interference and when dimensioning grounding systems.

Short-circuit current sources and short-circuit current waveforms

Depending on the short-circuit location, the following short-circuit current sources must be taken into consideration when calculating the short-circuit currents:

Short-circuit current sources

- ▷ system infeed (external infeed, emergency power generator);
- ▷ synchronous machines (generators, motors);
- ▷ asynchronous machines (generators, motors);
- ▷ converter-fed d. c. drives with temporary inverter operation.

Short-circuit current waveforms

The time characteristic of short-circuit currents depends on the point in time at which the short circuit occurs. The oscillograms are produced when the voltage is zero following the positive half-wave.

The waveform of the short-circuit current at the fault location largely depends on the associated system infeed or infeeds.

Fig. 1.2/2 shows the short-circuit current waveform for a three-phase short circuit re-

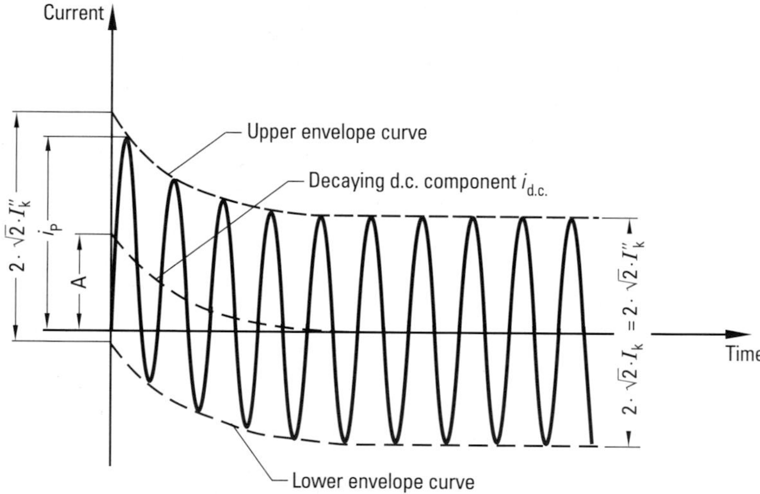

I''_k = Initial symmetrical short-circuit current

i_p = Maximum asymmetrical short-circuit current

I_k = Sustained short-circuit current

$i_{d.c.}$ = Decaying d.c. component of short-circuit current

A = Initial value of d.c. component $i_{d.c.}$

Fig. 1.2/2
Waveform of short-circuit current for three-phase short circuit remote from generator terminals with constant a.c. component (schematic characteristic)

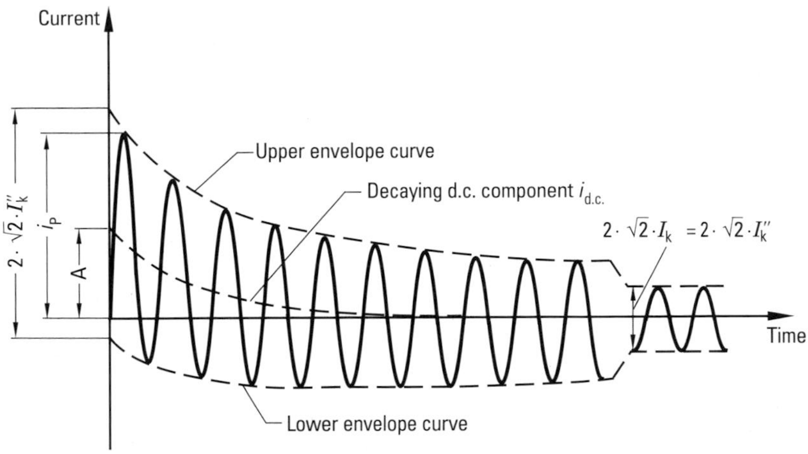

I_k'' = Initial symmetrical short-circuit current

i_p = Maximum asymmetrical short-circuit current

I_k = Sustained short-circuit current

$i_{d.c.}$ = Decaying d.c. component of short-circuit current

A = Initial value of d.c. component $i_{d.c.}$

Fig. 1.2/3
Waveform of short-circuit current for short circuit close to generator terminals with decaying a.c. component (schematic characteristic)

mote from the generator terminals occurring in the phase conductor at the least favorable switching instant for a three-phase short circuit. This waveform should be expected for short circuits which are supplied via system infeeds. The symmetrical short-circuit current does not vary with time (condition: $I_k'' = I_k$).

Fig. 1.2/3 shows the short-circuit current waveform for a synchronous generator occurring in the phase conductor at the least favorable switching instant for a three-phase short circuit. This short circuit is classified as being close to the generator terminals since the symmetrical short-circuit current decays (over time) from the initial value I_k'' to the sustained short-circuit current I_k.

If the short circuit is cleared before the sustained short-circuit current is reached, the symmetrical breaking current I_b is already lower than the initial value I_k'' (condition: $I_k'' > I_b \geq I_k$).

Fig. 1.2/4 shows the waveform for the three-phase short-circuit current of a l. v. asynchronous motor occurring in the phase conductor at the least favorable switching instant for a three-phase short circuit. This short circuit is close to the generator terminals and the symmetrical short-circuit current decays from the initial value I_k'' to zero within just a few cycles (condition: $I_k'' \gg I_b$; $I_k = 0$).

The short-circuit currents shown in Fig. 1.2/2 to 1.2/4 have one feature in common: in approximately 10 ms (one half-cycle at 50 Hz), they reach their maximum possible value, the maximum asymmetrical short-circuit current (i_p). The maximum asymmetrical short-circuit current is the determining factor in evaluating the dynamic stress in the event of a fault.

The significance of the other currents is considered in the calculation examples. These examples also show that not only one but a combination of several of the above-mentioned short-circuit current sources is involved. The short-circuit current at the fault location then consists of these transferred short-circuit currents. If the short circuits do not occur directly at the terminals of the synchronous or asynchronous machine, the decay process will not be as pronounced as depicted in Figs. 1.2/3 and 1.2/4.

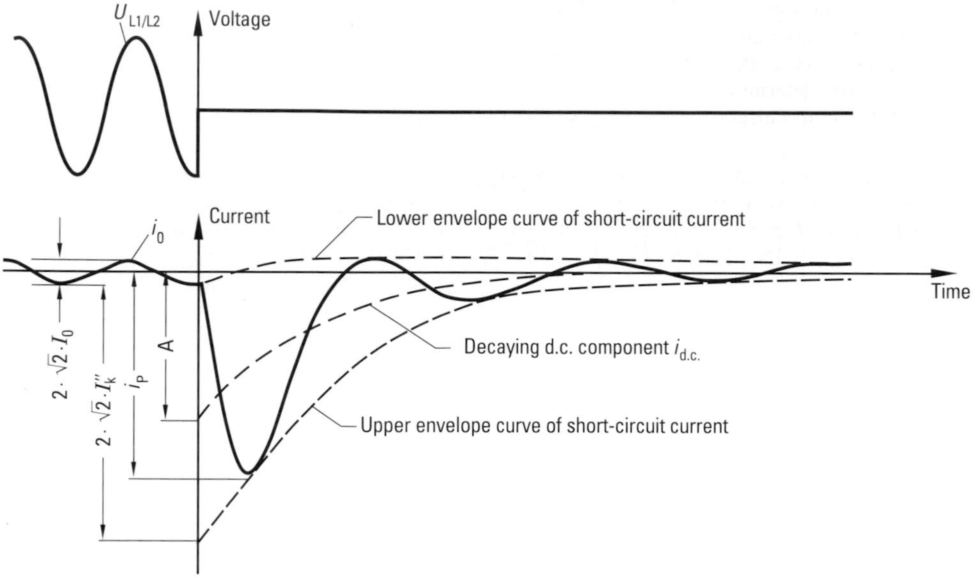

Fig. 1.2/4
Waveform of the short-circuit current of an l.v. asynchronous motor occurring in the phase conductor at the least favorable switching instant; i_0 no-load current (oscillogram for three-phase short circuit between terminals for no-load operation)

General factors to be considered when determining short-circuit currents

According to the relevant standards, maximum and minimum short-circuit currents must be calculated for the types of fault named on page 44 when dimensioning and selecting equipment.

If the circuit state of the system changes during the short circuit, additional calculations will be necessary.

Additional calculations are, however, not required in the majority of practical cases. Depending on the application, the r.m.s. value of the a.c. component and the peak value (i_p) of the short-circuit current are more significant. The highest maximum asymmetrical short-circuit current (i_p) depends on the time constant of the decaying d.c. component and the frequency (f), i.e. on the ratio R/X or X/R of the short-circuit impedance ($\underline{Z}_k$), and is reached if the short circuit occurs at voltage zero. i_p also depends on the decay of the a.c. component of the short-circuit current.

In meshed systems, a number of aperiodic time constants occur. It is, therefore, not possible to provide a simple method for calculating i_p and $i_{d.c.}$. Special methods for calculating i_p to a sufficient degree of accuracy are indicated in IEC 60909/DIN VDE 0102.

Dynamic stress

The three-phase short circuit is generally the most important in three-phase systems with regard to the making and breaking loads of the switching devices and the dynamic load placed on the equipment as it produces the highest short-circuit current (I_k'', i_p).

Thermal stress

The three-phase short circuit is important with respect to thermal stress of the equipment in the case of short circuits remote from the generator terminals. With short circuits close to the generator terminals, the thermal stress caused by the phase-to-phase fault current may be higher than that caused by the three-phase short circuit owing to the higher sustained short-circuit currents. The phase-to-ground fault current must also be taken into consideration for the thermal stress in equipment in systems with effective neutral grounding, if the sustained phase-to-ground fault current exceeds that of a three-phase short circuit.

Upstream fuses

If fuses are used for system protection, the short-circuit currents must initially be calculated as if the fuses were not present. The fuse let-through currents, which then represent the maximum asymmetrical short-circuit currents of the downstream equipment, must be determined from the calculated short-circuit currents and fuse characteristics.

Information on taking asynchronous motors into account when determining the short-circuit currents can be found in the standards IEC 60909/DIN VDE 0102.

Computer programs

As can be seen from the following examples, calculating the short-circuit currents with the aid of a pocket calculator is relatively complicated even for very simple radial systems. Calculations for more extensive and meshed systems are only possible using a computer.

For this reason, computer programs, such as SINCAL and NETOMAC, are used at Siemens depending on the type of fault and task to be undertaken.

The program *KUBS plus* (for short-circuit current calculation, back-up protection, and selectivity) can also be used for simple l.v. radial systems. L.v. three-phase a.c. motors are not taken into consideration here.

1.2.2 Calculation examples

Calculating short-circuit currents will now be illustrated using two examples. The method of calculation is that specified in the standards IEC 60909/DIN VDE 0102.

Diagrams, tables, and a number of equations, which provide information on the most important characteristic equipment data necessary for calculating the short-circuit currents, can be found on page 69 onwards. The values specified are mostly average values. In real applications, calculations must be based on the actual values for the equipment (provided they are known). If the data for calculating a new system is incomplete, the specified average values can be applied. Care should then be taken to ensure that the assumed values are within the narrowest possible tolerance range.

Example 1:

Low-voltage system with motor

The assumed system circuit with the characteristic data of the equipment is shown in Fig. 1.2/5.

The l.v. system is fed via a 400 kVA transformer from a 10 kV system. A number of sub-distribution boards are supplied from the 400 V busbar via four-core cables. The l.v. system is effectively grounded.

The following short-circuit currents must be calculated for the indicated fault locations F1. F2, and F3:

F1: $I''_{k3\,max}$, $i_{p3\,max}$, $I_{b3\,max}$

F2: $I''_{k2\,min}$

F3: $I''_{k1\,min}$

▷ ***Three-phase short-circuit currents***
Highest three-phase short-circuit currents

Fault location F1

Impedances of equipment and short-circuit current path:

System impedance:

$$Z_{(1)Q} = \frac{c \cdot U_n^2}{S''_{kQ\,max}} = \frac{1.1 \cdot 10^2\ \text{kV}^2}{150\ \text{MVA}} = 0.733\ \Omega.$$

Where $R_{(1)Q} = 0.1 \cdot X_{(1)Q}$:

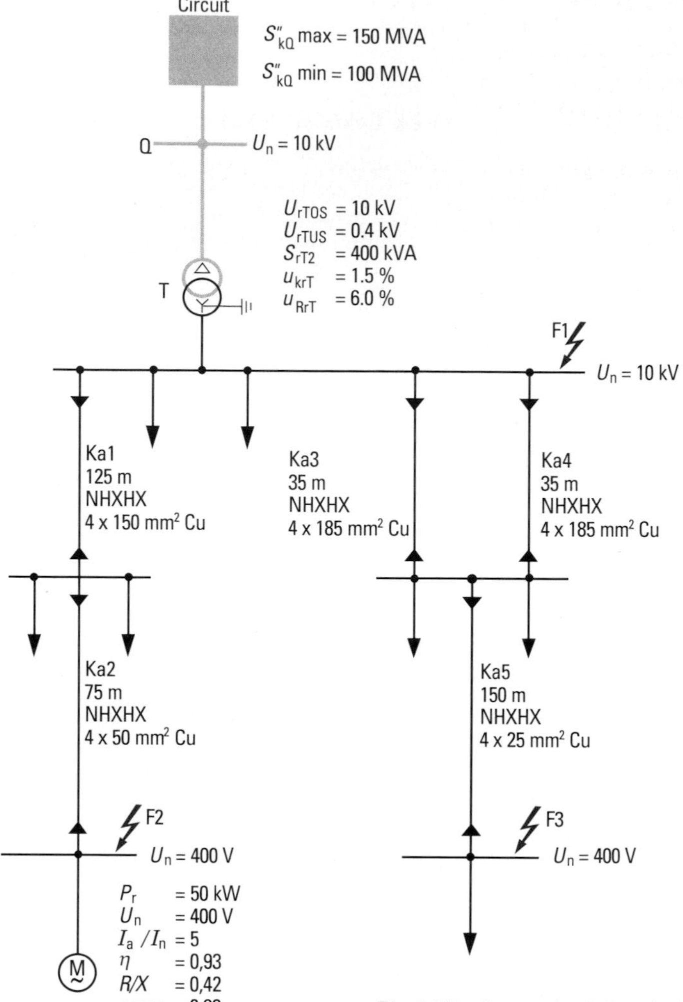

Fig. 1.2/5 System circuit for calculation example 1

$$X_{(1)\mathrm{Q}} = \frac{Z_{(1)\mathrm{Q}}}{1.005} = \frac{0.733\ \Omega}{1.005} = 0.729\ \Omega,$$

$$\underline{Z}_{(1)\mathrm{Q}} = (0.0729 + \mathrm{j}0.729)\ \Omega.$$

Impedance of transformator:

$$U_{\mathrm{XrT}} = \sqrt{u_{\mathrm{krT}}^2 - u_{\mathrm{RrT}}^2} = \sqrt{6^2 - 1.5^2}\% = 5.81\%,$$

$$R_{(1)\mathrm{T}} = \frac{u_{\mathrm{RrT}} \cdot U_{\mathrm{rTUS}}^2}{S_{\mathrm{rT}}} = \frac{1.5\% \cdot 0.4^2\ \mathrm{kV}^2}{100\% \cdot 0.4\ \mathrm{MVA}} = 0.006\ \Omega,$$

$$X_{(1)\mathrm{T}} = \frac{u_{\mathrm{xrT}} \cdot U_{\mathrm{rTUS}}^2}{S_{\mathrm{rT}}} = \frac{5.81\% \cdot 0.4^2\ \mathrm{kV}^2}{100\% \cdot 0.4\ \mathrm{MVA}} = 0.02324\ \Omega,$$

$$\underline{Z}_{(1)\mathrm{T}} = (0.006 + \mathrm{j}0.02324)\ \Omega$$

Impedance of system plus transformer:

$$\underline{Z}_{(1)\alpha} = \underline{Z}_{(1)Q}\left(\frac{U_{rTUS}}{U_{rTOS}}\right)^2 + \underline{Z}_{(1)T},$$

$$\underline{Z}_{(1)\alpha} = (0.0729 + j0.729)\,\Omega\left(\frac{0.4\,\text{kV}}{10\,\text{kV}}\right)^2 + (0.006 + j0.02324)\,\Omega,$$

$$\underline{Z}_{(1)\alpha} = (0.00612 + j0.0244)\,\Omega.$$

Cable impedances Ka1 and Ka2:

$$\underline{Z}_{(1)Ka} = l(r' + jx'),$$

$$\text{Ka1}: \; l = 0.125\,\text{km},\, r' = 0.124\,\Omega/\text{km},\, x' = 0.0804\,\Omega/\text{km},$$

$$\text{Ka2}: \; l = 0.075\,\text{km},\, r' = 0.387\,\Omega/\text{km},\, x' = 0.0848\,\Omega/\text{km},$$

$$\underline{Z}_{(1)Ka1} = l(r' + jx') = 0.125\,\text{km}\,(0.124 + j0.0804)\,\Omega/\text{km} = (0.0155 + j0.0101)\,\Omega,$$

$$\underline{Z}_{(1)Ka2} = l(r' + jx') = 0.075\,\text{km}\,(0.387 + j0.0848)\,\Omega/\text{km} = (0.029 + j0.0064)\,\Omega.$$

The specific cable impedances were taken from the Siemens publication "Power Cables and their Application" (see Section 7.3).

Impedance of motor:

$$S_{rM} = \frac{P_{rM}}{\eta_r \cos\varphi_r},$$

$$Z_{(1)M} = \frac{U_{rM}^2}{\left(\frac{I_a}{I_r}\right)\cdot S_{rM}} = \frac{0.4^2\,\text{kV}^2}{5\cdot\frac{0.05}{0.93\cdot 0.83}\,\text{MVA}} = 0.494\,\Omega,$$

$$\underline{Z}_{(1)M} = Z_{(1)M}\cdot\left(\frac{\frac{R}{X}}{\sqrt{1+\left(\frac{R}{X}\right)^2}} + j\frac{1}{\sqrt{1+\left(\frac{R}{X}\right)^2}}\right),$$

$$\underline{Z}_{(1)M} = 0.494\,\Omega\left(\frac{0.42}{\sqrt{1+0.42^2}} + j\frac{1}{\sqrt{1+0.42^2}}\right) = (0.191 + j0.455)\,\Omega.$$

Sum of impedances for cables Ka1, Ka2, and motor:

$$\underline{Z}_{(1)\beta} = \underline{Z}_{(1)Ka1} + \underline{Z}_{(1)Ka2} + \underline{Z}_{(1)M},$$

$$\underline{Z}_{(1)\beta} = (0.0155 + j0.0101)\,\Omega + (0.029 + j0.0064)\,\Omega + (0.191 + j0.455)\,\Omega,$$

$$= (0.2355 + j0.4715)\,\Omega,$$

$$Z_{(1)\beta} = \sqrt{0.2355^2 + 0.4715^2}\,\Omega = 0.527\,\Omega.$$

Initial symmetrical short-circuit current $I''_{k3\,max}$:

System component via transformer:

$$I''_{kN3\,max} = \frac{c\cdot U_n}{\sqrt{3}\cdot\underline{Z}_{(1)\alpha}} = \frac{1.0\cdot 0.4\,\text{kV}}{\sqrt{3}\cdot(0.00612 + j0.0244)} = (2.23 - j8.91)\,\text{kA},$$

$$I''_{kN3\,max} = \sqrt{2.23^2 + 8.91^2}\,\text{kA} = 9.18\,\text{kA}.$$

Motor component via cables Ka1 and Ka2:

$$I''_{kM3\,max} = \frac{c\cdot U_n}{\sqrt{3}\cdot\underline{Z}_{(1)\beta}} = \frac{1.0\cdot 0.4\,\text{kV}}{\sqrt{3}\,(0.2355 + j0.4715)\,\Omega} = (0.196 - j0.392)\,\text{kA},$$

$$I''_{kM3\,max} = \sqrt{0.196^2 + 0.392^2}\,\text{kA} = 0.438\,\text{kA}.$$

Sum:

$$\underline{I}''_{k3\,max} = \underline{I}''_{kN3\,max} + \underline{I}''_{kM3\,max},$$

$$\underline{I}''_{k3\,max} = (2.23 - j8.91)\text{ kA} + (0.196 - j0.392)\text{ kA} = (2.43 - j9.30)\text{ kA},$$

$$I''_{k3\,max} = \sqrt{2.43^2 + 9.30^2}\text{ kA} = \underline{9.61\text{ kA}}.$$

Maximum asymmetrical short-circuit current $i_{p3\,max}$:

System component:

$$i_{pN3\,max} = \kappa \cdot \sqrt{2} \cdot I''_{kN3\,max} = 1.48 \cdot \sqrt{2} \cdot 9.18\text{ kA} = 19.21\text{ kA},$$

$$\kappa = 1.48 \quad \text{where} \quad \frac{R_{1(\alpha)}}{X_{1(\alpha)}} = \frac{0.00612}{0.0244} = 0.25$$

(According to Fig. 1.2/13)

Motor component via cables Ka1 and Ka2:

$$i_{pM3\,max} = \kappa \cdot \sqrt{2} \cdot I''_{kM3\,max} = 1.24 \cdot \sqrt{2} \cdot 0.438\text{ kA} = 0.77\text{ kA},$$

$$\kappa = 1.24 \quad \text{where} \quad \frac{R_{1(\beta)}}{X_{1(\beta)}} = \frac{0.2355}{0.4715} = 0.50.$$

Sum:

$$i_{p3\,max} = i_{pN3\,max} + i_{pM3\,max} = (19.21 + 0.77)\text{ kA} = \underline{19.98\text{ kA}}.$$

Symmetrical breaking current $I_{b3\,max}$:

System component:

$I_{bN3\,max} = I''_{kN3\,max} = 9.18$ kA (remote from generator terminals).

Motor component via cables Ka1 and Ka2:

$$I_{bM3\,max} = \mu_{0.1} \cdot q_{0.1} \cdot I''_{kM3\,max} = 0.78 \cdot 0.13 \cdot 0.438\text{ kA} = 0.044\text{ kA}.$$

$$I_{rM} = \frac{P_r}{2 \cdot \cos\varphi_r \cdot \sqrt{3} \cdot U_{rM}} = \frac{50\text{ kW}}{0.93 \cdot 0.83 \cdot \sqrt{3} \cdot 0.4\text{ kV}} = 93.5\text{ A},$$

$$\frac{I''_{kM3\,max}}{I_{rM}} = \frac{0.438\text{ kA}}{0.0935\text{ kA}} = 4.68,$$

$$\mu_{0.1} = 0.62 + 0.72 \cdot e^{-0.32 \cdot 4.68} = 0.78$$

(According to Fig. 1.2/14)

$$m = \frac{P_r}{p} = \frac{0.050\text{ MW}}{2} = 0.025\text{ MW},$$

$$q_{0.1} = 0.57 + 0.12 \cdot \ln 0.025 = 0.13$$

(According to Fig. 1.2/15)

Sum:

$$I_{b3\,max} = I_{bN3\,max} + I_{bM3\,max} = (9.18 + 0.044)\text{ kA} = \underline{9.22\text{ kA}}.$$

<u>Fault location F2</u>

Impedances of equipment and short-circuit current path:

System and transformer plus cables Ka1 and Ka2:

$$\underline{Z}_{(1)\gamma} = \underline{Z}_{(1)\alpha} + \underline{Z}_{(1)\text{Ka1}} + \underline{Z}_{(1)\text{Ka2}},$$

$$\underline{Z}_{(1)\gamma} = (0.0061 + j0.0244)\,\Omega + (0.0155 + j0.0101)\,\Omega + (0.029 + j0.0064)\,\Omega,$$

$$\underline{Z}_{(1)\gamma} = (0.0506 + j0.0409)\,\Omega.$$

Initial symmetrical short-circuit current $I_{k3\,max}$:

System component via transformer and cables Ka1 and Ka2:

$$\underline{I}''_{kN3\,max} = \frac{c \cdot U_n}{\sqrt{3} \cdot \underline{Z}_{(1)\gamma}} = \frac{1.0 \cdot 0.4\,kV}{\sqrt{3}\,(0.0506 + j\,0.0409)\,\Omega} = (2.76 - j\,2.23)\,kA,$$

$$I''_{kN3\,max} = \sqrt{2.76^2 + 2.23^2} = 3.55\,kA.$$

Motor component:

$$I''_{kM3\,max} = \frac{c \cdot U_n}{\sqrt{3} \cdot \underline{Z}_{(1)M}} = \frac{1.0 \cdot 0.4\,kV}{\sqrt{3}\,(0.191 + j\,0.455)\,\Omega} = (0.18 - j\,0.43)\,kA$$

$$I''_{kM3\,max} = \sqrt{0.18^2 + 0.43^2} = 0.47\,kA.$$

Sum:

$$\underline{I}''_{k3\,max} = \underline{I}''_{kN3\,max} + \underline{I}''_{kN3\,max},$$

$$\underline{I}''_{k3\,max} = (2.76 - j2.23)\,kA + (0.18 - j0.43)\,kA = (2.94 - j2.66)\,kA,$$

$$I''_{k3\,max} = \sqrt{2.94^2 + 2.66^2}\,kA = \underline{3.97\,kA}.$$

Maximum asymmetrical short-circuit current $i_{p3\,max}$:

System component via transformer and cables Ka1 and Ka2:

$$i_{pN3\,max} = \kappa \cdot \sqrt{2} \cdot I''_{kN3\,max} = 1.04 \cdot \sqrt{2} \cdot 3.55\,kA = 5.22\,kA,$$

$$\kappa = 1.04 \quad \text{where} \quad \frac{R_{(1)\gamma}}{X_{(1)\gamma}} = \frac{0.0506}{0.0409} = 1.24 \text{ according to formula in Fig. 1.2/13.}$$

Motor component:

$$i_{pM3\,max} = \kappa \cdot \sqrt{2} \cdot I''_{kM3\,max} = 1.30 \cdot \sqrt{2} \cdot 0.47\,kA = 0.86\,kA,$$

$$\kappa = 1.30 \quad \text{where} \quad \frac{R_{(1)M}}{X_{(1)M}} = \frac{0.191}{0.455} = 0.42.$$

Sum:

$$i_{p3\,max} = i_{pN3\,max} + i_{pM3\,max}$$

$$i_{p3\,max} = (5.22 + 0.86)\,kA = \underline{6.08\,kA}.$$

Symmetrical breaking current $I_{b3\,max}$:

System component via transformer and cables Ka1 and Ka2:

$$I_{bN3\,max} = I''_{kN3\,max} = 3.55\,kA \text{ (remote from generator terminals).}$$

Motor component:

$$I_{bM3\,max} = \mu_{0.1} \cdot q_{0.1} \cdot I''_{kM3\,max} = 0.76 \cdot 0.13 \cdot 0.47\,kA = 0.05\,kA,$$

$$\frac{I''_{kM3\,max}}{I_{rM}} = \frac{0.47}{0.0935} = 5,$$

$$\mu_{0.1} = 0.62 + 0.72\,e^{-0.32 \cdot 5} = 0.76 \quad \text{According to formula in Fig. 1.2/14,}$$

$$m = \frac{P}{p} = \frac{0.050\,MW}{2} = 0.025\,MW,$$

$$q_{0.1} = 0.57 + 0.12 \cdot \ln 0.025 = 0.13 \quad \text{According to Fig. 1.2/15.}$$

Sum:

$$I_{b3\,max} = I_{bN3\,max} + I_{bM3\,max} = (3.55 + 0.05)\,kA = \underline{3.60\,kA}.$$

Fault location F3

Cable impedances for cables Ka3, Ka4 and Ka5:

$$\underline{Z}_{(1)\mathrm{Ka3}} = 0.035\ \mathrm{km}\ (0.0991 + \mathrm{j}0.0798)\ \Omega/\mathrm{km} = (0.0035 + \mathrm{j}0.0028)\ \Omega.$$

$$\underline{Z}_{(1)\mathrm{Ka4}} = \underline{Z}_{(1)\mathrm{Ka3}}\ .$$

$$\underline{Z}_{(1)\mathrm{Ka5}} = 0.15\ \mathrm{km}\ (0.727 + \mathrm{j}0.088)\ \Omega/\mathrm{km} = (0.1091 + \mathrm{j}0.0132)\ \Omega.$$

$$\underline{Z}_{(1)\vartheta} = \frac{\underline{Z}_{(1)\mathrm{Ka3}}}{2} + \underline{Z}_{(1)\mathrm{Ka5}} = \left(\frac{0.0035 + \mathrm{j}0.0028}{2} + 0.1091 + \mathrm{j}0.0132\right)\ \Omega,$$

$$\underline{Z}_{(1)\vartheta} = (0.111 + \mathrm{j}0.015)\ \Omega.$$

The specific cable impedances were taken from the Siemens publication "Power Cables and their Application" (see Section 7.3).

Parallel impedance of system via transformer and motor feeder with cable:

$$\underline{Z}_{(1)\varepsilon} = \frac{\underline{Z}_{(1)\alpha} \cdot \left(\underline{Z}_{(1)\mathrm{Ka1}} + \underline{Z}_{(1)\mathrm{Ka2}} + \underline{Z}_{(1)\mathrm{M}}\right)}{\underline{Z}_{(1)\alpha} + \underline{Z}_{(1)\mathrm{Ka1}} + \underline{Z}_{(1)\mathrm{Ka2}} + \underline{Z}_{(1)\mathrm{M}}},$$

$$\underline{Z}_{(1)\varepsilon} = \left[\frac{(0.00612 + \mathrm{j}0.0244) \cdot (0.0155 + \mathrm{j}0.0101 + 0.029 + \mathrm{j}0.0064 + 0.191 + \mathrm{j}0.455)}{(0.00612 + \mathrm{j}0.0244 + 0.0155 + \mathrm{j}0.0101 + 0.029 + \mathrm{j}0.0064 + 0.191 + \mathrm{j}0.455)}\right]\ \Omega.$$

$$\underline{Z}_{(1)\varepsilon} = (0.0061 + \mathrm{j}0.0232)\ \Omega.$$

Sum of parallel impedances of system via transformer and motor feeder plus cable impedance of Ka3 parallel to Ka4 and Ka5:

$$\underline{Z}_{(1)} = \underline{Z}_{(1)\varepsilon} + Z_{(1)\vartheta} = (0.0061 + \mathrm{j}0.0232)\ \Omega + (0.111 + \mathrm{j}0.015)\ \Omega,$$

$$\underline{Z}_{(1)} = (0.1171 + \mathrm{j}0.0382)\ \Omega.$$

Initial symmetrical short-circuit current $I''_{\mathrm{k3\,max}}$:

$$\underline{I}''_{\mathrm{k3\,max}} = \frac{c \cdot U_{\mathrm{n}}}{\sqrt{3} \cdot \underline{Z}_{(1)}} = \frac{1 \cdot 0.4\ \mathrm{kV}}{\sqrt{3} \cdot (0.1171 + \mathrm{j}0.0382)\ \Omega} = (1.79 - \mathrm{j}0.58)\ \mathrm{kA},$$

$$I''_{\mathrm{k3\,max}} = \sqrt{0.58^2 + 1.79^2}\ \mathrm{kA} = \underline{1.88\ \mathrm{kA}}.$$

Maximum asymmetrical short-circuit current $i_{\mathrm{p3\,max}}$:

$$i_{\mathrm{p3\,max}} = 1.15 \cdot \kappa \cdot \sqrt{2} \cdot I''_{\mathrm{k3\,max}} = 1.15 \cdot 1.02 \cdot \sqrt{2} \cdot 1.88\ \mathrm{kA} = \underline{3.12\ \mathrm{kA}},$$

$$\kappa = 1.02 \qquad \frac{R_{(1)}}{X_{(1)}} = \frac{0.1171}{0.0382} = 3.06 \quad \text{According to formula in Fig. 1.2/14.}$$

The factor 1.15 is a safety factor applied in the case of meshed systems and common impedance.

Symmetrical breaking current $I_{\mathrm{b3\,max}}$:

Since the motor component of $I''_{\mathrm{kN3\,max}}$ is much smaller than the system component and $I_{\mathrm{bN3\,max}} = I''_{\mathrm{kN3\,max}}$, $I_{\mathrm{b3\,max}} = I''_{\mathrm{k3\,max}} = 1.88$ kA can be applied.

▷ ***Phase-to-phase fault currents***
Minimum phase-to-phase fault currents

When calculating the minimum short-circuit currents, the following should be expected:

- Minimum symmetrical short-circuit power applied: $S''_{\mathrm{kQ\,min}} = 100$ MVA
- $c_{\mathrm{min}} = 0.95$
- Cable temperature = 80 °C.

Motors are assumed to be switched off.

Fault location F1

System impedance:

$$\underline{Z}_{(1)\mathrm{Q\,min}} = \frac{1.0 \cdot U_\mathrm{n}^2}{S''_\mathrm{kQ\,min}} = \frac{1.0 \cdot 10^2\,\mathrm{kV}^2}{100\,\mathrm{MVA}} = 1.0\,\Omega.$$

Where $R_{(1)\mathrm{Q\,min}} = 0.1 \cdot X_{(1)\mathrm{Q\,min}}$:

$$X_{(1)\mathrm{Q\,min}} = \frac{Z_{(1)\mathrm{Q\,min}}}{1.005} = \frac{1.0\,\Omega}{1.005} = 0.995\,\Omega.$$

$$\underline{Z}_{(1)\mathrm{Q\,min}} = (0.0995 + \mathrm{j}\,0.995)\,\Omega.$$

Impedance of system plus transformer:

$$\underline{Z}_{(1)\alpha\,\mathrm{min}} = \underline{Z}_{(1)\mathrm{Q\,min}} \left(\frac{U_\mathrm{rTUS}}{U_\mathrm{rTOS}}\right)^2 + \underline{Z}_{(1)\mathrm{T}},$$

$$\underline{Z}_{(1)\alpha\,\mathrm{min}} = (0.0995 + \mathrm{j}\,0.995)\,\Omega \left(\frac{0.4\,\mathrm{kV}}{10\,\mathrm{kV}}\right)^2 + (0.006 + \mathrm{j}\,0.02324)\,\Omega,$$

$$\underline{Z}_{(1)\alpha\,\mathrm{min}} = (0.00616 + \mathrm{j}\,0.0248)\,\Omega.$$

$$\underline{Z}_{(2)\alpha\,\mathrm{min}} = \underline{Z}_{(1)\alpha\,\mathrm{min}} \quad \text{(Negative-sequence impedance = positive-sequence impedance).}$$

Initial symmetrical short-circuit current $I''_\mathrm{k2\,min}$:

$$\underline{I}''_\mathrm{k2\,min} = \frac{c \cdot U_\mathrm{n}}{\underline{Z}_{(1)\alpha\,\mathrm{min}} + \underline{Z}_{(2)\alpha\,\mathrm{min}}} = \frac{0.95 \cdot 0.4\,\mathrm{kV}}{2 \cdot (0.00616 + \mathrm{j}\,0.0248)\,\Omega},$$

$$\underline{I}''_\mathrm{k2\,min} = (1.79 - \mathrm{j}\,7.22)\,\mathrm{kA},$$

$$I''_\mathrm{k2\,min} = \sqrt{1.79^2 + 7.22^2}\,\mathrm{kA} = \underline{7.44\,\mathrm{kA}}.$$

Fault location F2

Cable impedances for cables Ka1 und Ka2 at 80 °C:

$$\underline{Z}_{(1)\mathrm{Ka1e}} = (1.24 \cdot 0.0155 + \mathrm{j}\,0.0101)\,\Omega = (0.0192 + \mathrm{j}\,0.0101)\,\Omega,$$

$$\underline{Z}_{(1)\mathrm{Ka2e}} = (1.24 \cdot 0.029 + \mathrm{j}\,0.0064)\,\Omega = (0.0360 + \mathrm{j}\,0.0064)\,\Omega.$$

Impedance of system plus transformer and cables Ka1 and Ka2:

$$\underline{Z}_{(1)\,\mathrm{min}} = \underline{Z}_{(1)\alpha\,\mathrm{min}} + \underline{Z}_{(1)\mathrm{Ka1e}} + \underline{Z}_{(1)\mathrm{Ka2e}},$$

$$\underline{Z}_{(1)\,\mathrm{min}} = (0.00616 + \mathrm{j}\,0.0248)\,\Omega + (0.0192 + \mathrm{j}\,0.0101)\,\Omega + (0.036 + \mathrm{j}\,0.0064)\,\Omega$$

$$\underline{Z}_{(1)\,\mathrm{min}} = (0.06136 + \mathrm{j}\,0.0413)\,\Omega \quad \text{(Negative-sequence impedance = positive-sequence impedance).}$$

Initial symmetrical short-circuit current $I''_\mathrm{k2\,min}$:

$$\underline{I}''_\mathrm{k2\,min} = \frac{c \cdot U_\mathrm{n}}{2 \cdot \underline{Z}_{(1)\alpha\,\mathrm{min}}} = \frac{0.95 \cdot 0.4\,\mathrm{kV}}{2 \cdot (0.06136 + \mathrm{j}\,0.0413)\,\Omega} = (2.13 - \mathrm{j}\,1.43)\,\mathrm{kA},$$

$$I''_\mathrm{k2\,min} = \sqrt{2.13^2 + 1.43^2}\,\mathrm{kA} = \underline{2.57\,\mathrm{kA}}.$$

Fault location F3

Cable impedances for cables Ka3, Ka4 and Ka5 at 80 °C:

$$\underline{Z}_{(1)\mathrm{Ka3e}} = (1.24 \cdot 0.0035 + \mathrm{j}\,0.0028) = (0.00434 + \mathrm{j}\,0.0028)\,\Omega,$$

$$\underline{Z}_{(1)\mathrm{Ka4e}} = \underline{Z}_{(1)\mathrm{Ka3e}}$$

$$\underline{Z}_{(1)\mathrm{Ka5e}} = (1.24 \cdot 0.1091 + \mathrm{j}\,0.0132) = (0.13528 + \mathrm{j}\,0.0132)\,\Omega.$$

Sum of impedances for system, transformer, cable Ka3 parallel to Ka4 and Ka5:

$$\underline{Z}_{(1)\min} = \underline{Z}_{(1)\alpha\min} + \frac{\underline{Z}_{(1)\mathrm{Ka3e}}}{2} + \underline{Z}_{(1)\mathrm{Ka5e}},$$

$$\underline{Z}_{(1)\min} = (0.00616 + \mathrm{j}\,0.0248)\,\Omega + (0.00217 + \mathrm{j}\,0.0014)\,\Omega + (0.13528 + \mathrm{j}\,0.0132)\,\Omega,$$

$$\underline{Z}_{(1)\min} = (0.1436 + \mathrm{j}\,0.0394)\,\Omega.$$

Initial symmetrical short-circuit current $I''_{\mathrm{k2\,min}}$:

$$\underline{I}''_{\mathrm{k2\,min}} = \frac{c \cdot U_\mathrm{n}}{2 \cdot \underline{Z}_{(1)\alpha\min}} = \frac{0.95 \cdot 0.4\,\mathrm{kV}}{2 \cdot (0.1436 + \mathrm{j}\,0.0394)\,\Omega} = (1.23 - \mathrm{j}\,0.34)\,\mathrm{kA},$$

$$I''_{\mathrm{k2\,min}} = \sqrt{1.23^2 + 0.34^2}\,\mathrm{kA} = \underline{1.28\,\mathrm{kA}}.$$

▷ ***Phase-to-ground fault currents***
Minimum phase-to-ground fault currents

When calculating the minimum short-circuit currents, the following should be expected:

- Minimum symmetrical short-circuit power applied: $S''_{\mathrm{kQ\,min}} = 100\,\mathrm{MVA}$
- $c_{\min} = 0.95$
- Cable temperature = 80 °C.

Motors are assumed to be switched off.

Fault location F1

Positive-sequence impedance of transformer:

$$\underline{Z}_{(1)\mathrm{T}} = (0.006 + \mathrm{j}\,0.02324)\,\Omega.$$

Zero-sequence impedance of transformer:

$$R_{(0)\mathrm{T}} = R_{(1)\mathrm{T}} = 0.006\,\Omega,$$

$$X_{(0)\mathrm{T}} = 0.95 X_{(1)\mathrm{T}} = 0.95 \cdot 0.02324\,\Omega = 0.0221\,\Omega,$$

$$\underline{Z}_{(0)\mathrm{T}} = (0.006 + \mathrm{j}\,0.0221)\,\Omega.$$

Positive-sequence impedance of system plus transformer:

$$\underline{Z}_{(1)\alpha\min} = (0.00616 + \mathrm{j}\,0.0248)\,\Omega \quad \text{(Negative-sequence impedance = positive-sequence impedance)}.$$

Initial symmetrical short-circuit current $I''_{\mathrm{k1\,min}}$:

$$\underline{I}''_{\mathrm{k1\,min}} = \frac{\sqrt{3} \cdot c \cdot U_\mathrm{n}}{2 \cdot \underline{Z}_{(1)\alpha\min} + \underline{Z}_{(0)\mathrm{T}}} = \frac{\sqrt{3} \cdot 0.95 \cdot 0.4\,\mathrm{kV}}{2 \cdot (0.00616 + \mathrm{j}\,0.0248)\,\Omega + (0.006 + \mathrm{j}\,0.0221)\,\Omega}$$

$$\underline{I}''_{\mathrm{k1\,min}} = (2.20 - \mathrm{j}\,8.63)\,\mathrm{kA},$$

$$I''_{\mathrm{k1\,min}} = \sqrt{2.20^2 + 8.63^2}\,\mathrm{kA} = \underline{8.90\,\mathrm{kA}}.$$

Fault location F2

Zero-sequence impedances of cables Ka1 and Ka2 at 80 °C:

$$R_{(0)\mathrm{Ka1e}} = 3.45\,R_{(1)\mathrm{Ka1e}}, \quad X_{(0)\mathrm{Ka1}} = 3.95\,X_{(1)\mathrm{Ka1}},$$

$$R_{(0)\mathrm{Ka2e}} = 2.90\,R_{(1)\mathrm{Ka2e}}, \quad X_{(0)\mathrm{Ka2}} = 7.3\,X_{(1)\mathrm{Ka2}},$$

$$\underline{Z}_{(0)\mathrm{Ka1e}} = (3.45 \cdot 0.0192 + 3.95 \cdot \mathrm{j}\,0.0101)\,\Omega = (0.0662 + \mathrm{j}\,0.04)\,\Omega.$$

$$\underline{Z}_{(0)\mathrm{Ka2e}} = (2.9 \cdot 0.036 + 7.3 \cdot \mathrm{j}\,0.0064)\,\Omega = (0.1044 + \mathrm{j}\,0.0467)\,\Omega.$$

Zero-sequence impedances of transformer plus cables Ka1 and Ka2:

$$\underline{Z}_{(0)\,\mathrm{min}} = \underline{Z}_{(0)\mathrm{T}} + \underline{Z}_{(0)\mathrm{Ka1e}} + \underline{Z}_{(0)\mathrm{Ka2e}},$$

$$\underline{Z}_{(0)\,\mathrm{min}} = (0.006 + \mathrm{j}0.0221)\,\Omega + (0.0662 + \mathrm{j}0.04)\,\Omega + (0.1044 + \mathrm{j}0.0467)\,\Omega$$

$$\underline{Z}_{(0)\,\mathrm{min}} = (0.1766 + \mathrm{j}0.1088)\,\Omega.$$

Positive-sequence impedance of system plus transformer and cables Ka1 and Ka2:

$\underline{Z}_{(1)\alpha\mathrm{min}} = (0.06136 + \mathrm{j}0.0413)$ (Negative-sequence impedance = positive-sequence impedance).

Initial symmetrical short-circuit current $I''_{\mathrm{k1\,min}}$:

$$\underline{I}''_{\mathrm{k1\,min}} = \frac{\sqrt{3}\cdot c\cdot U_{\mathrm{n}}}{2\cdot \underline{Z}_{(1)\,\mathrm{min}} + \underline{Z}_{(0)\,\mathrm{min}}} = \frac{\sqrt{3}\cdot 0.95\cdot 0.4\,\mathrm{kV}}{2\cdot(0.06136 + \mathrm{j}0.0413)\,\Omega + (0.1766 + \mathrm{j}0.1088)\,\Omega},$$

$$\underline{I}''_{\mathrm{k1\,min}} = (1.57 - \mathrm{j}\,1.00)\,\mathrm{kA},$$

$$I''_{\mathrm{k1\,min}} = \sqrt{1.57^2 + 1.0^2}\,\mathrm{kA} = \underline{1.86\,\mathrm{kA}}.$$

Fault location F3

Zero-sequence impedances of cables Ka3, Ka4 and Ka5 at 80 °C:

$$R_{(0)\mathrm{Ka3e}} = 3.5\,R_{(1)\mathrm{Ka3e}},\quad X_{(0)\mathrm{Ka3}} = 3.7\,X_{(1)\mathrm{Ka}},$$

$$R_{(0)\mathrm{Ka5e}} = 2.25\,R_{(1)\mathrm{Ka5e}},\quad X_{(0)\mathrm{Ka5}} = 12.5\,X_{(1)\mathrm{Ka}},$$

$$\underline{Z}_{(0)\mathrm{Ka3e}} = (3.5\cdot 0.00434 + 3.7\cdot \mathrm{j}0.0028)\,\Omega = (0.0152 + \mathrm{j}0.0104)\,\Omega,$$

$$\underline{Z}_{(0)\mathrm{Ka4e}} = \underline{Z}_{(0)\mathrm{Ka3}},$$

$$\underline{Z}_{(0)\mathrm{Ka5e}} = (2.25\cdot 0.13528 + 12.5\cdot \mathrm{j}0.0132)\,\Omega = (0.3044 + \mathrm{j}0.165)\,\Omega.$$

Zero-sequence impedances of transformer plus cables Ka3 parallel to Ka4 and Ka5:

$$\underline{Z}_{(0)\,\mathrm{min}} = \underline{Z}_{(0)\mathrm{T}} + \frac{\underline{Z}_{(0)\mathrm{Ka3e}}}{2} + \underline{Z}_{(0)\mathrm{Ka5e}},$$

$$\underline{Z}_{(0)\,\mathrm{min}} = (0.006 + \mathrm{j}0.0221)\,\Omega + \frac{(0.0152 + \mathrm{j}0.0104)}{2}\Omega + (0.3044 + \mathrm{j}0.165)\,\Omega,$$

$$\underline{Z}_{(0)\,\mathrm{min}} = (0.318 + \mathrm{j}0.1923)\,\Omega.$$

Sum of impedances for system, transformer, cable Ka3 parallel to Ka4 and Ka5:

$\underline{Z}_{(1)\alpha\mathrm{min}} = (0.1436 + \mathrm{j}0.0394)\,\Omega$ (Negative-sequence impedance = positive-sequence impedance).

Initial symmetrical short-circuit current $I''_{\mathrm{k1\,min}}$:

$$\underline{I}''_{\mathrm{k1\,min}} = \frac{\sqrt{3}\cdot c\cdot U_{\mathrm{n}}}{2\cdot \underline{Z}_{(1)\,\mathrm{min}} + \underline{Z}_{(0)\,\mathrm{min}}} = \frac{\sqrt{3}\cdot 0.95\cdot 0.4\,\mathrm{kV}}{2\cdot(0.1436 + \mathrm{j}0.0394)\,\Omega + (0.318 + \mathrm{j}0.1923)\,\Omega},$$

$$\underline{I}''_{\mathrm{k1\,min}} = (0.91 - \mathrm{j}0.41)\,\mathrm{kA},$$

$$I''_{\mathrm{k1\,min}} = \sqrt{0.91^2 + 0.41^2}\,\mathrm{kA} = \underline{0.99\,\mathrm{kA}}.$$

Table of calculated initial symmetrical short-circuit currents and maximum asymmetrical short-circuit curent $i_{p3\,max}$

Fault location		F1	F2	F3
Max. three-phase	$I''_{k3\,max}$ (kA)	9.61	3.98	1.88
Min. phase-to-phase	$I''_{k2\,min}$ (kA)	7.44	2.57	1.28
Min. phase-to-ground	$I''_{k1\,min}$ (kA)	8.90	1.86	0.99
Max. asymmetrical short-circuit current	$i_{p3\,max}$ (kA)	19.98	6.08	3.12

Generally speaking, the initial phase-to-phase symmetrical short-circuit currents (I''_k) are $\sqrt{3}/2$ times the three-phase initial symmetrical short-circuit currents. The c factor for calculating the maximum values in 400 V systems is 1.0. The c factor for calculating the minimum values is 0.95. The product of 9.61 kA $\cdot \sqrt{3}/2 \cdot 0.95$ for fault location F1 is 7.91 kA. A further reduction to 7.44 kA results from the lower applied system symmetrical short-circuit power of 100 MVA and the omission of the motor component.

Example 2:

Short-circuit in low-voltage system with additional generator infeed

The assumed system circuit and the characteristic data of the equipment are shown in Fig. 1.2/6.

Fault location F1 has a multi-end single-phase infeed. Fault locations F2 and F3 are supplied via common impedances, i.e. they have a meshed infeed.

A central 400 V l. v. substation is supplied by a 20 kV system from the busbar Q via a NEKEBA cable and a 1 MVA transformer in vector group Dyn5. The substation is equipped with a generator with a rated output of 500 kVA, which operates in parallel with the external system. The neutrals of the transformer and generator have effective grounding. A number of main distribution boards are connected radially to the central 400 V busbar via four-core cables. Further radial cables branch off this main distribution board to individual sub-distribution boards.

The following short-circuit currents must be calculated for the indicated fault locations F1, F2 and F3:

Fault locations F1, F2 and F3:

$I''_{k3\,max}$, $i_{p3\,max}$, $I_{b3\,max}$, $I''_{k2\,min}$,
$I''_{k1\,max}$, $I''_{k1\,min}$.

▷ ***Three-phase short-circuit currents***
Maximum three-phase short-circuit currents

Fault location F1:

Impedances of equipment and short-circuit current path:

System impedance:

$$Z_{(1)Q} = \frac{c \cdot U_n^2}{S''_{kQ}} = \frac{1.1 \cdot 20^2\ \text{kV}^2}{500\ \text{MVA}} = 0.88\ \Omega.$$

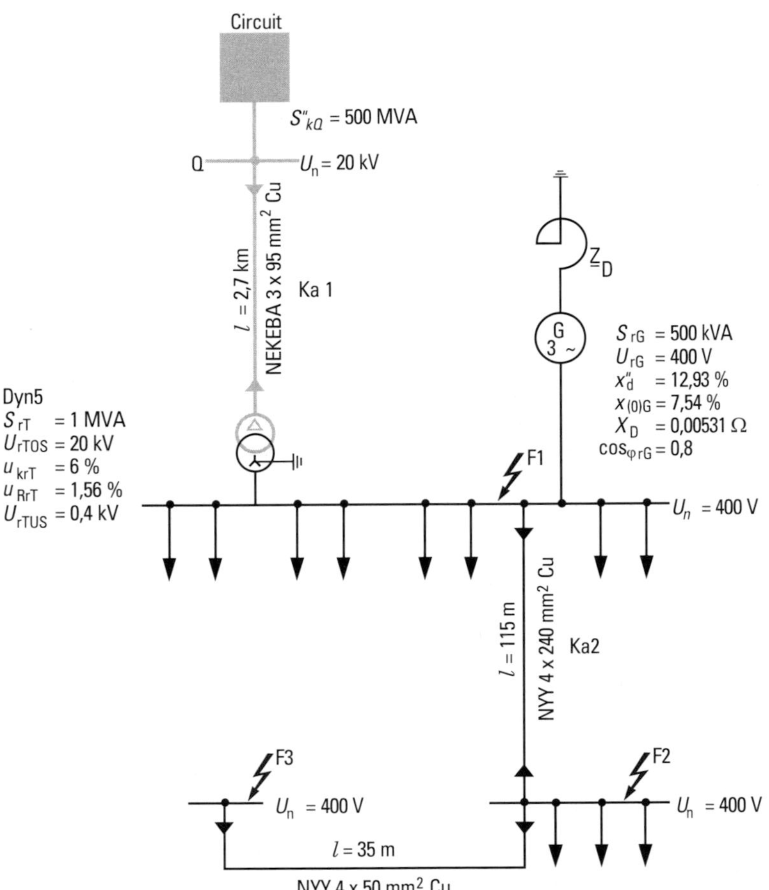

Fig. 1.2/6 System circuit for calculation example 2

Where $R_{(1)Q} = 0.1 \cdot X_{(1)Q}$:

$$X_{(1)Q} = \frac{Z_{(1)Q}}{1.005} = \frac{0.88\,\Omega}{1.005} = 0.8756\,\Omega,$$

$$\underline{Z}_{(1)Q} = (0.08756 + \mathrm{j}\,0.8756)\,\Omega.$$

Cable impedance of Ka1, 20 KV:

$$\underline{Z}_{(1)\mathrm{Ka1}} = l(r' + \mathrm{j}x'),$$

$$r' = 0.193\,\Omega/\mathrm{km},$$

$x' = 0.1285\,\Omega/\mathrm{km}$ (see Siemens publication "Power Cables and their Application")

$$\underline{Z}_{(1)\mathrm{Ka1}} = 2.7\,\mathrm{km}\,(0.193 + \mathrm{j}\,0.1285)\,\Omega/\mathrm{km} = (0.5211 + \mathrm{j}\,0.34695)\,\Omega.$$

System and cable impedance:

$$\underline{Z}_{(1)\alpha} = \underline{Z}_{(1)Q} + \underline{Z}_{(1)\mathrm{Ka1}} = (0.60866 + \mathrm{j}\,1.22255)\,\Omega,$$

$$\underline{Z}_{(1)\alpha t} = \underline{Z}_{(1)\alpha}\left(\frac{U_{\mathrm{rTUS}}}{U_{\mathrm{rTOS}}}\right)^2 = (0.60866 + \mathrm{j}\,1.22255)\,\Omega\left(\frac{0.4\,\mathrm{kV}}{20\,\mathrm{kV}}\right)^2 = (0.00024 + \mathrm{j}\,0.00049)\,\Omega.$$

Transformer impedance:

$$u_{\mathrm{XrT}} = \sqrt{u_{\mathrm{krT}}^2 - u_{\mathrm{RrT}}^2} = \sqrt{6^2 - 1.56^2}\% = 5.794\%,$$

$$R_{(1)\mathrm{T}} = \frac{u_{\mathrm{RrT}} \cdot U_{\mathrm{rTUS}}^2}{100\% \cdot S_{\mathrm{rT}}} = \frac{1.56\% \cdot 0.4^2\ \mathrm{kV}^2}{100\% \cdot 1\ \mathrm{MVA}} = 0.00249\ \Omega,$$

$$X_{(1)\mathrm{T}} = \frac{u_{\mathrm{XrT}} \cdot U_{\mathrm{rTUS}}^2}{100\% \cdot S_{\mathrm{rT}}} = \frac{5.794\% \cdot 0.4^2\ \mathrm{kV}^2}{100\% \cdot 1\ \mathrm{MVA}} = 0.00927\ \Omega,$$

$$\underline{Z}_{(1)\mathrm{T}} = (0.00249 + \mathrm{j}\,0.00927)\ \Omega.$$

System, cable, and transformer impedance:

$$\underline{Z}_{(1)\beta} = \underline{Z}_{(1)\alpha t} + \underline{Z}_{(1)\mathrm{T}} = (0.00273 + \mathrm{j}\,0.00976)\ \Omega.$$

$$Z_{(1)\beta} = Z_{\mathrm{kN}} = \sqrt{0.00273^2 + 0.00976^2}\ \Omega = 0.01013\ \Omega.$$

Generator impedance:

$$X_{(1)\mathrm{G}} = X_{\mathrm{d}}'' = \frac{x_{\mathrm{d}}'' U_{\mathrm{rG}}^2}{100\% \cdot S_{\mathrm{rG}}} = \frac{12.93\% \cdot 0.4^2\ \mathrm{kV}^2}{100\% \cdot 0.5\ \mathrm{MVA}} = 0.04138\ \Omega,$$

$$R_{(1)\mathrm{G}} = 0.15 \cdot X_{\mathrm{d}}'' = 0.15 \cdot 0.04138\ \Omega = 0.00621\ \Omega,$$

$$\underline{Z}_{(1)\mathrm{G}} = (0.00621 + \mathrm{j}\,0.04138)\ \Omega.$$

Correction for generator impedance according to IEC 60909/DIN VDE 0102 (see also Section 1.2.3):

$$K_{\mathrm{G}} = \frac{U_{\mathrm{n}}}{U_{\mathrm{rG}}} \cdot \frac{c_{\max}}{1 + x_{\mathrm{d}}'' \cdot \sin \varphi_{\mathrm{rG}}} = \frac{0.4\ \mathrm{kV}}{0.4\ \mathrm{kV}} \cdot \frac{1}{1 + 0.1293 \cdot 0.6} = 0.928,$$

$$\underline{Z}_{(1)\mathrm{GK}} = K_{\mathrm{G}} \cdot \underline{Z}_{(1)\mathrm{G}} = 0.928(0.00621 + \mathrm{j}\,0.04138)\ \Omega = (0.00576 + \mathrm{j}\,0.0384)\ \Omega,$$

$$Z_{(1)\mathrm{GK}} = \sqrt{0.00576^2 + 0.0384^2}\ \Omega = 0.03883\ \Omega.$$

Initial symmetrical short-circuit current $I_{\mathrm{k3\,max}}''$:

System component:

$$\underline{I}_{\mathrm{kN3\,max}}'' = \frac{c \cdot U_{\mathrm{n}}}{\sqrt{3} \cdot \underline{Z}_{\mathrm{kN}}} = \frac{0.4\ \mathrm{kV}}{\sqrt{3} \cdot (0.00273 + \mathrm{j}\,0.00976)\ \Omega} = (6.14 - \mathrm{j}\,21.95)\ \mathrm{kA},$$

$$I_{\mathrm{kN3\,max}}'' = 22.8\ \mathrm{kA}.$$

Generator component:

$$\underline{I}_{\mathrm{kG3\,max}}'' = \frac{c \cdot U_{\mathrm{n}}}{\sqrt{3} \cdot \underline{Z}_{(1)\mathrm{GK}}} = \frac{0.4\ \mathrm{kV}}{\sqrt{3} \cdot (0.00576 + \mathrm{j}\,0.0.0384)\ \Omega} = (0.88 - \mathrm{j}\,5.88)\ \mathrm{kA},$$

$$I_{\mathrm{kG3\,max}}'' = 5.95\ \mathrm{kA}.$$

Sum:

$$\underline{I}_{\mathrm{k3\,max}}'' = \underline{I}_{\mathrm{kN3\,max}}'' + \underline{I}_{\mathrm{kG3\,max}}'' = (6.14 - \mathrm{j}\,21.95 + 0.88 - \mathrm{j}\,5.88)\ \mathrm{kA} = (7.02 - \mathrm{j}\,27.83)\ \mathrm{kA},$$

$$I_{\mathrm{k3\,max}}'' = \underline{28.70\ \mathrm{kA}}.$$

Maximum asymmetrical short-circuit current $i_{\mathrm{p3\,max}}$:

System component:

$$i_{\mathrm{pN3\,max}} = \kappa \cdot \sqrt{2} \cdot I_{\mathrm{kN3\,max}}'' = 1.44 \cdot \sqrt{2} \cdot 22.8\ \mathrm{kA} = 46.43\ \mathrm{kA},$$

$\kappa = 1.44$ where $R_{(1)\beta}/X_{(1)\beta} = 0.00273/0.00976 = 0.28$

(according to Fig. 1.2.13).

Generator component:

$$i_{\text{pG3 max}} = \kappa \cdot \sqrt{2} \cdot I''_{\text{kG3 max}} = 1.64 \cdot \sqrt{2} \cdot 5.95\,\text{kA} = 13.80\,\text{kA},$$

$\kappa = 1.64$ where $R_{(1)GK}/X_{(1)\text{GK}} = 0.15$
(according to Fig. 1.2.13).

Sum:

$$i_{\text{p3 max}} = i_{\text{pN3 max}} + i_{\text{pG3 max}} = (46.43 + 13.80)\text{kA} = \underline{60.23\,\text{kA}}.$$

Symmetrical breaking current $I_{\text{b3 max}}$:

System component:

$$I_{\text{bN3 max}} = I''_{\text{kN3 max}} = 22.8\,\text{kA} \quad \text{(remote from generator terminals)}.$$

Generator component:

$$I_{\text{bG3 max}} = \mu_{0.05} \cdot I''_{\text{kG3 max}} = 0.755 \cdot 5.95\,\text{kA} = 4.49\,\text{kA},$$

$$I_{\text{rG}} = \frac{S_{\text{rG}}}{\sqrt{3} \cdot U_{\text{rG}}} = \frac{0.5\,\text{MVA}}{\sqrt{3} \cdot 0.4\,\text{kV}} = 0.722\,\text{kA},$$

$$\frac{I''_{\text{kG3 max}}}{I_{\text{rG}}} = \frac{5.95\,\text{kA}}{0.722\,\text{kA}} = 8.24,$$

$\mu_{0.05} = 0.755$ where $t_{\min} = 0.05\,\text{s}$
(according to Fig. 1.2.14).

Sum:

$$I_{\text{b3 max}} = I''_{\text{bN3 max}} + I''_{\text{bG3 max}} = (22.8 + 4.49)\,\text{kA} = \underline{27.29\,\text{kA}}.$$

Fault location F2:

Impedances of equipment and short-circuit current path:

Parallel impedance of system via transformer and generator:

$$\underline{Z}_{(1)\gamma} = \frac{\underline{Z}_{(1)\beta} \cdot \underline{Z}_{(1)\text{GK}}}{\underline{Z}_{(1)\beta} + \underline{Z}_{(1)\text{GK}}} = \frac{(0.00273 + \text{j}\,0.00976)(0.00576 + \text{j}\,0.0384)}{(0.00273 + \text{j}\,0.00976) + (0.00576 + \text{j}\,0.0384)}\,\Omega$$

$$= \frac{(-0.000359 + \text{j}\,0.00016)}{(0.008496 + \text{j}\,0.04816)}\,\Omega$$

$$= (0.00197 + \text{j}\,0.0078)\,\Omega,$$

$$Z_{(1)\gamma} = \sqrt{0.00197^2 + 0.0078^2}\,\Omega = 0.00805\,\Omega.$$

Cable impedance Ka2, 0.4 kV:

$$\underline{Z}_{(1)\text{Ka2}} = l(r' + \text{j}x') = 0.115\,\text{km}\,(0.0754 + \text{j}\,0.0798)\,\Omega/\text{km} = (0.00867 + \text{j}\,0.00918)\,\Omega,$$

$$\underline{Z}_{\text{k}} = \underline{Z}_{(1)\delta} = \underline{Z}_{(1)\gamma} + \underline{Z}_{(1)\text{Ka2}} = (0.01064 + \text{j}\,0.01698)\,\Omega,$$

$$Z_k = \sqrt{0.01064^2 + 0.01698^2}\,\Omega = 0.020\,\Omega.$$

Initial symmetrical short-circuit current $I''_{\text{k3 max}}$:

$$I''_{\text{k3 max}} = \frac{c \cdot U_{\text{n}}}{\sqrt{3} \cdot Z_{\text{k}}} = \frac{0.4\,\text{kV}}{\sqrt{3} \cdot 0.020\,\Omega} = \underline{11.55\,\text{kA}}.$$

Comprising:

Generator component:

$$I''_{\mathrm{kG3\,max}} = I''_{\mathrm{k3\,max}} \frac{Z_{(1)\gamma}}{Z_{(1)\mathrm{GK}}} = 11.55\,\mathrm{kA}\,\frac{0.00805\,\Omega}{0.03883\,\Omega} = 2.39\,\mathrm{kA}.$$

System component:

$$I''_{\mathrm{kN3\,max}} = I''_{\mathrm{k3\,max}} \frac{Z_{(1)\gamma}}{Z_{(1)\beta}} = 11.55\,\mathrm{kA}\,\frac{0.00805\,\Omega}{0.01013\,\Omega} = 9.18\,\mathrm{kA}.$$

Maximum asymmetrical short-circuit current $i_{\mathrm{p3\,max}}$:

$$i_{\mathrm{p3\,max}} = 1.15 \cdot \kappa_{\mathrm{k}} \cdot \sqrt{2} \cdot I''_{\mathrm{k3\,max}} = 1.15 \cdot 1.17 \cdot \sqrt{2} \cdot 11.55\,\mathrm{kA}$$
$$= \underline{21.98\,\mathrm{kA}},$$

$\kappa = 1.17$ where $R_{\mathrm{k}}/X_{\mathrm{k}} = 0.01064/0.01698 = 0.627$
(according to Fig. 1.2.13).

The factor 1.15 is a safety factor applied in the case of meshed systems and common impedance (IEC 60909 / DIN VDE 0102).

Symmetrical breaking current $I_{\mathrm{b3\,max}}$:

The following approximation applies in meshed systems (IEC 909 / DIN VDE 0102):

$$I_{\mathrm{b3\,max}} \approx I''_{\mathrm{k3\,max}} = \underline{11.55\,\mathrm{kA}}.$$

Fault location F3:

Impedances of equipment and short-circuit current path:

Parallel and cable impedance Ka2:

$$\underline{Z}_{(1)\delta} = (0.01064 + \mathrm{j}0.01698)\,\Omega.$$

Cable impedance Ka3:

$$\underline{Z}_{(1)\mathrm{Ka3}} = l\,(r' + \mathrm{j}x'),$$
$$r' = 0.387\,\Omega/\mathrm{km},$$
$$x' = 0.0848\,\Omega/\mathrm{km},$$
$$\underline{Z}_{(1)\mathrm{Ka3}} = 0.035\,\mathrm{km}\,(0.387 + \mathrm{j}0.0848)\,\Omega/\mathrm{km} = (0.01354 + \mathrm{j}0.00297)\,\Omega,$$

Parallel and cable impedances Ka2 + Ka3:

$$\underline{Z}_{\mathrm{k}} = \underline{Z}_{(1)\varepsilon} = \underline{Z}_{(1)\delta} + \underline{Z}_{(1)\mathrm{Ka3}} = (0.02418 + \mathrm{j}0.01995)\,\Omega,$$
$$Z_{\mathrm{k}} = \sqrt{0.02418^2 + 0.01995^2}\,\Omega = 0.03135\,\Omega.$$

Initial symmetrical short-circuit current $I''_{\mathrm{k3\,max}}$:

$$I''_{\mathrm{k3\,max}} = \frac{c \cdot U_{\mathrm{n}}}{\sqrt{3} \cdot Z_{\mathrm{k}}} = \frac{0.4\,\mathrm{kV}}{\sqrt{3} \cdot 0.03135\,\Omega} = \underline{7.37\,\mathrm{kA}}.$$

Maximum asymmetrical short-circuit current $i_{\mathrm{p3\,max}}$:

$$i_{\mathrm{p3\,max}} = 1.15 \cdot \kappa \cdot \sqrt{2} \cdot I''_{\mathrm{k3\,max}} = 1.15 \cdot 1.04 \cdot \sqrt{2} \cdot 7.37\,\mathrm{kA} = \underline{12.47\,\mathrm{kA}},$$

$\kappa = 1.04$ where $R_{\mathrm{k}}/X_{\mathrm{k}} = 0.02418/0.01995 = 1.21$

(according to Fig. 1.2.13).

The factor 1.15 is a safety factor applied in the case of meshed systems and common impedance (IEC 60909 / DIN VDE 0102).

Symmetrical breaking current $I_{b3\,max}$:

$$I_{b3\,max} \approx I''_{k3\,max} = \underline{7.37\ \text{kA}}.$$

▷ ***Phase-to-phase fault currents***
Minimum phase-to-phase fault currents

Fault location F1:

System impedance:

$$\underline{Z}_{(1)Q} = (0.08756 + \text{j}\,0.8756)\ \Omega.$$

It is assumed here that the minimum short-circuit power of the system (S''_{ka}) is also 500 MVA.

Cable impedance Ka1 at 80 °C, 20 kV:

$$\underline{Z}_{(1)\text{Ka1e}} = 1.24 \cdot R_{(1)\text{Ka1}} + \text{j}X_{(1)\text{Ka1}} = (1.24 \cdot 0.5211 + \text{j}\,0.34695)\ \Omega = (0.64616 + \text{j}\,0.34695)\ \Omega.$$

System + cable impedance Ka1:

$$\underline{Z}_{(1)\alpha} = \underline{Z}_{(1)Q} + \underline{Z}_{(1)\text{Ka1e}} = (0.73372 + \text{j}\,1.22255)\ \Omega,$$

$$\underline{Z}_{(1)\alpha t} = (0.73372 + \text{j}1.22255)\ \Omega \left(\frac{0.4\ \text{kV}}{20\ \text{kV}}\right)^2 = (0.00029 + \text{j}\,0.00049)\ \Omega = \underline{Z}_{(1)\alpha t}.$$

Transformer impedance:

$$\underline{Z}_{(1)T} = (0.00249 + \text{j}\,0.00927)\ \Omega.$$

System, cable, and transformer impedance:

$$\underline{Z}_{(1)\beta} = \underline{Z}_{(1)\alpha t} + \underline{Z}_{(1)T} = (0.00278 + \text{j}\,0.00976)\ \Omega,$$

$$Z_{(1)\beta} = \sqrt{0.00278^2 + 0.00976^2}\ \Omega = 0.01015\ \Omega,$$

$$\underline{Z}_{(1)\beta} = \underline{Z}_{(2)\beta},$$

$$\underline{Z}_{kN} = 2 \cdot \underline{Z}_{(1)\beta} = 2(0.00278 + \text{j}\,0.00976)\ \Omega = (0.00556 + \text{j}\,0.01952)\ \Omega,$$

$$Z_{kN} = \sqrt{0.00556^2 + 0.01952^2}\ \Omega = 0.0203\ \Omega.$$

Generator impedance, negative-sequence impedance = positive-sequence impedance:

$$\underline{Z}_{kGK} = \underline{Z}_{(1)GK} + \underline{Z}_{(2)GK} = 2 \cdot \underline{Z}_{(1)GK} = 2(0.00576 + \text{j}\,0.0384)\ \Omega = (0.01152 + \text{j}\,0.0768)\ \Omega,$$

$$Z_{kGK} = \sqrt{0.01152^2 + 0.0768^2}\ \Omega = 0.07766\ \Omega.$$

Initial symmetrical short-circuit current $I''_{k2\,min}$:

System component:

$$\underline{I}''_{kN2\,min} = \frac{c \cdot U_n}{\underline{Z}_{kN}} = \frac{0.95 \cdot 0.4\ \text{kV}}{(0.00556 + \text{j}\,0.01952)\ \Omega} = (5.13 - \text{j}\,18.00)\text{kA},$$

$$I''_{kN2\,min} = 18.71\ \text{kA}.$$

Generator component:

$$\underline{I}''_{kG2\,min} = \frac{c \cdot U_n}{\underline{Z}_{kGK}} = \frac{0.95 \cdot 0.4\ \text{kV}}{(0.01152 + \text{j}\,0.0768)\ \Omega} = (0.73 - \text{j}4.84)\ \text{kA},$$

$$I''_{kG2\,min} = 4.89\ \text{kA}.$$

Sum:

$\underline{I}''_{k2\,min} = (0.73 - j\,4.84 + 5.13 - j\,18.00) = (5.86 - j\,22.84)$ kA,

$I''_{k2\,min} = \underline{23.60\ kA}$.

<u>Fault location F2:</u>

Impedances of equipment and short-circuit current path:

Parallel impedance of system via transformer and generator:

$$\underline{Z}_{(1)\gamma} = \frac{\underline{Z}_{(1)\beta} \cdot \underline{Z}_{(1)GK}}{\underline{Z}_{(1)\beta} + \underline{Z}_{(1)GK}} = \frac{(0.00278 + j\,0.00976)(0.00576 + j\,0.0384)}{(0.00278 + j\,0.00976) + (0.00576 + j\,0.0384)}\,\Omega = (0.002 + j\,0.0078)\,\Omega,$$

$$Z_{(1)\gamma} = \sqrt{0.002^2 + 0.0078^2}\,\Omega = 0.00805\,\Omega,$$

Cable impedance Ka2 at 80 °C, 0.4 kV:

$$\underline{Z}_{(1)Ka2e} = 1.24 \cdot R_{(1)Ka2} + j\,X_{(1)Ka2} = (1.24 \cdot 0.00867 + j\,0.00918)\,\Omega = (0.01075 + j\,0.00918)\,\Omega,$$

(Negative-sequence impedance = positive-sequence impedance).

Parallel and cable impedance Ka2:

$$\underline{Z}_{(1)\delta} = \underline{Z}_{(1)\gamma} + \underline{Z}_{(1)Ka2e} = \underline{Z}_{(2)\delta} = (0.01275 + j\,0.01698)\,\Omega.$$

Total impedance:

$$\underline{Z}_k = \underline{Z}_{(1)\delta} + \underline{Z}_{(2)\delta} = 2 \cdot \underline{Z}_{(1)\delta} = 2(0.01275 + j\,0.01698)\,\Omega = (0.0255 + j\,0.03396)\,\Omega,$$

$$Z_k = \sqrt{0.0255^2 + 0.03396^2}\,\Omega = 0.04247\,\Omega.$$

Initial symmetrical short-circuit current $I''_{k2\,min}$:

$$I''_{k2\,min} = \frac{c \cdot U_n}{Z_k} = \frac{0.95 \cdot 0.4\ kV}{0.04247\,\Omega} = \underline{8.95\ kA}.$$

$\underline{Z}_{(1)GK} = \underline{Z}_{(2)GK}$ and $\underline{Z}_{(1)\beta} = \underline{Z}_{(2)\beta}$:

System component:

$$I''_{kN2\,min} = I''_{k2\,min}\frac{Z_{(1)\gamma}}{Z_{(1)\beta}} = 8.95\ kA\,\frac{0.00805\,\Omega}{0.01015\,\Omega} = 7.09\ kA.$$

Generator component:

$$I''_{kG2\,min} = I''_{k2\,min}\frac{Z_{(1)\gamma}}{Z_{(1)GK}} = 8.95\ kA\,\frac{0.00805\,\Omega}{0.03883\,\Omega} = 1.85\ kA.$$

<u>Fault location F3:</u>

Cable impedance Ka3 at 80 °C, 0.4 kV:

$$\underline{Z}_{(1)Ka3e} = 1.24 \cdot R_{(1)Ka3} + j\,X_{(1)Ka3} = (1.24 \cdot 0.01354 + j\,0.00297)\,\Omega = (0.01679 + j\,0.00297)\,\Omega.$$

Parallel and cable impedance Ka2 and Ka3:

$$\underline{Z}_{(1)\varepsilon} = \underline{Z}_{(1)\delta} + \underline{Z}_{(1)Ka3e} = (0.02954 + j\,0.01995)\,\Omega.$$

Total impedance, negative-sequence impedance = positive-sequence impedance:

$$\underline{Z}_k = \underline{Z}_{(1)\varepsilon} + \underline{Z}_{(2)\varepsilon} = 2 \cdot \underline{Z}_{(1)\varepsilon} = 2\,(0.02954 + j\,0.01995)\,\Omega = (0.05908 + j\,0.0399)\,\Omega,$$

$$Z_k = \sqrt{0.05908^2 + 0.0399^2}\,\Omega = 0.07129\,\Omega.$$

Initial symmetrical short-circuit current $I''_{k2\,min}$:

$$I''_{k2\,min} = \frac{c \cdot U_n}{Z_k} = \frac{0.95 \cdot 0.4\,kV}{0.07129\,\Omega} = \underline{5.33\,kA}.$$

▷ ***Phase-to-ground fault currents***
a) Maximum short-circuit currents with phase-to-ground fault

Fault location F1:

Impedances of equipment and short-circuit current path:

System, cable, and transformer impedance:

$$\underline{Z}_{(1)\beta} = (0.00273 + j\,0.00976)\,\Omega.$$

Generator impedance:

$$\underline{Z}_{(1)GK} = (0.00576 + j\,0.0384)\,\Omega.$$

Zero-sequence impedance of transformer:

$$R_{(0)T} = R_{(1)T} = 0.00249\,\Omega \quad \text{(see page 60)},$$
$$X_{(0)T} \approx 0.96 \cdot X_{(1)T} = 0.96 \cdot 0.00927\,\Omega = 0.0089\,\Omega \quad \text{(see page 60)}$$
$$\underline{Z}_{(0)T} = \left(R_{(0)T} + j X_{(0)T}\right) = (0.00249 + j\,0.0089)\,\Omega.$$

Zero-sequence impedance of generator:

$$X_{(0)G} = \frac{x_{(0)G} \cdot U_{rG}^2}{100\% \cdot S_{rG}} = \frac{7.54\% \cdot 0.4^2\,kV^2}{100\% \cdot 0.5\,MVA} = 0.02413\,\Omega,$$

$$R_{(0)G} = R_{(1)G} = 0.00621\,\Omega \quad \text{(see page 60)},$$
$$\underline{Z}_{(0)G} = 0.00621 + j\,0.02413\,\Omega,$$
$$\underline{Z}_{(0)GK} = K_G \cdot \underline{Z}_{(0)G} = 0.928(0.00621 + j\,0.02413)\,\Omega = (0.00576 + j\,0.02239)\,\Omega.$$

The neutral of the generator is connected to ground via a reactor to reduce the circulating currents. The reactor is dimensioned such that the phase-to-ground fault current of the generator does not exceed the three-phase short-circuit current in the event of a terminal short circuit. The reactor has an inductive reactance of $X_{rD} = 0.00531\,\Omega$. When calculating the phase-to-ground fault currents, this value must be multiplied by three (zero-sequence reactance).

$$\underline{Z}_{(0)D} = 3 \cdot j X_D = j\,3 \cdot 0.00531\,\Omega = j\,0.01593\,\Omega.$$

The equivalent impedances of the parallel connections can be derived in order to avoid having to perform calculations with symmetrical components.

Positive-sequence and negative-sequence impedance of parallel connection of system via transformer and generator:

$$\underline{Z}_{(1)\gamma} = (0.00197 + j\,0.0078)\,\Omega \quad \text{(see page 61)},$$
$$\underline{Z}_{(0)\gamma} = \frac{\underline{Z}_{(0)T} \cdot (\underline{Z}_{(0)GK} + \underline{Z}_{(0)D})}{\underline{Z}_{(0)T} + \underline{Z}_{(0)GK} + \underline{Z}_{(0)D}} = \frac{(0.00249 + j\,0.0089)\,[0.00576 + j\,(0.02239 + 0.01593)]}{(0.00249 + j\,0.0089) + 0.00576 + j\,(0.02239 + 0.01593)}\,\Omega$$
$$= (0.00184 + j\,0.00724)\,\Omega.$$

Total impedance, negative-sequence impedance = positive-sequence impedance:

$$\underline{Z}_k = \underline{Z}_{(1)\gamma} + \underline{Z}_{(2)\gamma} = \underline{Z}_{(0)\gamma} = 2 \cdot \underline{Z}_{(1)\gamma} + \underline{Z}_{(0)\gamma}$$
$$= [2 \cdot 0.00197 + 0.00184 + j(2 \cdot 0.0078 + 0.00724)]\,\Omega = (0.00578 + j\,0.02284)\,\Omega,$$
$$Z_k = \sqrt{0.00578^2 + 0.02284^2}\,\Omega = 0.02356\,\Omega.$$

Initial symmetrical short-circuit current I''_{k1max}:

$$I''_{k1max} = \frac{\sqrt{3} \cdot c \cdot U_n}{Z_k} = \frac{\sqrt{3} \cdot 1 \cdot 0.4\,\text{kV}}{0.02356} = \underline{29.41\,\text{kA}}.$$

Fault location F2:

Impedances of equipment and short-circuit current path:

Parallel and cable impedance of parallel connection of transformer and generator:

$$\underline{Z}_{(1)\delta} = \underline{Z}_{(2)\delta} = (0.01064 + \text{j}\,0.01698)\,\Omega$$ (Negative-sequence impedance = positive-sequence impedance).

Zero-sequence impedance of parallel connection of transformer and generator:

$$\underline{Z}_{(0)\gamma} = (0.00184 + \text{j}\,0.00724)\,\Omega.$$

Zero-sequence impedance of cable Ka2:

According to Fig. 1.2.8 and 1.2.10:

$$R_{(0)\text{EKa2}} = 3.55 \cdot R_{(1)\text{Ka2}} = 3.55 \cdot 0.00867\,\Omega = 0.03078\,\Omega,$$

$$X_{(0)\text{EKa2}} = 3.5 \cdot X_{(1)\text{Ka2}} = 3.5 \cdot 0.00918\,\Omega = 0.03213\,\Omega,$$

$$Z_{(0)\text{Ka2}} = (0.03078 + \text{j}\,0.03213)\,\Omega.$$

Zero-sequence impedance parallel connection + cable Ka2:

$$\underline{Z}_{(0)\delta} = \underline{Z}_{(0)\gamma} + Z_{(0)\text{Ka2}} = (0.03262 + \text{j}\,0.03937)\,\Omega.$$

Total impedance:

$$\underline{Z}_k = \underline{Z}_{(1)\delta} + \underline{Z}_{(2)\delta} + \underline{Z}_{(0)\delta} = [2 \cdot 0.01064 + 0.03262 + \text{j}\,(2 \cdot 0.01698 + 0.03937)]\,\Omega$$
$$= (0.0539 + \text{j}\,0.07333)\,\Omega,$$

$$Z_k = \sqrt{0.0539^2 + 0.07333^2}\,\Omega = 0.091\,\Omega.$$

Initial symmetrical short-circuit current I''_{k1max}:

$$I''_{k1max} = \frac{\sqrt{3} \cdot c \cdot U_n}{Z_k} = \frac{\sqrt{3} \cdot 1 \cdot 0.4\,\text{kV}}{0.091\ \Omega} = \underline{7.61\,\text{kA}}.$$

Fault location F3:

Impedances of equipment and short-circuit current path:

Parallel and cable impedances Ka2 + Ka3:

$$\underline{Z}_{(1)\varepsilon} = \underline{Z}_{(2)\varepsilon} = (0.02418 + \text{j}\,0.01995)\,\Omega$$ (see page 62).

Zero-sequence impedance, parallel connection + cable Ka2:

$$\underline{Z}_{(0)\delta} = (0.03262 + \text{j}\,0.03937)\,\Omega.$$

Zero-sequence impedance of cable Ka3:

According to Fig. 1.2.8 and 1.2.10:

$$R_{(0)\text{EKa3}} = 2.85 \cdot R_{(1)\text{Ka3}} = 2.85 \cdot 0.01354\,\Omega = 0.03859\,\Omega,$$

$$X_{(0)\text{EKa3}} = 7.1 \cdot X_{(1)\text{Ka3}} = 7.1 \cdot 0.00297\,\Omega = 0.02109\,\Omega,$$

$$\underline{Z}_{(0)\text{Ka3}} = (0.03859 + \text{j}\,0.02109)\,\Omega.$$

Zero-sequence impedance, parallel and cable impedances Ka2 + Ka3:

$$\underline{Z}_{(0)\varepsilon} = \underline{Z}_{(0)\delta} + \underline{Z}_{(0)\text{Ka3}} = (0.07121 + \text{j}\,0.06046)\ \Omega.$$

Total impedance:

$$\underline{Z}_{\text{k}} = \underline{Z}_{(1)\varepsilon} + \underline{Z}_{(2)\varepsilon} + \underline{Z}_{(0)\varepsilon} = (0.11957 + \text{j}\,0.10036)\ \Omega$$ (Negative-sequence impedance = positive-sequence impedance).

$$Z_{\text{k}} = \sqrt{0.11957^2 + 0.10036^2}\ \Omega = 0.1561\ \Omega.$$

Initial symmetrical short-circuit current $I''_{\text{k1 max}}$:

$$I''_{\text{k1 max}} = \frac{\sqrt{3} \cdot c \cdot U_{\text{n}}}{Z_{\text{k}}} = \frac{\sqrt{3} \cdot 1 \cdot 0.4\ \text{kV}}{0.1561\ \Omega} = \underline{4.44\ \text{kA}}.$$

b) Minimum short-circuit currents with phase-to-ground fault

Fault location F1:

$$\underline{Z}_{(1)\gamma} = \underline{Z}_{(2)\gamma} = (0.002 + \text{j}\,0.0078)\ \Omega \quad \text{(see page 64)},$$

$$\underline{Z}_{(0)\gamma} = (0.00184 + \text{j}\,0.00724)\ \Omega \quad \text{(see page 65)}.$$

Total impedance, negative-sequence impedance = positive-sequence impedance:

$$\underline{Z}_{\text{k}} = 2 \cdot \underline{Z}_{(1)\gamma} + \underline{Z}_{(0)\gamma} = (0.00584 + \text{j}\,0.02284)\ \Omega,$$

$$Z_{\text{k}} = \sqrt{0.00584^2 + 0.02284^2}\ \Omega = 0.02357\ \Omega.$$

Initial symmetrical short-circuit current $I''_{\text{k1 min}}$:

$$I''_{\text{k1 min}} = \frac{\sqrt{3} \cdot c \cdot U_{\text{n}}}{Z_{\text{k}}} = \frac{\sqrt{3} \cdot 0.95 \cdot 0.4\ \text{kV}}{0.02357\ \Omega} = \underline{27.92\ \text{kA}}.$$

Fault location F2:

$$\underline{Z}_{(1)\delta} = (0.01275 + \text{j}\,0.01698)\ \Omega \quad \text{(see page 64)},$$

$$\underline{Z}_{(0)\gamma} = (0.00184 + \text{j}\,0.00724)\ \Omega \quad \text{(see page 65)}.$$

Zero-sequence impedance of cable Ka2 at 80 °C:

$$\underline{Z}_{(0)\text{Ka2e}} = 1.24 \cdot R_{(0)\text{EKa2}} + \text{j}X_{(0)\text{EKa2}} = (1.24 \cdot 0.03078 + \text{j}\,0.03213)\ \Omega$$
$$= (0.03817 + \text{j}\,0.03213)\ \Omega.$$

Zero-sequence impedance of parallel connection of transformer and generator + cables Ka2 at 80 °C:

$$\underline{Z}_{(0)\delta} = \underline{Z}_{(0)\gamma} + \underline{Z}_{(0)\text{Ka2e}} = (0.04001 + \text{j}\,0.03937)\ \Omega.$$

Total impedance, negative-sequence impedance = positive-sequence impedance:

$$\underline{Z}_{\text{k}} = 2 \cdot \underline{Z}_{(1)\delta} + \underline{Z}_{(0)\delta} = (0.06551 + \text{j}\,0.07333)\ \Omega.$$

$$Z_{\text{k}} = \sqrt{0.06551^2 + 0.07333^2}\ \Omega = 0.09833\ \Omega.$$

Initial symmetrical short-circuit current $I''_{\text{k1 min}}$:

$$I''_{\text{k1 min}} = \frac{\sqrt{3} \cdot 0.95 \cdot 0.4\ \text{kV}}{0.09833\ \Omega} = \underline{6.69\ \text{kA}}.$$

Fault location F3:

$$\underline{Z}_{(1)\varepsilon} = (0.02954 + \mathrm{j}\,0.01995)\ \Omega \quad \text{(see page 64)},$$

Zero-sequence impedance of cable Ka3 at 80 °C:

$$\underline{Z}_{(0)\mathrm{Ka3e}} = 1.24 \cdot R_{(0)\mathrm{EKa3}} + \mathrm{j}\,X_{(0)\mathrm{EKa3}} = (1.24 \cdot 0.03859 + \mathrm{j}\,0.02109)\ \Omega$$
$$= (0.04785 + \mathrm{j}\,0.02109)\ \Omega.$$

Zero-sequence impedance of parallel connection of transformer and generator + cables Ka2 and Ka3 at 80 °C:

$$\underline{Z}_{(0)\varepsilon} = \underline{Z}_{(0)\delta} + \underline{Z}_{(0)\mathrm{Ka3e}} = (0.08786 + \mathrm{j}\,0.06046)\ \Omega.$$

Total impedance, negative-sequence impedance = positive-sequence impedance:

$$\underline{Z}_{\mathrm{k}} = 2 \cdot \underline{Z}_{(1)\varepsilon} + \underline{Z}_{(0)\varepsilon} = (0.14694 + \mathrm{j}\,0.10036)\ \Omega,$$
$$Z_{\mathrm{k}} = \sqrt{0.14694^2 + 0.10036^2}\ \Omega = 0.17794\ \Omega.$$

Initial symmetrical short-circuit current $I''_{\mathrm{k1\,min}}$:

$$I''_{\mathrm{k1\,min}} = \frac{\sqrt{3} \cdot 0.95 \cdot 0.4\ \mathrm{kV}}{0.17794\ \Omega} = \underline{3.7\ \mathrm{kA}}.$$

These currents are only relevant when dimensioning protective devices.

Table of calculated initial symmetrical short-circuit currents I''_{k} and maximum asymmetrical short-circuit current $i_{\mathrm{p3\,max}}$

Fault location		F1	F2	F3
Max. three-phase	$I''_{\mathrm{k3\,max}}$ (kA)	28.70	11.55	7.37
Min. phase-to-phase	$I''_{\mathrm{k2\,min}}$ (kA)	23.60	8.95	5.33
Max. phase-to-ground	$I''_{\mathrm{k1\,max}}$ (kA)	29.41	7.61	4.44
Min. phase-to-ground	$I''_{\mathrm{k1\,min}}$ (kA)	27.92	6.69	3.70
Max. asymmetrical short-circuit current	$i_{\mathrm{p3\,max}}$ (kA)	60.23	21.98	12.47

Generally speaking, the initial phase-to-phase symmetrical short-circuit currents I''_{k} are $\sqrt{3}/2$ times the three-phase initial symmetrical short-circuit currents. The difference between the maximum and minimum initial symmetrical short-circuit currents is due to the different c factors, the higher cable resistances resulting from temperature rises (factor 1.24), and possibly the different levels of initial symmetrical short-circuit power applied at the supply terminals.

1.2.3 Equipment impedance values

The impedance values of the individual items of equipment must be known before the symmetrical and asymmetrical short-circuit currents for the various fault types can be calculated. The impedance values can be determined from the equipment data which is provided by the respective manufacturer. If this data is not available at the project planning stage, empirical values should be used. In order to make this task easier for planners and designers, the principal formulas for determining the impedance values from the equipment data are specified below. They are followed by tables and diagrams, which provide information on the average values of equipment data as well as the linear resistance and inductive reactance of equipment.

The relevant operands in accordance with IEC 60909/DIN VDE 0102 are specified in Figs. 1.2/13 to 1.2/15.

Synchronous generators

Initial reactance

$$X''_d = \frac{x''_d \cdot U^2_{rG}}{100\% \cdot S_{rG}} \text{ in } \Omega/\text{phase}$$

S_{rG} Rated power of generator (MVA)

U_{rG} Rated voltage of generator (kV)

x''_d Specific initial reactance (%) (subtransient reactance)

$R_G = 0.05 \cdot X''_d$ for generators with $U_{rG} > 1$ kV and $S_{rG} \geq 100$ MVA

$R_G = 0.07 \cdot X''_d$ for generators with $U_{rG} > 1$ kV and $S_{rG} < 100$ MVA

$R_G = 0.15 \cdot X''_d$ for generators with $U_{rG} \leq 1000$ V

Approximate values for generators with $U_{rG} \leq 1000$ V are listed in Table 1.2/4 on page 76.

Synchronous motors and synchronous capacitors

In the case of generators without unit transformers where the generator is connected directly (e. g. in low-voltage systems), the generator impedance $\underline{Z}_G = (R_G + jX''_d)$ must be multiplied by a correction factor K_G.

$$K_G = \frac{U_n}{U_{rG}} \cdot \frac{c_{max}}{1 + x''_d \sin \varphi_{rG}}$$

c_{max} Voltage factor

U_n Rated voltage of system

U_{rG} Rated voltage of generator

φ_{rG} Phase angle between I_{rG} and $U_{rG}/\sqrt{3}$

When performing calculations, synchronous motors and synchronous capacitors should be treated in the same way as synchronous generators.

Asynchronous motors

Motors in public low-voltage systems can be disregarded.

Asynchronous motors and groups of motors can also be disregarded if the sum of their rated currents in the vicinity of the fault location is less than one percent of the initial symmetrical short-circuit current I''_k without taking the motors into consideration. In all other cases, please refer to IEC 60909/DIN VDE 0102 Section 2.

Two-winding transformers

Resistance

$$R_T = \frac{u^2_{Rr} \cdot U^2_{rT}}{100\% \cdot S_{rT}} \text{ in } \Omega/\text{phase}$$

Inductive reactance

$$X_T = \frac{u_{Xr} \cdot U^2_{rT}}{100\% \cdot S_{rT}} \text{ in } \Omega/\text{phase}$$

$$u_{Xr} = \sqrt{u^2_{kr} - u^2_{Rr}}$$

S_{rT} Rated power of transformer (MVA)

U_{rT} Rated voltage of transformer (kV) (mid-tap values)

u_{Rr} Rated resistive voltage drop (%)

u_{Xr} Rated reactive voltage drop (%)

u_{kr} Rated short-circuit voltage (%)

Approximate values for transformer characteristics are specified in Tables 1.2/5 and 1.2/6 (pages 76/77). The impedances for balanced operation, which are equal to the impedances of the positive-sequence system and the negative-sequence system, can be determined from these characteristic values:

$(\underline{Z}_T = \underline{Z}_{(1)T} = \underline{Z}_{(2)T})$.

Table 1.2/1
Impedance values for l.v. transformers

Transformer vector group	$R_{(0)T}$ / $R_{(1)T}$	$X_{(0)T}$ / $X_{(1)T}$
Dy	1	0.95
Dz, Yz	0.4	0.1
Yy [1)]	1	7 to 100 [2)]

[1)] The transformers Yy are not suitable for protection by automatic disconnection of the power supply in TN systems with overcurrent protective devices or RCDs.
[2)] H.v. neutral not grounded.

It is more difficult to predict the zero-sequence impedances of transformers which are required, for example, when calculating the phase-to-ground fault currents. The magnitude of the zero-sequence impedances depends to a great extent on the method of connection and the mechanical construction.

Table 1.2/1 provides an overview of the most frequently encountered impedance values for l. v. transformers as quotients of the resistances and inductive reactances in zero-sequence and positive-sequence systems where $f = 50$ Hz.

Overhead lines

The resistance values for copper, aluminum, Aldrey, aluminum-steel, and Aldrey-steel overhead-line conductors manufactured in accordance with DIN 48 201, 48 204, and 48 206 are specified in Tables 1.2/7 and 1.2/8.

Mean values for inductive reactance per unit length are specified in Table 1.2/2 and mean values for zero-sequence inductive reactance per unit length in Table 1.2/3.

Cables

The resistance per unit length r' in Ω/km and the inductance per unit length l' in mH/km can be found in the Siemens publication "Power Cables and their Application" (see Section 7.3). According to IEC 60909/DIN VDE 0102, the d.c. resistance value can be used in place of the resistance value.

The inductive reactance per unit length is calculated as follows:

$$x' = 2 \cdot \pi \cdot f \cdot 10^{-3} \cdot l' \; \Omega/\text{km}$$

f System frequency (1/s (Hz))

l' Inductance per unit length (mH/km)

For 50 Hz, $x' = 2 \cdot \pi \cdot 50 \cdot 10^{-3} \cdot l' \; \Omega/\text{km}$.

Table 1.2/2
Inductive reactance per unit length X'_L in positive-sequence systems for overhead-line conductors where $f = 50$ Hz

Nominal cross section q_n (mm^2)	Inductive reactance per unit length X'_L (Ω/km) with mean phase-to-phase spacing d (cm) 50	60	70	80	90	100
10	0.37	0.38	0.40	0.40	0.41	0.42
16	0.36	0.37	0.38	0.39	0.40	0.40
25	0.34	0.35	0.37	0.37	0.38	0.39
35	0.33	0.34	0.35	0.36	0.37	0.38
50	0.32	0.33	0.34	0.35	0.36	0.37
70	0.31	0.32	0.33	0.34	0.35	0.35
95	0.29	0.31	0.32	0.33	0.34	0.34
120	0.29	0.30	0.31	0.32	0.33	0.34

The following can be applied: $\frac{X'_L}{\Omega/\text{km}} = 2\pi\, 10^{-2} \left(0.25 + \ln \frac{d}{r}\right)$

d Mean phase-to-phase spacing
r Conductor radius

Table 1.2/3
Quotients of resistances and inductive reactances in zero-sequence and positive-sequence systems for l.v. overhead lines with four conductors of identical cross section where $f = 50$ Hz

$\frac{R_{(0)L}}{R_{(1)L}}$	2 when calculating maximum short-circuit current 4 when calculating minimum short-circuit current
$\frac{X_{(0)L}}{X_{(1)L}}$	3 when calculating maximum short-circuit current 4 when calculating minimum short-circuit current

The specified value for $R_{(0)L}/R_{(1)L}$ can be used for conductor temperatures of 20 °C and 80 °C.

$R_{(0)L}$ Resistance of zero-sequence system
$R_{(1)L}$ Resistance of positive-sequence system
$X_{(0)L}$ Reactance of zero-sequence system
$X_{(1)L}$ Reactance of positive-sequence system

Mean values for inductive reactance per unit length $x'_{(1)}$ per conductor where $f = 50$ Hz

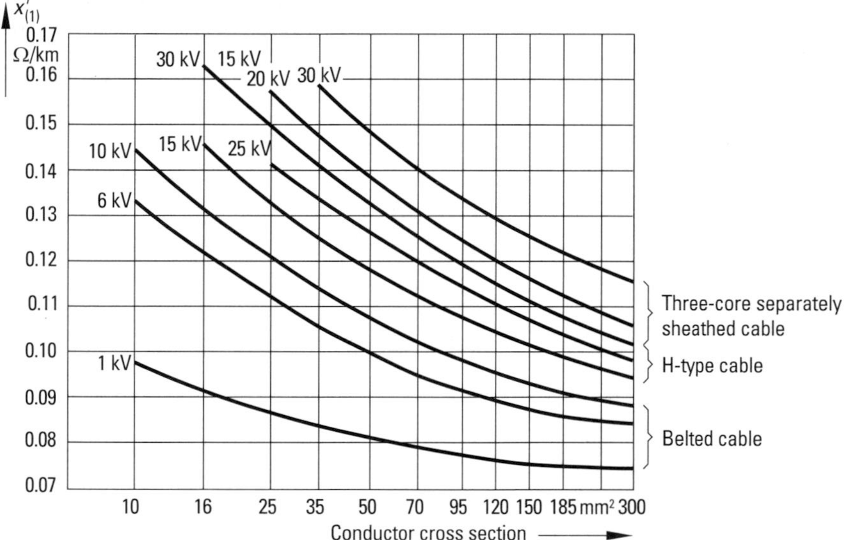

Fig. 1.2/7 Three-phase cable for 1 to 30 kV

Mean values for inductive reactance per unit length for the positive-sequence and negative-sequence system can be found in Fig. 1.2/7.

Mean values for resistance and inductive zero-sequence reactance are specified in Figs. 1.2/8 to 1.2/11.

Mean resistance ratios for the zero-sequence system relative to the positive-sequence system for 1 kV four-core cables per conductor where $f = 50$ Hz

$R_{(0)}$ with 4th conductor only as return conductor (Ω)
$R_{(0)M}$ 4th conductor and cable sheath as return conductor (Ω)
$R_{(0)E}$ 4th conductor and ground as return conductor (Ω)
$R_{(0)ME}$ 4th conductor, cable sheath, and ground as return conductor (Ω)

Positive-sequence resistance $R_{(1)}$ (Ω per conductor)

Specific soil resistivity $\rho = 100$ Ωm

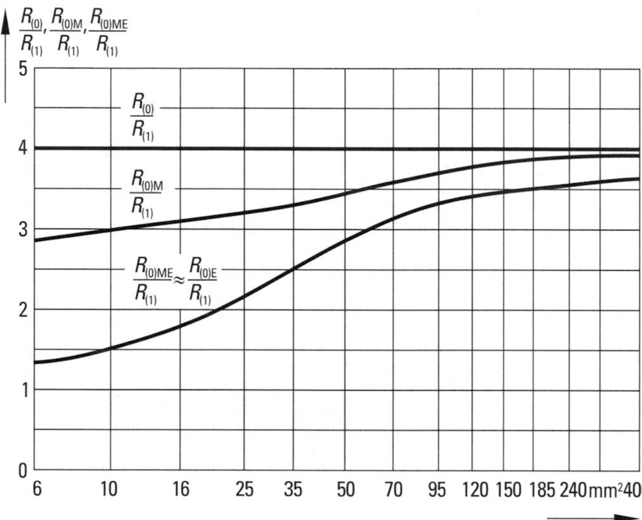

Fig. 1.2/8
Cable with copper conductors (NKBA and PROTODUR)

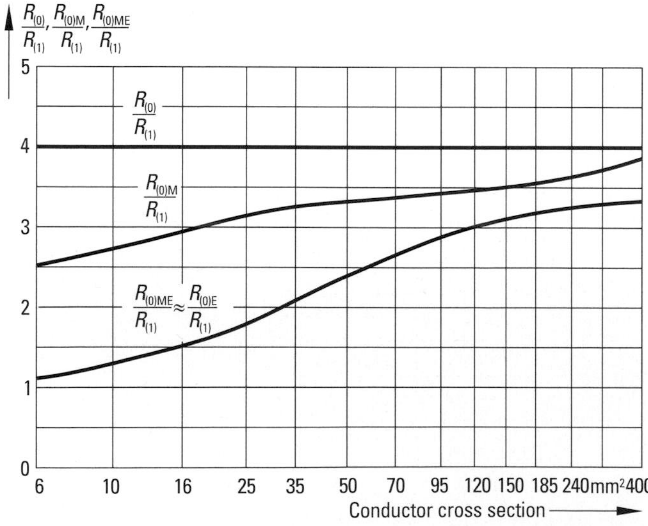

Fig. 1.2/9
Cable with aluminium conductors (NKBA and PROTODUR)

Mean inductive reactance ratios for the zero-sequence system relative to the positive-sequence system for 1 kV four-core cables per conductor where $f = 50$ Hz

$X_{(0)}$ with 4th conductor only as return conductor (Ω)
$X_{(0)M}$ 4th conductor and cable sheath as return conductor (Ω)
$X_{(0)E}$ 4th conductor and ground as return conductor (Ω)
$X_{(0)ME}$ 4th conductor, cable sheath, and ground as return conductor (Ω)

Inductive positive-sequence reactance $X_{(1)}$ (Ω per conductor)

Specific soil resistivity $\rho = 100\ \Omega$m

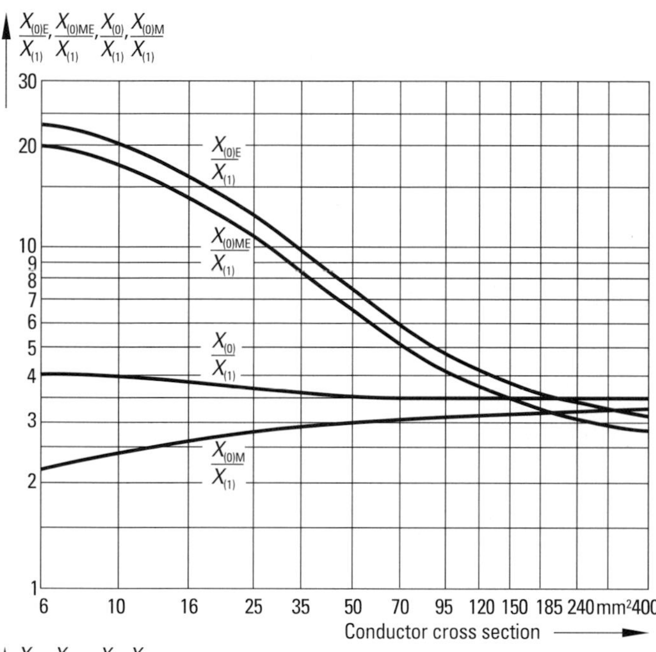

Fig. 1.2/10
Cable with copper conductors (NKBA and PROTODUR)

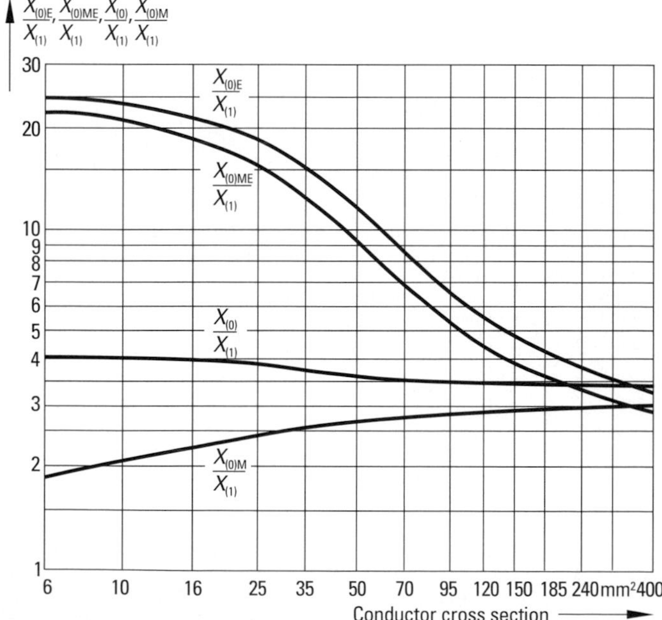

Fig. 1.2/11
Cable with aluminum conductors (NKBA and PROTODUR)

Busbars

Resistance can be disregarded in the case of large cross sections.

Mean values for the inductive reactance per unit length of busbars (rectangular-section conductors) can be found in Fig. 1.2/12.

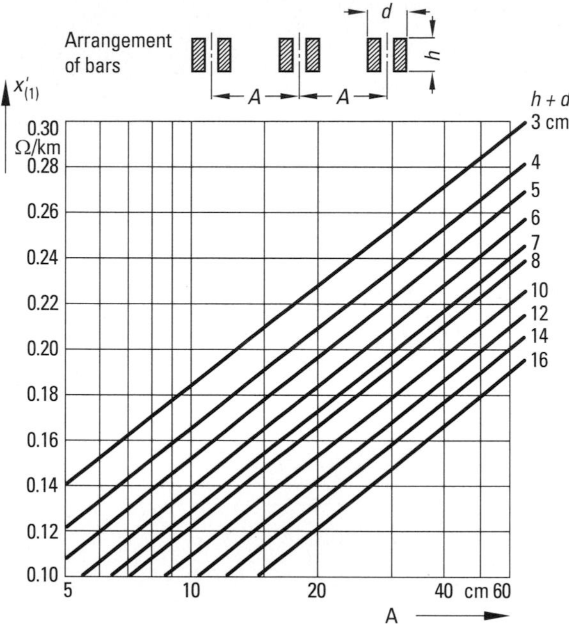

Fig. 1.2/12
Mean values for inductive reactance per unit length $x_{(1)}$ per phase conductor where $f = 50$ Hz

1.2.4 Calculation variables to IEC 60909 / DIN VDE 0102

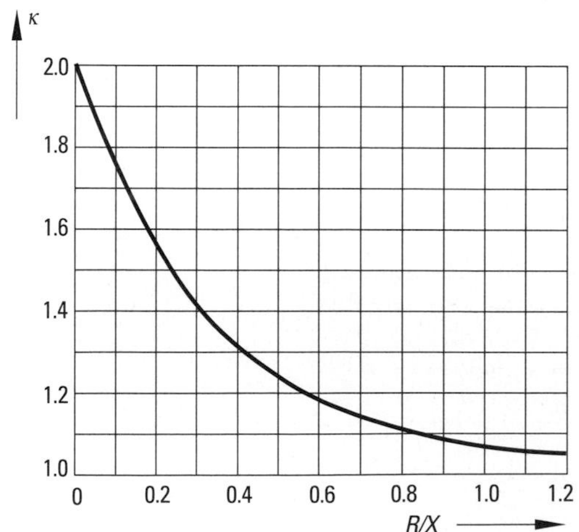

Fig. 1.2/13
Factor κ as function of R/X

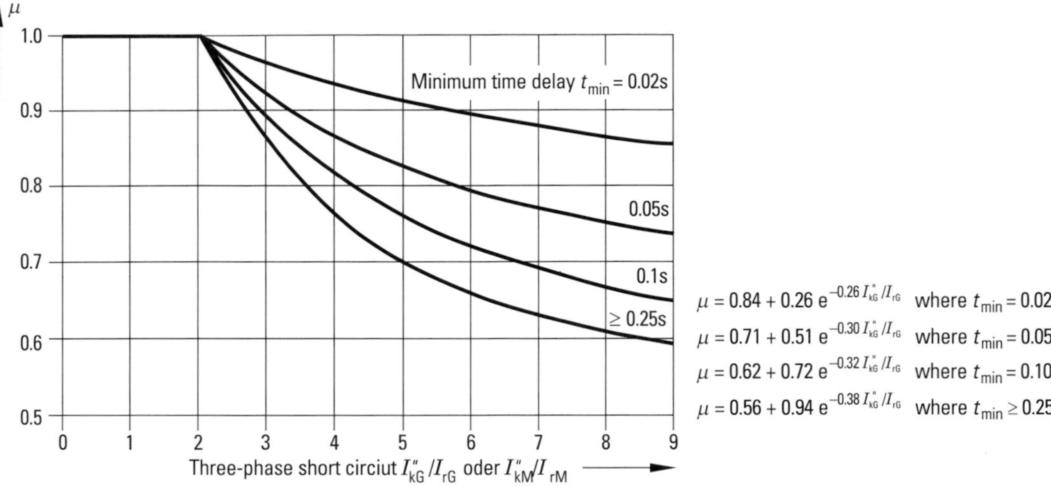

Fig. 1.2/14 Factor μ for calculating symmetrical breaking current I_b

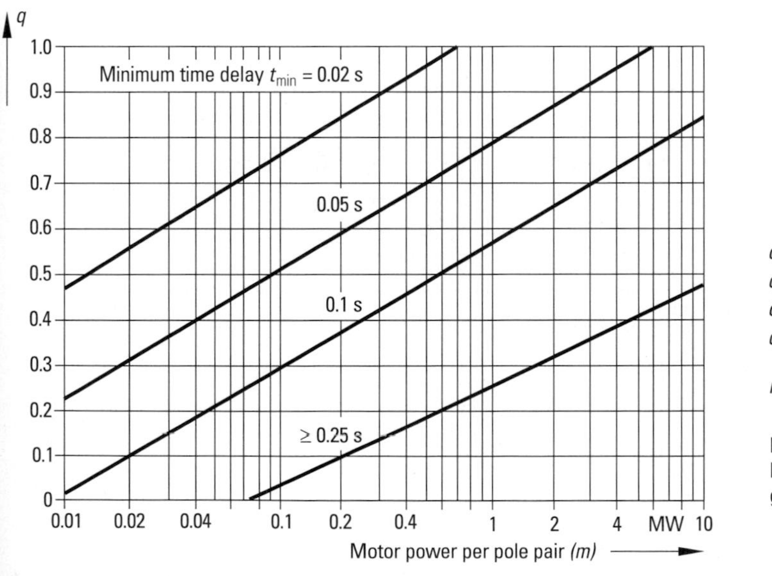

$q = 1.03 + 0.12 \ln m$ where $t_{min} = 0.02$ s
$q = 0.79 + 0.12 \ln m$ where $t_{min} = 0.05$ s
$q = 0.57 + 0.12 \ln m$ where $t_{min} = 0.10$ s
$q = 0.26 + 0.10 \ln m$ where $t_{min} \geq 0.25$ s

m Rated motor power per pole pair (MW)

If the calculation for *q* as specified in IEC 909/DIN VDE 0102 yields values greater than 1, *q* is given the value 1.

Fig. 1.2/15 Factor q for calculating the symmetrical breaking current I_b of asynchronous motors

Table 1.2/4 Characteristic values of l.v. synchronous generators

		Turbine-driven generators	Number of poles	Salient-pole generators	Number of poles
Rated power	kVA	40 to 1400		1600 to 3600	
Subtransient reactance x''_d (saturated)	%	10 to 15	4 to 14	10 to 12	2
				11 to 23	4
Transient reactance x'_d (saturated)	%	20 to 40	4 to 14	13 to 17	2
				26 to 36	4
Synchronous reactance x_d (unsaturated)	%	150 to 300	4 to 14	170 to 220	2
				260 to 300	4
No-load short-circuit ratio K_0		0.4 to 0.8	4 to 14	0.6 to 0.7	2
				0.4 to 0.5	4
Negative-sequence reactance $x_{(2)}$	%	$\approx x''_d$	4 to 14	$\approx x''_d$	2+4
Zero-sequence reactance $x_{(0)}$	%	$(0.4 \text{ to } 0.8) \cdot x''_d$	4 to 14	$(0.4 \text{ to } 0.6) \cdot x''_d$	2+4
Subtransient time constant T''_d	s	0.001 to 0.03	4 to 14	0.001 to 0.035	2+4
Transient time constant T'_d	s	0.04 to 1.0	4 to 14	0.04 to 1.2	2+4
Time constant of d.c. component T_g	s	0.01 to 0.25	4 to 14	0.03 to 0.25	2+4

Table 1.2/5
Characteristic values of three-phase distribution transformers
High-voltage rating $U_{rOS} = 3000$ to 24 000 V; rated impedance voltage $u_{kr} = 4\%$
According to DIN 42 500

S_{rT}	kVA		50	75	100	125	160	200	250	315	400	500	630
u_{Rr}	%		2.1	1.89	1.75	1.64	1.47	1.42	1.3	1.24	1.15	1.1	1.03
U_{rUS}	V	230	Yyn0									–	
	V	400	Yzn5						Dyn5				
	V	525	Yyn0										

According to DIN 42 503

S_{rT}	kVA		50	75	100	125	160	200	250	315	400	500	630
u_{Rr}	%		2.3	2.09	1.95	1.84	1.75	1.65	1.64	1.56	1.5	1.43	1.33
U_{rUS}	V	230	Yyn0									–	
	V	400	–						Dyn5				
	V	525	Yyn0										
S_{rT}	kVA		50	75	100	125	160	200					
u_{Rr}	%		2.5	2.27	2.15	2.0	1.94	1.8	–				
U_{rUS}	V	400	Yzn5										

(continued next page)

Table 1.2/5
Characteristic values of three-phase distribution transformers *(continued)*
According to DIN 42511

S_{rT}	kVA	250	315	400	500	630	800	1000	1250	1600
u_{Rr}	%	1.78	1.71	1.61	1.56	1.48	1.38	1.35	1.31	1.24
U_{rUS}	V 400	Dyn5								
	V 525	Yyn0								
S_{rT}	kVA	250	315	400	–					
u_{Rr}	%	1.84	1.73	1.62						
U_{rUS}	V 230	Yyn0								

High-voltage rating $U_{rOS} = 24000$ to 36000 V; rated impedance voltage $u_{kr} = 6\%$

According to DIN 42511

S_{rT}	kVA	100	125	160	200	250	315	400	500	630	800	1000	1250	1600
u_{Rr}	%	2.1	2.0	1.88	1.78	1.75	1.71	1.61	1.56	1.48	1.38	1.35	1.31	1.24
U_{rUS}	V 230	Yyn0			–									
	V 400	–			Dyn5									
	V 525	Yyn0												
S_r	kVA	100	125	160	200	250	315	400						
u_{Rr}	%	2.3	2.16	2.0	1.9	1.84	1.73	1.62						
U_{rUS}	V 230	–				Yyn0								
	V 400	Yzn5				–								

High-voltage rating for to 250 to 800 kVA: $U_{rOS} =$ 3000 to 12000 V
to 1000 to 1500 kVA: $U_{rOS} =$ 5000 to 12000 V
to 250 to 1600 kVA: $U_{rOS} =$ 12000 to 24000 V

Rated impedance voltage $u_{kr} = 6\%$

Table 1.2/6
Characteristic values of three-phase distribution transformers
Rated impedance voltage u_{kr}

High-voltage rating	kV	6 to 20	30	60	110
u_{kr}	%	3.5 to 8	6 to 11	9 to 12	9 to 15

Rated resistance voltage u_{Rr}

Rated power	MVA	0.1	0.32	1	3.2	10	32
u_{Rr}	%	1.8 to 2.1	1.5 to 1.8	1.3 to 1.5	0.8 to 1.0	0.5 to 0.7	0.4 to 0.6

Table 1.2/7
Resistances r' (Ω/km per conductor) at 20 °C with mean conductor temperature for overhead-line conductors, manufactured according to DIN 48 201 Part 1 (copper), Part 5 (aluminum), and Part 6 (E-AlMgSi)

Nominal cross section q_n mm²	Conductor diameter $d = 2r$ mm	Copper $r'_{20\,°C}$ Ω/km	Aluminum $r'_{20\,°C}$ Ω/km	E-AlMgSi (Aldrey) $r'_{20\,°C}$ Ω/km
10	4.10	1.806	–	–
16	5.10	1.139	1.802	2.090
25	6.30	0.746	1.181	1.370
35	7.50	0.527	0.834	0.967
50	9.00	0.366	0.579	0.671
70	10.50	0.276	0.437	0.507
95	12.50	0.195	0.309	0.358
120	14.00	0.155	0.246	0.285
150	15.80	0.124	0.196	0.227
185	17.50	0.100	0.159	0.184
240	20.30	0.075	0.119	0.138
300	22.50	0.061	0.097	0.112
400	26.00	0.046	0.072	0.084
500	29.10	0.037	0.058	0.067
625	32.60	–	0.046	0.054
800	36.90	–	0.036	0.042
1000	41.10	–	0.029	0.034

Table 1.2/8
Resistance per unit length r' (mean values at 50 Hz) for overhead-line conductors manufactured according to DIN 48 204 and 48 206 [from "Aluminium-Freileitungen" (Aluminum overhead lines), Aluminium-Verlag GmbH, Düsseldorf]

The resistance values are calculated for the actual cross section of the aluminum or Aldrey layers, allowing for the lay ratio referred to the electrical resistances (at 20 °C) of 0.02826 Ω mm²/m for aluminum and 0.0328 Ω mm²/m for Aldrey, where the conductivity of aluminum is 35.38 m/Ω mm² and that of Aldrey is 30.58 m/Ω mm²

Nominal cross sections [1)] q_n/q_{nS} mm²	Resistance per unit length r' for aluminum/steel Ω/km	for Aldrey/steel Ω/km	Nominal cross sections [1)] q_n/q_{nS} mm²	Resistance per unit length r' for aluminum/steel Ω/km	for Aldrey/steel Ω/km
16/2.5	1.8792	2.180	210/35	0.1380	0.1601
25/4	1.2027	1.395	210/50	0.1363	0.1581
35/6	0.8353	0.9689	230/30	0.1249	0.1449
44/32	0.6566	0.7616	240/40	0.1188	0.1378
50/8	0.5946	0.6898	265/35	0.1117	0.1269
50/30	0.5644	0.6547	300/50	0.09488	0.11006
70/12	0.4130	0.4791	305/40	0.0949	0.11009
95/15	0.3058	0.3547	340/30	0.0853	0.0989
95/55	0.2992	0.3471	380/50	0.0757	0.0879
105/75	0.2733	0.3170	385/35	0.0749	0.0869
120/20	0.2374	0.2754	435/55	0.0666	0.0772
120/70	0.2364	0.2742	450/40	0.0644	0.0747
125/30	0.2259	0.2621	490/65	0.0590	0.0684
150/25	0.1939	0.2249	550/70	0.0526	0.0610
170/40	0.1682	0.1952	560/50	0.0515	0.0597
185/30	0.1571	0.1822	680/85	0.0426	0.0494

1) q_n Nominal cross section of aluminum or Aldrey layers
q_{nS} Nominal cross section of steel core

1.3 System protection

System configurations

Star-type system configurations (radial systems, double spur systems) are normally used for building and industrial power systems. A number of switchgear stations and distribution boards are required for distributing power from the infeed to the load. The protection equipment of these devices is connected in series.

Objectives of system protection

The objective of system protection is to detect faults and to selectively isolate faulted parts of the system. It must also permit short clearance times to limit the fault power and the effect of arcing faults.

Mutual system interference

High power density, high individual power outputs, and the relatively short distances in industrial and building power systems mean that l. v. and h. v. systems are closely linked. Activities in the l. v. system (short circuits, starting current) also have an effect on the h. v. system. If the situation is reversed, the control state of the h. v. system affects the selectivity criteria in the secondary power system.

It is, therefore, necessary to adjust the power system and its protection throughout the entire distribution system and to coordinate the protective functions.

1.3.1 Terminology

Electrical installations in a power system are protected either by protection equipment allocated to the installation components or by combinations of these protective elements.

Protection in event of protection device failure

If a protection device fails, the upstream protection device must provide the necessary protection.

Rated switching capacity

The rated switching capacity is the maximum value of the short circuit that the protection device is able to clear correctly. The protection device may be used in power systems for rated switching capacities up to this value.

Back-up protection

If a short circuit, which is higher than the rated switching capacity of the protection device used, occurs at a particular point in the system, back-up protection must provide protection for the downstream installation component and for the protection device by means of an upstream protection device.

Selectivity

Selectivity is advisable for series-connected protection devices to ensure the greatest possible level of supply reliability for the unaffected feeders. System protection is regarded as "selective" if only the protection device closest to the fault location (relative to the direction of power flow) is triggered.

Partial selectivity

In certain situations, partial selectivity (up to a particular short-circuit current) is sufficient or selectivity may not be required at all. The probability of faults occurring and the effects of these on the load must then be considered for unfavorable scenarios.

1.3.1.1 Protection equipment – features

Low-voltage protection devices[1)]

Low-voltage high-breaking-capacity fuses

Low-voltage high-breaking-capacity (l. v. h. b. c.) fuses have a high breaking capacity. They fuse quickly to restrict the peak short-circuit current. The protective characteristic is determined by the selected utilization category of the l. v. h. b. c. fuse (e. g. full-range fuse for overload and short-circuit protection, or partial range fuse for short-circuit protection only) and the rated current (Fig. 1.3/1).

Low-voltage circuit-breakers

Circuit-breakers are distinguished according to their method of operation (high and low current limiting). The protective functions are determined by selectable electromechanical or electronic releases:

Releases

- ▷ Overload protection by means of inverse time-delay overload releases ("a" releases), e. g. bimetallic releases
- ▷ Releases
- ▷ Short-circuit protection by means of instantaneous overcurrent releases ("n" releases), e. g. solenoid releases
- ▷ Short-circuit protection by means of definite short-time-delay overcurrent releases ("z" releases) for selective grading or for I^2-dependent delay ($I^2 \cdot t = \text{constant}$)
- ▷ Ground fault protection by means of ground fault current releases ("g" re-

---

1) The Siemens technical publication "Switching, Protection and Distribution in Low-Voltage Networks" contains descriptions and methods of operation for l. v. switchgear and protection devices as well as terminology and definitions.

leases) with definite or I^2-dependent delay

▷ Fault current protection by means of residual current releases

Electronic releases also permit new tripping criteria which are not possible with electromechanical releases.

Protective characteristics

The protective characteristic is determined by the rated circuit-breaker current as well as the setting and operating values of the releases (see Table 1.3/5).

High-voltage protection equipment

High-voltage high-breaking-capacity fuses

High-voltage high-breaking-capacity (h.v. h.b.c.) fuses can only be used for short-circuit protection. They do not provide overload protection. A minimum short-circuit current is, therefore, required for correct operation. H.v. h.b.c. fuses restrict the peak short-circuit current. The protective characteristic is determined by the selected rated current (Fig. 1.3/2).

High-voltage circuit-breakers

Circuit-breakers can provide time-overcurrent protection (definite and inverse), time-overcurrent protection with additional directional function, or differential protection. Distance protection is seldom used in the distribution systems described here.

Protective characteristics

Owing to cost-related factors, higher expenditure is generally justifiable for protective devices in h.v. power systems. Secondary relays, the characteristic of which is also determined by the actual current transformation ratio, are used in most cases. Static digital protection devices are also being used to an increasing extent.

1.3.1.2 Low-voltage protection equipment assemblies

Protection equipment assemblies

With series-connected distribution boards, it is possible to arrange the following protection devices in series (relative to the direction of power flow):

▷ Fuse with downstream fuse

▷ Circuit-breaker with downstream miniature circuit-breaker

▷ Circuit-breaker with downstream fuse

▷ Fuse with downstream circuit-breaker

▷ Fuse with downstream miniature circuit-breaker

▷ Several parallel infeeds with or without coupler units with downstream circuit-breaker or downstream fuse

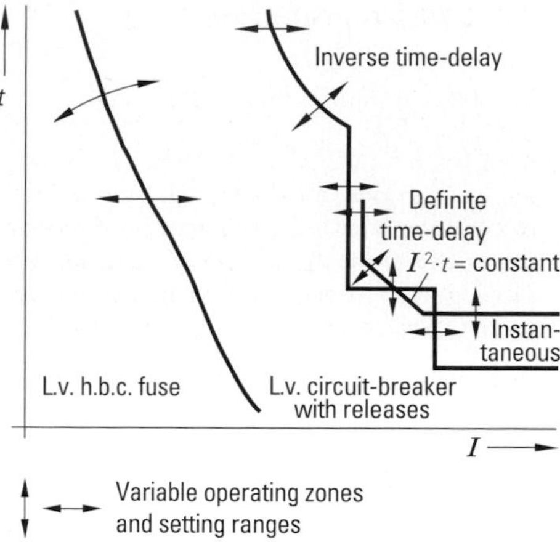

Fig. 1.3/1
Protective characteristic of l.v. h.b.c. fuse and l.v. circuit-breaker with releases

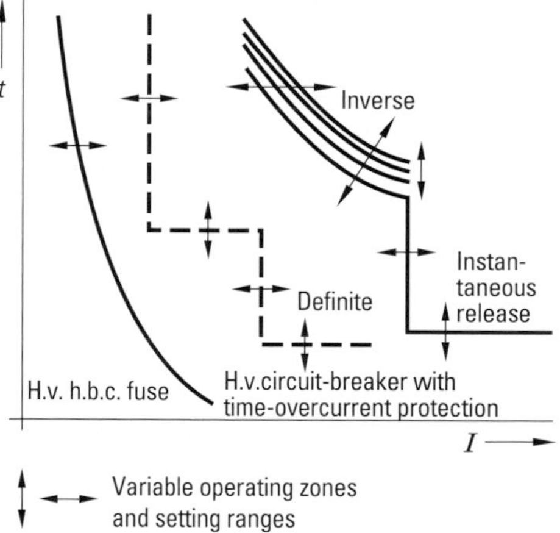

Fig. 1.3/2
Protective characteristic of h.v. h.b.c. fuse and h.v. time-overcurrent protection

Current selectivity must be verified in the case of meshed l.v. systems.

The high and low-voltage protection for the transformers feeding power to the l.v. system must be harmonized and adjusted to the additional protection of the secondary power system. Appropriate checks must be carried out to determine the effects on the primary h.v. system.

In h.v. systems, h.v. h.b.c. fuses are normally only installed upstream of the transformers in the l.v. infeed. With the upstream circuit-breakers, only time-overcurrent protection devices with different characteristics are usually connected in series. Differential protection does not or only slightly affects the grading of the other protection devices.

Electrical installation engineering is concerned, first and foremost, with the erection of electrical installations in l. v. power systems. For this reason, system protection focuses primarily on low-voltage systems.

1.3.1.3 Selectivity criteria

In addition to factors such as rated current and rated switching capacity, a further criterion that must be considered when implementing a protection device is selectivity. Selectivity is important because it ensures optimum supply reliability. The following criteria can be applied for selective operation of series-connected protection devices:

Time grading

▷ Time difference for clearance (time grading)

Current grading

▷ Current difference for operating values (current grading)

Inverse time grading

▷ Combination of time and current grading (inverse time grading)

Direction (directional protection), impedance (distance protection), and current difference (differential protection) are also used.

Requirements for selective behavior of protection devices

Protection devices can only behave selectively if both the highest and lowest short-circuit currents for the relevant system points are known at the project planning stage.

As a result:

▷ The highest short-circuit current determines the required rated switching capacity of the circuit-breaker.

▷ The lowest short-circuit current is important for setting the overcurrent release; the operating value of this release must be less than the lowest short-circuit current so that, in the event of a short circuit, the instantaneous overcurrent release can trigger correctly (it is not necessary to wait for the inverse-time overload release to respond).

▷ The observance of specified tripping conditions determines the maximum conductor lengths or their cross sections.

▷ Selective current grading is only possible if the short-circuit currents are known.

▷ In addition to current grading, partial selectivity can be achieved using combinations of carefully matched protection devices.

▷ The highest short-circuit current can be both the three-phase and the single-phase short-circuit current.

▷ With infeed into l. v. power systems, the single-phase fault current will be greater than the three-phase fault current if transformers with the Dy connection symbol are used (which is frequently the case). An $X_{(o)T}/X_T$ ratio of 0.95 is used for such transformers.

▷ The single-phase short-circuit current will be the lowest fault current if the damping zero phase-sequence impedance of the l. v. cable is active.

With large installations, it is advisable to determine all short-circuit currents using a special computer program.

Grading the operating currents with time grading

Grading of the operating currents is also taken into consideration with time grading, i.e. the operating value of the overcurrent release of the upstream circuit-breaker must be at least 1.25 times the operating value of the downstream circuit-breaker. Scattering of operating currents in definite-time-delay overcurrent releases (z) is thus compensated ($\leq \pm 10\%$).

Plotting the tripping characteristics of the graded protection devices in a grading diagram will help verify and visualize selectivity.

Time sequence for circuit-breakers

The time sequence of the breaking operation of the circuit-breakers must also be taken into consideration. Fig. 1.3/3 illustrates the individual time-related terms using two graded l. v. circuit-breakers as an example.

Grading time
Delay time

The grading time t_{st} is the interval required between the tripping characteristics of two series-connected protection devices to ensure correct operation of the protection device immediately upstream of the fault. The delay time to be set at the circuit-breaker (t_d) is obtained from the sum of the grading times.

1.3.1.4 Preparing grading diagrams

Preparation by hand

The following must be observed when plotting the tripping characteristics on log-log graph paper:

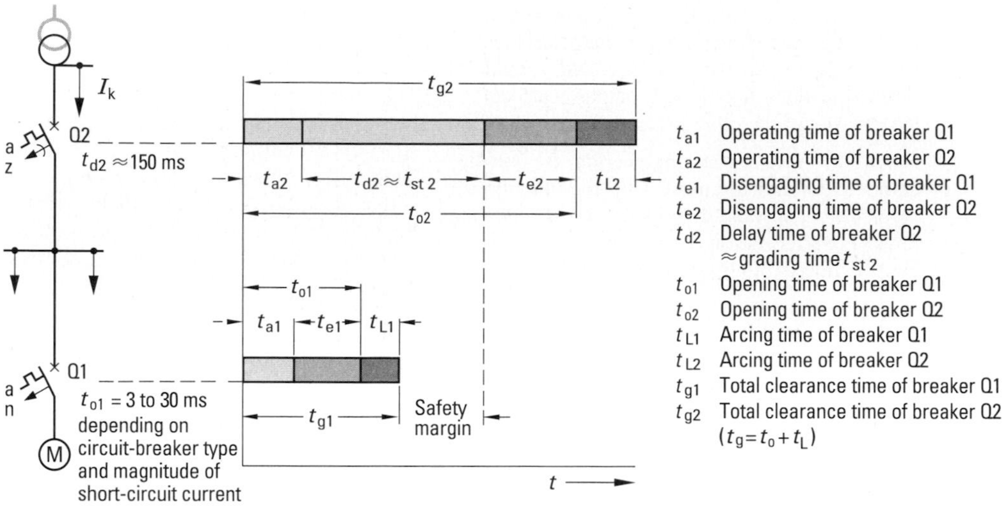

Fig. 1.3/3
Time sequence for the breaking operation of two graded l.v. circuit-breakers in the event of a short circuit

General recommendations

▷ To ensure positive selectivity, the tripping characteristics must neither cross nor touch.

▷ With *electronic* inverse-time-delay overcurrent releases, there is only one tripping characteristic as it is not affected by preloading. The selected characteristic must, therefore, be suitable for the motor or transformer at operating temperature.

▷ With *mechanical* (thermal) inverse-time-delay overload releases (a), the characteristics shown in the manufacturers' catalogs apply for cold releases. The opening times ($t_ö$) are up to 25% less at normal operating temperatures.

Tolerance range of tripping characteristics

▷ The tripping characteristics of circuit-breakers in the manufacturers' catalogs are usually only average values and must be extended to include the tolerance ranges (only in Figs. 1.3/4, 1.3/20, and 1.3/25).

▷ With *overcurrent releases* (I_i = n, I_d = z, $I_d + I_i$ = zn), the tolerance may, according to EN 60947-2/IEC 60947-2/VDE 0660 Part 101, be ± 20% of the current setting.

Significant tripping times

For the sake of clarity, only the delay time (t_d) is plotted for circuit-breakers with *definite-time-delay* overcurrent releases (z), and only the opening time (t_o) for circuit-breakers with *instantaneous* overcurrent releases (n).

Grading principle

Delay times and operating currents are graded in the opposite direction to the flow of power, starting with *distribution boards*

▷ *without fuses*, for the load breaker with the highest current setting of the overcurrent release,

▷ *with fuses*, for the fused outgoing circuit from the busbars with the highest rated fuse-link current.

Circuit-breakers are used in preference to fuses in cases where fuse links with high rated currents do not provide selectivity vis-à-vis the definite-time-delay overcurrent release (z) of the transformer feeder circuit-breaker, or only with very long delay times (t_d) (400 to 500 ms).

Procedure with two or more voltage levels

In the case of selectivity involving two or more voltage levels (Fig. 1.3/37 ff.), all currents and tripping characteristics on the high-voltage side are converted and referred to the low-voltage side on the basis of the transformation ratio.

Tools for preparing grading diagrams:

Grading diagram tools

▷ Standard forms with paired current values for commonly used voltages, e.g. 20/0.4 kV, 10/0.4 kV, 13.8/0.4 kV, and so on

▷ Templates for plotting the tripping characteristics

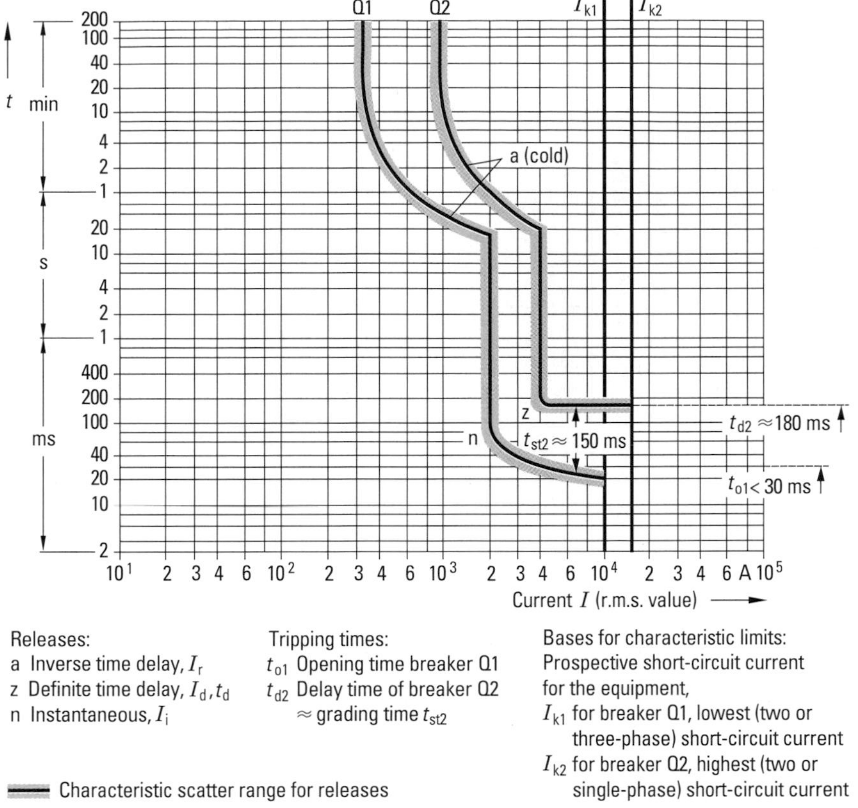

Fig. 1.3/4 Grading diagram with tripping characteristics for breakers Q1 and Q2 from Fig. 1.3/3

Fig. 1.3/4 shows a hand-drawn grading diagram with tripping characteristics for two series-connected circuit-breakers. The time sequence for the breaking operation illustrated in Fig. 1.3/3 was used here (time selectivity).

Preparing grading diagrams using a computer program

"Computer-Assisted Protective Grading" [1)]

Computer program for grading paths, grading diagram

This program allows grading paths involving several voltage levels to be generated in the form of block diagrams and then represented together with the associated grading diagram (interaction of the current-time characteristics) in just one chart (DIN A3) (Fig. 1.3/5).

[1)] Siemens computer program

Grading path for several voltage levels

A grading path diagram may comprise a maximum of ten series-connected system and protective elements (20 different elements are included in the program). The current scaling is adjusted if the grading path extends over several voltage levels. The number of decades in the log-log scale can be freely selected.

Separate short-circuit current calculations

The minimum and maximum short-circuit currents of the individual voltage levels are determined with separate short-circuit current calculations (e.g. using the program SINCAL = Siemens Network Calculation). The base points of these short-circuit current ranges are used to assign the ranges to the corresponding current scale.

The ranges of the protection device characteristics are limited at the respective voltage levels by the maximum short-circuit currents.

Grading diagram for sample system

I_n-MOTOR, BI-MOTOR W, BI-MOTOR K, I_a-MOTOR, S-MOTOR K, S-MOTOR W, AS-NSH, LS-T.1, I-0.4 kV min, SR-T.1, I-0.4 kV max, I-10 kV min, I-10 kV max

t: 1000 s … 0,001; 10 min … 1 min

10^1 2 5 10^2 2 5 10^3 2 5 10^4 2 5 10^5 5 10^6 0.40 kV

1 2 5 10 2 5 100 2 5 1000 2 5 10000 2 A 10.00 kV

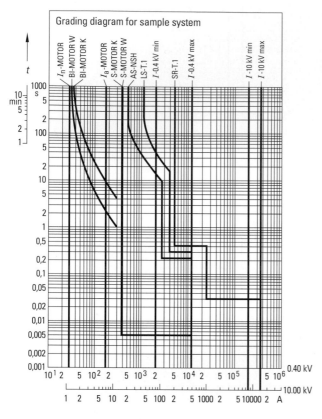

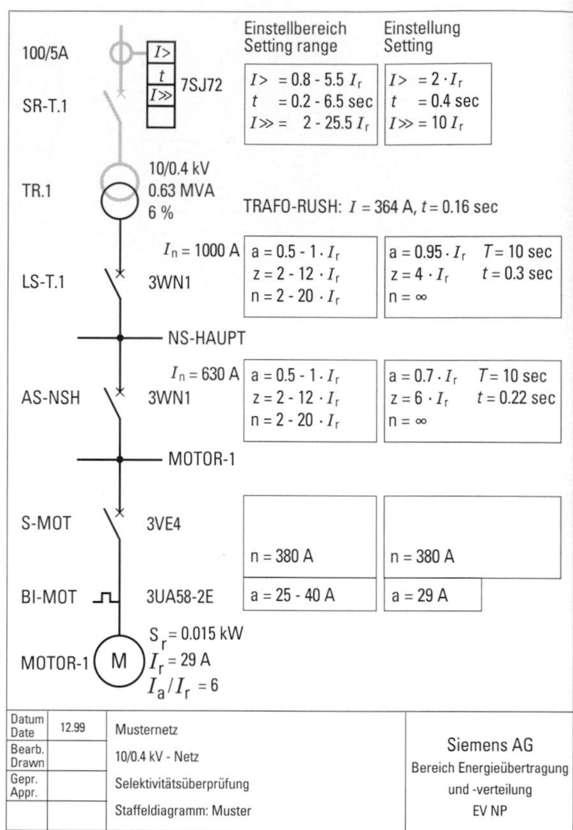

Fig. 1.3/5 Computer-generated grading diagram with grading flow

The complete characteristic of the protection device is plotted even if combinations of protection devices have been used (e. g. contactor with bimetallic and l. v. circuit-breaker).

Database

A comprehensive database, which can be expanded as required, has been created for protection equipment with an inverse-time characteristic curve (tripping characteristic).

L.v. h. b. c. fuses, overload relays (bimetallic), l. v. circuit-breakers, h. v. h. b. c. fuses, and inverse-time relays are stored in the database. The characteristics of time-overcurrent protection relays are converted and plotted using the setting data.

Checking selectivity

These characteristics allow selectivity to be checked interactively by the person processing the data. The program displays appropriate messages if characteristics intersect or the selected grading interval is incorrect.

Variable processing

The protection equipment settings can be adjusted, the protection relays can be interchanged, and the sequence or number of elements in the grading path can be changed.

Setting tables

Once selectivity has been checked, tables can be compiled listing the settings for all relays (releases) either in alphabetical order (each relay is assigned a unique name) or assigned to busbars (Fig. 1.3/6).

The SINCAL-UMZ program can be used to check the short-circuit conditions under which the protection equipment used responds and whether it has selective grading.

Low-voltage time grading

Grading and delay times

Only the grading time t_{st} and delay time t_d are relevant for time grading between several series-connected circuit-breakers or in conjunction with l. v. h. b. c. fuses (Fig. 1.3/7).

RELAISEINSTELLUNGEN DES PROJEKTES :
RELAY SETTING OF THE PROJECT : MUSTER

EINBAUORT LOCATION	SCHUTZELEMENT TYPE OF RELAY	EINSTELLBEREICH SETTING RANGE	EINSTELLWERT SETTING	DIAGRAMM DIAGRAM
AS-NSH	3WN1	a= 0.50- 1.00x I_r z= 2.00- 12.00x I_r n= 2.00- 20.00x I_r	In= 630. A a= 0.70x I_r z= 6.00x I_r t>=0.2200sec n=unendlich unlimited	muster
BI-MOT	3UA58-2E	a= 25.00- 40.00 A	a= 29.00 A	muster
LS-T. 1	3WN1	a= 0.50- 1.00x I_r z= 2.00- 12.00x I_r n= 2.00- 20.00x I_r	I_r= 1000. A a= 0.95x I_r z= 4.00x I_r t>=0.3000sec n=unendlich unlimited	muster
S-MOT	3VE4	n= 380. - 380. A	n= 380 A	muster
SR-T. 1	7SJ72	I>= 0.80- 5.50x I_r t>=0.2000-6.5000sec I>>= 2.00- 25.50x I_r	I>= 2.00x I_r t>=0.4000sec I>>= 10.00x I_r	muster

Fig. 1.3/6 Printout of relay settings (excerpt) in English and German using Fig. 1.3/5 as an example

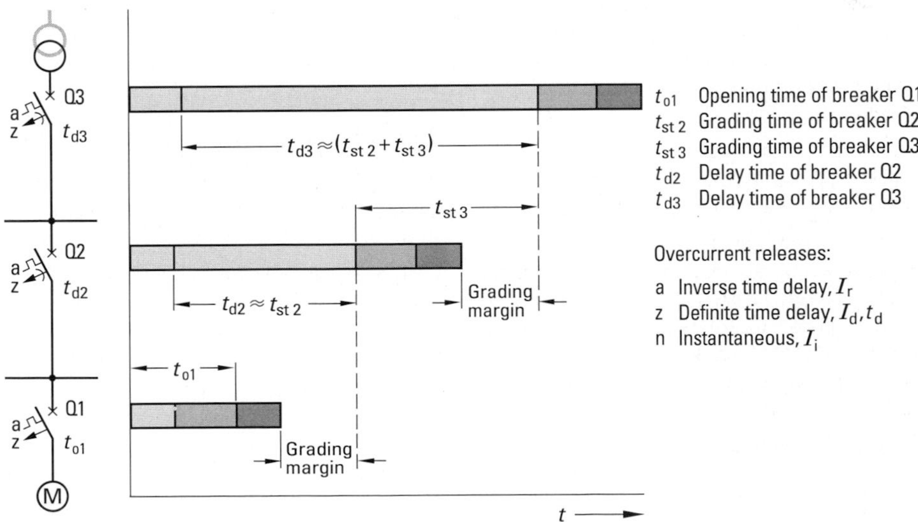

Fig. 1.3/7 Time grading for several series-connected circuit-breakers

The delay time t_{d2} of breaker Q2 can be equated approximately with the grading time t_{st2}; the delay time t_{d3} of breaker Q3 is calculated from the sum of the grading times $t_{st2} + t_{st3}$.

The resulting inaccuracies are corrected by the calculated grading margins. In the interests of simplicity, only the grading times are added.

Proven grading times t_{st}

Proven grading times

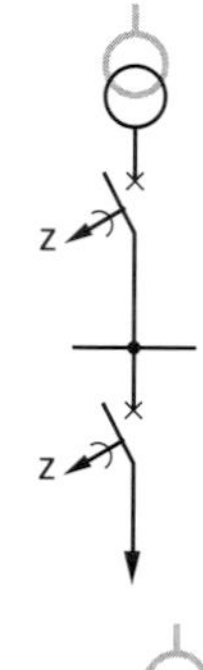

Series-connected circuit-breakers:

▷ With *electronic* overcurrent releases
between breaker Q1 and Q2 70 ms
between breaker Q2 and Q3 at least 100 ms
or
between breakers with predefined timers (3WN breakers) and between breakers with time-reduced selectivity control max. 50 ms

▷ With mechanical or separate time-delay relays 10 ms

▷ With *mechanical* overcurrent releases 150 ms

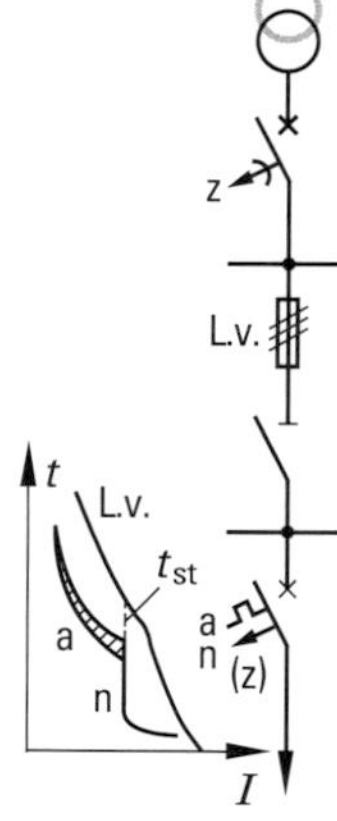

Irrespective of the type of "z" release (mechanical or electronic), a grading time of 70 ms to 100 ms is necessary between a *circuit-breaker* and a *downstream l. v. h. b. c. fuse.*

Between a *l. v. h. b. c. fuse* and a *downstream circuit-breaker*, a grading time t_{st} (safety margin) of at least 1 s must be maintained from the prearcing-time/current characteristic of the l. v. h. b. c. fuse to the point at which the tripping characteristics a and n or (z) intersect, in order to allow for the scatter band of the "a" release (see Fig. 1.3/25).

Back-up protection

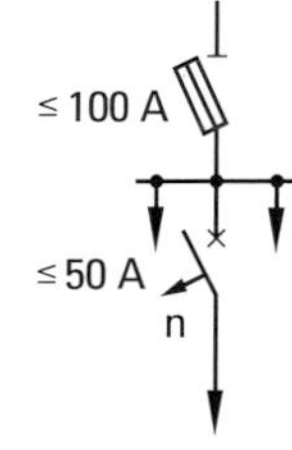

According to the Technical Supply Conditions of the public utilities (see Section 7.1), miniature circuit-breakers must be fitted with back-up fuses with a rated current of 100 A (max.) to prevent any damage being caused by short-circuit currents.

The DIN VDE and IEC standards also permit a switching device to be protected by one of the upstream protection devices with an adequate rated short-circuit switching capacity if both the feeder and the downstream protection device are also protected (back-up protection).

Literature

Literature on l. v. installations

Further information on l. v. switching and protection devices can be found in the Siemens publication "Switching, Protection and Distribution in Low-Voltage Networks".

High-voltage time grading[1)]

Operating and grading time

The following must be observed when determining the grading time (t_{st}) on the high-voltage side:

Once the protection device has been energized (Fig. 1.3/8), the set time must elapse before the device issues the tripping command to the shunt or undervoltage release of the circuit-breaker (operating time t_k).

The release causes the circuit-breaker to open. The short-circuit current is interrupted when the arc has been extinguished.

Only then does the protection system revert to the normal (rest) position (release time).

The grading time t_{st} between successive protection devices must be greater than the sum of the total clearance time t_g of the breaker and the release time of the protection system.

Since a spread of time intervals, which depends on a number of factors, has to be expected for the protection devices (including circuit-breakers), a grading margin is incorporated in the grading time.

Whereas grading times t_{st} of less than 400 to 300 ms were not possible with protection devices with mechanical releases, the more modern electronic and digital releases permit grading times of only 300 or 250 ms.

1.3.2 Protection equipment for low-voltage power systems

Tables 1.3/1 and 1.3/2 provide an overview of the protection equipment for l. v. systems dealt with in this manual. For the sake of completeness, the protection equipment in the h. v. system of outgoing transformer feeders has also been listed in Table 1.3/2.

Protecting conductors and cables against overheating, as defined in the DIN VDE specification "Erection of power installations with nominal voltages up to 1000 V"

[1)] See also Section 1.3.5.1

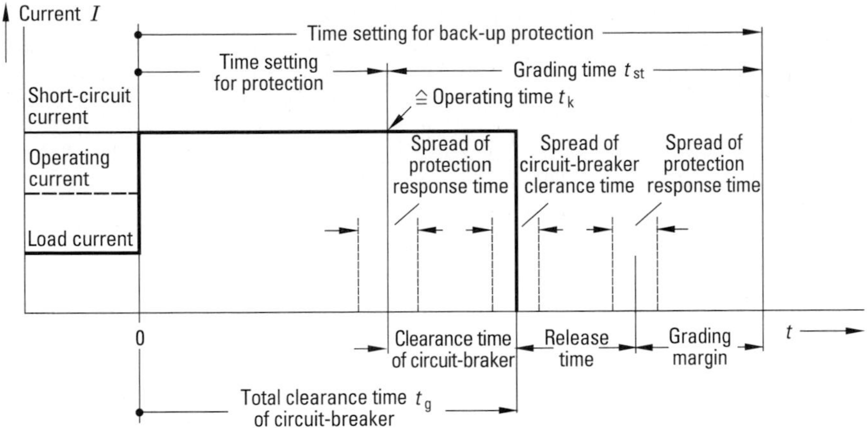

Fig. 1.3/8 Time grading in h.v. systems

Table 1.3/1
Overview of conductor and cable overcurrent protection devices discussed in this manual together with their protection ranges

Overcurrent protection devices	Standard	Overload protection	Short-circuit protection	See Section or Page
Fuses gL	EN 60269/IEC 269/ DIN VDE 0636	×	×	Section 4.1.1
Miniature circuit-breakers	EN 60898/IEC 898/ DIN VDE 0641-11	×	×	Section 4.1.2
Circuit-breakers with overload and overcurrent releases	EN 60947-2/IEC 947-2/ DIN VDE 0660-101	×	×	Page 89
Switchgear fuses aM	EN 60269/IEC 269/ DIN VDE 0636	–	×	Page 96
Switchgear assemblies with back-up fuse, utilization category gL or aM, and contactor with overload relay or starter circuit-breaker and contactor with overload relay	EN 60269/IEC 269/ DIN VDE 0636	–	×	Page 93 ff.
	EN 60947-4-1/IEC 947-4-1/ DIN VDE 0660-102	×	–	
	EN 60947-2/IEC 947-2/ DIN VDE 0660-101	–	×	Page 98
	EN 60947-4-1/IEC 947-4-1/ DIN VDE 0660-102	×	–	

× Protection provided
– No protection provided

DIN VDE 0100 Part 430, is dealt with in Section 1.7.

Overcurrent protection for conductors and cables

In accordance with this specification, overcurrent protection devices must be used to protect conductors and cables against overheating which may result from operational overloads or dead short circuits.

1.3.2.1 Circuit-breakers with protective functions

Protective functions of l. v. circuit-breakers

Circuit-breakers are used, first and foremost, for overload and short-circuit protection. In order to increase the degree of protection, they can also be equipped with additional releases, e.g. for clearance with undervoltage,

Table 1.3/2
Overview of protection grading schemes discussed in this manual for outgoing transformer and l.v. feeders

Protection devices					
	H.v.	Switch-disconnectors, h.v. h.b.c. fuses		Circuit-breakers, instrument transformers, time-overcurrent protection	Switch-disconnectors, h.v. h.b.c. fuses
	L.v.	Circuit-breakers or l.v. h.b..c. fuses	Tie breaker	Circuit-breakers	Network circuit-breakers and network master relays
Cost		Low	Justifiable	High	Low
High-voltage side Transformers with thermal release or full thermal protection Low-voltage side with various series-connected protection devices in radial systems, and parallel-connected l.v. h.b.c. fuses in interconnected systems		h.v. h.b.c. H.v. L.v. Individual and parallel operation customary Optional ≤ 630 A l.v. h.b.c. ≤ 50 A, ≤ 100 A		H.v. L.v. Individual and parallel operation customary	h.v. h.b.c. H.v. L.v. Only and parallel operation customary (Interconnected system)
Infeeds H.v. / L.v.	See page or Section	Page 118		Page 120	98 Q2
Outgoing feeders		Page 107 · Page 93 · Page 93 · Section 4.1.1 and 4.1.2	Page 114	Page 105 · Page 109 · Page 110 · Page 118	Section 5.6

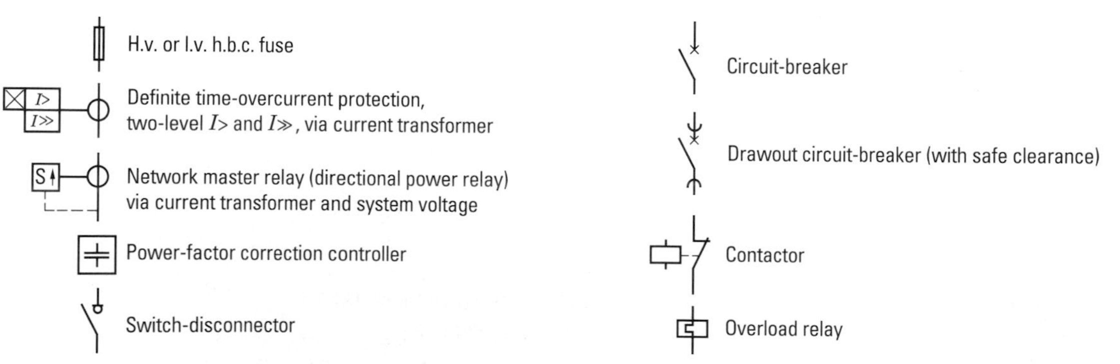

or with supplementary modules for detecting fault/residual currents.

The circuit-breakers are distinguished according to their protective function:

▷ Circuit-breakers for system protection to EN 60947–1 / IEC 60947–1 / DIN VDE 0660–100
and EN 60947–2 / IEC 60947–2 / DIN VDE 0660–101
▷ Circuit-breakers for motor protection to EN 60947–1 / IEC 60947–1 / DIN VDE 0660–100
and EN 60947–2 / IEC 60947–2 / DIN VDE 0660–101
▷ Circuit-breakers used in motor starters to EN 60947–1 / IEC 60947–1 / DIN VDE 0660–100 and EN 60947–4-1 / IEC 60947–4-1 / DIN VDE 0660–102
▷ Miniature circuit-breakers to EN 60898 / IEC 60898/DIN VDE 0641–11

Depending on their method of operation, circuit-breakers are available as:

Zero-current interrupters
Current limiters

▷ Zero-current interrupters (low current limiting) or
▷ Current limiters (fuse-type current limiting).

When configuring selective distribution boards, zero-current interrupters are more suitable as upstream protection devices and current limiters as downstream protection devices.

Overload and overcurrent protection

Tables 1.3/3 and 1.3/4 provide an overview of releases and relays in l. v. circuit-breakers. Table 1.3/5 contains the operating ranges of the overcurrent releases. According to the standards specified in Table 1.3/1, the operating value, at which the releases trigger, may deviate by ± 20% from the set value.

Overcurrent releases

The instantaneous electromagnetic overcurrent releases have either fixed or variable settings, whereas the electronic overcurrent releases used in Siemens circuit-breakers all have variable settings.

Modules

The overcurrent releases can be integrated in the circuit-breaker or supplied as separate modules for retrofitting or replacement. Possible exceptions are indicated in the manufacturers' specifications.

Overload releases

Mechanical (thermal) inverse-time-delay overload releases ("a" releases) are not always suitable for networks with a high harmonic content. Circuit-breakers with electronic overload releases must be used in such cases.

Short-circuit protection with "z" releases

In the case of older circuit-breakers with definite (short-)time-delay overcurrent releases (z) used for time-grading short-circuit protection, it should be noted that the circuit-breakers are designed for a specific maximum permissible thermal and dynamic load. If, in the event of a short circuit, the time delay results in this load being exceeded, an "n" release must also be used to ensure that the circuit-breaker is opened instantaneously with very high short-circuit currents. The information supplied by the manufacturer should be consulted when selecting the release type to be used.

Reclosing lockout after short-circuit tripping

A number of circuit-breakers can be fitted with a mechanical and/or electrical reclosing lockout which, after short-circuit tripping, prevents reclosing to the short-circuit.

The circuit-breaker can only be closed again after the fault has been eliminated and the lockout has been reset manually.

Fault-current/residual-current protection

The global importance of fault-current protection devices has grown in the field of protection technology due to the high level of protection they provide (protection of personnel and property) and their extended scope of protection (alternating and pulsating current sensitivity). Please also refer to Section 4.1.3 "Residual-current-operated protection devices".

Apart from residual-current-operated circuit-breakers, miniature circuit-breaker assemblies, e. g. miniature circuit-breakers with fault-current tripping, are being used to an increasing extent for domestic, commercial, and industrial applications.

MCBs with fault-current tripping

These circuit-breaker assemblies are available as compact factory-built devices or as assemblies comprising a miniature circuit-breaker (see Section 4.1.2) as the basic device and an additional add-on module.

Table 1.3/3 Symbols for releases according to protective function

Symbols for releases according to protective function

Protective function	Siemens symbol	Time-delay characteristics of release	Graphical symbol to EN 60617/ DIN 40713		
			Circuit diagram or		Block diagram
Overload protection	a	Inverse-time delay			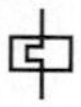
Selective short-circuit protection	z [1]	Definite-time delay by timing element or inverse-time delay			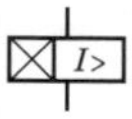 $I>$
Fault current/residual current/ground fault protection	g [1]	Definite-time delay or inverse-time delay	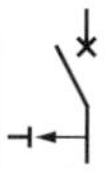		 $I \perp$
Short-circuit protection	n	Instantaneous			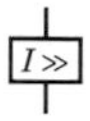 $I\gg$

[1] Also with "time-reduced selectivity control" for 3WN breakers

Release combinations will be referred to below by the letter symbols only, i.e. "an", "az", "azn", and "zn" releases, etc.

Type and range of releases and relays

Table 1.3/4
Circuit-breaker releases and relays with protective function

Function	Release	Relay
Overload protection	Overload release Inverse-time delay or electronic delay	Overload relay Thermal delay or electronic delay Thermistor protection release devices
Short-circuit protection	Overcurrent release Instantaneous electromagnetic or electronic release	Overcurrent relay Instantaneous electromagnetic release
Selective short-circuit protection	Overcurrent release Short electromagnetic or electronic delay	–

Table 1.3/5
Operating ranges of the overcurrent releases (to EN 60947/IEC 60947/DIN VDE 0660)

Applications (primarily for short-circuit current clearance)	Time-delay characteristic	Operating ranges of inverse-time-delay overcurrent release as multiple of set value I_r
Circuit-breaker for generator protection	Instantaneous or short-time delay	Approx. 3 to 6 · I_r
Circuit-breaker for line protection	Instantaneous	Approx. 6 to 12 · I_r
Circuit-breaker for motor protection	Instantaneous or short-time delay [1]	Approx. 8 to 15 · I_r

[1] Poss. short-time delay for rush current shunting

Miniature circuit-breakers with fault-current/ residual-current tripping

The assembly comprising a circuit-breaker and additional add-on module has established itself for circuit-breakers with rated currents (I_n) $\leq$ 400 A and fault-current/residual-current tripping.

Technical features

The additional add-on module for residual-current tripping used in system protection applications includes the following technical features:

▷ Rated residual current ($I_{\Delta n}$), adjustable in steps, e. g. 30 mA/100 mA/300 mA/1000 mA/3000 mA
▷ Tripping time (t_a), adjustable in steps, e. g. instantaneous 60 ms/100 ms/250 ms/ 500 ms/1000 ms
▷ Operation depends on system voltage
▷ Sensitivity, tripping with alternating and pulsating d. c. fault currents
▷ Reset button "R" for resetting after residual-current tripping
▷ Test button "T" for testing the miniature circuit-breaker assembly
▷ Status display for the current leakage/residual current (I_Δ) in the downstream circuit, e. g. by means of colored LEDs:
  - green: $I_\Delta \leq 0.25\, I_{\Delta n}$
  - yellow: $0.25\, I_{\Delta n} < I_\Delta \leq 0.5\, I_{\Delta n}$
  - red: $I_A < I_\Delta < 0.5\, I_{\Delta n}$
    $I_A \cong$ Tripping current of additional residual-current module
▷ Disconnection of the electronics overvoltage protection prior to insulation measurement in the installation
▷ "Remote tripping"
▷ "Auxiliary switch"

Interface to bus systems

The miniature circuit-breaker assemblies can be equipped with appropriate interfaces to bus systems to enable the exchange of information and interaction with other components in the electrical installation.

Miniature circuit-breaker assemblies sensitive to universal current

Miniature circuit-breaker assemblies, which are sensitive to universal current, are required for industrial applications for electrical installations in which smooth d. c. fault currents or currents with a low residual ripple occur in the event of a fault.

Standards

The standards EN 60947-2/IEC 60947-2/ DIN VDE 0660-101 apply for circuit-breakers with additional add-on fault-current or residual-current modules.

Selection criteria for circuit-breakers

When selecting the appropriate circuit-breakers for system protection, special attention must be paid to the following characteristics:

"KUBS" computer program

▷ Type of circuit-breaker and its releases according to the respective protective function and tasks
▷ Rated voltages
▷ Short-circuit strength and rated short-circuit making and breaking capacity
▷ Rated and maximum load currents
▷ The computer program "KUBS"[1)] (short-circuit current calculation, back-up protection, and selectivity) can provide assistance in selecting the most suitable circuit-breaker.

The system voltage and system frequency are crucial factors for selecting the circuit-breakers according to

▷ rated insulation voltage (U_i) and
▷ rated operating voltage (U_e).

Rated insulation voltage (U_i)

The rated insulation voltage (U_i) is the standardized voltage value for which the insulation of the circuit-breakers and their associated components is rated in accordance with HD 625/IEC 60664/DIN VDE 0110, Insulation Group C.

Rated operating voltage (U_e)

The rated operating voltage (U_e) of a circuit-breaker is the voltage value to which the rated short-circuit making and breaking capacities and the short-circuit performance category refer.

Short-circuit current

The maximum short-circuit current at the installation location is a crucial factor for selecting the circuit-breakers according to

▷ short-circuit strength as well as
▷ rated short-circuit making and breaking capacities.

Dynamic short-circuit strength

The dynamic short-circuit strength is the maximum asymmetric short-circuit current. It is the highest permissible instantaneous value of the prospective short-circuit current along the conducting path with the highest load.

Thermal fault withstand capability (1 s current)

The permissible thermal short-circuit strength is referred to as the rated short-time current (1 s current). It is the maximum current which the breaker is capable of with-

[1)] Siemens computer program

standing for 1 s without any damage occurring.

Rated switching capacity

The rated switching capacity of the circuit-breakers is specified as the rated short-circuit making capacity and rated short-circuit breaking capacity in conjunction with the short-circuit performance category P-1 or P-2 (see Table 1.3/7).

Rated short-circuit making capacity (I_{cm})

The rated short-circuit making capacity (I_{cm}) is the short-circuit current which the circuit-breaker is capable of making at the rated operating voltage + 10%, rated frequency, and a specified power factor. It is expressed as the maximum peak value of the prospective short-circuit current, and is at least equal to the rated short-circuit breaking capacity (I_{cn}), multiplied by the factor n specified in Table 1.3/6.

Rated short-circuit breaking capacity (I_{cn})

The rated short-circuit breaking capacity (I_{cn}) is the short-circuit current which the circuit-breaker is capable of breaking at the rated operating voltage + 10%, rated frequency, and a specified power factor (cos φ). It is expressed as the r.m.s. value of the alternating current component.

Switching capacity category

Switching capacity categories, which specify how often a circuit-breaker can switch its rated making and breaking current as well as the condition of the breaker after the specified switching cycle, are defined for circuit-breakers in EN 60947/IEC 60947/DIN VDE 0660 and in accordance with IEC 60157-1 (Table 1.3/7). The rated switching capacity is based on the test sequence O-t-CO-t-CO. The rated switching capacity can also be specified on the basis of the shortened switching sequence O-t-CO (see Table 1.3/7 for explanation of O, t, and C).

Table 1.3/6
Correlation between the rated short-circuit making capacity and the rated short-circuit breaking capacity I_{cn} (to EN 60 947-2/IEC 60 947-2/ DIN VDE 0660-1)

Rated short-circuit breaking capacity I_{cn} A	Power factor cos φ	Rated short-circuit making capacity I_{cm} (minimum) $n \cdot I_{cn}$
$I_{cn} \leq 1500$	0.95	$1.41 \cdot I_{cn}$
$1500 < I_{cn} \leq 3000$	0.9	$1.42 \cdot I_{cn}$
$3000 < I_{cn} \leq 4500$	0.8	$1.47 \cdot I_{cn}$
$4500 < I_{cn} \leq 6000$	0.7	$1.53 \cdot I_{cn}$
$6000 < I_{cn} \leq 10000$	0.5	$1.7 \cdot I_{cn}$
$10000 < I_{cn} \leq 20000$	0.3	$2.0 \cdot I_{cn}$
$20000 < I_{cn} \leq 50000$	0.25	$2.1 \cdot I_{cn}$
$50000 < I_{cn}$	0.2	$2.2 \cdot I_{cn}$

Table 1.3/7
Switching performance categories to DIN VDE 0660 and IEC 60 157-1

Switching capacity category	P-1	P-2
Switching cycle[1)]	O–t–CO	O–t–CO–t–CO
Condition of circuit-breaker after rated switching capacity test	Temperature-rise test with rated thermal current must be performed. Adjacent insulation material must not be damaged. Slight characteristic shifts are permissible for the overload release.	The circuit-breaker should be able to carry its rated thermal current without the need for maintenance (temperature-rise test only required in cases of doubt). Characteristic shifts are not permissible for the overload release.

1) O Opening
CO Opening and closing
t Interval

Rated circuit-breaker currents

The rated duty, e.g. continuous operation, intermittent operation, or short-time operation plays a decisive role in selecting the switchgear according to its rated currents.

The following rated currents are distinguished according to the thermal characteristics:

▷ Rated thermal current I_{th}
▷ Rated continuous current I_u
▷ Rated operating current I_e

Rated thermal current I_{th}, rated continuous current I_u

Rated thermal current, rated continuous current

The rated thermal current I_{th} or I_{the} for motor starters in enclosures is defined as an 8 h current in accordance with EN 60947-1, -4-1, -3/IEC 60947-1, -4-1, -3/DIN VDE 0660-100, -102, -107. It is the maximum current which can be carried during this time without the temperature limit being exceeded. The rated continuous current I_u can be carried for an unlimited time.

With adjustable inverse-time-delay releases and relays, the maximum current setting is the rated continuous current I_u.

Rated operating current I_e

Rated operating current

The rated operating current I_e is the current that is determined by the operating conditions of the switching device, the rated operating voltage and rated frequency, rated switching capacity, the rated duty, utilization category[1)], contact life, and the degree of protection.

Application examples for circuit-breakers with protection

Application examples and tripping characteristics

The principal application examples and typical tripping characteristics of modern circuit-breakers currently available from Siemens are specified in Table 1.3/8.

1.3.2.2 Switchgear assemblies

Switchgear assemblies

Switchgear assemblies are series-connected switching and protection devices which perform specific tasks for protecting a system component; the first device (relative to the flow of power) provides the short-circuit protection.

Switchgear assemblies are usually grouped into subassemblies (see Section 1.4), e. g. as

▷ non-withdrawable subassemblies or
▷ withdrawable MCC units.

Comparisons between the properties and protection characteristics of fuses and circuit-breakers can be found on page 101 ff.

Switchgear assemblies with fuses

Fuses and circuit-breakers

Fuses and molded-case circuit-breakers

If the prospective short-circuit current I_k exceeds the rated short-circuit breaking capacity I_{cn} of the circuit-breaker, the latter must be provided with upstream fuses (Fig. 1.3/9).

Protection and operating ranges

Defined protection and operating ranges are assigned to each device in the switchgear assembly. The "a" release monitors overload currents, while the "n" release detects short-circuit currents up to the rated short-circuit breaking capacity I_{cn} of the circuit-breaker.

The circuit-breaker provides protection against all overcurrents up to its rated short-circuit breaking capacity I_{cn} and ensures all-pole opening and reclosing.

The fuses are only responsible for short-circuit clearance with higher short-circuit currents I_k. In this case too, the circuit-breaker opens all its poles almost simultaneously via its "n" release, tripped by the let-through current I_D of the fuse. The fuse must, therefore, be selected such that its let-through current I_D is less than or equal to the rated short-circuit breaking capacity I_{cn} of the circuit-breaker.

Fuse, contactor, and thermal inverse-time-delay overload relay

The contactor is used to switch the motor on and off. The overload relay protects the motor, motor supply conductors, and contactor against overloading. The fuse upstream of the contactor and overload relay provides protection against short circuits. For this reason, the protection ranges and characteristics of all the components (Fig. 1.3/10) must be carefully coordinated with each other.

The switchgear assembly comprising contactor and overload relay is referred to as a motor starter or, if a three-phase a.c. motor is started directly, a direct-on-line starter.

Specifications for contactors and motor starters

The standards EN 60947-4-1/IEC 60947-4-1/DIN VDE 0660-102 apply for *contactors* and *motor starters* $\leq$ 1000 V for direct-on-line starting (with maximum voltage).

Degrees of damage for switchgear assemblies

When short-circuit current protection equipment is selected for switchgear assemblies, a distinction is made between various types of protection according to the permissible degree of damage (as defined in EN 60947-4-1/IEC 60947-4-1/DIN VDE 0660-102)[1)]:

Type "a" Destruction and replacement of individual components or complete switching device

[1)] The utilization category indicates the duty and load of the switching devices (refer to the device standards EN 60947/IEC 60947/DIN VDE 0660).

[1)] In EN 60947-4-1/IEC 60947-4-1/DIN VDE 0660-102, behavior in the event of a short-circuit has been modified as follows:
Coordination type "1":
Destruction of the contactor and overload relay is permitted. The contactor and/or overload relay must be replaced if required.
Coordination type "2":
The overload relay must not be damaged. Contact welding at the contactor is, however, permitted if this can be separated easily or if the contactor can be replaced easily.

Table 1.3/8
Application examples for modern Siemens circuit-breakers and their typical tripping characteristics

Circuit-breaker type	Rated current	Application example	Tripping characteristics
Circuit-breaker 3WN	250 to 6300 A	*Protection of distribution systems, motors, transformers, and generators* – High rated short-time current for time grading – Two series, 3WN1 and 3WN6, with high and medium rated switching capacity – Electronic, microprocessor-based overcurrent releases with integral power source – Time-reduced selectivity control with total delay time of 50 ms	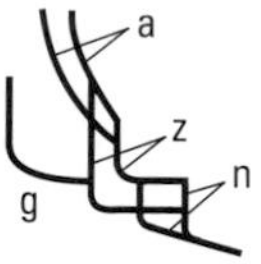 a, z, n, g
Current-limiting circuit-breaker 3VF	6.3 to 2500 A	– *Designed and tested to EN 60 947/IEC 60 947/ DIN VDE 0660. Possible applications:*	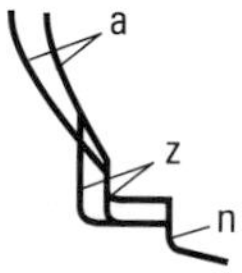 a, z, n
		System protection ≤ 2500 A – Optional adjustable overload and overcurrent releases: Carefully adapted to protection requirements	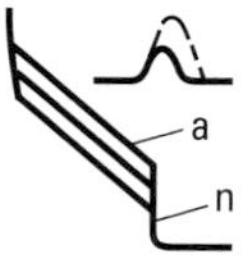 a, n
		Motor protection ≤ 500 A – Electronic overload releases with variable time-lag class: Effective protection when motor under full load	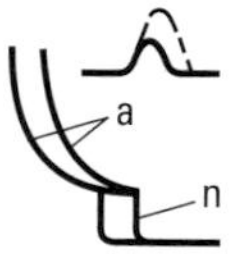 a, n
		Starter assemblies ≤ 500 A – Unsusceptible to inrush currents: Breaker not tripped by direct-on-line motor starting	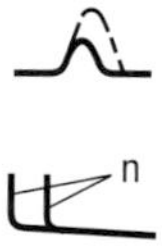 n
		As non-automatic circuit-breaker ≤ 2000 A – With integrated overcurrent releases, no back-up fuses required	 n
Circuit-breaker 3RV1	0.16 to 100 A	– 3RV1 circuit-breakers for motor protection with overload and overcurrent protection	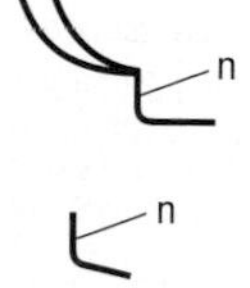 a, n n

a Overload tripping z Short-time-delay overcurrent tripping n Instantaneous overcurrent tripping
g Ground-fault tripping

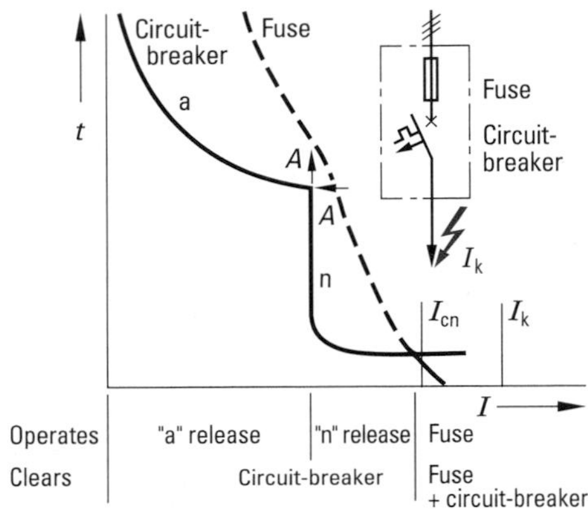

a Inverse-time-delay overload release
n Instantaneous electromagnetic overcurrent release
I_{cn} Rated short-circuit breaking capacity
I_k Prospective sustained short-circuit current
A Safety margins

Fig. 1.3/9
Switchgear assembly comprising fuse and circuit-breaker

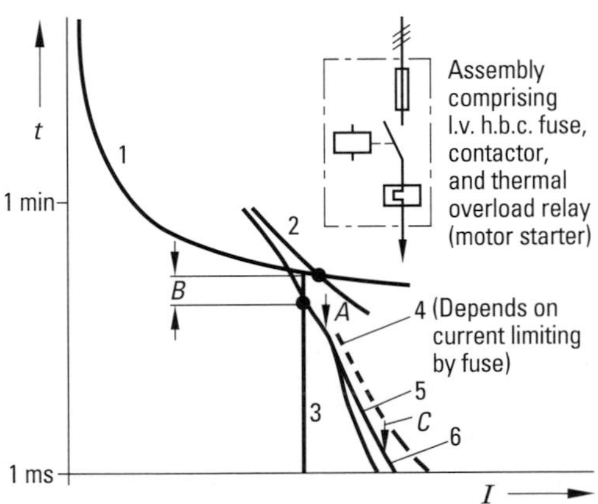

1 Tripping characteristic of (thermal) inverse-time-delay overload relay
2 Destruction characteristic of thermal overload relay
3 Rated breaking capacity of contactor
4 Characteristic of contactor for easily separable welding of contacts
5 Prearcing-time/current characteristic of fuse, utilization category aM
6 Total clearance-time characteristic of aM fuse

A, B, C Safety margins for reliable short-circuit protection

Fig. 1.3/10
Switchgear assembly comprising fuse, contactor, and thermal inverse-time-delay overload

Type "b" Welding of contacts and permanent change in characteristic values of overload relay

Type "c" Welding of contacts without permanent change to operating values of overload relay

Protection and operating ranges of equipment

Grading diagram for motor starter

The protection ranges and the relevant characteristics of the equipment constituting a switchgear assembly used as a motor starter are illustrated in the grading diagram in Fig. 1.3/10.

The fuses in this assembly must satisfy a number of conditions:

▷ The time-current characteristics of fuses and overload relays must allow the motor to be run up to speed.
▷ The fuses must protect the overload relay from being destroyed by currents approximately 10 times higher than the rated current of the relay.
▷ The fuses must interrupt overcurrents beyond the capability of the contactor (i.e. currents approximately 10 times higher than the rated operating current I_e of the contactor).
▷ In the event of a short-circuit, the fuses must protect the contactor such that any damage does not exceed the specified degrees of damage (see above). (Depending on the rated operating current I_e, contactors must be able to withstand motor starting currents of between 8 and 12 I_e without the contacts being welded.)

To satisfy these conditions, the following safety margins A, B, and C must be maintained between certain characteristics of the devices:

Protection of overload relay

In order to *protect the overload relay*, the prearcing-time/current characteristic of the fuse (an l. v. h. b. c. switchgear fuse of utilization category aM was used in this example; refer to the following section "Selecting fuses") must lie in *margin A* below the intersection of the tripping characteristic of the overload relay (1) with its destruction curve (2).

Protection of contactor

In order to *protect the contactor against excessively high breaking currents*, the prearcing-time/current characteristic of the fuse from the current value, which corresponds to the breaking capacity of the contactor (3), must lie in *margin B* below the tripping characteristic of the overload relay (1).

In order to *protect the contactor against contact welding*, time-current characteristics, up to which load currents can be applied, can be specified for each contactor

▷ without welding or
▷ easily separable welding (characteristic 4 in Fig. 1.3/10).

In both cases, therefore, the fuse must respond in good time. The total clearance time characteristic of the fuse (6) must lie in *margin C* below the characteristic of the contactor for easily separable contact welding (4) (total clearance time = prearcing time + extinction time).

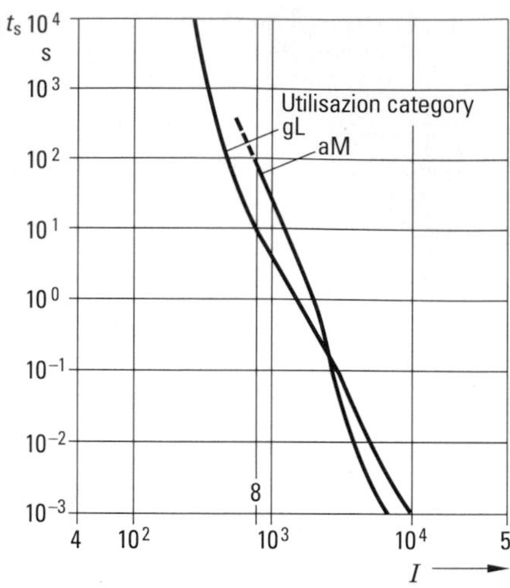

Fig. 1.3/11
Comparison of prearcing-time/current characteristics of l.v. h.b.c. fuses of utilization categories gL and aM, rated current 200 A

Selecting fuses

Fuses for motor starters are selected according to the aforementioned criteria.

L.v. h.b.c. switchgear fuses

Compared with l.v. h.b.c. fuses of utilization category gL used to protect conductors and cables, *l.v. h.b.c. switchgear fuses of utilization category aM* provide the advantage of weld-free short-circuit protection for the maximum motor power which the contactor is capable of switching.

Owing to their more effective current limiting abilities (as compared with those of line-protection fuses), they are very effective in relieving contactors of high peak short-circuit currents i_p since they respond more rapidly in the upper short-circuit range (see Fig. 1.3/11).

It is, therefore, preferable to use switchgear fuses rather than line-protection fuses with relay settings < 80 A at higher operating currents with correspondingly lower short-circuit current attenuation.

Table 1.3/9 shows the classification of the fuses based on functional features.

Classification of l. v. h. b. c. fuses and comparison of characteristics of gL and aM utilization categories

L.v. h. b. c. fuses are divided into functional and utilization categories in accordance with

Table 1.3/9
Classification of l.v. h.b.c. fuses based on their functional characteristics defined in EN 60269-1/IEC 60269-1/DIN VDE 0636-10

Functional category			Utilization category	
Designation	Rated continuous current ≤	Rated breaking current	Designation	Protection of
Full-range fuses				
g	I_n	$\geq I_{amin}$	gL/gG gR gB	Cables and conductors Semi-conductors Mining installations
Back-up fuses				
a	I_n	$\geq 4\ I_n$ $\geq 2.7\ I_n$	aM aR	Switchgear Semi-conductors

I_{amin} Minimum rated breaking current

their type. They can continuously carry currents up to their rated current.

Functional category g (full-range fuses)

Functional category g applies to full-range fuses which can interrupt currents from the minimum fusing current up to the rated short-circuit breaking current.

Utilization category gL

This category includes fuses of *utilization category gL* used to protect cables and conductors.

Functional category a (back-up fuses)

Functional category a applies to back-up fuses which can interrupt currents above a specified multiple of their rated current up to the rated short-circuit breaking current.

Utilization category aM

This functional category applies to switchgear fuses of *utilization category aM*, the minimum breaking current of which is approximately four times the rated current. These fuses are thus only intended for short-circuit protection. For this reason, fuses of functional category "a" must not be used above their rated current. A means of overload protection, e.g. thermal time-delay relay, must, therefore, always be provided.

Comparison of utilization categories gL and aM

The prearcing-time/current characteristics of l. v. h.b.c fuses of utilization category gL and aM for 200 A are compared in Fig. 1.3/11.

Switchgear assemblies without fuses (fuseless design)

Back-up protection (cascade-connected circuit-breakers)

Cascade connection

If two circuit-breakers with "n" releases of the same type are connected in series along one conducting path, they will open simultaneously in the event of a fault (K) in the vicinity of the distribution board (Fig. 1.3/12).

The short-circuit current is thereby detected by two series-connected interrupting devices and effectively extinguished. As a result, the downstream circuit-breaker with a lower rated switching capacity can be installed at a location where the prospective short-circuit current exceeds its rated switching capacity.

Protection and operating ranges of the circuit-breakers

Fig. 1.3/12a shows the single-line diagram and Fig. 1.3/12b the principle of a cascade connection. The rated current of the up-

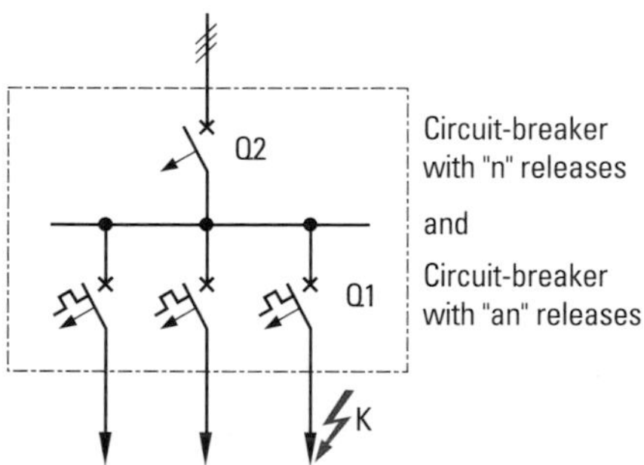

Fig. 1.3/12 a
Single-line diagram of a back-up circuit (cascade connection) in a sub-distribution board

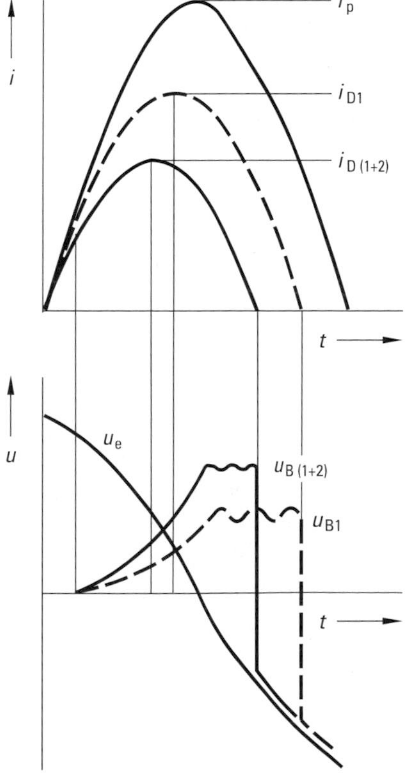

i_p	Maximum asymmetrical short-circuit current (peak value)
u_e	Source voltage (operating voltage)
u_{B1}	Arc voltage of outgoing circuit-breaker Q1
i_{D1}	Let-through current of outgoing circuit-breaker Q1
$u_{B(1+2)}$	Sum of arc voltages of upstream circuit-breaker Q2 and outgoing circuit-breaker Q1
$i_{D(1+2)}$	Actual let-through current (less than i_{D1})

Fig. 1.3/12 b
Principle of a back-up circuit (cascade connection)

stream circuit-breaker Q2 is selected in accordance with its rated operating current.

The circuit-breaker Q2 can, for example, be used as a main circuit-breaker or group circuit-breaker for several feeders in sub-distribution boards. Its "n" release is set to a very high operating current – if possible, to the rated short-circuit breaking capacity of the downstream circuit-breakers. The outgoing circuit-breaker Q1 provides overload protection and also clears (unaided) relatively low short-circuit currents which may be caused by short circuits to exposed conductive parts, insulation faults, or short circuits at the end of long conductors and cables. The upstream circuit-breaker Q2 only opens at the same time if high short-circuit currents occur as a result of a dead short circuit in the vicinity of the outgoing circuit-breaker Q1 (restricted selectivity).

Circuit-breakers with "an" releases and contactor

Protection and operating ranges of devices

The circuit-breaker provides overload and short-circuit protection (also for the contactor), while the contactor performs switching duties (Fig. 1.3/13). The requirements that must be fulfilled by the circuit-breaker are the same as those that apply to the fuse in switchgear assemblies comprising fuse, contactor, and overload relay (see Fig. 1.3/10).

Starter circuit-breaker with "n" release, contactor, and overload relay (a)

Overload protection is provided by the overload relay in conjunction with the contactor, while short-circuit protection is provided by the starter circuit-breaker. The operating current of its "n" release is set as low as the starting cycle will permit, in order to allow for low short-circuit currents in the instantaneous breaking range (Fig. 1.3/14). The advantage of this switchgear assembly is that it is possible to determine whether the fault was an overload or short circuit according to whether the contactor (via the overload relay) or the starter circuit-breaker has opened.

Readiness for reclosing

Further advantages of the starter circuit-breaker following short-circuit tripping are three-phase circuit interruption and immediate readiness for reclosing.

The switchgear assemblies with the starter circuit-breaker are becoming increasingly important in fuseless control units.

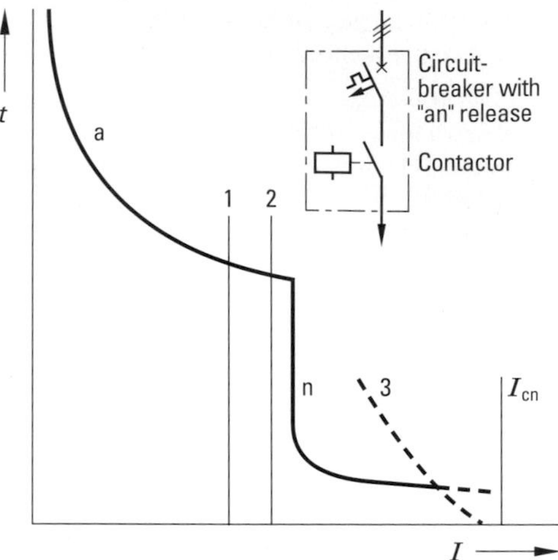

1 Rated breaking capacity of contactor
2 Rated making capacity of contactor
3 Characteristic of contactor for easily separable contact welding
a Characteristic of inverse-time-delay overload release
n Characteristic of instantaneous electromagnetic overcurrent release
I_{cn} Rated short-circuit breaking capacity of circuit-breaker

Fig. 1.3/13
Switchgear assembly comprising circuit-breaker and contactor

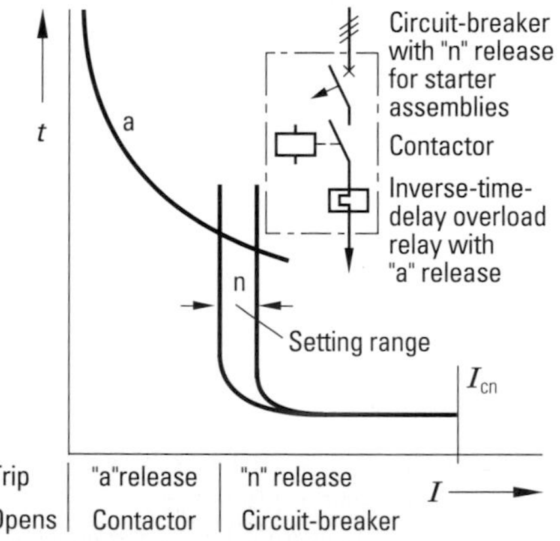

a Characteristic of (thermal) inverse-time-delay overload relay
n Characteristic of adjustable instantaneous overcurrent release

Fig. 1.3/14
Switchgear assembly comprising circuit-breaker, adjustable overcurrent release, contactor, and overload relay

Switchgear assemblies with thermistor motor-protection devices

Overload relays and releases cease to provide reliable overload protection when it is no longer possible to establish the winding temperature from the motor current. This is the case with:

▷ high switching frequencies
▷ irregular, intermittent duty
▷ restricted cooling and
▷ high ambient temperatures.

In these cases, switchgear assemblies with thermistor motor-protection devices are used. The switchgear assemblies are designed with or without fuses depending on the installation configuration.

Temperature sensor in motor winding

The degree of protection that can be attained depends on whether the motor to be protected has a thermally critical stator or rotor. The operating temperature, coupling time constant, and the position of the temperature sensor in the motor winding are also crucial factors. They are usually specified by the motor manufacturer.

Motors with thermally critical stators

Motors with thermally critical stators can be adequately protected against overloads and overheating by means of thermistor motor-protection devices *without* overload relays. Feeder cables are protected against short circuits and overloads either by fuses and circuit-breakers (Fig. 1.3/15a) or by fuses alone (Fig. 1.3/15b).

Motors with thermally critical rotors

Motors with thermally critical rotors (even if started with a locked rotor) can only be provided with adequate protection if they are fitted with an additional overload relay or release. The overload relay and release also protect the conductors against overloads (Fig. 1.3/15a, c, and d).

1.3.2.3 Selecting protection equipment

Short-circuit protection of branch circuits

Branch circuits in distribution boards and control units can be provided with short-circuit protection by means of fuses or by means of circuit-breakers without fuses. The level of anticipated current limiting, which in fuses with low rated currents is higher than in current-limiting circuit-breakers with the same rated current, may also be a crucial factor in selecting protection equipment.

Comparing the protection characteristics of fuses with those of current-limiting circuit-breakers

The following should be taken into consideration when comparing the protection characteristics of fuses and circuit-breakers:

▷ the rated short-circuit breaking capacity, which can vary considerably,
▷ the level of current limiting, which with fuses $\leq$ 400 A is always higher than for current-limiting circuit-breakers with the same rated current,
▷ the shape of the prearcing-time/current characteristics of fuses and the tripping characteristics of circuit-breakers,
▷ clearance conditions in accordance with HD 384.4.41/IEC 60364-4-41/DIN VDE 0100-410, Section 6.1.3 "Protection measures in TN systems" (see Chapter 2).

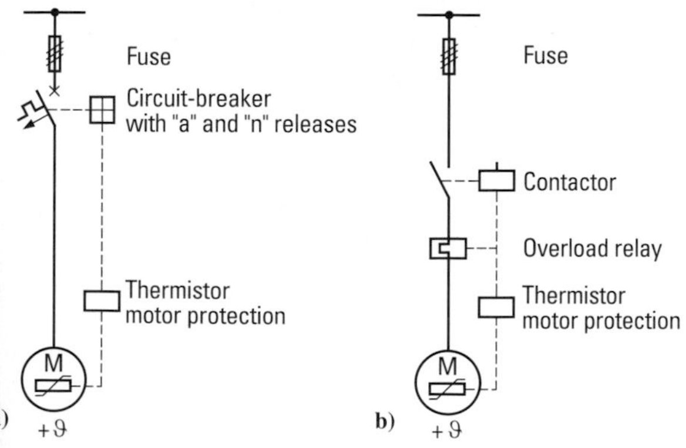

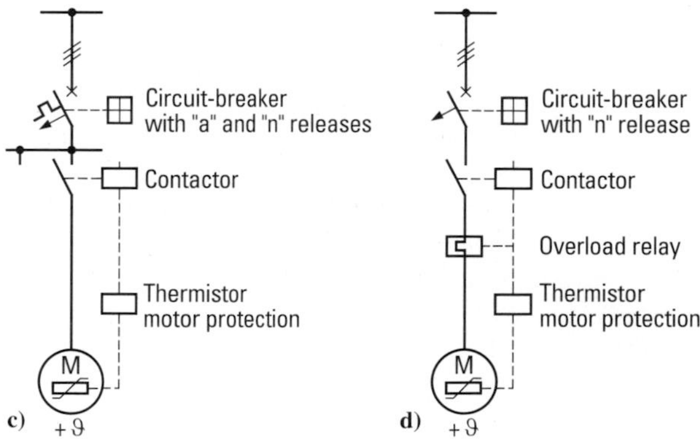

Fig. 1.3/15
Switchgear assemblies with thermistor motor-protection devices *plus* additional overload relay or release (block diagram)

Comparison of current-limiting characteristics

Current limiting with l. v. h. b. c. fuses and circuit-breakers

Fig. 1.3/16 shows the current-limiting characteristics of a circuit-breaker (rated continuous current 63 A, at 400 V and 50 Hz) compared to the l. v. h. b. c. fuses (type 3NA, utilization category gL, rated currents 63 A and 100 A). Owing to the high motor starting currents, however, the rated current of the fuse must be higher than the rated operating current of the motor, i.e. a circuit-breaker with a minimum rated current of 63 A or a fuse with a minimum rated current of 100 A is required for a 30 kW motor.

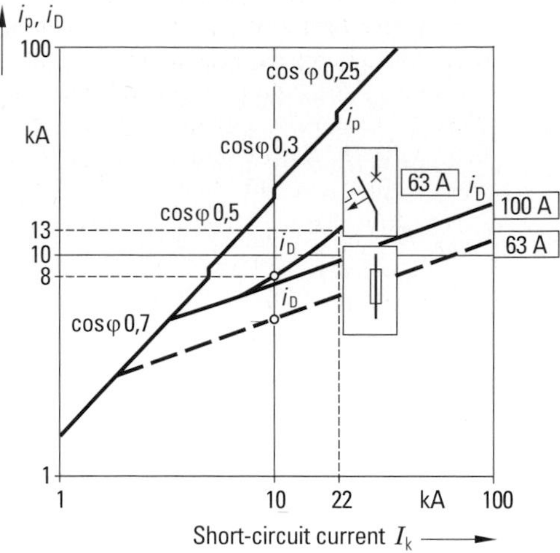

i_D Let-through currents
i_p Maximum assymmetrical short-circuit current

e.g. where I_k = 10 kA:

i_D Fuse (100 A) 7.5 kA
i_D Circuit-breaker 8 kA

Fig. 1.3/16
Current limiting characteristics of circuit-breaker (63 A) and l.v. h.b.c. fuses (63 and 100 A)

Comparison between the tripping characteristics and rated short-circuit breaking capacity of fuses with those of circuit-breakers with the same rated current and a high switching capacity

Tripping characteristics and rated short-circuit breaking capacity

The prearcing-time/current characteristic a of the 63 A fuse link (utilization category gL) and the "an" tripping characteristic b of a circuit-breaker are, by way of example, plotted in the time-current diagram Fig. 1.3/17. The current setting for the inverse-time-delay overload release of the circuit-breaker corresponds to the rated current of the fuse link.

Current limiting range (1)

The typical test range for fuse currents is between 1.3 and 1.6 times the rated current (A), while the test range for the limiting tripping currents of the overload release is between 1.05 and 1.2 times the current setting (B). The adjustable overload release enables the current setting and, therefore, the limiting tripping current to be matched more closely to the continuous loading capability of the equipment to be protected than would be possible with a fuse, the different current ratings of which only permit approximate matching. Although the limit current of the fuse is adequate for providing overload protection for cables and conductors, it is not sufficient for the starting current of motors where a fuse with the characteristic a' would be needed.

Overload range (2)

In the overload range (2), the prearcing-time/current characteristic of the fuse is steeper than the tripping characteristic of the overload release. This is desirable for overload protection of cables and conductors; the flatter tripping characteristic b is, however, required for the overload protection of motors.

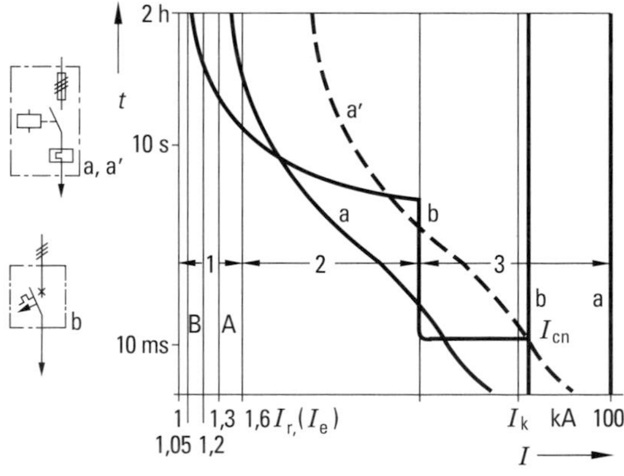

1 Current limiting range
2 Overload range
3 Short-circuit current range
A Test range for fuse currents
B Test range for limiting tripping currents of circuit-breaker
I_{cn} Rated short-circuit breaking capacity

Fig. 1.3/17
Characteristics and rated switching capacities of fuse (a) and circuit-breaker (b) with "an" releases

Short-circuit current range (3)

In the short-circuit current range (3), the instantaneous release of the circuit-breaker detects short-circuit currents above its operating value faster than the fuse. At higher currents, the fuse trips more quickly and, therefore, limits the short-circuit current more effectively than a circuit-breaker.

Extremely high rated switching capacity of l. v. h. b. c. fuses

This results in an extremely high rated breaking capacity for fuses of over 100 kA at an a.c. operating voltage of 690 V. The rated short-circuit breaking capacity I_{cn} of circuit-breakers, however, depends on a number of factors, e.g. the rated operating voltage U_e and the type.

A comparison between the protection characteristics of fuses, circuit-breakers, and their switchgear assemblies can be found in Tables 1.3/10 and 1.3/11.

Selecting circuit-breakers for distribution boards with and without fuses

Distribution boards and control units can be constructed with or without fuses.

Distribution boards with fuses

The standard design of distribution boards with fuses (Table 1.3/12) includes switchgear assemblies comprising circuit-breakers and fuses, whereby a specific task is allocated to each protection device.

The feeder circuit-breaker provides overload protection and selective short-circuit protection for the transformer and distribution board. The Siemens circuit-breaker 3WN is ideal for this purpose.

The switchgear assemblies comprising fuse and circuit-breaker, which provide system protection, protect the lines to the sub-dis-

Table 1.3/10 Comparison between the protective characteristics of fuses and circuit-breakers

Characteristic	Fuse	Circuit-breaker
Rated switching capacity (a.c.)	> 100 kA, 690 V	f (I_r, U_e, type[1])
Current limiting	f (I_r, I_k)	f (I_r, I_k, U_e, type[1])
Additional arcing space	None	f (I_r, U_e, I_k, type[1])
External indication of operability	Yes	No
Operational reliability	With additional costs[2]	Yes
Remote switching	No	Yes
Automatic all-pole breaking	With additional costs[3]	Yes
Indication facility	With additional costs[4]	Yes
Interlocking facility	No	Yes
Readiness for reclosing after		
clearing overload	No	Yes
clearing short-circuit	No	f (condition)
Interrupted operation	Yes	f (condition)
Maintenance costs	No	f (number of operations and condition)
Selectivity	No additional costs	With additional costs
Replaceability	Yes[5]	With unit of same make
Short-circuit protection		
Cable	Very good	good
Motor	Very good	good
Overload protection		
Cable	Adequate	good
Motor	Not possible	good

[1] The term "type" embraces: current extinguishing method, short-circuit strength through internal impedance, type of construction
[2] For example, by means of shockproof fuse switch-disconnectors with snap-action closing
[3] By means of fuse monitoring and associated circuit-breakers
[4] By means of fuse monitoring
[5] Due to standardization

Table 1.3/11
Comparison between the protective characteristics of different switchgear assemblies

Equipment to be protected and switching rate	*Protection devices with fuses*					
Fuse Circuit-breaker Contactor Overload protection Thermistor motor protection	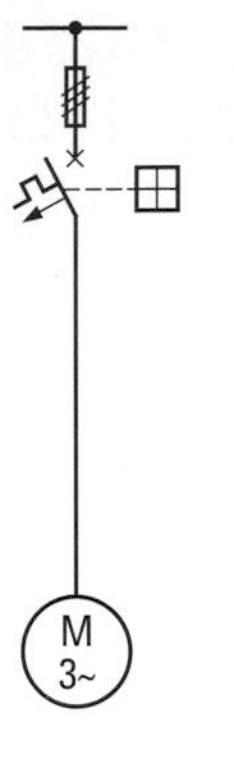 M 3~	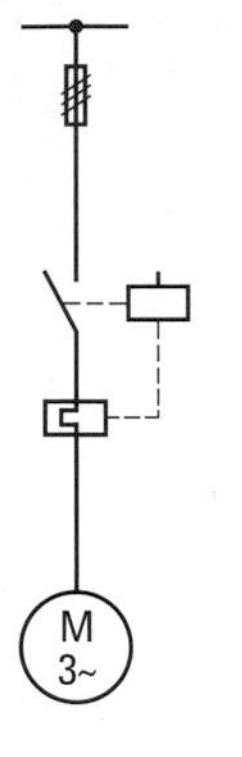 M 3~	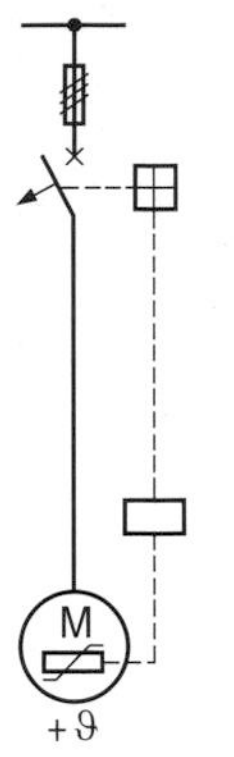 M +ϑ	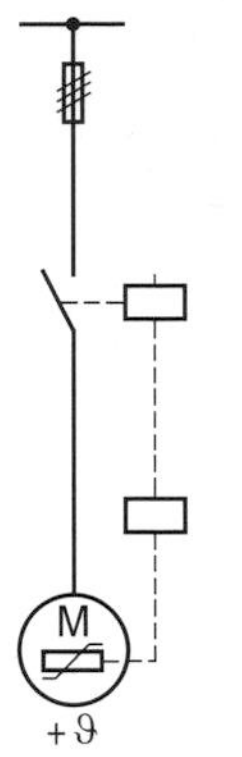 M +ϑ	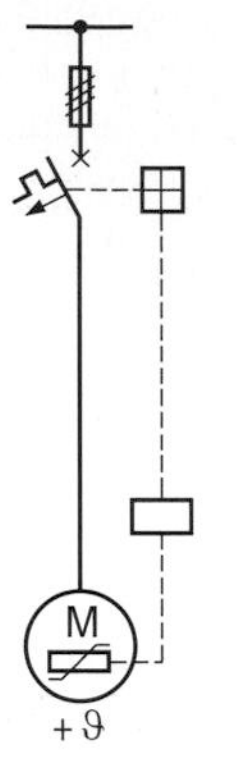 M +ϑ	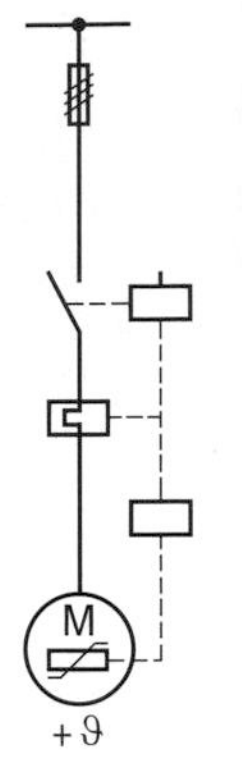 M +ϑ
Overload protection						
– Cable	++	++	+	+	++	++
– Motor (with thermally critical stator)	++[1)]	++	++	++	++	++
– Motor (with thermally critical rotor)	++[1)]	++	+	+	++	++
Short-circuit protection						
– Cable	++	++	++	++	++	++
– Motor	++	++	++	++	++	++
Switching rate	–	++	–	++	–	++

Equipment to be protected and switching rate	*Protection devices without fuses*					
– Circuit-breaker Contactor Overload protection Thermistor motor protection	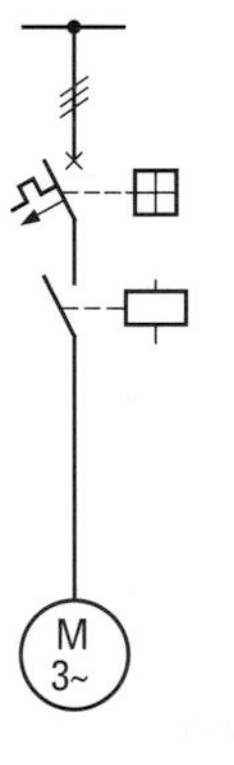 M 3~	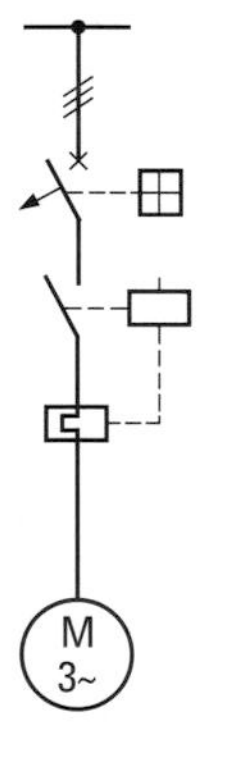 M 3~	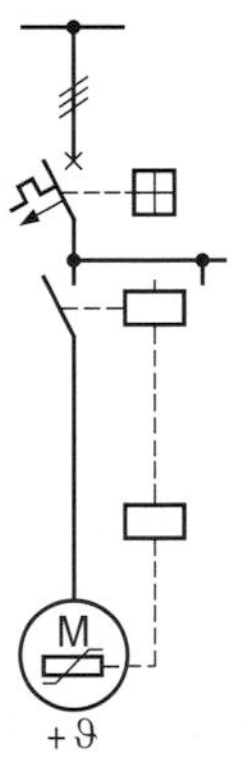 M +ϑ	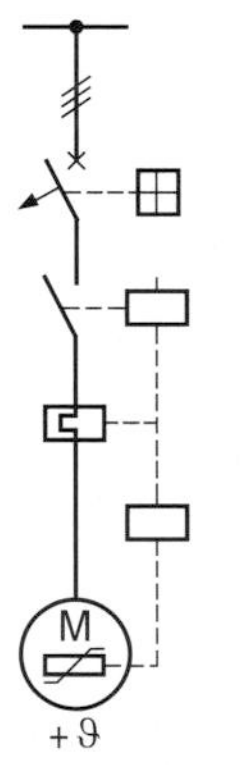 M +ϑ	 M 3~	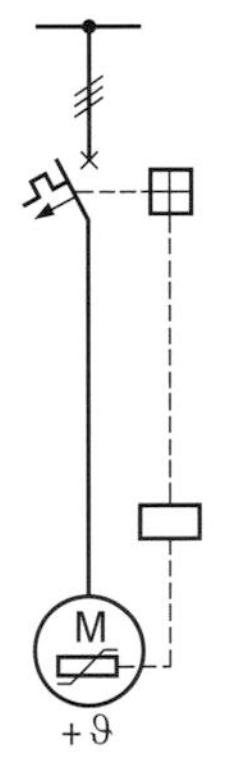 M +ϑ
Overload protection						
– Cable	++	++	++	++	++	+
– Motor (with thermally critical stator)	++[1)]	++	++	++	++[1)]	++
– Motor (with thermally critical rotor)	++[1)]	++	++	++	++[1)]	++
Short-circuit protection						
– Cable	++	++	++	++	++	++
– Motor	++	++	++	++	++	++
Switching rate	+	+	+	+	–	–

[1)] Protection with slight functional loss following failure of external conductor

++ Very good + Good – Poor

Table 1.3/12
Distribution boards with fuses *and* circuit-breakers

No.	Type of circuit-breaker	Type code	Rated short-circuit breaking capacity I_{cn}	Type of release/relay a: Adjustable	a: Fixed setting	z: Adjustable	n: Fixed setting	n: Adjustable	Back-up fuse I_{cn} > 100 kA	Tripping characteristic (↕ ↔ Adjustable release)
Feeder circuit-breaker										
1	Circuit-breaker for selective protection	3WN	$\geq I_{k1}$	×	–	×	–	×		
Distribution circuit-breaker										
2	Fuse and circuit-breaker for system protection	3NA 3VF	$\geq I_{k2}$ $\leq I_{k2}$	– –	– ×	– –	– –	– ×	× –	
Load circuit-breaker										
3	Fuse and circuit-breaker for motor protection	3NA 3RV1	$\geq I_{k3}$ $\leq I_{k3}$	– ×	– –	– –	– ×	– –	× –	
4	Fuse and direct-on-line starter	3NA 3ND 3TW	$\geq I_{k3}$ $\geq I_{k3}$ $\leq I_{k3}$	– – ×	– – –	– – –	– – –	– – –	× × –	

tribution board against overloads and short circuits. The switchgear assemblies comprising fuse and circuit-breaker, which provide motor protection, as well as fuses, contactor, and overload relay protect the motor feeder cable and the motor against overloads and short circuits.

Distribution boards without fuses (fuseless design)

In the case of distribution boards without fuses (Table 1.3/13), short-circuit protection is provided by circuit-breakers for system protection. In the case of load circuit-breakers, short-circuit protection is provided by circuit-breakers for motor protection only or for starter assemblies together with the contactor.

The protection ranges of the switchgear assemblies comprising circuit-breaker, contactor, and overload relay are specified in Section 1.3.2.2.

Further technical data can be found in the literature supplied by the manufacturer.

Table 1.3/13
Distribution boards with circuit-breakers *without* fuses

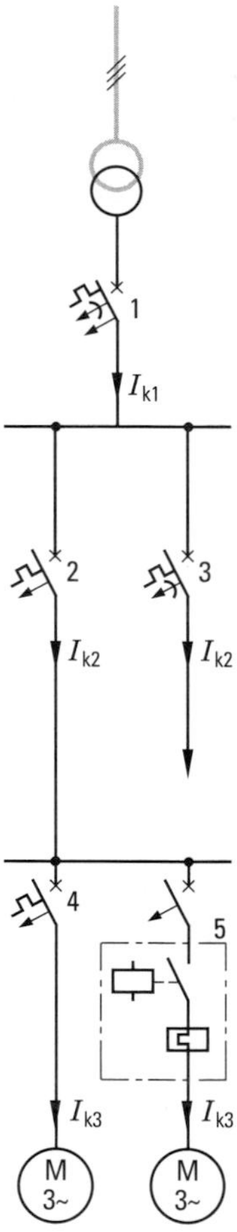

No.	Type of circuit-breaker	Type code	Rated short-circuit breaking capacity I_{cn}	Type of release/relay					Tripping characteristic
				a		z	n		↕ Adjustable ↔ release
				Adjus-table	Fixed setting	Adjus-table	Fixed setting	Adjus-table	
Feeder circuit-breaker									
1	Circuit-breaker for selective protection	3WN	$\geq I_{k1}$	×	–	×	–	×	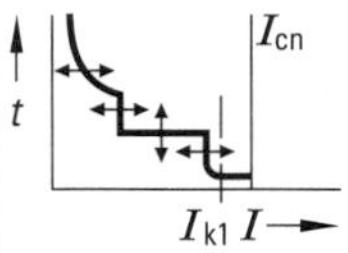
Distribution circuit-breaker									
2	Circuit-breaker for system protection	3VF	$\geq I_{k2}$	–	×	–	–	×	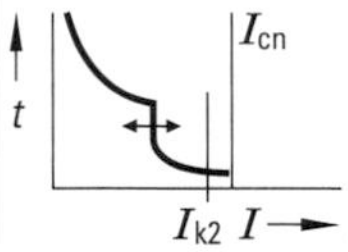
3	Circuit-breaker for selective protection	3WN	$\geq I_{k2}$	×	–	×	–	–	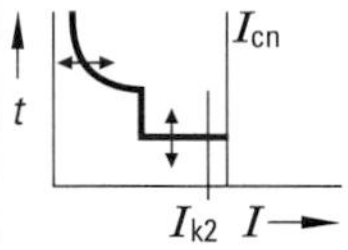
Load circuit-breaker									
4	Circuit-breaker for motor protection	3RV1	$\geq I_{k3}$	×	–	–	×	–	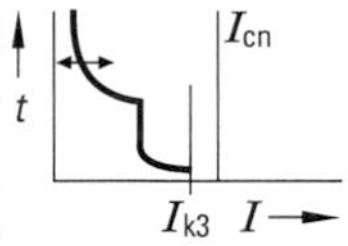
5	Circuit-breaker and direct-on-line starter	3RV1	$\geq I_{k3}$	–	–	–	×	–	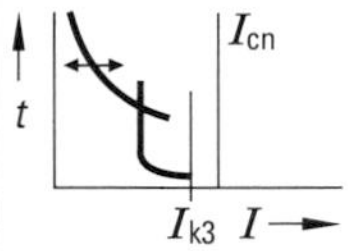
		3TW	–	×	–	–	–	–	

1.3.3 Selectivity in low-voltage systems

Selectivity and selectivity types

Selectivity is achieved with two series-connected protection devices if, when a fault occurs after the downstream protection device, only the downstream device is triggered.

Selectivity types A distinction is made between three types of selectivity:

▷ *Absolute selectivity:* the downstream protection device always trips alone (even with the lowest and highest short-circuit currents) up to its rated switching capacity.

▷ *Partial selectivity:* the downstream protection device trips (alone) more quickly up to a particular short-circuit current value (*selectivity limit*).With values above this limit, however, the let-through current pulse is sufficient to trip the upstream protection device as well. The selectivity limit can be determined either by compar- **Selectivity limit** ing characteristics or by carrying out appropriate measurements.

▷ *Full selectivity:* although the protection device assembly provides only partial selectivity, the maximum prospective short-circuit current at the downstream protection device is below the selectivity limit.

Caution! Fully selective protection device assemblies can become partially selective again due to system modifications and the resulting increase in short-circuit current.

Determining the selectivity type

Selective behavior

The selective behavior of two series-connected protection devices can be determined in two ways:

▷ by comparing characteristics
▷ by performing experimental selectivity measurements

Comparing characteristics

Two types of diagram can be used for comparing characteristics:

▷ the *time-current diagram*
▷ the $I^2 \cdot t$ *diagram*

Since the characteristics are compared over several orders of magnitude, they are usually plotted on log-log paper.

In the overload range, operating and total clearance times are approximately the same and can be plotted on one time-current diagram.

In order to assess selectivity in the event of a short-circuit, particularly with times < 100 ms, the operating values of the upstream protection device must be compared with the let-through values of the downstream protection device in the $I^2 \cdot t$ diagram.

All characteristics must – if not already specified by the manufacturer – be assigned a scatter band to enable selectivity to be determined reliably. In the case of switchgear, EN 60947–2/IEC 60947–2/DIN VDE 0660–101 specifies a scatter of $\pm 20\%$ for the instantaneous overcurrent release. The operating times, which are sometimes considerably shorter at normal operating temperatures, must be taken into consideration for electromechanical overload releases.

Selectivity measurement

All questions relating to selectivity can be answered by carrying out measurements in an appropriate system. These measurements are virtually indispensable, particularly when assessing selectivity in the event of a short circuit, owing to the extremely rapid switching operations when current-limiting protection equipment is used.

The measurements can, however, be very costly and complicated, which is why many manufacturers publish selectivity tables for their switchgears (see Table 1.3/14).

1.3.3.1 Selectivity in radial systems

Selectivity between series-connected fuses

The incoming feeder and the outgoing feeders of the busbar of a distribution board carry different operating currents and, therefore, also have different cross sections. Consequently, they are usually protected by fuses with different rated currents, which ensure selectivity on account of their different operating behaviors.

Selectivity between series-connected fuses with identical utilization categories

When fuses of the same utilization category (e.g. gL or gG) are used, selectivity is ensured across the entire overcurrent range up to the rated switching capacity (absolute selectivity) if the rated currents differ by a factor of 1.6 or higher (Fig. 1.3/18). The Joulean heat values ($I^2 \cdot t$ values) should be compared in the case of high short-circuit currents. In the example, a 160 A l. v. h. b. c. fuse would also have absolute selectivity with respect to a 100 A l. v. h. b. c. fuse.

Selectivity between series-connected fuses with different utilization categories

Characteristics only have to be compared if fuses with different utilization categories (e.g. gG and aM) are being checked for selectivity. Information regarding this as well as selectivity at other operating voltages (e.g. 500 V or 690 V) can be found in the literature supplied by the manufacturer.

Selectivity between series-connected circuit-breakers

Selectivity by grading the operating currents of instantaneous overcurrent releases (current grading)

Selectivity can be achieved by grading the operating currents of instantaneous overcurrent releases ("n" releases) (Fig. 1.3/19), provided that:

Current grading with different short-circuit currents

▷ either the short-circuit currents in the event of a short circuit at the respective locations of the circuit-breakers are sufficiently different,

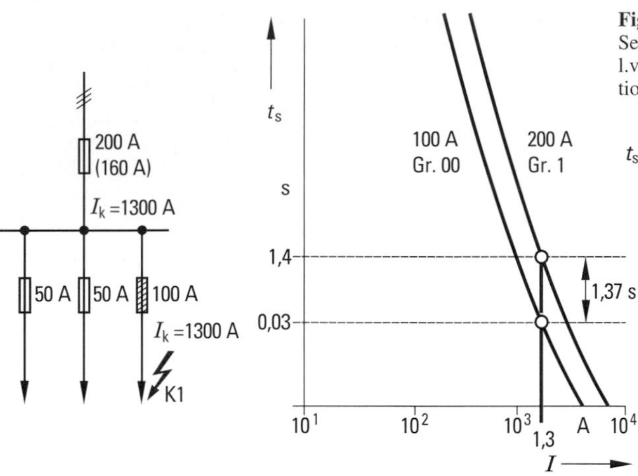

Fig. 1.3/18
Selectivity between series-connected l.v. h.b.c. fuses with identical utilization categories (example)

t_s Prearcing time

Current grading with differently configured "n" releases

▷ or the rated currents and, therefore, the "n" release values of the upstream and downstream circuit-breakers differ accordingly.

5-second breaking and line-protection conditions

In keeping with the 5-second breaking condition specified in HD 384.4.41/IEC 60364-4-41/DIN VDE 0100-410 or the 5-second line-protection condition specified in DIN VDE 0100-430 (if line protection cannot be provided in any other way), the "n" release must generally be set to 4000 A so that even very small short circuits are cleared at the input terminals of the downstream circuit-breaker Q1 within the specified time.

Only partial selectivity can be established by comparing characteristics for current grading since the increased appearance of broken lines in the characteristics in the range $\leq$ 20 ms which result from the complicated dynamic switching and tripping operations does not permit conclusions to be drawn with regard to selectivity.

Possible solution: dynamic selectivity

Dynamic selectivity

Selectivity through circuit-breaker coordination (dynamic selectivity)

Effect of dynamic processes

With high-speed operations, e. g. in the event of a short circuit, and the interaction of series-connected protection devices, the dynamic processes in the circuit and in the electromechanical releases have a considerable effect on selectivity behavior, particularly if current limiters are used.

Selectivity is also achieved if the downstream current-limiting protection device triggers so quickly that, although the let-through current does momentarily exceed the operating value of the upstream protection device, the "mechanically slow" release does not have time to trigger. The let-through current depends on the maximum asymmetrical short-circuit current and current limiting characteristics.

Selectivity limits of two series-connected circuit-breakers

A maximum short-circuit value – the *selectivity limit* – up to which the downstream circuit-breaker can open more quickly and alone (selective protection) can be determined for each switchgear assembly.

Table 1.3/14 shows an example of a selectivity table. The selectivity limit indicated in the table may be well above the operating value of the instantaneous overcurrent release in the upstream circuit-breaker (see Fig. 1.3/20).

Irrespective of this, it is important to check the selectivity in the event of an overload by comparing the characteristics and by means of tripping times in accordance with the relevant regulations.

Generally speaking, only partial selectivity is possible in the case of dynamic selectivity with short circuits. This may be sufficient (*full selectivity*) if the prospective maximum short-circuit current at the downstream protection device is lower than the established selectivity limit.

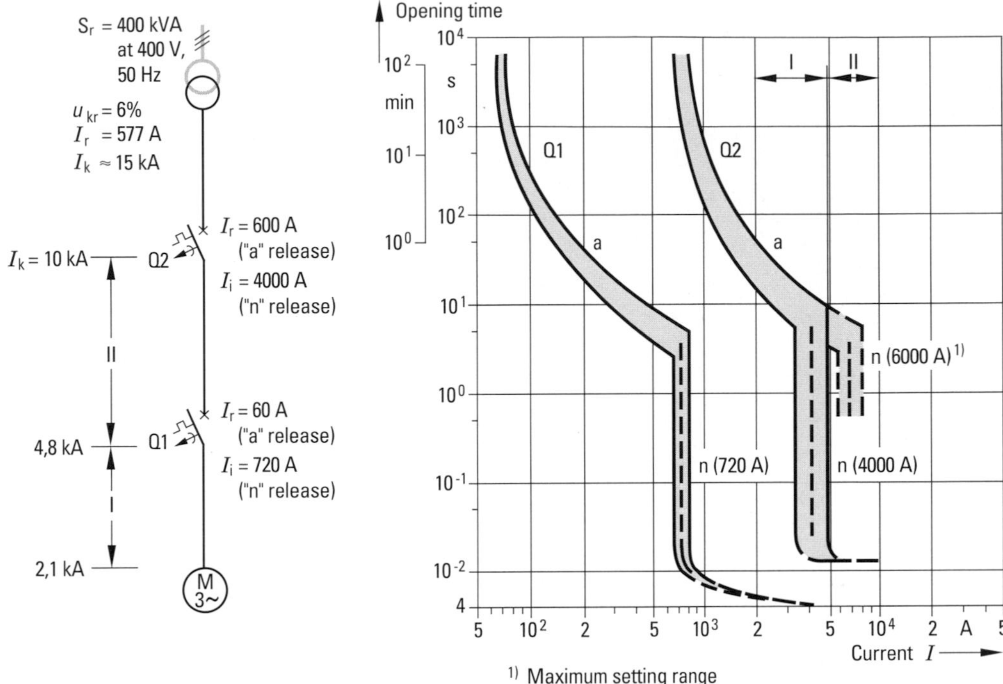

a) Single-line diagram

b) Tripping characteristics

Q1 Circuit-breaker for motor protection (current limiting)
Q2 Circuit-breaker (zero-current interrupter)

a Inverse-time-delay overload release
n Instantaneous electromagnetic overcurrent release

Fig. 1.3/19
Current selectivity for two series-connected circuit-breakers at different short-circuit current levels (example)

With partial selectivity, which usually arises with current grading owing to the clearance condition (see Fig. 1.3/19), consideration of dynamic selectivity provides a suitable opportunity for establishing full selectivity without having to use switchgear with short-time-delay overcurrent releases.

Selectivity by means of short-time-delay overcurrent releases (time grading)

Time grading by short-time-delay releases

If current grading is not possible on account of the requirements listed on page 105 and cannot be achieved by selecting the switchgear in accordance with the selectivity tables (dynamic selectivity), selectivity can be provided by time grading short-time-delay overcurrent releases. This requires grading of both the tripping delays and the appropriate operating currents.

The upstream circuit-breaker is equipped with short-time-delay overcurrent releases (z) so that, if a fault occurs, only the downstream circuit-breaker disconnects the affected part of the installation from the system. Time grading can be implemented to safeguard selectivity if the prospective short-circuit currents are almost identical. This requires grading of both the tripping delays and the operating currents of the overcurrent releases.

Time grading with virtually identical short-circuit currents

In addition to the diagram with the four series-connected circuit-breakers, Fig. 1.3/20 also contains the associated grading diagram. The necessary grading time, which allows for all scatter bands, depends on the operating principle of the release and the type of the circuit-breaker.

Electronic "z" releases

With electronic short-time-delay overcurrent releases ("z" releases), a grading time of approximately 70 ms to 100 ms from circuit-breaker to circuit-breaker is sufficient to allow for all scatter bands.

Table 1.3/14
Example of a selectivity table (selectivity limits) for two series-connected circuit-breakers (400 V AC, 50 Hz[1])

Downstream miniature circuit-breakers				Upstream circuit-breakers Overcurrent release with fixed setting Type													
				3VF3						3VF4				3VF5			
Type characteristic	Rated current (I_n) A			50	63	80	100	125	160	125	160	200	250	200	250	315	400
		Limit value of $I>$ release A		400	500	630	800	1000	1280	1250	1600	2000	2500	2000	2500	3150	4000
			I_{cu}[2] kA I_{cn} kA	25/40 70/100	25/40 70/100	25/40 70/100	25/40 70/100	25/40 70/100	25/40 70/100	40/70 100	40/70 100	40/70 100	40/70 100	45/70 100	45/70 100	45/70 100	45/70 100
	A	A	kA	Selectivity limits (kA)													
5SX2…-5 A	2	6	6	6	6	6	6	6	6	6	6	6	6	6	6	6	6
	10	30	6	2.5	4	4	4.5	4.9	6	6	6	6	6	6	6	6	6
	16	48	6	2.3	3.7	3.7	4.4	5	6	6	6	6	6	6	6	6	6
	32	96	6	1.8	3	3	3.5	3.7	6	6	6	6	6	6	6	6	6
	40	120	6	1.5	2	2	2.4	2.7	3.2	3.9	4.6	6	6	6	6	6	6
5SX2/5SX4…-6 B	6	30	6/10	3.2	6/10	6/9.7	6/10	6/10	6/10	6/10	6/10	6/10	6/10	6/10	6/10	6/10	6/10
	10	59	6/10	2.5	6/6.2	4.8	6/6.2	6/6.5	6/10	6/10	6/10	6/10	6/10	6/10	6/10	6/10	6/10
	13	65	6/10	2.3	4.6	3.8	4.6	5.1	6/8.9	6/10	6/10	6/10	6/10	6/10	6/10	6/10	6/10
	16	80	6/10	2.3	4.6	3.8	4.6	5.1	6/8.9	6/10	6/10	6/10	6/10	6/10	6/10	6/10	6/10
	20	100	6/10	2.3	4.6	3.8	4.6	5.1	6/8.9	6/10	6/10	6/10	6/10	6/10	6/10	6/10	6/10
	25	125	6/10	2.1	3.4	3	3.4	3.7	5.2	6/9.6	6/10	6/10	6/10	6/10	6/10	6/10	6/10
	32	160	6/10	2.1	3.4	3	3.4	3.7	5.2	6/9.6	6/10	6/10	6/10	6/10	6/10	6/10	6/10
	40	200	6/10	1.8	2.3	2.2	2.4	2.5	3.6	6	6	6	6	6	6	6	6/10
	50	250	6/10		2.3	2.2	2.4	2.7	3.6	5.1	5.9	6	6	6	6	6	6/10
5SX2/6SX4…-7 C	0.5	5	6/10	6/10	6/10	6/10	6/10	6/10	6/10	6/10	6/10	6/10	6/10	6/10	6/10	6/10	6/10
	1	10	6/10	6/10	6/10	6/10	6/10	6/10	6/10	6/10	6/10	6/10	6/10	6/10	6/10	6/10	6/10
	1.5	15	6/10	6/10	6/10	6/10	6/10	6/10	6/10	6/10	6/10	6/10	6/10	6/10	6/10	6/10	6/10
	2	20	6/10	6/10	6/10	6/10	6/10	6/10	6/10	6/10	6/10	6/10	6/10	6/10	6/10	6/10	6/10
	3	30	6/10	2.5	6/8.2	6/6.3	6/8.2	6/8.6	6/10	6/10	6/10	6/10	6/10	6/10	6/10	6/10	6/10
	4	40	6/10	2.5	6/8.2	6/6.3	6/8.2	6/8.6	6/10	6/10	6/10	6/10	6/10	6/10	6/10	6/10	6/10
	6	60	6/10	2.5	6/8.2	6/6.3	6/8.2	6/8.6	6/10	6/10	6/10	6/10	6/10	6/10	6/10	6/10	6/10
	8	80	6/10	2.3	3.7	3.3	3.8	4.6	6/9.4	6/10	6/10	6/10	6/10	6/10	6/10	6/10	6/10
	10	100	6/10	2.3	3.7	3.3	3.8	4.6	6/9.4	6/10	6/10	6/10	6/10	6/10	6/10	6/10	6/10
	13	130	6/10	2.1	3.9	3.2	3.8	4.4	6/7.5	6/10	6/10	6/10	6/10	6/10	6/10	6/10	6/10
	16	160	6/10	2.1	3.9	3.2	3.8	4.4	6/7.5	6/10	6/10	6/10	6/10	6/10	6/10	6/10	6/10
	20	200	6/10	2.1	3.9	3.2	3.8	4.4	6/7.5	6/10	6/10	6/10	6/10	6/10	6/10	6/10	6/10
	25	250	6/10	1.9	3	2.6	3	3.6	4.9	6/8	6/9.1	6/10	6/10	6/10	6/10	6/10	6/10
	32	320	6/10	1.9	3	2.6	3	3.6	4.9	6/8	6/9.1	6/10	6/10	6/10	6/10	6/10	6/10
	40	200	6/10	1.4	2.1	2	2.2	2.3	2.9	3.6	4.8	6/6.5	6/6.5	6/6.5	6/6.5	6/6.5	6/10
	50	500	6/10			1.9	2.1	2.2	2.9	3.6	4.8	6/6.2	6/6.2	6/6.2	6/6.2	6/6.3	6/10
	63	630	6			1.5	1.5	1.5	2.5	3	4	5	5	4	4	5	5
5SX2…-8 D	2	40	6	4.2	6	6	6	6	6	6	6	6	6	6	6	6	6
	6	120	6	2.3	4.1	3.8	4.2	4.3	6	6	6	6	6	6	6	6	6
	10	200	6	1.9	3.7	3.2	3.7	4	6	6	6	6	6	6	6	6	6
	16	320	6	1.7	3.3	2.9	3.3	3.5	4.7	6	6	6	6	6	6	6	6
	32	640	6				2.4	2.7	3.7	6	6	6	6	6	6	6	6
	40	800	6					1.5	3	4	4.9	6	6	6	6	6	6
	50	1000	6						2.6	4	4.8	6	6	6	6	3.9	6
5SX6/5SX7…-7 C	63	630	6/10				1	1.2	1.5	2.5	3	4	4	4	3	4	6
	80	800	6/10					1.2	1.5	1.5	2	3	3	3	2.5	3	6
	100	1000	6/10						1.5	1.5	2	3	3	3	2.5	3	5
5SX6/5SX7…-8 D	63	1230	6/10								2	4	4	3	2.5	4	6
	80	1600	6/10									3	3	2.5	2	3	5
	100	2000	6/10										2.5		2	3	5

1) In 240/415 V, 50 Hz networks the selectivity limits must be reduced by 10%.
2) Rated short-circuit breaking capacity I_{cu} for circuit-breakers to IEC 60947-2. Rated switching capacity I_{cn} for miniature circuit-breakers to EN 60898.

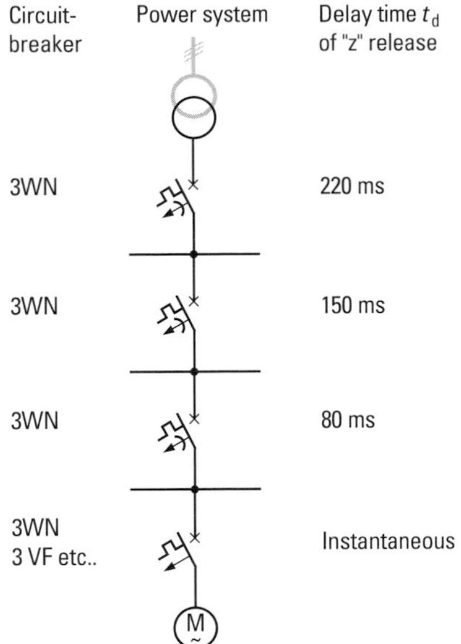

Fig. 1.3/20
Required delay time settings for electromagnetic short-time-delay "z" releases for selective short-circuit protection

Operating current

The operating current of the short-time-delay overcurrent release should be set to at least 1.45 times (twice per 20% scatter, unless other values are specified by the manufacturer) the value of the downstream circuit-breaker.

Additional "n" releases

In order to reduce the short-circuit stress in the event of a dead short circuit at the upstream circuit-breakers, these can be fitted with instantaneous electromagnetic overcurrent releases in addition to the short-time-delay releases (Fig. 1.3/21). The value selected for the operating current of the instantaneous electromagnetic overcurrent releases must be high enough to ensure that the releases only operate in the case of dead short-circuits and, under normal operating conditions, do not interfere with selective grading.

Time-reduced selectivity controller

Time-reduced selectivity controller

A microprocessor-controlled time-reduced selectivity controller has been developed for circuit-breakers to prevent excessively long tripping times when several circuit-breakers are connected in series. This control function allows the tripping delay to be reduced to 50 ms (max.) for the circuit-breakers located upstream of the short circuit.

The method of operation of the time-reduced selectivity controller is illustrated in Fig. 1.3/22. A short circuit at K1 is detected by Q1, Q3, and Q5. If the time-reduced selectivity controller is active, Q3 is temporarily disabled by Q1 and Q5 by Q3 by means of appropriate communications lines. Since Q1 does not receive any disabling signal, it trips after 10 ms. A short circuit at K2 is only detected by Q5; since it does not receive any disabling signal, it trips after 50 ms. Without time-reduced selectivity control, tripping would only occur after 150 ms.

Selectivity between circuit-breaker and fuse

When considering selectivity in conjunction with fuses, a permissible scatter band of ± 10% in the direction of current flow must be allowed for in the time-current characteristics.

Circuit-breaker with downstream fuse

Selectivity between "an" releases and fuses with very low rated currents

In the overload range up to the operating current I_i of the instantaneous overcurrent release, partial selectivity is achieved if the upper scatter band of the fuse characteristic does not touch the tripping characteristic of the fully preloaded instantaneous overcurrent release and maintains a safety margin $t_A \geq 1$ s (Fig. 1.3/23). A reduction in the tripping time of up to 25% must be allowed for at normal operating temperatures (unless the manufacturer states otherwise).

Absolute selectivity for circuit-breakers without short-time-delay overcurrent releases is achieved if the let-through current of the fuse I_D does not reach the operating current of the instantaneous overcurrent release (refer to current limiting diagram for l.v. h.b.c. fuses in Section 4.1.1). This is, however, only to be expected for a fuse, the rated current of which is very low compared with the rated continuous current.

Selectivity ratios between "az" releases and fuses with relatively high rated currents

Due to the dynamic processes that take place in electromagnetic releases, absolute selectivity can also be achieved with fuses, the I_D of which briefly exceeds the operating current (see dynamic selectivity on page 106). Once again, selectivity can only be verified by means of appropriate measurements with I_i. Absolute selectivity can be achieved by using circuit-breakers with short-time-delay overcurrent releases ("z" releases) if the safety margin for the operating current I_d

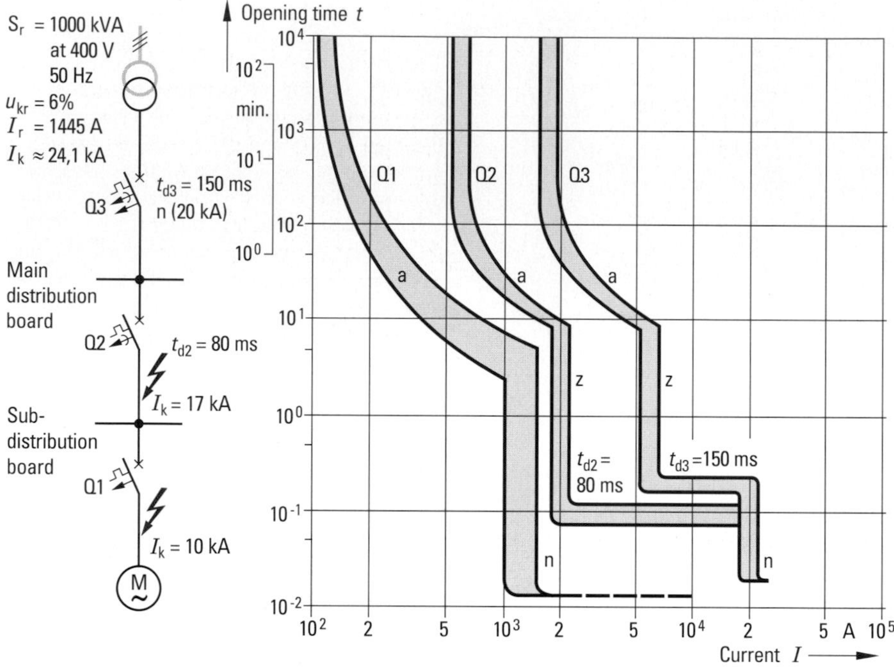

a) Single-line diagram b) Grading diagram

Q1	Circuit-breaker for motor protection
Q2	Circuit-breaker for system protection
Q3	Circuit-breaker
a	Inverse-time-delay overload release
n	Instantaneous electromagnetic overcurrent release
z	Short-time-delay overcurrent release
t_{d3}	Delay time of circuit-breaker Q3
t_{d2}	Delay time of circuit-breaker Q2
I_k	Max. short-circuit current

Fig. 1.3/21
Selectivity between three series-connected circuit-breakers with limitation of short-circuit stress by means of an additional "n" release in circuit-breaker Q3

between the upper scatter band of the fuse characteristic and the delay time of the "z" release t_d is selected so that $t_A \geq 100$ ms (Fig. 1.3/24).

Selectivity between fuse and downstream circuit-breaker

Selectivity ratios in the overload range

In order to achieve selectivity in the overload range, a safety margin of $t_A \geq 1$ s is required between the lower scatter band of the fuse and the characteristic of the inverse-time-delay overload release (Fig. 1.3/25).

In the case of short-circuits, it is important to remember that, after the releases in the circuit-breaker have tripped, the fuse continues to be heated during the arcing time. The selectivity limit lies approximately at the point where a safety margin of 70 ms between the lower scatter band of the fuse and the operating time of the instantaneous release or the delay time of the short-time-delay overcurrent release is undershot.

Short-circuit range

A reliable and usually relatively high selectivity limit for the short-circuit range can be determined in the $I^2 \cdot t$ diagram. In this diagram, the maximum let-through $I^2 \cdot t$ value of the circuit-breaker is compared with the minimum prearcing $I^2 \cdot t$ value of the fuse (Fig. 1.3/26). Since these values are maximum and minimum values, the scatter bands are not necessary.

Selectivity with parallel infeeds

Improving selectivity with parallel infeeds

With parallel infeeds to a busbar, the total short-circuit current ($I_{k\Sigma}$) that occurs in the faulted outgoing feeder comprises the partial

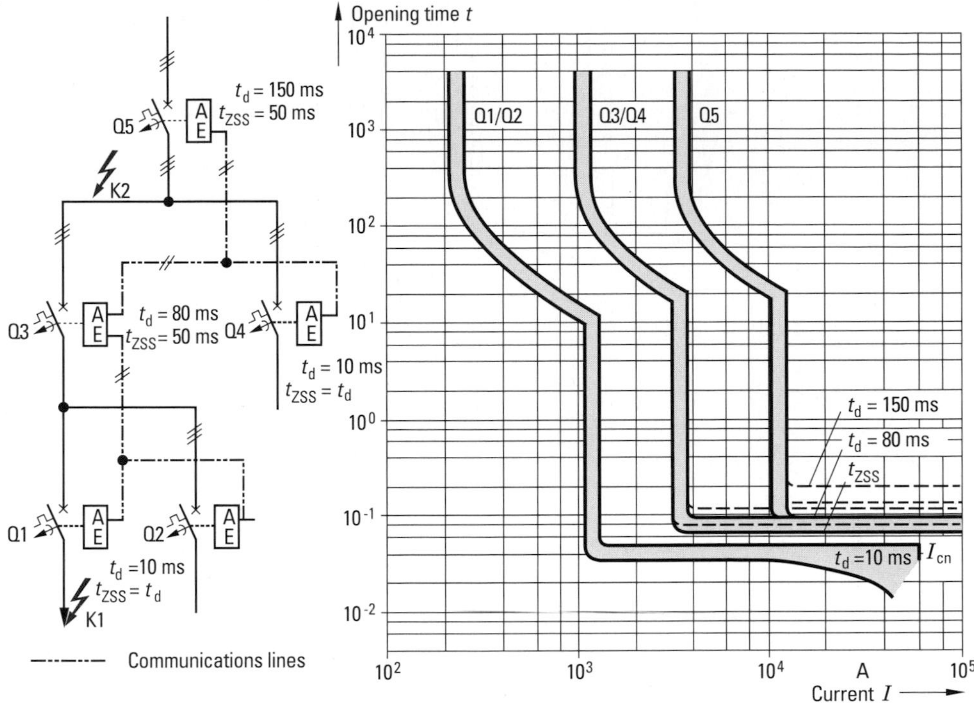

t_d	Max. operative time of disabling signal or delay time with inactive tim-reduced selectivity control
t_{ZSS}	Delay time of all circuit-breakers with active time-reduced selectivity control wich detect the short circuit and do not receive a disabling signal
A	Output sending disabling signal
E	Input receiving disabling signal
I_{cn}	Rated short-circuit breaking capacity

Fig. 1.3/22 Time-reduced selectivity control of series or parallel-connected circuit-breakers

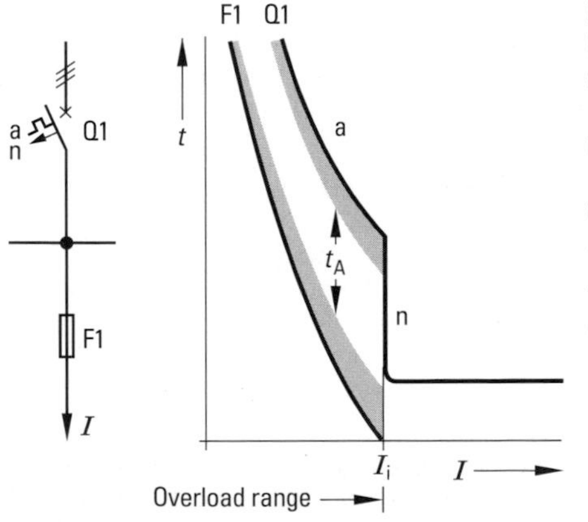

F1	Fuse
Q1	Circuit-breaker
a	Inverse-time-delay overload delay
n	Instantaneous electromagnetic overcurrent release
t_A	Safety margin
I_i	Operating current of "n" release

The time current characteristics (scatter bands) do not touch.

Fig. 1.3/23
Selectivity between circuit-breaker and downstream fuse in overload range

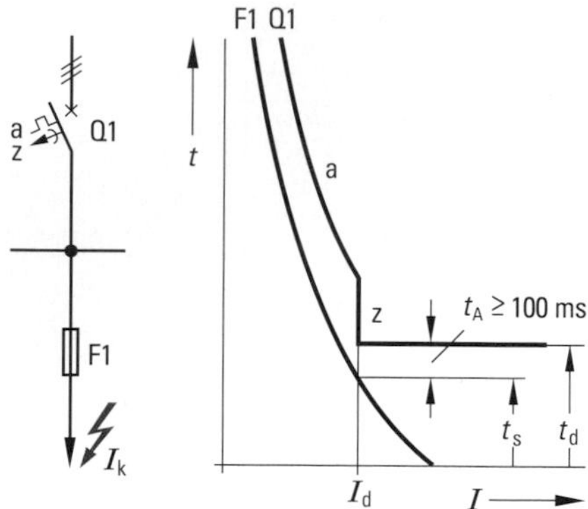

a Overload release
z Short-time-delay overcurrent release
t_A Safety margin
I_d Operating current of "z" release
t_s Prearcing time of fuse
t_d Delay time of "z" release

Fig. 1.3/24
Selectivity between circuit-breaker with "az" releases and downstream fuse; short-circuit current range

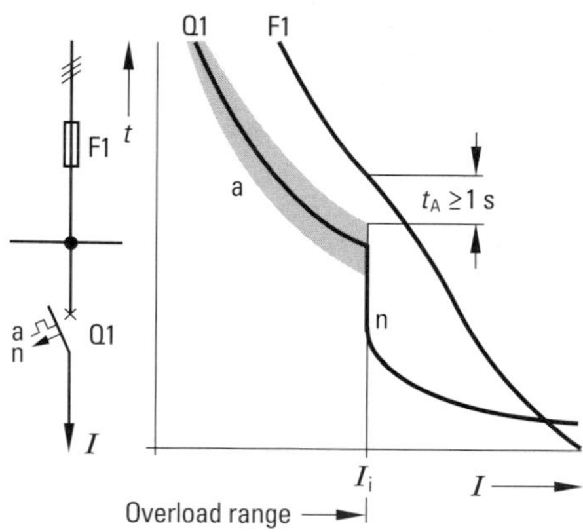

Q1 Circuit-breaker
F1 Fuse
a Inverse-time-delay overload release
n Instantaneous electromagnetic overcurrent release
t_A Safety margin
I_i Operating current of "n" release

The time-current characteristics (scatter bands) do not touch

Fig. 1.3/25
Selectivity between fuse and downstream circuit-breaker; overload range

short-circuit currents (I_{kPart}) in the individual infeeds and represents the base current in the grading diagram (Fig. 1.3/27). This is the case for all fault types.

Two identical infeeds

If a short circuit occurs in the outgoing feeder downstream of the circuit-breaker Q1, the total short-circuit current $I_{k\Sigma}$ of ≤ 20 kA, for example, flows via this, while the infeed circuit-breakers Q2 and Q3, with the outgoing feeder connected centrally to the busbars and incoming feeders of equal length, each carry only half this current, i.e. ≤ 10 kA.

Parallel operation permits additional current selectivity by means of a shift in the tripping characteristic of the "az" releases of the infeed circuit-breaker

Additional current selectivity with parallel transformer operation

In the grading diagram, the tripping characteristic of circuit-breakers Q2 and Q3 must, therefore, be considered in relation to the base current of the circuit-breaker Q1.

Characteristic displacement factor

Since the total short-circuit current is ideally distributed equally among the two infeeds (ignoring the load currents in the other outgoing feeders) with the outgoing feeder located at the center of the busbars, the tripping characteristic of circuit-breakers Q2 and Q3 can be shifted optimally to the right

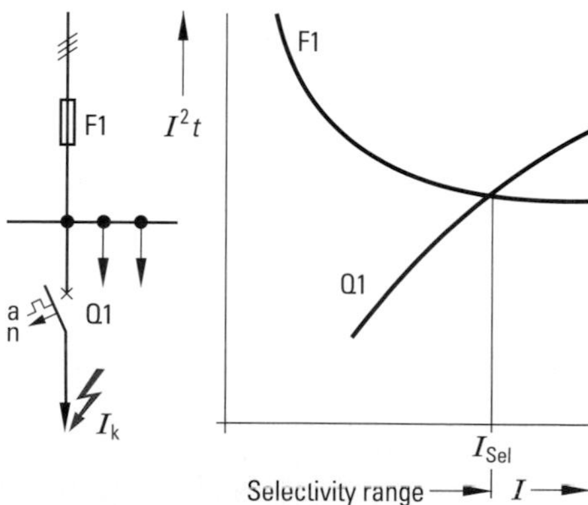

Q1 Circuit-breaker (max. let-through value)
F1 Fuse (min. prearcing value)
I_{Sel} Selectivity limit

Fig. 1.3/26
Selectivity between fuse and downstream circuit-breaker; short circuit

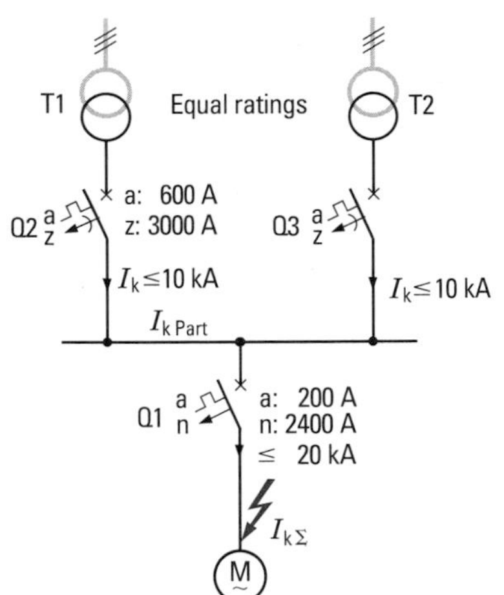

a) Single-line diagram

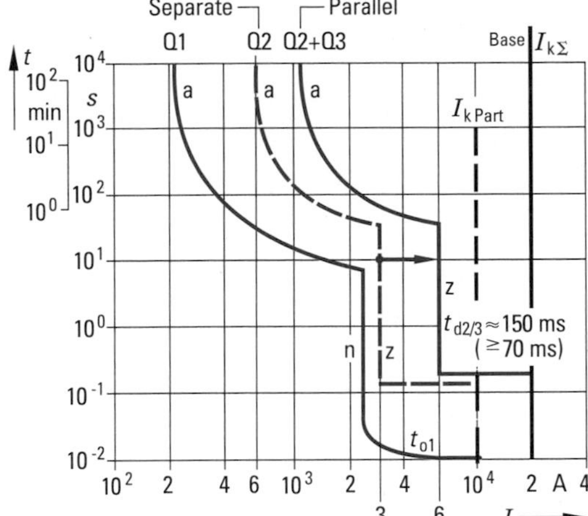

b) Grading diagram

Q2, Q3	Infeed circuit-breakers
Q1	Circuit-breaker
a	Inverse-time-delay overload release
z	Time-delay overcurrent release
n	Instantaneous overcurrent release

$t_{d2/3}$	Delay time of circuit-breakers Q2 and Q3 150 ms mechanical "z" release (≥70 ms elecronic "z" release)
t_{o1}	Opening time of circuit-breaker Q1
$I_{k\Sigma}$	Total short-circuit current
$I_{k\,Part}$	Partial short-circuit current

Fig. 1.3/27
Selectivity with two infeed transformers of the same rating and operating simultaneously. Example with outgoing feeder in the center of the busbars

along the current scale by a characteristic displacement factor of 2 up to the line $I_{k\Sigma}$, which represents the base current for this fault condition. The result of this is selectivity both with regard to time and current.

If the characteristic of the individual circuit-breaker is used instead of the shifted characteristic, the exact short-circuit current (distribution), which flows through the circuit-breaker, must be taken into consideration.

With asymmetrical configurations and with infeeds and outgoing feeders located in the busbars, short-circuit current distribution will differ according to the impedance along the incoming feeders.

Reduced selectivity with l.v. h.b.c. fuses with a rating of 630 to 1000 A near an infeed

This is particularly significant in the case of fused outgoing circuits with high current ratings, e.g. 630 to 1000 A. It is important to ensure that a safety margin of ≥ 100 ms between the tripping characteristic of the "z" release and the prearcing-time/current characteristic of the l.v. h.b.c. fuse is provided not only with parallel operation, but also with individual transformer operation.

When setting the releases of circuit-breakers Q1, Q2, and Q3, it must be ensured that selectivity is also achieved for operation with one transformer and for all short-circuit currents (single to three-phase).

For cost-related reasons, "z" releases for the feeder circuit-breakers must also be provided for low and medium rated fuse currents; "n" releases must also be used since the resulting current selectivity is not sufficient.

Three identical infeeds

With parallel operation of three transformers, the selectivity ratios will, owing to the additional current selectivity, be more favorable than with two units since the characteristic displacement factor is < 2 and < 3. Once again, "az" releases are required for the circuit-breakers in the infeeds in order to achieve unequivocal selectivity ratios.

Additional "n" release necessary!

Furthermore, it is necessary to provide additional "n" releases to allow a fault between the transformer and feeder circuit-breaker to be detected (see Fig. 1.3/28). For this purpose, the "z" releases of the circuit-breakers Q1 to Q3 must be set to a value $< I_k$ and the "n" releases to a value $> I_k$ but $< I_{k\Sigma}$. The highest and lowest fault currents are important here. Due to the "n" releases, only the faulted transformer infeed will trip on the high-voltage and low-voltage side. The circuit-breakers in the "healthy" infeeds remain operative.

Infeeds parallel-connected via tie breakers

Protective functions under fault conditions

Tie breakers must perform the following protective functions in fault situations:

▷ Instantaneous release with faults in the vicinity of the busbars
▷ Relieve outgoing feeders of the effects of high total short-circuit currents

Selecting the circuit-breakers

The type of device used in the outgoing feeders and the selectivity ratios depend primarily on whether circuit-breakers with current-zero cut-off, i.e. without current limiting, or with current limiting are used as tie breakers.

Instantaneous, current-limiting tie breakers relieve the outgoing circuits of the effects of high unlimited total peak short-circuit currents (i_p) and, therefore, permit the use of less complex and less expensive circuit-breakers.

Setting the overcurrent releases in tie breakers

Note on setting the overcurrent releases in tie breakers:

The values set for the overcurrent releases must be as high as possible in order to prevent operational interference caused by the tie breakers opening at relatively low short-circuit currents, e. g. in the outgoing feeders of the sub-distribution boards.

With two infeeds

With two infeeds and depending on the fault location (left or right busbar section or feeder), only the associated partial short-circuit current (e. g. $I_{k\,Part\,2}$) flows through the tie breaker Q3 (see Fig. 1.3/29).

With three infeeds and fault

With three infeeds, the ratios are different according to which of the outgoing feeders shown in Figs. 1.3/30 a and b is faulted.

In the center busbar section

If a fault occurs at the outgoing feeder of the center busbar section (Fig. 1.3/30 a), approximately equal partial short-circuit currents flow through the tie breakers Q4 and Q5.

In the outer busbar section

If a fault occurs at the outgoing feeder of the outer busbar section, (Fig. 1.3/30 b), two partial short-circuit currents flow through the tie breaker Q4.

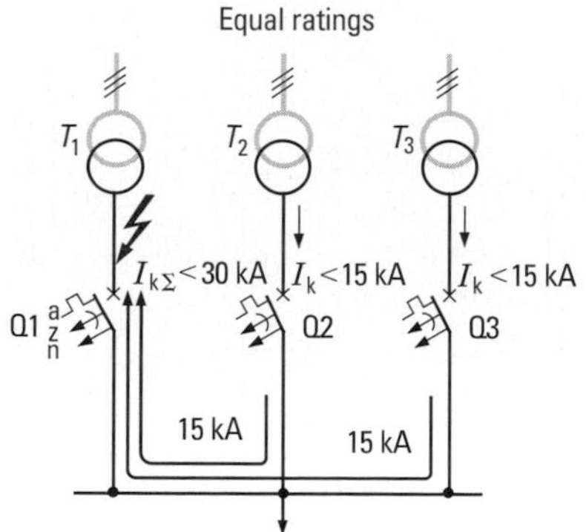

Fig. 1.3/28
Selectivity with three infeed transformers operating simultaneously

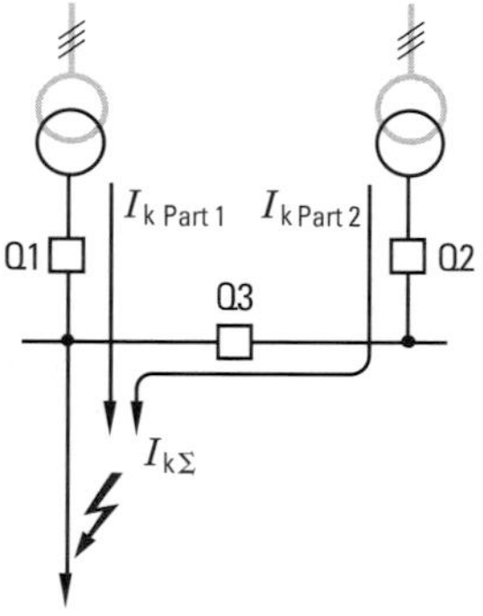

Fig. 1.3/29
Short-circuit distribution via the tie breaker Q3 with two infeeds Q1 and Q2

Computer-assisted selectivity check

Precise values for the short-circuit currents, which flow through the tie breakers, are required to permit optimum setting of the overcurrent releases. They provide information concerning selective characteristics with a large number of different fault currents, and are determined and evaluated with the aid of a computer program.

Selectivity and undervoltage protection

If a short circuit occurs, the system voltage collapses to a residual voltage at the short-circuit location. The magnitude of the residual voltage depends on the fault impedance. With a "dead" short circuit, the fault impe-

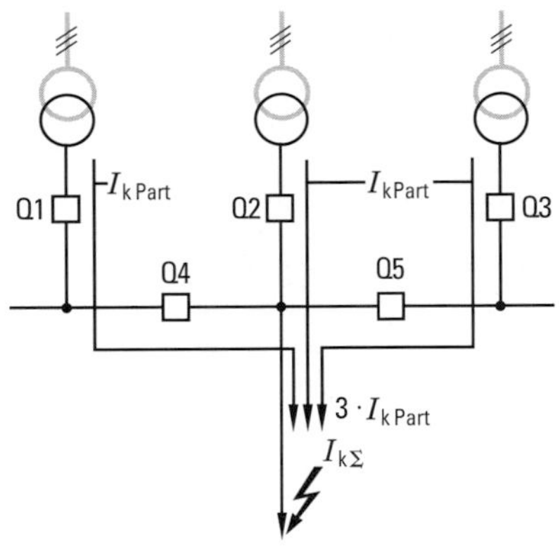

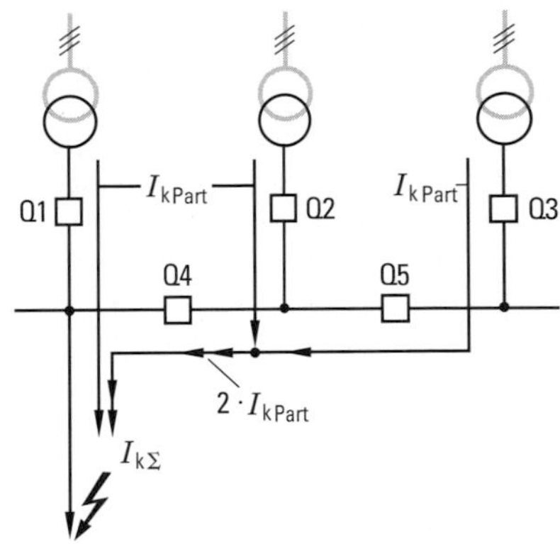

a) Fault at the outgoing feeder of the center busbar section b) Fault at the outgoing feeder of the outer busbar section

Fig. 1.3/30
Distribution of the short-circuit currents used in determining the settings for the overcurrent release in the tie breakers Q4 and Q5 with three infeeds and faults a and b in the outgoing feeders of different busbar sections

dance and, therefore, the voltage at the short-circuit location drop to almost zero. Generally speaking, however, arcs with arc-drop voltages between approximately 30 V and 70 V occur with short circuits. This voltage, starting at the fault location, increases proportionately to the intermediate impedance with increasing proximity to the power source.

Fig. 1.3/31 illustrates the voltage ratios in faulted l. v. switchgear with a "dead" short circuit.

If a short circuit occurs at K1 (Fig. 1.3/31a), the rated operating voltage U_e drops to $0.13 \cdot U_e$ at the busbar of the sub-distribution board and to $0.5 \cdot U_e$ at the busbar of the main distribution board. The next upstream circuit-breaker Q1 clears the fault. Depending on the size and type of the circuit-breaker, the total breaking time is ≤ 30 ms for zero-current interrupters and a maximum of 10 ms for current-limiting circuit-breakers.

If a short-circuit occurs at K2 (Fig. 1.3/31b), the circuit-breaker Q2 opens. It is equipped with a short-time-delay overcurrent release (z). The delay time is at least 70 ms. During this time, the rated operating voltage at the busbar of the main distribution board is reduced to $0.13 \cdot U_e$.

If the rated operating voltage drops to 0.7–0.35 times this value and the voltage reduction lasts longer than approximately 20 ms, all of the circuit-breakers with undervoltage releases open. All contactors also open if the rated control supply voltage collapses to below 75% of its rated value for longer than 5 ms to 30 ms.

Tripping delay for contactors and undervoltage releases

Undervoltage releases and contactors with tripping delay are required to ensure that the selective overcurrent protection is not interrupted prematurely. These are not necessary if current-limiting circuit-breakers, which have a maximum total clearing time of 10 ms, are used.

1.3.3.2 Selectivity in meshed systems

Selectivity functions in meshed systems

Two selectivity functions must be performed in meshed systems:

1. Only the short-circuited cable should be disconnected from the system.
2. If a short-circuit occurs at the terminals of an infeed transformer, only the faulted terminal must be disconnected from the system.

Node fuses

Node fuse

The nodes of a meshed l. v. system are normally equipped with cables with the same cross section and with l. v. h. b. c. fuses (utilization category gL) of the same type and rated current (Fig. 1.3/32).

If a short-circuit (K1) occurs along the meshed system cable, the short-circuit currents I_{k3} and I_{k4} flow to the fault location.

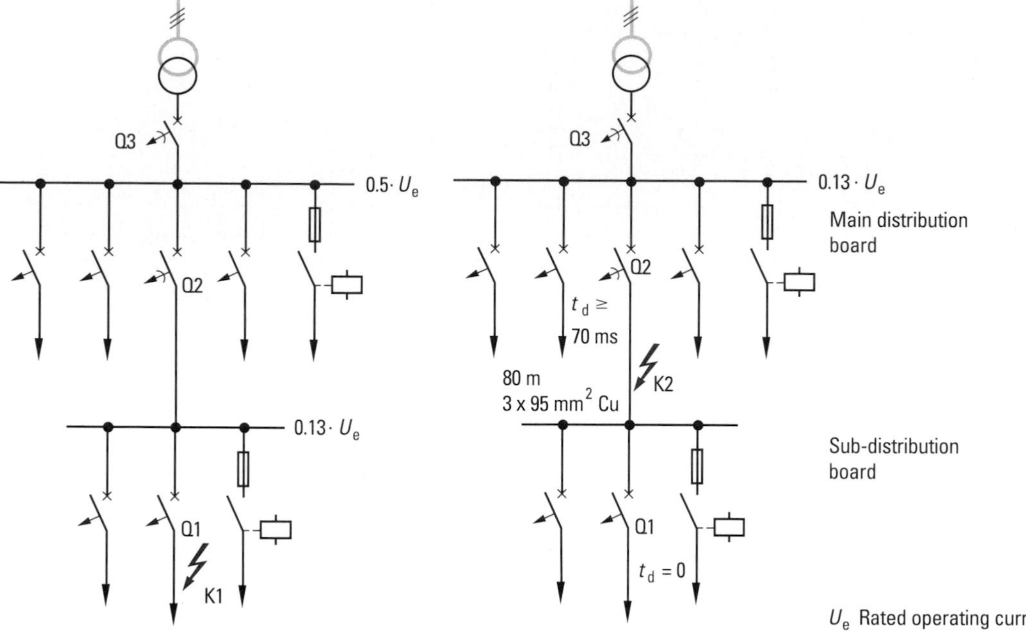

Fig. 1.3/31 Voltage ratios for short-circuited l.v. switchgear with a main and sub-distribution board

Short-circuit current I_{k3} from node a comprises the partial currents I_{k1} and I_{k2} which may differ greatly depending on the impedance ratios.

Permissible current ratio

Selectivity of the fuses at node a is achieved if fuse F3, through which the total current I_{k3} flows, melts and fuse F1 or F2, through which the partial short-circuit I_{k1} or I_{k2} flows, remains operative. In the case of Siemens l.v. h.b.c. fuses, the permissible current ratio $I_{k1}/(I_{k1} + I_{k2})$ for high short-circuit currents is 0.8.

Power transformers in meshed systems

Feeder circuit-breaker with network master relay

In multi-phase meshed systems (Fig. 1.3/33), i.e. infeed via several h.v. conductors and transformers, power feedback from the l.v. system to the fault location should be prevented if a fault occurs in a transformer substation or h.v. conductor. A network master relay (reverse power relay) and a "network circuit-breaker" are required for this purpose. This is a three-phase circuit-breaker, possibly without overcurrent release, but with a capacitor-delayed shunt release for network circuit-breaker (open-circuit shunt release with memory).

If a short circuit occurs on the h.v. side of the transformer (K1) or between the transformer and network circuit-breaker (K2) or along the cable (K3) (Fig. 1.3/34), the h.v. h.b.c. fuse operates on the h.v. side; on the l.v. side, power flows back to the fault location via the network circuit-breaker. The open-circuit shunt release receives the tripping pulse from the network master relay. The fault location is thus selectively disconnected from the power system.

Network circuit-breaker without "z" release

If the outgoing transformer feeders are protected by network master relays, no "z" release is provided or the value set for this is so high that the thermal overload capability of the transformer can be fully utilized.

Network master relays

Network master relays are used in conjunction with network circuit-breakers. In multi-feed l.v. power systems, they ensure fast, selective disconnection of a damaged h.v.

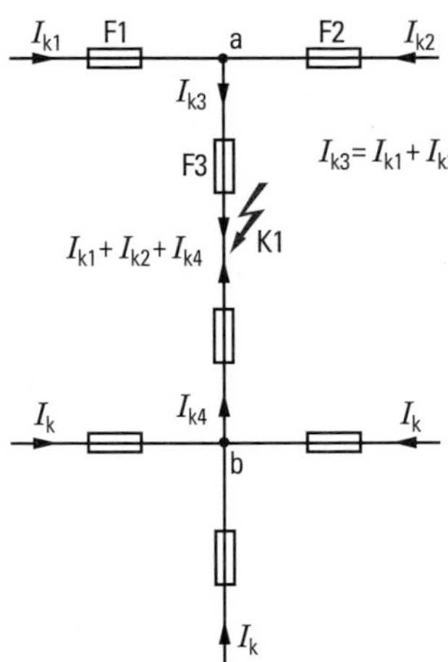

Fig. 1.3/32
Short-circuited cable with its two incoming feeder nodes a and b

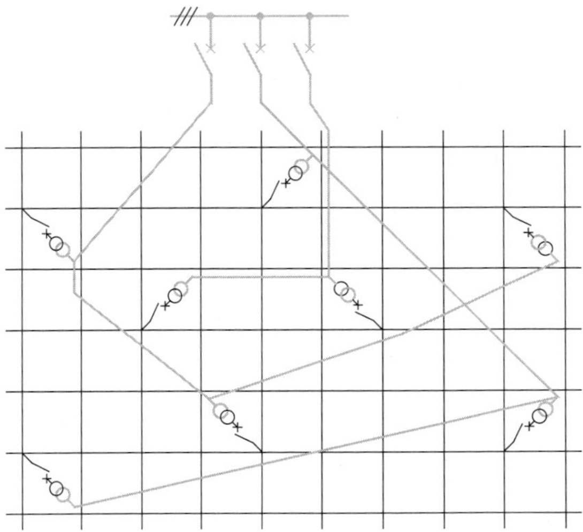

Fig. 1.3/33
Example of a meshed system with multi-phase infeed

cable from the connected transformer substations. The relay detects a reversal in the flow of power if, in the event of a short circuit in an h.v. feeder cable of the meshed system, high currents flow via the l.v. power system and the transformers of the damaged h.v. cable to the fault location.

To prevent errors, however, the network master relay permits circulating currents up to the same value as the rated current at the rated voltage (setting can be varied between 2 A and 6 A using the spring bias). Fig. 1.3/35 shows the tripping characteristic for the standard setting (6 A) and for various other voltages.

Network circuit-breakers

Selecting the network circuit-breaker

When selecting the network circuit-breaker and its rated switching capacity, it is important to remember that the highest short-circuit current must be expected in the event of a short circuit between the transformer terminals and the circuit-breaker. In this case, the total short-circuit current of all the infeed points flows through the meshed system and the circuit-breaker to the short-circuit location. The total short-circuit current may be higher than the short-circuit current of the relevant transformer.

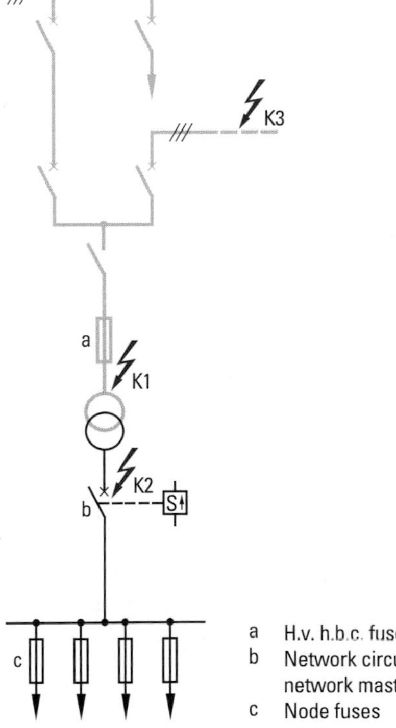

a H.v. h.b.c. fuses
b Network circuit-breaker with network master relay
c Node fuses

Fig. 1.3/34
Single-line diagram showing the infeed point of a meshed l.v. power system

Tripping characteristic of the network master relay

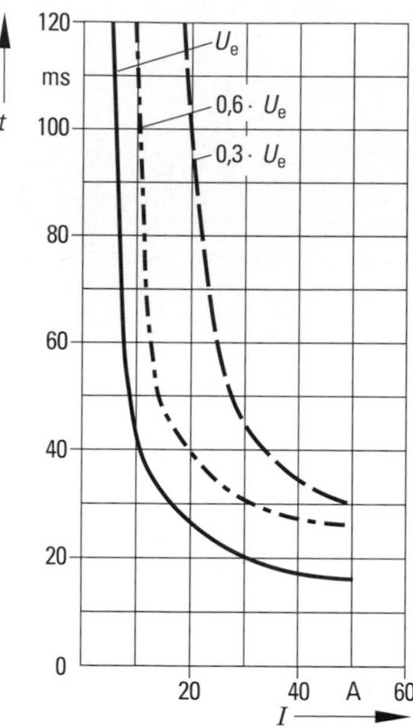

Fig. 1.3/35
Tripping characteristic of the network master relay 7RM with standard setting (6 A)

Technical details regarding network master relays and network circuit-breakers can be found in the literature supplied by the respective manufacturer.

1.3.4 Protection of capacitors

According to IEC 60358/VDE 0560 Part 4, capacitor units must be suitable for continuous operation with a current, the r.m.s. value of which does not exceed 1.3 times the current which flows with a sinusoidal voltage and rated frequency.

Owing to the above-mentioned dimensioning requirements, no overload protection is provided for capacitor units in the majority of cases.

Capacitors in systems with harmonic components

The capacitors can only be overloaded in systems with devices which generate high harmonics (e.g. generators and converter-fed drives). The capacitors, together with the series-connected transformer and short-circuit reactance of the primary system, form an anti-resonant circuit. Resonance phenomena occur if the natural frequency of the resonant circuit matches or is close to the frequency of a harmonic current generated by the power converter.

Reactor-connected capacitors

The capacitors must be provided with reactors to prevent resonance (see Section 1.6). An LC resonant circuit, the resonance frequency of which is below the lowest harmonic component (250 Hz) in the load current, is used instead of the capacitor. The capacitor unit is thus inductive for all harmonic currents that occur in the load current and can, therefore, no longer form a resonant circuit with the system reactance.

Settings for overload relay

If thermal time-delay overload relays are used to provide protection against overcurrents, the tripping value can be 1.3 to 1.43 times the rated current of the capacitor since, allowing for the permissible capacitance deviation, the capacitor current can be $1.1 \cdot 1.3 = 1.43$ times the rated capacitor current.

With transformer-heated overload relays or releases, a higher secondary current flows due to the changed transformation ratio of the transformers caused by the harmonic components. This may result in premature tripping.

Harmonics suppression by means of filter circuits

An alternative solution would be to use filter circuits to remove the majority of harmonics from the primary system (see Sections 1.6.3 and 1.6.4). The filter circuits are also series resonant circuits which, unlike the reactor-connected capacitors, are tuned precisely to the frequencies of the harmonic currents to be filtered. As a result, the impedance is almost zero.

Short-circuit protection

L.v. h.b.c. fuses with utilization category gL are normally used in capacitor units for short-circuit protection.

A rated fuse current of 1.6 to 1.7 times the rated capacitor current is required to prevent the fuses from tripping in the overload range and when the capacitors switch.

1.3.5 Protection of distribution transformers

The following are used as protection devices in high-voltage systems:

H.v. h.b.c. fuses

High-voltage high-breaking-capacity (h.v. h.b.c.) fuses usually used in conjunction with switch-disconnectors to protect radial feeders and transformers against short circuits.

Circuit-breakers with protection

Protection relays

▷ Protection relays connected to current transformers (protection core) can be used to perform all protection-related tasks irrespective of the magnitude of the short-circuit currents and rated operating currents of the required circuit-breakers.

Digital protection

▷ Modern protection equipment is controlled by microprocessors (digital protection) and supports all of the protective functions required for a high-voltage outgoing feeder.

Protection as component of substation control and protection system

▷ Digital protection also allows operating and fault data, which can be called up via serial data interfaces, to be collected and stored. Digital protection can, therefore, be incorporated in substation control and protection systems as an autonomous component.

Standards for protection relays

Static protection relays must comply with the standards IEC 60255 and DIN VDE 0435-303.

Current transformer rating for protection purposes

Current transformers are subject to the standards IEC 60185, IEC 60186, and DIN VDE 0414-1, -2, and -3. Current transformers with 5P or 10P cores must be used for connecting protection equipment.

The required rated output and overcurrent factor must both be determined on the basis of the information provided in the protection relay descriptions.

Time-overcurrent protection

▷ Time-overcurrent protection via current transformers for protecting cables and transformer feeders can be either two-phase or three-phase. The method of neutral-point connection is an important factor here.

Relay operating currents with emergency generator operation

▷ Care should be taken to ensure that the operating currents of the protection relays provided for normal system operation are also attained in the event of faults during emergency operation using generators with relatively low rated outputs.

Three-phase time-overcurrent protection

In the interests of upgradeability, it is advisable to configure the time-overcurrent protection as a three-phase system right from the very outset, irrespective of the method of neutral-point connection.

1.3.5.1 Protection with overreaching selectivity

Ideally, transformer feeders should be protected by:

H.v. h.b.c. fuses

▷ high-voltage high-breaking-capacity (h. v. h. b. c.) fuses used in conjunction with switch-disconnectors for rated transformer outputs ≤ 1600 kVA (approx.) for low switching rates, or

Circuit-breakers with protection

▷ circuit-breakers with protection (see page 125 ff.) ≥ 800 kVA (approx.) and for high switching rates; also when several circuit-breakers with "z" releases are arranged in series on the low-voltage side and selectivity is not possible with upstream h. v. h. b. c. fuses.

The anticipated selectivity ratios must, therefore, be checked before the protection scheme is chosen and dimensioned.

Protection by means of h. v. h. b. c. fuses

Dimensioning h. v. h. b. c. fuses

The rated current of the h. v. h. b. c. fuses specified by the manufacturers for the rated output of each transformer should be used when dimensioning the h. v. h. b. c. fuses. The lowest rated current is dictated by the rush currents generated when the transformers are energized and is 1.5 to 2 times the rated transformer current.

Minimum breaking current

In order to determine the maximum rated current, the minimum breaking current I_{amin} of the fuse must be exceeded in the event of a short circuit on the secondary side of the transformer up as far as the busbars of the installation. I_{amin} is normally 4 to 5 times the rated transformer current. The fuse link can be chosen between the specified limits according to the selectivity requirements.

Back-up protection with protection zones

H.v. h. b. c. fuses must provide sufficient back-up protection in the event of possible failure of the downstream protection device. Fig. 1.3/36 shows the necessary protection zone for three possible circuits. The reach of the back-up protection zone is inversely proportionate to the rated fuse current.

Safety margins between the prearcing-time/ current characteristic of h. v. h. b. c. fuses and the characteristics of other protection devices

Safety margins

The rated current of the h. v. h. b. c. fuse links must be selected such that there is a minimum safety margin amounting to 25% of

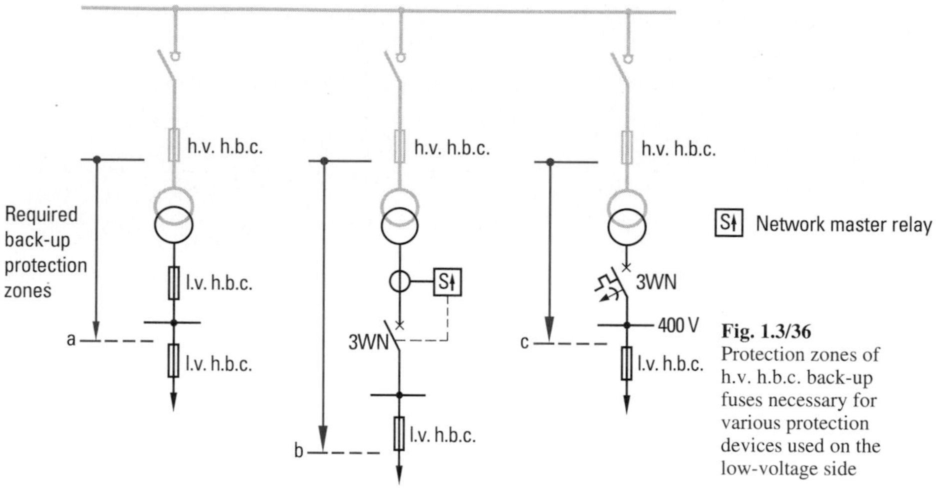

Fig. 1.3/36
Protection zones of h.v. h.b.c. back-up fuses necessary for various protection devices used on the low-voltage side

I_{amin} with respect to the short-circuit current I_k of the transformer between the calculated maximum current in the event of a short circuit in the vicinity of the busbar on the low-voltage side (converted to the high-voltage side) and the minimum breaking current I_{amin} (the circle in the prearcing-time/current characteristic) (see Figs. 1.3/37 to 1.3/40).

Further information on safety margins, e.g. for gradings as shown in Fig. 1.3/36, case b and c, is given below.

Grading of h. v. with l. v. h. b. c. fuses in infeed circuits

Grading with l. v. h. b. c. fuses

Example of a transformer with a rated output of 400 kVA (Fig. 1.3/37):

L.v. h. b. c. fuse switch-disconnectors or motor fuse-disconnectors (maximum rated current 630 A) are mainly used for transformers with rated outputs ≤400 kVA; circuit-breakers with overcurrent releases are used on the low-voltage side for rated outputs ≥500 kVA.

It is acceptable for the prearcing-time/current characteristics F2 (l. v. h. b. c.) and F3 (h. v. h. b. c.) – referred to 0.4 kV – to touch and the switch-disconnector to be tripped on the high-voltage side by the upstream h. v. h. b. c. fuse since both fuses protect the same system element and interruption will occur in all cases (restricted selectivity). H.v. h. b. c. fuses with higher rated currents (e.g. 160 A as shown in Fig. 1.3/38) would not be suitable here since their minimum breaking current I_{amin} is 12 kA, i.e. well above the short-circuit current I_k which the transformer can carry (max. 9.5 kA).

Grading of h. v. h. b. c. fuses with network circuit-breakers and downstream l. v. h. b. c. fuses

Selecting the h. v. h. b. c. fuse rating

In meshed systems with several transformers and parallel system operation, the l. v. feeder circuit-breakers are not fitted with overcurrent releases (az) but, instead, have separate 7RM19 network master relays which only respond to reverse currents. Given the absence of the "az" release as a grading element, the back-up protection range of the h. v. h. b. c. fuse must be extended as shown in Fig. 1.3/36, case b. In Fig. 1.3/38, this is achieved by selecting the h. v. h. b. c. fuse with the lower current rating.

The proximity of the characteristics F1 and F2 in this case does not have a detrimental effect on selectivity since the ring interconnections also function as infeeds in the event of a fault which means that selectivity is improved as a result of the total short-circuit current $I_{k\Sigma}$ in the feeder being distributed among the infeeds (with the two ring interconnections in Fig. 1.3/38). This is possible because a higher short-circuit current ($I_{k\Sigma}$) flows through the l. v. h. b. c. fuse F1 than through the h. v. h. b. c. fuse F2 (I_{kT}).

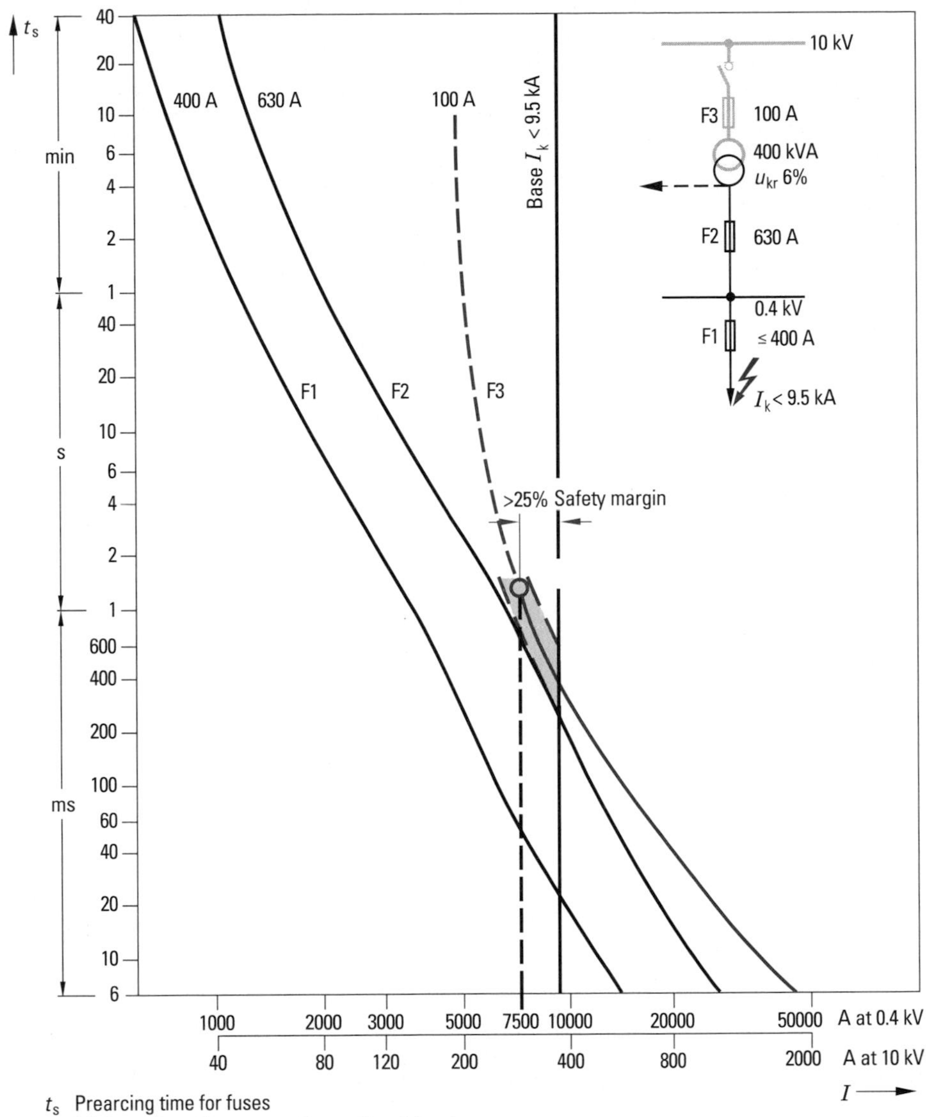

Fig. 1.3/37 Example showing grading of h.v. h.b.c. with l.v. h.b.c. fuses in infeed circuits

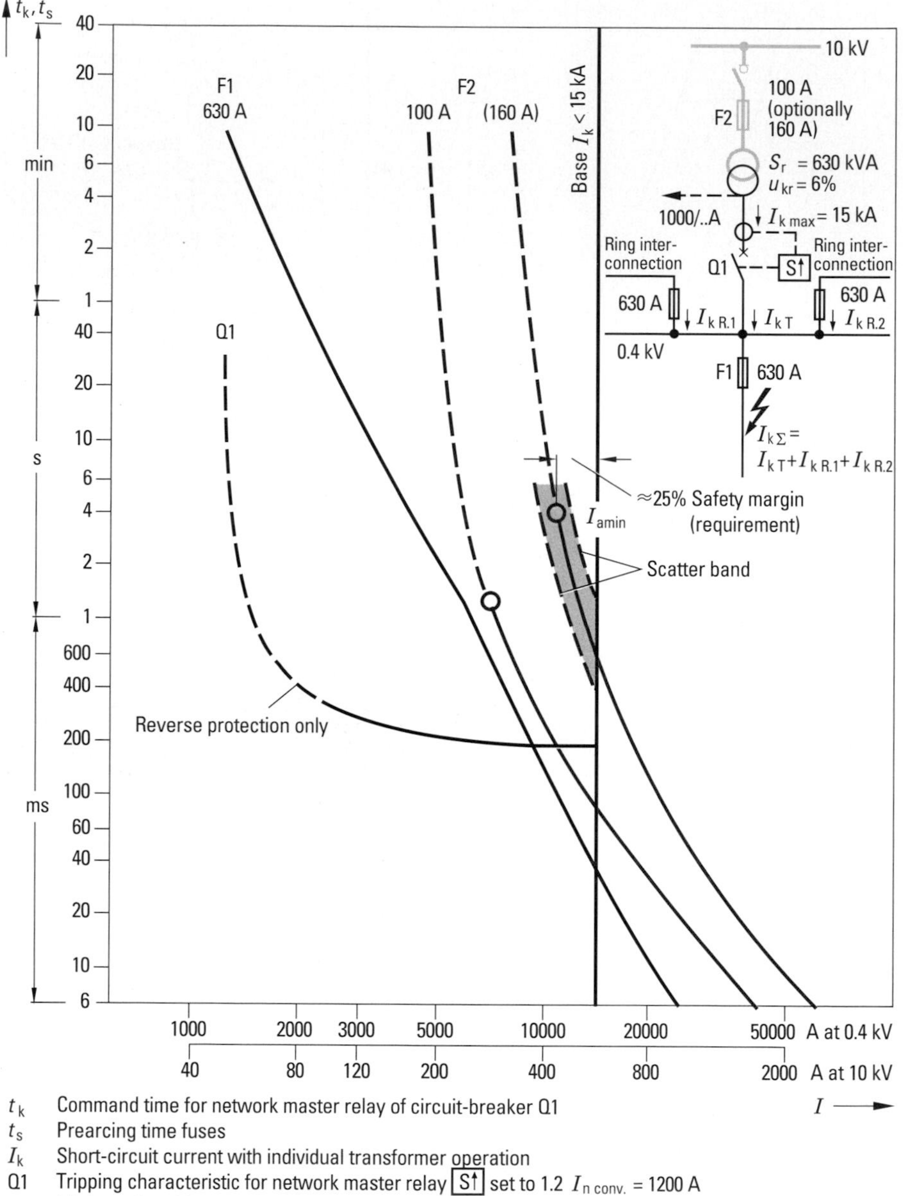

Fig. 1.3/38
Example showing grading with h.v. h.b.c. fuses – network master relay in the infeed – and l.v. h.b.c. fuses in the outgoing feeder; transformer rating 630 kVA

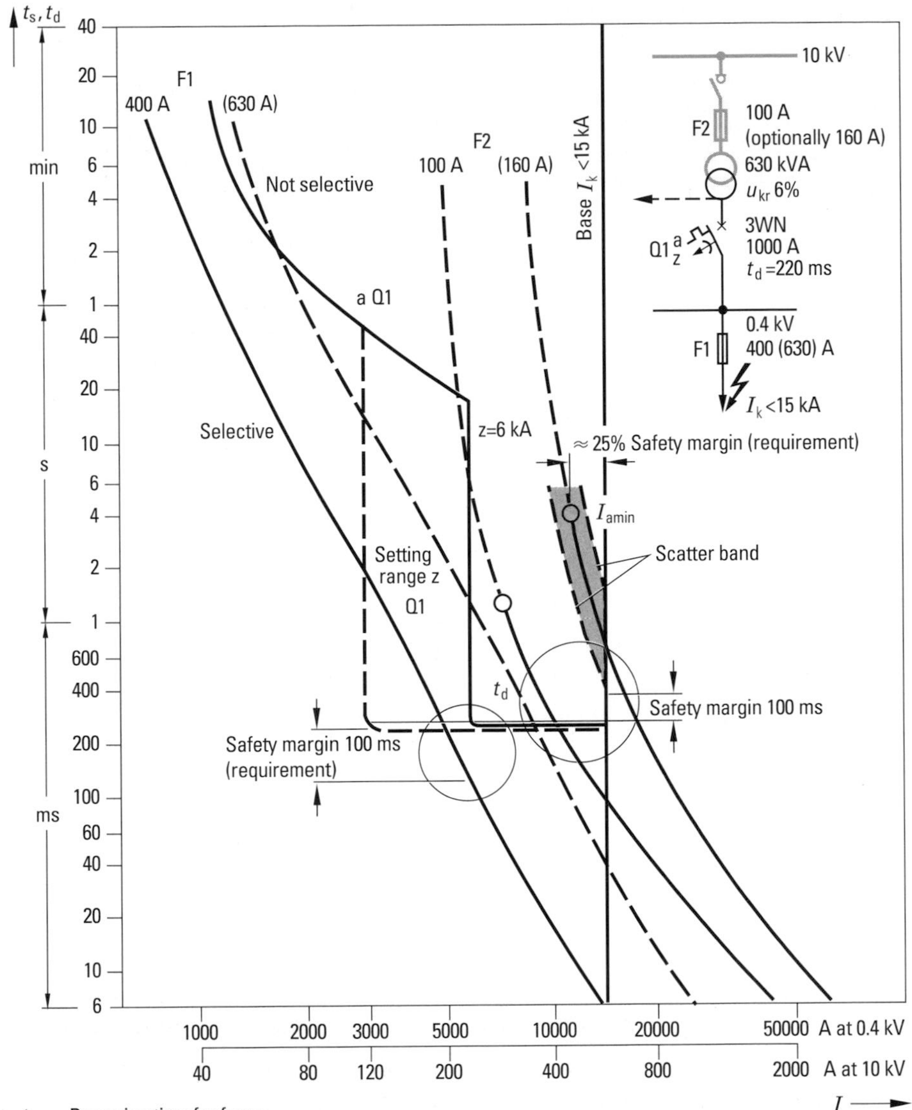

t_s Prearcing time for fuses
t_d Delay time for "z" release (Q1)
Minimum breaking current I_{amin} of h.v. h.b.c. fuse

Fig. 1.3/39 a
Example showing grading of h.v. h.b.c. fuses F2 with circuit-breaker Q1 and downstream l.v. h.b.c. fuse F1 in the outgoing feeder

Grading of h. v. h. b. c. fuses with low-voltage circuit-breakers and downstream l. v. h. b. c. fuses using a 630 kVA transformer as an example

Requirements

Selectivity is required between the protection devices of the feeders and those of the infeed, which together form a functional unit; a safety margin of at least 100 ms is necessary between the characteristic of an l. v. h. b. c. fuse and that of a "z" release (Fig. 1.3./39a).

Between l. v. h. b. c. fuses and "z" releases

Selectivity is achieved with the 400 A fuse link used in the example. The setting and delay time (t_d) must be adjusted with the "z" release (setting 6 kA).

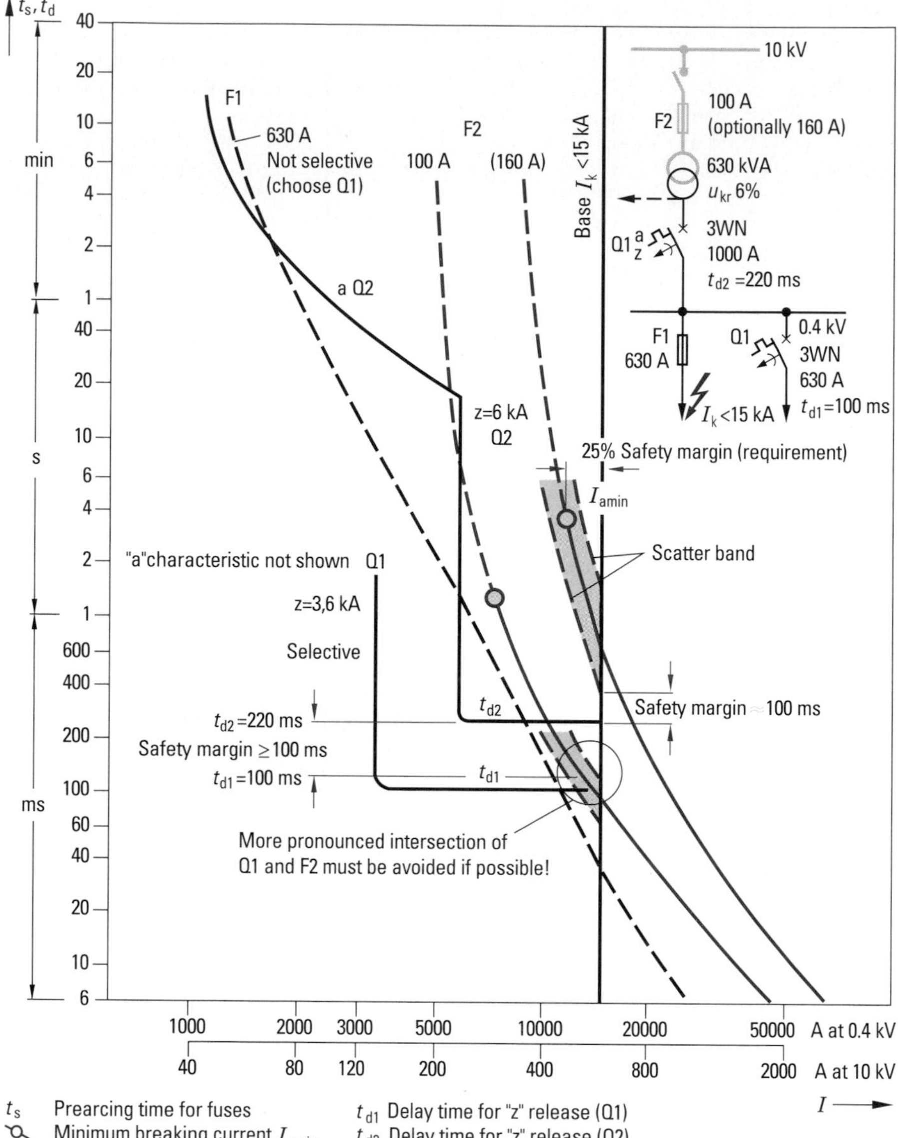

t_s Prearcing time for fuses
Minimum breaking current I_{amin}
t_{d1} Delay time for "z" release (Q1)
t_{d2} Delay time for "z" release (Q2)

Fig. 1.3/39 b
Example showing grading of h.v. h.b.c. fuses F2 with circuit-breaker Q2 and downstream circuit-breaker Q1 with "az" release in the outgoing feeder
Result: If circuit-breakers are used instead of l.v. h.b.c. fuses with a high current rating, selectivity can be achieved relatively easily. Intersection of the characteristics Q1 and F2 must be avoided if possible, otherwise nuisance tripping may occur. The h.v. h.b.c. fuse links with the higher current rating must be used in such cases.

In such cases, selectivity can be achieved more easily using downstream circuit-breakers, e.g. 3WN (Fig. 1.3/39b), or using a considerably more powerful transformer, the circuit-breaker of which can be set to a higher value on the "z" release.

Between h.v. h.b.c. fuses and "z" releases

Since the protection devices in the infeed form a functional unit, a restriction in selectivity in the upper short-circuit current range is accepted in the case of faults in the vicinity of the busbars (as indicated by the circle in the diagram for the 100 A h.v. h.b.c. fuse in Fig. 1.3/39b).

Safety margin between h.v. h.b.c fuse and "z" release

If, on the other hand, selectivity is required, e.g. with different switching priorities at the two voltage levels or in order to avoid the high-voltage switchgear having to be switched off, for example, when h.v. h.b.c. fuses are replaced, there should be a safety margin of approximately 100 ms on the base line I_k between the characteristic of the "z" release and the left-hand limit of the scatter band of the prearcing-time/current characteristic of the h.v. h.b.c fuse.

Scatter band of h.v. h.b.c. fuses

According to EN 60282-1/DIN VDE 0670-4, the scatter band width of h.v. h.b.c. fuse links can be ±20%. Siemens h.v. h.b.c. fuse links have a scatter band width of ±10%.

Protection by means of circuit-breakers with definite-time overcurrent protection (UMZ)

Grading with low-voltage protection devices

Requirements

The two infeed circuit-breakers (in Fig. 1.3/40) form a functional unit and require selectivity with respect to the protection devices on the low-voltage side.

Outgoing feeders with l.v. h.b.c. fuses

If low-voltage fuses are connected downstream, selectivity with circuit-breakers with mechanical releases (3WE) can only be achieved up to a certain maximum fuse current rating; in the example, Q2 with mechanical "z" releases (setting range 3 to 6 kA) ≤400 A for F1. Larger l.v. h.b.c. fuses are also selective if 3WN circuit-breakers with a "z" release range of 2 to 12 · I_r are used.

Outgoing feeders with mixed components

If outgoing feeders with mixed components are used, the safety margin of at least 100 ms relative to the largest permissible l.v. h.b.c. fuse link for F1 is an important factor in determining the setting for the "z" release of Q2. In the case of mechanical "z" releases with the highest current setting of 6 kA, this results in a delay time t_d of 220 ms for the smallest permissible safety margin of 100 ms. This determines the starting point for all subsequent upward and downward grading in the diagram.

Outgoing feeders with circuit-breaker

Since selectivity cannot be achieved using l.v. h.b.c. fuses with a higher current rating (see Fig. 1.3/39a), circuit-breakers with time or, if possible, current grading should be used.

Based on the assumption that verification of the short-circuit currents would show that current grading would be possible, a 630 A (Q1) distribution circuit-breaker with "an" releases was selected.

Intersection of the characteristics Q2 and Q3 in the middle short-circuit range is permissible because:

- ▷ the "a" release of the low-voltage circuit-breaker Q1 (not shown in Fig. 1.3/40) protects the transformer against overloading, which only occurs in the range 1–1.3 times the rated current of the transformer;
- ▷ there is a safety margin of ≥150 ms (≈300 ms in the example shown in Fig. 1.3/40) between the $I>$ tripping value of the UMZ protection and the l.v. h.b.c. fuse characteristic F1; selectivity is, therefore, achieved.

Higher rated transformer outputs and broader setting ranges for the "z" release of Q2 make it easier for the characteristic Q3 $I>$ to be shifted to the left of the characteristic Q2 z. This also provides a certain degree of back-up protection with respect to the "a" release of circuit-breaker Q2.

UMZ protection

Nowadays, digital devices are used to provide UMZ protection in practically all applications. They have broader setting ranges, allow a choice between definite-time (UMZ) and inverse-time (AMZ) overcurrent protection or overload protection, provide a greater and more consistent level of measuring accuracy, and are self-monitoring.

Selecting current transformers for UMZ protection

The following points should be observed when selecting current transformers for UMZ protection (these considerations are applied in the example shown in Fig. 1.3/40):

Current transformers with a rating of 40 to 200 A could be selected for rated currents of 36.4 A on the high-voltage side of the 630 kVA transformer, with the characteristic Q3 $I>$ at 200 A positioned at the abscissa for 10 kV, and with the broad setting ranges. Here, it is important to bear in mind the higher investment costs for current transformers with lower rated primary currents.

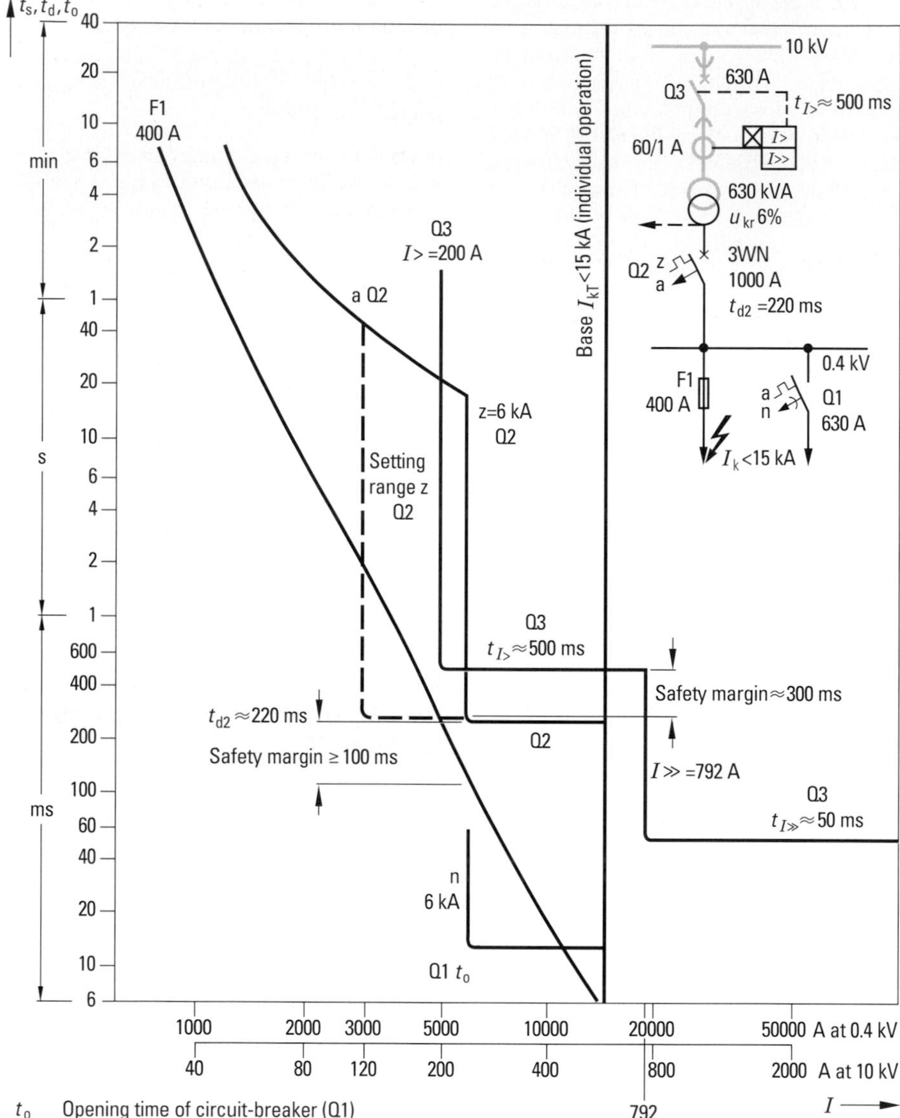

t_o Opening time of circuit-breaker (Q1)
t_{d2} Delay time of "z" release (Q2)
$t_{I>}$ Delay time of $I>$ release of UMZ protection (Q3)
$t_{I\gg}$ Delay time of I_0 release

For the sake of clarity, the "a" characteristic of circuit-breaker Q1 is not shown

Fig. 1.3/40
Example showing grading of circuit-breaker with UMZ protection (Q3), circuit-breaker 3WN, 1000 A with "az" releases (Q2) and downstream outgoing feeders, e.g. 400 A l.v. h.b.c. fuse (F1) and 630 A distribution circuit-breaker (Q1) in a 630 kVA transformer feeder

If, for example, 60/1 A current transformers are selected, the current sensors must be set as follows:

Setting the current sensors

Current sensor $I >$:

The setting for a selected operating value of 200 A is as follows:

$$I_p = \frac{200\ \text{A}}{60/1} = 3.3\ \text{A}.$$

Timing element for $I >$ excitation:
$t_{I>} = 0.5$ s

Current sensor $I \gg$:

Current sensor $I \gg$ should only respond to faults on the high-voltage side (in the shortest possible time).

Operating current I approximately $I_{kT} \cdot 1.20$ (safety margin relative to I_{kT})

$$I_{kT} = \frac{I_{nT} \cdot 100\%}{u_{kr}} = \frac{36.4\ \text{A} \cdot 100\%}{6\%} = 606.6\ \text{A},$$

Operating current $= I_{kT} \cdot 1.2 = 728$ A

Operating current (in secondary circuit) $=$

$$\frac{728\ \text{A}}{60/1} \approx 12.1\ \text{A}$$

1.3.5.2 Equipment for protecting distribution transformers (against internal faults)

The following signaling devices and protection equipment are used to detect internal transformer faults:

- ▷ Devices for monitoring and protecting liquid-cooled transformers (e.g. Buchholz protectors, temperature detectors, contact thermometers etc.)
- ▷ Temperature monitoring systems for GEAFOL resin-encapsulated transformers comprising:
  - temperature sensors in the low-voltage winding and
  - signaling and tripping devices in the incoming-feeder switchpanel
- ▷ The thermistor-type thermal protection protects the transformer against overheating resulting from increased ambient temperatures or overloading. Furthermore, it allows the full output of the transformer to be utilized irrespective of the number of load cycles without the risk of damage to the transformer.

These signaling and protection devices do not have to be included in the grading diagrams (e.g. Fig. 1.3/27).

1.4 Low-voltage switchboards and distribution systems

1.4.1 General

Low-voltage switchboards and distribution systems represent the links between the equipment for generating (generators), transporting (cables, overhead lines) and converting (transformers) electrical energy and the loads that consume this energy such as motors, solenoid valves as well as heating, lighting, air-conditioning and IT equipment.

Fig. 1.4/1 contains an overview of the complete range of Siemens low-voltage switchboards and distribution systems.

The selection criteria represented here are grouped into the following four categories:

Currents

Rated currents of busbars,
Rated currents of infeeds,
Rated currents of outgoing feeders,
Short-circuit strength of busbars.

Degree of protection and type of installation

Degree of protection,
Protection against electric shock (safety class),
Enclosure material,
Type of installation (against a wall, free-standing),
Number of operating faces

Equipment mounting type

Fixed-mounted,
Removable,
Withdrawable

Application

Ten different possible applications.

The low-voltage switchboards and distribution systems described here conform to the following standards:

EN 60439-1/ IEC 60439-1/ VDE 0660 Part 500	Low-voltage switchgear and controlgear assemblies; Type-tested and partially type-tested assemblies
EN 60439-2/ IEC 60439-2/ VDE 0660 Part 502	Low-voltage switchgear and controlgear; Particular requirements for busbar trunking system
EN 60439-3/ IEC 60439-3/ DIN VDE 0660-504	Particular requirements for low-voltage switchgear and controlgear assemblies intended to be installed in places where unskilled persons have access for their use.
EN 60439-4/ IEC 60439-4/ DIN VDE 0660-501	Low-voltage switchgear and controlgear assemblies; Particular requirements for assemblies for construction sites (ASC)

There are no significant differences between the wide range of construction types of low-voltage switchboards and distribution systems. For this reason, terms and designations are not always used consistently by the manufacturer and the user even for the same product. In most cases, it will be the application that determines the designation from the point of view of the user.

Main switchboard
Sub-distribution board

In order to overcome these difficulties regarding different terms and designations, only the terms "main switchboard" and "sub-distribution board" are used in the example of a low-voltage system in an industrial plant (Fig. 1.4/2).

The main switchboard is supplied directly via one transformer for each busbar section. The motor control centers, control systems, distribution boards for lighting, heating, air conditioning, workshops etc., which are connected downstream and thus supplied from the main switchboard, are classed as sub-distribution boards.

Point distribution boards

The term "point distribution board" (Fig. 1.4/3) is used to describe all switchboards and distribution boards which distribute power radially from a point source via cables and overhead lines to remote loads. The necessary switching, protective and measuring equipment is grouped centrally in the switchboard or distribution board.

Busbar trunking system

With "busbar trunking systems" (Fig. 1.4/4), on the other hand, the power is distributed via relatively long, enclosed busbars to the immediate locality of the loads. The loads are connected to the busbars via tap-off units with fuses, circuit-breakers or miniature circuit-breakers, and short stub lines or cables.

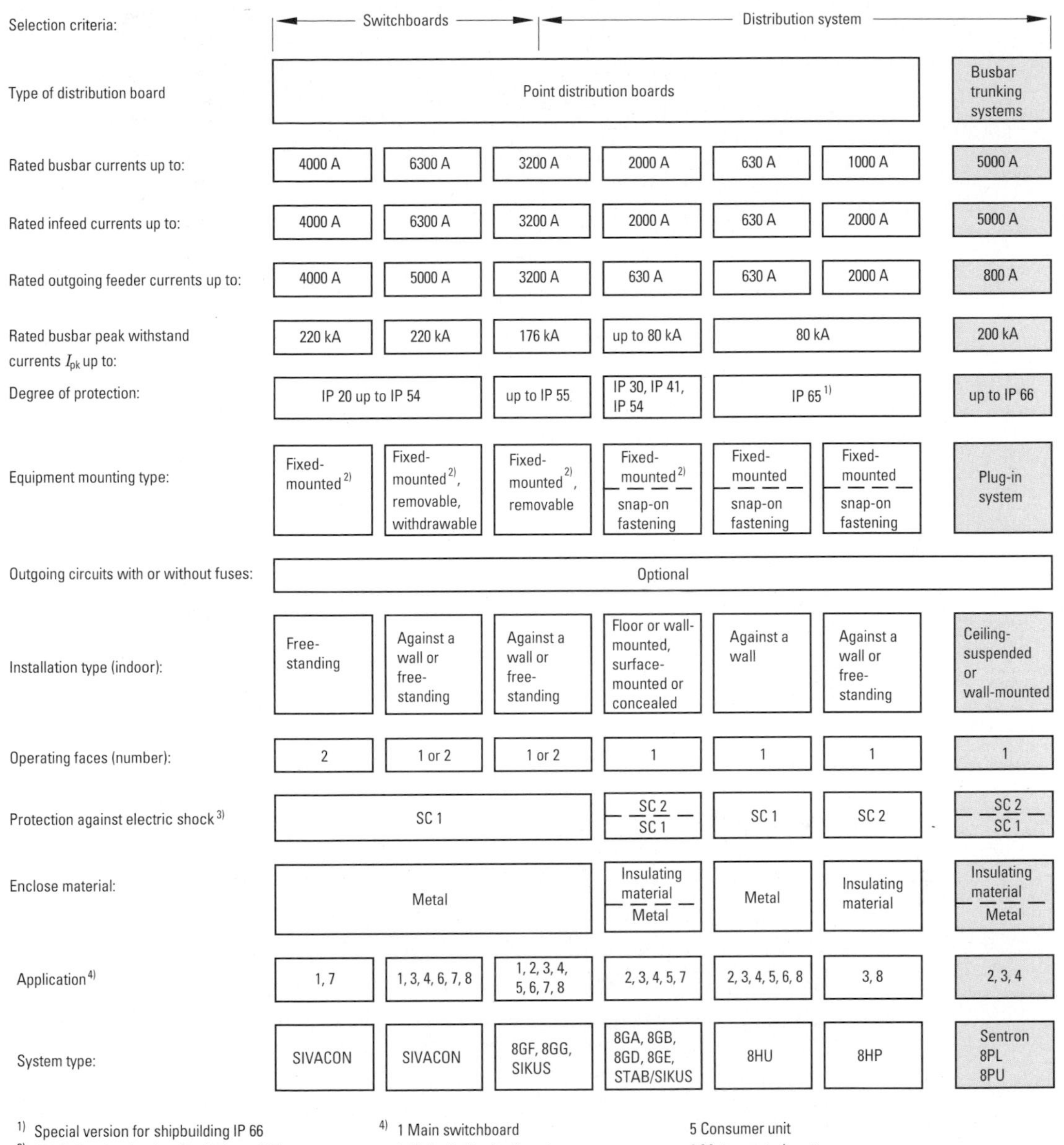

Selection criteria:	Switchboards	Switchboards	Distribution system	Distribution system	Distribution system	Distribution system	Distribution system
Type of distribution board	Point distribution boards						Busbar trunking systems
Rated busbar currents up to:	4000 A	6300 A	3200 A	2000 A	630 A	1000 A	5000 A
Rated infeed currents up to:	4000 A	6300 A	3200 A	2000 A	630 A	2000 A	5000 A
Rated outgoing feeder currents up to:	4000 A	5000 A	3200 A	630 A	630 A	2000 A	800 A
Rated busbar peak withstand currents I_{pk} up to:	220 kA	220 kA	176 kA	up to 80 kA	80 kA		200 kA
Degree of protection:	IP 20 up to IP 54		up to IP 55	IP 30, IP 41, IP 54	IP 65[1]		up to IP 66
Equipment mounting type:	Fixed-mounted[2]	Fixed-mounted[2], removable, withdrawable	Fixed-mounted[2], removable	Fixed-mounted[2] / snap-on fastening	Fixed-mounted / snap-on fastening	Fixed-mounted / snap-on fastening	Plug-in system
Outgoing circuits with or without fuses:	Optional						
Installation type (indoor):	Free-standing	Against a wall or free-standing	Against a wall or free-standing	Floor or wall-mounted, surface-mounted or concealed	Against a wall	Against a wall or free-standing	Ceiling-suspended or wall-mounted
Operating faces (number):	2	1 or 2	1 or 2	1	1	1	1
Protection against electric shock[3]	SC 1			SC 2 / SC 1	SC 1	SC 2	SC 2 / SC 1
Enclose material:	Metal			Insulating material / Metal	Metal	Insulating material	Insulating material / Metal
Application[4]	1, 7	1, 3, 4, 6, 7, 8	1, 2, 3, 4, 5, 6, 7, 8	2, 3, 4, 5, 7	2, 3, 4, 5, 6, 8	3, 8	2, 3, 4
System type:	SIVACON	SIVACON	8GF, 8GG, SIKUS	8GA, 8GB, 8GD, 8GE, STAB/SIKUS	8HU	8HP	Sentron 8PL 8PU

[1] Special version for shipbuilding IP 66
[2] Circuit-breaker (optional: withdrawable)
[3] Safety class: SC 1 = protective-conductor terminal, SC 2 = total insulation, SC 3 = extra-low voltage

[4] 1 Main switchboard
2 Main distribution board
3 Lighting and power distribution board
4 Sub-distribution board
5 Consumer unit
6 Motor control center
7 Power-factor correction
8 Control system

Fig. 1.4/1 Selection criteria for low-voltage switchboards and distribution systems

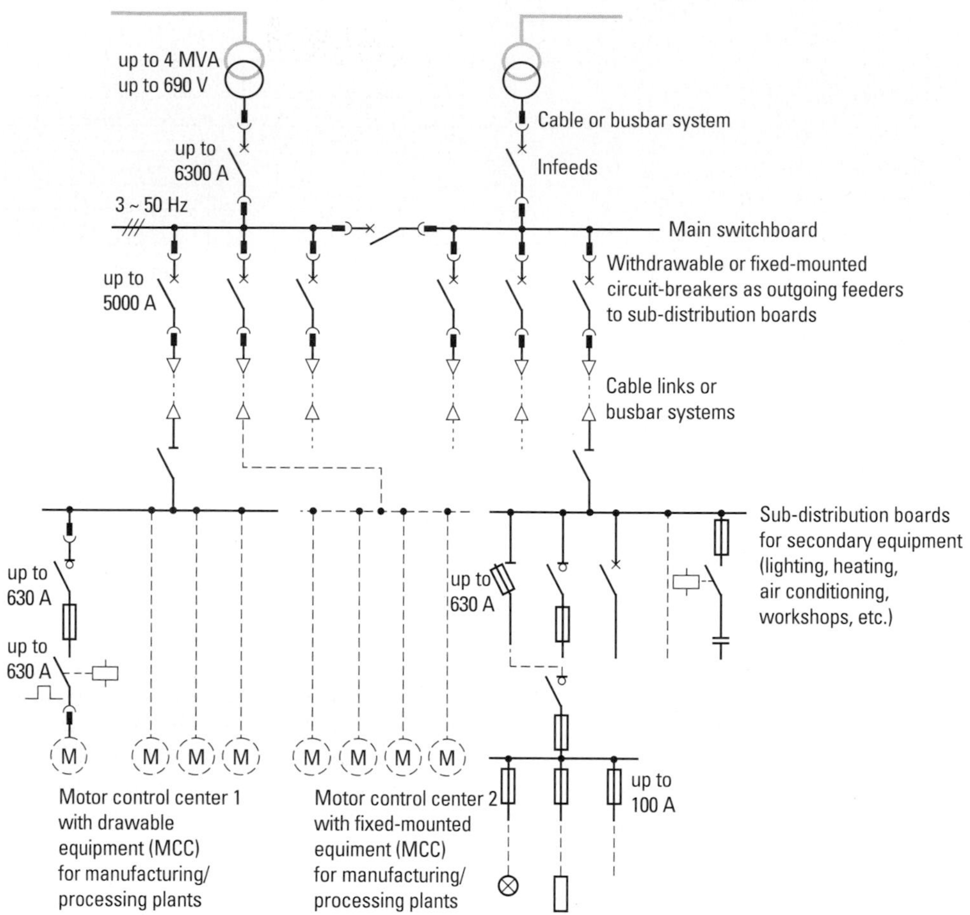

Fig. 1.4/2 Typical configuration of a low-voltage system in an industrial plant

Busbar trunking systems (with tap-off units of various sizes and in various positions) are used to supply work centers, machines etc. in large factories and laboratories. Tap-off units can be provided at practically any point in the busbar run which means that busbar trunking systems are particularly suitable for loads, the locations of which change on a frequent basis. They are also used as rising main busbars in high-rise buildings where they supply the storey distribution boards (e.g. STAB wall-mounting distribution boards).

Switchboards and distribution boards – types of construction

The aforementioned standards for low-voltage switchgear and controlgear assemblies describe the different types of construction. They also contain construction and testing guidelines for the manufacturer as well as installation instructions for the user.

Open-type assemblies

In open-type assemblies (Fig. 1.4/5), parts that may be live during operation, such as main busbars, vertical busbars, items of equipment, terminals and conductors, are accessible from all sides because the open frame is not fitted with any covers.

Open-type assemblies must only be installed in closed electrical operating areas.

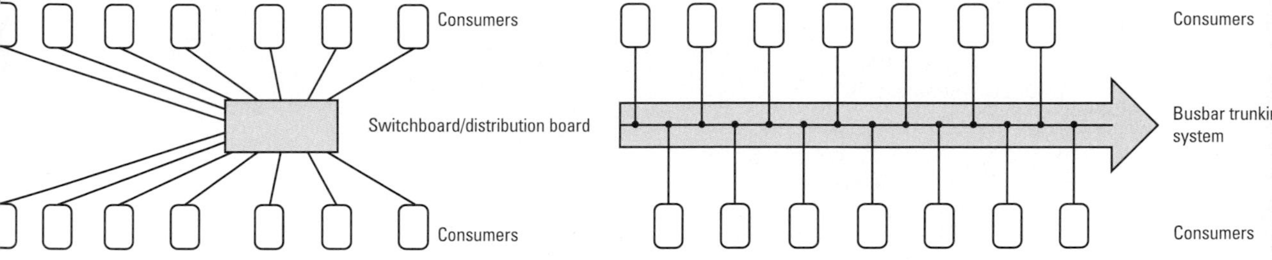

Fig. 1.4/3 Point distribution system

Fig. 1.4/4 Busbar trunking system

On the basis of the directives of the employers' liability insurance association which were drawn up for the operators of electrical equipment (accident prevention regulations, "Elektrische Anlagen und Betriebsmittel" [Electrical Installations and Equipment] (VBG 4)), this type of assembly is only of secondary importance. Non-enclosed contactor frames of this type are, however, still used in certain cases for switching drive motors.

Dead-front assemblies

In contrast to open-type assemblies, dead-front assemblies provide protection against contact with live parts on the operating side, but are not enclosed on the other sides which permit access.

This type of assembly must, therefore, also be installed in closed electrical operating areas. For the reasons mentioned above, dead-front assemblies are also used only very rarely.

Enclosed assemblies

Enclosed assemblies (Fig. 1.4/6), as the name suggests, are enclosed on all sides so that contact with parts which may be live during operation is prevented. Degree of protection IP 2X, at least, must be provided. Installation is permissible in generally accessible operating areas.

Enclosed assemblies are the most widely used nowadays as they offer users optimum protection with regard to personnel and equipment.

Switchboard assemblies usually comprise several cubicles. A group of cubicles (up to four) constitutes a transportable unit.

Fig. 1.4/5
Open-type low-voltage switchgear assembly

Fig. 1.4/6
Enclosed low-voltage switchgear assembly

Most of the cubicles in enclosed assemblies are fitted with full-height doors, as can be seen in Fig. 1.4/6. Versions with individual compartment doors are used mainly for items of equipment mounted in withdrawable units. With fixed-mounted units, the fronts of the cubicles are also covered with full-height doors.

Withdrawable units

A withdrawable unit is a replaceable assembly in which a number of items of equipment are grouped together and interconnected to form a functional unit. A withdrawable unit can be installed in such a way that one isolating gap is open (as per EN 60947-3/IEC 60947-3/DIN VDE 0660-107) whereas the unit itself remains connected (mechanically) to the switchgear assembly.

The withdrawable arrangement (Fig. 1.4/7) is invariably associated with the totally-enclosed cubicle construction.

This is further divided into individual compartments for the withdrawable units (outgoing feeder, infeed or coupling unit) and thus provides an extremely high level of operational and personal safety.

Box-type assemblies

Box-type distribution boards (Fig. 1.4/8), made of insulating material or sheet steel, consist of individual boxes which are permanently connected to each other and contain items of equipment such as main busbars, fuses, switches and contactors. Contact with parts that may be live during operation is prevented by means of enclosures and covers. Box-type distribution boards can, therefore, be installed in generally accessible operating areas.

Outdoor installation

By attaching a protective cowl and by providing the appropriate degree of protection (at least IP 65), this type of distribution board can, unlike the types already described, also be installed outdoors.

Recommendations for selection

Switchboards

In accordance with Fig. 1.4/1, a distinction is made between "switchboards" and "distri-

Fig. 1.4/7
Replaceable withdrawable unit

bution systems". Switchboards are characterized primarily by:

▷ High rated equipment currents ≤6300 A (approx.)
▷ High short-circuit strength ≤220 kA peak
▷ Sheet steel as enclosure material
▷ Overall height of at least 2000 mm
▷ Equipment mounting types: fixed, removable, withdrawable
▷ Degree of protection of enclosure between IP 20 and IP 55

Distribution systems

Distribution systems, on the other hand, have the following features:

▷ Rated currents of the installed equipment ≤3200 A (approx.)
▷ Short-circuit strength ≤176 kA peak
▷ Various enclosure materials (insulating material or sheet steel)
▷ Overall height < 2000 mm
▷ Equipment mounting type mostly fixed
▷ Degree of protection up to IP 66

The selection criteria in Fig. 1.4/1 should be used when choosing a switchboard or distribution board type for a particular application. When the bottom of the table is reached, the most suitable system type will have been selected almost automatically.

Fig. 1.4/8
Multi-box-type, totally insulated distribution board

Detailed descriptions of the selected types together with further technical data, equipment ranges, dimensions etc. can be found in the following sections:

1.4.2 Standard low-voltage switchboards
1.4.3 Box-type low-voltage distribution systems
1.4.4 Busbar trunking and busway systems
1.4.5 Power and domestic distribution boards

Section 1.4.6 "Planning low-voltage switchgear and distribution systems, software tools (P.I.S.A.A)" contains information on designing and dimensioning switchboards and distribution boards. It also provides an introduction to various planning tools.

Short-circuit strength

The prospective short-circuit current (I_k'' = initial symmetrical short-circuit current, i_p = peak short-circuit current) at the point at which the switchboard or distribution board is installed – i.e. between the infeed transformer on the one side and the cable-connected loads on the other side – must not be higher than the short-circuit strength values specified for the product by the manufacturer. This requirement can, if necessary, be fulfilled by connecting a current-limiting protective device (short-circuit current-limiting reactor, current-limiting circuit-breakers or l. v. h.b.c. fuses) on the line side (see Section 1.3.2). Details of the short-circuit strength can be provided by the manufacturer in accordance with EN 60439-1/IEC 60439-1/VDE 0660 Part 500 by specifying:

▷ the maximum permissible prospective rated short-circuit current at the input terminals of the switchgear assembly with integrated short-circuit protective device in the infeed,
▷ the rated short-time current and rated peak withstand current,
▷ the prospective rated short-circuit current together with the current-limiting switching device.

In the case of switchgear assemblies with several infeeds (which are not all in operation at the same time), the short-circuit strength may be specified for each of these. If, however, several infeeds are in operation at the same time, or if the switchgear assembly is equipped with an infeed and outgoing feeders for high-capacity rotating motors which can supply the short circuit, the short-circuit strength of the switchgear assembly must be determined and specified on the basis of the individual sustained short-circuit current values preselected by the user.

Calculating the short-circuit current

If these values are not available, a short-circuit current calculation must be performed in accordance with IEC 60909/DIN VDE 0102 (see Section 1.2), for example.

KUBS program

A suitable PC program such as KUBS (short-circuit calculation, backup protection and selectivity) can be used to help perform the calculation.

Degree of protection

Depending on the installation location and surrounding conditions, a switchboard or distribution board type should be chosen which provides the necessary degree of protection against contact and against the ingress of foreign bodies and water (EN 60 529/IEC 60529/DIN VDE 0470-1). Section 1.4.6 contains a list of the different degrees of protection and requirements that must be fulfilled by the enclosure of an item of equipment. According to the German standard DIN VDE 0100-731, switchboards and distribution boards with a degree of protection less than IP 2X must only be installed in closed electrical operating areas.

In rooms that are accessible to everyone, switchboards and distribution board types must be chosen that provide protection against accidental contact and against contact with parts that may be live during operation, i.e. the degree of protection must be at least IP 20. This requirement is fulfilled by all of the assembly types described.

Enclosure

According to EN 60439-1/IEC 60439-1/VDE 0660 Part 500, enclosures must be designed in such a way that all external surfaces provide protection to at least IP 2X. The distances between the external surfaces (if these are not made of insulating material) and active parts must, at the very least, conform to the values for creepage distances and clearances specified in EN 60439-1/IEC 60439-1/VDE 0660 Part 500.

Insulated enclosure

The enclosures of certain distribution system types can be made of metal or insulating material. The insulated enclosure offers extensive protection against corrosion and a higher degree of protection against contact.

Surfaces of metal enclosures

The enclosures or casings and metal parts of the mounting structures of all types of switchboard and distribution board are protected against corrosion by a high-quality surface finish and, therefore, only need to be repaired/reworked or maintained if they are damaged. Most of the switchboard enclosures are provided with a durable, highly-resistant epoxy resin powder coating.

Protection against electric shock

Switchgear assemblies must be designed in such a way that they provide the necessary protection against contact with live parts. It must also be possible to be able to implement appropriate measures to provide protection against shock (in accordance with HD 384.4.41/IEC 60364-4-41/DIN VDE 0100-410).

The specifications regarding protective measures are described in Chapter 2.

Cost-effective design

To allow the most cost-effective switchboard or distribution board to be chosen in each case, the characteristic features of the various systems should be compared before a decision is taken and before any structural work is carried out.

Features to be considered include:

Installation, access, delivery

Installation location

- ▷ Generally accessible operating areas,
- ▷ Electrical operating areas,
- ▷ Closed electrical operating areas.

Type of installation

- ▷ On the floor, against a wall,
- ▷ On the floor, free-standing in the room
- ▷ Fixed to a wall, in a stairwell or a recess
- ▷ Suspended from the ceiling
- ▷ On a mounting structure
- ▷ In a cavity wall

Type of access

- ▷ For operation by non-experts,
- ▷ On one or two sides (for operation),
- ▷ Top and rear access for installation of and maintenance inspections on busbars,
- ▷ Front and rear access for connection work and modifications.

Installation dimensions

- ▷ Overall height, width and depth.

Delivery

- ▷ Height and width of doors
- ▷ Elevator dimensions
- ▷ Where necessary, carrying capacity of cranes/elevators

Ambient conditions

When choosing and planning the switchgear assemblies, the ambient temperature, climatic conditions and altitude of the installation site must be taken into consideration as these factors determine the load carrying capacity of all the different items of equipment. Extreme changes in temperature and relative air humidity as well as outdoor installation affect the design of the switchgear assemblies (e. g. degree of protection, canopies, breathers and heating).

Equipment mounting types

Mounting types

Switchgear as well as control, protective and monitoring devices can be permanently installed in removable units or in withdrawable units in switchgear assemblies. The options for replacing items of equipment in the event of a fault and for modifying switchgear assemblies during operation are determined by the mounting type.

Special requirements

Special requirements, such as explosion protection, protection against corrosive atmospheres, induced-shock design (earthquakes, plane crashes) must be taken into consideration within the scope of additional agreements made between the manufacturer and the user.

1.4.2 Standard low-voltage switchboards

1.4.2.1 Introduction

Application

Low-voltage switchboards for rated currents $\leq$6300 A (approx.) are used primarily as main switchboards in industrial plants and large commercial establishments, in power stations and refineries as well as in high-rise buildings and large hospitals; in other words, wherever substantial power demands have to be catered for.

System configuration

The block diagram in Fig. 1.4/2 illustrates how an industrial power supply can be logically divided into a "main switchboard" (power center) supplied directly by transformers, whose infeed, outgoing and busbar coupling switches consist only of circuit-breakers, and the various "sub-distribution boards" for a wide range of different applications.

This separation between the high-current main switchboard and downstream sub-distribution boards has a number of advantages:

Main switchboard

▷ the main switchboard is installed in the immediate vicinity of the infeed transformers. The cable or busbar connections to the transformers are short,
▷ the circuit-breakers incorporated in it for the infeeds, outgoing feeders and busbar couplings are of one type only with current ratings between 630 A and 6300 A,
▷ if these circuit-breakers are designed as withdrawable units, the disconnecting switch for each circuit-breaker can be omitted (in contrast to fixed-mounted units),
▷ the two-source infeed, together with the busbar coupling switch – which is open under normal operating conditions – permits, in the event of one of the infeeds failing, a switchover to be made to the second infeed for all the downstream sub-distribution boards. This, in turn, reduces downtimes and increases supply availability for all loads.

Load shedding may be necessary depending on the degree of utilization of the transformers.

Sub-distribution boards

▷ the sub-distribution boards are located centrally in relation to their loads with the result that the cables that feed them are short,
▷ due to the relatively long cable or busbar connections between the main switchboard and sub-distribution board, sub-distribution boards with low short-circuit strength values can be used for busbars, vertical busbars etc.

Busbar and cable infeed

Switchboards in the immediate vicinity of the transformers can be supplied effectively via enclosed busbar ducts. Entry into the switchboard is then possible either from below (as with cables) or from above via additional extensions. With cable connections, it is important to bear in mind that, in the case of a 2 MVA transformer with a rated current of approximately 3000 A at 400 V, up to 14 parallel cables with a cross-sectional area of 240 mm² per external conductor must be secured and connected at the incoming-feeder panel. Installation outlay is, therefore, considerably higher.

Outgoing feeders

Outgoing feeders are always connected via cables. Depending on the distances involved, parallel cables are also used as of 250 A (approx.) due to the voltage drop and currents to be carried.

Cable racks False floors

The cables are usually routed downwards in this type of arrangement. Cables racks are suspended from the ceiling of the storey below in order to install the cables. In ground-floor factory areas where no cellar is available, false floors are often used so that the cables can be installed easily and accessibly in the space provided (height: approx. 0.5 m).

When dimensioning and installing cable racks and cable basements, it is important to make sure that the permissible bending radii are observed for connecting the cables.

Operating and maintenance procedures

The operating and maintenance procedures should be implemented in accordance with DIN VDE 0100-729. In the case of switchboards with withdrawable items of equipment – particularly if these have high current ratings and are, therefore, relatively large – it is also important to ensure that sufficient space is left on the operating side between the switchboard and the opposite surface (wall, the next switchboard, machines etc.) to allow access to personnel and any lifting trucks and trolleys required to replace the equipment (see Section 2.2).

Main switchboards perform an extremely important task in supplying all downstream sub-distribution boards and associated loads. It is, therefore, essential to ensure that the outgoing-feeder units can be maintained properly by providing sufficient space in the gangways.

Isolation facilities

If several transformers feed into a common busbar or into a number of busbar sections (which are connected by means of special coupling switches), provision should be made for an additional means of isolation between the incoming-feeder circuit-breaker and busbar so that the incoming-feeder circuit-breaker can be isolated for maintenance purposes or in the event of a fault.

This can be achieved by means of:

▷ a disconnector or switch-disconnector ($\leq$3000 A),

▷ l. v. h.b.c. fuse-bases with isolating links ($\leq$1250 A),

▷ l. v. h.b.c. fuse switch-disconnectors with isolating links ($\leq$ 630 A),

▷ withdrawable circuit-breakers ($\leq$6300 A).

The last option reduces not only costs but also the volume of the switchboard and is the type of isolation most frequently used. This also applies to sectionalizing points with circuit-breakers between two busbar sections.

Sectionalizing points

The purpose of sectionalizing points – and in certain cases bus sectionalizers with fixed-mounted switch-disconnectors – is to limit the short-circuit power with several infeeding transformers to the power rating of a single transformer. The switchboard must, however, be operating under normal operating conditions with the tie breaker open.

Sectionalizing points are also used to continue to supply part of the power required by the load feeders of a system infeed should this fail. The tie breaker must be closed for this purpose.

Cross-coupling

Switch-disconnectors and or circuit-breakers can also be used to connect two switchboards installed back-to-back, each with one busbar run. This type of connection is referred to as a cross-coupling.

Isolation of outgoing feeders

It must also be possible to isolate the individual outgoing feeders from the main or vertical busbar in a simple and straightforward manner to enable the circuits to be replaced as well as modifications and maintenance to be performed. In most cases, the l. v. h.b.c. fuse-links in l. v. h.b.c. fuse-bases or l. v. h.b.c. fuse switch-disconnectors, which are provided in any case for short-circuit protection, can be used for this purpose. The safest and most convenient way of maintaining and checking equipment is if the operating personnel are able to completely remove the equipment from its enclosure by means of two simple handles. Withdrawable switchboards offer this facility.

Type of busbar system

Main switchboards usually take the form of a so-called four-conductor system with the three phase conductors L1, L2 and L3 as well as a PE or PEN conductor bar laid in the lower part of the panels. This is due to the fact that unbalanced loads such as heating, lighting and single-phase motors are rarely connected to the main switchboards. The phase conductors are located either in the upper part of the switchboard or at the rear in a staggered formation. If necessary, a fifth, insulated neutral conductor bar can be incorporated in the horizontal busbar system and vertical busbar arrangements (removable and withdrawable-unit design). See Section 1.4.6 for further information.

Busbar load carrying capacity

In the case of switchboards manufactured to EN 60439-1/IEC 60439-1/DIN VDE 0660 Part 500, only the data supplied by the manufacturer is relevant for the load carrying capa-city of the busbars with regard to the rated current and rated operating current under normal operating conditions. The values specified in catalogs, product data sheets or other publications were determined on the basis of experiments and extrapolation performed at the testing stations in accordance with the type test defined in the relevant standards.

Factors affecting load carrying capacity

The rated currents and rated short-circuit currents of the busbars have different values according to the type of switchboard and depend on:

- ▷ the mounting position within the sections,
- ▷ the relationship of the conductors to each other,
- ▷ the cross-sectional area of the conductor material,
- ▷ the strength of the conductor material,
- ▷ the clearances between the supports,
- ▷ possible thermal effects of other components.

Cross-sectional area of PE conductor

The PE conductor which is conductively connected to the frame must be adequately dimensioned in order to discharge any short-circuit currents that may occur.

The cross-sectional areas of protective conductor bars (PE) must at least conform to the values specified in Table 4 in Section 7.4.3.1.7 or Table 4a in Section 7.4.3.1.10 in EN 60439–1/IEC 60439–1/DIN VDE 0660 Part 500 in the case of internal protective conductor connections (protective conductors for exposed conductive parts for which suitable protection cannot be provided by means of the fixing element).

Cross-sectional area of PEN conductor

The requirements for PEN conductor bars are the same as those for PE conductor bars except that higher unbalance currents may have to be taken into consideration.

The cross-sectional area of the PEN conductor bar should correspond to 25% of the phase conductor cross-sectional area in order to be realistically dimensioned. If current-limiting power converters are used in grounded-neutral systems, the PEN conductor bar must be dimensioned in the same way as the secondary phase conductors.

Position of infeed

An important aspect that must be considered when designing the busbars is the physical position of the infeed within a switchboard. Figure 1.4/9 indicates the possible arrangements of infeeds and outgoing feeders and the resulting stress values. In these cases, the busbars are not divided by sectionalizing points even when there are two or more transformer infeeds. This is nevertheless recommended due to the costs that can be saved when a switchboard is procured (equipment with low breaking capacity and lower short-circuit strength) and in the interests of operational reliability (limits the extent of the damage that can be caused in the event of a fault and reduces downtimes).

Busbar material

Busbars are nowadays usually made of electrolytic copper (strength F30). Aluminum and copper-clad aluminum are only of secondary importance in switchboards.

Marking busbars

The busbars are identified by means of heat-resistant adhesive strips labeled L1, L2 and L3 for the phase conductors and N for the neutral conductor. A green/yellow adhesive strip is used for the PE or PEN protective conductor (one for each section).

The earlier practice of painting the busbars yellow-green-violet is no longer appropriate and does not conform to DIN 43 671 or EN 60439–1/IEC 60439–1/VDE 0660 Part 500 which specify that all the phase conductors must be the same color.

Motivepower loads, increase in short-circuit current

Any relatively large motivepower loads supplied by the outgoing feeders of a switchboard increase the short-circuit current in the event of a short-circuit. This should be taken into consideration when dimensioning the main busbars. It is advisable to calculate the rated short-circuit currents using a suitable PC program (e.g. KUBS).

Rated values of transformers

Table 1.4/1 contains data on rated currents and rated short-circuit currents of transformers with rated outputs of between 50 kVA and 3150 kVA, rated short-circuit voltages u_{kr} of 4% and 6% as well as rated operating voltages at the transformer terminals of 400 V and 525 V. This data is used for dimensioning a switchboard.

Conversion at 690 V

In order to determine the values at 690 V, the values specified for the rated current and rated short-circuit current at 400 V for the same transformer output rating must be divided by 1.73 ($\sqrt{3}$).

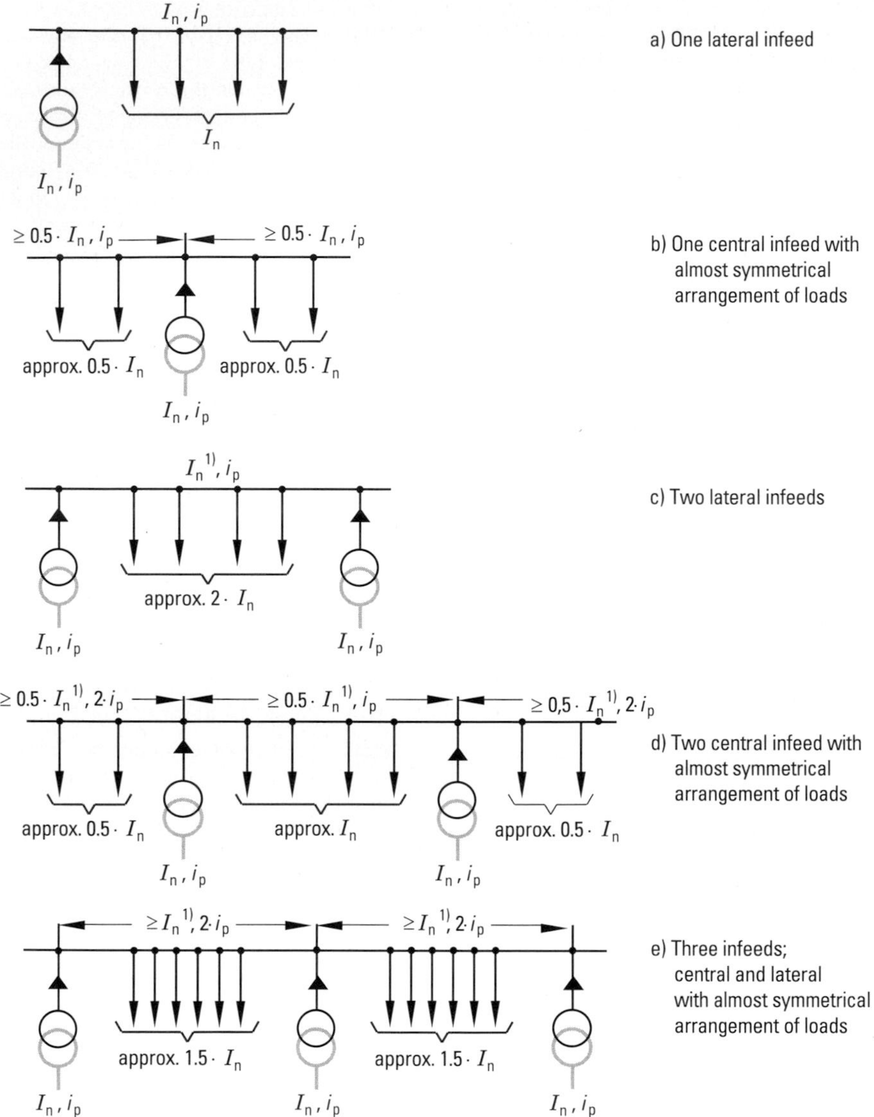

[1] With the transformers operating together and the busbars not divided

I_n Rated transformer current

i_p Peak short-circuit current via the transformer connected to a system with unlimited short-circuit power

Fig. 1.4/9 Distribution of busbar load according to the position of the infeeds

Table 1.4/1
Rated currents and rated short-circuit currents of transformers with rated outputs of 50 kVA to 3150 kVA at rated operating voltages of 400 V and 525 V

Rated transformer output S_{nr}	Rated operating voltage 400 V			Rated operating voltage 525 V		
	Rated current I_n	Rated short-circuit voltage: $u_{kr} = 4\%$ [1]	$u_{kr} = 6\%$ [2]	Rated current I_r	Rated short-circuit voltage: $u_{kr} = 4\%$ [1]	$u_{kr} = 6\%$ [2]
		Rated short-circuit current I_k [3]			Rated short-circuit current I_k [3]	
kVA	A	$A_{(r.m.s)}$	$A_{(r.m.s)}$	A	$A_{(r.m.s)}$	$A_{(r.m.s)}$
50	72	1 805	1 203	55	1 375	916
100	144	3 610	2 406	110	2 750	1 833
200	288	7 220	4 812	220	5 500	3 667
250	360	9 025	6 015	275	6 875	4 583
315	455	11 375	7 583	346	8 660	5 775
400	578	14 450	9 630	440	11 000	7 333
500	722	18 050	12 030	550	13 750	9 166
630	910	22 750	15 166	693	17 320	11 550
800	1 156	28 900	19 260	880	22 000	14 666
1 000	1 444	36 100	24 060	1 100	27 500	18 333
1 250	1 805	45 125	30 080	1 375	34 375	22 916
1 600	2 312	57 800	38 530	1 760	44 000	29 333
2 000	2 890	72 250	48 167	2 203	55 075	36 717
2 500	3 613	90 325	60 227	2 753	68 825	45 883
3 150	4 552	113 800	75 867	3 469	86 725	57 817

[1] To DIN 42 500 and 42 503 for rated outputs of 50 kVA to 630 kVA
[2] To DIN 42 511 for rated outputs of 100 kVA to 1600 kVA
[3] With the high-voltage side of the transformer connected to a stiff system, $I_k \approx I''_k$

Conversion to rated peak withstand current

The factor n is used to convert the r.m.s. value of the rated short-circuit current I_k to the rated peak withstand current I_{pk} for switchboards in accordance with EN 60439–1/IEC 60 439–1/VDE 0660 Part 500 unless otherwise specified by the customer. The rated peak withstand current (peak value of the first half cycle of the rated short-circuit current including the d. c. component), which is paramount from the point of view of electrodynamic loading, is determined by multiplying the r.m.s value by the factor n (Table 1.4/2).

Table 1.4/2
Factors n for determining the rated peak withstand currents for different rated short-circuit current r.m.s. values

Rated short-circuit current r.m.s value I_k kA	Power factor $\cos \varphi$	Factor n
up to 5	0.7	1.5
> 5 to 10	0.5	1.7
> 10 to 20	0.3	2.0
> 20 to 50	0.25	2.1
> 50	0.2	2.2

The variation of the power factor $\cos \varphi$ with the current values corresponds to the conditions encountered in most cases. If more severe conditions arise in particular situations, e. g. in the vicinity of large transformers, special agreements must be made between the manufacturer and customer.

1.4.2.2 SIVACON low-voltage switchboards

Requirements

Modern industrial and building services systems are planned and implemented in line with important requirements such as economic efficiency, flexibility, safety and reliability. These requirements must also be fulfilled by the switchboards for power supply and distribution.

SIVACON low-voltage switchboards fulfill these requirements as follows:

Economic efficiency

▷ Economic efficiency
  – Design based on a sophisticated modular concept with a large number of stan-

dardized components thus permitting efficient series production,
- Only those modules and components which provide the required functionality are used, whereby functional requirements determine the structural requirements,
- Computer-assisted engineering procedures which are made available directly to the consultant allow fast, time-saving planning and customizing to take account of operator-specific requirements;

Flexibility

▷ Flexibility
- Can be implemented as power centers, main and sub-distribution boards as well as motor control centers (MCC) thanks to different equipment mounting types
- Can be easily adapted to modified operating conditions thanks to appropriate cubicle and device modules
- Different functional units can be freely combined within a switchgear cubicle

Safety

▷ Safety and reliability
- Only type-tested assemblies (TTA) used
- Type and routine testing to EN 60439-1/IEC 60439-1/DIN VDE 0660 Part 500
- Complete quality assurance by means of comprehensive quality inspections

Design

Modular concept

A functional-based modular concept allows the following to be configured in the SIVACON low-voltage switchboard system:

Switchboard configurations

▷ Power centers ≤ 6300 A
▷ Main and sub-distribution boards ≤ 4000 A
▷ Motor control centers ≤ 2500 A

Based on the modular SIVACON structure (Fig. 1.4/10), a switchgear cubicle or switchboard consists of modular groups, which are made up of individual modules, and individual modules.

Standard outer size

All of the switchgear cabinets (switchgear cubicles) have a standard outer size which means that the rooms used to house the switchboards can be planned and designed systematically and efficiently.

Functional compartments

These are divided into functional compartments, which for the most part cannot be modified. This allows standardization, particularly where parallel requirements exist, e. g. in the busbar and cable connection compartments.

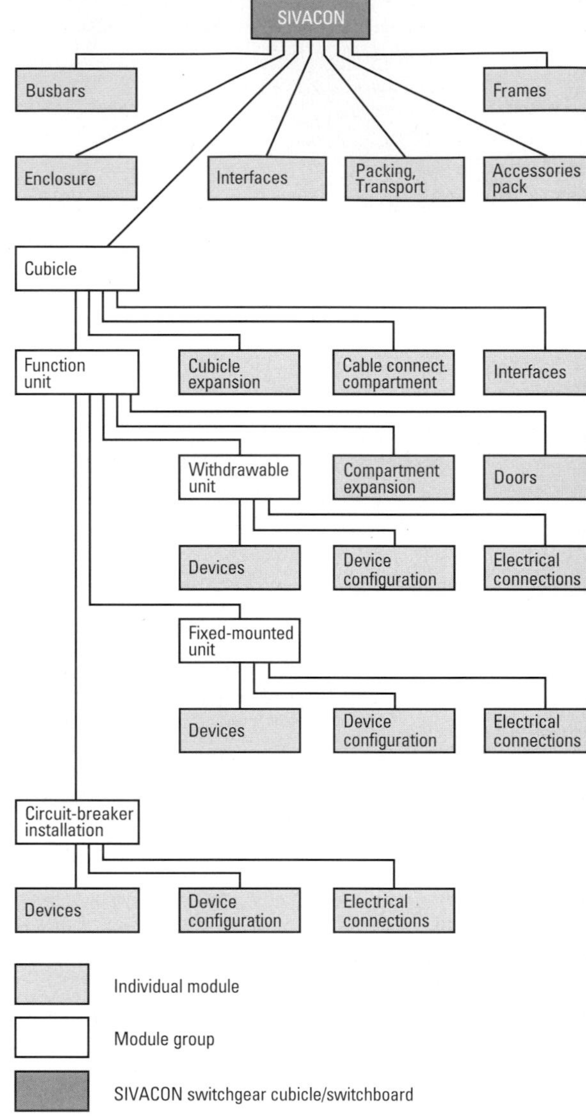

Fig. 1.4/10 SIVACON modular concept

Device compartment

The individual functional compartments can be configured as required thus providing the flexibility required by the individual functions. The device compartment can, for the most part, be freely configured in accordance with the needs of the user.

Equipment designs

The possibilities for configuring the device compartment are based on the following equipment designs:

▷ Circuit-breaker design
▷ Withdrawable-unit design
▷ Fixed-mounted design

These determine, first and foremost, efficiency and cost effectiveness as well as the degree of adaptability.

The device compartment itself always has the same dimensions.

Function unit

All function units can be integrated in the compartment irrespective of the busbar position and the installation techniques used.

Modular design

SIVACON's modular structure enables components to be combined in many different ways even if these have different designs.

Compartmentalization

The device compartment is divided into ten 175 mm modules (175 mm = 1 M). By inserting additional racks, the device compartment can be divided into compartments ranging from 1 M to 10 M depending on the module size.

Internal separation

According to EN 60439–1/IEC 60439–1/VDE 0660 Part 500, internal separation is possible up to type 4 (Fig. 1.4/11).

Packing density

Very high packing densities are possible, e.g. up to 40 withdrawable units per switchgear cubicle.

Circuit-breaker design

Requirements vis-à-vis operational reliability and operator safety

A large number of main and sub-distribution boards as well as high-power loads are usually connected downstream of low-voltage switchboards for large energy demands ("power centers"). The following requirements vis-à-vis long-term operational reliability and operator safety are, therefore, particularly important:

▷ The functions "Supply", "Coupling", and "Feed" must be assured a high degree of reliability over long periods.
▷ Maintenance and inspections should only require short downtimes.

These exacting requirements are met using circuit-breaker components with the following features:

▷ Function compartments with protective separation for
  – switchgear
  – auxiliary equipment and measuring instruments
  – cable and busbar connections
▷ Total separation from adjacent cubicles by means of solid-wall switchgear cabinets
▷ Circuit-breakers moved to the inspection and disconnected position with door closed
▷ Cabinet widths tailored to circuit-breaker sizes, e.g. only 400 mm wide for circuit-breakers ≤ 1600 A (Fig. 1.4/12)
▷ Variable use of circuit-breakers with rated currents of 630 A to 6300 A
▷ Sufficient space provided for auxiliary equipment and for various connection systems

3WN circuit-breakers

Siemens 3WN circuit-breakers are used for the entire rated current range from 630 A to 6300 A.

This means:

▷ Freely selectable infeed direction without restriction of the technical data
▷ High short-time current carrying capacity for time-grading short-circuit protection ≤ 500 ms
▷ Short-circuit protection with time-reduced selectivity control with full selectivity and very short delay times (50 ms)
▷ No ammeter or current transformer required if the operating current LCD is used
▷ Displays and operation with closed door
▷ No complex grounding or short-circuiting devices required for maintenance if a 3WN grounding switch is used

Auxiliary equipment compartment

The auxiliary equipment compartment is a separate compartment for the circuit-breaker. It also provides space for large control units and interlocks.

The auxiliary equipment module can be easily disconnected from the power circuit. Modifications resulting from service-related changes can thus be carried out outside the switchgear cubicle without any difficulty.

Cable connection compartment

The size of the cable connection compartment is matched to the rated breaker current. The compartment facilitates the connection of both cables and busbars. This considerably reduces assembly times and the bilateral cubicle separation from adjacent cubicles provides additional protection for personnel.

The cables and busbars can be fed in from below or above.

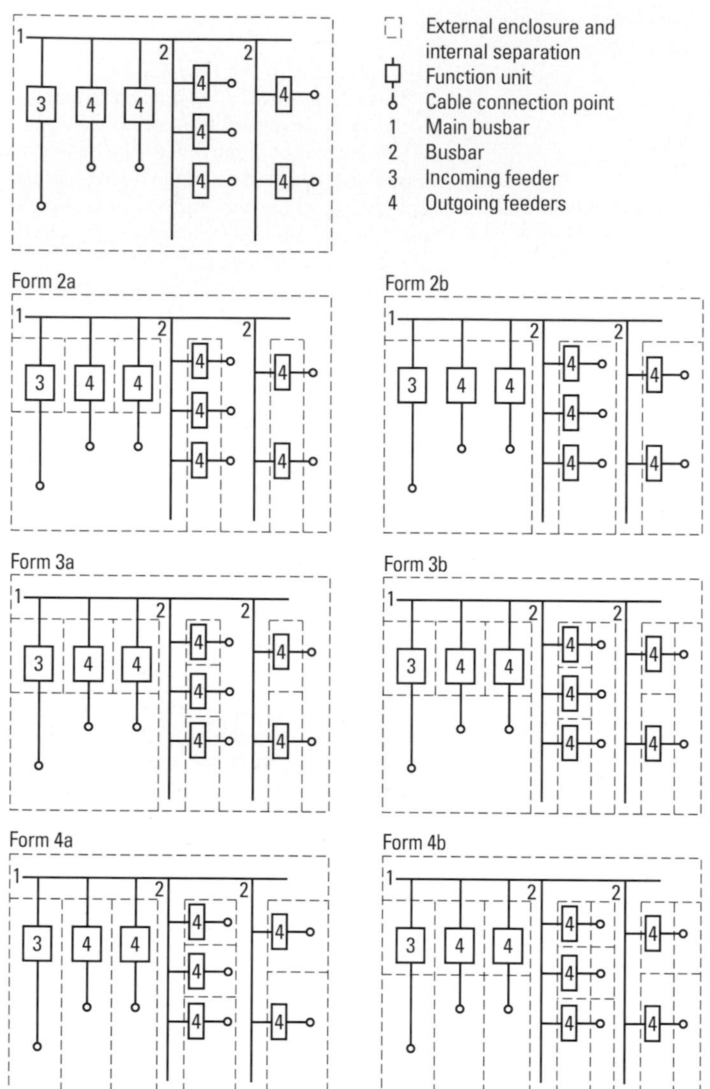

Fig. 1.4/11 Internal separation types to EN 60 439-1/IEC 60 439-1/VDE 0660 Part 500

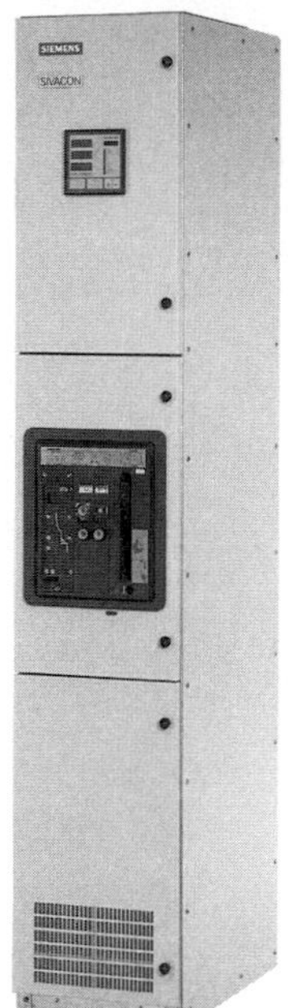

Fig. 1.4/12
SIVACON switchgear cubicle with 3WN withdrawable circuit-breaker, 1600 A, cubicle width 400 mm

Fig. 1.4/13
SIVACON switchgear cubicle with withdrawable units
Left: Size 1 withdrawal units
Right: Cable connection compartment

Withdrawable-unit design

Flexibility through withdrawable-unit design

Technical installations and equipment in power stations, industrial plants, hospitals, and office complexes require an adequate source of power which is available on a continuous basis. Changing operational requirements, e.g. changes to required motor outputs or the connection of new loads, mean that low-voltage switchboards must be able to provide a high degree of flexibility. The SIVACON switchboard system has achieved this by means of a new, ergonomic withdrawable-unit design (Fig. 1.4/13).

The SIVACON withdrawable-unit design ensures easy handling, requires only short assembly times, and provides a high level of switchboard availability during operation.

Withdrawable-unit size

The withdrawable units are also designed in accordance with the 175 mm modular grid. As a result, their height can vary between 1 M and 4 M depending on the required function and load current ($\leq$ 630 A max.).

The withdrawable units with a height of 1 M are available for the total width of the device compartment in the following sizes: size 1 for rated load currents $\leq$ 125 A, size ½ (half device compartment width), and size ¼ (¼ device compartment width) for small loads, e.g. solenoid valves, actuators, low-power motors, etc. up to a rated current of 35 A.

Advantages

This type of withdrawable-unit design offers users the following advantages:

- ▷ test and disconnected position with closed door,
- ▷ high packing density with minimum base area (up to 40 ¼-size withdrawable units in one switchgear cubicle),
- ▷ standardized user interface for all withdrawable units,
- ▷ visible isolating gap in the incoming and outgoing circuit of withdrawable unit (see Fig. 1.4/16),
- ▷ smooth insertion of withdrawable units; no insertion force necessary,
- ▷ load feeders can be easily adapted to modified operating conditions without the switchgear cubicle having to be disconnected,
- ▷ separate functional compartments in cubicle (type of internal compartmentalization must be at least 3 b).

Plug-on bus system

The plug-on bus system is embedded in split plastic casing shells with phase separation and is protected against internal arcs (Fig. 1.4/14).

Fig. 1.4/14
Plug-on bus system in a SIVACON switchgear cubicle for withdrawable units

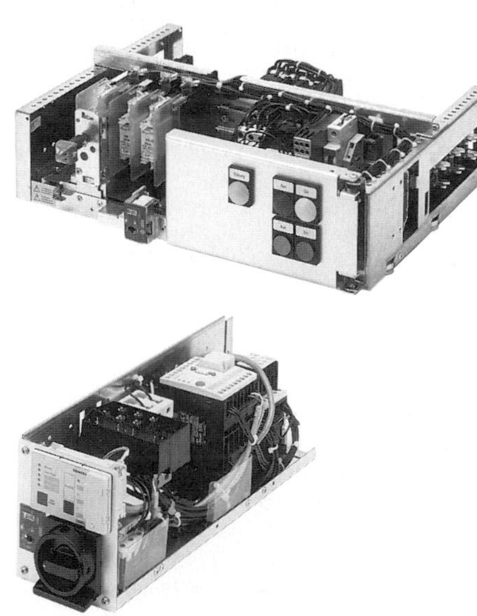

Fig. 1.4/15
Top: Size 1 withdrawable unit for Delta-wye connection ≤ 37 kW at 400 V AC
Bottom: Size ¼ withdrawable unit for direct contactor ≤ 15 kW and SIMOCODE-DP motor protection and control device

Three and four-pole version

This type of construction provides protection against accidental contact (see Section 2.3) with live plug-on buses without the need for additional guards (e. g. shutters). The plug-on bus system is arranged vertically on the left-hand side of the switchgear cubicle. It can be configured as a three-pole or four-pole version.

Standard operation for all ratings

The withdrawable units, which are operated and handled in the same way for all power ratings, offer a higher level of safety and reliability for day-to-day use.

All parts – including the isolating contacts – are located within the confines of the withdrawable units. This virtually eliminates the risk of any damage occurring even outside the switchgear cubicle.

A size-1 withdrawable unit is shown in Fig. 1.4/15.

Reliable operation

Specially designed isolating contact systems allow the "test" and "disconnected position" to be set without the withdrawable units needing to be moved. The degree of protection of the switchboard remains unchanged (Fig. 1.4/16). The visible isolating gap and the clearly indicated position of the withdrawable unit ensure reliable operation.

The withdrawable unit always passes through the "disconnected position" when it is moved to the "test" and "connected position".

Main switch with motor switching capacity, door interlock

Each withdrawable unit has a mains switch with motor switching capacity (6 to 8 · I_n) and disconnector characteristic. In size 1 – 4 withdrawable units, the operating mechanism of the mains switch together with the door interlock ensures that the compartment door in front of the withdrawable unit cannot be opened until the mains switch has been switched off.

Front panel

In size ¼ and ½ withdrawable units, the operating mechanism of the mains switch is installed in the front panel of the withdrawable units. The front panels also form the plates covering the front of the withdrawable units in the switchgear cubicle. These withdrawable units, therefore, have no compartment doors (Fig. 1.4/15, bottom).

Once the mains switch has been switched off, the withdrawable unit is de-energized and the mechanical interlock is released so that the isolating contacts can be moved.

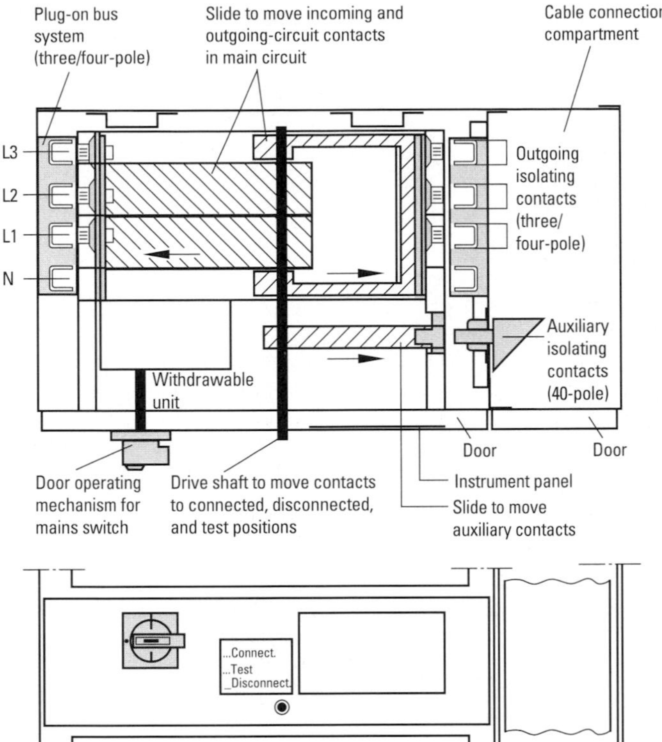

Fig. 1.4/16
Withdrawable unit in disconnected position

Two isolating gaps per withdrawable unit

The withdrawable units of the SIVACON low-voltage switchboard provide two-fold protection for operating personnel by means of:

▷ mains switch with disconnector characteristic and door interlock and interlock for the isolating contact system,
▷ movable isolating contact system with a disconnected position in both the incoming and outgoing circuits of the withdrawable unit to permit maintenance.

Modifying the withdrawable-unit compartments

A range of simple options for converting the withdrawable units from one size to another allow the switchboard to be adapted quickly to new or changing operating requirements.

If size ¼ and ½ withdrawable units are used, adapter plates with integrated wiring also permit modification without the switchboard needing to be disconnected.

Cable connection compartment

Cables are connected in the ergonomic cable connection compartment at the side of the withdrawable unit. This compartment has a width of up to 400 mm.

Fixed-mounted design

In a number of cases, e.g. building installations, it is not necessary to replace components when the system is in operation and brief downtimes do not result in any significant reduction in follow-on costs.

Fixed-mounted design

In such cases, the fixed-mounted design (Fig. 1.4/17) provides adequate flexibility with a high level of efficiency and reliability by means of:

▷ freely combinable function modules,
▷ simple replacement of function modules when the switchboard is de-energized,
▷ short conversion and downtimes thanks to universal cubicle busbars at the side of the unit,
▷ add-on modules allowing user-specific separation and compartmentalization.

Function modules

The function modules permit flexible and cost-effective installation, especially if operational changes or modifications in line with new load specifications become necessary. In addition to equipping subracks with any desired combination of devices or device

Fig. 1.4/17
SIVACON switchgear cubicle, fixed-mounted design

Fig. 1.4/18
Universal cubicle busbar for SIVACON fixed-mounted switchgear cubicle with connection facilities for busbars, cable lugs, and terminals

components, the user can also freely combine the function modules within a switchgear cubicle. When the function modules are inserted in the switchgear cubicle, they are first mounted in the available slots and screwed to the switchgear cubicle. This retaining system simplifies installation and permits "one-man-assembly".

Add-on modules

Add-on modules to meet individual requirements regarding convenience and safety are available, e. g.:

- ▷ plates between the function modules for internal separation,
- ▷ individual doors for functional compartments which may have been created as a result of internal separation,
- ▷ insulating barriers covering live connections or parts.

Universal cubicle busbar, connection facilities

The universal cubicle busbar (Fig. 1.4/18) offers a range of different connection facilities and simplifies the connection of cables, conductors, and busbars. As a result, a function module or individual device can be connected where it is required without the need for drilling or punching. The connections are clearly visible when the switchgear cubicle is opened and can be easily inspected. Re-

Fig. 1.4/19
SIVACON switchgear cubicle, in-line plug-in and strip design

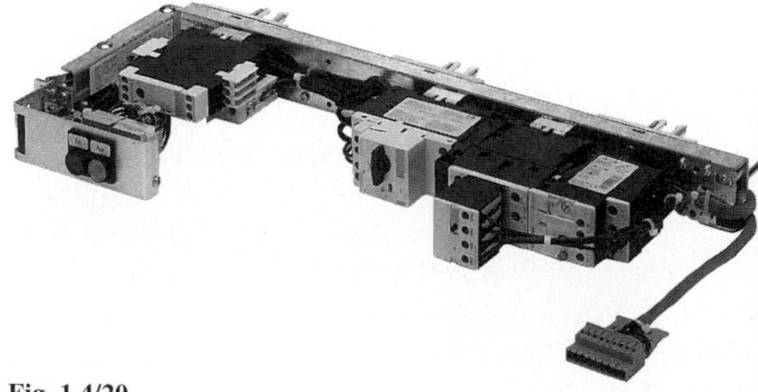

Fig. 1.4/20
In-line plug-in module, 11 kW, with overall height of 50 mm

moving the transparent barrier provides easy access to the cubicle busbar connections and permits modifications or expansions to be carried out quickly.

witching evices in strip rm, in-line lug-in modules

One of the most important features of switching devices in strip form for outgoing cable units ≤ 630 A and in-line plug-in modules primarily for motor feeders ≤ 45 kW is their compact design which permits modular cubicle assembly with a high packing density (Fig. 1.4/19).

Both the switchable strips and the in-line plug-in modules are fitted with incoming plug-in contacts (Fig. 1.4/20).

This design allows the switching devices and switching device assemblies to be converted quickly and replaced easily without the switchboard having to be disconnected.

The cubicle busbar is located at the rear of the cubicle and is fitted with a touch guard, which has tapping openings in the 50 mm grid.

For top-mounted main busbar system

For rear-mounted main busbar system

Fig. 1.4/21
SIVACON solid-wall switchgear cabinets

Frame and enclosure

rame onstruction

The frames used for the SIVACON low-voltage switchboard are constructed using solid-wall technology. The single-section side panels are prepunched (with all the necessary knockouts), beveled, and connected together with braces to form a dimensionally-accurate and sturdy frame. The frame is riveted together which means that it requires no maintenance, welding, bolting, or re-tightening (Fig. 1.4/21). Removable frame and panel sections facilitate busbar assembly – the busbars are inserted in insulated busbar holders.

ole matrix ystem

Hole matrix systems, which allow subsequent expansion, ensure both uncomplicated electrical disassembly of the switchgear cabinets and simple retrofitting of new function modules.

orrosion rotection/ urface rotection

All of the sheet-steel sections used are sendzimir-galvanized and ensure durable corrosion protection and correct grounding connections.

The powder-coated panels (color: RAL 7032, pebble gray) provide additional surface protection and give the switchboard an attractive appearance.

oors, hinges, ocks

The doors are attached to the frame construction by means of hinges (opening angle 120°).

Pressure compensation

Spring-loaded locks prevent the doors from opening inadvertently and also ensure safe pressure compensation in the event of overpressure caused by arcing faults. In addition, the top plates are fitted with pressure-relief valves.

Main busbar system

Position of busbar

"Safe busbar"

The main busbar system with rectangular busbar cross-sections for rated currents ≤ 6300 A can be assembled in various configurations (Fig. 1.4/22) and consists of the three phase conductors L1, L2, and L3 as well as the PE, N, and PEN conductors. A second busbar system, e.g. "safe busbar", can be installed in any switchboard.

Busbar connections

The busbar connections at the shipping split of a switchboard are freely accessible from the front thus allowing the switchboard joins to be bolted together much more easily on site. The bolted connections in the busbar system have been used successfully in Siemens switchboards for decades and do not require any maintenance.

Installation

The main busbar system can be inserted and removed from above in the case of motor control centers ≤ 2500 A and from the rear

Side view

2200 G 400

Position of main busbar system: top
Busbar current: $I_n \leq 2500$ A, $I_{pk} \leq 50$ kA
Installation: wall-mounted or free-standing
Cable entry: from below
Application: motor control centers, sub-distribution boards

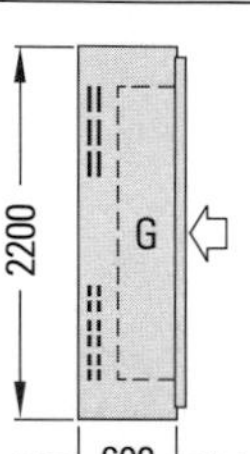

Position of main busbar system: rear (top and/or bottom)
Busbar current: $I_n \leq 4000$ A, $I_{pk} \leq 100$ kA
Installation: wall-mounted or free-standing
Cable entry: from above and/or below
Application: main distribution/sub-distribution boards and integrated motor control centers

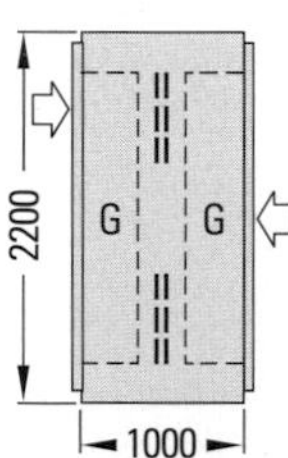

Position of main busbar system: center (top and/or bottom)
Busbar current: $I_n \leq 4000$ A, $I_{pk} \leq 100$ kA
Installation: free-standing double-fronted
Cable entry: from below and/or above
Application: main distribution/sub-distribution boards and integrated motor control centers in double-fronted design

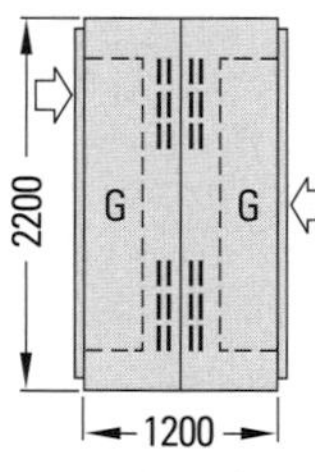

Position of main busbar system: rear (top and/or bottom)
Busbar current: $I_n \leq 4000$ A, $I_{pk} \leq 100$ kA
Installation: free-standing back-to-back
Cable entry: from below and/or above
Application: main distribution/sub-distribution boards and integrated motor control centers in back-to-back design with completely separated busbar systems

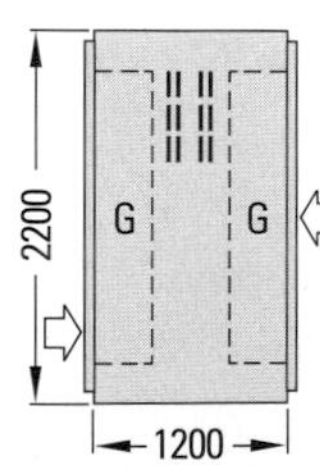

Position of main busbar system: center top
Busbar current: $I_n \leq 6300$ A, $I_{pk} \leq 100$ kA
Installation: free-standing double-fronted
Cable entry: from below
Application: power centers

Dimensions in mm

G Device installation compartment   Operating side   II Busbars

Fig. 1.4/22
Different configurations for SIVACON low-voltage switchboards resulting from the variable position of the main busbar system

side of the switchgear cubicle in the case of main and sub-distribution boards. This means that on-site modification of the main busbar system (as a result of adaptation to changed power demands) or subsequent installation in the switchboard compartment across the entire length of a transport unit is possible in cases where the installation space is restricted.

Four-pole switchable configuration

The main busbar system can also be configured as a four-pole system in cases where individual loads or functions require four-pole switching. In such cases, all conductors (L1, L2, L3, N) are installed along the same busbar run.

Switchgear cubicle arrangement with SIVACON

Freely arranged switchgear cubicles

Circuit-breaker, withdrawable-unit, and fixed-mounted switchgear cubicles, the main busbars of which are located in the same position, can be arranged alongside each other in any sequence. When the front faces and back panels of all the items are arranged in the same way, all the switchgear cubicles have the same depth.

Double-fronted arrangement

The double-fronted arrangement is also possible for all three configurations.

The wide range of possible combinations permits the switchgear cubicles to be assembled in accordance with specific individual requirements and the respective switching tasks.

Table 1.4/3 contains the basic technical data for the SIVACON low-voltage switchboard.

1.4.2.3 Communication in SIVACON low-voltage switchboard systems via PROFIBUS-DP

Application

Modern industrial production plants require high-performance programmable controllers to control and monitor their increasingly complex process cycles. Since the supply of power to the various electrical mechanisms in a production plant is an important factor in ensuring that processes are performed correctly, it is essential that the low-voltage switchboard is incorporated in process control and monitoring.

Process reliability and a high level of plant availability can only be ensured if extensive data exchange takes place between the low-

Table 1.4/3 Basic technical data of the SIVACON low-voltage switchboard

Standards	EN 60439-1 IEC 60439-1 VDE 0660 Part 500 DIN VDE 0106-100 IEC 1641 for testing behavior in the event of internal faults (arcing fault)		
Rated voltages			
Rated impulse strength (U_{imp}) Overvoltage category Pollution severity	8 kV III 3		
Rated insulation voltage U_i Rated operation voltage U_e	1000 V 690 V		
Main busbars, three and four-pole, horizontal			
Rated current U_n Rated peak withstand current U_{pk} Rated short-time withstand current I_{cw}	≤6300 A ≤ 220 kA ≤ 100 kA		
Connecting, plug-on, and cubicle busbars, three and four-pole, vertical	Circuit-breaker design	Withdrawable-unit design	Fixed-mounted design Plug-on design
Rated current I_n Rated peak withstand current i_{pk} Rated short-time withstand current I_{cw}	≤6300 A ≤ 220 kA ≤ 100 kA	1000 A 143 kA 65 kA	2000 A 110 kA 50 kA
Rated device currents I_n			
Circuit-breakers Outgoing cable units Motor control units	≤6300 A ≤ 800 A ≤ 630 A		
Degree of protection to EN 60529/ IEC 60529, DIN 40050	IP 20 to IP 54		
Surface protection (Coating to DIN 43656) Frame sections Panels Doors	 Sendzimir-galvanized/powder-coated Sendzimir-galvanized/powder-coated Powder-coated		
Color (standard)	RAL 7032 (pebble gray)		
Dimensions of switchgear cubicles (Preferred dimensions to DIN 41488)			
Height mm	2200		
Width mm	400, 500, 600, 800, 1000		
Depth			
with single-fronted design mm	400, 600		
with double-fronted design mm	1000		
with power centers mm	1200		

voltage switchboard and the programmable controllers via serial bus systems.

The SIVACON low-voltage switchboard (see Section 1.4.2.2) supports this advanced technology in the form of a standardized solution as it incorporates both power distribution and motor feeder control in a fully-integrated bus concept (Fig. 1.4/23).

PROFIBUS-DP

The transparent, manufacturer-independent PROFIBUS-DP, which is certified to EN 50170, is used for fast, reliable, and cyclic data exchange. It is a universal system, which provides a straightforward means of networking a large number of cubicle devices and allows signals to be transmitted within milliseconds, even over distances of several kilometers.

Technical data

Technical data of the PROFIBUS-DP:

▷ Manufacturer-independent, standardized to EN 50 170
▷ Two-wire copper conductors or optical fibers
▷ Bus system can be expanded ≤ max. 100 km (if optical fibers are used)
▷ Maximum transmission rate: 12 Mbaud
▷ Up to 124 bus devices
▷ Various frames: cyclic data exchange, acyclic data exchange, diagnostic data exchange, parameter data exchange
  - if a device fails, all other devices continue to operate,
  - integrated solution for networking to PROFIBUS-FMS and AS-interfaces.

Information exchange

In order to make the best possible use of this exchange of detailed information between the low-voltage switchboard and the automation or process control system, all data relevant for the low-voltage switchboard, such as control commands, check-back signals, operating and diagnostic data can be integrated in the appropriate process displays of an operator station and thus made available to control-room personnel.

Transfer time intervals

In order to reduce transfer times, important data, such as control commands, is transmitted very quickly and cyclically while diagnostic data, for example, is only transmitted if changes occur. Parameter data is

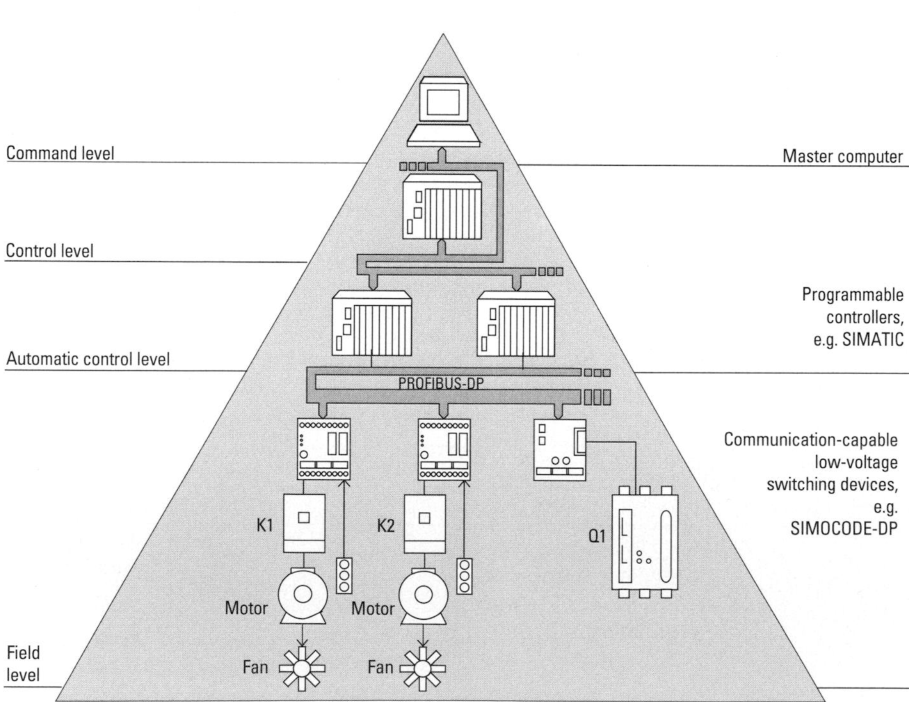

Fig. 1.4/23 Diagram of plant with SIMOCODE-DP

only transmitted acyclically or when the communication process is initiated.

This ensures that the transmission capacity of the bus is used to optimum effect and prevents the cycle time in the central processing units of the programmable controllers from being increased unnecessarily.

Communication-capable low-voltage switching devices

Control of power flow and motor feeders

Communication-capable low-voltage switching devices with a PROFIBUS-DP interface must be used at the field level (see Fig. 1.4/23) to control the power flow and motor feeders and to collect all important operating, diagnostic, and statistical data. The 3WN6 circuit-breaker is available for power distribution and the communication-capable motor protection and control device SIMOCODE-DP for controlling and monitoring motor feeders.

3WN6 circuit-breaker with DP/3WN6 PROFIBUS-DP interface module

3WN6 circuit-breaker

Power distribution is controlled and monitored from the superordinate automation level via the DP/3WN6 PROFIBUS-DP interface module (Fig. 1.4/24) of the 3WN6 circuit-breaker. The DP/3WN6 PROFIBUS-DP interface module is linked to the 3WN6 circuit-breaker via a point-to-point connection. This connection is matched to the specific environment (e. g. high currents, sliding contacts in the case of the withdrawable-unit version, etc.) of the circuit-breaker. Conversion to the PROFIBUS-DP protocol takes place in the DP/3WN6 interface module.

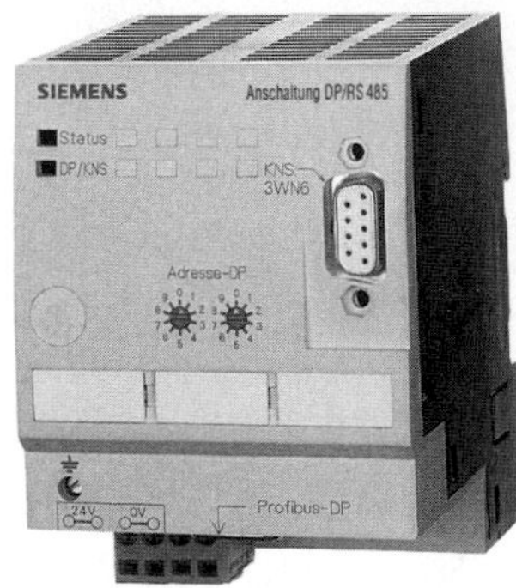

Fig. 1.4/24
DP/3WN6 PROFIBUS-DP interface module for 3WN6 circuit-breakers

Functions

The DP/3WN6 PROFIBUS-DP interface module performs the following functions:

- ▷ Analog measured values:
  e. g. phase currents, ground-fault current, cos φ
- ▷ Event information:
  e. g. type of tripping, load shedding, phase asymmetry
- ▷ Operating states:
  e. g. switch in on/off position, in test position, ready for closing
- ▷ Set-up (parameterization):
  e. g. current setting of overload release, time-lag class, operating value of short-circuit release
- ▷ Diagnostic data:
  e. g. current value at last trip, peak current value of last 15 minutes
- ▷ Remote control:
  e. g. switch on/off.

System start-up

Only the PROFIBUS-DP address has to be preset in order to start up the DP/3WN6 PROFIBUS-DP interface module. The baud rate and other bus parameters are set by the master module; the interface module is adjusted automatically to these settings. The communications link with the circuit-breaker is established automatically when the supply voltage is connected.

Parameterization

The DP/3WN6 PROFIBUS-DP interface module is configured either using the COM PROFIBUS parameterization software in the case of SIMATIC S5, using STEP 7 for SIMATIC S7, or suitable software from other manufacturers.

The circuit-breaker can be parameterized either using the rotary switch/keypad on the overcurrent release, the hand-held controller of the circuit-breaker, Win3WN6, or the application program.

SIMOCODE-DP motor protection and control device

SIMOCODE-DP motor protection and control device

The motor feeders of the SIVACON low-voltage switchboard are integrated in the process automation system via the communication-capable motor protection and control device SIMOCODE-DP.

SIMOCODE-DP is a signal-processing I/O module which supports all the functions for motor protection and for controlling, interlocking, and monitoring the motor feeder in the low-voltage switchboard. It also supplies detailed operating, diagnostic, and statistical data. SIMOCODE-DP is, therefore, a system that turns the switchboard into an intelligent low-voltage switchboard.

The SIMOCODE-DP system is required for every motor feeder. It consists of the following components:

Components

- ▷ Basic unit
- ▷ Expansion module
- ▷ Operator control module
- ▷ Current transformer for analog current value display
- ▷ Win-SIMOCODE-DP/Professional software
- ▷ Manual

Basic unit

Basic unit

The SIMOCODE-DP basic unit (Fig. 1.4/25) with its four inputs and four outputs supports all of the necessary protection and control functions and establishes the link to the automation level via the PROFIBUS-DP. The four inputs are supplied with power by the internal 24 V DC voltage source. The expansion module, operator control module, or a personal computer can be connected via the system interface. The basic unit is available in three different versions for the control supply voltages 24 V DC / 115 V AC / 230 V AC.

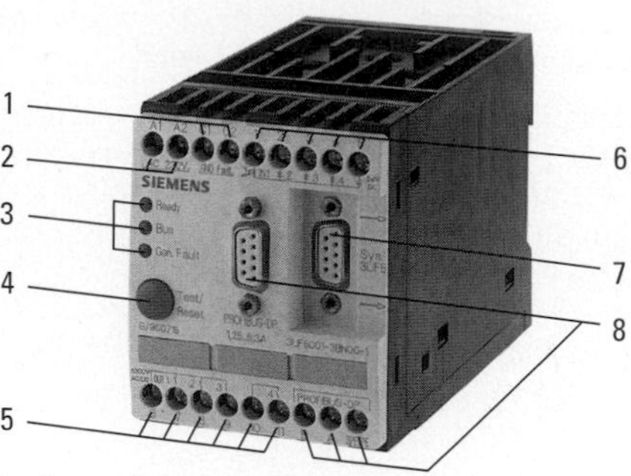

1 Connection of thermistor sensor circuit for thermistor-type motor protection
Option: Connection of summation current transformer for ground-fault detection
2 Connection of control supply voltage 24 V DC / 115 V AC / 230 V AC
3 3 LEDs
4 Device test, manual reset:
Configurable automatic reset, remote reset via bus or input
5 3 + 1 relay outputs:
Configurable function assignment. NO/NC contact behavior by means of signal conditioning
6 4 opto-coupled inputs:
Internal 24 V DC power supply
Configurable function assignment
7 System interface: Connection of expansion module, operator control module, or personal computer
8 Bus port, PROFIBUS-DP standard 9-pole SUB-D socket. Terminals for withdrawable units

Fig. 1.4/25 SIMOCODE-DP basic unit

Expansion module

Expansion module

The SIMOCODE-DP-E expansion module (Fig. 1.4/26) provides the system with an additional eight inputs and four outputs. The expansion module itself is supplied by the basic unit. The eight inputs must be connected to an external power supply. The three different voltage versions available are 24 V DC / 115 V AC / 230 V AC. The connections to the basic unit and to the operator control module or personal computer are established via system interfaces.

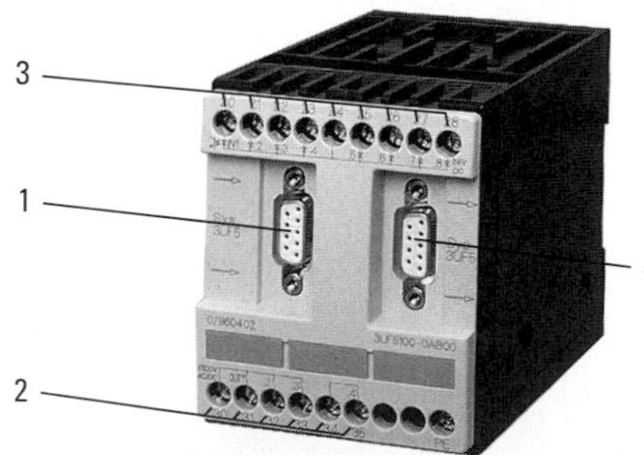

1 System interface:
Interface connection to basic unit
Power supplied from basic unit
2 3 + 1 relay outputs:
Configurable function assignment
NO/NC contact behavior by means of signal conditioning
3 8 opto-coupled inputs:
External 24 V DC / 115 V AC / 230 V AC power supply
Configurable function assignment
4 System interface: Connection of operator control module or personal computer

Fig. 1.4/26 SIMOCODE-DP expansion module

Operator control module

Operator control module

The SIMOCODE-DP operator control module (Fig. 1.4/27) is used to control a drive mechanism from the switchgear cabinet. It can be connected to the basic unit or to the expansion module. The module is powered by the basic unit. It is also possible to connect a personal computer to the system interface. The module is installed in the front pa-

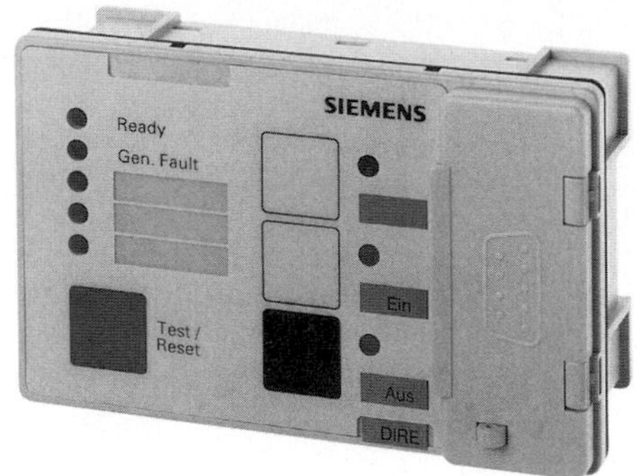

Fig. 1.4/27
SIMOCODE-DP operator control module

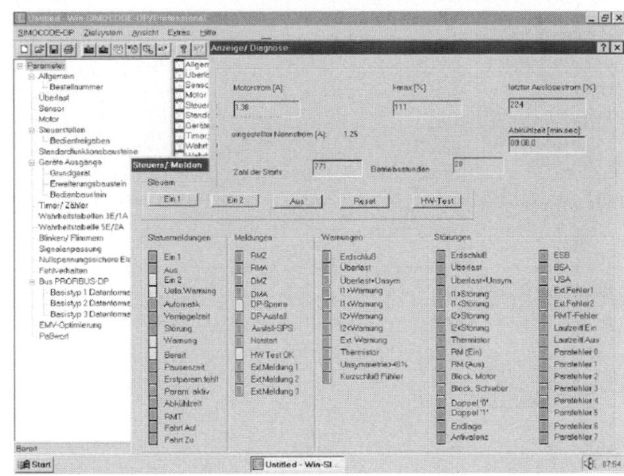

Fig. 1.4/28
Parameterization, operation, monitoring, and testing with Win-SIMOCODE-DP/Professional

nel of a distribution board or in the switchgear cabinet door.

The operator control module is protected to IP 54. It features three operator buttons and six LEDs.

Current transformer for analog current value display

Analog current value display

The SIMOCODE-DP system features an external current transformer to allow current values to be displayed on a pointer instrument in the switchgear cabinet door. This also enables the current flowing through the motor feeder to be read on site on the low-voltage switchboard.

Win-SIMOCODE-DP/Professional software

Software

The Win-SIMOCODE-DP/Professional software allows the motor feeder to be parameterized, operated, monitored, and tested "on-line" via the PROFIBUS-DP or point-to-point via the RS 232 system interface (Fig. 1.4/28). "On-line" means that parameter, control, signaling, and statistical data can be written or read acyclically via a class-2 PROFIBUS-DP communications processor installed in the PC/PG (e.g. CP5412 to SIMOCODE-DP) parallel to the usual cyclic data exchange with the automation/process control system.

Win-SIMOCODE-DP/Professional can be run under Windows '95.

Manual

Manual

The manual contains a detailed description of the motor protection and control functions and of the PROFIBUS-DP communications interface (Order No. 3UF5700-0AA00-0 German, 3UF5700-0AA00-1 English).

Motor protection functions

Motor protection functions

Similar to an electronic overload relay, the SIMOCODE-DP system measures the motor current flowing through all three phase conductors via the current transformer integrated in the basic unit. Current setting ranges between 0.25 A to 820 A are possible with just six device variants.

Tripping class

Tripping class

The tripping class can be varied between CLASS 5 and CLASS 30 in six steps. As a result, the disconnecting time can be adjusted extremely accurately to the starting inertia, which permits highly efficient utilization of the motors. With tripping class 10, for example, tripping occurs within 10 s for three-pole symmetrical loads with 7.2 times the set current under cold-start conditions, and within 30 s for tripping class 30. The tripping current in the maximum current range is between 1.10 and 1.20 times the set current (Fig. 1.4/29). The maximum release time deviation in the overload range is ± 10%. The SIMOCODE-DP system, therefore, complies with the requirements specified in EN 60947 / IEC 60947 / DIN VDE 0660.

Asymmetrical loads

Asymmetrical loads

Three-phase asynchronous motors respond to slight system voltage asymmetry with a relatively high asymmetrical current consumption. The resulting losses caused by the generated negative-sequence field lead to an increase in temperature in the stator and rotor winding.

The SIMOCODE-DP system also measures the asymmetrical current consumption of the motor to be monitored and, in the case of motor current asymmetry of approximately 40%, trips prematurely in accordance with the characteristic for two-pole loads (Fig. 1.4/30). This prevents the motor from overloading even in the event of a phase failure.

Single-pole loads

Protection of single-phase motors

The SIMOCODE-DP system is designed for protecting three-phase asynchronous motors. If it is used to protect single-phase motors, it is essential to ensure that the microprocessor is programmed for one phase only. For this purpose, the SIMOCODE-DP system must be set to "single-phase motor" during parameterization.

Thermistor motor protection

Temperature monitoring

In addition to overload protection, the SIMOCODE-DP system can also monitor the temperature of the motor windings if a PTC or NTC thermistor detector circuit is connected. This will ensure "thermistor motor protection" under severe operating conditions, e.g. restricted coolant flow or extreme temperature fluctuations in the motor. Apart from the familiar binary thermistor detectors, SIMOCODE-DP can also evaluate analog (KTY) thermistor detectors. Analog thermistor detectors are characterized by a resistance curve which rises more linearly with temperature. The advantage of this is that an alarm and tripping threshold can be parameterized in the SIMOCODE-DP system.

The thermistor connection has an open-circuit interlock, i.e. the device trips if an open terminal or open circuit is detected along the supply line.

After tripping, the device is ready for closing again as soon as the winding temperature at the mounting location of the thermistor drops 5 K below the operating temperature of the thermistor.

Ground-fault detection

Ground-fault detection

The SIMOCODE-DP system allows the following types of ground-fault detection:

▷ Internal ground-fault detection via the current transformers integrated in the basic unit. Apart from thermistor motor protection, they also permit detection of fault currents in the case of operation at nominal values 30% greater than the set current I_e.

▷ Ground-fault detection via an external summation current transformer. The transformer allows fault currents of 0.3 A, 0.5 A, and 1 A to be detected. A suitable basic unit must be selected.

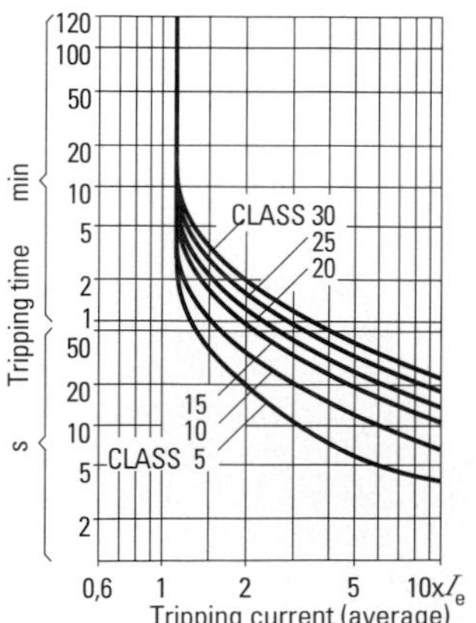

Fig. 1.4/29
Tripping characteristics for 3-pole symmetrical loads, CLASS 5 to CLASS 30

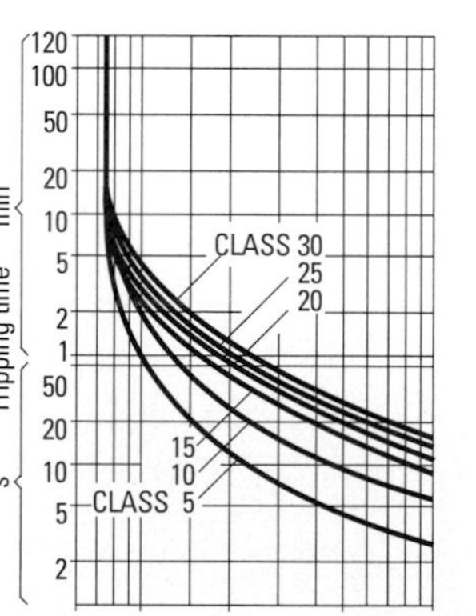

Fig. 1.4/30
Tripping characteristics for 2-pole symmetrical loads, CLASS 5 to CLASS 30

Current limits

Undershooting and overshooting current limits

The variable current limits inform the process engineer whether the production equipment has entered a critical state. For example, a ruptured belt (e.g. on a rolling train) could be indicated by the message "Lower current limit undershot" and an overload (e.g. in a stirring machine) by the message "Upper current limit overshot". These messages can prevent the machine from malfunctioning because they prompt the process engineer to take appropriate measures, e.g. to reduce the material mass being processed by the stirring machine.

Locking protection

Locking protection

Motor locking protection for operation at nominal values involves a comparison with a variable current limit. Tripping is instantaneous.

Manual, remote, and automatic reset

Manual, remote reset

The SIMOCODE-DP system is reset using the Test/Reset button. A remote reset is performed by connecting a pushbutton to one of the inputs, via the PROFIBUS-DP, or using the Test/Reset button on the operator control module.

Automatic reset

Automatic resetting is possible by means of appropriate parameterization.

Recovery time

Recovery time

The purpose of the recovery time is to give the motor time to cool down after an overload. The recovery time is 5 minutes and can be parameterized to a maximum of 60 minutes.

Setting protective functions to alarm/emergency start

In highly automated plants, the production process is frequently more important than the state of an individual motor because the failure of a motor feeder along a sequence of several consecutive processes can inflict heavy financial losses on the owner.

Alarm

To prevent such failures, all of the motor protection functions of the SIMOCODE-DP system can be set to "Alarm".

Emergency start

The SIMOCODE-DP system can also be set to "Emergency start" if the system has to be started up quickly after overload tripping. If the emergency start function is activated, the thermal memory in the SIMOCODE-DP system is erased and the motor feeder is reconnected immediately.

Control functions

Control functions

In addition to overload protection, the SIMOCODE-DP system can also support the following control functions:

▷ Direct-on-line starter
▷ Reversing starter
▷ Star-delta starting
▷ Dahlander starter
▷ Pole-changing switch
▷ Solenoid valve control
▷ Slide control, actuator
▷ SIKOSTART (soft starter)

Continuous or inching mode

The continuous or inching mode can also be set for all control functions.

All control functions can be implemented using the basic unit, provided that no more than four inputs and four outputs are required. If necessary, the expansion module could also provide an additional eight inputs and four outputs.

Permanently stored interlocks

All of the interlocks and logic operations required for the respective control function (e.g. for slide control cycles, actuators, instantaneous shutdown of the motor following operation of the limit or torque limit switches) are permanently stored in the SIMOCODE-DP system. This offers the following advantages:

▷ Reduces the workload of the application program in the programmable controller, i.e. fast configuration, fewer possible faults, shorter program cycle times
▷ Time-critical functions of the motor feeder do not depend on the signal propagation times at the automation level and along the communications link.

Propagation times for motor start-up and run-down and for slide control between the limit positions can also be set.

Circuit diagrams for control functions

The circuit diagrams for the most important control functions are shown in Figs. 1.4/31, 1.4/32, and 1.4/33.

Manual/automatic changeover and operator releases

Automatic mode Manual mode

In automatic mode, the control commands (e.g. ON, OFF) are passed onto the SIMO-

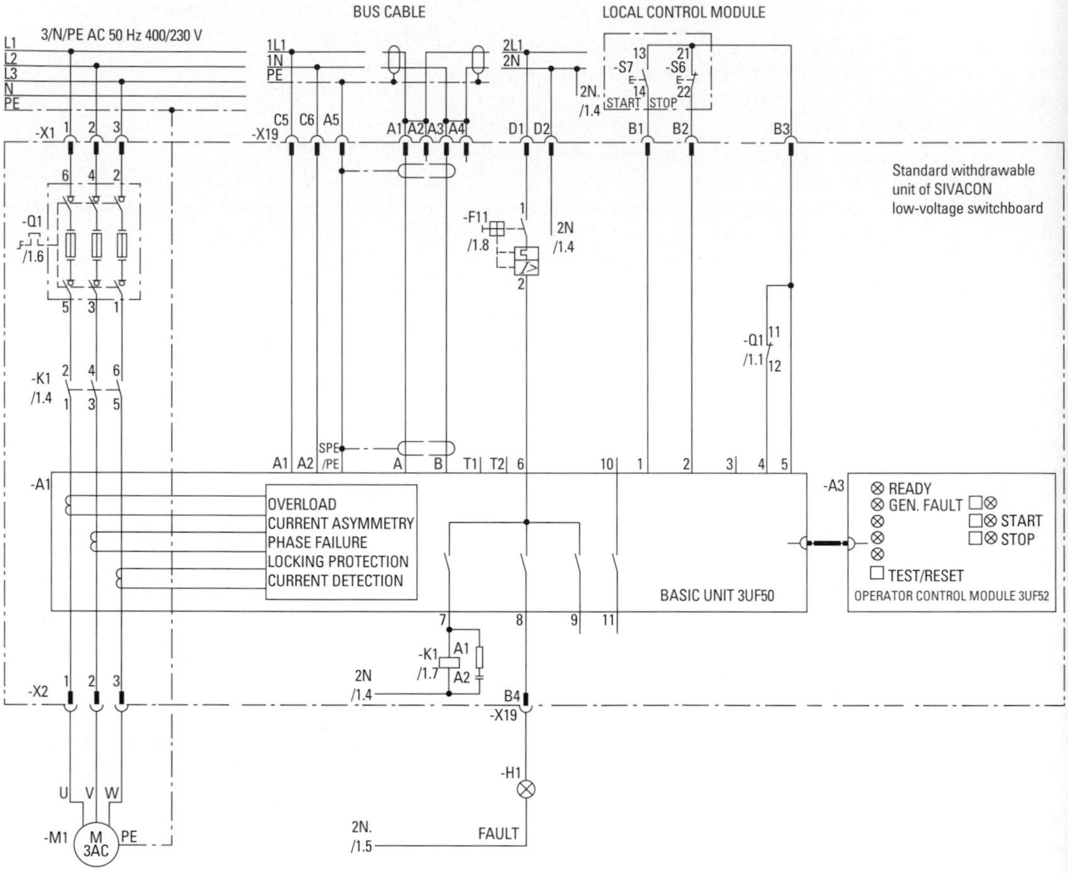

Fig. 1.4/31
Circuit for direct-on-line starting of three-phase motors with communication-capable SIMOCODE-DP motor protection and control device

Manual mode

CODE-DP system via the PROFIBUS-DP. In this mode, control commands from the switchgear cabinet (e. g. from the operator control module) or from the local control station are only permitted if the appropriate operator releases have been parameterized. If, however, the automatic control level changes over to manual mode or command is transferred to the local control station by means of a manual/automatic key-operated switch, control commands via the bus are ignored and the feeder can only be controlled via the operator control module or the local control stations.

Distributed signal logic, delay, and conditioning

Function modules

The high degree of flexibility of the SIMOCODE-DP system is characterized, in particular, by the following function modules integrated in the system:

- ▷ Four truth tables (three each with three inputs and one output; one with five inputs and two outputs)
- ▷ Two timers
  - variable setting (0.5 s to 60 min)
  - output behavior: ON delay, OFF delay, passing make contact
- ▷ Two counters, 0.....65535
- ▷ Four signal processing modules
  - level inverted without memory
  - edge rising with memory
  - edge falling with memory

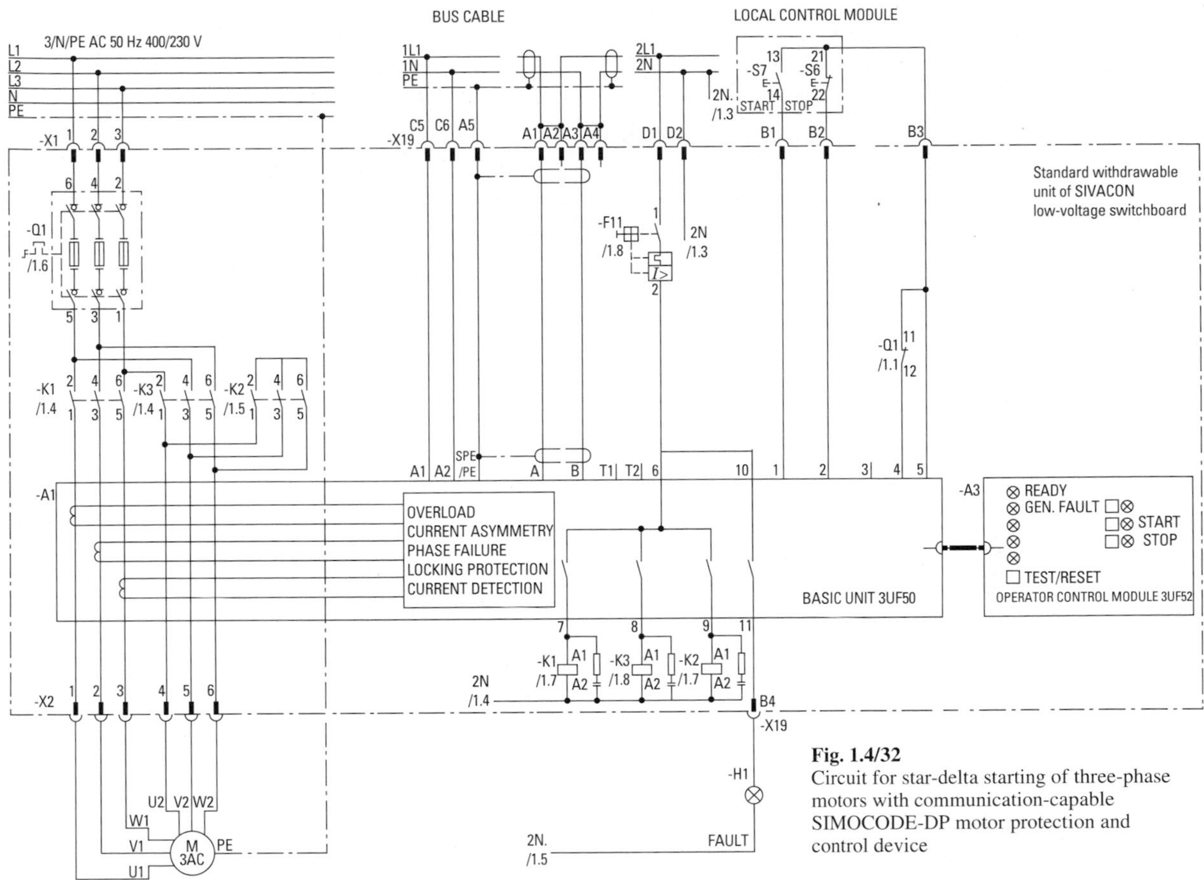

Fig. 1.4/32
Circuit for star-delta starting of three-phase motors with communication-capable SIMOCODE-DP motor protection and control device

- ▷ Two elements non-resetting on voltage failure
  - current information remains stored even if the control voltage fails
- ▷ Six signal conditioning modules
  - three flashing, three flickering

The inputs and outputs of the system can be assigned not only to the predefined control functions but also to the truth tables, timers, and counters. The control functions can, therefore, be adapted to suit the user's requirements. Output behavior (NC/NO contact) is determined by the signal processing modules.

Graded start-up

Time-graded start-up

Reconnection of the motor feeders after a power failure can be time graded, thus enabling the system to be started up quickly and selectively.

Diagnostics

Operating data

The SIMOCODE-DP system determines a large amount of operating, diagnostic, and statistical data in the motor feeder. This data can also be read directly from the display elements. The following data is available:

- ▷ Operating data: current level (%), ON, OFF, right, left, fast, slow, slide open, slide closed, slide moving, alarm, fault, interlock time running, recovery time running
- ▷ Diagnostic data: overload alarm/tripping, thermistor motor protection alarm/tripping, ground-fault alarm/tripping, locking protection, upper/lower current limit reached, time-out, exclusive-OR error at slide, etc. **Diagnostic data**
- ▷ Statistical data: number of motor starts, operating hours, number of trips, current at last overload trip (%), thermistor detector value, counter value **Statistical data**

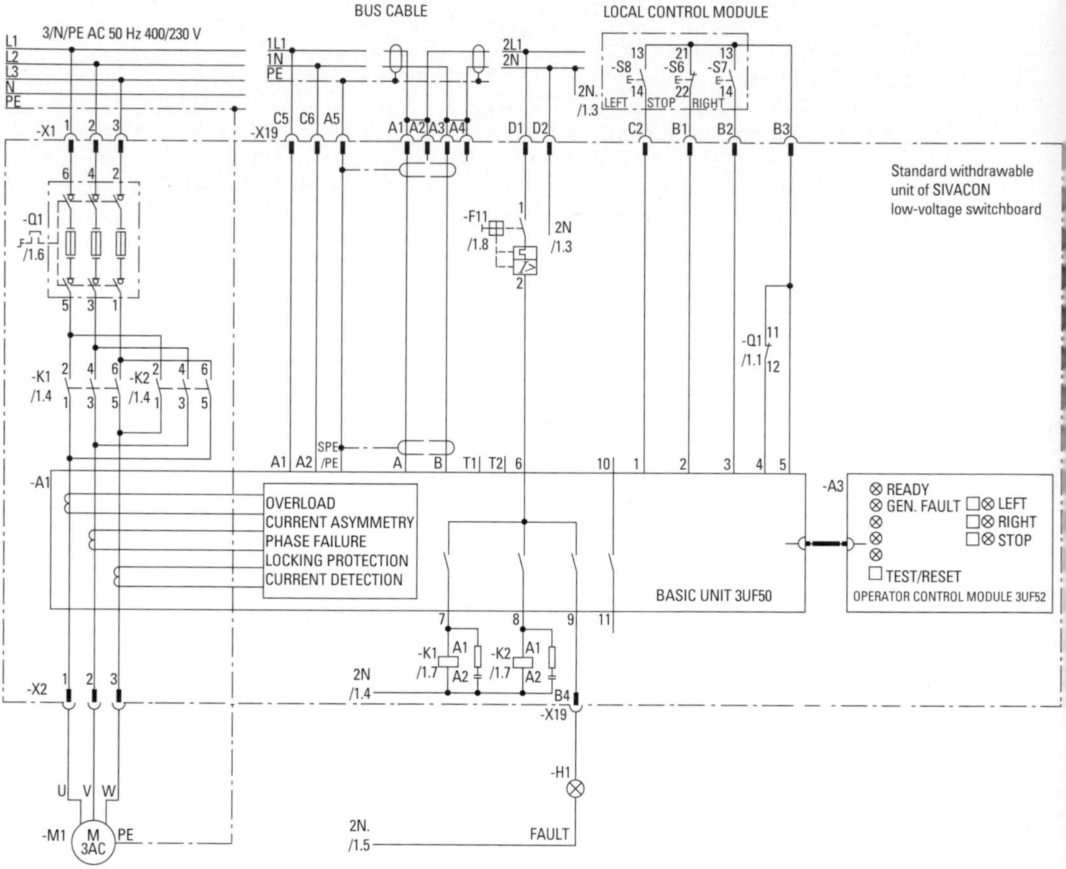

Fig. 1.4/33
Circuit for reversing direction of three-phase motors with communication-capable SIMOCODE-DP motor protection and control device

Retrieval

All data can be retrieved:

- ▷ directly at the switchgear cabinet via the personal computer using the Win-SIMOCODE-DP/Professional software,
- ▷ via the PROFIBUS-DP: the data must be preprocessed by the application program and then passed on to a suitable operator interface, for example,
- ▷ via the PROFIBUS-DP/DPV1 using a communications processor installed in the PC/PG together with the Win-SIMOCODE-DP/Professional software.

Data communication via PROFIBUS-DP

Application program

The communications processor (CP/IM) installed in the programmable controller (AG) is responsible for managing the SIMOCODE-DP systems connected to the PROFIBUS-DP and forms the interface to the application program. Further processing of the SIMOCODE-DP data, i.e. integration in cross-plant control and subsequent preprocessing, is carried out by the application program.

Data communication

The PROFIBUS-DP has a number of different data channels to ensure short transfer times. As a result, four bytes are continuously sent from the SIMOCODE-DP system to the automation level and back again. The 20-byte diagnostic data is only collected from the automation level if an appropriate request is made by the SIMOCODE-DP system, i.e. the diagnostic channel is event-driven. The 213-byte parameter data is sent to the SIMOCODE-DP system when the programmable controller is started. In addition to cyclic data communication, all parameter data, control commands, operating, diagnostic, and statistical data can also be acyclically read/written from/to the SIMOCODE-DP system via the PROFIBUS-DP by means of a class 2 PROFIBUS-DP communications processor installed in a PC/PG, for example.

Fig. 1.4/34 illustrates the basic structure of the automatic control level with the SIMOCODE-DP system and the PROFIBUS-DP.

Parameterization

Parameterization The SIMOCODE-DP system can be easily integrated in a PROFIBUS-DP network using the parameterization and service program COM PROFIBUS SIMATIC S5 or STEP 7 SIMATIC S7.

The COM PROFIBUS parameterization and service program can be called up via a PC, for example. The master system is opened via File "NEW" after which the individual slaves, e.g. a SIMOCODE-DP system, are assigned. Double-clicking the SIMOCODE-DP icon calls up the parameter list in which all feeder-specific parameters can be set.

The address and baud rate for the SIMOCODE-DP system are always set via the system interface using the Win-SIMOCODE-DP software. The same software can be used to set all the parameters via the system interface or on-line via the PROFIBUS-DP (see Fig. 1.4/34).

Other functions

Autonomous operation

Autonomous operation If system faults occur along the communications channel, the SIMOCODE-DP system automatically switches to manual mode. Depending on the selected parameters, either the load feeder is disconnected or the operating state is retained. All subsequent control operations are performed manually.

Standard function modules

Standard function modules The SIMOCODE-DP system provides the user with a number of standard function modules. Standard function modules are self-contained units and allow, for example, time-graded restarting of drive mechanisms following a system failure. The standard function modules operate independently of the selected control function and their use is optional. They also make it possible to reduce the workload of the automation/process control system and to assign functions to the distributed, autonomous SIMOCODE-DP system.

Test function

Test function The combined Test/Reset button can be used to test whether the SIMOCODE-DP system is functioning correctly. If the button is pressed for less than one second, a lamp test is carried out. If it is pressed for up to two

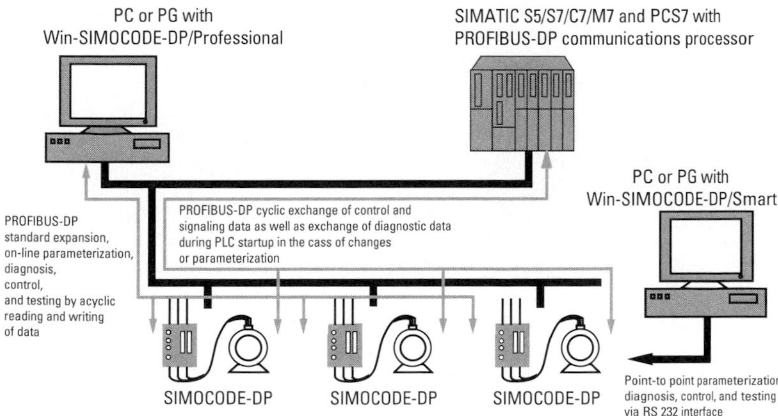

Fig. 1.4/34 Data communication via PROFIBUS-DP

seconds, the device hardware, current detection function, and the thermistor or ground-fault input are tested. By pressing the button for up to five seconds, the current transformers, load, and microprocessor are tested without first having to disconnect the motor feeder. If the button is pressed for longer than five seconds, the output relay is de-energized and the complete SIMOCODE-DP system is checked. If the main circuit is de-energized, the current transformers and load are not tested.

Self-monitoring

Self-monitoring

The SIMOCODE-DP processor is monitored continuously. If a fault occurs, SIMOCODE-DP switches the load to the parameterized safe state ("Off" or "Retain operating state" – monostable or bistable behavior of the output relay).

Tripping behavior of auxiliary contacts in event of voltage failure

Tripping behavior of auxiliary contacts

The tripping behavior of the auxiliary contacts in the event of a supply voltage failure differs depending on the version of the SIMOCODE-DP system:

▷ If the control supply voltage fails, the auxiliary contacts assume the "tripped" position and can be controlled again when the voltage supply is restored;
A brief voltage failure ≤ 200 ms does not cause the state of the auxiliary contacts to change. These SIMOCODE-DP devices are used in systems in which the control supply voltage is not monitored separately.

▷ If the control supply voltage fails, the state of the auxiliary contacts does not change. The auxiliary contacts only switch in the event of overloads, provided that the supply voltage is available. These SIMOCODE-DP devices are used in systems in which the control supply voltage is monitored separately.

Straight-through design

Straight-through current transformers

In devices with a current setting range ≤ 100 A, the main circuits do not have to be connected to the overload relay, but instead simply pass through the current transformers integrated in the compact device. This prevents any power loss at the contact resistors of the terminal connections. The bushing openings are dimensioned in such a way that they can accommodate all standard types of insulated copper conductor (10 mm diameter ≤ 25 A current setting and 15 mm ≤ 100 A).

Harmonic-insensitive current detection

Harmonic-insensitive current detection

The SIMOCODE-DP system can protect motors in outgoing feeders with soft-starters as current detection is insensitive to harmonics.

Temperature range

Temperature range

The SIMOCODE-DP system is designed to be used at ambient temperatures of between −25 °C and + 60 °C.

Communication-capable switching devices in SIVACON low-voltage switchboards

Withdrawable-unit design

The communication-capable switching devices are intended primarily for use in "withdrawable-unit" low-voltage switchboards (see Section 1.4.2.2). Together, these devices represent an ideal combination, provide maximum reliability and availability, and support a range of monitoring and diagnostic functions.

Installing 3WN6 circuit-breakers with DP/3WN6 PROFIBUS-DP interface module

3WN6 circuit-breakers

3WN6 circuit-breakers are installed in the SIVACON low-voltage switchboard in the customary manner (see Section 1.4.2.2).

The DP/3WN6 PROFIBUS-DP interface module is located in the auxiliary equipment and measuring compartment of the circuit-breaker cubicle.

Control and display elements

All of the circuit-breaker control and display elements are located on the control console of the circuit-breaker and can be accessed from the front of the switchboard. In the case of the overcurrent releases, which are integrated in the circuit-breaker control console, a digital display indicating all three phase conductor currents or the phase conductor carrying the highest load is also available in addition to the LEDs to indicate the reason why the circuit-breaker trips. This means that the ammeters previously incorporated in the instrument panel of the circuit-breaker cubicle are no longer necessary.

Installing the SIMOCODE-DP system

SIMOCODE-DP system

If the SIMOCODE-DP system is used, all conventional control and display elements required for the motor feeder, e.g. pushbuttons, LEDs, ammeters, and elapsed-hour meters, are no longer necessary as they are already integrated in the SIMOCODE-DP system.

The basic unit, expansion and operator control module are interconnected with shielded cables and can, therefore, be installed at physically separate locations. It is also possible to install the basic unit and expansion module side by side and to connect them using a shielded plug connector. With this configuration, the operator control module of each motor feeder is installed in the front panel of the switchboard and can, therefore, be operated from the front. The purpose of this is to allow control commands to be entered, the motor feeder to be parameterized, and switchboard data collected by the SIMOCODE-DP system to be displayed with the switchboard door closed.

Bus installation in switchgear cubicle

Two-wire bus system

The two-wire bus system, the PROFIBUS-DP for the communication-capable switching devices is connected in each switchgear cubicle in accordance with the principle of permanent wiring, i.e.:

- ▷ in the circuit-breaker cubicle at the control terminals in the auxiliary equipment compartment,
- ▷ in the motor control panel at each of the control connectors in the withdrawable-unit compartments.

Fig. 1.4/35 illustrates how the bus cable is installed in the switchgear cubicle and how the withdrawable units of a motor control panel are connected.

If this wiring and connection method is used, the PROFIBUS-DP connection will not have to be interrupted and the downstream motor feeders will not have to be disconnected from the PROFIBUS-DP should a communications device or withdrawable unit have to be replaced.

Shielding

Bus cable shielding

If the bus cable is shielded, no faults will occur while data is being exchanged between programmable controllers, the connected SIMOCODE-DP systems, and the DP/3WN6 PROFIBUS-DP interface modules for the 3WN6 circuit-breaker, during normal operation or in the event of a short circuit in the low-voltage switchboard. A wiring compartment which is separate from the busbar is used for cubicle-to-cubicle bus installation in order to eliminate any effects caused by arcing faults.

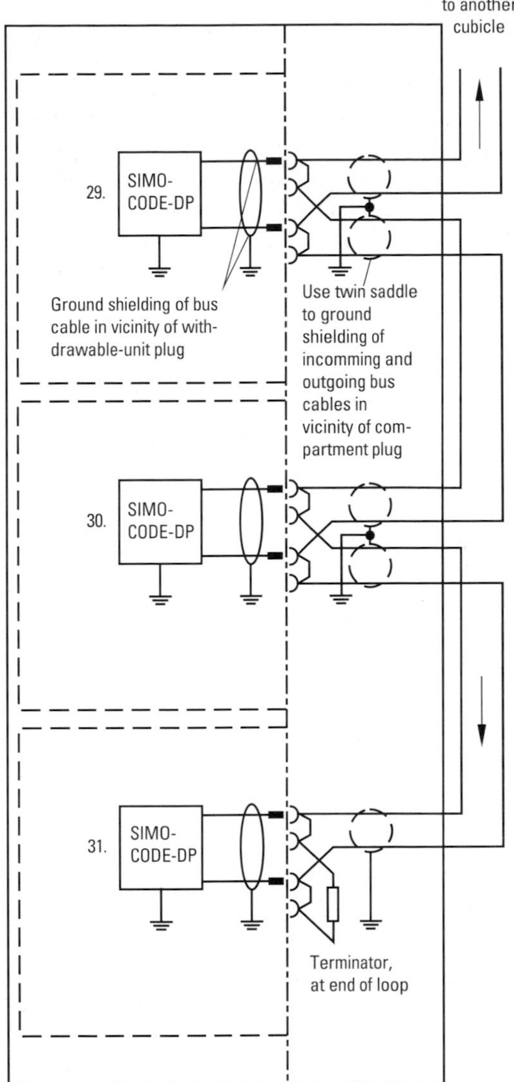

Fig. 1.4/35
Diagram showing installation of bus cable in switchgear cubicle and connections to control connectors of withdrawable units

Advantages of using communication-capable switching devices

Advantages

Communication-capable switching devices offer many different advantages with regard to planning, commissioning, and operating low-voltage switchboards:

- ▷ simplified planning and shorter delivery times because a detailed protection and control concept is no longer required when placing orders for low-voltage switchboards (adaptation by means of parameterization, not by replacing devices),
- ▷ less documentation thanks to fixed-programmed, yet freely selectable switching operations and the use of standardized circuit diagrams,
- ▷ short installation times on site, fewer potential sources of error because the complex parallel control wiring and terminal boards are no longer required on account of the two-wire PROFIBUS-DP cable,
- ▷ less time and expense involved in commissioning as SIMOCOD-DP system can be parameterized centrally,
- ▷ simplified troubleshooting using the detailed diagnostic data,
- ▷ extensive process monitoring with the aid of detailed information sent to the control room,
- ▷ unnecessary downtimes can be avoided because specific preventive maintenance is possible using the statistical data,
- ▷ no replacement time relays or display, operator, alarm, and coupling relays have to be stocked as the functions performed by these are integrated in the SIMOCODE-DP system.

1.4.3 Box-type low-voltage distribution systems

Definition

Box-type low-voltage distribution systems consist of a range of system boxes with different, yet carefully-matched dimensions. They can be combined to form complex distribution boards which can be of any size and which can perform any desired function.

Components

The following components, which are matched to the individual box modules, enable fast and efficient assembly:

- ▷ fuse assemblies for all standard low-voltage switching devices and electrical installation elements (e.g. DIAZED fuse-base),
- ▷ mounting plates for any device configuration,
- ▷ busbar systems,
- ▷ extensive range of connection facilities and accessories.

Standards

All components of the distribution system comply with the following standards:

- ▷ EN 60 439-1 / IEC 60439-1 / VDE 0660 Part 500 “Type-tested and partially type-tested assemblies”,
- ▷ DIN VDE 0100 “Erection of power installations with rated voltages below 1000 V”,
- ▷ HD 625.1.S1 / IEC 60664-1 / DIN VDE 0110-1 “Insulation co-ordination for equipment within low-voltage systems”,
- ▷ EN 60 439-3 / IEC 60439-3 / DIN VDE 0660-504 “Particular requirements for low-voltage switchgear and controlgear assemblies intended to be installed in places where unskilled persons have access for their use – distribution boards”,
- ▷ DIN VDE 0603-1 / 0606 “Consumer units and meter panels AC 400 V”.

Testing

If type-tested components are used in order to assemble a low-voltage distribution board, the majority of the tests usually required are no longer necessary. Only a routine test has to be carried out after the distribution board has been fully assembled. This involves checking the wiring, electrical functions, insulation resistance, and verifying the required protective measures in accordance with EN 60 439-1 / IEC 60439-1 / DIN VDE 0660-500.

Variable configurations

An important advantage of box-type low-voltage distribution systems is that they can be configured in accordance with specific structural conditions (box size and arrangement can be varied, e.g. optimum utilization of the “wasted” space under a staircase).

System boxes which have been adapted to the mounting depth of the installation and switching devices used means compact units with minimum space requirements. This results in lower construction costs (“converted space”). Furthermore, useful living space or space for industrial purposes is made available.

Simple expansion of the distribution board by adding new box modules also means that space for subsequent installation does not have to be provided which results in lower capital outlay.

Enclosure material

Depending on the prevailing requirements and the service environment, box-type distribution systems are made of insulating material or sheet steel (also stainless steel).

The following low-voltage distribution systems are described below:

- ▷ 8HP insulation-enclosed distribution system,
- ▷ 8HU sheet-steel-enclosed distribution system.

1.4.3.1 8HP insulation-enclosed distribution system

Application

The type-tested (TTA) 8HP insulation-enclosed distribution system is used for low-voltage main and sub-distribution boards in industrial and residential/functional buildings and as a carrier system for open and closed-loop control systems.

The high degree of protection – IP 65 (special version IP 66) – means that the systems can be used in damp and dusty environments, e.g. on board ships, or as worksite distribution boards, e.g. in steel works or quarries. Owing to its resistance to corrosive atmospheres, the system is predestined for use in the chemical industry, in papermills, or sewage treatment plants.

Since the distribution boards provide a high level of protection against shock currents (class 2 equipment), they can be safely used by personnel who have not received specialist training.

Construction

The modular system (Figs. 1.4/36 to 1.4/38) features five basic modules (box sizes). Based on box size 2, they have a ground area ratio of 0.5 : 1 : 1.5 : 2 : 4 (see Fig. 1.4/38).

This permits the compact assembly of contiguous complex distribution and control systems.

The different cover heights (standard 147 mm) and intermediate frames allow the mounting depth to be optimized to the built-in devices used.

The side panels of the boxes have removable knockouts for:

- ▷ the connecting flanges so that two system boxes can be permanently connected both to each other and in the distribution board assembly,
- ▷ cable glands (heavy-gauge threaded connections).

Transparent and non-transparent covers are available for all distribution boxes.

Advantages

The design of the insulation-enclosed distribution system provides the user with a range of technical advantages with respect to distribution systems for building installation and industrial power distribution:

- ▷ a high level of operator protection by means of total insulation,
- ▷ can be used in damp and dusty environments thanks to a high degree of protection (IP 65, special versions with IP 66),
- ▷ corrosion and pollutant-resistant,
- ▷ high degree of mechanical strength thanks to high-quality materials (e.g. glass-fiber-reinforced polyester),
- ▷ resistance to heat and UV rays,
- ▷ halogen-free (therefore no consequential damage resulting from the effects of fires that occur outside the system),
- ▷ flame-retardant, self-extinguishing,
- ▷ suitable for export to the USA (UL approval of insulation encapsulation),
- ▷ aseismic design,
- ▷ tested for use in civil defense installations.

Usage of 8HP insulation-enclosed distribution boards is facilitated by:

- ▷ simple planning by the customer (see Section 1.4.6),
- ▷ lower-cost storage of spare parts thanks to the modular system design,
- ▷ efficient assembly and simple finish of the system boxes,
- ▷ no maintenance required (reduced costs),
- ▷ "just-in-time" due to short delivery times.

Constructional advantages

Handling of the distribution units and the components of the system when the system is in operation, when the distribution boards are being assembled, and when the boards are connected on site, is facilitated by a series of constructional features, three of which will be described in more detail.

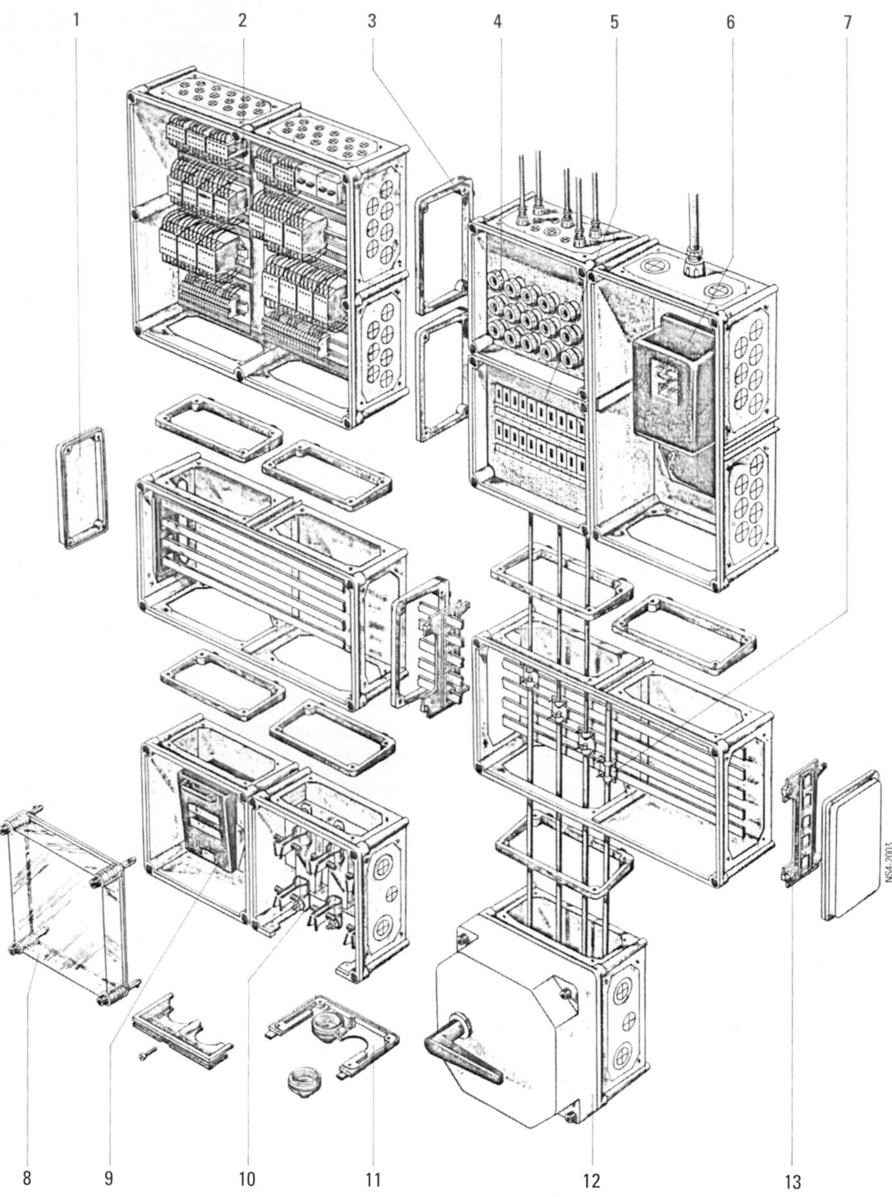

1 Cover plate
2 Box with control devices
3 Gasket
4 DIAZED fuse box
5 Snap-on devices
6 Meter box
7 Busbar run components
8 Cover
9 L.v. h.b.c. fuse switch-disconnector
10 L.v. h.b.c. fuse-bases
11 Cable-entry plate
12 Switch-disconnector
13 Busbar holder

Fig. 1.4/36 Construction of an 8HP-type low-voltage distribution board

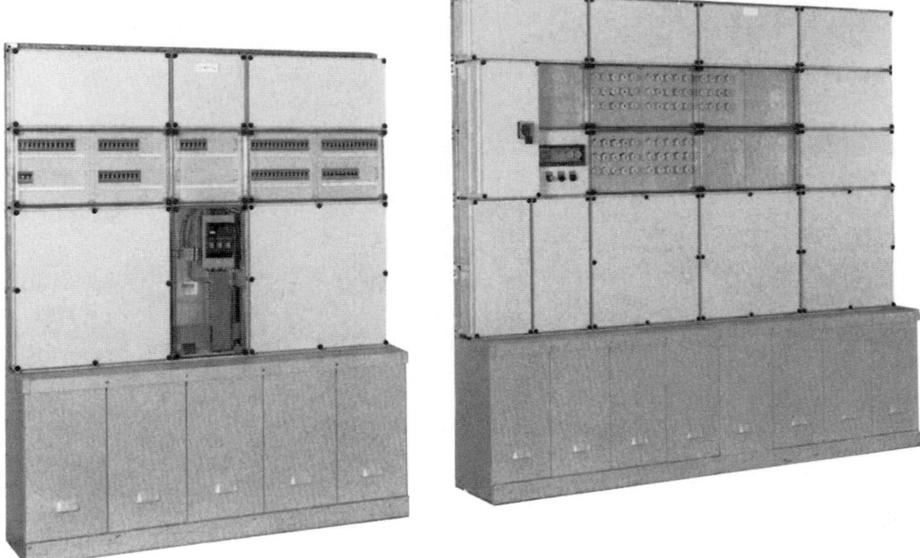

Fig. 1.4/37 a Insulation-enclosed distribution board with cable compartment covers

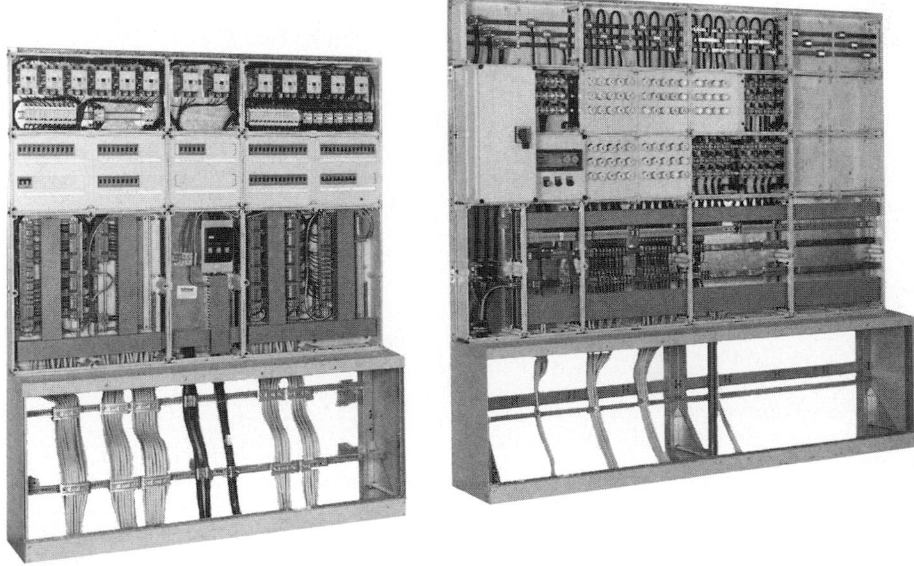

Fig. 1.4/37 b Insulation-enclosed distribution board (with covers removed)

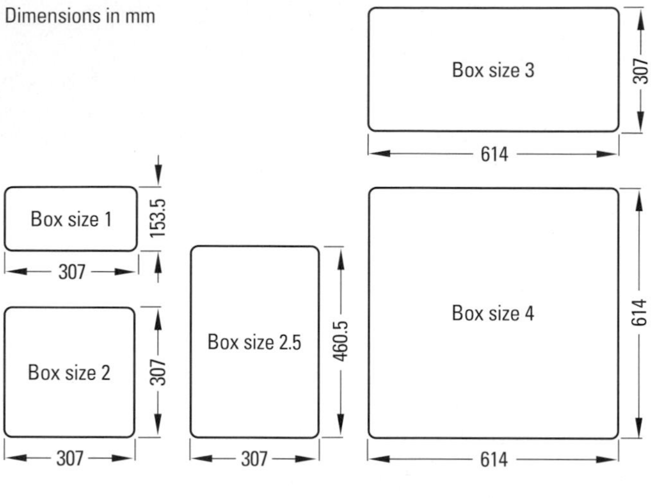

Fig. 1.4/38 Basic modules of the 8HP system (dimensions)

Securing covers

All of the covers have sealable quick-release locks, which are closed by applying thumb pressure, but can only be opened using a tool such as a screwdriver (Fig. 1.4/39). They can be replaced by manually-operated locks. An indicator clearly shows whether the cover has been properly secured (to ensure the degree of protection).

It is also possible to convert a standard cover into a hinged cover by inserting two hinge-screws.

Finish

The use of glass-fiber-reinforced polyester for the bases of the boxes and for the non-transparent covers and polycarbonate for transparent covers avoids any problems when installing the equipment. Holes and openings for manually-operated mechanisms, indicator lights, and flaps can be provided without any difficulty as standard metal machining tools, e. g. twist drills, hole saws, pad saws, or hole cutters, can be used.

Cable connection

In order to facilitate the connection of cables on site, a split cable-entry plate, which can be fitted to any of the boxes in the system (Fig. 1.4/40), has been developed for cables with large cross sections. All cables can thus be installed quickly and inexpensively from the front; cable glands are no longer required.

Cables fed in from below can be concealed by means of a cable-compartment cover (see Fig. 1.4/37).

Outdoor installation

The 8HP distribution system can also be installed outdoors. In order to protect the system against direct solar radiation (excessive heating), rain, and snow, it should be installed at a suitably sheltered location or be fitted with a special canopy.

Supply program

The system can be supplied in three different ways:

1) Complete distribution board supplied by the manufacturer, ready for connection,

2) Disassembled distribution board components for self-assembly by the user,

3) Distribution units for individual assembly, e. g. fuse boxes, boxes containing circuit-breakers, boxes for control systems.

The distribution units named in point 3) are available as boxes with:

▷ mounting plates, e. g. for control systems with the Siemens mini programmable controller logo,
▷ snap-on DIN-rail-mounted devices, e. g. miniature circuit-breakers,
▷ DIAZED and NEOZED fuses,
▷ l. v. h. b. c. fuses (NH00 to NH3),
▷ fuse switch-disconnectors ($\leq$ 630 A),
▷ switch-disconnectors $\leq$ 800 A and switch-disconnectors with fuses $\leq$ 250 A,
▷ load transfer switches $\leq$ 630 A and automatic synchronizers $\leq$ 1000 A,
▷ circuit-breakers $\leq$ 630 A.

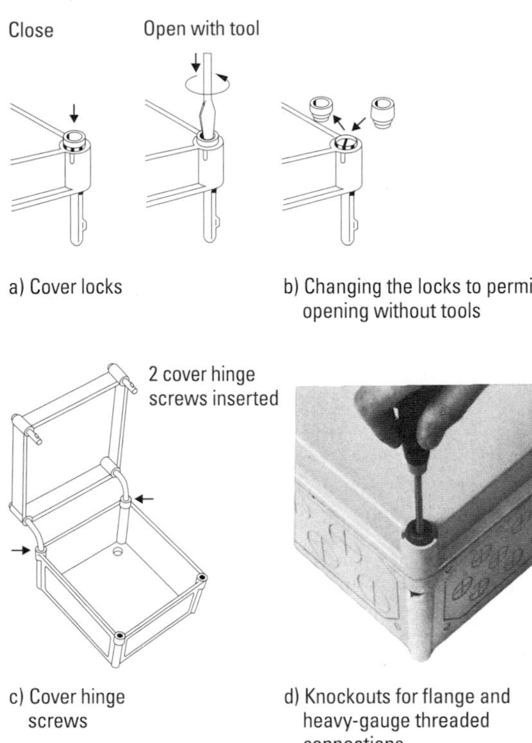

Fig. 1.4/39
Alternative cover fastenings and knockouts at the sides of the boxes

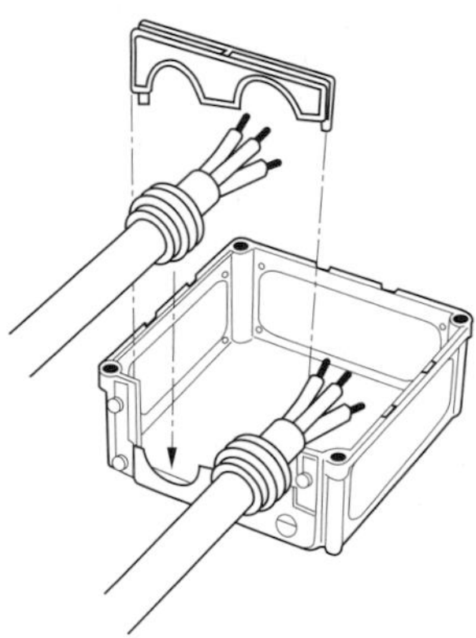

Fig. 1.4/40 Split cable-entry plate

▷ power-factor correction capacitors,
▷ circuit-breakers for transformer infeeds ≤ 1250 kVA.

Technical data

The technical data of the 8HP distribution system together with the rated currents and short-circuit strength of the busbars can be seen in Table 1.4/4.

1.4.3.2 8HU sheet-steel-enclosed distribution system

Application

8HU sheet-steel-enclosed distribution boards are particularly suitable for applications in mechanically hazardous locations or in "hostile" operating areas since their boxes are made of 2.5 mm thick sheet steel.

The sheet-steel boxes can be installed individually or as part of distribution unit assemblies.

The 8HU distribution system uses the same fuse assemblies as those of the 8HP distribution system. Mounting plates, mounting rails, and busbars ≤ 630 A are available for installing freely configured equipment assemblies.

As a result, the 8HU distribution system can be used as a main and sub-distribution board as well as a control system.

Construction

Six boxes (Fig. 1.4/41) with two different heights, but with the same width grading, form the basic modules of the system. The three 248-mm high boxes can, therefore, also be assembled edgewise with the three 315-mm high boxes. The standard mounting depth of the boxes is 150 mm. The two largest boxes are also available with a depth of 250 mm for larger devices (e. g. incoming-feeder disconnectors).

Assembly

The boxes have flange openings on all four sides which are used to join the boxes together. Adapter flanges can be used to join small flange openings to large flange openings. Gaskets, which ensure that protection is provided to IP 65, must be fitted between the assembled boxes.

Unused flange openings must be sealed with covers.

Cable entries

The side panels of the boxes or the connecting flange must be drilled to provide holes for cable entries. Pre-drilled connecting flanges with between one and four openings for cables with outer diameters of 4 mm to 19 mm are available. Cables with outer dia-

Table 1.4/4
Technical data for the 8HP distribution system together with the rated currents and short-circuit strength of the busbars

Rated voltage		AC 690 V, DC 600 V
Insulation group I	To HD 625.1 S1 / EN 60664-1 / DIN VDE 0110-1 und -20	600 ≤ CTI [1)], built-in devices are subject to specifications from manufacturer
Impulse voltage withstand level/pollution severity		8 kV/3
Minimum clearances Minimum creepage distances		8 mm 12.5 mm
Rated current	Busbars Built-in devices Infeed	250 A, 400 A, 630 A, 1000 A up to 2000 A up to 2000 A
Degree of protection	DIN 40050/ EN 60529/IEC 529	IP 65 IP 66 as special version for shipbuilding industry
Color	Encapsulation elements Transparent cover Cable compartment cover	RAL 7035, light gray Colorless RAL 7035, light gray
Ambient temperature		−40 °C to +55 °C

[1)] Comparative Tracking Index; guide value for tracking resistance to IEC 0112 and DIN VDE 0303-1

Rated busbar currents/busbar dimensions and infeeds

Rated busbar current	Busbar dimensions		Infeed	
			unilateral	central
	Phase conductor L1, L2, L3	N and PE conductors		
A	mm	mm	A	A
250	12 × 5	12 × 5	250	400
400	20 × 8	20 × 8	400	800
630	2 × 20 × 8	20 × 8	630	1000
1000	2 × 30 × 10	30 × 10	1000	1805

Short-circuit strength of busbars

Rated busbar current	Distance between busbar holders			
	307 mm		614 mm	
	I_{th} (1 s)	i_{pk}	I_{th} (1 s)	i_{pk}
A	kA	kA	kA	kA
250	10	40	10	40
400	20	70	20	70
630	30	70	30	70
1000	40	80	30	60

meters of between 37 mm and 60 mm can be fed through the connecting flanges with cable entries or through cable glands.

Covers

The covers are attached using four M6 hexagon bolts for box widths ≤ 375 mm and six bolts for box widths of 500 mm. They can also be fitted with two or three hinges along one of their longest sides. Boxes with hinged covers can be fitted with lockable knob-type quick-release locks instead of threaded joints (Fig. 1.4/42). The degree of protection, however, is then reduced to IP 54.

PE conductor connection

All covers must be connected to the boxes with a flexible equipment-grounding (PE) conductor via a latching-type connector. Each box contains PE terminal studs (identified accordingly) which are provided to permit equipment-grounding connections between boxes.

Two C-profiles (20 mm × 8 mm) are attached to each box base to allow equipment to be secured. The mounting plates made of sendzimir-galvanized sheet-steel (which belong to the system program), mounting rails, or fuse assemblies can be attached to these profiles using four M5 non-slip sliding nuts.

Securing built-in equipment

Installation components

The installation components available include:

▷ Components for busbar runs ≤ 630 A
▷ Fuse assemblies with DIAZED, NEOZED, and l. v. h. b. c. fuses
▷ Fuse assemblies with l. v. h. b. c. fuse switch-disconnectors ≤ 400 A
▷ Fuse assemblies for snap-on devices (e. g. circuit-breakers)
▷ 160 A to 630 A switch-disconnectors and 160 A to 400 A switch-disconnectors with fuse-links and lockable operating mechanisms (already installed in 8HU boxes).

General technical data

The general technical data for the 8HU sheet-steel-enclosed distribution system can be found in Table 1.4/5.

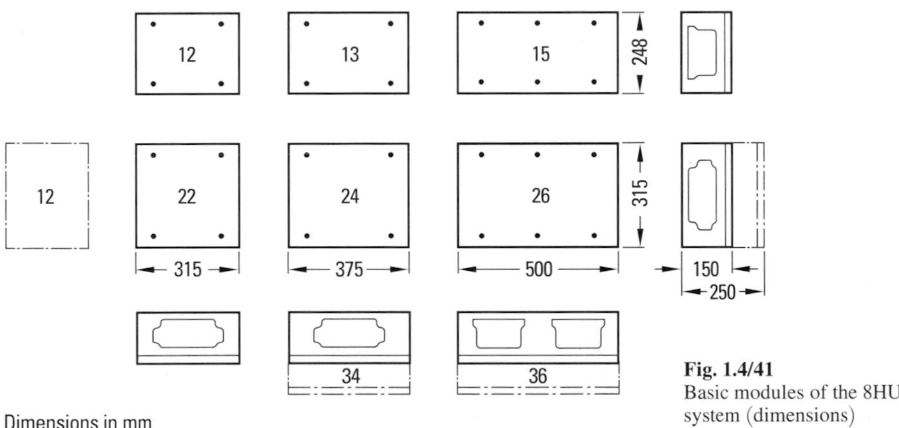

Fig. 1.4/41
Basic modules of the 8HU system (dimensions)

Table 1.4/5
General technical data for the 8HU-type steel-sheet-enclosed distribution system

Rated voltage		AC 690 V, DC 600 V
Rated current	Busbars	630 A
	Built-in devices	630 A
	Infeed	630 A
Short-circuit strength of busbars		40 kA to 70 kA, depending on version (peak withstand current I_{pk})
Material		Welded sheet steel (thickness 2.5 mm) (stainless steel version also available)
Surface		Weather-proof powder coating (polyester-resin-based)
Color		RAL 7031, blue-gray
Degree of protection	EN 60 529/IEC 60 529/ DIN VDE 0470-1	IP 65 (IP 54 with knob-type quick-release locks)

Fig. 1.4/42
8HU-type box, size 13, with hinged cover

1.4.4 Busbar trunking and busway systems (overhead busway systems)

1.4.4.1 8PL busbar trunking system and Sentron Busways with variable tap-offs

Design, standards

Busbar trunking systems are manufactured as TTAs (type-tested switchgear and controlgear assemblies) and PTTAs (partially type-tested switchgear and controlgear assemblies) in accordance with EN 60 439-1 / IEC 60439-1 / VDE 0660 Part 500 and EN 60439-2 / IEC 60439-2 / VDE 0660 Part 502. They consist of enclosed busbars, which have a large number of openings along their entire length to which tap-off units can be attached (Fig. 1.4/43).

The tap-off units can be simply plugged and locked into any of the tap-off openings to allow power to be drawn by the connected electrical loads. The busbars do not necessarily have to be de-energized beforehand.

Applications

Busbar trunking systems which are installed horizontally have a wide range of applications, e.g. on productions sites in the mechanical engineering and automobile industry, the timber, paper, and textile industry, the food industry, as well as in workshops, experimental and research laboratories, exhibition and trade-fair stands (Fig. 1.4/44).

If installed vertically, they can also be used as rising main busbars in high-rise office, residential, and functional buildings (Fig. 1.4/45).

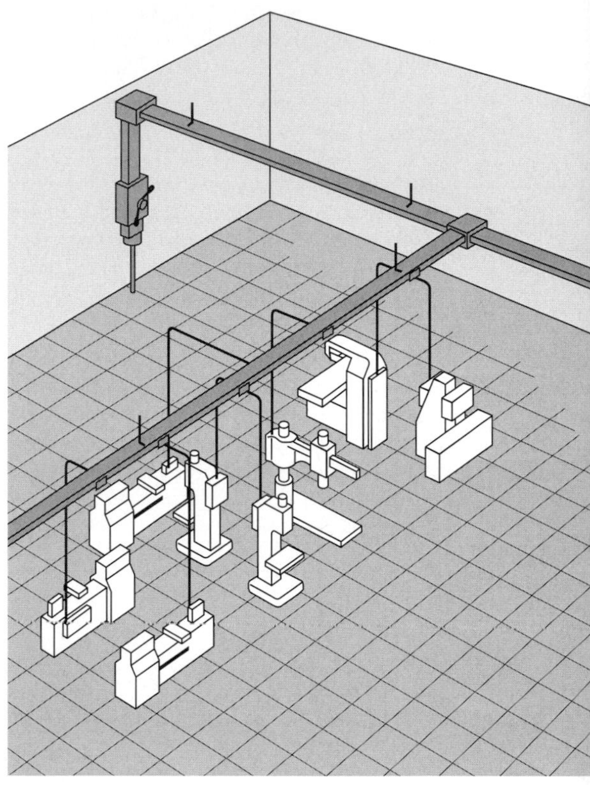

Fig. 1.4/43
Illustration showing an 8PL busbar trunking system with variable tap-offs

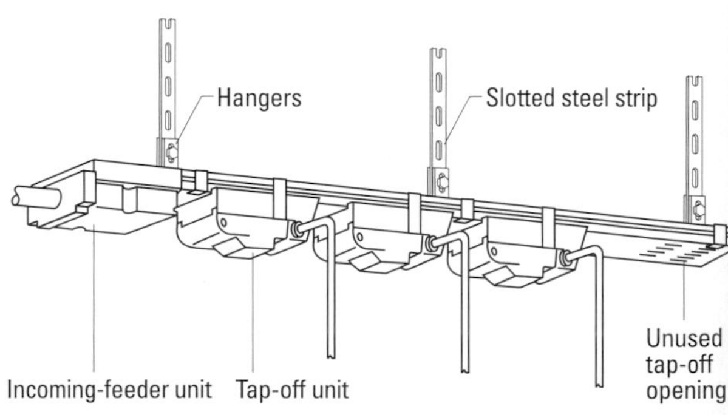

Fig. 1.4/44
Busbar trunking system installed horizontally

The busbar trunking systems are machines, motors, and electrical equipment – which are continuously regrouped due to changes in the production process – that are connected to the busbar trunking systems (Fig. 1.4/46).

Alternatively, busbar trunking systems can be used for supplying distribution boards directly.

Components

The range of busbar trunking systems includes type-tested trunking units, trunking unit connections, fittings and end covers, incoming-feeder units, adapter units, expansion units, and base supports as well as tap-off units in a wide variety of different versions and assembly variants including accessories such as mounting components for brackets and hangers, etc.

System-specific and tested fire-protection barriers complete the product range.

Copper, which is highly conductive and has ideal electrical properties, is normally used for the busbars and contacts. The contacts are silver or tin-plated.

Installation

The busbar components and tap-off units of the busbar trunking system are easy to assemble and install on site.

The trunking units, which are available in different lengths, can be assembled to any length using the trunking-unit connections. They can also be connected in parallel. Appropriate fittings are available to change the direction of the busbar run, e. g. angled, T-shaped, and cross units. Internal and external angles are formed using bend sections to allow the busbar to be adapted to the structural features and fittings of the room in which it is to be installed. Tap-off openings are spaced at regular intervals along the busbar (the distance between the openings varies depending on the busbar trunking system). This allows the tap-off units to be installed very closely together. The end of the busbar is sealed with an end cover.

Suspension

The busbar trunking system can be suspended directly from the ceiling or girders or attached directly to walls using simple or sliding mounting brackets. Alternatively, they can be suspended at a distance from the ceiling using hangers.

Luminaire and cable mounts

With the 8PL system, luminaire hangers or cable holders can be simply clipped on and secured using two screws (Fig. 1.4/47).

Incoming-feeder units

The incoming-feeder units are available for a wide range of different currents. The cables and/or conductors can be connected to a terminal strip.

Tap-off units

Tap-off units for rated currents between 16 A and 5000 A and different component configurations are available for connecting the electrical loads. The rated peak withstand current (I_{pk}) is 200 kA (max.) for Sentron Busway tap-off units. An interrupter is only fitted to de-energize the system when mounting and removing the tap-off units and to allow the fuse-links to be changed safely.

Fig. 1.4/45
Busbar trunking system installed vertically as a rising main busbar

Fig. 1.4/46 Sentron Busway trunking system

Table 1.4/6 Technical data for 8PL busbar trunking systems and Sentron Busways

Type	8PL	Sentron low-amp Busways 8PM1	Sentron Busways 8PM2	SMS Universal 5VE
Rated current (A)	160/250/400	100/160/200/250/315/400/500/630/800/1000	200/400/630/800/1000/1200/1350/1600/2000/2500/3150/4000/5000	16
Short-circuit strength of busbars (kA)	24–50	6–50	28–200	0.18 (1.5 mm^2) 0.30 (2.5 mm^2)
Rated voltage (V)	500 AC 600 DC	690 AC	690 AC	250/400 AC
Degree of protection	IP 43	IP 40–66	IP 40–66	IP 20–43
Protection against electric shock	Class 2	Class 1	Class 2	Class 3
Color of busway casing	RAL 7030	RAL 7042	RAL 7042	RAL 9011

Tap-off units consist of a base section with wiring compartment, a cable entry fitting with strain-relief clamp for the cable to be connected, and an integrated shockproof contact assembly for connection to the busbars of the trunking unit. Depending on the requirements of the load feeders, the tap-off units can be equipped with socket outlets, miniature circuit-breakers, r.c.c.b. devices, circuit-breakers, or other switching devices.

Fig. 1.4/46 shows a possible application for the Sentron Busway trunking system.

Advantages

The advantages of using busbar trunking systems are:

- ▷ unrestricted assembly and relocation of machines,
- ▷ clearly arranged cable runs,
- ▷ high connection density,
- ▷ high degree of operational reliability thanks to fuse-protected, interlocking tap-off units and fuse changes with the system de-energized,
- ▷ large connection compartments to facilitate wiring,
- ▷ safe mounting and removal of tap-off units using one hand.

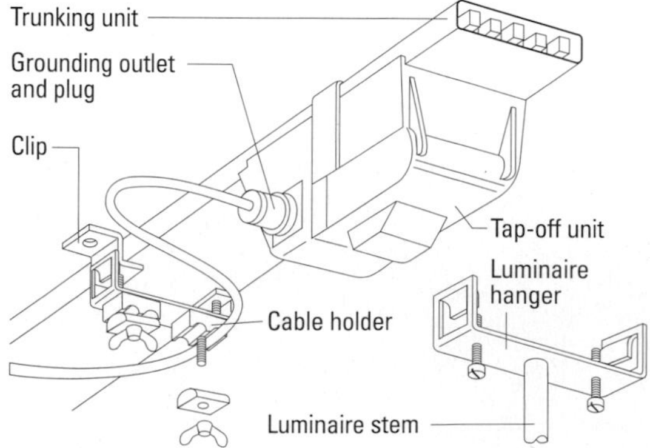

Fig. 1.4/47
Cable and luminaire hangers of the 8PL busbar trunking system

Technical data

Table 1.4/6 contains the technical data for the Siemens busbar trunking systems and Sentron Busways.

1.4.4.2 8PU busway system

Application

Busway systems are used as an alternative to parallel cables to establish electrical connections between transformers and low-voltage switchboards as well as between switchboards and other switchboards.

Enclosure

The 8PU sheet-steel-enclosed and resin-encapsulated busway systems are two easily interconnectable variants for indoor and outdoor use.

Degree of protection

The degrees of protection provided fulfill the most stringent requirements. The degree of protection for the sheet-steel-enclosed version is IP 42. This increases to IP 54 (in compliance with EN 60529 / IEC 60529 / DIN VDE 0470-1) if gaskets are used to seal the joints between the busbar units. The resinencapsulated version is protected to IP 54 for indoor applications or, with gaskets fitted in the joints between the busbar units, to IP 66 for outdoor use.

Busbars

The copper busbars of both versions (the number of conductors can be 3, $3^1/_2$, 4, or 5) are tightly sandwiched together and are electrically isolated from each other by means of five-layer insulating material and from the sheet-steel enclosure or resin encapsulation by means of three-layer insulating material.

Short-circuit strength

The design principle saves space and provides a high level of short-circuit strength and optimum stability.

By arranging the busbars next to each other in the sequence L1, L2, L3, N, PE, the magnetic fields generated during operation largely cancel each other out so that the reactance is reduced to a minimum and the voltage drop depends only on the ohmic resistance of the copper busbars.

Low voltage drop

As a result, the voltage drop is reduced to less than 2% of the applied operating voltage per 100 m of busbar.

Standards

The 8PU busway system is suitable for use in low-voltage systems $\leq$ 750 V AC, 50 to 60 Hz, and is designed for rated currents of 500 A to 5000 A. As a type-tested switchgear and controlgear assembly (TTA), it complies with the standards EN 60439-1 / IEC 60439-1 / VDE 0660 Part 500 and EN 60439-2 / IEC 60439-2 / VDE 0660 Part 502.

Components

A comprehensive range of components is available thus enabling the 8PU busway system to be used in transformer and switchboard rooms with very different structural features.

The component range includes:

▷ linear sections:
straight busbar sections, connecting bar sections, expansion units, connecting elements;

▷ elements for changing direction:
flat right-angle sections, on-edge right-angle sections, Z-sections, bends, T-bar sections;

▷ wall, ceiling, and floor mounting:
ceiling suspension units, wall-mounting, floor-mounting, fixing accessories and mounts (rigid, spring-loaded, or sliding);

▷ wall and ceiling openings:
wall and ceiling bushings, fire-protection barriers;

▷ for connection to transformers:
flange plates, laminated connection accessories, expansion straps;

▷ for connection to switchboards:
solid terminal brackets.

Detailed component descriptions, technical data, and dimensions together with planning information can be found in the Siemens Catalog I 2.36 "8PU Busway System".

Fig. 1.4/48 shows an example of a busbar connection between a transformer and low-voltage switchboard. Fig. 1.4/49 shows the components required to establish the connection.

1.4.5 Power and domestic distribution systems

Application and use

Power and domestic distribution systems can be divided into permanently-installed and portable distribution boards.

Permanently-installed power and domestic distribution systems include:

▷ small, wall-mounting, and floor-mounting distribution boards,
▷ modular distribution-board systems,
▷ meter and meter distribution cabinets.

Portable power and domestic distribution systems include:

▷ cable distribution cabinets.
▷ worksite distribution boards,
▷ fairground and campsite distribution boards,
▷ multi-outlet distribution boards.

Depending on their design, power and domestic distribution systems can be used as main and sub-distribution boards for supplying power to electrical loads, e. g. in residential and office buildings, in commercial and industrial plants, at markets and fairgrounds, as well as on construction sites and campsites.

Planning

The intended application must be known before power and domestic distribution systems can be planned and configured. A list of specifications, a block diagram, or a circuit diagram should be available.

Planning tools

The planning of Siemens power and domestic distribution boards (see Section 1.4.6) is facilitated considerably if the following tools are used:

▷ computer programs for planning and calculating, as well as for generating location and circuit diagrams (P.I.S.A.A, see Section 1.4.6),
▷ PC program "KUBS" (short-circuit current calculation, back-up protection, and selectivity).

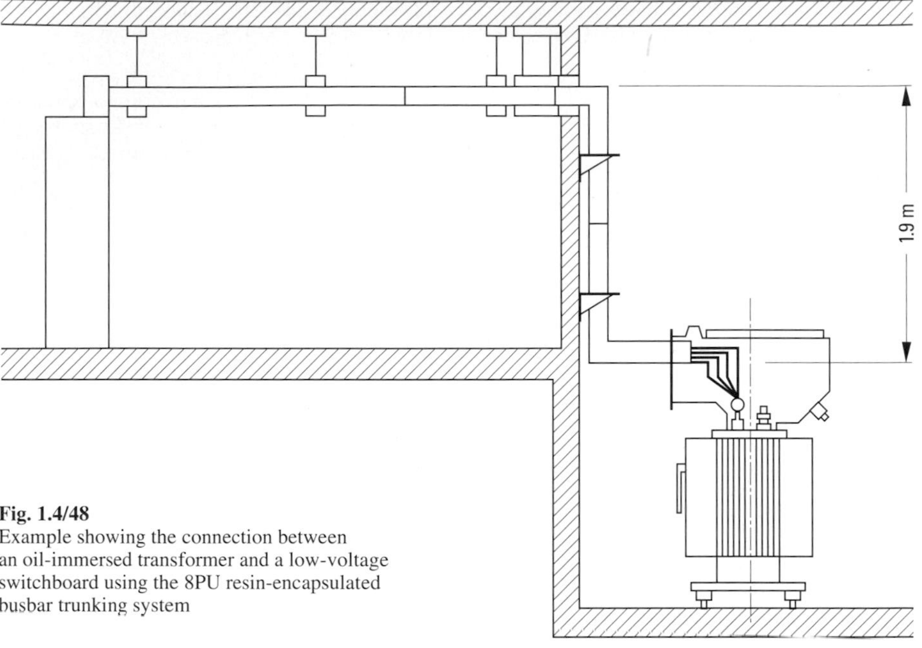

Fig. 1.4/48
Example showing the connection between an oil-immersed transformer and a low-voltage switchboard using the 8PU resin-encapsulated busbar trunking system

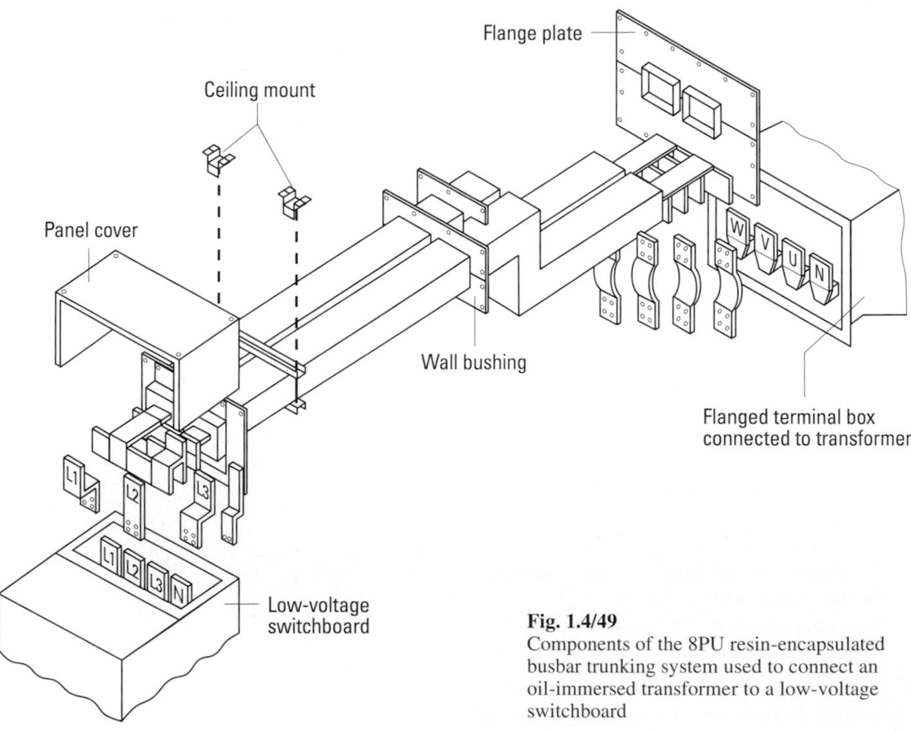

Fig. 1.4/49
Components of the 8PU resin-encapsulated busbar trunking system used to connect an oil-immersed transformer to a low-voltage switchboard

Standards

The relevant standards must be observed when planning Siemens power and domestic distribution boards, e. g.:

▷ DIN VDE 0603–1
"Consumer units and meter panels AC 400 V",

▷ EN 60439–1 / IEC 60439–1 / DIN VDE 0660–500
"Low-voltage switchgear and controlgear assemblies – Type-tested and partially type-tested assemblies",

▷ EN 60439–3 / IEC 60439–3 / DIN VDE 0660–504
"Low-voltage switchgear and controlgear assemblies – Particular requirements for low-voltage switchgear and controlgear assemblies intended to be installed in places where unskilled persons have access for their use – Distribution boards",

▷ EN 60439–4 / IEC 60439–4 / DIN VDE 0660–501
"Low-voltage switchgear and controlgear assemblies – Particular requirements for assemblies for construction sites",

▷ IEC 60364 / DIN VDE 0100
"Erection of power installations with rated voltages below 1000 V",

▷ HD 625.1 S1 / IEC 60664–1 / DIN VDE 0110–1
"Insulation co-ordination for equipment within low-voltage systems",

▷ EN 60204 / IEC 60204 / DIN VDE 0113
"Electrical equipment of machines",

▷ TAB
"Technische Anschlußbedingungen der örtlichen Energie-Versorgungs-Unternehmen (EVU)" (Technical supply conditions of the local public utility; applicable in Germany only),

▷ DIN 43871
"Installationskleinverteiler für Einbaugeräte bis 63 A" (Small distribution boards for built-in devices ≤ 63 A),

▷ DIN 43880
"Installationseinbaugeräte" (Installation devices).

Short-circuit strength

The short-circuit strength of the power and domestic distribution boards largely depends on the cross section of the busbars and the distance between the busbar holders of the installed busbar system (see Fig. 1.4/50).

Siemens power and domestic distribution systems

Table 1.4/7 provides an overview of the power and domestic distribution systems available from Siemens together with the associated technical data.

See Table 1.4/8 for information on the STRATUM distribution system.

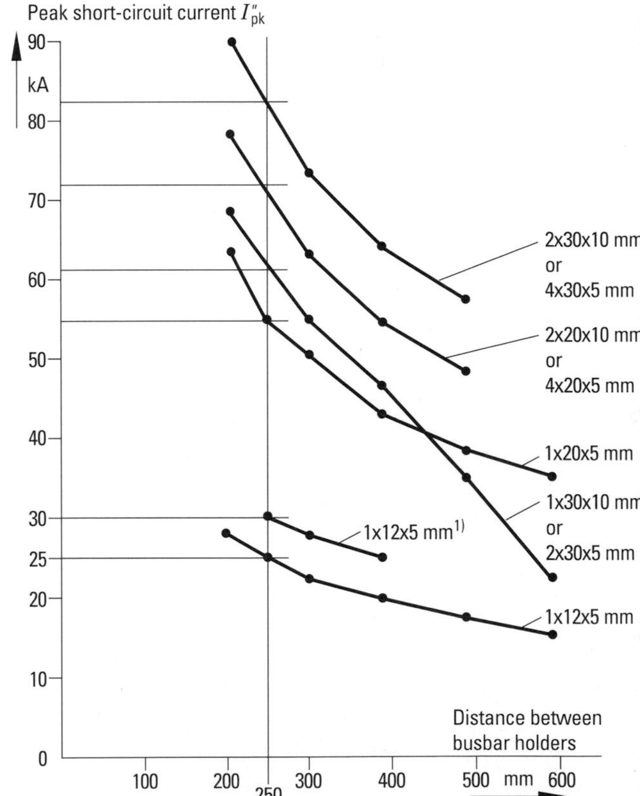

[1)] Reinforced busbar holder

Fig. 1.4/50
Short-circuit strength of the busbar systems from the 8GA SIKUS floor-mounting distribution board range as a function of the cross section of the busbars and the distance between the busbar holders

Table 1.4/7 Overview and technical data of Siemens power and domestic distribution systems SIMBOX, STAB, SIKUS, SIPRO

Fixed distribution boards

Designation	SIMBOX 63- Small distribution boards	STAB 160- Wall-mounting distribution boards	STAB 400- Wall-mounting distribution boards	STAB- Wall-mounting distribution boards Flush and surface-type	SIKUS 630- Floor-mounting distribution boards	SIKUS- Floor-mounting distribution boards	SIKUS 3200- Modular distribution-board system	STAB Universal- Wall-mounting distribution boards, Floor-mounting distribution boards	SIKUS Universal- Modular distribution-board system	SIPRO Wall-mounting distribution boards, Meter cabinets, Meter distribution cabinets
Type	8GB	8GE1	8GE2	8GD	8GE3	8GA	8GG	8GF	8GF	8GR
Compliance with:	DIN VDE 0603-1, DIN 43 880	EN 60439-1/-3, IEC 60439-1/-3, DIN VDE 0660-500/ -504	EN 60439-1/-3, IEC 60439-1/-3, DIN VDE 0660-500/ -504	EN 60439-1/-3, IEC 60439-1/-3, DIN VDE 0660-500/ -504	EN 60439-1, IEC 60439-1, DIN VDE 0660-500	EN 60439-1, IEC 60439-1, DIN VDE 0660-500	EN 60439-1, IEC 60439-1, DIN VDE 0660-500	NF, CEI; EN 60439-1/-3, IEC 60439-1/-3, DIN VDE 0660-500/ -504	NF, CEI; EN 60439-1, IEC 60439-1, DIN VDE 0660-500	EN 60439-1, IEC 60439-1, DIN VDE 0660-500, DIN VDE 0603-1
Application	Residential buildings Functional buildings	Residential buildings Functional buildings	Residential buildings Functional buildings Industrial buildings	Functional buildings Industrial buildings	Functional buildings Industrial buildings	Functional buildings Industrial buildings	Functional buildings Industrial buildings	Residential buildings Functional buildings Industrial buildings	Functional buildings Industrial buildings	Residential buildings Functional buildings
Rated current (A)	63	160	400	400	630	2000	3200	630	3150	400
Degree of protection	IP 30	IP 43	IP 43	IP 30, IP 41, IP 54	IP 55	IP 30, IP 41, IP 54	IP 30, IP 55	IP 30, IP 55	IP 30, IP 55	IP 31, IP 54
Protection against electric shock	Class 2	Class 2	Class 1 and 2	Class 1 and 2	Class 1 and 2	Class 1 and 2	Class 1	Class 1	Class 1	Class 2
Cabinet depth, ext. (mm)	Surface type 72.5 Surface type[1] 100 Flush-type[2] 70 Flush-type 97	125	216	136, 160, 220	225	250, 480	400, 600, 800, 1000	200, 235	400, 600, 800, 1000	210
Cabinet height, ext. (mm)	Surface-type 272, 397, 542, 667 Flush-type 311, 436, 581, 706	665, 815, 965, 1115	965, 1115, 1265, 1415	451, 750, 1125	2000	1605, 1975	1600, 1800, 2000, 2200	666, 866, 1066, 1266, 1466, 1866	1700, 1900, 2100	790, 940, 1090, 1240, 1390
Cabinet width, ext. (mm)	Surface-type 342 Flush-type 361	315, 565, 815, 1065	315, 565, 815, 1065, 1315	300, 550, 800, 1050, 1300	600, 850, 1100	350, 600, 850, 1100, 1350	400, 600, 800, 850, 1000, 1100	850, 1000, 1100	400, 700, 1000	290, 540, 790, 1040, 1290
Device tiers	1 to 4	4 to 7	6 to 9	3, 6, 8	12	12, 15	–	3 to 8	–	5 to 12
Panels	–	1 to 4	1 to 4	1 to 4	2 to 4	1 to 5	–	–	–	1 to 5

[1] Hood-type distribution board
[2] For installation equipment with depth of 55 mm

able 1.4/8 Overview and technical data of Siemens STRATUM permanently-installed distribution

esignation	STRATUM 200- Wall-mounting distribution boards, Plastic housing, Surface-type	STRATUM 200- Wall-mounting distribution boards, Metal housing, Surface-type Flush-type	STRATUM 400- Wall-mounting distribution boards, Metal housing, Surface-type Flush-type	STRATUM 601- Wall-mounting and floor-mounting distribution boards Metal housing	STRATUM 603- Wall-mounting and floor-mounting distribution boards Metal housing	STRATUM 3200- Modular distribution-board system, Metal housing
ompliance with:	EN 60439-3, BS EN 60439-3, IEC 60439-3	EN 60439-3, BS EN 60439-3, IEC 60439-3	EN 60439-1, BS 5486 Teil 13, IEC 60439-1	EN 60439-1, BS EN 60439-1, IEC 60439-1	BN 60439-1, BS EN 60439-1, IEC 60439-1	EN 60439-1, BS EN 60439-1, IEC 60439-1
pplication	Residential buildings Functional buildings	Residential buildings Functional buildings	Residential buildings Functional buildings Industrial buildings	Functional buildings Industrial buildings	Functional buildings Industrial buildings	Functional buildings Industrial buildings
ated current A)	100	100	200	630	1250	3200
egree of protection	IP 30	IP 30	IP 30, IP 54	IP 30	IP 30	IP 30, IP 55
rotection against lectric shock	Class 2	Class 1	Class 1	Class 1	Class 1	Class 1
abinet depth, ext. nm)	110	102, 117	117	172, 188	310	400, 600, 800, 1000
abinet height, ext. nm)	210, 232	257, 296	496, 696, 896, 1096	822, 500, 1422, 1722, 2022	2305	2000
abinet width, ext. nm)	220, 292, 370, 474	259, 359, 403	395	814	1000	400, 600, 800, 1000
eeders	4 to 16	6 to 18	4 to 24	–	5 to 18	–

1.4.5.1 Permanently-installed power and domestic distribution systems

pplication

SIMBOX 63 small distribution boards

Small distribution boards ≤ 63 A are suitable as sub-distribution boards in all domestic electrical installations. Owing to their small overall depth, they are used in both residential and functional buildings and are usually installed in the vicinity of the respective load centers. They are connected downstream of the meter cabinet and the main or storey distribution board.

Equipment with a depth of 55 to 70 mm conforming to DIN 43 880 and clip-on mountings for DIN rails (35 mm × 7.5 mm) conforming to DIN EN 50 022 can be installed in the units. The distance between the DIN rails (tier to tier) in the small distribution board is 125 mm. Since the equipment is dimensioned in pitch units, each DIN rail tier can be fitted with devices of up to 12 pitch units. The width of one pitch unit is 18 mm.

instabus EIB data busses, which enable communication between *instabus EIB* devices, can be inserted and secured in the DIN rails.

Special version

An extremely flat special version of the SIMBOX 63 small distribution board is available with an overall depth of just 70 mm. This can be mounted in recesses of standard single-brick cemented walls to DIN 105.

N-type equipment with an overall depth of 55 mm (e.g. residual-current-operated circuit-breakers, miniature circuit-breakers, MINIZED low-capacity fuse switch-disconnectors, and *instabus EIB* equipment) can be installed in this special version.

Basic program, dimensions

One to four-tier flush-type, surface-type, and hood-type distribution boards for brick and concrete walls and one to four-tier small distribution boards for cavity walls are available as Class 2 units (total insulation) depending on the area of application.

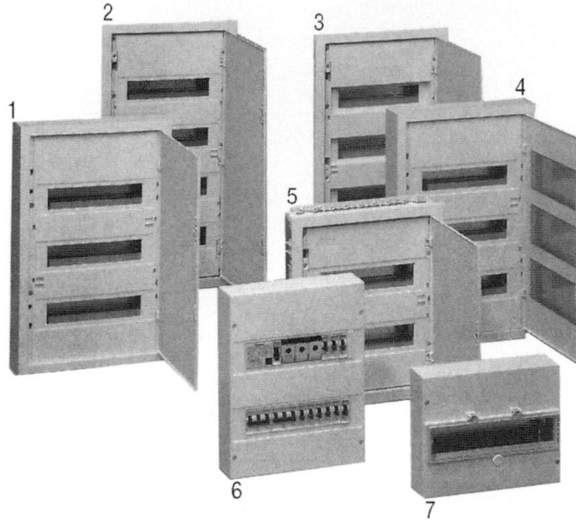

1 Small distribution boards, surface-type, three-tier, overall depth 72.5 mm
2 Small distribution boards, flush-type, four-tier, overall depth 70 mm
3 Small distribution boards, flush-type, four-tier, overall depth 97 mm
4 Small distribution boards, surface-type, three-tier, overall depth 100 mm with transparent windows in door
5 Cavity-wall distribution boards, two-tier, overall depth 70 mm
6 Hood-type distribution boards, two-tier, overall depth 100 mm
7 Hood-type distribution boards, single-tier, overall depth 100 mm with transparent door

Fig. 1.4/51
Selection of SIMBOX 63 small distribution boards from the basic program

Fig. 1.4/51 shows a number of these distribution boards. Table 1.4/7 indicates their various areas of application and dimensions.

Connection and wiring compartments

All SIMBOX 63 small distribution boards have large cable connection and wiring compartments. Cable entry is facilitated by special snap-locking sliding flanges with flexible plastic teeth (Fig. 1.4/52).

An N-conductor and a PE-conductor terminal with strain-relief clamp are provided for each circuit.

Large surfaces ensure sufficient heat dissipation for the installed equipment.

Housings, color

Test temperatures, thermal stability

The housings are made of impact-resistant polystyrene (color: RAL 9010, pure white). The flush-type units satisfy the glow-wire test with test temperatures of 650 °C and the surface-type units with test temperatures of 750 °C. The housings for cavity-wall installation are, however, made of flame-retardant polystyrene and are thus able to withstand a test temperature of 960 °C. All of the distribution boards are designed to be resistant to temperatures of 100 °C.

In order to increase the degree of mechanical protection, while at the same time retaining Class II protection, all small distribution boards (except the hood-type unit) are fitted with a sheet-steel door with baked enamel coating (color: RAL 9010, pure white). Every Siemens small distribution board is tested to DIN VDE 0603-1 and bears the VDE symbol.

Fig. 1.4/52
Cable entry in SIMBOX 63 small distribution boards

Flush-type version

The flush-type version consists of three components:

▷ wall-recess box with N-conductor and PE-conductor terminals,
▷ equipment rack with shockproof cover,
▷ trim frame with sheet-steel door and bolt lock.

Door and lock

The door can be rotated 180° to allow it to be hinged on either the right or left-hand side. The bolt lock can be easily replaced at a later stage by a socket lock. The spring-loaded hinges allow the door to be opened to an angle of 180°.

Adjusting equipment racks and trim frames

The equipment racks together with the trim frames and doors of wall-recess boxes which are not mounted horizontally can be adjusted at a later stage by means of the adjustable mounting slots.

The trim frame can be removed to allow walls to be papered without impairing the degree of shock protection.

All components can be installed quickly and easily by means of the ¼-turn, +/– quick-release screws.

Bushes, which allow wall-recess and cavity-wall boxes to be connected horizontally, make it possible to install several distribution boards side by side.

STAB and SIKUS distribution systems

Application

The STAB distribution systems 8GE1, 8GE2, and 8GD and the SIKUS distribution systems 8GE3 and 8GA listed in Table 1.4/7 are modular in design and are available in different sizes, depending on individual requirements and the respective application.

Special assemblies for Siemens switching devices, e.g. miniature circuit-breakers with a rated current of 0.5 A and circuit-breakers with a rated current of 2000 A and for a wide range of equipment assemblies, are available thus enabling the distribution systems to be configured in accordance with individual requirements.

Standards

All STAB and SIKUS distribution systems have been tested as "type-tested switchgear and controlgear assemblies (TTA)" in compliance with the standards EN 60 439-1/-3 / IEC 60439-1/-3, and DIN VDE 0660-500/ -504.

System components

Depending on the type of distribution board, they consist of the following system components:

▷ housing,
▷ jumpering (wiring) units,
▷ base frame,
▷ assemblies,
▷ busbars,
▷ accessories.

Individual system components

Individual system components are also available so that customer-specific distribution boards can be constructed. A wide range of different TTA configurations are possible even with just a few system components.

The well-conceived system design allows the following versions to be constructed in different sizes and with different degrees of protection and safety classes (Fig. 1.4/53):

▷ flush-type distribution boards,
▷ surface-type distribution boards,
▷ floor-mounting distribution boards.

Expanding distribution boards

The panels are suitable for mounting all types of assembly, busbar run, and installation equipment. Assemblies with keyhole mounts and DIN rails standardized to DIN EN 50 022 with a tier spacing of 125 mm or 150 mm are used. Individual parts such as terminal strips, extra-deep racks, mounting plates, panel covers are also available to allow low-voltage switchgear and equipment to be installed in accordance with customer requirements. Individual parts and assemblies are designed in such a way that they can be installed using only a few tools.

Degrees of protection, safety classes

Depending on the distribution system, the housings are available with degree of protection IP 30, IP 31, and IP 41 to IP 55 in accordance with EN 60529 / IEC 60529 / DIN VDE 0470-1 and with safety class 1 (grounding terminal) or 2 (total insulation).

Housings, color

All distribution-board housings with safety class 1 are made of galvanized, lacquered sheet steel (color: RAL 7035, light gray). The totally insulated housings also have a plastic lining and SIKUS floor-mounting distribution boards are made entirely of halogen-free plastic (polyurethane integral skin foam, color: RAL 9011, graphite black).

Doors and locks

All housing versions have sheet-steel doors with a twist or espagnolette lock with double-bit lock and three or four-point locking mechanism, which can also be fitted with different lock types at a later stage (e.g. for senate locks). Housings with safety class 2 provide total insulation for all lock variants.

A range of flanges is available to seal housing openings.

Fig. 1.4/53 STAB/SIKUS distribution system

Construction

Fig. 1.4/54 shows an example of a flush-type STAB wall-mounting distribution board (8GD) and surface-type STAB 160 wall-mounting distribution board (8GE1).

Fig. 1.4/55 shows the structure of the SIKUS floor-mounting distribution board 8GA. The SIKUS 630 floor-mounting distribution board 8GE3 has a similar structure.

Panel sizes and mounting dimensions

Figs. 1.4/56 and 1.4/57 illustrate the different possible internal structures with grid dimensions to DIN 43 870. The mounting dimensions comply with DIN 43 880.

Overall depths

The overall depths of the distribution boards must be selected in accordance with the switching devices and equipment to be installed as well as the required rated currents.

STAB Universal distribution boards

Application

STAB 8GF Universal distribution boards are used as main and sub-distribution boards in residential, functional, and industrial buildings.

Construction

STAB 8GF Universal distribution boards cater for the installation practices common to southern European countries, i.e. a large tier spacing of up to 200 mm between the DIN rails standardized to DIN EN 50 022.

Equipment conforming to DIN 43 880 with snap-on mountings and measuring up to 24 pitch units can be mounted on these DIN rails.

Horizontally-mounted circuit-breakers (MCCBs) with rated currents of up to 630 A

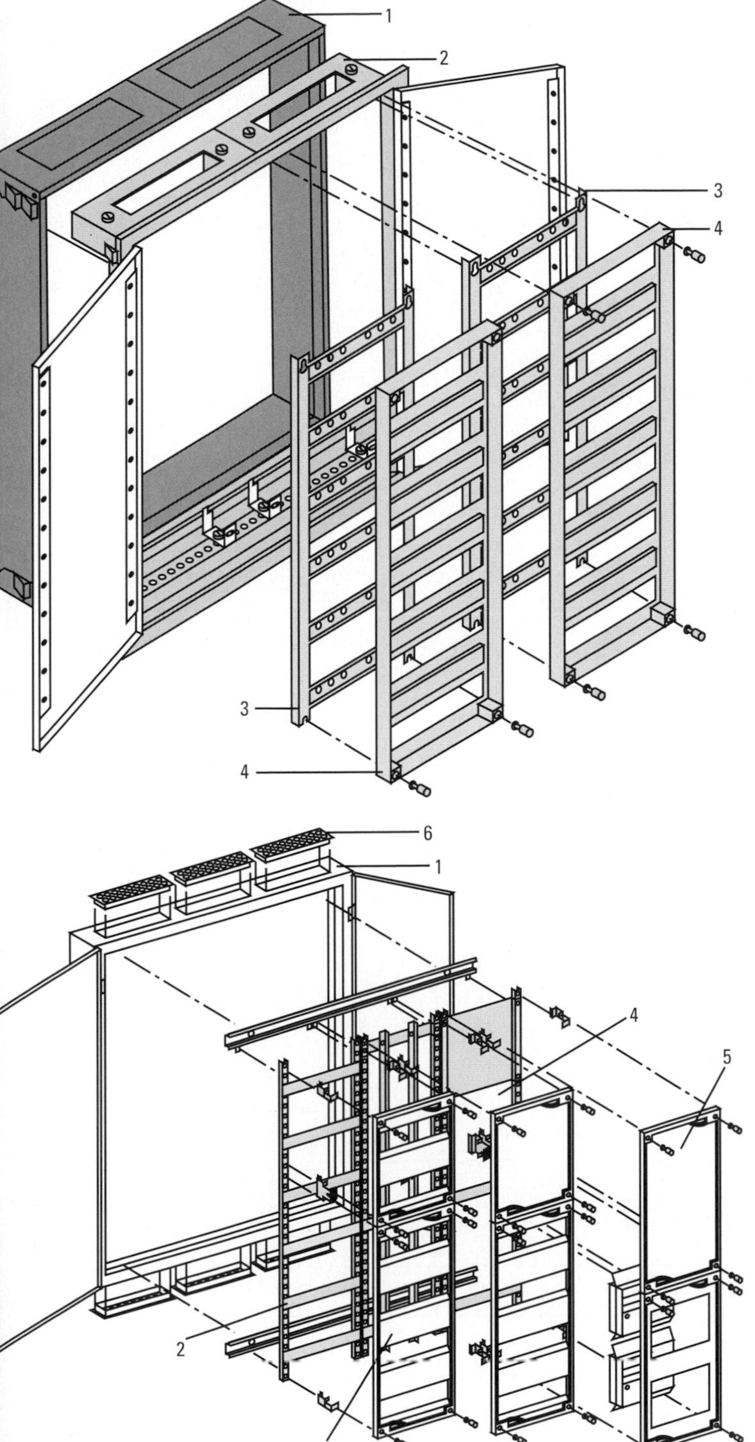

STAB wall-mounting
distribution board 8GD

Flush-type design
Overall depth 130 mm
Bracket-mounted

1 Sheet-steel wall-recess box
2 Flush-type distribution board
3 + 4 Distribution board panel assembly consisting of equipment rack (3) and panel cover (4)

STAB 160 wall-mounting
distribution board 8GE1

Surface-type design
Overall depth 125 mm

1 Surface-type distribution board including cable flange plate (6)
2 +3 Distribution board panel assembly consisting of equipment rack (2) and panel cover (3)
4 + 5 Distribution board panel assembly consisting of equipment plate (4) and panel cover (5)
6 Cable flange plate (6)

'ig. 1.4/54
tructure of different STAB wall-mounting distribution boards

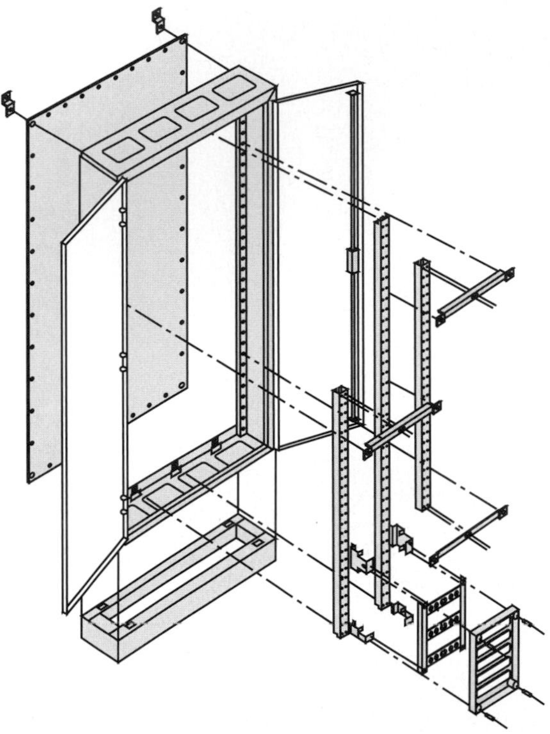

Fig. 1.4/55
Structure of a SIKUS floor-mounting distribution board 8GA with degree of protection IP 41 and safety class 2

can be installed as incoming-feeder circuit-breakers.

Equipment and circuit-breakers are covered with accurately fitting cubicle covers made of sheet steel.

A suitable assembly is available for measuring instruments.

Housing varian

The housings for STAB 8GF Universal distribution boards are designed as single cabinets. A version which allows several housings to be installed side by side is, however, also available.

Sheet-steel doors can be used to cover the front of the unit. Glass doors made of toughened safety glass, which were designed by the Italian designer Giugiaro, are also available to enhance the appearance of domestic distribution boards.

The housings are available for both wall-mounting and floor-mounting distribution boards.

Busbars

Floor-mounting distribution boards are fitted with a busbar compartment on one side in which a 4-pole busbar system with a rated current of 160 A, 250 A, 400 A, or 630 A can be installed.

System construction

In order to facilitate installation in the workshop, the entire distribution system can be assembled (devices mounted and wiring installed) outside the housing. The cable ducts for the wiring are screwed onto the stays of the supporting structure.

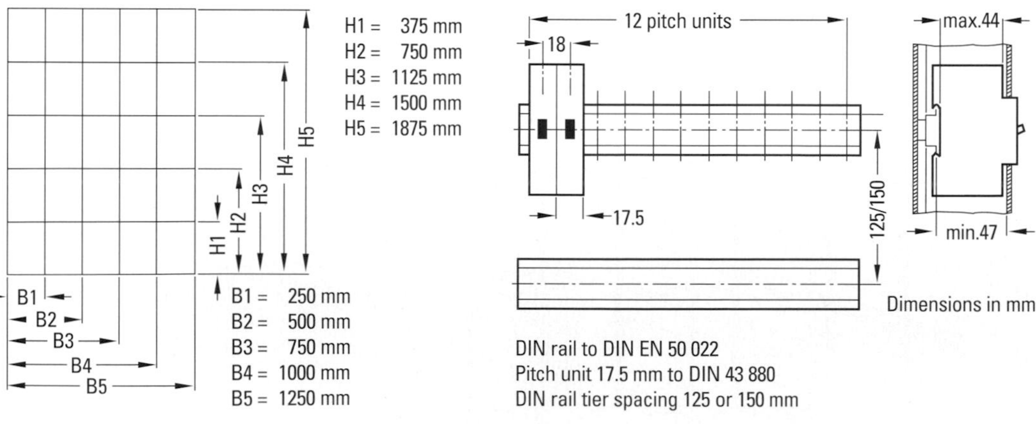

Fig. 1.4/56
Internal structure of SIKUS floor-mounting distribution board 8GA and STAB wall-mounting distribution board 8GD with grid dimensions (H × W) 375 mm × 250 mm and mounting dimensions

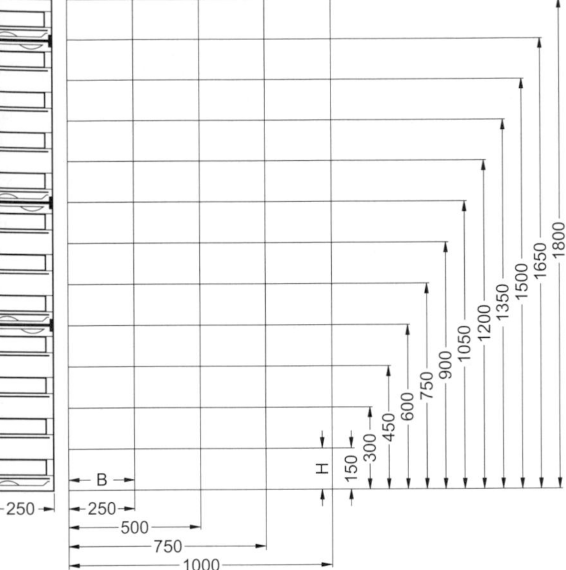

Fig. 1.4/57
Internal structure of SIKUS 630 floor-mounting distribution boards, as well as STAB 160 and STAB 400 wall-mounting distribution boards with grid dimensions (H × W) 150 mm × 250 mm

Fig. 1.4/58
STAB 8GF Universal floor-mounting distribution board

Fig. 1.4/58 is a detailed illustration of a disassembled STAB 8GF Universal floor-mounting distribution board.

Standards

STAB Universal distribution boards comply with the standards NF, CEI, and EN 60 439-1/-3 / IEC 60439-1 / DIN VDE 0660-500 and -504.

Degree of protection, protection against electric shock

STAB Universal distribution boards without doors are protected to IP 30. Boards with sheet-steel or designer glass doors have degree of protection IP 55.

All STAB Universal distribution boards conform to the requirements of safety class 1 (grounding terminal).

SIKUS 3200 and SIKUS Universal modular distribution-board systems

Application

The SIKUS 3200 (type 8GG) and SIKUS Universal (type 8GF) modular distribution-board systems consist of distribution board frames which can be used individually or as assemblies for a wide variety of different applications. They can be expanded using the appropriate system components so that they can be used, for example, as power or domestic distribution boards, for process, control, and automation applications, in network engineering for computer systems, as fire alarm systems, data and communication systems, or as control centers for heating, air-conditioning, and ventilation systems.

Construction

The distribution-board frame consists of a frame structure made of perforated stays. The frame is mounted on a base element and fitted with a top cover. A number of side panels, modular doors, and designer glass doors are available as side enclosure elements. Doors with an opening angle 180° can be fitted to all four sides of the frame structure. The doors can still be fully opened even if the cabinets are mounted side by side. The hinges can be fitted to the left or right-hand side of the doors and can be changed at any time.

Installation

The carefully-matched assemblies including the blanked modular door and panel covers facilitate assembly. Since the distribution boards are available in a large number of different overall widths (up to 1100 mm) and depths (up to 1000 mm) (see Fig. 1.4/61), switch-disconnectors and in-line switch-disconnectors with and without fuses can also be mounted in the SIKUS 3200 in addition to the circuit-breakers. A large mounting plate for installing equipment, DIN rail assemblies, and 19″ fixed or hinged frames completes the system program.

The modular distribution-board systems SIKUS 3200 and SIKUS Universal can be used for a wide range of applications enabling devices such as circuit-breakers with rated currents of up to 3200 A, small devices such as miniature circuit-breakers, fault protection devices, electromechanical and electronic DIN-rail-mounted devices, *instabus EIB* devices, SIMATIC programmable controllers, or SIMODRIVE motor controllers to be installed.

Fig. 1.4/59 shows one possible configuration of a SIKUS Universal 8GF modular distribution-board system (typical configuration for all modular distribution-board systems).

Types of construction

Modular distribution-board systems are available primarily as components for self-assembly or as pre-assembled unequipped and partially equipped cabinets (Fig. 1.4/61).

An appropriate range of accessories is available to satisfy international and national standards and certain installation practices. For example, sheet-steel covers with horizontal module spacing (200 mm grid system) satisfy the requirements of the NF and CEI standards in the case of the SIKUS Universal system.

In order to facilitate assembly of the SIKUS 3200 modular distribution-board system, pre-assembled distribution-board frames can be ordered with suitable assemblies for the most important switching devices (i.e. partially equipped). This applies in particular to incoming-feeder panels with installed circuit-breakers designed for rated currents ≤3200 A (Fig. 1.4/60).

Both types of construction support a fully-integrated grounding concept with safety class 1 and degree of protection up to IP 55.

Standards

Low-voltage power and domestic distribution boards, which are assembled using unequipped cabinets or with appropriate components in accordance with individual requirements, comply with the standards:

EN 60439-1 / IEC 60439-1 / DIN VDE 0660-500 as "partially type-tested switchgear and controlgear assemblies (PTTA)".

Partially-equipped SIKUS 3200 modular distribution-board cabinets comply with the same standards, but as "type-tested switchgear and controlgear assemblies (TTA)". This ensures:

- ▷ conformance with the specified temperature-rise limits,
- ▷ insulation resistance,
- ▷ short-circuit strength,
- ▷ effectiveness of the equipment grounding conductor,
- ▷ conformance with the specified clearances and creepage distances,
- ▷ IP degree of protection,
- ▷ correct mechanical operation.

The standards EN 60439-3 / IEC 60439-3 / DIN VDE 0660-504 and EN 60 204-1 / IEC 204-1 / VDE 0113 Part 1 also apply.

The fully-integrated distribution-board system from Siemens, ranging from the flat STAB 160 wall-mounting distribution boards to the SIKUS 3200 modular distribution-board cabinet, in conjunction with the appropriate assemblies for the different Siemens devices allow low-voltage power and domestic distribution boards to be assembled for any application in accordance with DIN 43 870.

STRATUM distribution-board system

STRATUM power and domestic distribution boards are used mainly in Great Britain, India, and Australia as main and sub-distribution boards in residential, functional, and industrial buildings.

An overview of the different versions and applications can be seen in Table 1.4/8.

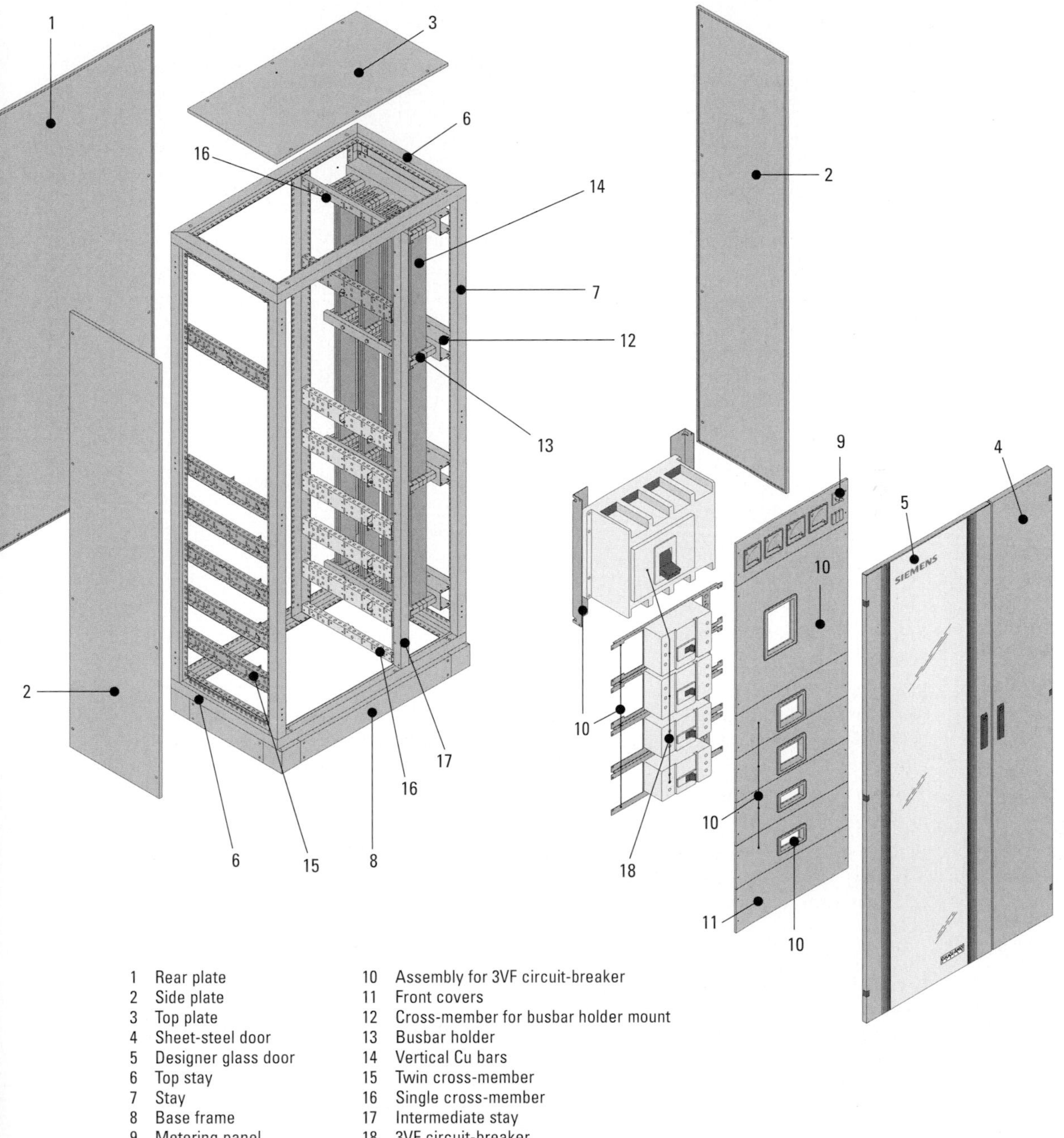

Fig. 1.4/59
Structure of a SIKUS Universal modular distribution board

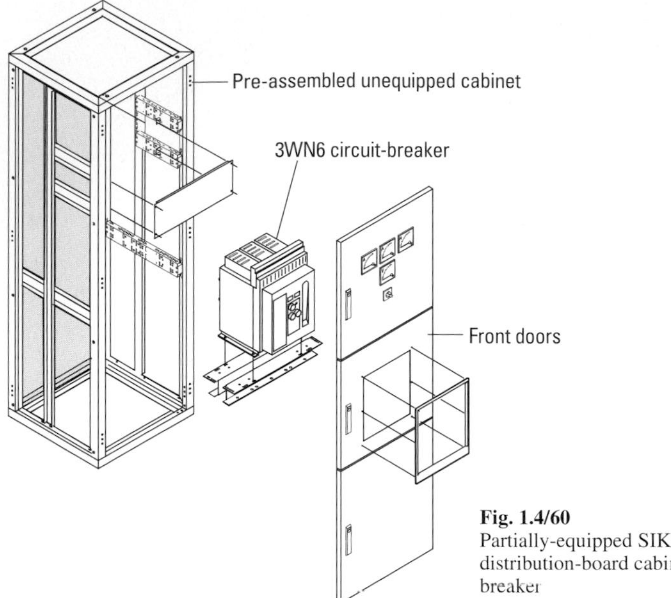

Fig. 1.4/60
Partially-equipped SIKUS 3200 modular distribution-board cabinet for a 3WN6 circuit-breaker

STRATUM 200 wall-mounting distribution boards

Application, construction

STRATUM 200 wall-mounting distribution boards are used as main and sub-distribution boards in residential and functional buildings.

The housings are available in either plastic or sheet steel.

In order to cater for British installation practices, the distribution boards are already equipped with a residual-current protective device (RCD) or a two-pole mains switch (*isolator*) to protect the infeed.

Cable and line protection is provided by miniature circuit-breakers, e.g. from the N system, which can be connected directly to a copper busbar with forked terminal lugs for the necessary power supply. In this way, between 4 and 16 load feeders can be provided (Fig. 1.4/62).

Sufficient space is available for cable entry and for devices to be connected.

Covers attached to the front of the unit have flaps which allow access to the operating elements of the installed equipment and ensure a high degree of protection against direct contact with live parts. The flaps can also be made of transparent, smoked plastic so that the control states of the installed equipment can be determined quickly.

Standards

STRATUM 200 wall-mounting distribution boards are manufactured and tested in compliance with the standards EN 60 439-3 / BS EN 60 439-3 and IEC 60 439-3.

STRATUM 400 wall-mounting distribution boards

Application, construction

STRATUM 400 wall-mounting distribution boards are used mainly in functional buildings and industrial plants as main and sub-distribution boards for operating currents $\leq$200 A.

The three-pole busbar system is mounted vertically in the center of the distribution board so that N-type miniature circuit-breakers (MCBs) from the 5SX2 series can be positioned to the right and left.

The busbar system and the mains switch with a rated current $\leq$250 A are pre-installed in all distribution boards (Fig. 1.4/63).

The sheet-steel housings, covers, and doors provide reliable protection against electric shock (safety class 1).

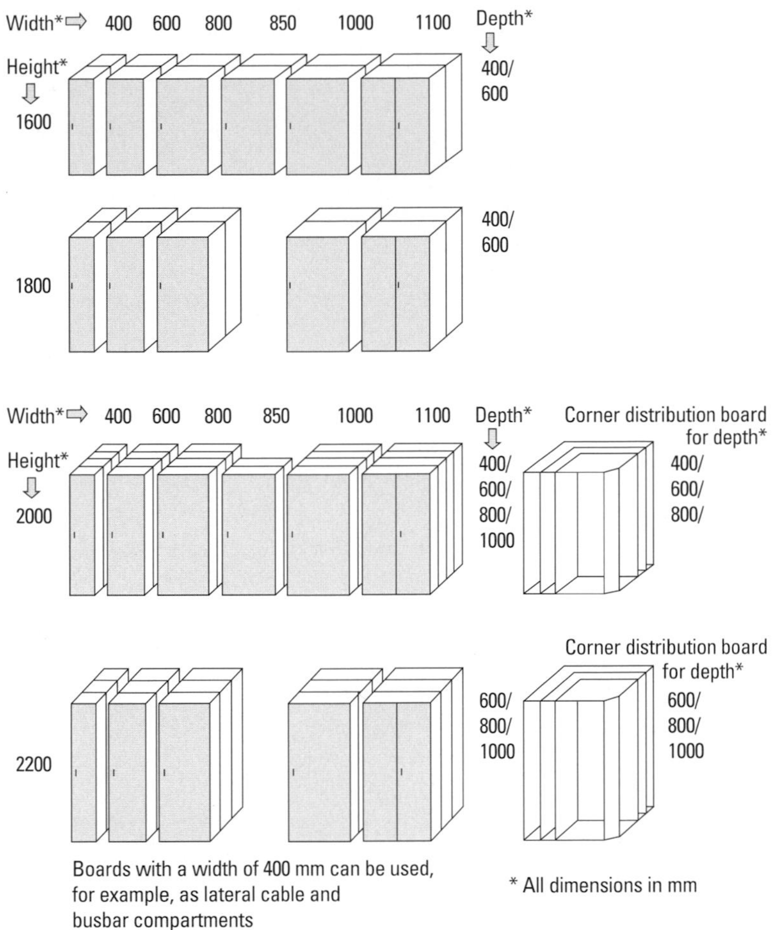

Fig. 1.4/61 SIKUS 3200 modular distribution-board system made of sheet steel, safety class 1, dimensions

STRATUM 400 wall-mounting distribution boards are available in different sizes for either flush-type or surface-type installation for between 4 and 24 load feeders (see Table 1.4/8).

tandards

STRATUM 400 wall-mounting distribution boards are manufactured and tested in compliance with the standards EN 60 439-1 / IEC 60 439-1 and BS 5486 Part 13.

STRATUM 601 and STRATUM 603 distribution boards

pplication, onstruction

STRATUM 601 and STRATUM 603 distribution boards are used in functional buildings and industrial plants as main and sub-distribution boards for rated currents of 250/400/630 A and 800/1250 A.

3VF circuit-breakers (MCCBs) with rated currents $\leq$ 250 A and 630 A can be connected to the left and right of the three-pole busbar system, which is mounted vertically in the center of the distribution board. In this way, it is possible to install MCCBs for between 5 and 18 load feeders. Owing to their different breaking capacities, the circuit-breakers can be adapted at relatively low cost to the prospective short-circuit currents in the load feeders. The different tripping characteristics of the MCCBs mean that they can be used for system or motor protection, in starter assemblies, or as non-automatic circuit-breakers.

The sheet-steel housings, covers, and doors provide reliable protection against electric

Fig. 1.4/62
STRATUM 200 wall-mounting distribution board

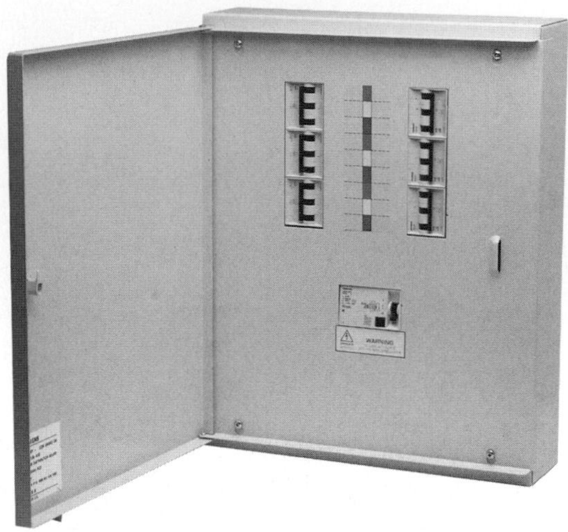

Fig. 1.4/63
STRATUM 400 wall-mounting distribution board

shock (safety class 1, degree of protection IP 30). The distribution boards are available as wall-mounting or floor-mounting boards. The example shown in Fig. 1.4/64 is a STRATUM 603 floor-mounting distribution board with front cover.

Standards

The STRATUM 601 and STRATUM 603 distribution boards are manufactured and tested in compliance with the standards EN 60439-1/BS EN 60439-1 and IEC 60439-1.

STRATUM 3200 modular distribution-board system

Application, construction

The STRATUM 3200 modular distribution-board system consists of cabinets which can be assembled side by side. They are used primarily in industrial applications as power centers and main distribution boards.

The sheet-steel housings of the cabinets are divided into two, three, or four compartments each with its own door. They can be equipped with incoming-feeder circuit-breakers (ACBs) with a rated current ≤3200 A, outgoing-feeder circuit-breakers (MCCBs), or fuses with a rated current ≤800 A (Fig. 1.4/65).

Instrumentation and control equipment can be installed in the cabinets.

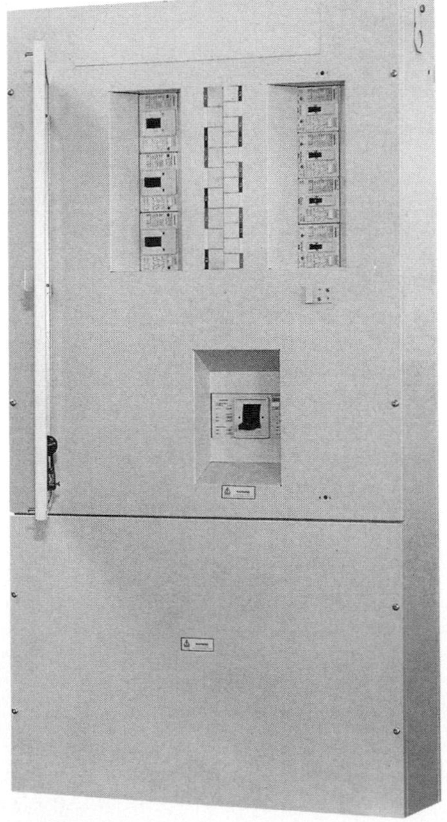

Fig. 1.4/64
STRATUM 603 floor-mounting distribution board

Fig. 1.4/65 STRATUM 3200 power center

hort-circuit trength

The busbar system is dimensioned for peak short-circuit currents ≤50 A lasting 3 s or 80 kA lasting 1 s.

Cables can be connected from above or below and can be accessed from the front or the rear of the unit.

tandards

STRATUM 3200 distribution boards are manufactured and tested as type-tested low-voltage switchgear and controlgear assemblies (TTA) in compliance with the standards EN 60 439-1 / BS EN 60 439-1 and IEC 60 439-1.

SIPRO wall-mounting distribution boards, meter cabinets, meter distribution cabinets, distribution cabinets, and floor-mounting distribution boards

pplication, anning

All electrical power imported from the local public utility system for domestic, agricultural, and industrial use passes through an incoming main feeder box and a meter cabinet system with metering units.

tandards

Meter cabinet systems in Germany must be installed by an authorized electrician in accordance with the relevant regional technical supply conditions of the responsible public utility. For this reason, it is advisable to check the suitability of the selected meter cabinet type including its equipment and installation location with the responsible public utility as early as the planning stage. Sufficient space should be left in the meter cabinet for expanding the consumer's installation at a later stage.

Construction

The cabinets of the SIPRO wall-mounting distribution-board system consist of plastic sections reinforced with metal rails. The plastic sections are connected to each other by means of corner joints and a rear panel made of insulating material or sheet steel. The inner surfaces of sheet-steel cabinets are lined with insulating material.

Snap-on assembly kit

The sheet-steel doors can be attached on both the right and left-hand side using snap-on hinges. Reinforcing stays in the doors increase their strength and protect the installed equipment against damage. The profiles of the side, top, and base panels of plastic SIPRO distribution cabinets are fitted with two parallel profile slots on their inner side, which are used to attach snap-on assembly kits and equipment cover panels. The snap-on assembly kits are installed by simply pushing the kit mounting bases into the lower profile slot, pushing the assembly kit into the cabinet, and securing it in place using two captive screws in the profile slot in the top of the cabinet. The equipment cover panels of the upper and lower connection compartments are even simpler to mount and require just one captive screw.

Sealable cable connection box

The lower connection compartment can be extended, e. g. for larger cable and conductor cross sections, by flange-mounting sealable cable connection boxes.

Add-on housings

Base frames

Add-on housings, which can be mounted on system-specific base frames under the SIPRO meter and meter distribution cabinets, are available to enlarge the equipment compartment and connection compartment. This converts a wall-mounting cabinet into a floor-mounting distribution board.

Protection against electric shock, degree of protection

SIPRO wall-mounting distribution boards can be used for flush, surface, and partially-recessed installation. All cabinets have safety class 2 (total insulation) and degree of protection IP 31 for basic doors and IP 54 for sealed doors.

Cabinet program

The following cabinet configurations are available:

▷ unequipped cabinets with separate snap-on assembly kits and equipment components,
▷ unequipped cabinets with pre-fitted snap-on assembly kits,
▷ unequipped cabinets with pre-fitted and pre-wired snap-on assembly kits,
▷ partially-equipped cabinets with pre-fitted snap-on assembly kits (meter or distribution panel) and space to allow equipment packages to be fitted by the customer and the public utility (e.g. busbars, main branch circuit terminals, l. v. h. b. c. fuse-bases, or l. v. h. b. c. fuse switch-disconnectors),
▷ complete cabinets, equipped according to customer specifications or the technical supply conditions of the local public utility.

Fig. 1.4/66 shows a SIPRO meter cabinet with snap-on assembly kits, fully-equipped with public utility equipment packages.

Snap-on assembly kits

The snap-on assembly kits consist of two U-section stays, which are connected by means of the N/PE-conductor terminal strip, an equipment rack, and the meter mounting plate. The U-section stays have a hole matrix for 4 mm self-tapping screws which allows a wide variety of equipment configurations.

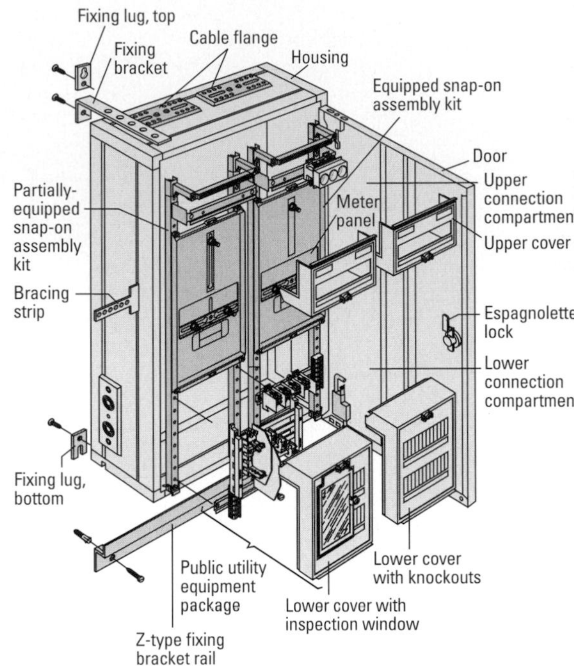

Fig. 1.4/66
SIPRO meter cabinet with snap-on assembly kits and public utility equipment packages

Public utility assembly kits

Public utility assembly kits for busbar runs, main conductor branch terminals, l. v. h. b. c. fuse switch-disconnectors, l. v. h. b. c. fuse-bases, flush-type circuit-breakers, control-line terminals, series cross-members, etc. are installed in the lower connection compartment.

The meters or rate changeover devices are installed in the middle section on standardized meter mounting plates (DIN 43 870). All installation devices approved by the public utility, such as residual-current circuit-breakers, NEOZED fuses, MINIZED D02 fuse switch-disconnectors, N-type miniature circuit-breakers, and the N/PE-conductor terminal kits, are installed in the upper connection compartment using DIN rails (DIN EN 50022).

1.4.5.2 Portable domestic distribution boards for outdoor use

Applications

If, for technical or cost-related reasons, it is not possible to erect permanently-installed domestic distribution boards indoors, or if meters cannot be readily accessed and read (e. g. isolated houses, leisure and holiday resorts which have no electrical connection facilities, building sites, fairgrounds and markets, public lighting systems, and traffic signals), cable distribution boards, distribution and meter connection cabinets are installed outdoors.

Design, construction

The outdoor cabinets of portable distribution-board systems (see Table 1.4/9) are made of either hot-galvanized and baked-enameled sheet steel (safety class 1 or safety class 2 if lined with plastic) or glass-reinforced polyester, type 833.5 to DIN 16 913, using the SMC (sheet molding compound) technique (safety class 2, total insulation). They comply with EN 60439-1 / IEC 60 439-1 / DIN VDE 0660-500 and are type-tested as TTAs.

Outdoor cabinets are available in different series and designs ranging from modular systems (the housing and base are assembled from individual parts and bolted together) to complete cabinets (the housing and base form a single unit). The cabinets are constructed in such a way that sufficient ventilation is provided between the base and cabinet and underneath the top cover. All ventilation channels are labyrinthal to prevent foreign bodies from entering the cabinet. The front panels are fitted with quick-release locks and allow damaged cabinet sections to be replaced easily. All molded parts used in the plastic version are colored and the top covers also have a UV-resistant, hardened two-component polyurethane coating. This permits the cabinets to be used in locations with severe environmental pollution. Cabinets which are destined for installation in desert, coastal, or mountainous regions can be fully coated.

Cable entry and equipment installation

In the case of cable distribution cabinets (see Fig. 1.4/67), the incoming-feeder cables are inserted from below and connected to an incoming main feeder box for l. v. h. b. c. fuses installed in or around the building which is to be supplied with electricity.

With fairground/market and worksite distribution cabinets (Fig. 1.4/68), the incoming-feeder cables are also inserted from below and connected to l. v. h. b. c. fuse-bases in the public utility termination panel of the distribution board. The bases of the fairground and market distribution boards have an amply dimensioned entry shaft with strain-relief fittings at the front to allow the consumer cables to be inserted.

Meters, time switches, and AF receivers are usually installed in totally-insulated plastic housings with transparent covers (degree of protection IP 54). Space is also provided for additional switching devices in front of the meter and for consumer outgoing feeders with DIAZED or NEOZED fuses, l. v. h. b. c. fuse blocks, miniature circuit-breakers, residual-current circuit-breakers, grounding or CEE outlets, etc.

Table 1.4/9
Overview and technical data of the Siemens portable power and domestic distribution systems

Designation	Cable distribution cabinets	Worksite distribution boards	Fairground/market distribution boards	Campsite distribution boards	Multi-outlet distribution boards
Type	8MB	8MM2	8MM4	8MM6	8MM7
Rated current (A)	1000	630	400	40	40
Cabinet depth (mm)	320, 335	360, 480	330	330	120 to 190
Base assembly	yes	yes	yes		–
Degree of protection	IP 43, IP 54	IP 43, IP 54	IP 43	IP 43	IP 43
Protection against electric shock	Safety class 1 and 2	Safety class 1 and 2	Safety class 2	Safety class 1 and 2	Safety class 2

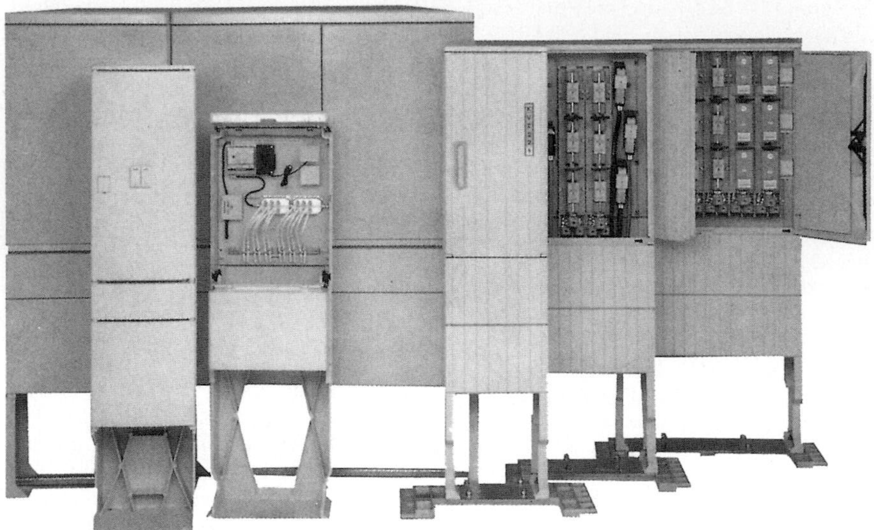

Fig. 1.4/67 Examples of 8MB portable cable distribution cabinets

Degree of protection

All outdoor cabinets have a minimum degree of protection of IP 43 in accordance with EN 60529 / IEC 60529 / DIN VDE 0470-1. The doors are attached to the side panels of the cabinet by means of sturdy pin-type hinges. They are fitted with three-point locking mechanisms, which can be modified to allow a double-locking system to be installed at a later stage.

1.4.6 Planning low-voltage switchgear and distribution systems, software tools (P.I.S.A.A)

In order to be able to supply all of the consumers in a residential building, office block, or factory with electrical energy, low-voltage switchgear and distribution systems must be configured for all network nodes in such a way that they fulfil the requirements of their respective installation sites and permit the connected consumers, cables, and wiring to be switched, protected, and monitored.

Fig. 1.4/2 shows an example of a low-voltage system installed in an industrial plant.

The owner of the system defines the local operating and environmental conditions, if necessary, in conjunction with the manufacturer, and also supplies the latter with all the electrical data of the switchgear or distribution system and network at the installation site, so that a system which is both optimally matched to the prevailing technical conditions and which is as cost-effective as possible can be configured for each individual network node. Various items of information are required for this purpose.

Ambient and installation conditions

- ▷ Mechanical loads
- ▷ Degree of protection (EN 60529 / IEC 60529 / DIN VDE 0470-1): shock hazard, protection against dust and water
- ▷ Ambient temperature and climatic conditions
- ▷ Corrosion resistance
- ▷ Type of installation and mounting (wall-mounting, free-standing)
- ▷ Open-type assembly, cabinet-type or box-type distribution board, busbar trunking system
- ▷ Covers and doors, transparent or non-transparent
- ▷ Maximum permissible dimensions of the switchgear or distribution board for transport and erection at the installation location
- ▷ Cable entry (cable duct, cable raceways)

Worksite distribution board

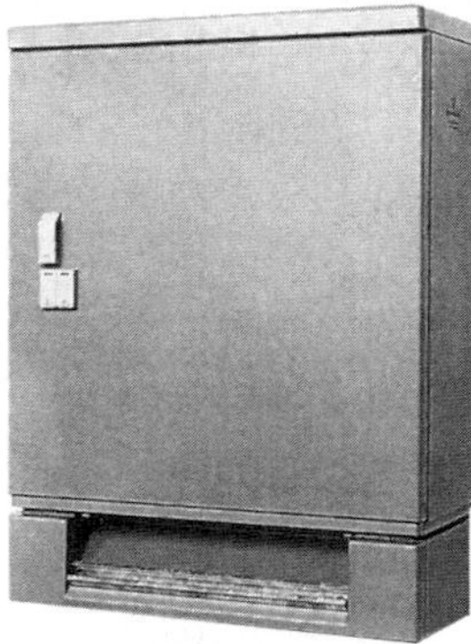

Fairground and market distribution board

Fig. 1.4/68
Examples of other portable distribution boards

▷ Base covering, if required
▷ Equipment mounting:
fixed-mounted
removable or withdrawable
quickly replaceable
adapted to busbars

Electrical conditions and data

A block diagram should be provided for planning and configuration purposes. The following electrical data should also be known:

▷ Applicable standards
▷ Applicable protective measures to HD 60384.4.41 / IEC 60384-4-41 / DIN VDE 0100-410:
protection against direct contact
protection against indirect contact
(see Chapter 2 for protective measures)
▷ Rated operating voltage and frequency
▷ Busbars:
current intensity, number of conductors (3, 4, 5 conductors)
▷ Short-circuit current at the installation location
▷ Position of the incoming-feeder cables (from above, below, side), type and cross section, number of cables and cores
▷ Position of the outgoing-feeder cables (upwards, downwards, sideways), type and cross section, number of cables and cores
▷ Number of outgoing feeders and details of their component configurations (contactors, circuit-breakers, fuses, etc.), rated output, rated and operating current, setting range of the thermal releases, etc.
▷ Outgoing feeders with control devices: details of control voltage and frequency, as well as location of the control stations
▷ If applicable, coincidence and load factor of the outgoing feeders or distribution board
▷ If applicable, communication via serial bus systems.

The more specific and complete this data is, the faster and more reliably the most suitable solution will be found for the application in question.

Detailed explanations pertaining to the mechanical and electrical data for low-voltage switchgear and distribution systems as well as the low-voltage switching devices used in them can be found in the Siemens publication "Switching, Protection and Distribution in Low-Voltage Networks".

Project planning

The two stages involved in planning low-voltage switchgear, distribution boards, and control units are referred to as *implementation planning* and *erection*. Implementation planning is usually carried out by designers and engineering consultants, the result of which is the specifications list. This forms the basis for erecting the equipment which is generally the responsibility of the fitters and construction personnel.

Project planning tools

In order to facilitate project planning, the manufacturers of low-voltage switchgear and distribution boards use a large number of tools. These tools are also available to customers, owners, planning consultants, etc.

The most important project planning tools are:

▷ lists and catalogs,
▷ special project planning documentation,
▷ computer programs,
▷ product data, symbols, and tender specifications in electronic form.

Many of these project planning tools are also available on the Internet.

Implementation planning

Implementation planning

Once all of the above-mentioned information has been provided, the designer can select and dimension the most cost-effective and technically suitable equipment for the application in question. The specifications list compiled during this process forms the basis for erecting the equipment.

Planning electrical equipment

During project planning, all electric circuits and outgoing feeders as well as the entire system must be dimensioned taking the following into consideration:

▷ overload protection,
▷ short-circuit protection,
▷ protection by disconnection of supply,
▷ voltage drop,
▷ selectivity.

Computer-assisted project planning methods

Engineering consultants, software companies, and manufacturers employ different computer-assisted project planning methods. They are used first and foremost to perform dimensioning calculations (e. g. KUBS for calculating short-circuits in low-voltage systems), but can also be used to generate circuit diagrams, location diagrams, and the specifications list itself.

Manufacturers support these project planning methods by providing diagrams as well as tender specifications in the most commonly used data formats (e. g. also available on the Internet).

Project planning example

On the basis of the block diagram (Fig. 1.4/69), the following example illustrates the steps (Fig. 1.4/70) involved in planning an 8HP distribution board manually.

The distribution board is "assembled" step by step beginning with the incoming-feeder unit.

Procedure for planning distribution boards

Incoming-feeder unit

Distribution board planning always begins with the incoming-feeder unit. This is particularly important because the position of the incoming-feeder unit is a major factor in determining the cross section of the busbars. The most cost-effective solution is always achieved when the current from the incoming-feeder terminals branches in two directions (Fig. 1.4/71).

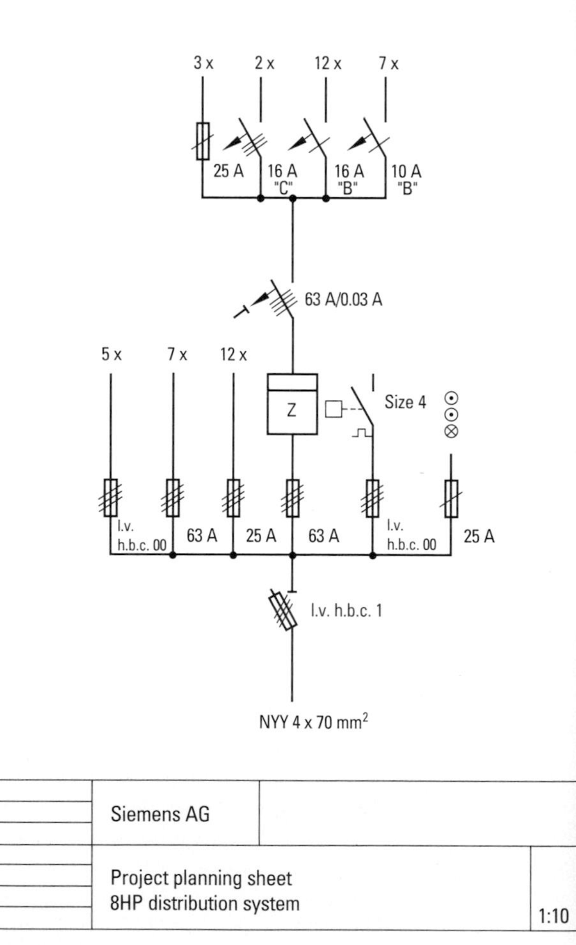

Fig. 1.4/69
Block diagram of a planned 8HP distribution board

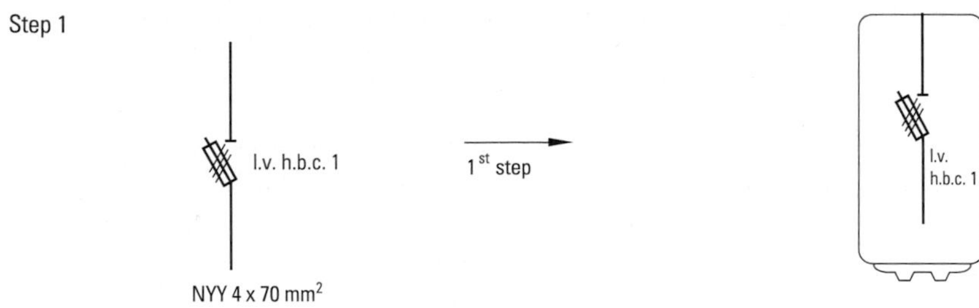

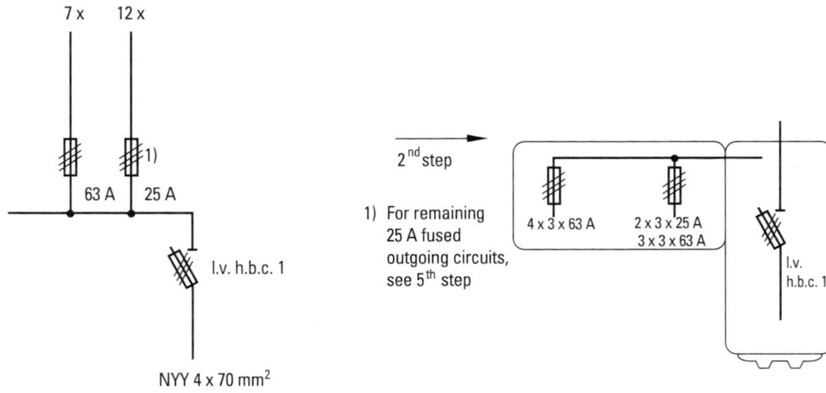

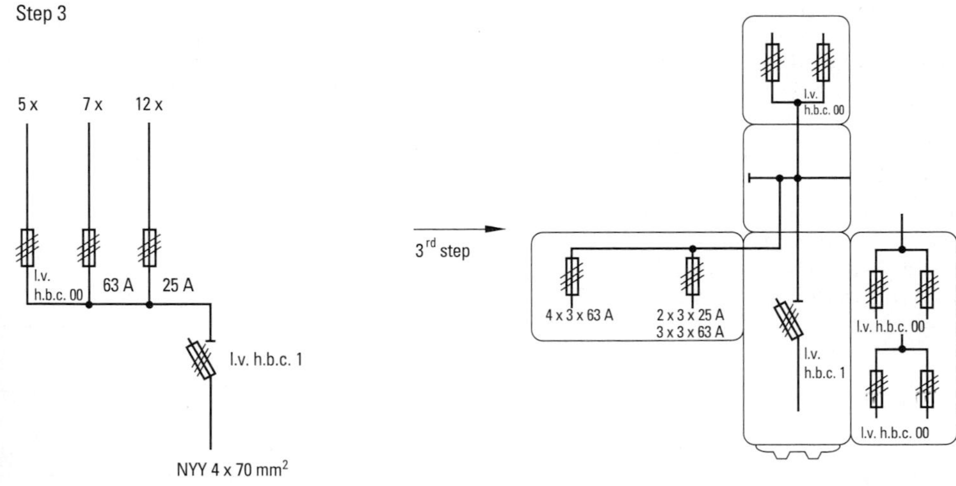

Fig. 1.4/70 Steps in planning a low-voltage distribution board

STEP 4 and STEP 5 see next page

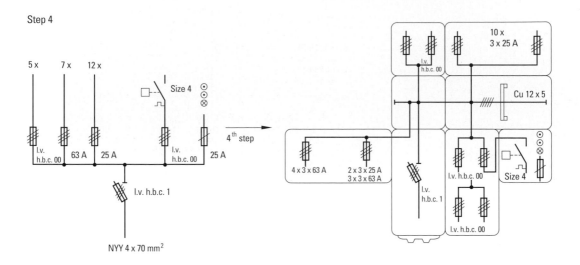

Step 4

5 x 7 x 12 x

Size 4

l.v. h.b.c. 00 63 A 25 A l.v. h.b.c. 00 25 A

l.v. h.b.c. 1

NYY 4 x 70 mm²

4th step

10 x 3 x 25 A

l.v. h.b.c. 00

Cu 12 x 5

4 x 3 x 63 A 2 x 3 x 25 A 3 x 3 x 63 A

l.v. h.b.c. 1

l.v. h.b.c. 00

Size 4

l.v. h.b.c. 00

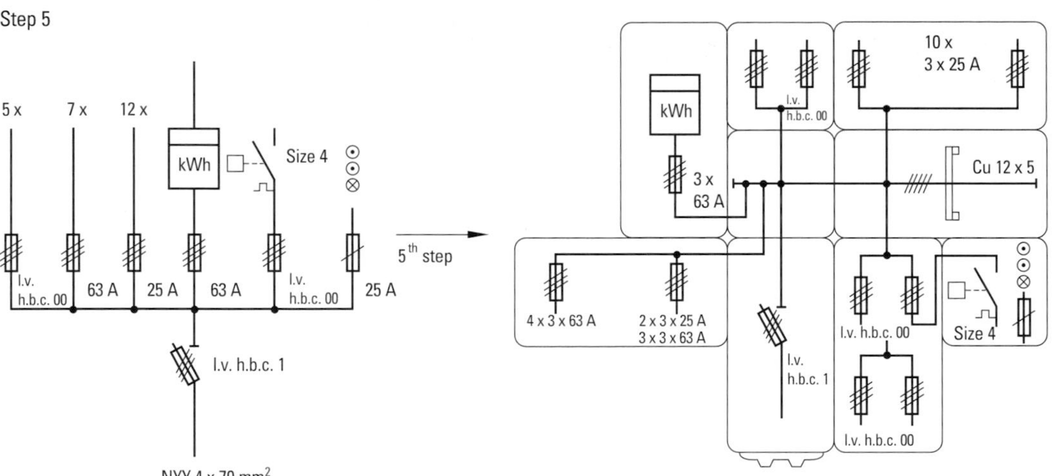

The corresponding types of infeed that can be used for switchboards are described in detail in Section 1.4.2. This also applies to box-type distribution systems (see Section 1.4.3) as well as power and domestic distribution systems (see Section 1.4.5).

The devices and their housings or installation compartments are selected and dimensioned using the aforementioned project planning methods or by the designer with the aid of appropriate catalogs.

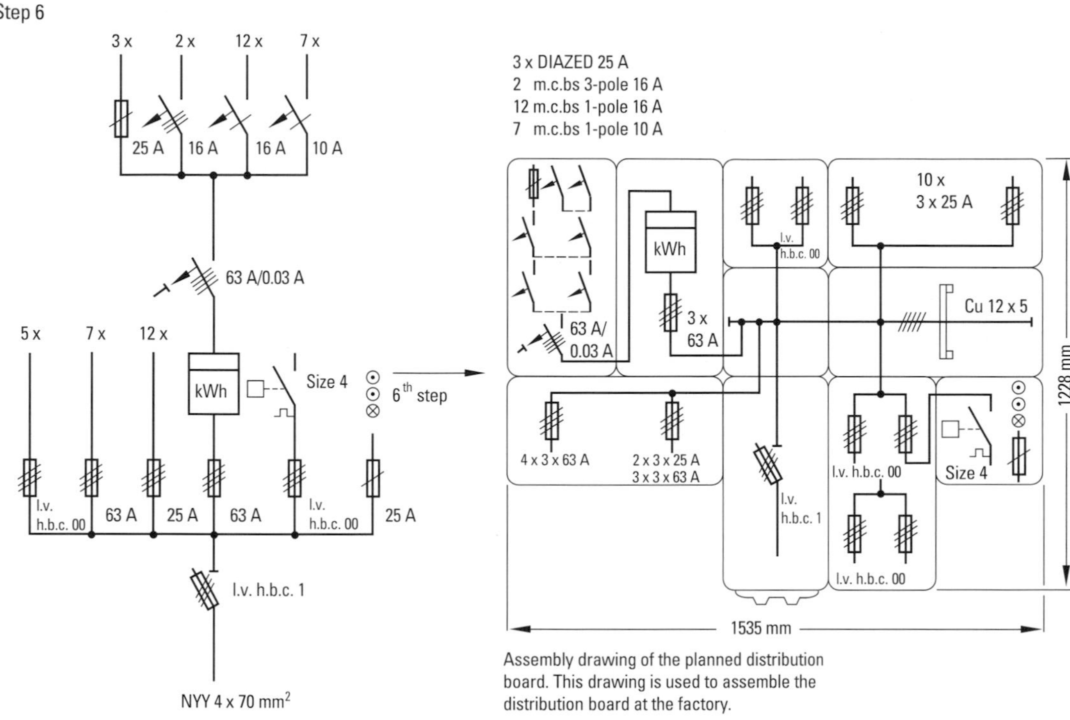

Fig. 1.4/70 Steps in planning a low-voltage distribution board *(continued)*

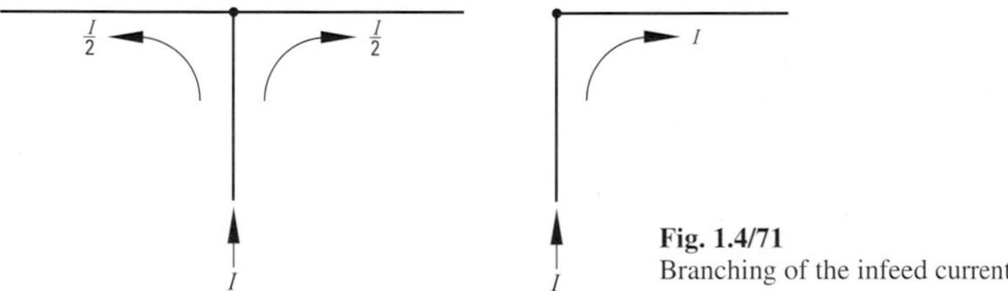

Fig. 1.4/71
Branching of the infeed current

The following aspects must be taken into consideration:

Short-circuit strength

The prospective short-circuit current at the installation location must not be greater than the values specified for the short-circuit strength of the switchgear and distribution board. This requirement can also be satisfied by connecting a current-limiting protective device on the line side (see Section 1.3.2). Information on calculating the short-circuit current at the installation location can be found in Section 1.2.

Overcurrent protective devices for incoming-feeder cables

It is generally advisable to connect the overcurrent protective devices required for the switchgear and distribution boards upstream of the incoming-feeder cables so that these are also protected (see Section 1.7). Only one disconnecting device (e. g. isolating link, load interrupter switch) is usually necessary in the infeed circuit inside the switchgear or distribution board. Control systems conforming to EN 60204/IEC 60204/DIN VDE 0113 must be fitted with a mains switch; in all other cases, this will depend on the costs involved and other operational criteria.

Selectivity will not be provided if fuse-links with the same current rating are connected at the beginning and end of the incoming-feeder cable. If selectively graded fuse-links and circuit-breakers are installed, care must be taken to ensure that the incoming-feeder cables are not overloaded. It is, therefore, advisable to provide fuse-links with different current ratings at the beginning of the cable for protection against short circuits and at the end for protection against overloads (see Section 1.7).

Bearing in mind the desired clearing times in the low-voltage system, which should be as short as possible, it is inadvisable to rely on just one protective device (circuit-breaker, h.v. h.b.c. fuse) connected on the high-voltage side of a transformer. Fuses or circuit-breakers on the low-voltage side always provide better protection thanks to shorter tripping times (see Section 1.3.2).

Fig. 1.4/72 shows the position of overcurrent protective devices used to protect the incoming-feeder cables and downstream distribution boards as well as the position of switch-disconnectors in the incoming-feeder units.

Overcurrent and overload protection for outgoing cables and installed equipment

In addition to the criteria specified in HD 60384 / IEC 60364 / DIN VDE 0100, a further distinction must be made for switchboards and distribution boards between:

- ▷ overcurrent and overload protection of outgoing cables, for which the fuse-link must be selected in accordance with the cross section of the conductor,
- ▷ overcurrent protection for installed equipment, which is intended to prevent short-circuit currents from exceeding any of the following equipment ratings:
- ▷ rated switching capacity (e. g. for motor-protection switches and circuit-breakers),
- ▷ resistance to welding (e. g. for contactors, auxiliary contactors, etc.),

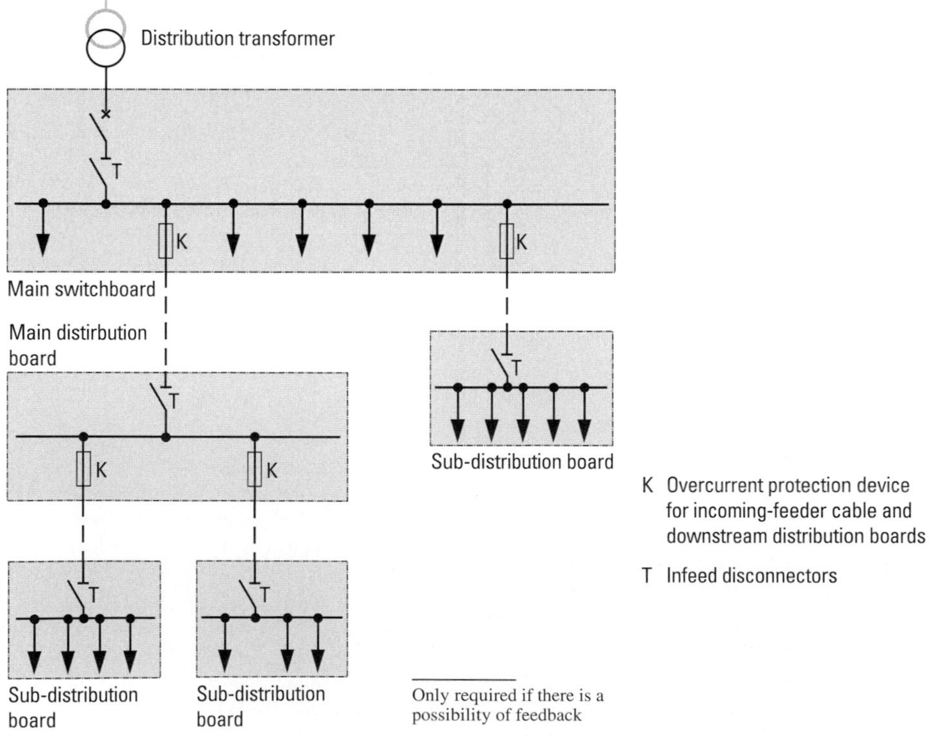

Fig. 1.4/72
Arrangement of overcurrent protection devices and disconnectors in switchboards and distribution boards

- ▷ thermal loading capacity (e.g. for bimetallic relays),
- ▷ dynamic stress (e.g. with short-circuit currents > 50 kA r.m.s.).

Fuse-links

The selected fuse-links must have a sufficiently high rating so that they can withstand the starting current of downstream motors.

The ratings of back-up fuses specified in the equipment data sheets have been determined by the manufacturers in accordance with these criteria and must never be exceeded.

Contactors

Contactors, bimetallic relays, and auxiliary switches must always be protected using back-up fuses, the current ratings of which do not exceed those specified by the manufacturer.

Motor-protection switches and circuit-breakers

Motor-protection switches and circuit-breakers can be used without back-up fuses provided that the prospective short-circuit current at the installation location is not greater than the rated switching capacity of the equipment.

In the case of motor-protection switches with a very low rated tripping current, short-circuit currents are reduced by the high internal impedance to such an extent that back-up fuses are unnecessary (short-circuit calculation: see Section 1.2).

Protection against direct contact

Protection against direct contact with live parts (see Section 2.2) can be achieved by (see HD 60384.4.41 / IEC 60364-4-41 / DIN VDE 0100-410, IEC 64 (CO) 196 / DIN VDE 0106, EN 60439-1 / IEC 60439-1/ DIN VDE 0660-500):

- ▷ providing an appropriate enclosure (at least IP 2X) or
- ▷ installing unprotected equipment (IP 00, IP 10) in closed electrical operating areas (in cabinets, if necessary) in compliance with the relevant specifications in HD 60384 / IEC 60364 / DIN VDE 0100.

According to HD 60384.4.41 / IEC 60364-4-41 / DIN VDE 0100-410, it must only be possible to remove barriers, which provide protection against accidental contact with live metal parts in distribution boards installed in generally accessible operating areas, by means of a key or special tool.

Protection against indirect contact

Protection against indirect contact can be achieved by additional protective measures (see Section 2.2). It is essential to take the protective measures of the infeed supply system into consideration here.

Information concerning operator-controllable equipment

Quick-release covers, which can be opened without tools, must only be used for the boxes of a distribution board, e.g. fuse boxes, which have an internal insulating lining for protection against accidental contact. This does not apply in closed electrical operating areas. Since fuses occasionally have to be replaced during operation, it is advisable to group contactors and their fuses, for example, in separate boxes; the fuse boxes, unlike the contactor boxes, are then fitted with quick-release covers.

Total insulation

If totally insulated switchboards or distribution boards are to be used, it is essential to ensure that the factory-fitted protective insulation is not penetrated by any conductive metal parts, such as switch shafts, metal cable glands, screws, etc.

Inactive metal parts within the insulation itself, such as base plates and equipment housings, must never be connected to the PE or PEN-conductor, even if they have a PE connection terminal.

Insulation enclosures provide optimum protection against corrosion and do not require any maintenance.

Metal enclosures

Enclosures and structural parts made of metal are generally protected against corrosion by a high-quality protective coating. Suitable protection must be provided for metal parts of switchboards and distribution boards (with the exception of totally insulated distribution boards) by means of a PE-conductor (see Section 2.2).

Space for equipment

When planning special equipment in enclosed switchgear assemblies and distribution boards (particularly in box-type distribution boards), adequate space must be provided for the following in addition to the basic space requirements of the equipment:

- ▷ voltage clearance (clearance in air) with respect to the enclosure,
- ▷ heat dissipation from the individual devices,
- ▷ arcing in switching devices (where appropriate),
- ▷ wiring,
- ▷ connection of incoming and outgoing feeder cables (connection compartment).

The equipment must, therefore, never be packed as closely as its external dimensions and the internal dimensions of the enclosure would theoretically allow.

In certain cases, it is necessary to observe manufacturer-specific restrictions if an item of equipment is used in an enclosure, e.g. with respect to rated current and switching capacity.

Heat dissipation

With open-type assemblies, the heat generated by all the devices is dissipated via the air freely flowing around the equipment. With enclosed switching devices, this heat can only be dissipated by means of heat exchange between the surface of the enclosure and the surrounding air. This explains why many enclosed devices have a reduced current rating.

The specified maximum power loss must not be exceeded as a result of too many heat-producing devices being installed or as a result of these being arranged too closely together.

Connection compartments

The internal or external space provided for connecting outgoing-feeder cables is a critical factor which determines the ease and efficiency with which connection work can be carried out after the switchboards have been erected.

Although a small enclosure may at first seem a cost-effective solution, it can, however, reveal itself as a false economy due to the time and expense involved in the initial and subsequent connection of cables and leads as a result of the limited amount of space. It is, therefore, essential to ensure that adequate connection space is provided when selecting enclosed switchboards and distribution boards.

Identification in circuit diagrams

It is important to be able to identify the various components of the equipment installed in switchboards and distribution boards in accordance with a number of different criteria. According to EN 60439-1 / IEC 60439-1 / DIN VDE 0660-500, it must be possible to identify the different circuits and their overcurrent protective devices clearly within the switchgear assembly. The identifying markings used for the individual equipment in the switchgear assembly must correspond to the information in the circuit diagrams and related documents supplied with the switchgear assembly.

Identification of cables and busbars

Cables and busbars must be marked in such a way that PEN-conductors as well as PE and N-conductors can be clearly distinguished from one another and from the phase conductors. With the busbars, which should if possible be bare, this can be achieved by means of either different shaped sections or different arrangements of the individual bars, or indeed a combination of the two.

Color coding using adhesive tape is recommended for bare busbars. The markings are then as follows (1 × per panel or box):

4-conductor systems
the phase conductors with L1, L2, L3,
the PEN-conductor with green/yellow striped adhesive tape,

5-conductor systems
the phase conductors with L1, L2, L3,
the PE-conductor with green/yellow striped adhesive tape,
the N-conductor with light blue adhesive tape.

Outgoing feeders from the neutral conductor in auxiliary circuits (e.g. return lines from contactor coils) have no special markings.

The terminal for the external PE-conductor must be identified by the ⏚ symbol, unless the PE-conductor is to be connected to a bar clearly identified by the green/yellow marking.

Identification of circuits

Apart from the identifying markings to be applied by the manufacturer of the switchgear and distribution boards, there are certain markings, which can only be applied by the installer once the assembly has been erected and connected, e.g. identification of the outgoing circuits.

Type of installation, accessibility

In order to ensure that the most cost-effective installation type is selected in all cases, the characteristic features of the switchgear and distribution boards should be compared before any structural measures are defined. These features include, for example:

- ▷ open or enclosed design (type of operating area),
- ▷ type of installation:self-supporting: free-standing in a room, against a wall, or in a wall recess; not self-supporting: for attachment to a wall, a rack, or in a wall recess,
- ▷ accessibility, e.g. for installation, maintenance, and operation,
- ▷ dimensions (height, depth),
- ▷ recommendations with respect to structural measures.

Special requirements

Possible special requirements, such as explosion protection, protection against corrosive atmospheres, vibration, consistent use of a particular PE-conductor, and the scope of the documentation to be supplied (e. g. circuit diagrams, parts lists, location diagrams, test reports) must be considered in the light of applicable standards or within the framework of additional agreements.

The following standards must be observed when installing low-voltage switchgear and distribution boards:

▷ EN 60 439–1/IEC 60439–1/DIN VDE 0660-500, "Low-voltage switchgear and controlgear assembles; Requirements for type-tested (TTA) and partially type-tested (PTTA) assemblies",
▷ VDE 0106 Part 100, "Protection against electric shock; Actuating members positioned close to parts liable to shock".

Further standards also exist for special cases, e. g.:

▷ EN 60204–1/IEC 60204–1/VDE 0113, "Electrical equipment of machines; General requirements",
▷ EN 60439–2/IEC 60439–2/DIN VDE 0660-502, "Low-voltage switchgear and controlgear; Particular requirements for busbar trunking system",
▷ EN 60439–3/IEC 60439–3/DIN VDE 0660-504, "Low-voltage switchgear and controlgear assemblies; Particular requirements for low-voltage switchgear and controlgear assemblies intended to be installed in places where unskilled persons have access for their use",
▷ EN 60439–4/IEC 60439–4/DIN VDE 0660-501, "Low-voltage switchgear and controlgear assemblies; Particular requirements for assemblies for construction sites",
▷ DIN VDE 0603, "Consumer units and meter panels AC 400 V",
▷ EN 60439/IEC 60439–5/DIN VDE 0660-503, "Cable distribution cabinets",
▷ DIN VDE 0660-505, "House connection boxes".

As with all other electrical equipment, all low-voltage switchgear and distribution boards must be installed and connected in accordance with the DIN VDE 0100 standards.

Erection

The most important erection requirements are described in the specifications list. An offer is first obtained from the fitters and construction personnel who are responsible for erecting the system.

Offer

The offer must contain prices for all of the services quoted in the specifications list. Deviations from technical specifications are only permitted in well-founded cases. Alternatives are also frequently requested to allow comparisons to be made (higher/lower prices). Location diagrams must also be included with the offer.

Order, execution

All quantities and technical specifications are then defined in the order and all materials and products, which are not currently in stock, must be ordered. Assembly and production lists must be compiled for parts manufactured internally. The system is then erected and commissioned, and the necessary documentation is handed over to the customer.

P.I.S.A.A software tools

Various programs for compiling offers and processing orders are available from engineering consultants, software companies, and manufacturers.

The most important steps and tasks in planning distribution boards can be carried out using P.I.S.A.A (Siemens planning tools for installation engineering, for compiling offers and processing orders) (Fig. 1.4/73).

Compiling offers with P.I.S.A.A

P.I.S.A.A can be used to compile offers quickly and clearly for industrial plants, residential or functional buildings as well as all other applications involving STAB, SIKUS, SIKUS Universal, SIMBOX 63, STRATUM domestic distribution boards or 8HP insulation-enclosed distribution boards and SIPRO meter cabinets.

The desired installation equipment is selected according to functional criteria. The program then suggests the appropriate types of distribution board and assembly kits. Space for future upgrades can also be included. Various discounts as well as assembly times are taken into consideration. The program also supports easy-to-use functions for preparing location and circuit diagrams.

SIEMENS
P.I.S.A.A
Version 2.1 für Windows
Planungshilfen für Installationstechnik von Siemens zur Angebotserstellung und Auftragsbearbeitung

Fig. 1.4/73 Siemens planning tool P.I.S.A.A

Meter cabinets are selected according to the specifications of the public utility by entering the place name and postal code of the installation location. If the public utility is not known, the cabinets can be selected on the basis of the technology described. A view diagram is created automatically.

All important key data, such as customer and supplier addresses, as well as calculation parameters (e. g. hourly rate and discounts), can be entered beforehand. "Best-sellers" can be easily defined using P.I.S.A.A to speed up the selection process even further. You can also perform calculations based on your own fixed prices and assembly times. Additional equipment from the Siemens range or from other manufacturers can, of course, be included in calculations.

Each offer can also be created with a separate cover letter and then printed out or saved as a file and processed further.

Order processing with P.I.S.A.A

Using P.I.S.A.A, an order can be based on an offer created previously or simply replanned. The appropriate assembly kits for the distribution boards are selected graphically. Parts, which were included in the offer as flat-rate charges or which are required as additional components, can also be added. Detailed calculations can be performed to check that all your data is correct. The program then outputs complete material and order lists for the different suppliers and warehouses.

Electrical domestic distribution boards can be fully planned and configured using the P.I.S.A.A project planning tool. The program saves the product data selected, creates view, location and circuit diagrams, and calculates prices along with the total price in the appropriate national currency or in EUROs. The user interface of the program can be set to the following languages – German, English, French, Italian, Portuguese, or Spanish.

Fig. 1.4/74 shows the sequence of steps when planning distribution boards with P.I.S.A.A.

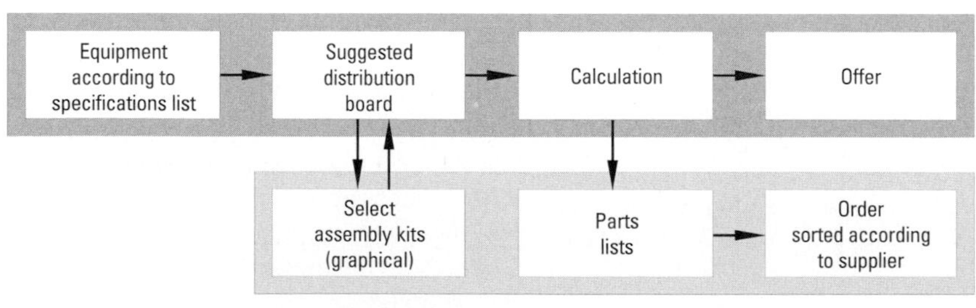

Fig. 1.4/74 Planning distribution boards with P.I.S.A.A

1.5 Grounding systems

Grounding principle

The main mass of the earth can be considered as an invariant reference potential (zero potential). This zero potential is applied to all parts conductively connected to ground. In practice, the exposed conductive parts of electrical equipment are connected to this reference potential directly. In the event of a short circuit to an exposed conductive part, the location surrounding the ground connection (grounding system) from where a person or animal can touch the exposed conductive parts of electrical equipment will, therefore, also have approximately the same potential. A hazardous touch voltage is thus prevented. The area in which these locations and the exposed conductive parts of the electrical equipment have approximately the same potential is, as a result of equipotential bonding, extended beyond the immediate vicinity of the grounding electrode to include the entire area occupied by an electrical consumer installation. Furthermore, the grounding system must also ensure that protective devices, such as an RCD (residual current protective device) or overcurrent protective device, are triggered in the event of a short circuit to an exposed conductive part (for example, in TT systems). The type and quality of the connection between ground and the electrical installations concerned are crucial factors here.

Operational and protective reasons

Grounding may be used for both operational reasons (functional or operational grounding) or protective reasons (protective grounding).

Standards/ specifications

The basic terms and characteristic values as well as the requirements concerning the scope and type of the equipment are defined in the relevant standards and specifications. These include: prEN 50179 / DIN VDE 0141, HD 384.4.41 / IEC 60364-4-41 / DIN VDE 0100-410, HD 384.5.54 / IEC 60364-5-54 / DIN VDE 0100-540, prEN 50114 / DIN VDE 0151, ENV 61024-1 / IEC 61024-1 / DIN VDE 0185-1, and IEC 60364-5-548 / DIN VDE 0800-2.

1.5.1 Basic requirements for grounding in electrical installations

Grounding in electrical power installations ≥ 1 kV

Operational and protective grounding in h.v. installations

Requirements concerning grounding in electrical installations > 1 kV are defined in prEN 50179 / DIN VDE 0141. Additional conditions stipulated by the responsible public utility may also apply in particular cases. Since operational grounding in h.v. installations is usually the responsibility of the respective public utility, only protective grounding will be dealt with in the following.

Grounding metallic parts in electrical equipment

Requirement specified in standard:

Grounding metallic parts in electrical equipment

Metallic parts in all electrical equipment with rated voltages > 1 kV, which are not part of the main circuit, must be grounded. If these metallic parts have a permanent and electrically conductive connection to grounded baseplates or metallic supporting structures, they do not require additional grounding (i.e. they do not require an additional ground connection).

> *Note*
> This requirement applies for all electrical devices to which the high voltage is applied directly. In certain cases, e.g. switchgear, these devices are integrated in switchgear assemblies at the factory or workshop. The design of these assemblies is such that the requirements are fulfilled when they are assembled. By way of contrast, the exposed conductive parts (housing) of switchgear, transformers, capacitors, generators, and motors must be connected to ground on site (i.e. to grounding electrode conductors which are connected to a grounding system).

Grounding metallic parts which are not part of electrical equipment

Requirement specified in standard:

Grounding external metallic parts

Metallic parts must be grounded if, in the event of a fault, there is a danger of them connecting with live parts by means of direct contact or electric arcs.

Metallic parts do not require additional grounding if they are bonded electrically with other grounded structural parts (e.g. if they are mounted on these parts, or attached to them via hinges, cradle mounts, etc.).

Note
No information concerning the distances which can be bridged by an electric arc, for example, is given in the standard. In practice, a distance of 1.50 m between the bare, live parts and the metallic housing elements (e.g. doors, windows, railings, covers) is generally used for open indoor installations, such as transformers with a primary voltage of up to 30 kV. Windows embedded in reinforced concrete walls, or metal grids, for example, which are mounted in large grounded frames, do not require an additional ground connection.

Clearance to barrier

The safe clearances specified in prEN 50179 / DIN VDE 0141 can be used for external, conductive components in rooms containing encapsulated indoor switching stations, the housings of which form part of the protective grounding system.

This clearance must be at least 115 mm for installations up to 10 kV and at least 215 mm for installations up to 20 kV.

Grounding h.v. cables and fittings

Armoring, metallic cladding, and shields

Requirement specified in standard:
Armoring, metallic cladding, or shields must be grounded at one end at least.

Note
Grounding at both ends should always be provided to increase the conductance of the shield for carrying the ground fault current in the event of a ground fault and to utilize the grounding continuity, for example, from station to station.

Grounding cable racks for h.v. cables

Cable racks

Requirement specified in standard:

Cable racks do not need to be grounded if the metallic shields or cladding of the cables which they support can carry the ground fault current.

Note 1
This always applies if the cross section of the shield or conductive cladding of the cable is dimensioned in accordance with the cable standard and is grounded at both ends.

Note 2
If, in exceptional cases, the shield can be grounded at one end only, the cable racks must also be grounded. The conductive surge continuity is not subject to any qualitative requirements.

Grounding in electrical power installations ≤ 1000 V

Requirements concerning grounding in electrical installations ≤1000 V are defined in the standards HD 384.4.41 / IEC 60364–4-41 / DIN VDE 0100–410 and HD 384.5.54 / IEC 60364–5-54 / DIN VDE 0100–540.

Functional and protective grounding in l.v. installations

Grounding in installations ≤1000 V must satisfy the combined requirements of functional and protective grounding, whereby protective grounding is of primary importance since operating personnel come into direct contact with l.v. devices.

The type of grounding used in an electrical installation and the specified minimum values for the grounding resistance depend on the type of ground connection for the selected network system and the protective devices.

Grounding in TN, TT, and IT systems

Grounding is required in the following network systems:

TN systems (see Section 2.4, Fig. 2.4/3):

▷ central grounding of a system point, usually the neutral point of the transformer or generator, and the equipment grounding conductor (PE or PEN conductor),
▷ grounding of main equipotential bonding;

TT systems (see Section 2.4, Fig. 2.4/2):

▷ central grounding of a system point, usually the neutral point of the transformer or generator,
▷ grounding of main equipotential bonding,
▷ common or separate grounding of the exposed conductive parts in the consumer installation;

IT systems (see Section 2.4, Fig. 2.4/1):

▷ grounding of main equipotential bonding,
▷ common or separate grounding of the exposed conductive parts in the consumer installation.

Interconnection of grounding systems for h.v. and l.v. installations

Interconnection of grounding systems

It is advisable to connect all operational and protective grounding systems to one common grounding system if the h.v. and l.v. system in question is installed in an area with building complexes or in an industrial plant. The reason for this is that total separation of the grounding systems is not possible in these situations.

This applies, in particular, to electrical building services management systems which explains why a *common* grounding system should always be installed.

Grounding in telecommunications systems

Functional and protective grounding in telecommunications systems

Requirements concerning the grounding of telecommunications systems are defined in IEC 60364-5-548 / DIN VDE 0800. Both functional and protective grounding are important in telecommunications systems.

Grounding electrodes for telecommunications systems

In addition to an independent grounding system, these standards also permit the following as grounding electrodes for a telecommunications system:

▷ foundation grounding electrodes of the building,
▷ lightning-protection grounding system in compliance with ENV 61024–1 / IEC 61024–1 / DIN VDE 0185–1,
▷ grounding in compliance with prEN 50179 / DIN VDE 0141.

Coordination of trades

These grounding electrodes must ensure that the systems remain operational. It is important that the different trades involved in the construction activities and the specified requirements are coordinated.

Grounding of lightning-protection systems

Grounding of lightning-protection systems

Requirements concerning the grounding of lightning-protection systems are defined in the standards ENV 61024-1 / IEC 61024-1 / DIN VDE 0185-1.

In addition to an independent grounding system, it is also possible to use the following as grounding electrodes for lightning-protection systems:

▷ foundation grounding electrodes of the building,
▷ armoring in reinforced concrete foundations,
▷ grounding in compliance with prEN 50179 / DIN VDE 0141.

Reduction of pace and touch voltages by means of potential grading in the vicinity of electrical installations

Reduction of pace and touch voltages

Potential grading is required for certain electrical installations to reduce pace and touch voltages which, in the event of a fault, can become dangerously high. The principle behind potential grading is illustrated in Fig. 1.5/1.

Potential grading in installations > 1 kV

The areas in which potential grading may become necessary for *h.v. installations* > 1 kV are described in prEN 50179 in Appendix D / DIN VDE 0141 Section 4.4.2. These areas include:

▷ the contact zone around stations with metallic outer walls,
▷ the operating locations in and around indoor installations,
▷ the contact zone around metal conductive fences surrounding outdoor installations,
▷ the operating locations in and around outdoor installations.

The following measures are recommended:

Conductor grounding electrodes for potential grading

Contact zones:

A conductor grounding electrode laid at a depth of approximately 0.5 m and at a distance of approximately 1 m from the objects. The electrode must be connected to the grounding system.

At operating locations:

Mesh-type conductor grounding electrodes or structural steel mesh laid in the building foundations or in the soil at a depth of approximately 0.20 m. The interwoven, metallically conductive mesh or matting must be connected to the grounding system.

Minimum cross sections

The minimum cross sections of the conductor grounding electrodes are specified in Table 1.5/2.

Ground resistance

No special ground resistance is required for potential grading.

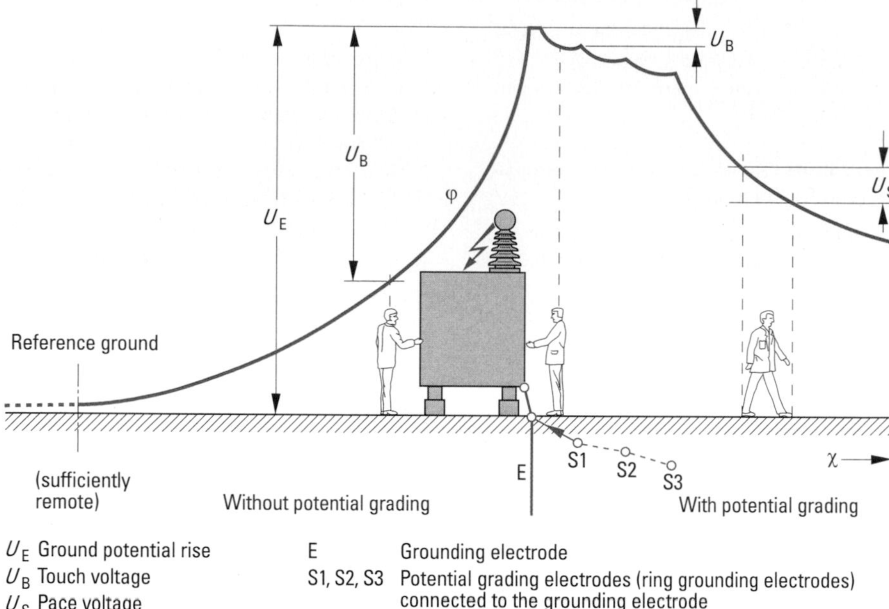

Fig. 1.5/1
Example showing the variation of ground-to-electrode potential and the voltages for a current-carrying grounding electrode

Insulation of operating location

As an alternative to potential grading, it is possible to use insulating mats to *insulate* the operating location, for example, in front of indoor installations.

Potential grading in installations <1 kV

In the case of *l.v. installations*, national standards for special applications, such as indoor and outdoor swimming pools, fountains, and agricultural facilities for livestock, specify measures for potential grading in the hazardous areas of these structural installations (see Section 5.9). These standards focus on the effect of hazardous pace voltages on wet or damp surfaces in the protection zones surrounding electrical installations.

Laying structural steel mats or mesh-type conductor grounding electrodes is recommended for potential grading.

Potential grading must be connected to the equipment grounding conductor via equipotential bonding conductors.

1.5.2 Determining the permissible ground resistance

Ground resistance

The maximum permissible resistance of a grounding system is determined by the requirements of the electrical installation to be protected. Whether this involves purely functional or protective grounding or a combination of both is a prime consideration.

Ground resistance of protective grounding in installations >1 kV

Verification with public utility

A resistance of 5 Ω is usually sufficient. This value should, however, always be verified with the public utility responsible for the h.v. supply.

Ground resistance (R_B) of central system point in TN and TT systems (operational grounding)

Total ground resistance in TN and TT systems

The standards HD 384.4.41 / IEC 60364-4-41 / DIN VDE 0100-410 do not specify a total ground resistance value for grounding the central system point (neutral point of the transformer or generator), except for special cases where a fault may occur between an external conductor and ground (for example, with overhead power lines). In such cases, the following condition must be fulfilled for the operational ground:

$$\frac{R_B}{R_E} \leq \frac{50\text{ V}}{U_0 - 50\text{ V}}$$

R_B Total ground resistance of all operational grounding electrodes

R_E Lowest assumed ground contact resistance of the external conductive parts, not connected to an equipment grounding conductor, via which a ground fault can occur

U_0 Rated voltage with respect to grounded conductors

Grounding main equipotential bonding

Main equipotential bonding

Main equipotential bonding for each building as specified in HD 384.4.41 / IEC 60364-4-41 / DIN VDE 0100-410 is used to improve the central grounding of the system point and equipment grounding conductor. This is achieved by connecting all the natural ground electrodes of a building and all metallic parts connected directly to ground together at each of these points. The central grounding system is, therefore, enhanced because all of the grounding electrodes are also connected to each other in parallel. This ensures that, for each building, the grounding quality of the equipment grounding conductor is considerably better than the ground contact resistance of a possible fault in the building system.

The grounding resistance of the main equipotential bonding is *not* subject to any specific requirements. The foundation grounding electrodes, for example, required in buildings are perfectly adequate as grounding electrodes.

Ground resistance of exposed conductive parts (R_A) in TT systems

Common protective device/ equipment grounding conductors

In TT systems, all exposed conductive parts of electrical equipment which are protected by a common protective device must be connected to a common grounding electrode via an equipment grounding conductor. All exposed conductive parts which can be touched at the same time must be connected to the same grounding electrode. The following condition applies for the ground resistance of the exposed conductive parts R_A:

$$R_A \cdot I_a \leq 50\text{ V}$$

R_A Resistance of the grounding electrode and equipment grounding conductor of the exposed conductive parts

I_a Current which causes the protective device to trigger automatically within 5 s. If an RCD is used, I_a corresponds to the rated residual current $I_{\Delta n}$

Example

1) Protective device: *Residual-current-operated circuit-breaker* (RCCB)

Rated residual current $I_{\Delta n} = 30$ mA

$$R_A \leq \frac{50\text{ V}}{I_{\Delta n}} \leq \frac{50\text{ V}}{30\text{ mA}} = 1.7\text{ k}\Omega$$

2) Protective device: *Miniature circuit-breaker*

Rated current 10 A, B characteristic $I_a \leq 5 \cdot I_n$

$$R_A \leq \frac{50\text{ V}}{I_a} \leq \frac{50\text{ V}}{5 \cdot 10\text{ A}} = 1\ \Omega.$$

In practice, the low ground resistance of 1 Ω is often very difficult to achieve. For this reason, RCDs with a rated residual current $I_{\Delta n}$ of 30 mA are usually used as protective devices.

Ground resistance of exposed conductive parts (R_A) in IT systems

Equipment grounding conductors

In IT systems, all of the exposed conductive parts must be connected to a grounded equipment grounding conductor individually, in groups, or all together.

Requirements at initial fault

The following condition must be fulfilled to prevent a touch voltage greater than ~50 V from being generated and sustained when an initial permissible fault occurs between the grounded exposed conductive part and ground:

$$R_A \cdot I_d \leq 50\text{ V}$$

R_A Resistance of the grounding electrode and equipment grounding conductor of the exposed conductive parts

I_d Fault current for the initial fault with negligible impedance between an external conductor and the equipment grounding conductor or an exposed conductive part to which it is connected. The value of I_d allows for the leakage currents and the total impedance of the electrical installation with respect to ground.

Capacitive leakage current

The fault current I_d corresponds to the sum of all leakage currents in the IT system. This primarily concerns the capacitive leakage currents of the conductor system. For example, a conductor system (NYM cable) measuring 100 m in length with conductor cross sections of 2.5 mm² would, therefore, have a total capacitive leakage current of 1.4 mA. In this case, a ground resistance value of 2 Ω would normally be adequate.

Requirements after initial fault

If not all exposed conductive parts in an IT system are connected to a common equipment grounding conductor but, instead, are grounded individually or in groups, the TT system requirements relating to the grounding of exposed conductive parts must be fulfilled since the situation here is similar to that in a TT system when an initial fault occurs (see page 207).

1.5.3 Dimensioning and types of grounding electrode

Dissipation resistance

The currents generated by an electrical installation and in the event of a fault (ground fault, short-circuit to ground) are carried to and from the "ground" conductor via the grounding electrode conductors and the grounding electrode. Depending on the requirements of the respective electrical installation, this presupposes minimum values for the dissipation resistance of the grounding electrodes and minimum cross sections for grounding electrode conductors.

Soil resistivity ϱ_E

Soil resistivity

The size and complexity of a grounding system are primarily determined by the "soil resistivity" of the earth in which the grounding system is to be embedded. It varies greatly according to the type of soil, its stratification, granular structure, and dampness. Seasonal fluctuations in dampness must also be taken into consideration.

Reference values (Table 1.5/1), which are also specified in prEN 50179 / DIN VDE 0141, can provide assistance during project planning.

If no site-specific data is available, measurements must be carried out to predetermine the soil resistivity (see Section 1.5.7).

Table 1.5/1
Reference values for the soil resistivity of various soil types

Soil type	Soil resistivity ϱ_E in Ωm
Marshy ground	5 to 40
Loam, clay, humus	20 to 200
Sand	200 to 2500
Gravel	2000 to 3000
Weathered rock	usually less than 1000
Sandstone	2000 to 3000
Granite	to 50000
Morainic debris	to 30000

Types of grounding electrode

A distinction is made between two typical types of grounding electrode:

- ▷ conductor grounding electrodes
- ▷ buried grounding electrodes

Conductor grounding electrodes

Conductor grounding electrodes are laid horizontally at a depth of between 0.5 m and 1 m. They can be made of strip, round-section, or stranded material and can be installed in radial, ring, or mesh configurations.

Buried grounding electrodes

Buried grounding electrodes are generally installed vertically at a considerable depth. They can be made of tubular, round-section, or sectioned material and can be installed as individual grounding electrodes or in radial, ring, or mesh configurations by establishing active ground conductor connections.

Natural grounding electrodes

Natural grounding electrodes are structural elements of a building or its foundations, or separate metallically conductive parts which are in permanent contact with the ground or water over a large area.

These elements include sheet piling, concrete pile reinforcements, steel building elements, metallic piping.

Foundation grounding electrodes

Foundation grounding electrodes are a special type of conductor grounding electrode. They are laid as closed rings in the exterior wall foundations of buildings. In the event of current flow to ground, the area within the ring has approximately the same potential which eliminates the risk of potential differences. The current raises the whole object to grounding-electrode potential relative to the surrounding environment.

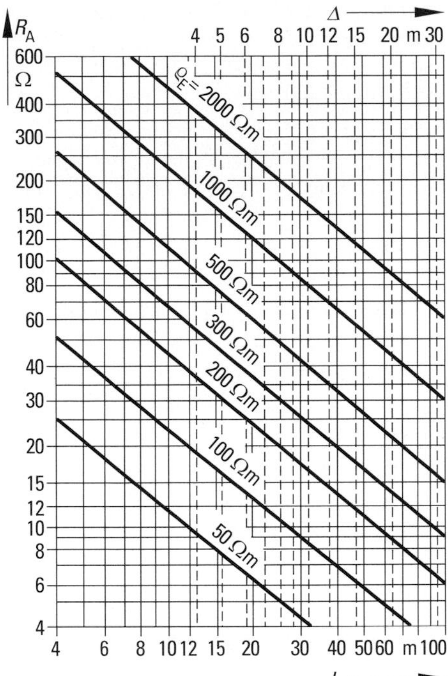

Strip grounding electrode: $R_A = \frac{\varrho_E}{\pi L} \cdot \ln \frac{2L}{d}$,

Ring grounding electrode: $R_A = \frac{\varrho_E}{\pi^2 D} \cdot \ln \frac{2\pi D}{d}$,

L Length of strip grounding electrode

D $\left(= \frac{L}{\pi}\right)$ Diameter of ring grounding electrode

d Diameter of grounding cable and round wire, or half width of grounding strip (in this case: 1.5 cm)

ϱ_E Soil resistivity in Ωm

Fig. 1.5/2
Dissipation resistance of conductor grounding electrodes (strip, round-section, or stranded material) laid straight or as ring in homogeneous soil

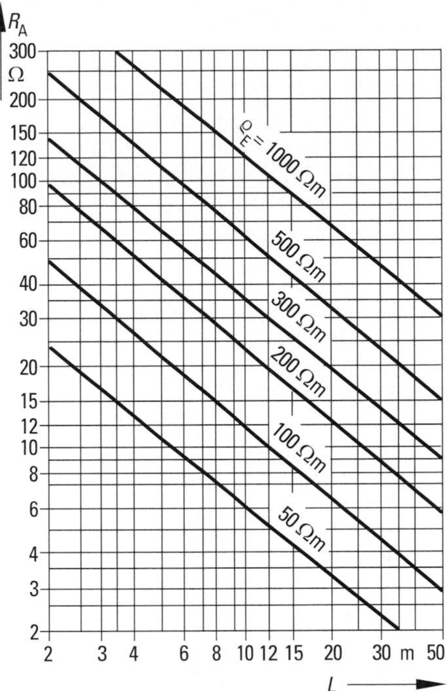

$R_A = \frac{\varrho_E}{2\pi L} \cdot \ln \frac{4L}{d}$

L Length of buried grounding electrode

d Diameter of grounding conductor bar (in this case: 2 cm)

ϱ_E Soil resistivity in Ωm

Fig. 1.5/3
Dissipation resistance of buried grounding electrodes arranged vertically in homogeneous soil

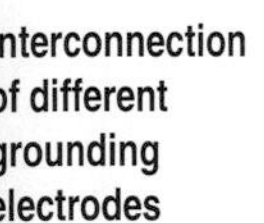
Interconnection of different grounding electrodes

The interconnection of different grounding electrode configurations to form a single grounding system is common practice and is permitted. It is, however, important to ensure that the different materials are compatible with each other (see page 213 ff.).

Dissipation resistance of grounding electrodes

Dissipation resistance

The dissipation resistance of a grounding electrode largely depends on the soil resistivity and the type of grounding electrode.

The standard prEN 50179 / DIN VDE 0141 contains graphs (Figs. 1.5/2 and 1.5/3), which allow the expected dissipation resistance to be determined for the two most commonly used types of grounding electrode. The graphs are based on the specified formulae.

Conductor grounding electrodes

In practice, the following approximate calculations are usually applied for conductor grounding electrodes:

Strip grounding electrode

Strip grounding electrode: $R_A = \frac{2 \cdot \varrho_E}{L}$,

Ring grounding electrode

Ring grounding electrode: $R_A = \frac{2 \cdot \varrho_E}{3 \cdot D}$,

Mesh grounding electrode

Mesh grounding electrode: $R_A = \frac{\varrho_E}{2 \cdot D}$,

ϱ_E Soil resistivity in Ωm

L Length of strip grounding electrode in m

D Diameter of enclosed area (ideal circle with mesh grounding electrode) in m

The calculated dissipation resistance values R_A can, however, only be used as rough values since they are lower than the values calculated in accordance with prEN 50179 / DIN VDE 0141.

Conductor grounding electrodes have an optimum length. This means that the dissipation resistance ceases to drop as of a specific electrode length. This, however, only applies to electrodes ≥ 500 m. With radial configurations, the dissipation resistance drops if the angles between the radial electrode arms are less than 60°.

Foundation grounding electrodes are regarded as conductor grounding electrodes. Experience has shown that very high, stable dissipation resistance values can be achieved with foundation grounding electrodes. Measurements carried out over a period of many years reveal, for example, that dissipation resistances of between 2 Ω and 15 Ω can be achieved with electrode lengths ≥ 40 m.

As a rule, conductor grounding electrodes should be laid at a depth of between 0.5 m and 1 m, provided that the soil conditions permit this.

Stones and coarse gravel in the immediate vicinity of the grounding electrode increase the dissipation resistance. In such cases, the grounding electrode should be surrounded with homogeneous soil and the soil then compacted.

Buried grounding electrodes

Buried grounding electrodes should, as far as possible, be driven into the ground perpendicular to the surface. They are of particular benefit in cases where the soil resistivity is inversely proportionate to the depth.

If several buried grounding electrodes are needed to achieve the required dissipation resistance, the distance between the electrodes should be twice the active length of a single electrode.

It is important to note that, with high soil resistivity (for example, in the upper soil layers), buried grounding electrodes are not active along their entire length.

In practice, the following approximate calculation is also used to determine the dissipation resistance of buried grounding electrodes:

$$R_A = \frac{\varrho_E}{L}$$

L Active length of buried grounding electrodes in m

As with the approximate calculations for conductor grounding electrodes, lower values are obtained with this formula than those shown in Fig. 1.5/3.

It is, therefore, essential to verify the results by carrying out appropriate measurements (see page 217).

1.5.4 Dimensioning grounding electrode conductors

Ground fault current/disconnecting time/ minimum cross section

The operating or fault currents to be carried and the disconnecting times relevant for fault currents must be taken into consideration when dimensioning grounding electrode conductors (connection between the installation parts to be grounded and the grounding electrode). Furthermore, the values specified for the minimum cross sections must also be observed (see Table 1.5/4). If common grounding systems for installations ≤ 1000 V and > 1 kV and a common grounding electrode conductor are used, the grounding electrode conductor must be dimensioned for the maximum possible single requirement. It is not necessary to presuppose a scenario where two faults occur simultaneously.

Ground fault currents in installations > 1 kV

The following play a decisive role when dimensioning the grounding electrode conductor of the *protective grounding* for installations > 1 kV:

Double-phase-to-ground fault current I''_{k2E}

▷ double-phase-to-ground fault I''_{k2E} with h.v. systems with insulated neutral point (auxiliaries systems in industrial installations and power stations) and with ground-fault compensation (overland and urban systems of public utility),

ingle-phase-to-round fault urrent I''_{k1E}

▷ single-phase-to-ground fault current I''_{k1E} with h.v. systems with low-resistance neutral point (systems with rated voltages >110 kV).

These currents can be calculated approximately using the respective system fault level S''_k, the rated system voltage U_n, as well as the short-circuit positive-phase-sequence impedance and zero-phase-sequence impedance of the system ($\underline{Z}_{(1)}$ and $\underline{Z}_{(0)}$):

double-phase-to-ground fault current:

$$I''_{k2E} = \frac{S''_k}{2 \cdot U_n},$$

single-phase-to-ground fault current:

$$I''_{k1E} = \frac{1.1 \cdot U_n}{\sqrt{3}\,(2 \cdot \underline{Z}_{(1)} + \underline{Z}_{(0)})}.$$

Since the high voltage for electrical installations in buildings is usually supplied from systems with rated voltages < 110 kV (i.e. from systems with insulated neutral point or with ground fault compensation), only the double-phase-to-ground fault current needs to be determined, whereby the disconnecting time for the fault t_F is assumed to be no more than 1 s.

Example

S''_k = 500 MVA (system fault level)

U_n = 20 kV (rated system voltage)

$$I''_{k2E} = \frac{S''_k}{2 \cdot U_n} = \frac{500\ \text{MVA}}{2 \cdot 20\ \text{kV}} = 12.5\ \text{kA}.$$

Ground fault currents in installations ≤1000 V

Key factor: single-phase-to-ground fault current

The single-phase-to-ground fault current plays a decisive role when dimensioning the grounding electrode conductor of the *functional and protective grounding* in installations ≤1000 V with direct grounding in TN and TT systems. For transformers with delta-wye connections (Dy), the single-phase-to-ground fault current can be calculated approximately using the rated system voltage U_n and the sum of the upstream system impedances $\underline{Z}_k$ (in h.v. systems) and the transformer impedance $\underline{Z}_T$:

$$I''_{k1E} = \frac{c \cdot U_n}{\sqrt{3}\,(2 \cdot \underline{Z}_k + \underline{Z}_T)}.$$

A disconnecting time of approximately 0.5 s can be used here (time within which the first overcurrent protective device in the l.v. installation usually triggers in the event of a fault). Bearing in mind the selectivity requirements, this time may, in certain cases, be considerably shorter.

> *Note*
> It is assumed here that the unprotected cable run between the transformer and transformer circuit-breaker is designed to be inherently ground-fault-resistant and short-circuit-proof and that a transformer fault will be detected by automatic monitoring devices and cleared on the h.v. side.

Example

U_n = 400 V (rated system voltage)

S''_k = 500 MVA (system fault level of h.v. system)

u_{kr} = 6% (short-circuit voltage of transformers)

S_T = 2 · 630 kVA (transformer output)

System impedance

$$\underline{Z}_Q = \frac{1.1 \cdot U_n^2}{S''_k} = \frac{1.1\,(400\ \text{V})^2}{500\ \text{MVA}} = 0.352\ \text{m}\Omega.$$

Transformer impedance

$$\underline{Z}_T = \frac{u_{kr} \cdot U_n^2}{S_T} = \frac{6\% \cdot (400\ \text{V})^2}{1260\ \text{kVA} \cdot 100\%} = 7.6\ \text{m}\Omega$$

$$I''_{k1E} = \frac{c \cdot U_n}{\sqrt{3}\,(2 \cdot \underline{Z}_Q + \underline{Z}_T)}$$

$$= \frac{400\ \text{V}}{\sqrt{3}\,(2 \cdot 0.352\ \text{m}\Omega + 7.6\ \text{m}\Omega)} = 27.8\ \text{kA}.$$

With a rated system voltage of 400 V, $c = 1$.

(More detailed information on calculating the short-circuit current can be found in Section 1.2.)

Determining the cross sections of grounding electrode conductors

Cross section of grounding electrode conductor

To allow the cross section of the grounding electrode conductor to be determined, prEN 50179 / DIN VDE 0141 specifies the permissible short-circuit current density of a number of different materials (see Fig. 1.5/4). The permissible final temperature for the respective material when the disconnecting time t_F is reached is used as a basis here.

The required cross section A is calculated as follows:

$$A = \frac{I''_k}{G}$$

G Permissible short-circuit current density

Example

For h.v. protective grounding:

$I''_{k2E} = 12.5$ kA

$t_F = 1$ s

For l.v. functional and protective grounding:

$I''_{k1E} = 27.8$ kA

$t_F = 0.5$ s

Material: Galvanized steel.

Short-circuit current density

According to Fig. 1.5/4, the permissible short-circuit current density for galvanized steel (line 4) is:

G (where $t_F = 1$ s) = 70 A/mm²,

G (where $t_F = 0.5$ s) = 100 A/mm².

The cross section of the grounding electrode conductor is calculated as follows:

for the h.v. protective grounding:

$$A = \frac{12.5 \cdot 10^3 \text{ A}}{70 \text{ A/mm}^2} = 180 \text{ mm}^2$$

for the l.v. functional and protective grounding:

$$A = \frac{27.8 \cdot 10^3 \text{ A}}{100 \text{ A/mm}^2} = 278 \text{ mm}^2.$$

1.5.5 Requirements for grounding electrode components

Grounding electrode materials

Grounding electrode materials

The following materials have proven suitable for grounding electrodes and are prescribed for this purpose in prEN 50179 / DIN VDE 0141:

▷ Steel, hot-galvanized,
Strip, round, shaped cross sections and tubes for conductor and buried grounding electrodes;

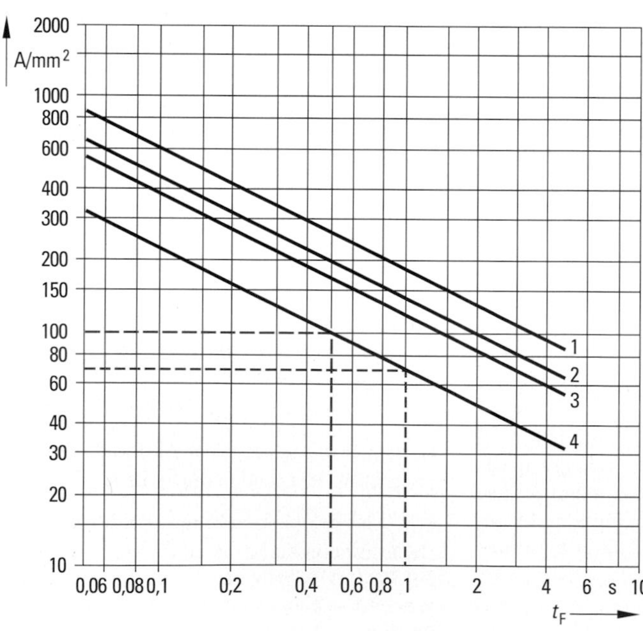

1 Copper, bare or galvanized
2 Copper, galvanized or lead-cladded
3 Aluminum, for grounding electrode conductors only
4 Steel, galvanized

Fig. 1.5/4
Short-circuit current density *G* for main ground buses, grounding electrode conductors, and grounding electrodes as a function of the duration of the fault current t_F. Lines 1, 3, and 4 apply for a final temperature of 300 °C. The dashed line 2 applies for a final temperature of 150 °C.

- ▷ Steel, lead-cladded,
Round wire for conductor grounding electrodes;
- ▷ Steel, copper-cladded,
Round bar for buried grounding electrodes;
- ▷ Copper, bare,
All electrode shapes;
- ▷ Copper, tin-cladded,
Stranded materials for conductor grounding electrodes;
- ▷ Copper, galvanized,
Strip cross sections for conductor grounding electrodes;
- ▷ Copper, lead-cladded,
Stranded materials and round cross sections for conductor grounding electrodes.

The following corrosion hazards should be taken into consideration when selecting the material for the grounding electrode.

Minimum dimensions for grounding electrodes

Minimum dimensions

Since grounding electrodes are subject to mechanical loads and corrosion, minimum dimensions are defined for the electrodes which ensure a reasonable service life (Table 1.5/2). These minimum dimensions are specified in tables in the standards prEN 50170/ DIN VDE 0141, prEN 50114 / DIN VDE 0151, HD 384.5.54 / IEC 60364-5-54 / DIN VDE 0100-540.

Electrolytic corrosion of grounding electrodes

Electrolytic corrosion

The risk of electrolytic corrosion of the grounding electrodes and subsequent ineffectiveness depend on the inherent corrosion of the electrode components (e.g. with individual grounding electrodes without any metallically conductive connections to other grounding electrodes) and the formation of corrosive elements (e.g. if grounding electrodes made of different materials – materials with different chemical potential values – are connected together).

Inherent corrosion

Inherent corrosion

The properties of the individual electrode materials and their compatibility with the different soil types must be taken into consideration to prevent inherent corrosion. Information regarding this can be found in prEN 50114 / DIN VDE 0151 (a summary is given below).

Hot-galvanized steel is extremely resistant in almost all soil types. The reason for this is the heterogeneous iron-zinc alloy layers and the outer layer of pure zinc which tends to form a covering layer. The prerequisite for a reasonable service life is a sufficiently thick, non-porous, flawless zinc coating.

Hot-galvanized steel is also suitable for embedding in concrete.

Lead-cladded round steel wire

Lead tends to form a good covering layer in soil and is, therefore, resistant to many soil types. There is, however, a risk of corrosion in extremely alkaline environments (pH value ≤ 10). Lead must, therefore, never be in direct contact with concrete.

In soil, the steel core may corrode if the lead cladding is damaged. Although the risk of corrosion is negligible in well aerated soils (e.g. in sand), it can be much greater in poorly aerated soils (e.g. in loam and clay).

Steel with extruded copper cladding and electrolytically copper-plated steel

Damage to the copper cladding will considerably increase the risk of corrosion to the steel core. The copper layer must, therefore, be seamless. At points where multi-section buried grounding electrodes are coupled, copper cladding or copper coating layers must be seamless and interconnected with the same conductance.

Bare copper is generally extremely resistant in soil.

Tin-plated or galvanized copper is – similar to bare copper – generally very resistant in soil. Tin-plated copper is currently only used for stranded grounding electrodes, and galvanized copper only for strip grounding electrodes.

Formation of corrosive elements

Formation of corrosive elements through combination of different metals

Direct, metallically conductive connections between grounding electrodes made of different materials and the interconnection of grounding electrodes with other subterranean installations (tanks, containers, cables, conductive foundations, etc.) may result in the formation of corrosive elements. Depending

Table 1.5/2
Materials for grounding electrodes and their minimum dimensions with regard to corrosion and mechanical strength (source: prEN 50179 / DIN VDE 0141)

Material		Soil type	Minimum dimensions				
			Conductor			Coating/cladding	
			Diameter mm	Cross section mm^2	Thick-ness mm	Individual values m	Average values m
Steel	Hot-galvanized	Strip[2]		90	3	63	70
		Shaped section (incl. plates)		90	3	63	70
		Tubes	25		2	47	55
		Round bar for buried grounding electrode	16			63	70
		Round wire for conductor grounding electrode	10				50
	Lead-cladded[1]	Round wire for conductor grounding electrode	8			1000	
	Extruded copper cladding	Round bar for buried grounding electrode	15			2000	
	Electrolytically copper-plated	Round bar for buried grounding electrode	14.2			90	100
Copper	Bare	Strip		50	2		
		Round wire for conductor grounding electrode		25[3]			
		Stranded	1.8 *	25			
		Tubes	20		2		
	Tin-plated	Stranded	1.8 *	25		1	5
	Galvanized	Strip		50	2	20	40
	Lead-cladded[1]	Stranded	1.8 *	25		1000	
		Round wire		25		1000	

* for individual wire
1) Not suitable for direct embedding in concrete
2) Strip, rolled or cut, with rounded edges
3) May be used under exceptional circumstances if experience has shown that the risk of corrosion is extremely low. 16 mm²

on the metal/soil open-circuit potential of the respective material and the area ratios of the individual metallic components, which may, under certain circumstances, influence each other (area of the cathodic material S_k in relation to the area of the anodic material S_a), there may be a risk of corrosion to metals with the higher negative potential due to the formation of an effective elemental potential.

The information shown in Table 1.5/3 (from prEN 50114 / DIN VDE 0151) is based on experience gained in connecting grounding electrodes made of different materials with area ratios of $S_k : S_a \geq 100 : 1$.

It is, therefore, evident that problems may arise not only with combinations of very different metals, such as steel with copper, but also when connecting steel grounding elec-

able 1.5/3
nformation based on experience in connecting grounding electrodes made of different materials with area ratios of $S_k : S_a \geq 100 : 1$ (from prEN 0114 / DIN VDE 0151)

aterial with small surface area	Material with large surface area								
	Galvanized steel	Steel	Steel in concrete	Galvanized steel in concrete	Stainless steel	Copper	Tin-plated copper	Galvanized copper	Lead-cladded copper
Galvanized steel	+	+ Zinc corrosion	–	+ Zinc corrosion	–	–	–	+	+
Steel	+	+	–	+	–	–	–	+	+
Steel in concrete	+	+	+	+	+	+	+	+	+
Lead-cladded steel	+	+	–	+	–	–	+	+	+
Cu-cladded steel	+	+	+	+	+	+	+	+	+
Stainless steel	+	+	+	+	+	+	+	+	+
Copper	+	+	+	+	+	+	+	+	+
Tin-plated copper	+	+	+	+	+	+	+	+	+
Galvanized copper	+	+ Zinc corrosion	+ Zinc corrosion	+ Zinc corrosion	+ Zinc corrosion	+ Zinc corrosion	+ Zinc corrosion	+	+ Zinc corrosion
Lead-cladded copper	+	+	+ Lead corrosion	+	+ Lead corrosion	+ Lead corrosion	+	+	+

+ Compatible
– Non-compatible

trodes to "steel in concrete". This is especially the case with grounding electrodes which are connected to the foundation steel of buildings, for example (e.g. via the main equipotential bonding of the building).

Corrosion protection measures

Further corrosion protection measures, which are particularly important when connecting grounding electrodes to grounding electrode conductors, are specified in Section 4 of prEN 50144 / DIN VDE 0151. The most important measures are as follows:

▷ On account of increased susceptibility at the transition zone between soil and air, galvanized steel must be protected against corrosion at least 0.3 m both above and below the surface of the soil. Thin coatings are not sufficient. Coatings, which have good bonding characteristics and do not absorb any moisture, provide adequate protection.

▷ Connections can be established by means of bolts, welding (e.g. exothermic welding), soldering, and using crimped and pressure connectors. In concrete, it is also possible to use wedge-type connectors.

▷ Two bolts (at least M 8) or one bolt (at least M 10) must be used to connect flat strips to each other and to steel constructions.
Lead cladding covering copper and steel conductors must be removed before the bolted connections are established.

▷ Connecting points in soil must have the same corrosion characteristics as the material of the grounding electrodes. Connecting points which, as a result of being worked, when installed or for production-related reasons, do not have the same corrosion protection must be provided with a corrosion-proof coating after they have been assembled.

- ▷ Connecting points in concrete between individual reinforcing rods and between steel reinforcement elements and galvanized steel do not require any corrosion protection. Connecting points, exposed copper surfaces, and any lead cladding at connections between steel reinforcement elements and copper conductors must, however, be coated.
  Coating the lead cladding of lead-cladded steel conductors is sufficient for connections between steel reinforcement elements and the steel conductors.
- ▷ When filling in pits and ditches in which grounding electrodes have been laid, care must be taken to ensure that no slag, coal, or rubble comes into direct contact with the grounding electrode material.

1.5.6 Requirements for the components of grounding electrode conductors

Materials for grounding electrode conductors

Proven grounding electrode conductors

In practice, the following materials have proven suitable for grounding electrode conductors. They are also prescribed for this purpose in the relevant standards:

- ▷ Copper, aluminum, steel
  for insulated, mechanically protected installation (a "mechanically protected" grounding electrode conductor is laid, for example, in a tube of sufficient strength or as a cable)
- ▷ Copper, steel
  for insulated, mechanically unprotected installation
- ▷ Copper, hot-galvanized steel
  for exposed installation

Minimum dimensions for grounding electrode conductors

Mechanical strength, current carrying capacity

Minimum dimensions of grounding electrode conductors are necessary to make sure that their mechanical strength and current carrying capacity are sufficient to withstand ground faults.

Minimum cross sections where $U_n > 1$ kV

The following minimum dimensions specified in prEN 50179 / DIN VDE 0141 must be complied with for electrical power installations with a rated voltage > 1 kV:

Copper	16 mm²
Aluminum	35 mm²
Steel	50 mm²

With grounding electrode conductors which are laid *bare* in soil, the minimum dimensions are the same as those specified for grounding electrodes, since grounding electrode conductors *laid bare in soil* are regarded as grounding electrodes.

The actual cross sections to be used for grounding electrode conductors in installations > *1 kV* must, however, always be determined using the calculations on page 211 ff.

Minimum cross sections for underground installation where $U_n \leq 1000$ kV

The following minimum cross sections specified in HD 384.5.54 / IEC 60384-5-54 / DIN VDE 0100-540 must be complied with for electrical power installations with a rated voltage ≤ *1000 V*:

For installation with protection against corrosion and mechanical stress:

Copper	2.5 mm²
Aluminum	4 mm²
Steel	7.5 mm²

For installation without protection against corrosion and mechanical stress:

Copper	16 mm²
Steel	16 mm²

For bare installation without protection against mechanical stress:

Copper	25 mm²
Hot-galvanized steel	50 mm²

> *Note*
> If steel is selected as the grounding electrode conductor material, the cross section must be approximately three times that for copper. With copper conductors, the cross section needed to achieve the desired mechanical strength must sometimes be greater than that required for the specified current carrying capacity.

Overground installation

No minimum cross sections are specified for overground installation. The standards HD 384.5.54 / IEC 60364-5-54, Table 54 F / DIN VDE 0100-540, Table 2 (Table 1.5/4) and the section "Determining the cross sections of grounding electrode conductors"

Table 1.5/4
Assignment of equipment grounding conductor cross sections to external conductor cross sections (Table 54 F from HD 384.5.54 / IEC 60384–5-54 and DIN VDE 0100-540, Table 2)

Cross section of external conductor of installation	Minimum cross section of appropriate equipment grounding conductor
S mm^2	S_P mm^2
$S \leq 16$ $16 < S \leq 35$ $S > 35$	S 16 $\frac{S}{2}$

(see page 211 ff.) must, however, be taken into consideration when determining the actual cross sections to be used.

It should be noted that the requirements with regard to the cross sections of grounding electrode conductors are the same as those for equipment grounding conductors (PE and PEN).

Installation notes for grounding electrode conductors

Installation notes

Information concerning the installation of grounding electrode conductors can be found in prEN 50179 / DIN VDE 0141. Information concerning corrosion protection measures required for bare grounding electrode conductors laid in soil can be found in prEN 50114 / DIN VDE 0151.

It should be emphasized that overground grounding electrode conductors must be installed so that they are visible or behind special covers and protected against mechanical and chemical destruction. They may also be installed in concrete, provided that their connecting points are easily accessible.

Reinforcing steel elements in reinforced concrete structures may be used as grounding electrode conductors provided that they have the specified minimum cross sections, and are welded together or are connected in some other way so that they are secure and have the same electrical conductance.

The values in the table above are only valid if the equipment grounding conductor is made of the same metal as the external conductor. If this is not the case, the cross section of the equipment grounding conductor must be such that the conductance is the same as that specified in the table.

1.5.7 Measurements and tests on grounding systems

Measurements/ tests

In order to obtain information concerning grounding characteristics before erecting an electrical installation, it is advisable to carry out measurements to establish the soil resistivity (see also Section 2.4).

The dissipation resistance for a newly erected grounding system and its current operational status can be verified by performing appropriate tests. These tests include measuring the dissipation resistance and visual inspections.

Measuring the soil resistivity

Soil resistivity

The following two measuring techniques have proven successful in practice:

▷ the 4-probe method
▷ measurement at a grounding rod

Both the specific surface resistance values and those of the deeper soil layers can be determined using both measuring techniques to provide information concerning the type of grounding electrode to be used.

4-probe method

With the 4-probe method, the individual layers are measured by varying the distance between the auxiliary probes (intervals of 0.5 m).

Measurement at grounding rod

Measurements are taken at a grounding rod using an additional auxiliary grounding electrode and a measuring probe. The first measurement is made with the grounding rod at a depth of 1 m. The rod is then driven deeper and deeper. The specific resistance values of the lower soil layers are determined by performing measurements at the different depths.

If a conductor grounding electrode is to be installed, it is particularly advisable to carry out several separate measurements at different positions and to calculate an average value. The seasonal fluctuations in soil moisture content must also be taken into consideration.

The measurements must be carried out using grounding measuring bridges.

Measuring the dissipation resistance

Dissipation resistance

Initial testing

The grounding system and, in particular, the dissipation resistance must be tested after installation and before commissioning in accordance with prEN 50179 / DIN VDE 0141 and HD 384.6.61 / IEC 60384–6-61 / DIN VDE 0100–610. Initial testing must be carried out by means of appropriate measurements.

Measuring the dissipation resistance is important for those parts of a grounding system which are installed to produce a specific grounding resistance (e.g. the grounding of the main equipotential bonding does not need to be measured). Two measuring techniques have proven successful in practice:

▷ the indirect-acting measuring technique
▷ the ammeter-voltmeter measuring technique

Indirect-acting measuring technique

An external a.c. voltage is used for the *indirect-acting measuring technique*. The measured grounding resistance can be read directly from the measuring instrument scale.

Ammeter-voltmeter measuring technique

Measurements using the *ammeter-voltmeter measuring technique* are based on the system voltage. The resistance must be determined by calculation (quotient derived from the measured voltage and measured current).

A measuring probe and an auxiliary grounding electrode are required for both techniques.

Layout plan

In order to ensure that measurement results are not influenced by other components, the position of the grounding electrodes and grounding electrode conductors should be recorded in a layout plan. This is particularly advisable for large grounding systems.

State checks

State checks

According to prEN 50179 / DIN VDE 0141 and EN 50110 / DIN VDE 0105-1, the grounding systems within a network must be inspected at a number of locations (for example, by excavating at various points). The state of the grounding components should be assessed by means of visual inspection (e.g. checking for signs of corrosion). Five years is considered a reasonable interval for these checks.

1.6 Power-factor correction and harmonic filtering

1.6.1 Introduction

Active power
Reactive power

Many electrical loads draw not only on useful active power P but also reactive power Q, which is not involved in the transfer of energy.

Voltage drop
Losses

If the reactive power is supplied by the power-supply system, it gives rise to an increased voltage drop and additional losses in the generators, transformers, switchgear, overhead lines and cables.

Power-factor correction

It is possible to stabilize the system voltage and reduce the transmission losses by means of power-factor correction equipment, which provides the reactive power directly at the loads. The load on the equipment for generating and transmitting power is reduced, and the level of active power that can be supplied is, thus, much higher (Fig. 1.6/1).

Public utility requirements

The public utilities, therefore, require their customers to ensure that the ratio of active power P to apparent power S does not fall below a certain value.

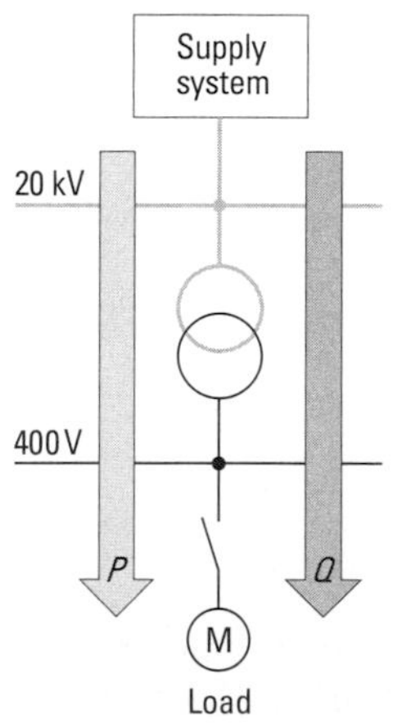

a) Uncorrected system

b) Corrected system

P Active power
Q Reactive power

Fig. 1.6/1 Power flow in low-voltage systems

1.6.2 Power-factor correction of linear loads by shunt capacitors

General principles

Current

Linear loads, such as motors and reactors, draw a current which is almost sinusoidal from the supply system.

Since inductive reactive power is necessary to generate their magnetic fields, the current lags the supply voltage by a phase angle φ (Fig. 1.6/2).

Apparent power

These loads, therefore, consume an apparent power S which is always greater than the required active power, and is calculated from the vector sum of the active and reactive power.

$S = \sqrt{P^2 + Q^2}$ (kVA).

Power factor

The ratio of active power to apparent power is referred to as the power factor ($\cos\varphi$).

$\cos\varphi = \frac{P}{S}$.

Shunt capacitors

In most cases, fundamental-frequency reactive power is corrected by means of shunt capacitors. These can be associated with individual loads or groups of loads, or may be installed centrally to correct a complete system.

Required capacitor power

In order to correct a given power factor $\cos\varphi_1$ to an improved power factor $\cos\varphi_2$, a capacitor rating Q_C of

$Q_C = P \cdot (\tan\varphi_1 - \tan\varphi_2)$

is required.

Fig. 1.6/3 shows the power diagram for an uncorrected and a corrected system.

Desired $\cos\varphi_2$

A lagging power factor of between 0.9 and 0.98 should, if possible, be used for a corrected system.

Overcorrection

Public utilities frequently stipulate a power factor greater than 0.9.

Overcorrection ($Q_c > Q_1$) should, on the whole, be avoided in order to prevent the transmission of capacitive reactive power, which can result in an increase in the system voltage.

Determining the capacitor rating

Planning new installations

In order to determine the capacitor rating for the purposes of planning a new installation, the reactive-power consumption of the individual loads must be added with an allowance for an appropriate coincidence factor a.

The calculation must be based on the active power values and power factors encountered during operation. In the case of drives, these can depart from the rated values.

Rough estimate

It is often sufficient to make a rough estimate of the capacitor rating. As a rule of thumb, it can be assumed that

$Q_C = 0.3 \cdot a \cdot S$ (kvar)

a coincidence factor
S installed apparent load power

This is based on a correction to $\cos\varphi_2 = 0.9$ and a mean load power factor $\cos\varphi_1 = 0.75$.

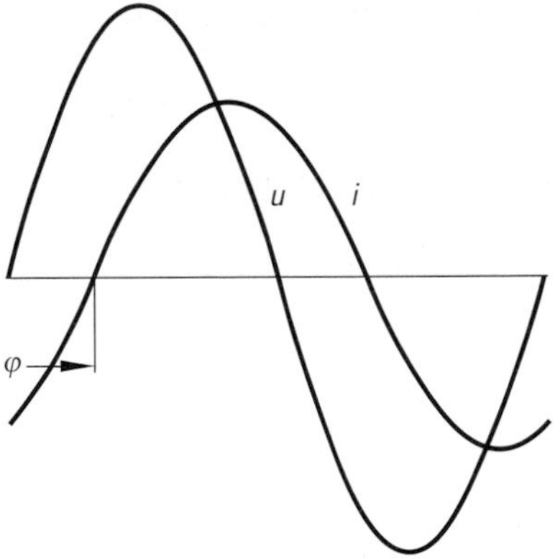

Fig. 1.6/2
Phase displacement φ between current i and voltage u with a linear, inductive load.

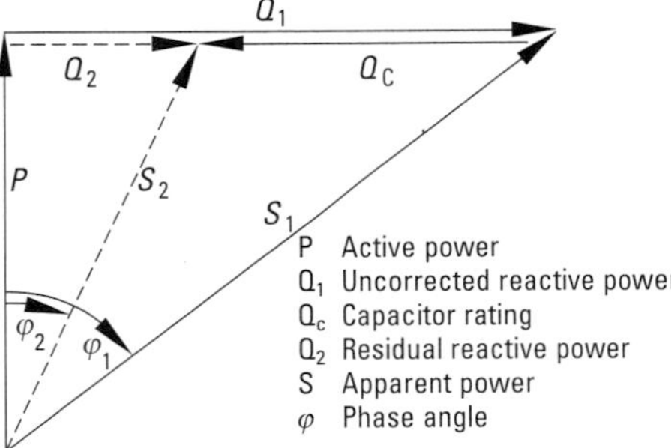

Fig. 1.6/3
Power diagram for an uncorrected (index 1) and a corrected (index 2) system

Retrospective correction

In the case of installations that are already in operation, the required capacitor rating can be determined by means of appropriate measurements.

Measurements with meters

If active and reactive energy-meters are installed, the capacitor rating requirement can be determined from the monthly electricity bill.

The required calculation is:

$$Q_C = \frac{W_b - W_w \cdot \tan \varphi_2}{t} \text{ (kvar)}.$$

W_b	Reactive power (kvarh)
W_w	Active energy (kWh)
t	Operating time (h)
$\tan \varphi_2$	Corresponds to required power factor $\cos \varphi_2$

Supply-system analyzers

The reactive-power requirements and existing power factor $\cos \varphi_1$ can also be easily determined for network segments using portable or permanently-installed supply-system analyzers (e.g. from the SIMEAS product range). For this purpose, only 3 or 4 phase connections must be established for the currents or voltages. All supply-system data is calculated on the basis of these and is either stored in the memory for the required period or can be called up online via a modem.

Table 1.6/1 Low-voltage shunt capacitors

Model	In cylindrical aluminum casing with SIGUT terminals
Type MKK	4RB5
Design	For single-phase and three-phase circuits, 5 kvar to 25 kvar per capacitor, for nominal system voltages of 50/60 Hz 230 V to 690 V AC, dry. Impregnating agent: nitrogen
Mounting position	Any
Degree of protection	IP 20 (with finger protection)
Current carrying capacity	Up to $1.5 \times I_n$
Dimensions	Diameter 122 mm and 142 mm Height 204 mm and 240 mm (depending on rating per capacitor)
Temperature class	– 25/D
Power loss	0.2 W/kvar in dielectric
Service life	100 000 h
	4RB5 with 25 kvar

Shunt capacitors

Capacitors are manufactured for single-phase or three-phase circuits (see Table 1.6/1).

Single-phase capacitors

The following formula applies for *single-phase capacitors*

$$Q_C = U^2 \cdot \omega \cdot C \cdot 10^{-3}$$

$$I_C = \frac{Q_C}{U}.$$

Three-phase capacitors

Three-phase capacitor elements can be connected internally in star (Y) or delta (Δ).

The capacitor rating and capacitor current are calculated for:

Star connection

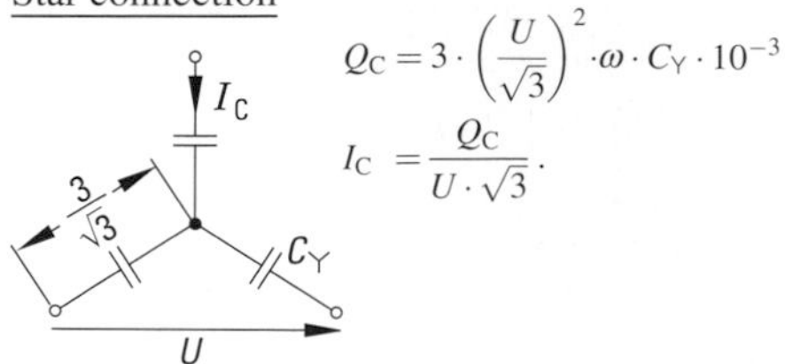

$$Q_C = 3 \cdot \left(\frac{U}{\sqrt{3}}\right)^2 \cdot \omega \cdot C_Y \cdot 10^{-3}$$

$$I_C = \frac{Q_C}{U \cdot \sqrt{3}}.$$

Delta connection

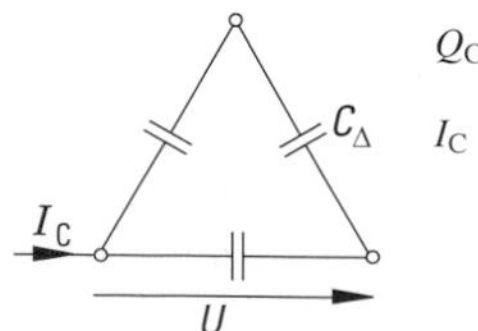

$$Q_C = 3 \cdot U^2 \cdot \omega \cdot C_\Delta \cdot 10^{-3}$$

$$I_C = \frac{Q_C}{U \cdot \sqrt{3}}.$$

Q_C Capacitor rating (kvar)
U Capacitor operating voltage (kV)
I_C Capacitor current (A)
C Capacitance (µF)
ω Angular frequency ($2\,\pi f$) (s^{-1})
f System frequency (Hz).

In internal delta circuits, only 1/3 of the capacitance is required for the same capacitor rating.

In the case of higher system voltages (≥500 V), however, an internal star connection is more reliable and economical.

'oltage and requency

It is important to ensure that the rated voltage of the capacitor corresponds to the operating voltage of the system at the point of installation.

If the operating voltage and frequency differ from the rated voltage and frequency of the capacitor, the power output by the capacitor changes.

The following formula applies:

$$Q_2 = Q_1 \cdot \left(\frac{U_2}{U_1}\right)^2 \cdot \frac{f_2}{f_1}$$

Index 1 Capacitor rating
Index 2 Power output by the capacitor for different operating values

The nominal system voltage should never exceed the rated voltage of the capacitors at the point at which they are installed.

Power-factor correction configurations

Loads can be corrected *individually*, *in groups* or *centrally* (Fig. 1.6/4).

The choice of configuration must be considered from both an economic and technical point of view.

Individual correction

Evaluation criteria

Individual correction is recommended when
▷ large loads with
▷ constant power factors
▷ switched on for long periods

have to be corrected.

Advantages

One advantage of this type of correction is that the load on the supply cable to the loads is reduced. In many cases, the capacitors can be connected directly to the terminals of the individual loads and activated and deactivated using common switchgear.

Disadvantages

Care must be taken when using individual correction for motors. In the case of pole-changing motors, or motors that are connected via star-delta starters, the correction capacitor must not be momentarily disconnected from the supply system (danger of phase opposition). This also applies to motors that are operated intermittently. In this case, the capacitor must always be discharged sufficiently before the motor is started (<10 % of its rated voltage). In order to prevent dangerous self-excitation, the capacitor rating connected directly to the motor terminals should be <90 % of the no-

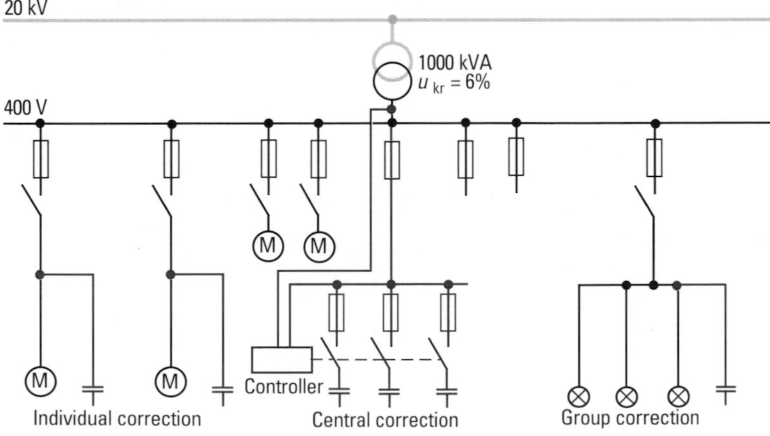

Fig. 1.6/4 Methods of power-factor correction

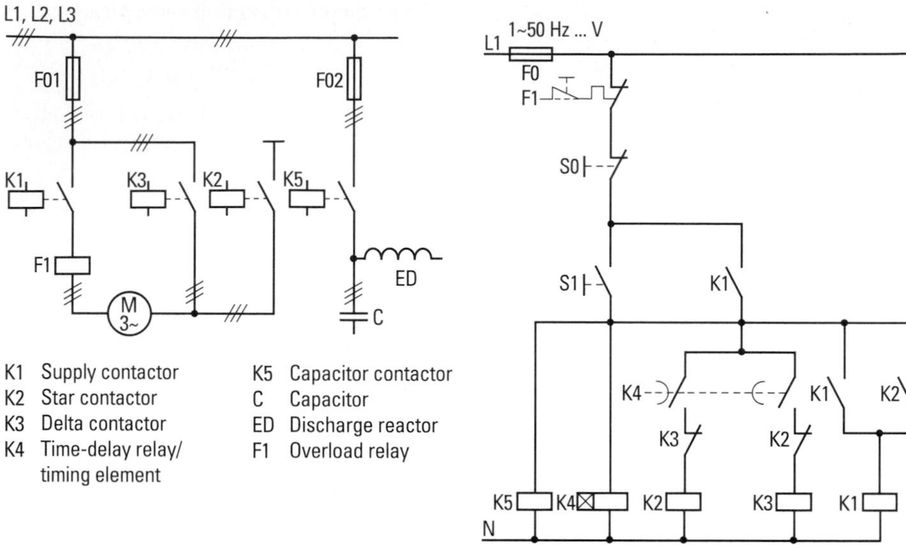

Fig. 1.6/5 Preferred capacitor connection in a star-delta-contactor assembly

load reactive-power consumption. One solution here is to connect the capacitor via a separate contactor which is integrated in the motor controller (Fig. 1.6/5).

Group correction

In a group correction system, one correction unit is associated with each load group. This may consist of motors or fluorescent lamps connected to the system via a common contactor or switch. In this arrangement, as with individual correction, separate switchgear is often not necessary for switching the capacitors.

Central correction

Power-factor correction units are used for central correction. These are directly associated with a main or sub-distribution board.

Evaluation criteria

This type of correction is particularly suitable when

▷ a large number of loads with
▷ different power requirements
▷ switched on for varying periods

are connected to the system.

Advantages

Further advantages of central correction:

▷ the correction equipment can be easily checked because of its centralized arrangement,
▷ retro-installation or extension is relatively simple,
▷ the capacitor rating is always matched to the reactive power requirements of the loads and
▷ with regard to the coincidence factor, a lower capacitor rating than would be required for individual load correction is often sufficient.

Power-factor correction units

Power-factor correction units (Fig. 1.6/6) consist of a controller and a power section including:

Power section

▷ shunt capacitors,
▷ contactors for capacitor switching,
▷ fuses for the capacitor circuits,
▷ elements for discharging the capacitors when they are disconnected from the system.

A power-factor correction unit is characterized by its power rating, which comprises the sum of the output values of the branch circuits, its step function ratio and number of steps.

Branch circuits

A branch circuit represents the capacitor power that is regulated by a controller output.

Step function ratio

The step function ratio is the total number of branch circuits expressed as a ratio. The step

Fig. 1.6/6
Power-factor correction unit 4RY with 500 kvar (10 × 50 kvar)

function ratio of five identical branch circuits, for example, is 1:1:1:1:1.

The step function ratio is freely selectable and can be modified by changing the branch circuit ratings.

Number of steps

The number of steps is determined by the step function ratio and the number of branch circuits. The number of steps is calculated by adding together the numbers of the step function ratio.

Example

Control unit 500 kvar
Step function ratio 1:1:2:2:2:2
(50 kvar: 50 kvar: 100 kvar: 100 kvar: 100 kvar: 100 kvar)
Number of steps:
$1 + 1 + 2 + 2 + 2 + 2 = 10$.

The total power of the control unit (500 kvar) can be switched in ten steps, although only six controller outputs are used.

In order to achieve sufficiently accurate control and, at the same time, avoid operating the contactors too frequently, the number of steps chosen should be at least 1 and not more than 12.

Controller

The controller measures the reactive power at the infeed point via current and voltage transformers. If the measured values deviate from the specified setpoint, it issues control commands to the capacitor switching contactors and thus connects or disconnects the capacitor power in steps.

There are two main types of controller:

▷ circuit controllers,
▷ controllers with freely selectable step function ratios

Circuit controllers

Circuit controllers were developed in order to distribute the switching cycles carried out by the correction unit evenly among all the contactors. This, however, resulted in the following disadvantages:

▷ only branch circuits with the same rating are possible,
▷ the six or twelve switching steps that are used most require a corresponding number of output relays in the controller,
▷ the number of steps is, therefore, limited to six or twelve,
▷ the step function ratio is limited to 1:1:1:1:1:1 or 1:1:1:1:1:1:1:1:1:1:1:1,
▷ if the capacitor rating is increased, the ratings of all the branch circuits must be increased accordingly,
▷ the step function ratio cannot be freely defined.

Modified circuit controllers with selectable switching sequences, e.g. 1:2:2... or 1:2:2:3:3, only behave in the same way as standard circuit controllers if the step (capacitor) ratings are the same.

Controller with freely selectable step function ratio

The Siemens SIMEAS C (Fig. 1.6/7) is a state-of-the-art, intelligent controller with a multifunctional display. It offers the following advantages:

▷ any capacitor ratings can be connected to 5 of the 6 output relays (only the lowest capacitor rating must be connected to the first controller output in order for the C/k values to be set automatically),
▷ only 6 output relays for up to 63 steps,
▷ existing controllers can be expanded without difficulty,
▷ optimized switching operations through automatic determination of the capacitor rating for each controller output and immediate connection of the controller outputs that are most suitable for the required capacitor rating (intelligent control).

Fig. 1.6/7 SIMEAS C controller

▷ multifunctional display shows all important data at a glance. Additional information on the supply system, e.g. system voltage, load current, total harmonic distortion of voltage and current, active power, reactive power and apparent power,

▷ simple operation and communication with PC via optical RS232 interface.

C/k value

The C/k value is the operate value of the controller and is determined from the connected capacitor rating of the first controller output and the transformation ratio of the incoming-supply current transformer to which the controller is connected. Modern controllers determine this value automatically within the first few minutes after the correction unit has been started up.

Capacitors

Only shunt capacitors should be used in power-factor correction units (see Table 1.6/1). Shunt capacitors are self-healing, usually three-phase and have a large capacitance (high rating).

Self-healing means that weak points in the dielectric, which lead to disruptive discharges with occasional voltage peaks in the supply system, can be insulated again (healed) by the material that evaporates in the electric arc.

The evaporation process also extinguishes the electric arc.

The reduction in capacitance caused by these fault locations is negligible.

Shunt capacitors are also characterized by their high pulse carrying capacity and capacitive stability.

An integrated safety mechanism deactivates the capacitor if it is overloaded or at the end of its service life and disconnects it from the supply system. Shunt capacitors have an extremely long service life.

Capacitor switching contactors

Inrush making currents $>100 \cdot I_{Cn}$ occur when modern, extremely low-loss shunt capacitors are energized.

Contactors that frequently energize and de-energize the capacitors in power-factor correction units must be designed and tested for these high inrush making currents. Siemens 3TK4 or SIRIUS 3RT16 capacitor switching contactors are equipped with precharging resistors that are energized via leading auxiliary contacts before the main contacts close and the precharging resistors are bridged. In this way, the inrush making current is reduced to a non-critical value.

Discharge reactors

Before shunt capacitors are energized again, they must be discharged to a non-critical value. The discharge time must be shorter than the response time of the controller. Discharge reactors are mainly used for this purpose in Siemens power-factor correction units. They have a low d.c. resistance, which results in short discharge times, but a high a.c. resistance which means that losses that occur when the capacitors are energized are negligible.

1.6.3 Correction of converter-fed non-linear loads

With the continuing development of power electronics, the number of converter-fed loads has steadily increased. Converters draw lagging reactive power from the three-phase system with a non-sinusoidal current.

System perturbation due to three-phase bridge converters

System perturbation caused by converters is explained below with reference to the three-phase bridge converter. This is the converter used most frequently for large loads.

Fundamental-frequency reactive power

Phase-control reactive power

Commutation reactive power

The output voltage of a converter can be varied continuously by means of delay-angle control. As a result, the valve currents, with increasing delay angle α, are displaced further into the "inductive" region. The converter draws reactive power from the supply, which is often referred to as phase-control reactive power. This is accompanied by the commutation reactive power, which results from the finite rate-of-rise of the valve currents, and is also drawn by uncontrolled converters. With an overlap angle u, a converter draws a fundamental-frequency reactive power

$$Q_{(1)} = P_{(1)} \tan (\alpha + u/2).$$

from the supply.

Harmonic currents and resonances

Fundamental component

The supply-side converter current is not sinusoidal. Using the Fourier analysis it can, however, be broken down into sinusoidal components as follows:

▷ a fundamental component which is characterized by the fact that it oscillates with the system frequency,

▷ harmonics that are designated a harmonic number v.

Harmonics

Harmonics have a frequency of

$$f_v = v \cdot f_{(1)}$$

v harmonic number

$f_{(1)}$ system frequency

Under ideal conditions, the amplitude of a particular harmonic is

$$I_{(v)} = \frac{1}{v} \cdot I_{(1)}$$

$I_{(1)}$ Amplitude of the fundamental component.

Fourier analysis

Fig.1.6/8 illustrates how a converter current is broken down according to the Fourier analysis.

The three-phase bridge converter gives rise to harmonics of the orders

$$v = 6 \cdot k \pm 1 \ (k = 1, 2, 3, \dots)$$

i.e. the 5th harmonic, the 7th, 11th, 13th, 17th, 19th, etc.

Amplitudes of the harmonic currents

The amplitudes of the harmonic currents depend on their orders, the reactances in the commutation loops, smoothing of the direct current and the delay angle, as well as the magnitude of the fundamental current.

In practice, the following values can be assumed:

$I_{(5)} = 0.25\, I_{(1)}, I_{(7)} = 0.13\, I_{(1)},$
$I_{(11)} = 0.09\, I_{(1)}$ and $I_{(13)} = 0.07\, I_{(1)}.$

While the amplitude of the fifth harmonic can assume substantially higher values with less d.c. smoothing, the higher-order harmonic currents are generally less significant.

Supply-voltage distortion

The harmonic currents distort the system voltage, give rise to losses and can provoke resonance phenomena.

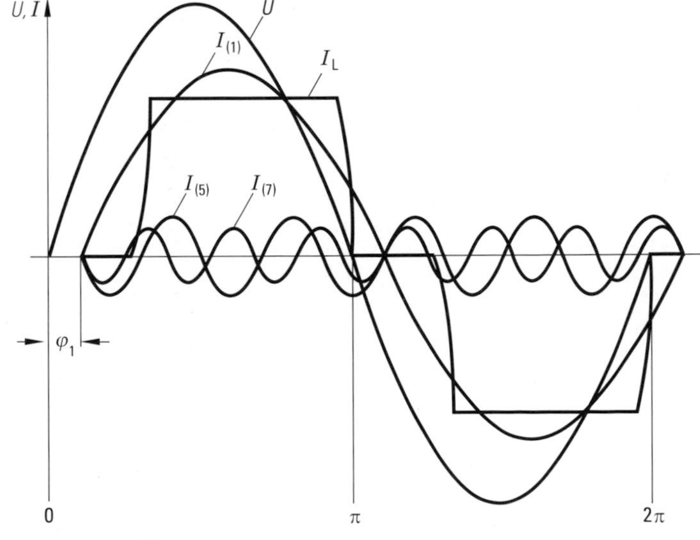

I_L Total load current
$I_{(1)}$ Fundamental component
$I_{(5)}$ 5th harmonic
$I_{(7)}$ 7th harmonic
U System voltage (phase-to-earth voltage)
φ_1 Phase displacement between system voltage and fundamental component

Fig.1.6/8
Breaking down the converter current into fundamental and harmonic components.

Resonances

Care should be taken when shunt capacitors are used for power-factor correction in systems with converters (Fig. 1.6/9). The capacitors, with the reactance of the supply system, form a resonant circuit in which individual harmonic currents can be considerably amplified depending on the rating of the capacitors connected.

Reactor-connected capacitors

The capacitors used for power-factor correction in systems with converters are connected via series reactors. The result is a series-resonant circuit which is tuned in such a way that its resonant frequency is below the 5th harmonic. This means that the correction unit is reactive for all the harmonics present in the conductor current and cannot introduce any other resonances.

Avoiding resonance effects

Correction for lagging reactive power and harmonics

Since a proportion of the harmonic currents, particularly the 5th harmonic, flows into the correction unit, the supply system is relieved not only of lagging reactive power, but also of harmonics.

This applies to supply systems in which a large part of the total load is formed by non-linear loads (current converters).

Secondary distribution systems, in which this load component is very small, but whose primary h.v. systems have a high harmonic content are, however, also quite common. In these cases, harmonics from the primary supply system are blocked or filtered.

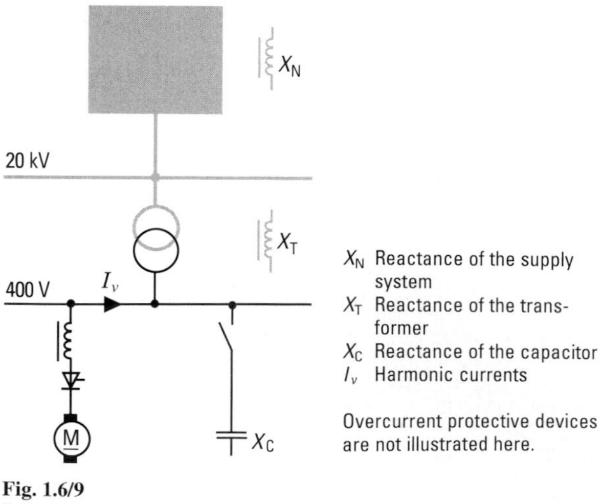

Fig. 1.6/9
Resonance effects resulting from correction by capacitors in supply systems with converter-fed loads

Reactor/capacitor ratio *p*

Reactor-connected capacitors with different reactor/capacitor ratios are used for these different tasks.

The reactor/capacitor ratio p is specified in %. The reactance X_L (of the reactor) is calculated from the capacitor reactance X_C for the system frequency $f_{(1)}$ multiplied by the reactor/capacitor ratio p.

Center frequency f_R

With the reactor/capacitor ratio p, it is possible to determine the center frequency f_R of the series-resonant circuit formed by the reactor and capacitor using the following formula:

$$f_R = f_{(1)} \cdot \frac{1}{\sqrt{p}}$$

With a reactor/capacitor ratio of $p = 7\%$, the center frequency can be calculated as follows:

$$f_R = 50\text{ Hz} \cdot \frac{1}{\sqrt{0.07}} = 189\text{ Hz}.$$

Reactor-connected controllers

Reactor-connected controllers are available with the following reactor/capacitor ratios: $p = 5\%$, 5.67%, 7%, 8%, 12.5% and 14%.

The lower the value for p, the closer the center frequency of the reactor-connected controllers to 250 Hz. The 5th and 7th harmonics from the supply system are then filtered accordingly.

As the value for p increases, the center frequency of the reactor-connected controllers moves away from 250 Hz.

The blocking effect on harmonics from the primary h.v. system, or audio-frequency remote control signals, is then correspondingly high.

In order to choose the correct reactor-connected controller, it is important to consider the load structure in the low-voltage system, as well as the different factors in the primary h.v. system (harmonics, audio-frequency remote control operation).

Component loading

The reactor and capacitor are subject to different loads due to the different reactor/capacitor ratios and the different levels of harmonic filtering that result from this.

Quality requirements

The higher the level of harmonic filtering, the greater the demands made on the components in reactor-connected capacitor units.

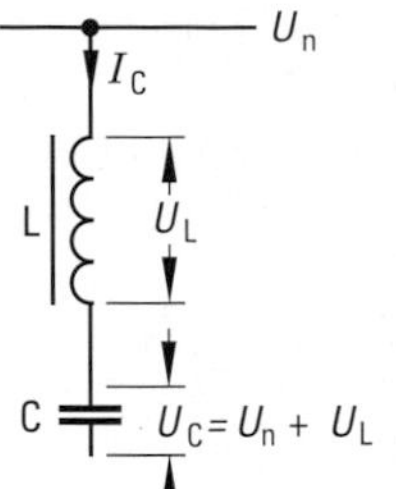

Fig. 1.6/10
Voltage stress on reactor-connected capacitors

For example, a reactor-connected capacitor must always be designed for a higher voltage, since the series circuit (comprising the reactor and capacitor) results in an increase in voltage at the capacitor terminals (Fig. 1.6/10).

The capacitor must also be capable of absorbing the harmonic currents, in addition to the fundamental-frequency current. In order to prevent the selected center frequency from being displaced into critical areas, a reactor-connected capacitor must have a very high capacitive stability.

This also applies to the reactor, which, in addition to the fundamental-frequency current, must also absorb the higher-frequency harmonics. The behavior of the reactor must not be affected by these, i.e. its reactance must remain linear over a wide current range.

Selection

Reactor-connected correction equipment is chosen, first and foremost, in accordance with the required capacitor rating.

Reactor-connected controllers should always be used in low-voltage systems with a share of more than 20% non-linear loads (converter drives, UPS systems, discharge lamps, welding machines, arc furnaces, etc). When choosing the reactor/capacitor ratio, it is important to consider any audio-frequency remote control systems that may be connected to the supply system and the desired degree of partial harmonic filtering of low-frequency current harmonics (see Table 1.6/2).

5.67% reactor-connected controllers are recommended if there are no audio-frequency remote control systems present and if the required capacitor rating is ≥200 kvar.

For lower ratings, different types of 7% or 14% reactor-connected controllers are available.

Audio-frequency remote control systems (AF remote control)

Different public utilities use audio-frequency remote control systems to carry out switching operations in their supply systems. The supply system is superimposed with audio-frequency pulses which control the receiver relays distributed throughout the supply system. These trigger the required switching operations, such as rate changing, load shedding, switching lighting systems, etc.

Audio frequency

If a public utility uses audio-frequency remote control systems, an audio frequency of

Table 1.6/2
Overview of audio-frequency blocking circuits and harmonic filtering of different power-factor correction units.

Type of power-factor correction unit	Reactor/capacitor ratio p % with centre frequency f_R at system frequency 50 Hz	Filtering of the 5th harmonic	Adequate blocking effect on audio frequencies
Without reactors 4RY	–	–	> 250 Hz, consult with public utility for higher capacitor ratings
Passive, tuned filter circuit 4RY	4 % with $f_R = 245$ Hz	High, up to 90 %	> 550 Hz
With reactors 4RF16	5.67 % with $f_R = 210$ Hz	Good, up to 50 %	> 350 Hz
With reactors 4RF17	7 % with $f_R = 189$ Hz	Good, up to 30 %	> 250 Hz
With reactors 4RF18	8 % with $f_R = 177$ Hz	Up to 20 %	> 210 Hz
With reactors 4RF19-Z	12.5 % with $f_R = 141$ Hz	Low, up to 15 %	> 175 Hz
With reactors 4RF19	14 % with $f_R = 134$ Hz	Low, up to 10 %	> 160 Hz
Combined reactor-connected 4RF34	5.67 % and 12.5 % parallel connection	Good, up to 40 %	> 160 Hz up to 190 Hz

between 160 Hz and 2000 Hz is normally selected as a carrier signal for the telegrams.

Loading by capacitors

Since the resistance of capacitors is very low due to their frequency-dependant reactance for high-frequency signals, high audio-frequency currents can occur in capacitors even at low audio-frequency voltages. This increases the loading on the transmitters and can cause the remote control system to malfunction.

Audio frequency blocking circuits

Capacitors without reactors in supply systems with predominantly linear loads must, therefore, be provided with suitable audio-frequency blocking circuits (parallel blocking circuits tuned to the audio frequency). Reactor-connected capacitors have an adequate blocking effect on specific audio frequencies, depending on the reactor/capacitor ratio.

Table 1.6/2 shows an overview of the correction units available (without impermissible loading by capacitors).

Fig. 1.6/11
Tuned, passive filter circuit 2 × 175 kvar for 504 A harmonic current

Tuned, passive filter circuits

If it is necessary to selectively filter self-generated current harmonics, tuned passive filter circuits can be used (Fig. 1.6/11).

Filter circuits

Filter circuits consist of high-quality, series-connected shunt capacitors with high capacitive stability and appropriately dimensioned filter reactors. These series-resonant circuits are precisely tuned to the current harmonics to be filtered and present very low impedances at these frequencies. They absorb the harmonic currents so that these flow predominantly in the filters and not in the supply system.

Filter circuits are available for the 5th and 7th harmonics. A common filter circuit is planned for the 11th and 13th harmonics.

If filter circuits for different harmonic numbers are installed in a supply system, they must be connected upwards starting with the lowest harmonic number. They are disconnected in the reverse order.

In many cases, a filter circuit for the 5th harmonic, which also partially filters the 7th harmonic, is sufficient.

Tuned, passive filter circuits are capacitive at the fundamental frequency and contribute to power-factor correction with their capacitor rating.

Filter circuits are, however, not connected or disconnected via controllers. This takes place via special processes together with the non-linear loads in the supply system.

Filter circuits can be combined with reactor-connected correction units.

1.6.4 Dynamic power-factor correction and active harmonic filtering

Restrictions posed by conventional solutions

Conventional solutions for power-factor correction and harmonic filtering were described in Sections 1.6.1 to 1.6.3.

Problems with dynamic load changes

Conventional solutions, however, cause problems in many supply systems, or are simply unsuitable for certain tasks. The overall system impedance changes when capacitors are connected. Even when reactor-connected capacitors are used, resonance occurs which can cause problems with even-order harmonics, interharmonic components or with the third harmonic.

If the supply system configuration and load structure change after the power-factor correction units or passive filter circuits have been installed, the units may be overloaded and malfunction.

Due to their delayed switching times (capacitor discharge time), conventional units are too slow for dynamic load changes (hoisting gear, lifts, welding machines etc.).

In some cases, flicker compensation (an extremely fast type of power-factor correction with response times in the millisecond range) can be used to reduce the load current and the resultant voltage drops in the system impedances. In the case of conventional units, this is a problem that cannot be solved. When modern converter drives are used, high harmonic currents are often generated; the reactive-power consumption, however, is almost zero. A high level of harmonic filtering (without connecting the capacitor rating to the supply system, as this is undesirable here) is impossible with conventional filter circuits. Problems that occur when passive, tuned filter circuits are used in supply systems with audio-frequency remote control systems also mean that new solutions have to be found.

SIPCON P – Siemens Power Conditioner

Correcting system perturbations

SIPCON P (P = parallel coupling) is an active filter that is used to correct the system perturbations of any load in three-phase systems (Fig. 1.6/12).

SIPCON P is equipped with an IGBT (insulated gate bipolar transistor) frequency converter that is regulated by a specially developed software package and controller. SIPCON P generates five discrete voltages with different frequencies and phase angles. The entire setup, which is connected to the supply system via an LCL filter, can be regarded as a power source for a fundamental-frequency current and four harmonic currents of the orders 5, 7, 11 and 13. If the phase angle and amplitude of the currents supplied by SIPCON P are correct, the inductive load reactive current and the respective harmonic currents are partially or fully extinguished at the point of common coupling (Fig. 1.6/13).

SIPCON P controller

Like a conventional power-factor correction controller, the SIPCON P controller measures the currents and voltages of the three-phase system via current and voltage transformers and uses the measured values to calculate the active and the reactive components as well as the harmonic components of the load current. The controller compares the actual values with the setpoints defined by the user and ensures that the SIPCON currents are supplied to the points of common coupling in the appropriate manner.

SIPCON P does not cause any interference in audio-frequency remote control systems, since wide-band filtering does not take place; only the desired reactive and harmonic currents are corrected here. Even a high level of 5th harmonic filtering with

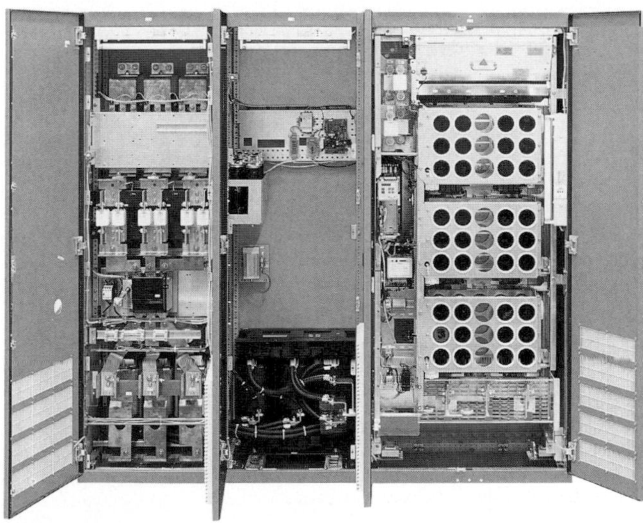

Fig. 1.6/12
SIPCON P controller with a rated power of 610 kvar

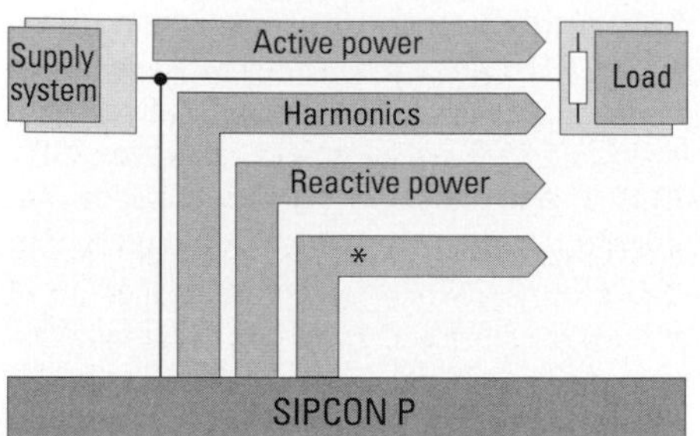

Fig. 1.6/13
SIPCON removes undesirable load current components from the supply system

exceed the rated power of SIPCON P, the device will still be safe with 100% of its rated power connected to the power-supply system.

- No capacitive power is supplied. There is, therefore, no danger of overcompensation at the mounting position, provided that the power factor of the non-linear loads is already > 0.9.
- Discrete operation, no interference in audio-frequency remote control systems.
- The user can parameterize a priority list in accordance with the rated power of the device for the harmonics that are to be filtered and change this at any time.

250 Hz influences non-adjacent ripple control frequencies with 210 Hz, 217 Hz, 228 Hz or 270 Hz.

All existing conventional power-factor correction units and filter circuits for low-voltage systems can be replaced by SIPCON P. This is due to the rated power of the converter and the considerable progress made in developing IGBT power semiconductors.

Stepless, dynamic power-factor correction, flicker compensation

SIPCON P is used for stepless and dynamic power-factor correction and is, therefore, also suitable for rapidly changing loads and flicker compensation.

It can correct both inductive and capacitive reactive power.

No deviations whatsoever occur from the defined target power factor cos φ_2 and no system perturbations arise as a result of switching operations.

Advantages

When SIPCON P is used, the system impedance does not change and there is no danger of resonance phenomena. Audio-frequency remote control signals are not affected at all.

Active harmonic filtering

SIPCON P is used for selective, active harmonic filtering of up to four discrete harmonics (orders 5, 7, 11 and 13). Fig. 1.6/14 shows SIPCON P connected to the supply system and load via a current measuring point, as well as active filtering of harmonics of the orders 5, 7, 11 and 13.

The most important advantages of active filtering are:

- Overloading is not possible. If the harmonic currents at the mounting position

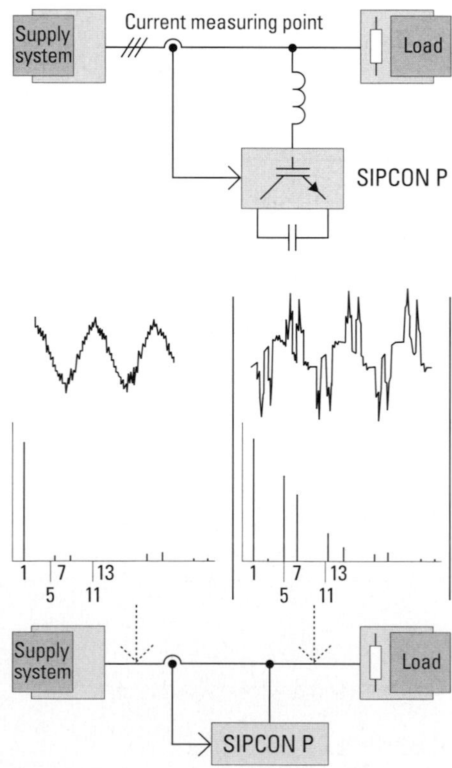

Fig. 1.6/14
SIPCON P, filtering of harmonics of the orders 5, 7, 11 and 13.

Harmonic measurement

Harmonic currents

A knowledge of the harmonic currents at the mounting position is an important prerequisite when planning passive or active filter circuits. The most reliable method for determining the harmonic currents is to measure these as part of a supply system analysis.

Supply-system analyzers

The Siemens SIMEAS and OSZILLOSTORE product range includes different supply-system analyzers for permanent installation or portable use.

The user must carry out three or four-phase current and voltage measurements at the measuring point.

It is useful to carry out measurements of all relevant system data over an extended period of time (e.g. 1 week) according to EN 50 160 and IEC 1000-2-2 in order to determine the maximum harmonic values accurately.

Special software packages, such as OSCOP P and OSCOP Q, simplify the evaluation of system analyses and supply the required system data in either numeric or graphic form.

Once the maximum harmonic currents have been determined (in particular the 5th, 7th, 11th and 13th order), the passive or active filters can be dimensioned and rated accordingly.

1.7 Power cables and their application

In April 1999, Publicis MCD Verlag (Erlangen, Germany) published the fifth, extensively revised and expanded edition of the manual "Kabel und Leitungen für Starkstrom". Based on the practical requirements of planners and installation engineers, this manual takes an in-depth look at power cable characteristics, criteria for selecting cables, as well as the relevant standards.

In view of the scope of the manual, this section will cover only topics of particular importance for building installations.

Readers requiring more detailed information on power cables are advised to consult the manuals "Kabel und Leitungen für Starkstrom", fifth extensively revised and expanded edition, 1999 (cited as [1] in this section), and "Power Cables and their Application, Part 2", fourth revised edition, 1989 (cited as [2] in this section), both distributed by Publicis MCD Verlag (Erlangen).

1.7.1 Insulated wires and flexible cables

Guidelines for selecting cables

Only types of cable that conform to the relevant standards may be used in wiring systems for electrical installations.

These include the following standards:

- ▷ IEC 60 050/DIN VDE 0289
  Definitions for cables, wires, and flexible cords for power installation
- ▷ HD 186, HD 308, HD 402, IEC 60 446, DIN VDE 0293
  Identification of cores in cables and flexible cords used in power installations with nominal voltages up to 1000 V
- ▷ HD 383/IEC 60 228/DIN VDE 0295
  Conductors of cables, wires and flexible cords for power installation
- ▷ HD 21/IEC 60 227/DIN VDE 0281
  PVC cables, wires, and flexible cords for power installation with nominal voltages up to and including 450/750 V
- ▷ DIN VDE 0250
  Cables, wires and flexible cords for power installation.
- ▷ HD 22/IEC 60 245/DIN VDE 0282
  Rubber-insulated cables of rated voltages up to and including 450/750 V
- ▷ HD 603/IEC 60 502-1/DIN VDE 0276-603
  Distribution cables of nominal voltages U_0/U 0.6/1 kV
- ▷ HD 620/DIN VDE 0276-620
  Distribution cables of nominal voltages U_0/U 3.6/6 (7.2) kV to 20.8/36 kV
- ▷ EN 60 811/IEC 60 811/DIN VDE 0473
  Insulating and sheathing materials of electric cables; common test methods.

These standards stipulate cable types for fixed installation and portable electrical equipment. They also contain specifications regarding cable construction, characteristics, and tests, as well as information for use.

Harmonized standards

The standards HD 21 / DIN VDE 0281 and HD 22 / DIN VDE 0282 are harmonized specifications elaborated by CENELEC[1]. These standards apply in countries affiliated to CENELEC, and will become increasingly important within the European single market.

Type designations

The types of cable specified in these standards:

▷ PVC-insulated cables (HD 21 / DIN VDE 0281),
▷ Rubber-insulated cables (HD 22 / DIN VDE 0282)

have type designations which are the same in all CENELEC countries. They are intended only for cables that conform with harmonized standards, and are subdivided into three parts, as shown in Table 1.7/1.

Examples of type designations

1. PVC-insulated single-core non-sheathed cable, 1.5 mm², with blue core **H07V-U1.5BU**
2. 07RN rubber-insulated three-core flexible cable for moderate mechanical stress, 2.5 mm², with green/yellow core (protective conductor) **H07RN-F3G2.5**
3. 03VV PVC-insulated two-core sheathed flexible cable for low mechanical stress, 0.75 mm², without green/yellow core (protective conductor) **H03VV-F2X0.75**

The initial letter H indicates that the type of cable complies in all respects with the harmonized standards.

The initial letter A is used to designate supplementary types of cable (e.g. with non-standard numbers of cores and conductor cross-sectional areas). This letter indicates that a cable does, in principle, conform to the harmonized standards, but is approved for use in one particular country only (approved national supplementary type).

[1] European Committee for Electrotechnical Standardization.

DIN VDE 0250 contains all the types of cable not covered by the harmonized standards. These include all variants with rated voltages of $U_0/U \geq 0.6/1$ kV, as well as cables for specific applications. The former national designation (beginning with "N") remains unchanged for cables of this type, i.e. with multi-core cables a distinction is made between:

▷ cables with green/yellow protective conductor (the letter "J" preceded by a hyphen is appended to the appropriate type designation, e.g. NYM-J),
▷ cables without green/yellow protective conductor (the letter "O" preceded by a hyphen is appended to the type designation, e.g. NYM-O).

VDE mark of conformity, HAR

To verify that cables conform to the above standards to DIN VDE, the VDE Testing Agency permits the VDE mark of conformity to be used in the form of the symbol ◁VDE▷ or an identification thread. The letters ◁HAR▷ also have to be added for types of cable that conform with harmonized standards.

Identification thread

The color coding of the identification thread is black/red for types to national standards and black/red/yellow for types to harmonized standards.

Manufacturer's name/symbol

Cables also have to carry an identification thread indicating their origin, or the name of the manufacturer on at least one core or on the sheath.

Core identification

Core identification for insulated cables is specified in DIN VDE 0293 (see Table 1.7/2).

Color designations

As part of the harmonization process (HD 308), the new designations for core colors have been incorporated in the DIN VDE standards (Table 1.7/3).

Colored cores must be used as follows:

▷ The green/yellow core must be used exclusively as a protective conductor (PE or PEN). This core must not be used for any other purpose.
▷ The blue core is used as a neutral conductor. This core can be used as required, but not as a protective conductor.

Table 1.7/1 Key to types of designation (extract)

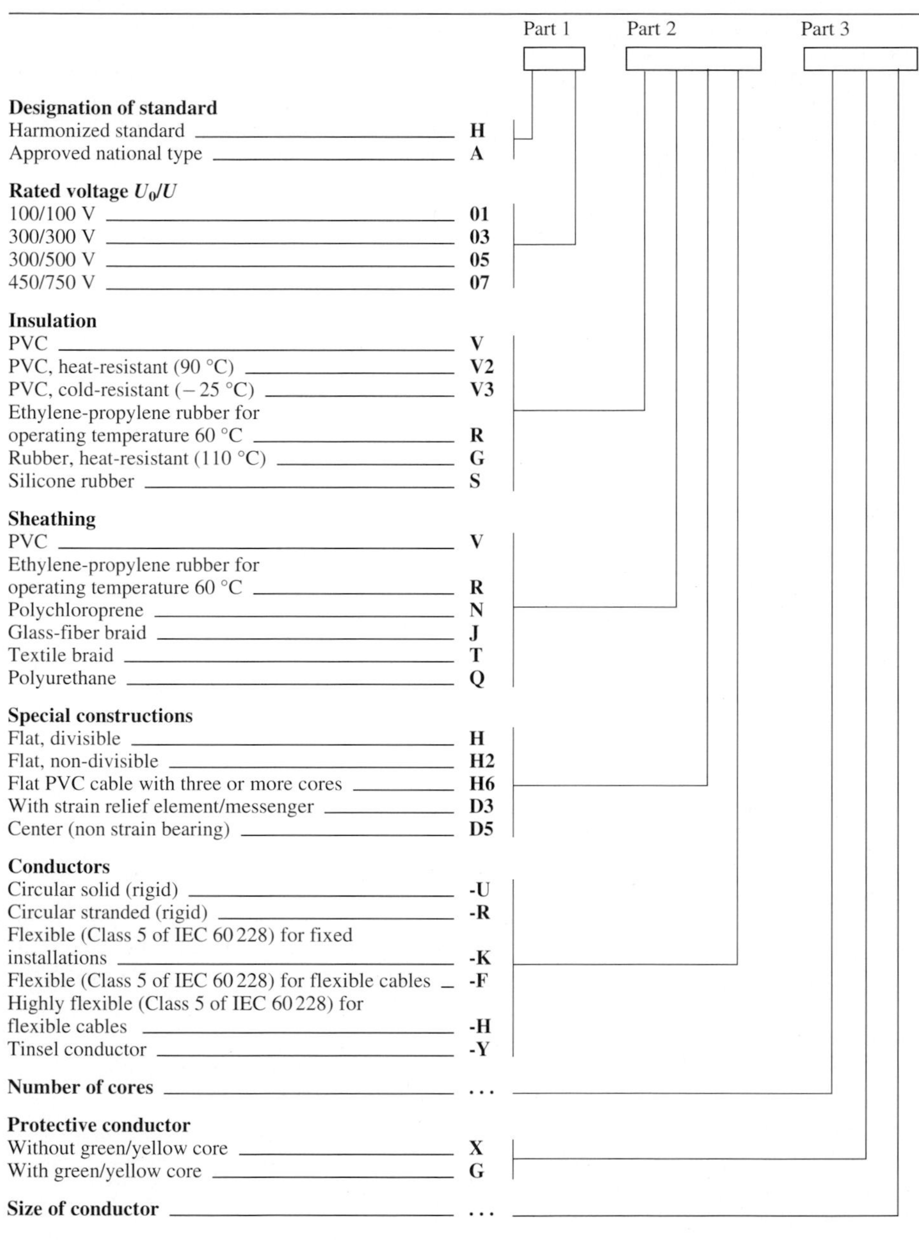

	Code	Part
Designation of standard		Part 1
Harmonized standard	**H**	
Approved national type	**A**	
Rated voltage U_0/U		Part 1
100/100 V	**01**	
300/300 V	**03**	
300/500 V	**05**	
450/750 V	**07**	
Insulation		Part 2
PVC	**V**	
PVC, heat-resistant (90 °C)	**V2**	
PVC, cold-resistant (– 25 °C)	**V3**	
Ethylene-propylene rubber for operating temperature 60 °C	**R**	
Rubber, heat-resistant (110 °C)	**G**	
Silicone rubber	**S**	
Sheathing		Part 2
PVC	**V**	
Ethylene-propylene rubber for operating temperature 60 °C	**R**	
Polychloroprene	**N**	
Glass-fiber braid	**J**	
Textile braid	**T**	
Polyurethane	**Q**	
Special constructions		Part 2
Flat, divisible	**H**	
Flat, non-divisible	**H2**	
Flat PVC cable with three or more cores	**H6**	
With strain relief element/messenger	**D3**	
Center (non strain bearing)	**D5**	
Conductors		Part 2
Circular solid (rigid)	**-U**	
Circular stranded (rigid)	**-R**	
Flexible (Class 5 of IEC 60228) for fixed installations	**-K**	
Flexible (Class 5 of IEC 60228) for flexible cables	**-F**	
Highly flexible (Class 5 of IEC 60228) for flexible cables	**-H**	
Tinsel conductor	**-Y**	
Number of cores	. . .	Part 3
Protective conductor		Part 3
Without green/yellow core	**X**	
With green/yellow core	**G**	
Size of conductor	. . .	Part 3

Designations for color identification of cable cores are contained in Table 1.7/3
For examples of type designations, see page 232

Table 1.7/2 Core identification for insulated cables

Number of cores	Cables *with* green/yellow core	Cables *without* green/yellow core
Cables for fixed installation		
1	green/yellow	black
2	green/yellow, black[1)]	black, blue
3	green/yellow, black, blue	black, blue, brown
4	green/yellow, black, blue, brown	black, blue, brown, black
5	green/yellow, black, blue, brown, black	black, blue, brown, black, black
6 and over	green/yellow, additional cores black and numbered	black and numbered
Flexible cables		
1	–	black
2	–	brown, blue
3	green/yellow, brown, blue	black, blue, brown
4	green/yellow, black, blue, brown	black, blue, brown, black
5	green/yellow, black, blue, brown, black	black, blue, brown, black, black
6 and over	green/yellow, additional cores black and numbered	black and numbered

1) According to DIN VDE 0100-540, this 2-core variant is only permissible for conductor cross-sectional areas ≥ 10 mm² copper

▷ If power supply cables are used in telecommunications systems to DIN VDE 0800, the green/yellow core must also be used exclusively as a protective conductor.

Metal sheaths

Metal sheaths, as well as any bare sheath wires must not be used alone as either current-carrying conductors for normal operation, or as protective conductors (HD 384/ IEC 60364/DIN VDE 0100).

Selecting cables

In addition to the relevant DIN VDE standards for cables, the following must also be observed for the purpose of selecting cables in Germany: the DIN VDE installation specifications, and (where applicable) the special regulations of the public utility or inspection authorities (e.g. factory and shop inspectorates, regional mines inspectorates).

Table 1.7/3
Designations for color identification of cable cores to DIN VDE 0293 compared with the old designations to DIN 47002

Color	Old designation to DIN 47002	New designation DIN VDE 0293
Black	sw	BK
Brown	br	BN
Red	rt	RD
Orange	or	OG
Yellow	ge	YE
Green	gn	GN
Blue	bl	BU
Violet	vi	VT
Gray	gr	GY
White	ws	WH
Pink	rs	PK
Turquoise	tk	TQ

Selecting the rated voltage

The rated and operating voltages of cables are stipulated by the definitions given in HD 516 S2/DIN VDE 0298–300.

Rated voltage

The rated voltage of an insulated power cable is the voltage on which its construction and testing of its electrical characteristics are based. The rated voltage is expressed by two AC voltage values U_0/U where:

U_0 = the r.m.s. value between an external conductor and ground (non-insulating environment),

U = the r.m.s. value between any two phase conductors in a multi-core cable or a system of single-core cables.

In systems with AC voltage, the rated voltage of a cable must be at least equal to the nominal voltage of the system in which it is used. This also applies to the values U_0 and U. In systems with DC voltage, the nominal voltage of the system must not exceed the rated voltage of the cable by more than a factor of 1.5.

Operating voltage

The operating voltage is the voltage present between the conductors or between a conductor and ground in a power installation under normal conditions.

Continuously permissible operating voltage

Cables with rated voltages $U_0/U \leq 0.6/1$ kV are suitable for use in three-phase AC, single-phase AC, and DC systems, in which the maximum continuously permissible operating voltage does not exceed the rated voltage of the cables by more than

▷ 10 % for cables with rated voltages $U_0/U \leq 450/750$ V

▷ 20 % for cables with rated voltages $U_0/U = 0.6/1$ kV.

Cables with rated voltages $U_0/U > 0.6/1$ kV are suitable for use in three-phase and single-phase AC systems, in which the maximum operating voltage $U_{b\,max}$ does not exceed the rated voltage of the cables by more than 20 %.

The cables may be used:

a) in three-phase and single-phase AC systems with an effectively grounded neutral point;

b) in three-phase and single-phase AC systems where the neutral point is not effectively grounded, provided that individual ground faults are sustained for no longer than 8 hours, and the total ground-fault time in any year does not exceed 125 hours. If these conditions cannot be complied with, cables with a higher rated voltage must be selected to prolong the service life of the cabling.

Cables in DC systems

If cables are used in DC systems, the continuously permissible DC operating voltage between the conductors must not exceed the permissible AC operating voltage by more than a factor of 1.5. In DC systems with single-phase grounding, the value must be multiplied by 0.5.

Selecting conductor cross-sectional areas

Current-carrying capacity

The conductor cross-sectional areas selected must ensure that the current load under normal operating conditions does not exceed the current-carrying capacity of the conductor, and that no part of the conductor is at any time heated above the maximum permissible operating temperature. The temperature rise or current-carrying capacity of a cable with a particular cross-sectional area depends on its construction, its material properties, and the particular operating conditions.

The following requirement must always be fulfilled, whatever the particular operating conditions:

current load $I_b \leq$ current-carrying capacity I_z.

Cables for fixed installation

DIN VDE 0298-4 (February 1988) was the first German national standard to specify comprehensive recommended values for the current-carrying capacity of cables for fixed installation and of flexible cables. This standard subsumes the former German division of installation types (three groups) under the following internationally agreed basic types of installation:

Installation type A
Installation in thermally-insulated walls,

Installation type B
Installation of cables in conduit or duct on or in a wall,

Installation type C
Installation of cables directly on or in a wall/under plaster

Installation type E
Installation of cables in free air.

The current-carrying capacities specified in DIN VDE 0298-4 for PVC-insulated cables were taken from Report R 64-001 for the

Harmonization Document HD 384.5.523 S 1, and, where necessary, coordinated with national requirements.

This coordination was required because the current-carrying capacities in Report R 64-001 were determined on the basis of the most unfavorable installation conditions. For example, all the values specified for Installation Types B and C apply to installation on wooden walls. In Germany, however, cables are usually installed on wall surfaces with more favorable heat dissipation than wood. Depending on the construction of the wall, the current-carrying capacity of cables in these types of installation can be higher than the values specified in the standard. In certain cases, where cables are installed on masonry or plaster, the current-carrying capacity was, therefore, increased slightly on the basis of the measuring results obtained, enabling the tried-and-tested allocation of protective equipment to be retained (see Section 1.7.3).

Operating conditions

Values for current-carrying capacity are always based on clearly defined operating conditions, representative examples of which can be found in Table 1.7/4. The current-carrying capacities for cables installed in free air or buildings apply, for example, to individually installed single-core and multi-core cables operated continuously at a defined ambient temperature in three-phase AC systems (if three conductors are loaded), or in single-phase AC systems (if two conductors are loaded). These operating conditions presuppose that no other non-loaded conductors are involved.

Installation types

Numerous other methods can be used to install cables in buildings, in addition to the installation types presented in Table 1.7/4. A representative selection of additional installation types can be found in Table 1.7/5, together with references to the internationally agreed installation types that specify the current-carrying capacity.

able 1.7/4 Operating conditions for cables for fixed installation

eference operating conditions ɔr determining the rated current I_n	*Other (site) operating conditions,* and calculation of current-carrying capacity $I_n = I_n \cdot \Pi f$
ype of operation	
ontinuous operation at the values for current-carrying capacity Table 1.7/6	Current-carrying capacity with intermittent operation to Section 1.7.2
ıstallation conditions	
ıstallation type A1, A2[1)] ıstallation in thermally-insulated walls single-core non-sheathed cables in conduit (A1) multi-core cable in conduit (A2) multi-core cable in wall (A2)	Conversion factors for grouping to Table 1.7/9 for multi-core cables to Table 1.7/8
ıstallation type B1, B2 ıstallation in conduit or ducts single-core non-sheathed cables in conduit on wall[2)] (B1) single-core non-sheathed cables in conduit on wall (B1) single-core non-sheathed cables, single-core light-sheathed cable, or multi-core cable in conduit in wall or under plaster (B1) multi-core cable in conduit on wall[2)] or on floor (B2) multi-core cable in conduit on wall or on floor (B2)	Conversion factors for grouping to Table 1.7/9 for multi-core cables to Table 1.7/8
ıstallation type C[3)] irect installation multi-core cable on wall or floor single-core light-sheathed cable on wall or floor multi-core cable in wall or under plaster flat webbed cables under plaster	Conversion factors for grouping to Table 1.7/9 for multi-core cables to Table 1.7/8
ıstallation type E ıstallation in free air, i.e. unhindered heat dissipation is ensured by: clearance from wall to Tables 1.7/5 and 1.7/6 clearance (≥ 2 × cable diameter) between cables installed side by side clearance (≥ 2 × cable diameter) between cable runs above one another	Conversion factors for grouping to 1.7/9 for multi-core cables to Table 1.7/8
mbient conditions	
mbient temperature 30 °C	Conversion factors for site ambient temperatures to Table 1.7/7
ufficiently large and ventilated rooms, in which the ambient temperaıre is not noticeably increased by heat loss from the cables	
rotection against direct solar radiation, etc.	See [1], Section 18.4.2
perating frequency 50 to 60 Hz	

Thermally-insulated walls comprise an external weatherproof board, thermal insulation, and an internal board made of wood or similar material. The thermal resistivity of the internal board is 0.1 K · m/W. A conduit or multi-core cable is installed in the wall so that it is close to the internal board, but does not necessarily touch it. It is assumed that the heat loss of the cable is dissipated only via the internal board. The conduit may be made of metal or plastic.

No distinction is made between single-core and multi-core cables in conduits (see also Note on page 244).

Conduits are installed on walls in such a way that the clearance between the conduit and wall is less than 0.3 × the diameter of the conduit.

The cables are installed in such a way that the clearance between them and the surface of the wall is less than 0.3 × the outer diameter of the cables.

Table 1.7/5 Common fixed installations and corresponding installation types

Installation	Description	Installation type for determining current-carrying capacity
Room	Single-core non-sheathed cables in conduit in thermally-insulated wall[1)]	A1 (A)[2)]
Room	Multi-core cable or multi-core light-sheathed cable in conduit in thermally-insulated wall[1)]	A2 (A)[2)]
Room	Multi-core cable or multi-core light-sheathed cable installed directly in thermally-insulated wall[1)]	A2 (A)[2)]
	Single-core non-sheathed cables, single-core cables or light-sheathed cables in conduit on (wooden) wall, or with clearance of < 0.3 × outer diameter of conduit	B1
	Multi-core cable or multi-core light-sheathed cable in conduit on (wooden) wall, or with clearance of < 0.3 × outer diameter of conduit	B2
	Single-core non-sheathed cables, single-core cables, or light-sheathed cables in enclosed conduit on (wooden) wall	B1
	Multi-core cable or multi-core light-sheathed cable in enclosed conduit on (wooden) wall	B2
	Single-core non-sheathed cables, single-core cables, or light-sheathed cables in conduit in masonry/concrete with thermal resistivity ≤ 2 K · m/W	B1
	Multi-core cable or multi-core light-sheathed cable in conduit in masonry/concrete with thermal resistivity ≤ 2 K · m/W	B2

1) The maximum thermal resistivity of the inside of the wall is 0.1 K · m/W
2) To DIN VDE 0298-4, see also Note on page 244

Table 1.7/5 Common fixed installations and corresponding installation types *(continued)*

Installation	Description	Installation type for determining current-carrying capacity
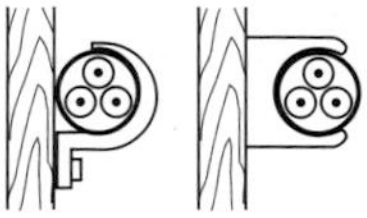	Single-core or multi-core cable(s) or light-sheathed cable – installed on (wooden) wall or with clearance < 0.3 × outer diameter of cable between cable and wall	C
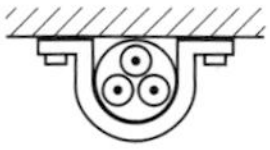	– installed under (wooden) ceiling	C
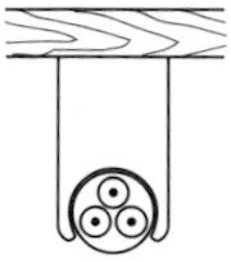	– installed with clearance > 0.3 × outer diameter of cable between cable and ceiling	E
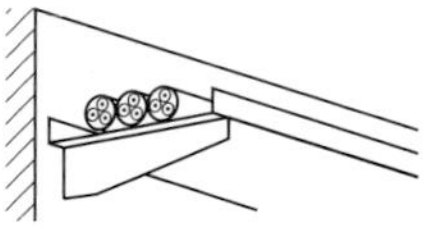	– on cable tray (blank)	C or E
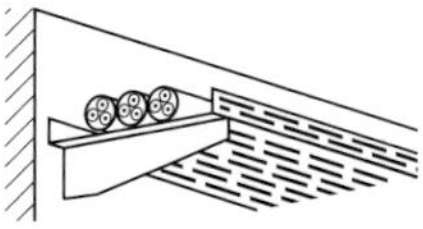	– on cable tray (perforated), horizontal or vertical	E[1)]
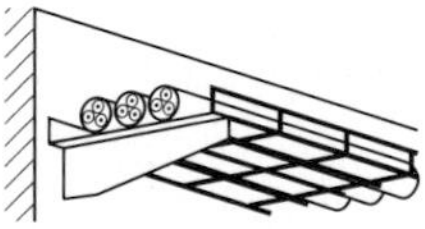	– on cable consoles	E[1)]
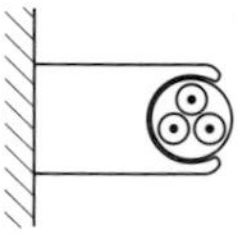	– with clearance > 0.3 × outer diameter of cable between cable and wall	E[1)]
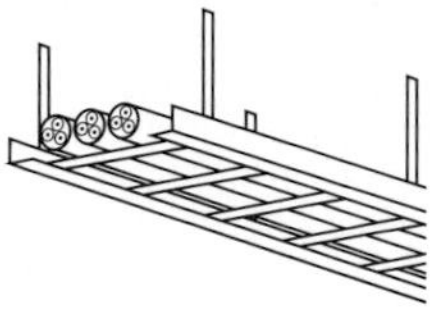	– on cable ladder	E[1)]
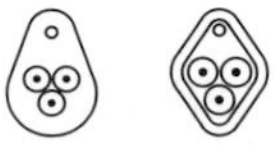	Single-core or multi-core cable(s) or light-sheathed cable, suspended on messenger or with built-in messenger	E

[1)] Particular attention must be paid to cables installed vertically with restricted ventilation. The ambient temperature at the upper end of vertical installations can rise considerably, making it necessary to reduce the current-carrying capacity.

Table 1.7/5 Common fixed installations and corresponding installation types *(continued)*

Installation	Description	Installation type for determining current-carrying capacity
	Single-core or multi-core cable(s), or light-sheathed cable in building cavity [1) 2) 4)]	$1.5\ d \le V < 5\ d$ installation type B2 $5\ d \le V < 50\ d$ installation type B1
	Single-core non-sheathed cables in conduit in building cavity [1) 3) 4)]	$1.5\ d \le V < 20\ d$ installation type B2 $20\ \mathrm{d} \le V$ installation type B1
	Single-core or multi-core cable(s), or light-sheathed cable in conduit in building cavity [4)]	B2
	Single-core non-sheathed cables in enclosed conduit in building cavity [1) 3) 4)]	$1.5\ d \le V < 20\ d$ installation type B2 $20\ \mathrm{d} \le V$ installation type B1
	Single-core or multi-core cable(s), or light-sheathed cable in enclosed conduit in building cavity [4)]	B2
	Single-core non-sheathed cables in enclosed conduit in masonry/concrete with thermal resistivity ≤2 K · m/W [1) 2) 4)]	$1.5\ d \le V < 5\ d$ installation type B2 $5\ d \le V < 50\ d$ installation type B1
	Single-core or multi-core cable(s), or light-sheathed cable in enclosed conduit in masonry/concrete with thermal resistivity ≤ 2 K · m/W [4)]	B2
	Single-core or multi-core cable(s) or light-sheathed cable – in ceiling cavity [1) 2)] – in raised floor [1) 2)]	$1.5\ d \le V < 5\ d$ installation type B2 $5\ d \le V < 50\ d$ installation type B1

[1)] *V* is the smaller measurement; or the diameter of a building duct/cavity; or the internal height of a rectangular, enclosed conduit, floor or ceiling cavity.
If *V* is greater than 50 *d*, the relevant rated values for installation types C or E must be used.
The internal height of the cavity is more important than the width.

[2)] *d* is the outer diameter of a multi-core cable or a multi-core light-sheathed cable,
d is 2.2 × the outer diameter if three single-core cables or single-core light-sheathed cables are grouped in a trefoil arrangement.
d is 3 × the outer diameter if single-core cables or single-core light-sheathed cables are installed flat in one plane.

[3)] *d* is the outer diameter of a conduit, or the height of an enclosed conduit.

[4)] Particular attention must be paid to cables installed vertically with restricted ventilation. The ambient temperature at the upper end of vertical installations can rise considerably, making it necessary to reduce the current-carrying capacity.

Table 1.7/5 Common fixed installations and corresponding installation types *(continued)*

Installation	Description	Installation type for determining current-carrying capacity
	Single-core non-sheathed cables, single-core cable or light-sheathed cable in conduit on wooden wall – horizontal installation[1]	B1
	– vertical installation[1] [2]	B1
	Multi-core cable(s) or light-sheathed cable in conduit on wooden wall – horizontal installation[1]	B2
	– vertical installation[1] [2]	B2
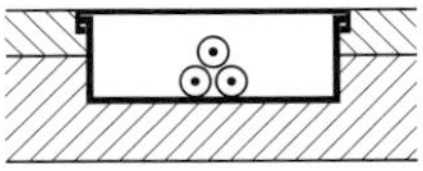	Single-core non-sheathed cables in duct for underfloor installation[1]	B1
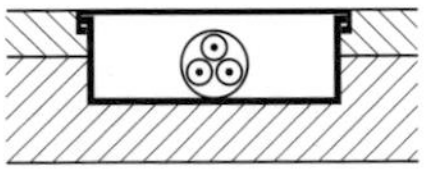	Multi-core cable(s) or light-sheathed cable in conduit for underfloor installation [1]	B2
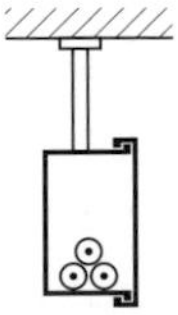	Single-core non-sheathed cables in suspended conduit[1]	B1
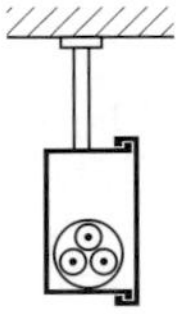	Multi-core cable(s) or light-sheathed cable in suspended conduit[1]	B2

1) The current-carrying capacities for installation types B1 and B2 in Table 1.7/6 apply to an individual electric circuit.
If more than one electric circuit is installed in a conduit, the conversion factors to Table 1.7/9 must be used, irrespective of whether the conduit is equipped with internal compartmentalization or isolation.

2) Particular attention must be paid to cables installed vertically with restricted ventilation. The ambient temperature at the upper end of vertical installations can rise considerably, making it necessary to reduce the current-carrying capacity.

Table 1.7/5 Common fixed installations and corresponding installation types *(continued)*

Installation	Description	Installation type for determining current-carrying capacity
	Single-core non-sheathed cables in conduit with non-ventilated duct (horizontal or vertical installation)[2) 3)]	$1.5\,d \leq V < 20\,d$ installation type B2 $20\,d \leq V$ installation type B1
	Single-core non-sheathed cables in conduit with ventilated duct in floor[1) 4)]	B1
	Single-core or multi-core cable or light-sheathed cable in an open or ventilated duct (horizontal or vertical installation)[4)]	B1
	Single-core or multi-core cable(s) or light-sheathed cable installed directly in masonry/concrete with thermal resistivity ≤ 2 K · m/W[5)] – without additional mechanical protection	C
	– with additional mechanical protection	C

[1)] If a multi-core cable or a multi-core light-sheathed cable is installed in a conduit or ventilated cable duct in the floor, the current-carrying capacity for installation type B2 must be used.

[2)] d is the outer diameter of the conduit,
V is the internal height of the cable duct.
The internal height of the cable duct is more important than the width.

[3)] Particular attention must be paid to vertical, non-ventilated ducts.
The air temperature at the upper end of cavities can rise considerably, making it necessary to reduce the current-carrying capacity.

[4)] It is recommended that these installation types be used only where access is restricted to authorized persons, and reductions in current-carrying capacity, as well as fire hazards due to accumulated refuse can be avoided.

[5)] The current-carrying capacity can be higher for cables with a nominal conductor cross-sectional area ≤ 16 mm^2.

Table 1.7/5 Common fixed installations and corresponding installation types *(continued)*

Installation	Description	Installation type for determining current-carrying capacity
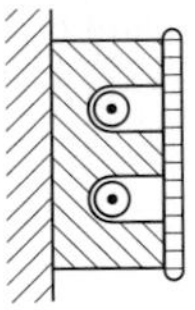	Single-core non-sheathed cables in molded ducts/sections[1]	A1 (A)[2]
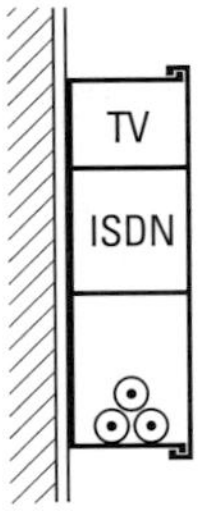 TV / ISDN	Single-core non-sheathed cables, single-core cables or light-sheathed cables in skirting board duct	B1
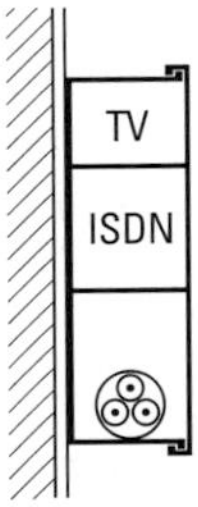 TV / ISDN	Multi-core cable(s) or light-shielded cable in skirting board duct	B2
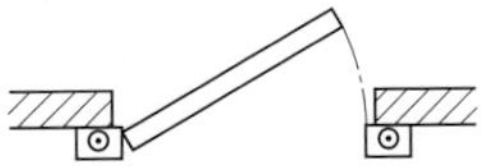	Single-core non-sheathed cables in conduits or single-core or multi-core cable(s) or light-sheathed cable in door panels[1]	A1 (A)[2]
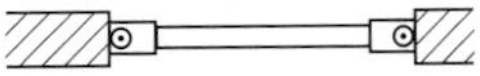	Single-core non-sheathed cables in conduits or single-core or multi-core cable(s) or light-sheathed cable in window frames[1]	A1 (A)[2]

[1] The thermal conductivity of the enclosure is assumed to be very poor because of the material used and possible air pockets. Installation types B1 or B2 may be used with variants corresponding to one of the installations in conduits on a wooden wall.

[2] To DIN VDE 0298-4, see also Note on page 244

Current-carrying capacity: rated value

The current-carrying capacities of cables with a permissible operating temperature of 70 °C for fixed installation are listed in Table 1.7/6 for installation types A1, A2, B1, B2, C, and E. The values specified here apply for a maximum ambient temperature of 30 °C, which is usual in Germany.

Site operating conditions, conversion factors

When cables are actually installed, the appropriate conversion factors (e.g. to Tables 1.7/7 – 1.7/9) may have to be used to take account of site operating conditions deviating from the reference values (e.g. different ambient temperatures, grouping).

Installation types A1 and A2

> *Note*
> To date, DIN VDE 0298-4 does not differentiate between installation type A 1 (single-core cables in conduits) and A2 (multi-core cables in conduits), and thus departs from standard international practice. This national difference has been provisionally retained in Table 1.7/5.

Short-time and intermittent operation

Cables that are not operated continuously at constant load may have current-carrying capacities higher than those specified in Table 1.7/6. The criteria and procedure for these types of operation are discussed in detail in [1], Sections 18.6.2 and 18.6.3.

DC loading

Where cables are loaded with direct current, the same current-carrying capacities (corresponding to the number of loaded cores) may be used.

Current-carrying capacity

The current-carrying capacity I_z for the actual operating conditions is calculated by multiplying the rated value I_r (Table 1.7/6) by the product of all the relevant conversion factors f (Tables 1.7/7 to 1.7/9 or DIN VDE 0298-4) using the following equation:

$$I_z = I_z \cdot \Pi f \quad \text{(A)}$$

Πf product of all conversion factors

The current-carrying capacity must be determined for:

- ▷ normal operating conditions
- ▷ operation under fault conditions (short circuit).

Short-circuit rating

The short-circuit rating must be calculated in accordance with [1] Section 12.2, and is not considered here.

Voltage drop

Overload protection

In low-voltage systems in particular, the conductor cross-sectional area must be dimensioned according to the specified voltage drop Δu (see page 275); furthermore, suitable protective equipment must be used to prevent cables from being thermally overloaded (see Section 1.7.3). The relevant standards for the particular system must also be adhered to.

Maximum length of cables

Checks should also be carried out in accordance with DIN VDE 0100 Supplementary Sheet 5 to ensure compliance with maximum permissible cable lengths, specifically regarding:

- ▷ protection against indirect contact,
- ▷ protection in the event of short circuits,
- ▷ specified voltage drop.

Terms and definitions

When the cross-sectional areas of cables are determined, the terms and definitions laid down in DIN VDE 0298-4 apply. These and other specifications are discussed extensively in Section 1.7.2, and should also be applied to insulated wires and flexible cables.

Cables for fixed installation

The current-carrying capacity values specified in Table 1.7/6 also apply to cables installed in the same way as wiring and flexible cables (e.g. NYY, NYKY).

able 1.7/6 Current-carrying capacity. Cables for fixed installation. Installation types A1, A2, B1, B2, C, and E

sulation	PVC											
ype designation[1)]	NYM, NYBUY, NHYRUZY, H07V-U, H07V-R, H07V-K, NHXMH[2)]											
aximum permissible erating temperature	70 °C											
mbient temperature	30 °C											
o. of loaded nductors	2	3	2	3	2	3	2	3	2	3	2	3
stallation type[3)]	A1		A2		B1		B2		C		E	
	In thermally-insulated walls				On or in walls or under plaster in conduits or ducts				Installed directly		In free	
	Single-core non-sheathed cables in conduit		Multi-core cable in conduit		Single-core non-sheathed cables in conduit on wall		Multi-core cable in conduit on wall or floor		Multi-core cable on wall or floor		Multi-core cable, conforming to specified clearances (≥ 0,3 d)	
			Multi-core cable in wall		Single-core non-sheathed cables in duct on wall		Multi-core cable in duct on wall or floor		Single-core light-sheathed cable on wall or floor			
					Single-core non-sheathed, single-core light-sheathed, and multi-core cables in conduit in masonry				Multi-core cable, flat webbed cable in wall or under plaster			
ated cross-sectional ea of copper conictor in mm²	Current-carrying capacity in A											
1.5	15.5	13.5	15.5	13.0	17.5	15.5	16.5	15.0	19.5	17.5	22	18.5
2.5	19.5	18.0	18.5	17.5	24	21	23	20	27	24	30	25
4	26	24	25	23	32	28	30	27	36	32	40	34
6	34	31	32	29	41	36	38	34	46	41	51	43
10	46	42	43	39	57	50	52	46	63	57	70	60
16	61	56	57	52	76	68	69	62	85	76	94	80
25	80	73	75	68	101	89	90	80	112	96	119	101
35	99	89	92	83	125	110	111	99	138	119	148	126
50	119	108	110	99	151	134	133	118	168	144	180	153
70	151	136	139	125	192	171	168	149	213	184	232	196
95	182	164	167	150	232	207	201	179	258	223	282	238
20	210	188	192	172	269	239	232	206	299	259	328	276

For type designations, see Table 1.7/1 and [1]
Insulation made of cross-linked polyethylene compound
For installation types not listed here, see Table 1.7/5

Example 1

Installation on wall

A total of 50 H07V-U single-core non-sheathed cables, each with a conductor cross-sectional area of 6 mm^2 are to be installed in a conduit on the wall of a cellar. The single-core non-sheathed cables form 10 three-phase circuits, each with the required PE and neutral conductors. The ambient temperature is 40 °C. The current-carrying capacity now has to be determined for the specified conditions.

Operating conditions (continuous load):

Maximum permissible operating temperature	70 °C
Ambient temperature	40 °C

Result:

Rated value I_r to Table 1.7/6 (installation type B1, $\vartheta_U = 30\,°C, n = 3$)	36 A
Conversion factor for	
– ambient temperature to Table 1.7/7	0.87
– grouping to Table 1.7/9	0.48
Total conversion factor Πf	0.417
This yields a current-carrying capacity I_z of	15.0 A.

Example 2

Installation under ceiling

Two NYM-J 3 × 2.5 light-sheathed cables for single-phase alternating current (2 loaded cores) and two NYM-J 4 × 1.5 for three-phase loading (3 loaded cores) are installed under a ceiling. All four cables are installed side by side so that they touch each other. The maximum ambient temperature is 35 °C. The current-carrying capacity now has to be determined.

Operating conditions (continuous load):

Maximum permissible operating temperature	70 °C
Ambient temperature	35 °C

Result:

conductor cross-sectional area	2.5 mm^2	1.5 mm^2
number of loaded cores	$n = 2$	$n = 3$
Rated value I_r to Table 1.7/6 (installation type C, $\vartheta_U = 30\,°C$)	27 A	17.5 A
Conversion factors for		
– ambient temperature to Table 1.7/7	0.94	0.94
– grouping to Table 1.7/9	0.68	0.68
Total conversion factor Πf	0.64	0.64
Current-carrying capacity I_z under the given operating conditions	17.2 A	11.2 A.

Example 3

Installation in duct

An NYM 10 × 2.5 10-core light-sheathed cable is used to control a small machine tool. The cable is routed in a duct on a wall and on the machine. The ambient temperature is 40 °C.

Operating conditions (continuous load):

Maximum permissible operating temperature	70 °C
Ambient temperature	40 °C

Result:

Rated value I_r to Table 1.7/6 (installation type B2, $\vartheta_U = 30\,°C, n = 3$)	20 A
Conversion factors for	
– ambient temperature to Table 1.7/7	0.87
– multi-core cable to Table 1.7/8	0.55
Total conversion factor Πf	0.48

When all 10 cores are loaded simultaneously under the given operating conditions, the 10-core light-sheathed cable has a current-carrying capacity I_z of 9.6 A.

Example 4

Installation on cable rack

At a nominal voltage of $U_n = 400$ V, 60 kVA of three-phase power has to be transmitted by means of a maximum of 6 NYBUY lead-sheathed cables. The cables are installed side by side on cable racks. The racks are located in a large room with an ambient temperature of 45 °C. What conductor cross-sectional area has to be selected?

The operating current at 60 kVA is

$$I_b = \frac{60 \cdot 10^3}{\sqrt{3} \cdot 400\ \text{V}} = 86.6\ \text{A}$$

Table 1.7/7
Conversion factors for ambient temperatures other than 30 °C

Insulation	PVC
Maximum permissible operating temperature	70 °C
Ambient temperature °C	Conversion factor
10	1.22
15	1.17
20	1.12
25	1.06
30	**1.00**
35	0.94
40	0.87
45	0.79
50	0.71
55	0.61
60	0.50
65	0.35

Table 1.7/8
Conversion factors for multi-core cables with rated cross-sectional area ≤ 10 mm²

Number of loaded cores	Conversion factor
5	0.75
7	0.65
10	0.55
14	0.50
19	0.45
24	0.40
40	0.35
61	0.30

Conversion factors for
- ambient temperature to Table 1.7/7 0.79
- grouping to Table 1.7/9 0.79

Total conversion factor Πf 0.62

When converted to the "reference operating conditions" in accordance with Table 1.7/4, the "fictitious load current" (for ambient temperature $\vartheta_u = 30\,°C$ and individual installation) is, therefore, at least

$$I_{bf} = \frac{86.6\text{ A}}{0.62} = 139.7\text{ A}.$$

The required minimum current-carrying capacity for six cables installed in parallel is, thus, 23.3 A per conductor.

Table 1.7/6 shows that the appropriate installation type (E) with three loaded cores requires a minimum conductor cross-sectional area of 2.5 mm² with a current-carrying capacity of 25 A. Determining the cross-sectional area shows that six NYBUY-J 4 × 2.5 lead-sheathed cables are required.

Guidelines for laying cables

Protection against mechanical damage

Cables must either be equipped with a covering or positioned to HD 384.5.52/DIN VDE 0100-520 to provide adequate protection against mechanical damage. Within normal arm's reach, a covering is always required to protect cables against mechanical damage. The covering of light-sheathed cables, for example, is considered adequate for this purpose.

At points subject to particular hazards, such as floor bushings (Fig. 1.7/1), all cables (including light-sheathed cables) must be provided with additional protection, e.g. by means of slip-over plastic or steel conduits, or other coverings, which must be fixed securely.

Cable routing

Cables in and under plaster, as well as behind wall paneling, must be routed vertically or horizontally, or parallel to the edges of the room (Fig. 1.7/2) (HD 384.5.52/DIN VDE 0100-520 and DIN 18015 Parts 1 and 3).

Shared core covering in an electric circuit

With single-core non-sheathed cables in conduits or ducts, only the conductors of one main circuit (including related auxiliary circuits) may be laid in the same conduit/duct. This requirement does not apply to electrical and closed electrical operating areas.

Several main circuits (including related auxiliary circuits) may be combined in one multi-core cable/flexible cable, provided this is not prohibited by other standards.

If auxiliary circuits are installed separately from the main circuits, several auxiliary circuits may be combined in a multi-core cable. The same applies to single-core non-sheathed cables in a conduit or duct.

Installation in cast or tamped concrete

Flexible cables may only be laid directly in concrete that has been shaken or tamped if they are installed in conduits in accordance with type "AS" to DIN VDE 0605 (see Section 5.2). Light-sheathed cables, however, may be installed in recesses and covered with concrete in a similar way to underplaster installation.

Table 1.7/9 Conversion factors for grouping

Arrangement	Number of multi-core cables or number of single-phase or three-phase AC electric circuits comprising single-core cables (2 or 3 current-carrying conductors)														
	1	2	3	4	5	6	7	8	9	10	12	14	16	18	20
Trefoil arrangement directly on wall, on floor, in conduit or duct on or in wall (installation types A to E)	1.00	0.80	0.70	0.65	0.60	0.57	0.54	0.52	0.50	0.48	0.45	0.43	0.41	0.39	0.3[illegible]
One layer on wall or on floor, touching (installation type C)	1.00	0.85	0.79	0.75	0.73	0.72	0.71	0.70	0.70	0.70	0.70	0.70	0.70	0.70	0.70
One layer on wall or on floor, with spacing equal to outer diameter of cable *d* (installation type C)	1.00	0.94	0.90	0.90	0.90	0.90	0.90	0.90	0.90	0.90	0.90	0.90	0.90	0.90	0.90
One layer under ceiling, touching (installation type C)	0.95	0.81	0.72	0.68	0.66	0.64	0.63	0.62	0.61	0.61	0.61	0.61	0.61	0.61	0.6[illegible]
One layer under ceiling, with spacing equal to outer diameter of cable *d* (installation type C)	0.95	0.85	0.85	0.85	0.85	0.85	0.85	0.85	0.85	0.85	0.85	0.85	0.85	0.85	0.85
One layer on cable tray (perforated), horizontally or vertically, touching (installation type E) — ≥20 mm	1.00	0.88	0.82	0.79	0.77	0.76	0.76	0.73	0.73	0.73	0.73	0.73	0.73	0.73	0.73
	1.00	0.88	0.82	0.78	0.75	0.73	0.73	0.73	0.72	0.72	0.72	0.72	0.72	0.72	0.72
≥225 mm	1.00	0.88	0.81	0.76	0.73	0.71	0.71	0.71	0.70	0.70	0.70	0.70	0.70	0.70	0.70
One layer on cable rack or on cable clamps, etc. touching (installation type E) — ≥20 mm	1.00	0.87	0.82	0.80	0.80	0.79	0.79	0.78	0.78	0.78	0.78	0.78	0.78	0.78	0.78

⊙ Symbol for single-core or multi-core cable

Notes

When these conversion factors are applied to the values in Table 1.7/6, the number of loaded cores, type of cable, and installation type must conform.

When multi-core cables and two or three loaded cores are grouped together, the conversion factor for the total number of grouped cables must be selected and applied to the current-carrying capacity for cables with two or three loaded cores.

If n cables in a group of single-core cables are loaded, the conversion factor must be determined for $n/2$ or $n/3$ electric circuits and applied to the current-carrying capacity of two or three loaded cores.

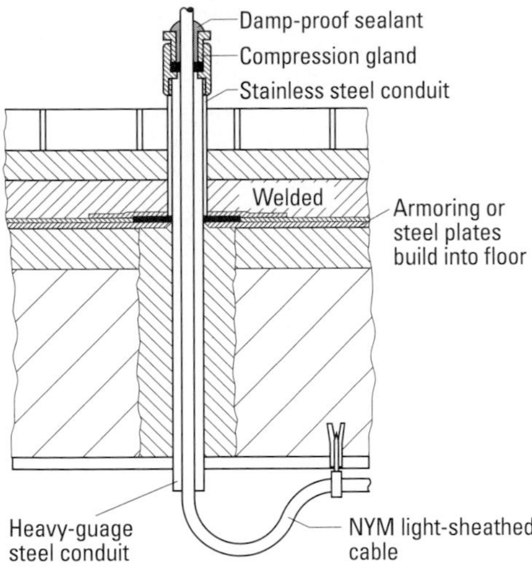

Fig. 1.7/1
Floor bushing comprising a stainless steel conduit with welded high-strength steel conduit

Installation in the ground

Flexible cables must not, under any circumstances, be laid in the ground or in inaccessible underground ducts outside buildings. Only cables may be used for applications of this type.

Installation in water

The suitability of rubber-insulated flexible cables (e.g. HYDROFIRM-cables) for continuous operation in water must be verified by means of special tests. This type of application is not taken into account by the tests for standard cable types.

Fixing cables

The type and form of cables must be taken into consideration when the appropriate fixing method is selected. Flat webbed cables, for example, can be fixed by means of gypsum plaster, cable clamps made of insulating material or metal with an insulated intermediate layer, steel nails with washers made of insulating material, or adhesives.

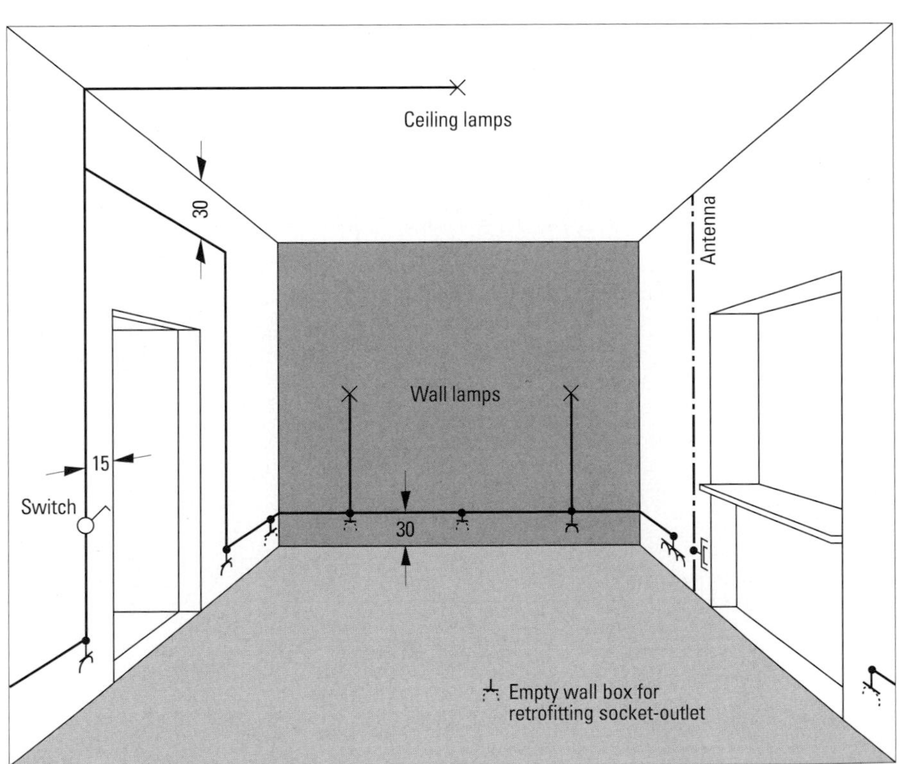

Fig. 1.7/2 Cables installed under plaster in a room to DIN 18015 (measurements in cm)

Damp-proof cables must not be fitted using hook-nails even if they are installed under plaster. Nails may, however, be attached loosely to hold the cables in place but must be removed after the gypsum plaster has been applied. Section 1.7.4 contains further guidelines on materials for reducing installation time.

Permissible bending radii

The factors that are important for determining for the minimum permissible bending radii are the type and outer diameter of cables, as well as the installation type and operating conditions.

To avoid damage to cables, the minimum permissible bending radii listed in Table 1.7/10 must be strictly adhered to. These bending radii (taken from the standards HD 516 S2/DIN VDE 0298-300) have been tried and tested in practical applications.

Strain relief

Cables for portable electrical equipment have to be relieved of stress and strain at the points of connection. The protective conductor must be longer than the other cables so that it is loaded after these in the event of the strain relief failing.

To prevent cables from buckling at the cable entries of devices, the entries must be rounded off, or sleeves must be used (Fig. 1.7/3). It is not permissible to relieve strain by knotting cables, e.g. where they enter devices on building sites.

Table 1.7/10 Minimum permissible bending radii for power cables

Type of cable	Minimum permissible bending radius Rated voltage ≤ 0.6/1 kV				> 0.6/1 kV
Cables for fixed installation	Outer diameter of cable *d* in mm				
	≤ 10	> 10 ≤ 25	> 25		
Fixed installation	4 *d*	4 *d*	4 *d*		6 *d*
Formed bend	1 *d*	2 *d*	3 *d*		4 *d*
Flexible cables	Outer diameter of cable *d* in mm				
	≤ 8	> 8 ≤ 12	> 12 ≤ 20	> 20	
Fixed installation	3 *d*	3 *d*	4 *d*	4 *d*	6 *d*
Free moving	3 *d*	4 *d*	5 *d*	5 *d*	10 *d*
At cable entry	3 *d*	4 *d*	5 *d*	5 *d*	10 *d*
Forced guiding[1)] e.g. Drum operation	5 *d*	5 *d*	5 *d*	6 *d*	12 *d*
Cable wagon operation	3 *d*	4 *d*	5 *d*	5 *d*	10 *d*
Drag chain operation	4 *d*	4 *d*	5 *d*	5 *d*	10 *d*
Roller guides	7.5 *d*	7.5 *d*	7.5 *d*	7.5 *d*	15 *d*

[1)] Suitability for this type of operation must be ensured by means of special constructional characteristics.

Notes:
d = outer diameter of cable or thickness of flat cable.
With cable types that can be used for several applications, it may be necessary to consult the manufacturer

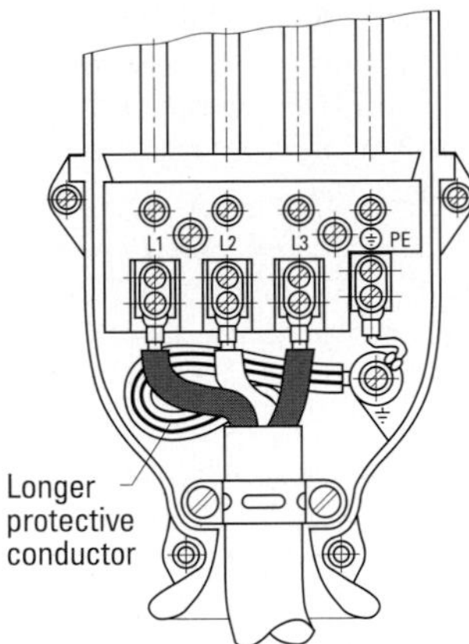

Fig. 1.7/3
Example of strain relief, grip, and funnel-shaped entry (to HD 384.5.52/DIN VDE 0100-520, Section 11.7).

1.7.2 Power cables for voltages up to 30 kV

The types of cable that may be used for cable systems in electrical installations must conform to the relevant European, international, or national standards.

Cable markings

Cables bear the following markings, which show that they have been manufactured to meet the relevant specifications, and which manufacturer produced them:

▷ identification marking on the outer sheath (thermoplastic-insulated cables), and
▷ an identification strip in the cable (paper-insulated cables).

The following information appears on the sheath of thermoplastic-insulated cables for low and medium voltage:

▷ manufacturer's type designation or symbol, and year of manufacture
▷ type designation and rated voltage U
▷ the VDE mark of conformity
▷ length marking for round cables with outer diameters ≥10 mm (to facilitate determination of length).

Examples of markings for thermoplastic-insulated cables[1]:

"1999 PROTODUR NAYY–J 1 kV ◁VDE▷ 0276"
as well as continuous length marking,

"1999 PROTOTHEN X NA2XS(F)2Y 20 kV ◁VDE▷ 0276"
as well as continuous length marking.

Cables with paper insulation and a metal sheath have an identification strip under the metal sheath, which indicates their origin, e.g.

"1999 'manufacturer' ◁VDE▷ 0276".

Colors of outer sheaths

The colors used for the outer sheaths of power cables can be found in Table 1.7/11.

Outer sheaths made of polyethylene (PE) should always be black to ensure better durability.

Under normal environmental conditions (e.g. effects of weather or of substances usually found in soil), the colors of outer sheaths must not change to the extent that they can no longer be differentiated clearly or become unrecognizable.

It should be noted, however, that colored PVC sheaths turn black if exposed to the effects of sulfur compounds – especially hydrogen sulfide. The sulfur compounds that occasionally occur in soil result from the bacterial decomposition of organic substances under the exclusion of air (e.g. fecal matter, effluent, and manufactured gases).

Core identification $U_0/U \leq 0.6/1$ kV

Core identification of power cables is regulated by HD 186, HD 308, HD 402/IEC 60446/DIN VDE 0293.

Table 1.7/11 Colors of outer sheaths

Power cables	Color of outer sheath
Rated voltage ≤ **0.6/1 kV**	black
but	
– for underground mines	yellow
for intrinsically safe installations in locations with explosion hazard	blue
– for cables with functional endurance E30/E90	orange
Rated voltage > **0.6/1 kV**	
– with PVC sheath	red
– with PE sheath	black

[1] PROTODUR cable = cable with PVC (polyvinylchloride) insulation,
PROTOTHEN X-cable = cable with XLPE (cross-linked polyethylene) insulation:

The relevant specifications can be found in Tables 1.7/12 and 1.7/13; color designations and allocations are explained in Table 1.7/14.

The cores of single-core cables are always black. The core ends must be green/yellow, black, blue, or brown, depending on their purpose.

If cables have a core with a reduced conductor cross-sectional area, this core must be marked either GNYE (in variants with protective conductor) or BU (in variants without protective conductor).

The cores of single-core paper-insulated cables are beige. In addition to color identification, the cores of paper-insulated cables

Table 1.7/12 Core identification for PROTODUR and PROTOTHEN-X cables

Number of cores	With protective conductor[1)] (suffix "G", in Germany also "J")	Without protective conductor [1)] (suffix "X", in Germany also "O")	With concentric conductor (suffix "C") Used as neutral conductor (N), protective conductor (PE), or PEN conductor
2 3 4 5	GNYE/BK GNYE/BK/BU GNYE/BK/BU/BN GNYE/BK/BU/BN/BK	BK/BU BK/BU/BN BK/BU/BN/BK BK/BU/BN/BK/BK	BK/BU BK/BU/BN BK/BU/BN/BK BK with printed numbers
6 and over	GNYE/other BK cores with printed numbers on the inside beginning with 1, and ending with GNYE as the outermost core	BK cores with printed numbers on the inside beginning with 1	BK cores with printed numbers on the inside beginning with 1

[1)] Protective conductor = PE, PEN conductor

Table 1.7/13 Core identification for paper-insulated cables

Number of cores	With protective conductor[1)] (suffix "G", in Germany also "J")	Without protective conductor[1)] (suffix "X", in Germany also "O")	With Al sheath (metal sheath). Used as neutral conductor (N), protective conductor (PE) or PEN conductor
2 3 4 5	– GNYE/BK/BU GNYE/BK/BU/BN GNYE/BK/BU/BN/BK	BK/BU BK/BU/BN BK/BU/BN/BK –	BK/BU BK/BU/BN BK/BU/BN/BK –

[1)] Protective conductor = PE, PEN conductor

Table 1.7/14 Explanation of color designations and use of cores

Color desig.	Colors	Use
GNYE	green/yellow, green/beige (with paper-insulated cables)	PE and PEN conductors. This core must not be used for any other purpose. The only exception applies to public utility supply networks in Germany, where this core is also permissible for grounded neutral conductors (N) with conductor cross-sectional areas >6 mm^2
BK	Black	External conductors
BU	Blue	Neutral conductors (N), also permissible for external conductors
BN	brown, beige (with paper-insulated cables)	External conductors

that do not function as protective conductors (PE, PEN) can also be identified by means of printed numbers.

Core identification $U_0/U \leq 0.6/1$ kV

According to the relevant standards, the cores of cables with rated voltages of over 0.6/1 kV are not identified, i.e. the insulation of thermoplastic-insulated cables of this type is not colored, and the paper insulation has no colored layers of covering paper. With cable joints, this means that cross-bonding is no longer required to connect cores of the same color. Installation personnel should be instructed to verify that the phase sequence is correct.

Core identification to national standards

Core identification of cables can vary considerably between national standards. When cables are installed and connected in accordance with national standards, it is essential, therefore, that the appropriate standards are taken into consideration and complied with.

Allocating voltages

A distinction must be made between the rated voltages of cables (equipment) and the nominal voltages of the systems in which they are used.

Tables 1.7/15 and 1.7/16 show how these voltages are allocated, and contain values for maximum continuously permissible operating voltages, as well as rated lightning impulse voltages.

Table 1.7/15 Allocation of nominal system voltages (between external conductors) to rated cable voltages

Rated voltages of cable U_0/U	Nominal voltage U_n of system, i.e. of external conductors in		
	Three-phase systems	Single-phase systems	
	$U_n \leq U = \sqrt{3} \cdot U_0$	External conductor insulated $U_n \leq 2\, U_0$	One external conductor grounded $U_n = \leq U_0$
kV	kV	kV	kV
0.6/1	1	1.2	0.6
1.8/3[1)]	3	–	–
3.6/6	6	7.2	3.6
6/10	10	12	6
8.7/15[1)]	15	–	–
12/20	20	24	12
18/30	30	36	18
To: IEC 60071-1 IEC 60183 DIN VDE for cables	To: DIN VDE 0101 DIN VDE 0111	To: DIN VDE for cables	

Table 1.7/16 Maximum permissible voltage U_m of cable and rated lightning impulse voltage

Rated voltages of cable U_0/U	Max. permissible voltage U_m of cable for three-phase systems	Max. permissible voltage U_m of cable for single-phase systems		Rated lightning impulse voltage U_{rB}
		Both external conductors insulated	One external conductor grounded	
kV	kV	kV	kV	kV
0.6/1	1.2	1.4	0.7	–
1.8/3[1)]	3.6	–	–	40
3.6/6	7.2	8.3	4.2	60
6/10	12	14	7	75
8.7/15[1)]	17.5	–	–	95
12/20	24	28	14	125
18/30	36	42	21	170
To: IEC 60071–1 IEC 60183 DIN VDE for cables	DIN VDE 0101 DIN VDE 0111	DIN VDE for cables		IEC 60071-1 DIN VDE 0101 DIN VDE 0111

[1)] This voltage range to IEC 60071-1 and IEC 60183 is no longer included in the DIN VDE standards for cables

Voltage distribution for cables is illustrated in Fig. 1.7/16.

Nominal system voltage (U_n) The nominal system voltage U_n is the voltage that characterizes the power supply system or a part of it. The actual voltage can vary within the permissible tolerances of U_n.

Operating voltages ($U_b/U_{b\,max}$) The instantaneous value of a voltage is the operating voltage U_b. The upper limit of this voltage is determined by the maximum permissible operating voltage $U_{b\,max}$. The operating voltage (e.g. of a three-phase system) is the r.m.s. value of the highest phase-to-phase voltage that can occur at any time or at any point in the system under normal operating conditions.

Rated voltage of cables (U_0/U) The rated voltage of cables is specified by the voltages U_0/U, where U_0 is the voltage between an external conductor and a metal sheath or ground, and U is the voltage between the external conductors of a three-phase system: $U = \sqrt{3} \cdot U_0$.

Maximum voltage for equipment (U_m) The maximum voltage for equipment U_m is the r.m.s. value of the highest phase-to-phase voltage for which an item of equipment is rated. With cables, the voltage is $U_m = 1.2 \cdot U$, and can be found in Table 1.7/16.

Rated lightning impulse voltage (U_{rB}) Cables must conform to the values U_{rB} specified in Table 1.7/16, i.e. using this value to test a new cable must not cause a disruptive discharge.

Voltage conditions The following voltage conditions must be adhered to when cables are selected (taking the existing power supply system into consideration):

$$U_n \leq U_0 \text{ or } U,$$

$$U_{b\,max} \leq \frac{1}{\sqrt{3}} U_m.$$

Special applications In addition to the specified applications, 0.6/1 kV cables may be used in:

- ▷ DC systems with maximum operating voltage $U_{b\,max}$ not exceeding 1.8 kV phase-to-phase, or 1.8 kV phase-to-ground.
- ▷ In three-phase systems with maximum phase-to-phase voltage $U_{b\,max} = 3.6$ kV, if the conductor cross-sectional area is ≥ 240 mm^2 or the cable is equipped with concentric conductors or armor.

Ground fault, short circuit to ground Depending on how the neutral point is treated, a distinction is made between the following operating conditions in single-phase and three-phase systems:

- ▷ In systems with a low-resistance-grounded neutral point, ground faults are disconnected immediately, i.e. within one second.
- ▷ In systems with an insulated neutral point or with ground fault compensation, individual ground faults must not be sustained for longer than 1 hour, provided that the standards for the particular cable type do not specify longer times.
- ▷ Short circuits to ground comprise double ground faults as well as phase-to-phase and three-phase grounded faults. These are disconnected within seconds.

Current-carrying capacity (terms, definitions, and regulations)

DIN VDE 0276–1000 specifies the standard terms, definitions, and regulations applicable to the current-carrying capacity of cables.

Current-carrying capacity Current-carrying capacity denotes the permissible current I_z under specified operating conditions.

In addition to the above definition, the following must be explained:

Operating conditions, rated values The specified values for current-carrying capacity under "reference operating conditions" are "rated values" to DIN 40200, and should be designated rated current I_r. In a similar way, the reference operating conditions (to DIN VDE 0276-1000, operating conditions for calculating rated currents) are "ratings".

The following equation applies for current-carrying capacity under given operating conditions:

$$I_z = I_r \cdot \Pi f,$$

Conversion factor where Πf is the product of all the relevant conversion factors, such as the influence of grouped adjacent cables.

Load Load is the short designation for current load. Load designates the currents a cable may be required to carry during a particular type of operation, or in the event of a fault. The operating current I_b is calculated from the maximum permissible operating voltage $U_{b\,max}$ in kV and the effective power P for transmission in kW:

- ▷ For direct current:

$$I_b = \frac{P}{U_{b\,max}} \quad \text{in A}$$

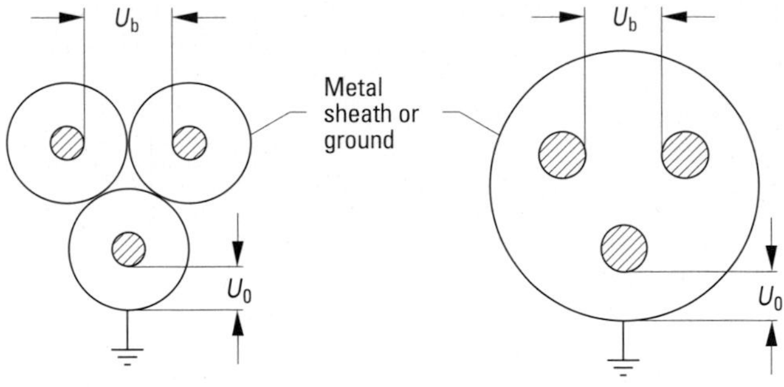

a) Three single-core cables

b) Three-core cable

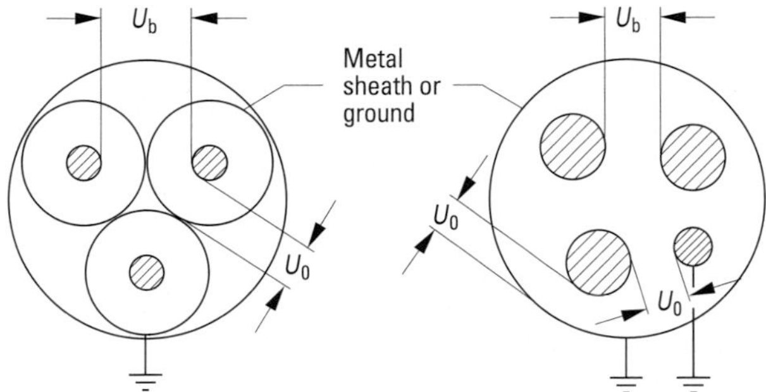

c) Three-core cable with individual core shields

d) Four-core cable with grounded protective or neutral conductor

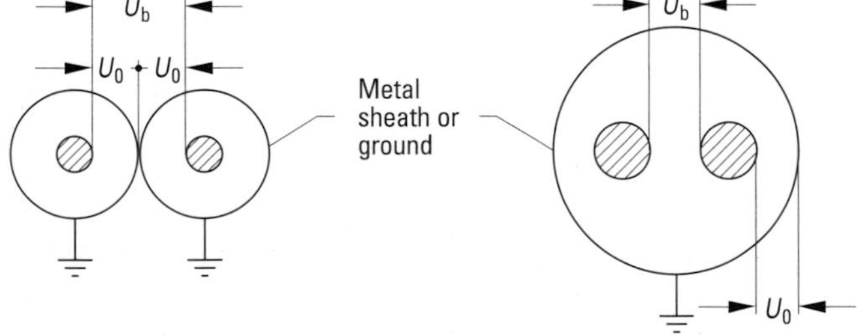

Two single-core cables

Two-core cable

e) Single-phase alternating current, both external conductors insulated: $U_b = 2 \cdot U_0$

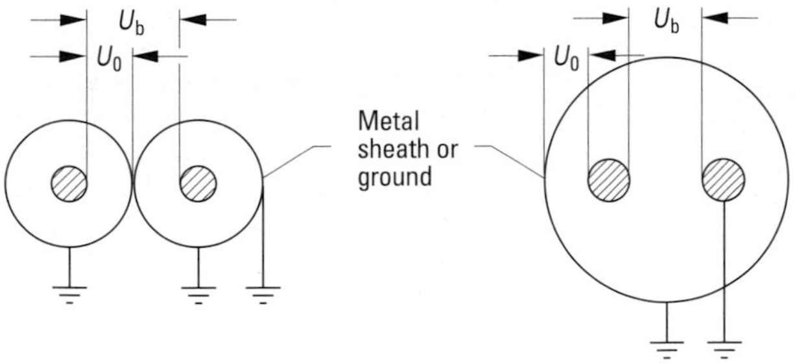

Two single-core cables

Two-core cable

f) Single-phase alternating current, one external conductor grounded: $U_b = U_0$

Fig. 1.7/4
Voltage distribution for cables

▷ For single-phase alternating current:

$$I_b = \frac{P}{U_{b\,max} \cdot \cos\varphi} \quad \text{in A}$$

▷ For three-phase alternating current:

$$I_b = \frac{P}{\sqrt{3} \cdot U_{b\,max} \cdot \cos\varphi} \quad \text{in A.}$$

Operating current

Under normal operating conditions, the load is the operating current I_b. With public utility operation or other cyclical types of operation, the operating current is the maximum load value.

Permissible operating temperature

The permissible operating temperature ϑ_{Lr} is the maximum permissible temperature of the conductor during normal operation. This temperature is used to calculate the current-carrying capacity during normal operation, and is specified in accordance with the service life in the relevant construction specifications.

Selecting conductor cross-sectional area

The selected conductor cross-sectional area must ensure that, under normal operating conditions, the current load does not exceed the current-carrying capacity of a conductor, and that no part of the conductor at any time exceeds the permissible operating temperature. The temperature rise or current-carrying capacity of a cable with a particular conductor cross-sectional area depends on its construction, material properties, and the specific operating conditions. The following condition must always be fulfilled, whatever the particular operating conditions:

Current load $I_b \leq$ current-carrying capacity I_z.

The most unfavorable operating conditions that can occur during operation and along the cable route are crucial here. Taking these conditions into consideration ensures that no part of the conductor at any time exceeds the permissible operating temperature.

Permissible temperature rise

The temperature rise of a cable depends on its construction, material properties, and operating conditions. Additional temperature rises (caused by grouping with other cables, heating lines, direct solar radiation, etc.) must be taken into account. Impermissibly high operating temperatures and temperature rises cause cables to age more rapidly.

The permissible temperature rise of a cable – excluding other heat sources – is determined by means of the permissible operating temperature and the ambient temperature (see Table 1.7/17).

Table 1.7/17 Permissible operating temperatures

Type of cable	Standards	Permissible operating temperature °C
XLPE-insulated cable (PROTOTHEN-X)	HD 603/IEC 60502-1/ DIN VDE 0276-603, HD 604/DIN VDE 0276-604, HD 602/IEC 60502-2/ DIN VDE 0276-620, HD 622/DIN VDE 0276-622	90
PVC-insulated cable (PROTODUR)	HD 603/IEC 60502-1/ DIN VDE 0276-603; IEC 60502/ DIN VDE 0271	70
Paper-insulated mass-impregnated cable (belted cable)[1)] 6/10 kV	HD 621/IEC 60055/ DIN VDE 0276-621	65
single-core cable, three-core separately-sheathed cable 6/10 kV		70
12/20 kV		65
18/30 kV		60

1) The cores, which are stranded together, are enclosed by shared insulation – belt insulation – which is covered by the metal sheath, e.g. NKBA, NAKBA

Temperature rise

The temperature rise $\Delta\vartheta_L$ of a conductor for any load current I_b is calculated as follows at a constant ambient temperature ϑ_U and without taking changes in resistance due to the conductor temperature and dielectric losses into account:

$$\Delta\vartheta_L = \vartheta_{Lr} - \vartheta_U = \Delta\vartheta_r \left(\frac{I_b}{I_r}\right)^2 \text{ in K with}$$

$$\Delta\vartheta_r = \vartheta_{Lr} - 30\ °\text{C in K.}$$

Ambient temperature

Since the ambient temperature ϑ_U cannot always be determined by means of measurements, it often has to be estimated. If no other values based on measurements or experience are available, the following ambient temperatures can be assumed for central European conditions:

▷ Cables installed in free air:

unheated cellars 20 °C

rooms without air-conditioning (not heated in summer) 25 °C

factory bays, work rooms, etc. 30 °C

▷ Cables installed in the ground:

with installation depth 0.7 m to 1.0 m 20 °C

Air temperatures over 30 °C occur, for example:

▷ where there is insufficient protection against direct solar radiation,
▷ in poorly ventilated rooms,
▷ with machines or installations with high heat dissipation.

In certain cases, heat loss from the cable itself can increase the ambient temperature. This applies mainly to cable ducts (see example on page 267).

Ground temperatures below 20 °C should be used in calculations only if verified by measurements taken over a prolonged period of time. During the summer months, the temperature at a depth of 1 m below a concrete or asphalt covering exposed to solar radiation can, for example, exceed 25 °C.

If the ambient temperature ϑ_U of a cable installation in air deviates from the reference value of 30 °C (e.g. if air temperatures are high for long periods, or if heating cables are routed parallel to cables), the current-carrying capacity I_z or required conversion factor f_ϑ can be determined with sufficient accuracy

$$I_z = I_r \cdot \sqrt{\frac{\Delta\vartheta_L}{\Delta\vartheta_r}} = I_r \cdot \sqrt{\frac{\vartheta_{Lr} - \vartheta_U}{\vartheta_{Lr} - 30\ °C}} = I_r \cdot f_\vartheta \text{ in A}$$

ϑ_U ambient temperature
$\Delta\vartheta_L$ temperature rise ($\Delta\vartheta_L = \vartheta_{Lr} - \vartheta_U$)
f_ϑ required conversion factor

Normal operation

Normal operation denotes any type of operation (e.g. continuous, short-time, intermittent, cyclical, and public utility operation), during which the permissible operating temperature determined by the insulation is not exceeded.

Overcurrents

Overcurrents include overload currents as well as short-circuit currents. Over a limited period of time, these give rise to conductor temperatures that exceed the permissible operating temperatures.

Overcurrent protection devices must be used to protect cables against detrimental temperature rises (see Section 1.7.3). It may be necessary to dimension the conductor cross-sectional area in accordance with the specifications for thermal short-circuit ratings of the short-circuit load as well.

Overload currents

Overload currents are caused by overloading during normal operation of fault-free circuits. The permissible temperatures for conditions of this type have not yet been standardized. These temperatures depend on the duration/number of overloads, which, in turn, depend on the heat/pressure performance characteristics of cables, and the effects of accelerated aging.

Short-circuit currents

Short-circuit currents are caused by faults between live conductors with negligible impedance, which have different potentials during normal operation. The permissible short-circuit temperatures (see Table 1.7/24) are restricted to a maximum short-circuit duration of 5 seconds.

In isolated-neutral and compensated systems, the phase-to-ground fault current is referred to as the ground-fault current. Since ground faults cause increased voltage stress in the fault-free conductors, temperatures exceeding the permissible operating temperature cannot be permitted in these cases.

Type of operation

The type of operation designates the characteristics of the current, current-carrying capacity, and load as a function of time.

Continuous operation

Continuous operation is a type of operation with constant current, the duration of which is sufficient for the cable to reach a thermally stable condition, but is otherwise not restricted with respect to time.

Public utility operation

Public utility operation is a common type of operation in public electricity supply networks. This type of operation is characterized by a 24-hour load cycle with a distinctive maximum load and load factor. The load factor m is the ratio of the average and maximum loads in a 24-load cycle. A common load factor value is $m = 0.7$.

Guidelines for dimensioning

Cables must always be dimensioned to ensure that, even when all relevant conversion factors f have been applied, the current-carrying capacity does not fall below the specified load; i.e. the following equation always applies:

$$I_b \leq I_z = I_r \cdot \Pi f.$$

In a technical manual such as this, the project planning and design guidelines for cable installations can only be described briefly. Readers requiring more detailed information on operating types, definitions, and characteristics of cables are advised to consult the fifth, extensively revised and expanded edition of the manual "Kabel und Leitungen für Starkstrom", 1999 [1] (see Section 7.3).

1.7.2.1 Guidelines for project planning and design

Installation in free air

Reference operating conditions

The following are considered to be reference operating conditions:

- ▷ operating type,
- ▷ installation conditions,
- ▷ ambient conditions.

These reference operating conditions form the basis for the current-carrying capacity specifications, which are discussed in greater detail in Part 2 of the Siemens manual "Power Cables and their Application", fourth revised edition, 1989. This manual is referred to as [2] throughout the rest of this section.

The reference operating conditions for installation in free air can be found in Table 4.2.9 in [2], together with guidelines for other (site) operating conditions. These specifications form the basis for the rated values in the current-carrying capacity Tables (also in [2]). When cables are installed in free air, the temperature rise caused by heat loss must be dissipated from the surface of the cable only by means of unimpeded natural convection and radiation, excluding other heat sources, and must not cause the ambient temperature to increase noticeably (infinitely high thermal capacity of the environment, e.g. large room or hall). This is ensured if the following conditions are fulfilled:

- ▷ The clearance between cables and the wall, ceiling, or floor must be at least 0.3 × the diameter of the cable. The spacing between cables arranged side by side in one layer must be at least 2 × the diameter of the cable.
- ▷ With cables arranged side by side and in several vertical layers, the additional vertical clearance between the layers (cable trays or racks) must be greater than 30 cm.
- ▷ Protection against direct heating by sunlight, and similar.
- ▷ Rooms must be sufficiently large and adequately ventilated to prevent the ambient temperature from increasing due to heat loss from cables.

If a cable is installed directly on a wall or on the floor, the current-carrying capacity must be reduced by a factor of 0.95 in comparison with installation in free air.

Ambient temperature, grouping

The reference ambient temperature for cables installed in free air is 30 °C. Other ambient temperatures, as well as groupings, must always be taken into account using the appropriate conversion factors specified in [2].

Multi-core cables

The current-carrying capacity of multi-core cables (control cables) can be calculated using conversion factors specified in [1]. Table 1.7/7 is also applicable for cables installed in air.

Cables installed in buildings

In cases where PVC-insulated cables are installed in buildings in the same way as flexible cables, the current-carrying capacity specifications in Table 1.7/5 can be used.

Where XLPE-insulated cables with a permissible operating temperature of 90 °C are installed in buildings, the current-carrying capacities specified in Table 1.7/8 must be used. The operating conditions listed in Table 1.7/4 can also be applied to the installation types considered here (A1, A2, B1, B2, C, and E); the operating conditions specified in Table 1.7/5 can be applied to other common types of installation in buildings.

The values specified in Table 1.7/18 apply to XLPE-insulated cables for fixed installation. They were derived from Report R 64-001 to HD 384.5.523 S 1.

Example 1

Installation on racks

The entire electrical power supply for an office installation on the sixteenth floor of a high-rise building is to be provided by means of N2XY wiring cables in a vertical duct. In this example, the permissible operating temperature for wiring cables at 90 °C will be applied, since the design of the building is such that these cables are not near other types of cable with lower permissible operating temperatures. Preliminary considerations

Table 1.7/18 Current-carrying capacity. Cables for fixed installation. Installation types A1, A2, B1, B2, C, and E

Insulation	XLPE											
Type designation[1]	N2XY, N2XSY, N2XH, N2XCH											
Permissible conductor operating temperature	90 °C											
Ambient temperature	30 °C											
Number of loaded cores	2	3	2	3	2	3	2	3	2	3	2	3
Type of installation[2]	A1		A2		B1		B2		C		E	
	Directly in thermally-insulated walls, or in conduits				On or in walls, or under plaster in conduits or ducts				Installed directly		In free air	
	Single-core cables in conduit				Single-core cables in conduit on wall		Multi-core cable in conduit on wall or on floor		Multi-core cable on wall or on floor		≥ 0,3 d ≥ 0,3 d Multi-core cable installed conforming to specified clear-ances	
	Multi-core cable in wall		Multi-core cable in conduit		Single-core cables in conduit on wall		Multi-core cable in conduit on wall or on floor		Single-core cables on wall or on floor			
					Single-core cables, multi-core cables in conduit in masonry				Multi-core cable in wall or under plaster			
Copper conductor: Nominal cross-sectional area in mm²	Current-carrying capacity in A											
1.5	19.0	17.0	18.5	16.5	23	20	22	19.5	24	22	26	23
2.5	26	23	25	22	31	28	30	26	33	30	36	32
4	35	31	33	30	42	37	40	35	45	40	49	42
6	45	40	42	38	54	48	51	44	58	52	63	54
10	61	54	57	51	75	66	69	60	80	71	86	75
16	81	73	76	68	100	88	91	80	107	96	115	100
25	106	95	99	89	133	117	119	105	138	119	149	127
35	131	117	121	109	164	144	146	128	171	147	185	158
50	158	141	145	130	198	175	175	154	209	179	225	192
70	200	179	183	164	253	222	221	194	269	229	289	246
95	241	216	220	197	306	269	265	233	328	278	352	298
120	278	249	253	227	354	312	305	268	382	322	410	346

1) For type designations, see 157 ff in [1]
2) For installation types not listed here, see Table 1.7/5

Table 1.7/19
Conversion factors for ambient temperatures other than 30 °C

Permissible operating temp.	90 °C
Ambient temperature °C	Conversion factor
10	1.15
15	1.12
20	1.08
25	1.04
30	**1.00**
35	0.96
40	0.91
45	0.87
50	0.82

regarding the voltage drop have yielded the following cross-sectional areas:

6 N2XY 4 × 35 RM cables for $I_b = 100$ A per system
8 N2XY 4 × 6 RE cables for $I_b = 40$ A per system
18 N2XY 4 × 2.5 RE cables for $I_b = 20$ A per system

The 32 cables are installed side by side on two cable racks, which are to be installed back to back in the vertical duct with a clearance ≥225 mm. Sufficiently dimensioned ventilation ensures that, even at the upper end of the duct, an air temperature of 35 °C is not exceeded. The current-carrying capacity of the cables now has to be checked.

The first step towards solving this planning and design problem is to determine the required conversion factors:

Conversion factor for

- ambient temperature to Table 1.7/19 0.96
- grouping to Table 1.7/9 0.70

Total conversion factor Πf 0.67

The basic value for the above conversion factors can be found in Table 1.7/18, column E, under the installation type "In free air" and for three loaded cores. The current-carrying capacity for the individual cables is, therefore:

N2XY 4×35 RM $I_r = 158$ A $\Rightarrow I_z = 105.9$ A $> I_b$

N2XY 4× 6 RE $I_r = 54$ A $\Rightarrow I_z = 36.2$ A $> I_b$

N2XY 4×2.5 RE $I_r = 32$ A $\Rightarrow I_z = 21.4$ A $> I_b$

This means that the wiring cables with cross-sectional areas of 35 mm^2 and 2.5 mm^2 are sufficiently dimensioned; instead of N2XY 4 × 6 RE, however, 4 × 10 RM must be installed. With these cables, the current-carrying capacity $I_r = 75$ A $\Rightarrow I_z = 50.3$ A, and is, therefore, sufficient for this installation.

Example 2

Cables in lamp installation

To supply lamps built into a suspended ceiling (cavity height: approximately 30 cm), special fire-prevention requirements make it necessary to install a halogen-free type of cable with improved characteristics in the event of fire (SIENOPYR cable N2XH). Like wiring cables, this type also has a permissible operating temperature of 90 °C. The total wattage of all the connected lamps is 5 kW at a supply voltage of 400 V. The required cross-sectional area of the cable now has to be determined for a maximum ambient temperature of 35 °C.

According to Table 1.7/18, this installation corresponds to installation type B1, since the clearance of the ceiling cavity is $V = 30$ cm, and, therefore, far larger than the probable outer diameter d of the cable.

The conversion factor for the ambient temperature 35 °C is 0.96, referred to the rated current-carrying capacity value in Table 1.7/18 at the specified ambient temperature of 30 °C. The AC load of 5 kW yields an operating current of:

$$I_b = \frac{5000\ \text{W}}{400\ \text{V}/\sqrt{3}} = 21.7\ \text{A}$$

Converted to $\vartheta_U = 30$ °C, this would yield a minimum rated current-carrying capacity value of:

$$I_z = \frac{21.7\ \text{A}}{0.96} = 22.6\ \text{A}$$

Table 1.7/18 shows that a cross-sectional area of 1.5 mm^2 is sufficient for the required type of cable (N2XH), since the two loaded cores have a current-carrying capacity $I_r = 23$ A. This means that the SIENOPYR cable N2XH 3 × 1.5 is dimensioned sufficiently to supply all the lamps with 5 kW.

Installation in the ground

The reference operating conditions for installation in the ground (together with guidelines for site operating conditions) specified in [1], Tables 18.5, 18.7, 18.9, and 18.11 apply as the rated values for cables installed in the ground.

Type of operation

The type of operation is assumed to be the public utility supply load, which is characterized by a defined 24-hour load cycle with a distinctive maximum load and a load factor $m = 0.7$.

Installation conditions

The installation conditions assume an installation depth of approximately 0.7 m. Only one system is considered in each case (direct current, single-phase and three-phase alternating current).

Grouping

Reciprocal temperature rise in grouped cables must always be taken into account using the products of the two conversion factors f_1 and f_2, which are determined by ambient conditions and grouping.

Bedding in soil

In the ground, cables are usually embedded in a layer of sand or sieved soil, and are covered, for example, with bricks.

Ambient conditions

The characteristic values for the ambient conditions are:

▷ ground temperature at installation depth: $\vartheta_E = 20$ °C,
▷ soil-thermal resistivity
  – moist: $\rho_E = 1.0$ K · m/W,
  – dry: $\rho_x = 2.5$ K · m/W.

The above thermal-resistivity values are required to calculate the current-carrying capacity using the two-layer model. Details of this calculation method can be found in [1], Section 18.

Soil-thermal resistivity in moist areas

The reference soil-thermal resistivity of 1 K · m/W in moist areas applies to sandy soil with a normal moisture content in a temperate climate with a maximum soil temperature of approximately 25 °C. Lower values can occur during the colder seasons (if precipitation is sufficient) and with more favorable types of soil. Higher values must be selected in regions with higher soil temperatures, extended dry periods, or complete lack of precipitation.

Soil-thermal resistivity in dry areas

Depending on capacity utilization, XLPE-insulated cables are especially prone to drying out the soil in a particular area directly surrounding the cable, which means that all the moisture in the soil immediately surrounding the cable is displaced. The fact that heat dissipation from the cable is inhibited as a result of this is taken into account by selecting a soil-thermal resistivity of 2.5 K·m/W for dry areas.

Rubble, ash slag, waste

The values for soil-thermal resistivity are very high in ground that contains rubble, ash, slag, waste, or organic matter. In such conditions, it may be necessary and more economical to replace the soil around the cable. The exact amount of soil that must be replaced is determined by means of measurements and calculations.

Cables or cable routes with high loads can dry out soil. In the Tables in [1], therefore, the rated values I_r (i.e. the current-carrying capacities under the reference operating conditions) were determined by distinguishing schematically between a dry and a moist area surrounding cables (two-layer model). The effects of the dry area on the current-carrying capacity were taken into account when the conversion factors f_1 and f_2 were determined.

Installation depth

The installation depth of cables has only a very slight influence on their rated values. The deeper the cables are installed, however, the lower the ambient temperature (and, generally, the soil-thermal resistivity). This is because the deeper areas of the soil are usually more moist and remain more uniformly moist than the upper layers. When low-voltage and high-voltage cables are laid individually at the usual installation depth (0.7 m to 1.2 m) at an ambient temperature of 20 °C and soil-thermal resistivity of 1 K · m/W, it is not necessary to convert the rated values specified in the tables.

Protective covers

In conduits, duct blocks, as well as in badly compacted soil beneath covers, air pockets can result in additional thermal resistivity. If this is the case, and clearances are not increased (e.g. with parallel cables/cable systems), the current-carrying capacity of the cables must be reduced.

Crossing cable routes

Problems can arise where cable routes cross, particularly if a large number of cables are involved. Cables must, therefore, be installed with sufficient clearances in these cases. Heat dissipation must also be improved by using the most favorable bedding material possible. If cables that cross are also grouped more densely, excessive temperature rises can be prevented by means of a sufficiently large brick-lined duct at the crossing point, which allows the cables to be crossed in air.

Crossing district heating lines

Where cables are installed close to or cross district heating lines, dangerous additional temperature rises can occur, particularly if the heating lines are poorly insulated. The continuous heat loss from the heating pipes can cause the soil to dry out considerably. For this reason, adequate clearances should be maintained between cables and heating lines (the clearances between the cables may also have to be increased), and heating lines must be insulated on all sides accordingly. Installing insulation between a heating line and cables insulates the cables against heat from the pipes; measures of this type, however, also inhibit heat dissipation from the cables, and are not, therefore, recommended.

Example 1

Installation in the ground, grouping

Three NYFGY 3 × 185 SM 3.6/6 kV PROTODUR cables are installed in parallel. The transmission power now has to be determined for the following conditions:

Operating conditions:

Permissible operating temperature	70 °C
Cables in ground side by side, clearance between cables	7 cm
Covering: bricks	
Public utility supply load, load factor m	0.7
Soil-thermal resistivity (moist) ρ_E	1.5 K m/W
Ambient temperature	30 °C
Resulting operating values:	
Rated value I_r to [1], Table 18.7	395 A
Conversion factors	
– f_1 to [1], Table 18.14	0.80
– f_2 to [1], Table 18.18	0.77
– Covering: bricks	1.00
Total conversion factor Πf	0.62

The three cables can be loaded with

$I_z = 3 \cdot 395\ \text{A} \cdot 0.62 = 735\ \text{A}$

The transmission power of the cable joint is, therefore:

$S = \sqrt{3} \cdot U \cdot I = \sqrt{3} \cdot 6\ \text{kV} \cdot 735\ \text{A} = 7.64\ \text{MVA}.$

Example 2

Installation in the ground, grouping

Using PROTOTHEN-X cables, 5 MVA has to be transmitted at an operating voltage of 10 kV. Since the cables are installed in a limited area close to the switchgear station, a conductor cross-sectional area of 70 mm^2 must not be exceeded. The cables are installed in a trefoil arrangement.

Operating conditions:

Permissible operating temperature	90 °C
Cables in ground side by side, clearance between systems	7 cm
Covering: bricks	
Continuous load, load factor m	1.0
Soil-thermal resistivity (moist) ρ_E	1.5 K · m/W
Ambient temperature	20 °C

The desired transmission power of 5 MVA yields a load current of:

$$I_b = \frac{S}{\sqrt{3} \cdot U} = \frac{5\ \text{MVA}}{\sqrt{3} \cdot 10\ \text{kV}} = 289\ \text{A}.$$

A preliminary estimate assumes that two cable systems are required. This results in the following conversion factors:

– f_1 to [1], Table 18.14	0.86
– f_2 to [1], Table 18.18	0.72
– Covering: bricks	1.00
Total conversion factor Πf	0.62

This means that, when converted to the "reference operating conditions", the "fictitious load current" I_{bf} for each cable system is:

$$I_{bf} = \frac{289\ \text{A}}{2 \cdot 0.62} = 233\ \text{A}.$$

Table 18.7 in [1] shows that it is necessary to select two cable systems of the type N2XS2Y 1 × 70 RM/16, 6/10 kV, each with a copper-conductor cross-sectional area of 70 mm^2. The rated value of this cable (current-carrying capacity under the reference operating conditions) is $I_r = 268$ A, and is, therefore, sufficient for the "fictitious load current" of 233 A.

Additional calculations for N2XS2Y 1×35 RM/16, 6/10 kV reveal that three systems are required for a conductor cross-sectional area of 35 mm^2 at a rated current of $I_r = 187$ A. The following conversion factors apply:

– f_1 to [1], Table 18.14	0.86
– f_2 to [1], Table 18.18	0.62
– Covering: bricks	1.00
Total conversion factor Πf	0.53

This results in the following current-carrying capacity for the three cable systems connected in parallel with a conductor cross-sectional area of $35\ \text{mm}^2$:

$I_z = 3 \cdot I_r \cdot \Pi f = 3 \cdot 187\ \text{A} \cdot 0.53 = 297\ \text{A}.$

Example 3

Installation in air

A total of 33 multi-core cables to Fig. 1.7/5 are installed on cable racks in a cable basement (see Fig. 1.7/5). The individual cable types and their load currents are specified in Table 1.7/20. The cable basement is sufficiently large to ensure that, even when the cables are loaded, the temperature does not exceed the ambient temperature $\vartheta_U = 40\ °C$ (infinitely high thermal capacity).

The requirement

$I_b \leq I_z = I_r \cdot \Pi f$

results in the condition

$\frac{I_b}{I_r} \leq \Pi f.$

Grouping up to 5 cables on a total of 7 cable racks yields a reduction factor of $f_H \approx 0.91$ (see [1], Table 18.22). The conversion factor for other air temperatures f_ϑ is determined using [1], Table 18.19.

The results in Table 1.7/20 show that the ratio of current load to current-carrying capacity (I_b/I_r) is never greater than the product of the conversion factors $f_H \cdot f_\vartheta$. The cables can, therefore, be used under the assumed operating conditions without their service lives being restricted.

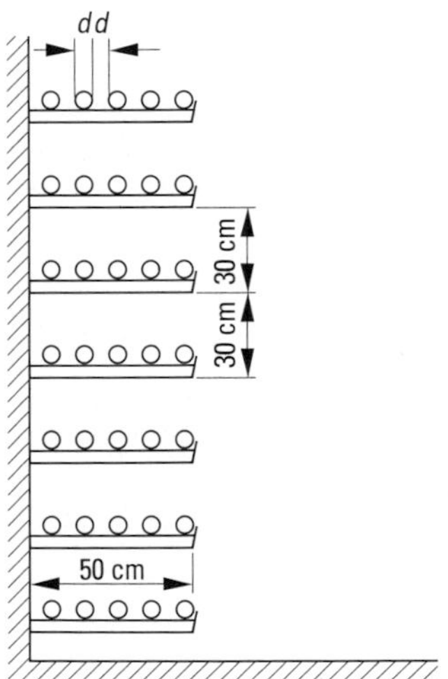

Fig. 1.7/5
Arrangement of cables to Table 1.7/20 on cable racks in a cable basement

Installation in ducts

Non-ventilated and covered ducts

In non-ventilated and covered ducts, most of the heat loss generated by the cables is dissipated only via the walls, top, and bottom of the duct. Natural ventilation is generally prevented by the compartmentalization prescribed for ducts. The resulting heat accumulation raises the temperature of the air surrounding the cables in ducts, which results in a current-carrying capacity lower than that for installation in free air.

The temperature rise of the air in the duct depends on the dimensions of the duct and on the magnitude of the heat loss of all the cables. The number of cables that generate the heat loss and their distribution have no influence on the temperature rise.

Ducts in the ground

Ducts in buildings

If a duct is located in the ground, the heat generated in the duct is dissipated outwards into the soil via the walls of the duct by means of thermal conduction. The calculation procedure for this type of duct is described in detail in [1] and, will not, therefore, be discussed here. If, however, the duct is inside a building (e.g. industrial buildings, power stations), the simple calculation method outlined below can be applied. Before this method is discussed, however, some general guidelines on installing cables in ducts will be described.

Guidelines for installing cables

Cables can either be fixed directly to the walls of the duct (e.g. by means of cable clamps) or installed on cable racks, ladders, or trays. The vertical clearance between racks depends on their width, but should not be less than 30 cm. This clearance must also be maintained when installing the innermost cables. Where cables are installed on racks and ladders, or are fixed directly to the walls, the spacing between the cables should be equal to the diameter of the cable, in order to minimize heat transfer from cable to cable. This is particularly important where the current load is high.

Dimensions of ducts

The height of accessible cable ducts should be no less than 2.0 m. The width must be selected to ensure a passageway clearance of at least 60 cm to 80 cm. Where cable racks are used with a vertical clearance of approximately 30 cm between the racks, the width

Table 1.7/20 Technical data for installing cables in a cable basement (Example 3)

Type designation Number of cores and conductor cross-sectional area U_0/U kV	NYFGY 3 × 120 SM 3.6/6	NYCWY 4 × 240 SM 0.6/1	NEKEBA 3 × 95 RM 12/20	NA2XY 4 × 95 SE 0.6/1
Number of cables Load I_b A Load factor $m = 1.0$ Ambient temperature $\vartheta_U = 40$ °C	13 200	7 320	6 175	7 200
Rated values I_r A to [1], Table Permissible operating temperature ϑ_{Lr} °C	274 18.8 70	443 18.6 70	243 18.10 65	308 18.6 90
I_b/I_r	0.73	0.72	0.72	0.65
f_ϑ to [1], Table 18.19 $f_H \cdot f_\vartheta$ ($f_H = 0.91$)	0.87 0.79	0.87 0.79	0.85 0.77	0.91 0.83

of the racks should be limited to approximately 50 cm to facilitate installation.

Temperature rise of air in ducts

Fig. 1.7/6 shows the temperature rise of the air in the duct as a function of the heat loss per meter of the duct with the size of the duct as a variable. This graph can be used in conjunction with the rated values I_r for cables in free air in [1] either to determine the conductor cross-sectional areas required to transmit a given power, or (where the number of cables, conductor cross-sectional area, current load, and dimensions of the duct are known) to calculate the temperature rise that will occur.

When the size of the duct is determined, only surfaces via which heat can actually be dissipated must be taken into account. For example, walls or tops of ducts adjoining warm machine rooms, transformer cells, or similar areas must not be included in the calculation.

Planning and design

An installation can be planned and designed as follows:

The first step involves selecting the approximate conductor cross-sectional area for each cable, which should be roughly 30 % higher than the value required for installation in free air. With high currents, it may be necessary to use several cables for each cable run. The next step involves drawing a sketch of the duct, including the required height, width, number of racks, and arrangement of the cables, following the rules mentioned above.

In accordance with the selected arrangement of the cables in the completed sketch, the conversion factor f_H for groups installed in free air can now be determined to [1], Tables 18.20 to 18.23. The total heat loss of all the cables in the duct and the resulting temperature rise of the air in the duct to Fig. 1.7/6 can then be determined. The resulting value must be added to the temperature of the air in the duct with unloaded cables, and the conversion factor f_ϑ for increased ambient temperature (to [1], Table 18.19) determined for this. If these two conversion factors are multiplied, the product must not be less than the ratio of the desired load I_b to the rated value I_r:

$$\frac{I_b}{I_r} \leq \Pi f = f_H \cdot f_\vartheta$$

If this is not the case, either the number of cables, the conductor cross-sectional area, or the size of the duct must be increased.

r.m.s value

The temperature-rise time constant of a duct is very high in comparison to that of the cables (see page 269, section on short-time and intermittent operation). The temperature rise of the air in the duct can, therefore, be determined using the heat loss calculated from the r.m.s. value I_q of the currents over 24 hours:

$$I_q = \sqrt{\frac{I_{b1}^2 \cdot t_1 + I_{b2}^2 \cdot t_2 + \ldots + I_{bi}^2 \cdot t_i}{t_1 + t_2 + \ldots + t_i}}$$

$$t_1 + t_2 + \ldots + t_i = 24\ \text{h.}$$

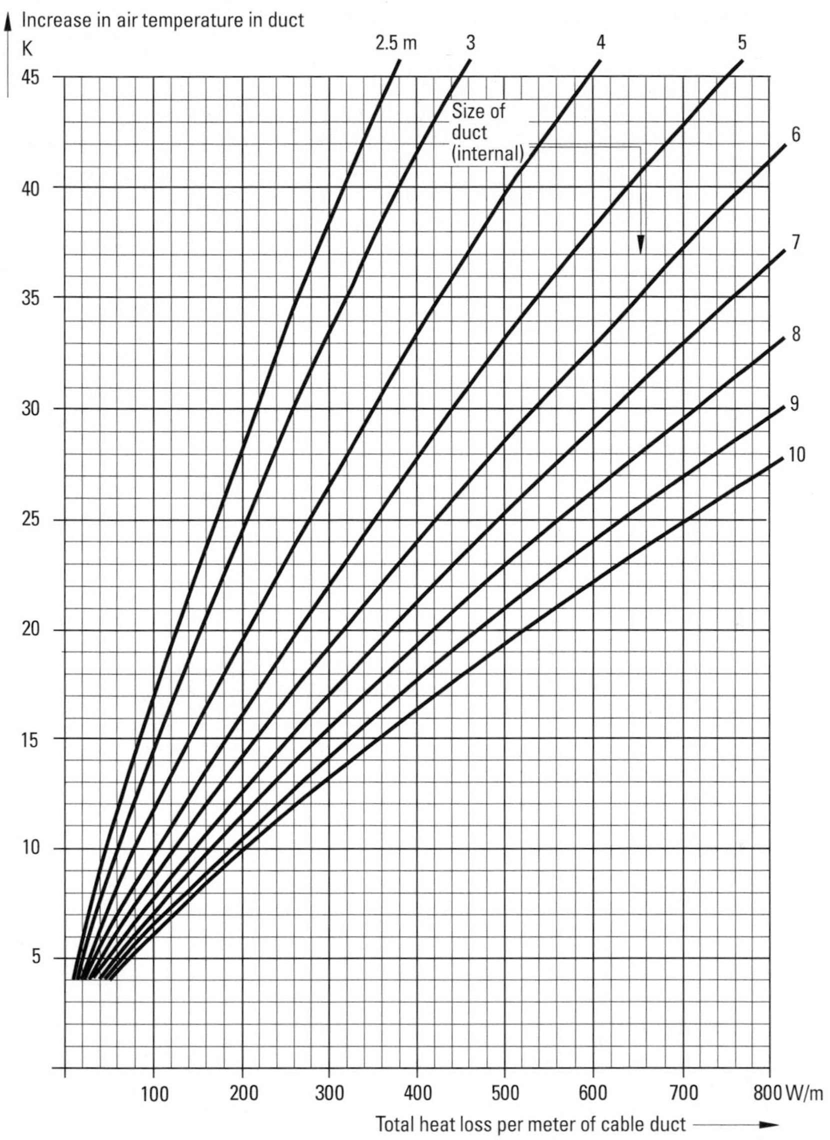

Fig. 1.7/6 Temperature rise of air in duct as a function of heat loss per meter of cable duct

where I_{b1}, I_{b2}, . . ., I_{bi} are the currents that flow during the times t_1, t_2, . . ., t_i.

Grouping a larger number of cables

If the number of cables grouped is larger than allowed for in the relevant Tables in [1], the reduction factors to Fig. 1.7/7 can be applied. These also apply to single-core cables if a circuit comprises the required number of grouped single-core cables instead of a multi-core cable. In these cases, the reduction factors refer to the current-carrying capacity I_r of a grouped cable system.

If more than six cable racks are arranged one above the other, the conversion factor for six cable racks can be used. If there is a clearance of more than 2 cm between the wall to which the racks are fixed and the first cables on the racks, the conversion factor derived from Fig. 1.7/7 for racks loaded with one layer of cables can be multiplied by a factor of 1.05.

The approximate reduction factor for grouping in air f_H can be derived from the reduction factor f_{Hh} (horizontal component) for grouping on a cable rack to Fig. 1.7/7a and the reduction factor f_{Hv} (vertical component) for grouping cable racks with similar loads one above the other.

This is:

$f_H = f_{Hh} \cdot f_{Hv}$

with

$f_{Hv} = 0.95$ for two cable racks one above the other,

$f_{Hv} = 0.93$ for three cable racks one above the other,

$f_{Hv} = 0.90$ for six and more cable racks one above the other.

In the case of cable racks with one layer of cables, the value for f_H as the product of f_{Hh} and f_{Hv} can be taken directly from the curves in Fig. 1.7/7b.

If the number of cables installed in a duct and their loads are not known, the conductor cross-sectional areas must be dimensioned for an assumed total reduction factor. If the reduction factor originally used is insufficient, a final review must be carried out to determine whether forced ventilation is necessary.

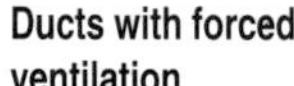

Ducts with forced ventilation

Forced ventilation must be provided if natural heat dissipation via the walls of the duct is too low (i.e. the air in the duct becomes too hot, and the conductor temperature exceeds the permissible value), and if no other solution is possible, e.g. increasing the size of the duct (the cooling surface).

In such cases, the calculation is usually based on the total heat loss generated in the duct, and does not take into account the fact that heat continues to be dissipated via the walls of the duct. This ensures that the fans are dimensioned adequately and that sufficient reserves remain available for subsequent expansions.

The required air rate Q depends on the total heat loss $\Sigma P'$ generated by the cables in the duct, the length of the duct l and the temperature rise of the cooling air $\Delta\vartheta_{co}$ between the duct inlet and outlet. This is calculated as follows:

$$Q = \frac{\Sigma P' \cdot l}{c_p \cdot \Delta\vartheta_{co}} \quad \text{in} \quad \frac{\text{m}^3}{\text{s}}.$$

$\Sigma P' = \Sigma (P'_i + P'_d)$ takes into account both the heat loss due to current P'_i and, if neces-

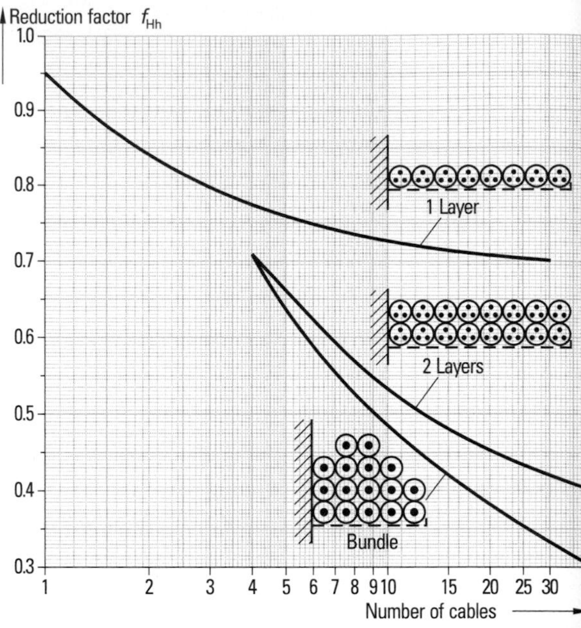

a) As a function of the arrangement and number of cables on a cable rack: reduction factor f_{Hh}

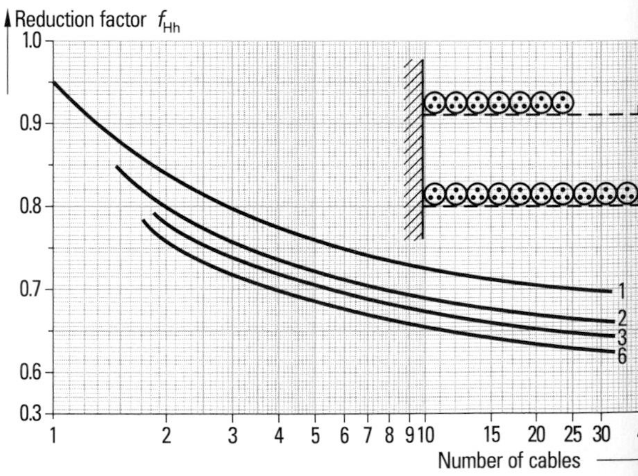

b) As a function of the number of cable racks with approximately the same load with one layer of cables: reduction factor f_H

Fig. 1.7/7
Reduction factors with grouped multi-core cables (or bundled single-c[...] cables belonging to one circuit) on cable racks

sary, the dielectric losses P'_d of all the cables in the duct. c_p is the specific thermal capacity of the air at constant pressure, and depends on the temperature as well as the humidity of the air; approximate calculations are made on the basis of $c_p = 1.3\ \text{kJ/Km}^3$.

The air velocity v is calculated using the cross-sectional area yielded by the height and width of the duct:

$$v = \frac{Q}{A} \quad \text{in} \quad \frac{\text{m}}{\text{s}}.$$

If noise disturbance is to be prevented, the air velocity must not exceed 5 m/s.

The temperature rise of the air in the duct must be selected taking account of the temperature of the available cooling air at the inlet and the permissible temperature at the outlet. The temperature of the cooling air before it is heated is usually identical to the ambient temperature ϑ_a selected for project planning and design. Taking the maximum permissible conductor temperature ϑ_{Lr} into account, the cable with the highest (relative) temperature rise yields the following for the temperature rise of the cooling air:

$$\Delta\vartheta_{co} \leq \vartheta_{Lr} - \vartheta_a - \Delta\vartheta_L$$

with

$$\Delta\vartheta_L = \Delta\vartheta_r \left(\frac{I_b}{I_r}\right)^2 = (\vartheta_{Lr} - 30\ °\text{C}) \left(\frac{I_b}{I_r}\right)^2 \quad \text{in K.}$$

Since the moving air improves the heat dissipation of cables considerably, the conversion factor for grouping f_H need not be used here.

Example

Installation in a duct in a building

The cables listed in Table 1.7/21 are to be installed in a 20 m long duct with a cross-sectional area of 2.2 m × 1.5 m, which can dissipate heat via all its surfaces. The cables are loaded with the operating currents specified in the same table.

The operating time is initially planned for eight hours daily at full load. Operation at full load for 16 hours should also be possible; forced ventilation can be provided for this purpose. The assumed ambient temperature with unloaded cables is 35 °C. Fig. 1.7/8 shows the planned arrangement of the cables in the duct.

A conversion factor of $f_H \approx 0.91$ must be used for groups of five cables on a total of seven racks (cf. [1], Table 18.22).

With eight-hour operation, the resulting r.m.s. value of the currents of the cables in Table 1.7/21, column 1, is:

$$I_q = \sqrt{\frac{I_{b1}^2 \cdot t_1}{t_1 + t_2}} = 205\ \text{A} \cdot \sqrt{\frac{8\ \text{h}}{24\ \text{h}}} = 118\ \text{A}.$$

The heat loss is calculated using the resistances per unit length specified in the tables in the manual "Power Cables and their Application", Part 2 (fourth edition), e.g. for the cables in Table 1.7/21, column 1:

$$\begin{aligned} P' &= 3 \cdot I_q^2 \cdot R'_w \\ &= 3 \cdot (118\ \text{A})^2 \cdot 0.151\ \Omega/\text{km} = 6.31\ \text{W/m} \end{aligned}$$

and

$$\sum P' = 14 \cdot 6.31 = 88.3\ \text{W/m}.$$

The heat loss of the cables in columns 2, 3, and 4 was determined in the same way, and entered in Table 1.7/21. The sum of the heat losses of all the cables is, therefore:

$$\begin{aligned} \sum P' &= 88.3 + 43.3 + 27.2 + 34.9 \\ &= 193.7\ \text{W/m} \end{aligned}$$

The air temperature in the duct increases in accordance with Fig. 1.7/6 by approximately 12 K, rising to 35 + 12 = 47 °C.

For the cables in Table 1.7/21, column 1, the conversion factor for temperature is obtained from the equation on page 257:

$$f_\vartheta = \sqrt{\frac{\Delta\vartheta_L}{\Delta\vartheta_r}} = \sqrt{\frac{70 - 47}{70 - 30}} = 0.76.$$

Comparing the ratio I_b/I_r with the product of the conversion factors $f_\vartheta \cdot f_H$ shows that the permissible operating temperature is not exceeded for any of the cables.

Using the cables in Table 1.7/21, column 1, as an example, the permissible temperature rise of the cooling air now has to be determined.

The following applies:

$$\Delta\vartheta_L = 40\ \text{K} \left(\frac{205}{313}\right)^2 = 17.2\ \text{K}$$

$$\Delta\vartheta_{co} \leq 70 - 35 - 17.21 \leq 8\ \text{K}.$$

With the transition to 16-hour operation:

$$I_q = 205\ \text{A} \cdot \sqrt{\frac{16\ \text{h}}{24\ \text{h}}} = 167\ \text{A}$$

and

$$P' = 3 \cdot (167\ \text{A})^2 \cdot 0.151\ \Omega/\text{km} = 12.6\ \text{W/m}.$$

Table 1.7/21 Technical data for installing cables in a duct in a building

Column		1	2	3	4
Type designation Number of cores and conductor cross-sectional area U_0/U	 kV	NYFGY 3 × 150 SM 3.6/6	NYSEY 3 × 240 RM 6/10	NEKEBA 3 × 70 RM 12/20	NEKEBA 3 × 120 RM 12/20
Number of cables Load I_b Load factor $m = 1.0$ Ambient temperature $\vartheta_U = 35$ °C	 A	14 205	8 240	6 120	7 165
Current-carrying capacity I_r to [1], Table Permissible operating temperature ϑ_{Lr} Resistance per unit length R'_w	A °C Ω/km	313 18.8 70 0.151	423 18.8 70 0.0934	200 18.10 65 0.318	279 18.10 65 0.184
I_b/I_r		0.65	0.57	0.60	0.59
8-hour operation I_q P' $\Sigma P'$	 A W/m W/m	 118 6.31 88.3	 139 5.41 43.3	 69 4.54 27.2	 95 4.98 34.9
f_ϑ $f_H \cdot f_\vartheta$ ($f_H = 0.91$)		0.76 0.69	0.76 0.69	0.72 0.66	0.72 0.66
$\Delta\vartheta_L$ $(\Delta\vartheta_{co})_{max}$	K K	17.2 17.8	12.9 22.1	12.6 17.4	12.2 17.8

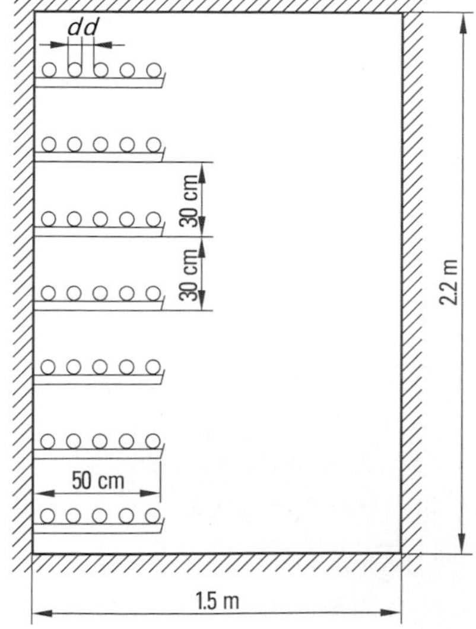

Fig. 1.7/8
Arrangement of cables to Table 1.7/21 on cable racks in a duct in a building

The sum of the heat losses in the duct is, therefore, doubled in comparison to 8-hour operation:

$\Sigma P' = 2 \cdot 193.7 = 387.4$ W/m,

and the air temperature in the duct increases by approximately 20 K (see Fig. 1.7/6) to $35 + 20 = 55$ °C. It is, therefore, necessary to ventilate the duct.

As shown earlier, a maximum temperature rise of the air in the duct of $(\Delta\vartheta_{co})_{max} =$ 18 K would be permissible for the cables in Table 1.7/21, column 1.

The values for the remaining cables can be found in Table 1.7/21.

A maximum permissible temperature rise of the cooling air of $\Delta\vartheta_{co} = 10$ K is selected for the example. With a duct length of 20 m and a duct cross-sectional area of $1.5 \text{ m} \times 2.2 \text{ m} = 3.3 \text{ m}^2$, the resulting air rate required is:

$$Q = \frac{\Sigma P' \cdot l}{c_p \cdot \Delta\vartheta_{co}} \text{ in } \frac{\text{m}^3}{\text{s}}; \quad c_p = 1.3 \frac{\text{kJ}}{\text{K} \cdot \text{m}^3}$$

$$= \frac{387.4 \text{ W/m} \cdot 20 \text{ m}}{1.3 \cdot 10^3 \text{ Ws/m}^3 \cdot 10 \text{ K}} = 0.60 \frac{\text{m}^3}{\text{s}}$$

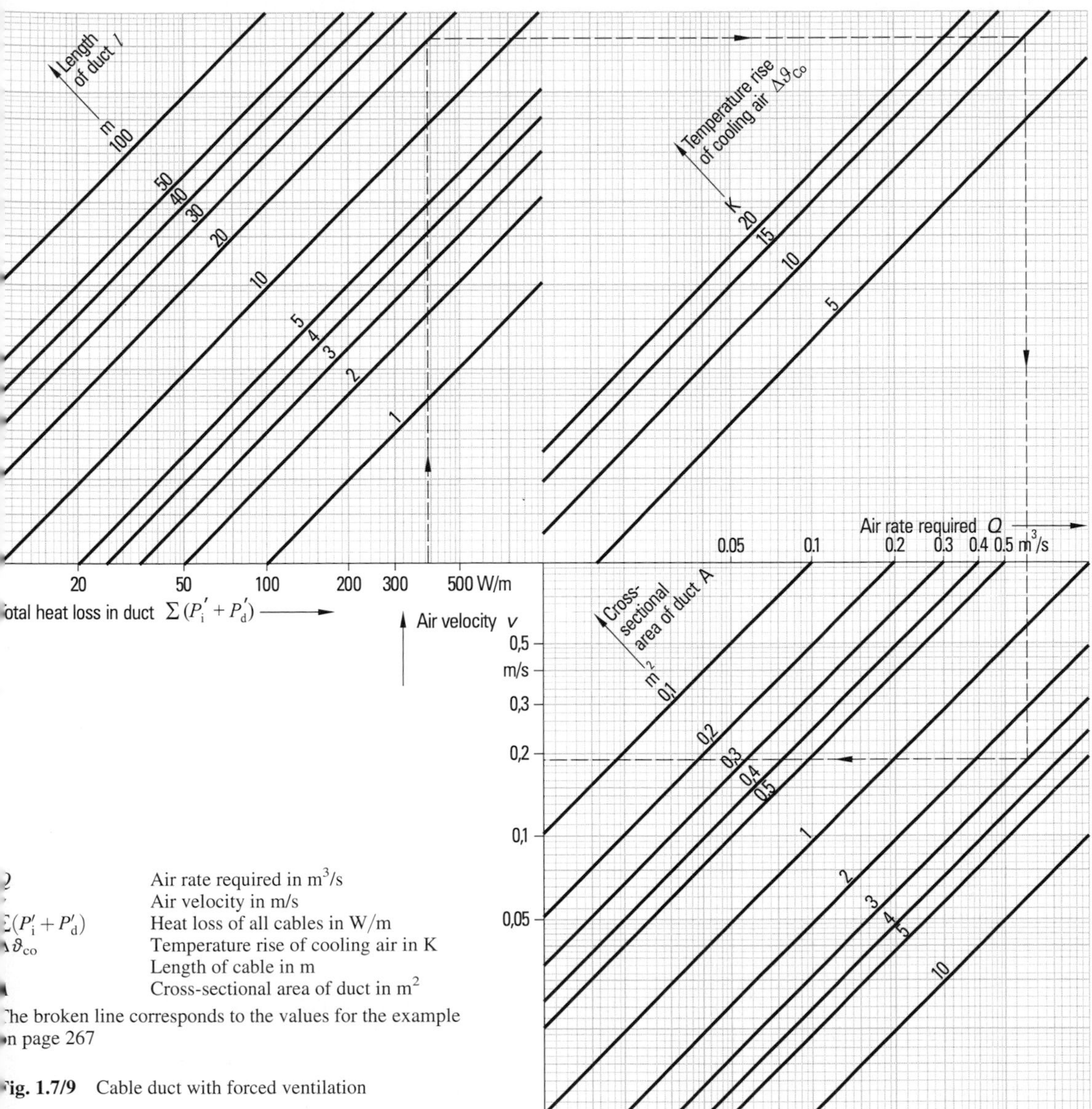

Q Air rate required in m^3/s
v Air velocity in m/s
$\Sigma(P'_i + P'_d)$ Heat loss of all cables in W/m
$\Delta\vartheta_{co}$ Temperature rise of cooling air in K
l Length of cable in m
A Cross-sectional area of duct in m^2

The broken line corresponds to the values for the example on page 267

Fig. 1.7/9 Cable duct with forced ventilation

with an air velocity of

$$v = \frac{0.60\ \mathrm{m^3/s}}{3.3\ \mathrm{m^2}} = 0.182\ \frac{\mathrm{m}}{\mathrm{s}}.$$

Results with approximately the same values can also be found in Fig. 1.7/9.

Short-time and intermittent operation

Calculation method

At a constant current load, the temperature of a cable rises until, after a certain period, its final temperature is reached. The temperature rises (approximately) in accordance with an exponential function or several combined exponential functions. Precise calculations can only be performed with considerable time and effort. Calculations using a minimum time value, the r.m.s. value, and adiabatic heat rise are both simple and yield realistic results.

With cables installed in the ground, the thermal capacity of the soil surrounding the cables plays a significant role, which cannot easily be calculated. Since the thermal capacity of soil is always higher than that of air, calculations using the method described below will always yield reliable results.

Minimum time value

The minimum time value τ of a cable is the value equal to a fifth of the time taken for the cable to almost reach its permissible temperature rise at a constant current load. For installations in free air, this value is determined primarily by the material properties of the cable.

Fig. 1.7/10 specifies the calculation rules required to derive the conversion factors for determining the current-carrying capacity (f_{KB} for short-time operation, and f_{AB} for intermittent operation):

$I_{KB} = f_{KB} \cdot I_r$ or $I_{AB} = f_{AB} \cdot I_r$.

The minimum time values τ for 1-kV PVC-insulated cables can be found in Fig. 1.7/11.

r.m.s. value

The calculation method using the r.m.s. value takes into account the "average" thermal effect of the current. The r.m.s. value can, however, only be calculated in order to determine the conductor cross-sectional area if the cyclic duration factor of the peak current remains below the values specified in Table 1.7/22.

The conductor temperature obtained from this calculation has a mean value corresponding to the permissible temperature, but exceeds this value at the end of the load period, and falls below it at the end of the no-load period. The longer the load time t_b and duty cycle t_s in relation to the time value τ, and the smaller the cyclic duration factor (ED), the greater the temperature variation of the conductor, since the current during the load period exceeds the r.m.s. value.

The r.m.s. value of a sequence of currents of the magnitude I_{bi} that flow during any time interval is:

$$I_q = \sqrt{\frac{I_{b1}^2 \cdot t_1 + I_{b2}^2 \cdot t_2 + \ldots + I_{bi}^2 \cdot t_i}{t_1 + t_2 + \ldots + t_i}}$$

with $t_1 + t_2 + \ldots + t_i = 24$ h.

Table 1.7/22
Permissible cyclic duration factor for cables as a function of nominal conductor cross-sectional area

Nominal conductor cross-sectional area in mm^2	Permissible cyclic duration factor in s
≤ 6	4
10 ≤ 25	8
35 ≤ 50	15
70 ≤ 150	30
185 and greater	60

The following applies for this intermittent operation:

$I_{AB} = I_z \leq I_q$.

Adiabatic heat rise

The current-carrying capacity I_{KB} for operating times in the seconds range (e.g. starting currents of electric machines) can also be calculated according to the rules for the short-circuit rating (adiabatic heat rise) with the conductor cross-sectional area q and the short-time current density J_{th} for $t_b = 1$ s to Table 1.7/23:

$$I_{KB} = q \cdot J_{th} \sqrt{\frac{1\text{ s}}{t_b}}.$$

The short-time current density J_{th} specified here for determining the permissible current during short-time operation is such that the particular permissible operating temperature ϑ_{Lr} is not exceeded. This exploits the fact that, prior to the increased load, the cable under consideration is either not loaded or is loaded only slightly. The short-time current density J_{th} as a function of the initial temperature ϑ_a (conductor temperature prior to increased load) can be found in Table 1.7/23.

If the cable is not loaded before the short-time load begins, for example, then:

$I_0 = 0, \quad \vartheta_a = \vartheta_U.$

If, however, the cable is loaded with I_0 before the short-time load, the following applies:

$$I_0 \neq 0, \vartheta_a = \vartheta_U + (\vartheta_{Lr} - 30\ °C) \left(\frac{I_0}{I_r}\right)^2.$$

Example 1

Short-time operation

A 4-core NYY 4 × . . . SM 0.6/1 kV PROTODUR cable is to be used to connect a motor. The rated output of the three-phase AC motor is $S_n = 175$ kVA; according to the manufacturer's specifications, $I_{an} = 7 \cdot I_n$ ap-

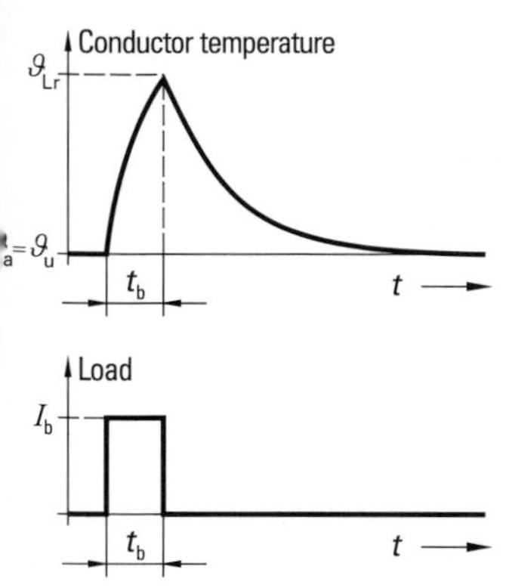

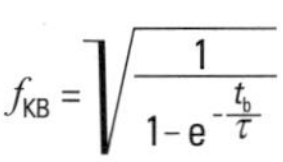

$$f_{KB} = \sqrt{\frac{1}{1-e^{-\frac{t_b}{\tau}}}}$$

without previous load
$I_0 = 0$

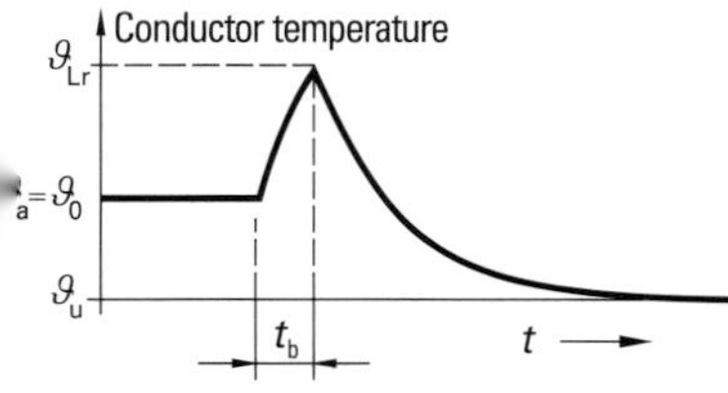

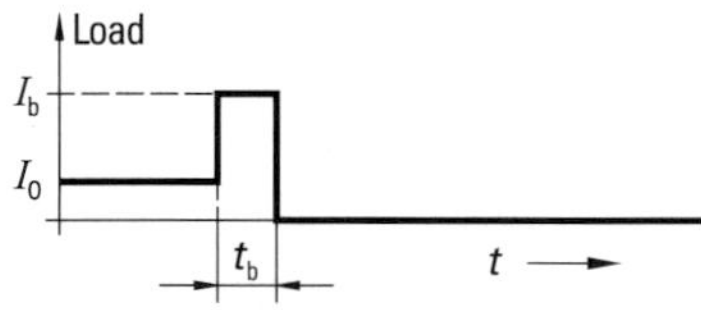

$$f_{KB} = \sqrt{\frac{1-\left(\frac{I_0}{I_b}\right)^2 e^{-\frac{t_b}{\tau}}}{1-e^{-\frac{t_b}{\tau}}}}$$

with previous load
$I_0 < I_b$
$I_0 \neq 0$

a) Short-time operation

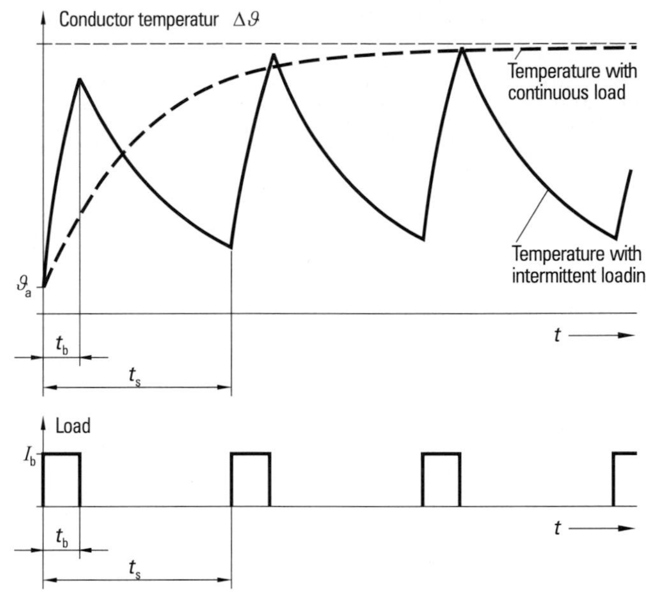

$$f_{AB} = \sqrt{\frac{1-e^{-\frac{t_s}{\tau}}}{1-e^{-\frac{t_b}{\tau}}}} = \sqrt{\frac{1-e^{-\frac{t_s}{\tau}}}{1-e^{-\frac{t_b}{\tau}\frac{ED}{100}}}}$$

$$ED = \frac{t_b}{t_s} 100 \ (\%)$$

ϑ_{Lr}	Permissible operating temperature at
ϑ_a	Initial temperature of conductor
ϑ_u	Ambient temperature
t_b	Load time
t_s	Duty cycle
τ	Minimum time value of cable (see Table in [2], Section 5)
I_b	Load current
I_0	Previous load current
ED	Cyclic duration factor (%)

Fig. 1.7/10
Determining the conversion factors f_{KB} for short-time and f_{AB} for intermittent operation

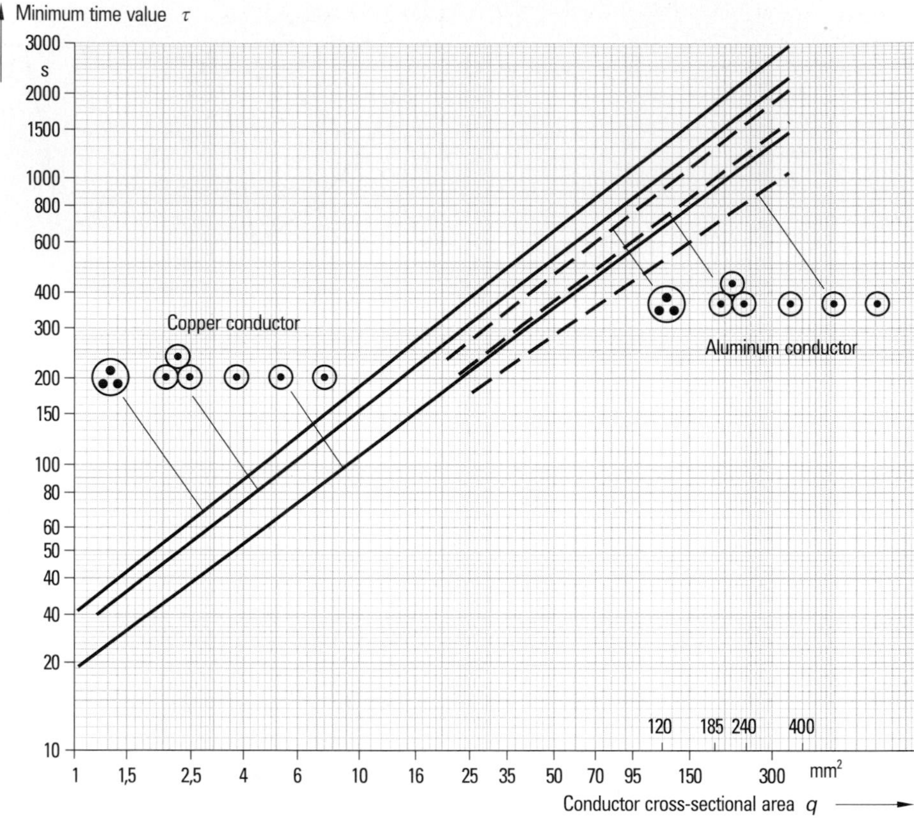

Fig. 1.7/11
Minimum time values τ as a function of conductor cross-sectional area q; PVC-insulated cable for 0.6/1 kV

Table 1.7/23
Short-time current densities J_{th} for calculating current with short-time operation in the seconds range

Type of cable	Permissible operating temp. ϑ_{Lr} °C	Conductor temperature ϑ_a prior to loading								
		90 °C	80 °C	70 °C	65 °C	60 °C	50 °C	40 °C	30 °C	20 °C
		Short-time current density (1 s) in A/mm²								
Copper conductor										
XLPE-insulated cable	90	–	39.9	56.6	63.9	70.3	81.8	92.3	102	111
PVC-insulated cable	70	–	–	–	29.0	41.2	58.8	72.7	84.7	95.6
Aluminum conductor										
XLPE-insulated cable	90	–	26.5	37.7	42.4	46.6	54.3	61.3	67.7	73.8
PVC-insulated cable	70	–	–	–	19.3	27.4	39.0	48.2	56.2	63.5

plies for the starting current; start-up is completed after 30 s.

The following, therefore, applies:

Rated current of motor:

$$I_n = \frac{S_n}{\sqrt{3} \cdot U_n} = \frac{175 \cdot 10^3}{\sqrt{3} \cdot 400} = 253 \text{ A},$$

Starting current:

$$I_{an} = 7 \cdot 253 = 1771 \text{ A}.$$

According to [2] Table 5.1.5, at a specified ambient temperature of $\vartheta_U = 30$ °C, a conductor cross-sectional area of 120 mm² is sufficient for the rated current of the motor

(253 A). A minimum time value $\tau = 1221$ s is specified for this.

The conversion factor for short-time operation (KB) is, therefore:

$$f_{KB} = \sqrt{\frac{1}{1-e^{-\frac{t_b}{\tau}}}} = \sqrt{\frac{1}{1-e^{-\frac{30}{1221}}}} = 6.42.$$

The current-carrying capacity of the NYY 4 × 120 SM 0.6/1 kV PROTODUR cable is specified in [2], Table 5.1.5, as $I_r = 276$ A; for the starting time of $t_b = 30$ s, a short-time current of:

$$I_{KB} = f_{KB} \cdot I_r = 6.42 \cdot 176 = 2772 \text{ A}$$

is permissible; the conductor cross-sectional area is thus sufficiently dimensioned.

In comparison with the minimum-time-value method, the r.m.s. value method with a time of 24 h = 86400

$$I_q = \sqrt{\frac{(1771)^2 \cdot 30 + (253)^2 \cdot 86370}{86400}}$$

yields a mean current load of $I_q = 255$ A, which is less than the current load $I_r = 276$ A.

Example 2

Intermittent operation

A crane motor is periodically loaded with a constant load I_b during the time interval t_b and a subsequent break during a duty cycle of t_s. The cyclic duration factor (ED) is calculated as follows for a load period t_b and a duty cycle t_s:

Cyclic duration factor

$$\text{ED} = \frac{t_b}{t_s} \cdot 100\,\%.$$

The following values apply for the crane motor: cyclic load factor (ED) = 20 %, load current $I_b = 575$ A, duty cycle $t_s = 360$ s, ambient temperature $\vartheta_u = 40$ °C. An initial estimation is based on the cable type NYY 4 × 150 SM 0.6/1 kV; according to [2], Table 5.1.5, this cable has a rated value of $I_r = 315$ A for installation in air, and a minimum time value $\tau = 1465$ s. The following applies for intermittent operation:

$$f_{AB} = \sqrt{\frac{1-e^{-\frac{t_s}{\tau}}}{1-e^{-\frac{t_s}{\tau}}}} = \sqrt{\frac{1-e^{-\frac{t_s}{\tau}}}{1-e^{-\frac{t_s}{\tau}\frac{ED}{100}}}}$$

$$= \sqrt{\frac{1-e^{-\frac{360}{1465}}}{1-e^{-\frac{360}{1465}0.2}}} = 2.13$$

and

$$I_{AB} = I_r \cdot f_\vartheta \cdot f_{AB}$$
$$= 315 \text{ A} \cdot 0.87 \cdot 2.13 = 584 \text{ A}$$

with $f_\vartheta = 0.87$ to [2], Table 4.2.12, due to the higher ambient temperature. The selected cable type with a conductor cross-sectional area of 150 mm^2 is, therefore, sufficiently dimensioned for the load current $I_b = 575$ A with intermittent operation.

Stress in the event of short circuits

When cable installations are being planned, it is essential to check that the cables and cable accessories selected are able to withstand the dynamic and thermal short-circuit stress encountered.

The most important variables here are the peak short-circuit current I_p (for dynamic stress), and the thermally equivalent short-circuit current I_{th} (for thermal stress).

The following variables are derived from the initial symmetrical short-circuit power S_k'':

initial symmetrical short-circuit current

$$I_k'' = \frac{S_k''}{\sqrt{3} \cdot U_n},$$

peak short-circuit current (peak value)

$$I_p = \kappa \cdot \sqrt{2} \cdot I_k''.$$

The initial symmetrical short-circuit power S_k'' is a characteristic of the system under consideration; the peak value κ depends on the ratio R/X of the system impedance, and can be determined in accordance with DIN VDE 0102 (see Section 1.2).

Thermally equivalent current

The thermally equivalent current I_{th} takes the actual short-circuit current flow into consideration. Throughout the entire duration of a short circuit, the current determined in this way would generate the same heat as the short-circuit current that actually flows, the time characteristic of which is determined by a decaying DC component and an AC component (see Section 1.2).

The magnitude of the short-circuit currents that occur depends on the system upstream of the point at which the short circuit occurs, and can be determined either by asking the operator of the system, or by means of network calculations. For further details, see Section 1.2 "Calculating short-circuit currents in three-phase systems".

Peak value

The following section considers the effects of short-circuit currents. If no information on the system is available apart from details of the initial symmetrical short-circuit current, the peak value $\kappa = 1.8$ must be selected.

Dynamic stress

The forces that occur during a short circuit are proportional to the square of the peak short-circuit current. As a result, these forces can cause mechanical stress in cables and at sealing ends, even with moderately high short-circuit currents. In armored multi-core cables, the short-circuit forces that occur within the cable are taken up by the stranding, sheath, and armor.

A peak short-circuit current of up to 40 kA is permissible for multi-core low-voltage cables to HD 603/IEC 60502-1/DIN VDE 0276-603 and HD 604/DIN VDE 0276-604. In low-voltage installations, peak short-circuit currents are not as significant as in medium-voltage installations, since these currents are limited to values below 40 kA (e.g. by means of the fuses used, which blow during the current rise within the first half-wave). For this reason, additional binding bands for low-voltage cables and checks for greater resistance to sudden short circuits are uncommon.

Fixing cables and cable accessories

Cable accessories nevertheless have to be fitted so that they resist impact. Single-core cables not installed in the ground must be fitted to their respective supports. To prevent additional temperature rises, cable clamps made of non-magnetic material, or steel clamps, in which the magnetic circuit is not closed, must be selected.

In addition to the magnitude of the short-circuit currents, the clearance between the short-circuited conductors is a crucial factor with regard to the short-circuit forces that occur. To ensure that cables and accessories are fixed safely and reliably, it is necessary to:

- ▷ ensure that the maximum displacement of cables between the fixing points does not exceed 5 % of the distance between fixings (with thermoplastic-insulated cables),
- ▷ adhere to the maximum permissible surface pressure of the conductor on the insulation (depending on material) in the vicinity of the fixing point,
- ▷ ensure that the strength of the clamp/strap is not exceeded, and that the number of turns per unit length is selected accordingly.

Permissible distance between fixings

The clearance between any two fixing points that satisfies all three requirements must be used as the permissible distance between fixings for planning and design purposes.

The individual steps for determining the permissible distance between fixings are described in detail in [1]. Specifying this distance is an extremely complex procedure, and will not, therefore, be discussed any further here.

Installation on rising cable runs

Irrespective of how cables are dimensioned for mechanical short-circuit forces, clearances of 1.5 m must not be exceeded with cables installed vertically and fixed to rising cable runs by means of clamps. Care should be taken to ensure that a suitable type of cable clamp is selected (e.g. with a cover shell made of non-magnetic material).

Thermal stress

When cables are selected with regard to their thermal short-circuit rating, one of the most important factors to be taken into account is the stress on the conductor; in some cases, however, the type of sheath or shield may also have to be considered.

With phase-to-phase ungrounded faults and three-phase faults, only the external conductors are thermally stressed.

The phase-to-phase ungrounded initial symmetrical short-circuit current is $\sqrt{3}/2\times$ the three-phase initial symmetrical short-circuit current. With phase-to-ground faults, phase-to-phase grounded faults, and double ground faults, the shields or metal sheaths of the cables are also adversely affected.

The greatest stress on the conductors generally occurs in the event of a three-phase short circuit. The short-circuit current in the event of phase-to-ground and phase-to-phase grounded faults can exceed the current that occurs with three-phase short circuits only in systems with low-resistance neutral-point grounding, in which the requirements for effective neutral-point grounding to DIN VDE 0111 are met.

Determining conductor cross-sectional areas for short circuits

The following condition must be fulfilled, if the conductor cross-sectional area is to be determined for the short circuit:

$I_{th} \leq I_{thz}$

I_{th} thermally equivalent short-circuit current (load)

I_{thz} thermally equivalent short-circuit rating.

The thermally equivalent short-circuit rating I_{thz} is derived by multiplying the rated short-time current density J_{thr} from Table 1.7/24 by the nominal conductor cross-sectional area q_n, and using the short-circuit duration t_k in the following equation:

$$I_{thz} = q_n \cdot J_{thr} \sqrt{\frac{t_{kr}}{t_k}} \quad \text{where} \quad t_{kr} = 1 \text{ s}.$$

The short-circuit duration t_k is specified by means of the characteristic of the protective device used, or by the response time of the protective device – taking the operating time of the switch and protective device into consideration.

If the cable is operated with a current less than its current-carrying capacity, a lower initial conductor temperature at the beginning of the short circuit ϑ_a can be assumed for the calculation.

The short-circuit rating of the copper-wire screens of thermoplastic-insulated cables is specified in Fig. 1.7/12; the short-circuit rating of the metal sheaths of paper-insulated cables can be found in [1], if required.

Example

Determining conductor cross-sectional areas for short circuits

A check is to be carried out to determine whether the conductor cross-sectional area $q = 120$ mm^2 of a 10-kV XLPE-insulated cable with copper conductor is sufficiently dimensioned for a thermally equivalent short-circuit current $I_{th} = 26$ kA and a short-circuit duration of $t_k = 0.5$ s. The load prior to the start of the short circuit generates a conductor temperature of 60 °C.

Table 1.7/24 specifies the following rated short-time current density for the conductor temperature $\vartheta_a = 60$ °C:

$J_{thr} = 159$ A/mm^2.

The required minimum conductor cross-sectional area for the specified load $I_{th} = 26$ kA is, therefore:

$$q \geq \frac{I_{th}}{J_{thr}} \cdot \sqrt{\frac{t_k}{t_{kr}}} = \frac{26\text{ kA}}{159\text{ A/mm}^2} \cdot \sqrt{\frac{0.5\text{ s}}{1\text{ s}}}$$

$$= 115.6 \text{ mm}^2.$$

The N2XS2Y 1 × 120 RM/16, 6/10 kV cable is, thus, sufficiently dimensioned for the conductor short circuit.

In addition to the above, the load on the shield must be specified for a short-circuit duration of $t_k = 0.5$ s. Fig. 1.7/12 shows that the copper shield with a cross-sectional area of 16 mm^2 with $t_k = 0.5$ s has a thermal rating of approximately 4.5 kA. Under normal system conditions, this corresponds to a permissible initial symmetrical short-circuit current of approximately 4.4 kA.

Determining the voltage drop

Voltage drop in low-voltage systems

In low-voltage systems in particular, checks must be carried out to determine whether the conductor cross-sectional area selected with regard to the current-carrying capacity fulfils the requirements relating to the voltage drop.

Voltage drop in high-voltage systems

In high-voltage systems with very long cable runs, it is also advisable to carry out this type of check, particularly if the capacitive current of PVC-insulated cables has to be taken into consideration (see [1], Chapter 20).

When a cable[1] of length l is operated in a three-phase system with a current I_b, the effective resistance per unit length R'_w and reactance per unit length X'_L cause a longitudinal voltage drop ΔU at the end of the cable of:

$$\Delta U = \sqrt{3} \cdot I_b \cdot l (R'_w \cdot \cos\varphi + X'_L \cdot \sin\varphi)$$

$$= U_a - U_e.$$

In this equation, φ is the phase angle between the voltage U_e and the load current I_b that lags behind it (inductive load).

With the leading load current I_b (capacitive load), the sign of X'_L must be reversed, which means that ΔU can also be negative.

The percentage voltage drop relative to the nominal system voltage U_n is calculated as follows:

$$\Delta u = \frac{\Delta U}{U_n} \cdot 100\,\%.$$

[1] The term "cable" is used here in its general sense to denote an electrical conductor (e.g. insulated cable, insulated conductor, or overhead line)

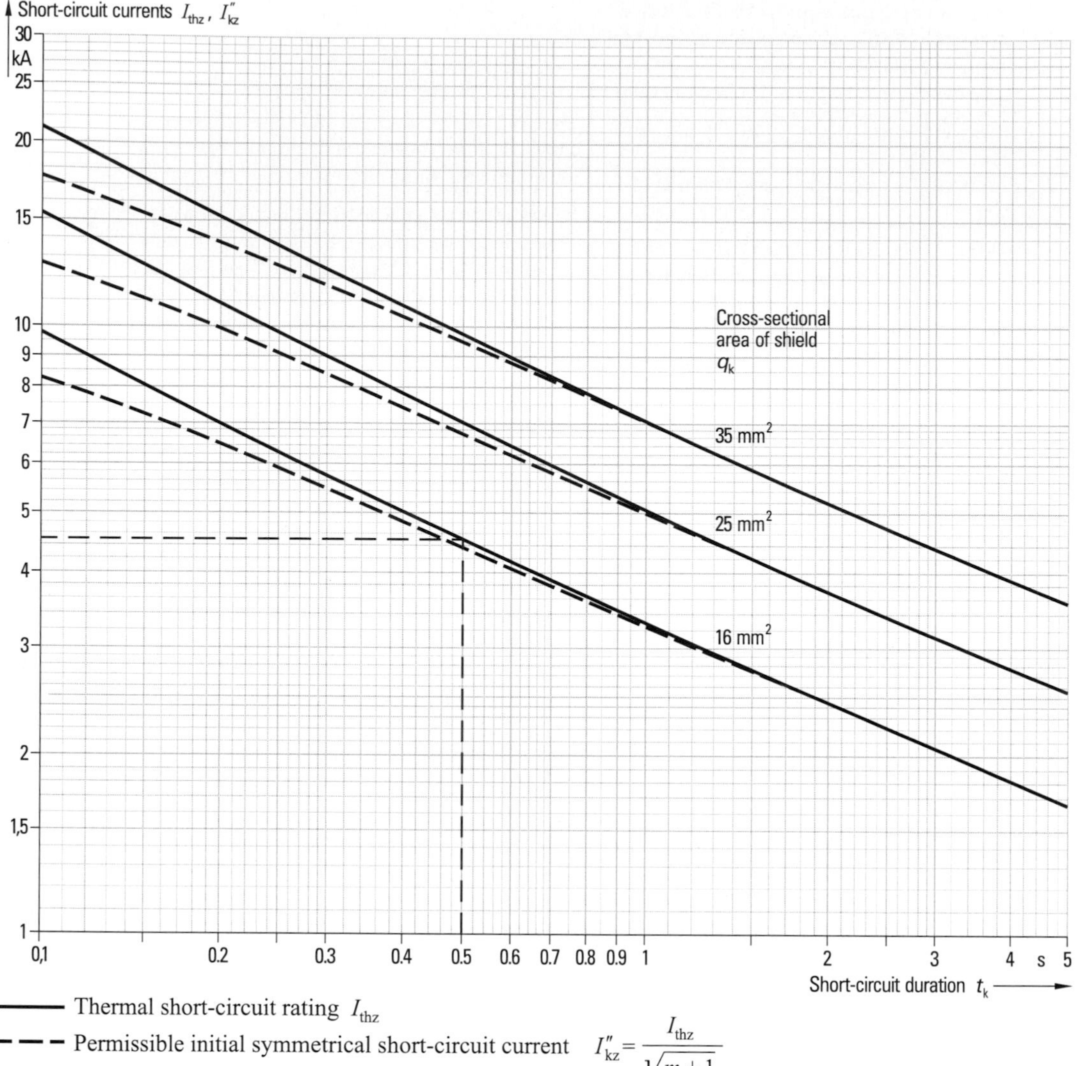

——— Thermal short-circuit rating I_{thz}

– – – Permissible initial symmetrical short-circuit current $I''_{kz} = \dfrac{I_{thz}}{\sqrt{m+1}}$

Fig. 1.7/12 Short-circuit rating of copper-wire screens in thermoplastic-insulated cables

U_a and U_e are the voltages (phase-to-phase in a three-phase AC system, between the external conductors) at the supply end and loaded end of the cable (see Table 1.7/25).

When the voltage drop in cables in DC systems, and in cables with a conductor cross-sectional area of up to 16 mm^2 in single-phase and three-phase AC systems is calculated, only the DC resistance per unit length of the cable at operating temperature has to be taken into account. For cables with conductor cross-sectional areas greater than 16 mm^2 in single-phase AC and three-phase AC systems, however, the resistance per unit length and reactance per unit length must be considered. With non-armored cables and especially with insulated wiring cables and flexible cables, the limit is considerably higher.

Table 1.7/25 contains all the formulae for calculating the voltage drop for a cable supplied from one end with one consumer at the load end.

able 1.7/24
ermissible short-circuit temperature ϑ_e and rated short-time current density J_{thr}
ables with copper conductors

ype of cable sulation	Permissible operating temperature ϑ_{Lr} °C	Permissible short-circuit temperature ϑ_e °C	Conductor temperature ϑ_a at beginning of short circuit								
			90 °C	80 °C	70 °C	65 °C	60 °C	50 °C	40 °C	30 °C	20 °C
			Rated short-time current density J_{thr} in A/mm^2 for $t_{kr} = 1$ s								
oft-soldered ints	–	160	100	107	115	119	122	129	136	143	150
LPE-insulated ables	90	250	143	148	154	157	159	165	170	176	181
VC-insulated ables											
≤ 300 mm^2	70	160	–	–	115	119	122	129	136	143	150
> 300 mm^2	70	140	–	–	103	107	111	118	126	133	140
aper-insulated ables 6/10 kV	65	170	–	–	–	124	127	134	141	147	154
hree-core eparately-sheathed ables											
12/20 kV	65	170	–	–	–	124	127	134	141	147	154
18/30 kV	60	150	–	–	–	–	117	124	131	138	145

ables with aluminum conductors

ype of cable sulation	Permissible operating temperature ϑ_{Lr} °C	Permissible short-circuit temperature ϑ_e °C	Conductor temperature ϑ_a at beginning of short circuit								
			90 °C	80 °C	70 °C	65 °C	60 °C	50 °C	40 °C	30 °C	20 °C
			Rated short-time current density J_{thr} in A/mm^2 for $t_{kr} = 1$ s								
LPE-insulated ables	90	250	94	98	102	104	105	109	113	116	120
VC-insulated ables											
≤ 300 mm^2	70	160	–	–	76	78	81	85	90	95	99
> 300 mm^2	70	140	–	–	68	71	73	78	83	88	93
aper-insulated ables 6/10 kV	65	170	–	–	–	82	84	89	93	97	102
hree-core eparately-sheathed ables											
12/20 kV	65	170	–	–	–	82	84	89	93	97	102
18/30 kV	60	150	–	–	–	–	77	82	87	91	96

Table 1.7/25 Cable supplied from one end with one consumer

	Transmission power[1]	Voltage drop $(U_a - U_e)$ V	Transmission losses kW
Direct current 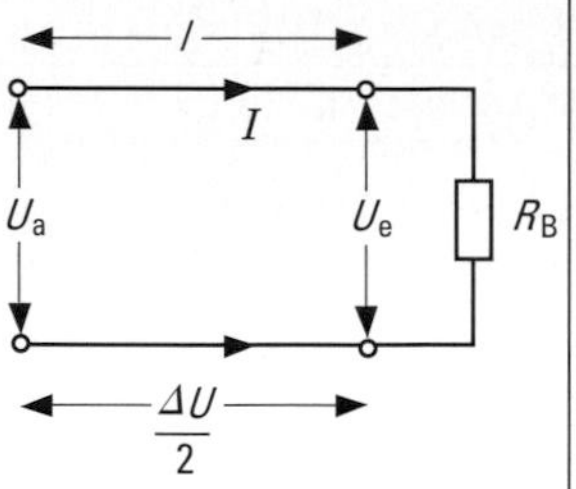	$P = I \cdot U$ $P = I^2 \cdot R_B \cdot 10^{-3}$	$\Delta U = 2 \cdot l \cdot I \cdot R'_\vartheta$	$V = 2 \cdot l \cdot I^2 \cdot R'_\vartheta \cdot 10^{-3}$
Single-phase alternating current 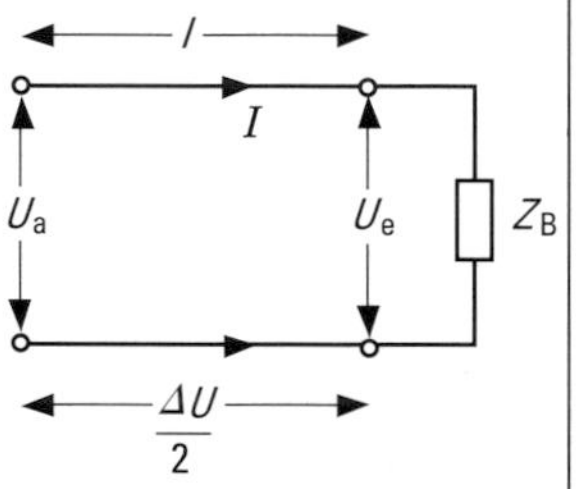	$S = I \cdot U$ $P = I \cdot U \cdot \cos\varphi$ $Q = I \cdot U \cdot \sin\varphi$ $S = I^2 \cdot Z_B \cdot 10^{-3}$	$\Delta U = 2 \cdot l \cdot I \cdot Z'_{(q)}$	$V = 2 \cdot l \cdot I^2 \cdot R'_w \cdot 10^{-3}$
Three-phase alternating current 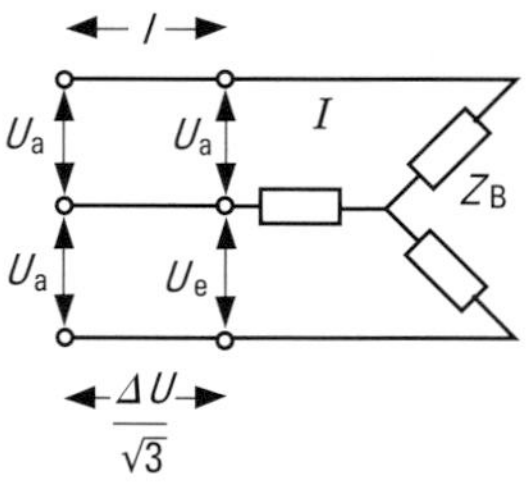	$S = \sqrt{3} \cdot I \cdot U$ $P = \sqrt{3} \cdot I \cdot U \cdot \cos\varphi$ $Q = \sqrt{3} \cdot I \cdot U \cdot \sin\varphi$ $S = 3 \cdot I^2 \cdot Z_B \cdot 10^{-3}$	$\Delta U = \sqrt{3} \cdot l \cdot I \cdot Z'_{(q)}$	$V = 3 \cdot l \cdot I^2 \cdot R'_w \cdot 10^{-3}$

[1] To calculate the transmission power, the nominal voltage U_n of the systems or the rated voltage of the devices may be selected

S Apparent power in kVA
P Effective power in kW
Q Reactive power in kVAr
U Voltage in kV[1]
U_a Voltage at supply end of cable
U_e Voltage at load end of cable
I Current in A
l Length of cable in km
X'_L Reactance of cable in Ω
R_B Ohmic load resistance in Ω/km
X_B Inductive load resistance in Ω
R'_ϑ DC resistance per unit length of cable at operating temperature in Ω/km

R'_w Resistance of cable per unit length at operating temperature in Ω/km
$R'_{(q)}$ Sum of impedances as a function of cross-sectional area in Ω/km
$Z_B = \sqrt{R_B^2 + X_B^2}$ Impedance in Ω/km
$\cos\varphi = \frac{R_B}{Z_B}$ Power factor
$\sin\varphi = \frac{X_B}{Z_B}$

Table 1.7/26 Cables supplied from one or two ends with evenly distributed load

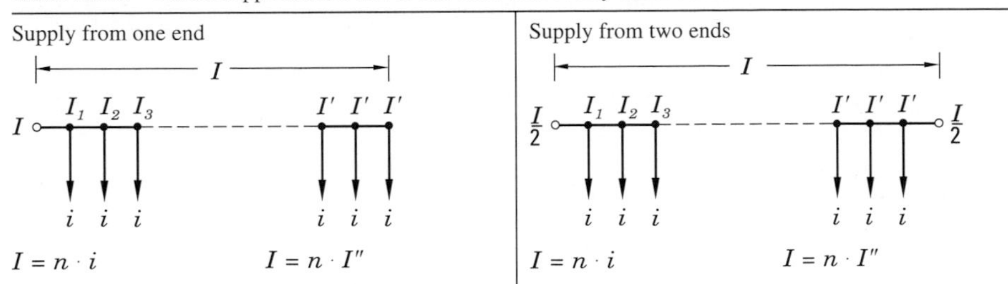

Three-phase alternating current

Supply from one end	Supply from two ends
$\Delta U = \frac{\sqrt{3}\cdot n(n+1)}{2}\, i\cdot l'\cdot Z'_{(q)}$	$\Delta U = \frac{\sqrt{3}\cdot n(n+2)}{8}\, i\cdot l'\cdot Z'_{(q)}$
$= \frac{\sqrt{3}(n+1)}{2n}\, I\cdot l\cdot Z'_{(q)}$	$= \frac{\sqrt{3}(n+2)}{8n}\, I\cdot l\cdot Z'_{(q)}$
$V = \frac{n(n+1)(2n+1)}{2}\, i^2\cdot l'\cdot R'_w\cdot 10^{-3}$	$V = \frac{n(n+1)(n+2)}{4}\, i^2\cdot l'\cdot R'_w\cdot 10^{-3}$
$= \frac{(n+1)(2n+1)}{2n^2}\, I^2\cdot l\cdot R'_w\cdot 10^{-3}$	$= \frac{(n+1)(n+2)}{4n^2}\, I^2\cdot l\cdot R'_w\cdot 10^{-3}$
for very high n:	for very high n:
$\Delta U = \frac{\sqrt{3}\cdot n^2}{2}\, i\cdot l'\cdot Z'_{(q)}$	$\Delta U = \frac{\sqrt{3}\cdot n^2}{8}\, i\cdot l'\cdot Z'_{(q)}$
$= \frac{\sqrt{3}}{2}\, I\cdot l\cdot Z'_{(q)}$	$= \frac{\sqrt{3}}{8}\, I\cdot l\cdot Z'_{(q)}$
$V = n^3\cdot i^2\cdot l'\cdot R'_w\cdot 10^{-3}$	$V = \frac{n^3}{4}\cdot i^2\cdot l'\cdot R'_w\cdot 10^{-3}$
$= I^2\cdot l\cdot R'_w\cdot 10^{-3}$	$= \frac{1}{4}\, I^2\cdot l\cdot R'_w\cdot 10^{-3}$

With direct current and single-phase alternating current:

ΔU the right side of the equation must be multiplied by $\frac{2}{\sqrt{3}}$

V the right side of the equation must be multiplied by $\frac{2}{3}$

ΔU Voltage drop in V

V Transmission losses in kW

Table 1.7/26 summarizes the formulae for calculating the voltage drop for cables supplied from one or two ends with several identical consumers. These formulae can be used to carry out calculations for radial and ring systems. The calculation method is somewhat simplified in these cases by introducing the sum of impedances as a function of the conductor cross-sectional area:

$$Z'_{(q)} = R'_w \cdot \cos\varphi + X'_L \cdot \sin\varphi$$

The resistances per unit length can be taken directly from [2]; the reactances per unit length can be calculated using the inductance specifications in [2].

Example

Three-phase AC power of $P = 75$ kW is to be transmitted over a length of $l = 0.15$ km with $\cos\varphi = 0.9$, a nominal voltage of $U_n = 400$ V and $\Delta u = 2\%$. The following applies:

$$\Delta U = \frac{\Delta u \cdot U_a}{100\%} = \frac{2\% \cdot 400\ \text{V}}{100\%} = 8.0\ \text{V}.$$

In accordance with Table 1.7/25:

$$I_b = \frac{S}{\sqrt{3} \cdot U_a} = \frac{P}{\sqrt{3} \cdot U_a \cdot \cos \varphi}$$

$$= \frac{75 \cdot 10^3 \text{ W}}{\sqrt{3} \cdot 400 \text{ V} \cdot 0.9} = 120 \text{ A}.$$

The following condition applies for the sum of impedances:

$$Z'_{(q)} = (R'_w \cdot \cos \varphi + X'_L \cdot \sin \varphi) \leq \frac{\Delta U}{\sqrt{3} \cdot I_b \cdot l}$$

$$= \frac{8 \text{ V}}{\sqrt{3} \cdot 120 \text{ A} \cdot 0.15 \text{ km}} = 0.257 \; \Omega/\text{km}.$$

Using the impedance values from lines 34 and 44 from Table 5.1.5 in [2], together with $\cos \varphi = 0.9$ and $\sin \varphi = 0.436$, for a four-core PROTODUR cable, the following values are obtained for the given conductor cross-sectional areas:

70 mm²: $R'_w = 0.321 \; \Omega/\text{km}$
$L'_b = 0.262 \text{ mH/km}$
$X'_L = 0.0823 \; \Omega/\text{km}$
$(R'_w \cdot \cos \varphi + X'_L \cdot \sin \varphi) = 0.325 \; \Omega/\text{km}$

95 mm²: $R'_w = 0.232 \; \Omega/\text{km}$
$L'_b = 0.261 \text{ mH/km}$
$X'_L = 0.0820 \; \Omega/\text{km}$
$(R'_w \cdot \cos \varphi + X'_L \cdot \sin \varphi) = 0.245 \; \Omega/\text{km}$

These results clearly show that only the NYY-J 4 × 95 SM 0.6/1 kV cable may be used for the maximum permissible voltage drop Δu of 2 %.

Note
Since technical terms, characteristic electrical values, and guidelines on installing cables are discussed in detail in [1] and [2], they are not repeated here.

1.7.3 Protecting cables against excessive temperature rises due to overcurrents

Rule Overcurrent protective devices must be used to protect cables against excessive temperature rises, which can occur as a result of overloads during normal operation, as well as dead short circuits.

Overcurrent protective devices Depending on the type selected, overcurrent protective devices provide protection against overloads and short circuits, or one of these conditions only (Table 1.7/27).

Standards The relevant standards for these applications are HD 384.4.43/IEC 60364-4-43/DIN VDE 0100-430 and HD 384.4.473/IEC 60364-4-473.

Location Overcurrent protective devices for providing protection against overloads and/or short

Table 1.7/27 Use of overcurrent protective devices

Overcurrent protective devices	To Standard	Overload protection	Short-circuit protection
NEOZED/DIAZED/l.v. h.b.c. fuses, utilization category gL/gG (see Section 4.1.1)	EN 60269/IEC 60269/ DIN VDE 0636	×	×
Partial-range fuses for device protection aM (see Section 4.1.1)	EN 60269/IEC 60269/ DIN VDE 0636		×
Miniature circuit-breakers B, C and D (see Section 4.1.2)	EN 60898/IEC 60898/ DIN VDE 0641–11	×	×
Circuit-breakers with overcurrent releases and instantaneous overcurrent releases	EN 60947/IEC 60947/ DIN VDE 0660	×	×
Switchgear and controlgear with inverse-time release only	EN 60947/IEC 60947/ DIN VDE 0660	×	
Circuit-breakers with instantaneous over-current release only	EN 60947/IEC 60947/ DIN VDE 0660		×
Thermistor-type motor protection, e.g. for motor circuits	DIN VDE 0110 and EN 60947/IEC 60947/ DIN VDE 0660	×	

× suitable for protection

circuits must be installed at the origin of each electric circuit, as well as at points in the circuit where the current-carrying capacity or short-circuit rating is reduced. This can occur, for example, with reduced conductor cross-sectional areas, or when cable installation conditions are changed and the upstream protective device can no longer provide adequate protection.

Overload protection

The following conditions must be fulfilled when overcurrent protective devices are allocated to the rated conductor cross-sectional areas of cables to protect cables against excessive temperature rises in the event of an overload:

$I_b \leq I_n \leq I_z$

$I_2 \leq 1.45 \cdot I_z$.

I_b Prospective operating current of electric circuit

I_n Rated current of overcurrent protective device

> *Note*
> With settable overcurrent protective devices, I_n corresponds to the setting value

I_z Current-carrying capacity of cables to IEC 60183/DIN VDE 0276 (installed in the ground), and HD 21.1/DIN VDE 0276 (fixed installation in buildings)

I_2 Current that causes the overcurrent protective device to trip under the conditions defined in the device specifications (conventional tripping current)

Consequently, the rated current I_n can be equal to the current-carrying capacity I_z when overload protective devices with $I_z \leq 1.45 \cdot I_n$ are used. This applies, for example, to miniature circuit-breakers to EN 60898/IEC 60898/DIN VDE 0641-11 with tripping characteristics B, C, and D.

Conventional tripping current

The overcurrent protective devices specified in Table 1.7/28 have a maximum conventional tripping current of 1.45 × their rated current, i.e. with the installation types stipulated in the Table, these overcurrent protective devices can be selected with a rated current up to the magnitude of the current-carrying capacity of the cables (the occasionally lower values for I_n result from the specified rated current steps of the protective devices).

If the selected overcurrent protective device has a different conventional tripping current, the above no longer applies, and the following relationship between the rated current and current-carrying capacity of the device applies:

$$I_n \leq \frac{1.45}{k} \cdot I_z$$

k Factor for determining the conventional tripping current $I_2 = k \cdot I_n$

In such cases, the overcurrent protective devices must be matched with the nominal conductor cross-sectional area of the cable in accordance with the above condition, as the following example shows.

Example

According to Table 1.7/28, a multi-core cable with two loaded cores of 10 mm^2 installed in a conduit on a wall can be allocated an overcurrent protective device with the rated current $I_n = 50$ A, if the conventional tripping current of the device is $I_2 \leq 1.45 \cdot I_n$.

If, however, an overcurrent protective device with a conventional tripping current $I_2 \leq 1.6 \cdot I_n$ is to be used, according to the relationship

$$I_n \leq \frac{1.45}{k} \cdot I_z$$

$$\leq \frac{1.45}{1.6} \cdot 52 \text{ A}$$

$$\leq 47 \text{ A}$$

the rated current of the device selected must not exceed 47 A, i.e. it must not be greater than the rated current step of 40 A.

Settable overcurrent releases

With overcurrent protective devices equipped with settable overcurrent releases (such as circuit-breakers and protective circuit-breakers to EN 60947/IEC 60947/DIN VDE 0660), I_n corresponds to the setting value of the release.

These devices can be set on the basis of the current-carrying capacity values in Table 1.7/6 to IEC 60183/DIN VDE 0276 for cables and HD 21.1/ DIN VDE 0276 for flexible cables.

If the conditions for the cables to be protected differ from those in Table 1.7/28 and

Table 1.7/28

Current-carrying capacity of cables for fixed installation. Installation types A1, A2, B1, B2, C, as well as installation type E in free air, and allocation of overcurrent protective devices for overload protection

Insulation	PVC																							
Type designation	H07V-U, H07V-R, H07V-K, NYIF, NYIFY, NYM, NYMZ, NYMT, NHYRUZY, NYBUY, NYDY, NHXMH, NHMH, NYY, NYCY																							
Maximum permissible operating temperature	70 °C																							
Ambient temperature	30 °C																							
Number of loaded cores	2		3		2		3		2		3		2		3		2		3		2		3	
Installation type	A1				A2				B1				B2				C				E			
	In thermally-insulated walls								On or in walls or under plaster in conduit or duct								Installed directly				In free air			
	Single-core non-sheathed cables in conduit				Multi-core cables or light-sheathed cables in conduit				Single-core non-sheathed cables in conduit on wall				Multi-core cables or light-sheathed cables in conduit on wall or floor				Multi-core cables or light-sheathed cables on wall or floor				Multi-core cables or light-sheathed cables installed conforming to specified clearances (≥ 0,3 d)			
					Multi-core cables or light-sheathed cables directly in wall				Single-core non-sheathed cables in conduit on wall				Multi-core cables or light-sheathed cables in conduit on wall or floor				Single-core cables or light-sheathed cables on wall or floor							
									Single-core non-sheathed cables, single-core or multi-core cables, or light-sheathed cables in conduit in masonry								Multi-core cables or light-sheathed cables, flat webbed cable in wall or under plaster							
Rated cross-sectional area of copper conductor in mm²	Current-carrying capacity I_z in A and rated current I_n of overcurrent protective device, which must have conventional tripping current $I_2 \leq 1.45\ I_n$																							
	I_z	I_n	I_z	I_n	I_z	I_n	I_z	I_n	I_z	I_n	I_z	I_n	I_z	I_n	I_z	I_n	I_z	I_n	I_z	I_n	I_z	I_n	I_z	I_n
1.5	15.5	13	13.5	13	15.5	13	13.0	13	17.5	16	15.5	13	16.5	16	15.0	13	19.5	16	17.5	16	22	20	18.5	16
2.5	19.5	16	18	16	18.5	16	17.5	16	24	20	21	20	23	20	20	20	27	25	24	20	30	25	25	25
4	26	25	24	20	25	25	23	20	32	25	28	25	30	25	27	25	36	35	32	25	40	35	34	25
6	34	25	31	25	32	32	29	25	41	35	36	35	38	35	34	32	46	40	41	35	51	50	43	40
10	46	40	42	40	43	40	39	35	57	50	50	50	52	50	46	40	63	63	57	50	70	63	60	50
16	61	50	56	50	57	50	52	50	76	63	68	63	69	63	62	50	85	80	76	63	94	80	80	80
25	80	80	73	63	75	63	68	63	101	100	89	80	90	80	80	80	112	100	96	80	119	100	101	100
35	99	80	89	80	92	80	83	80	125	100	110	100	111	100	99	80	138	125	119	100	148	125	126	125
50	–		–		–		–		–		–		–		–		–		–		180	160	153	125
70	–		–		–		–		–		–		–		–		–		–		232	200	196	160
95	–		–		–		–		–		–		–		–		–		–		282	250	238	200
120	–		–		–		–		–		–		–		–		–		–		328	315	276	250

from the conditions on which the Tables in Section 1.7.1 are based, e.g.:

▷ other environmental conditions,
▷ other ambient temperatures,
▷ grouped cables,
▷ cables installed in the ground,

the applicable current-carrying capacity of the cables under these conditions must first be determined.

The overcurrent protective devices are then selected on the basis of the value determined for the permissible current-carrying capacity under the differing operating conditions I_b, in the same way as described above for I_z.

Overload protection of conductors in parallel

Conductors connected in parallel can be protected either individually or by means of a shared protective device.

Protecting conductors individually allows the conductor cross-sectional area to be precisely matched to the protective device. If a cable run fails, the remaining parallel conductors can continue to operate. This is best ensured by providing individual protection at the source and end of the electric circuit.

If conductors connected in parallel are protected by means of a shared protective device, the following conditions must be fulfilled:

▷ all the conductors must have the same electrical characteristics (type, cross-sectional area, length, and type of installation),
▷ the conductors must not be tapped at any point.

When the shared overload protective device is selected, the current-carrying capacity I_z is the sum of the current-carrying capacities of all the conductors, taking possible reduction factors into account (e.g. grouping).

Dispensing with overload protection

In the following cases, it is permissible, in accordance with HD 384.4.43/IEC 60364-4-43/DIN VDE 0100-430, to dispense with an overload protective device:

▷ in cable joints, in which the occurrence of overload currents does not have to be taken into account. This is subject to the requirement that the cable runs contain neither taps nor plug-and-socket devices.
▷ in connecting cables between electrical machines, starters, transformers, rectifiers, batteries, switchgear or similar parts of systems,
▷ in auxiliary circuits,
▷ in public utility networks comprising cables installed in the ground or overhead lines.

Overload and short-circuit protective devices should not be used if interrupting the electric circuit could give rise to a hazardous situation, e.g.

▷ in the field circuits of rotating machines,
▷ in the supply circuits of solenoids,
▷ in the secondary circuits of current transformers,
▷ in electric circuits used for safety purposes, such as fire-extinguishing, safety-lighting, smoke and heat extraction systems, as well as fire service elevators.

The cable runs in these electric circuits should be designed so that damaging temperature rises due to overload currents does not have to be taken into account.

Short-circuit protection

Short-circuit protective devices are designed to interrupt the short-circuit currents in the conductors of an electric circuit, before they can cause a temperature rise that would endanger the conductor insulation, connecting points and junctions, as well as the surroundings of the cables.

The breaking capacity of the protective device must, therefore, be at least equal to the highest current in the event of a dead short circuit at the point of installation.

An overcurrent protective device selected in accordance with the paragraph “Overload protection” (see page 281), with a breaking capacity not less than the current in the event of a dead short circuit at the point of installation, ensures that the downstream cable run is protected not only in the event of an overload, but also in the event of a short circuit.

If overload protection is not required for a particular electric circuit, or if, for special reasons, a shared overcurrent protective device is not necessary, the short-circuit protective device is selected in accordance with the conductor cross-sectional area to be protected, the circuit length, and the loop impedance of the system on the supply side.

The following relationship applies for the permissible time until the required disconnection of a short-circuit current at any point of an electric circuit:

$$t = \left(k \cdot \frac{q}{I}\right)^2$$

Where:

t = permissible break-time in the event of a short circuit in s

I = r.m.s. value of the current with an assumed dead short circuit in A

k = material coefficient of the conductor material, for example,
- 115 As/mm^2 with PVC-insulated copper conductors
- 141 As/mm^2 with rubber-insulated copper conductors

q = conductor cross-sectional area in mm^2.

In general, a permissible break-time t of 5 s can be assumed for short circuits.

The current in the event of a dead short circuit (point at which dead short circuit occurred) or the loop impedance of the short-circuit loop can be determined by means of:

▷ calculation, e.g. to DIN VDE 0102,
▷ simulations using system models,
▷ system measurements,
▷ appropriate inquiries made to the public utility.

Tables 1.7/29 to 1.7/32[1)] below provide information for selecting the required overcurrent protective device in accordance with the selected conductor cross-sectional area, the circuit length, and the magnitude of the short-circuit current (represented by means of the loop impedance upstream of the overcurrent protective device to be selected). The assumed break-time is 5 s.

Example

With a loop impedance of 500 mΩ upstream of the overcurrent protective device (in this case, fuses), and a PVC-insulated cable with a length of 120 m and conductor cross-sectional area 2.5 mm^2 Cu connected to the device, a fuse with a maximum rated current I_n of 16 A is permissible for short-circuit protection (see Table 1.7/29). If the length of the cable were only 115 m, a fuse with a rated current of I_n = 20 A could be selected.

Dispensing with short-circuit protection

In the following cases, it is permissible, in accordance with HD 384.4.43/IEC 60364-4-43/DIN VDE 0100-430, to dispense with a short-circuit protective device:

▷ in public utility networks comprising cables installed in the ground or overhead lines,
▷ for connecting cables between electrical machines, starters, transformers, rectifiers, batteries, and associated switchgear, where the protective devices are located in the switchgear.

Lighting circuits

Overcurrent protective devices for protecting lighting circuits must not exceed 25 A.

Fluorescent lamps and circuits comprising tubular fluorescent lamps, as well as lighting circuits with E 40 lamp holders to EN 60238/IEC 60238/VDE 0616 Part 1 can be protected by overcurrent protective devices with higher values. In such cases, consideration should be given to the permissible loading of the cable and the wiring accessories.

Overcurrent protective devices for protecting lighting circuits in domestic installations must not exceed 16 A.

Socket-outlet circuits

To protect electric circuits with socket-outlets, the protective devices must be matched not only with the permissible loading of the cables, but also with the rated current of the connected socket-outlets. The lower of these two values is the more important.

[1)] The values in these tables are from Supplementary Sheet 5 of DIN VDE 0100

Table 1.7/29
Permissible length of cables with copper conductors, PVC or rubber insulation, and fuses, utilization category gL/gG to EN 60269/IEC 60269/DIN VDE 06360

Rated voltage of system: 400 V AC, 50 Hz

Disconnection after 5 s, or after permissible short-circuit temperature has been reached

Nominal conductor cross-sectional area	Rated current of protective device	Minimum short-circuit current	Loop impedance upstream of protective device in mΩ								
			10	50	100	200	300	400	500	600	700
			Maximum permissible length l_{max} in m								
mm²	A	A									
1.5	6	27	270	269	267	264	261	258	255	252	249
1.5	10	47	155	154	152	149	146	143	140	137	134
1.5	16	65	112	111	109	106	103	100	97	94	91
1.5	20	126	58	57	55	52	49	46	43	40	36
1.5	25	135	54	53	51	48	45	42	39	36	32
2.5	10	47	253	251	249	244	239	234	229	224	219
2.5	16	65	183	181	178	173	169	164	159	154	148
2 5	20	85	139	138	135	130	125	120	115	110	105
2.5	25	110	108	106	103	98	93	88	83	78	73
2.5	32	165	72	70	67	63	57	52	47	42	36
4	16	65	297	294	290	282	274	266	258	250	241
4	20	85	227	224	220	212	204	196	187	179	171
4	25	110	175	172	168	160	152	144	135	127	118
4	32	150	128	125	121	113	105	96	88	79	71
4	40	190	101	98	94	86	77	69	60	51	42
4	50	280	68	65	61	53	45	36	27	18	8
6	20	85	342	337	331	319	307	294	282	270	257
6	25	110	264	259	253	241	229	216	204	191	178
6	32	150	193	188	182	170	158	145	132	119	106
6	40	190	152	147	141	129	116	104	91	77	64
6	50	260	111	106	100	87	75	62	48	35	20
6	63	330	87	82	76	64	57	38	24	10	0
10	25	110	441	433	423	403	382	361	340	319	298
10	32	150	323	315	305	284	264	242	221	199	178
10	40	190	255	246	236	216	195	173	152	130	107
10	50	260	185	177	167	146	125	103	81	58	34
10	63	320	150	142	132	111	89	67	44	20	0
10	80	440	108	100	90	69	46	23	0	0	0
16	32	150	512	499	483	450	417	384	350	315	280
16	40	190	404	391	374	341	308	274	240	205	169
16	50	260	294	281	265	231	198	163	127	91	54
16	63	320	238	225	209	175	141	106	69	32	0
16	80	440	172	159	143	109	73	37	0	0	0
16	100	580	130	117	100	65	29	0	0	0	0

Table 1.7/30
Permissible length of cables with copper conductors, PVC or rubber insulation, and miniature circuit-breakers to EN 60898/IEC 60898/DIN VDE 0641-11, tripping characteristic B

Rated voltage of system: 400 V AC, 50 Hz

Disconnection after 0.1 s, or after permissible short-circuit temperature has been reached
Length values are the same for break-times 0.4 s and 5 s

Nominal conductor cross-sectional area	Rated current of protective device	Minimum short-circuit current	Loop impedance upstream of protective device in mΩ								
			10	50	100	200	300	400	500	600	700
			Maximum permissible length l_{max} in m								
mm²	A	A									
1.5	6	30	243	242	240	237	234	231	288	225	222
1.5	10	50	145	144	143	140	137	134	131	128	125
1.5	16	80	91	89	88	85	82	79	76	73	70
1.5	20	100	72	71	70	67	64	61	57	54	51
1.5	25	125	58	57	55	52	49	46	43	40	36
2.5	10	50	238	236	233	229	224	219	214	209	204
2.5	16	80	148	146	144	139	134	129	124	119	114
2 5	20	100	118	116	114	109	104	99	94	89	84
2.5	25	125	95	93	90	85	80	75	70	65	60
2.5	32	160	74	72	69	64	59	54	49	44	38
4	16	80	241	238	234	226	218	210	202	193	185
4	20	100	193	190	186	178	169	161	153	145	136
4	25	125	154	151	147	139	131	122	114	106	97
4	32	160	120	117	113	105	96	88	80	71	62
4	40	200	96	93	89	80	72	64	55	46	37
6	20	100	290	285	279	267	255	243	230	218	205
6	25	125	232	227	221	209	197	184	172	159	146
6	32	160	181	176	170	158	145	133	120	107	94
6	40	200	144	139	133	121	109	96	83	70	56
6	50	250	115	110	104	92	79	66	53	39	25
6	63	315	91	86	80	68	55	41	28	14	0
10	25	125	388	380	370	350	329	308	287	265	244
10	32	160	303	295	284	264	243	222	201	179	157
10	40	200	242	234	223	203	182	160	139	116	94
10	50	250	193	185	175	154	132	111	88	66	42
10	63	315	153	144	134	113	92	69	47	23	0
16	32	160	480	467	451	418	385	351	317	282	247
16	40	200	383	370	354	321	288	253	219	184	148
16	50	250	306	293	277	243	210	175	140	104	67
16	63	315	242	229	213	179	145	110	73	36	0

Table 1.7/31
Maximum permissible length of cables with copper conductor, PVC or XLPE insulation, and fuses, utilization category gL/gG to EN 60269/IEC 60269/DIN VDE 0636

Rated voltage of system: 400 V, 50 Hz

Disconnection after 5 s, or after permissible short-circuit temperature has been reached

Nominal conductor cross-sectional area	Rated current of protective device	Minimum short-circuit current	Loop impedance upstream of protective device in mΩ				
			10	50	100	200	300
			Maximum permissible length l_{max} in				
mm^2	A	A	m	m	m	m	m
25	63	320	374	354	328	275	221
25	80	440	271	250	224	170	115
25	100	580	204	183	157	102	46
25	125	750	157	136	109	54	0
25	160	930	125	104	77	21	0
35	80	440	372	343	307	233	157
35	100	580	280	251	215	140	62
35	125	750	215	186	149	73	0
35	160	930	172	143	106	28	0
35	200	1350	116	87	49	0	0
35	250	1600	97	67	29	0	0
50	100	580	376	337	288	187	83
50	125	750	289	249	200	97	0
50	160	930	231	191	141	38	0
50	200	1350	156	116	65	0	0
50	250	1600	130	90	39	0	0
70	125	750	408	352	281	136	0
70	160	930	326	270	199	53	0
70	200	1350	220	164	92	0	0
70	250	1600	184	127	54	0	0
70	315	2200	130	73	0	0	0
95	160	930	438	361	265	70	0
95	200	1350	296	219	122	0	0
95	250	1600	246	169	72	0	0
95	315	2200	174	97	0	0	0
95	400	2750	135	58	0	0	0
120	200	1350	362	267	148	0	0
120	250	1600	302	207	88	0	0
120	315	2200	213	118	0	0	0
120	400	2750	165	70	0	0	0
150	200	1350	426	314	174	0	0
150	250	1600	355	243	103	0	0
150	315	2200	250	139	0	0	0
150	400	2750	195	83	0	0	0
150	500	3900	129	17	0	0	0

Table 1.7/32
Maximum permissible length of cables with copper conductors, PVC or XLPE insulation and fuses, utilization category gL/gG to EN 60269/IEC 60269/DIN VDE 0636

Rated voltage of system: 400 V, 50 Hz

Disconnection after 5 s, or after permissible short-circuit temperature has been reached

Nominal conductor cross-sectional area	Rated current of protective device	Minimum short-circuit current	Loop impedance upstream of protective device in mΩ				
			10	50	100	200	300
			Maximum permissible length l_{max} in				
mm²	A	A	m	m	m	m	m
25/16	63	320	291	275	255	214	172
25/16	80	440	211	195	174	133	90
25/16	100	580	159	143	122	80	36
25/16	125	750	122	106	85	42	0
25/16	160	930	0	0	0	0	0
35/16	80	440	235	217	195	148	100
35/16	100	580	177	159	136	89	40
35/16	125	750	136	118	95	47	0
35/16	160	930	0	0	0	0	0
50/25	100	580	265	238	203	132	59
50/25	125	750	203	176	141	69	0
50/25	160	930	162	135	100	27	0
50/25	200	1350	0	0	0	0	0
70/35	125	750	282	244	195	95	0
70/35	160	930	225	187	138	37	0
70/35	200	1350	152	114	64	0	0
70/35	250	1600	127	88	38	0	0
95/50	160	930	303	251	185	49	0
95/50	200	1350	205	152	85	0	0
95/50	250	1600	170	118	51	0	0
95/50	315	2200	120	67	0	0	0
120/70	200	1350	274	203	113	0	0
120/70	250	1600	229	157	67	0	0
120/70	315	2200	161	90	0	0	0
120/70	400	2750	125	54	0	0	0
150/70	200	1350	292	216	120	0	0
150/70	250	1600	243	167	71	0	0
150/70	315	2200	172	95	0	0	0
150/70	400	2750	133	57	0	0	0

1.7.4 Materials for installing and fixing cables

As a result of the increasing use of electrical power, as well as developments in modern communication technologies, ever more extensive wiring systems are required in both functional and residential buildings. Whether buildings are constructed for industrial, commercial, administrative, or residential purposes, installing and fixing cables accounts for a considerable amount of installation work. Reducing installation times not only allows work to be carried out more cost effectively, it often considerably facilitates the installation process. The following section describes the most important materials for installing and fixing cables.

For individual cables

onduits to IN VDE 0605

The conduits selected depend on the type of installation and the particular conditions under which cables are installed. Plastic, special plastic, steel, and aluminum conduits are available.

The following types of conduit are distinguished to DIN VDE 0605:

▷ conduits for high mechanical stress, for installation in concrete, and on, under, and in plaster (designation "AS"),
▷ conduits for moderate mechanical stress, for installation on, under, and in plaster (designation "A"),
▷ conduits for low mechanical stress, only for installation under and in plaster (designation "B"),
▷ conduits with particular electrical characteristics (designation "C"),
▷ conduits with flame-retardant characteristics (designation "F"),
▷ conduits that are thermally stable up to 105 °C (designation "105").

lastic or metal onduits

Plastic conduits prevent accidental energization due to cable insulation faults. By contrast, metal conduits that are not internally insulated must be incorporated in additional protective measures against excessively high touch voltages if they contain cables with basic insulation (e.g. H07V, etc.).

lechanical tress

Whether an insulating or a high-strength conduit is selected depends on the particular mechanical stress involved.

hermal stress

If plastic conduits are exposed to abnormal thermal stress due to cold (e.g. if construction work is carried out in winter) or heat (e.g. when panels are assembled for use in concrete construction methods; see Section 5.2), special variants, such as polypropylene conduits, must be used.

Installation type

Flexible conduits, which can easily be bent manually, are preferred for underplaster installation. Rigid conduits are more expedient for installation on plaster, since they require fewer fixing points.

Pull-in tool

Problems often arise during the time-consuming process of pulling cables into conduits, since cables can become skewed or lodged in the conduit. This tends to occur particularly where cables have to be pulled into cable conduits or ducts already in use.

The KATI-Blitz pull-in tool (Fig. 1.7/13) solves these familiar problems. The highly flexible spiral spring locator with its rounded head fits into any bend down to a radius of 30 mm. The 3 mm thick PC-glass-fiber rod is just as flexible as the spiral spring locator, but its inner stability means that it is as robust as a rod. These characteristics enable any resistance or bend to be overcome easily. The tool action of KATI-Blitz can be improved even further by using a cable lubricant. This reduces frictional resistance by approximately 50 %, preventing wear or

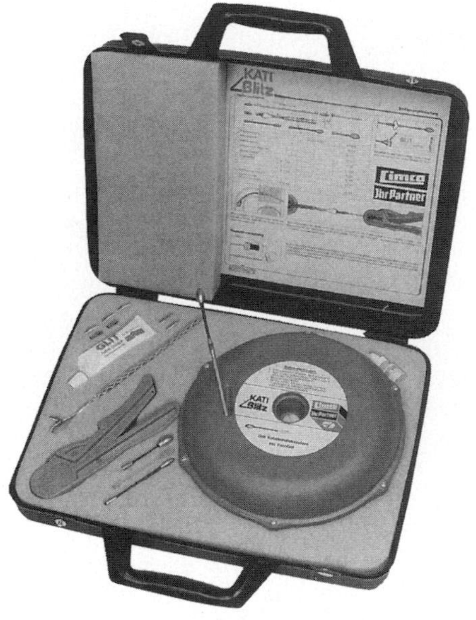

Fig. 1.7/13 KATI-Blitz pull-in tool

Fig. 1.7/14
Retainer clip with integrated nail

Fig. 1.7/15
Multiple pressure saddle with oblong hole, screw thread and retaining base; suitable for butt-mounting

Fig. 1.7/16
Flexible clip for high-strength plastic, steel, and aluminum conduits

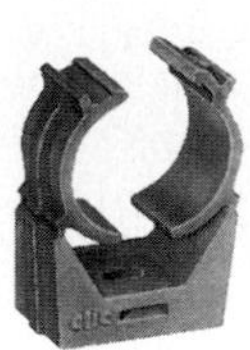

Fig. 1.7/17
Snap-in clip for conduits

damage to cables, and the short circuits that this can lead to. The cable lubricant leaves a durable waxy film on the outer sheath, which also protects the installed cables against breakage and drying out.

Insulating cable clamps

A wide range of clamps are available for fixing individual cables or conduits. Figs. 1.7/14 to 1.7/17 show a selection of these clamps.

Cable clamps

Cable clamps are intended primarily for installation on anchor rails; they can also be used, however, with all standard mounting rails, and can be attached directly on sheet steel, wood, or masonry.

For several cables

Selection criteria

The great variety of installation materials available, and the fact that they can often be used in a wide range of applications means that only the most commonly used installation materials can be described here. The particular installation conditions must always be considered when the materials are selected. The following factors are extremely important:

- ▷ number and outer diameter of cables,
- ▷ installation type, with regard to mechanical, thermal, and chemical stress,
- ▷ installation type, with regard to appearance (concealed installation, e.g. in false ceilings; visible installation, e.g. in electrical operating areas, offices, and hospitals),
- ▷ easy installation of additional cables,
- ▷ cost of installation materials and costs for installing them,
- ▷ required separation of power installations and communications systems.

Multiple saddles

Multiple saddles are generally used to install no more than five cables, especially if the installation point is visible. Variants are available in graded sizes for outer diameters from approximately 5 mm to 25 mm, and 8 mm to 38 mm. The saddles are fixed to hollow rails made of plastic or hot-galvanized sheet steel (Fig. 1.7/18).

Cable clamps

Cable clamps permit any number of cables to be installed, even above one another in several layers. The clamps are available in graded sizes for outer diameters of 10 mm to 100 mm.

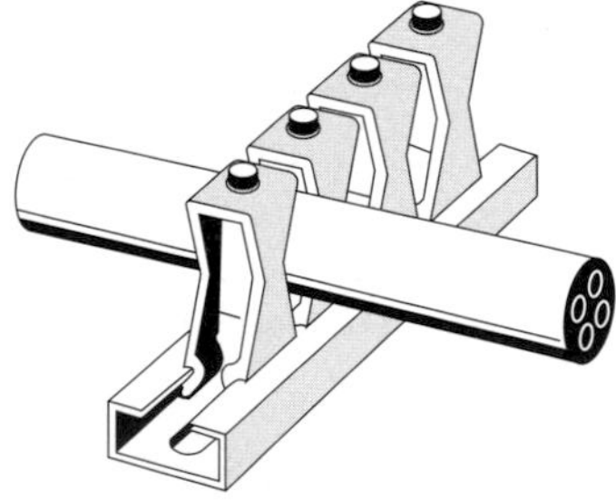

Fig. 1.7/18
Insulating hammering saddles for screwless snap-on fixing

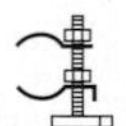

Single clamp with one lug

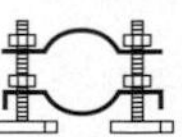

Single clamp with two lugs and Al cover shell

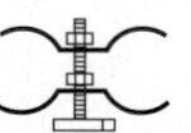

Double clamp

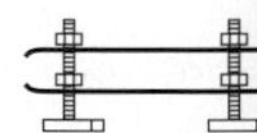

Multiple clamp

Fig. 1.7/19 Cable clamps for one or more cables

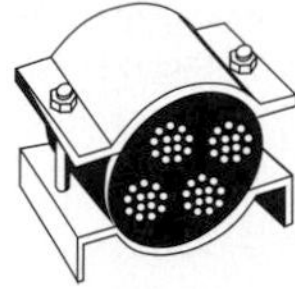

Fig. 1.7/20
Single clamp with two lugs installed with cable

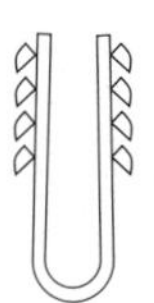

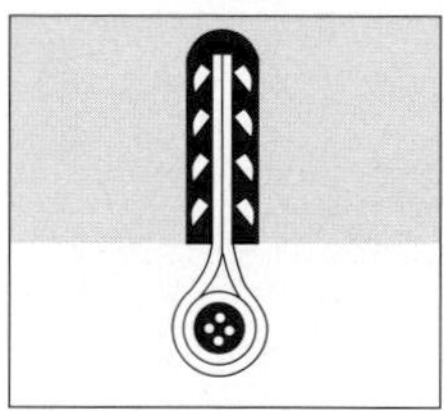

Fig. 1.7/23
Plug clip for individual cables, cable bundles, and flexible conduits

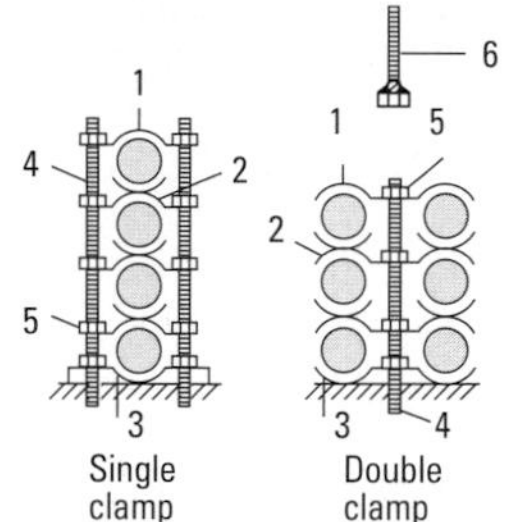

1 Cover shell
2 Intermediata shell
3 Back shell
4 Fixing element
5 Hexagonal nut
6 Extension bolt

Fig. 1.7/21
Several cables installed one above the other

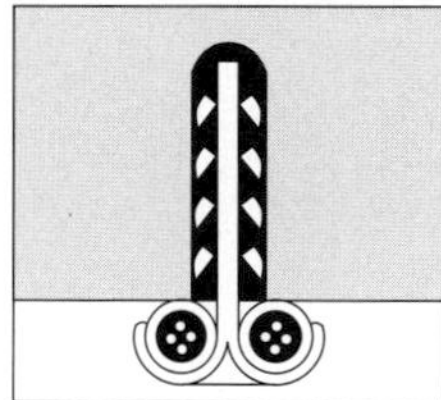

Fig. 1.7/24
Twin push-in clip for two individual cables in parallel

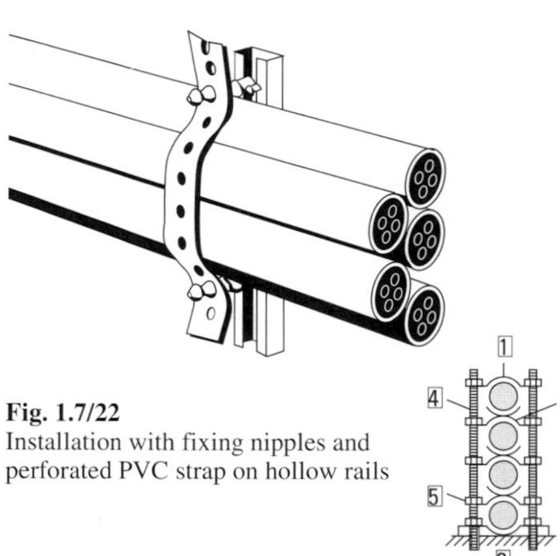

Fig. 1.7/22
Installation with fixing nipples and perforated PVC strap on hollow rails

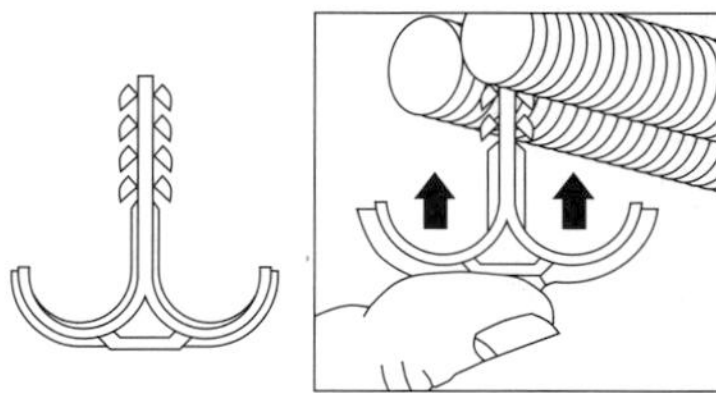

Fig. 1.7/25
Twin push-in clip for two parallel rigid conduits or cable bundles

They are supplied as single clamps with one or two lugs, as double clamps, and as multiple clamps (Figs. 1.7/19 to 1.7/21).

Cable clamps comprise a back shell, and cover shell or back shell with a fixing element (bolts or cleats with nuts). For fixing single-core cables using single clamps with two lugs, a variant is available with an aluminum cover shell to prevent temperature rises due to eddy currents.

Al cover shell

The form of the clamps and the back shells, which must be screwed tightly, help prevent even pressure-sensitive cables from being deformed. These cable clamps can also be used to fix conduits.

Extension bolts permit additional clamps to be fitted to cable clamps already installed.

Plastic straps

Various types of plastic strap are used primarily in concealed installations, e.g. for bundling cables in false ceilings.

Fixing nipples on hollow rails allow cables to be secured (even in layers) using perforated PVC straps (Fig. 1.7/22).

One of the most effective means of fixing cables, conduits and ducts is the plug-in system shown in Figs. 1.7/23 to 1.7/31.

Cable ducts

Hard PVC cable ducts are particularly expedient where cables are grouped, and in systems in which cables often have to be rerouted or added (Fig. 1.7/32). These can be left open (concealed installation), or used with a form-fit cover (visible installation). The ducts can be compartmentalized by means of slot-in partitions to accommodate

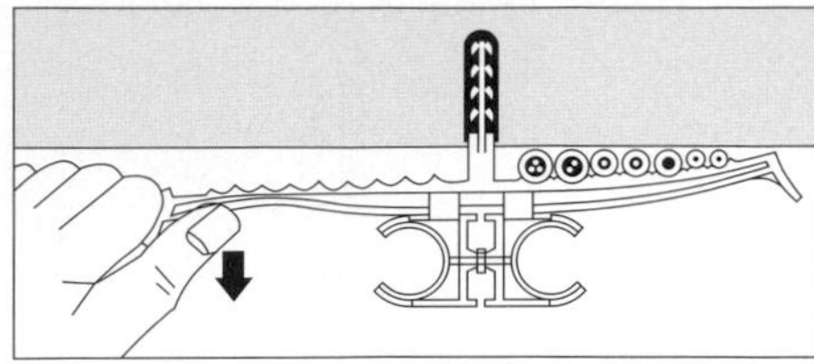

Fig. 1.7/26
Two-sided bracket clip for cables, with clips on bottom for conduits; left arm of bracket bent downwards

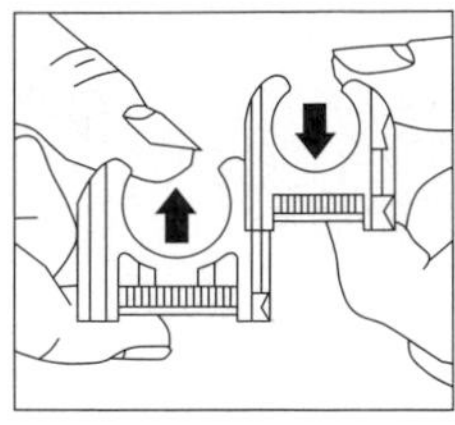

Fig. 1.7/29
Euro clip: buttable and pluggable; can be fixed with clip-on plug-in dowels, screw-type dowels, and impact dowels

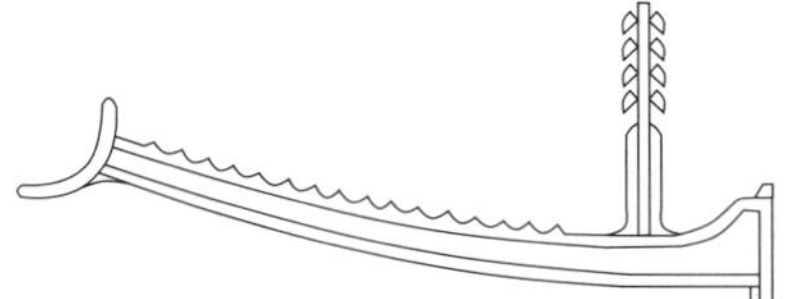

Fig. 1.7/27 One-sided bracket clip for cables

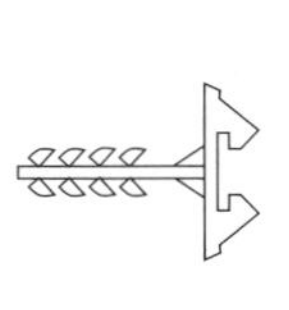

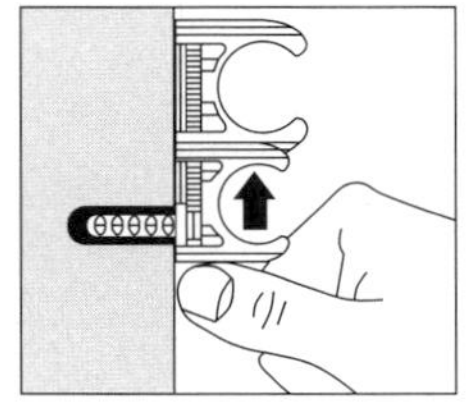

Fig. 1.7/30
Clip-on plug-in dowel for Euro clip and multi-binder

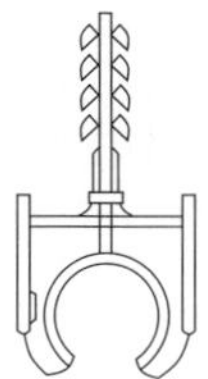

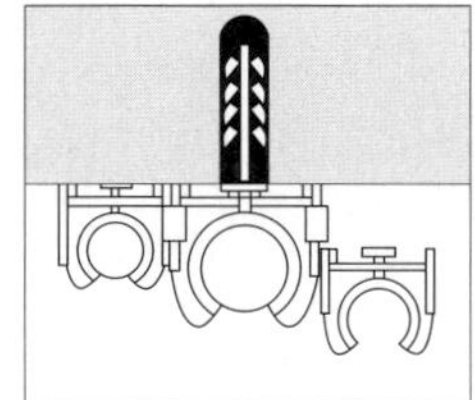

Fig. 1.7/28
One-part buttable conduit clips for rigid conduits

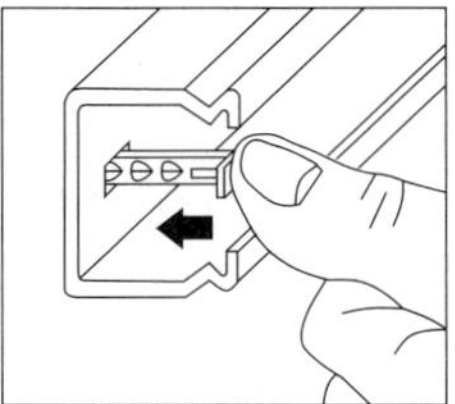

Fig. 1.7/31 Dowel pin

The range of sizes and accessories available enables ducts of all types to be implemented.

Sill-type ducts

Switches, socket-outlets, and similar devices (Fig. 1.7/33) can be installed in special variants, e.g. sill-type ducts (see Section 5.1.4.6).

Skirting board ducts

Skirting board ducts (Fig. 1.7/34) are particularly well suited to the concrete construction method and for modernizing older buildings. A box unit enables a switch, a socket-outlet, or similar devices to be installed at any point on the duct (see Section 5.1.4.3). The range of products is complemented by molded sections for changes in direction.

Floor-mounted ducts

Floor-mounted ducts are used to install power and communication cables between the wall terminals and desks, e.g. in offices (Fig. 1.7/35).

Larger groups of cables are installed on supports, in cable gutters, in cable raceways, or on cable racks.

Flexible cable supports

Flexible cable supports made of galvanized round steel bars are particularly easy to install (Fig. 1.7/36). They can be bent into any required form on site.

The supports are open on all sides and permit any type of cable distribution. This open construction prevents dust deposits, and ensures that the cables are well cooled.

Since the brackets of the support are used not only to hold the cables, but also for fixing, no special cross-members are required.

Cable gutters, cable raceways

Cable gutters or raceways (see Section 5.1.4.3) are available in different variants made of plastic or hot-galvanized sheet steel.

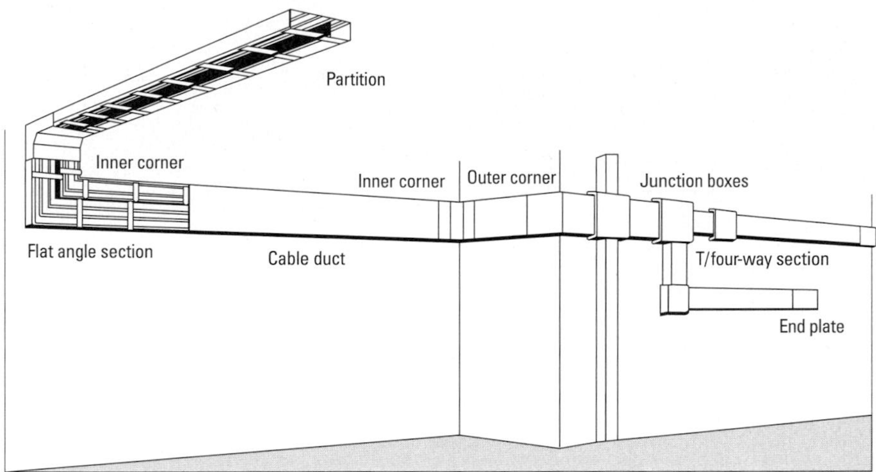

Fig. 1.7/32 Installation of ducts for cables

The accessories include bends, connectors, etc. Cable gutters and raceways are fixed on wall cross-members or on special fixing elements (e.g. hangers and clamping brackets), which are mounted on the ceiling.

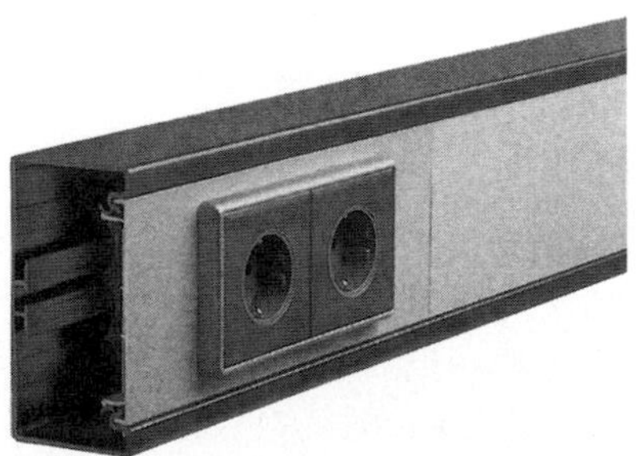

Fig. 1.7/33
Sill-type duct with socket-outlets for power systems

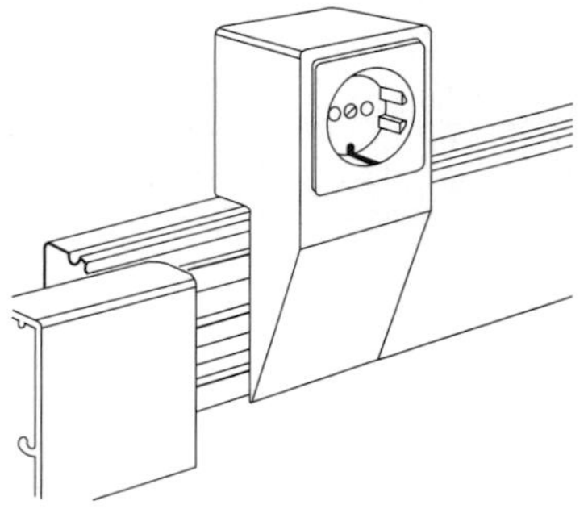

Fig. 1.7/34 Skirting board duct with box unit

As a rule, cables are simply inserted into gutters and raceways. If cables have to be fixed, e.g. onto vertical sections, they are secured by means of plastic boot clamps or notched straps (Fig. 1.7/37).

Cable racks

Cable racks (Fig. 1.7/38) are available in different variants from specialist companies. They generally comprise prefabricated modular components, and can be free standing, suspended, or mounted on walls vertically or horizontally. Planning and installation is also usually carried out by the manufacturer. The installed cables are fixed in the same way as described for cable gutters and raceways.

Siemens special box-type rollers

Siemens special box-type rollers enable cables to be installed efficiently on cable racks. These allow the cables to be installed after the rack has already been mounted on the wall or ceiling. This means that the cables no longer have to be laid out on the floor prior to installation on the racks. The cables are removed by swinging out the upper roller.

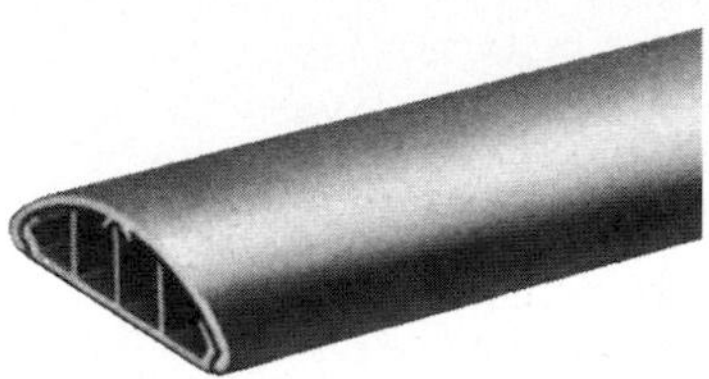

Fig. 1.7/35 Floor-mounted duct with partitions

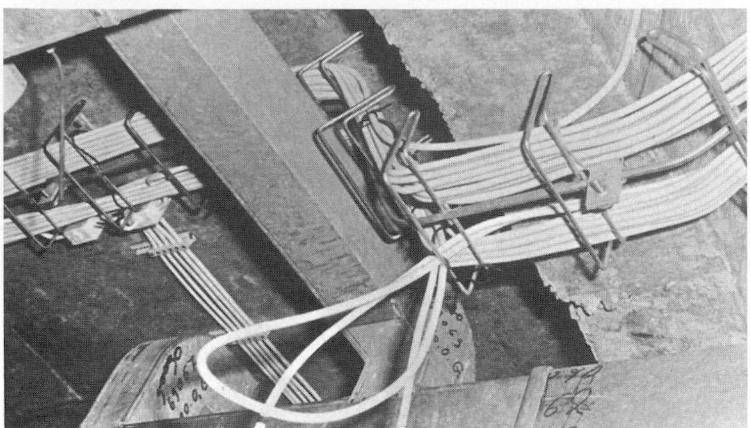

Fig. 1.7/36
Flexible cable supports made of galvanized round steel bars

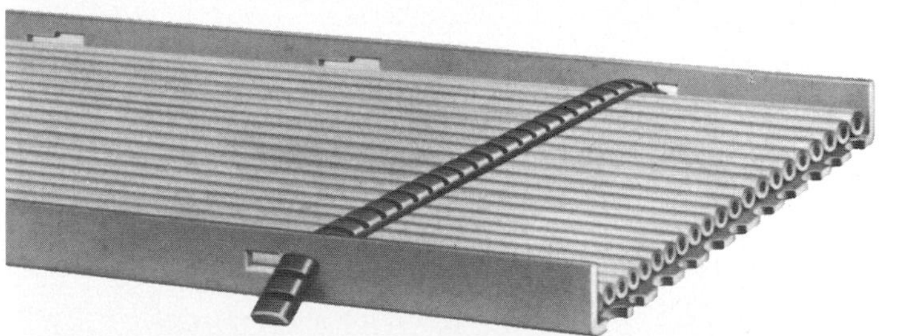

Fig. 1.7/37
Cables fixed with a notched strap in a cable gutter

Fig. 1.7/38
Cable racks in a cable duct

1.8 Supply systems for safety services

1.8.1 Introduction

Scope, standards

According to IEC 60050(826) :06.98/DIN VDE 0100-200, a supply system for safety services is a power-supply system, the purpose of which is to maintain the operability of personal-safety (current-using) equipment. This power-supply system includes the power source and circuits up to the terminals of the current-using equipment.

Establishing whether or not this equipment is indeed necessary, and, if so, as of which limit values and with which degree of protection it is to be implemented, is, however, not the task of electrotechnical standards. This, together with the actual safety equipment required for a particular application, is determined on the basis of regional or national statutory requirements, occupational safety and health regulations as well as the appropriate insurance-related stipulations.

Required safety equipment and its use

Required safety equipment

The safety equipment that is essential for personal safety includes:

▷ Emergency lighting for escape routes and dangerous work areas,
▷ Fire alarm systems,
▷ Equipment for alerting visitors and employees and issuing them with appropriate instructions,
▷ Equipment used to supply water for fire fighting,
▷ Elevators for fire fighters,
▷ Smoke and heat extraction equipment,
▷ CO_2 alarm systems.

This equipment must continue to function correctly, even when the general power supply is disrupted.

When and where ?

The aforementioned standards and statutory requirements specify that required safety equipment must be installed in buildings or facilities in which one or more of the following conditions apply:

▷ Problematic escape routes,
▷ Visitors not familiar with the building/area,
▷ Escape difficult/impossible, e.g. because of illness,
▷ Large number of employees,
▷ Dangerous materials are processed and stored,
▷ Flammable materials are stored.

These situations can be encountered in:

- High-rise buildings
- Meeting places/assembly points
- Department stores and office buildings
- Exhibition centers
- Hotels and guesthouses
- Schools
- Hospitals and homes
- Workrooms and storage areas
- Large garages

1.8.2 Power-supply system design

Safety power supply

Safety power supply

If supply systems for safety services are prescribed by law, it is important to ensure that the standard HD-384.5.56/IEC 60364-5-56/DIN VDE 0100-560 is observed.

As already mentioned in the definition in IEC 60050 (826)/DIN VDE 0100-200, the supply system must include a power source, which is independent of the general power supply, as well as the distribution and secondary distribution system up to the current-using equipment to be supplied. The requirements specified in the relevant standards regarding the individual components of the supply system and the installation requirements are outlined briefly below:

Power sources

Mains-independent power sources

The following are classified as mains-independent power sources according to HD 384.3S1/ IEC 60364-3-35/DIN VDE 0100-560:

▷ rechargeable batteries,
▷ primary cells,
▷ generators with drive motors that are independent of the general power supply,
▷ an additional incoming supply from the general power supply which is independent of the normal incoming supply from the mains (appropriate measures must be taken to ensure that both incoming supplies cannot fail at the same time).

Selection

The independent power source is selected on the basis of the requirements of the current-using equipment to be supplied. The most important criteria are the *ON delay* and the *operating time.*

General power supply

Safety power supply

As shown in Fig. 1.8/1, the required safety equipment is usually supplied by the general power-supply system. Only in the event of a mains failure is the distribution system of the safety power supply automatically switched over to the independent power source.

The power source that is most suitable for a particular application depends on the maximum open-circuit time that is permitted for the required safety equipment and the necessary supply time.

The *primary cell* is normally used to supply directly assigned current-using equipment, since (unlike a battery) it cannot be recharged.

Additional independent incoming supply

An additional independent incoming supply from the general power supply is only encountered in exceptional cases due to the considerable mains-specific requirements that have to be fulfilled. Rechargeable batteries and generators with mains-independent drive motors *(generating sets)* are, therefore, the most common power sources.

Generating sets

The advantage of rechargeable batteries is their short ON delay. Generating sets have the advantage of a long operating time.

Installation

According to Sections 562.1 to 562.3 of the aforementioned standards, central, independent power sources must be permanently installed at a suitable location that is only accessible to qualified personnel. The room must be designed in such a way that gas or smoke can be extracted and fresh air supplied. It is, therefore, advisable to install the power source in a separate room. The walls and ceilings of this room must be made of non-flammable materials to ensure that the power source is not endangered if a fire breaks out in the adjacent rooms.

Permitted protective measures

Protective measures

According to Section 561.2 of the aforementioned standards, preference should be give to protective measures that do not cause an automatic shutdown when the first error occurs.

According to HD 384.4.41/IEC 60364-4-41/DIN VDE 0100-410, these protective measures include:

▷ Safety Extra Low Voltage (SELV) and Protective Extra Low Voltage (PELV),
▷ Protection in the IT system,
▷ Protection provided by Class 2 equipment,
▷ Protection provided by non-conductive rooms,
▷ Protection provided by ungrounded, local equipotential bonding,
▷ Protective separation.

These measures are recommended because they ensure that short circuits to exposed conductive parts do not inevitably cause the system to shut down automatically. It is important to remember, however, that this is only a recommendation. The extent to which the protective measures can be implemented in each individual case also depends on the

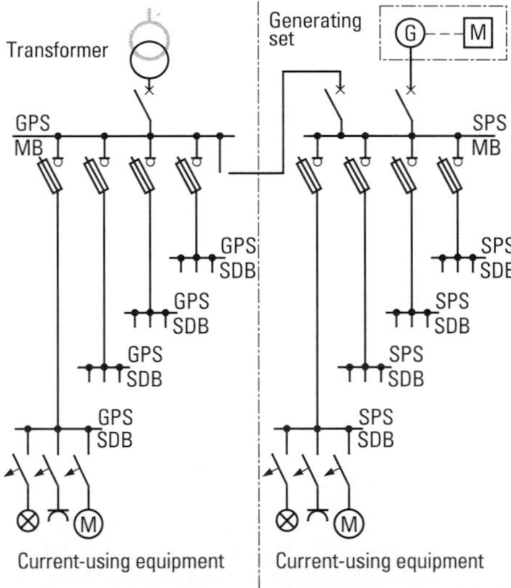

Fig. 1.8/1
Separate general power supply and safety power supply with a generating set.

operating conditions of the loads and the rooms themselves.

This applies, in particular, to situations in which the safety extra low voltage is limited to 50 V AC/120 V DC and to the special requirements regarding protection provided by non-conductive rooms and ungrounded, local equipotential bonding.

Protection in the TN system

If protection is provided in the TN system, it is essential to ensure that a fault in a circuit does not affect parallel circuits as well. This can be achieved by designing the distribution and secondary distribution systems in such a way that, in the event of a fault of negligible impedance at any point between the outer conductor and the equipment grounding conductor, or the exposed conductive part connected to this, the overcurrent protective device which is immediately upstream of the fault location switches off selectively and automatically within the specified time interval. Operation from both the independent power source and the normal mains supply should be considered here. If this system configuration is chosen, the circuit is disconnected in the event of a short circuit to an exposed conductive part, but only within the affected area. The extent to which this is acceptable, and therefore the extent to which the TN system can be implemented, must be considered on a case-to-case basis.

Wiring system

Wiring system

Appropriate requirements for the wiring systems of supply systems for safety services are specified in Section 563 of the standards. An independent distribution and secondary distribution system, which is separate from the general power-supply system, must be used to supply the current-using equipment of the supply system for safety services. This is illustrated in Fig. 1.8/1. With regard to this separation, Section 563.1 of the standards states that electrical faults, modifications, or any type of manipulation in a system must not interfere with the operational reliability of the other systems. This may require fire-resistant separation measures, separate cable paths or special enclosures.

The circuits of supply systems for safety services must, therefore, have their own electric lines or cables and be routed through separate conduits. Combining several circuits in the same enclosure is not permitted. Furthermore, circuits must not be routed through areas subject to explosion hazards or fire hazards. If the circuits are routed through areas subject to fire hazards, the electric lines or cables, should, at the very least, be provided with non-flammable coverings.

Overload protection

Overload protection

According to HD 384.4.47/IEC 60364-4-47/ DIN VDE 0100-470, overload protection for the circuits is not mandatory. If no overload protection measures are implemented, however, it is important to make sure that possible overloads are taken into account when the cross section of the circuits is determined.

Short-circuit protection

Short-circuit protection

With regard to the short-circuit protection used for the circuits, overcurrent protective devices should be selected and installed with due care and attention in order to ensure that the overcurrent in one circuit does not adversely affect the operational reliability of the upstream circuits or circuits connected in parallel to this. As with the TN system, it is important to verify that the overcurrent protective device is tripped selectively and at the appropriate time irrespective of whether the system is supplied by an independent power source or the normal power supply. This verification should be carried out on the basis of appropriate calculations.

1.8.3 Measures for operation in the event of fire

Maintaining operability in the case of fire

Maintaining operability in the case of fire

According to Section 561.1.2 of the standards, the design and installation location of the equipment of a supply system for safety services (which must continue to operate even in the event of a fire) must be such that it continues to function for a reasonable period of time after the fire has broken out. The standards do not, however, specify how long the equipment must continue to operate.

The following values can be used as a guide:

▷ Operability must be maintained for 30 minutes for fire detectors and alarms, and to allow buildings to be evacuated.

▷ Operability must be maintained for 90 minutes for fire-fighting equipment in buildings.

The standards specify that the operability of the equipment can be maintained by choosing the appropriate

- design and
- installation location.

Protection through design or installation location

Protection with independent power source and main distribution board

In the case of the central, independent power source and main distribution board of the safety power supply, it is the choice of location at which this equipment is installed that is paramount. Each piece of equipment should be installed in a separate room. The walls and ceilings of these rooms should be made of non-flammable materials and be capable of withstanding fire for at least the aforementioned periods. Both protection options can be implemented for the cables and electric lines that lead away from the main distribution board:

Cables/electric lines

▷ Cables and electric lines covered with insulating and sheathing material which retains its dielectric properties for a specific period of time when exposed to fire, for example, the types specified in IEC 60331/DIN VDE 0472-814 installed in support systems that have been tested accordingly,

▷ Busbar systems, the components of which are non-flammable,

▷ Cables and electric lines laid in ducts and vertical raceways made of non-flammable materials, the fire resistance of which has been tested accordingly,

▷ Cables and electric lines laid, for example, in the ground or between the unprepared floor and screed with the result that these are not exposed to fire at all.

Fire compartments

The final circuits which lead to the current-using equipment should, as far as possible, always be limited to the fire compartment that is to be supplied. The sub-distribution board should also be located in the fire compartment and have at least one non-flammable covering.

Sub-distribution boards

Current-using equipment

Current-using equipment, such as lights, loudspeakers and fire alarms, must be installed at appropriate locations so that it can perform its respective tasks correctly. The extent to which this equipment functions in the event of a fire is, however, also restricted by its design. Remedial measures here could take the form of redundant configurations, i.e. two circuits or ring feeders for each section of the building (fire compartment) among which the current-using equipment is distributed alternately.

Fire-fighting water systems and elevator systems

As with independent power sources, fire-fighting water systems and elevator systems should be located in separate rooms with a level of fire protection that is at least capable of maintaining the operability of the equipment.

2 Protective measures

2.1 Introduction

Standards

Protective measures are dealt with in detail in Chapter/Group 4 of HD 384/IEC 60364/ DIN VDE 0100.

Chapter/Group 4 contains specifications concerning the following protective measures:

IEC/ HD Chap.	DIN VDE Part	Protective measure
41	410	Protection against electric shock
42	420	Protection against thermal effects
43	430	Protection of cables and cords against overcurrent
44	440	Protection against overvoltage
45	450	Protection against undervoltage
46	460	Isolation and switching
47	470	Application of protective measures
48	480	Choice of protective measures

"Protection against electric shock" is of particular importance and is, therefore, the primary focus of this chapter.

Requirements

The basic requirements for "Protection against electric shock" are described primarily in HD 384.4.41 / IEC 60364-4-41 / DIN VDE 0100-410. The standards HD 384.5.54/ IEC 60364-5-54 / DIN VDE 0100-540 must also be observed.

The application of protective measures is dealt with in HD 384.4.47/IEC 60364-4-47/ DIN VDE 0100-470. Protection against electric shock is discussed in Part 471 of DIN VDE 0100-470.

The choice of protective measures according to external influences is dealt with in IEC 60364-4-481. The corresponding standards for HD and DIN VDE are still in preparation. Restrictions with respect to the protective measures that can be applied are specified in the various parts of Chapter/Group 7 of HD 384 / IEC 60364 / DIN VDE 0100 (see Section 5.9).

HD 384 / IEC 60364 / DIN VDE 0100 are standards for erecting power installations with nominal voltages[1)] ≤1000 V (frequencies 15–1000 Hz), and d.c. voltages ≤1500 V. The applicable parts of HD 384 / IEC 60364 / DIN VDE 0100 still do not contain any special information concerning d.c. voltage systems. In these cases, HD 384.4.41 / IEC 60364-4-41 / DIN VDE 0100-410 apply.

Hazardous touch voltage

The measures specified in HD 384.4.41 / IEC 60364-4-41 / DIN VDE 0100-410 are intended to prevent the occurrence, or at least the persistence of a hazardous touch voltage (U_B). Since it is the flow of current and not the voltage which constitutes the criterion for a hazard, it was agreed upon at an international level that touch voltages $U_B >$ 50 V AC or $>$ 120 V DC must be considered as hazardous.

In contrast to this voltage U_B which can actually pass through people or animals (also referred to as U_T, derived from the word "touch"), the conventional limit for permissible continuous touch voltages is denoted by U_L.

Permissible continuous touch voltage U_L

The limit for permissible continuous touch voltages U_L is 50 V for a.c. voltage and 120 V for d.c. voltage. In special cases or under certain environmental conditions, Chapter/Group 7 of HD 384 / IEC 60364 / DIN VDE 0100 defines lower values, e.g. 25 V AC and 60 V DC for agricultural operating areas.

Fault voltage U_F, touch voltage U_B

The touch voltages U_B and U_L are not always identical to the fault voltage U_F (Fig. 2.1/1). If the equipment grounding conductor is intact, the touch voltage is consider-

1) See Section 7.2 for explanations

ably lower than the fault voltage. If the equipment grounding conductor is interrupted or not fitted, the touch voltage U_B can be approximately equal to the nominal system voltage (with respect to ground) in the event of a short circuit to an exposed conductive part. This case is similar to direct contact with a live conductor (see Note).

> *Note*
> *According to the agreement, the touch voltage is only the voltage that can occur in the event of a short circuit to an exposed conductive part. Voltages that occur when live parts are touched do not constitute touch voltages as defined for protection against electric shock.*

Measures for protection against electric shock

In accordance with HD 384.4.41 / IEC 60364-4-41 / DIN VDE 0100-410, protection against electric shock is provided by:

Basic protection

▷ protection against electric shock under normal conditions (also referred to as protection against direct contact or basic protection).
This measure is intended to prevent all contact with live parts;

Fault protection

▷ protection against electric shock under fault conditions (also referred to as protection in the event of indirect contact or fault protection).
This measure is intended to prevent a hazardous touch voltage from occurring or persisting over a long period of time following failure of the basic protection.

Since, in a number of cases, the fault protection becomes ineffective when the basic protection fails (e.g. damage to the reinforced insulation used for electrical equipment of safety class 2), additional protection by means of RCDs is possible and is a requirement for particularly hazardous situations.

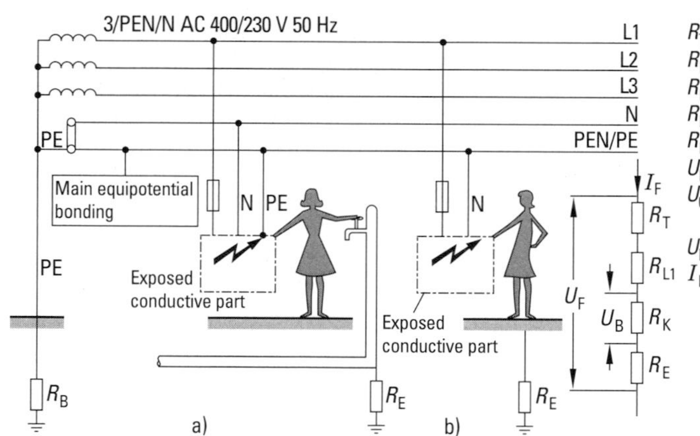

R_T Resistance of transformer in Ω
R_{L1} Resistance of all interconnecting cables in Ω
R_K Resistance of person (hand to foot) in Ω
R_E Ground contact resistance of location in Ω
R_B Operational ground resistance in Ω
U_B Touch voltage in V
U_0 Nominal voltage with respect to grounded conductor in V
U_F Fault voltage in V
I_F Fault current in A

Case a): Intact equipment grounding conductor

Until disconnection by protective device:
where cross section of equipment grounding conductor = cross section of external conductor $U_B \leq 0.5 \cdot U_0$,
with supplementary grounding of equipment grounding conductor $U_B \approx 0.25 \cdot U_0$ (depending on distance of grounded point, voltage balance)

where cross section of equipment grounding conductor = 1/2 cross section of external conductor
$U_B \leq 0.66 \cdot U_0$, with supplementary grounding of equipment grounding conductor $U_B \approx 0.25 \cdot U_0$
(depending on distance of grounded point, voltage balance)

Case b): No equipment grounding conductor

$R_{ges} = R_T + R_{L1} + R_K + R_E + R_B$ in Ω

$R_{ges} = 0.015\,\Omega + 0.5\,\Omega + 1000\,\Omega + 20\,\Omega + 2\,\Omega$

$R_{ges} = 1022.515\,\Omega$

$$I_F = \frac{U_0}{R_{ges}} = \frac{230\text{ V}}{1022.515\,\Omega} = 0.2249\text{ A}$$

$$U_B = R_K \cdot I_F = 1000\,\Omega \cdot 0.2249\text{ A} = 224.9\text{ V}$$

$$U_F = (R_T + R_{L1} + R_K + R_E) \cdot I_F = 1022.515\,\Omega \cdot 0.2249\text{ A} = 229.5\text{ V}$$

Fig. 2.1/1 Fault and touch voltage in TN systems

upplementary rotection

This measure is also referred to as supplementary protection.

This results in:

▷ measures which provide protection against direct and in the event of indirect contact or
▷ separate measures which can be applied simultaneously
  • for protection against direct contact and
  • protection in the event of indirect contact.

Protection against electric shock is always independent of the magnitude of the nominal voltage, i.e. appropriate measures are necessary even if the nominal voltages are very low. Additional measures, such as warning signs, should only be provided in exceptional cases and under certain conditions defined in Chapter 47 and Part 470.

rotection n the event of ndirect contact: xceptions

Protection in the event of indirect contact is not necessary for:

▷ the bases of overhead line insulators and all metal parts connected to these bases,
▷ reinforced concrete columns, the reinforcement of which is inaccessible,
▷ metal tubes or other metal enclosures for protecting electrical equipment of safety class 2,
▷ exposed conductive parts of electrical equipment if these are small (< 50 mm × 50 mm), if contact with the surface is not possible or the conductive parts cannot be grasped due to their arrangement. Protection may only be omitted if an equipment grounding conductor would be difficult to connect or unreliable, e.g. with bolts, rivets, rating plates.

In accordance with the National Appendix (Germany) of DIN VDE 0100-470 for public distribution systems, protection against indirect contact is also unnecessary for service entry masts / steel towers and reinforced concrete columns with accessible reinforcement.

Protection against direct contact: exceptions

Protection against direct contact is obligatory, except in the case of measures which provide protection against direct and indirect contact (see Section 2.2.1).

2.2 Protection against direct and in the event of indirect contact

leasures

Measures that provide protection against both direct and indirect contact are described in HD 384.4.41 / IEC 60364-4-41 / DIN VDE 0100-410, Section 411.

These measures are:

▷ protection by means of extra-low voltage: SELV and PELV.
  Abbreviations:
  SELV: Safety Extra Low Voltage
  PELV: Protective Extra Low Voltage
  FELV (Functional Extra Low Voltage) is also specified in IEC 60364-4-41 as a measure for providing protection against direct and indirect contact. In HD 384 and DIN VDE 0100, FELV is intended for special cases where protection by automatic disconnection of the power supply is required.
▷ protection by limiting the level of discharge energy.

Note
Closer examination reveals that the specifications in Section 411 of HD 384.4.4 / IEC 60364-4-41 / DIN VDE 0100-410 are only correct under certain conditions since supplementary protection against direct contact is always required for nominal voltages AC > 25 V and DC > 60 V even in the case of SELV (safety extra-low voltage). In Group 700 of DIN VDE 0100, supplementary protection against direct contact is, in certain cases, required as of 0 volts.

2.2.1 Protection by extra-low voltage: SELV and PELV

2.2.1.1 Protection by SELV

Protection by SELV is permitted for AC ≤ 50 V and DC ≤ 120 V if a power source with safety separation or a power source with a similar measure is used (see Fig. 2.2/2).

These power sources include:

Power sources with safety separation

▷ transformers with safety separation to EN 60742 / IEC 60742 / DIN VDE 0551-1 with appropriate identification (Fig. 2.2/1),

> *Note*
> The protective measure for the primary circuit must be taken into consideration in the case of transformers with protective shielding.

▷ power sources which provide the same level of safety as the above-mentioned transformers, e.g. motor-generator sets with separate windings to IEC 2/915 CDV / VDE 530 Part 1 or diesel-generator sets;

▷ electrochemical power sources, e.g. electrochemical batteries to DIN VDE 0510, storage batteries, or other galvanic elements. If electrochemical batteries are used, it is essential to ensure that devices, which also have safety separation from the supply system, are used for floating operation.
During charging, high d.c. voltages (max. 19% above U_n) are permitted; steps must, however, be taken to ensure that a.c. voltages higher than 50 V resulting from faults, e.g. in the rectifier, cannot occur;

▷ other power sources which do not depend on higher-voltage electric circuits, e.g. generators;

▷ certain electronic devices where the voltage at the output terminals and with respect to ground is not higher than 50 V AC and 120 V DC (or lower if lower values have been agreed upon) even if a fault occurs in the device.
Higher voltages are not permitted even if the voltages are limited to values below 50 V AC and 120 V DC or the power source is disconnected instantaneously and immediately when live parts or exposed conductive parts of faulty electrical equipment are touched.

Furthermore, it must also be ensured that:

Conditions in SELV circuits

▷ live parts (external conductors, middle conductors and, where applicable, neutral conductors) of SELV circuits are not connected to ground or equipment grounding conductors,

▷ exposed conductive parts in SELV circuits are not "inadvertently" connected to exposed conductive parts of other electrical equipment or to ground.

> *Note*
> "Inadvertently" connected: the exposed conductive surface of electrical equipment of safety class 3 is placed temporarily on a grounded component. If, however, an electrical device of safety class 3 is permanently attached to grounded metallic components, e.g. baseframe, braces etc., the requirements for applying the "SELV" protective measure are no longer satisfied.

Safety separation for electrical equipment

▷ safety separation is not only required for the power sources, it is also a requirement for the entire SELV circuit.

This means that:

- separation is also required for electrical equipment to which SELV circuits and circuits with other voltages are connected,
- conductors for SELV circuits must, if possible, be installed so that they are separate from conductors for high-voltage circuits (additional insulation by means of non-metallic jackets or

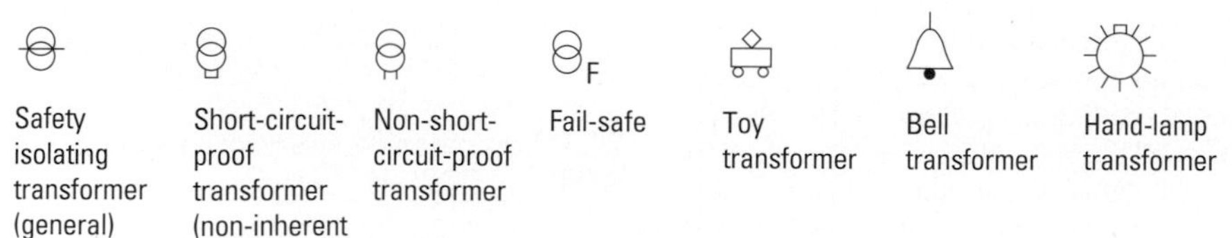

Fig. 2.2/1
Identification symbols for transformers with safety separation to EN 60742 / IEC 60742 / DIN VDE 0551-1

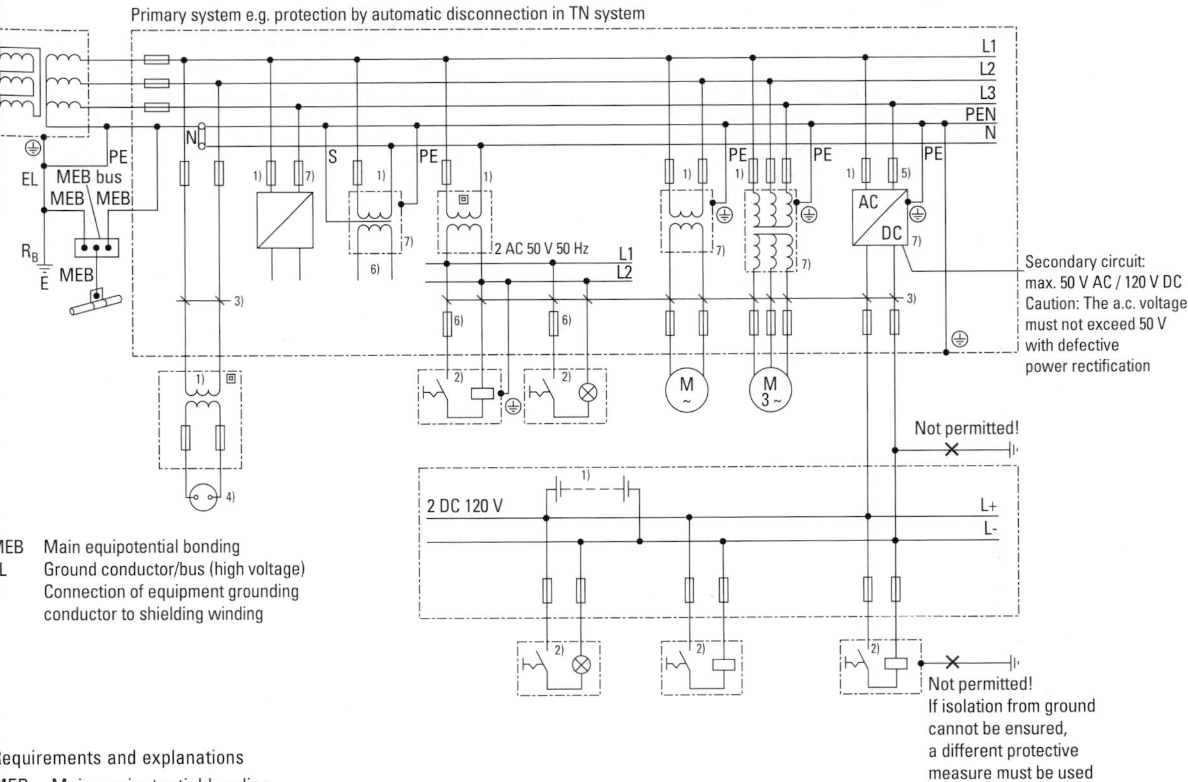

equirements and explanations

IEB Main equipotential bonding
L Ground conductor/bus (high voltage)
Connection of equipment grounding conductor to shielding winding

) Power sources: safety isolating transformers to EN 60742-1/IEC 60742-1/ DIN VDE 0551-1, or motor-generator sets to IEC 2/915 CDV/VDE 0530 Part 1, or storage batteries to DIN VDE 0510, or devices with an equivalent level of safety, or electronic devices in which voltages higher than 50 V AC / 120 V DC cannot occur after the 1st fault.
) Electrical equipment must have the same safety separation (electrical isolation) as the power sources if circuits with a higher voltage are also connected.
) Cables must be installed separately from circuits with a higher voltage or must be provided with insulation for the highest possible operating voltage.
) Plug-and-socket devices for SELV must not be used in circuits in which other protective measures are implemented. Sockets must not have ground contacts. It should only be possible to connect plugs to circuits with the same or a lower SELV.
) Charging rectifiers must be equipped with a safety isolating transformer.
) Control circuits, which are supplied via transformers, require single-pole protection only.
) Electrical equipment is usually safeguarded by its mounting attachments, i.e. equipment grounding conductors do not have to be connected.

SELV is a protective measure against direct contact ≤ 25 V AC, 60 V DC and a protective measure against indirect contact ≤ 50 V AC, 120 V DC without additional requirements. Restrictions for special cases. Protection against direct contact for voltages > 25 V AC, > 60 V DC

- by insulating live parts (the insulation must withstand a test voltage of 500 V AC for 1 minute)

  or

- by concealing or enclosing live parts to degree of protection ≥ IP 2X

ig. 2.2/2 Protection by SELV

equivalent enclosures must be provided as a minimum requirement).
Conductors for SELV circuits can, however, be contained in multi-core cables, conductors, or conductor bundles together with conductors of other circuits provided that they are sufficiently insulated to withstand the maximum permissible operating voltage. Conductors may, however, still have to be installed separately for other reasons associated with electromagnetic compatibility (EMC) (see Section 3).

▷ protection against direct contact with live parts must be provided for SELV > 25 V AC and 60 V DC,

Plug-and-socket devices

▷ plug-and-socket devices for SELV must not fit the plug-and-socket devices which are used for circuits with other protective measures. Similarly, it must not be possible to use plug-and-socket devices intended for SELV 25 V AC in plug-and-socket devices for SELV 50 V AC,

▷ SELV and PELV plug-and-socket devices cannot be interchanged,

▷ plugs and sockets for SELV circuits do not have ground contacts.

2.2.1.2 Protection by PELV

Differences to SELV

PELV is subject to the same requirements as SELV, with the following differences (Fig. 2.2/3):

▷ in PELV circuits, a live part and/or the exposed conductive parts can be connected to an equipment grounding conductor/ground,

▷ plugs and sockets can have a ground contact,

▷ protection against direct contact is only nonessential ≤ 6 V AC (r.m.s. value) or 15 V DC (harmonic free), except for electrical equipment ≤ 25 V AC or 60 V DC which is usually only used in dry rooms or locations and where full contact with the surface of live parts is not to be expected.

2.2.2 Protection by functional extra-low voltage (FELV) without safety separation

Standards

FELV is a voltage ≤ 50 V AC and ≤ 120 V DC; it is described in Section 411 of IEC 60364-4-41 and is, consequently, regarded as a protective measure. According to HD 384 and DIN VDE 0100, FELV is a special form of the protective measure for the primary system and is, therefore, also described in HD 384.4.47 / DIN VDE 0100-470.

Electrical isolation

With FELV, a transformer must have electrical isolation (basic isolation), i.e. basic isolation must be used to separate an FELV circuit from a circuit with a higher voltage. As a result, the FELV circuit including all electrical equipment for the primary voltage must be insulated.

Owing to the electrical isolation, disconnection in the FELV circuit to provide protection against electric shock is not required until the 2nd fault (3rd fault in the case of primary IT systems).

Requirements

Protection against direct contact must be provided by:

▷ barriers or enclosures such as those specified in Section 412 of IEC 60364-4-41,

▷ insulation which is suitable for the test voltage on the primary side.

Protection in the event of indirect contact must be provided by:

▷ connecting the exposed conductive parts in the FELV circuit to the equipment grounding conductor on the primary side,

▷ disconnection in the event of a fault via the primary protective device.

It must not be possible to interchange the plugs and sockets with the plugs and sockets of other circuits.

Live parts of the FELV circuit can be connected to the equipment grounding conductor/ground. This connection is not necessary for protection in the event of indirect contact.

General information concerning SELV, PELV, FELV

Interconnection of circuits for safety and functional extra-low voltage

SELV, PELV, and FELV circuits must not be interconnected. SELV circuits may, however, be connected to SELV circuits, PELV circuits to PELV circuits and FELV circuits to FELV circuits. It is important to ensure that the voltages do not accumulate as a result (Fig. 2.2/5).

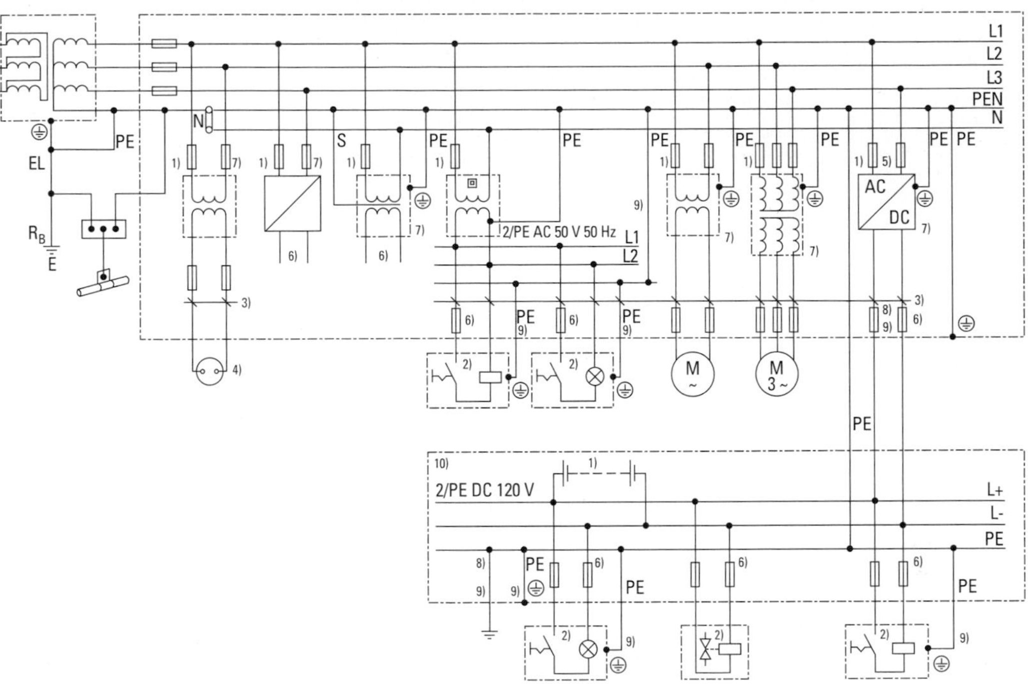

Requirements and explanations

1) Power sources: safety isolating transformers to EN 60742-1/IEC 60742-1 / DIN VDE 0551-1, or transformers to DIN VDE 0804, or motor-generator sets to IEC 2/915 CDV / VDE 0530 Part 1, or storage batteries to DIN VDE 0510, or devices with an equivalent level of safety, or electronic devices in which voltages higher than 50 V AC / 120 V DC cannot occur after the 1st fault.

2) Electrical equipment must have the same safety separation (electrical isolation) as the power sources if circuits with a higher voltage are also connected.

3) Cables must be installed separately from circuits with a higher voltage or must be provided with insulation for the highest occurring operating voltage.

4) Plug-and-socket devices for PELV must not be used in circuits in which other protective measures are implemented. It should only be possible to connect plugs to circuits with the same or a lower voltage.

5) Charging rectifiers must be equipped with a safety isolating transformer.

6) 2-pole protection is required if the external conductor is not grounded. Control circuits, which are supplied via transformers, require single-pole fuse protection only.

7) Electrical equipment is usually safeguarded by its mounting attachments, i.e. equipment grounding conductors do not have to be connected.

8) Connect the equipment grounding conductor bar to the equipment grounding conductor of the TN system. Alternatively, ground directly on site (see also 9)), the equipment grounding conductor is then referred to as a functional grounding conductor.

9) These connections are not equipment grounding conductors in the usual sense. They are only required for disconnection in the case of double faults.

10) Notation for equipment grounding conductors incorporated in the cable: 2/PE AC 50 Hz 50 V and 2/PE DC 120 V.

The following supplementary protective measures against direct contact must be used for voltages > 6 V AC and > 15 V DC, or > 25 V AC and > 60 V DC in dry locations:

- protection by insulating live parts (the insulation must withstand a test voltage of 500 V AC for 1 minute)

  or

- protection by concealing or enclosing live parts to degree of protection ≥ IP 2X

Fig. 2.2/3 Protection by PELV with safety separation

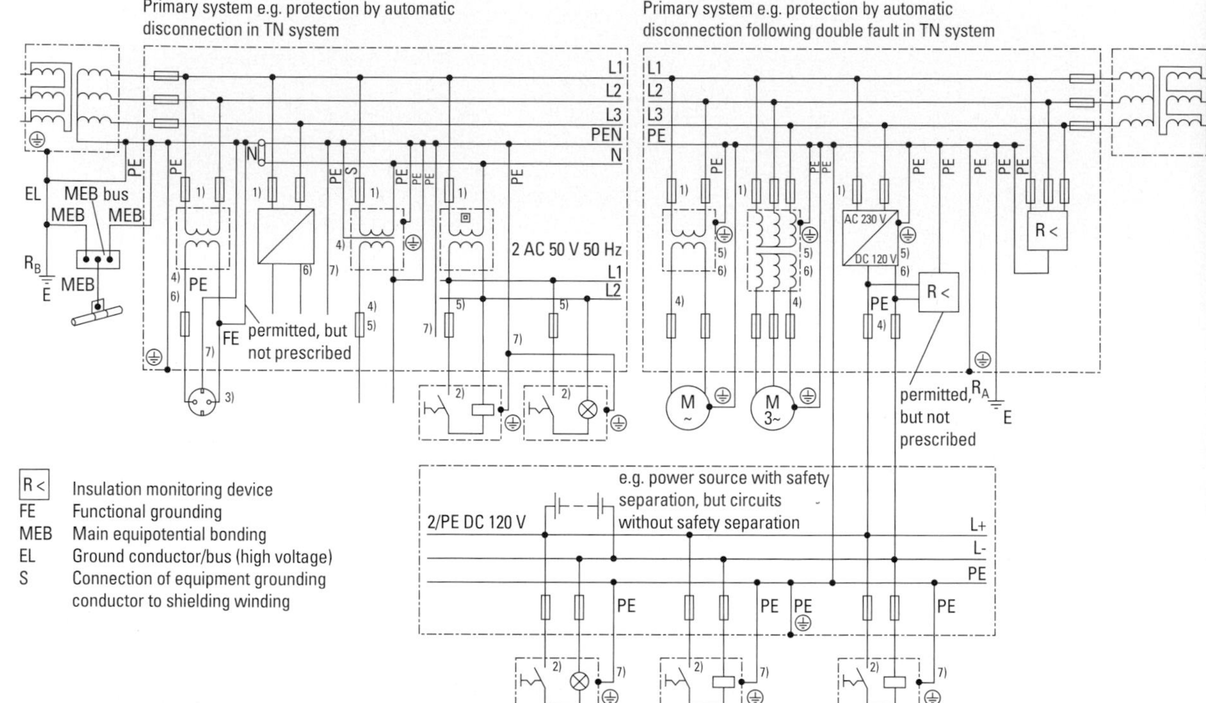

Requirements and explanations

Disconnection in the event of a fault is only required at voltages of 50 V AC / 120 V DC, i.e. after 2 faults, e.g. short circuit to exposed conductive part and overspill of the primary voltage to the FELV circuit.
Disconnection by means of primary fuses, for example, is sufficient.

1) Power sources: control-power transformer to IEC 989 / DIN VDE 0550-3 or better, or devices with an equivalent level of safety.
2) Electrical equipment does not have to have safety separation (electrical isolation).
3) It must not be possible to use plug-and-socket devices for FELV in circuits in which other protective measures are implemented. It should only be possible to connect plugs to circuits with the same or a lower voltage (no grounding-type plug-and-socket devices; Norvo plug-and-socket devices, for example, must be used instead).
4) Fuses on the secondary side are not part of the protective measure "protection by disconnection". They are required for short-circuit protection and for disconnecting double faults with hazardous movements[1)].
5) Control circuits, which are supplied via transformers, require single-pole protection only. Protection of non-grounded conductors is sufficient in grounded circuits. All-pole protection is required in all other cases, e.g. if power is supplied from storage batteries.
6) Electrical equipment is usually safeguarded by its mounting attachments, i.e. equipment grounding conductors do not have to be connected.
7) With FELV, the equipment grounding conductor on the primary side must be incorporated in the cable and connected to the exposed conductive parts For example, the equipment grounding conductor from the power system of the charging rectifier must be used in the case of storage batteries.

The following must be implemented as a protective measure against direct contact:
- protection by insulating live parts (the insulation must withstand a test voltage of 500 V AC for at least 1 minute)
  or
- protection by concealing or enclosing live parts to degree of protection ≥IP 2X

1) Hazardous movements as defined in the relevant German accident prevention regulation (VBG 5[1)]) are movements made by
- parts of the powered equipment,
- tools of the equipment or parts thereof,
- workpieces or parts thereof,

or
- other components

where the moving parts constitute hazards or potential hazards.

Fig. 2.2/4 Protection by FELV without safety separation

1) VBG 5: German standards for "powered equipment" (**V**orschriftenwerk der **B**erufs**g**enossenschaft)

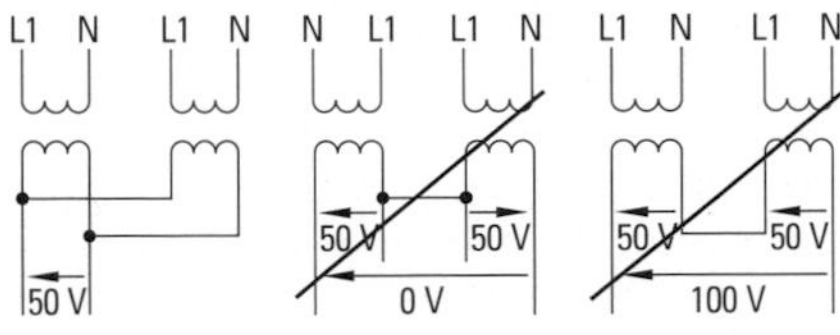

Fig. 2.2/5
Permitted and non-permitted parallel connection of SELV, PELV, and FELV circuits

2.2.3 Protection by limiting steady-state touch current and charge

Steady-state touch current and charge

The measure "protection by limiting the steady-state touch current and charge" is not yet adequately described in HD 384.4.41 / IEC 60364-4-41 / DIN VDE 0100-410. At present, protection against direct contact is not obligatory if the discharge energy does not exceed 350 mJ (see also VBG 4[1)], § 8 No. 1).

2.3 Protection against electric shock under normal conditions (protection against direct contact or basic protection)

Protection against direct contact is intended to prevent contact with parts which are energized (live) under normal operating conditions.

A distinction is made between the following:

▷ full protection against direct contact,
▷ partial protection against direct contact.

Application

According to HD 384.4.41 / IEC 60364-4-41 / DIN VDE 0100-410, Section 412, there are no restrictions for partial protection against direct contact.

In Germany, partial protection is subject to the restriction specified in the standard National Appendix NB.2.

In addition to these measures,

▷ supplementary protection by means of RCDs (see Section 2.3.3),

represents an additional protective measure which cannot be used on its own.

2.3.1 Full protection against direct contact

Full protection against direct contact can be achieved by:

Basic insulation

▷ insulating live parts, whereby the live parts must be fully covered with insulation (basic insulation). The insulation must be attached in such a way that it can only be removed by destroying it,

Barriers/ enclosures

▷ barriers or enclosures, whereby the live parts behind the barriers or enclosures must be arranged in such a way that at least degree of protection IP 2X (openings < 12.5 mm) to EN 60529 / IEC 529 / DIN VDE 0470-1 can be achieved. It must only be possible to remove the barriers and enclosures using a tool or key.

Degree of protection

Note
In accordance with EN 60529 / IEC 529 / DIN VDE 0470-1, protection against direct contact can also be indicated by the additional 3rd character in the IP code.

Full protection against direct contact can then be expressed by the current degree of protection IP 2X or by the new designation IP XXB. To permit this, the alternative degrees of protection must be specified in the respective standards for the equipment (constructional requirements). The alternative degree of protection is included in HD 384 / IEC 60364 / DIN VDE 0100.

The letter at the end of the IP code means:

A safe from touch by the back of the hand,
B safe from finger-touch,
C straight conductors, diameter 2.5 mm,
D straight conductors, diameter 1 mm.

Protection by insulation

Operational/ functional insulation

Protection by insulation is achieved by fully covering live parts with basic insulation (Fig. 2.3/1).

[1)] VBG 4: German standards for "electrical installations and equipment" (Vorschriftenwerk der Berufsgenossenschaft)

The basic insulation is more than the operational/functional insulation which is required to allow the electrical equipment to function correctly (Fig. 2.3/2).

For example, construction components or exposed conductive parts (barrier or enclosure).

Protection by barriers or enclosures

Barriers or enclosures can be made of insulating material or metal. Barriers/enclosures made of insulating material must not/cannot satisfy the requirements for protection by insulation. They must, however, be sufficiently robust so that they do not rupture under normal stress. Metal barriers or enclosures must not sag to such an extent that they come into contact with uninsulated live parts.

Test finger

With degree of protection IP 2X, openings in barriers and enclosures, which satisfy the specifications for full protection against direct contact, must not be larger than 12.5 mm. With degree of protection IP XXB, larger openings are permitted provided that contact with live parts when a straight test finger is inserted into the opening is prevented, e.g. by means of a prod guard. In accordance with EN 60529 / IEC 60529 / DIN VDE 0471-1, it is theoretically possible to insert the test finger up to a depth of 80 mm into openings measuring 12.5 mm. Contact with live parts must, however, be prevented (Fig. 2.3/3).

Exceptions

Larger, temporary openings, which can arise when certain parts are replaced (e.g. bulbs, screw-in fuse-links), do not contradict the requirements for full protection against direct contact.

Fig. 2.3/1
Protection by insulation (basic insulation)

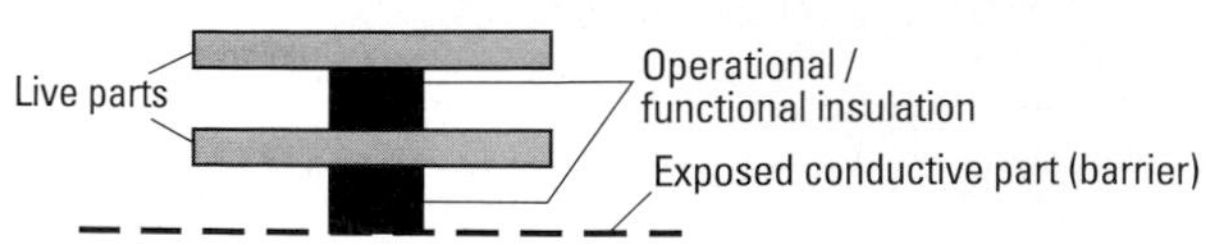

Fig. 2.3/2 Operational/functional insulation

Exceptions also apply to items of equipment which, according to their constructional requirements, require larger openings to permit correct operation. The necessary degree of protection IP 2X or IP XXB (see page 307) applies for all surfaces (this usually also includes the floor), but not for horizontal surfaces which are easily accessible. A degree of protection of at least IP 4X in accordance with EN 60529 / IEC 60529 / DIN VDE 0471-1 or IP XXD must be provided here.

Note
IP 3X and IP XXC provide adequate protection against direct contact with horizontal surfaces for equipment and equipment assemblies manufactured in accordance with applicable constructional requirements (e.g. to EN 60439-1 / IEC

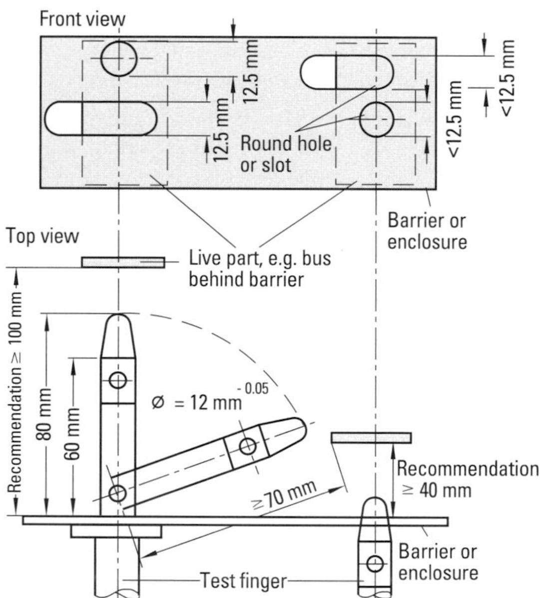

Insertion depth of test finger into openings and minimum clearances of live parts behind openings:

Round hole	12 to 12.5 mm:	80.0 mm
or	11.0 mm:	16.5 mm
slot	10.0 mm:	13.0 mm
	9.0 mm:	8.0 mm
	8.0 mm:	4.0 mm

Fig. 2.3/3
Minimum requirements for the layout of uninsulated live parts behind barriers or enclosures (observe recommendations)

60 439-1 / DIN VDE 0660-500) until this is regulated by an international standard. The higher degree of protection for horizontal surfaces is necessary to prevent objects such as necklaces from entering the openings.

ttaching and emoving barriers nd enclosures

Barriers and enclosures must be attached securely. It must only be possible to remove or open these:

- ▷ with a key or tool,
- ▷ after the power supply to the live parts, for which the barrier or enclosure is to provide protection against direct contact, has been switched off (disconnected). It should not be possible to switch the voltage back on again until the barriers or enclosures have been refitted in their original positions. An interlock with the disconnecting device is required. It must, however, be possible to bypass the interlock when service work is carried out.
  In a number of countries, e.g. France, the neutral conductor must also be disconnected; this does not apply to SELV, PELV, and FELV circuits with nominal voltages ≤ 25 V AC and 60 V DC,
- ▷ if an intermediate cover with degree of protection IP 2X or IP XXB prevents contact with live parts and the intermediate cover can only be removed using a key or tool.

If equipment, which is to restore a specific function (e.g. miniature circuit-breaker or overload relay), is installed behind barriers or enclosures or behind intermediate covers, which can only be removed using a key or tool, DIN VDE 0106 Part 100[1)] must also be observed.

2.3.2 Partial protection against direct contact

Full protection against direct contact is not always necessary. According to HD 384.4.41 / IEC 60 364-4-41 / DIN VDE 0100-410, partial protection against direct contact is also permitted. Unfortunately, the fact that partial protection against direct contact may only be applied in special cases is no longer immediately apparent (see also Section 2.3).

Partial protection can be achieved:

- ▷ by means of obstacles,
- ▷ by means of suitable clearances.

Restriction to electrical operating areas

According to DIN VDE 0100-731 (see Section 5.9.11), these measures are only permitted in electrical operating areas (HD/IEC stipulations in preparation).

Protection by means of obstacles

Since obstacles provide only partial protection against direct contact (Fig. 2.3/4), intentional contact, e.g. by deliberately bypassing obstacles, does not need to be prevented.

Obstacles must be provided in order to prevent:

- ▷ accidental contact with live (hazardous) parts, e.g. if wooden barriers, rails, or grilles are used,

or

- ▷ accidental contact with live parts when the equipment is being used under normal operating conditions, e.g. by implementing measures as specified in DIN VDE 0106-100.

The obstacles must be attached in such a way that they cannot be inadvertently removed. It should, however, be possible to remove the obstacles without using a key or tool.

Protection by means of clearances

Normal arm's reach

Protection by means of suitable clearances is achieved by ensuring that no parts that can

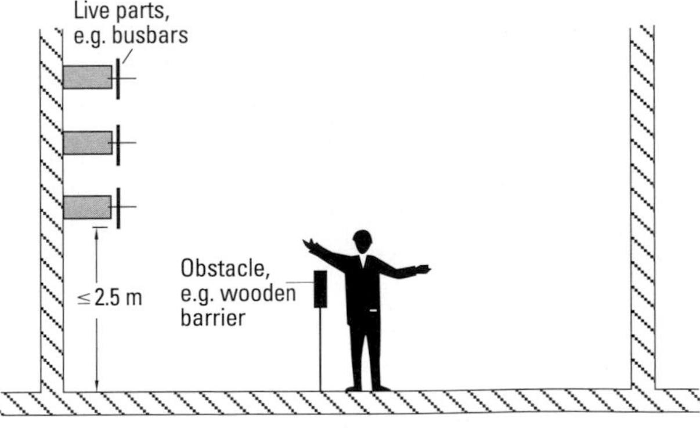

Fig. 2.3/4 Protection by means of obstacles

[1)] European and international standards currently in preparation

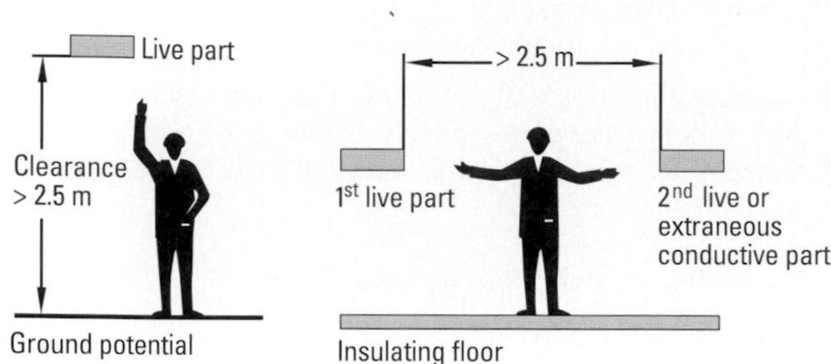

Fig. 2.3/5 Protection by means of clearances

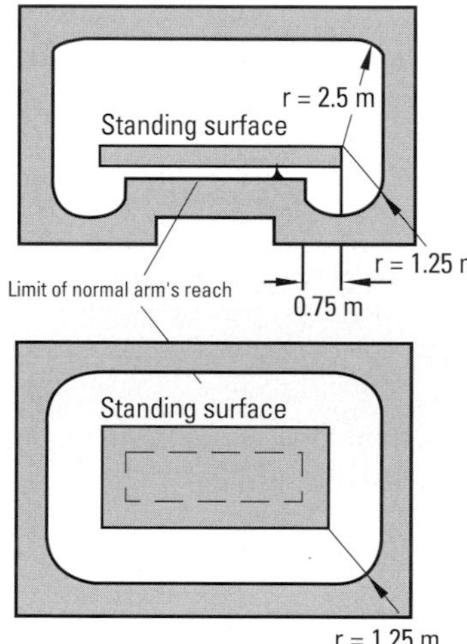

Fig. 2.3/6 Normal arm's reach

be accessed simultaneously (Fig. 2.3/5) with different electric potential are located within normal arm's reach (Fig. 2.3/6). Live parts which are located at a distance of ≤2.5 m from each other are defined as being simultaneously accessible. Clearances must be increased accordingly at locations where bulky or long, conductive objects are handled. If, however, the standing surface is limited due to an obstacle, e.g. meshed grille, normal arm's reach begins and ends at this obstacle. Normal arm's reach ends at a height of 2.5 m irrespective of any obstacles.

2.3.3 Supplementary protection by RCDs

Supplementary protection by means of RCDs is also referred to as "protection against direct contact". According to Section 412.5 of HD 384.4.41 / IEC 60364-4-41 / DIN VDE 0100-410, the requirements for this type of protection can, however, only be satisfied if residual-current protective devices with a rated residual current ($I_{\Delta n}$) of ≤30 mA are used (see Sections 4.1.3 and 5.9). Other alternative measures could, however, also ensure protection against direct contact, e.g. SELV (safety extra-low voltage) (see Section 2.2.1).

Third protection level

In practice, supplementary protection by means of RCDs has proved to be an effective protective measure against faults caused by defective basic insulation or faults in totally-insulated equipment. It is, therefore, also referred to as the third protection level. This description is, however, inappropriate since the 3rd protection level is seldom implemented. In the majority of cases, a residual-current protective device (RCD) is used both for protection in the event of indirect contact and for supplementary protection.

2.4 Protection against electric shock under fault conditions (previously Protection in the event of indirect contact or Fault protection)

:andards

Protection against electric shock under fault conditions is described in HD 384.4.41 / IEC 60364-4-41 / DIN VDE 0100-410 Protection against electric shock. According to HD 384.4.47 / IEC 60364-4-47 / DIN VDE 0100-470 "Application of protective measures", all electrical equipment must be provided with protection in the event of indirect contact or protected by means of measures, such as those described in Sections 411 ("Protection by SELV and PELV") and 413 ("Protection in the event of indirect contact") of HD 384 / IEC 60364 / DIN VDE 0100. Exceptions to this are parts for which, according to HD 384.4.47 / IEC 60364-4-47 / DIN VDE 0100-471, this protection is not necessary (see also Section 2.1).

rotection by utomatic isconnection of ower supply

The fault protective measure "protection by automatic disconnection of the power supply" must always be applied. The following can also be used for certain parts of a system if protection by automatic disconnection of the power supply is impracticable or undesirable (because the power supply is indispensable for these parts):

- protection by means of non-conducting locations (see Section 2.4.9)

or

- protection by means of ungrounded, local equipotential bonding (see Section 2.4.10)

Furthermore, the following can also be used in any electrical installation for certain equipment or for parts of the electrical installation:

- protection by means of extra-low voltage: SELV and PELV (see Section 2.2.1)

or

- class 2 equipment or
- equipment with safety separation (see Section 2.4.11).

2.4.1 Protection by automatic disconnection of power supply

leasures for utomatic isconnection of ower supply

Protection by automatic disconnection of the power supply must be coordinated with respect to the type of ground connection (Systems according to type of ground connection, see Section 2.4.2) and the properties of the equipment grounding conductors and protective devices.

The disconnecting times specified in HD 384.4.41 / IEC 60364–4-41 / DIN VDE 0100–410 are based on IEC 60479 / DIN VDE 0140-479.

The power supply must be disconnected automatically by means of:

▷ RCDs (residual-current protective devices, see also the following explanations).
If RCDs are connected in series, "S"-type time-delay RCDs can also be used to provide the required selectivity.
The type of equipment connected is all the more important if RCDs are used.
Particularly in the case of motors supplied via frequency converters (the most commonly encountered type of motor nowadays), has it become necessary to use "universal-current-sensitive" RCDs instead of the usual "pulse-current-sensitive" RCDs.

or by means of:

▷ overcurrent protective devices (fuses, circuit-breakers, miniature circuit-breakers).

The power supply of the circuit or equipment to be protected must be disconnected as quickly as possible in order to minimize the risk of physical injury to personnel.

Explanations relating to RCDs

RCD (Residual Current protective Device) is a generic term for:

- devices **with** an auxiliary power source, also referred to as differential-current protective devices

and for:

- devices **without** an auxiliary power source, also referred to as fault-current protective devices.

No European standards exist for devices with an auxiliary power source. These devices are, therefore, not permitted for providing protection in the event of indirect contact or for supplementary protection against direct contact (provided that such protection is required).

Under certain circumstances, a disconnecting time of 5 seconds is permitted depending on the type of ground connection. The disconnecting time may be exceeded if the prospective touch voltage (U_B) is not higher than the conventional permissible touch voltage (U_L) (e.g. by using an additional (local) equipotential bonding conductor (see Section 2.4.7)) or if disconnection is not necessary on account of the protection against electric shock.

Furthermore, it is also necessary to connect the exposed conductive parts of the electrical equipment to equipment grounding conductors under the conditions specified for each system according to the type of ground connection. Exposed conductive parts, which are accessible simultaneously, must be connected to the same grounding system (equipment grounding conductor system with common grounding).

Cross sections of equipment grounding conductors

The cross sections for all equipment grounding conductors (irrespective of the system's grounding connection) must either be selected:

- according to Table 2.4/1 (identical to Table 54F in HD 384.5.54/IEC 60364-5-54 / DIN VDE 0100-540) – Earthing arrangements, protective conductors, equipotential bonding conductors

or

- calculated using the following formula:

$$S \leq \frac{\sqrt{I^2 \cdot t}}{k}$$

where:

I Maximum possible fault current in A (at least the current which causes the protective device to disconnect within the required time, max. 5 s),

t Operating time of protective device in s with max. possible fault current,

k Material coefficient, dependent on conductor and insulating material (usually PVC-insulated copper conductors, k = 115),

S Cross section of equipment grounding conductor in mm.

If the formula or Table 2.4/1 yields a non-standard cross section, the next largest standard cross section must be selected. The use of cables/conductors 150/70 mm^2 is, however, generally accepted.

Determining the cross sections of the equipment grounding conductor using these two methods does not mean that the conditions for protection by automatic disconnection of the power supply do not have to be checked. With PEN conductors, the current flowing through the neutral conductor must also be taken into consideration when dimensioning the cross sections.

Minimum cross section of equipment grounding conductors

Irrespective of the cross section of the external conductors, the minimum cross section of the equipment grounding conductors outside an enclosure (e.g. cable, conductor, conduit) shared with the external conductors is:

- 2.5 mm^2 for mechanically protected installation,
- 4 mm^2 for mechanically unprotected installation.

Table 2.4/1
Cross sections of equipment grounding conductor in relation to cross section of external conductor

Cross section S of associated external conductor in mm^2	Minimum cross section S_p of equipment grounding conductor in mm^2
$S \leq 16$	S (i.e. same as external conductor)
$16 < S \leq 35$	16
$S > 35$	$S/2$ (i.e. ½ external conductor)

2.4.2 Systems according to type of ground connection

Systems according to type of ground connection

According to HD 384.3 / IEC 60364-3 / DIN VDE 0100-300 "Assessment of general characteristics of installations", the following systems exist (classified according to the type of ground connection):

▷ IT system (Fig. 2.4/1),
▷ TT system (Fig. 2.4/2),
▷ TN system (Fig. 2.4/3) with its three variants:
- TN-C system,
- TN-C-S system,
- TN-S system.

- The first letter (I or T) refers to the grounding characteristics of the power source (i.e. primarily to the grounding of

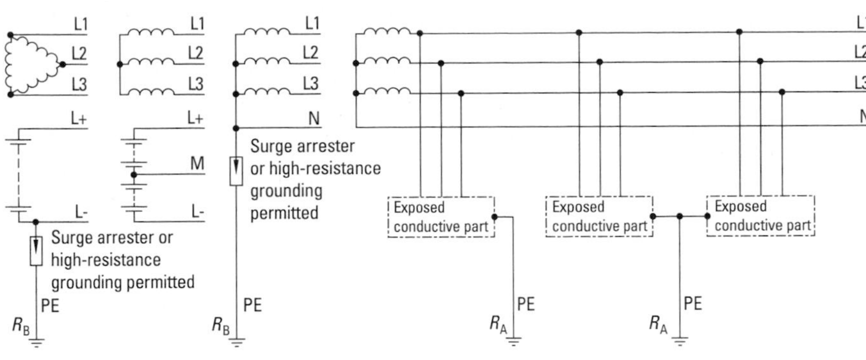

R_A Contact resistance of frame ground
R_B Operational ground resistance

If neutral conductors are used: Observe the dielectric strength of the equipment used! Overcurrent detection is required in the neutral conductor. Overcurrent detection must disconnect all of the external conductors including the neutral conductor. The neutral conductor must not be disconnected before the external conductors or connected after the external conductors.

Fig. 2.4/1 IT system
I Isolation, point in system not directly grounded
T Terre (ground), exposed conductive parts not directly grounded (supplementary equipotential bonding may be necessary)

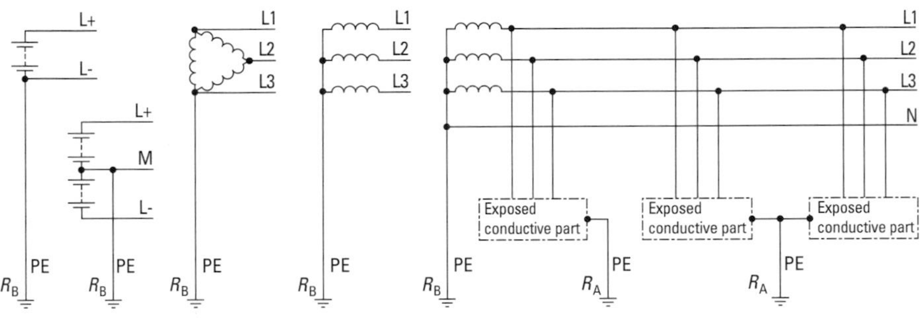

R_A Contact resistance of frame ground, dependent on operating current of protective device
R_B Operational ground resistance, no fixed value, recommendation for $R_B \leq 2\,\Omega$

If neutral conductors are used: Overcurrent detection is also required in the neutral conductor in the case of "protection by disconnection" with overcurrent protective devices. The neutral conductor must not be disconnected or connected before the external conductors.

Fig. 2.4/2 TT system
T Terre (ground), direct grounding of point in system required
T Terre (ground), exposed conductive parts directly grounded (supplementary equipotential bonding may be necessary)

the public utility power system or of an industrial system).

- The second letter (T or N) refers to the grounding characteristics of the exposed conductive parts.
- The supplementary letters in the TN system (C, C-S, S) refer to the combination or separation of the equipment grounding conductor and neutral conductor (see Fig. 2.4/3).

2.4.3 Main equipotential bonding

An important requirement for applying protection by automatic disconnection of the power supply is that main equipotential bonding exists in every building, irrespective of the type of power supply system (except in buildings in which only SELV, PELV, or safety separation is used) (see Fig. 2.4/4 to 2.4/6).

The following must be interconnected:

Which components have to be included?

▷ main equipment grounding conductor (equipment grounding conductor along the supply cable of the TN system or equipment grounding conductor of a TT or IT system leading off from a grounding electrode (e.g. foundation grounding electrode))

▷ main grounding conductor (equivalent to the above-mentioned main equipment grounding conductor),

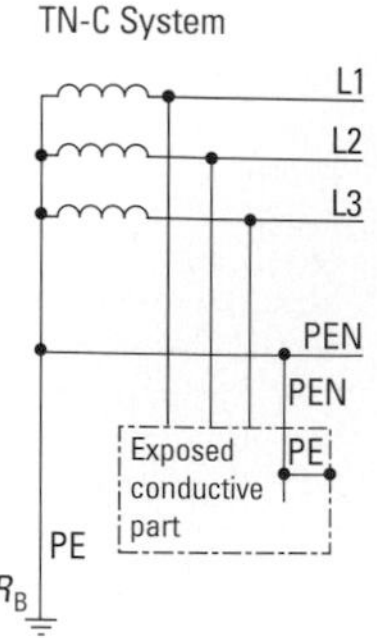

TN-C System

C: Neutral and equipment grounding conductors installed as PEN throughout.

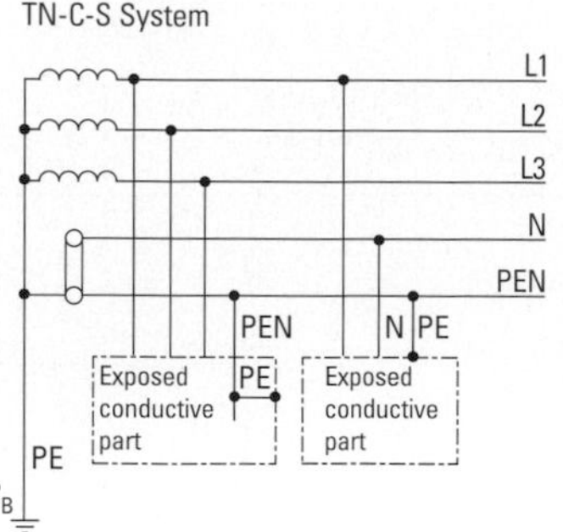

TN-C-S System

C-S: Neutral and equipment grounding conductors partly installed together as PEN, partly installed separately.

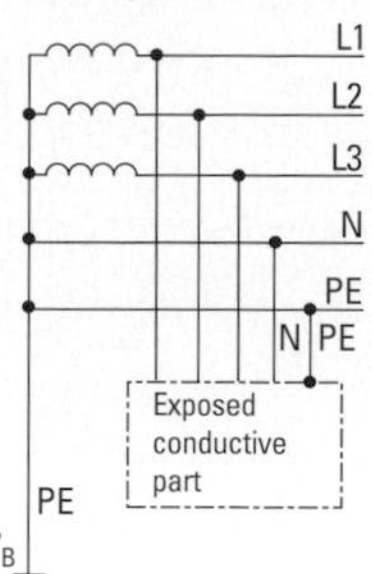

TN-S System

S: Neutral and equipment grounding conductors installed separately throughout.

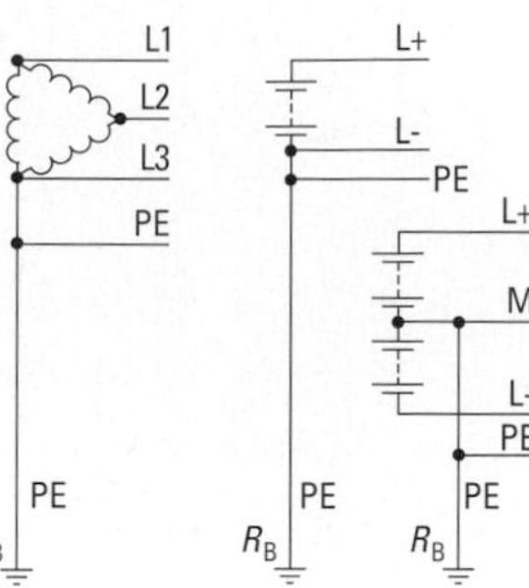

Only the TN-S system is possible with a grounded external conductor.
A grounded external conductor must not be used as a PEN conductor, i.e. separate external and equipment grounding conductors are required throughout.

R_B Operational ground resistance, no fixed value, recommendation for $R_B \leq 2\ \Omega$

Fig. 2.4/3 TN system
T Terre (ground), direct grounding of point in system required
N Neutral, exposed conductive parts connected to grounded point in system via equipment grounding conductor (PE) and/or PEN conductor (supplementary equipotential bonding may be necessary)

▷ main ground terminal or bus (equipotential bonding strip),
▷ extraneous conductive parts such as:
- metal piping of supply systems, e.g. for gas and water,
- metal parts of building construction, central heating, and air conditioning,
- main metal reinforcement elements of building construction, e.g. armored concrete (if the armor is accessible).

Connection to main ground terminal or bus

If the above-mentioned parts originate from outside the building, they must be connected to each other and to the main ground terminal or bus as closely as possible to the point at which they enter the building.

Metal enclosures of telecommunications cables are an exception here. They should only be included in the main equipotential bonding if the owner/operator has given his consent, otherwise responsibility for preventing all potential hazards lies with the owner/operator.

What other components are necessary ?

In addition to those parts mentioned in the standards, the following components must also be included:

▷ lightning protection grounding electrodes,
▷ outdoor antenna installations,
▷ surge arresters.

Cross section of main equipotential bonding conductor

The cross sections of the main equipotential bonding conductors must be selected according to the cross section of the main equipment grounding conductor. The following points must be observed:

▷ the cross section must be at least half the cross section of the equipment grounding conductor (the Cu conductor must have a minimum cross section of 6 mm^2),
▷ restriction to 25 mm^2 Cu or conductors with identical conductivity are permitted.

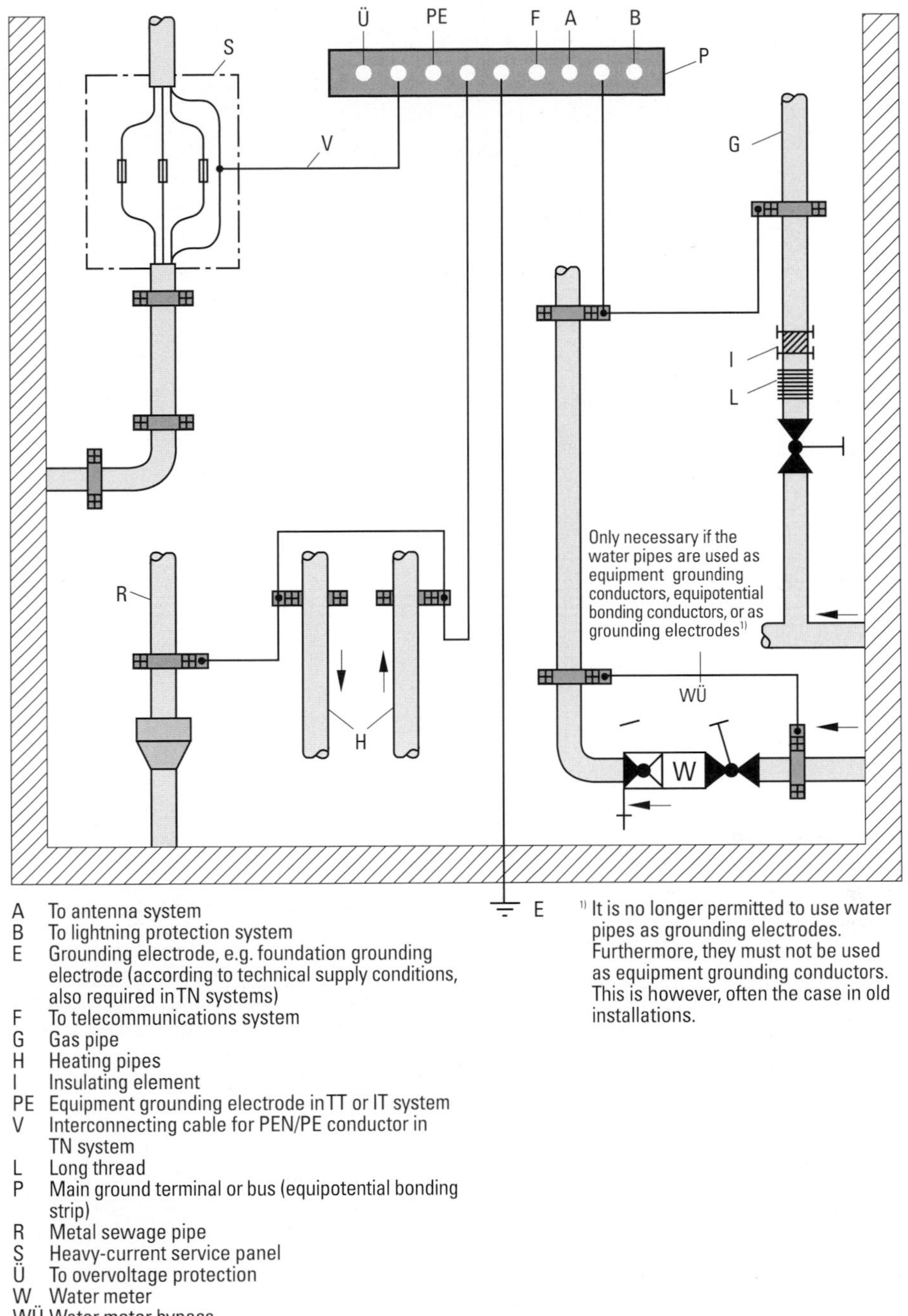
Ü
PE
F
A
B
P
S
V
G
I
L
R
H
W
WÜ
E
Only necessary if the water pipes are used as equipment grounding conductors, equipotential bonding conductors, or as grounding electrodes[1)]
A To antenna system
B To lightning protection system
E Grounding electrode, e.g. foundation grounding electrode (according to technical supply conditions, also required in TN systems)
F To telecommunications system
G Gas pipe
H Heating pipes
I Insulating element
PE Equipment grounding electrode in TT or IT system
V Interconnecting cable for PEN/PE conductor in TN system
L Long thread
P Main ground terminal or bus (equipotential bonding strip)
R Metal sewage pipe
S Heavy-current service panel
Ü To overvoltage protection
W Water meter
WÜ Water meter bypass
1) It is no longer permitted to use water pipes as grounding electrodes. Furthermore, they must not be used as equipment grounding conductors. This is however, often the case in old installations.

Fig. 2.4/4 Example of layout for main equipotential bonding

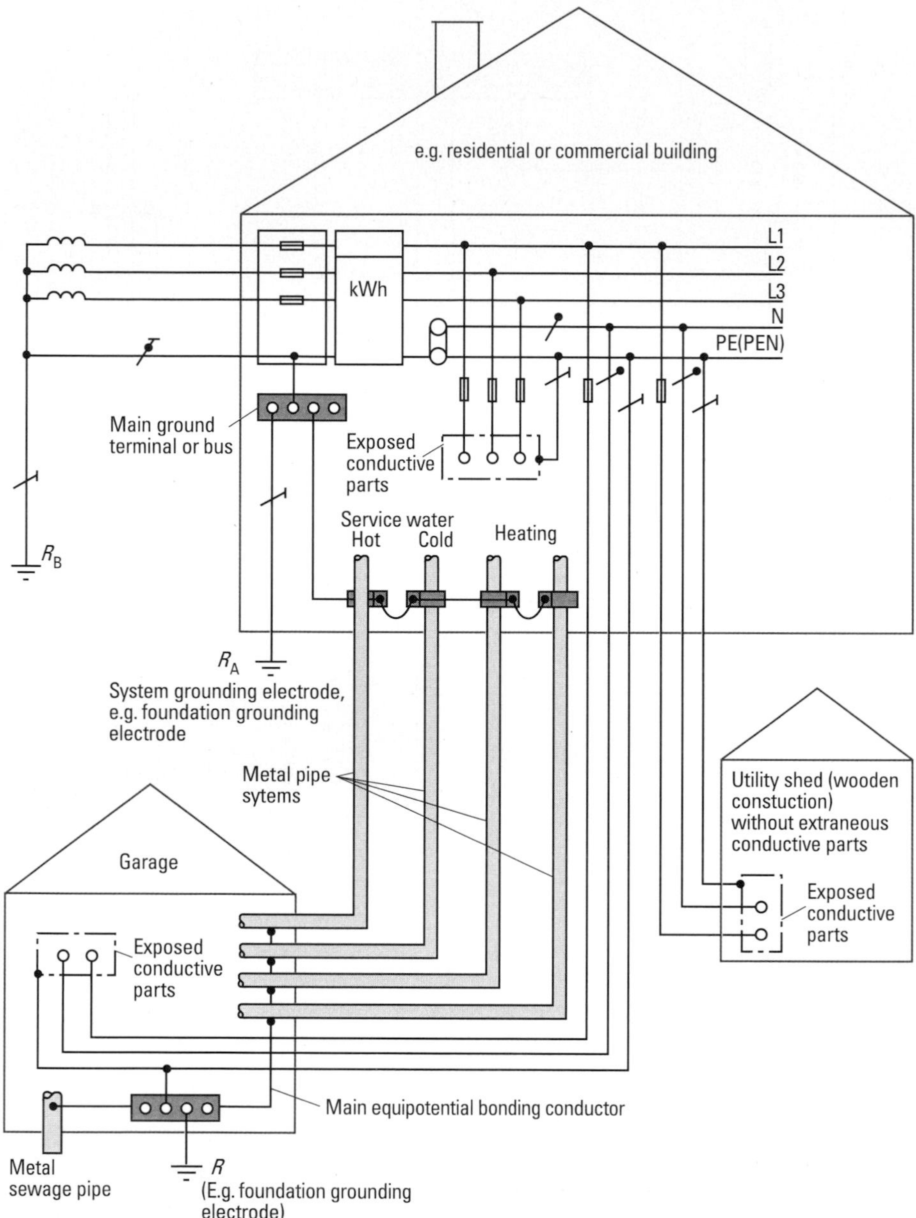

Fig. 2.4/5 Main equipotential bonding in TN system (each building with an electrical installation)

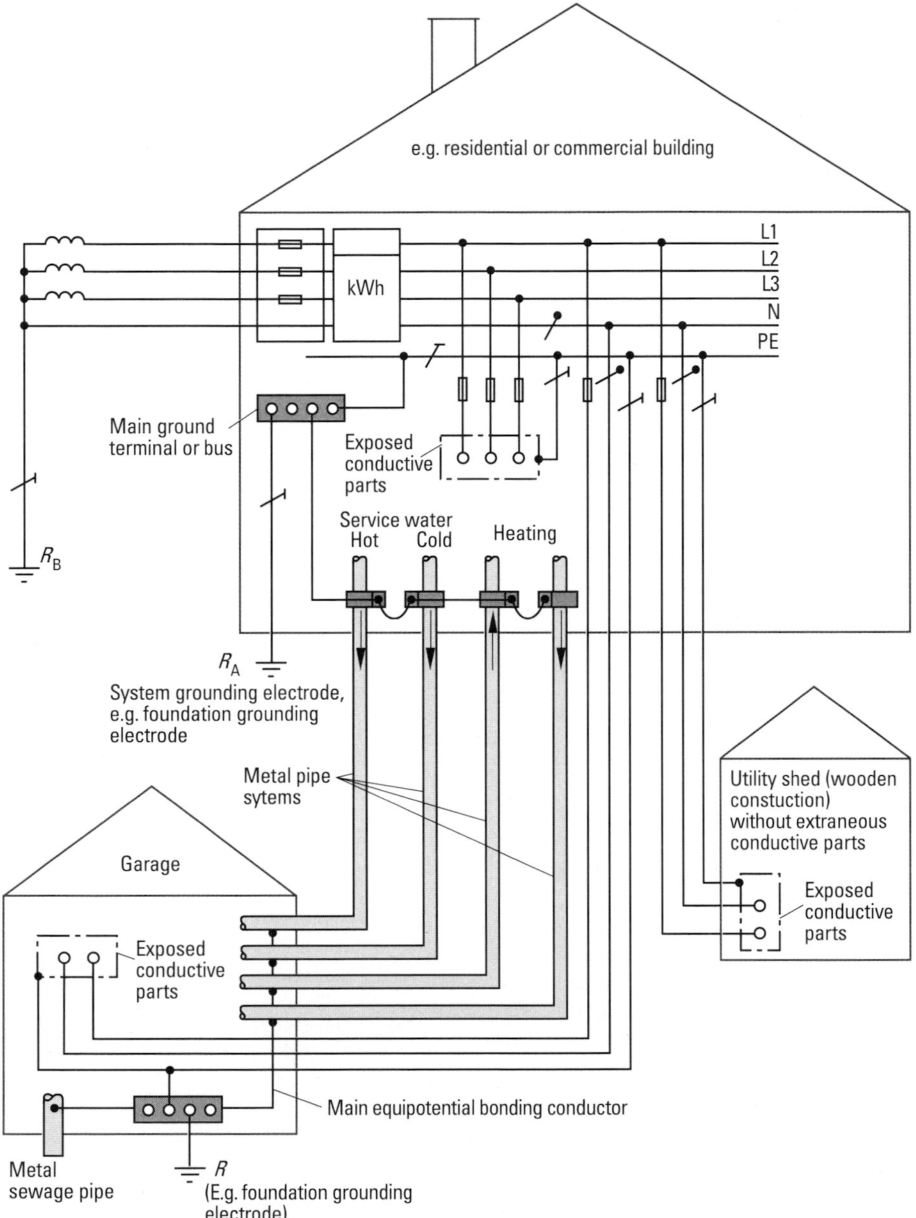

Fig. 2.4/6
Main equipotential bonding in TT system (each building with an electrical installation). Also applicable for the IT system.

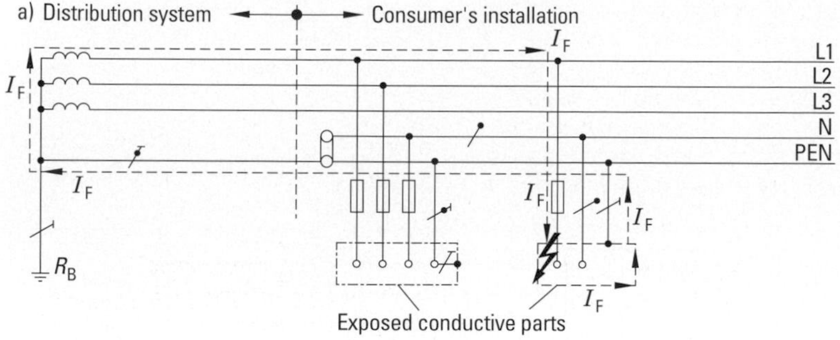

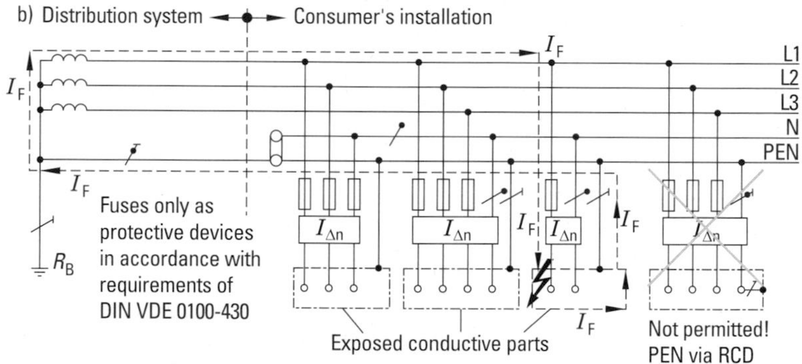

Fig. 2.4/7
Protection by automatic disconnection of power supply in a TN-C-S system
a) with overcurrent protective devices as disconnecting devices
b) with RCDs as disconnecting devices (PEN must not be routed via the RCD)

2.4.4 TN system

Requirements for TN system

The requirements for the three TN system variants (TN-C, TN-C-S, TN-S system) are virtually identical. The same protective devices can also be used (RCDs and overcurrent protective devices). Exception: RCDs must not be used in the TN-C section (see also Fig. 2.4/7).

PEN conductors which have been split into equipment grounding conductors and neutral conductors are not subject to any restrictions.

The following requirements must be satisfied in the TN system:

Requirements for grounding (operational grounding R_B)

▷ All exposed conductive parts of the electrical installation must be connected to the grounded point (usually the neutral point or, if this is inaccessible, an external conductor) of the supply system. The grounding connection (operational grounding R_B) must be established at or near (i.e. ≤50 m) the associated transformer or generator. Requirements for grounding electrodes are specified in HD 384.5.54 / IEC 60364-5-54 / DIN VDE 0100-540.

No value is specified for the resistance of the operational grounding electrode R_B. The following condition must, however, be fulfilled if an abnormal fault occurs between an external conductor and ground (e.g. an overhead power line falls onto a crash barrier):

$$\frac{R_B}{R_E} \leq \frac{50\ \text{V}}{U_0 - 50\ \text{V}},$$

With 230/400 V, this results in a ratio of 1/3.6,
where:

R_B Total ground resistance of all parallel grounding electrodes (including those of the power supply system);

R_E Lowest ground contact resistance of extraneous conductive parts not connected to an equipment grounding conductor via which a fault between the external conductor and ground can occur;

U_0 Nominal a.c. voltage (r.m.s. value) with respect to ground in volts;

50 V Limit for permissible continuous touch voltage (with a.c. voltage). The specified 50 V can also be used for d.c. voltage; it is, however, advisable to use the permissible touch voltage U_L specified for the d.c. voltage.

Requirements for equipment grounding/ neutral conductors

▷ In permanently installed cable systems, the equipment grounding conductor, which connects the exposed conductive parts to the neutral point of the power source, can also be used as the neutral conductor. A combined neutral conductor and equipment grounding conductor, which is referred to as a PEN conductor, can only be used for Cu conductors with a cross section ≥ 10 mm² and Al conductors ≥ 16 mm². In the case of cables with concentric conductors, the cross section of the PEN conductor can be reduced to 4 mm² if the connecting points have two screw clamps per terminal.

▷ With cross sections < 10 mm² Cu or < 16 mm² Al, or if the requirements for a TN-S system have to be fulfilled, the equipment grounding conductor and neutral conductor must always be configured as separate conductors. Furthermore, equipment grounding conductors and neutral conductors should not be connected to each other again once they have been separated, and the neutral conductor should no longer be grounded.

Note
In buildings with large telecommunications installations, it is advisable not to use PEN conductors (TN-C system), where possible, in order to prevent electrical interference. Due to the possibility of electrical interference, a TN-S system is also required in certain areas of hospitals, irrespective of the cross section.

Multiple infeed in TN system

Caution
Problem-free installation of TN-S systems is only possible at locations where two or more infeeds are not required simultaneously (see Fig. 2.4/8). Problems may also be encountered with UPS installations.

Separation into equipment grounding and neutral conductors

▷ The equipment grounding and neutral conductor can be separated at terminals or busbars; each conductor must have its own connecting point. There may, however, only be one PEN bus to which any number and sequence of equipment grounding conductors, neutral conductors, and PEN conductors can be connected (see Fig. 2.4/9).

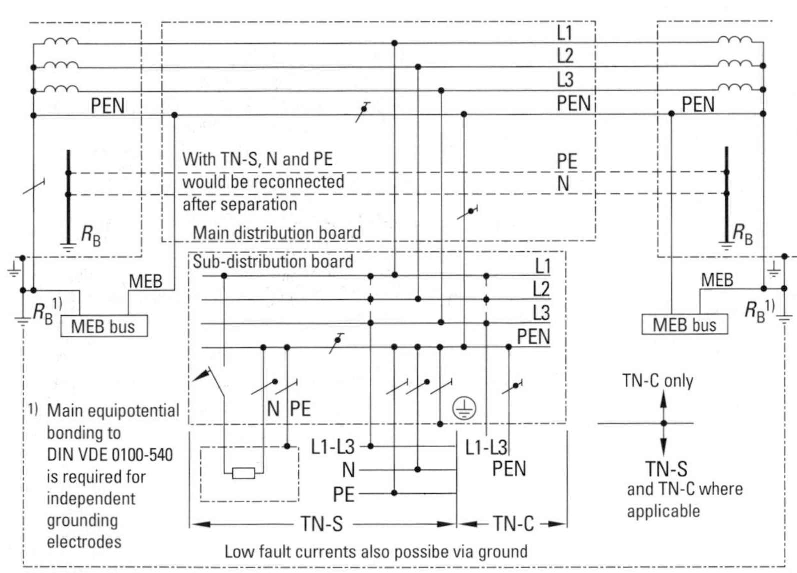

Fig. 2.4/8 Example of multiple infeed in TN system

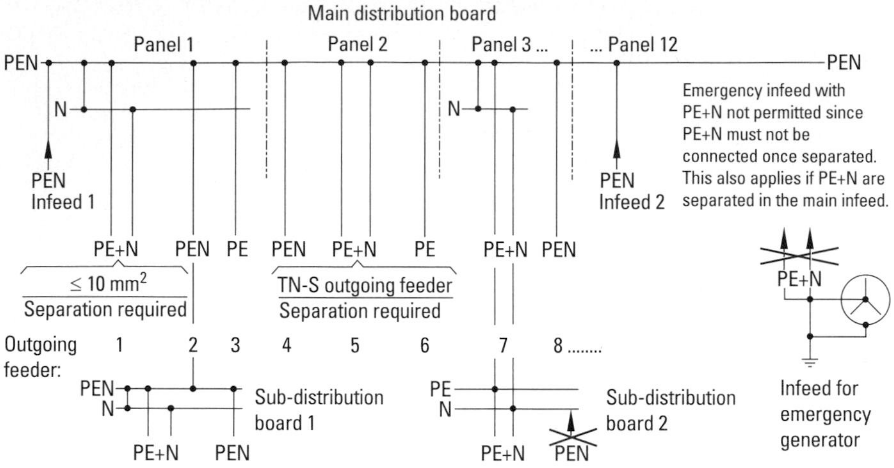

Main distribution board
Panel 1 | Panel 2 | Panel 3 ... | ... Panel 12
PEN — PEN
N
PEN Infeed 1
Emergency infeed with PE+N not permitted since PE+N must not be connected once separated. This also applies if PE+N are separated in the main infeed.
PEN Infeed 2
PE+N PEN PE PEN PE+N PE PE+N PEN
≤ 10 mm²
Separation required
TN-S outgoing feeder
Separation required
PE+N
Outgoing feeder: 1 2 3 4 5 6 7 8........
PEN N Sub-distribution board 1 PE+N PEN
PE N Sub-distribution board 2 PE+N PEN
Infeed for emergency generator

Fig. 2.4/9
4 or 5 conductors in TN system, necessary and advisable separation into N and PE

Isolated neutral conductor busbars or terminals are necessary and advisable if the insulation resistance of neutral conductors with respect to ground must be tested repeatedly, e.g. at locations exposed to fire hazards, because only one connection (tab between the PEN busbar and neutral conductor busbar) has to be released to allow the required measurements to be performed.

Disconnecting times

▷ The characteristic values of the protective devices and the loop impedance must be matched so that, if a fault with negligible impedance occurs at any point in the electrical installation, the power source will be disconnected automatically within the time specified in Table 2.4/2 (see below) which is identical to Table 41B in HD 384.4.41 / IEC 60364-4-41 / DIN VDE 0100-410.

This requirement is satisfied by the following conditions:

$$Z_S \cdot I_a \leq U_0 \quad \text{or} \quad Z_S \leq \frac{U_0}{I_a}$$

where:

Z_S Impedance of ground-fault loop, consisting of the external conductor and equipment grounding conductor of the circuit between the power source and fault location;

Note
Impedance encompasses ohmic and inductive resistance values which can be measured using modern instruments; calculations often only take the ohmic resistance into consideration. This error should, therefore, be allowed for when the result is evaluated.

I_a Current which causes automatic disconnection of the circuit within the time specified in Table 2.4/2 (identical to Table 41B in HD 384.4.41 / IEC 60364-4-41 / DIN VDE 0100-410) or, where applicable, within 5 seconds; the rated residual current $I_{\Delta n}$ must be taken into consideration if RCDs are used. The required disconnecting currents for fuses used as disconnecting devices or overcurrent protective devices are specified in Table 2.4/3.

For circuit-breakers applies:
Value of the short-circuit release plus 20%.

For miniature circuit-breakers applies:

- Tripping characteristic A: $3 \cdot I_n$,
- Tripping characteristic B: $5 \cdot I_n$,
- Tripping characteristic C: $10 \cdot I_n$,
- Tripping characteristic D: $20 \cdot I_n$.

The tripping currents specified for the circuit-breakers and miniature circuit-breakers are such that tripping/disconnection occurs within less than 100 ms. The disconnecting times specified in the table are thus always achieved. In accordance with the relevant standards, RCDs must trip within max. 200 ms; the disconnecting times of most RCDs are, however, shorter.

U_0 Nominal a.c. voltage (r.m.s. value) with respect to ground.

▷ The disconnecting times (Table 2.4/2) must be taken into consideration in the TN system:

The disconnecting times specified in Table 2.4/2 apply for branch circuits which supply hand-held equipment or Class 1 portable equipment via socket-outlets or non-detachable connections.

The disconnecting time for distribution circuits (also inside buildings) can be longer than the values specified in the table. A time of 5 seconds must, however, not be exceeded. A maximum disconnecting time of 5 s is also sufficient for branch circuits, which are used solely for supplying non-detachable loads, provided that no circuits for which the above table is applicable are supplied from the same distribution board. If the table has to be used for socket-outlet circuits, for example, but the disconnecting time of max. 5 seconds is to be permitted for non-detachable loads, one of the following two conditions must be satisfied:

- the impedance of the equipment grounding conductor between the distribution board and the point at which the equipment grounding conductor is connected to the main equipotential bonding conductor must not exceed the following value:

$$R_{PE} = \frac{50\text{ V}}{U_0} \cdot Z_S$$

where:

U_0 Nominal a.c. voltage (r.m.s. value) with respect to ground;

Z_S Impedance of ground-fault loop, consisting of the external conductor and equipment grounding conductor of the circuit between the power source and every possible fault location;

50 V Limit for permissible continuous touch voltage (with a.c. voltage)

or

- a local equipotential bonding conductor, which connects the same extraneous conductive parts as the main equipotential bonding conductor and also fulfills the other requirements for main equipotential bonding, must be installed in the distribution board (Fig. 2.4/10).

If the disconnecting conditions cannot be satisfied when overcurrent protective devices are used, RCDs should be used instead. Alternatively, a supplementary equipotential bonding conductor can be provided (see Section 2.4.7).

Table 2.4/2
Maximum permissible disconnecting times in TN system

Nominal voltage U_0 of electrical installation with respect to ground in V	Disconnecting time in s
230 ($U_0/U = 230/400$ V)	0.4
400 ($U_0/U = 400/690$ V)	0.2
> 400	0.1

Note 1: The disconnecting time for the associated nominal voltage applies to voltages within the tolerance range specified in IEC 38.
Note 2: With intermediate voltage values, the next highest voltage from the table must be used.

2.4.5 TT system

Requirements for TT system

In TT systems, the exposed conductive parts of the electrical equipment are not connected to the grounded neutral point or a grounded external conductor as is the case with TN systems. Instead, the exposed conductive parts are connected to ground individually, in groups, or all together (grounded equipment grounding conductors) (Fig. 2.4/11).

As with the TN system, the neutral point or, if the neutral point is inaccessible, an external conductor is connected to ground in the TT system too (see Fig. 2.4/2). No resistance value is specified for operational grounding. In contrast to the TN system, however, no additional requirement (with respect to R_B/R_E) has to be fulfilled.

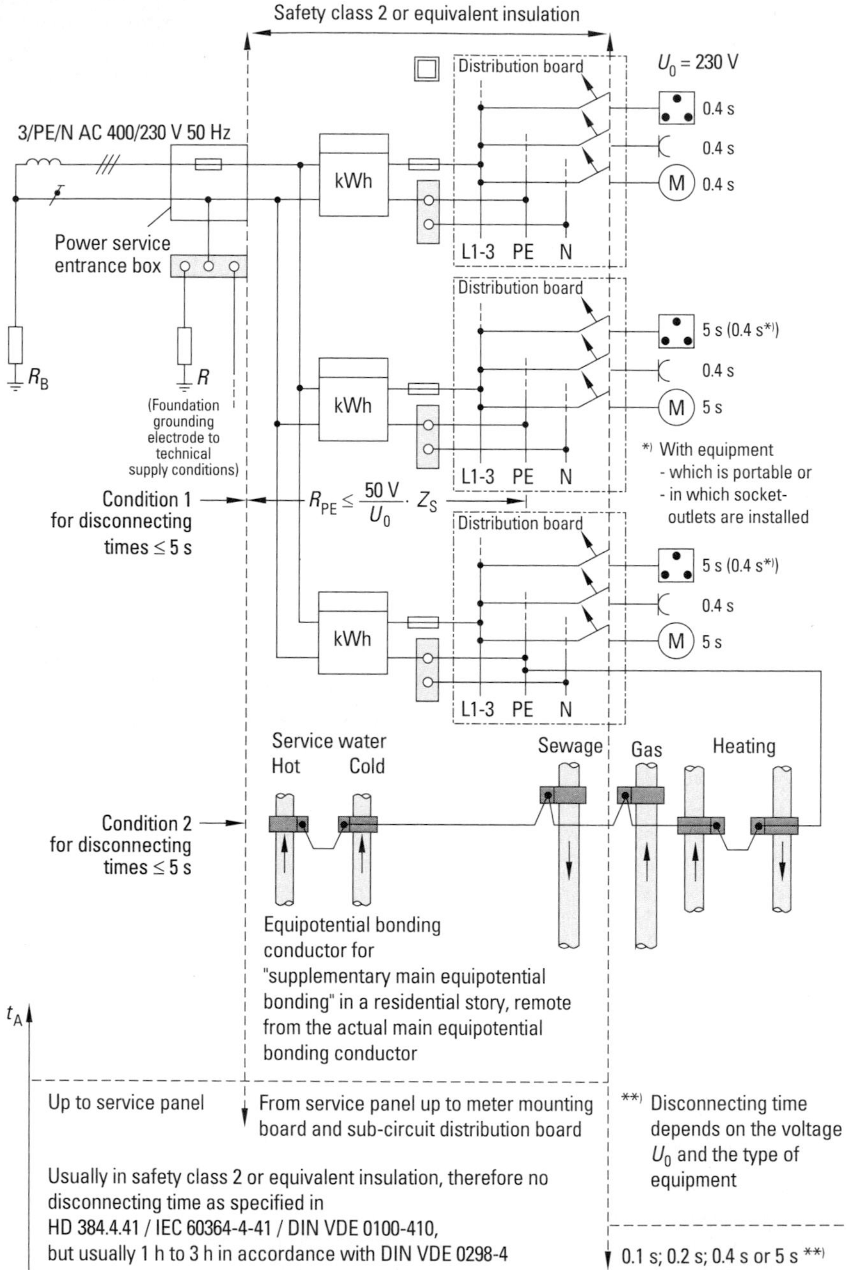

Fig. 2.4/10 Maximum permissible disconnecting times in TN system (single-line representation)

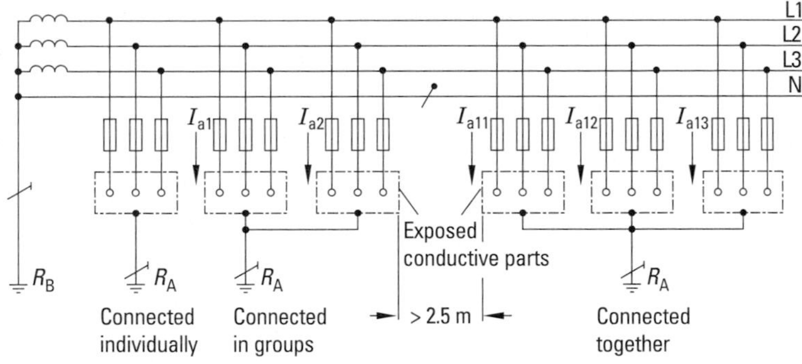

Fig. 2.4/11 Frame grounding in TT system

The following requirements must be satisfied in the TT system:

Requirements for grounding

▷ All exposed conductive parts, which are protected by the same protective devices, must be connected to a common grounding electrode via equipment grounding conductors.

▷ The following condition must be satisfied for the ground resistance of the exposed conductive parts:

$$R_A \leq \frac{50\ \text{V}}{I_a}$$

where:

R_A Sum of resistance values of the grounding electrode and equipment grounding conductor between the exposed conductive part and ground (see Fig. 2.4/12);

I_a Current which causes automatic disconnection of the circuit within 5 seconds, without additional discrimination as in the TN system; the rated residual current $I_{\Delta n}$ must be taken into consideration if RCDs are fitted. The required disconnecting currents for overcurrent protective devices are specified in Table 2.4/3;

50 V Limit for permissible continuous touch voltage (with a.c. voltage). The specified 50 V can also be used if U_L is limited to 25 V. It is, however, advisable to use the permissible touch voltage U_L limited to 25 V.

Protective devices

▷ RCDs and, in certain cases, fuses can be used as protective devices. Since, due to the required low ground resistance, the disconnecting conditions in TT systems

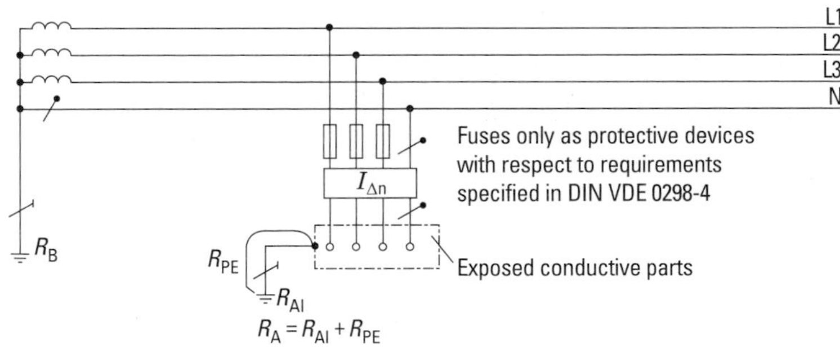

Fig. 2.4/12
Protection by automatic disconnection of power supply with RCDs in TT system (determination of R_A)

Table 2.4/3

Required short-circuit currents I''_{kerf} (disconnecting current) for line-protection fuses to EN 60629 / IEC 60629 / DIN VDE 0636-10, utilization category gL/gG, gM, with regard to the critical thermal short-circuit load (160 °C) of PVC-insulated cables with copper conductors and with regard to the disconnecting times of 5 s; 0.4 s; 0.2 s, and 0.1 s to HD 384.4.41 / IEC 60364-4-41 / DIN VDE 0100-410

Rated cross section q	Rated current I_n in A																												
	2	4	6	10	16	20	25	32	35	40	50	63	80	100	125	160	200	224[1)]	250	300[1)]	315	355[1)]	400	425[1)]	500	630	800	1000	1250
mm²	Required short-circuit current $I''_{k\,erf}$ in A																												
0.5	9	19	28	76	180	400																							
0.75			28	55	115	190	370	350	890										Unprotected cross section range										
1				48	93	137	250	290	550	520																			
1.5					70	100	175	200	360	350	860	1880																	
2.5						86	115	150	225	220	520	890	1750	3200															
4									130	200	330	570	950	1600	3100	6100													
6											250	400	640	1050	1900	3200	6800	4800		8400									
10												330	430	640	1080	1800	3450	3200	5400	6300	10 700								
16														580	715	1140	2100	2000	3100	5000	5700	6000	11 100	8800	22 000				
25																950	1380	1350	2000	3700	3500	4300	6100	6800	10 100	28 000			
35																	1250	1250	1700	2300	2450	3100	4100	4900	7150	16 000			
50																				2050	2050	2000	2950	3200	5000	10 050	21 000	39 000	90 000
70																							2800	2900	3500	7100	13 500	22 500	51 000
95																										5150	9400	15 500	35 500
120																										5000	7400	12 000	25 000
150																											6600	9400	19 000
185																												8800	15 000
240																													11 000
300																	Protected cross section range referred to protection provided in the event of a short circuit, assuming that the required short-circuit current $I''_{k\,erf}$ can flow.										↓	↓	↓
400																													
500																											6600	8800	11 000
2 × 50																											9000	14 500	32 000
2 × 70																											6600	10 500	19 500
2 × 95 to																												8500	12 300
2 × 500																												8500	15 000
3 × 50																												9400	19 000
3 × 70 to																												8800	13 000
3 × 500																													11 000
4 × 50 to 4 × 500	↓ 9.5	↓ 19	↓ 28	↓ 48	↓ 70	↓ 86	↓ 115	↓ 150	↓ 130	↓ 200	↓ 250	↓ 330	↓ 430	↓ 580	↓ 715	↓ 950	↓ 1250	↓ 1250	↓ 1700	↓ 2050	↓ 2050	↓ 2000	↓ 2800	↓ 2900	↓ 3500	↓ 5000	↓ 6600	↓ 8800	↓ 11 000
5 s	9.5	19	28	48	70	86	115	150	130	200	250	330	430	580	715	950	1250	1250	1700	2050	2050	2000	2800	2900	3500	5000	6600	8800	11 000
0.4 s	17	32	50	80	120	150	210	250	200	300	460	610	800	1050	1300	1800	2200	2200	3000	3200	3900	3700	5100	5600	6600	9900	12 000	17 000	22 000
0.2 s	20	40	60	95	140	180	250	320	270	410	570	720	980	1200	1700	2100	2800	2800	3800	4400	4700	4500	6200	6400	8000	11 000	14 000	20 000	27 000
0.1 s	23	48	73	120	180	220	300	390	440	500	690	900	1100	1400	2000	2500	3300	3500	4300	5100	5700	6100	7800	8200	10 000	13 000	18 000	23 000	32 000

[1)] Fuse not specified in EN 60269 / IEC 60269 / DIN VDE 0636, but available from Siemens

Note:

The last four lines refer to the disconnecting times for protection against electric shock: 0.4 s for socket-outlet circuits with 230 V; 0.2 s for socket-outlet circuits with 400 V; 0.1 s for socket-outlet circuits with >400 V

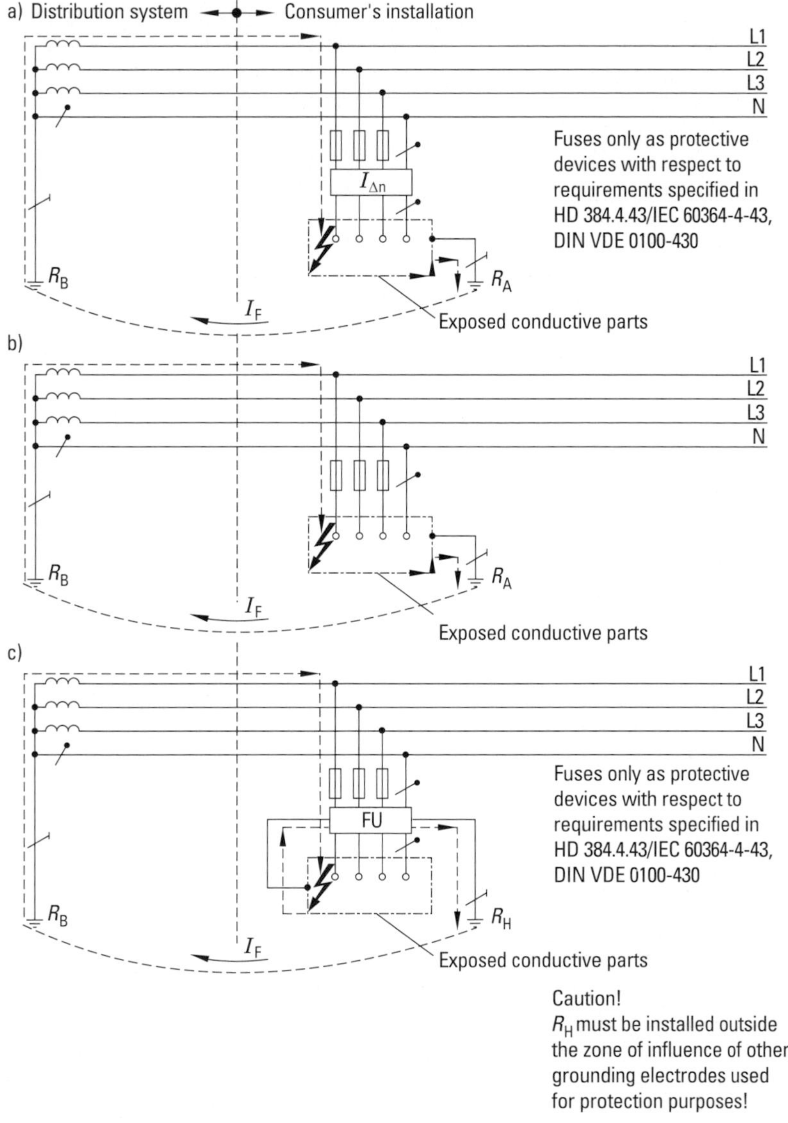

Fig. 2.4/13
Protection by automatic disconnection of power supply in TT system
a) with RCD as protective device
b) with overcurrent protective device as protective device
c) with fault-voltage-operated protective device as protective device

with overcurrent protective devices can only be fulfilled in the lower rated current range (e.g. with fuses up to approximately 16 A, because in this case R_A must be less than 0.7 Ω), it is advisable to use RCDs. Alternatively, supplementary equipotential bonding can also be provided (see Section 2.4.7).

▷ In special cases, fault-voltage-operated protective devices which, however, are no longer standardized in Germany, can also be used. See also Fig. 2.4/13a to c.

2.4.6 IT system

Requirements in IT system

As in TT systems, the exposed conductive parts of electrical equipment in IT systems are connected to ground (grounded equipment grounding conductors) individually, in groups, or all together.

Live parts not grounded or only via high impedance

In contrast to TN and TT systems, all live parts including the neutral conductor must be insulated with respect to ground. If necessary, they can be grounded by means of a sufficiently high impedance (no impedance values are specified). The impedance can be located between the neutral point of the system and ground or between an artificial neutral point and ground. It can also be located between an external conductor and ground. Overcurrent protective devices can also be installed to damp overvoltages.

Since the fault current at the first fault is very low owing to the ungrounded system and the sufficiently high impedance to ground (Fig. 2.4/14), disconnection after the first fault is not necessary.

In IT systems, the following requirements must be satisfied:

Ground resistance of exposed conductive parts

▷ The ground resistance of the exposed conductive parts must be sufficiently low to allow the following condition (see Fig. 2.4/14) to be satisfied:

$R_A \cdot I_d \leq 50\ \text{V}$ or

$$R_A \leq \frac{50\ \text{V}}{I_d}$$

where:

R_A Sum of resistance values of the grounding electrode and of the equipment grounding conductor between the exposed conductive part and ground;

I_d Fault current occurring at the first fault with negligible impedance between an external conductor and an exposed conductive part. Both the leakage currents and the total impedance of the electrical installation with respect to ground are included in this value.

50 V Limit for permissible continuous touch voltage (with a.c. voltage). The specified 50 V can also be used if U_L is limited to 25 V. It is, however, advisable to use the permissible touch voltage U_L limited to 25 V.

Insulation monitoring required

▷ An insulation monitoring device is required which issues an acoustic and/or visual signal to indicate the first fault between a live part (i.e. including any neutral conductors which may be installed) and an exposed conductive part or with respect to ground. Insulation monitoring is not a requirement of IEC 60364.

Disconnecting conditions for 2nd fault

▷ Once the first fault has occurred, the requirement for automatic disconnection of the power supply must be satisfied in the event of a second fault.
The conditions for disconnection of the power supply when the second fault oc-

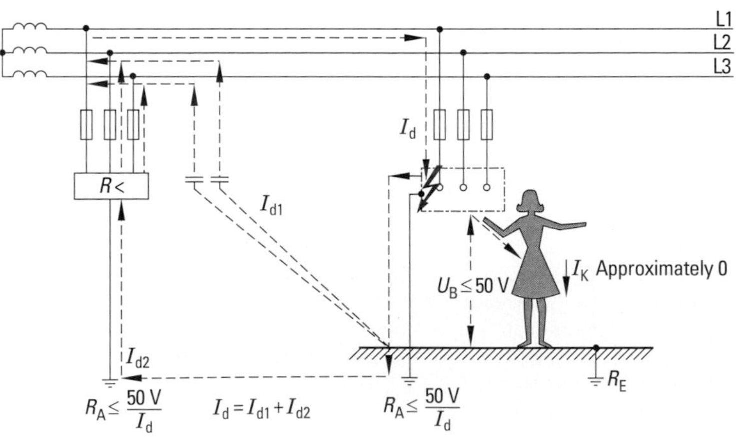

Fig. 2.4/14 Fault current at first fault in IT system

curs depend on the grounding type of the exposed conductive parts (individual/ group grounding or common grounding).

The following apply:

- If the exposed conductive parts are connected to ground individually or in groups via equipment grounding conductors, the disconnecting conditions for the TT system must be satisfied.
- If the exposed conductive parts are connected to ground together via equipment grounding conductors, the disconnecting conditions for the TN system with the following variation must be satisfied:

• If a neutral conductor **is not** fitted

$$Z_S \leq \frac{U}{2 \cdot I_a}$$

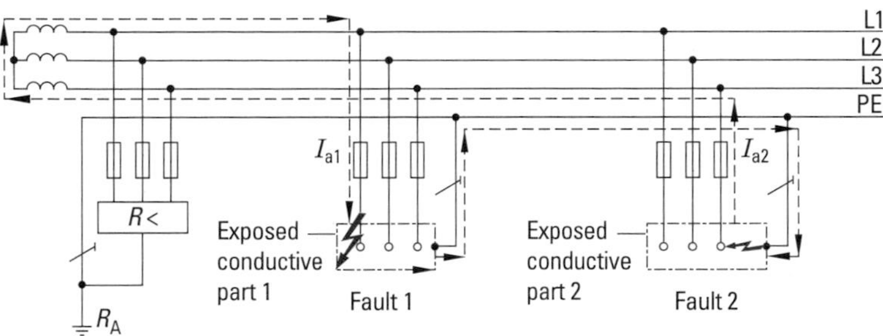

a) Actual fault loop with two faults in an IT system in which all exposed conductive parts are connected together to a grounded equipment grounding conductor.

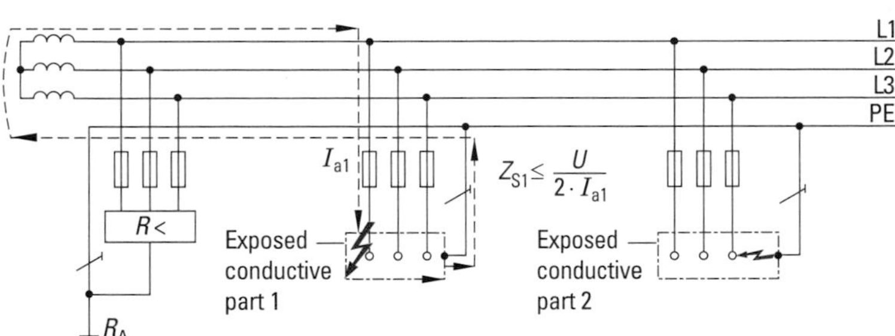

b) Fault loop to be taken into consideration with respect to exposed conductive part 1, with two faults in an IT system, in which all exposed conductive parts are connected together to a grounded equipment grounding conductor.

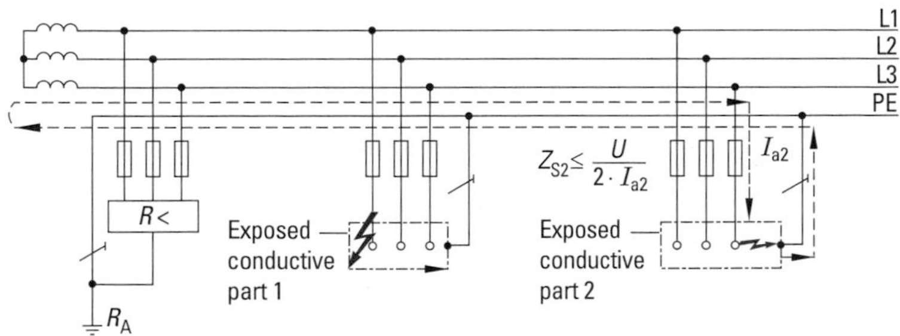

c) Fault loop to be taken into consideration with respect to exposed conductive part 2, with two faults in an IT system, in which all exposed conductive parts are connected together to a grounded equipment grounding conductor.

Fig. 2.4/15 a–e
Actual ground-fault loop and ground-fault loops to be taken into consideration in IT system

L1
L2
L3
N
PE
$R<$
R_A
I_{a1}
I_{a2}
Exposed conductive part 1
Exposed conductive part 2

d) Actual fault loop with two faults in an IT system with distributed neutral conductor, in which all exposed conductive parts are connected together to a grounded equipment grounding conductor.

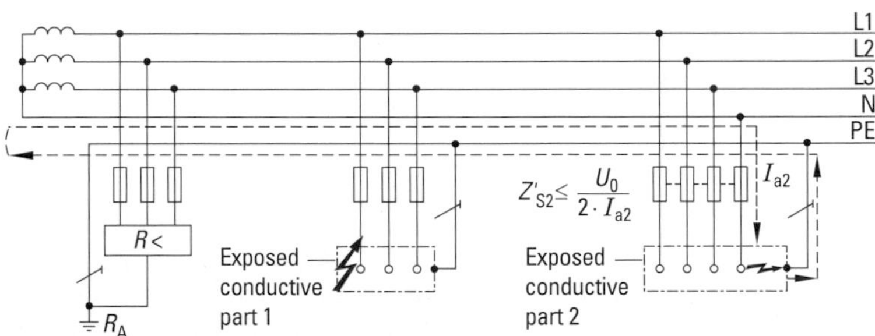

e) Fault loop to be taken into consideration with two faults in an IT system distributed neutral conductor, in which all exposed conductive parts are connected together to a grounded equipment grounding conductor.

- If a neutral conductor **is** fitted

$$Z'_S \leq \frac{U_0}{2 \cdot I_a}$$

where:

- U_0 Nominal a.c. voltage (r.m.s. value) between external conductor and neutral conductor
- U Nominal a.c. voltage (r.m.s. value) between external conductors
- Z_S Impedance of fault loop, consisting of the **external conductor** and **equipment grounding conductor** of the circuit (see Fig. 2.4/15a to 15c);
- Z'_S Impedance of fault loop, consisting of the **neutral conductor** and **equipment grounding conductor** of the circuit (Fig. 2.4/15d and 15e.);
- I_a Current which causes disconnection of the circuit within the time t specified in Table 2.4/4 (identical to Table 41B in HD 384.4.41 / IEC 60364-4-41 / DIN VDE 0100-410), where applicable, or within 5 seconds for all other circuits provided that this disconnecting time is permitted, e.g. in distribution circuits.

Disconnecting times

The following disconnecting times must be achieved:

Table 2.4/4
Disconnecting times in IT system with and without neutral conductor

Nominal voltage of electrical installation in V U_0/U	Disconnecting time in s	
	Neutral conductor not distributed	Neutral conductor distributed
230/400	0.4	0.8
400/690	0.2	0.4
580/1000	0.1	0.2

Note: 1. The disconnecting time for the associated nominal voltage applies for voltages which are within the tolerance range to IEC 38.

Note: 2: The next highest voltage value in the table must be used for intermediate voltage values.

2.4.7 Supplementary equipotential bonding

Reason for use, type of application

If, depending on the type of ground connection, the conditions for automatic disconnection of the power supply cannot be satisfied throughout the entire installation or in part of an installation, not even if RCDs are fitted, a local equipotential bonding conductor (also referred to as supplementary equipotential bonding) can be used.

This may be the case in d.c. systems, for example, because it is also not possible to use universal-current-sensitive RCDs directly in the d.c. voltage system. In cases where disconnection is required for reasons other than protection against electric shock, e.g. for fire protection, equipotential bonding may not be sufficient.

Supplementary equipotential bonding can be used throughout the entire installation, in part of an installation, or just for one individual item of equipment.

In the case of non-detachable electrical equipment, the electric circuits of which cannot satisfy the disconnecting conditions, all simultaneously accessible exposed conductive parts must be included in the supplementary equipotential bonding. This also applies to all extraneous conductive parts which can be touched at the same time as these exposed conductive parts (Fig. 2.4/16).

If possible, all key metal building constructions, including metal elements in reinforced concrete, should also be connected to the equipotential bonding.

The equipotential bonding must also be connected to the equipment grounding conductor systems of all equipment, including the equipment grounding conductors of socket-outlets.

Conditions for effectiveness

The effectiveness of supplementary equipotential bonding (conductors) must be verified by carrying out appropriate measurements. The following condition must be satisfied:

$$R \leq \frac{50\ \mathrm{V}}{I_a}$$

where:

R Resistance of the equipotential bonding connection between two exposed conductive parts or an exposed conductive part and an extraneous conductive part;

I_a Current which would cause the respective protective device to disconnect,
- for RCDs: rated residual current $I_{\Delta n}$
- for overcurrent protective devices: current which would cause the protective device to disconnect with 5 seconds;

50 V Limit for permissible continuous touch voltage U_L (with a.c. voltage).

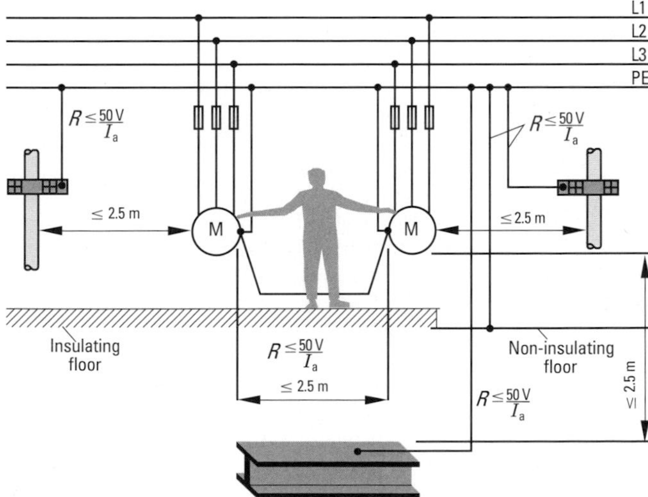

Fig. 2.4/16
Local supplementary equipotential bonding as substitute for unfulfilled disconnecting condition

The specified 50 V can also be used if the permissible touch voltage U_L is limited to 25 V. It is, however, advisable to use the permissible touch voltage U_L limited to 25 V.

Apart from the above-mentioned condition, the equipotential bonding conductors must have the following minimum cross sections:

- between two exposed conductive parts: cross section of the smaller equipment grounding conductor,
- between an exposed conductive part and an extraneous conductive part: half the cross section of the equipment grounding conductor relative to the exposed conductive part.

Irrespective of these minimum cross sections, equipotential bonding conductors must satisfy the minimum cross section requirements for equipment grounding conductors if they are installed outside an enclosure which they share with the external conductors.

The following applies:

▷ mechanically protected: min. 2.5 mm^2 Cu,

▷ mechanically unprotected: min. 4 mm^2 Cu.

Since equipotential bonding conductors are regarded as equipment grounding conductors, the requirements for equipment grounding conductors, especially with respect to their color and minimum cross sections, must also be satisfied.

In addition to the supplementary equipotential bonding described here, requirements for supplementary equipotential bonding are also specified in the special Group 700 provisions in HD 384 / IEC 60 364 / VDE 0100; the requirements in this case differ slightly (see Section 5.9.1).

2.4.8 Protection by means of Class 2 equipment or equipment with equivalent insulation (total insulation)

Protection by means of Class 2 equipment or equipment with equivalent insulation (total insulation) is achieved as follows:

▷ the equipment has double or reinforced insulation

or

▷ the equipment is factory-assembled (e.g. to EN 60439-1 / IEC 60439-1 / DIN VDE 0660-500) and is marked with an appropriate symbol (see 417-IEC 5172 and Reg. No. 00154 to DIN 30600).

Cables and leads are not marked with the symbol. Conductors with basic insulation which have an additional enclosure made of insulating material (cables/leads) can be used without additional fault protection.

Basic insulation, supplementary insulation

Other equipment, which only has basic insulation, can be upgraded by fitting supplementary insulation, in exceptional cases by fitting reinforced insulation.

All of the following conditions must be satisfied:

▷ All conductive parts of electrical equipment, which are separated from live parts by basic insulation only, must be surrounded by an insulating covering with degree of protection IP 2X or IP XXB (at least) to EN 60529 / IEC 60529 / DIN VDE 0470-1.

▷ The insulating covering must withstand the mechanical, electrical, and thermal loads which may occur at the installation site. Paints, lacquers, and similar coatings are not usually sufficient to satisfy these requirements.

▷ If the insulating covering was not tested beforehand and its effectiveness is uncertain, a test to HD 384.6.61 / IEC 60364-6-61 / DIN VDE 0100-610 must be carried out.

> *Note*
> Since these standards do not currently contain any specific requirements, it is advisable to perform the customary test with 4000 V for 1 minute.

▷ No conductive parts, via which voltages can be conducted inadvertently out of the system, should be incorporated in the insulating covering. The protection must not be impaired if mechanical joints or connections have to be incorporated in the insulating covering. Insulating adapters, for example, must be provided in this case. Plug-and-socket devices can be incorporated in the outer insulating covering. Screws made of insulating material

Protection is provided:

- by using Class 2 equipment bearing the symbol ⧈

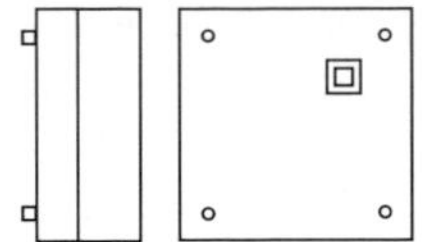

or

- by attaching supplementary insulation ≥ IP 2X or IP XXB to equipment with basic insulation

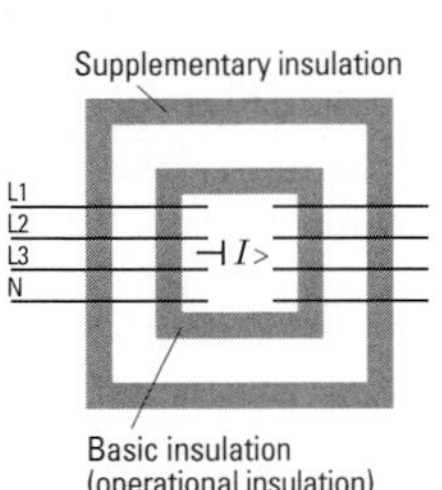

or

- by attaching reinforced insulation to uninsulated live parts

Requirements and explanations

1) If terminal block mounting rails are used as equipment grounding conductor or PEN bars, these bars must be installed with appropriate insulation.

2) Metal threaded conduit joints are not permitted. Furthermore, fixing screws and other exposed conductive parts must not project out of the covering, unless they are insulated against internal live parts and exposed conductive parts (as per safety class 2).

3) Terminals for equipment grounding conductors are not permitted.

Supplementary protection against direct contact is not required.

Fig. 2.4/17 Protection by means of total insulation

should only be used in the outer covering if the insulation is not impaired when these screws are replaced by metal screws.

▷ If covers or doors can be opened without the aid of tools, live parts behind the covers or doors must be concealed by a panel made of insulating material ≥ IP 2X or ≥ IP XXB. Materials to IP 3X or IP XXC must be used for domestic distribution boards (devices suitable for operation by untrained personnel). It should only be possible to remove the covers concealing these devices with the aid of tools.

▷ The symbol ⧈ should be attached at visible points on the outside and inside of the covering.

Reinforced insulation

With total insulation, DIN VDE 0106-100 must also be observed for actuating elements in the immediate vicinity of live parts. EN and IEC standards are in preparation.

Conditions for equipment grounding conductors

Conductive (de-energized) parts within the total insulation must not be connected to equipment grounding conductors unless this is expressly permitted in the relevant standards for the equipment in question.

Connecting points for equipment grounding conductors leading to downstream equipment can be provided. Equipment grounding conductors and their connecting points within the total insulation must be treated as live parts, i.e. they must, for example, be insulated against exposed conductive parts.

No equipment grounding conductors should be fitted along moving connecting cables. Old cables can be replaced by cables with equipment grounding conductors. These equipment grounding conductors should only be connected at the plug and not directly at/in totally insulated equipment. In the case of non-detachable connections, looped-in equipment grounding conductors are permitted for downstream equipment.

Conversion to safety class 1

If conductive parts are connected to an equipment grounding conductor, a Class 2 device becomes a Class 1 device. The symbol ▣ must, therefore, be permanently concealed (e.g. using adhesive tape). Conductive (de-energized) parts within the insulation then become exposed conductive parts and must be connected to the equipment grounding conductor. An equipment grounding conductor connecting point bearing the symbol ⏚ must be provided.

The total insulation and insulating covering must not have a negative effect on the operation of the equipment incorporated therein. The way in which equipment is attached and cables and conductors are connected must not impair the specified degree of protection.

See Fig. 2.4/17 for a further explanation of total insulation.

2.4.9 Protection by non-conducting locations

Requirements

This protective measure is intended to prevent simultaneous contact with exposed conductive parts of different equipment which could assume a different electric potential if the basic insulation fails.

All of the following requirements must be satisfied (Fig. 2.4/18):

▷ The exposed conductive parts must be arranged in such a way that, under normal conditions, personnel cannot touch
  - two exposed conductive parts simultaneously

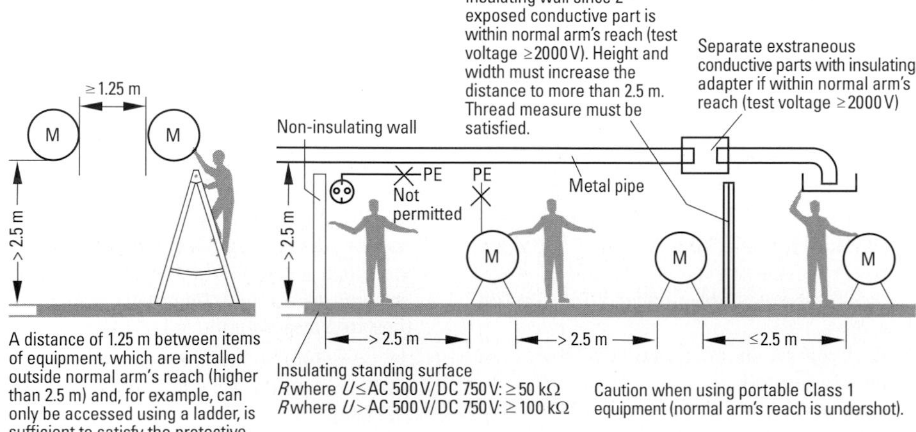

Fig. 2.4/18 Protection by means of non-conducting locations

or

- an exposed conductive part and an extraneous conductive part simultaneously,

if there is a possibility that these parts will have different electric potentials (almost always the case) if the basic insulation fails.

▷ No equipment grounding conductors must be installed at a non-conducting location. As a result, no equipment grounding conductors can or may be connected to Class 1 equipment. In the past, Class 1 equipment was forbidden in Germany and did not usually have any equipment grounding conductor connection.

The requirement specifying that simultaneous contact with exposed conductive parts, and with exposed conductive parts and extraneous conductive parts must be prevented is satisfied by one or more of the following measures:

▷ The location concerned must have an insulating floor and insulating walls.

▷ Exposed conductive parts and extraneous conductive parts must be installed at a distance of more than 2.5 m from each other. At heights in excess of 2.5 m, the distance between the parts can be reduced to 1.25 m.

Obstacles

▷ Obstacles can be positioned between exposed conductive parts and between exposed conductive parts and extraneous conductive parts in order to increase the distance between the parts to the specified values. Such obstacles should preferably be made of insulating material. If they are made of metal, they must not be connected to grounded parts or exposed conductive parts.

▷ Insulation or insulated arrangement of extraneous conductive parts. This type of insulation must have sufficient mechanical strength and be tested with a test voltage of at least 2 kV.
Any leakage current should not exceed 1 mA.

Resistance values of insulating floors and walls

▷ The resistance of insulating floors and walls must not be lower than **50 kΩ** at voltages ≤ 500 V AC / 750 V DC and **100 kΩ** at voltages >500 V AC / 750 V DC at any point at which a measurement is taken. If these values are not achieved, the respective areas are classified as extraneous conductive parts and must be covered with insulating material.

▷ The implemented measures must be permanent and must not be rendered ineffective.

▷ Steps must be taken to ensure that extraneous conductive parts cannot carry voltages inadvertently out of the system from the location in question (non-conducting locations).

2.4.10 Protection by ungrounded, local equipotential bonding

Requirements

Provided that the following requirements are fulfilled (Fig. 2.4/19), ungrounded, local equipotential bonding prevents hazardous touch voltages from occurring:

▷ Simultaneously accessible exposed conductive parts and extraneous conductive parts must be connected by means of equipotential bonding conductors. The cross section of the ungrounded equipotential bonding conductors is not specified. The cross section should be selected as with grounded equipotential bonding conductors. According to HD 384.5.54 / IEC 60364–5-54 / DIN VDE 0100–540, a cross section equivalent to the cross section of the smaller equipment grounding conductor must be used between two exposed conductive parts, and a cross section equivalent to half the cross section of the equipment grounding conductor between exposed conductive parts and extraneous conductive parts.

▷ Local equipotential bonding must not be connected to ground via exposed conductive parts or extraneous conductive parts.

▷ Measures must be taken to ensure that personnel cannot pick up hazardous touch voltages when entering a location with the protective measure "protection by means of ungrounded, local equipotential bonding". This applies in particular if a conductive floor, which is insulated against ground potential, is connected to the equipotential bonding.

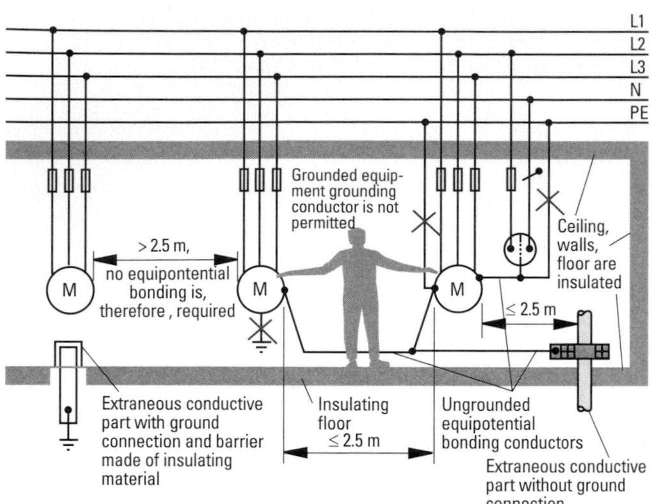

Fig. 2.4/19
Protection by means of ungrounded, local equipotential bonding

2.4.11 Protection by safety separation

Protection by means of safety separation can be used for the power supply

▷ of just one item of current-using equipment

or

▷ of several items of current-using equipment.

In both cases, a safety-separated power source prevents hazardous touch voltages from being picked up at the exposed conductive part(s) of the equipment if the basic insulation fails (Fig. 2.4/20).

Conditions for applying both variants

The following conditions must be satisfied for both variants:

▷ The product of voltage (in volts) multiplied by the length of the cables to all connected items of equipment (in meters) should not exceed the value 100,000 and the total cable length should not exceed 500 m (250 m at 400 V and 435 m at 230 V).

▷ The following must be used:
- transformers with safety separation to EN 60742 / IEC 60742 / DIN VDE 0551–1, bearing the symbol ⊖

or

- motor-generator sets with separate windings to IEC 2/915 CDV / DIN VDE 0530-1

or

- other power supply units with an identical level of electrical safety.

▷ Portable transformers with safety separation must be designed to safety class 2 or have an equivalent degree of insulation. Permanently-installed transformers with safety separation must also be designed to safety class 2, have an equivalent degree of insulation, or be constructed such that insulation, which conforms to safety class 2, separates the output from the input and from the conductive housing of the transformer.

▷ Live parts must not be connected to other circuits nor to ground/equipment grounding conductors.

▷ Exposed conductive parts in the secondary circuit must not be connected to ground or to exposed conductive parts for which other protective measures are used.

▷ Secondary circuits must be safety separated from other circuits. This separation must correspond to that between the primary and secondary circuit (only applies to cables/conductors in certain cases). Wherever possible, circuits with safety separation should not be installed together with circuits without safety separation in one cable/conductor. If this is not possible, all conductors must be insulated to withstand the highest possible voltage.

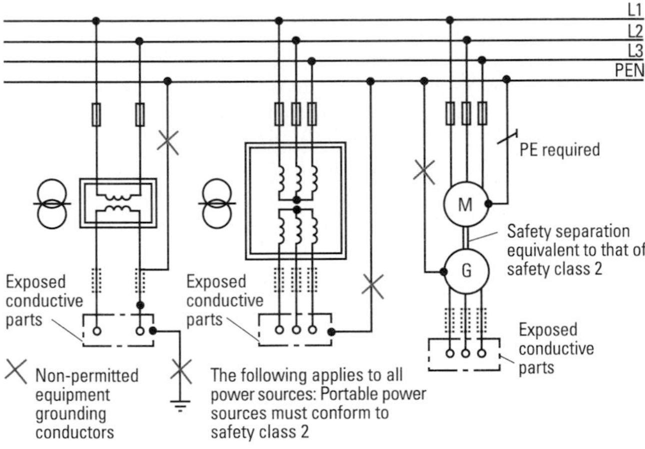

Fig. 2.4/20
Protection by means of safety separation

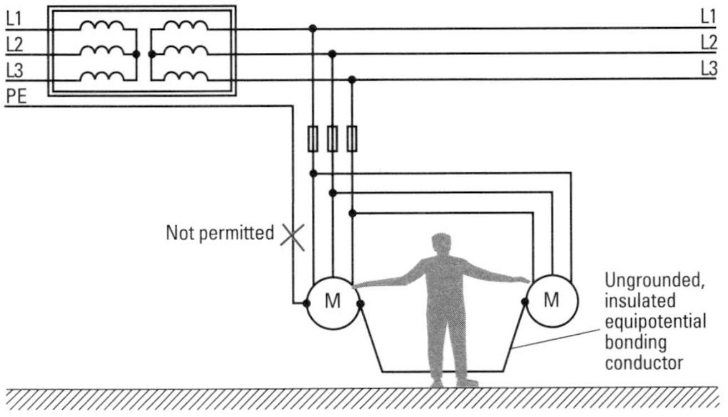

Fig. 2.4/21
Protection by means of safety separation with ungrounded equipotential bonding and several items of current-using equipment connected to a transformer

▷ The voltage of a circuit with safety separation must not exceed 500 V.

▷ Only one circuit (alternating or three-phase current) should be formed.

Additional requirement for several items of equipment

The following requirements must also be satisfied for safety separation with several items of current-using equipment:

▷ All exposed conductive parts downstream of a power source must be connected to each other using an ungrounded, insulated equipotential bonding conductor (see Fig. 2.4/21) which must not be connected to equipment grounding conductors or exposed conductive parts for which other protective measures are used, or a metal enclosure encasing the power source. Class 2 equipment can be used without an equipotential bonding conductor.

▷ Socket-outlets and plugs must be provided with a ground contact which must be connected to the equipotential bonding conductor.

▷ If two faults occur between different external conductors and the equipotential bonding conductor, a protective device must disconnect the power supply within the time specified for the TN system (0.1 s for voltages to ground >400 V; 0.2 s for 400 V; 0.4 s for 230 V; *but not 5 s*).

Safety separation with one item of equipment only

Safety separation must be used for the power supply of just one item of current-using equipment if this is expressly specified in the relevant sections of Chapter/Group 7 of HD 384 / IEC 60364 / DIN VDE 0100.

2.5 Verifying protective measures in the event of indirect contact

Standards

The tests are specified in the section entitled "Initial verification" of the relevant standards prHD 384.6.61 / IEC 60364-6-61 / DIN VDE 0100-610 since it is not only protection in the event of indirect contact, i.e. fault protection, but also overall protection against electric shock which must be verified when an electrical system is installed and put into operation. Furthermore, testing must also be carried out in accordance with the requirements specified in the other parts of the standards from the HD 384 / IEC 60364 / DIN VDE 0100 series.

In addition to the tests specified in prHD 384.6.61 / IEC 60364-6-61 / DIN VDE 0100–610, more extensive tests, which are not described here, may also be necessary, e.g. to EN 60204-1 / IEC 60204-1 / DIN VDE 0113-1.

> *Note*
> The relevant European standard for initial verification is still in preparation. The most recent draft (prHD 384.6.61) is, therefore, also taken into consideration in the following.

2.5.1 General

Verification by testing

According to prHD 384.6.61, IEC 60364-6-61, and DIN VDE 0100-610, the system fitter or an authorized inspection agency must verify that the conditions regarding the protection of personnel, livestock (the generic term "animals" is sometimes used), and property are satisfied, preferably while electrical installations (specified in the German standard "Electrical power installations") of up to 1000 V AC and 1500 V DC (only 1000 V AC installations are specified in the title of the German standard) are being installed in buildings or, at the latest, when they are commissioned for the first time. The requirements must also be satisfied if electrical systems in buildings are modified, repaired, or expanded.

Visual inspection, trial/measurement

These tests include visual inspections and trial/measurement (testing). In a number of cases, calculations or simulations using system models can be performed instead of the measurements.

It is important to ensure that personnel and livestock cannot be endangered and that property cannot be damaged during testing. The tester must, therefore, be a trained electrician as defined in IEC 60050 (195) (identical to the specifications in EN 50110-1 / DIN VDE 0105-100; no equivalent IEC publication is currently available) who can judge the potential hazards that may arise during testing. A suitably trained person can be commissioned to carry out simple testing tasks.

The tests described below are primarily based on the IEC publication IEC 60364-6-61 because, as already mentioned, an equivalent HD is still in preparation. Considerable differences may, therefore, exist with the national standards of other European countries. Furthermore, test requirements are sometimes inconsistent with the regulations for electrical installation, particularly in the case of DIN VDE and prHD. An attempt will, therefore, be made to point out the most important differences.

Test report

The IEC publication states, for example, that a test report must be compiled once testing has been completed. It does not, however, specify the type and scope of the report. A test report should always contain the following information (a standard form is currently being discussed):

- name and address of customer,
- name and address of contractor,
- order code or job number,
- type and scope of installed system,
- designation of distribution board(s),
- numbers/designations of circuits,
- measuring instruments used,
- visual inspections documented in form of checklists ("OK" or "N/A"),
- tests/measurements documented ("OK" or "N/A"),
- signature and date.

A list of measured values can be omitted on account of the time and effort involved. This does not, however, mean that the values do not have to be recorded during the tests.

2.5.2 Verification

2.5.2.1 Verification by means of visual inspection

Condition of system

Visual inspection involves performing a visual check on the electrical system (the inspection only covers electrical equipment that is permanently installed) in order to establish whether the system is intact and complete. Visual inspection must be carried out prior to trials/measurements, usually after the entire system has been shut down. The visual inspection deals first and foremost with the system configuration which is compared with the design specifications. At this stage, it is possible to detect serious design faults.

In addition, the following must also be established:

- ▷ whether the electrical equipment complies with the safety requirements of the relevant equipment standards. This is possible, for example, by checking the equipment markings or equipment certificates (if available);
- ▷ whether the electrical equipment has been selected and installed in accordance with the regulations for electrical installation and the manufacturer's specifications;
- ▷ whether any damage occurred during installation as this could have a detrimental effect on safety.

The following must also be established (where applicable) by means of visual inspection:

- ▷ whether protection against electric shock is provided by measuring the required clearances, e.g. with protection by means of obstacles or protection by positioning the equipment outside of normal arm's reach. Here, it is important to remember that the protective measure "protection by means of non-conducting locations" can only be verified if the electrical system only contains equipment that is permanently installed. Technically speaking, this means that no plug-and-socket devices must be present. These requirements are not specified in the relevant sections (e.g. 413.3) of HD 384.4.41 / IEC 60364-4-41 / DIN VDE 0100-410;
- ▷ whether fire barriers and other precautionary measures to prevent fires spreading are provided, and whether protection against thermal effects (e.g. maximum permissible surface temperatures under normal operating conditions, as defined in HD 384.4.42 / IEC 60364-4-42 / DIN VDE 0100-420) exists (see Section 527.2 to 527.4). Where necessary, checks must also be carried out to establish whether the requirements for locations exposed to fire hazards according to Part 482 have been taken into consideration;
- ▷ whether cables, conductors, and conductor bars have been selected in accordance with HD 384.4.43 / IEC 60364-4-43 / DIN VDE 0100-430 and HD 384.5.523 / IEC 60364-5-523 / DIN VDE 0298-4 with respect to current-carrying capacity and voltage drops (see also Section 2.5.2.9 for voltage drops);
- ▷ whether the protective and monitoring devices have been selected and set in accordance with the circuit documentation and the system specifications (see also prHD 384.5.53 / IEC 60364-5-53 / DIN VDE 0100-530);
- ▷ whether the required and appropriate isolating and switching devices are fitted and have been installed correctly (see also HD 384.4.46 / IEC 60364-4-46 / DIN VDE 0100-460 and HD 384.5.537 / IEC 60364-5-537 / DIN VDE 0100-537);
- ▷ whether the equipment and necessary protective measures have been selected taking external influences into consideration (see Section 512.2 of HD 384.5.51 / IEC 60364-5-51 / DIN VDE 0100-510);
- ▷ whether alphanumeric and/or color coding has been provided for neutral conductors and equipment grounding conductors (including equipotential bonding conductors) and whether additional identification has been provided at the ends of PEN conductors (blue at the ends of conductors with green/yellow insulation and green/yellow at the ends of PEN conductors with blue insulation) (see Section 514.3 of HD 384.5.51 / IEC 60364-5-51 / DIN VDE 0100-510);
- ▷ whether circuit documentation, warnings, and similar information have been provided (see Section 514.5 of HD 384.5.51/ IEC 60364-5-51 / DIN VDE 0100-510), particularly if more than one distribution board/ switchgear unit is installed in the same system;

- ▷ whether circuits, protective devices, switches, terminals, etc. are labeled correctly (see Section 514 of HD 384.5.51 / IEC 60364-5-51 / DIN VDE 0100-510);
- ▷ whether the connections between cables and conductors are correct. In cases of doubt, this can be verified by means of appropriate measurements. The resistance measured at the connecting point should not be greater than the resistance of a 1 m length of conductor of the appropriate cross section (resistance values can be found in Table 2.5/2 on page 341);
- ▷ whether the system components are easily accessible to permit identification (e.g. equipment designations, set values, conductor designations), operation, and maintenance. This applies, in particular, to emergency stop devices (according to Section 537.3) and shutdown devices for mechanical maintenance (according to Section 537.4 of HD 384.5.51 / IEC 60364-5-51 / DIN VDE 0100-510).

2.5.2.2 Verification by means of trial/measurement

Fulfillment of purpose

Trial/measurement involves carrying out trials (e.g. actuating test buttons) and measurements (e.g. loop impedance measurement) in electrical systems in order to determine whether the electrical system fulfils its purpose.

The primary objective of this type of testing is to establish the characteristics of the electrical system which cannot be determined by means of visual inspection, e.g. using measuring instruments.

Requirements for measuring instruments

The HD and IEC publications do not contain any information concerning the measuring instruments required for the tests. They merely include a note that indicates that the requirements for measuring instruments and display devices are to comply with the standards from the IEC 61557 series "Electrical safety in low voltage distribution systems up to 1000 V AC and 1500 V DC – Equipment for testing, measuring or monitoring of protective measures". Corresponding requirements are, however, specified in the National Appendix of the German standard (see Table 2.5/1).

It is essential that measuring errors caused by the measuring instrument itself as well as by the measuring method used are taken into consideration for all measurements.

Measuring instruments can have a measuring error of ± 30 % which is particularly significant in the case of borderline results.

Sequence of trials and measurements

With trial/measurement, it is particularly important to ensure that the trials and measurements listed below are carried out, provided that they are relevant to the installation work

Table 2.5/1 Standards for measuring instruments

Measuring task	Standards[2)]
Voltage[1)] and current	IEC 60051, DIN 43780, DIN 43751 Part 1 to 3, DIN VDE 0410 and 0411
Fault current, fault voltage, and touch voltage	DIN VDE 0413-6
Insulation resistance	DIN VDE 0413-1
Loop impedance (loop resistance)	DIN VDE 0413-3
Resistance of grounding conductors, equipment grounding conductors, and equipotential bonding conductors	DIN VDE 0413-4
Ground resistance	
– Potentiometer method	DIN VDE 0413-5
– Voltmeter-ammeter method	DIN VDE 0413-3 and -7
Phase sequence	DIN VDE 0413-9
Resistance of floors and walls with	
– DC voltage	DIN VDE 0413-1
– AC voltage	DIN VDE 0413-5 or -7
High-voltage test	DIN VDE 0432-2 and -3

1) The internal resistance should not be below 0.7 kΩ/V and should not be above 500 kΩ. With DC voltages above 500 V up to 1500 V, an internal resistance of 1.5 MΩ should not be exceeded.

2) The standards from the EN 50197 series are only available in draft form for a number of measuring instruments.

that has been performed. If possible, the tests should be carried out in the following sequence as this can reduce potential hazards which may arise during testing as a result of faults in the system. The relevant requirements for the tests are described in the following sections.

The following trials and/or measurements must be carried out:

▷ continuity of the equipment grounding conductors, the connections at the main equipotential bonding and supplementary equipotential bonding (see Section 2.5.2.2.1);
▷ insulation resistance of the electrical system (see Section 2.5.2.2.2);
▷ safety separation of the circuits in the case of SELV, PELV, and protection by electrical separation (see Section 2.5.2.2.3);
▷ resistance of insulating floors and walls (see Section 2.5.2.2.4);
▷ protection by automatic disconnection of the power supply (see Section 2.5.2.2.5);
▷ functional test on residual-current protective devices (see Section 2.5.2.2.6)
▷ voltage polarity (see Section 2.5.2.3);
▷ dielectric strength (see Section 2.5.2.4);
▷ functional test (see Section 2.5.2.5);
▷ thermal effects. This test is carried out by means of visual inspection (see Section 2.5.2.1) which means that no further testing is required here;
▷ voltage drop (see Section 2.5.2.6).

Other methods

Other methods, e.g. calculation or system simulation, can also be used instead of the measurements listed here.

This is particularly relevant for determining the loop impedance, which is required in the case of protection by automatic disconnection of the power supply in TN systems, and for determining the voltage drop.

If faults are detected during one of the tests, the tests that may have been affected by the fault must be repeated once the fault has been rectified.

Testing methods

The testing methods described below are intended as examples. Other methods are also possible provided that they yield adequate test results.

2.5.2.2.1 Measuring the continuity of equipment grounding conductors, connections at main equipotential bonding and supplementary equipotential bonding; measuring the resistance of equipment grounding conductor connections

Continuity of equipment grounding conductor

Measuring the continuity of the equipment grounding conductor is no longer stipulated in IEC 60364-6-61. This requirement has been deleted in Amendment 2 of IEC 60364-6-61. In TN systems, continuity of the equipment grounding conductor must be determined by measuring the loop impedance. In TT systems, it is determined by measuring R_A, whereby the connections between the equipment grounding conductors and the exposed conductive parts must also be included. The requirements for TT and TN systems also apply to IT systems, depending on how the exposed conductive parts are connected to ground (individually, in groups, or all together); see also page 347 ff.

Measuring resistance

Since this measurement is still stipulated in a number of national standards and verification of the resistance of the equipotential bonding connection is specified for equipotential bonding conductors under Section 413.1.6 of HD 384.4.41 / IEC 60364-4-41 / DIN VDE 0100-410, it is necessary to demonstrate how verification can be provided. Furthermore, a note appearing in Appendix D of Amendment 1 of IEC 60364-6-61 recommends that the continuity of the equipment grounding conductors be measured before the loop impedance. This measurement is intended to prevent a hazardous touch voltage from occurring while the loop impedance is being measured, if an equipment grounding conductor connection is defective. The measurement for the resistance of the equipment grounding conductor connection is approximately equivalent to the measurement for the continuity of the equipment grounding conductor connection.

Power source

A current of at least 0.2 A from a power source, the no-load voltage of which is between 4 V and 24 V (DC or AC), should be used for the measurement. A number of other standards specify a current of at least 10 A. In the case of equipment grounding conductors, the resistance determined in this way must be less than or equal to the value which is derived from the following condition for the resistance of the equipment grounding conductor.

Resistance of equipment grounding conductor

Since only one half of the fault loop is measured when the resistance of the equipment grounding conductor is measured, this is taken into consideration in the following condition by including the resistance of the external conductor in relation to the cross section of the equipment grounding conductor:

a) With negligible system impedance:

For *TN systems*: $$R \leq \frac{m}{m+1} \cdot \frac{U_0}{I_a}.$$

For *IT systems*, in which the neutral conductor is not distributed:

$$R \leq \frac{m}{m+1} \cdot \frac{U}{2\,I_a}.$$

For *IT systems*, in which the neutral conductor is distributed:

$$R \leq \frac{m}{m+1} \cdot \frac{U_0}{2\,I_a}.$$

b) With non-negligible system impedance:

For *TN systems*: $$R \leq \frac{0.8\,m}{m+1} \cdot \frac{U_0}{I_a}.$$

where:

R Resistance of equipment grounding conductor between exposed conductive part and next point along main equipotential bonding in Ω

U_0 Nominal AC voltage (r.m.s. value) with respect to ground in V

U Nominal AC voltage between two conductors in V

I_a Current which causes automatic disconnection of the circuit within the time specified for TN systems in Table 41A and for IT systems in Table 41B of HD 384.4.41 / IEC 60364-4-41 / DIN VDE 0100-410, or within 5 s according to the conditions in Section 413.3.5 of the above-mentioned standards

m Ratio between resistance of equipment grounding conductor connection and resistance of associated external conductor

0.8 is a specified factor which can be replaced by known system impedance values.

System impedance

In certain cases (e.g. in building installation), the system impedance cannot be omitted. The value can either be obtained from the responsible public utility or the formula with the factor 0.8 must be used.

System impedance could be omitted in large-scale industrial applications if the resistance of the equipment grounding conductor in a branch circuit has to be verified for a socket-outlet or a lighting element.

Resistance of equipotential bonding conductor

Condition for resistance

When the resistance of the equipotential bonding conductor is checked (as a substitute for non-fulfillment of the disconnection condition and for systems in particularly hazardous locations, e.g. as stipulated in HD 384.7.701 / IEC 60364-7-701 / DIN VDE 0100–701), the determined resistance of the equipotential bonding conductor must be less than or equal to the value derived from the condition:

$$R \leq \frac{50\text{ V}}{I_a}$$

where:

R Maximum permissible resistance of equipotential bonding and equipment grounding conductor connection to be verified by measurement, in Ω

50 V Limit for permissible continuous touch voltage (with AC voltage). The specified 50 V can also be used for DC voltage. It is, however, advisable to use the permissible touch voltage for DC voltage $U_L = 120$ V

I_a Current which causes or would cause automatic disconnection of the protective devices within the time specified in Table 41A and 41B of HD 384.4.41 / IEC 60364-4-41 / DIN VDE 0100-410 (see also Section 2.4.4, Table 2.4/3). With supplementary equipotential bonding, I_a is

- *with unfulfilled disconnection condition*: the disconnecting current which would have to flow to allow disconnection by the protective device within the specified time;
- *with supplementary equipotential bonding in particularly hazardous locations*: the disconnecting current of the protective device in circuits with the highest required disconnecting current in this range.

alculation

The resistance of the equipotential bonding connection both for the "unfulfilled disconnection condition" and for "particularly hazardous locations" can also be determined by calculation, provided that the connection can be traced visually and the length can be determined. The calculation is performed by multiplying the appropriate resistance value from Table 2.5/2 by the length. The value calculated in this way must be less than or equal to the value which is calculated using the formula given above.

alculation ia voltage drop

Since the international, regional and most of the national standards do not currently contain any information on how to carry out the measurement or how to interpret the result when checking the continuity and resistance of the main equipotential bonding connections, the voltage drop that occurs at the specified measured current should be measured and used to determine the resistance of the equipment grounding conductor connection. A measuring instrument, which indicates the resistance immediately, can also be used. The measured values must not exceed the resistance values specified in the table below, depending on the cross section and length of the equipment grounding conductor connections.

Table 2.5/2
Ohmic resistance R of copper conductors per meter at an ambient conductor temperature of 30 °C

Conductor cross section mm²	Conductor resistance $R_{30\,°C}$ mΩ/m
1.5	12.5755
2.5	7.5661
4	4.7392
6	3.1491
10	1.8811
16	1.1858
25	0.7525
35	0.5467
50	0.4043
70	0.2817
95	0.2047
120	0.1632
150	0.1341
185	0.1091

The conductor resistances R_x can be calculated for other temperatures x using the following equation:

$R_x = R_{30\,°C}\,[1 + \alpha \cdot (x - 30\,°C)]$
α = Temperature coefficient (for copper $\alpha = 0.00393\ K^{-1}$)

Performing test

No information is provided with regard to how the test is to be performed, whether the equipment grounding conductor connections have to be disconnected during the measurement (which is not recommended since it is very easy to forget to reconnect them), or whether the measurement should be carried out while the equipment grounding conductor is still connected. The latter can result in more favorable values if parallel circuits are routed via structural parts, equipotential bonding conductors, or shields, for example. This does not have any negative effects unless the "parallel circuits" can be removed easily. Such cases must be taken into consideration when analyzing the measuring results.

2.5.2.2.2 Measuring the insulation resistance of electrical systems

The insulation resistance must be measured in all electrical systems irrespective of the magnitude of the voltage (Figs. 2.5/1 and 2.5/2). The measurement must be carried out:

Measurement between live conductors

▷ between each live conductor and the other interconnected live conductors (it is important to remember that the neutral conductor is also a live conductor). This measurement is not specified in all national standards and may lead to problems if current-using equipment is connected. The current-using equipment should, therefore, be disconnected if this measurement is required.

Measurement between live conductors and ground

▷ between each live conductor, including the neutral conductor, with respect to ground or grounded equipment grounding conductors.
To simplify matters, the *insulation resistance of all interconnected live conductors* (in several or all circuits) can also be measured with respect to ground, preferably with respect to grounded equipment grounding conductors. The simplified measuring method must not be used for locations exposed to fire hazards, i.e. in these cases, live conductors must also be measured with respect to each other (if necessary, disconnect any connected current-using equipment) and with respect to ground/equipment grounding conductors. In TN-C systems, it is also permissible to carry out measurements with respect to a

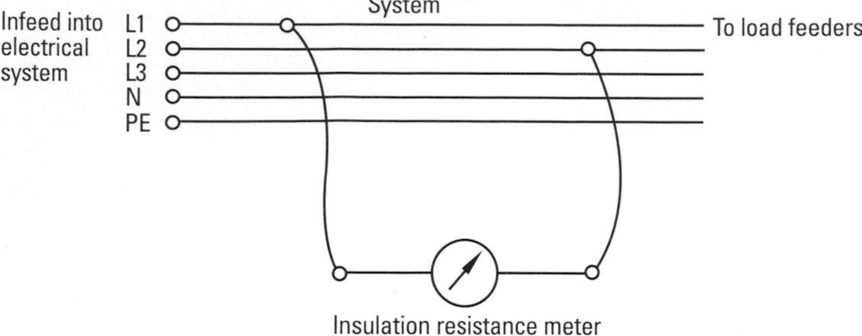

Fig. 2.5/1 Measuring insulation resistances between each live conductor

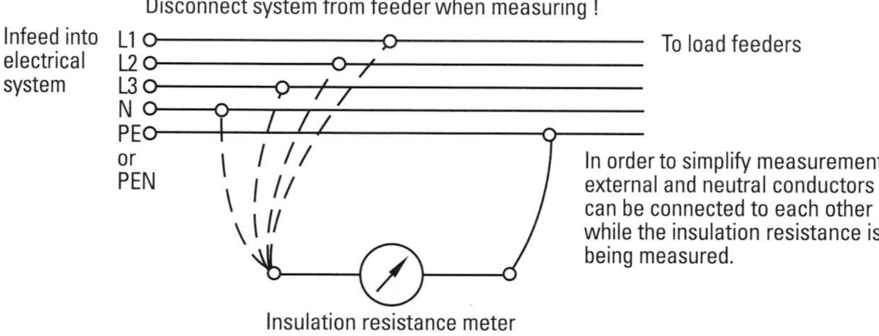

Fig. 2.5/2 Measuring insulation resistance between each live conductor and ground

PEN conductor (since it is also a grounded conductor). The neutral conductor(s) must be disconnected from the PEN busbar/PEN conductor if such measurements are carried out.

Points to remember during measurement

The measurement must be carried out at the infeed point of the electrical system. The simplest way of performing the measurement is during installation *when no electrical equipment has yet been connected.* Otherwise, the measurement can be carried out at feeder terminals on the infeed side of the distribution board (all live conductors, including the N-conductor links, if fitted). The individual circuits must not be interrupted by fuses, switching devices, or any other interrupt devices, i.e. when measurements are carried out, fuses must be inserted and switching devices must be closed in the individual circuits. If necessary, measurements must also be carried out at the output of the distribution board. This is, however, relatively complicated due to the connections that are required. Since, in control circuits, it may not always be possible to close all contacts simultaneously, it is advisable to check only the outgoing cables/feeders in control circuits. Current-using equipment can remain connected. Furthermore, if the insulation resistance is being measured in circuits which contain electronic equipment, measurements should only be made between the interconnected external and neutral conductors with respect to ground in order to prevent the equipment from being destroyed. Similar steps must be taken downstream of control-power transformers. In this case, any connections between an external conductor and equipment grounding conductor should be disconnected and both conductors measured together with respect to ground.

inimum values r insulation sistance

Table 2.5/3
Measuring DC voltage and minimum values for insulation resistance depending on nominal voltage of circuit (identical to Table 61A of prHD 384.6.61 / IEC 60364-6-61 / DIN VDE 0100-610)

Type and nominal voltage of circuit	Measuring DC voltage V	Insulation resistance MΩ
SELV and PELV circuits	250	≥ 0.25
Circuits ≤ 500 V FELV circuits Circuits with safety protection	500	≥ 0.5
Circuits > 500 V	1000	≥ 1.0

leasuring oltage, easuring nstrument

The measurements must be performed using a DC voltage (the magnitude of which must be selected from Table 2.5/3 according to the type and magnitude of the nominal voltage in the circuit) and an instrument, e.g. a megger to DIN VDE 0413–1 (see also Table 2.5/1), which can supply a measuring current of at least 1 mA. The results of this measurement should be better (higher insulation resistance) than or equal to the values specified in Table 2.5/3.

oints to emember uring neasurement

The system must be switched off (de-energized) during the measurement. As already mentioned, the current-using equipment can remain connected. This is, however, not possible if conductor-to-conductor measurements are required. If the minimum value specified in Table 2.5/3 cannot be attained while the current-using equipment is connected, the equipment can be disconnected. If disconnection of the current-using equipment does not yield the required values, the live conductors can also be measured individually with respect to ground. If the measurement then only yields a value close to that specified in the table, the system should be checked since a much higher value than the value in the table is normally attained if the system has been installed correctly and is functioning properly.

Other standards, e.g. EN 60204-1 / IEC 60204-1 / DIN VDE 0113-1, may specify insulation resistance values higher than those in Table 2.5/3, e.g. 1 MΩ even for circuits <500 V. Standards may, however, also specify lower values. For example, EN 60439-1 / IEC 60439-1 / DIN VDE 0660-500 specifies a value of just 230 kΩ referred to a rated voltage of 230/400 V.

2.5.2.2.3 Measuring insulation resistance for protection by safety separation of circuits

Measuring circuits with respect to other circuits

Safety separation must be verified for the following circuits with respect to other circuits and, in some cases, to ground (not with PELV) by measuring the insulation resistance:

- ▷ SELV circuits,
- ▷ PELV circuits,
- ▷ circuits with safety separation.

Safety separation with SELV circuits

SELV circuits

In the case of SELV circuits (circuits with safety extra-low voltage), the insulation resistance between all live parts of the SELV circuits and all live parts of other circuits, as well as with respect to ground or equipment grounding conductors in higher-voltage circuits, must be measured to verify that the insulation resistance is greater than or equal to the values specified in Table 2.5/3.

Safety separation with PELV circuits

PELV circuits

With PELV circuits (grounding of live parts and/or exposed conductive parts is permissible), the insulation resistance between all live parts of the PELV circuit and all live parts of other circuits, as well as with respect to ground or equipment grounding conductors in higher-voltage circuits, must be measured to verify that the insulation resistance is greater than or equal to the values in Table 2.5/3. In grounded PELV circuits, measurements only have to be carried out at the ungrounded conductors; if, however, there is a danger of the connected electrical equipment being damaged, the connection to ground/

equipment grounding conductor can be disconnected to allow both conductors to be measured with respect to ground. The ungrounded equipotential bonding conductor must also be checked in the case of safety separation with several items of current-using equipment.

Safety separation with safety-separated circuits

Safety-separated circuits

With safety-separated circuits, the insulation resistance between all live parts of the circuit with safety separation and other circuits, as well as with respect to ground or equipment grounding conductors in circuits without safety separation, must be measured to verify that the insulation resistance is greater than or equal to the values in Table 2.5/3.

2.5.2.2.4 Measuring resistance of insulating floors and walls

Insulating floors and walls

Strictly speaking, this test should be called "Insulation resistance of insulating floors and walls" since it is the insulation resistance that is verified by the measurement.

At least three measurements

At locations where the protective measure "Protection by means of insulating floors and walls" or more correctly "Protection by means of non-conductive locations" is applied, at least three measurements must be carried out to verify that the insulation resistance of floors and walls matches or exceeds the values specified in Table 2.5/3. The three measurements to be carried out in each instance must be taken at every surface of the room or area in question. Protection is the primary objective here, which means that ceilings, doors/door frames, and window frames should also be checked. Since this is not specified in the standard, large surfaces which are located within normal arm's reach should also be checked as a minimum requirement.

If accessible extraneous conductive parts are installed in the room in question, at least one of the three measurements must be performed for the surface concerned at a distance of 1 m from the accessible extraneous conductive parts.

The resistance must be measured between the test electrode and an equipment grounding conductor of the remaining electrical system or with respect to ground. Where possible, the surfaces to be measured should not yet have been painted or coated in any way, unless this paint/coating is provided for insulation purposes.

Test electrodes

Two different test electrodes can be used to perform the measurements. Testing with test electrode 1 is recommended.

Testing with test electrode 1

Test electrode 1 consists of a square metal plate (each side measuring 250 mm) and a square of moist, water-permeable paper or fabric (each side measuring 270 mm) which has been shaken to remove any excess water droplets. The paper or fabric is placed between the metal plate and the surface to be measured. During the measurement, the test electrode must be pressed against the floor with a force of 750 N and against the walls with 250 N.

The measuring setup for measurements with test electrode 1 is illustrated in Figs. 2.5/4 and 2.5/5.

Testing with test electrode 2

Test electrode 2 (Fig. 2.5/3) consists of a metal tripod. The parts of the tripod which come into contact with the surface to be tested form the points of an equilateral triangle. Each of the contact surfaces must have an area of approximately 900 mm^2 and the resistance should be less than 5000 ohms. As with test electrode 1, the surfaces must also be moistened or covered with a damp cloth. During the measurement, the same forces must be applied as for measurements with test electrode 1.

Performing measurement

Measurements can be performed in the following way using either test electrode 1 or test electrode 2:

Power source

Where possible, measurements should be taken with respect to ground/equipment grounding conductor using the local nominal voltages and nominal frequencies, i.e. the grounded supply system at the measuring location should be used. If this is not possible, a transformer with separate windings or a different independent power source can be used instead. In such cases, one of the conductors of these power sources must be grounded or connected to the equipment

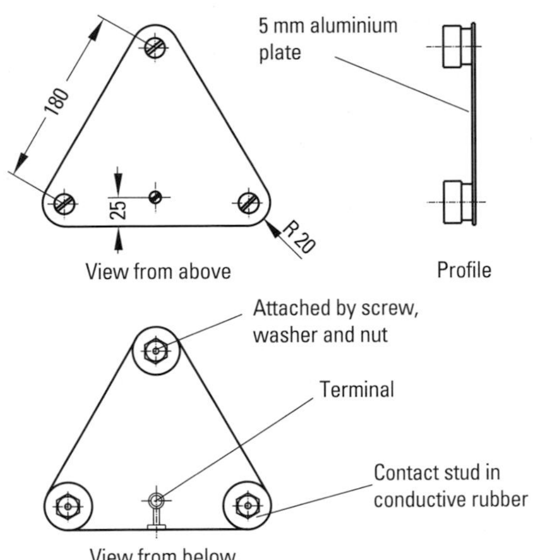

Fig. 2.5/3 Test electrode 2

grounding conductor while the measurement is being taken.

The resistance between the loaded metal plate and ground/equipment grounding conductor, i.e. the insulation resistance of the floor or walls to be measured, is calculated using the following formulas:

Voltage divider method

Determining resistance using voltage divider method (Fig. 2.5/4):

The resistance of the floor or surface to be measured is calculated using the following formula:

$$R_x = R_i \left(\frac{U_0}{U_x} - 1 \right)$$

where:

R_x Resistance of floor or wall with respect to ground/equipment grounding conductor in Ω

R_i Internal resistance of voltmeter in Ω
The internal resistance of the voltmeter must not be below the lower limit value specified in footnote[1)] of Table 2.5/1 (since hazardous shock currents can occur even at relatively low internal resistances if the metal plate is touched) and should not exceed the upper limit values;

U_0 Measured voltage with respect to ground in V

U_x Measured voltage with respect to metal plate (test electrode) in V.

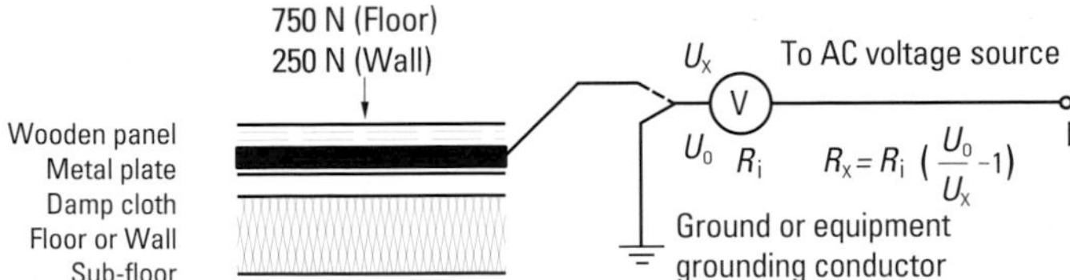

Fig. 2.5/4 Measuring setup for measuring resistance of floors and walls using voltage divider method

Electrostatic charge

If a DC voltage measuring instrument is used to measure the insulation resistance, the resistance can be read off directly at the measuring instrument. Instances where the discharge capacity for electrostatic charges has to be measured must be considered separately. The German standard DIN 51 953 can be applied in such cases.

Ammeter-voltmeter method

Determining resistance using ammeter-voltmeter method (Fig. 2.5/5):

With this measurement, an additional resistance of approximately 100 kΩ must be inserted between the voltage source and the measuring setup in order to prevent hazardous voltages.

The impedance Z_X of the floor or surface to be measured is calculated using the following formula:

$$Z_x = \frac{U_x}{I}$$

where:

Z_x Impedance of floor or surface in Ω
U_x Voltage measured at metal plate (test electrode) with respect to ground in V
I Current flowing to ground via floor or surface in A

2.5.2.2.5 Verifying protection by automatic disconnection of power supply

Mandatory tests

These tests must be carried out in all electrical systems since, according to Section 413 of HD 384.4.41 / IEC 60364-4-41 / DIN VDE 0100-410, protection by automatic disconnection of the power supply must be implemented in all electrical systems (provided that this is feasible). The insulation resistance of the electrical system must always be tested (as described in Section 2.5.2.2.2) for protection by automatic disconnection of the power supply, irrespective of the type of system grounding (system configuration).

Visual inspection recommended

Wherever protection by automatic disconnection of the power supply is implemented, it is advisable to carry out the following visual inspections (if relevant to the respective system), even if no visual inspections other than those listed in Section 2.5.2.1 are specified in the IEC publication:

The purpose of the inspections is to ensure that:

- ▷ equipment grounding conductors, grounding electrode conductors, and equipotential bonding conductors have the minimum cross section specified in HD 384.5.54 / IEC 60364-5-54 / DIN VDE 0100-540 (in some cases, this can also be determined by means of the specified measurements),
- ▷ equipment grounding conductors, grounding electrode conductors, and equipotential bonding conductors are installed correctly (e.g. equipment grounding conductors, and particularly PEN conductors, in TN systems should be incorporated in the same cable as the external conductors or, where possible, installed directly next to the external conductors), and that the connecting points are locked to prevent accidental loosening and, where necessary, protected against corrosion (e.g. different conductor materials for outdoor installation),
- ▷ equipment grounding and PEN conductors are not confused with the outer conductors,
- ▷ equipment grounding conductors and neutral conductors are not confused,
- ▷ the specifications for equipment grounding and PEN conductors with regard to connecting and isolating points have been observed (disconnection only possible with tools; connecting and isolating points can only be disconnected individually; one connecting point for each conductor),

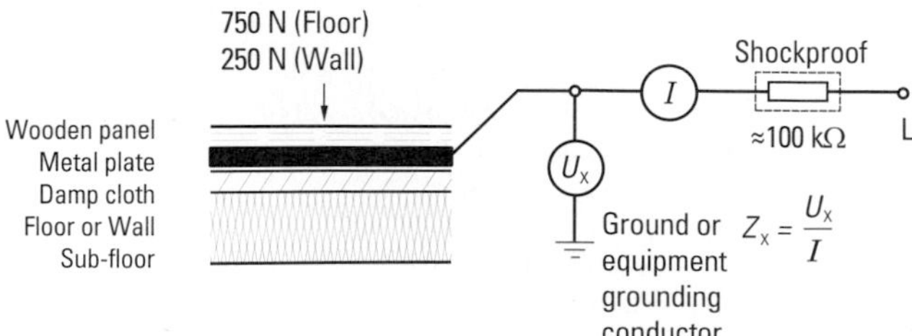

Fig. 2.5/5 Measuring setup for measuring impedance Z_X of floors and walls using ammeter-voltmeter method

- ▷ the ground contacts of the plug-and-socket devices can function properly (not bent or dirty, and not coated with paint),
- ▷ no *overcurrent protective devices* are fitted in equipment grounding or PEN conductors, and equipment grounding and PEN conductors do not contain any switching devices (interrupters). Devices, which can only be removed with a tool, are an exception here,
- ▷ the number and sizes of the protective devices fitted, such as overcurrent and residual-current protective devices, insulation monitoring devices, surge arresters, correspond to the relevant standards for installation. It should, however, be pointed out that surge arresters are not always required.

Tests in TN systems

The following tests are required in TN systems:

a) verify total grounding resistance R_B of all operational grounding electrodes;
b) verify loop impedance;
c) check tripping characteristics of associated protective devices.

Re a):

Total ground resistance R_B

The total grounding resistance R_B of all operational grounding electrodes only has to be determined if an abnormal fault (see Section 413.1.3.7 of HD 384.4.41 / IEC 60364-4-41/ DIN VDE 0100-410) is anticipated. This type of fault could, for example, be encountered if in the case of overhead power lines there is a possibility of a fault occurring between an external conductor and ground, e.g. if an overhead power line falls onto a "well-grounded" crash barrier. In this case, verification must be provided that the equipment grounding conductor and the exposed conductive parts connected to it do not carry a voltage (with respect to ground) in excess of 50 V. This can be verified by means of the following condition:

$$\frac{R_B}{R_E} \leq \frac{50\ \text{V}}{U_0 - 50\ \text{V}}$$

where:

R_B Total grounding resistance of all parallel grounding electrodes (including those of power supply system), in Ω;

R_E Lowest ground contact resistance of extraneous conductive parts not connected to an equipment grounding conductor via which a fault between the external conductor and ground can occur, in Ω;

U_0 Nominal AC voltage (r.m.s. value) with respect to ground, in V.

50 V Limit for permissible continuous touch voltage (with AC voltage). The specified 50 V can also be used for DC voltage. It is, however, advisable to use the permissible touch voltage U_L of 120 V specified for the DC voltage.

In all other cases, verification of the total grounding resistance R_B of all operational grounding electrodes is not an essential requirement.

Verification for conditions to be fulfilled

Verification can be provided by measuring R_B and R_E, as illustrated in Fig. 2.5/6. R_E does not have to be measured if a value of $< 2\ \Omega$ is measured for R_B, since a value of $< 7\ \Omega$ for R_E does not occur in practice. This means, therefore, that the condition in the equation is satisfied.

Measuring R_B

The measurement as illustrated in Fig. 2.5/6 is performed as follows:

A constant alternating current must be passed through the grounding electrode T and auxiliary grounding electrode T_1. The distance between the two electrodes must be such that mutual interference is prevented (a distance of approximately 20 m should be sufficient).

A second auxiliary electrode T_2, e.g. a metal spike driven into the ground, must be positioned halfway between T and T_1. The voltage drop between T and T_2 is measured.

The resistance of the grounding electrode is then the voltage between T and T_2, divided by the current which flows between T and T_1, on condition that there is no mutual interference between the grounding electrodes.

Two further measurements must be carried out using the auxiliary electrode T_2. This is positioned 6 m closer to and then 6 m further away from the electrode T. If the values resulting from the three measurements are approximately the same, the average of the three measurements must be used. If this is not the case, further measurements must be taken.

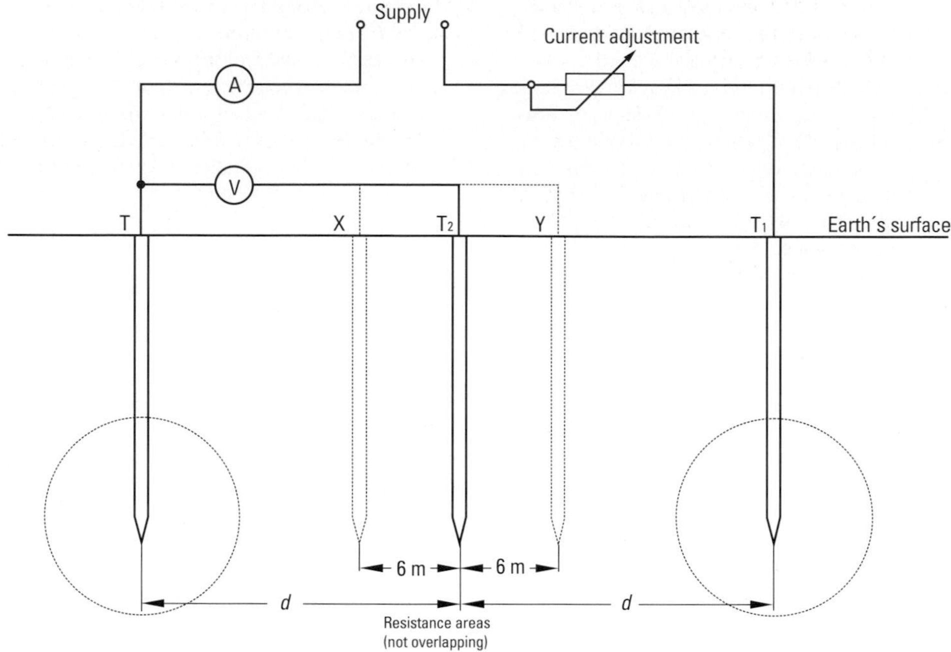

T : earth electrode under test, disconnected from all other sources of supply
T_1 : auxiliary earth electrode
T_2 : second auxiliary earth electrode
X : alternative position of T_2 for check measurement
Y : further alternative position of T_2 for the other check measurement
d : appr. 10 m

Fig. 2.5/6
Measuring setup for measuring total grounding resistance R_B, ground contact resistance R_E, and frame grounding resistance R_A

If the measurements are taken at the nominal system frequency, the internal resistance of the voltmeter must be at least 200 Ω/V.
The power source for the test must be isolated from the general power supply system, for example by means of a transformer with separate windings.

Re b):

Measuring loop impedance

The loop impedance must be measured as follows. Alternatively, verification is possible by:

- calculation or
- by performing a system simulation using a system model.

As already mentioned in Section 2.5.2.2.1, verification by measuring the resistance of the equipment grounding conductor (between the neutral point of the power source and the individual exposed conductive parts) is no longer specified. To increase the level of safety, it is, however, advisable to carry out the resistance measurement *before measuring the loop impedance.*

Furthermore, measuring the resistance of the equipment grounding conductor can also be used as an alternative to measuring the loop impedance. Section 2.5.2.2.1 describes how this measurement is performed.

The loop impedance must be measured at the same frequency as the nominal frequency of the general power supply system. The determined loop impedance must comply with the requirements specified in HD 384.4.41 / IEC 60364-4-41 / DIN VDE 0100-410, i.e.:

$$Z_S \cdot I_a \leq U_0 \quad \text{or} \quad Z_S \leq \frac{U_0}{I_a}$$

where:

Z_S Impedance of ground-fault loop, in Ω;
I_a Current which causes automatic disconnection of circuit within required time, in A;
U_0 Nominal AC voltage (r.m.s. value) with respect to ground, in V.

Measuring methods

The following two measuring methods can be used for measuring the loop impedance. It is important to note that the result obtained from the measurement is better than that in a fault scenario, since the conductors do not heat up while the measurement is being carried out. In the event of a fault, the conductors can reach a maximum temperature of 160 °C (e.g. with PVC insulation) before the circuit is disconnected. This results in a considerable increase in conductor resistance which, in turn, can increase the disconnecting time in borderline cases. This error must, therefore, be taken into consideration when analyzing the measuring results.

To simplify matters, the error can be taken into consideration if, instead of the condition

1st condition

$$Z_S \leq \frac{U_0}{I_a},$$

the condition

2nd condition

$$Z_S\,(\mathrm{m}) \leq \frac{2}{3} \cdot \frac{U_0}{I_a}$$

is satisfied.

where:

$Z_S(\mathrm{m})$ Impedance of measured ground-fault loop (external conductor, equipment grounding conductor or PEN conductor) in Ω;
I_a Current which causes automatic disconnection of circuit within specified time, in A;
U_0 Nominal AC voltage (r.m.s. value) with respect to ground, in V.

In the second condition, the system impedances are taken into consideration by multiplying the result by 2/3 so that they do not have be measured separately.

Furthermore, a slightly inaccurate result is obtained for the measurements described below due to the inductive and capacitive components.

Measuring method 1

Verification of loop impedance by measuring voltage drop

With this method, a hazardous situation may arise if the equipment grounding conductor is interrupted during the measurement or the equipment grounding conductor is not connected to an exposed conductive part. To remedy this problem, it is, therefore, advisable to measure the resistance of the equipment grounding conductor beforehand as described in 2.5.2.2.1.

Measuring voltage drop

The voltage drop in the circuit, for which verification is required, is determined by carrying out measurements with and without a variable load resistor (see measuring setup in Fig. 2.5/7) and calculating the loop impedance using the following formula:

$$Z = \frac{U_1 - U_2}{I_R}$$

where:

Z Impedance of ground-fault loop in Ω
I_R Current via load resistor R in A
U_1 Measured voltage without load resistor in V
U_2 Measured voltage with load resistor in V

When the measurement is carried out, the resistance should first be adjusted so that the difference between U_1 and U_2 is considerable.

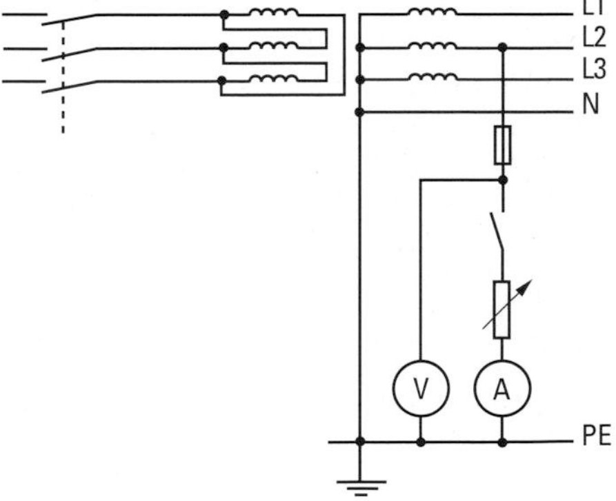

Fig. 2.5/7
Measuring setup for verification of loop impedance by measuring voltage drop

Measuring method 2

Verifying loop impedance by measuring with system-independent power source

The system transformer must be disconnected on the input side and the windings short-circuited before the measurement is performed. The secondary circuit is supplied with power via a separate power source (see the measuring setup in Fig. 2.5/8) and the loop impedance is calculated using the following formula:

$$Z = \frac{U}{I}$$

where:

Z Impedance of ground-fault loop in Ω
I Measured test current in A
U Measured test voltage in V.

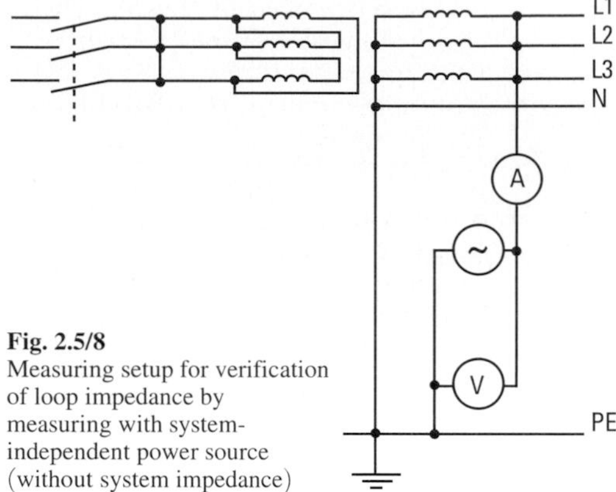

Fig. 2.5/8
Measuring setup for verification of loop impedance by measuring with system-independent power source (without system impedance)

Re c):

The tripping characteristics of the protective devices must comply with the relevant standards for these devices.

The following methods of inspection are used:

- RCDs: visual inspection and testing (see Section 2.5.2.2.6),
- overcurrent protective devices: visual inspection, e.g. release settings of circuit-breakers, rated current, and type of miniature circuit-breaker, rated current of fuses.

Tests in TT systems

In addition to the visual inspection specified in Section 2.5.2.1 and the visual inspection recommended on page 352, it is also advisable to carry out the following visual inspections (which are specified in DIN VDE but not in IEC 60364-6-61).

Additional recommended visual inspections

It is important to ensure that:

▷ all exposed conductive parts, which can be accessed simultaneously (normal arm's reach) or are connected downstream of a common protective device, have a common grounding electrode or grounded equipment grounding conductor,

▷ any installed overcurrent protective devices are configured in such a way that the neutral conductor is never disconnected before the external conductors and is not connected after the external conductors.

Additional recommended tests

The following tests should also be carried out:

Measuring R_A

a) The frame grounding resistance R_A must be measured in the same way as the total grounding resistance R_B (as illustrated in Fig. 2.5/6). The frame grounding resistance including the equipment grounding and grounding conductors between the exposed conductor part and grounding electrode must be low enough so that the disconnecting current required for the disconnecting time of 5 s flows. The following condition must be satisfied taking the permissible continuous touch voltage U_L into consideration:

$$R_A \leq \frac{50\ \text{V}}{I_a}$$

where:

R_A Sum of resistance values of grounding electrode (including grounding conductor) and equipment grounding conductor between exposed conductive part and ground, in Ω
I_a Current in A, which causes automatic disconnection of circuit within 5 s; the rated residual current $I_{\Delta n}$ must be taken into consideration if RCDs are fitted. The required disconnecting currents for overcurrent protective devices are specified in Table 2.4/3 in Section 2.4.4
50 V Limit for permissible continuous touch voltage (with AC voltage).

The specified 50 V can also be used if the permissible touch voltage U_L is limited to 25 V. It is, however, advisable to use the conventional touch voltage U_L = 25 V.

Preferred testing method

In TT systems, the frame grounding resistance R_A can also be determined using the following testing method (Fig. 2.5/9). The advantage of this method is that only *one* auxiliary grounding electrode is required.

b) The tripping characteristics of the associated protective devices must be checked:

- RCDs: visual inspection and testing (see page 352),
- overcurrent protective devices: visual inspection, e.g. release settings of circuit-breakers, rated current and type of miniature circuit-breaker, rated current of fuses.

The continuity of equipment grounding conductors must also be determined by visual inspection (see Section 2.5.2.1).

The IEC publication incorrectly refers to Section 612.1 of HD 384.6.61 / IEC 60364–6-61 / DIN VDE 0100–610 at this point. This section does not, however, deal with testing by visual inspection, but rather testing by measurement. In TT systems, measurements should also be carried out according to Section 2.5.2.2.1, especially if visual inspection of the equipment grounding conductors is difficult.

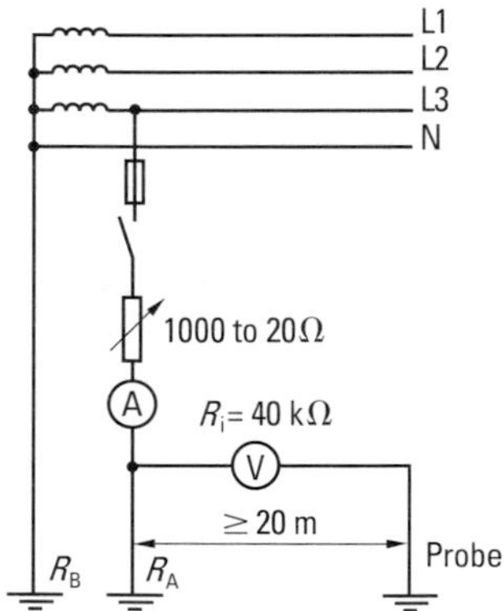

Fig. 2.5/9
Example showing measurement of frame grounding resistance R_A

Tests in IT systems

Additional visual inspections/trials

In addition to the visual inspection specified in Section 2.5.2.1 and the visual inspection recommended in Section 2.5.2.2.5, it is also advisable to carry out the following tests – which are not specified in IEC 60364-6-61 – by means of visual inspection and/or trials.

It is important to ensure that:

a) live conductors are not grounded directly (if necessary, via an impedance),
b) the exposed conductive parts are connected to a grounded equipment grounding conductor individually, in groups, or all together,
c) an acoustic signal and a visual signal are output when the test device of the insulation monitoring device is actuated or when an insulation fault is simulated in the system (by looping a resistor between an external conductor and ground or the grounded equipment grounding conductor),

It should be possible to acknowledge the acoustic signal. It should, however, not be possible to acknowledge the visual signal until the simulated fault has been rectified.

> *Note*
> Since the insulation monitoring device, which provides protection against electric shock, is not prescribed in the IEC publication, this test is only necessary if the insulation monitoring device is required for other reasons, e.g. for a fault-tolerant power supply.

Additional tests

The following tests are also necessary:

Fault current/ leakage current I_d

▷ The maximum possible fault current/leakage current I_d at the first fault must be calculated. In cases of doubt, a measurement must be performed or the value must be estimated on the basis of planning documents.
The calculation/measurement is not necessary if all exposed conductive parts of the electrical equipment are connected to a high-resistance system or a system grounded via an impedance.

Measurements should only be performed instead of calculations if calculations are not possible. If a measurement is performed, the associated system grounding may pose a hazard if a second fault occurs simultaneously.

Testing must verify that the product $R_A \cdot I_d$ does not exceed the limit for the permissible continuous touch voltage U_L (50 V is specified in HD 384.4.41 and DIN VDE 0100-410).

The following conditions must also be verified, irrespective of whether the exposed conductive parts are grounded individually, in groups, or all together:

Ground resistance R_A

▷ if the exposed conductive parts are connected to ground/equipment grounding conductor individually or in groups, the grounding resistance R_A (of the exposed conductive parts/equipment grounding conductors) must be determined, as specified in a) for TT systems (see page 350).

or

Loop resistance

▷ if the exposed conductive parts are connected to ground/equipment grounding conductor all together, the loop resistance must be determined, as specified in b) for TN systems (see page 348). For this measurement, an impedance-free connection must be established between the neutral point of the system and the equipment grounding conductor at the power source.

2.5.2.2.6 Visual inspection and testing of residual-current protective devices (RCDs)

With RCDs, the following must be determined by visual inspection:

Selection of correct RCD

▷ the correct RCD has been selected, i.e. that it is at least pulse-current-sensitive (in some countries, AC type RCDs (i.e. only alternating-current-sensitive) are still permitted), or a universal-current-sensitive RCD has been selected, for example, for converter-fed circuits;

▷ the magnitude of the rated residual current $I_{\Delta n}$ corresponds to the stipulated value.

Testing effectiveness

The effectiveness (tripping at specified time, maximum permissible touch voltage) of the RCDs must be verified by means of one of the three testing methods listed below. Equivalent testing methods can, however, also be used:

Testing method 1

Simulating residual current between live conductor and exposed conductive part/equipment grounding conductor

Fig. 2.5/10 illustrates a testing method in which a residual current I_Δ is simulated with a variable resistor R_P (the rating of which depends on the rated residual current $I_{\Delta n}$) which is connected between a live conductor on the load side of the RCD and the exposed conductive parts downstream of the RCD. A variable value between 200 Ω and 2000 Ω is sufficient for all standard rated residual currents $I_{\Delta n}$.

The RCD must trip at the rated residual current $I_{\Delta n}$ at the latest. Measurements in accordance with this method can be used for TN-S, TT, and IT systems. In IT systems, a point on the power source (e.g. the neutral point or an external conductor on the transformer) must be grounded or connected to the equipment grounding conductor of the exposed conductive parts while the test is being carried out. In the case of measurements with testing method 1, a check is performed at the same time to establish whether a (grounded) equipment grounding conductor is connected to the exposed conductive part. If, however, no such connection exists, a relatively high touch voltage may, in certain cases, occur at the exposed conductive part. The voltmeter on which the touch voltage can be read must, therefore, be monitored during the measurement.

Testing method 2

A variable resistor R_P, the rating of which depends on the rated residual current $I_{\Delta n}$, is

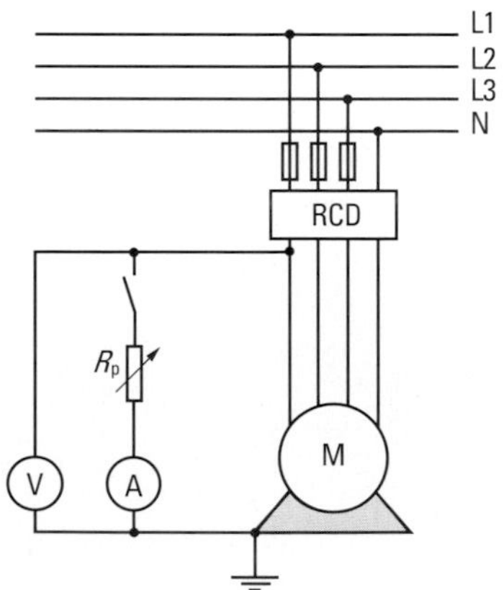

Fig. 2.5/10
Testing method 1 for TN-S, TT, and IT systems

Simulating residual current between two external conductors

also used for testing method 2 (Fig. 2.5/11). With this testing method, however, the resistor R_P is connected between an external conductor on the supply side of the RCD and another external conductor or the neutral conductor on the load side of the RCD. Current-using equipment, which is connected downstream of the RCD, must be disconnected during testing.

The RCD must trip at the rated residual current $I_{\Delta n}$ at the latest. The advantage of this testing method is that no excessively high touch voltages are generated if an equipment grounding conductor is interrupted, since the measurement is not carried out with respect to an exposed conductive part/equipment grounding conductor. The disadvantage, however, is that the continuity of the equipment grounding conductor cannot be tested at the same time.

This testing method can be used in all systems according to the type of ground connection. Furthermore, in IT systems it is not necessary to ground a point on the power supply system.

Testing method 3

Measuring method with auxiliary grounding electrode

An independent auxiliary grounding electrode and a variable resistor R_P (the rating of which depends on the rated residual current $I_{\Delta n}$) are required for testing method 3 (Fig. 2.5/12). The resistor is connected in series with an ammeter between a live conductor on the load side of the RCD and the exposed conductive parts. In addition, a voltmeter is also connected between the exposed conductive parts and the auxiliary grounding electrode. The voltmeter is used to measure the voltage between the exposed conductive part and the auxiliary grounding electrode when a residual current I_Δ – which must not, however, exceed the rated residual current $I_{\Delta n}$ – is generated with the resistor. The measured voltage U corresponds to the touch voltage U_L which would occur in the event of a fault.

The RCD must trip at the rated residual current $I_{\Delta n}$ at the latest. This condition is represented by the following formula:

$$U \leq U_L \cdot \frac{I_\Delta}{I_{\Delta n}},$$

where U_L is the conventional permissible continuous touch voltage and U is the voltage at which tripping occurs.

This testing method can be used for all systems according to the type of ground connection. In IT systems, an external conductor or the neutral point must be connected to ground during the measurement. A measurement can, however, only be performed if an auxiliary grounding electrode is used.

2.5.2.3 Voltage polarity

Voltage polarity

In cases where the standards prohibit the installation of single-pole switches in neutral

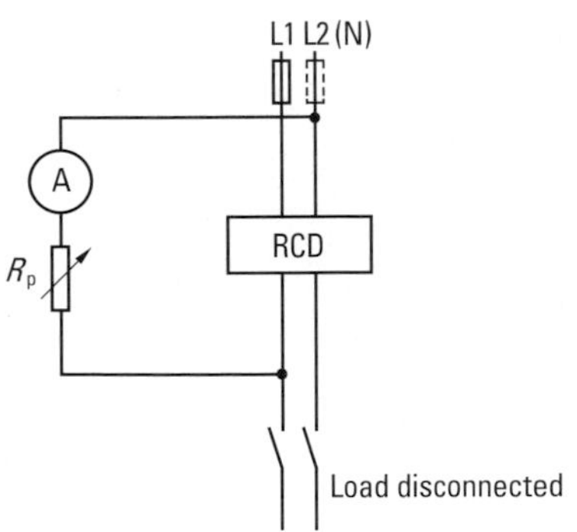

Fig. 2.5/11
Testing method 2 for TN-S, TT, and IT systems

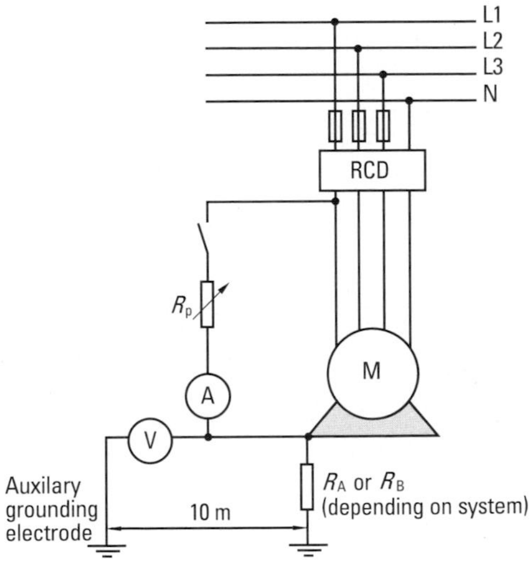

Fig. 2.5/12
Testing method 3 for TN-S, TT, and IT systems, only in conjunction with auxiliary grounding electrode

conductors, the voltage polarity must be determined by means of an appropriate test. The purpose of this test is to verify that single-pole switches are installed in the respective external conductors and not in the neutral conductor.

2.5.2.4 Dielectric strength

No detailed information is currently available on this test. It should be pointed out that the dielectric strength of equipment that has been installed on site and has not been type-tested must be tested in accordance with the standards EN 60439 / IEC 60439 / DIN VDE 0660.

Use of type-tested equipment

This very complicated and difficult test is unnecessary if type-tested equipment is used.

2.5.2.5 Functional test

Functional testing of equipment

All assemblies, such as low-voltage switchgear assemblies (distribution boards), drive mechanisms, actuators, interlocks, etc., must undergo a functional test. As with visual inspections and trials, the primary objective of this test is to determine whether the equipment has been attached, installed, and configured correctly (according to the specifications in the relevant equipment standards).

This test should also verify that the RCDs installed in the system function correctly and that the permissible touch voltage is not exceeded at the rated residual current $I_{\Delta n}$. The tests are described in Section 2.5.2.2.5.

2.5.2.6 Voltage drop

The required standards are still in preparation.

Limits for voltage drop

It is, therefore, advisable to determine the voltage drop in a three-wire three-phase system on the basis of Fig. 2.5/13 and to compare the results with the specified values (e.g. with the recommendations in HD 384.5.52 / IEC 60364-5-52 / DIN VDE 0100-520 which state that the voltage drop should not exceed 4%).

The characteristics, i.e. the maximum length of Cu cables/conductors, are based on a voltage drop of 1%. The cable/conductor lengths for other voltage drops can be determined by simply multiplying the values by the specified voltage drop percentage.

The determined cable/conductor lengths must be divided by two for alternating current.

If aluminum conductors are used, these lengths must be converted by dividing the

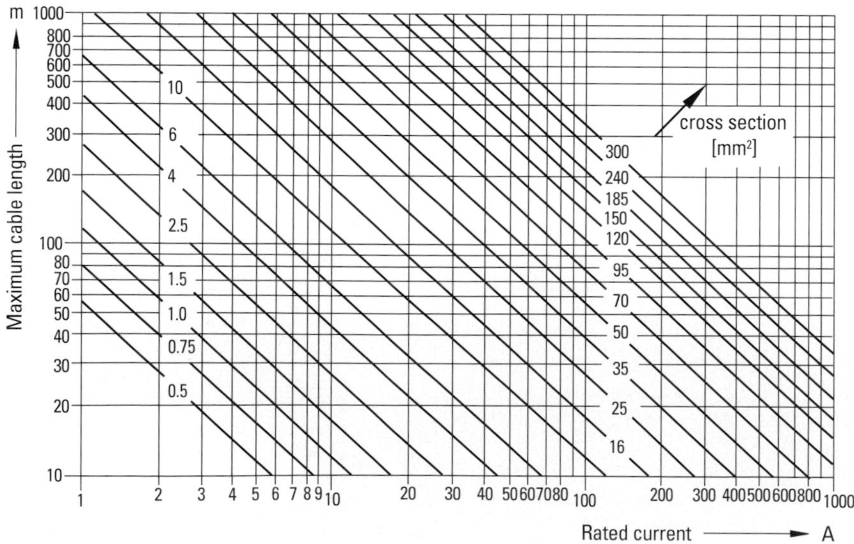

Fig. 2.5/13
Determining permissible cable/conductor length with voltage drop of 1% in three-phase system

determined length by 1.6 for three-phase current and by 3.2 for alternating current.

The characteristics in Fig. 2.5/13 are based on the following simplified calculations:

Three-wire three-phase system
(e.g. where $U_n = 400$ V)

$$\Delta u = \frac{I_n \cdot l \cdot \cos\varphi \cdot \sqrt{3}}{U_n \cdot q \cdot \kappa} \cdot 100\%.$$

Two-wire AC system
(e.g. where $U_n = 230$ V)

$$\Delta u = \frac{I_n \cdot l \cdot \cos\varphi \cdot 2}{U_n \cdot q \cdot \kappa} \cdot 100\%.$$

where:

I_n Rated current in A
l Length of single cable/conductor in m
U_n Nominal voltage in V
q Cross section of conductor in mm^2
κ Conductivity of conductor material

2.5.2.7 Repeat tests

Measures in preparation

Repeat tests are specified in Appendix F of Amendment 2 of IEC 60364-6-61. The necessary measures for these tests are still in preparation.

2.6 Protection against overvoltages of atmospheric origin or switching overvoltages in low-voltage systems

Origin of overvoltages

In low-voltage installations, overvoltages can occur as a result of atmospheric conditions (particularly if power is supplied via overhead power line systems), when circuits with long cables or motors are switched, and in circuits with low-voltage fluorescent lamps with central p.f. correction. A distinction is made between protection against lightning stroke current (high energy) and protection against overvoltage (e.g. remote lightning strikes and switching overvoltages, which propagate at low power levels along the cables).

A distinction is made between the following types of lightning protection:

Protection against lightning stroke current

▷ *External protection against lightning stroke current*
- Protection of buildings and annexes against destruction by fire.
- Protection of buildings and annexes against mechanical destruction caused by the explosive vaporization of water and by the electrodynamic force of the lightning strike (non-incentive lightning).

▷ *Internal protection against lightning stroke current / lightning protection equipotential bonding*
- Protection of personnel within buildings against the effects of the lightning strike.
- Protection of the electrical installation against destruction caused by the uncontrolled penetration of partial lighting stroke currents.

A complete lightning protection system must incorporate both elements (external and internal protection against lightning stroke current). See Fig. 2.6/1.

Arresters

Requirement categories

The lightning protection system can be divided into air terminals (lightning arresters) and devices for limiting transient overvoltages, referred to as arresters of requirement categories A, B, C, and D to DIN VDE 0675-6 (draft). General principles with regard to protection against electromagnetic lightning pulses are defined in IEC 61312-1. Other international standards are currently in preparation.

Arrangement in overhead system

Arresters of requirement category A for system voltages 250 V AC and 500 V AC (external conductor to ground), which can be installed directly in the overhead lines, are available for installation in low-voltage overhead systems.

They are installed in low-voltage overhead systems at intervals of approximately 1000 m or at intervals of approximately 500 m in areas with a high isokeraunic level.

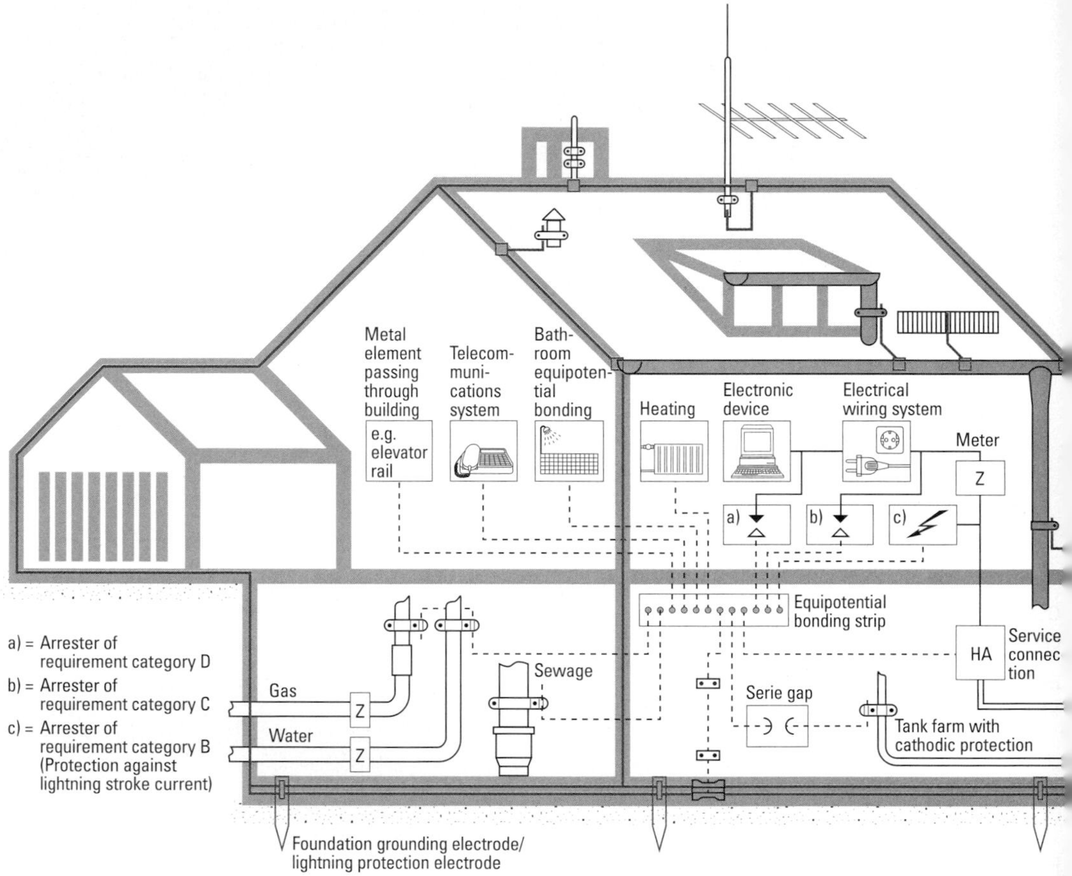

Fig. 2.6/1
External (red) and internal (blue) protection against lightning stroke current with arresters of requirement categories B, C, and D

To provide optimum protection, arresters must be installed at the following locations along the overhead system:

▷ if possible, at every junction,
▷ at the end of each long dead-end feeder,
▷ if possible, close to a service entrance, and
▷ at the cable entrance fittings in the case of intermediate cables.

Level of protection

The level of protection increases with the number of arresters used.

The arresters provide optimum protection if the feeder cables are kept as short as possible and the grounding resistance is kept as low as possible ($\leq 5\ \Omega$). Existing grounding electrodes, e.g. lightning protection, foundation grounding electrodes, and so on, can be used for grounding the arresters.

In TN systems, the electrodes used for grounding the arresters must be connected to the neutral conductors.

In TT systems, an arrester can also be fitted between the neutral conductor and ground.

In IT systems, an arrester can be fitted between the neutral point of the power source and ground (insulation of power source from ground).

Overvoltage protection in electrical installations

Fig. 2.6/2 contains a diagram of an electrical installation (requiring overvoltage protection) which has been divided into different overvoltage categories. The illustration

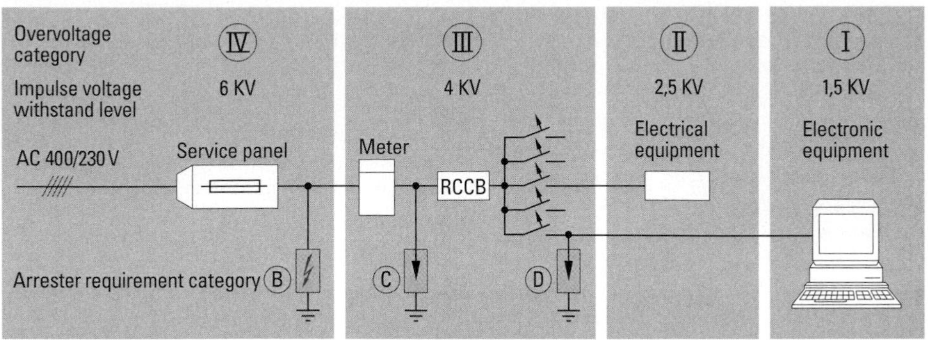

I Equipment which requires a particularly high level of protection
II Equipment to be connected to the permanently installed wiring system
III Items of equipment which are elements of the permanently installed wiring system
IV Equipment at the infeed of the wiring system

Fig. 2.6/2 Overvoltage category (IEC 60664-1 / VDE 0110 Part 1) and impulse voltage withstand level

Overvoltage categories

shows that the permissible overvoltage must be limited by means of suitable surge arresters with increasing proximity to the electrical equipment. Arresters of requirement categories B, C, and D are designed for this purpose.

The following technical data is crucial for installing a system:

Rated voltage

- Rated voltage U_n :
  Maximum voltage permitted at the terminals of the arrester.

Discharge capacity

- Discharge capacity:
  The discharge capacity is specified as a lightning impulse current of waveform 10/350 for B arresters and as nominal discharge capacity of waveform 8/20 for C and D arresters.

  Typical values are as follows:

Arrester of requirement category	B	C	D
Discharge capacity in kA	75	15	3

Protection level

- Protection level U_p :
  The protection level is the voltage value to which overvoltages are restricted by the arrester.

Arrester of requirement category	B	C	D
Protection level U_p in kV	4	1.5	1.5

Dimensioning

When the system is dimensioned, arresters of requirement class B are generally sufficient to discharge overvoltages resulting from direct or indirect lightning strikes, and arresters of requirement category C to discharge overvoltages resulting from switching operations.

Arresters of requirement category D must be selected to protect individual, high-quality electronic devices.

When the system is dimensioned, it is important to choose the correct back-up fuses for the individual arresters.

Back-up fuses

The required back-up fuses must be specified by the manufacturer of the arresters.

Installation location

Surge arresters of requirement categories B and C must always be installed as closely as possible to the service connection and, therefore, upstream of residual-current protective devices (RCDs). This prevents residual-current protective devices from being triggered inadvertently when overvoltages and any associated or subsequent follow currents are discharged.

Arrester types

The Siemens product range includes the following arresters:

▷ Arresters of requirement category B (non-expulsion type) for installation in distribution boards (Fig. 2.6/3).
▷ Arresters of requirement category C (standard type) for installation in distribution

boards, with contacts for remote display and plug-in arrester module both with and without contacts for remote display (Fig. 2.6/4).

▷ Arresters of requirement category D as supplementary module for simple and subsequent mounting on the DELTA grounding-type socket-outlet insert (see Section 4.3.1).

Various accessories, such as busbars and interaction limiting reactors, are available for the arresters.

instabus EIB

Using binary inputs and the arresters with contacts for remote display, it is possible to integrate the overvoltage protection system in building system engineering applications with *instabus EIB* (see Section 4.4).

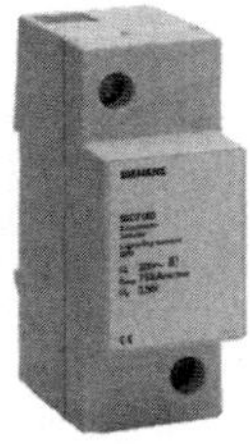

Fig. 2.6/3
Lightning stroke current arrester of requirement category B to DIN VDE 0675-6 (draft)

Fig. 2.6/4
Surge arrester of requirement category C to DIN VDE 0675-6 (draft)

3 Electromagnetic compatibility (EMC)

3.1 Introduction and terminology

Introduction

Electricity has established itself as an important power source in almost all areas of human life. This, together with an increase in the installation density of electrical and electronic devices and systems leads to greater susceptibility to mutual interference and an increase in the extent to which the devices and systems affect and are affected by their environment.

Limit values and test procedures had to be defined to prevent or restrict reciprocal susceptibility and interference.

This gave rise to the term "electromagnetic compatibility (EMC)" together with its regulations, directives, and test procedures. These will be discussed in the following sections. Equipment design which complies with EMC requirements[1)] will also be illustrated using the new control bus system <u>*instabus*</u> *EIB* as an example.

Definition of EMC

EMC is the capability of an electrical device to function satisfactorily in an electromagnetic environment, to which other devices also belong, without causing unacceptable interference to this environment.

This gives rise to the following terms:

EMC terminology

▷ Emitted interference
▷ Interference immunity

Emitted interference, interference suppression

In German, the term "interference suppression" is used as a synonym for the term "emitted interference". Interference suppression in Germany dates back to the high-frequency equipment law of 1927. This law,

Laws/directives

amended in 1949, was one of the first laws to be passed in the new Federal Republic of Germany and continues to serve as the basis for the interference suppression of electrical and electronic equipment. The law was supplemented with the necessary implementation regulations by the Directives of the Federal Ministry for Post and Telecommunications which were published in the Official Gazette (Fig. 3/1).

Standards

The limit values, according to which interference suppression must be implemented, are specified in the Directives of the telecommunications provider (Deutsche Telekom) and in the following standards:

▷ DIN VDE 0871 Industrial, scientific and medical equipment
▷ EN 55013/DIN VDE 0872 Radio and television broadcasting equipment
▷ EN 55104/DIN VDE 0875 Household appliances
▷ DIN VDE 0876 Radio-interference measuring apparatus

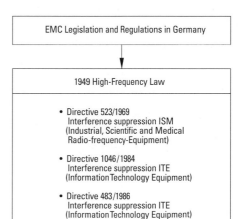

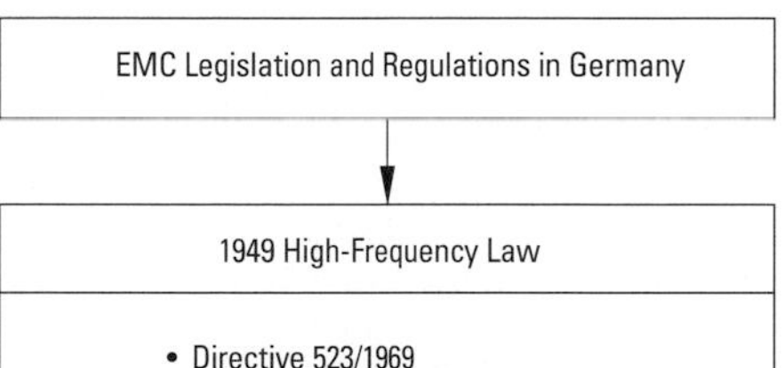

Fig. 3/1
Legislation and Directives for EMC in Germany

[1)] Further information can be found in the following Siemens technical publications:
"EMV und Blitzschutz leittechnischer Anlagen", Franz Pigler, Munich: Siemens 1990;
"CE Conformity Marking", Anton Kohling, Erlangen: Publicis MCD 1996.

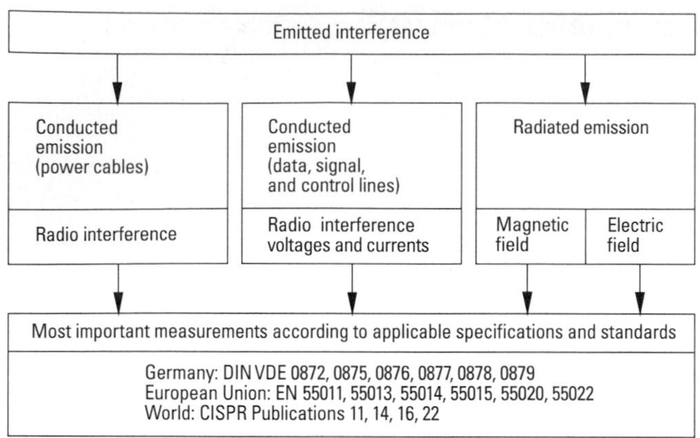

Fig. 3/2 Emitted interference and important measurements in compliance with applicable standards

- ▷ DIN VDE 0877 Methods of measurement of radio interference
- ▷ DIN VDE 0878 Information technology equipment and telecommunications devices and systems
- ▷ DIN VDE 0879 Motor vehicles

In addition, there may also be a number of other industry or user-specific requirements and regulations (e.g. chemical industry, telecommunications, property insurance companies) which have to be complied with.

3.1.1 Emitted interference

Types of emitted interference

Emitted interference is grouped into three categories (Fig. 3/2) in accordance with which all relevant devices and systems must be tested using appropriate methods of measurement.

The tested devices and systems can be divided into limit classes B or A according to their location.

Limit class B

Limit class B includes household appliances, such as domestic electrical devices, audio and television broadcasting devices, home computers, etc. Such appliances must comply with the relevant interference limits. The manufacturer must check these limits and ensure that they are complied with during series production.

Limit class A

Limit class A includes non-domestic products (e.g. industrial machine controllers, automation systems, etc.). The advantage of this class is that the interference limits are higher than for limit class B. The disadvantage, however, is that the products that belong to this class have a limited scope of application.

3.1.2 Interference immunity

Methods of measurement, limit values

The methods of measurement and limit values for interference immunity can be found in DIN EN 61 000-4-X / IEC 61 000-4-X (Fig. 3/3) and in the standards EN 50 082-1 and -2.

Since interference immunity is a crucial characteristic with regard to the quality of the aforementioned devices and systems, it forms an important part of the new EMC legislation for the European Union.

3.1.3 Generic EU directives for EMC

Generic EU directives for EMC

A generic EU directive, which was published in the Official Gazette of the European Union on May 3, 1989 and which must be converted to the domestic law of the respective EU member states, currently applies for EMC. The technical EMC standards, which this directive contains, are defined in European standards (EN) (Fig. 3/4).

The EU directive must be applied for the following electrical/electronic devices and equipment:

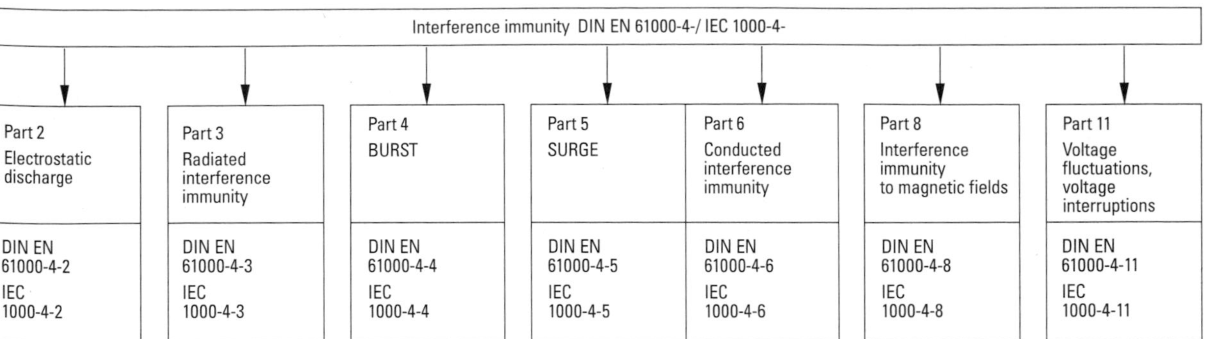

Fig. 3/3
Standards for interference immunity of electrical/electronic devices and systems

- ▷ Private audio and television broadcasting receivers
- ▷ Industrial equipment
- ▷ Mobile radio equipment
- ▷ Commercial, mobile radio and radiotelephone equipment
- ▷ Medical and scientific apparatus and equipment
- ▷ Information technology equipment
- ▷ Household appliances and electronic household equipment
- ▷ Radio equipment for aviation and shipping
- ▷ Electronic teaching and instruction equipment
- ▷ Telecommunication networks and equipment
- ▷ Transmitters for audio and television broadcasting
- ▷ Lights and fluorescent lamps

EMC legislation and standards in EU

Council Directives of May 3, 1989 to harmonize domestic EMC laws of member states

European EMC standards

Fig. 3/4
EMC legislation and standards in the European Union (EU)

Protection-specific objectives

The devices and equipment listed above must be designed and manufactured in such a way that the following protection-specific objectives are fulfilled:

- ▷ The generation of electromagnetic interference must be restricted such that radio and telecommunications equipment and other electrical/electronic devices can function correctly.
- ▷ The devices must be sufficiently immune to electromagnetic interference so that correct operation is possible.

The following European standards have so far been published:

- ▷ EN 50081-1 / VDE 0839 Part 81-1
  Electromagnetic compatibility – Generic emission standard – Part 1: Residential, commercial and light industry
- ▷ EN 50081-2 / VDE 0839 Part 81-2
  Electromagnetic compatibility – Generic emission standard – Part 2: Industrial environment
- ▷ EN 50082-1 / VDE 0839 Part 82-1
  Electromagnetic compatibility – Generic immunity standard – Part 1: Residential, commercial and light industry
- ▷ EN 50082-2 / VDE 0839 Part 82-2
  Electromagnetic compatibility – Generic immunity standard – Part 2: Industrial environment
- ▷ EN 50090-2-2 / VDE 0829 Part 2-2
  Home and building electronic systems (HBES) – Part 2-2: System overview – General technical requirements
- ▷ EN 55011 / VDE 0875 Part 11
  Limits and methods of measurement of

radio disturbance characteristics of industrial, scientific and medical (ISM) radio-frequency equipment

- ▷ EN 55013 / VDE 0872 Part 13
  Limits and methods of measurement of radio disturbance characteristics of broadcast receivers and associated equipment
- ▷ EN 55014-1 / VDE 0875 Part 14-1
  Electromagnetic compatibility – Requirements for household appliances, electric tools, and similar apparatus – Part 1: Emission – Product family standard
- ▷ EN 55014-2 / VDE 0875 Part 14-2
  Electromagnetic compatibility – Requirements for household appliances, electric tools, and similar apparatus – Part 2: Immunity – Product family standard
- ▷ EN 55015 / VDE 0875 Part 15-1
  Limits and methods of measurement of radio disturbance characteristics of electrical lighting and similar equipment
- ▷ EN 55020 / VDE 0872 Part 20
  Electromagnetic immunity of broadcast receivers and associated equipment
- ▷ EN 55022 / VDE 0878 Part 22
  Limits and methods of measurement of radio disturbance characteristics of information technology equipment
- ▷ EN 61000-3-2 / IEC 1000-3-2 / VDE 0838 Part 2
  Electromagnetic compatibility (EMC) – Part 3: Limits – Section 2: Limits for harmonic current emissions (equipment input current up to and including 16 A per phase)
- ▷ EN 61000-3-3 / IEC 1000-3-3 / VDE 0838 Part 3
  Electromagnetic compatibility (EMC) – Part 3: Limits – Section 3: Limitation of voltage fluctuations and flicker in low-voltage supply systems for equipment with rated currents up and including 16 A
- ▷ EN 61131-2 / IEC 1131-2 / VDE 0411 Part 500
  Programmable controllers – Part 2: Equipment requirements and tests

Fig. 3/5
CE mark of conformity for devices which comply with the EMC requirements specified in the EN standards.

CE mark and declaration of conformity

Devices, which are manufactured and tested in accordance with these standards, must be provided with the CE mark of conformity (Fig. 3/5). The manufacturer must also issue a CE declaration of conformity for these devices.

Occurrence and effects of interference

When electrical installations and devices are in operation, electric and magnetic fields as well as conducted voltages and currents of different magnitudes occur which can cause internal system interference and also influence other sensitive installations and devices. This interference is caused, in particular, by switching operations, higher-frequency electrical signals, and by the strong fields generated when lightning strikes. Electromagnetic radiation from transmitters and radio equipment can also have a severe, negative effect. Electronic devices, which function with low voltages and currents and which can, therefore, be influenced by low interference levels, are particularly susceptible.

Fig. 3/6 illustrates the correlation between electromagnetic susceptibility and interference emission.

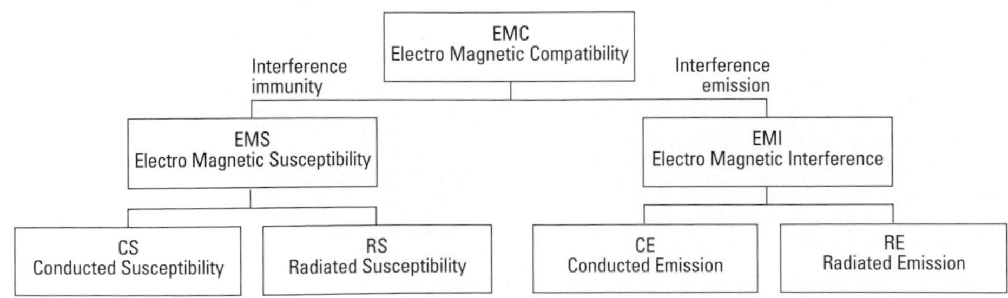

Fig. 3/6 Electromagnetic compatibility (EMC): correlation and terminology

3.2 EMC using building system engineering with *instabus EIB* as an example

The aforementioned susceptibility and interference emission are important factors to be considered, for example, with regard to electrical building system engineering (see Section 4.4). The *instabus EIB* (European Installation Bus) control bus system and associated equipment in conjunction with microelectronics and digital data transmission provide flexible service management in residential and functional buildings. Stringent EMC requirements for the bus systems and bus devices were fulfilled as early as the development phase to ensure reliable, interference-free operation in all areas of application. In addition, simple installation regulations must also be observed when systems are installed.

Simple shielding

With the *instabus EIB* control bus system, information is transmitted between the individual bus devices in the form of telegrams (see Section 4.4.1.1). A shielded MC[1] cable, e.g. PYCYM 2 x 2 x 0.8 mm, is used for the bus line; a twisted-pair core is used both to supply the bus devices with ungrounded safety extra-low voltage (SELV) and for data transmission. The DIN rail in the distribution boards, along which the bus line/data bus is laid, also acts as a shield (Fig. 3/7).

The electrical bit pulses are injected into the bus line and extracted symmetrically via the bus device transformers.

Efficient transmission

This transmission method has the following advantages:

▷ High common-mode rejection
▷ Low radiated interference
▷ Optimum transient protection
▷ Low-resistance line coupling with low interference susceptibility

Low harmonic content

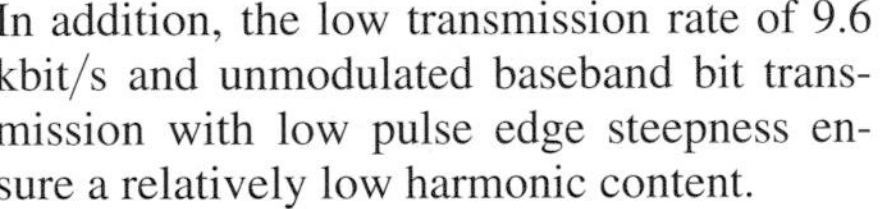

In addition, the low transmission rate of 9.6 kbit/s and unmodulated baseband bit transmission with low pulse edge steepness ensure a relatively low harmonic content.

[1] Measuring and control
[2] 1 character ≅ 11 bits;
8 bits of data information
3 bits of check and control information

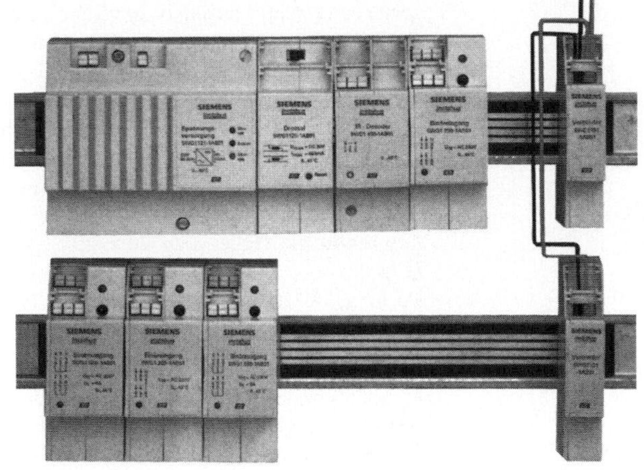

Fig. 3/7
Modular devices of *instabus EIB* control bus system on DIN rail (to DIN EN 50022) with data bus

In the bus devices themselves, multiple sampling is used to identify an incoming useful pulse as a valid data bit and to distinguish it from a spurious pulse. Furthermore, each telegram is safeguarded with the parity bits in each character and by a checksum which is one character long[2].

3.3 EMC equipment design

Design and development

It is important to observe the following points when designing and developing products which are to comply with EMC requirements:

▷ Determine the equipment functions and the EMC environment.
▷ Determine the EMC standards to be fulfilled and the resulting equipment requirements.
▷ Plan the layout and wiring, and elaborate the mechanical design of the device. Any shielding that may be required must also be taken into consideration.
▷ If necessary, run an EMC simulation on a PC or workstation. Whether or not this

serves any useful purpose should be established beforehand. It is nowadays possible, for example, to simulate the EMC characteristics of digital circuits (internal EMC in PCBs), relatively accurately. The EMC characteristics of analog circuits and the overall external EMC behavior of the device are extremely difficult to simulate using the technology currently available.

▷ Perform EMC measurements for each equipment design.

The internal design of electronic devices (this also applies to the *instabus EIB* devices) always consists of one or more PCBs, any shielding that may be required, and the actual housing which can also be manufactured with integrated shielding.

Equipment design

It is, therefore, essential to observe the following points when designing devices which are to comply with EMC requirements:

▷ Arrange preamplifiers immediately next to the sensors.
▷ Select the highest possible operating voltage.
▷ Select low impedances.
▷ Ensure that analog functional modules and digital functional modules are physically separate.
▷ Provide shielding by implementing suitable measures (e.g. grounding wires on the PCB). Attach large grounding surfaces (shielding surfaces) at critical points.
▷ Use short printed conductors and avoid conductor loops e.g. by using multilayers.
▷ Fit inputs/outputs with interference suppression elements.
▷ Connect additional capacitors to the power supply lines of integrated circuits.
▷ Ensure that the surface of the shielding material is in direct contact with other shielding elements or with the PCB.
▷ Improve interference immunity by means of sample and hold, integration, modulation (decoupling of spurious and useful signal).

3.4 Compliance with EMC installation rules

Installation of bus line

Bus lines that have been installed correctly can help ensure interference-free operation of an installation. Large-area loops, which can be formed by the bus line together with other grounded parts, are frequently the cause of EMC interference. If magnetic fields are generated (as a result of lightning striking, for example), overvoltages which can cause flashovers are induced in these loops.

When a system is installed, the following rules must, therefore, be observed in addition to the installation specifications in DIN VDE 0100:

▷ To prevent induced voltages, the bus line and power lines must be laid as closely as possible to each other. This also applies to other grounded parts.
▷ Line ends should be as far away from grounded parts as possible (to reduce coupling capacitances).
▷ The distance between the external lightning protection system and any other components must be as great as possible.
▷ Overvoltage protection devices (gas-filled overvoltage arresters) which are suitable for the bus system must also be used in buildings with lightning protection systems.

4 Electrical installation equipment and systems

4.1 Protection equipment for load circuits

4.1.1 Fuse systems

General

Fuses are high-quality technical devices which, despite their small size, are able to clear even the highest short-circuit currents reliably. For this reason, they are used to protect lines and cables against unacceptable thermal and dynamic effects. On account of their diverse tasks and different operating conditions, they must also ensure a high level of operating safety, low power loss, optimum selectivity ratios, resistance to ageing, and a high level of current limiting.

Standards

Fuse systems are categorized according to their area of application and design:

▷ l.v. h.b.c. (low-voltage high-breaking-capacity) fuses,
▷ D (DIAZED) fuses,
▷ D0 (NEOZED) fuses,
- NEOZED fuses VBG 4,
- MINIZED switch-disconnectors >N< D01,
- MINIZED switch-disconnectors >N< D02,

▷ cylindrical fuses.

The requirements relevant to fuse systems are specified in the standards EN 60269 / IEC 60269 / DIN VDE 0636:

The following are also applicable:

VDE 0638 for D0 fuse combination units (MINIZED switch-disconnectors >N<),

VDE 0680 Part 4 for l.v. h.b.c. detachable fuse handle, and
VDE 0680 Part 7 for gauge piece wrench.

Application, utilization categories

Full-range fuses provide protection against unacceptable overload and short-circuit currents. Fuses in the gL/gG utilization category to EN 60269 / IEC 60269 / DIN VDE 0636 are adapted to the load characteristics of cables and lines and are, therefore, intended primarily for cable and line protection.

Furthermore, gL/gG fuses can also be used to protect motor circuits and to protect switching devices, e.g. contactors and circuit-breakers, against short circuits.

Fuse types

Fuses are divided into plug-in and screw-in systems.

Plug-in systems (l.v. h.b.c. fuses) are intended for operation by specially-trained personnel and do not provide any shock protection or rated current polarization. Insulating covers and barriers for protection against direct contact are, however, available for the majority of plug-in systems.

Screw-in systems (D and D0 fuses) can be operated by untrained personnel which means that shock and rated current polarization protection must be ensured.

L.v. h.b.c. (low-voltage high-breaking-capacity) fuses

Fuse types

L.v. h.b.c. fuses (Fig. 4.1/1) consist of the l.v. h.b.c. fuse-base and the l.v. h.b.c. fuse-link, which is inserted into or removed from the l.v. h.b.c. fuse-base using the l.v. h.b.c. detachable fuse handle. The l.v. h.b.c. detachable fuse handle is not required if the l.v. h.b.c. fuse-base has a hinge attachment.

As with the l.v. h.b.c. fuse switch-disconnectors, the hinge attachment is used to hold and insert the l.v. h.b.c. fuse-link into the contact clips of the l.v. h.b.c. fuse-base (Fig. 4.1/2). L.v. h.b.c. fuse blocks, l.v. h.b.c. fuse switch-disconnectors, or motor fuse-disconnectors can also be used instead of the l.v. h.b.c. fuse-base.

L.v. h.b.c. fuse-bases are secured in place by means of screws. Certain fuse types, however, can also be snapped onto DIN rails to

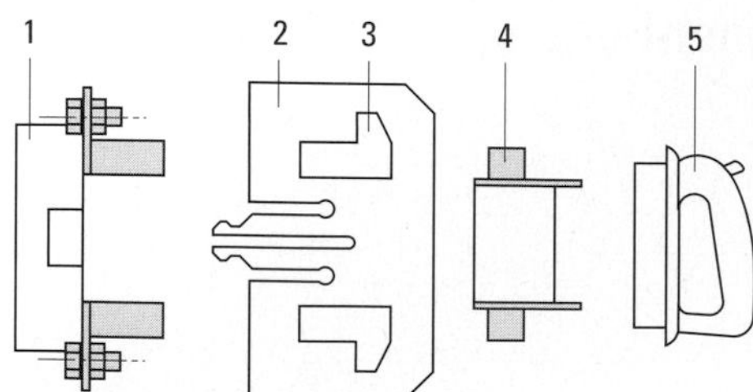

1 L.v. h.b.c. fuse-base
2 Barrier
3 Insulating cover
4 L.v. h.b.c. fuse-link
5 Detachable handle

Fig. 4.1/1 Components of an l.v. h.b.c. fuse

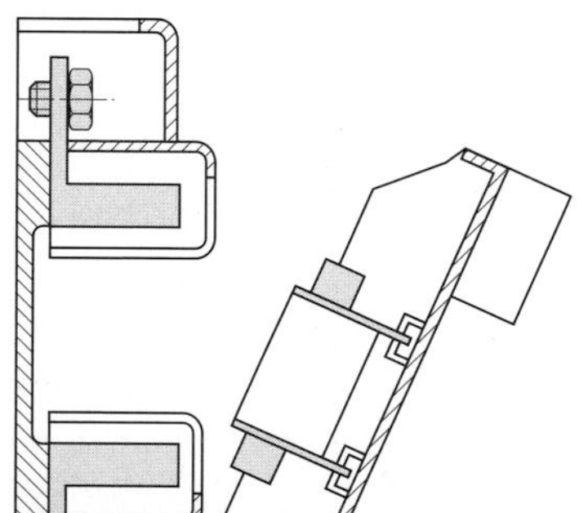

Fig. 4.1/2
L.v. h.b.c. fuse-base with hinge attachment and inserted l.v. h.b.c. fuse-link

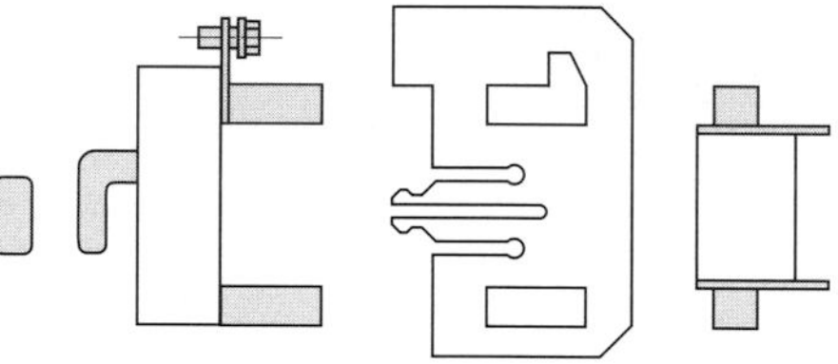

Fig. 4.1/4
L.v. h.b.c. bus-mounting fuse-base for snap-on mounting and locking or plug-on mounting and screwing to Cu busbars

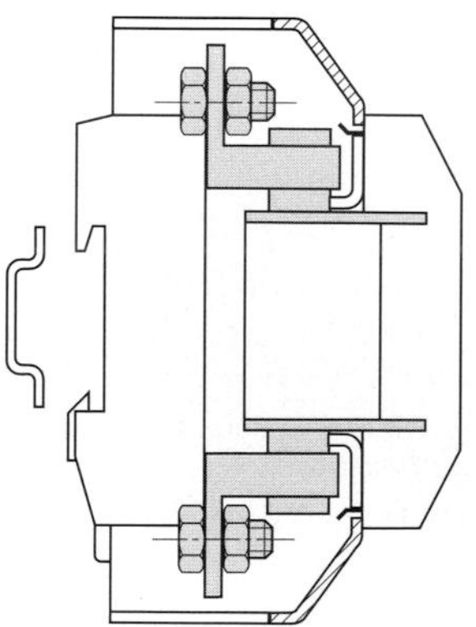

Fig. 4.1/3
Snap-on l.v. h.b.c. fuse-base with shock protection guard, for mounting on DIN rails to DIN EN 50022

DIN EN 50022 (4.1/3). In this case, it is advisable to use the more sturdy 15 mm high DIN rail.

L.v. h.b.c. bus-mounting fuse-bases can be plugged onto busbars (Fig. 4.1/4).

L.v. h.b.c. fuse-links with insulated or live puller lugs (Fig. 4.1/5) are fitted with an indicator which clearly identifies l.v. h.b.c. fuse-links that have been tripped by an overcurrent or short-circuit current.

The indicator can be attached to the front or end of the fuse body. L.v. h.b.c. fuse-links with a combination indicator, which comprises an end and central indicator, facilitate status reading.

These two indicators are connected to each other outside the ceramic cartridge. This avoids holes having to be drilled in the ceramic cartridge which would make it less stable (Fig. 4.1/6).

The l.v. h.b.c. detachable fuse handle can be used with or without a leather sleeve.

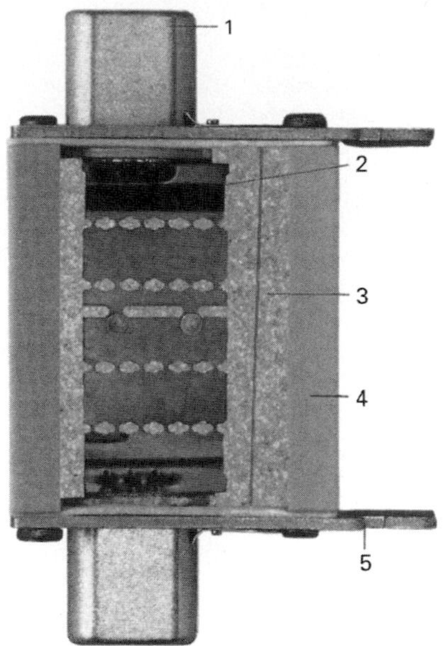

1 Contact pin
2 Fuse-element
3 Quartz sand
4 Steatite body
5 Puller lug, live

Fig. 4.1/5
Cross section of an l.v. h.b.c. fuse-link with live puller lug

Fig. 4.1/6
L.v. h.b.c. fuse-link, size 00, with combination indicator (end and central indicator)

Sizes

L.v. h.b.c. fuses for cable and line protection are available in eight sizes (000, 00, 0, 1, 2, 3, 4, and 4 a) for rated currents between 2 A and 1250 A with a rated voltage of 500 V AC (Table 4.1/1) and five sizes (000, 00, 1, 2, and 3) for rated currents between 2 A and 500 A with a rated voltage of 690 V AC.

The rated currents of these different sizes overlap. This makes planning easier and facilitates subsequent expansion of the electrical installation.

Size 000/00 l.v. h.b.c. fuse-links ≤250 V DC and size 0 to 4 a l.v. h.b.c. fuse-links ≤440 V DC are suitable for use in direct voltage installations.

D (DIAZED) fuses

Fuse types

D or DIAZED fuses (Fig. 4.1/7) consist of a fuse-base (fuse-mount), shock-protection guard, gauge piece, fuse-link, and fuse-carrier (screw cap). The DIAZED fuse-link is screwed into and out of the fuse-base with the fuse-carrier. The gauge piece can take the form of an adapter screw, or an adapter ring in the case of DIAZED bus-mounting fuses. It provides the rated current polarization protection for the fuse-link specified during project planning.

DIAZED fuse-bases can, depending on their type, be screw-mounted, snapped onto DIN rails to DIN EN 50022 (width 35 mm), or plugged onto busbars.

The base-contact bar of the fuse-base is designed to hold the gauge piece. The gauge piece together with the base-contact studs of

Table 4.1/1
Assignment of l.v. h.b.c. fuse-links to l.v. h.b.c. fuse-bases for rated voltages 500 V AC and 690 V AC

Size	Rated current of fuse-bases A	for fuse-links A
000	160	≤ 2 to 100
00	160	≤ 6 to 160
0	160	6 to 160
1	250	≤ 16 to 250
2	400	≤ 35 to 400
3	630	≤ 200 to 630
4	1250	≤ 500 to 1250
4a	1250	630 to 1250

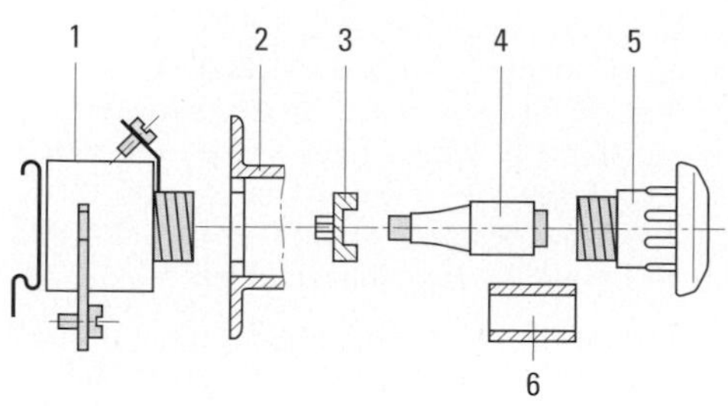

1 Built-in fuse-mount
2 Shock-protection guard
3 Gauge piece (adapter screw)
4 Fuse-link
5 Fuse-carrier (screw cap)
6 Liner (required if the next smallest fuse-link size is used)

Fig. 4.1/7 Components of a DIAZED fuse

the DIAZED fuse-link, which have different diameters depending on the rated current, ensures rated current polarization. The term "rated current polarization" means that fuse-links with a rated current higher than that intended cannot be used. Consequently, the connected cables and lines are reliably protected against unacceptable overloads.

Each rated current is assigned a color so that it can be identified quickly. Both the gauge piece and the indicator of the fuse-link bear this color (Table 4.1/2).

Sizes

DIAZED fuses for cable and line protection are available in four sizes (ND_z, D II, D III, and D IV) for rated currents between 2 A and 100 A with a rated voltage of 500 V AC (Table 4.1/3) and in one size (D III) for rated currents between 2 A and 63 A with rated voltages of 690 V AC and 600 V DC. Although the rated current intensities of the individual sizes do not overlap, liners can be used in the fuse-carriers to allow size D II DIAZED fuse-links, for example, to be used in size D III DIAZED fuse-bases.

Size ND_z with E 16 thread for rated currents between 2 A and 25 A with a rated voltage of 500 V AC/DC is an exception.

D0 (NEOZED) fuses

Fuse types

The design of D0 or NEOZED fuses (Fig. 4.1/8) is identical to that of the DIAZED fuses. NEOZED fuses are, however, better suited to today's installation conditions since the dimensions of the fuse-bases are matched to the 18 mm module size of miniature circuit-breakers and DIN-rail-mounted devices.

Size D01 and D02 NEOZED fuse-bases have the same module size as well as the same width and fixing dimensions. The module size is 27 mm (1.5 modular units), i.e. 1.5 times that of the miniature circuit-breaker >N< (Fig. 4.1/9). D03 fuse-bases, however, have a module size of 45 mm, which corresponds to 2.5 times that of the single-pole miniature circuit-breaker >N<.

Table 4.1/2
Identification colors of the gauge pieces and indicators of DIAZED and NEOZED fuse-links

DIAZED fuse-links Size	Rated current intensity A	Identification color	NEOZED fuse-links Size
D II	2 4 6 10 16	Pink Brown Green Red Gray	D01
	20 25	Blue Yellow	D02
D III	35 50 63	Black White Copper	
D IV	80 100	Silver Red	D03

Table 4.1/3
Assignment of DIAZED fuse-links to DIAZED fuse-bases for rated voltage 500 V AC/DC

Size	Thread	Rated current of fuse-bases A	For fuse-links A
ND_z	E 16	25	2 to 25
D II	E 27	25	2 to 25
D III	E 33	63	35 to 63
D IV	R 1¼"	100	80 to 100

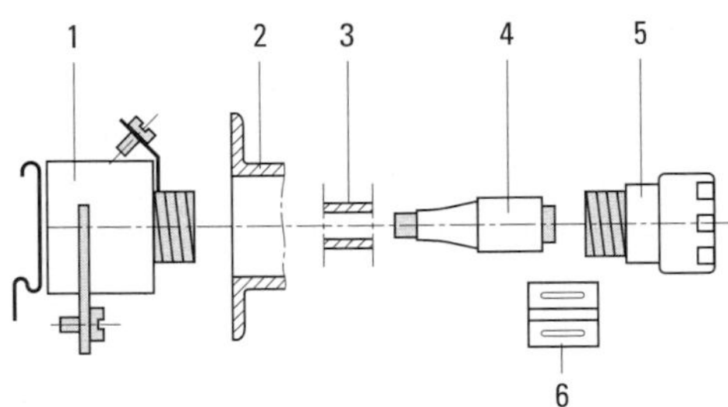

1 Built-in fuse-mount
2 Shock-protection guard
3 Adapter sleeve
4 Fuse-link
5 Fuse-carrier (screw cap)
6 Retaining spring (required if the next smallest fuse-link size is used)

Fig. 4.1/8 Components of a NEOZED fuse

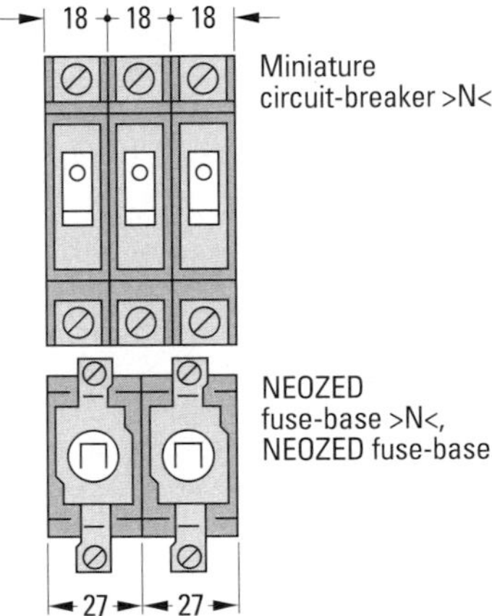

Fig. 4.1/9
Module sizes of the built-in fuse-bases D01 and D02, referred to the miniature circuit-breakers >N<

Table 4.1/4
Assignment of NEOZED fuse-links to NEOZED fuse-bases for rated voltages 400 V AC and 250 V DC (also applies to NEOZED fuse-links and NEOZED fuse-bases >N<)

Size	Thread	Rated current of fuse-bases A	For fuse-links A
D01	E 14	16	2 to 16
D02	E 18	63	20 to 63
D03	M 30 × 2	100	80 to 100

NEOZED fuse bases >N<

The extra-low NEOZED fuse-bases >N< were designed for the very flat N-type small distribution boards, currently known as SIMBOX 63 small distribution boards (see Section 1.4.5).

Sizes

NEOZED fuses for cable and line protection are available in three sizes (D01, D02, and D03) for rated currents between 2 A and 100 A with rated voltages of 400 V AC and 250 V DC (Table 4.1/4).

NEOZED fuses VBG 4

VBG 4-type NEOZED fuses are shock-protected in accordance with the accident prevention regulations of the German employers' liability insurance association (VBG). VBG 4-type NEOZED fuses comply with the standard EN 60269 / IEC 60269 / DIN VDE 0636 and bear the German VDE symbol.

These fuses are a more advanced version of the standard NEOZED fuses. Their design and dimensions, conforming to DIN 43880, have been adapted to the miniature circuit-breakers and, in particular, to the MINIZED switch-disconnectors >N< D02.

Construction

The VBG 4-type NEOZED fuses consist of a split, heat-resistant plastic housing (degree of protection IP 20), the fuse-link, the adapter sleeve to ensure the rated current polarization of the selected fuse-link, and the fuse-carrier (screw cap).

Sizes

They are available in the sizes D01 and D02 with one or three poles (Fig. 4.1/10).

The plastic housing contains the contact elements, the multiple thread gland, the spring-loaded base contact to ensure good contacting, and the snap-on slide for fast installation.

To allow conductors to be connected, the ends of the contacts are fitted with modern box elevator terminals with +/– terminal screws (suitable for connection work using a battery-powered screwdriver). They are designed for conductor cross sections ≤35 mm². The copper cross sections of the contact elements for both sizes are designed for a rated current of 63 A. The input terminals are suitable for connecting coupling busbars and individual conductors.

The housing halves are designed in such a way that, once they are joined together, a screwdriver can still be inserted in the recesses in which the terminal screws are located (this also means that the screws cannot be lost).

The fuse-bases are mounted on DIN EN 50022 rails, and can be flush-mounted with other DIN-rail-mounted devices (see Section 4.2).

Sizes D01 and D02 are designed as single and three-pole versions for rated voltages 400 V AC and 250 V DC, as well as for 440 V AC for the shipbuilding industry. NEOZED fuse-links with a rated current between 2 A and 63 A and the appropriate adapter sleeves can be used.

The mounting depth is 70 mm. The module width is 27 mm (1.5 modular units) for both sizes. Accessories, such as single busbars and three-phase busbars with single and double-input and output terminals, complete the product range.

MINIZED switch-disconnectors >N< D01 and D02

Design

The MINIZED switch-disconnector >N< is a NEOZED fuse combination unit in which a switching element is in series with a NEOZED fuse. The MINIZED switch-disconnector >N< provides maximum shock protection since all accessible metal parts are de-energized when the power supply is disconnected so that the NEOZED fuse-link can be replaced without any risk of injury.

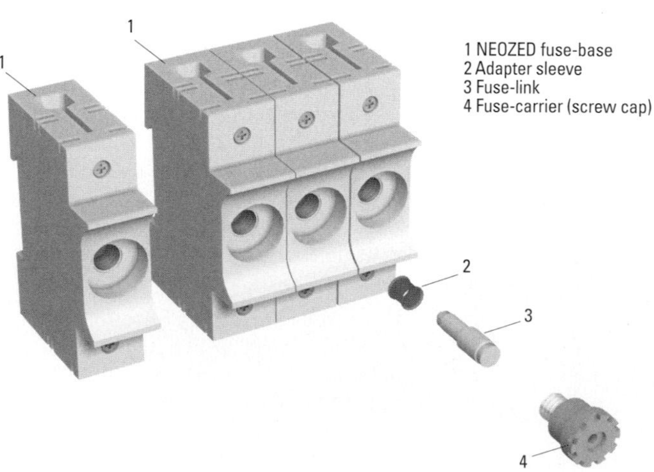

Fig. 4.1/10 VBG 4-type NEOZED fuse, one and three-pole

The MINIZED switch-disconnectors >N< D01 and D02 satisfy the requirements of utilization category AC 22 in compliance with VDE 0638, i.e. switching of mixed resistive and inductive loads including slight overloads.

ated breaking apacity

The rated breaking capacity when used together with NEOZED fuse-links is 50 kA.

INIZED switch-isconnector >N< D01

The MINIZED switch-disconnector >N< D01 is suitable for size D01 NEOZED fuse-links with a maximum rated current of 16 A.

odule size 8 mm per pole

The MINIZED switch-disconnector >N< D01 with a width of 18 mm (1 modular unit) per pole is intended specifically for D01 NEOZED fuse-links between 2 A and 16 A.

A mechanical lock-out prevents the system from being energized if the NEOZED fuse-links are not inserted correctly.

The system can only be energized if the NEOZED fuse-links are inserted and contact correctly in the withdrawable unit. NEOZED fuse-links and the switch-disconnector are connected in series in a single structural unit.

Once the power supply has been switched off, all accessible metal parts are de-energized so that the fuse-link can be changed safely by the operator. The circuit states can be seen in Fig. 4.1/11. No adapter sleeves are available for the MINIZED switch-disconnector >N< D01 for the different rated currents of the NEOZED fuse-links as they are not necessary.

MINIZED switch-disconnector >N< D02

The MINIZED switch-disconnector >N< D02 can only be energized if the NEOZED fuse-link or the dummy insert included in the scope of supply is screwed in correctly so that the closing lock-out is released (Fig. 4.1/12). This means that there must always be sufficient contact force acting on the fuse-link without intervention by the operator.

Module size 27 mm per pole

The MINIZED switch-disconnector >N< D02 has a snap-on mounting and, with a mounting depth of 55 mm, corresponds to the dimensions of the N system (see Section 4.2.1). It has a module size of 27 mm per pole and can, therefore, be installed instead of a size D01 or D02 NEOZED fuse-base. It is particularly easy to mount the switch-disconnector on 12 mm × 5 mm busbars with a centering spacing of 40 mm using the busbar adapter, which is simply plugged on and secured with a screw. The MINIZED switch-disconnector >N< D02 is designed for D02 NEOZED fuse-links 20 A – 63 A, but is also suitable for D01 NEOZED fuse-links 2 A – 16 A provided that retaining springs and gauge pieces are used.

Since the MINIZED switch-disconnector >N< D02 is protected against unauthorized power tapping, it can be used freely in distri-

Infeed from below

Energized

De-energized

De-energized
Off-circuit replacement of fuse-link

Fig. 4.1/11 Circuit states of the MINIZED switch-disconnector >N< D01

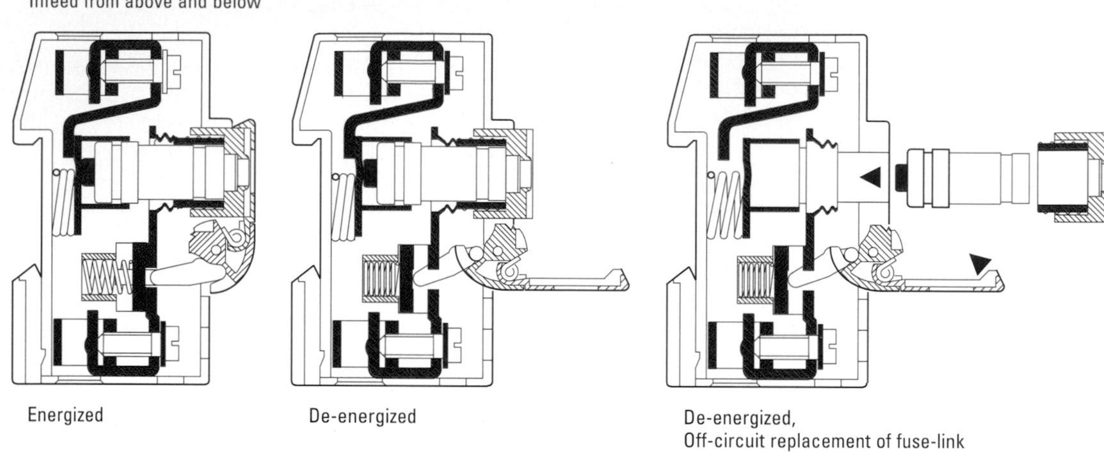

Fig. 4.1/12 Cross section of the MINIZED switch-disconnector >N< D02

bution systems and for applications involving intermediate meters.

Cylindrical fuses

Compact design Cylindrical fuses are similar in shape and size to the DIN-rail-mounted devices (see Section 4.2). They have a smaller module size than NEOZED and DIAZED fuses so that they require even less space when installed in distribution boards. They are snapped onto DIN rails in compliance with DIN EN 50022.

Sizes Cylindrical fuses consist of a fuse housing closed on all sides (fuse-mount) with an integrated snap-on mounting, a hinge attachment which holds the fuse-link, and a cylindrical fuse-link (Fig. 4.1/13). The mounts for cylindrical fuses are available in four sizes. The smallest mount is 8.5 × 31.5 (the first number indicates the diameter and the second number the length of the fuse-link in millimeters). The other sizes are:

10 × 38, 14 × 51, and 22 × 58.

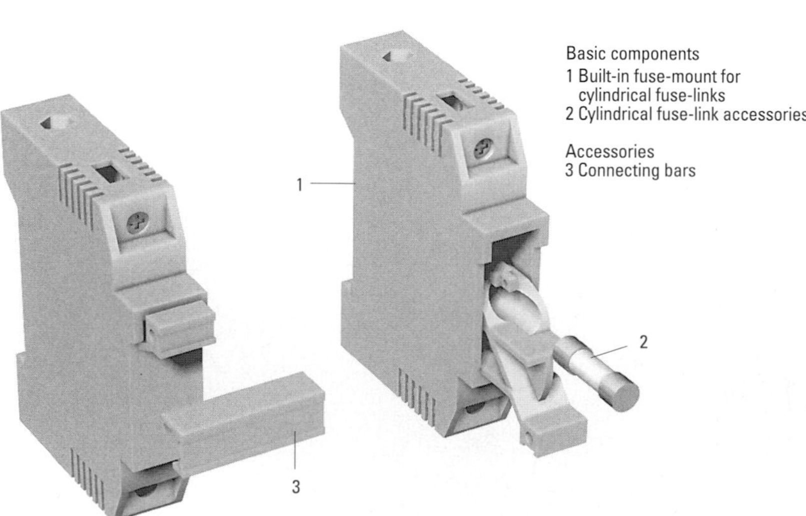

Fig. 4.1/13 Construction of a cylindrical fuse

The unit widths are adapted to the module size 18 mm (1 modular unit) and multiples thereof (Table 4.1/5). The following versions are available for each size: 1-pole, 1-pole with N conductor, 2-pole, 3-pole, and 3-pole with N conductor.

Table 4.1/6 shows the different fuse-link/ fuse-mount assignments.

The fuse-mount has ventilation slots for heat dissipation. This prevents the maximum permissible temperature increase of 70 K (as specified in HD 630.2.1 S2 / IEC 60629-2-1) from being exceeded in the mount.

In accordance with HD 630.2.1 S2 / IEC 60269-2-1, gauge pieces which ensure polarization of fuse-links with different rated currents are not provided.

use-mount

The fuse-mount consists of two plastic shell halves, which have two contact clips and a hinge attachment on the inside to hold the fuse-link.

The hinge attachment allows the fuse-link to be changed by hand without tools. The fuse-link in the hinge attachment should only be swung in or out when de-energized. This should be carried out by a trained technician only. Once the hinge attachment has been swung out, the operator can safely change the fuse-link.

Conductor connection

The conductors are connected by means of box elevator terminals with +/– screws (posi-drive). Since sizes 14 × 51 and 22 × 58 are not symmetrical (due to space and design-related reasons), they extend well below the DIN rail once mounted. When planning domestic distribution boards, it is, therefore, essential that the tier spacing between the DIN rails is at least 175 mm for these two sizes.

Table 4.1/5 Module widths and mounting depths of cylindrical fuse-mounts

Size	Module width					Mounting depth
∅× length mm	1-pole mm	1-pole +N mm	2-pole mm	3-pole mm	3-pole +N mm	mm
8.5 × 31.5	18	36	36	54	72	70
10 × 38	18	36	36	54	72	70
14 × 51	26	52	52	78	104	76
22 × 58	35	70	70	105	140	80

Table 4.1/6
Assignment of cylindrical fuse-links to cylindrical fuse-mounts as a function of the rated voltages 400 V AC and 500 V AC

Size ∅× length mm	Rated current Fuse-mount A	Fuse-link A	Rated voltage V	Rated switching capacity kA
8.5 × 31.5	20	2, 4, 6, 10, 16, 20	400 AC	20
10 × 38	32	2, 4, 6, 8, 10, 12, 16, 20, 25, 32	500 AC 500 AC 400 AC	100
14 × 51	50	4, 6, 8, 10, 12, 16, 20, 25, 32, 40, 50	500 AC 500 AC 400 AC	100
22 × 58	100	8, 10, 12, 16, 20, 25, 32, 40, 50, 63, 80, 100	500 AC 500 AC 500 AC 400 AC	100

Standards

Cylindrical fuses are manufactured and tested in accordance with the standards IEC 60269–2, IEC 60629–2-1, EN 60269–2, NF C63 210, C63211 and C60200, NBN C63269-2 and 2-1. They are primarily intended for the European market, in particular for the countries France, Italy, Portugal, Spain, and Belgium.

Application

Cylindrical fuse-links in utilization category gG are used as full-range fuses for cable and line protection and those in utilization category aM as partial-range fuses for switchgear and motor protection.

If required, individual fuse-mounts can be connected via the opening handle of the hinge attachment and using the connecting bar (included in accessories) so that they can operate in unison. A fixture is provided on the opening handle and connecting bar to allow a marking tag to be attached.

Cylindrical fuse-links

Cylindrical fuse-links consist of a ceramic cartridge, fuse-element, high-grade quartz sand, ring caps, and contact caps (Fig. 4.1/14). The fuse-element is embedded in chemically pure sand of a selected grain size, runs parallel to the interior wall of the ceramic cartridge, and is welded or secured at both ends with the ring caps. The attached contact caps seal off the fuse body.

Fuse-links with a rated current ≤ 6 A have fuse-elements made of a thin Cu round-section wire. Fuse-links ≥ 8 A have fuse-elements made of a bare or silver-plated Cu board with punch-outs. In utilization category gG, the links have precision-soldered "bottleneck" sections.

Technical data for the cylindrical fuse-links can be found in Tables 4.1/5 and 4.1/6. The prearcing-time/current characteristics are shown in Fig. 4.1/15.

Technical data on fuse-links

Fuse-links, functional description

Fuse-links for cable and line protection are designed in such a way that, together with the arc extinguishing medium (quartz sand), they operate reliably from the lowest fusing current to the rated breaking current. They therefore protect the connected cables and lines against unacceptable overloads (Fig. 4.1/16).

Prearcing-time current characteristic

The prearcing-time/current characteristics (Figs. 4.1/13, 4.1/17a, 4.1/17b, 4.1/17c, 4.1/22, and 4.1/23) illustrate how the fuses operate. They indicate the virtual time as a function of the prospective short-circuit current, i.e. the prearcing time.

The prearcing-time/current characteristics shown in the diagrams were recorded at a temperature of 10 °C and apply to non-preloaded fuse-links. Since deviations resulting from manufacturing tolerances are unavoidable, the diagrams show average prearcing-time/current characteristics, the maximum deviation of which may, according to the relevant standards, be ± 10% along the current axis (prearcing-time/current ranges to EN 60269/IEC 60269/DIN VDE 0636). Siemens fuse-links deviate by ± 5% (max.) which enhances selective properties (Fig. 4.1/18).

Selectivity

Selectivity between series-connected fuses always exists if only the fuse, which is closest to the fault source, trips in the event of a fault. Fuses connected upstream should, therefore, not trip so that the fault-free system components continue to function normally (Fig. 4.1/19).

In radial systems, fuses with different rated currents can be assigned easily using selectivity tables. If no selectivity tables are avail-

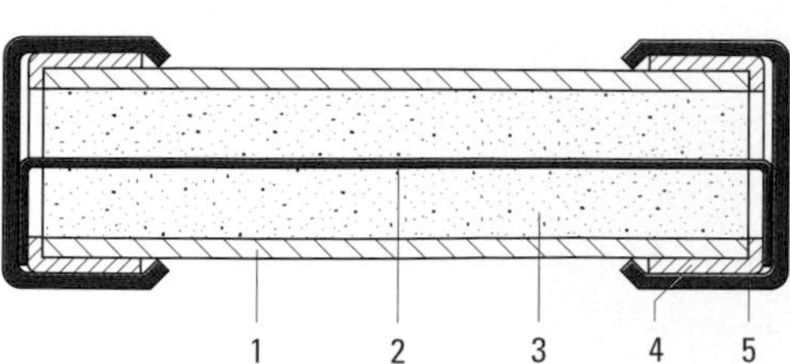

1 Ceramic cartridge
2 Fuse-element
3 High-grade quartz sand
4 Ring cap
5 Contact cap

Fig. 4.1/14 Cross section of a cylindrical fuse-link, size 10 × 38

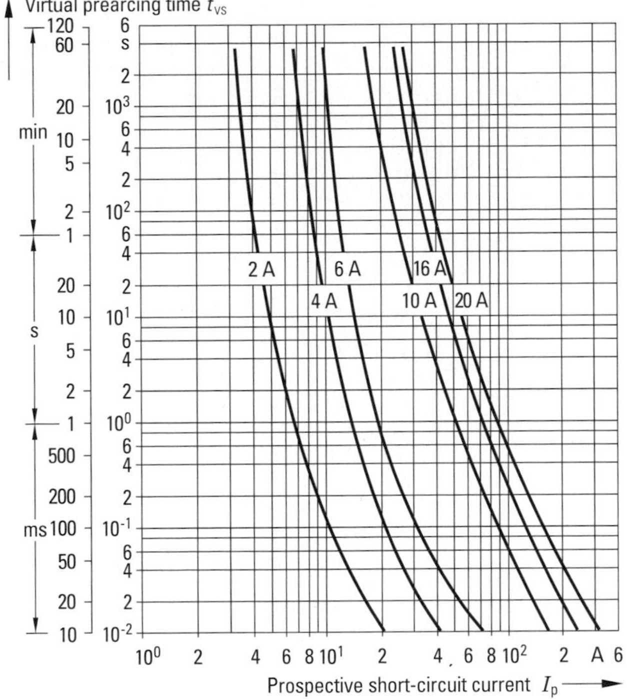

Fig. 4.1/15 Prearcing-time/current characteristics of cylindrical fuse-links, utilization category gG

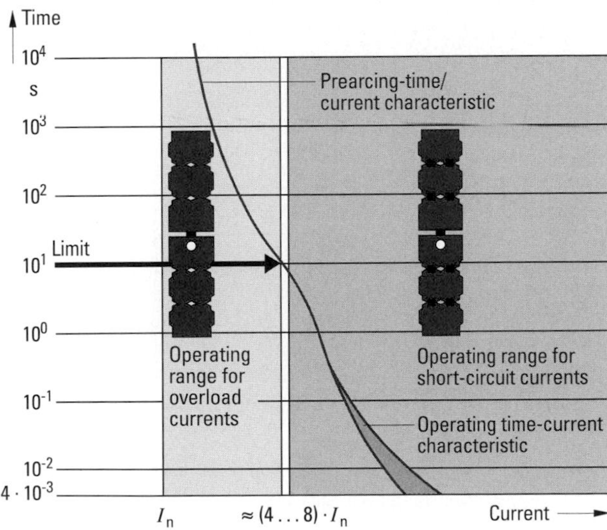

Fig. 4.1/16 Operating ranges of fuse-links in utilization category gL/gG

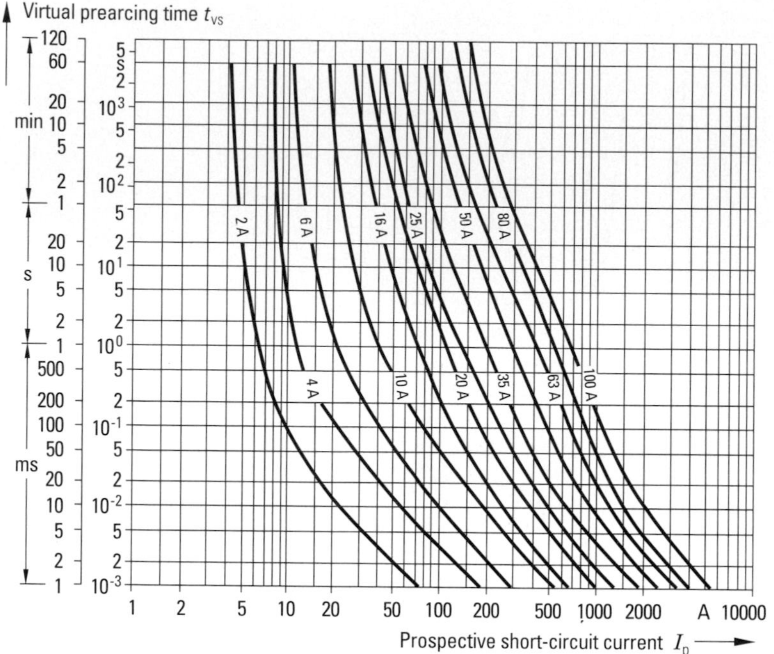

Fig. 4.1/17 a
Prearcing-time/current characteristics of 400 V AC and 250 V DC NEOZED fuse-links, utilization category gL/gG

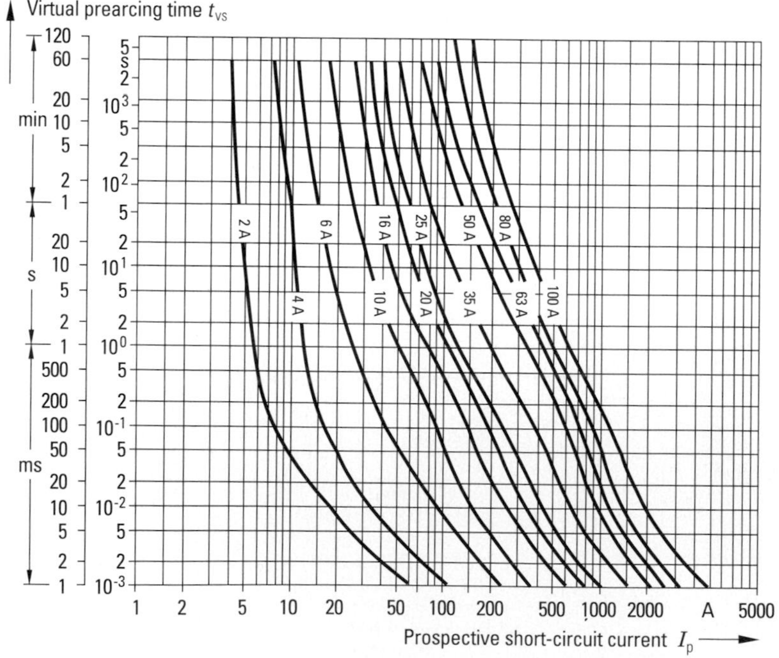

Fig. 4.1/17 b
Prearcing-time/current characteristics of 500 V AC/DC DIAZED fuse-links, utilization category gL/gG

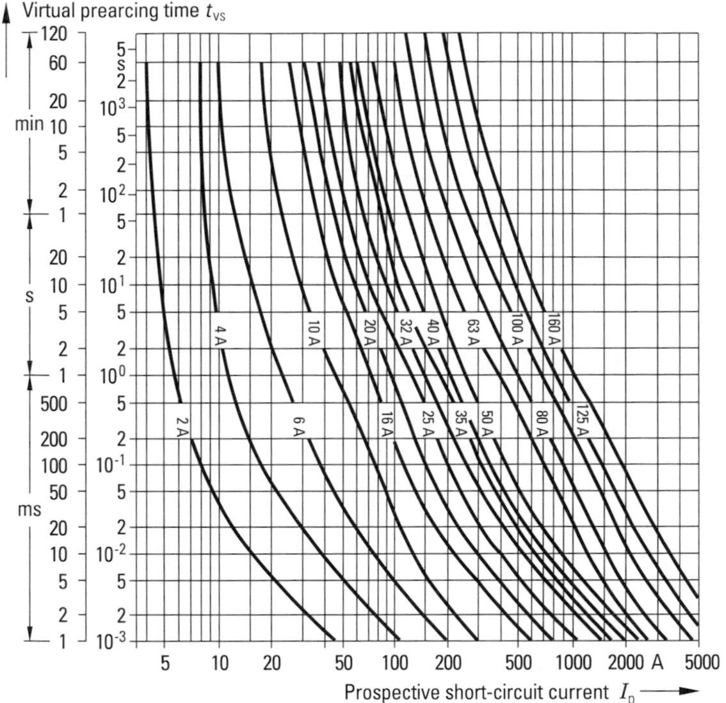

Fig. 4.1/17 c
Prearcing-time/current characteristics of size 00, 500 V AC and 250 V DC l.v. h.b.c. fuse-links, utilization category gL/gG

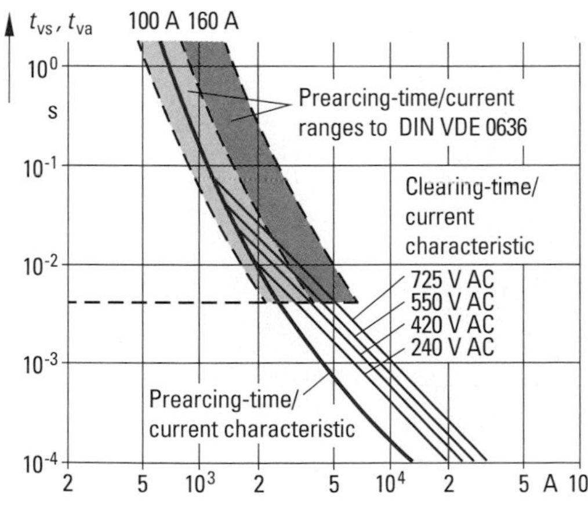

Fig. 4.1/18
Clearing-time/current characteristics of various test voltages using an l.v. h.b.c. fuse-link, size 00/100 A, as an example

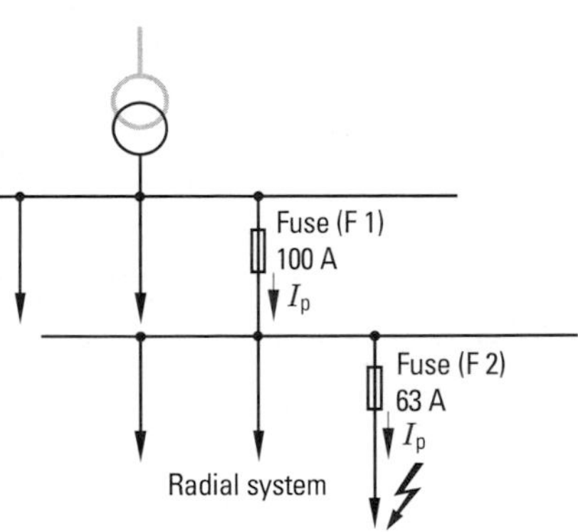

Fig. 4.1/19
Selectivity means that only the fuse, which is closest to the system fault (in this case, only the 63 A fuse), trips in the event of a fault.

able, the selectivity ratio of 1 : 1.6 (e.g. 63 A : 100 A) can be applied to achieve selective grading with fuses (see Section 1.3.3) which comply with the relevant standards (Fig. 4.1/20).

Power loss

Although the power losses associated with fuses are unavoidable, they should be kept as low as possible to increase cost efficiency, reduce power consumption, and eliminate unacceptable temperature rises. Siemens l.v. h.b.c., DIAZED, NEOZED, and cylindrical fuse-links are characterized by extremely low power loss levels which are below the acceptable values specified in EN 60269 / IEC 60269 / DIN VDE 0636.

Breaking capacity

Siemens l.v. h.b.c., DIAZED, NEOZED, and cylindrical fuse-links are characterized by their high level of switching reliability. With overcurrent breaking operations, the fusing temperature causes the breaking operation to occur at the point at which the cross section of the fuse-element is at its narrowest. With short-circuit currents, the fusing temperature is reached abruptly at all points along the fuse-element at which the cross section narrows.

In order to ensure reliable breaking even at the highest possible short-circuit current, the rated breaking capacity of 50 kA with alternating current, as specified in EN 60269 / IEC 60269 / DIN VDE 0636, is in some cases more than doubled by Siemens l.v. h.b.c., DIAZED, NEOZED, and cylindrical fuse-links.

In addition to a high rated breaking capacity, high current limiting is also important for protecting electrical installations reliably against excessively high short-circuit currents (Fig. 4.1/21). Thanks to their excellent technical characteristics, Siemens fuse-links have a very high level of current limiting (Fig. 4.1/22).

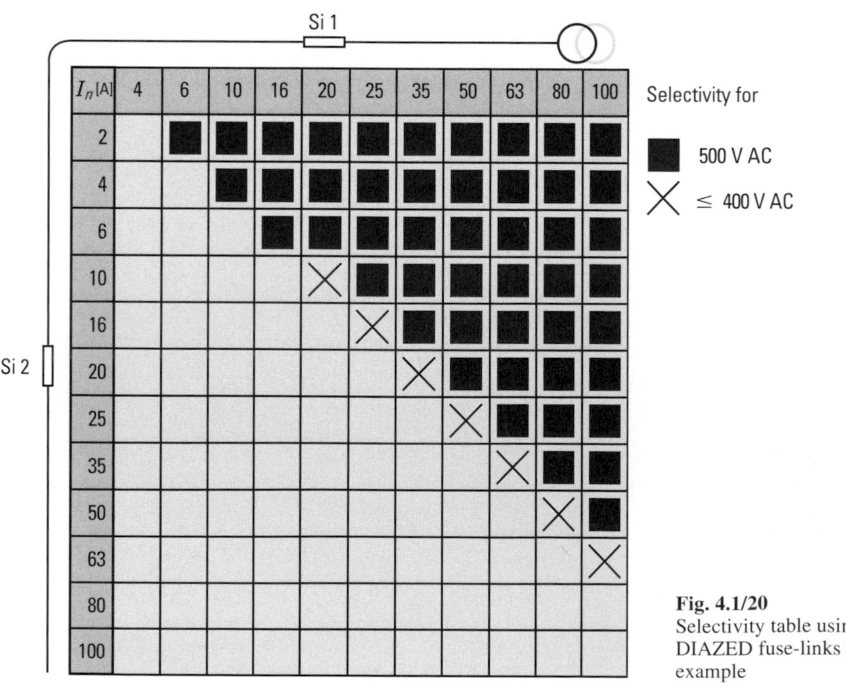

I_n [A]	4	6	10	16	20	25	35	50	63	80	100
2		■	■	■	■	■	■	■	■	■	■
4			■	■	■	■	■	■	■	■	■
6				■	■	■	■	■	■	■	■
10					X	■	■	■	■	■	■
16						X	■	■	■	■	■
20							X	■	■	■	■
25								X	■	■	■
35									X	■	■
50										X	■
63											X
80											
100											

Fig. 4.1/20
Selectivity table using DIAZED fuse-links as an example

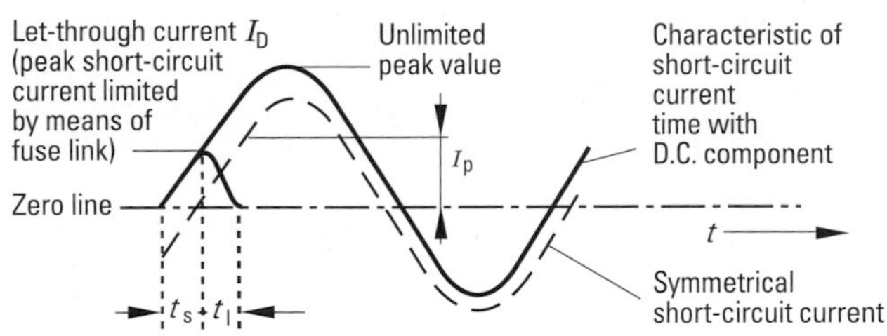

t_s Prearcing time
I_p Prospective short-circuit current at installation location (kA) (r.m.s. value)
t_l Arc extinction time

Fig. 4.1/21
Current limiting by means of l.v. h.b.c. fuse-link

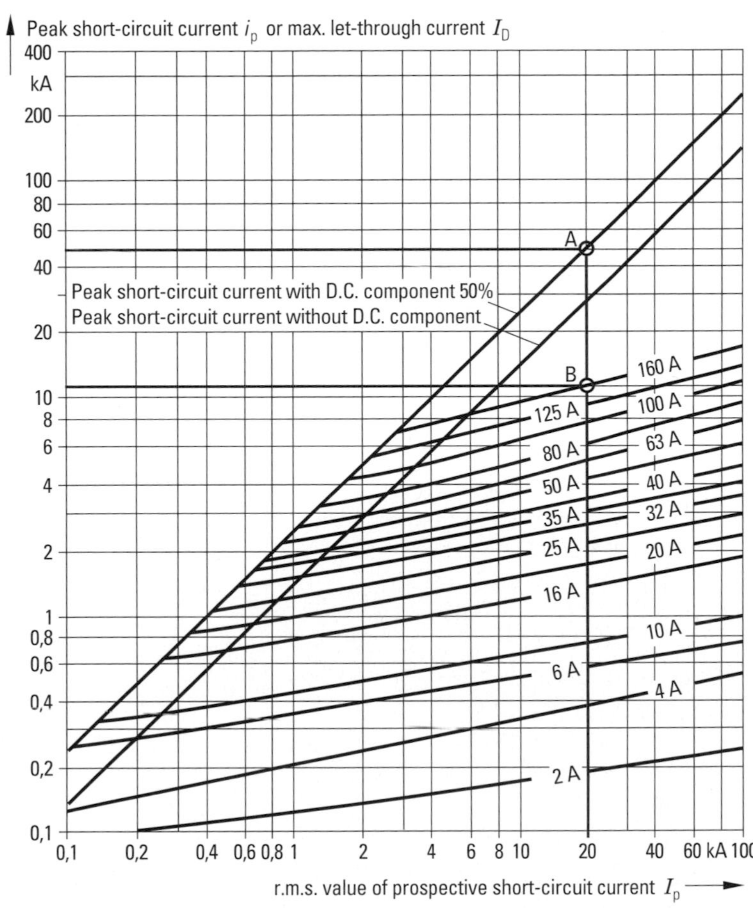

Fig. 4.1/22
Let-through current characteristics of l.v. h.b.c. fuse-links, size 00 / 500 V AC and 250 V DC, utilization category gL/gG

Example

L.v. h.b.c. fuse-links for 160 A rated current.
Prospective short-circuit current $I_p = 20$ kA (calculated r.m.s. value),
D.C. component = 50 % (assumed).

At which short-circuit current will the limited current peak (see Fig. 4.1/22) be reached when the l.v. h.b.c. fuse-links respond and how high would the peak short-circuit current i_p be if there were no l.v. h.b.c. fuse-links?

Result:

With l.v. h.b.c. fuse-links (160 A), the limited current peak is reached at 12 kA (the vertical line $I_p = 20$ kA intersects the characteristic of the 160 A l.v. h.b.c. fuse-link at point B).

Without l.v. h.b.c. fuse-links, the peak short-circuit current i_p is 50 kA (the vertical line $I_p = 20$ kA intersects the 50%-line at point A).

Ambient temperature

With Siemens fuse-links, the ambient temperature of 20 °C for fuses, as specified in the relevant standards, permits lengthy overload currents up to 1.15 times the rated current without any signs of deterioration. The special design of the fuse-elements also allows Siemens l.v. h.b.c., DIAZED, NEOZED, and cylindrical fuse-links to be used at ambient temperatures between – 5 °C and + 45 °C without reduction factors being taken into consideration.

In addition to fuses in utilization category gL/gG for cable and line protection, SILIZED fuse-links are also available which, with their extremely fast clearing characteristic, are matched to the load characteristics of the Siemens power thyristors and diodes (Fig. 4.1/23).

Busbar system for space-saving fuse-base arrangements

60 mm busbar system

The 60 mm busbar system is available to allow space-saving NEOZED, DIAZED, and l.v. h.b.c. fuse-base arrangements to be installed in incoming panels of distribution boards (see Section 1.4.5).

The system is modular in design and consists of the following components:

busbar holders, Cu conductor bars, input terminal boards, input and output terminals, as well as quick-assembly shock-protection components. The system also includes a range of rail-adaptable switchgear, fusegear, and installation equipment (Fig. 4.1/24).

The following can be assembled using the busbar system components:

- ▷ open busbar runs without shock-protection components (Fig. 4.1/25a)
- ▷ enclosed busbar runs with shock-protection components (Fig. 4.1/25b),
- ▷ equipped busbar runs (Fig. 4.1/25c).

With the 60 mm busbar system and the rail-adaptable installation equipment, it is possible to install a larger number of devices in distribution boards or to choose smaller distribution boards.

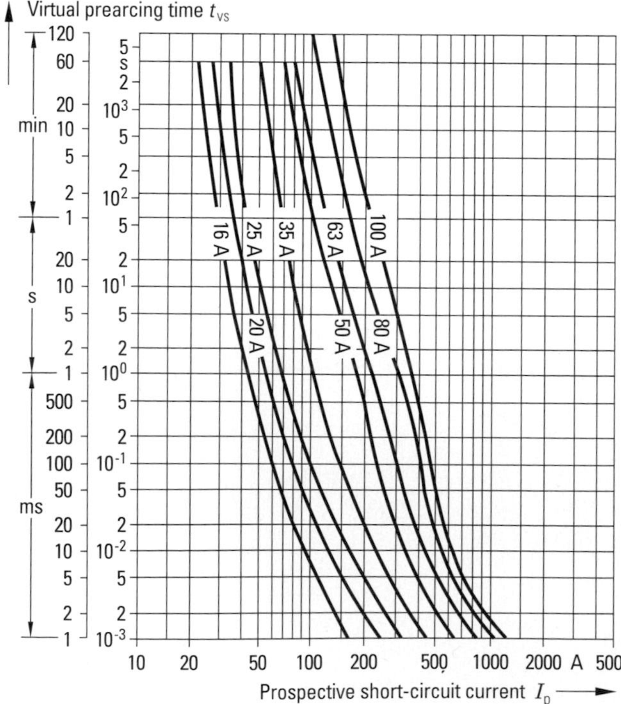

Fig. 4.1/23
Prearcing-time/current characteristics for SILIZED fuse-links

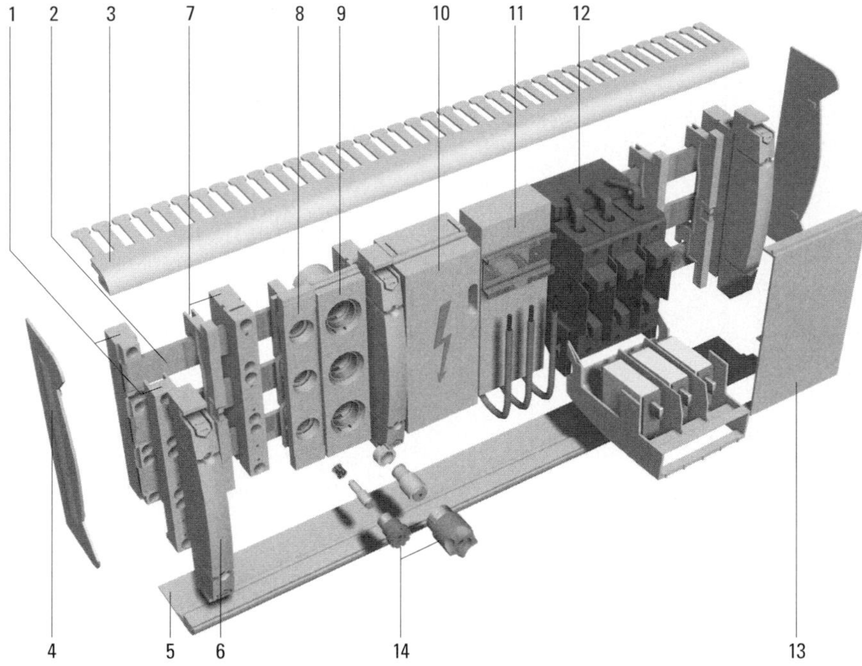

1 Busbar holder, 3-pole
2 Cu busbar
3 Holder for partitioning panel
4 Partitioning panel, slotted
5 Partitioning panel, closed
6 End closure
7 Spare cubicle rack
8 Spare cubicle cover
9 NEOZED bus-mounting fuse-base, 3-pole
10 DIAZED bus-mounting fuse-base, 3-pole
11 Adapter sleeve or adapter screw, fuse-links and screw caps
12 Terminal connection board with cover
13 Busbar adapter
14 L.v. h.b.c. fuse switch-disconnector, swung out

Fig. 4.1/24
60 mm busbar system fitted with rail-adaptable switchgear, fusegear, and installation equipment

Advantages

The busbar system also offers the following advantages:

▷ just one modular system for rated currents between 200 A and 630 A,
▷ small busbar cross sections resulting from center infeed or two end infeeds,
▷ fast mounting of rail-adaptable installation equipment since mechanical mounting and electrical contacting are carried out in one single operation,
▷ direct contacting on the Cu busbars,
▷ fewer points of contact and less contact resistance,
▷ less space required because devices can be flush-mounted on DIN rails (see Section 4.2),
▷ optimum utilization of space in front of Cu busbars,
▷ shock-protection components can be installed quickly by simply plugging them in,
▷ fast installation of supply leads using box elevator terminals,
▷ high level of operational reliability.

Design

The busbar runs consist of adjustable three and four-pole busbar holders for the main

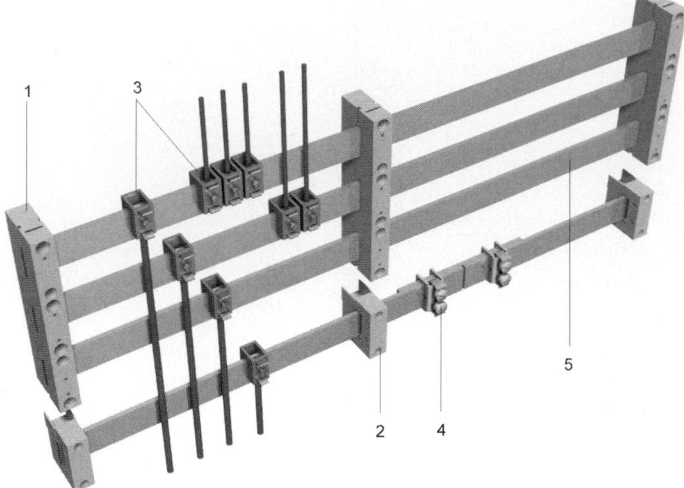

1 Busbar holder, 3-pole
2 N/PE busbar holder
3 Incoming/outgoing connection terminals
4 Extension terminal
5 Cu busbar

Fig. 4.1/25 a Open busbar run

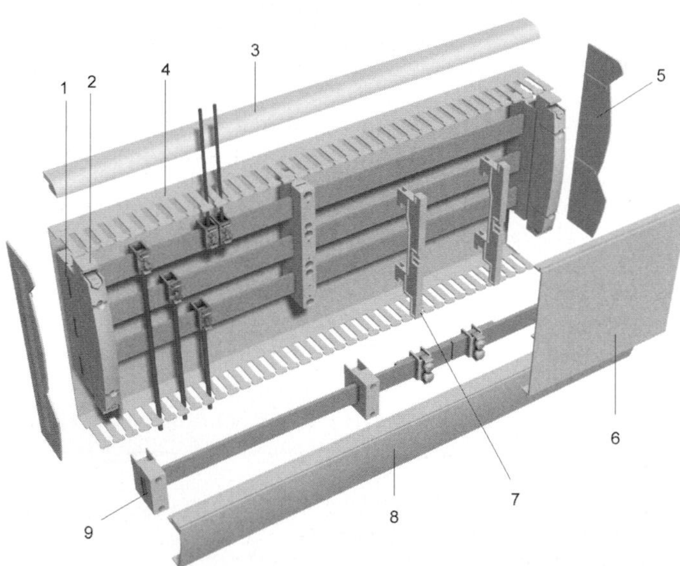

1 Holder for edge section
2 Base
3 Edge section
4 End closure
5 Spare cubicle rack
6 Spare cubicle cover
7 Busbar holder, 3-pole
8 N/PE busbar holder

Fig. 4.1/25 b Enclosed busbar run

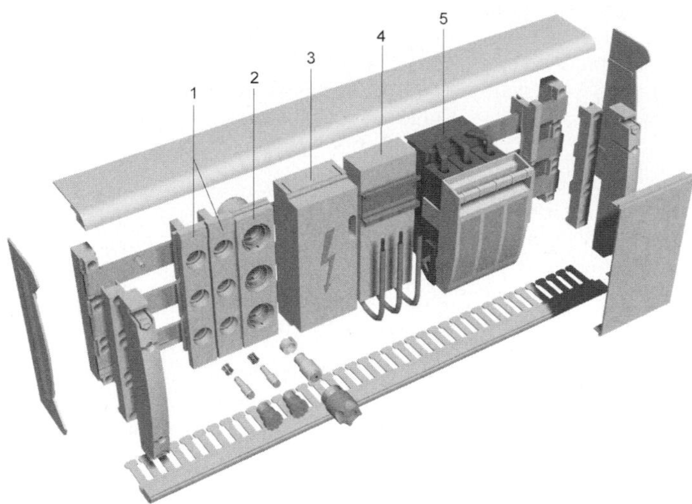

1 NEOZED bus-mounting fuse-base D02
2 DIAZED bus-mounting fuse-base D III
3 Terminal connection board
4 Busbar adapter
5 L.v. h.b.c. fuse switch-disconnector, size 00 (can be matched to busbar)

Fig. 4.1/25 c Equipped busbar run

busbars L1, L2, L3 or L1, L2, L3, N, individual holders for N, PE, or PEN conductors, the Cu busbars, as well as system-specific terminals for connecting the conductors to the infeed or to other outgoing feeders.

The busbar holders can be easily adjusted to the required Cu busbar width (12 mm, 15 mm, 20 mm, 25 mm, and 30 mm) and the busbar thicknesses (5 mm and 10 mm) using a detachable element. The set busbar dimensions can be read on the scale affixed to the side of the busbar holder.

The upper and lower parts of the busbar holder are made of high-quality, glass-fiber-reinforced polyester. The material is PVC and halogen-free, flame-retardant to UL 94 (V0), and has outstanding electrical, mechanical, and chemical properties.

The mounting holes of the busbar holders are always accessible from the outside so that the entire busbar system can be assembled with the installation equipment outside the distribution cabinet.

The centerline spacing from conductor bar to conductor bar is 60 mm and facilitates the mounting of rail-adaptable installation equipment designed for this dimension.

The equipment range includes:

▷ three-pole terminal connection boards for incoming feeders with conductor cross sections ≤ 120 mm²,

▷ three-pole NEOZED bus-mounting fuse-bases D02/E18 for adapter sleeves as well as input terminals with clamp-type connectors and locking locating springs for the Cu conductor bars measuring 12 mm × 5 mm to 30 mm × 10 mm and output terminals with box elevator terminals,

▷ three-pole DIAZED bus-mounting fuse-base D II/E27 and D III/E33 for adapter screws or ring gauge pieces as well as input terminals with clamp-type connectors and locking locating springs for Cu conductor bars measuring 12 mm × 5 mm to 30 mm × 10 mm and output terminals with box elevator terminals,

▷ three-pole l.v. h.b.c. fuse-bases, size 00 and 1, with shock protection,

▷ busbar adapters and busbar equipment racks with one or two DIN rails (DIN EN 50022) for motor feeders, circuit-breakers, and any component configuration,

▷ three-pole l.v. h.b.c. fuse switch-disconnectors, size 000, 00, 1, and 2.

Type-tested installation kits designed for the 60 mm busbar system (Figs. 4.1/26 and 4.1/27) are available to enable optimum mounting of this device range in STAB wall-mounting/SIKUS floor-mounting distribution systems.

Input and output terminals

The input and output terminals are designed as claw-type terminals and can thus be attached to Cu busbars (12 mm × 5 mm to 30 mm × 10 mm) at a later stage. They have captive recessed head screws and clamping members, as well as a flat spring which serves as a mounting aid. The cross sections range from 16 mm², 35 mm², 70 mm² to 120 mm² and from 150 mm² to 240 mm² for round conductors.

Rated dynamic short-circuit strength

The electrodynamic load placed on the busbar run depends on the magnitude of the short-circuit current, the length of the busbar, the support spacing between the busbar holders, and the distance between the busbars themselves.

Since a protective device (e.g. an l.v. h.b.c. fuse) is connected upstream of the busbars, the maximum possible let-through current (i_D) is the current to which a potential system short-circuit current (I_p) is limited by the inserted l.v. h.b.c. fuse-link.

The possible let-through current values of the l.v. h.b.c. fuse-link are specified by manufacturers in the form of current-limiting

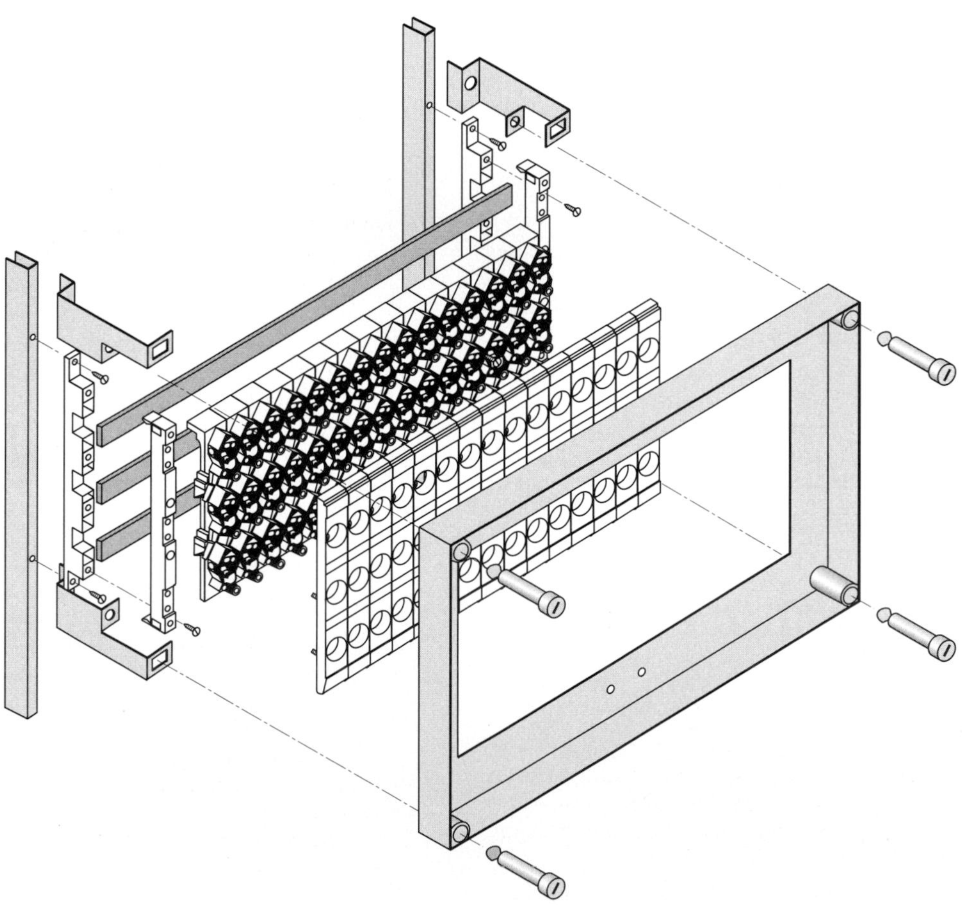

Fig. 4.1/26
Type-tested installation kits for the 60 mm busbar system with open busbar run and 15 three-pole NEOZED bus-mounting fuse-bases, size D01

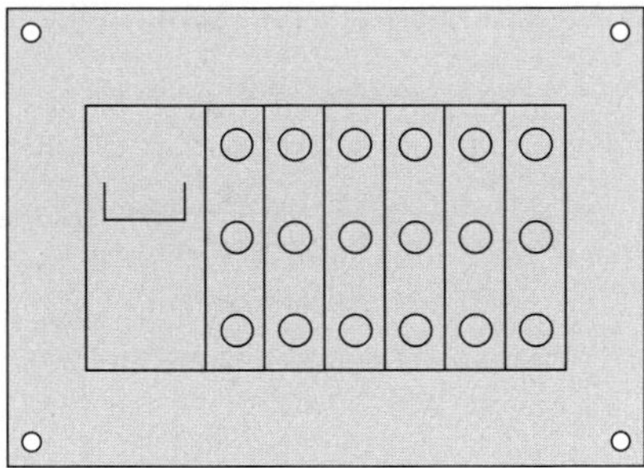

L.v. h.b.c. fuse switch-disconnector, size 00, three-pole DIAZED and NEOZED bus-mounting fuse-bases

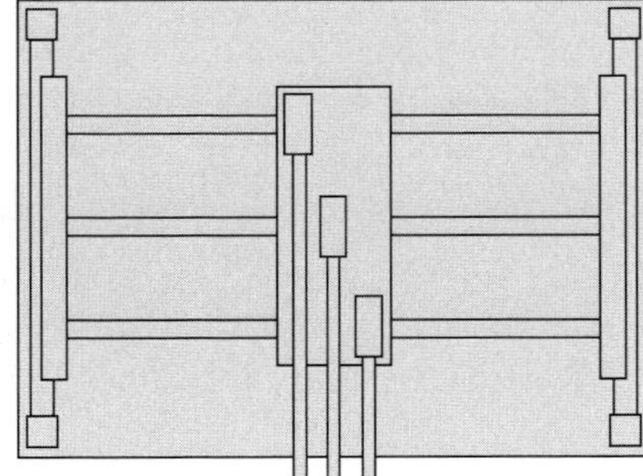

Terminal as center terminal without panel cover

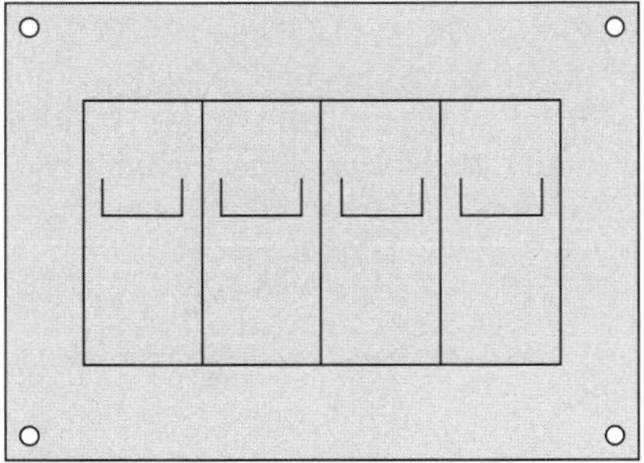

L.v. h.b.c. fuse switch-disconnectors, size 00

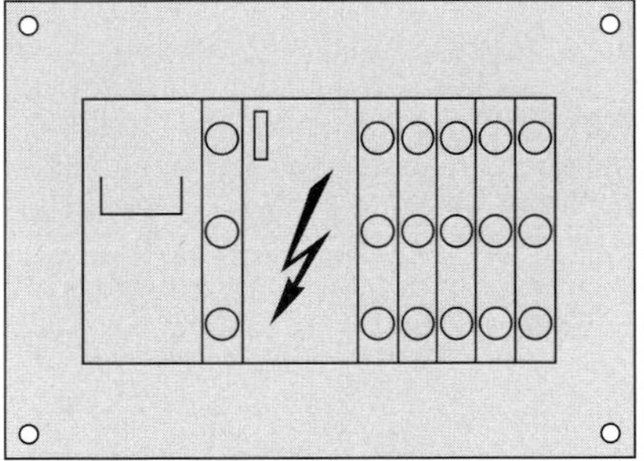

L.v. h.b.c. fuse switch-disconnectors, size 00, three-pole DIAZED and NEOZED bus-mounting fuse-bases together with terminal connection board

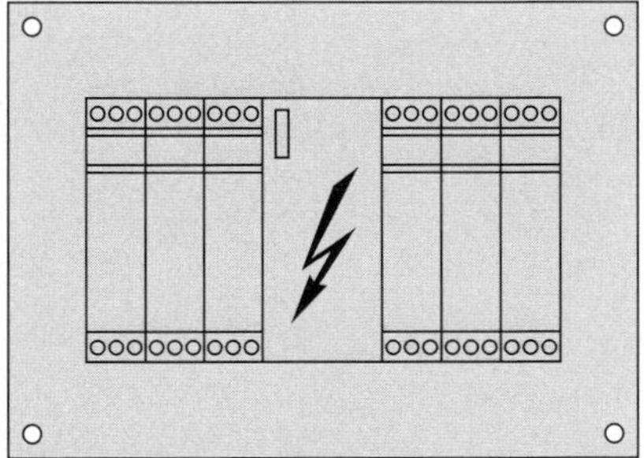

Terminal connection board and busbar adapter

Fig. 4.1/27
Equipment examples with the type-tested installation kit for 60 mm busbar system

diagrams as a function of the prospective short-circuit current I_p (r.m.s. value of the possible system short-circuit current) (Fig. 4.1/28).

In the case of busbar holders with busbars measuring 12 mm × 5 mm to 30 mm × 10 mm, the distance between two busbar holders should be adapted to the support spacing between the braces in the distribution boards and, if possible, should not exceed 250 mm. If busbars measuring 25 mm × 5 mm to 40 mm × 10 mm are used, the spacing can be greater. It should, however, be remembered that dynamic stability decreases as the support spacing increases. Furthermore, it is essential to ensure that the permissible current carrying capacity of the individual busbars is not exceeded.

A center infeed or infeed from both ends of the busbar must be provided in cases where the power values are approaching the permissible current carrying capacity.

4.1.2 Miniature circuit-breakers (MCBs)

Purpose

Miniature circuit-breakers are used, first and foremost, to protect cables and leads against overloads and short circuits. In doing so, they protect electrical equipment against overheating in line with the relevant standards, e.g. DIN VDE 0100-430.

In certain situations, miniature circuit-breakers in TN systems also provide protection against electric shocks in the event of excessively high touch voltages caused by insulation faults, e.g. in line with HD 384.4.41/ IEC 364–4-41/DIN VDE 0100-410.

Application

Miniature circuit-breakers can be used in all types of distribution system, in residential and functional buildings as well as for industrial applications. The different versions and wide range of accessories (auxiliary and fault signal contacts, open-circuit shunt releases, etc.) enable miniature circuit-breakers to fulfil the different requirements of the various areas of application.

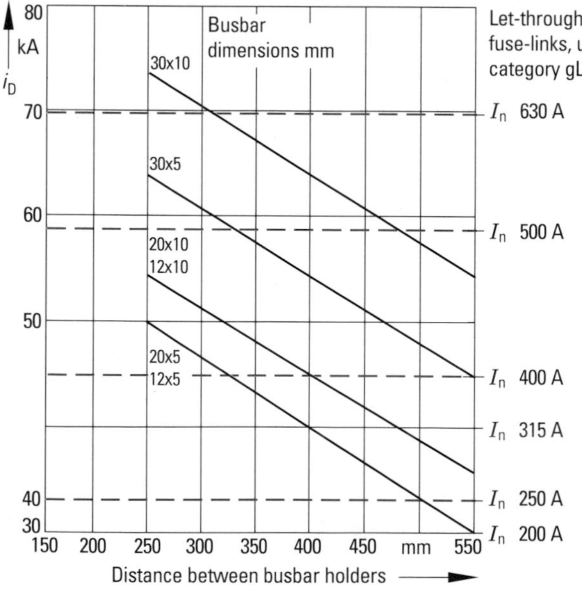

i_D: Let-through values (kA) of l.v. h.b.c. fuse-links, utilization category gL/gG with rated current 200 A to 630 A and prospective short-circuit current I_p = 120 kA

Sample reading for I_p 120 kA

Protective device:	L.v. h.b.c. fuse-link, 630 A, gL/gG (limited to i_D 74 kA (approx.))
Busbar:	30 mm x 10 mm
Rated dynamic short-circuit strength:	≤ 74 kA (approx.) with busbar holder spacing 250 mm or ≤ 57 kA (approx.) with busbar holder spacing 500 mm
Protective device:	L.v. h.b.c. fuse-link, 500 A, gL/gG (limited to i_D 55 kA (approx.))
Busbar:	20 mm x 10 mm
Rated dynamic short-circuit strength:	≤ 55 kA (approx.) with busbar holder spacing 250 mm or ≤ 45 kA (approx.) with busbar holder spacing 400 mm

Fig. 4.1/28 Diagram showing the rated dynamic short-circuit strength of the busbars

Four tripping characteristics (A, B, C, and D) are available for the equipment connected in the circuit to be protected.

Tripping characteristics

▷ *Tripping characteristic A* is particularly suitable for protecting transformers in measuring circuits and for electric circuits with long leads which, according to HD 384.4.41 S2/IEC 60364–4-41/DIN VDE 0100-410, have to be disconnected within 0.4 s,

▷ *Tripping characteristic B* is the standard characteristic for socket-outlet circuits in residential and functional buildings,

▷ *Tripping characteristic C* is useful for equipment with considerably high making currents, such as lamps and motors,

▷ *Tripping characteristic D* is suitable for pulse-generating equipment, such as transformers, solenoid valves, and capacitors.

Method of operation

Miniature circuit-breakers are designed for manual operation with an overcurrent trip-free mechanism (thermal instantaneous release) (Fig. 4.1/29). Multi-pole devices are connected both externally via the handles and internally via the releases.

Standards

The relevant international standard for miniature circuit-breakers is IEC 60898 which forms a basis for the European standard EN 60868 and the national standard DIN VDE 0641–11. The different circuit-breaker sizes are described in DIN 43880.

The disconnection conditions specified in the applicable standards, e.g. HD 384.4.41 S2/IEC 60364-4-41/DIN VDE 0100–410, must be observed with regard to operator protection.

Versions

Miniature circuit-breakers are available with one (Fig. 4.1/29), two, three, or four poles, and with a switched neutral conductor (1 pole + N and 3 pole + N).

Fig. 4.1/29
Single-pole miniature circuit-breaker

In line with the preferred series to IEC 60898, miniature circuit-breakers can have the following rated currents:

▷ Devices with overall depth of 55 mm – 0.3 A to 63 A,

▷ Devices with overall depth of 70 mm – 40 A to 125 A.

Auxiliary contacts, fault signal contacts, open-circuit shunt releases, undervoltage releases, and residual-current protective device units (residual current unit/RC unit) can be retro-fitted, depending on the device design.

Auxiliary contacts indicate the status of the miniature circuit-breaker but provide no information on whether the circuit-breaker was opened manually or automatically. Fault signal contacts indicate that the circuit-breaker was opened as a result of an overload or a short circuit. Open-circuit shunt releases are suitable for remote opening of circuit-breakers.

If the auxiliary contact and fault signal contact are connected to an *instabus EIB* binary input, the signals can also be read in an *instabus EIB* system. The miniature circuit-breaker can also be opened remotely via *instabus EIB* by means of the open-circuit shunt release in the *instabus EIB* binary output.

Depending on the design, Siemens miniature circuit-breakers also have the following characteristics:

▷ Terminals are safe from touch by fingers and the back of the hand in line with VDE 0106-100 (VBG4),

▷ Combination terminals for simultaneous connection of busbars and supply conductors,

▷ Master switch characteristics in line with EN 60204/IEC 60204/VDE 0113,

▷ Separate position indicator.

Alternating current miniature circuit-breakers are suitable for all a.c. and three-phase systems up to a voltage of 240/415 V and for all d.c. systems up to 60 V (1 pole) and 120 V (2 pole).

The rated voltage of the miniature circuit-breaker is 230/400 V AC.

Universal-current miniature circuit-breakers can also be used for 220 V DC (1 pole) and 440 V DC (2 pole).

To prevent the conductor insulation from being damaged in the event of a fault, certain values have been defined for maximum permissible temperatures. For PVC insulation, these values are 70 °C or 160 °C (continuous) for a maximum of 5 s (in the event of a short circuit).

Miniature circuit-breakers normally have two independent trip elements to protect the lines against overcurrents. In the event of an overload, a bimetallic element trips after a delay in accordance with the current intensity. If a certain threshold value is exceeded in the event of a short circuit, however, an electromagnetic overcurrent release trips immediately. The setting range (time/current limit range) of the miniature circuit-breakers to EN 60898/IEC 60898/DIN VDE 0641-11 is defined via the characteristic values I_1 to I_5 (Fig. 4.1/30). The characteristic values I_b, I_z of the lead correlate with these.

The international standard IEC 60898 has introduced new characteristics (B, C, and D) which have also been incorporated in EN 60898 and DIN VDE 0641-11.

The new tripping conditions of the miniature circuit-breakers facilitate assignment to the conductor cross-sections. The following conditions are listed in the relevant standards, e.g. DIN VDE 0100-430:

1st condition

$I_b \leq I_n \leq I_z$ (rated current rule),

2nd condition

$I_2 \leq 1.45 \cdot I_z$ (tripping current rule).

Since the second condition is fulfilled automatically with the new characteristics ($I_z = I_n$), the miniature circuit-breaker merely needs to be chosen according to the relationship $I_n \leq I_z$.

As a result, the rated currents of miniature circuit-breakers and conductor cross-sections can be reassigned in accordance with Table 4.1/7 referred to an ambient temperature of 30 °C (see DIN VDE 0100-430, Supplement 1) and in accordance with the type and cable grouping.

Siemens miniature circuit-breakers are available with tripping characteristics B, C, and

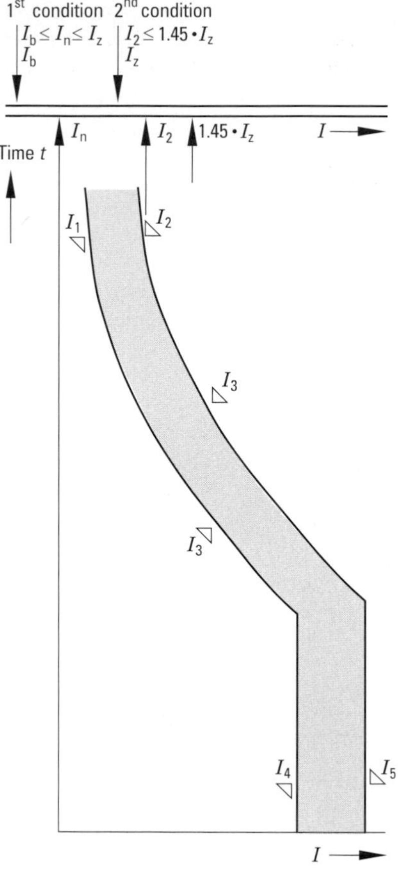

I_b	Anticipated operating current, i.e. the current determined by the load under uninterrupted operating conditions
I_z	Permitted continuous load current for a conductor with which the continuous temperature limit of the insulation is not exceeded
$1.45 \cdot I_z$	Maximum permissible overload current, restricted with respect to time, at which momentary overshooting of the continuous temperature limit does not affect the safety-related properties of the insulation
I_n	Rated current, i.e. the current for which the miniature circuit-breaker is designed and to which other rated variables refer (set value)
I_1	Conventional non-tripping current, i.e. the current which, under defined conditions, does not cause the circuit-breaker to trip
I_2	Conventional tripping current, i.e. the current which, under defined conditions, causes the circuit-breaker to trip within one hour ($I_n \leq 63$ A)
I_3	Tolerance limits
I_4	Holding current of the instantaneous electromagnetic overcurrent release (short-circuit release)
I_5	Tripping current of the instantaneous electromagnetic overcurrent release (short-circuit release)

Fig. 4.1/30
Reference values of electric lines and protection equipment

Table 4.1/7
Assignment of miniature circuit-breakers to conductor cross-sections
Example: Flat webbed cable, multi-core cable on or in the wall, installation type C[1)] at an ambient temperature of 30 °C

Rated cross-section q_n	Rated current I_n of the miniature circuit-breaker when protecting		I_z (line) Permissible continuous load current with	
	2 loaded conductors	3 loaded conductors	2 loaded conductors	3 loaded conductors
mm^2	A	A	A	A
1.5	16	16	19.5	17.5
2.5	25	20	26	24
4	32	32	35	32
6	40	40	46	41
10	63	50	63	57
16	80	63	85	76
25	100	80	112	96
35	125	100	138	119

[1)] Installation type C in line with DIN VDE 0298-4 and DIN VDE 0100-430, Supplement 1. The cables are secured in such a way that the distance between them and the wall surface is less than 0.3 × outer diameter of the cable

D and with the VDE sign based on the CCA procedure (**C**ENELEC **C**ertification **A**greement).

All of the tripping characteristics can be seen in Fig. 4.1/31. Due to the position of the tripping bands,

▷ the current pulse strength increases,
▷ the permitted cable lengths for operator protection decrease

from characteristics A to D.

Ambient temperature sensitivity

The release characteristics are defined in line with the relevant standards on the basis of an ambient temperature of + 30 °C. At higher temperatures, the thermal tripping characteristic in Fig. 4.1/31 shifts to the left. At lower temperatures, it shifts to the right. This means that the trip functionality is already active at lower currents (high temperature) or is not active until the current is relatively high (low temperature).

This should be noted in particular in very warm rooms when the circuit-breakers are installed in enclosed distribution boards (in which the I^2R loss of the installed devices results in considerable temperature rises) or in distribution boards located outdoors. Miniature circuit-breakers can be used at temperatures between −25 °C and + 55 °C with a relative humidity of up to 95%.

Resistance to climatic changes

Siemens miniature circuit-breakers are climate proof to IEC 68-2-30. They have been successfully tested with six climatic cycles.

Degree of protection

Since the circuit-breakers are mainly installed in distribution boards, their degree of protection must fulfil the requirements of the room type in question. Non-encapsulated miniature circuit-breakers with appropriate terminal covers are protected to IP 30 in line with EN 60529/IEC 60529/DIN VDE 0470-1.

Mounting

A snap-on mounting allows miniature circuit-breakers to be mounted quickly on 45 mm wide DIN rails to DIN EN 50022 (Fig. 4.1/32). Some versions can also be screwed onto mounting plates.

In addition to the accuracy of their tripping characteristics, another important feature of miniature circuit-breakers is their rated breaking capacity. This is classified in accordance with EN 60898/IEC 60898/DIN VDE 0641-11 into different breaking capacity categories and provides information on the levels up to which short-circuit currents can be disconnected (Table 4.1/8).

Rated breaking capacity

Depending on the version, Siemens miniature circuit-breakers support rated breaking capacity values of up to 25,000 A with VDE certification.

Current limiting classes

In order to provide the necessary selectivity with respect to series-connected fuses, miniature circuit-breakers with characteristics B and C ≤ 32 A are divided into three current limiting classes, according to their degree of current limitation.

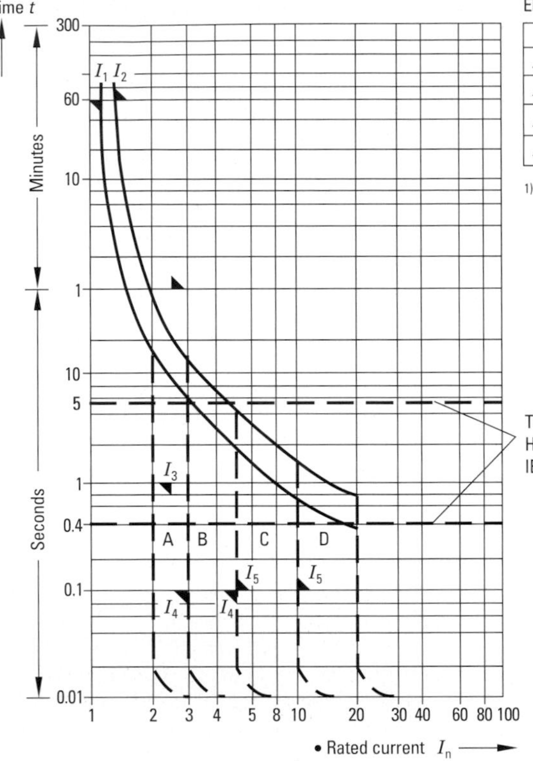

Miniature circuit-breaker
Tripping characteristics A, B, C, D to
EN 60898 / IEC 60898 / DIN 0641 ... 11

	A[1]	B	C	D
I_1 (t> 1h)	–	1.13 • I_n	1.13 • I_n	1.13 • I_n
I_2 (t< 1h)	–	1.13 • I_n	1.45 • I_n	1.45 • I_n
I_4 (t> 0,1s)	2 • I_n	3 • I_n	5 • I_n	10 • I_n
I_5 (t< 0,1s)	3 • I_n	5 • I_n	10 • I_n	20 • I_n

[1] Fulfils the conditions stipulated in DIN VDE 0100-410

Fig. 4.1/31
Time/current limit ranges of miniature circuit-breakers

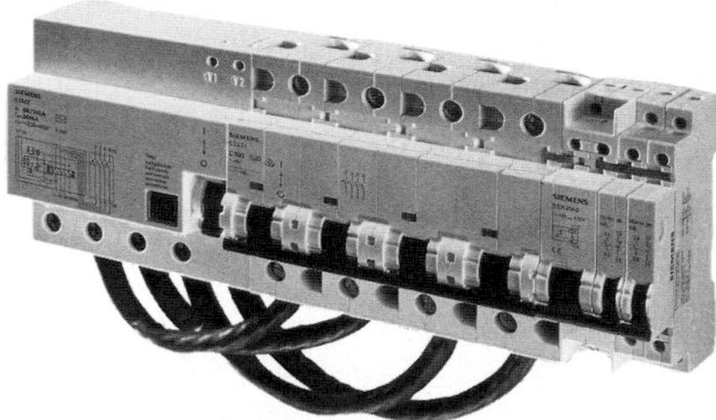

Fig. 4.1/32
High-current range of Siemens miniature circuit-breakers with built-on supplementary devices

The permitted let-through I^2t values are specified in the standards EN 60898/IEC 60898/DIN VDE 0641-11.

According to the technical supply conditions (TAB) of the German public utilities, only miniature circuit-breakers with a rated breaking capacity of at least 6,000 A from current limiting class 3 may be installed downstream of the meter in distribution boards of residential and functional buildings in order to provide the necessary selectivity.

The devices must carry the marking 6000 / 3.

Table 4.1/8
Rated breaking capacity categories for miniature circuit-breakers

Standard	Categories
EN 60898/IEC 60898/ DIN VDE 0641-11	1,500 A 3,000 A 4,500 A 6,000 A 10,000 A 15,000 A 20,000 A 25,000 A

Selectivity

Selectivity means that, in the event of a fault, only the protective device that is closest to the fault location along the current path trips. The power flow can thus be maintained in parallel circuits. The current characteristic of the tripping operation is illustrated in Fig. 4.1/33a with reference to current limiting classes. The Siemens miniature circuit-breaker B16 limits the current to significantly lower values than those specified for current limiting class 3.

Fig. 4.1/33b shows the selectivity limits of miniature circuit-breakers with different current limiting classes through the intersection of the circuit-beaker tripping characteristic with the melting characteristic of the fuse. The extremely effective current limitation capability of the miniature circuit-breaker also affects the high selectivity of the circuit-breakers with respect to the series-connected fuse.

Back-up protection

If the short-circuit current at the installation location of the miniature circuit-breaker exceeds its rated breaking capacity, a further short-circuit protective device has to be connected on the line side. Without impairing the operational reliability of the miniature circuit-breaker in cases such as these, the breaking capacity of the combination "miniature circuit-breaker B16 plus fuse 100 A gL" is increased to 35 kA.

Certain countries are using line-side circuit-breakers to an increasing extent instead of low-voltage high-breaking-capacity fuses. Depending on the type of circuit-breaker used, the overall breaking capacity is reduced considerably.

Although the circuit-breakers have a high rated breaking capacity, their current limiting capabilities within the limit switching capacity range of miniature circuit-breakers (6 kA/10 kA) are not yet sufficient to provide adequate support. Miniature circuit-breakers with rated currents of 6 to 32 A are, therefore, only protected by line-side circuit-breakers (types 3VF1 to 3VF6 and 3WN1/3WN5) up to a maximum of 15 kA (back-up protection).

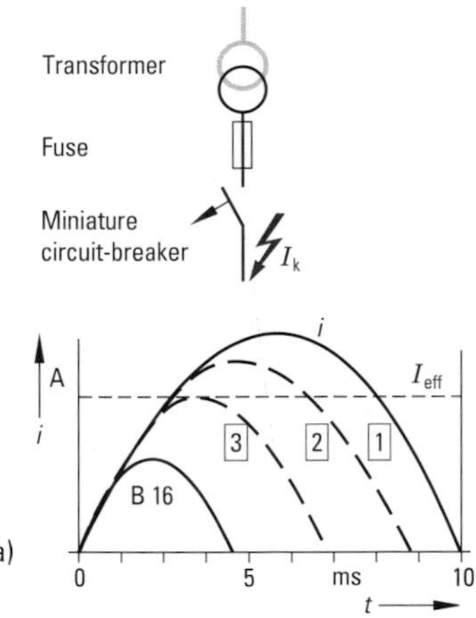

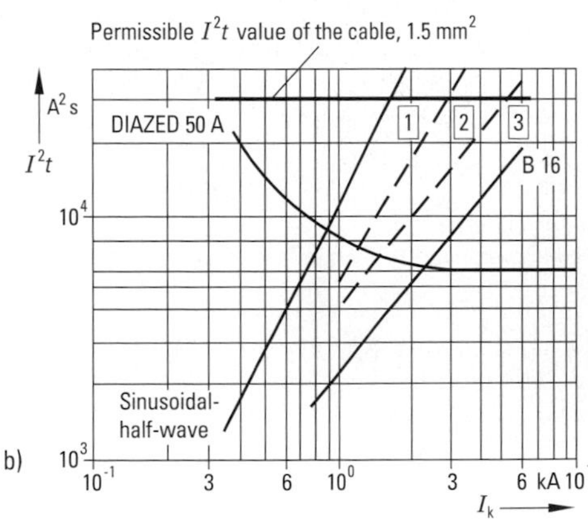

Fig. 4.1/33
Selectivity of miniature circuit-breakers of current limiting classes 1, 2 and 3 with respect to back-up fuses. Curve B16 applies to the Siemens 16 A circuit-breaker, tripping characteristic B

Further product information on Siemens miniature circuit-breakers can be found in the Siemens Catalog "Protective Switching and Fuse Systems, Building Management Systems with *instabus EIB*", Order Number: E20002-K8210-A101-A3-7600, and "Discrimination and back-up protection in fuseless low-voltage feeders", Order Number: E20001-P285-A649-V1-7400.

4.1.3 Residual-current protective devices (RCCBs)

Protective measures

Due to an increase in the consumption of electricity in households, as well as in commerce and industry, providing adequate protection for people and their property as well as for live-stock has become a matter of vital importance. Suitable protective measures include residual-current protective devices (Fig. 4.1/34) which are becoming increasingly widespread on account of the high degree and scope of protection they offer.

Basic insulation

Electrical accidents can generally be prevented by ensuring that the devices and installations are designed and constructed correctly – in other words by providing adequate basic insulation. Damage to the basic insulation can lead to faults which require further protective measures against excessively high shock currents (operator protection). These protective measures are stipulated in the standards HD 384/IEC 60364-3/DIN VDE 0100 which focus on protection against indirect contact and, in certain cases, additional protection against direct contact.

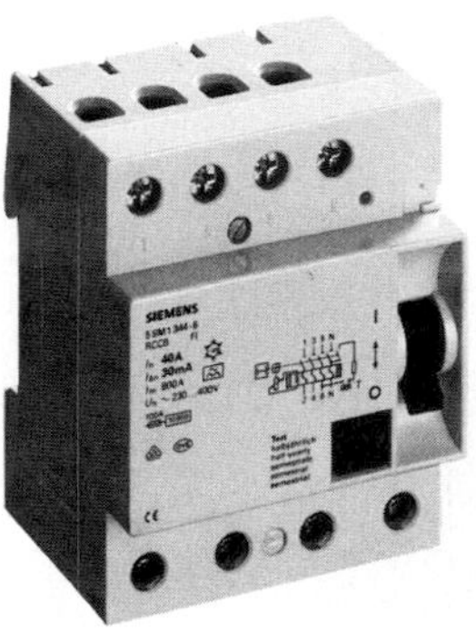

Fig. 4.1/34
Siemens residual-current protective devices, 3 pole + N for AC and pulsating, DC fault currents [symbol], can be used up to – 25 °C [symbol -25]

Insulation faults and their consequences

Insulation faults can have the following consequences:

- ▷ Short circuits to exposed conductive parts,
- ▷ Short circuits,
- ▷ Ground faults.

Any one of these can occur as a dead or high-resistance (i.e. resistive) electric-arc fault. Short circuits to exposed conductive parts are "accident risks" whereas short circuits and ground faults are "fire risks".

In order to provide adequate protection for people and their property, as well as for livestock against faults of this kind, appropriate protective measures have to be implemented. Residual-current protective devices are particularly suitable for this purpose (see also Sections 2.3 and 2.4).

4.1.3.1 Use of residual-current protective devices in different systems

Systems, permissible protective devices

The systems (TN, TT, IT systems) are defined in IEC 364-3/DIN VDE 0100-300 and the protective devices that may be used in them in HD 384.4.41/ IEC 60364-4-41/ DIN VDE 0100-410 (Table 4.1/9).

Residual-current protective devices can be used in all protective systems of an alternating or three-phase system and at ambient temperatures of + 45 °C to – 5 °C. The symbol [symbol -25] indicates that these devices can be used up to an ambient temperature of – 25 °C.

Fig. 4.1/35 illustrates how residual-current protective devices are installed in different systems.

Table 4.1/9
Requirements of German installation regulations for DIN VDE 0100 for using residual-current protective devices in accordance with EN 61008-1/IEC 1008

DIN VDE 0100	Title (short version)	Requirement (short version)
Part 410/A1/03.86	Protection against electric shock	Residual-current protective device
Part 559/03.83 and Part 559/A1: 1997-4 (IEC 64/911/CD: 1996)	Luminaires and lighting equipment	Residual-current circuit-breakers $I_{\Delta n} \leq 30$ mA for light projection stands
Part 701/05.84 (prHD 384.7.701/ IEC 364-7-701: 1984)	Locations containing a bath tub or shower basin	Socket outlets (zone 3) must be protected with residual-current circuit-breakers $I_{\Delta n} \leq 30$ mA
Part 702/06.92 (IEC 64/906/FDIS:1997)	Roofed swimming pools (swimming baths) and outdoor swimming facilities	Within the protection zones, residual-current circuit-breakers $I_{\Delta n} \leq 30$ mA
Part 704/11.87 (IEC 64/889/CD: 1986)	Construction site installations	Residual-current circuit-breakers $I_{\Delta n} \leq 30$ mA for socket outlets up to 16 A, otherwise $I_{\Delta n} \leq 500$ mA
Part 705/10.92	Agricultural and horticultural premises	Fire protection: residual-current circuit-breakers $I_{\Delta n} \leq 500$ mA for socket outlet circuits with residual-current circuit-breakers $I_{\Delta n} \leq 30$ mA
Part 706/06.92	Restrictive conductive locations	Residual-current circuit-breakers $I_{\Delta n} \leq 30$ mA
Part 708/10.93	Electrical installations in caravan parks and caravans	Socket outlets must be protected with residual-current circuit-breakers $I_{\Delta n} \leq 30$ mA
Part 720/03.83	Locations exposed to fire hazards	Ground-fault fire protection via residual-current circuit-breakers $I_{\Delta n} \leq 500$ mA
Part 721/04.84	Caravans, boats and yachts, as well as power supply thereof at camping sites and berths	The socket outlets at berths and in parking areas have to be protected with residual-current circuit-breakers $I_{\Delta n} \leq 30$ mA
Part 722/05.84	Temporary buildings, vehicles for travelling exhibitions and caravans	Residual-current circuit-breakers with $I_{\Delta n} \leq 500$ mA
Part 723/11.90	Classrooms with experimental desks	Residual-current circuit-breakers with $I_{\Delta n} \leq 30$ mA
Part 728/03.90	Stand-by power supply installations and other electrical supply installations for temporary operation	A residual-current protective device is also required as an alternative protective measure
Part 737/11.90	Humid/wet areas and locations, outdoor locations	Residual-current circuit-breakers $I_{\Delta n} \leq 30$ mA for socket outlets up to 32 A to connect equipment operated outdoors
Part 738/04.88	Fountains	Residual-current circuit-breakers for certain areas
Part 739/06.89 (Guideline)	Additional protection against direct contact	Residual-current circuit-breaker $I_{\Delta n} \leq 30$ mA for socket outlets up to 32 A in apartments

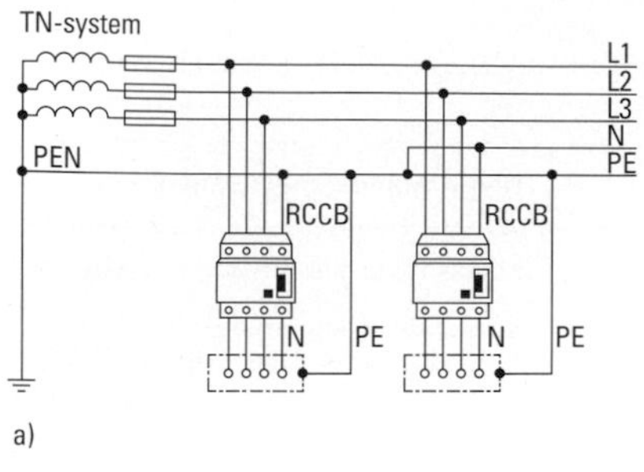

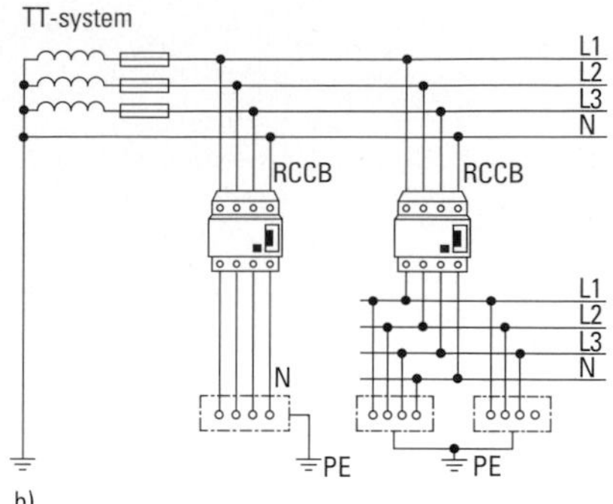

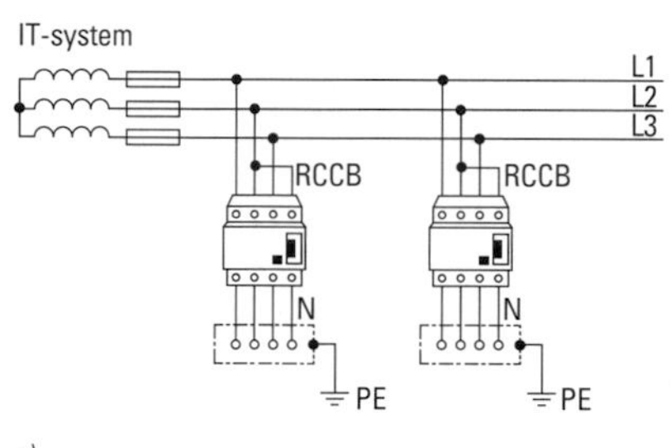

Fig. 4.1/35
Installation of residual-current protective devices in:
a) TN system,
b) TT system,
c) IT system.

4.1.3.2 Indirect and direct contact

Indirect contact

In the case of protection against dangerous short circuits to exposed conductive parts, a distinction is made between indirect and direct contact. With indirect contact (Fig. 4.1/36a), the residual current flows to ground via the equipment grounding conductor (PE conductor) if the insulation of the device is defective. Anyone who touches the faulty device when the fault occurs is connected in parallel to the fault circuit. The majority of the current will flow via the equipment grounding conductor due to the resistance ratio *equipment grounding conductor/person*.

Direct contact

The situation is, however, different in the case of unintentional, direct contact with operational, live parts or with external, non-grounded, conductive parts that conduct a current in the event of a fault (Fig. 4.1/36b). Here, the person acts as the equipment grounding device. The fault current that flows through the person's body can have fatal consequences.

The measures selected should, as far as possible, provide protection against indirect *and* direct contact.

Maximum protection

Residual-current protective devices with a rated residual current of no more than 30 mA offer the highest degree of protection of all comparable protective measures. They provide protection against indirect contact and, for the most part, against direct contact. The residual-current protective device trips as soon as a dangerous current flows through the human body.

It is then only the resistance of the human body that determines the current that flows through the person concerned. Investigations have shown that the resistance of the human body depends greatly on the condition of the skin at the point of contact and on the path of the current.

A resistance of approximately 1200 Ω was measured for a current that flowed from hand to hand or from hand to foot, for example.

Taking into consideration that the fault voltage is 230 V, the current that flows from hand to hand or from hand to foot is approximately 190 mA.

It is also important to consider the physiological reactions that can occur when these currents flow through the human body.

To clarify the correlation between the current magnitude and physiological reactions,

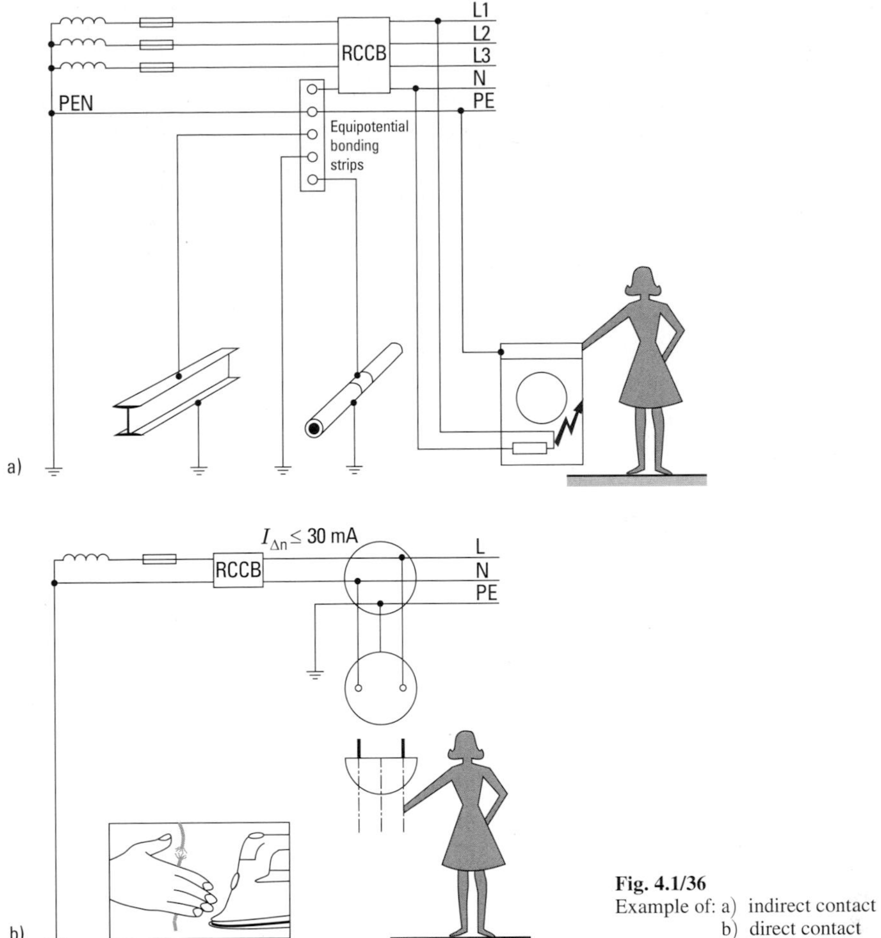

Fig. 4.1/36
Example of: a) indirect contact
b) direct contact

a large number of different electrical accidents were evaluated and measurements performed on animals.

Current intensity ranges to IEC 60479

The results of these evaluations and measurements are presented in the form of current intensity ranges (Fig. 4.1/37) in IEC 60479. The current/time values have been broken down into four ranges by means of curves.

Current/time values in range 4 are dangerous as they may cause the heart to beat irregularly, disturbing its pumping rhythm and impairing circulation, especially to the brain. This can have fatal consequences.

The trip scatter bands of the residual-current protective devices with rated residual currents of 10 mA and 30 mA are adapted to these ranges.

The excellent protective properties of residual-current protective devices are described in the following using direct contact as an example.

Fig. 4.1/38 illustrates an everyday situation in which an accident can occur.

High-resistance faults/faults to ground

An insulated metal frame on a wooden floor damages the connecting lead of the load (in this case a drill) and a voltage of 230 V passes through the metal frame which is on a

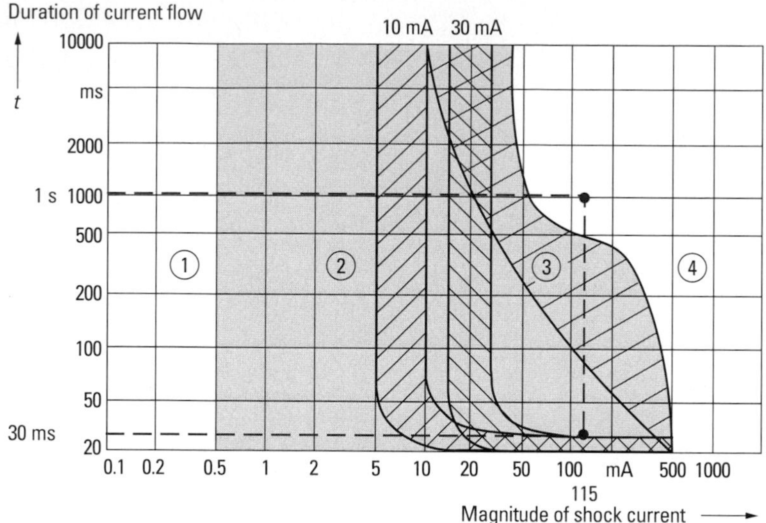

Fig. 4.1/37
Current intensity ranges for alternating current 50/60 Hz to IEC 60479 and trip scatter bands of the residual-current protective devices with rated residual current $I_{\Delta n} \leq 30$ mA and 10 mA

Range 1 Normally no noticeable consequences
Range 2 Normally no harmful consequences
Range 3 Muscle cramps, possible irregular heartbeat
Range 4 Danger of irregular heartbeat

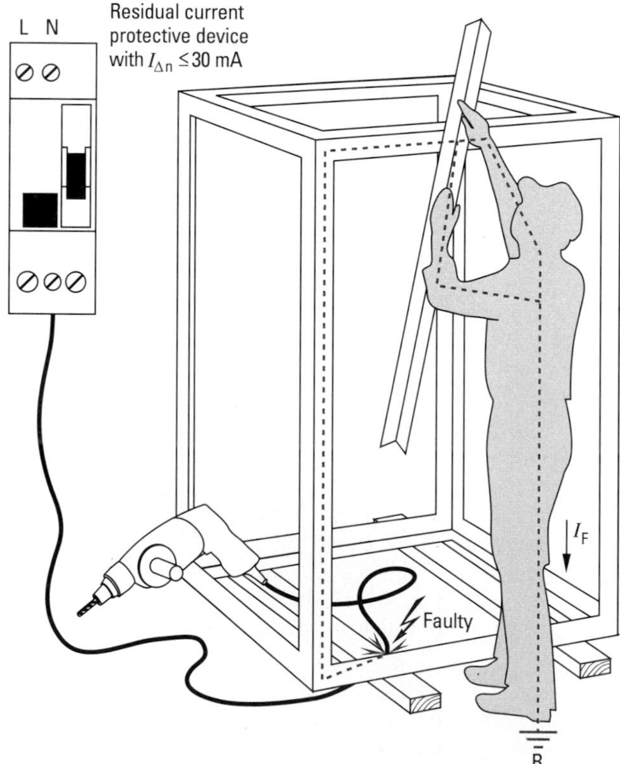

Fig. 4.1/38
Everyday example of direct contact:
- Without residual-current protective device, life threatening
- With residual-current protective device, not life threatening

$I_F = U_0/R = 230\ \text{V}/2000\ \Omega$
$= 115$ mA

U_0 Voltage to ground
R "Body-to-location" resistance of person

non-insulated floor (ground). A person touches the live frame (direct contact) and, assuming a "body-to-location" resistance of $R \approx 2000\ \Omega$, the following fault current flows through the human body:

$$I_F = \frac{230\ \text{V}}{2000\ \Omega} = 115\ \text{mA}.$$

If this current value of 115 mA is mapped on the current compatibility curve (Fig. 4.1/38) and a current flow duration of > 1 s is assumed, the current/time value is in range 4. It is highly likely that the person involved in the accident will die.

The current flow duration is > 1 s with 115 mA when a 16-A miniature circuit-breaker is used, for example, as this does not trip at this point.

This shock current of 115 mA is only detected and disconnected after 30 ms if a residual-current protective device with a rated residual current of $I_{\Delta n} \leq 30$ mA is used. The current/time value here is in current intensity range 2 (see Fig. 4.1/37), in which harmful effects do not normally occur.

4.1.3.3 Preventive measures for fire protection

High-resistance faults/faults to ground

Short circuits or ground faults are most likely to cause fire if relatively high resistances occur in the fault circuit at the electric-arc location (high-resistance faults/ faults to ground). Due to the fact that the fault currents are, in certain cases, well below the rated currents of the overcurrent protective device, it is not necessary for the fault to be disconnected by means of series-connected overcurrent protective devices such as fuses or miniature circuit-breakers.

Currents that are only slightly higher than the rated current of the overcurrent protective device are only disconnected after a relatively long period of time. For this reason, limits are set to provide protection against fires caused by ground-fault currents.

Faults that can cause ground-fault currents and form electric arcs include:

- ▷ Damage to the insulation of a lead or a piece of equipment,
- ▷ Charring of the area surrounding terminals of devices or motors,
- ▷ Interturn faults caused by overloading of motors or aging of reactors,
- ▷ Dampness or condensed water in equipment or parts of the installation,
- ▷ Conductive dust particles or dust accumulation in electrical equipment.

Faults such as these can cause high-resistance faults (to ground) and fires.

Thermal power: fire risk

Thermal power in excess of 100 W that occurs as a result of these faults represents a fire risk if it is released on a relatively small surface area of a few square millimeters. The time period during which this thermal power builds up is also important.

Protection against fires caused by ground-fault currents is, therefore, of prime importance.

Of the protective devices that trip in the event of a fault, it is only the residual-current protective device that offers extensive protection as it operates in accordance with the ground-fault monitoring principle. This method of fire protection by means of residual-current protective devices is stipulated in the relevant standards, e.g. DIN VDE.

Fire protection with residual-current protective devices

DIN VDE 0100 Part 720 (locations exposed to fire hazards) therefore specifies that suitable measures must be implemented to provide protection against fires that are caused by insulation faults. Only residual-current protective devices with a maximum rated residual current of 0.5 A are recommended. To provide even greater fire protection, residual-current protective devices with rated residual currents of ≤ 0.3 A are normally used. This is also recommended by property insurers.

Fig. 4.1/39 shows an example of an electrical installation with an insulation fault, which represents a fire risk. If the maximum possible continuous currents I_{ISO} are compared with the thermal values P_{ISO} (specified in Table 4.1/10) at the fault location when overcurrent or residual-current protective devices are used, it becomes apparent that only a residual-current protective device $I_{\Delta n} \leq 0.3$ A will provide reliable protection.

Table 4.1/10
Comparison of maximum possible continuous currents and thermal output values at the fault location when various overcurrent or residual-current protective devices are used

Protective devices	Max. possible continuous current I_{ISO}	P_{ISO} at $U_n = 230$ V
Fuse 10 A	15 A	3.45 kW
Miniature circuit-breaker B/C 16 A	18 A	4.14 kW
RCCB with $I_{\Delta n} = 0.5$ A	0.5 A	115 W
RCCB with $I_{\Delta n} = 0.3$ A	0.3 A	69 W
RCCB with $I_{\Delta n} = 30$ mA	0.03 A	6.9 kW

Minimum output required for a fire $P \approx 100$ W

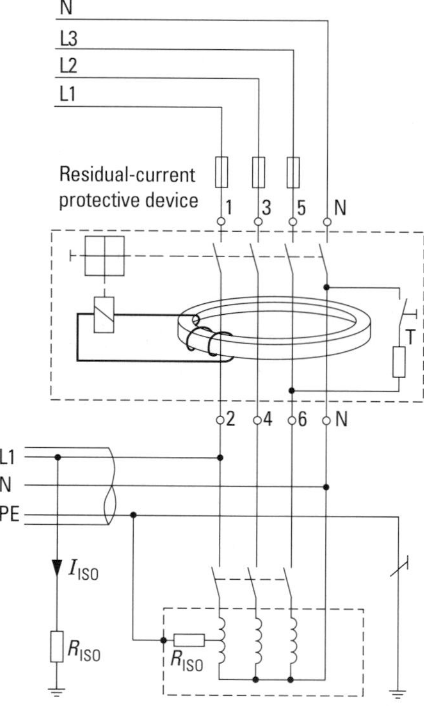

Fig. 4.1/39
Fire protection with residual-current protective devices

4.1.3.4 Structure and functional description of residual-current protective devices

The structure of a residual-current protective device is illustrated in Fig. 4.1/40. All current-carrying conductors of an installation are routed via the highly-permeable core of the summation current transformer W.

Tripping independent of line voltage and auxiliary power supply

If the installation is operating correctly, the vectorial sum of the ingoing and outgoing currents is zero. If a fault current is generated in the electrical installation due to faulty insulation, the current equilibrium in the summation current transformer is disturbed. Depending on the magnitude of the fault current, the core of the transformer is magnetized and a voltage is induced in the secondary winding n, which generates a current through the excitation winding of trip element A. When tripping current I_a of the residual-current protective device is reached, trip element A operates and disconnects the hazardous part of the installation. This trip principle functions independently of the line voltage or auxiliary power supply. Residual-current protective devices are fitted with a test device T to make sure that they are functioning correctly. This device should be actuated once every six months.

The usefulness of a protective measure for tripping in a TN or TT system is mainly determined by

▷ the reliability or failure rate of the residual-current protective device itself and
▷ the reliability or failure rate of the residual-current protective device in conjunction with the supply system, i.e. the protective measure.

For this reason, adequate protection can nowadays only be provided by means of tripping independent of the line voltage, as stipulated in EN 61 008/IEC 1008/DIN VDE 0664.

Differential current protection equipment

Differential current protection equipment with tripping *independent of the line voltage* cannot offer comparable protection as the electronics in the devices cease to function

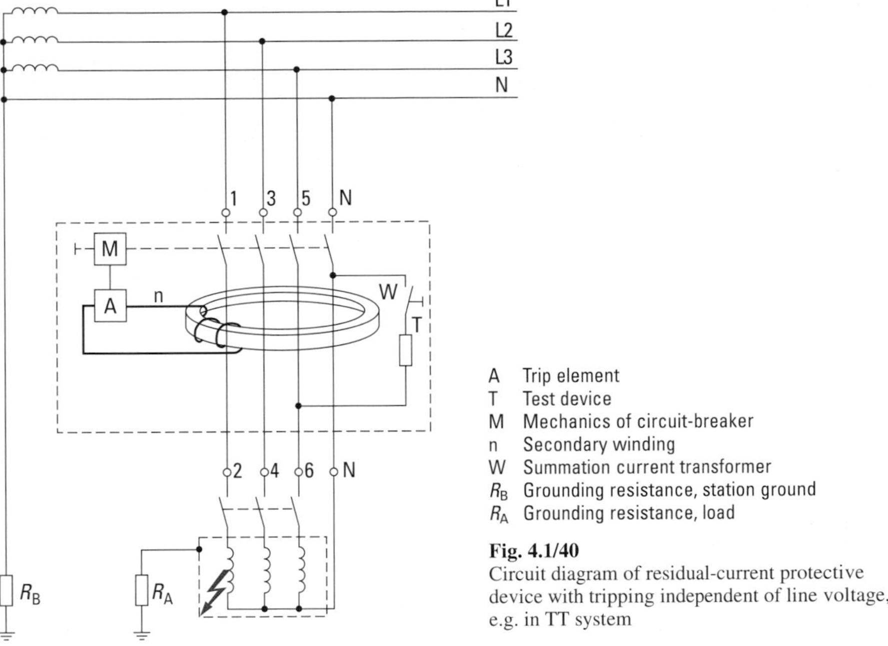

A Trip element
T Test device
M Mechanics of circuit-breaker
n Secondary winding
W Summation current transformer
R_B Grounding resistance, station ground
R_A Grounding resistance, load

Fig. 4.1/40
Circuit diagram of residual-current protective device with tripping independent of line voltage, e.g. in TT system

in the event of partial system failures. If necessary, checks should be carried out to determine the principle in accordance with which the evaluation electronics are supplied.

4.1.3.5 Residual-current protective devices for AC and pulsating DC fault currents to EN 61008/IEC 1008/DIN VDE 0664

New requirements

The residual-current protective devices used in the past only tripped in the event of sinusoidal AC fault currents in line with the requirements valid at that time. The increase in the number of electronic components in current-using devices for expanding the functional scope and/or for reducing power consumption necessitated a new generation of residual-current protective devices which trip both with AC and pulsating DC fault currents.

Fig. 4.1/41 shows the form of the load current and fault current of circuit types that generate fault currents of this kind in the event of a fault. Table 4.1/11 shows the permissible tripping current ranges for alternating and pulsating-current-sensitive residual-current protective devices according to EN 61 008/IEC 1008/DIN VDE 0664.

4.1.3.6 Universal-current-sensitive residual-current protective devices for industrial applications

Smooth DC fault currents

Circuit types in which smooth DC fault currents or currents with minor residual ripples occur in the event of a fault are being used to an increasing extent. Fig. 4.1/42 illustrates this using three-phase rectifier circuits as an example. AC or pulsating-current-sensitive residual-current protective devices cannot detect these DC fault currents and do not, therefore, trip. Current-using equipment that generates fault currents such as these in the event of a fault must, therefore, only be operated with universal-current-sensitive residual-current protective devices.

4.1.3.6.1 Structure and functional description

Supplementary unit for smooth DC fault currents

Universal-current-sensitive residual-current protective devices are based on pulsating-current-sensitive protective devices with tripping independent of the line voltage and a supplementary unit for detecting smooth DC fault currents.

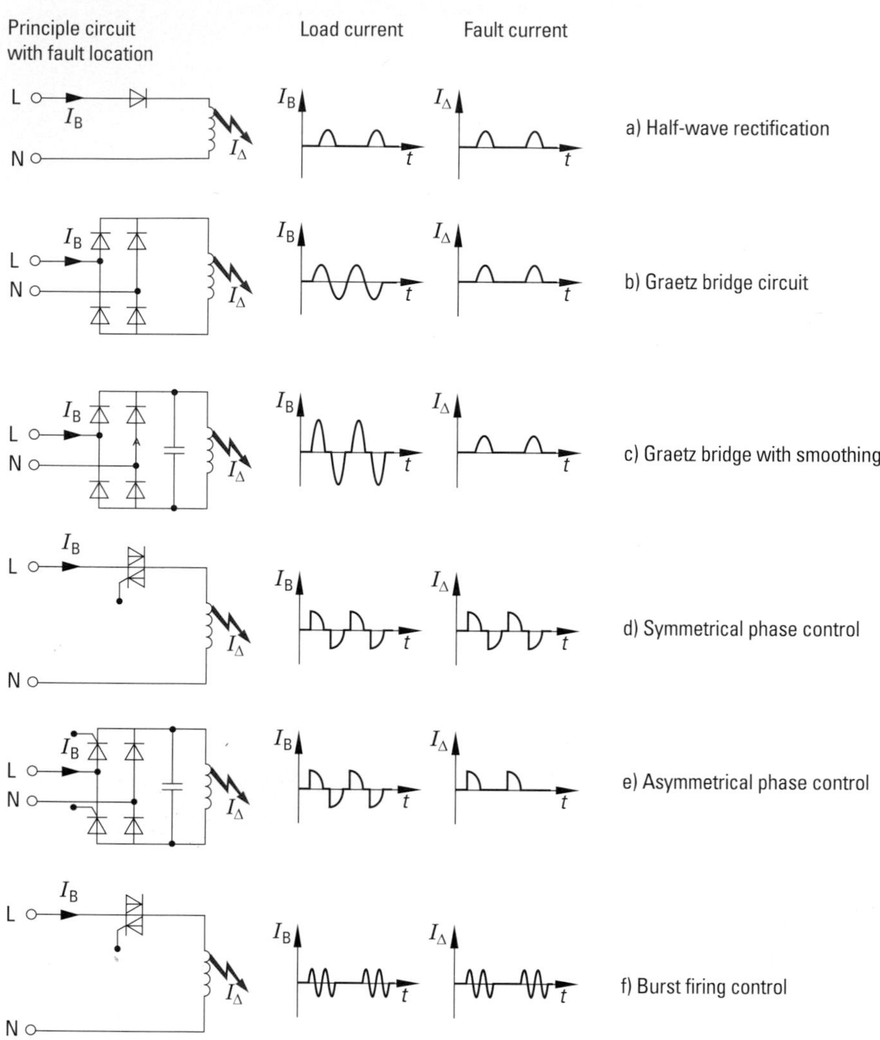

Fig. 4.1/41
Form of load current I_B and fault current I_Δ of circuit types that generate AC or pulsating DC fault currents in the event of a fault

The basic structure of the universal-current-sensitive residual-current protective device is shown in Fig. 4.1/43.

The summation current transformer W1 monitors the installation for AC and pulsating DC fault currents. The summation current transformer W2 detects the smooth DC fault currents in accordance with the principle of controlled inductance and, in the event of a fault, issues the trip command to trip element A via an electronic unit E which consists of a power supply unit and an evaluation module.

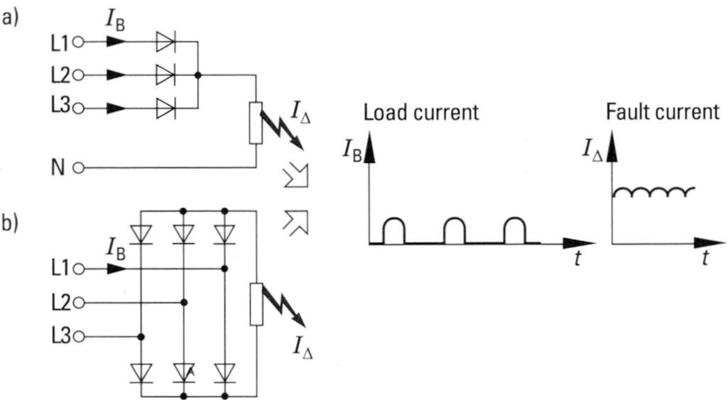

Circuit with fault location
a) Three-phase star connection
b) Three-phase bridge circuit, six coil

Fig. 4.1/42
Form of load current I_B and fault current I_Δ of circuit types which can generate DC fault currents with minor residual ripples in the event of a fault

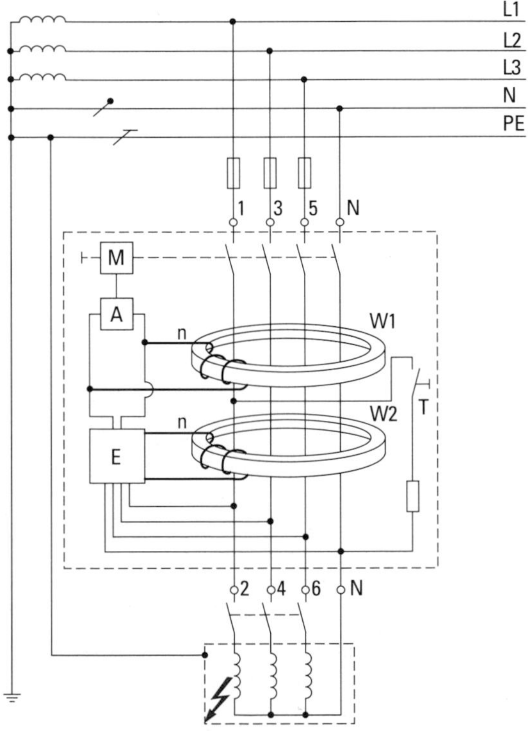

A Trip element
M Mechanics of the protective device
E Electronics for tripping with smooth DC fault currents
T Test device
W1 Summation current transformer for detecting the AC and pulsating DC fault currents
W2 Summation current transformer for detecting the smooth DC fault currents
n Secondary winding

Fig. 4.1/43
Circuit diagram of universal-current-sensitive residual-current protective devices, e.g. in TN S system

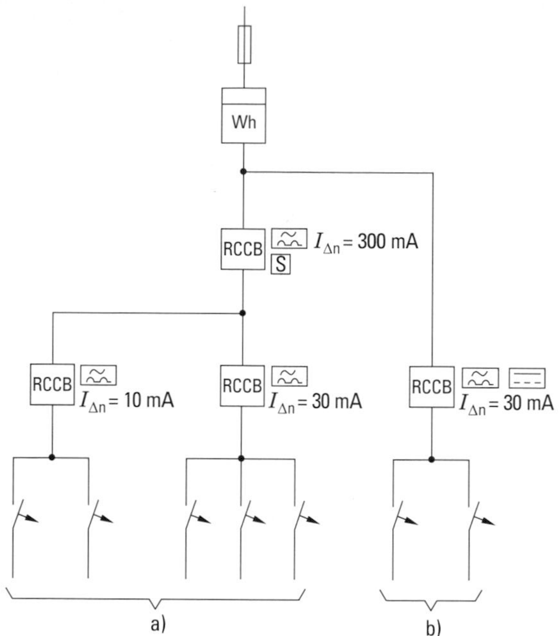

a) Electric circuits with current-using equipment in which alternating and/or pulsating DC fault currents can occur in the event of a fault
b) Electric circuits with current-using equipment in which alternating and/or pulsating DC fault currents and/or smooth DC fault currents can occur in the event of a fault
S Residual-current protective devices with time delay

Fig. 4.1/44
Use of universal-current-sensitive residual-current protective devices

4.1.3.6.2 Planning and erecting electrical installations with universal-current-sensitive residual-current protective devices

Important: When planning and erecting electrical installations, it should be noted that current-using equipment, which generates smooth DC fault currents in the event of a fault, is assigned to its own electric circuit with a universal-current-sensitive residual-current protective device (Fig. 4.1/44). As specified in EN 61 008/IEC 1008/DIN VDE 0664, branching of electric circuits such as these to pulsating-current-sensitive residual-current protective devices is not allowed as smooth DC fault currents ≥ 6 mA interfere with the tripping action of the protective devices.

According to EN 61008/IEC 1008/DIN VDE 0664, universal-current-sensitive residual-current protective devices are just as suitable for use in alternating or three-phase systems as the pulsating-current-sensitive residual-current protective devices.

4.1.3.6.3 Marks of conformity and areas of application

Universal-current-sensitive residual-current protective devices are provided with a special sign from the VDE Inspection and Certification Institute in the form of a VDE registration number.

These are used in electrical circuits with connected current-using equipment, which generates AC and/or pulsating DC fault currents as well as smooth DC fault currents in the event of a fault.

Examples of current-using equipment include:
▷ medical devices (such as X-ray machines, computer tomography equipment),
▷ frequency converters for lift control and variable-speed drives,
▷ Special heating cables that prevent the liquids being pumped through pipelines from freezing,
▷ uninterruptible power systems (UPS).

4.1.3.7 Tripping current ranges for residual-current protective devices

The tripping current ranges for alternating and pulsating-current-sensitive residual-current protective devices are, as shown in Table 4.1/11, stipulated in EN 61 008/IEC 1008/DIN VDE 0664.

As regards tripping in the event of smooth DC fault currents, Table 4.1/11 has been extended according to the requirements of IEC 60 479.

Table 4.1/11
Tripping current ranges for residual-current protective devices

Type of fault current	Pictorial marking	Tripping current ranges
AC fault currents		$0.50 \ldots 1 \cdot I_{\Delta n}$
Pulsating DC fault currents (pos. and neg. half-waves) Half-wave current		$0.35 \ldots 1.4 \cdot I_{\Delta n}$
Phase-angle-controlled half-wave currents Phase angles 90° el 135° el		 $0.25 \ldots 1.4 \cdot I_{\Delta n}$ $0.11 \ldots 1.4 \cdot I_{\Delta n}$
Half-wave current superimposed on a smooth DC fault current of 6 mA		max. $1.4 \cdot I_{\Delta n} + 6$ mA
Smooth DC fault current		$0.50 \ldots 2 \cdot I_{\Delta n}$

The sensitivity of the residual-current protective device is indicated by means of the following two symbols:

Symbols for sensitivity

Tripping with AC and pulsating DC fault currents,

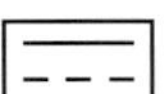
Tripping with smooth DC fault currents.

The first symbol is used with pulsating-current-sensitive protective devices. The second symbol has been taken from DIN 40900 Part 2.

The universal-current-sensitive residual-current protective devices fulfil the international standards IEC 755 Amendment 2 "General requirements for residual current operated protective device – Type B".

4.1.3.8 Selective and short-time-delay residual-current protective devices

Residual-current protective devices normally support instantaneous tripping. This means that series-connected residual-current protective devices of this type for selective tripping in the event of a fault will not function. To achieve selectivity, the devices connected in series must be graded, both with regard to the release time and the rated residual current.

Selective residual-current protective devices

Symbol S

Selective residual-current protective devices bear the symbol S.

The series-connected selective residual-current protective devices with instantaneous devices are used in large-scale industrial installations, for agricultural and horticultural applications, as well as in domestic installations so that protection is also provided for the part of the installation between the main distribution and sub-distribution boards.

Fig. 4.1/45 illustrates selective tripping with possible grading of the residual-current protective devices using an industrial application as an example.

Short-time-delay residual-current protective devices

In the case of current-using equipment that causes momentary high leakage currents when switched on (e.g. transient fault currents that flow via suppression capacitors between the outer conductor and PE conductor), instantaneous residual-current protective devices may trip inadvertently if the leakage current exceeds the rated residual current $I_{\Delta n}$ of the residual-current protective devices.

In cases such as these, in which it is not or only partly possible to remove these sources of interference, short-time-delay residual-current protective devices can be used.

Release time

These devices have a minimum release time of 10 ms, i.e. they must not trip with a fault-current pulse lasting 10 ms. The release requirements here are observed in line with

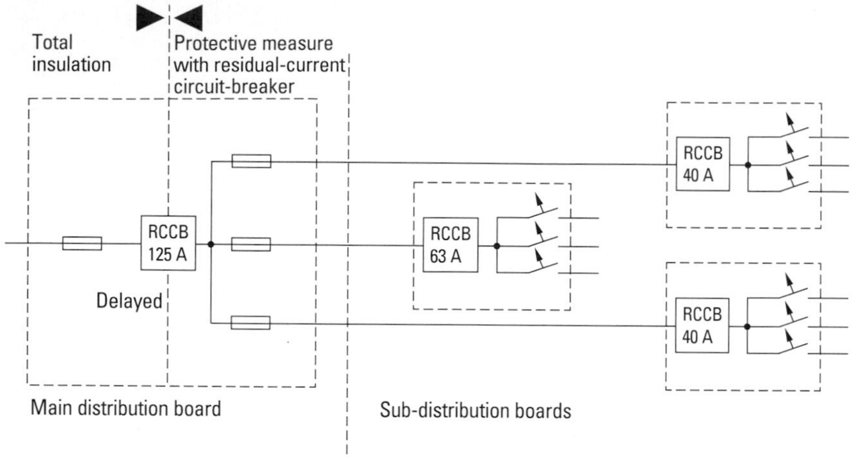

With delay [S]	Without delay
$I_{\Delta n}$ = 0.3 A; I_n = 63 A	$I_{\Delta n}$ = 10 mA; I_n = 16 A $I_{\Delta n}$ = 30 mA; I_n = 25, 40 and 63 A
$I_{\Delta n}$ = 0.5 A; I_n = 125 A	$I_{\Delta n}$ = 10 mA; I_n = 16 A $I_{\Delta n}$ = 30 mA; I_n = 25, 40, 63, 80, 100 and 125 A $I_{\Delta n}$ = 0.3 A; I_n = 25, 40, 63, 80, 100 and 125 A
$I_{\Delta n}$ = 1 A; I_n = 125 A	$I_{\Delta n}$ = 10 mA; I_n = 16 A $I_{\Delta n}$ = 30 mA; I_n = 25, 40, 63, 80, 100 and 125 A $I_{\Delta n}$ = 0.3 mA; I_n = 25, 40, 63, 80, 100 and 125 A and 0.5 A

Fig. 4.1/45
Possible grading of residual-current protective devices for selective tripping; example: industrial application

EN 61 008/IEC 1008/DIN VDE 0664. These

Surge strength devices bear the VDE sign and have a surge strength of 3 kA over and above the requirements specified in EN 61 008/IEC 1008/ DIN VDE 0664.

Symbol [K] Short-time-delay residual-current protective devices bear the symbol [K].

4.1.3.9 Rated breaking capacity and short-circuit strength of Siemens residual-current protective devices

Fault currents can also reach values approaching those of short-circuit currents, e.g. if there is an insulation fault in a well-grounded device such as a warm water storage container.

A ground fault can also cause a short circuit, e.g. due to the effect of an electric arc. For this reason, residual-current protective devices must have a sufficiently high rated breaking capacity and must be short circuit proof.

Short-circuit-type fault currents can also occur when using the neutral conductor as an equipment grounding conductor in TN systems. In this case, residual-current protective devices, together with the back-up fuse, must have an appropriate rated breaking capacity and sufficient rated short-circuit strength. The rated short-circuit strength of the combination must be specified on the devices.

mbol
00 A
10000

Together with an appropriate back-up fuse, the Siemens residual-current protective devices have a rated short-circuit strength of 10,000 A. According to EN 61 008/IEC 1008/DIN VDE 0664, this is the highest possible rated short-circuit strength.

4.1.3.10 Use of residual-current protective devices in building control system

mediate fault dication

All residual-current protective devices can also be used in building system engineering via the binary inputs of a bus system, e.g. *instabus EIB*, and the auxiliary contacts that are available for the residual-current protective devices (see Section 4.4). The signals can be read into the *instabus EIB* system by connecting the auxiliary contact to an *instabus EIB* binary input. Using an *instabus EIB* binary output, the status of the residual-current protective device is interrogated via the auxiliary contact and faults are indicated immediately.

4.1.3.11 Type range of Siemens residual-current protective devices

Siemens manufactures various types of residual-current protective devices. The following types can be used in residential and functional buildings and for industrial applications:

2 pole,
▷ Rated current
$I_n =$ 16 A, 25 A, 40 A, 63 A, 80 A
▷ Rated residual current
$I_{\Delta n} =$ 10 mA, 30 mA, 100 mA, 300 mA

4 pole,
▷ Rated current
$I_n =$ 25 A, 40 A, 63 A, 80 A, 100 A, 125 A
▷ Rated residual current
$I_{\Delta n} =$ 30 mA, 100 mA, 300 mA, 500 mA, 1000 mA.

Auxiliary contacts with the following functions can be retrofitted:
▷ 1 NO + 1 NC
or
▷ 2 NO
or
▷ 2 NC

Residual-current circuit-breakers with integrated overcurrent protection

Siemens also manufactures combined residual-current and miniature circuit-breakers, called residual-current circuit-breakers with integrated overcurrent protection in line with EN 61 009/IEC 1009/DIN VDE 0664:

2 pole and 4 pole
▷ Rated current
$I_n =$ 6 A, 10 A, 13 A, 20 A, 32 A
▷ Rated residual current
$I_{\Delta n} =$ 10 mA, 30 mA, 300 mA
▷ Characteristics B and C
(see Section 4.1.2).

RCCB modules for retrofitting on miniature circuit-breakers

RCCB modules can be retrofitted on Siemens miniature circuit-breakers.
▷ Rated current
$I_n =$ 40 A, 63 A, 80 A, 100 A
▷ Rated residual current
$I_{\Delta n} =$ 30 mA, 100 mA, 300 mA, 500 mA.

Protected socket outlets

If a residual-current protective device for protection against electric shock cannot be installed, a protected socket outlet is available to increase the level of protection (Fig. 4.1/46) e.g. in bathrooms or workshops, in line with the German standard DIN VDE 0662.

The socket outlets are available for:
▷ Rated current
$I_n =$ 16 A
▷ Rated residual current
$I_{\Delta n} =$ 10 mA and 30 mA.

Fig. 4.1/46
Protected socket outlet DELTA profile 16 A

4.1.3.12 Practical troubleshooting tips when using residual-current protective devices

Troubleshooting

If, during the planning stage of an electrical installation, the circuits are divided up appropriately among several protective devices and the installation is then commissioned correctly, no problems will normally be encountered later on when the installation is in operation. If a residual-current protective device in an electrical installation does, however, trip, troubleshooting can be carried out in one single step using the diagram in Fig. 4.1/47.

By carrying out appropriate visual inspections (see Section 2.5), special care must be taken to ensure that there is no electrical connection between

- ▷ the neutral conductor N and the equipment grounding conductor PE
- ▷ and/or
- ▷ the neutral conductors N of two or more protective devices

downstream of the residual-current protective device.

Unwanted tripping

Unwanted tripping of residual-current protective devices can also occur occasionally as a result of lightning surges or switching overvoltages in the electrical installations. These trips, however, mainly occur in current-using installations in which

- ▷ no overvoltage protection measures or no suitable measures are implemented (see also Section 2.6) and/or
- ▷ no current-using devices that comply with EMC requirements are fitted (see also Section 3.3).

Extensive investigations regarding unwanted tripping revealed that most occurrences could have been avoided by means of pulsating-current-sensitive residual-current protective devices which, for the most part, are surge-proof.

Surge arresters

If the electrical installation is also fitted with surge arresters, care must be taken to arrange these upstream of the residual-current protective device (with respect to ground). This should avoid unwanted tripping of the residual-current protective device caused by defective surge arresters.

Bundling of operational leakage curren[ts]

Unwanted tripping of residual-current protective devices can also occur if the installation is not planned and configured by qualified personnel; in other words the size of the installation and a high number of current-using devices cause bundling of operational leakage currents. Depending on the operational status of the installation, this can reach or exceed the tripping current of the series-connected residual-current protective device. The leakage current in the installation can be measured quite simply using a leakage current measuring device which is connected in series to the residual-current protective device in order to perform the measurement. The leakage current I can be determined in mA on the basis of the measured voltage U via a calibration curve that is supplied with the leakage current measuring device. This measuring instrument has proved to be useful for troubleshooting purposes and for monitoring the insulation condition in electrical installations, for example in traffic light systems.

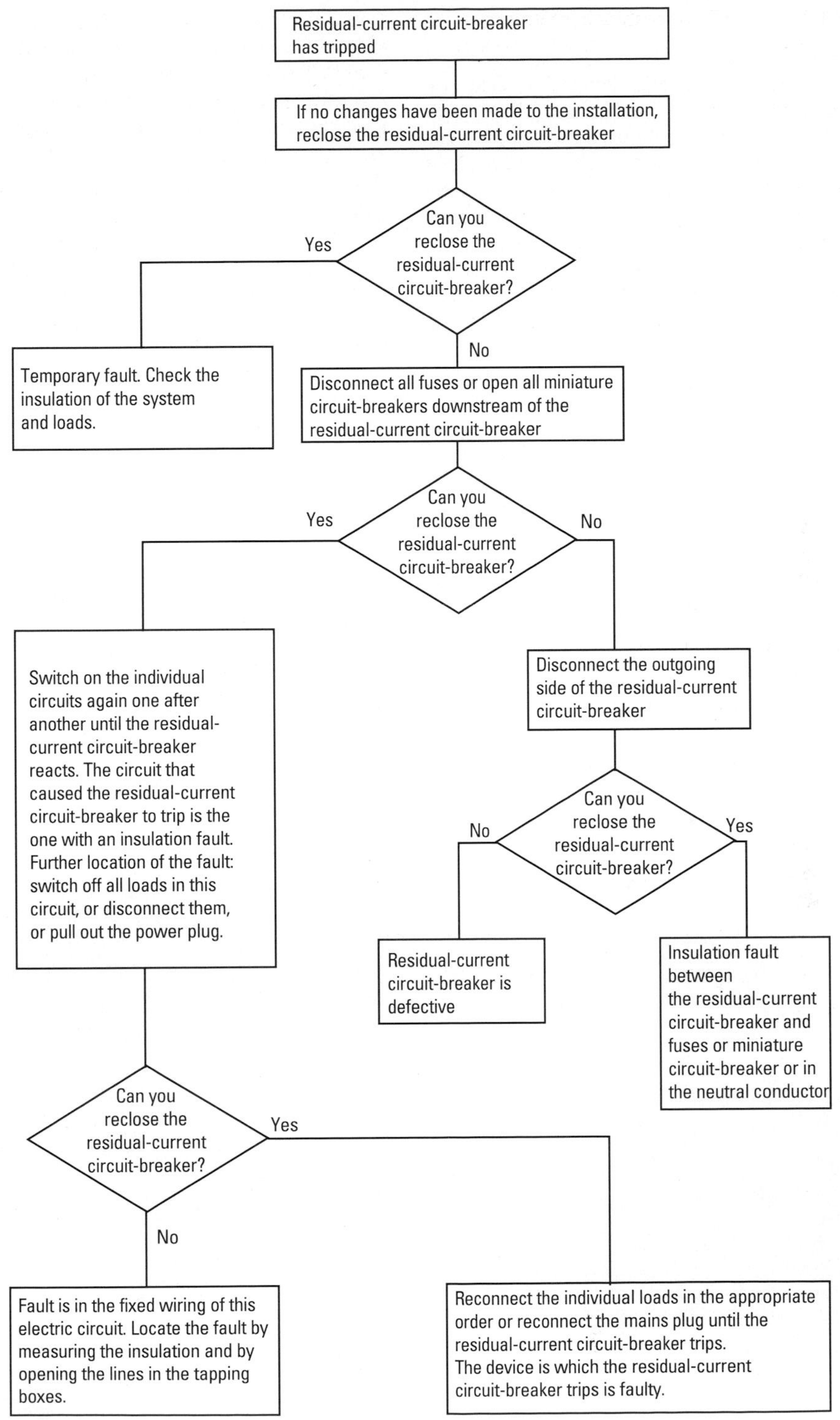

Fig. 4.1/47
Troubleshooting in an electrical installation if the residual-current protective device operates

4.1.4 Insulation monitoring devices

Effectiveness and application

In IT systems, the protective measure "protection against indirect contact" is achieved by means of insulation monitoring (see also Section 2.4.6). Insulation monitoring should be performed using an insulation monitoring device which monitors the insulation resistance of the electrical installation on a continuous basis. If an insulation fault occurs between an active part (including a neutral conductor, if fitted) and an exposed conductive part or with respect to ground, an acoustic and/or visual signal is issued immediately. The equipment is, however, not disconnected. This means that electrical equipment that is vital for sustaining patients' lives in hospitals and doctor's practices, for example, (see also Section 5.5) can continue to be used safely until it is possible to resolve the fault.

Standards

In Germany, DIN VDE 0107 stipulates that a.c. voltage systems (230 V AC, 50/60 Hz) in rooms used for medical purposes (such as operating theaters, intensive care units, examination rooms, and observation rooms) must be fitted with insulation monitoring devices.

Operating theater lights operated with 24 V AC can also be monitored.

To increase operational reliability in rooms used for medical purposes, load monitoring must also be carried out for the supply transformer.

Design

Fig. 4.1/48 shows a Siemens insulation monitor which conforms to the design and dimensions of the modular devices to DIN 43880 (see Section 4.2) and the CENELEC Report R023-001. The monitor is installed in domestic distribution boards (see Section 1.4.5) on DIN rails to EN 50022.

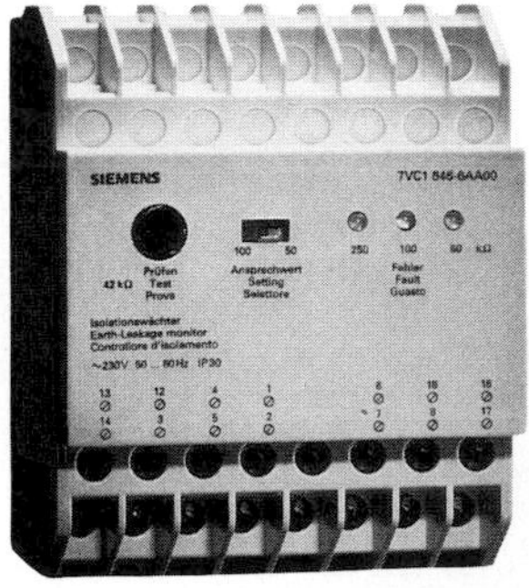

Fig. 4.1/48
Siemens insulation monitor 7VC1 646

Method of operation

The insulation monitor is connected to the ungrounded system to be monitored via a high-resistance resistor which is integrated in the device itself. The current generated is evaluated by means of a high-resistance resistor via a superimposed 24-V d.c. measuring voltage. The operate values 50 kΩ or 100 kΩ of the signaling relay are set with the changeover switch. An insulation fault that occurs at the minimum resistance of 50 kΩ stipulated in DIN VDE 0107 is thus indicated.

Light-emitting diodes provide a reliable means of indicating a drop in the insulation resistance to below 250 kΩ, 100 kΩ or 50 kΩ.

Measuring insulation resistance

A d.c. voltage is generated in the insulation monitor by means of a rectifier for measuring the insulation resistance. One pole is connected to ground and the other pole is connected to any one of the outer conductors of the system to be monitored by means of an electronic measuring circuit (Fig. 4.1/49). Depending on the how the electrical installation is insulated between the outer conductor and ground, a leakage current flows, the magnitude of which is evaluated by the insulation monitor. An LED row (calibrated to kΩ) indicates the insulation resistance. When the set operate value is reached, a visual and/or acoustic signal is issued via the measuring circuit. The acoustic signal can be switched off. The visual signal, however, continues to be output until the fault has been rectified.

Functional chec

The test button at the front enables the measuring circuit to be checked via an integrated 42 kΩ test resistance in line with DIN VDE 0107, Point 4.3.5.2.

In order to check whether the equipment is ready for operation, a ground fault is simulated via the test resistance between an outer conductor and ground. The insulation monitor then responds.

Important

Only *one* insulation monitor must be used for monitoring parts of the installation that are connected electrically. If a second insulation monitor is connected, the leakage current of the system is distributed among

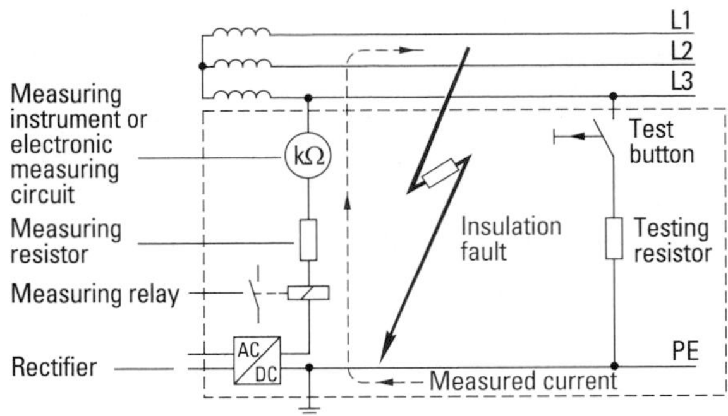

Fig. 4.1/49 Measuring circuit of the insulation monitor

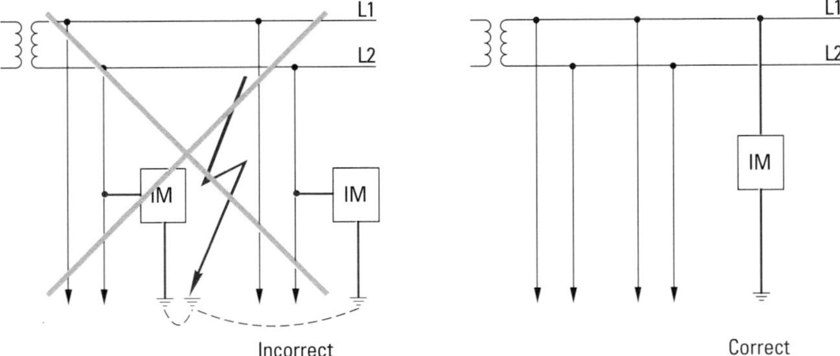

Fig. 4.1/50 Only install one insulation monitor IM in an electrical installation!

both monitors which would cause both of these to measure incorrect insulation values (Fig. 4.1/50).

A test and signal combination unit can be connected to the insulation monitor (Fig. 4.1/51) to allow faults to be indicated and the insulation monitor to be tested at locations other than the installation location.

The combination is fitted with a green indicator light which indicates that the installation is ready for operation, a yellow indicator light and a buzzer which are activated if an insulation fault occurs in the installation, a test button for performing a functional check on the insulation monitor, and a further key that can be used to cancel the acoustic signal. The visual fault signal, however, continues to be output until the fault has been rectified.

Fig. 4.1/51
Test and signal combination unit 7XV9 306

Power supply

The transformer (24 V AC / 1.2 VA) integrated in the insulation monitor supplies the test and signal combination unit. The transformer is designed to supply two test and signal combination units.

Installation

The combination is installed in a switch and socket box with a fixing dimension of 60 mm.

Technical data

Table 4.1/12 contains the technical data of the Siemens insulation monitor. The technical data for the test and signal combination unit can be found in Table 4.1/13.

Table 4.1/12 Technical data of the Siemens insulation monitor 7VC1 646

Monitored nominal voltage U_n in the IT system	24 ... 230 V AC
Rated voltage Frequency Power input	230 V AC + 10% / – 20% 50 ... 60 Hz < 5 VA at 230 V AC
Measuring d.c. voltage Measuring direct current D.c. internal resistance A.c. internal resistance	24 V DC < 0.2 mA 120 kΩ 120 kΩ
Insulation group (DIN VDE 0110) Noise immunity (DIN VDE 0843 T.4) Resistance to vibration (DIN IEC 60068-2-6)	C Severity 4 10 ... 150 Hz, 20 cycles, 5 gn
Maximum permitted external d.c. voltage	200 V DC (without damaging the device, any polarity)
Auxiliary supply	24 V AC 50 mA (sufficient for 2 test and signal combination units)
Signaling contacts Maximum switching voltage Continuous switching current Maximum switching capacity	2 changeover contacts 250 V AC / 300 V DC 10 A AC / 5 A DC 1250 VA AC / 250 W DC
System capacitance to ground Operate values Fault indication Fault pre-indication	< 10 μF 100 kΩ / 50 kΩ < 100 kΩ / < 50 kΩ < 250 kΩ
Degree of protection to EN 60529/IEC 60529/ DIN VDE 0470-1	IP 30 (built-in components) IP 10 (terminals)
Ambient temperature	–5 °C ... +50 °C
Connecting capacity of terminals Weight Specifications Dimensions (H × W × D)	≤ 1.5 mm^2 0.3 kg DIN VDE 0100, 0107, 0110, EN 61557-2/IEC 61557-2/VDE 0413 Part 2 90 × 72 × 53 mm

Table 4.1/13 Technical data of Siemens test and signal combination unit 7XV9 306

Control elements	Test button Delete key
Status indicator Fault indication	LED (green) LED (yellow) and buzzer
Degree of protection to EN 60529/IEC 60529/ DIN VDE 0470-1	IP 20 IP 44 with device insert seal (flush-type only)
Ambient temperature	–5 °C ... +50 °C
Connecting capacity of terminals Weight	≤ 1.5 mm^2 0.1 kg
Insertable labeling strips	Languages available: German, English, Spanish, French, Italian, Dutch
Rated voltage Power input	24 V AC + 10% / – 20% < 1.2 VA at 24 V AC

4.2 Mechanical, electromechanical, and electronic modular devices, timers

4.2.1 General

Modular devices

Modular device is a generic term for installation equipment which is used for switching, monitoring, display, control, regulation, and signaling purposes. In conjunction with *instabus EIB* built-in devices, this equipment supports an extremely wide range of functions when used in low-voltage switchgear assemblies as well as in power and domestic distribution systems. In Germany, the design of the modular devices (and their dimensions) is standardized to DIN 43 880 and is also defined in the CENELEC Report R 023-001. This standardization has simplified the planning and assembly of switchgear and distribution systems considerably and has also boosted development activities in this area. Consequently, new devices with new functions are being developed on a continuous basis and made available in line with these design specifications.

Standardization

The limitations posed by the extremely compact design, which permits space-saving mounting on DIN rails to DIN EN 50022, are the generation of specific heat and the dissipation of heat from switchgear assemblies and distribution boards.

N system

All of the mechanical, electromechanical, and electronic modular devices described in this section as well as the miniature circuit-breakers, residual-current protective devices, ground-leakage monitors, and some of the fusible links described in Sections 4.1.1 to 4.1.4 are constructed in line with the modular design of the *N system*; their dimensions comply with DIN 43 880 and the CENELEC Report R 023-001 in which the size limits for this modular design are specified (Fig. 4.2/1).

The device width is specified in modular units, e.g. $\frac{1}{2}$, 1, or 2 MU. One modular unit (MU) is 18 mm.

The devices of the *N system* with a mounting depth of 55 mm are identified by the letter >N<. Other device mounting depths are 70 mm and 92 mm.

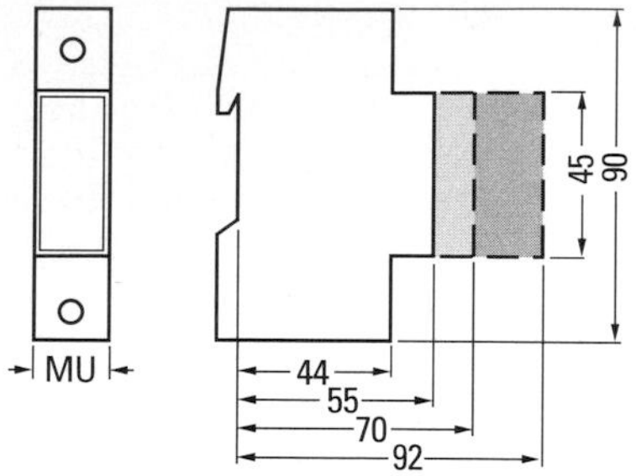

MU

Dimensions in mm

MU = 18 mm, or a multiple of 18 mm

Fig. 4.2/1
Overall dimensions for the housings of modular devices to DIN 43 880 and CENELEC Report R 023-001

Power and domestic distribution boards, such as Simbox 63, STAB wall-mounting and floor-mounting distribution boards, SIKUS modular distribution-board systems, are designed to accommodate any modular device (see Section 1.4.5).
The modular devices are clipped onto 35 mm DIN rails (conforming to DIN EN 50022) without tools being required.

***instabus EIB* built-in devices**

The dimensions of the *instabus EIB* built-in devices for installation in distribution boards are also based on DIN 43 880 and the CENELEC Report R 023-001.

Table 4.2/1 lists the functions, applicable standards, purpose, and applications of Siemens modular devices.

Table 4.2/1 Overview of functions, applicable standards, and applications of Siemens modular devices

Device	Standards	Function	Application		
			Functional	Residential	Industrial
Switches					
Switch-disconnector 16 A, 25 A	DIN VDE 0632-101 DIN VDE 0660-107	Switching loads or system components and for switching operations in control systems	•	•	•
Switch-disconnector 32 A	EN 60 947-3 DIN VDE 0660-107		•		•
Switch-disconnector 40 A, 63 A, 100 A	EN 60 947-3 DIN VDE 0632-101 DIN VDE 0660-107		•	•	•
Switch-disconnector 125 A	EN 60 947-3		•		•
Switch-disconnector 100 A, 125 A, 160 A, 200 A	EN 60 947-3 UL 508		•	•	•
Pushbutton	DIN VDE 0632		•		•
Switching devices					
Remote switch	DIN VDE 0637 DIN VDE 0632	Switching lighting systems	•	•	
Remote switch with central function			•		
Contactor (Insta contactors)	EN 60 947 IEC 60 947 DIN VDE 0660 Safety regulation: ZN 1/457	Switching lighting systems, ohmic and inductive loads	•	•	•

Table 4.2/1 (continued)

Device	Standards	Function	Application: Functional	Application: Residential	Application: Industrial
Switching devices					
Switch relay	DIN VDE 0435	Switching low loads and for contact multiplication in control systems		•	•
Soft-starting device	IEC 60947.4.2 DIN VDE 0660	Protecting motorized, mechanical drive units and pumps			•
Timers					
Stairwell lighting timer	EN 60669	Saving energy in stairwell lighting systems	•	•	
Pre-warning time switch for stairwell lighting	EN 60669 DIN VDE 0632 DIN 18015	Issuing warnings prior to stairwell lighting in multiple dwelling units being switched off	•	•	
Illuminated timer Energy-saving timer	IEC 60669 DIN VDE 0632	Saving energy in rooms which are used infrequently or for periods of varying duration	•	•	
Timer and control switch for ECG Dynamic	IEC 60669 DIN VDE 0632	Stairwell lighting timer for fluorescent luminaires with ECG Dynamic	•		
Timer for ventilation fans (operating time switch)	DIN VDE 0637	Overtravel for ventilation fans in washrooms	•	•	
Timer for industrial applications	IEC 60255 EN 61812-1 DIN VDE 0435-2021	Controlling time sequences in control systems			•
Mechanical time switch	IEC 60730 EN 60730 DIN VDE 0633	Switching in 30-minute steps (daily or weekly time sequences)	•		•
Digital time switch	IEC 60255 EN 60730 DIN VDE 0435-2021	Switching in 1-minute steps (daily, weekly, or twelve-month time sequences)	•	•	•
Programming time switches via PC		Creating, modifying, and documenting switching programs	•		•
Monitoring devices					
Undervoltage relay	IEC 60255 DIN VDE 0435-303 DIN VDE 0108	Monitoring power supplies for emergency lighting in public buildings	•		
Undervoltage relay, under/overvoltage relay	IEC 60255 DIN VDE 0435-303	Monitoring power supplies for saving operating parameters for devices or system components			•
Neutral conductor monitor	DIN VDE 0633	Monitoring neutral conductors for open-circuit	•		•

Table 4.2/1 (continued)

Device	Standards	Function	Application		
			Functional	Residential	Industrial
Monitoring devices					
Fuse monitor	IEC 60255 DIN VDE 0435-303	Monitoring fuses of all types	•		•
Short-time voltage relay		Monitoring power supplies for short-time failures of 20 ms			•
Phase display Phase monitor		Monitoring power supplies			•
Phase sequence display Phase sequence monitor		Monitoring phase sequence of a supply system			•
Counter, pulse counter	DIN VDE 0435	Monitoring operating hours and making operations of devices and systems	•		•
Current relay	IEC 60255 DIN VDE 0435-303	Monitoring emergency/alarm lighting, and motors	•		•
Thermistor motor protection relay	IEC 60255 DIN VDE 0632	Thermal protection of motor windings			•
Level-sensing relay	IEC 60255 DIN VDE 0435-303	Regulating fluid levels in containers	•		•
Twilight switch	EN 60669	Controlling lighting systems according to daylight intensity	•		

Switching loads

Modular devices are primarily switching-type devices and are used to switch different loads on and off. The effects of the increased making current of different loads on the switching contacts when a device is switched on are often underestimated.

Ohmic load

Ohmic loads, e.g. electric heaters, do not produce an increased making current (Fig. 4.2/2).

Filament lamp load

When filament lamps or halogen lamps are switched on, an 8 to 10-fold making current is generated since the filaments have low impedance when cold. The rated operating current does not stabilize for approximately 10 ms (Fig. 4.2/3).

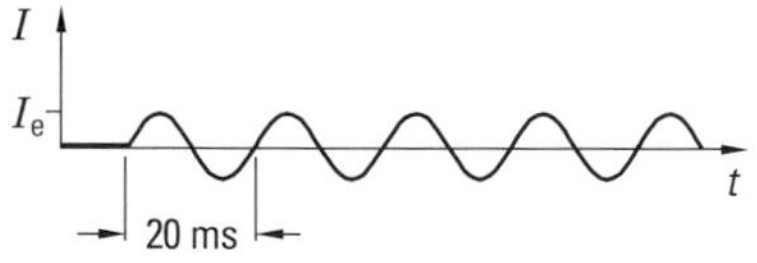

I_e : Rated operating current

Fig. 4.2/2
Current curve when ohmic load is switched on

Fluorescent lampload without p.f. Correction

Fluorescent lamps without p.f. correction behave inductively. When they are switched on, there is no increase in making current. Instead, the heating current, which is increased by the rated operating current after

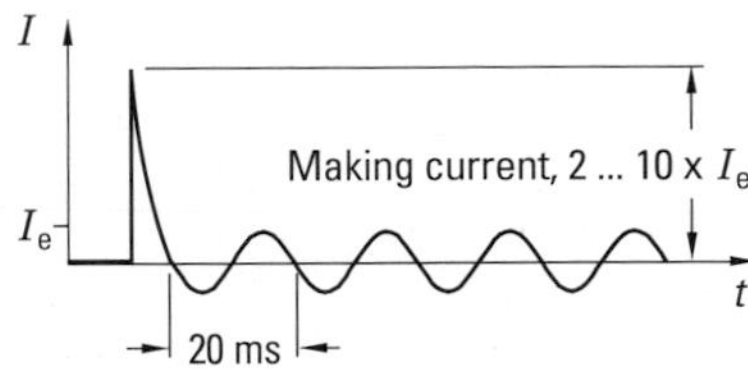

I_e : Rated operating current

Fig. 4.2/3
Current curve when filament lamp load is switched on

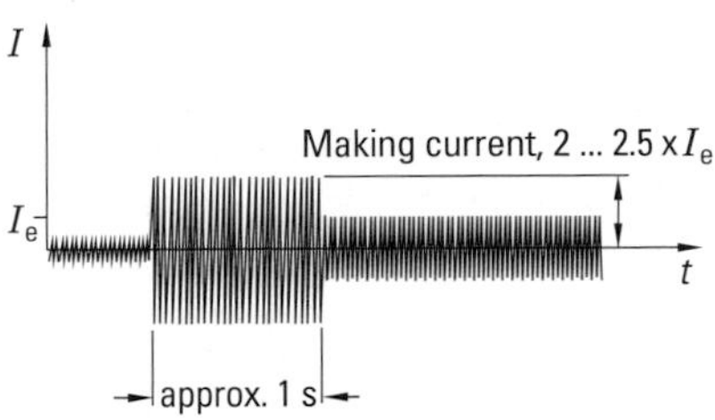

I_e : Rated operating current

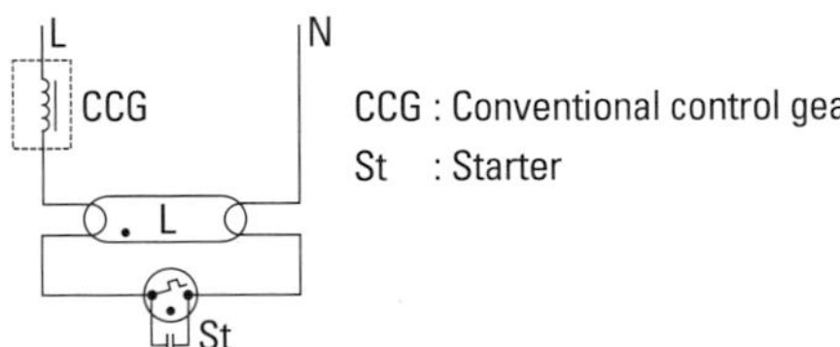

Fig. 4.2/4
Current curve when fluorescent lamp load without p.f. correction is switched on

ignition, flows first. An increased current then flows for a number of cycles (initial flickering). The time scale, therefore, extends over a relatively long period of time (Fig. 4.2/4).

Fluorescent lamp load with parallel p.f. correction

The capacitor, which is connected in parallel to the L/N input terminals, corrects the actual inductive load. It causes, however, an additional, extremely high making current (approximately 13 times higher) when the load is switched on (Fig. 4.2/5).

Twin-lamp circuit

The series capacitor in a lamp circuit provides the desired p.f. correction. An increased making current is prevented by the series connection with the reactor. Fluorescent lamps in twin-lamp circuits behave in a similar way to fluorescent lamps without p.f. correction (Fig. 4.2/6).

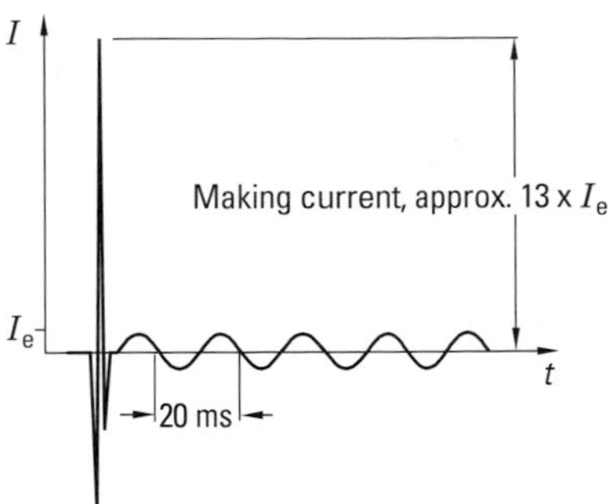

I_e : Rated operating current

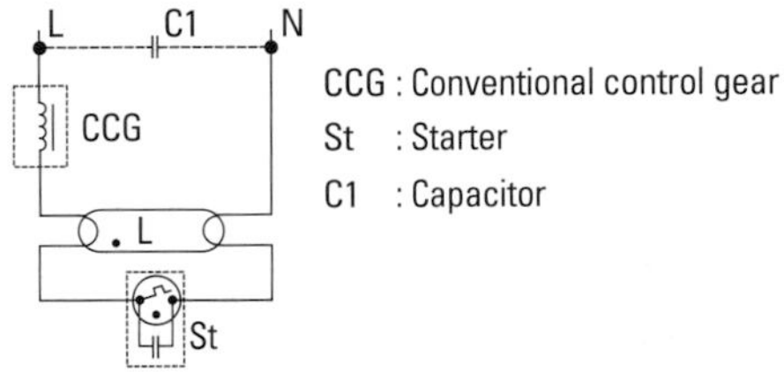

Fig. 4.2/5
Current curve when fluorescent lamp load with parallel p.f. correction is switched on

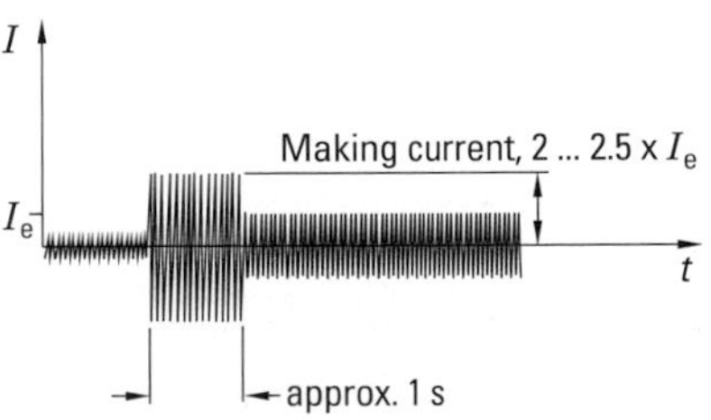

I_e : Rated operating current

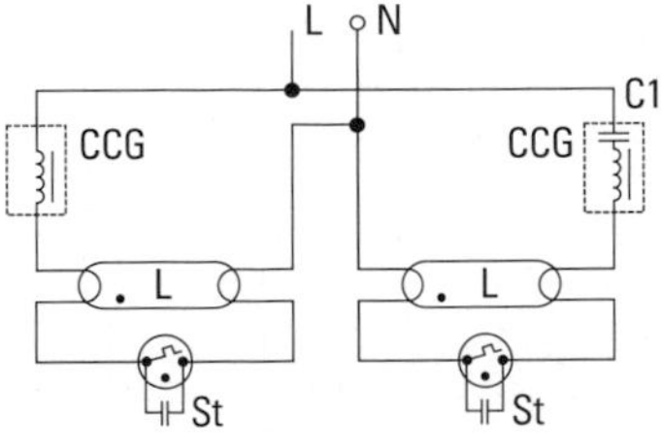

CCG : Conventional control gear
St : Starter
C1 : Capacitor

Fig. 4.2/6
Current curve when fluorescent lamp load of twin-lamp circuit is switched on

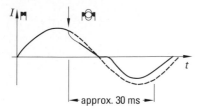

Fig. 4.2/7
Contact opening (zero-current interruption)

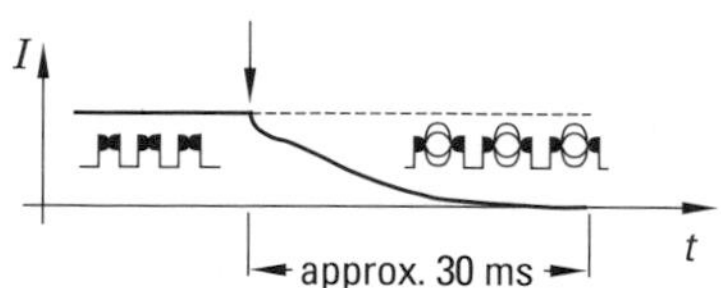

Fig. 4.2/8
Current curve when direct currents are switched off

Zero-current interruption

If a current-carrying contact opens, an electric arc always ignites above 24 to 30 V. The electric arc depends on the voltage, size of the isolating gap, contact speed, switching angle, and current intensity. According to the principle of zero-current interruption, the electric arc is quenched at the zero crossing after a maximum of $1\frac{1}{2}$ half-waves (30 ms) (Fig. 4.2/7). Additional quenching aids or current-limiting devices, such as miniature circuit-breakers, are not available.

Switching direct currents

None of the modular devices are specifically designed for switching direct currents. With d.c. voltages, there is no zero crossing at which the electric arc can be quenched. In order to allow sizeable currents to be switched, however, contacts are connected in series and the isolating gap is thus increased (Fig. 4.2/8). Planning data for switching direct currents is specified for a number of switching devices in the catalogs supplied by the manufacturer. This ensures that the electric arc is quenched after approximately 30 ms in order to eliminate the risk of fire.

Requirements for switching devices

All switching devices must be able to switch a wide variety of different loads. In addition to this, they must function reliably, and have a long service life and low operating costs. Their switching contacts must be able to cope with the various time/current characteristics and current loads that occur.

Types of contact

For this reason, a range of very different contacts has to be used in modular devices, e.g.:

- switches which are actuated manually:
  switching contacts with contact gap > 3 mm,
- contactors:
  contactor contacts with contact gap > 3 mm,
- relays and remote switches:
  switch contacts with contact gap > 3 mm,
- electronic devices used on printed circuit boards:
  μ-contacts with contact gap > 0.5 mm.

Planning switching devices

Selecting appropriate switching device

The rated operating currents, to utilization category AC-1 (ohmic load), which are normally used as a basis for selecting switching devices, do not permit an adequate level of planning reliability on account of the high making currents. In the case of contactors and switch relays for activating fluorescent lamps with parallel p.f. correction, for example, it becomes evident that the 16 A contact of the switch relay 5TT3 081 would be better for this type of load than a 20 A contact of a contactor (Table 4.2/2). Contact welding as a result of excessively high making currents can only be prevented by selecting appropriate devices that are designed for different lighting equipment.

Table 4.2/2
Example of a selection table for contactors and switch relays for activating fluorescent lamps with parallel p.f. correction

Fluorescent lamps with parallel p.f. correction with:			Maximum number of lamps with 230 V AC, 50 Hz				
Lamp type, rated output in W			S11	L18	L24	L36	L58
Capacitor capacity in μF			4.5	4.5	4.5	4.5	7.0
Suitable contactor / switch relay							
Type	Conducting paths	Rated operating current AC-1 per conducting path A					
5TT3 081	1	16	20	20	20	20	13
5TT3 8..	2	20	6	6	6	5	4
5TT3 9..	4	20	15	15	15	15	10
5TT3 835	3 (4)	21	–	33	–	33	21
5TT3 8..	1	25	6	6	6	5	4
5TT3 8..	4	24	8	8	8	8	5
5TT3 8..	4	40	45	45	45	45	25
5TT3 8..	4	63	70	70	70	70	43

Creepage distances and clearances

With rated operating voltages $U_c < 50$ V, insulated cables can be used for IT systems (see Section 1.7) if safety separation from the 230/400 V AC system can be provided in accordance with HD 384.4.41S2 / IEC 60364-4-41 / DIN VDE 0100-410.

Creepage distances and clearances

The device standards for remote switches (DIN VDE 0637) and for bell transformers (EN 60742 / IEC 60742 / DIN VDE 0551), therefore, specify creepage distances and clearances a = 8 mm and a rated impulse withstand voltage of ≥ 4 kV.

This means that the operating voltage of a device must be isolated from the connected voltage by means of a creepage distance and clearance of 8 mm for ≥ 4 kV if one of the voltages is a safety extra-low voltage (Fig. 4.2/9).

Safety extra-low voltage (SELV)

Safety extra-low voltages (SELV), such as those generated by bell transformers, are suitable for doorbell installations in residential buildings and for switching contactors, e.g. for lighting systems. This is due to the inexpensive cables that can be used as well as the less complex total insulation and lower safety class of the button (doorbell) installed outside the building. If a safety extra-low voltage is used, the button does not have to satisfy the same requirements as it would for 230/400 V AC systems.

When planning systems containing devices such as contactors, remote switches, and switch relays, it is important to remember that "safe separation" is ensured by means of creepage distances and clearances of 8 mm and a rated impulse withstand voltage of ≥ 4 kV. The protection concept itself, however, must not be interfered with by any device from these types of system. Switch relays, which can be used to isolate the circuits (Fig. 4.2/10), provide a solution to this problem.

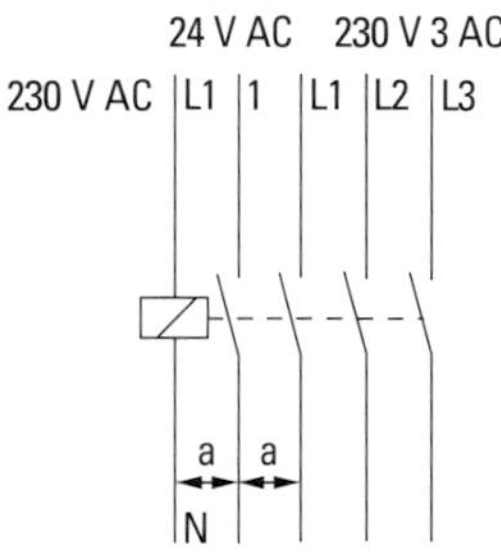

Fig. 4.2/9
Example: Creepage distances and clearances *a* in contactors, remote switches, and switch relays

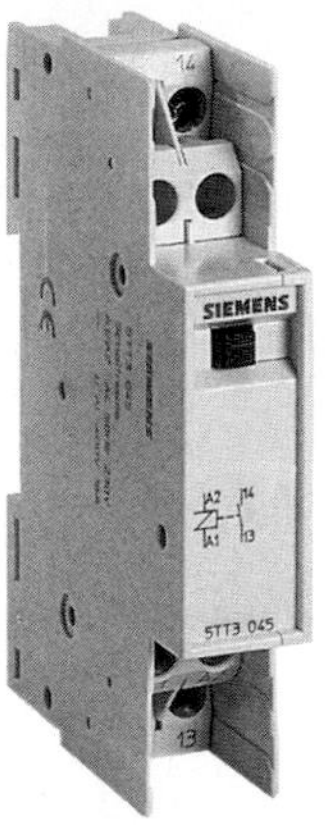

Fig. 4.2/10
Switch relay 5TT3 045 with 8 mm creepage distances and clearances, and rated impulse withstand voltage of ≥ 4 kV

4.2.2 Switching with switch-disconnectors

Using switch-disconnectors

Manually-operated switch-disconnectors are indispensable devices in electrical installations. They permit personnel to intervene in the technical processes of a system at any time. They can be used to connect and disconnect loads (e.g. lighting), to start, end, or select processes, or to isolate system components so that maintenance work can be carried out.

The different requirements that switch-disconnectors are expected to fulfill have led to a wide variety of functions and associated terms which do not always appear in the relevant standards.

Terms

Terms, such as ON switch, OFF switch, two-way switch, changeover switch, center-position switch, three-way switch, and isolating switch, describe the different types and characteristics of contacts. Control switch, operating switch, switch-disconnector, and isolator refer to the function of the contacts; pushbuttons and switches refer to the way in which the contact is actuated.

Selection

Switch-disconnectors with normally-open (NO) contacts are selected according to rated currents, rated operating voltages, and type of contact (Table 4.2/3).

Fig. 4.2/11
16 A switch-disconnector 5TE7 112, 2 pole

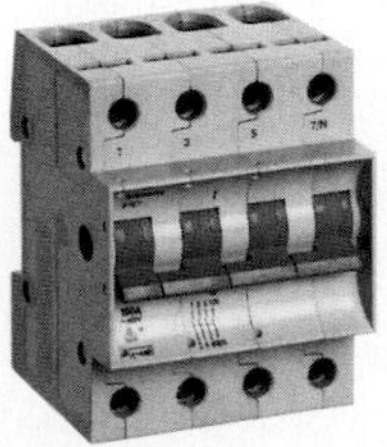

Fig. 4.2/12
100 A switch-disconnector 5TE7 714, 4 pole

Table 4.2/3 Selection table showing switch-disconnector types with NO contacts

Rated voltage, number and type of contacts [1)] Rated current, utilization category AC-1 A	230 V	400 V	690 V	1 NO	2 NO	3 NO	3 NO + N	3 NO + N link	4 NO
16	•			•					
		•			•	•	•		
25		•				•	•		
32	•			•					
		•			•	•	•		
40	•			•					
		•			•	•	•		
63	•			•					
		•			•	•	•		
80	•			•					
		•			•	•	•		
100	•			•					
		•			•	•	•		
			•		•	•		•	•
125		•							
			•		•	•		•	•
160			•		•	•		•	•
200			•		•	•		•	•

1) "3NO + N" means: Switch-disconnector with 3 NO contacts and one other leading contact for the N-conductor

"3NO + N link" means: Switch-disconnector with 3 NO contacts and terminal link for the N-conductor of the incoming and outgoing cable; N is not switched.

Type and use

Switch-disconnectors with rated currents between 16 A and 32 A are used as control switches and switch-isolators. They have a width of 1 MU and 2 MU (for three or more NO contacts) and are switched via a toggle (Fig. 4.2/11). Switch-disconnectors with rated currents between 40 A and 125 A are 1 MU wide per contact, can be rail-mounted on the input or output side, and have a toggle (Fig. 4.2/12).

Special type for public utilities

Three-pole 63 A types with a mechanism which can be locked in the ON or OFF position, are used in meter panels in residential buildings and are sometimes prescribed by the public utilities. Another variant can be disabled with a special key, i.e. it can not be actuated once in the OFF position and its input terminals are no longer accessible (Fig. 4.2/13). It is also possible to use a lock to prevent these switch-disconnectors from being actuated.

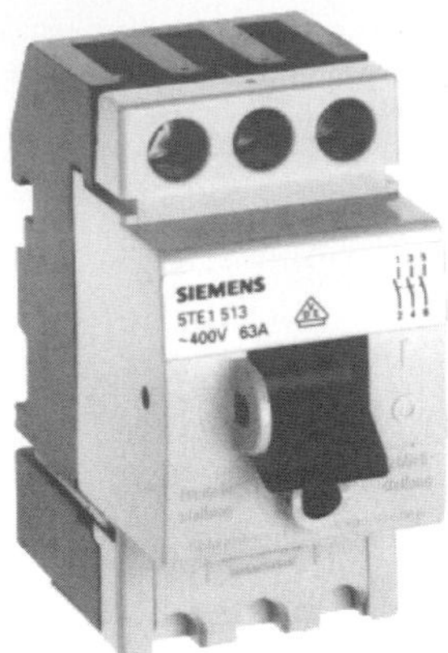

Fig. 4.2/13
63 A switch-disconnectors 5TE1 513, 3 pole, special type for public utilities

Isolating switches

The 690 V switches with 100 A to 200 A, which overlap with the 100 A and 125 A toggle-type switches, are actuated via a finger-grip knob. They have an 8 mm isolating gap and are used as isolating switches to isolate system components, sub-distribution boards or busbar trunking systems. The housing front is transparent so that the isolating gap is visible. A lock can be used to immobilize the switch in the OFF position in order to prevent unauthorized access.

Emergency switches

Emergency switches with identical dimensions (to EN 60204 / IEC 60204 / DIN 0113), 3 or 4 NO contacts and the signal colors "red" for the finger-grip knob and "yellow" for the housing complete this range of switches (Fig. 4.2/14).

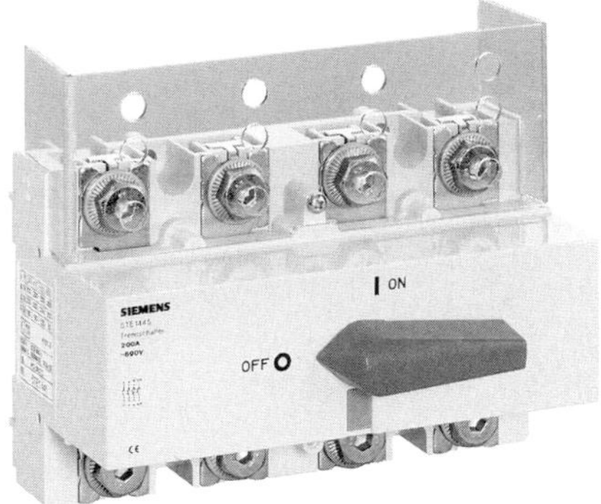

Fig. 4.2/14 Emergency switch

Accessories

Auxiliary circuit switches with 6 A contacts and one or two changeover contacts – which can be installed on the left, right, or both sides – permit remote indication.

Switch extensions (Fig. 4.2/15) allow the switches to be actuated even when the distribution board is closed.

Control switches, pushbuttons

All control switches and pushbuttons are designed for a rated current of 16 A. The following types are available:

- ▷ switch with 1 or 2 changeover contacts (1 C or 2 C); width: 1 or 2 MU; knob color: gray,
- ▷ switch with 1 NO contact (1 NO) and 230 V neon pilot lamp, for activation up to a cable length of 5 m or 150 m; width: 1 MU; knob color: gray (Fig. 4.2/16),
- ▷ two-way switch with two OFF positions and center position, each with 1 or 2 NO contacts (1 NO or 2 NO) for OPEN and CLOSED; width: 1 MU; knob color: gray,
- ▷ pushbutton with 1 NO contact and 1 NC contact (1 NO + 1 NC); pushbutton colors: gray, red, green, yellow, blue; width: 1 MU (Fig. 4.2/17),
- ▷ pushbutton with 1 NO contact or 1 NC contact (1 NO or 1 NC) in gray and with 230 V neon pilot lamp, for activation up to a cable length of 5 m or 150 m; width: 1 MU; knob color: gray.

Fig. 4.2/15
Extension shaft 5TE9 010

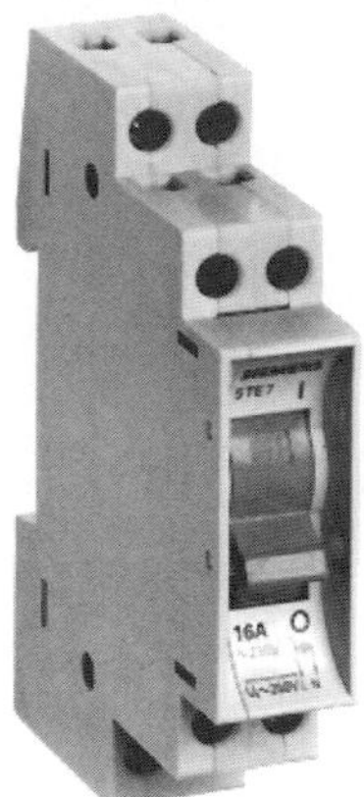

Fig. 4.2/16
Switch 5TE7 101 with 230 V neon pilot lamp

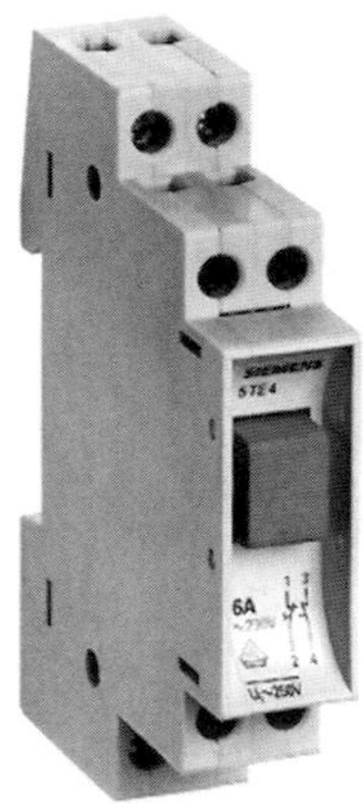

Fig. 4.2/17
Pushbutton 5TE4 700 with 1 NO and 1 NC contact

4.2.3 Switching with remote switches

Function

Remote switches (Fig. 4.2/18) are activated by means of a pushbutton. A minimum pulse time of 30 ms is sufficient. The minimum pulse time, which is triggered by manual activation is between 100 and 200 ms.

Activation by means of current pulses causes a locking mechanism to be triggered in the remote switch. Whenever the switch is actuated, the contact is closed and remains closed, or is opened and remains open. The coil of the remote switch is, therefore, de-energized when the contact is open or closed.

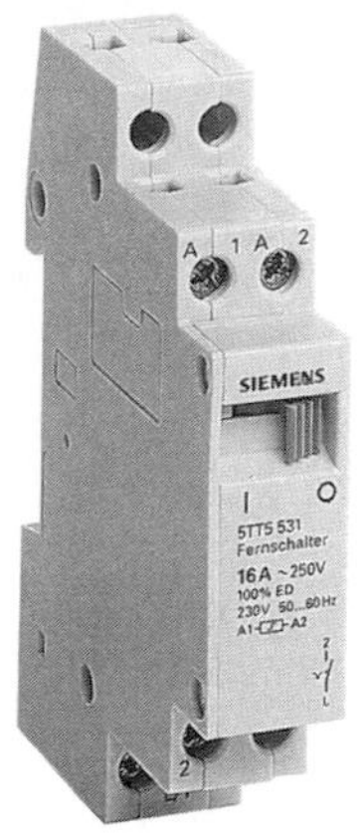

Fig. 4.2/18
Remote switch 5TT5 531, single pole

Crossover circuit

The crossover circuit, which, in the past, was often used for corridors with several switching points, for example, has been almost entirely replaced in residential and functional buildings by remote switches (Fig. 4.2/19).

Nowadays, the crossover circuit seems very complex, particularly because 3 and 4-core cables are required. The large number of connecting points in the flush-mounted splitting boxes makes it difficult to allocate the wires.

Momentary-contact circuit with remote switches

The momentary-contact circuit with remote switches has a more transparent structure and is less complicated. Load and actuation cables are separated. Since the momentary contact switches are connected in parallel, a 2-core cable is sufficient.

Protection against incorrect pushbutton operation

If a pushbutton jams, the coil of the remote switch is permanently energized. Remote switches are equipped with a protective PTC thermistor resistor (Fig. 4.2/20) in order to withstand this (i.e. continuous 230 V voltage). After approximately 5 s of continuous operation, the resistance increases and protects the coil. A cooling time of approximately 1 minute after the pushbutton has been released to allow the protective resistor to cool down is acceptable for this protective function.

If neon lamps are used in the pushbuttons, the neon lamp current continuously flows through the coil of the remote switch. The cable capacitor of the pushbutton cable amplifies this current.

Compensation of neon lamp current

Malfunctions may occur if too many neon lamps are used and the cables are too long. The neon lamp current then prevents the ar-a: Crossover circuit

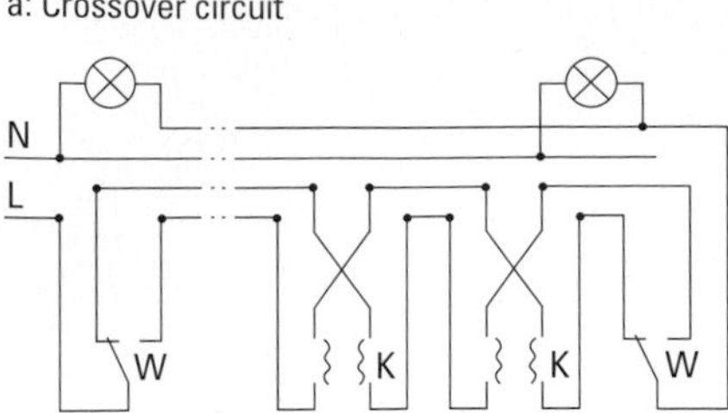

W: Two-way switch
K: Crossover switch
● Connecting points

b: Momentary-contact circuit with remote switch

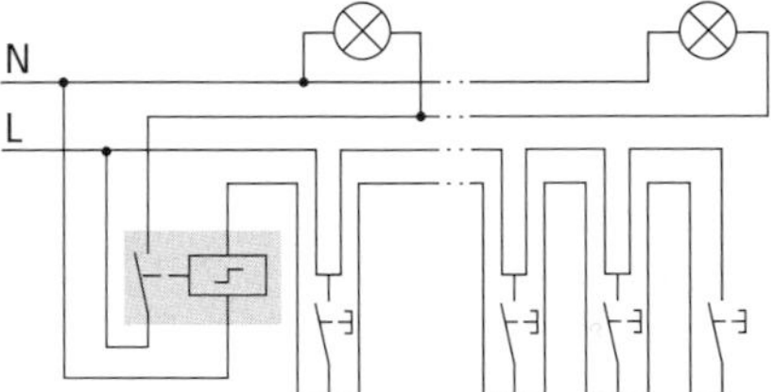

Fig. 4.2/19
Crossover circuit (a) and momentary-contact circuit with remote switches (b)

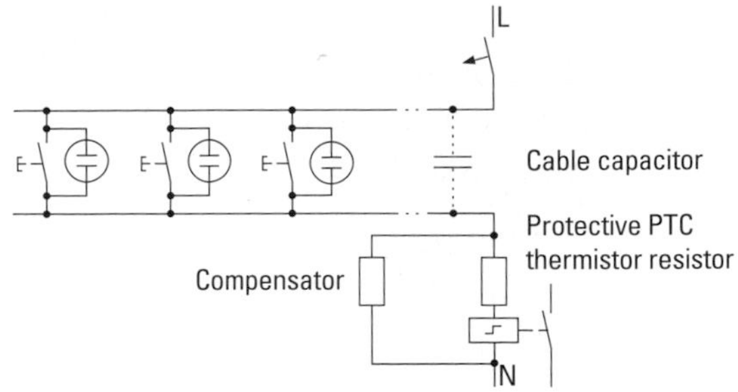

Fig. 4.2/20
Momentary-contact circuit with remote switch and compensator

Fig. 4.2/21 Compensator 5TG8 230

mature from returning to its original position with the result that the locking mechanism does not operate when the pushbutton is pressed. A 5TG8 230 compensator provides a solution to this problem (Figs. 4.2/20 and 4.2/21).

A resistor/PTC thermistor assembly carries the current past the coil of the remote switch; the coil has a lower initial load and a higher neon lamp load can be applied to the remote switch. Several compensators can be used together for one coil. Table 4.2/4 shows that the permissible neon lamp current increases with the different remote switches and compensators used.

Figs. 4.2/22 to 4.2/24 show a 3-pole remote switch with compensator as well as examples of circuits with remote switches.

Central circuit

Apart from pushbutton actuation of the remote switch to change the circuit state from ON to OFF or OFF to ON, remote switches are also available with a "Central ON" (ZE) and "Central OFF" (ZA) function.

Table 4.2/4
Permissible neon lamp current if compensators are used

With remote switch type	Permissible neon lamp current for following situations		
	Without compen-sator mA	1 com-pensator mA	2 com-pensators mA
5TT5 553	10	30	50
5TT5 6..	5	20	35
5TT5 15.	4	14	24
5TT5 16.	4	26	48

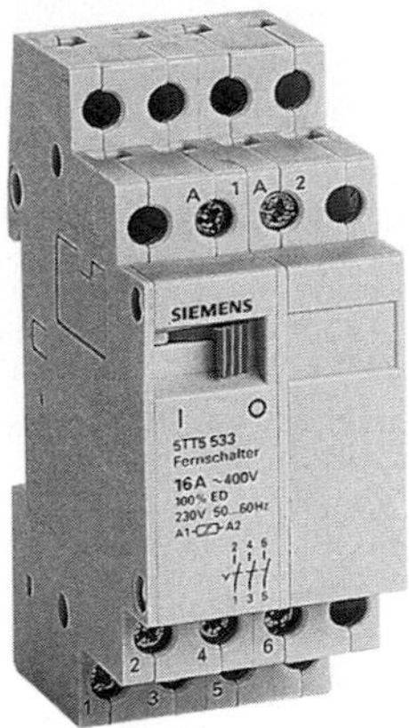

Fig. 4.2/22 Remote switch 5TT5 533, 3 pole

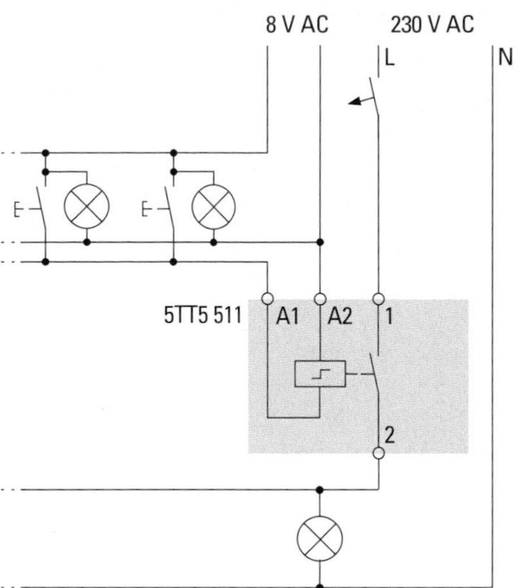

Fig. 4.2/24
Circuit diagram (1 pole) for lighting system with remote switch 5TT5 511 and pushbuttons with locating lamp, operating voltage 8 V AC

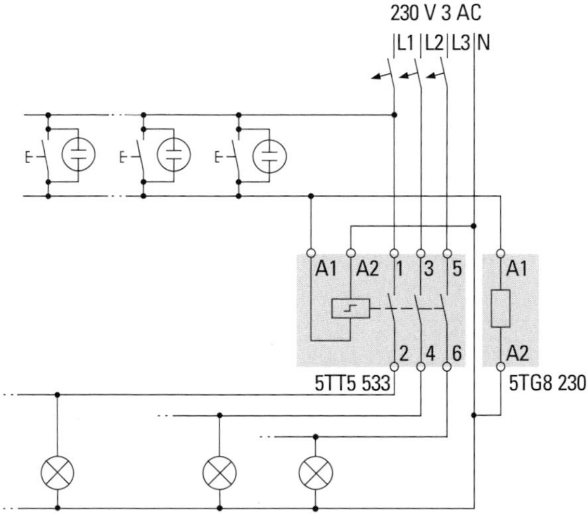

Fig. 4.2/23
Circuit diagram (3 pole) for lighting system with remote switch 5TT5 533 and compensator 5TG8 230, operating voltage 230 V AC

They allow system components to be switched on and off manually and centrally via a pushbutton or timer. Fig. 4.2/25 shows an example of such a circuit in which any number of remote switches can be switched on or off centrally. The user of the local system can switch "his light" on again after the system has been switched off centrally.

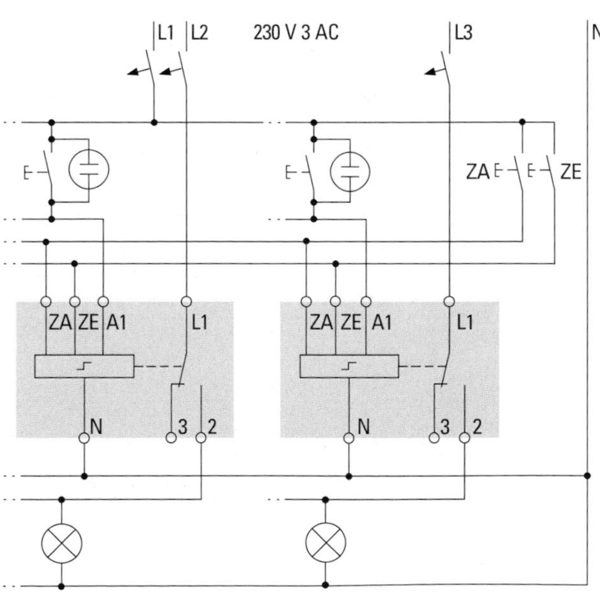

Fig. 4.2/25
Example of circuit allowing remote switches to be switched ON/OFF manually and centrally via a pushbutton

4.2.4 Switching with contactors

Magnetic system

Contactors (Fig. 4.2/26) are available as 1, 2, and 4-pole devices. They are always equipped with a d.c. magnetic system in order to prevent humming, particularly in residential and functional buildings. Less complicated a.c. magnetic systems were rejected for this reason.

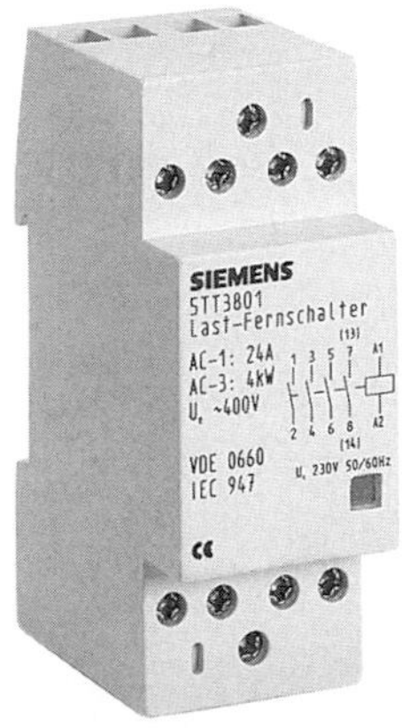

Fig. 4.2/26
Contactor 5TT3 801

Types

Table 4.2/5 shows a selection of the different types in accordance with utilization category AC-1 (switching of ohmic loads) and AC-3 (switching of three-phase asynchronous machines), as well as different operating voltages, and contact types.

Auxiliary contacts

An auxiliary contact (Fig. 4.2/27) with two contacts (variants: 2 NO contacts and 1 NC contact / 1 NO contact) can be attached to the four-pole contactors.

Temperatures

The d.c. magnetic system requires a relatively high number of coil windings compared to the a.c. magnetic system. Small sizes, high switching capacities, and a long service life are, however, required. The temperature limits (surface temperatures) are specified in the standards EN 60947-1 / IEC 60947-1 / DIN VDE 0660-100 and EN 60947-4-1 / IEC 60947-4-1 / DIN VDE 0660-102 as follows:

An increase in temperature of 50 K is permitted for parts which do not have to be touched under normal operating conditions and which are not made of metal. In other words, this is a temperature of 90 °C based

Table 4.2/5 Contact ratings, operating voltages, and contact types of contactors

Utilization category	Contact rating	AC-1 A AC-3 kW	20 1.3	20 4.0	24 4.0	24 4.0	25 1.3	40 11	63 15
Operating voltages		24 V AC	•	•			•		
		24 V DC		•					
		24 V AC/DC			•			•	•
		115 V AC		•		•			
		110 V DC				•			
		230 V AC	•	•	•		•	•	•
		220 V DC			•			•	•
Types of switching contact		1 NO					•		
		1 NC					•		
		2 NC	•						
		2 NO	•						
		1 NO + 1 NC	•						
		4 NO		•	•	•		•	•
		3 NO + 1 NC		•	•			•	•
		2 NO + 2 NC			•			•	•
		4 NC			•			•	•

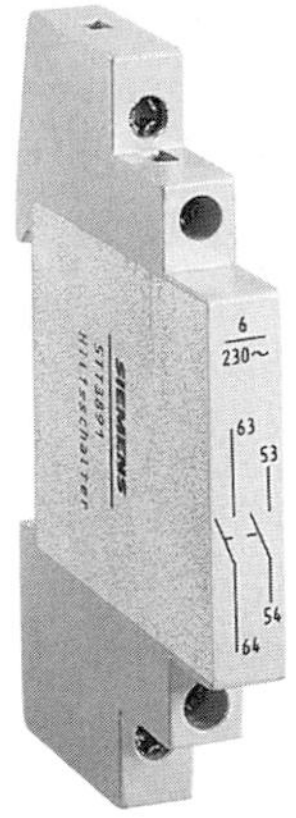

Fig. 4.2/27
Auxiliary contact with 2 NO contacts for 24 A, 40 A, and 63 A contactors

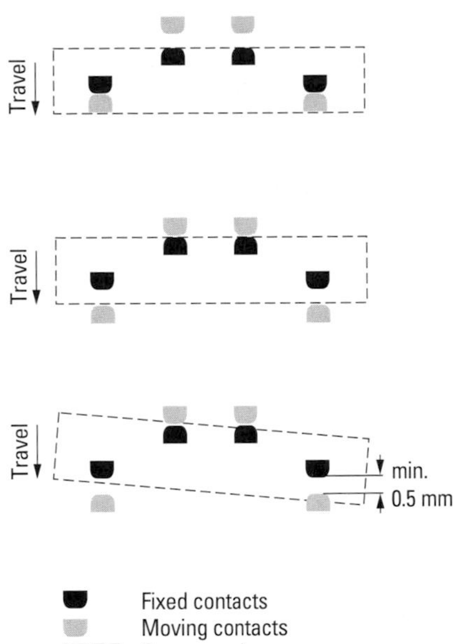

Fig. 4.2/28
Minimum requirement for positively-driven contacts

on a maximum ambient temperature of 40 °C in distribution boards!

Maximum surface temperatures of 70 °C can occur in practice. If a large number of contactors are used, the distribution board should be designed for a maximum internal ambient temperature of 40 °C. If several contactors are connected side by side, it is advisable to insert a spacer with a width of 0.5 MU after every second contactor in order to enlarge the heat-dissipating surface.

Positively-driven contacts

In accordance with the safety regulations ZH 1/457 issued by the German employers' liability insurance association, positively-driven contacts must be used for controlling powered metal-working presses to allow "emergency shutdown", where prescribed, and for similar applications. Emergency shutdowns may also be required in electrical building installations, e.g. for medical equipment in doctors' surgeries.

The term "positively-driven operation" is used if the switching contacts are mechanically connected in such a way that it is impossible for NC and NO contacts to be closed at the same time. The NO contact should only close after the NC contact has opened properly.

There must be a minimum contact gap of 0.5 mm between the NC contacts before the NO contacts close (Fig. 4.2/28). This gap must be ensured in all operating states under rated operating conditions throughout the electrical and mechanical service life of the contacts. The manufacturer is responsible for carrying out tests to verify that his products satisfy this requirement.

EMC (electromagnetic compatibility)

An electromagnetic switching device, which creates an isolating distance by means of switching contacts, is always a potential source of electromagnetic interference (see Section 3.3).

Since, however, contactors are only switched occasionally and the interference lasts for just a few milliseconds, the frequency, magnitude, and effects of this interference are negligible with respect to the requirements specified in the standards IEC 60947-1, EN 60947-1, DIN VDE 0660-100 and must be regarded as normal electromagnetic ambient conditions.

Noise voltage limiters

In the case of contacts with a rated current of 24 A, 40 A, and 63 A, the conducted radiated noise caused by the bridge rectifier of the d.c. magnetic system can be reduced reliably by means of a noise voltage limiter (Fig. 4.2/29) to a level below the permissible limits specified in EN 55014/VDE 874-14. Noise voltage limiters are connected in parallel to the contactor coil.

Fig. 4.2/29
Noise voltage limiter
5TT3 893

4.2.5 Switching with switch relays

Application

Switch relays (Fig. 4.2/30) are used to switch low loads, neon lamps, and fluorescent lamps. They are also used as contact multipliers in control systems and as coupling relays between safety extra-low voltages (SELV) and 230/400 V AC system voltages since they have creepage distances and clearances of 8 mm and a rated impulse withstand voltage of ≥ 4 kV.

Types

Switch relays are available for 8 V AC, 12 V AC, 24 V AC, 110 V AC, and 230 V AC, 50 Hz as well as for 12 V DC, 24 V DC, and 110 V DC. The rated operating current of the contacts is 16 A for utilization category AC-1. Switch relays can be supplied with 1 NO contact, 2 NO contacts, 1 changeover contact, or 2 changeover contacts. Manual actuation makes it possible for maintenance work to be done.

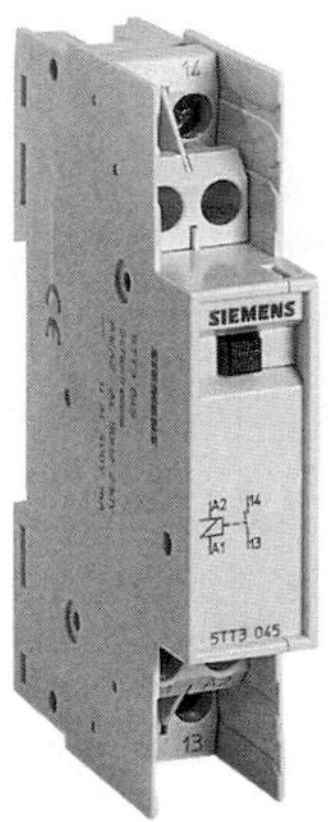

Fig. 4.2/30
Switch relay 5TT3 075
with changeover contacts

4.2.6 Switching with timers

Applications

Timers are used in residential buildings, mainly for switching off lighting after a time delay (e.g. stairwell lighting, courtyard, garden, and cellar lighting as well as lighting systems for rooms which are used for short periods of varying duration) and ventilation fans. They are also used as delay, wiper, flashing, and off-delay timers in control units.

Table 4.2/6 provides an overview of the Siemens timers together with their most important technical data, the standards to which they are constructed and tested, as well as their applications.

Timers for lighting systems

Timers for stairwell lighting, 4-conductor circuit

Various circuits are possible for stairwell lighting systems if the stairwell lighting timer is used (Fig. 4.2/31). The 4-conductor circuit with phase conductor switching (L switching) is used most frequently (Fig. 4.2/32).

The cables for the switching circuit (pushbuttons) and the load circuit (lights) are installed separately. The timer can be "repressed" (restarted) at any time. A standard lighting assembly in the loft can be operated independently by means of a switch (DA). A supplementary switch (DE) allows permanent lighting to be activated, e.g. if someone is moving into or out of an apartment.

Circuit with pre-warning timer

As a minimum requirement for electrical systems in residential buildings, the German industrial standard DIN 18015 (Section 4.3, Point 3) stipulates that timer circuits for stairwell lighting systems must have a prewarning function, e.g. dimming, to prevent sudden darkness.

The pre-warning time switch (Figs. 4.2/33 and 4.2/34) fulfills this requirement. It warns people using the stairwell that the stairwell lighting is about to be deactivated by switching to "semi-darkness" for 10 to 30 s. Elderly and infirm persons are thus given enough time to press the light switch again. This prevents them having to search for the light switch in total darkness and avoids the associated risk of injury. Since the semi-darkness phase has scarcely any significance with respect to the safety requirements, the operating time of the stairwell lighting timer can be minimized so

Table 4.2/6 Technical data, standards, and applications of Siemens timers

Device designation	Rated operating current A	Rated operating voltage U_c	Standard	Applications
Timer and pre-warning time switch for stairwell lighting	10	230 V AC, 50 Hz 230 V AC, 50/60 Hz	EN 60669 IEC 60669 DIN 18015	Limiting ON time for stairwell lighting and issuing warnings prior to lighting being switched off
Energy-saving timer	10		EN 60669 IEC 60669 DIN VDE 0632	Controlled deactivation of lighting systems, e.g. courtyard, garden, garage, and cellar lighting
Illuminated timer	10			Time-delayed deactivation of lighting systems in rooms which are used for short periods of varying duration
Timer for electronic control gear (ECG Dynamic)	10			Time-delayed deactivation of lighting systems with fluorescent lamps
Timer for ventilation fans	10		DIN VDE 0637	Time-delayed deactivation of ventilation fans
Delay timer	3	230 V AC, 50 Hz 230 V AC, 50/60 Hz	EN 60812–1 IEC 61812–1 IEC 60225 VDE 0435–2021	Time-oriented control operations
Wiper timer	3			
Flashing timer	3			
Off-delay timer	3	230 V AC, 50/60 Hz 24 .... 240 V AC/DC, 50 .... 400 Hz		
Multi-functional timer	1			Time-oriented sequential control

that safety and efficiency are perfectly balanced.

3-conductor circuit

The 3-conductor circuit with L switching of the stairwell lighting timer 5TT1 310 (Fig. 4.2/ 35a) is often encountered, but far less convenient. The load and switching circuits are not isolated due to the costs involved. If the light is switched on, the user must wait until the timer for the stairwell lighting switches off before it can be switched on again by pressing the light switch. This circuit has considerable disadvantages, particularly because it does not permit the pre-warning time switch 5TT1 313 to be used.

Fig. 4.2/31
Stairwell lighting timer 5TT1 311

Although the 3-conductor circuit with N switching (Fig. 4.2/35b) is no longer permitted in Germany on account of DIN VDE 0100-460, it is still widespread in other European countries.

Energy-saving timers

The energy-saving timer 5TT1 300 with a variable operating time of between 3 and 60 minutes is switched on when the pushbutton is pressed for the first time. Similar to a remote switch, it is switched off when the pushbutton is pressed a second time. If the pushbutton is not pressed a second time, the timer switches off once the set operating time has elapsed. Applications for this switch range from stairwell lighting to lighting systems for lofts, cellars, or communal laundry rooms which are rarely used for periods longer than 60 minutes.

Illuminated timers

The illuminated timer 5TT1 301 supports an additional function. If the pushbutton is pressed briefly, the switch functions in the same way as the standard timer 5TT1 311. The operating time is, however, reduced to a

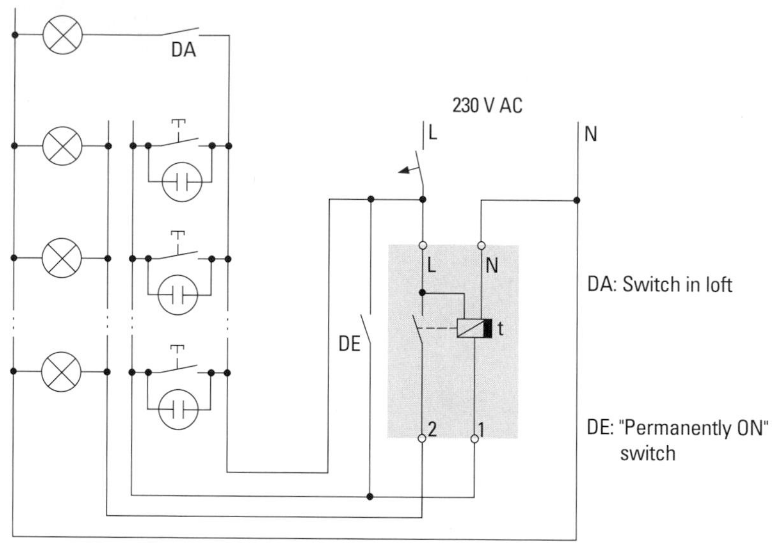

Fig. 4.2/32 4-conductor circuit with L switching

maximum of 5 minutes. If the pushbutton is pressed for longer than 1 second, the operating time of the switch is four times longer, whereby it is always the last pushbutton actuation that determines the operating time. This means that a long operating time is shortened again if the last pushbutton actuation was less than 1 second.

This illuminated timer can be fitted in rooms which are used for short periods of varying duration, e.g. for lighting systems in stairwells and garages.

Fig. 4.2/33
Pre-warning time switch 5TT1 313

Energy-saving lamps

Energy-saving lamps (fluorescent lamps with integrated electronic control gear) with screw bases do not respond to the reduced voltages of the pre-warning time switch 5TT1 313. The different control gear technologies currently used by various manufacturers do not permit a universal, inexpensive solution to be developed for controlling this type of energy-saving lamp. These energy-saving lamps can be used for all other timer circuits, except for the one shown in Fig. 4.2/34.

Timers for fluorescent lamps with ECG Dynamic

Fluorescent lamps are not normally used in timer circuits for stairwell lighting because frequent energization of the filaments to ignite the flow of gas reduces the service life of the lamps. Using electronic control gear (ECG), which starts fluorescent lamps more gently than conventional reactor circuits, does not improve matters since ECG is not suitable for frequent switching either.

However, in response to the demand for fluorescent lamps and their higher luminous efficiency as compared with incandescent lamps, a special timer, which considerably reduces the switching rate, has been developed for fluorescent lamps with ECG Dynamic (Figs. 4.2/36 and 4.2/37).

The timer for ECG Dynamic 5TT1 302 satisfies all the requirements vis-à-vis modern

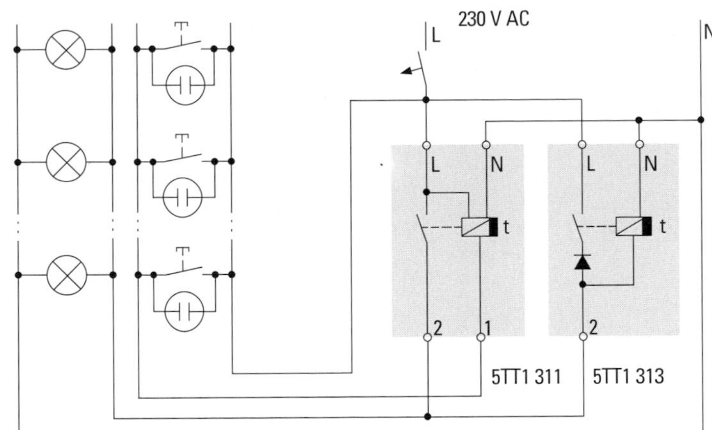

Fig. 4.2/34
4-conductor circuit for timer 5TT1 311 and pre-warning time switch 5TT1 313 with L switching

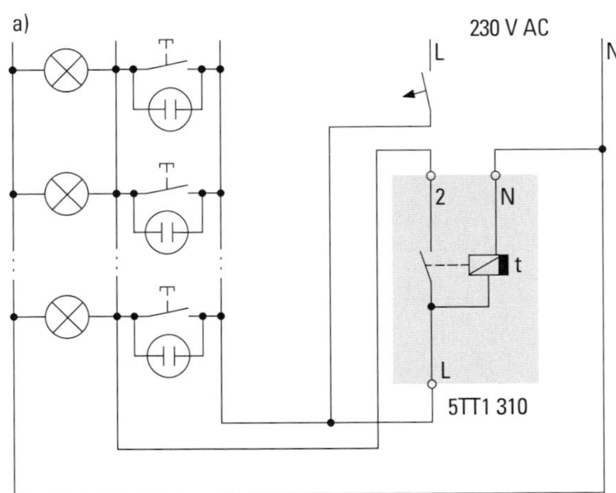

stairwell lighting systems. As with the standard stairwell lighting timer, the ECG is switched on when the pushbutton is pressed for the first time. Once the operating time has elapsed, the fluorescent lamps are dimmed (the degree of dimming can be adjusted). If the pushbutton is pressed again within 30 minutes after the pushbutton was first pressed, the lamps are switched on again to their full brightness for the set operating time of between 5 and 10 minutes. The lamps are not switched off completely until 30 minutes after the pushbutton was last pressed.

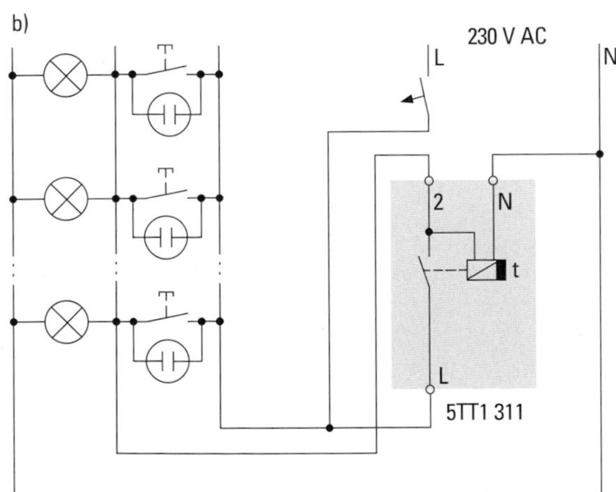

Fig. 4.2/35
3-conductor circuit for stairwell lighting timers with
a) L switching (5TT1 310), b) N switching (5TT1 311)

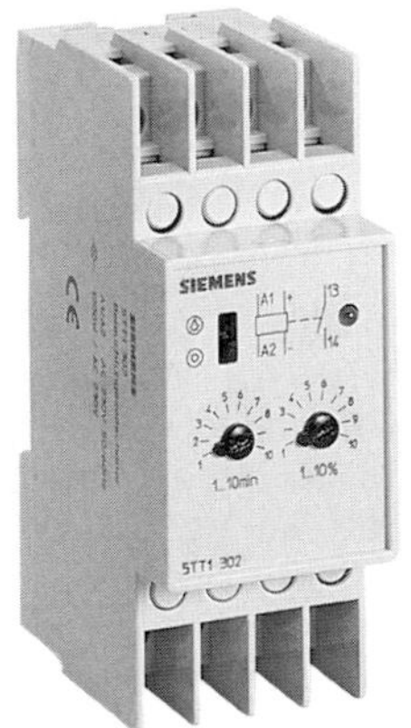

Fig. 4.2/36
Timer 5TT1 302 for fluorescent lamps with ECG Dynamic

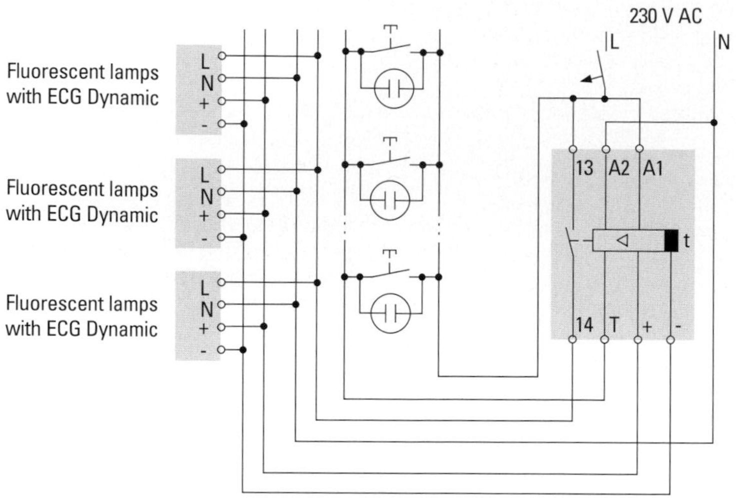

Fig. 4.2/37
Circuit for stairwell lighting system with timer 5TT1 302 for fluorescent lamps with ECG Dynamic

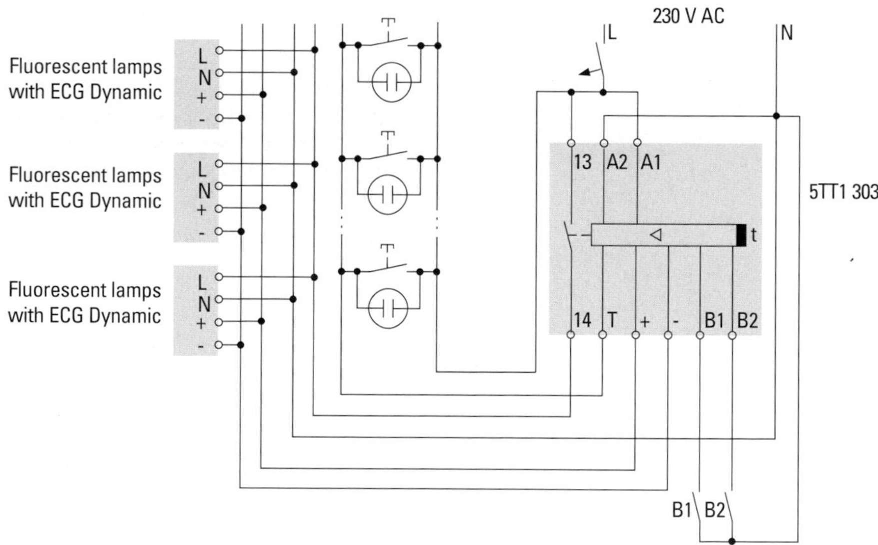

Fig. 4.2/38
Circuit for lighting system with control switch for fluorescent lamps with ECG Dynamic 5TT1 303

Control switches for fluorescent lamps with ECG Dynamic

In addition to the timer function for fluorescent lamps with ECG Dynamic, the dimming and maximum brightness time for the ECG Dynamic control switch can be switched to continuous operation by means of a switch or time switch (Fig. 4.2/38). If contact B1 closes, the infinitely continuous maximum brightness phase is activated; if contact B2 closes, the infinitely continuous dimming phase is activated.

This device is used for stairwell and corridor lighting, e.g. in homes for the elderly and hospitals.

In homes for the elderly, a time switch can be used to switch the corridor lighting to maximum brightness mode at meal times between 5pm and 7pm (contact B1) and to continuously dimmed mode between 7pm and 10pm (contact B2), whereby the lighting can always be switched to maximum brightness mode for the set time via a corridor pushbutton. The continuous lighting phase ends at 10pm. The lighting can then only be switched on when required using the corridor pushbuttons.

In hospitals, the lighting is switched to maximum brightness during the day (principal working times, lunch time, visiting times, shift changes, doctors' rounds). Dimming mode is activated during the rest periods in the afternoon and at night. A patient can always switch the lighting to maximum brightness mode for the set time by means of the corridor pushbutton. In emergencies, nurses can activate the "emergency operation" mode by means of a switch (contact B1), i.e. to continuous maximum brightness mode (infinitely continuous maximum brightness phase).

Timers for ventilation fans

Timers for ventilation fans are mainly used in sanitary installations and, owing to the high switching frequency, are equipped with a Triac switch. Fig. 4.2/39 shows the circuit of a room lighting system, which is simultaneously used to activate a ventilation fan. The switch (S) switches on the lighting system. The ventilation fan is then switched on via a timer (also referred to as an operating time switch) after a delay of approximately 1 minute. Once the lighting has been switched off, the ventilation fan continues to run for the time set on the timer.

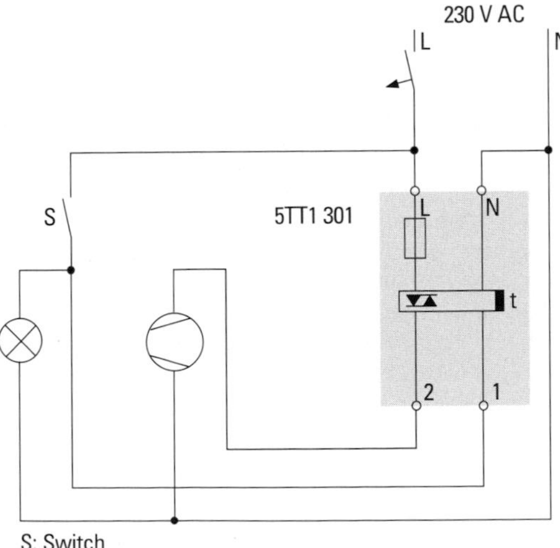

Fig. 4.2/39
Circuit of timer for ventilation fan 5TT1 301

Timers for industrial control systems

Single-function devices

The following single-function devices are available for use in control units:

- ▷ delay timers
- ▷ wiper timers
- ▷ flashing timers
- ▷ off-delay timers

Multi-functional timers

The multi-functional timer 5TT3 180 (Fig. 4.2/40) can be used to provide an extensive range of functions for time-oriented sequential control systems.

The following functions can be programmed using a rotary switch on the front of the device:

- ▷ ON/OFF delay,
- ▷ Passing make function,
- ▷ Clock-pulsed delay,
- ▷ Clock pulse generator, starting with a pulse,
- ▷ Drop-out delay,
- ▷ Pulse converter,
- ▷ Passing break function,
- ▷ Response and reset delay.

The required time ranges can be selected using the upper rotary switch (Table 4.2/7). A remotely-connected potentiometer enables the time ranges to be adjusted from a remote location.

The rated operating voltage of 24...240 V AC/DC and the multi-functionality of this timer mean that it can be used for an extremely broad range of applications.

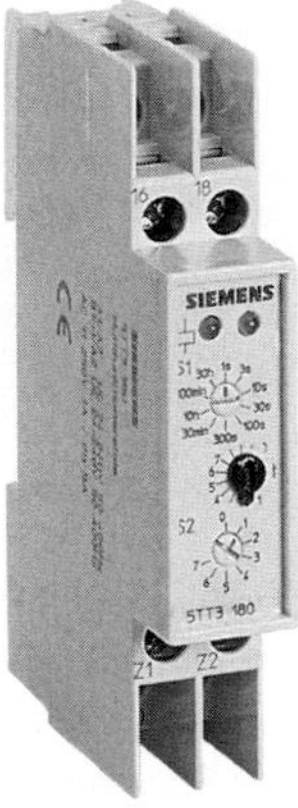

Fig. 4.2/40
Multi-functional timer
5TT3 180

Table 4.2/7
Settable time ranges of multi-functional timer 5TT3 180

Switch setting	Time ranges t
0	0.05– 1 s
1	0.15– 3 s
2	0.5– 10 s
3	1.5– 30 s
4	5–100 s
5	15–300 s
6	1.5– 30 min
7	5–100 min
8	0.5– 10 h
9	1.5– 30 h

Fig. 4.2/41
Synchronous time switch 7LS1 004

4.2.7 Switching with time switches

Applications

Time switches are used to activate systems and system components, for example:

irrigating systems, greenhouses, garden equipment, swimming pools, filtering systems, awning control units, interval bell systems, bell-ringing systems, shop-window lighting, neon signs, sports hall lighting, traffic signal control systems, street lighting, illuminated signs, office lighting, stairwell and entrance lighting, spotlight systems, preheating systems for industrial furnaces, spraying machines, baking ovens, heating systems, air-conditioning systems and devices, ventilation fans and systems, circulating pumps of heating systems, saunas, aquariums, fountains.

Types

A distinction is made between the following types of time switch:

- ▷ synchronous time switches
- ▷ quartz time switches
- ▷ digital time switches

Synchronous time switches

Synchronous time switches (Fig. 4.2/41) are driven by a synchronous motor and, therefore, require a system frequency of 50 Hz or 60 Hz. A switching disk, which rotates once every 24 hours, is driven by a gear unit. 48 control riders actuate special snap-action switches. The minimum switching interval is, therefore, half an hour. A time switch with a repeat switching sequence lasting 24 hours is referred to as a one-day time switch.

If the gear unit is modified so that the switching disk rotates once every 7 days, the minimum switching interval increases to 3.5 hours. The time switch is then referred to as a seven-day time switch.

Time switches with a one-hour disk have a minimum switching interval of 1.2 minutes and are used for cyclic switching operations. If this type of time switch is connected in series to a one-day time switch, the ventilation fan in a washroom can be switched on once every hour for 6 minutes, e.g. between 8am and 6pm.

Quartz time switches

Quartz time switches do not require a specific system frequency. A quartz resonator and an electronic module are used to supply the synchronous motor with a system-independent frequency. Rechargeable NiCd batteries provide the quartz time switch with a power reserve of more than 100 hours so that it does not have to be reset if the system supply is disconnected for long periods.

Power reserve

The minimum switching interval of 3.5 hours for seven-day operation is too long in the majority of cases. A one-day time switch with a minimum switching interval of 0.5 hours can, therefore, be connected in series. This combination allows the weekend to be "skipped" and the power supply to be switched on, for example, from Monday to Friday between 8am and 4pm.

Fig. 4.2/42 shows a weekly switching sequence with a seven-day time switch and two one-day time switches – the switching sequence for Monday to Friday is different to that for Saturday and Sunday.

Digital time switches

Digital time switches (Fig. 4.2/43) are electronic devices with an LCD and buttons for

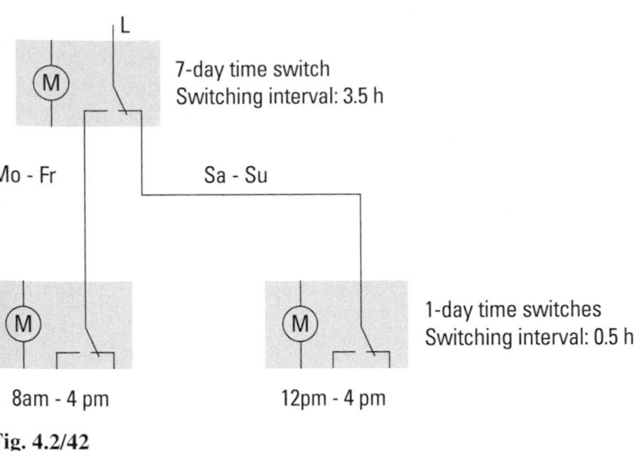

Fig. 4.2/42
Weekly switching sequence with 3 mechanical time switches

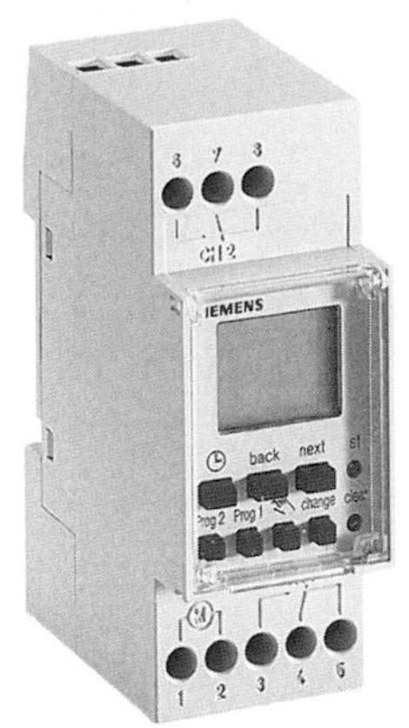

Fig. 4.2/43
Digital time switch
7LF4 120

entering the time and switching times. They do not depend on a specific system frequency and can be set in steps of 1 minute.

The programmed switching times are not deleted if the power supply fails since they are stored in an EEPROM. The time progresses normally during the power reserve mode.

Switching point

The switching point consists of the following parameters:

▷ the time,
▷ the weekday(s) which can be programmed as desired,
▷ the type of switching command, ON or OFF.

Daily and weekly program

The difference between the daily and weekly program is whether the switching operations are to take place on all or only certain days of the week. If days are "skipped" (any combination of days can be selected), this is referred to as a weekly program (Fig. 4.2/44). If a different switching sequence is to be used on the "skipped" days, a second weekly program is entered.

Date program

If, however, a third daily or weekly program is to be used on a certain day or at a certain time of the year, a start date and end date must also be entered. The third weekly program is then executed on the basis of the dates entered and has priority over the first and second weekly programs (Fig. 4.2/45).

The data is entered using the buttons and a detailed display. Flashing numbers and symbols prompt the user to enter the appropriate data (Fig. 4.2/46).

Summer/winter changeover

The summer/winter time is switched automatically at the appropriate weekends (Saturday night/Sunday morning) at the end of March and October.

If required, the following supplementary options are also available:

▷ H = semi-automatic:
The changeover is always performed on the entered date. If no date has been entered, no changeover is made.
▷ C = calculated:
The changeover is always performed on the same weekday during a specific week of the year.

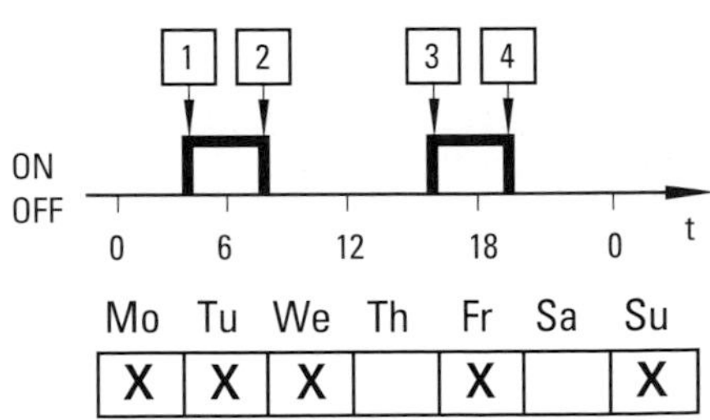

Fig. 4.2/44
Time sequence for standard weekly program

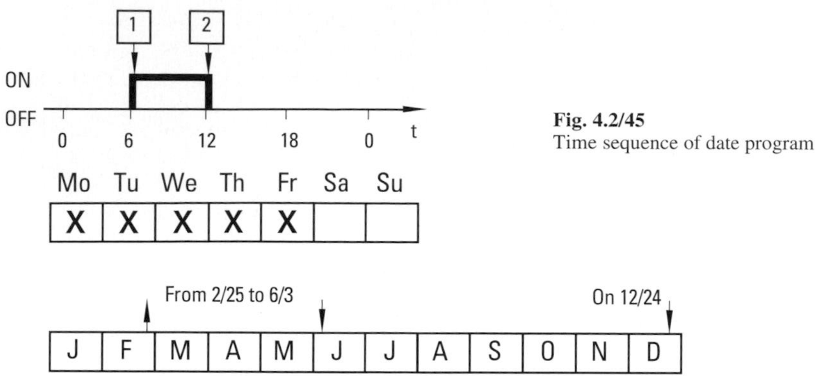

Fig. 4.2/45
Time sequence of date program

a) Display and buttons

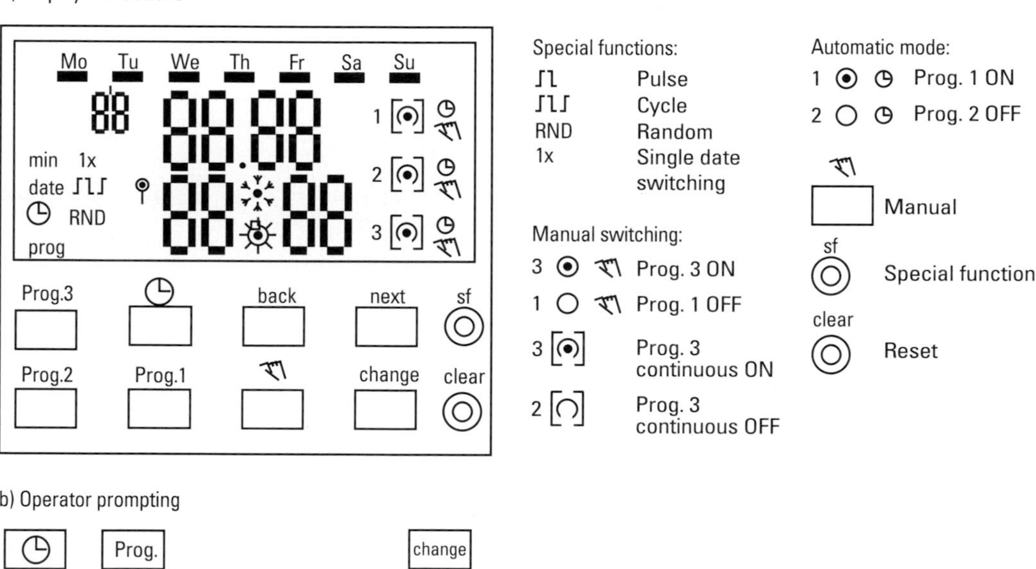

b) Operator prompting

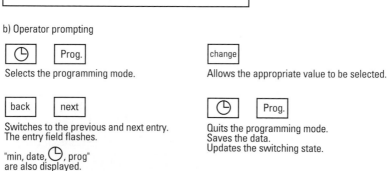

Fig. 4.2/46 Display and buttons (a) with operator prompting (b) of twelve-month time switch 7LF4 152

Special functions Digital time switches have the following special functions:

▷ *Single date switching*
A switching point with a date is deleted automatically after execution because it is not required in the following years, e.g. Easter vacation.

▷ *Cycle switching*
If, for example, a ventilation fan is to run for 5 minutes every hour between 8am and 6pm, 20 switching points (10 × ON and 10 × OFF) would usually have to be programmed. With cycle switching, the cycle is started with a "cycle ON" command instead of the individual ON and OFF commands, and the ON and OFF time is entered in minutes (between 1 and 59 minutes). This cycle is then repeated continuously until it is stopped by a "cycle OFF" command. If the cycle switching function is used, only four switching points have to be programmed: cycle ON, ON time, OFF time, cycle OFF (Fig. 4.2/47).

▷ *Pulse program for short switching times*
If switching intervals of less than one minute are required, a "pulse ON" command followed by a "pulse time" command of between 1 and 59 seconds can be programmed instead of an ON command and OFF command. This means that key commands for remote switches or self-latching circuits can be automated. Two switching points are required for a pulse program (Fig. 4.2/48).

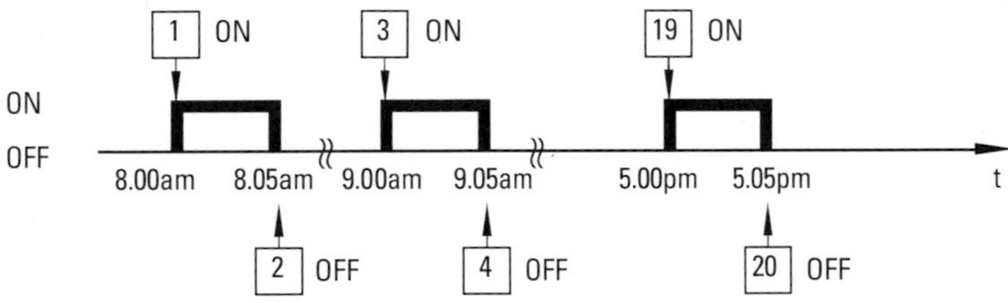

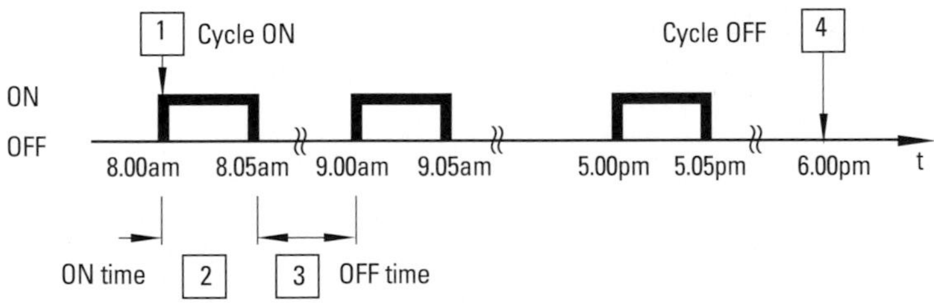

Fig. 4.2/47 Time sequence of cycle switching program compared to conventional ON/OFF switching

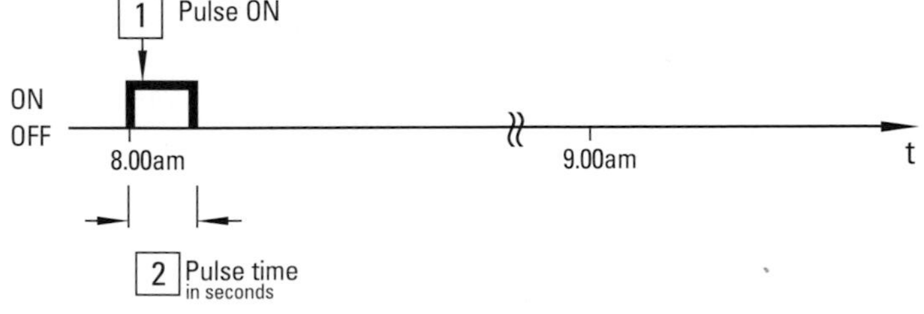

Fig. 4.2/48
Time sequence of pulse program

▷ *Random switching (RND)*
An ON or OFF command is delayed by a randomly programmed time of between 1 and 59 minutes before it is carried out.

Example:
The ON command is set for 5:20 am, the random time to 40 minutes. The system or device will be switched on at a randomly selected time between 5:20 am and 6 am. This switching mode is used for distributing the loads generated when operating processes are started (Fig. 4.2/49).

Creating time-switching programs using PC

Owing to the large number of date entries and switching commands, time-switching programs can sometimes be very complicated. The mechanically confined design of a digital time switch permits only standard programming in such cases. Complex switching programs can be created more quickly, more transparently, and more conveniently via the user interface of a PC (Fig. 4.2/50).

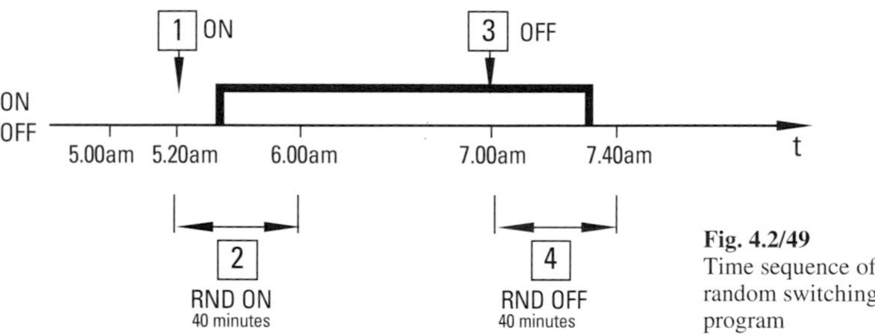

Fig. 4.2/49
Time sequence of random switching program

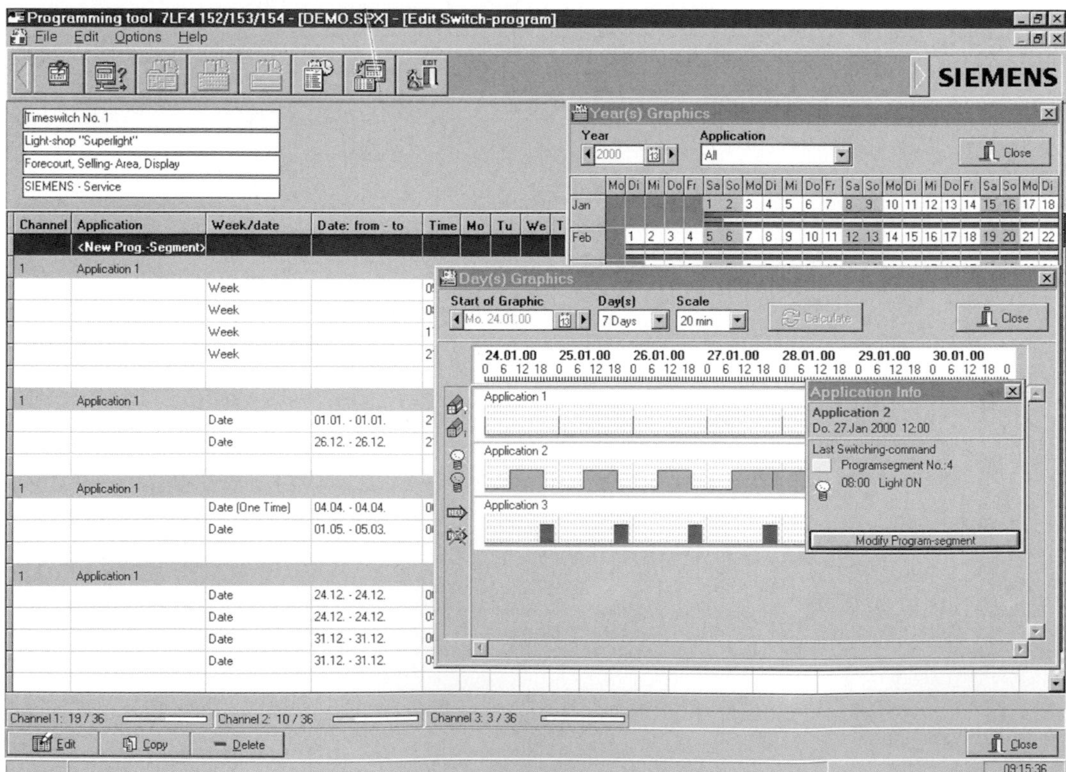

Fig. 4.2/50 PC user interface for programming time switching programs

The power supply unit 7LF4 144 with IR interface (Fig. 4.2/51) is required for transferring the switching programs, which have been created on the PC, to digital time switches from the 7LF4 15 series. The switching programs are downloaded to a portable IR programmer (Fig. 4.2/52) via a COM interface. The transmitter is then used to transfer the programs to the time switches installed in the system or to read out programs created manually on the time switch.

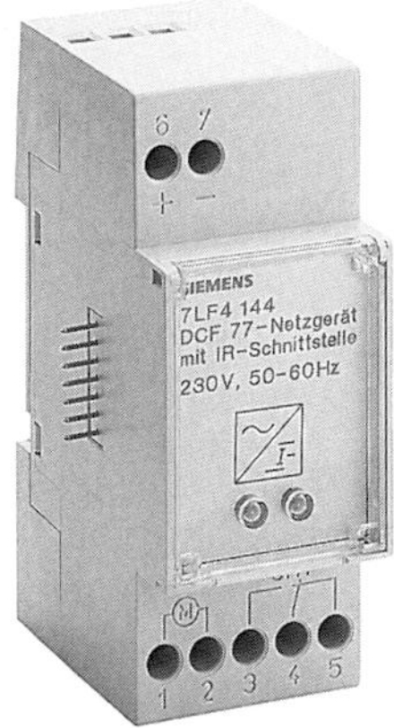

Fig. 4.2/51
Power supply unit 7LF4 144 with IR interface

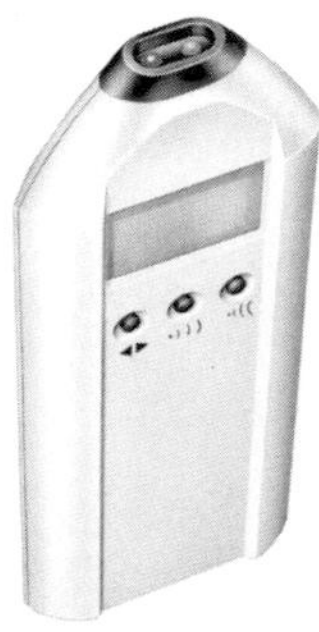

Fig. 4.2/52
IR programmer 7LF4 148

4.2.8 Monitoring with voltage relays

Undervoltage and overvoltage relays

Voltage relays are used to monitor the system voltage. A distinction is made between undervoltage and overvoltage relays. They protect systems and devices, the rated operating voltages (U_c) of which must not exceed or fall below the permitted limits by an unacceptable margin, e.g. +10%, −15%. They also switch on the emergency power supply if the system power supply fails.

All voltage relays can be used for 1, 2, or 3-phase monitoring. All 3 inputs of the voltage relay must, however, be assigned. A 1 and 2-phase connection renders asymmetry detection or neutral conductor monitoring ineffective (Fig. 4.2/53).

Application in accordance with DIN VDE 0108

Switchover to safety power supply

Applications for undervoltage relays are subject to the requirements of DIN VDE 0108 "Power installations and safety power supply in communal facilities", i.e. all major electrical loads must be switched over to a safety power supply if the general power supply is faulty (Fig. 4.2/54) (see also Sections 5.3 and 5.4).

A fault exists if the voltage of the general power supply drops by > 15 % (relative to the operating voltage) for 0.5 seconds (i.e. 195 V in the case of an operating voltage of 230 V).

A safety lighting system must be switched over to a safety power supply after 0.5 to 15 seconds depending on the use of the building. A battery system or a quick-starting standby generating set can be used as safety power supplies.

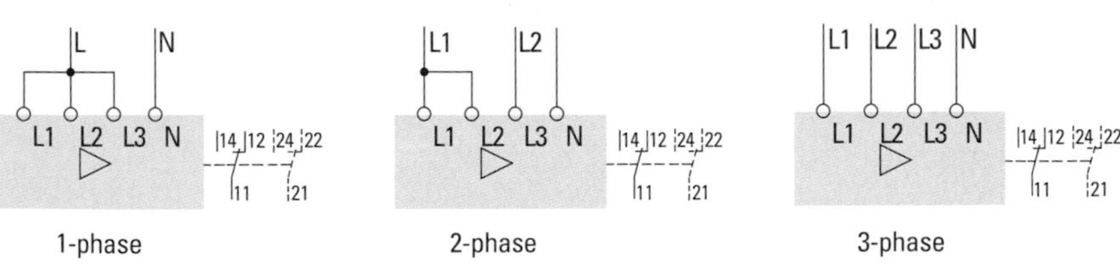

Fig. 4.2/53 Example of circuits for 1, 2, and 3-phase operation

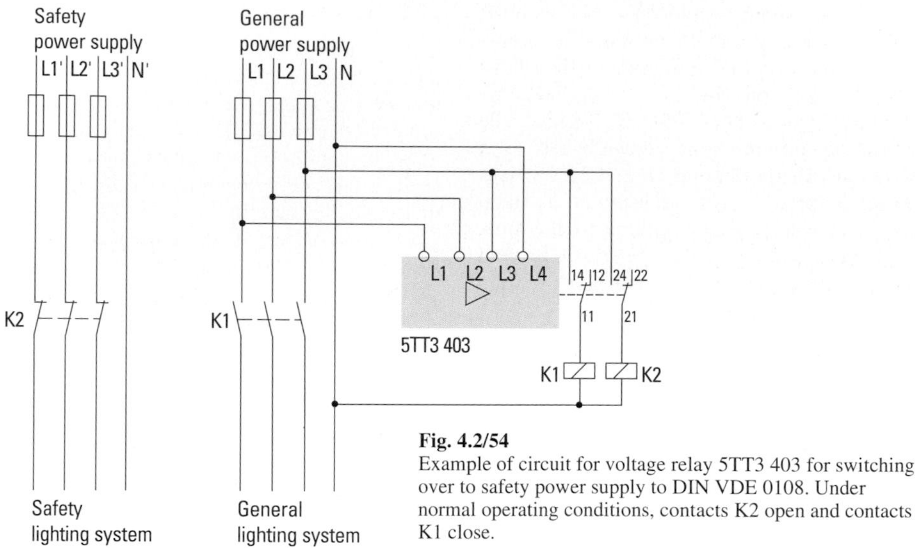

Fig. 4.2/54
Example of circuit for voltage relay 5TT3 403 for switching over to safety power supply to DIN VDE 0108. Under normal operating conditions, contacts K2 open and contacts K1 close.

Using undervoltage, overvoltage, and asymmetry detection in three-phase systems

Possible uses of voltage relays

The voltage relay 5TT3 408 (Fig. 4.2/55) is suitable for monitoring the operating voltage in three-phase systems to detect undervoltage, overvoltage, or asymmetry.

The voltage relays can be configured as follows:

▷ overvoltage from 0.9 to 1.3 · U_c ,
▷ undervoltage from 0.7 to 1.1 · U_c

with 4 % hysteresis in each case.

Voltages in three-phase system

The system configuration generally used in low-voltage systems is the three-phase system, which is made up of three a.c. voltages, each of which is out of phase by 120° (Fig. 4.2/56).

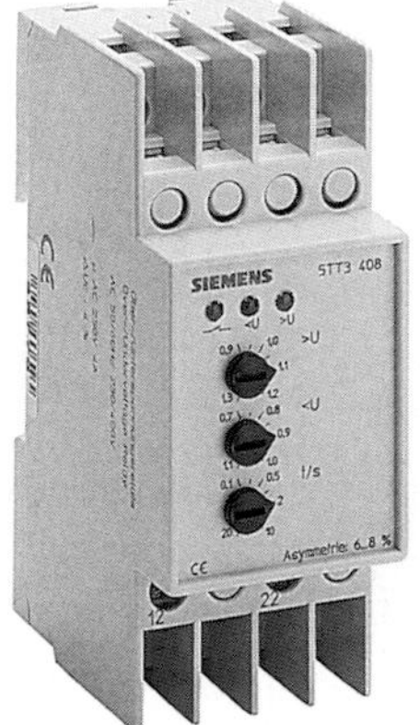

Fig. 4.2/55
Undervoltage/overvoltage relay 5TT3 408

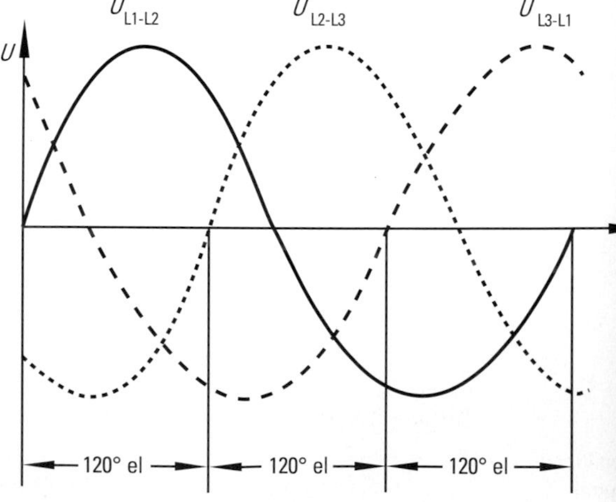

Fig. 4.2/56 A.c. voltages, out of phase by 120°

The phase conductors L1, L2, L3 form the three phase-to-phase voltages $U_{L1\text{-}L2}$, $U_{L2\text{-}L3}$, $U_{L3\text{-}L1}$. These voltages, which are represented geometrically in a phasor diagram, produce an equilateral triangle (Fig. 4.2/57). The three voltages to the neutral point N produce the phase-to-neutral voltages $U_{L1\text{-}N}$, $U_{L2\text{-}N}$, $U_{L3\text{-}N}$

symmetry

Under normal circumstances, the triangle is symmetrical, all voltages are equal, and all angles are 120°. Any deviations from this are referred to as asymmetry.

A distinction is made between two cases of asymmetry:

Case 1: With a "rigid" system, i.e. with constant phase-to-phase voltages, the phase-to-neutral voltages at the load can vary without the outer symmetry undergoing any change (Fig. 4.2/58).
This is the case with asymmetrical wye-connected loads and interrupted neutral conductors, i.e. with an open neutral point.

Case 2: If, however, the phase-to-phase voltages change, this will always result in a change in the phase-to-neutral voltages (Fig. 4.2/59).
This is the case, for example, with motor-driven equipment and if a phase conductor (phase) fails. The motor windings then induce a voltage in the "open" winding (feedback). Although this voltage is no longer equal to the original system voltage, it does allow the three-phase system downstream of the fuses to become symmetrical again at the monitoring point.

In order to detect asymmetry in the system, the three phase-to-phase voltages $U_{L1\text{-}N}$, $U_{L2\text{-}N}$, $U_{L3\text{-}N}$ must be measured and compared in the first case. Asymmetry is calculated as follows:

$$\text{Asymmetry} = \left(\frac{\text{Highest phase-to-phase voltage}}{\text{Lowest phase-to-phase voltage}} - 1\right) \cdot 100\ (\%)$$

Monitoring the neutral conductor

onitoring eutral onductor

In three-phase systems, single-phase loads are connected to the neutral conductor. If the neutral conductor is interrupted in a three-

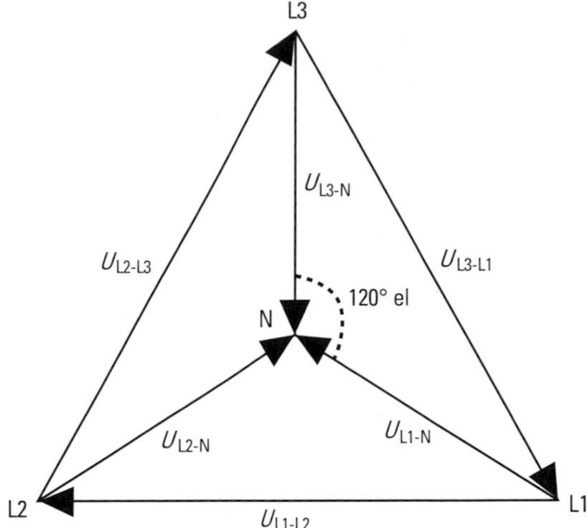

Fig. 4.2/57 Phasor diagram of a.c. voltages

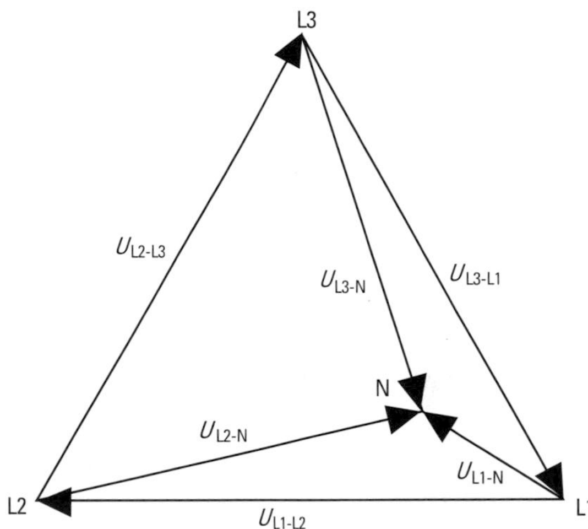

Fig. 4.2/58 Asymmetrical phase-to-neutral voltages

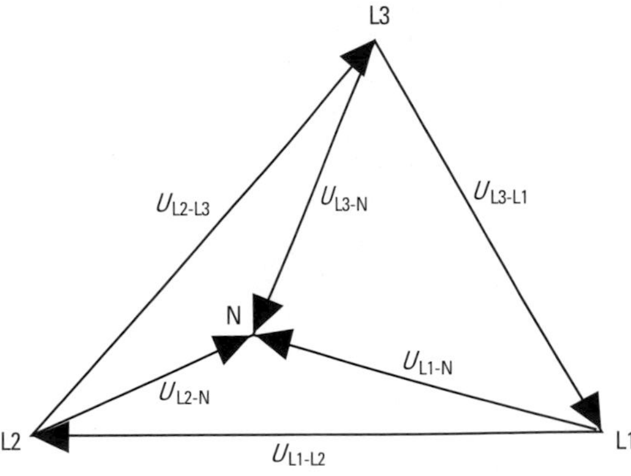

Fig. 4.2/59 Asymmetrical phase-to-phase voltages

phase system, the phase-to-neutral voltages become dangerously unbalanced as a result of the asymmetrical system load. The single-phase loads could then be destroyed by overvoltages or rendered inoperative by undervoltages.

The *neutral conductor monitor* 5TT3 410, which supports the following functions, is used to monitor the neutral conductor:

- ▷ detection of a missing neutral conductor in the system,
- ▷ detection of an interrupted neutral conductor in the device power supply cable,
- ▷ detection of an interchanged neutral conductor and phase conductor.

Fig. 4.2/60 shows a circuit diagram with the neutral conductor monitor.

Asymmetry and neutral conductor monitoring

Monitoring voltages

The following voltage relays are available for protecting medical appliances, laboratory equipment, and emergency lighting devices to DIN VDE 0108 against hazardous voltage fluctuations in the event of a broken neutral conductor:

- ▷ voltage relay 5TT3 405, non-adjustable, releases at 0.85 U_c and picks up again at 0.95 U_c,
- ▷ voltage relay 5TT3 406, adjustable between 0.7 and 0.95 U_c with 4% hysteresis, specially designed for medical equipment.

Fuse monitor

Monitoring fuses

Although fusible links do have indicators which show the functional state of the fuse (see Section 4.1.1), they do not have any contacts via which a fault signal could be issued.

The fuse monitor measures the voltage upstream and downstream of the fusible link, detects reverse voltages from the system, and can indicate exactly which fuse has failed.

Fig. 4.2/61 shows an example of a circuit for monitoring the fuses of a motor using the fuse monitor 5TT3 170.

Voltage relays for detecting short-time interruptions

Short-time interruptions >20 ms occur in the case of lightning strokes or internal switchover operations in the public utility system.

With sensitive technical processes in consumer installations, it is then often impossible to tell whether the process was interrupted by the public utility switchover or by another fault.

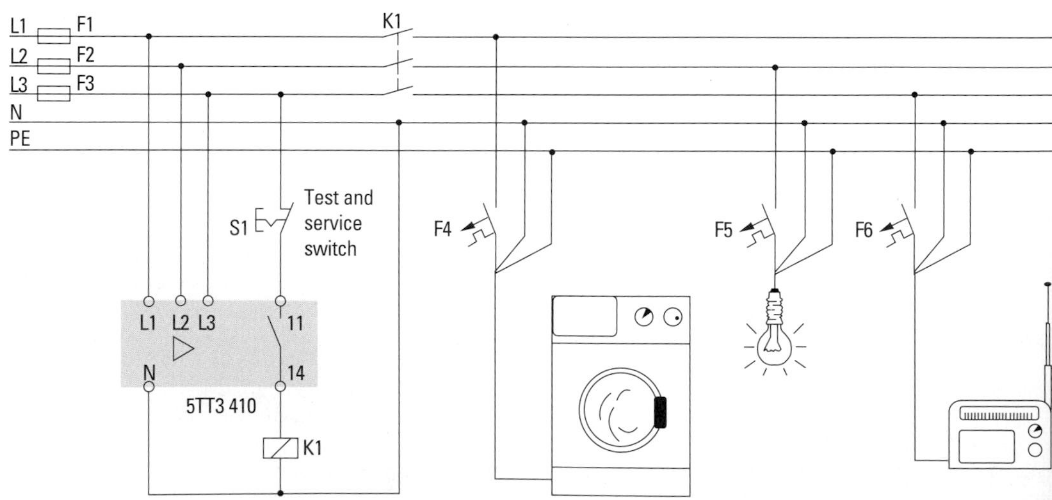

Fig. 4.2/60 Circuit diagram of neutral conductor monitor 5TT3 410 used to protect single-phase loads

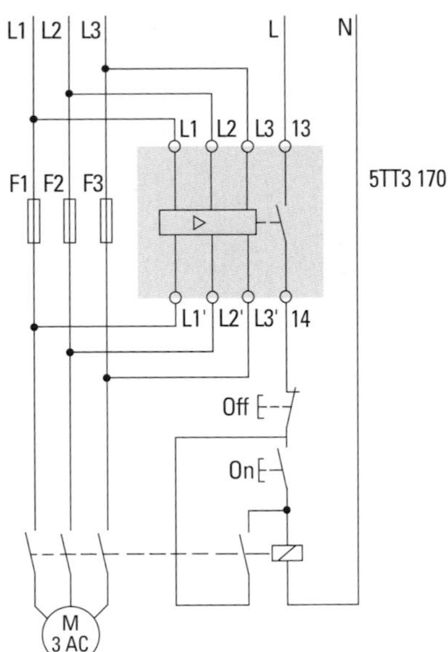

Fig. 4.2/61
Example of circuit for fuse monitor 5TT3 170

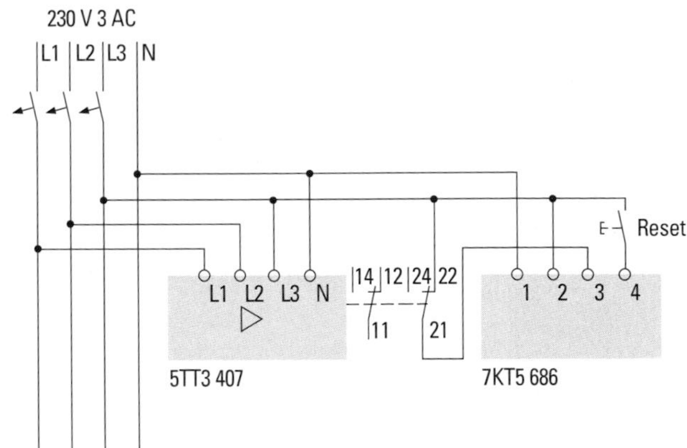

Fig. 4.2/62
Example of circuit with short-time voltage relay 5TT3 407 and supplementary pulse counter 7KT5 686

Short-time voltage relays

If a short-time voltage relay is used, a short-time interruption is clearly indicated as a result of the power supply being disconnected. The power supply can be switched on again by pressing a reset button.

Supplementary pulse counter

In certain cases, it is sufficient for a short-time interruption to be merely registered without the power supply being disconnected. In this case, the short-time voltage relay can be equipped with a supplementary pulse counter which simply counts the short-time interruptions (Fig. 4.2/62). The pulse counter can be reset if required.

Fig. 4.2/63
Current relay 5TT6 120 for undercurrent and overcurrent monitoring for three-phase connection, setting range 0.5 to 5 A

4.2.9 Monitoring with current relays

Current relays allow the operating currents of devices and systems to be monitored and compared with the prospective currents. If the operating current rises or falls in relation to the prospective current, a fault signal is issued.

Types

Current relays (Fig. 4.2/63) for monitoring single-phase and three-phase currents in a.c. and three-phase systems are available in two different versions with one setting range from 1 A to 10 A (15 A) and four setting ranges from 0.1 A to 1 A; 0.5 A to 5 A; 1.0 A to 10 A; and 1.5 A to 15 A (20 A). They are suitable for direct connection or connection to current transformers. The values in brackets (15 A) and (20 A) indicate the overrange of the current relays.

Delay time

Current relays have an adjustable delay time of 0.1 s to 20 s. Internal switchover operations with short-time current peaks do not, therefore, cause the current relays to operate. Depending on the type of relay, they can be overloaded by up to 20 A or 30 A for 3 seconds. This means that, if used in combination with a miniature circuit-breaker, which has a considerably shorter response time, they are provided with adequate protection against short circuits.

Short-circuit protection

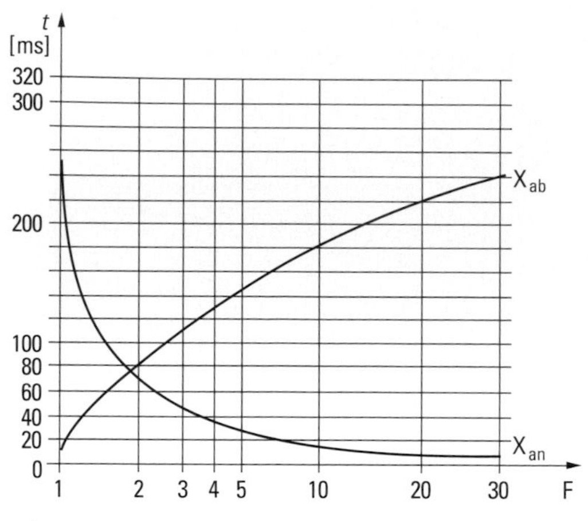

Fig. 4.2/64
Device-specific response time of all 5TT6 1.. current relays

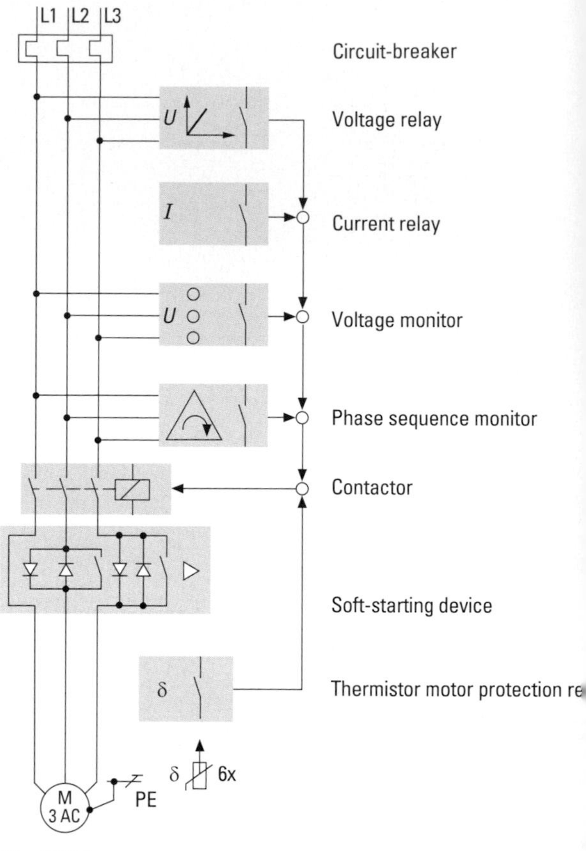

Fig. 4.2/65
Overview of protective and switching functions in motor feeder

Response time

The actual response time is calculated by adding together the set delay time and device-specific response time (Fig. 4.2/64).

Example:

The current relay is set to a delay time of 0.1 s and an operating current of 2 A. Short-time currents of 20 A are, however, applied to the current relay resulting in an overload factor F = 10 (20 A : 2 A). Fig. 4.2/64 indicates a device-specific response time of approximately 200 ms for the overrange current X_{ab} and factor 10. The total response time is then 0.3 s (0.1 s + 0.2 s).

Applications

Current relays are used, for example, for monitoring:

▷ emergency lighting with filament lamps,
▷ beacons, e.g. on high buildings along flight paths,
▷ refrigerators and heating systems.

4.2.10 Switching and protecting motors

Modular devices are also used for switching and protective purposes in motor feeders (see Fig. 4.2/65). Short-circuit protection can be provided by a circuit-breaker or fuse.

Voltage relays with asymmetry detection

A reverse voltage occurs if a phase fails and a three-phase load is connected. Owing to the different phase angle as compared with normal operation, the voltage relay detects the asymmetry and switches off the load at values between 6 % and 8 % (a difference of approximately 14 V to 18 V between the highest and lowest voltage).

Asymmetry may also be caused by a broken neutral conductor, in which case the voltage relay disconnects.

Voltage monitors

If a phase fails, the motor is likely to malfunction during two-phase operation. The voltage monitor disconnects as a preventive measure.

Phase sequence monitors

The phase sequence monitor determines the phase sequence. Fig. 4.2/66 shows a circuit diagram for automatic phase sequence correction for construction site distribution boards for which a clockwise phase se-

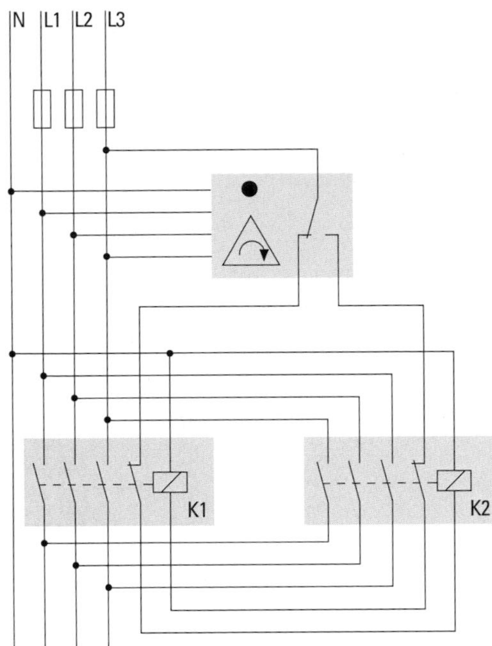

Fig. 4.2/66
Automatic phase sequence correction for construction site distribution boards

quence does not always exist because of the plug-and-socket connections.

Current relays

The current relay (Fig. 4.2/63) is used predominantly for measuring currents to protect motors and systems. If, for example, a fault occurs and currents different to those set are measured, a warning signal is issued prematurely before the circuit-breaker disconnects the electric circuit of the motor. The motor then has to be restarted once the cause of the fault has been eliminated.

Applications

Hard objects in *spiral conveyors*, e.g. in sewage treatment works, often cause the conveyor device to jam. With the appropriate settings, the current relay sends a signal via its contacts to warn the operator that a dangerous situation has arisen and that the motor may jam.

Current relays can also be used for motor feeders in:

▷ stirring machines,
▷ dust extraction systems in wood processing machines,
▷ crane control units.

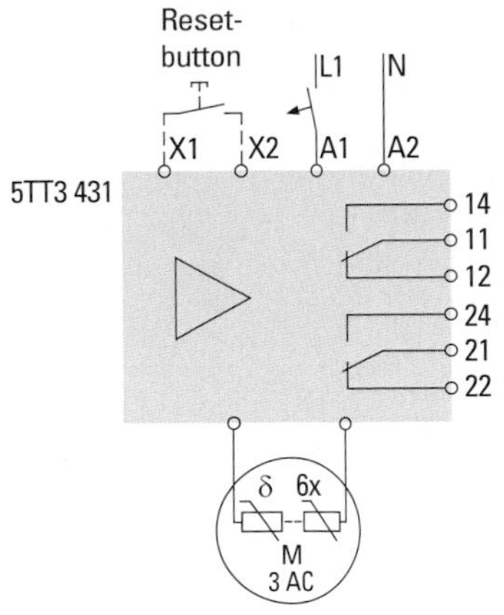

Fig. 4.2/67
Thermistor motor protection relay 5TT3 431

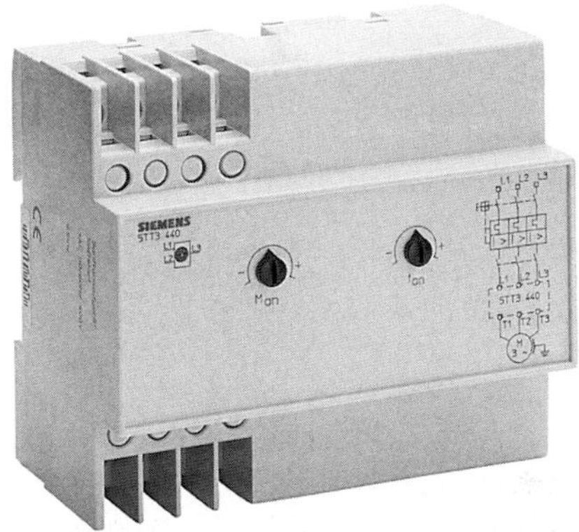

Fig. 4.2/68 Soft-starting device 5TT3 440

Thermistor motor protection

Motors with thermistors (PTC or NTC thermistor detectors in the windings to monitor the temperature of the motor) are required for this type of protection. If one of the thermistors (up to six detectors are possible) reaches the operating temperature, the thermistor motor protection relay (Fig. 4.2/67) switches off the motor.

Soft-starting devices

Soft-starting devices (Fig. 4.2/68) are electronic control units for soft-starting single-phase and three-phase asynchronous machines. The r.m.s. value of the voltage is increased continuously by means of a phase control function. The current and motor torque increase as the voltage increases. The drive mechanism starts smoothly with a minimum starting current.

The mechanical drive elements are thus protected against excessive wear and can be designed to operate more efficiently. Apart from a significant reduction in starting noise, the conveyed materials no longer topple or slip.

In order to prevent power losses in the device after the starting phase, the power electronics are shunted out with relay contacts during continuous operation.

The starting time for the starting torque and the acceleration time can be adjusted on the soft-starting device in order to match the motor forces and mechanical frictional forces. Fig. 4.2/69 shows the usual "starting ramp" of a soft-starting device.

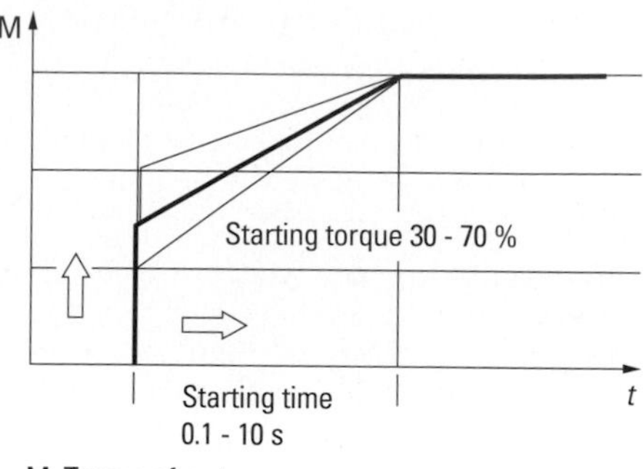

Fig. 4.2/69
Starting behavior of soft-starting devices

Excessively high motor temperatures At high switching rates, it is advisable to use the thermistor motor protection device to monitor the permissible motor temperature because every soft-start operation (reduction of the power input to two phases) represents an unbalanced load start, which leads to excessively high motor temperatures. The soft-starting device must not be operated with a capacitive load.

Types Soft-starting devices are available in 300 W – 5500 W versions for three-phase asynchronous machines and up to 1.5 kW for single-phase asynchronous machines.

Fig. 4.2/70 shows an example of a circuit for a soft-starting device.

Contactors Contactors are described in Section 4.2.4.

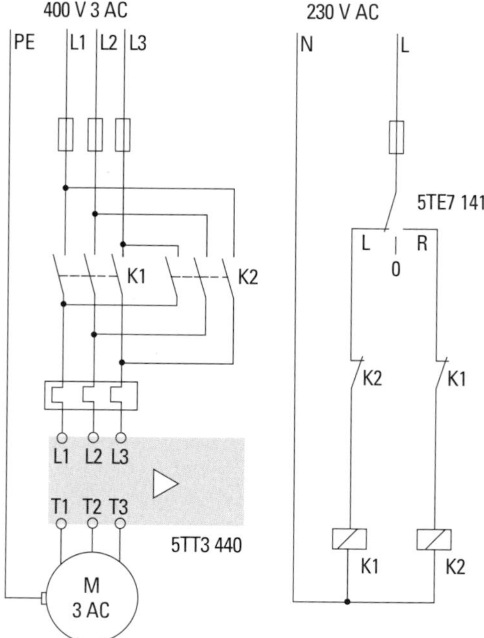

Fig. 4.2/70
Example of circuit for soft-starting device 5TT3 440 with changeover switch 5TE7 141 for rotation reversal

4.2.11 Monitoring with counters and pulse counters

Machinery and production equipment failures can result in considerable financial losses. Regular maintenance to prevent faults and malfunctions is, therefore, essential. Operating times, work cycles, the manufactured number of units, or making operations form the basis for maintenance schedules. Counters and pulse counters supply the information required to plan cost-effective maintenance schedules.

Electro-mechanical counters Electromechanical counters (Fig. 4.2/71) consist of a synchronous motor and a drum-type register and are available for 24 V AC, 115 V AC, and 230 V AC for 50 Hz and 60 Hz. They can count up to 99,999 operating hours to one decimal place.

Electro-mechanical pulse counters

Electromechanical pulse counters (Fig. 4.2/72) consist of a solenoid which drives the drum-type register. They register up to 999,999 making operations and are available in a 24 V AC and 230 V AC version for 50 Hz and 60 Hz.

If several counters or pulse counters are required, versions with 2 counting systems in one 2-MU housing can be used to save space. These double units also have a horizontal counter display.

Counter/pulse counter combination

Versions, in which a counter and pulse counter are combined (Fig. 4.2/73), are available to allow average running times, i.e. running time per making operation, to be determined.

Versions for direct current

Versions for direct current with a clock generator are based on the quartz resonator principle.

Electronic time and pulse counters

Electronic counters and pulse counters (Fig. 4.2/74) have a 5-mm high, 6-digit, 7-segment LED display. They are more efficient, can be reset after maintenance, and begin counting again from zero. Suppression of the leading zeros makes these counters easier to read. The data is stored in an EEPROM so that it is not lost in the event of a power failure.

Applications

The resettable electronic pulse counter is ideal, for example, for counting the daily output of a production line.

Fig. 4.2/71
Electromechanical counter 7KT5 745–3

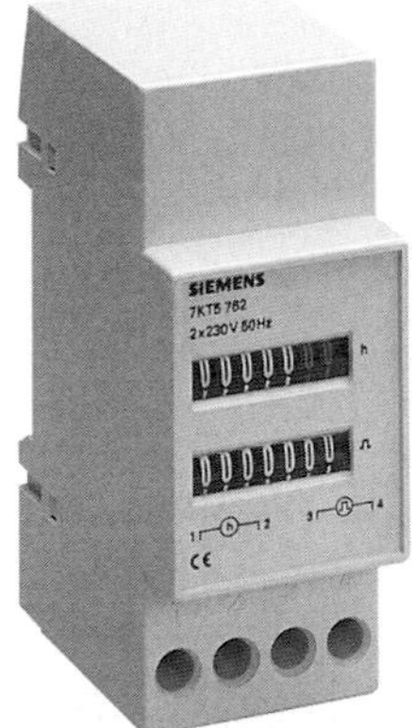

Fig. 4.2/73
Counter/pulse counter combination 7KT5 762

Fig. 4.2/72 Pulse counter 7KT5 753

Fig. 4.2/74 Electronic pulse counter 7KT5 780

Counting operating hours or making operations alone is often not sufficient to determine the required maintenance times, as is the case with large compressors for air-conditioning systems, for example. Since these systems are often operated at full load during the summer months, the oil must be changed more frequently at oil temperatures in excess of 50 °C. In addition to counting the operating hours, it is also advisable to take the oil sump temperature into consideration when determining the maintenance times. Fig. 4.2/75 shows a circuit used for determining operating hours for compressors in air-conditioning systems which measures the total number of operating hours and the number of full-load operations.

4.2.12 Monitoring and regulating with level-sensing relays

Monitoring fluid levels

Level-sensing relays are used for monitoring fluid levels. They are ohmmeters with a high resistance measuring range of up to 450 kΩ and can, therefore, be used for a wide range of applications:

- ▷ monitoring and regulating the level of conductive fluids and powders, e.g. maximum and minimum level,
- ▷ overflow and dry-running protection,
- ▷ monitoring and regulating the mixture ratio of conductive fluids,
- ▷ general resistance monitoring tasks, e.g. temperature limit determination with PTC resistors (**P**ositive **T**emperature **C**oefficient)

The broad resistance setting range, however, means that it is also possible to make a distinction between fluids and foam. To permit this with a minimum time delay, a sufficiently high operate value must be set so that the relay operates reliably if the COM and MAX electrode is covered with fluid, but not with foam.

Level regulation

Fig. 4.2/76 illustrates the principle of level regulation for an open container with free fluid admission and controlled pump drainage, as used for water treatment in sewage works or for cellar drainage.

Submersible electrodes

The submersible COM electrode for level measurement is installed at the lowest point. If the container holding the fluid is made of conductive material, it can be used as the reference electrode.

The MAX electrode is set to the maximum fluid level.

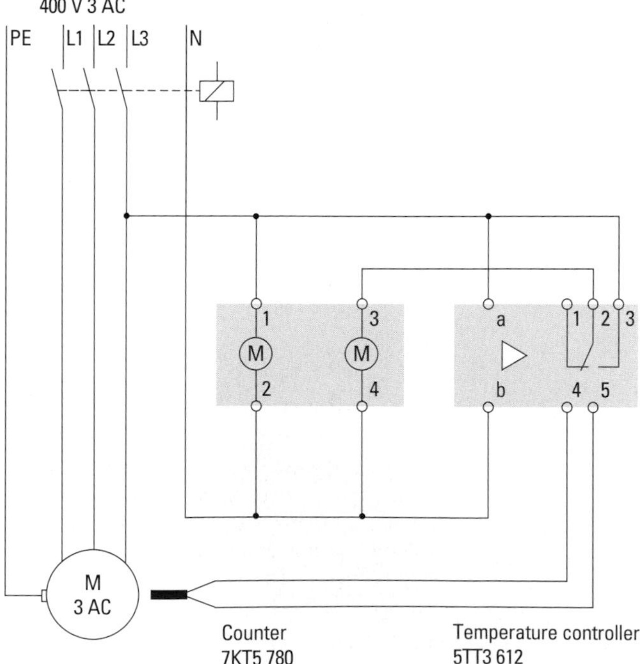

Fig. 4.2/75
Determination of operating hours for air-conditioning compressors by measuring total number of operating hours and number of full-load operations

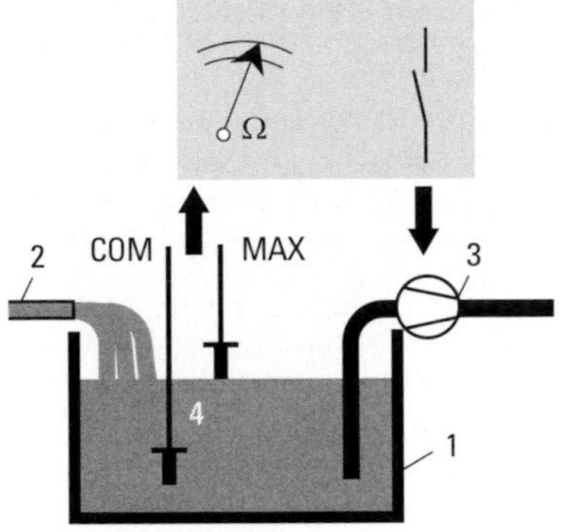

1 Open container
2 Free fluid admission
3 Controlled pump drainage
4 Resistance measurement

Fig. 4.2/76 Diagram illustrating level regulation

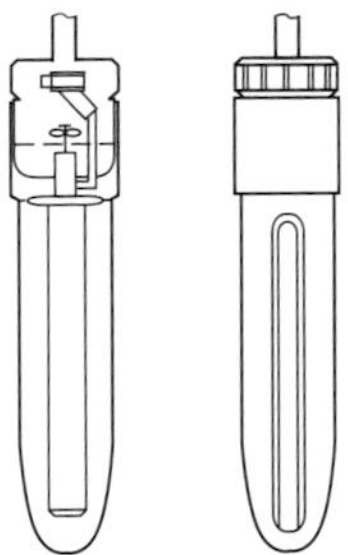

Fig. 4.2/77
Submersible electrode 5TG8 223

Submersible electrodes (Fig. 4.2/77) are available as stainless-steel versions which are resistant to temperatures of up to 90 °C, with a terminal connection or a 1-m long H07 RN-F connecting cable with high-quality insulation.

The level-sensing relay 5TT3 435 (Fig. 4.2/78) has two output relays for activating pumps. It can be used in two different operating modes, namely one-stage and two-stage level control. LEDs indicate the current operating status.

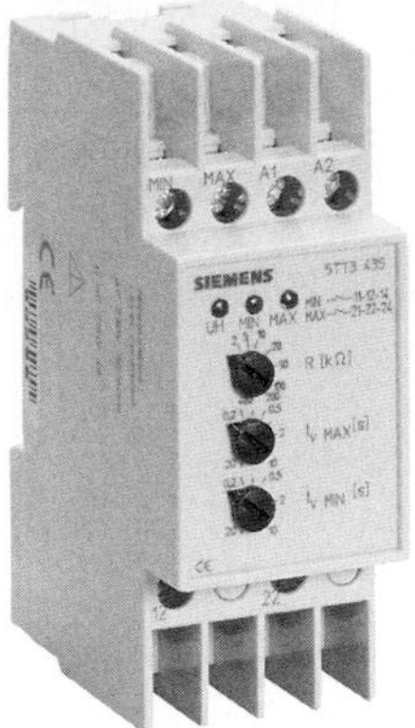

Fig. 4.2/78 Level-sensing relay 5TT3 435

Universal application

The level-sensing relay can also be used to monitor contacts, which must not be loaded with high currents (e.g. contacts in float switches, position transmitters, and limit switches), and Reed contacts as well as to check their event status (ON/OFF).

4.2.13 Monitoring and control with twilight switches

Operation, application

The light sensor of the twilight switch measures daylight intensity. If this reaches the intensity value set on the twilight switch, the device, e.g. lighting, is switched on.

An ON delay of at least 50 seconds and OFF delay of at least 50 seconds as well as a hysteresis 1.3 times the set value prevent continuous switching. If, for example, 10 lux is set and the device has switched on, it only switches off again at 13 lux if at least 50 seconds have elapsed.

The light sensor of the twilight switch must be installed in such a way that it cannot detect the activated lighting, otherwise the system may be switched every 50 seconds.

If used in conjunction with a timer, twilight switches are also suitable for activating lighting systems used to illuminate shop windows, billboards and highways, as well as spotlight systems.

Types

Twilight switches are available as 1-channel and 2-channel versions. Each channel can be preset to between 2 and 300 lux for dimming and to between 200 and 20,000 lux for illumination. Each setting range has a separate setting scale. The 2-channel version (Fig. 4.2/79) requires just one light sensor (Fig. 4.2/80). Both controller versions can be used for all lamp types.

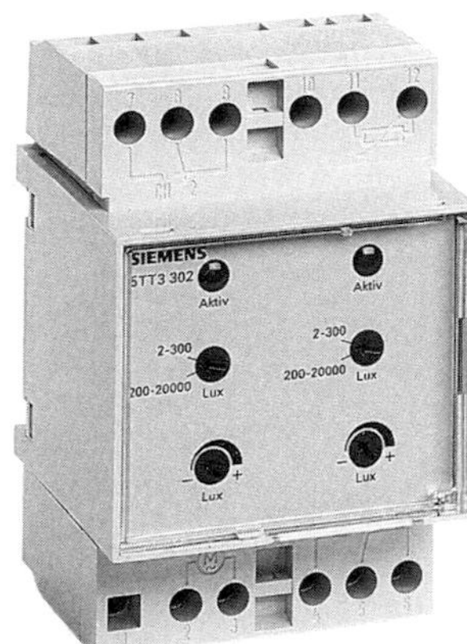

Fig. 4.2/79
Twilight switch 5TT3 301 for 2 channels

Fig. 4.2/80
Light sensor with wall-mounting bracket

4.2.14 Use in building control system with *instabus EIB*

Modular devices with interfaces

Mechanical, electromechanical, and electronic modular devices with switching-type contact outputs are conventional devices. They do not have a communicative interface and are, therefore, not "bus compatible".

Components for building control system with *instabus EIB* (see Section 4.4), however, do have interfaces which allow the modular devices described in Section 4.2 to be integrated in an *instabus EIB* installation. It is mostly the "sensory" capabilities of these conventional devices that are utilized in *instabus EIB* installations.

4.3 Operator communication, switching, control, and signaling systems, information and monitoring systems

4.3.1 Operator communication, switching, control and signaling systems

Jse

The use of electrical equipment and devices in residential, public, commercial and industrial buildings has opened up a whole new world of highly flexible solutions for operator communication, switching, controlling and signaling applications, as well as for information and monitoring systems in electrical building installations.

nfluences on device design

Actuators and devices with the required supplementary functions are designed in accordance with the various systems and engineering principles and practices in the countries concerned. The following criteria influence the design of devices:

▷ Voltage level,
▷ Connector configuration,
▷ Device box dimensions,
▷ Spacing for layout in assemblies,
▷ Conductor connection,
▷ Supplementary functions,
▷ Considerations of appearance/finish.

Standards

The technical requirements governing this area are laid down in the standards EN 60669-1/IEC 60669-1/DIN VDE 0632-1 (for switches) and IEC 60884-1/DIN VDE 0624-1 and DIN VDE 0620 (for socket-outlets). These specify standard performance characteristics, while at the same time permitting country-specific rated currents, types, designs, and versions.

Types

The following definitions are provided in these standards with regard to types:

Type A: This ensures that the connected conductors can be neither detached nor moved during work on the mounting surface.

Type B: During work on the mounting surface, the connected conductors are moved, or must be detached.

National systems

In view of the various connector configurations, a variety of systems have been developed worldwide.

Circular box system, spacing 71 mm

In Europe, the most commonly encountered system is the circular box system (Fig. 4.3/1), which is based on a device box with a diameter of 60 mm and spacing of 71 mm for layout in assemblies.

Modular system

In Southern Europe, a modular system (Fig. 4.3/2) is used, in which the individual contact blocks are butt-mounted in a supporting frame with a cover or guard. With this system, manufacturer-specific contact blocks and country-specific device boxes with various dimensions are available.

Monobloc system

Monobloc systems are used primarily in South East Europe and contain complete devices (which can also comprise several switching or connector functions) which are housed in device boxes. The device dimensions are not governed by standards, and are designed only for country-specific device boxes (Fig. 4.3/3).

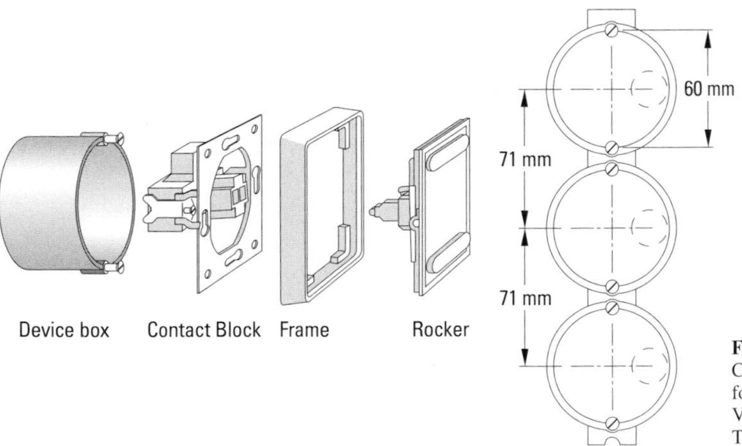

Fig. 4.3/1
Circular box system for devices, VDE design, Type A

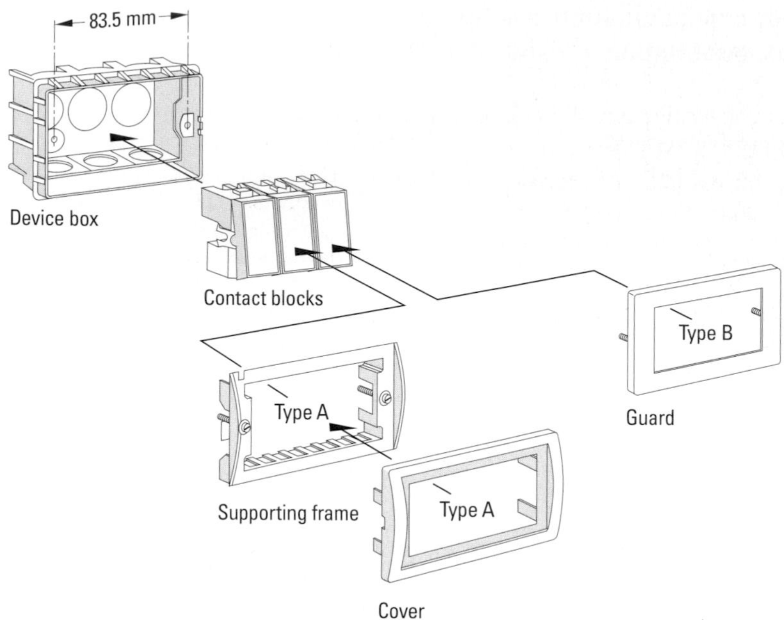

Fig. 4.3/2 Modular system for devices in Southern Europe, Type A and Type B

Type A

Users' requirements are continually changing, giving rise to constant changes in the appearance and finish of actuators. Type A (standardized at an international level) takes this into account, and ensures that the covers, which are an important factor in design, can be replaced without the connected conductors having to be detached.

Standardized contact blocks

The DELTA ranges from Siemens (Fig. 4.3/4) satisfy these changing requirements. This is ensured by standardized contact block (Fig. 4.3/5). The small base dimensions of the contact blocks make them extremely easy to install, since more room is available for the wiring in the 60-mm diameter device box. A screwless terminal connection system, as well as a circuit diagram with topographical pin designations make it even easier to connect conductors.

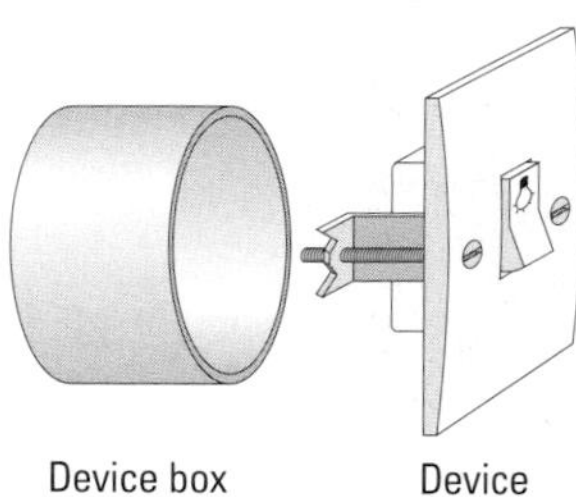

Fig. 4.3/3
Country-specific monobloc system for devices, Type B, with one switching function

The contact blocks are equipped with all-round protection against electric shock, i.e. live parts are safe from finger-touch.

In addition to this, the contact blocks are equipped with recoil claws, and are fitted in the device box with captive +/− screws. Extension claws are available for installing the contact blocks in device boxes fitted deeper in the wall. These are simply clipped on to the existing claws.

Voltage tests can be carried out from the front of the contact blocks after they have been installed.

Adjustment to plaster surfaces

The design of the actuators and switching device contact blocks allow any necessary adjustments to be made in line with plaster surfaces, ensuring that the devices switch reliably even when contact blocks are installed at an angle (Fig. 4.3/6).

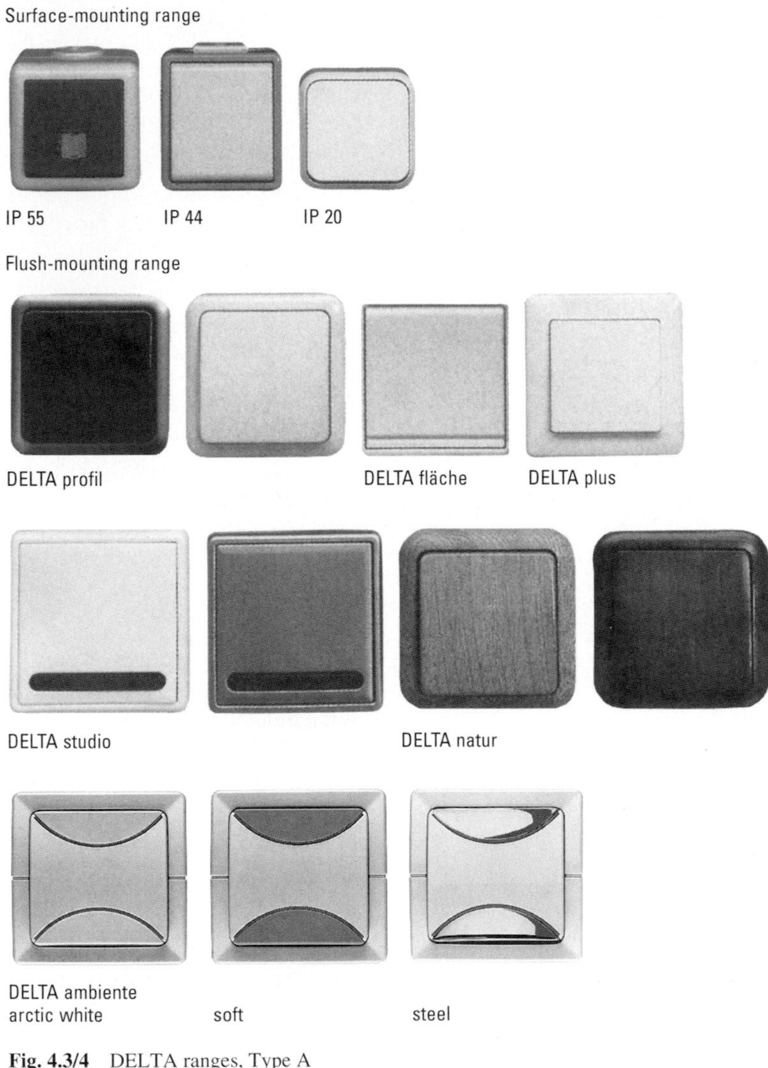

Fig. 4.3/4 DELTA ranges, Type A

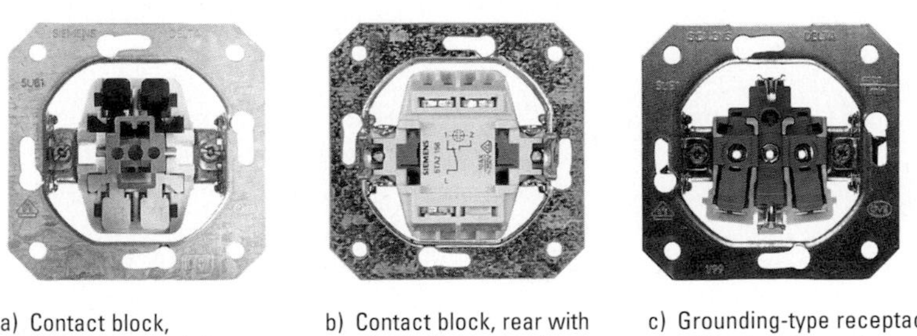

a) Contact block, changeover switch
b) Contact block, rear with circuit diagram and topographical pin designations
c) Grounding-type receptacle without design part

Fig. 4.3/5 DELTA ranges, contact block and grounding-type receptacle (without design part)

Snap-on system: Fast mounting

Frame and rocker with bearing block

Switching reliability: problem-free operation with angled contact blocks

Adjustment to plaster surfaces: 3 mm

1 mm

3 mm

Frame: can be fitted horizontally or vertically for flush-mounting boxes and integration in trunking

Molding material: thermoplastic, fracture resistant

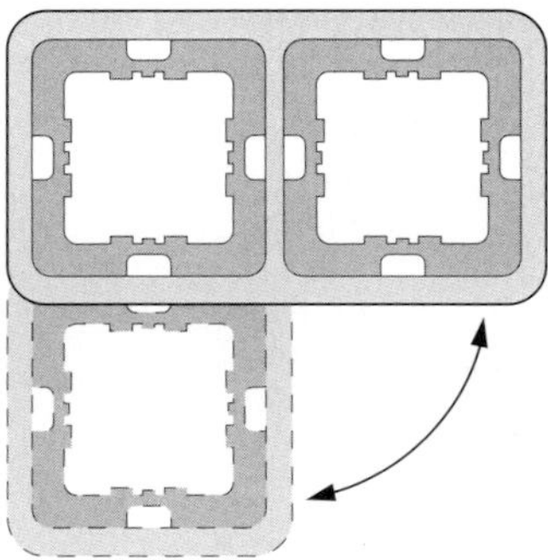

Fig. 4.3/6
Adjustment to plaster surfaces and switching reliability with DELTA contact blocks. Snap-on system for fast mounting of frames

Rocker, bearing block

The rocker and bearing block are designed as a single unit so that the rocker alone cannot be removed inadvertently.

Frames

The frame with the rocker is snapped on to the contact block either horizontally or vertically (see Fig. 4.3/6).

Locating lamp

The DELTA contact blocks are designed in such a way that a locating or pilot lamp can be retrofitted without the contact blocks having to be removed. An inspection window in the rocker provides optimum luminosity, and also complies with the requirements in the regulations for functional buildings that apply in Germany.

Environmentally compatible materials

The contact materials of the DELTA contact blocks are free of cadmium and nickel; the electro-plated coatings are free of passivations containing chrome VI. All plastics are halogen free, and free of pigments containing heavy metals.

Types of contact block

The contact blocks for the various device versions in the Siemens DELTA ranges are available as:

▷ Pushbuttons:
- Pushbuttons for various applications,
- Double pushbuttons for various applications,
- Louver blind pushbuttons;

▷ Switching devices:
- On-off switches, 1, 2, and 3 pole,
- Changeover switches,
- Twin changeover switches,
- Intermediate switches,
- Two-circuit switches,
- Switches with pilot lamp,
- Blind control switches,
- Key-operated blind control switches,
- Time switches,
- Operating time switches;

▷ Control devices:
- Rotary dimmers for various applications,
- Sensor dimmers for various applications,
- Step switches for various applications,
- Volume controls,
- Speed regulating rheostats,
- Room temperature controllers,
- Motion detectors;

▷ Signaling units:
- Visual signals,
- Information displays;

▷ Communication devices:
- Aerial sockets for various applications,
- Telephone sockets for various applications,
- Loudspeaker sockets,
- Telecommunication line units;

▷ Plug-and-socket devices for power supply:
- Grounding-type receptacles for various applications,
- Grounding-type receptacles with status indication,
- Grounding-type receptacles with overvoltage protection,

- Grounding-type receptacles with increased protection against electric shock (shutter),
- Grounding-type receptacles with integrated fault current protection and increased protection against electric shock (shutter),
- Socket-outlets with center grounding contact, 2 pole to CEE 7,
- Socket-outlets, 2 pole to American standard C73,
- Socket-outlets, 2 pole to American standard NEMA 5,
- Socket-outlets, 2 pole to British standard 1363,
- Socket-outlets, 2 pole to Italian standard CEI 23-5/16;

▷ Plug-and-socket devices for data and speech networks (see Section 4.3.2):

- D subminiature, Twinax, Western, BNC, BNC/TNC, IBM-LAN, CA, ACO, Western Nevada Omni, ICCS, and optical fibers.

Fig. 4.3/7
Grounding-type receptacle with status indication

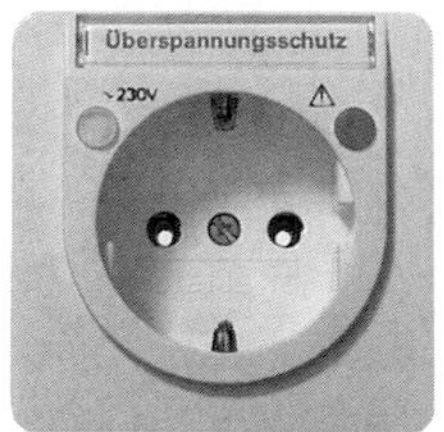

Fig. 4.3/8
Grounding-type receptacle with integrated overvoltage protection (see Section 4.1.4)

Versatility

The modular design of the devices in the DELTA contact block system for flush mounting enables them to be used in an extremely wide range of applications. For example, the functionality of DELTA grounding-type receptacles can be enhanced by adding modular accessories (Figs. 4.3/7 to 4.3/9). This does not affect the standard overall depth of 32 mm.

DELTA ranges

The DELTA ranges offer the high level of quality required for interior design purposes, and are available in various materials, forms and dimensions, with different surface finishes and colors.

Fig. 4.3/9
Grounding-type receptacle with integrated fault current protection and increased protection against electric shock (shutter) (see Section 4.1.3)

The actuating rockers of the DELTA ranges can also be connected to the DELTA bus coupling unit *instabus EIB* (Fig. 4.3/10) (see Section 4.4.2). Depending on the particular use and application, various functions can be controlled via the twisted-pair and Powerline versions of the *instabus EIB*.

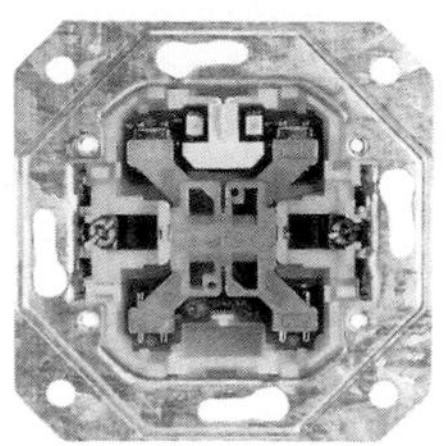

Fig. 4.3/10 DELTA bus coupling unit

DELTA studio

Design

The DELTA studio range is available in the colors light bronze and titanium white.

The covers and frames in the DELTA studio light bronze range are made of anodized aluminum. This means that the surfaces are insensitive to oils, chemicals, fiber-tipped pens, and perspiration.

The covers and frames in the DELTA studio titanium white range are powder coated. Replaceable tandem-rocker assemblies are available, allowing the color of devices and accessories (red, green, violet, blue, and lime green) to be matched to the particular environment. Fig. 4.3/11 shows some of the devices and accessories available in the DELTA studio range.

Frame assemblies with the basic dimensions 75 mm and 80 mm can be used with the devices of the DELTA studio range.

Hands-free speech systems

Hands-free speech systems are also available as a special supplementary function

DELTA ambiente

Design

The attractive design of the modern DELTA ambiente range is based on the results of studies carried out in cooperation with ergonomics experts into lighting (location aids) and the use of switches and pushbuttons by people of all ages.

Distinctive touch areas

To facilitate operation of the switching devices, a raised touch area has been integrated in the operating rocker (dimensions 65 mm × 65 mm). This touch area is available in various materials to meet different interior design requirements. In the DELTA ambiente arctic white range, the plastic touch areas are available in the color arctic white, or coated with gray soft-touch varnish, or in shiny stainless steel. The touch areas in the DELTA ambiete royal blue range are made of wood, and those in DELTA ambiente cosmos gray range of CORIAN[1)].

A segment-shaped lighting element acts as a guide for the person using the switch. The devices in the range can be fitted in two different frame assemblies made of die-cast aluminum (basic dimension 83 mm) with the forms "Contour" and "Convex". In this range, the socket-outlets are only equipped with additional protection against electric shock (shutter). Fig. 4.3/12 shows some of the devices and accessories in the DELTA ambiente range.

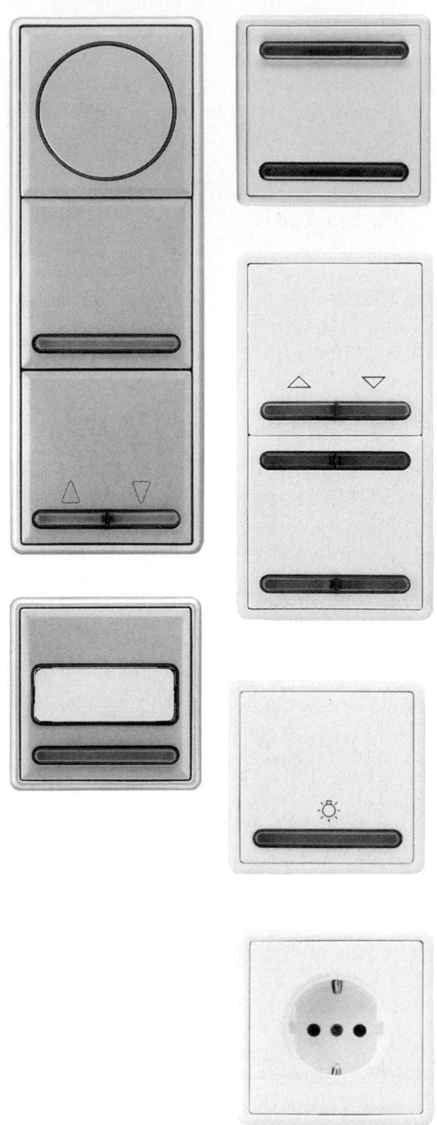

a) DELTA studio light bronze, from top to bottom: dimmer, pushbutton, louver blind pushbutton, pushbutton with labeling panel symbol

b) DELTA studio titanium white, from top to bottom: changeover switch, louver blind pushbutton, two-circuit switch, pushbutton with "light", grounding-type receptacle

Fig. 4.3/11
Some of the devices and accessories in the DELTA studio light bronze and titanium white ranges

[1)] CORIAN is a registered trademark. The material is a combination of 2/3 stone and 1/3 high-quality acrylic

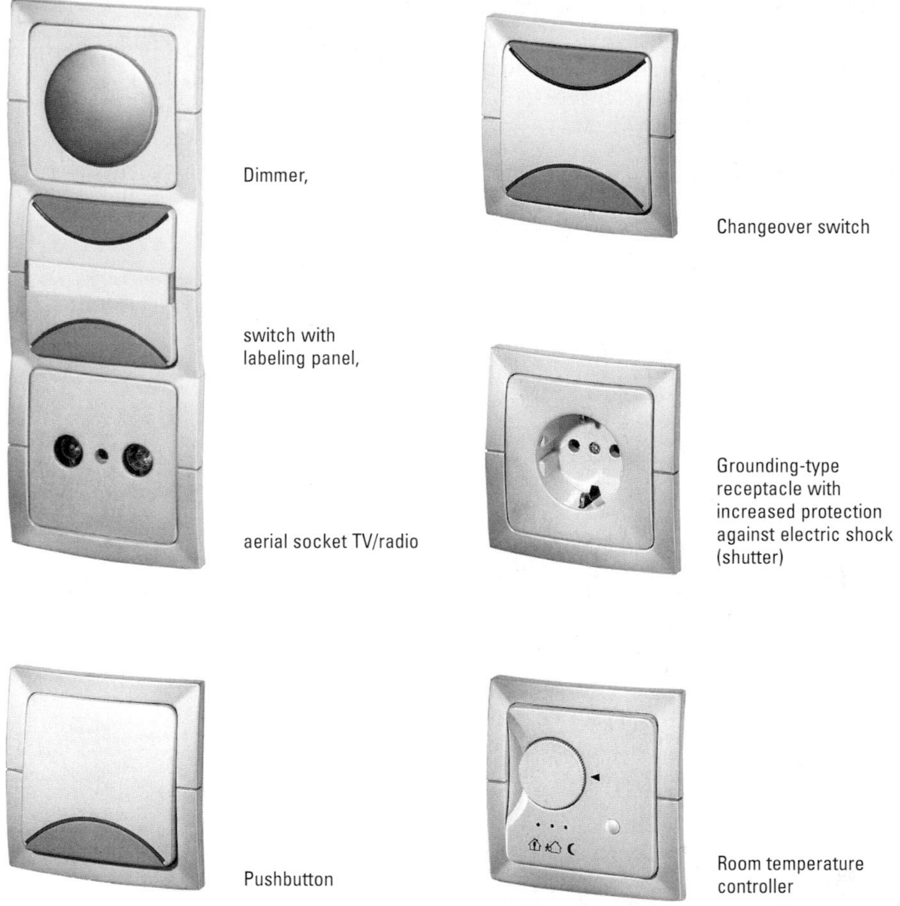

Fig. 4.3/12 Some of the devices and accessories in the DELTA ambiente range

DELTA natur

Real wood range: oak/maple

DELTA natur is a range of real wood devices and accessories for surface mounting (rocker dimensions 62 mm × 62 mm) with a decorative wood finish and a matching wooden frame assembly (basic dimension 80 mm). The clear-cut contours and distinctive profiles of this range contribute to its attractive appearance, especially when the devices and accessories are harmonized with the interior decor. The wood for the light oak, dark oak, and maple red versions is selected and worked in accordance with specialist manufacturing specifications (Fig. 4.3/13).

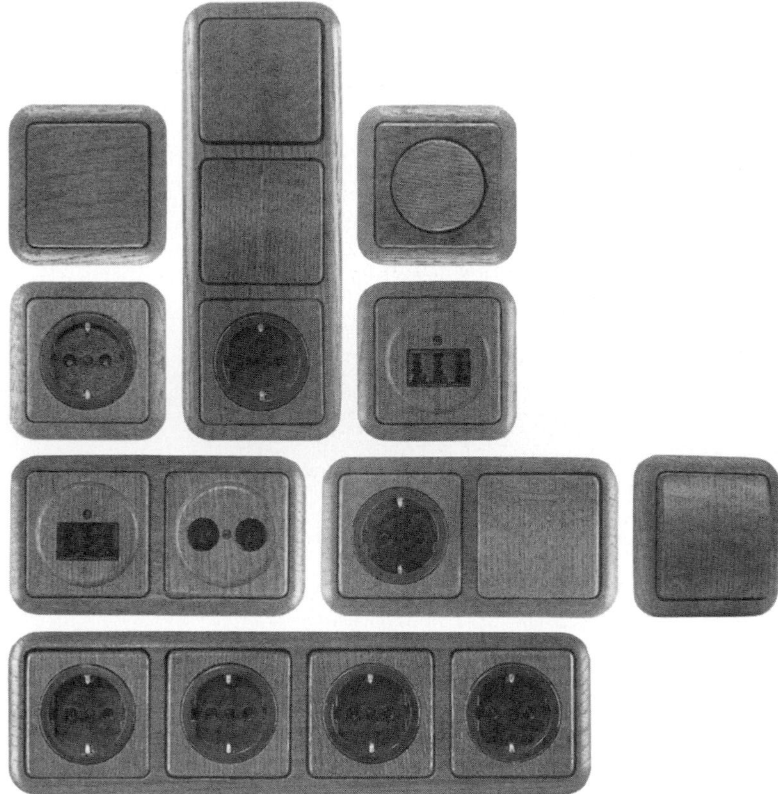

Fig. 4.3/13 Some of the devices and accessories in the DELTA natur range

DELTA profil

DELTA profil is a standardized range comprising rockers and covers (65 mm × 65 mm) and a matching frame assembly (basic dimension 83 mm). The design of the surfaces in the colors titanium white, pearl gray, safari green, taiga beige, sapphire blue, and silver contributes to the attractive appearance of the range (Fig. 4.3/14).

Surface-mounting housings

All the devices in the range can also be installed in surface-mounting housings for one and two contact blocks.

Fig. 4.3/14
Some of the devices and accessories in the DELTA profil anthracite range with silver frame assembly

DELTA fläche

Large-surface range, satin-frosted

This large-surface range (rocker dimension 70 mm × 70 mm) is available in the colors titanium white, sand, and electrical white with a satin-frosted surface and is particularly suitable for low-noise switching. The high-quality thermoplastic material ensures fracture-resistant mounting.

Surface-mounting housings

All the devices in the range can be installed in frame assemblies with the basic dimension 75 mm (Fig. 4.3/15) in surface-mounting housings for up to three contact blocks.

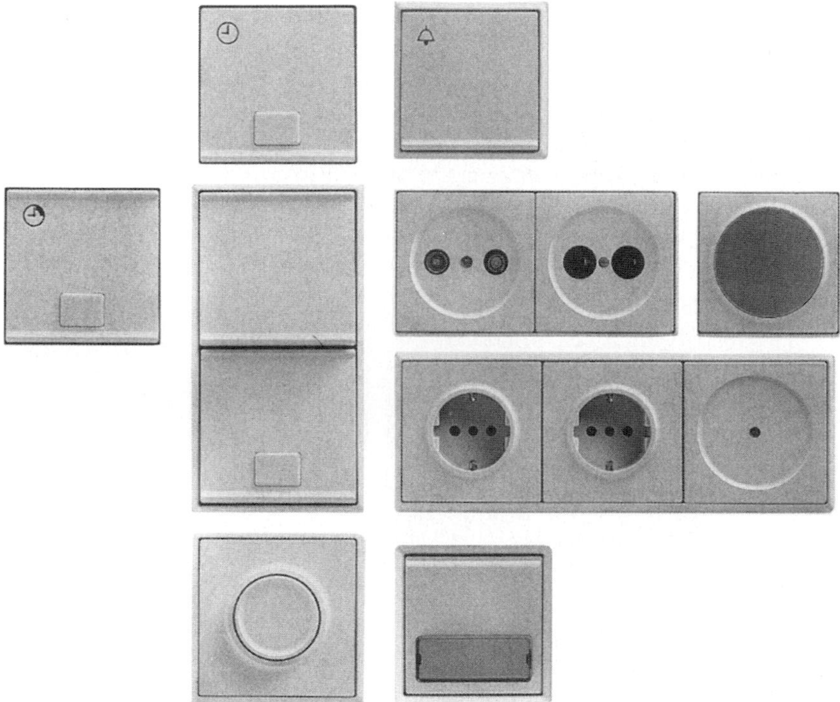

Fig. 4.3/15 Some of the devices and accessories in the DELTA fläche range

DELTA plus

Small-surface range

The DELTA plus range in the colors titanium white and electrical white is the basic range. It is also known as the small-surface range with its rocker dimensions of 55 mm × 55 mm. The frames (basic dimension 80 mm), however, are broader (Fig. 4.3/16). The thermoplastic material used in the frames is extremely durable and has an easy-care surface.

Surface-mounting housings

All the devices in the range can be installed in surface-mounting housings for one and two contact blocks.

Luminous call systems

A wide variety of functions for luminous call systems (see Section 4.3.2.3) are supported with this DELTA range.

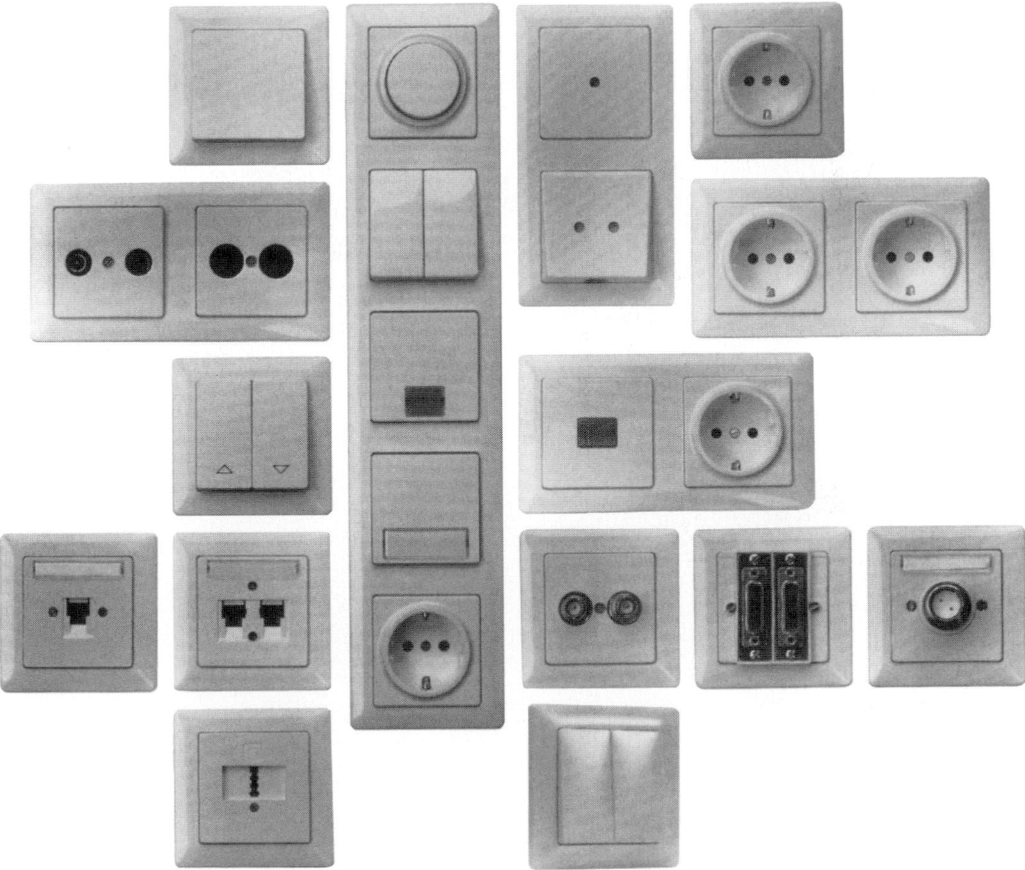

Fig. 4.3/16 Some of the devices and accessories in the DELTA plus range

Conversion to degree of protection IP 44

The DELTA ranges for flush mounting are also available in versions for use in damp and wet rooms, e.g. in washrooms, kitchens, laboratories, or in unheated and unventilated rooms, such as garages.

The contact blocks and actuators in the DELTA plus and DELTA fläche ranges can be converted to degree of protection IP 44 by means of a seal assembly and a special frame assembly (Fig. 4.3/17).

Until about 1990, the construction of residential buildings in Eastern-bloc countries was characterized by industrialized building techniques using prefabricated and standardized components. When paneled structure buildings of this type are renovated, it is therefore necessary to use compatible device boxes for flush mounting with a diameter of 40 mm (Fig. 4.3/18).

DELTA plus shallow design

Shallow devices and accessories with a standard design for the required pushbuttons/switches in the DELTA plus range are available for this type of application (Fig. 4.3/19).

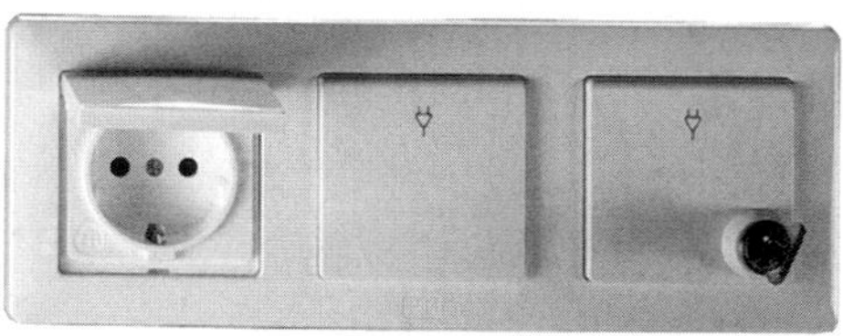

Fig. 4.3/17
DELTA plus, degree of protection IP 44

Fig. 4.3/18
Built-in socket for flush mounting diameter 40 mm

4.3/19 DELTA plus shallow pushbutton

Surface-mounting range

Surface-mounting range

The range of devices for surface mounting is used in those areas of residential and functional buildings where flush mounting is either not required (e.g. cellars, lofts, and technical operating areas) or is not possible for architectural or heat-insulation reasons.

Designs for extreme loads

Surface-mounting designs with degree of protection IP 44 (splash proof) and IP 55 (jet proof) – see Fig. 4.3/20 – are available for special operating conditions and extreme loads (e.g. in commercial and industrial applications).

Range for modular installation systems

Type A

The DELTA futura range is based on modular contact block technology and comprises Type A devices.

This range is used primarily in Southern Europe. In addition to various colored covers, it comprises the country-specific plug-and-socket devices (to Italian standard CEI), as well as supplementary electronic functions (Fig. 4.3/ 21).

Range for monobloc systems

Type B

Monobloc systems are used primarily in South East Europe. Two ranges are available (Type B) for applications of this type. Fig. 4.3/22 shows a switch from the super range and a pushbutton with “light“ symbol from the DELTA arte range. Country-specific plug-and-socket devices as well as various supplementary functions are also available for these ranges.

Fig. 4.3/20
Some of the devices and accessories in the DELTA range for surface mounting

Fig. 4.3/21
Some of the devices and accessories in the DELTA futura range

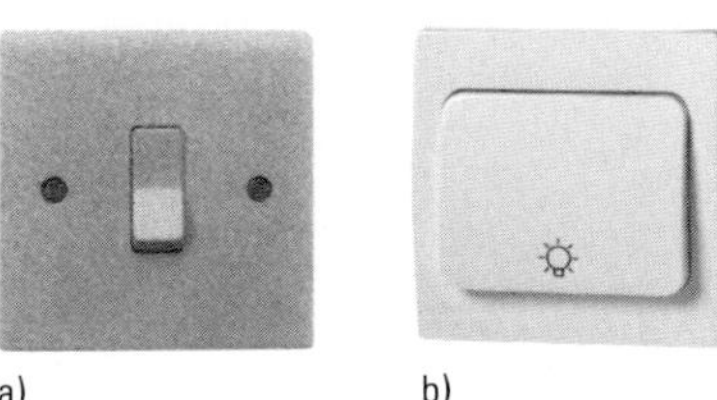

Fig. 4.3/22
Examples of devices and accessories from the ranges
a) super
b) DELTA arte

4.3.1.1 Electronic controllers for roller shutters and louver blinds

Use

While conventional, manually operated electro-mechanical shutters and blinds continue to be used, time-controlled electronic devices for opening and closing roller shutters and louver blinds are becoming increasingly important. These types of controller are used predominantly for applications in residential buildings.

Tasks

Electronic controllers operate motor-driven roller shutters and louver blinds, and can perform the following tasks automatically and reliably:

▷ Open and close roller shutters and louver blinds at programmed times by day, by night, and when the user is not present,
▷ Close shutters and blinds when sunlight is strong,
▷ Close shutters and blinds at a desired twilight level, which can be set by the user.

For house or apartment owners, the advantages of having these tasks performed by the electronic controller include greater convenience, a certain level of security against intruders, and protection for plants, furniture, and carpets against strong sunlight.

In the winter, the electronic controllers also help to save energy by closing and opening blinds/shutters at the most appropriate times in the evening and morning.

Controller for motor-driven roller shutters and louver blinds

DUOMATIC motor-driven roller shutters/louver blinds

The DUOMATIC electronic controller for motor-driven roller shutters/louver blinds is available in the following three versions:

▷ Version 1:
Automatic time control and infinite manual adjustment via pushbutton. Blind louvers can be infinitely adjusted via jogging mode;
▷ Version 2:
Automatic time control with sun tracking and infinite manual adjustment via pushbutton. Blind louvers can be infinitely adjusted via jogging mode;
▷ Version 3:
Automatic time control with sun and twilight tracking, and infinite manual adjustment via pushbutton. Blind louvers can be infinitely adjusted via jogging mode.

Design and dimensions

Fig. 4.3/23 shows a DUOMATIC blind control switch. All the DUOMATIC versions can be used in device boxes (diameter 60 mm), and in conjunction with installation switches from the DELTA ambiente, DELTA studio, DELTA profil, DELTA fläche and DELTA plus ranges, in either individual or multiple frames.

Technical data

The DUOMATIC blind control switch can be used in supply systems with 230 V AC, 50 Hz and can continue operating for 8 hours in the event of a power supply failure. The switching capacity is 6 A.

Operation, display, check

The following operation, display, and check functions are supported (Fig. 4.3/24):

▷ Operation by means of 4 keys, the functions of which are indicated by the symbols,
▷ Times to "LOWER" and "RAISE" the roller shutters/louver blinds may be programmed for each day,
▷ Forward and backward programming of the time (of day) and automatic timing,
▷ Infinite adjustment in manual mode,
▷ Switching from "automatic" to "manual" by means of a slide switch; mode is indicated on the LCD,
▷ Programmed automatic times can be displayed and checked on the LCD by briefly pressing a key (∇, Δ),
▷ Sensitivity of the sunlight sensor can be adjusted by means of a potentiometer,
▷ Sensitivity of the twilight sensor can be adjusted in a similar manner,
▷ Angle of blind louvers can be infinitely adjusted via jogging mode,
▷ Programmed automatic times can be varied by means of a random-check generator,
▷ Summer/winter time changeover.

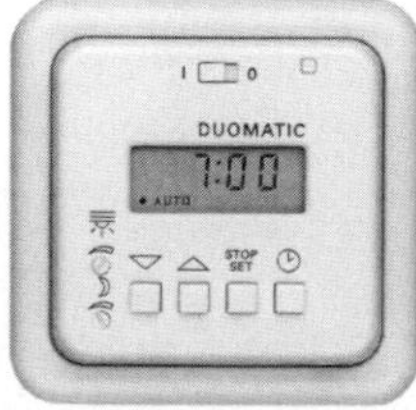

Fig. 4.3/23
DUOMATIC blind control switch Version 1, DELTA profil range

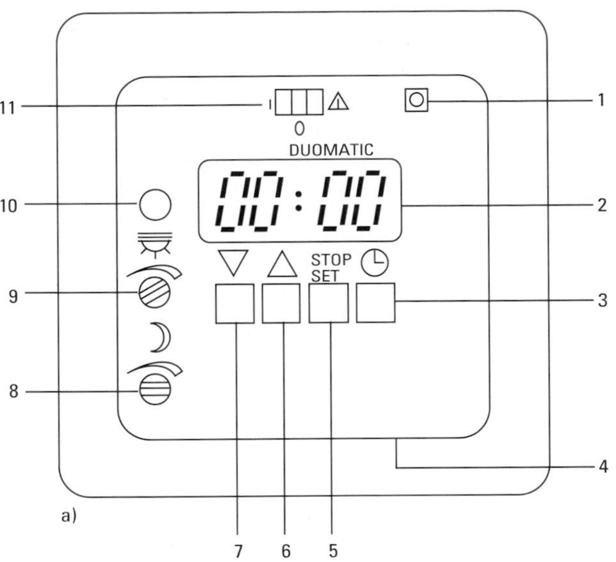

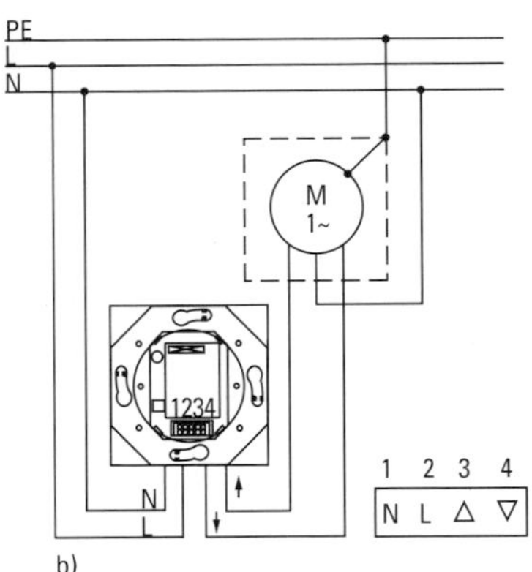

1 RESET key
2 LCD
3 Pushbutton for setting time
4 Socket for sunlight sensor
5 STOP/SET pushbutton
6 RAISE pushbutton
7 LOWER pushbutton
8 Adjusting screw for twilight sensivity
9 Adjusting screw for sunlight sensivity
10 Twilight sensor for indoors
11 Switch for setting:
  0 Manual mode
  I Automatic mode
  ⚠ Automatic mode + random-check generator (vacation)

Fig. 4.3/24
a) Operating, display and check functions of the DUOMATIC blind switch (version 3)
b) Connection diagram

Sunlight sensor

The sunlight sensor (Fig. 4.3/25) can be positioned anywhere on the windowpane by means of a suction-type attachment. A built-in phototransistor measures the intensity of the sunlight.

After ten minutes of intensive sunlight, the roller shutter/louver blind is lowered to the level of the sunlight sensor. If, however, the sunlight sensor is in the shade for more than two seconds during this ten-minute period, the scanning period (10 minutes) begins again when the sun reappears.

Lowering of the roller shutter/louver blind is stopped if the sunlight sensor is in the shade for approximately 0.75 s. If there is approximately 40 minutes of uninterrupted shade after lowering, the roller shutter/louver blind is raised. If, however, the sun shines again for only 1 s, the scanning period (40 minutes) begins again when the sun disappears. Raising of the roller shutter/louver blind is stopped if the sun shines for longer than 1 s. The roller shutter/louver blind is stopped by a limit switch at the "RAISE" or "LOWER" limit positions.

If the intensity of the sunlight exceeds the programmed sensitivity, a symbol (asterisk) on the LCD indicates that the sunlight program is active.

Twilight sensor

The twilight sensor works in conjunction with the "LOWER" command (programmed automatic timing). The roller shutter/louver blind is lowered when the programmed time for lowering is reached and the twilight sensor has registered the twilight for approximately 15 s. The sensitivity of the twilight sensor can be adjusted by means of a potentiometer.

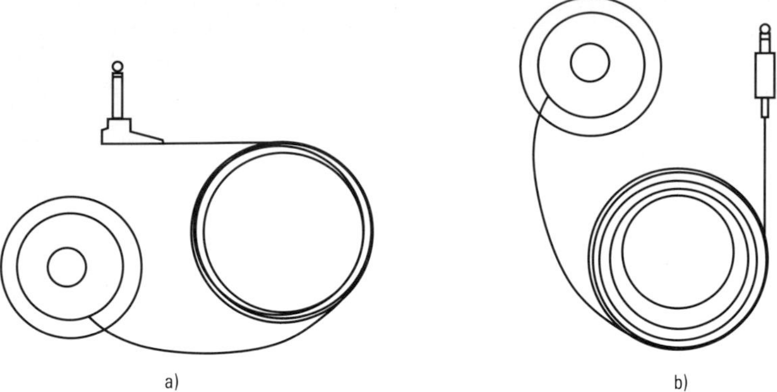

Fig. 4.3/25 a) Sunlight sensor for versions 1 and 2, with 1.5 m long two-core cable and radial plug
b) Sunlight/twilight sensor for version 3, with 3 m, 5 m, or 10 m long three-core cable and radial plug

Example

Programmed time for lowering:	7.30 p.m.,
Programmed twilight level:	8.30 p.m.,
Automatic lowering:	8.30 p.m. + 15 seconds.

The jogging mode is used to infinitely adjust the angle of blind louvers, Adjustment is carried out by briefly pressing the UP or DOWN key.

Random-check generator

The random-check generator varies the automatic opening and closing times. These switching commands are arbitrarily (randomly) delayed within a range of 0 to 15 minutes.

Conversion to mechanical system

Electronic blind controllers can be converted back to the existing mechanical system without any difficulty if a neutral conductor is available. A connection example is shown in Fig. 4.3/24.

Controlling several roller shutters/ louver blinds

To control several roller shutters/louver blinds using one DUOMATIC blind pushbutton or electronic blind control switch, it is necessary to assign a DUO plus control relay to each roller shutter/louver blind motor (Fig. 4.3/26). The control relay is available

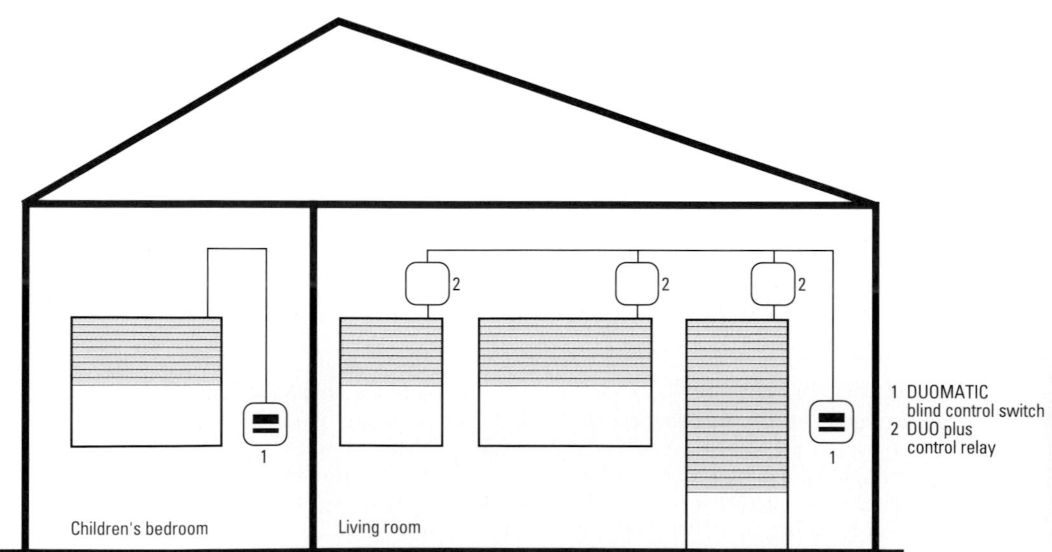

Fig. 4.3/26 DUO plus electronic control relay, 250 V AC, 50 Hz, 5 A Example of application

for surface mounting and for installation in device boxes for flush mounting (diameter 60 mm). It is rated at 250 V AC, 50 Hz and has a switching capacity of 5 A.

Using DUOMATIC blind pushbuttons or electronic blind control switches, it is possible, for example, to control several roller shutters/louver blinds either from a central location or individually. Electronic isolation and interlocking are ensured by the DUO plus control relay where roller shutters/louver blinds are controlled together. Central control has priority over individual control. Fig. 4.3/27 contains a number of typical circuit diagrams.

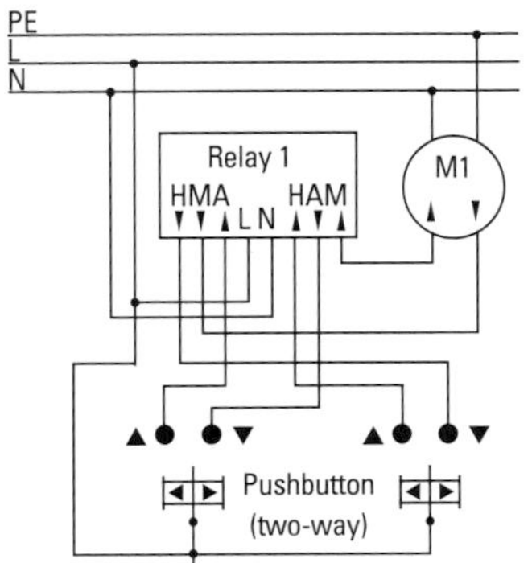

Control of one drive,
two-way of two blind pushbuttons.

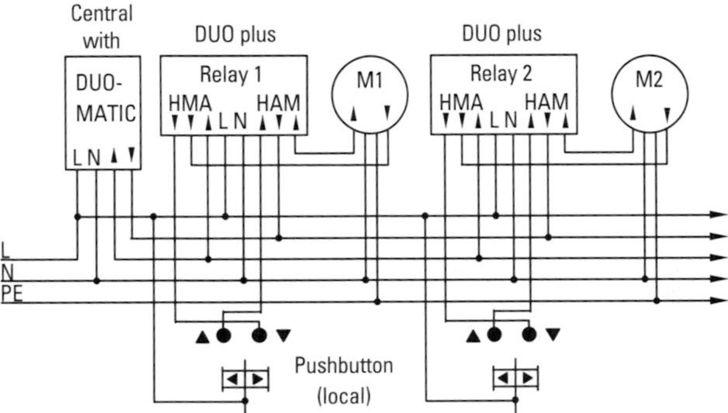

Control of several drives from a central location
DUOMATIC with individual local controls.

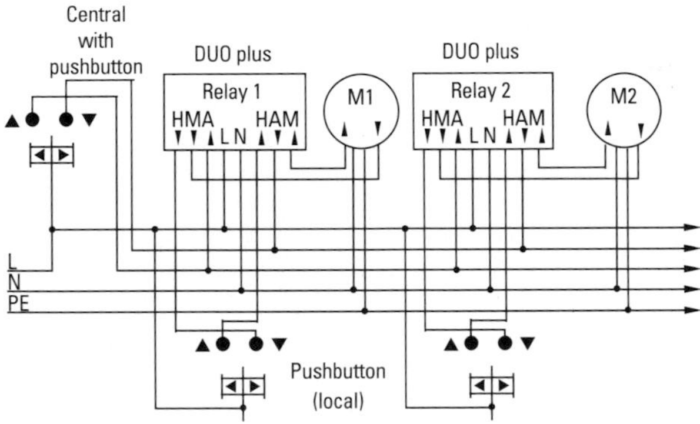

Control of several drives from a central location
Central blind pushbutton with individual local controls.

Fig. 4.3/27
Typical circuit diagrams

4.3.1.2 Connecting components for data and speech networks

General

The growing use of computers at the work place has led to an increasing demand for computer literacy among employees, as well as for reliable data transfer systems.

Each VDU requires not only a socket to connect it to the electrical distribution network, but also interfaces and connection elements for the high-quality cabling that connects it to the other computers in a network and to internal or external data processing centers.

Planning

Networks must be planned with great care and in close consultation with the user, since decisions regarding the performance and potential expansions of the system are of the utmost importance.

Every user needs an installation tailored to his specific requirements. Depending on these requirements, it may be appropriate to install either a central computer connected to numerous terminals, or a network of intelligent terminals that can also function independently of the central computer. Many users are also expressing a need for networked personal computers (PCs). In this type of system, network administration and access is usually controlled by one PC.

It is essential, therefore, not only to determine the user's specific requirements, but also the required network structure, taking into account the volume of data to be transferred. Computers and auxiliary equipment with various levels of performance and from different manufacturers can be interconnected without any difficulty within an office, a building or group of buildings by means of a data network.

Data networks

The most commonly encountered types of data network are:

▷ Ring networks,
▷ Star networks,
▷ Bus networks.

Ring networks

As the name suggests, in ring networks (Fig. 4.3/28a), users (network nodes) are connected in a ring. Data is transferred from network node to network node in one direction. The system is expanded by simply opening the ring and adding another user.

Star networks

In star networks (Fig. 4.3/28b), all the users are connected to a central computer, which also controls the system. This computer relays data to all the users for whom it is intended. This network structure enables operation in multi-channel mode, whereby a large number of connections can be made simultaneously. Additional users can simply be connected to the central computer, as required.

Bus networks

In bus networks (Fig. 4.3/28c), all the users are connected directly to a bus line. The data transmitted from one network node is initially received by all the others simultaneously. The data, however, is ignored by these - with the exception of the intended destination, of course. The network is expanded by tapping directly into the bus, enabling new users to be connected.

Connectors, connection units and sockets

System-specific connectors, connection units, and sockets for flush mounting are available for connecting and distributing the cables, and for connecting the terminals.

Flush-mounted devices

These components for data processing and data communication, as well as for telecommunications and electro-acoustic systems, are available in different versions in the DELTA ambiente, DELTA studio, DELTA profil, DELTA fläche and DELTA plus ranges (see Section 4.3.1). The components are designed for installation in device boxes (diameter 60 mm) and sill-type trunking (see Section 5.1.4.6) with a choice of vertical or angled outlets (Fig. 4.3/29).

Figs. 4.3/29 to 4.3/34 show some of the connecting components available.

Examples

In the following, a number of components for data processing and telecommunications will be used as examples of the extensive range available. Covers and frames from the DELTA ambiente, DELTA studio, DELTA profil, DELTA fläche and DELTA plus ranges can be used with all of the different interfaces (see Section 4.3.1).

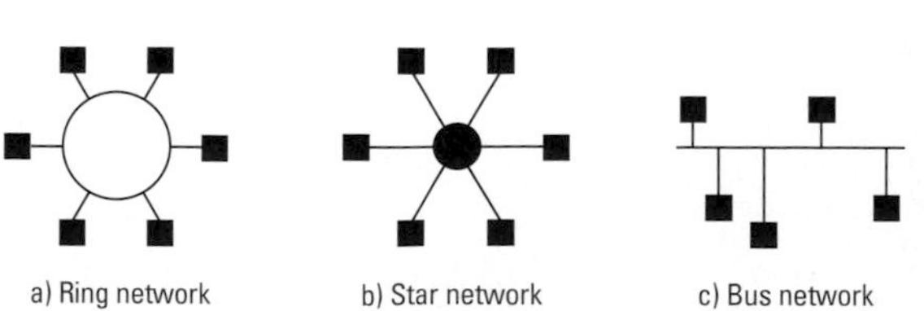

Fig. 4.3/28 Data networks

a) Vertical interface in the DELTA plus range

b) Angled interface in the DELTA fläche range

Fig. 4.3/29
D subminiature interface (for flush mounting)

Fig. 4.3/30 D subminiature connector

a) BNC interface with two coaxial cable sockets in the DELTA plus range

b) BNC connector for coaxial cable

Fig. 4.3/31
BNC plug-and-socket connector with bayonet lock

a) WE interface in the DELTA plus range

b) WE connector

Fig. 4.3/32
8-pole western plug-and-socket connector for terminal connection

D subminiature plug-and-socket connectors

D subminiature plug-and-socket connectors are used primarily to connect pluggable computer terminals and data transfer devices, as well as measuring and control equipment. The interface has a 9, 15 or 25-pole trapezoidal connector pin or a socket connector for connecting one device, or two connectors if two devices are to be connected. Fig. 4.3/29 a shows a flush-mounted interface with a vertical socket connector (in the DELTA plus range), and Fig. 4.3/29 b an angled socket connector in the DELTA fläche range. Fig. 4.3/30 shows the corresponding D subminiature connector.

BNC/TNC plug-and-socket connectors

BNC[1)]/TNC[2)] plug-and-socket connectors are used in the high-frequency range in coaxial cable networks. The standard version can be used for frequencies of up to 4 GHz, and is available for coaxial cable impedances of 50 Ω and 70 Ω.

With BNC plug-and-socket connectors, the connector is plugged into the socket of the interface by means of a bayonet lock (Fig. 4.3/31); with vibration-resistant TNC plug-and-socket connectors, this connection is established by means of a screw lock.

Western (WE) plug-and-socket connectors

Western (WE) plug-and-socket connectors are used in data processing (e.g. as twisted pair terminals[3)]) and in telecommunications (e.g. as ISDN[4)] basic access points).

The WE interfaces are available with one or two 6 or 8-pole WE sockets, which, when partially equipped with contacts, provide 4, 6 or 8-pole socket variants (Fig. 4.3/32).

Twinax plug-and-socket connectors

Twinax plug-and-socket connectors (Fig. 4.3/33) are used first and foremost in local data networks (LANs) with IBM terminals. The two primary conductors of the twinaxial cable are soldered to the built-in twinax

1) BNC **B**ayonet **N**orm **C**onnector
2) TNC **T**hreaded **N**orm **C**onnector
3) Two wires twisted together
4) ISDN **I**ntegrated **S**ervices **D**igital **N**etwork

socket of the interface and on the connector, and the metal braiding is clamped. When the connector has been plugged into the interface, it is secured by means of a union nut.

TLU line units

TLU telecommunication line units (Fig. 4.3/34) are used exclusively for connecting telecommunications devices, such as telephones, fax machines, and teleprinters. The adapter (connector) is available with different lengths of connection cable, or without a cable. The interfaces, which have one to three sockets, and the adapters are prepared for telephone connection (F coding), or for connection to supplementary devices (fax machines, etc.) and data terminal devices (video text) (N coding).

Optical fiber interfaces

Optical fiber interfaces in the DELTA profil range are tailored to meet the requirements of a modern cabling system with glass-fiber cables.

These systems ensure fast and reliable data transmission for the communication needs of the future. The many applications for which these systems can be used include:

▷ Telephone (speech, fax),
▷ Data transfer (client/server, computing, mail, Internet, virtual LAN),
▷ Multimedia (integration of speech, data, and video).

a) Interface in the DELTA plus range

b) Connector

Fig. 4.3/33 Twinax plug-and-socket connector

a) TLU interface in the DELTA natur range

b) TLU connector

Fig. 4.3/34 Telecommunication Line Units (TLUs)

a) Optical fiber interface

b) Optical fiber connector

Fig. 4.3/35 Optical fiber connection unit

4.3.2 Information and monitoring systems

4.3.2.1 Infrared motion detectors

Application

Infrared motion detectors are automatic switches that can be used to save energy, and increase safety and convenience (but are unsuitable for use in alarm systems). They are primarily used to control lighting systems in residential and functional buildings in accordance with demand.

Functional principles

Motion detectors are electronic ON/OFF switches that react to the movement of persons or other heat sources. Optical lenses sense the invisible infrared thermal radiation, and the motion detector then switches on the current-using device or devices (e.g. luminaires) connected to it. If no new movement is detected, the current-using devices are switched off again after a preset on time. The preset on time starts anew with every movement detected.

Brightness sensor

An integrated brightness sensor triggers the switching operation only when the brightness falls below a set level.

Time switches for maintained lighting

The "brightness" and "on time" functions are infinitely adjustable. A number of versions are also equipped with manually adjustable timers for maintained lighting.

Infrared motion detectors are available for flush mounting (Fig. 4.3/36 a) and surface mounting (Fig. 4.3/36 b).

Flush-mounted version

Motion detectors for flush mounting are designed to match the DELTA plus and DELTA profil ranges (see Section 4.3.1), and can be installed in device boxes to DIN 49073 (diameter 60 mm).

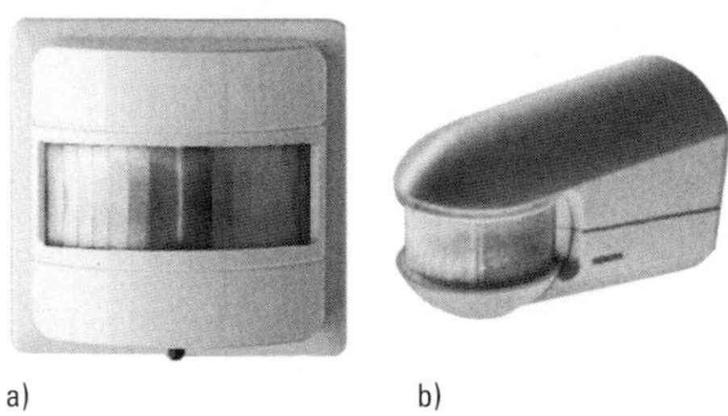

Fig. 4.3/36
a) DELTA profil 180i flush-mounted motion detector, titanium white, with degree of protection IP 20
b) DELTA matic surface-mounted motion detector with degree of protection IP 44

Installation height

The installation height should be approximately 0.9 m to 1.30 m.

Angle of detection

The angle of detection of the 180i motion detector for flush mounting is 180°. This can be reduced by means of two adjustable aperture wings.

Fig 4.3/37 show the angle of detection of the 180i motion detector for flush mounting.

New installations and rewiring

Since the 180i motion detectors for flush mounting have to be suitable both for installation in new buildings and in existing electrical installations, device versions are available for the circuits illustrated in Fig. 4.3/38.

Permissible lamp load

The permissible lamp load (switching capacity) can be seen in Table 4.3/1. Attention must be paid to the different values of the 2-wire/3-wire methods of connection.

Surface-mounted version

Where motion detectors are used in functional buildings and outdoors, the DELTA matic version for surface mounting is considered more suitable (see Fig. 4.3/36).

Degree of protection
Installation height

The degree of protection of this version corresponds to IP 44 in accordance with EN 60529/IEC 60529/DIN VDE 0470-1. The preferred installation height is 2.5 m.

Angles of detection

Depending on the application, the DELTA matic motion detectors for surface mounting can be used for the following angles of detection:

▷ 130° (see Fig. 4.3/39),
▷ 230° (see Fig. 4.3/40),
▷ 270° (see Fig. 4.3/41).

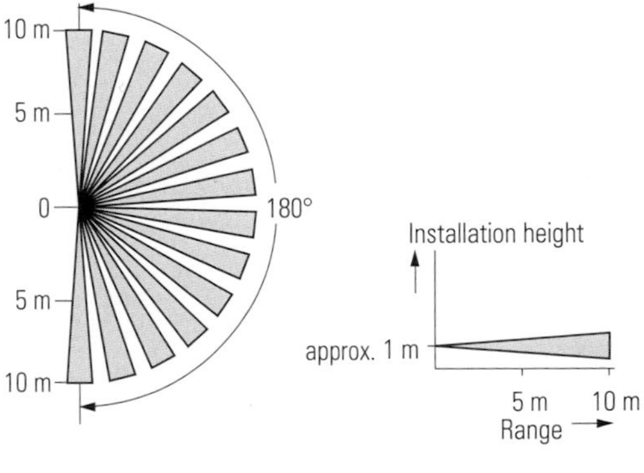

Fig. 4.3/37
Angle of detection of the 180i motion detector for flush mounting

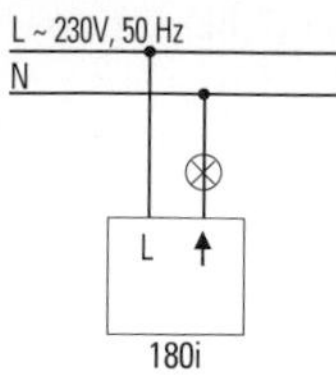

Motion detector connected to mains
N-conductor not required
2-wire connection

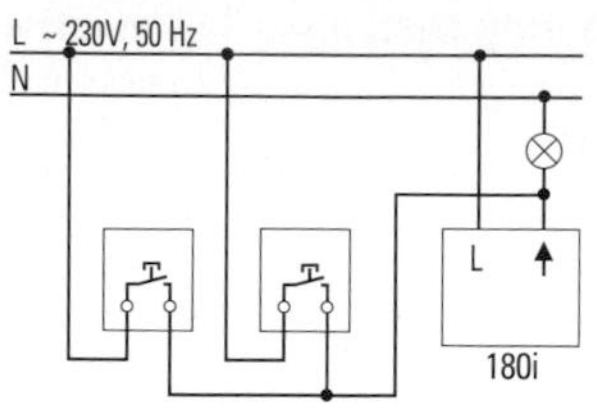

Motion detectors with pushbuttons
When the pushbutton is pressed, the motion detector is switched on for the preset time.
N-conductor not required
2-wire connection

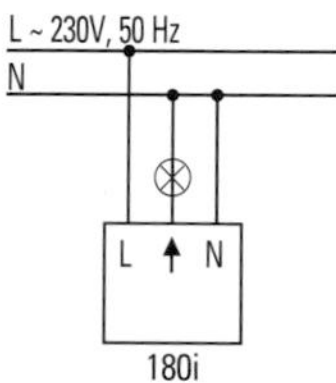

Motion detector connected to mains
N-conductor required
3-wire connection

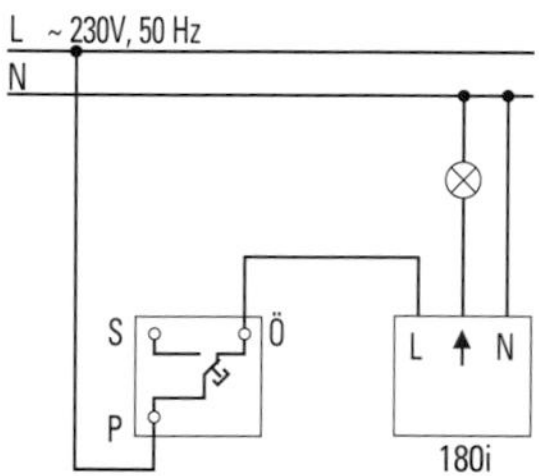

Motion detector with pushbutton, 1-pole NC contast.
When the pushbutton is pressed, the motion detector is switched on for the preset time.
N-conductor required
3-wire connection

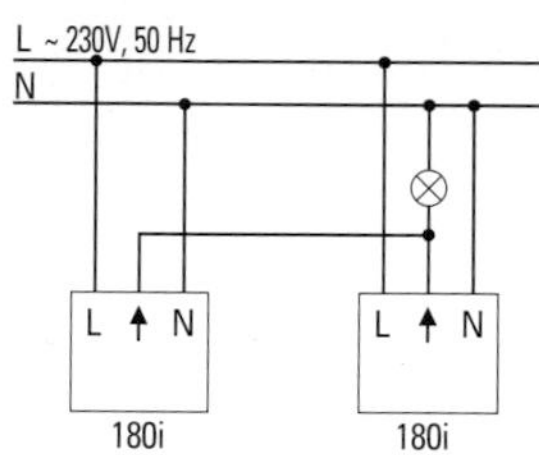

Two motion detectors in parallel
N-conductor required
3-wire connection

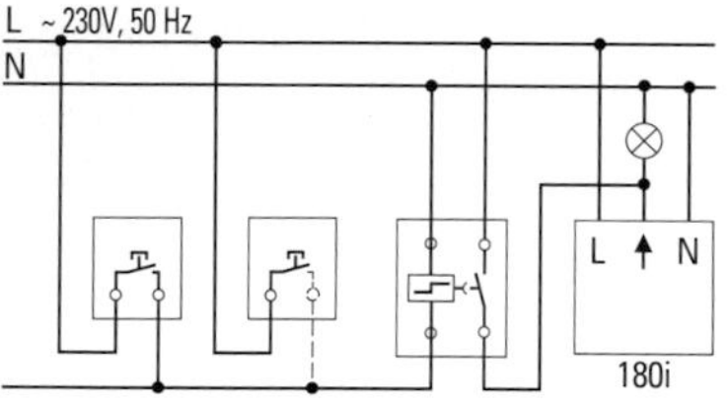

Parallel operation stairwell lighting timer with motion detector
N-conductor required
3-wire connection

Fig. 4.3/38 Typical circuits for the 180i flush-mounted motion detector

Table 4.3/1 Technical data for the 180i motion detector for flush mounting

Rated voltage	230 V AC, 50 Hz
Rated current	10 A; cos $\varphi = 0.5$ for devices with 3-wire connection
Switching capacity	2-wire connection: incandescent lamps 40 W to 400 W 3-wire connection: incandescent lamps 2300 W fluorescent lamps (p.f. uncorr.) 900 W parallel p.f. corr. 320 W Twin-lamp circuit with max. two ECGs or energy-saving lamps 1800 W Halogen lamps for 230 V AC 500 W *Important:* With ECGs and energy-saving lamps, note the high making currents
Time interval	Approx. 5 s to 5 mins (infinitely adjustable), retriggerable
Light sensor	Adjustable from 5 lux to 1000 lux
Angle of detection	180° (can be reduced by means of aperture wings)
Range	Approx. 10 m
Installation height	0.9 m to 1.3 m
Optics	Fresnel lens (1 level with 14 segments)
Parallel connection	3 to 4 devices with 3-wire connection
Contacts	1 make contact element, non-floating with 3-wire connection
Radio interference level	To DIN VDE 0875 interference level N
Permissible ambient temperature	0 °C to 35 °C (the switching capacity must be reduced for ambient temperatures ≥ 35 °C)

The 230° angle of detection can be divided into two separate zones, and the 270° angle of detection into three separate zones (Fig. 4.3/42). It is also possible to adjust the angle of detection as well as the range by tilting the spherical lens unit. Tilting the spherical lens unit (e.g. so that it points downwards) enables the range to be reduced (Fig. 4.3/43).

Masking angle of detection

If required, clip-on slats (half size) can be used to mask part of the sensing optics, thereby suppressing potential sources of interference (such as trees or bushes moved by the wind) that could cause the motion detector to switch inadvertently (Fig. 4.3/44). Using clip-on slats (full size), the angle of detection can be reduced or masked according to individual requirements.

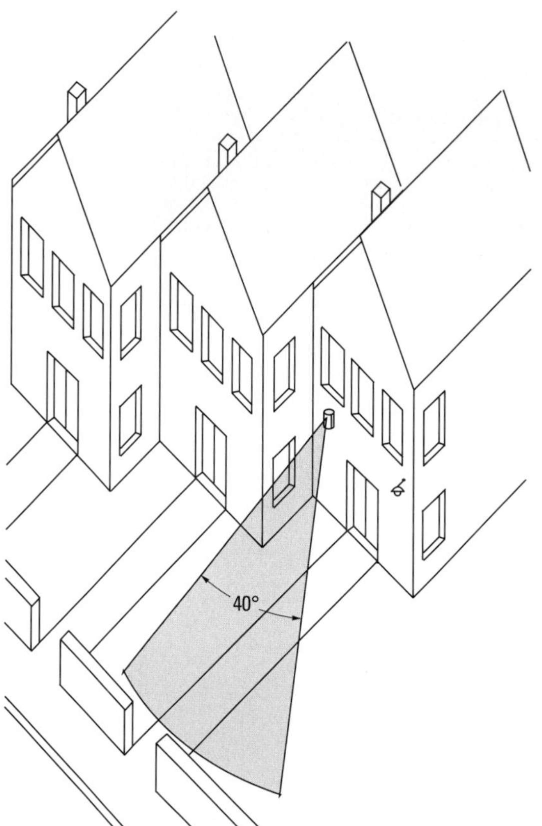

Fig. 4.3/39 Angle of detection 130° (reduced to 40°)

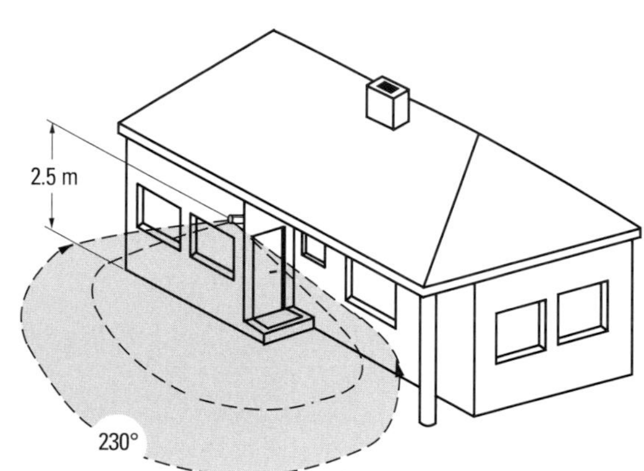

Fig. 4.3/40 Angle of detection 230°

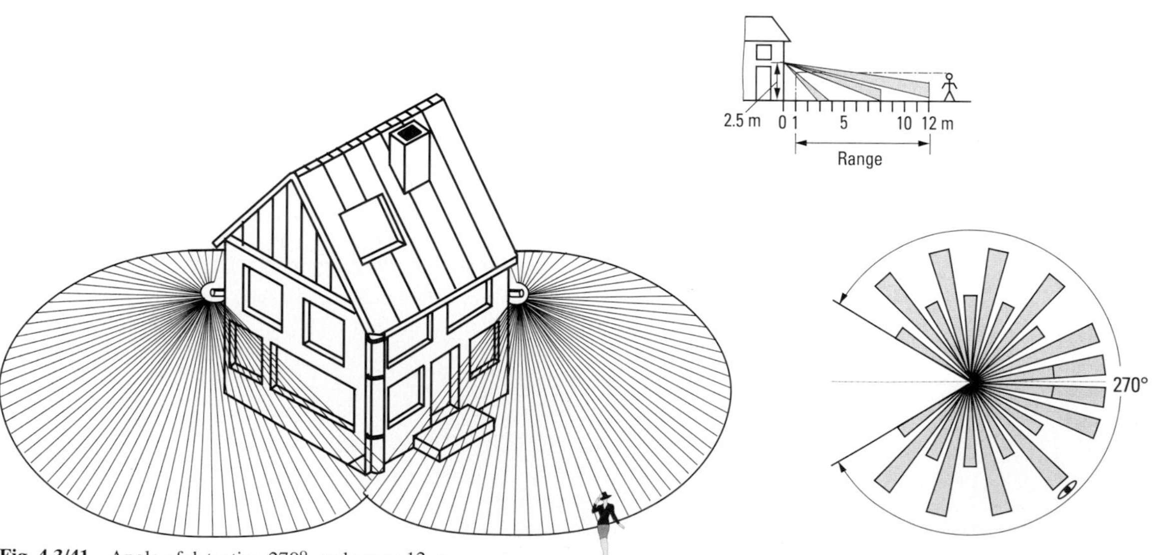

Fig. 4.3/41 Angle of detection 270°, and range 12 m

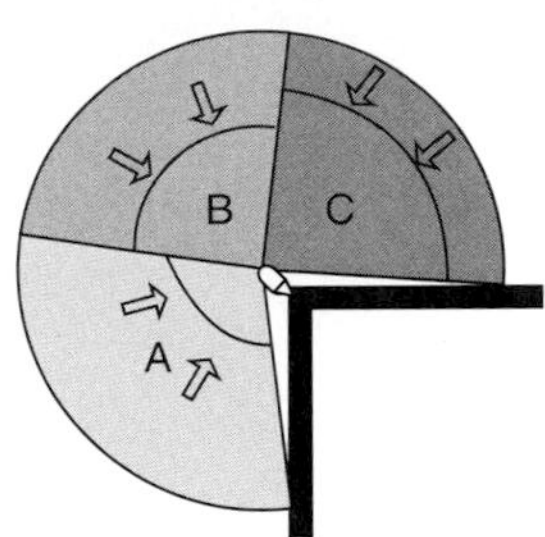

Fig. 4.3/42
Division of the 270° angle of detection into three separate zones

Permissible lamp loads

Table 4.3/2 contains the permissible lamp loads for the DELTA matic surface-mounted motion detector.

Fig. 4.3/45 illustrates typical circuits in which the DELTA matic surface-mounted motion detector can be used. A neutral conductor is required in all cases.

The built-in all-or-nothing relay of the DELTA matic surface-mounted motion detector is dimensioned to switch all types of lamp, such as incandescent lamps, fluorescent lamps, halogen lamps, mercury-vapor lamps, etc.

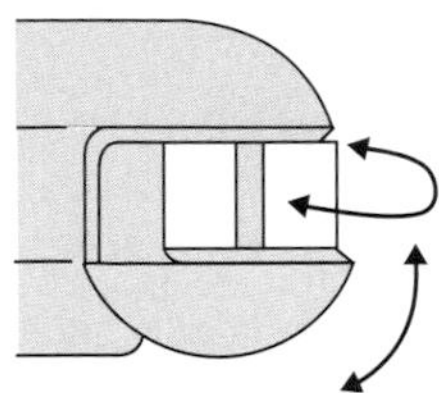

Fig. 4.3/43
Setting the angle of detection by tilting (horizontal axis); setting the range by tilting (vertical axis)

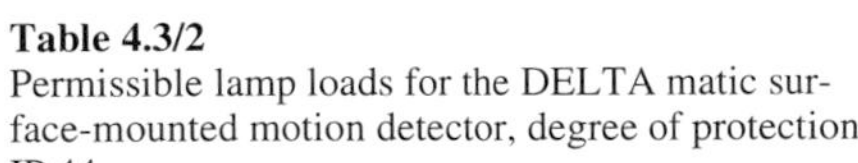

Table 4.3/2
Permissible lamp loads for the DELTA matic surface-mounted motion detector, degree of protection IP 44

Type of lamp	Maximum load W
Incandescent lamps	1600
Fluorescent lamps:	1200
p.f. uncorrected	650
parallel p.f. corrected twin-lamp circuit	2 × 1200
Halogen lamps (230 V AC):	1200
Low-voltage halogen lamps with transformer:	500
Mercury-vapor lamps:	
p.f. uncorrected	1000
parallel p.f. corrected	1000
Sodium-vapor lamps (high-pressure):	
p.f. uncorrected	1000
parallel p.f. corrected	1000
Mixed-light lamps:	2000
Dulux lamps:	
p.f. uncorrected	800
parallel p.f. corrected	560

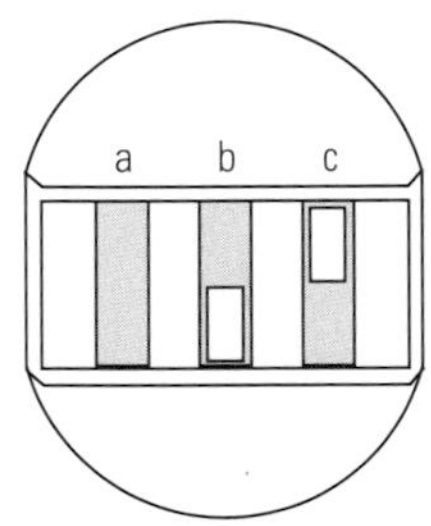

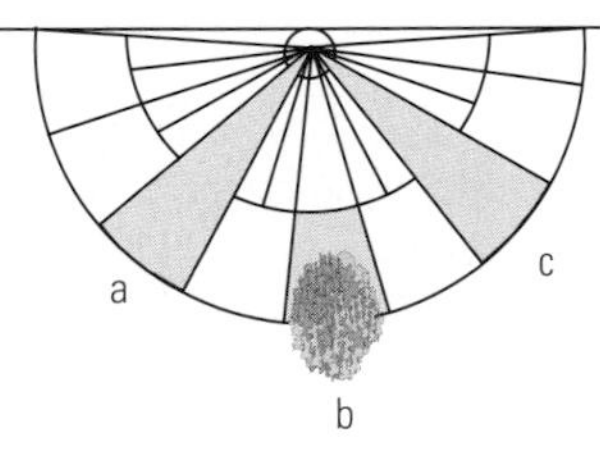

Fig. 4.3/44
Use of detachable clip-on slats to mask specific angles of detection (dark areas) of the sensing optics (230° angle of detection)

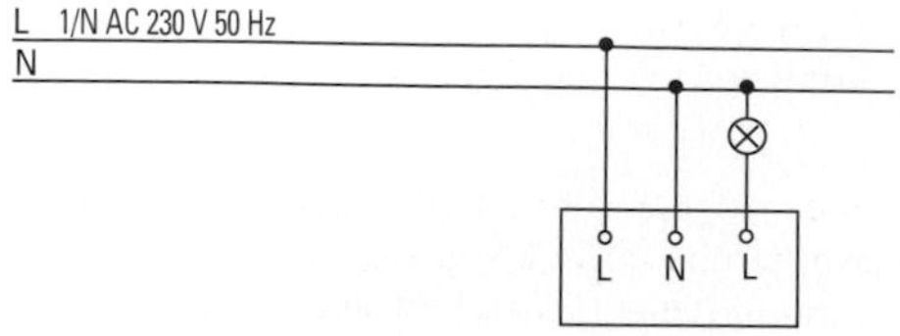

a) Individual device
Constant availability
of DELTA matic
(breaking operation)

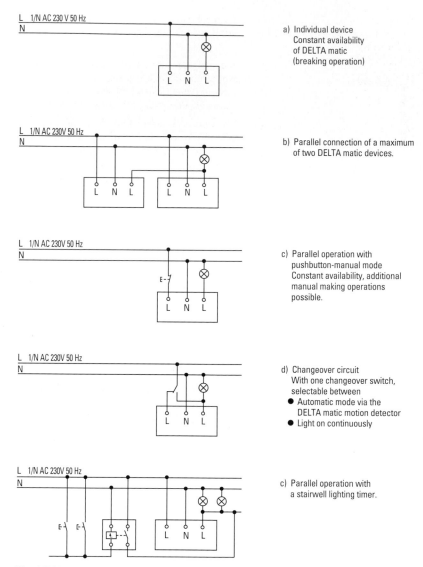

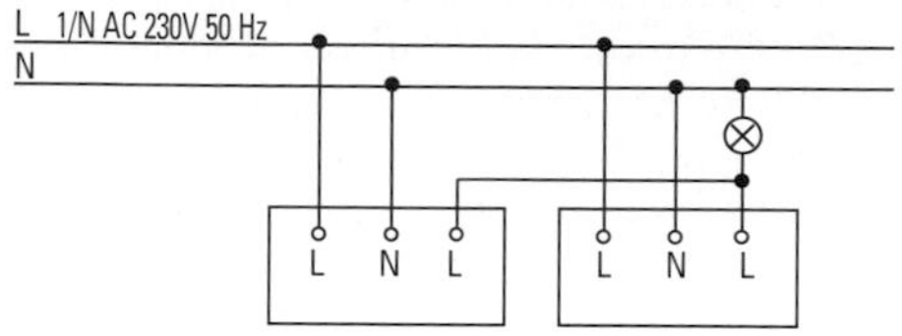

b) Parallel connection of a maximum
of two DELTA matic devices.

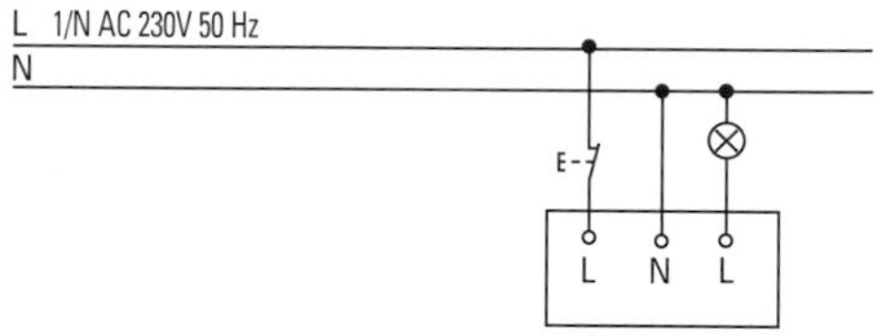

c) Parallel operation with
pushbutton-manual mode
Constant availability, additional
manual making operations
possible.

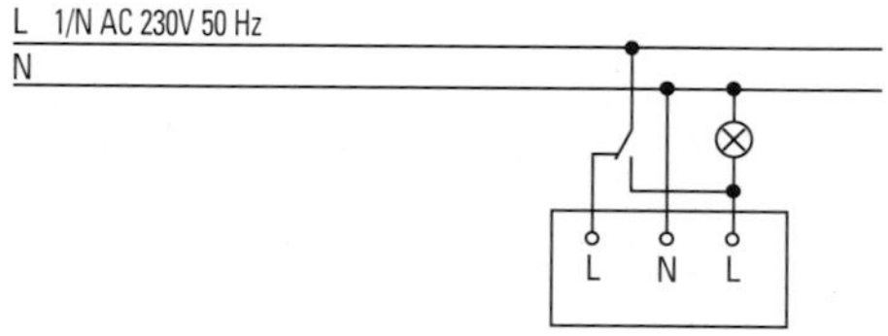

d) Changeover circuit
With one changeover switch,
selectable between
- Automatic mode via the DELTA matic motion detector
- Light on continuously

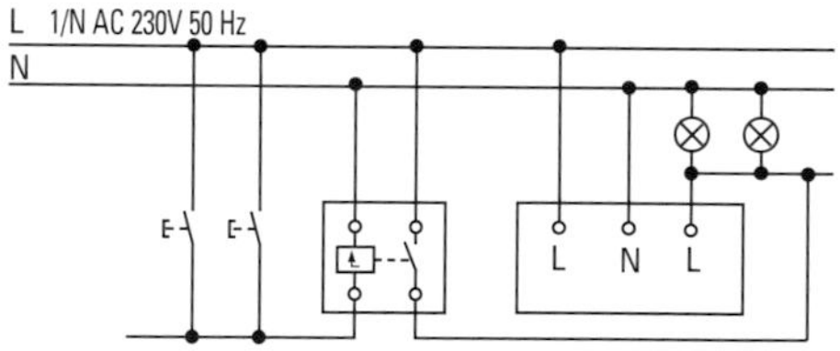

c) Parallel operation with
a stairwell lighting timer.

Fig. 4.3/45 Typical circuits for the DELTA matic surface-mounted motion detector

4.3.2.2 DELTA-FERN infrared system and infrared remote-control systems, DELTA FERN RF radio remote-control system

Infrared systems Infrared remote-control systems offer considerable advantages over other wireless remote-control systems, such as those based on ultrasound. They are not adversely affected by interference, room reflections, Doppler effects, or intermodulation noise.

To distinguish the IR signal from natural infrared light, it is encoded, and can be transmitted with a time or frequency code or in digitized form, according to application and use.

The high propagation rate of infrared light allows switching functions to be performed very quickly and reliably. The control voltages of other electrical devices do not affect the function of the infrared system.

Radio systems

In comparison with infrared remote-control systems, the advantage of remote-control systems based on radio is that they operate across a number of rooms. This allows functions to be executed remotely without wiring: an example of this would be a central OFF switch located at the exit of an apartment, a residential or functional building, which can be used to switch off the entire lighting system centrally. Louver blinds or roller shutters, and projection screens can also be controlled using a system of this kind.

Radio remote-control systems can also be linked to the *instabus EIB* control system (see Section 4.4).

4.3.2.2.1 DELTA-FERN infrared system

Use

DELTA-FERN is an infrared remote-control system for use in electrical installations. Equipped with four channels, the system is simple to operate, easy to install, highly flexible, and extremely user friendly.

Using only a wall or hand-held transmitter and the appropriate receivers as switches, dimmers, or pushbuttons, it is possible to convert conventional ON/OFF circuits (for example, in a living area, hall, or stairwell) to changeover, dimmer, or pushbutton circuits, without making changes to the existing wiring system.

This means that DELTA-FERN can be used when a residential or functional building is rewired (renovation) or when new wiring systems are installed (new building).

Components

Range of components

The DELTA-FERN infrared system comprises the following components (Fig. 4.3/46):

▷ 4-channel hand-held transmitter, all 4 channels fixed program;

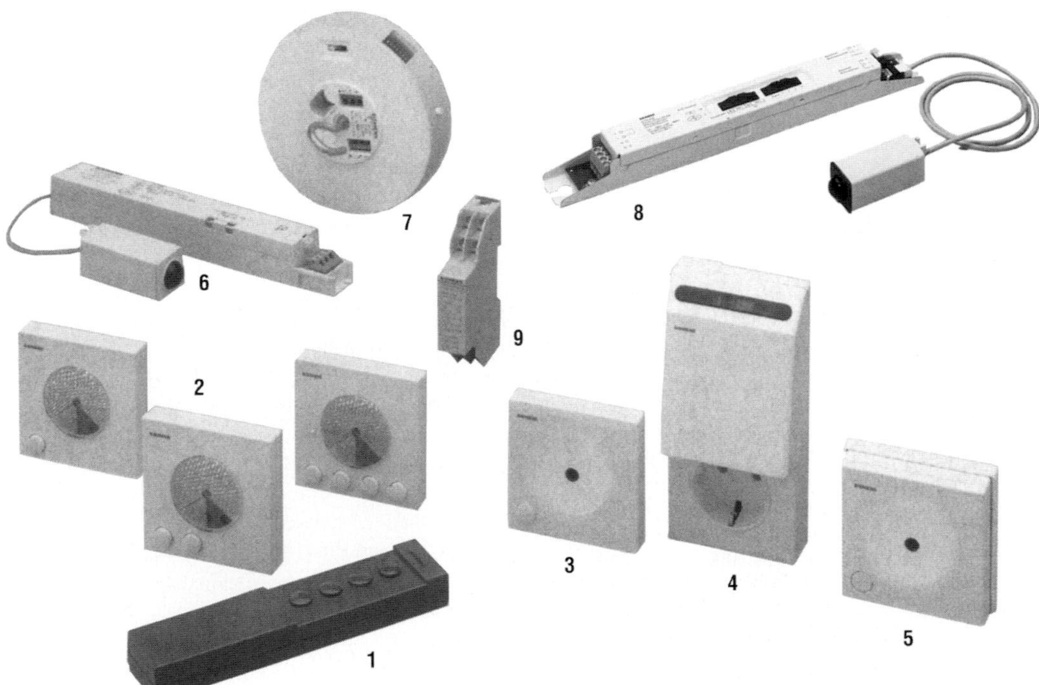

1 4-channel hand-held transmitter
2 1, 2, and 4-channel wall-mounted transmitter
3 Switchbox receiver
4 Plug-in socket receiver
5 Splitting box receiver
6 Built-in receiver
7 Ceiling-rose receiver
8 Switch/dim module
9 Link module

Fig. 4.3/46 Components of the DELTA-FERN infrared system for use in residential and functional buildings

- ▷ 4-channel wall-mounted transmitter, all 4 channels fixed program;
- ▷ 2-channel wall-mounted transmitter, switchable from channel 1 and 4 to channel 2 and 3;
- ▷ 1-channel wall-mounted transmitter, switchable to all 4 channels;
- ▷ Switchbox receiver as switch, dimmer or pushbutton, fits 60-mm diameter device box;
- ▷ Splitting box receiver as switch, dimmer or pushbutton;
- ▷ Built-in receiver as switch, "central OFF" switch, and pushbutton switch for fluorescent luminaires with conventional as well as electronic control gear;
- ▷ Switch/dim module as switch, "central OFF" switch for switching dimming fluorescent luminaires with dimmable electronic control gear;
- ▷ Ceiling-rose receiver as switch or dimmer;
- ▷ Plug-in socket receiver as switch or dimmer;
- ▷ Link module as power supply for remote-control switching of the "pushbutton" switchbox receiver or stairwell lighting timer.

Transmitters with 9 V monobloc battery

The transmitters do not require a mains connection, as the necessary power is supplied by an internal 9 V monobloc battery. The transmitters are suitable for wall-mounted, hand-held, or table-top operation. A current-using device or group of current-using devices can be controlled independently with each channel by means of the appropriate IR receiver (switch, dimmer or pushbutton).

Transmission range

The transmission range of the system is 15 m.

The wall-mounted transmitters can be bonded or screwed directly to the wall, or alternatively attached to the device box. The transmitters fit the frames of the DELTA studio and DELTA fläche switch/plug-in socket ranges.

For functional buildings, wall-mounted transmitters are available with 230 V AC, 50...60 Hz pilot lighting, in accordance with the German ordinance for working and business premises.[1]

[1] In Germany, the ordinance for working and business premises is published by the "Bundesanstalt für Arbeitsschutz" (Federal Institute for Occupational Safety).

Receivers

All IR receivers are designed for a mains voltage of 230 V AC, 50 . . . 60 Hz, and have 4 channels, which can be set by means of a slide switch. The permanently installed IR receivers can be operated manually in the usual way, or controlled remotely by means of IR transmitters.

The switchbox and splitting box receivers for installation as flush-type switches, dimmers, or pushbuttons can be used in two-wire or three-wire systems. A neutral conductor is not required for two-wire switching. This allows existing light switches and pushbuttons to be replaced with IR receivers (switches, dimmers, or pushbuttons). Where IR receivers are used, additional switching points can be retrofitted at any time.

Installation

Switchbox receivers

Switchbox receivers fit device boxes with a diameter of 60 mm, and can be used with the frames of the DELTA studio and DELTA fläche switch ranges in the DELTA studio and DELTA fläche surface-type housings.

Splitting box receivers

Splitting box receivers are installed in splitting boxes with a diameter of 70 mm. If there are too many wires, an additional splitting box must be connected.

Ceiling-rose receivers

Ceiling-rose receivers are connected between luminaires and the ceiling outlet box, and are attached to the latter by means of ceiling hooks.

Built-in receivers, switch/dim module

Built-in receivers are suitable for installation in luminaires and false ceilings, and the switch/dim module for fitting in luminaires. They are primarily used for installations in functional buildings.

Plug-in socket receivers

Plug-in socket receivers (as switches or dimmers) are installed quickly and easily. They are plugged into existing grounding-type receptacles, between the receptacles themselves and the current-using equipment.

Use in residential buildings

DELTA-FERN for rewiring

DELTA-FERN devices can be used to considerably enhance the functionality of an existing wiring system. Light switches are replaced by switchbox, splitting box, and ceiling-rose receivers. These are controlled by means of any number of wall-mounted transmitters, which can be installed in any desired configuration.

Typical circuit diagrams

Figs. 4.3/47 to 4.3/55 show typical circuit diagrams for rewiring residential buildings using switchbox receivers, splitting box receivers, and ceiling-rose receivers. No additional wiring is required.

Link module

A link module must be used to switch (4-wire) stairwell lighting timers (Fig. 4.3/50), if more than one switchbox receiver is used, and the stairwell is equipped with more than ten mechanical pushbuttons with "bright" (1.35 mA) neon lamps.

This also applies to installation remote-control systems (see Section 4.2). In this case, the remote switch must be connected by means of a link module. This is not necessary, however, if only one switchbox receiver and two mechanical pushbuttons with "bright" (1.35 mA) neon lamps (as in Fig. 4.3/51) are used.

Rewiring with switchbox receivers

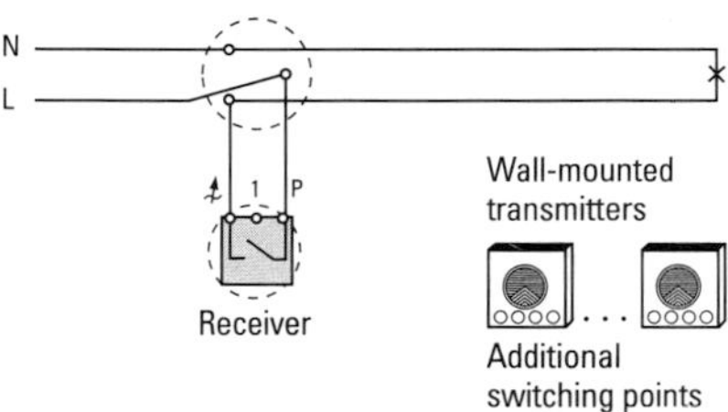

Fig. 4.3/47 ON/OFF circuit

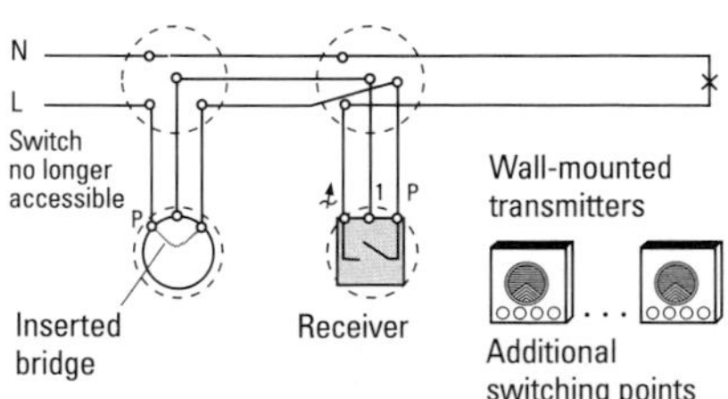

Fig. 4.3/48 Changeover circuit

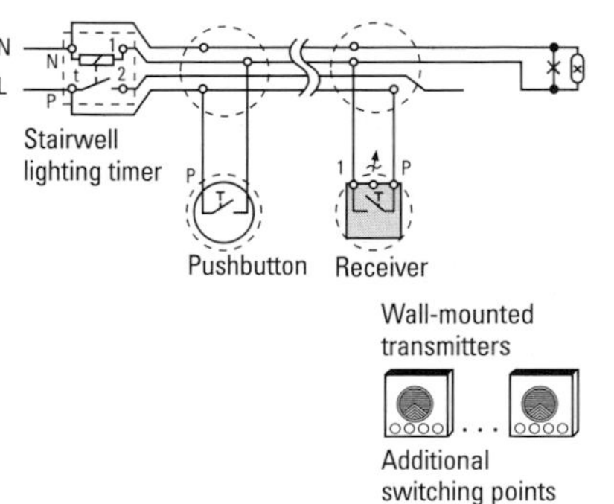

Fig. 4.3/49 Stairwell lighting timer, (4-wire)

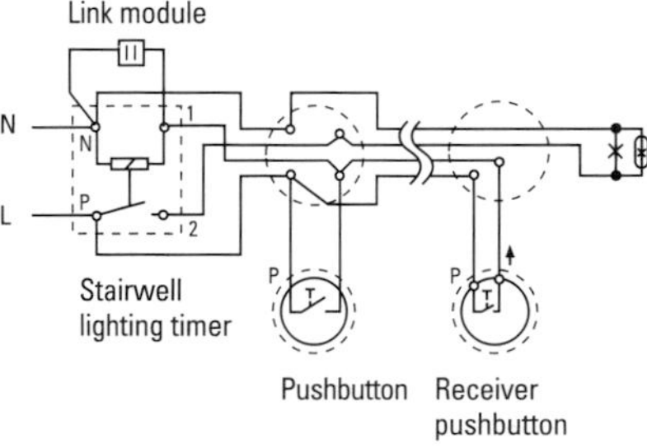

Fig. 4.3/50
Stairwell lighting timer with link module

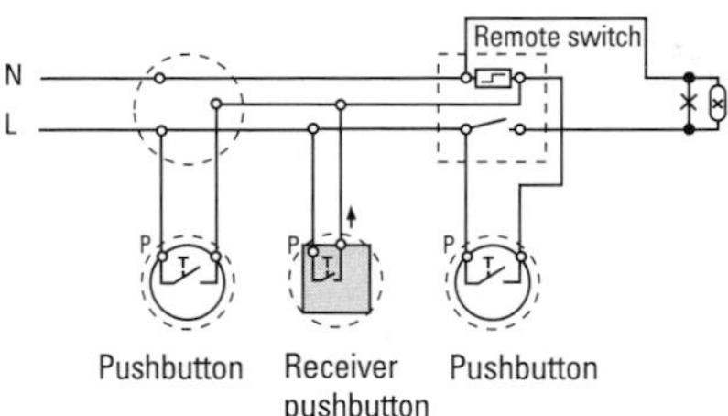

Fig. 4.3/51
Installation remote-control system without link module

Rewiring with splitting box receiver

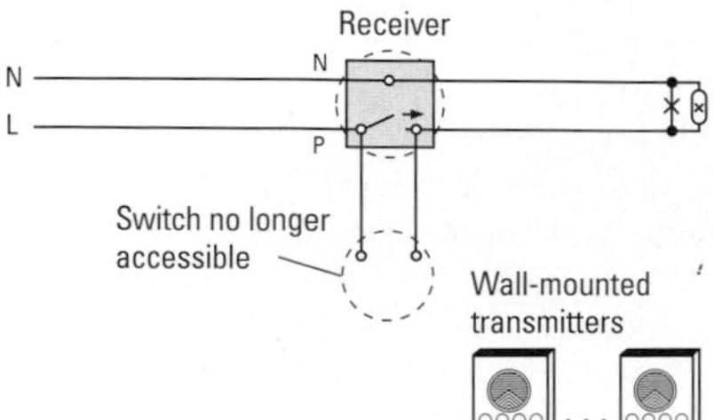

Fig. 4.3/52 ON/OFF circuit

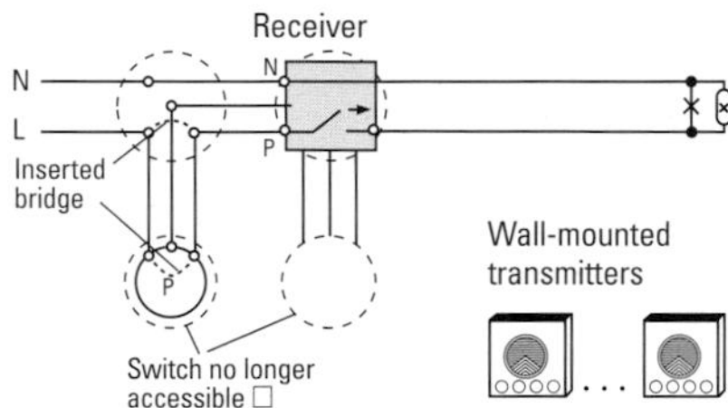

Fig. 4.3/53 Changeover circuit

Rewiring with ceiling-rose receiver

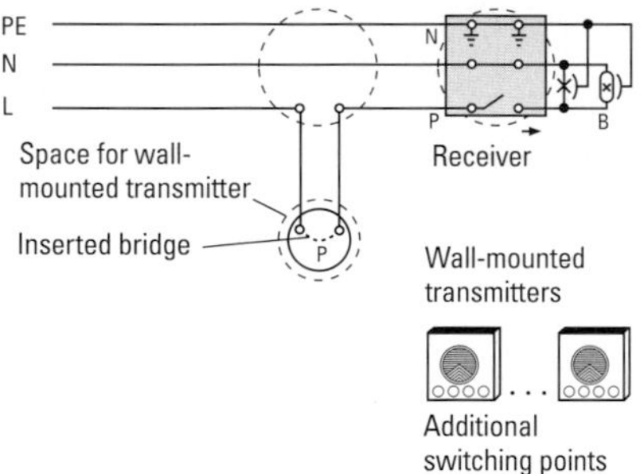

4.3/54 ON/OFF circuit

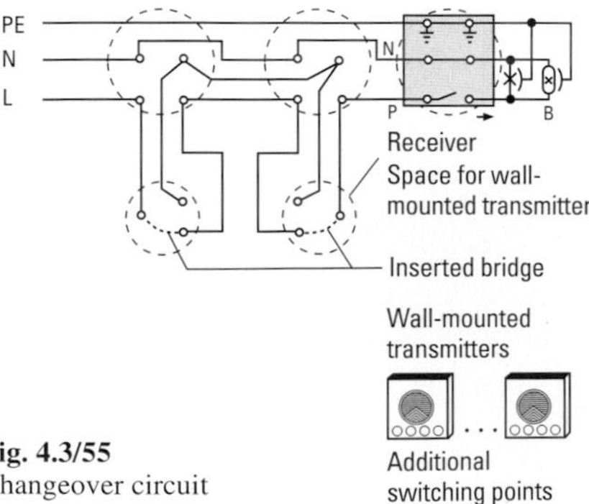

Fig. 4.3/55
Changeover circuit

DELTA-FERN and renovations

When people move into a rented or private apartment, they often find that there is only a very small number of switching points. During renovation, however, DELTA-FERN can be used to upgrade these minimal facilities to a much more convenient level. Figs. 4.3/56 and 4.3/57 illustrate how DELTA-FERN components can be used to significantly increase the level of convenience in an apartment, without new wiring and device boxes having to be installed. Fitting DELTA-FERN components is straightforward, requires little time and effort, and does not dirty the apartment in any way.

DELTA-FERN in new buildings

The DELTA-FERN infrared system offers particular advantages when installed in a new building, since time-consuming preplanning of switching points is no longer necessary.

Only the luminaires and the connection points for the other current-using devices are fitted in the cable run of the electrical installation. When building work is complete and the apartment has been fitted out and furnished, the wall and hand-held transmitters as well as the plug-in socket receivers can be positioned according to the tenants' or owners' individual requirements (see Fig. 4.3/58).

This type of electrical installation has proven to be particularly expedient in residential buildings with fair-faced masonry, exposed concrete, or large glass walls.

DELTA-FERN for controlling louver blinds/roller shutters

Louver blinds and roller shutters for windows can also be controlled using DELTA-FERN.

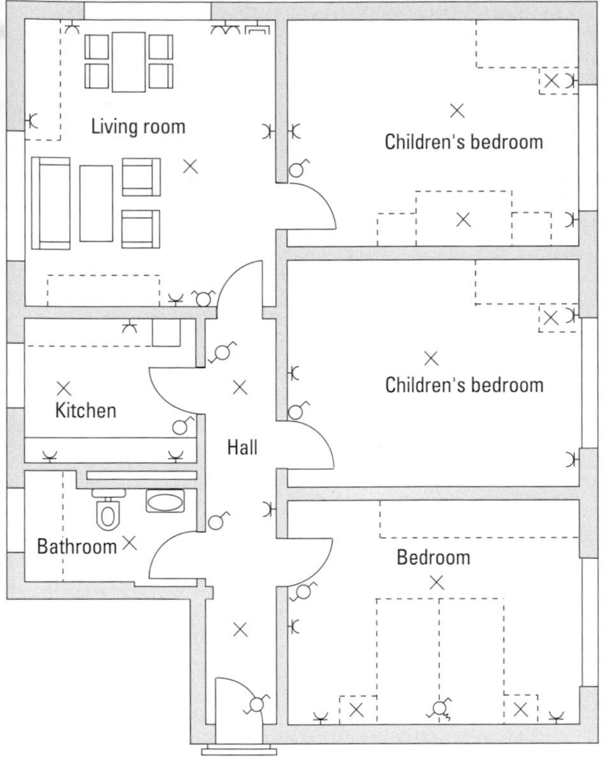

Fig. 4.3/56
Conventional wiring system in an apartment with one luminaire and (generally) only one switching point in each room

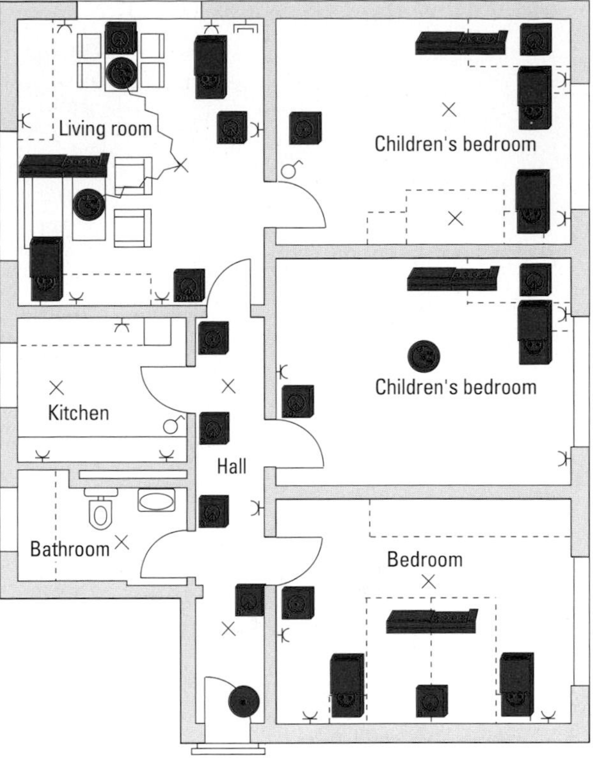

DELTA-FERN components

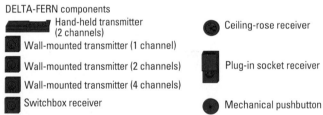

- Hand-held transmitter (2 channels)
- Wall-mounted transmitter (1 channel)
- Wall-mounted transmitter (2 channels)
- Wall-mounted transmitter (4 channels)
- Switchbox receiver
- Ceiling-rose receiver
- Plug-in socket receiver
- Mechanical pushbutton

Fig. 4.3/57
Existing electrical installation upgraded using DELTA-FERN components, allowing each luminaire in the rooms to be switched from several points

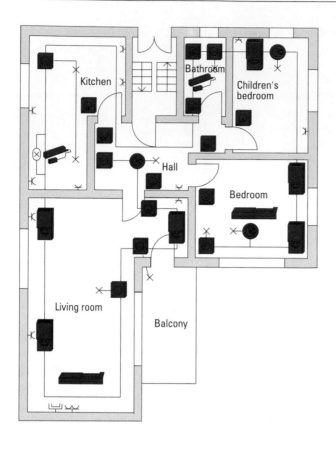

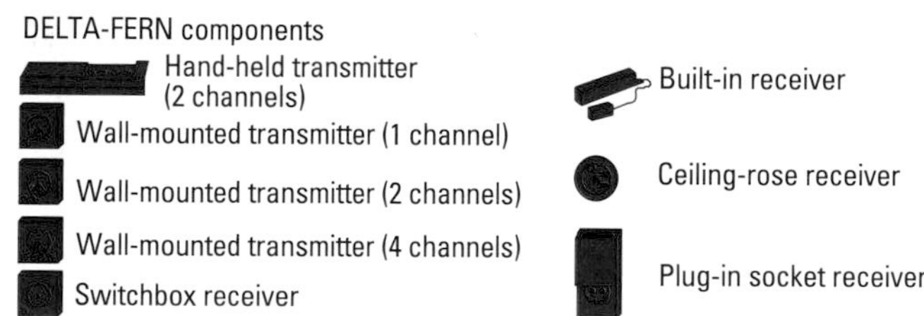

Fig. 4.3/58 Electrical installation fitted with the DELTA-FERN infrared system in a new apartment

Use in functional buildings

Use

In conjunction with preassembled, pluggable and programmable wiring systems, the DELTA-FERN infrared installation system is used mainly in buildings with complex lighting and louver blind systems.

Advantages

Using DELTA-FERN in this kind of environment has advantages for energy management, and offers flexibility with regard to subsequent changes in the use of a room or its structure. A distributed lighting power supply system (Fig. 4.3/59) also enables fire loads to be considerably reduced.

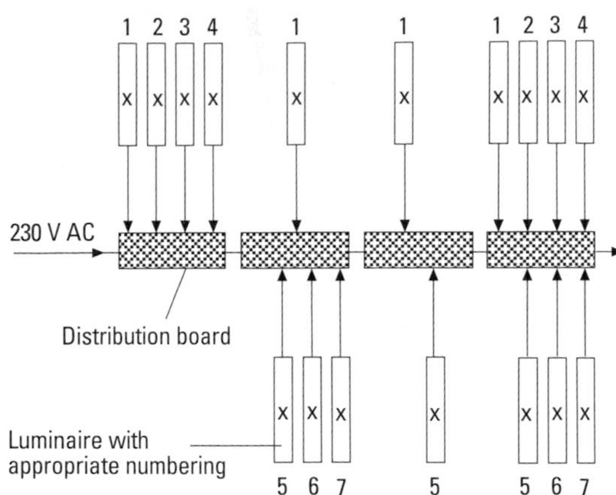

Fig. 4.3/59 Star connection

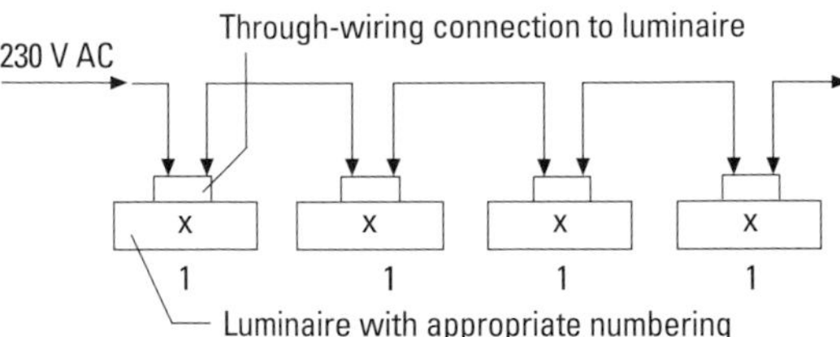

Fig. 4.3/60 Through connection

Planning

When planning electrical installations for lighting and louver blind systems, the following points relating to the installation have to be specified in advance:

▷ Building management:
- Usage, e.g. as an office, department store, manufacturing area, etc.,
- Lighting requirements for various room structures and visual tasks at the work place,
- Lighting circuitry, e.g. individual luminaires, continuous rows of luminaires, groups of luminaires, etc.;

▷ Energy management:
- Energy savings depending on the external/internal luminous intensity at the work place,
- Energy savings depending on centrally controlled, time-dependent functions;

▷ Louver blind control:
- manual,
- by means of programmable blind control units.

It is then necessary to specify the type of wiring system to be used.

A distinction is drawn here between:

▷ Star connection:
each luminaire is supplied separately via a terminal board (see Section 1.4.5) (Fig. 4.3/59);

▷ Through connection:
each luminaire is supplied by means of a through connection via the preceding luminaire (Fig. 4/60).

Selecting systems

Selecting and combining flexible infrared remote-control and installation systems is described below.

With the conventional type of installation, the management tasks set out above require an extensive wiring system.

Lighting system

Fig. 4.3/61 shows a room with a lighting system tailored to provide maximum flexibility with regard to possible divisions of the room axis. Each luminaire here can represent a separate functional unit.

Building management

The basic organizational characteristics of a room (building management) are determined by the x and y-axes. These permit a room to be subdivided into a large number of minimal spatial units, each with one luminaire, or into larger spatial units with several luminaires. Partitioning (e.g. using easily erected partitions) is usually carried out on the x-axis ("window axis").

Energy management could be carried out on the y-axis – for example, by switching off the continuous rows of luminaires near the window when sufficient daylight enters the room. It is also possible, however, to switch off both continuous rows of luminaires on the y-axis at specific times.

The following Figs. use the room represented in Fig. 4.3/61 to illustrate the possible ways of installing a lighting system.

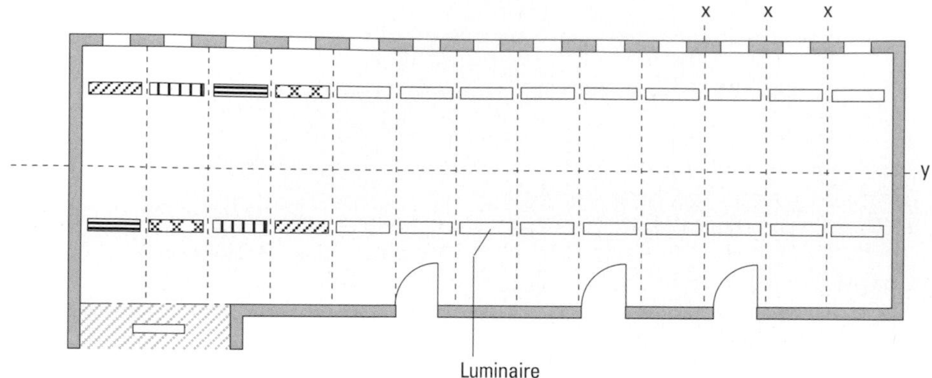

Fig. 4.3/61 Basic organizational characteristics of a room (building management)

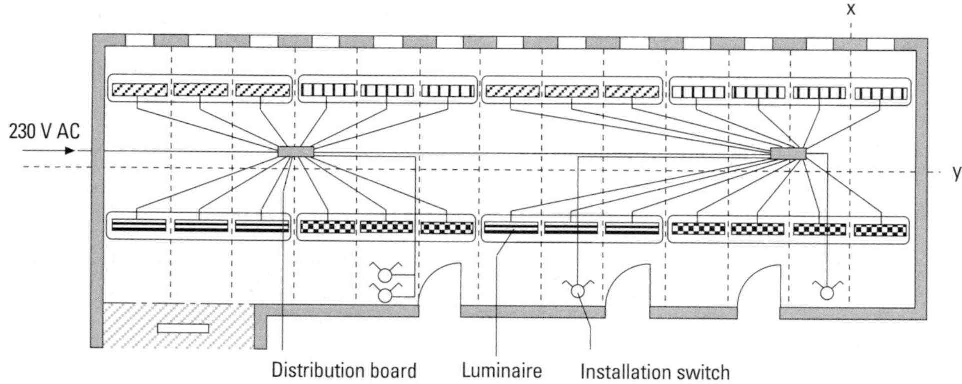

Fig. 4.3/62 Lighting system with conventional type of installation

Star connection

In Fig. 4.3/62, the connecting lead of each individual luminaire is routed to an assigned terminal board, where it is connected to terminals (star connection). The luminaires are connected to installation switches in accordance with the luminaire groups to be formed. Rewiring the terminal board enables the partitioning of the room to be changed on the *x*-axis. In this type of installation, energy management on the *y*-axis is not possible.

Through connection

Through connection is a variant that further reduces the amount of installation work required when changes are made to rooms. Fig. 4.3/63 shows this type of wiring with the DELTA-FERN infrared system. In this example, through connection is carried out on the luminaire itself; in other words, the main IR distribution board is located in the luminaire.

"Central programming point"

The DELTA-FERN infrared receiver is also relocated to the "central programming point" in the luminaire. This allows the system to be reprogrammed without opening the ceiling.

Building and energy management is possible on both the *x* and the *y*-axis.

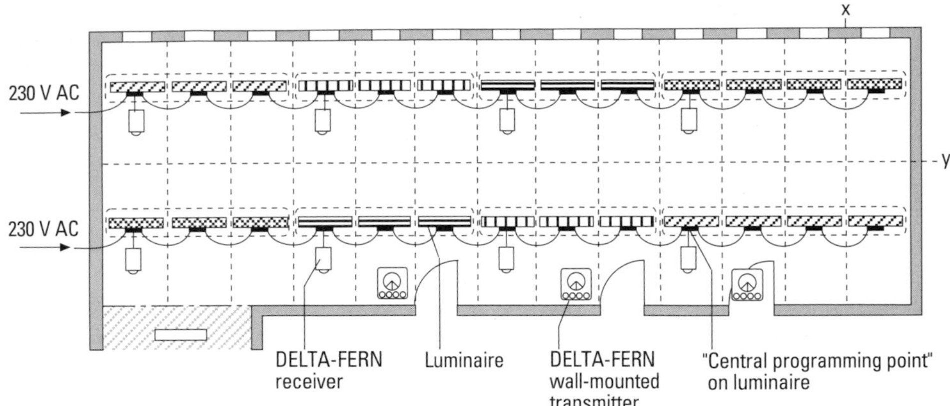

Fig. 4.3/63 Creation of a "central programming point" in the luminaire using through connection

Louver blind control

Switching functions

The following three switching functions are required for a louver blind control system:

▷ UP,
▷ STOP,
▷ DOWN.

These three switching functions also include adjustment of the louvers. Projection screens can be controlled in the same way.

Permanently installed mains control systems are equipped with standard louver blind knobs or rocker switches. These are operated manually at the point of installation. This type of technology can be used for small, individual rooms; its lack of flexibility, however, makes it unsuitable for large offices with many windows. In these larger environments, it is expedient to use the DELTA-FERN infrared installation system.

If the three mechanical switching functions (UP, STOP, DOWN) were integrated directly in the DELTA-FERN infrared installation system, they would occupy three channels (three keys), leaving only one channel (one key) for the lighting system. In order to avoid this, a 1-key solution has been developed.

1-key solution

The "1-key solution" converts the direct linear control of the mechanical switching functions ("UP, STOP, DOWN") to a circular configuration.

Fig. 4.3/64 contains a simplified representation of the switching functions of a mechanical and an electronic louver blind switch.

Mechanical switching functions always return to the STOP position when the switch knob is released in the UP or DOWN position. With electronic switching functions, the switch moves to the next switching function in the same way as a scanner (clock), whenever a control pulse is present.

Switching function as circular configuration

Group switches for installation remote-control are particularly well suited for converting switching functions to a circular configuration (see Section 4.2).

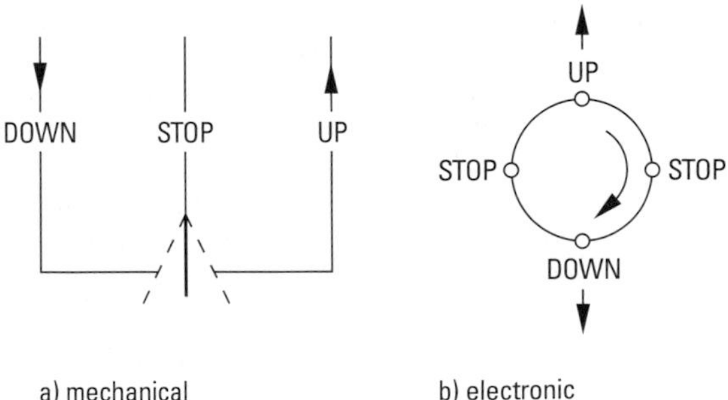

Fig. 4.3/64
Switching functions of a mechanical and an electronic louver blind switch

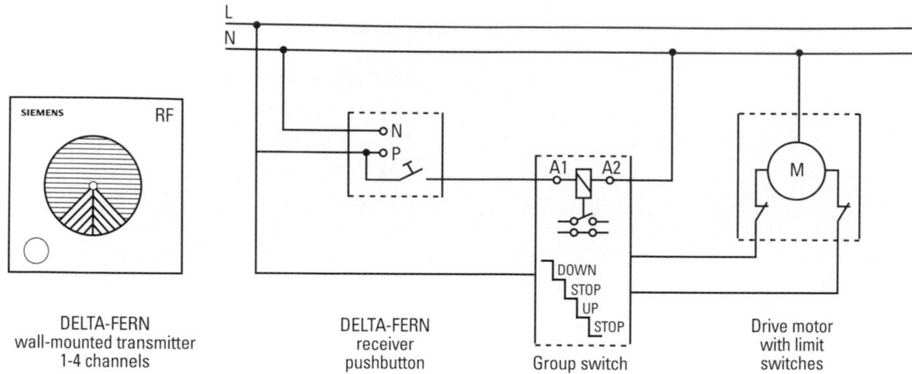

Fig. 4.3/65
Circuit diagram of a louver blind control system integrating components from the DELTA-FERN infrared system and a group switch (remote control) in a "1-key solution"

The "1-key solution" is implemented by integrating a DELTA-FERN receiver with momentary-contact function and a group switch (remote control) (Fig. 4.3/65).

When the IR light transmitted by an activated DELTA-FERN transmitter is received by a DELTA-FERN receiver with momentary-contact function, the output contact of the receiver closes for approximately 90 ms, applying a voltage to the A1/A2 relay coil of the group switch.

The contacts of the group switch, which are assigned in accordance with the switching step, are closed and mechanically interlocked. This switching state (e.g. 3–4 closed corresponds to "DOWN") reverts to "STOP" only if a new command is transmitted by the DELTA-FERN transmitter (Fig. 4.3/66).

Limit switches

In addition to the louver blind knob/rocker switches described above and the DELTA-FERN receivers, limit switches are provided for each motor unit.

The limit switches switch off the motor automatically if, for example, the blind has reached the "TOP/UP" or "BOTTOM/DOWN" limit. When the switch is in the "TOP" limit position, the group switch has to be operated again by actuating the DELTA-FERN transmitter key twice in order to access the "DOWN" function via the "STOP" position of the group switch (see Fig. 4.3/66).

Louver adjustment

When the blind is being lowered, it can be stopped at any point by means of the "STOP" function. The louvers of the blind can be adjusted by briefly actuating the "STOP" function.

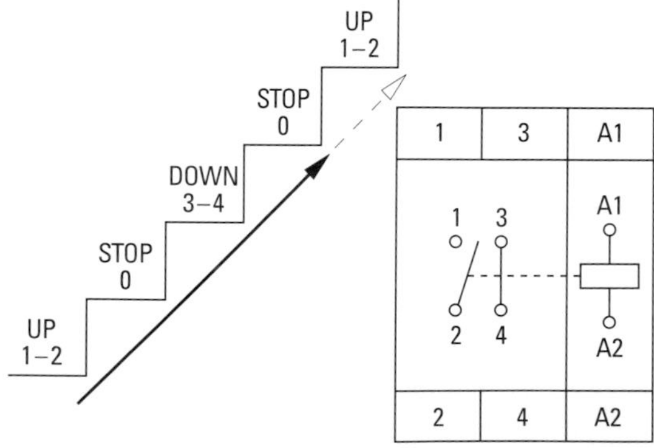

Fig. 4.3/66
Function and contact position of the group switch in the "1-key solution"

Lighting control

Introduction

Modern lighting concepts are characterized in particular by efficiency (Fig. 4.3/67) and convenience. To a large extent, this is due to the success of fully electronic control gear (ECG). Uniform light, around 25% less power demand, constant operational performance (despite mains fluctuations), and extremely low noise levels are just some of the advantages that have led to the ECG being used even in specialized environments such as film and music studios.

ECG Dynamic

A more advanced version of the tried-and-tested ECG is ECG Dynamic.

The broad dimming range (from 100% to 1% luminous flux) and "powerless" brightness control via a low-voltage control signal open up new fields of application in lighting technology for ECG Dynamic.

Significant energy saving

When fluorescent luminaires are operated with conventional series reactors, the power absorbed by the system is around 25% higher than with ECG. In addition to this, further power is required for conventional dimming.

Not only the high losses caused by the series reactors have to be taken into account. Consideration also has to be given to the constant power requirements of the auxiliary electronics and the heater transformer for heating the fluorescent coil, the power loss in the phase crossover dimmer, as well as the necessary base load.

As a result, the power used to operate an L58W fluorescent luminaire at full power, for example, increases to 80 W, before the power loss of the phase crossover dimmer and the base load are taken into consideration.

By contrast, when an ECG Dynamic is used, the system uses only 56 W to operate an L58W fluorescent luminaire at full power – a reduction of 30% with virtually the same luminous flux.

This is because, depending on the position of the dimmer, the ECG Dynamic heats the coil of the lamp only as much as is necessary for safe and reliable operation. In addition to this, no power is required to control brightness, and a base load is unnecessary.

Fig. 4.3/68 shows the operator panels and control units that can be used to manually control lighting systems with dimmable electronic control gear. Fig. 4.3/69 shows the control units suitable for non-mains as well as mains control.

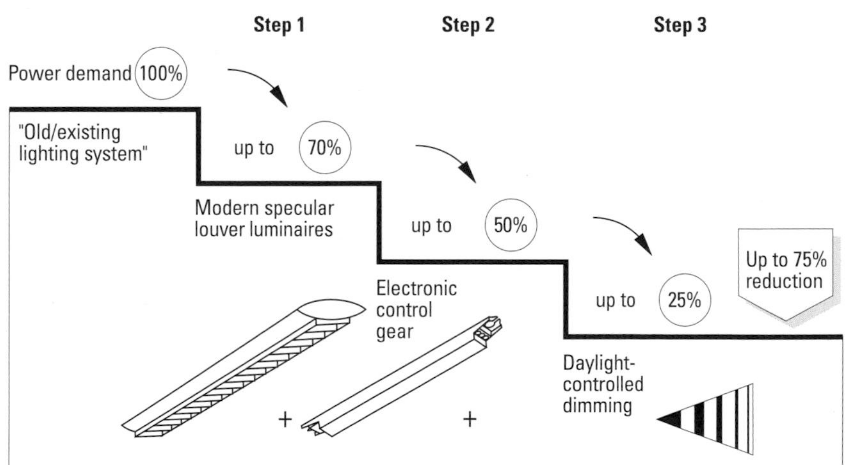

Fig. 4.3/67 Power demand of modern lighting concepts

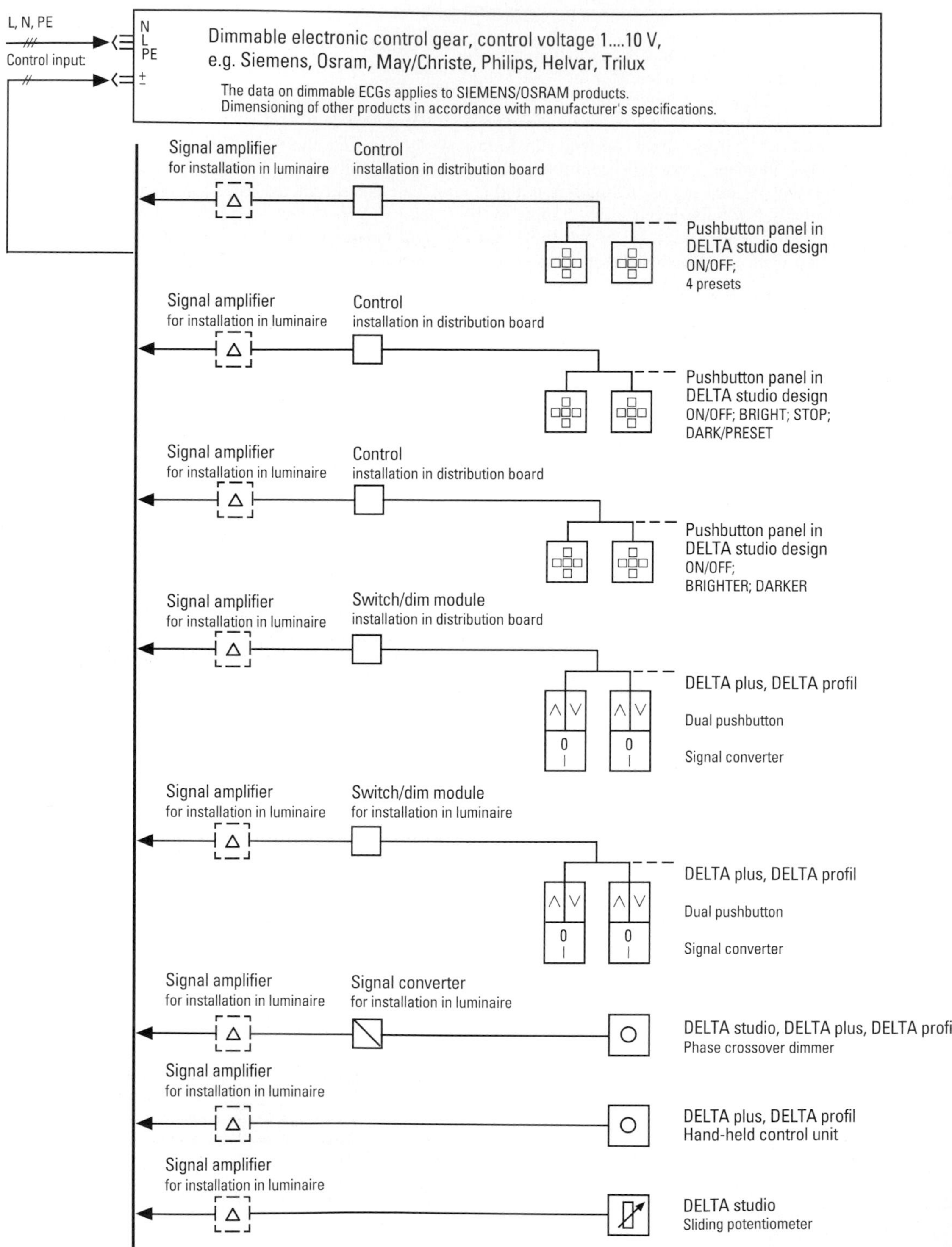

Fig. 4.3/68 Operator panels and control units for manual control

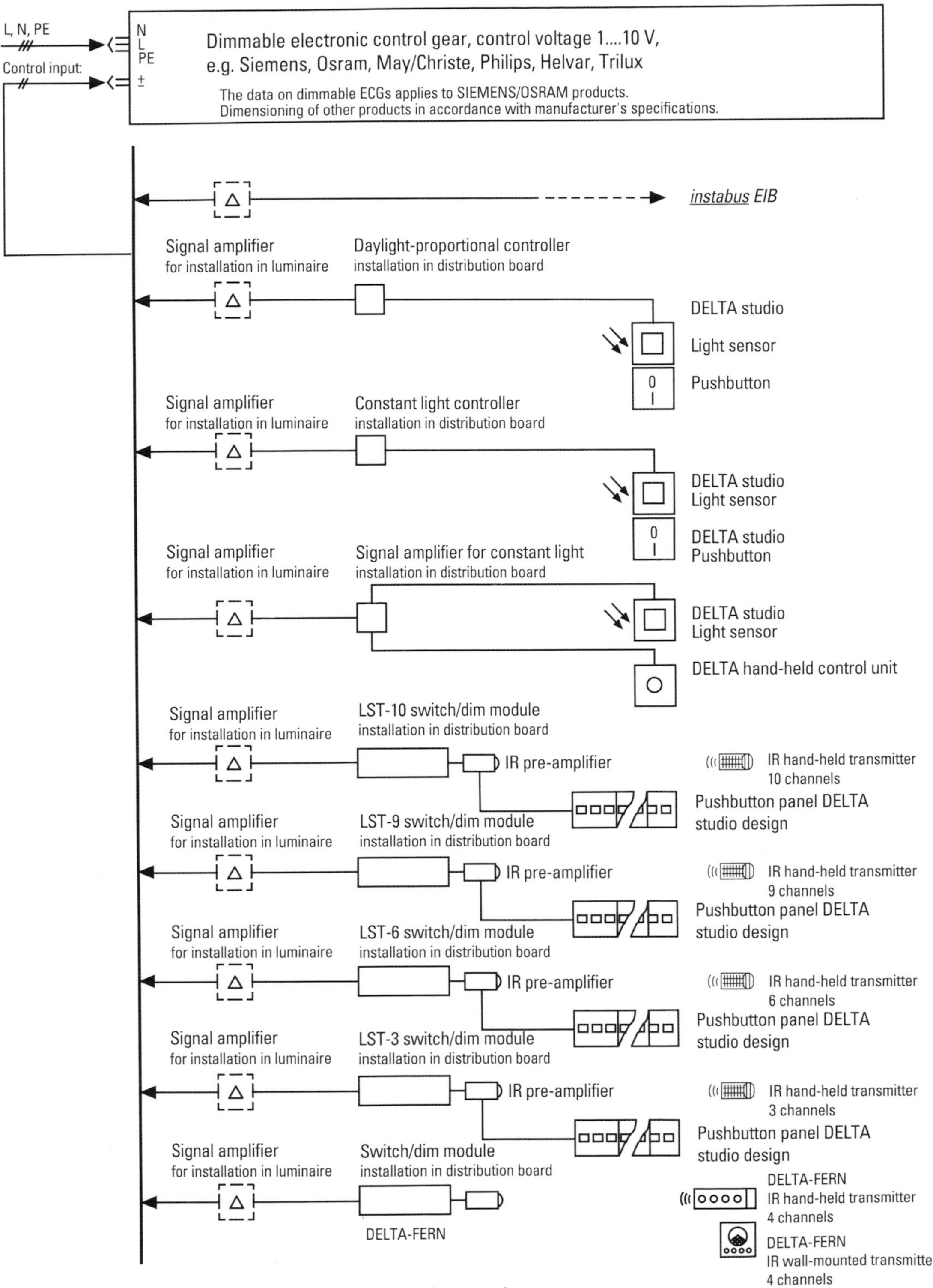

Fig. 4.3/69 Control units for non-mains and mains control

4.3.2.2.2 IR-64K infrared remote-control system

Introduction

In many electrical installation applications, modern infrared technology offers convenience, reliability, and a reduction of the costs involved in installing systems. This technology enhances and changes existing installations, making work places more user-friendly, flexible, and convenient.

Immunity to interference

The modular design and handling of the system is straightforward and offers the highest levels of immunity to interference when transmitting control signals using infrared light.

Transmission range

The IR-64K infrared remote-control system comprises a range of modules and devices (Fig. 4.3/70) for all types of remote-control task. It has a transmission range of 25m to 50m, and offers a broad spectrum of device combinations and expansion options. The high propagation rate of the infrared signal prevents interference problems from arising.

The 9-bit pulse code modulation (PCM) used in the IR-64K infrared remote-control system has a very high level of interference immunity in comparison with other remote-control systems.

The range of devices includes encodable hand-held transmitters with up to 64 channels, receiver/pre-amplifiers, decoders, power electronic assemblies with four or eight outputs, and the corresponding power supply units.

Compact systems with up to 64 switching functions are also available.

Assembly

The modules permit easy assembly of remote-control systems for simple switching functions through to complex applications. The modules can be integrated in devices or in systems for external installation, and are also suitable as device accessories.

Standards

All the components of the IR-64K infrared remote-control system conform to the standards EN 50081-1, EN 50082-1, IEC 60801-2, IEC 60801-4, and have BZT[1)] certification G105 376 C/IW, which is required in Germany.

Application

Fig. 4.3/71 shows some of the many potential applications for which the versatile IR-64K infrared remote-control system can be used.

Components

Selection

There are no restrictions regarding the ways in which components of the IR-64K can be combined with each other. The components selected depend essentially on the type of application, and the particular local conditions (Fig. 4.3/72).

The various hand-held transmitters are distinguished by differences in design, number of keys, and type of command encoding.

Hand-held transmitters

Encodable hand-held transmitters are available with up to 64 channels and degree of protection IP 30 and IP 54 to EN 60529/IEC 60529/DIN VDE 0470-1.

Receiver/pre-amplifiers

The receiver/pre-amplifiers can be operated with all the decoders and compact systems of the IR-64K infrared remote-control system. In selecting the appropriate pre-amplifier, the reception conditions, as well as installation, operational, and environmental factors should be taken into account. Receiver/pre-amplifiers are available with degree of protection IP 30 and IP 65.

Decoders

The decoders are distinguished by differences in design, number of receiver/pre-amplifier inputs, type of signal output, and operating voltage. All the modules can be encoded via DIL switches.

Additional expansion boards are available, which can be used to expand the system to accommodate up to 64 channels.

Booster modules

Booster modules are required to increase the switching capacity at the outputs of the decoders and their expansion boards. The booster modules are distinguished by their different switching capacities, number of outputs (four or eight) and operating voltage.

Compact system

In compact systems (Fig. 4.3/73) all the essential components of an IR-64K infrared remote-control system (such as the power supply unit, decoder with encoding switches, and the power relay) are integrated on a printed-circuit board in a housing with degree of protection IP 65.

Depending on the particular version, the compact systems have either two or four re-

[1)] BZT **B**undesamt für **Z**ulassung in der **T**elekommunikation (Federal Bureau for Approval in Telecommunications)

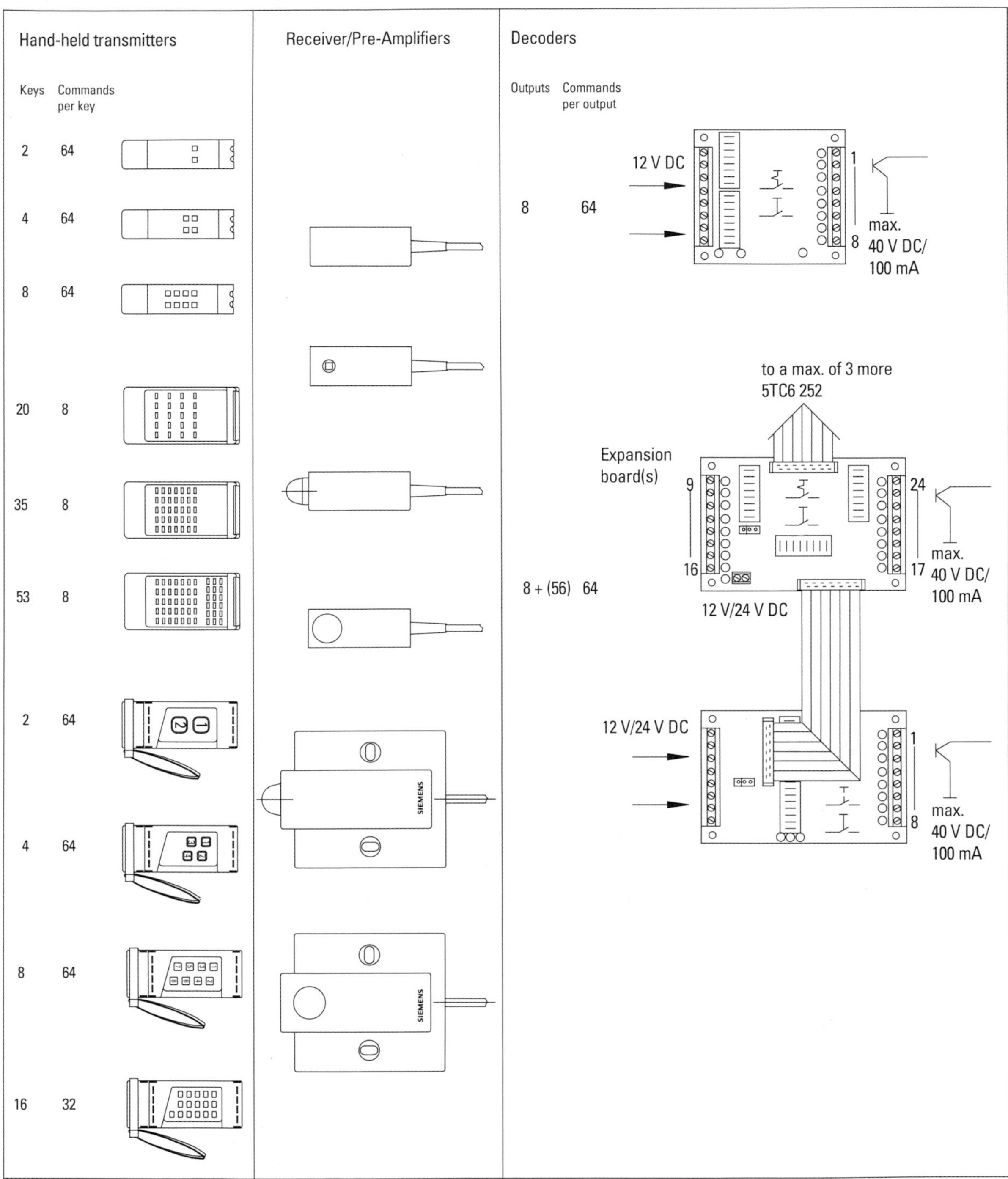

Fig. 4.3/70 System overview of the IR-64K infrared remote-control system

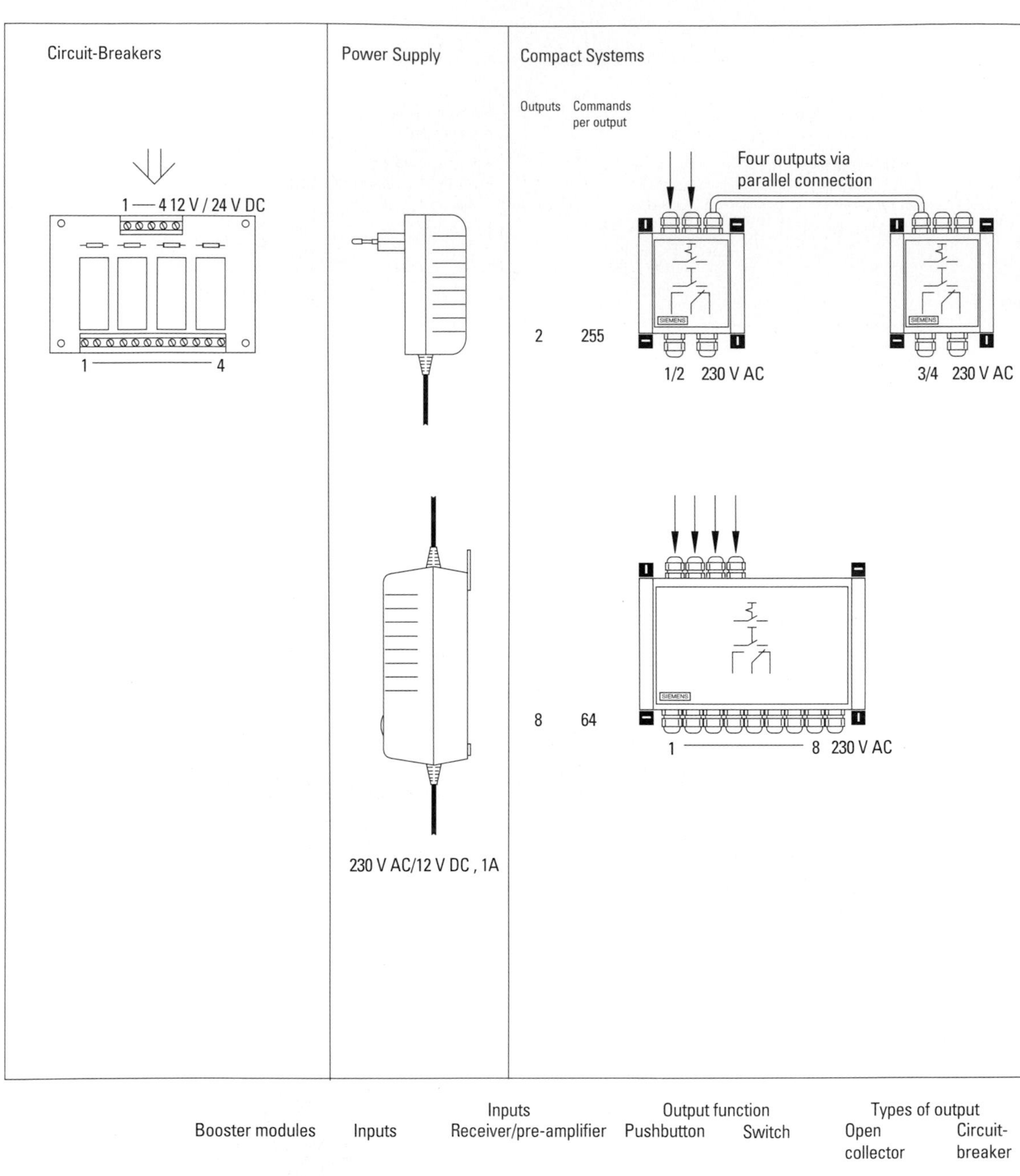

Fig. 4.3/70 (continued)

IR-64 K infrared remote-control system

Exterior lighting

- Residential complexes
- Stadiums
- Sports grounds

Systems, general

- Installation in damp rooms
- Voltageless switching
- Call systems for the catering trade

Consumer electronics

- Radios and TVs
- Sound studios
- Rotary aerials
- Projection screens
- Multimedia
- Lecture rooms

Application-oriented IR remote-control

- Planning and design of IR remote-control systems to user specifications

Control systems

- Motor control units, general (fans, louver blinds, curtains, gates, roof windows)
- Lifting platforms

Security systems

- Call and signal systems in homes for the elderly, nursing homes, and hospitals
- Medical equipment (general)

Industrial control systems

- Automotive electronics
- Test stands
- Mining
- Mechanical engineering
- Elevators
- Crane trolleys
- Crane systems
- Loading equipment

Leisure time and hobbies

- Toys
- Sports equipment (automatic firing ranges, ball pitching machines)
- Swimming pools (lighting, covers)

Fig. 4.3/71 Applications for the IR-64K infrared remote-control system

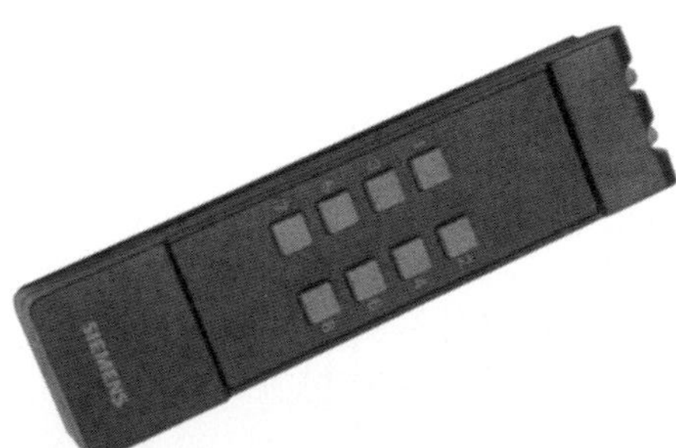

Hand-held transmitter

Decoder

Receiver/pre-amplifier

Booster module

Fig. 4.3/72 Some of the modules of the IR-64K infrared remote-control system

Fig. 4.3/73
Compact system with eight outputs and four receiver/pre-amplifier inputs

ceiver/pre-amplifier inputs, and two or eight power relay outputs with voltageless change-over contacts, which can be loaded up to 8 A or 16 A, depending on the load connected.

By connecting a maximum of four compact systems (each with eight outputs) in parallel, an IR remote-control system can be expanded to 32 outputs.

The encoding switches of the decoder are used to encode commands, and set the key or switch function of the decoder and expansion boards. Depending on the version, up to 255 different command numbers can be programmed for each output.

In the event of mains failure, the compact systems with eight outputs can be supplied with a controlled voltage of 12 V DC (e.g. stand-by power supply) via the 2-pole terminal. An additional terminal allows an external key to be connected, which can be used to reset all switching states simultaneously.

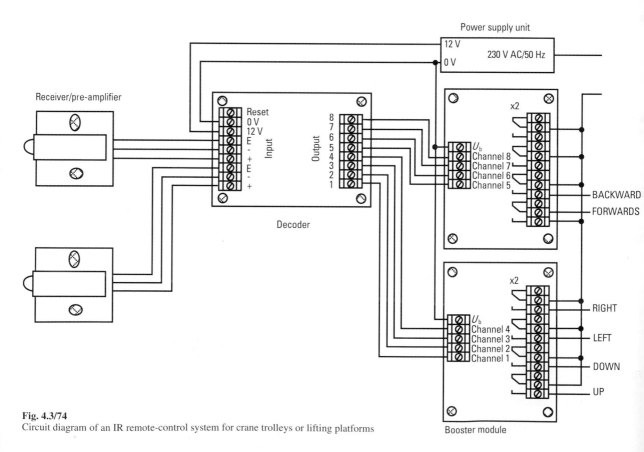

Fig. 4.3/74
Circuit diagram of an IR remote-control system for crane trolleys or lifting platforms

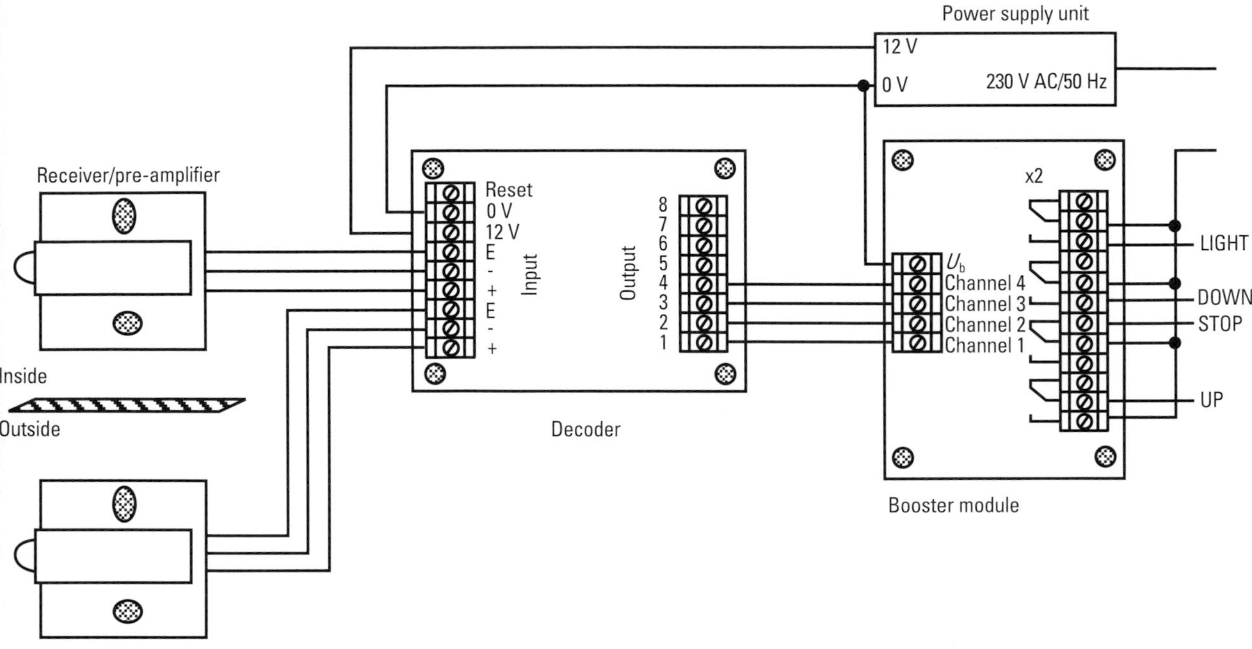

Fig. 4.3/75 Circuit diagram of an IR remote-control system for a roll-up gate

As described above, the receiver/pre-amplifiers must be selected to match the particular location in which they are to be used.

Power supply unit

Power for the modules is provided by power supply units, which supply a controlled voltage of 12 V DC with a load of 500 mA or 1 A.

Figs. 4.3/74 and 75 show examples of circuit diagrams for two different applications.

4.3.2.2.3 DELTA-FERN RF radio remote-control system

Use

DELTA-FERN RF is a radio remote-control system for use in electrical installations. The system is equipped with four function keys, and is particularly easy to use and install, highly flexible, and extremely convenient.

The system can be used when a residential or functional building is rewired (renovation) or when new wiring systems are installed (new building).

Rewiring

When rewiring existing systems, additional switching points can be installed for conventional ON/OFF, two-way, and intermediate switch circuits with any number of flat wall-mounted transmitters for surface mounting together with the corresponding receivers as switches, dimmers, or pushbuttons. This can be carried out without having to lay extra cables. The transmitters can also be installed to function across a number of rooms.

New installations

DELTA-FERN RF is ideal for use in new wiring systems. The highly flexible system is suited to a wide range of applications and uses, and requires very little pre-planning.

The switching points can be installed after the user has moved into the building, when they can be located at the points most suitable from an operational point of view. Since the switching type can be freely selected and wiring types or switch wiring do not have to be taken into account, the question of whether to use intermediate switch circuits, changeover or two-way switches, or pushbuttons does not arise.

The radio transmitters that function as light switches are operated using a standard 9 V monobloc battery, thus allowing several transmitters to be installed wherever they are required, irrespective of the location.

Switching operations can also be carried out using hand-held transmitters.

Components

Range of components

The DELTA-FERN RF radio remote-control system comprises the following components (Fig. 4.3/76):

▷ Hand-held transmitter with 4 function keys, key designation 1 to 4;
▷ Wall-mounted transmitter with 1 function key, key designation 1;
▷ Wall-mounted transmitter with 2 function keys, key designation 1 and 2;
▷ Wall-mounted transmitter with 4 function keys, key designation 1 to 4;
▷ Switchbox/splitting box receiver as switch, dimmer or pushbutton;
▷ Plug-in socket receiver as switch or dimmer; fits 60-mm diameter device box;
▷ Ceiling-rose receiver as switch or dimmer,
▷ Built-in receiver as switch, pushbutton, or as "central ON/OFF" switch for fluorescent luminaires with conventional or electronic control gear,
▷ Built-in receiver switch/dim module for switching luminaires with dimmable electronic control gear.

RF transmitters

The RF transmitters are operated using a 9 V monobloc battery and do not require a mains connection.

Fig. 4.3/76
Some of the components of the DELTA-FERN RF radio remote-control system

Transmission range

Each function key can be used independently of the others to control an item/group of current-using equipment via the corresponding RF receiver. The indoor transmission range is approximately 30 m.

Three wall-mounted transmitters with 1, 2 and 4 function keys, and a hand-held transmitter with 4 function keys are available.

The keys of the transmitter are designated 1 to 4. Each key uses its own transmission channel, which means that each transmission key has a unique function.

Depressing the keys carries out the following functions:

▷ short key depression: "ON" and "OFF",
▷ longer key depression: "Dimming".

The wall-mounted transmitters can be bonded or screwed directly to the wall, or attached to 60-mm diameter device boxes (with or without surface-mounting frames). The transmitters are designed as DELTA fläche transmitters, color: titanium white, and fit the 80 mm × 80 mm and 75 mm × 75 mm frames from the DELTA fläche titanium white range (see Section 4.3.1).

The RF transmitters permit additional switching points to be retrofitted at any time.

Receivers

The receivers can be used as required to switch or dim incandescent lamps, various ohmic loads, and fluorescent luminaires, as well as to operate remote switches or stairwell lighting timers (Fig. 4.3/77).

The receivers require a power supply of 230 V AC, 50/60 Hz; the switch/dim module for fluorescent luminaires requires a control voltage of 1 to 10 V DC. The receiver is available with 2-wire and 3-wire circuits. The 2-wire circuit variant allows the existing switch to be replaced by the "switch/dim" switchbox receiver. No N-conductor is required (receivers for the 2-wire circuit variants can be used only with incandescent lamps). This enables existing switches to be replaced with "switch/dim" switchbox receivers. The power for the electronics is supplied via the filament of the incandescent lamp.

A 3-wire circuit variant is required to switch fluorescent luminaires. The power for all the electronics in the 3-wire circuit variant is supplied directly from the mains. An N-conductor is required in this case. This circuit

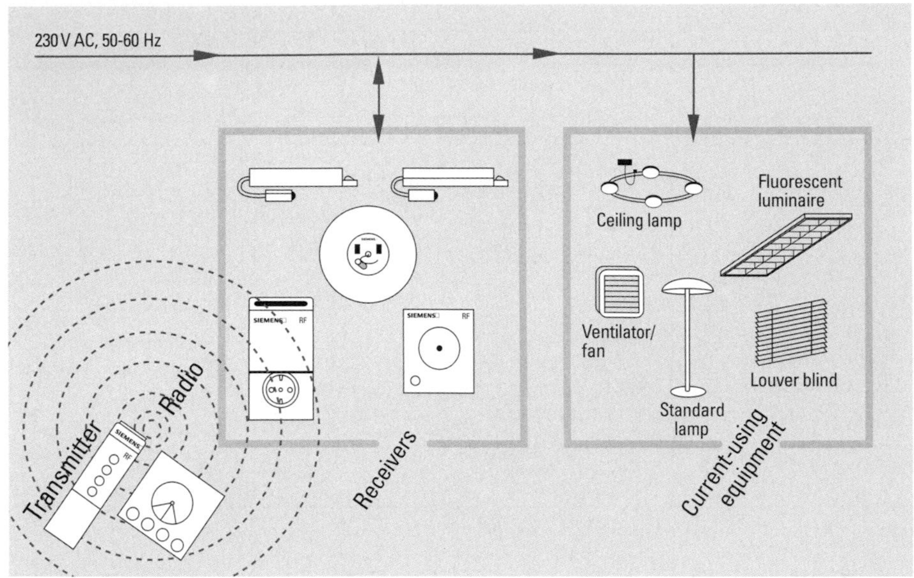

Fig. 4.3/77 Block diagram DELTA-FERN RF radio remote-control system, DELTA fläche titanium white

variant can also be used to switch incandescent lamps ≤ 1000 W. A built-in receiver with a switch/dim module (control voltage 1 to 10 V DC) is available for dimming fluorescent luminaires with electronic control gear (ECG Dynamic).

Assigning transmitters to receivers

The transmitters are assigned to receivers by means of a teach-in procedure.

The receivers have a slide switch with the position indicators "Operate" and "Learn".

When this switch is in the "Learn" position, the transmitter can be assigned to a particular receiver by depressing a key. The slide switch is then set in the "Operate" position, and the system is available for operation.

Up to 8 transmitters can be assigned to one receiver.

Clearing channels

If a transmitter is incorrectly programmed, the channel that was entered can be cleared by setting the slide switch of the receiver to "Learn" and depressing the key of the transmitter a second time.

Installation

Switchbox receivers

Switchbox receivers can be used in 60-mm diameter device boxes with the frames of the DELTA fläche range, as well as in the DELTA fläche surface-type housings.

Ceiling-rose receivers

Ceiling-rose receivers are connected between the luminaire and the ceiling outlet box, and attached to the ceiling by means of ceiling hooks.

Built-in receivers

Built-in receivers are suitable for installation in luminaires and false ceilings, and are primarily used for installations in functional buildings.

Switch/dim modules

The switch/dim module for fluorescent luminaires and luminaires with dimmable electronic control gear is suitable for installation in luminaires, and is primarily used for installations in functional buildings.

Plug-in socket receivers

Plug-in socket receivers (as switches or dimmers) are quickly and easily installed. They are plugged into existing grounding-type receptacles, between the receptacles themselves and the current-using equipment.

Use in residential buildings

DELTA-FERN RF for rewiring

The DELTA-FERN RF radio remote-control system allows the functionality of existing wiring systems to be considerably enhanced.

Light switches are replaced by switchbox receivers, which can be controlled by means of any number of wall-mounted transmitters. which can be installed in any desired configuration (Fig. 4.3/78).

Switchbox receivers

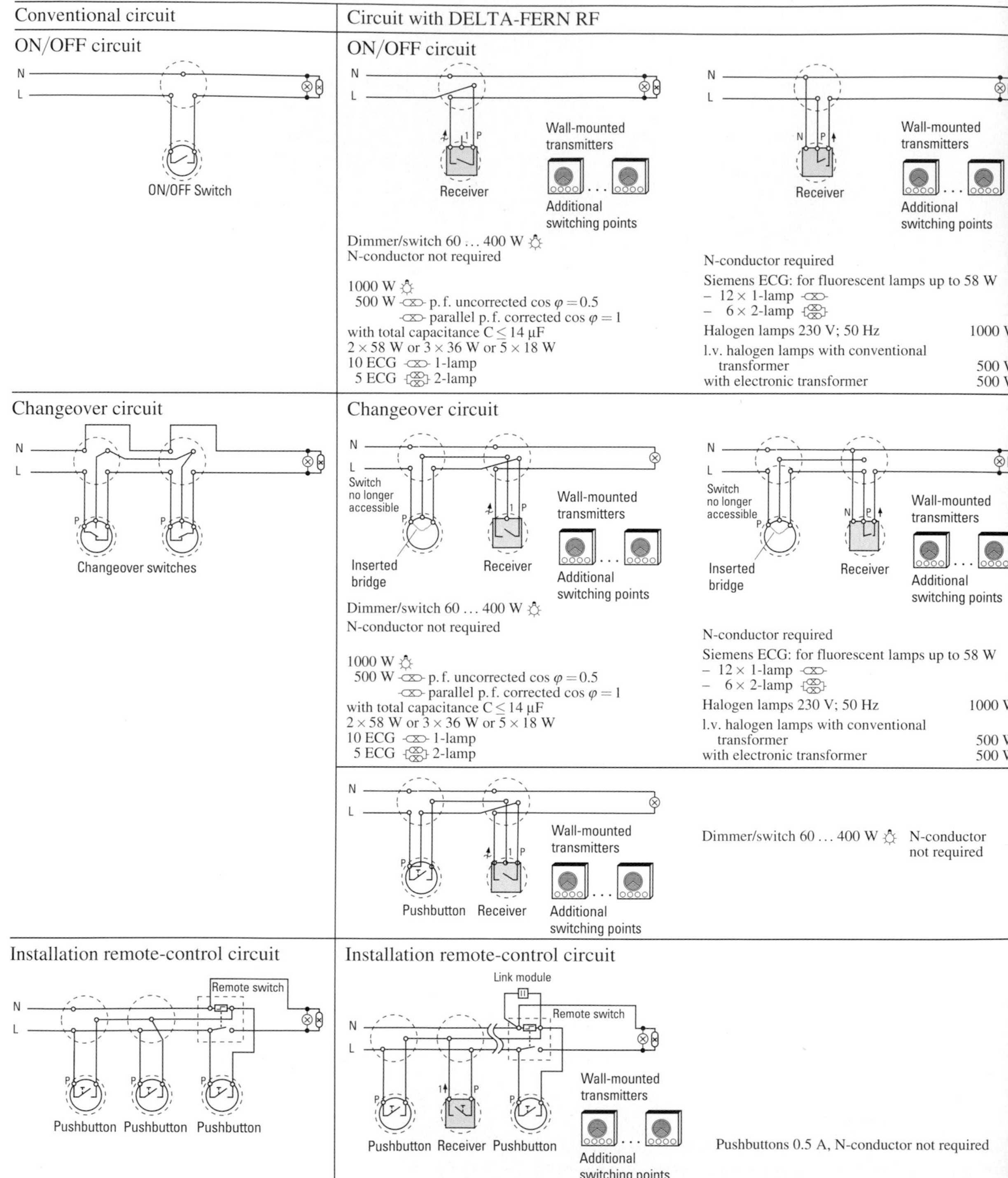

Fig. 4.3/78 Typical circuit diagrams for rewiring using the DELTA-FERN RF radio remote-control system

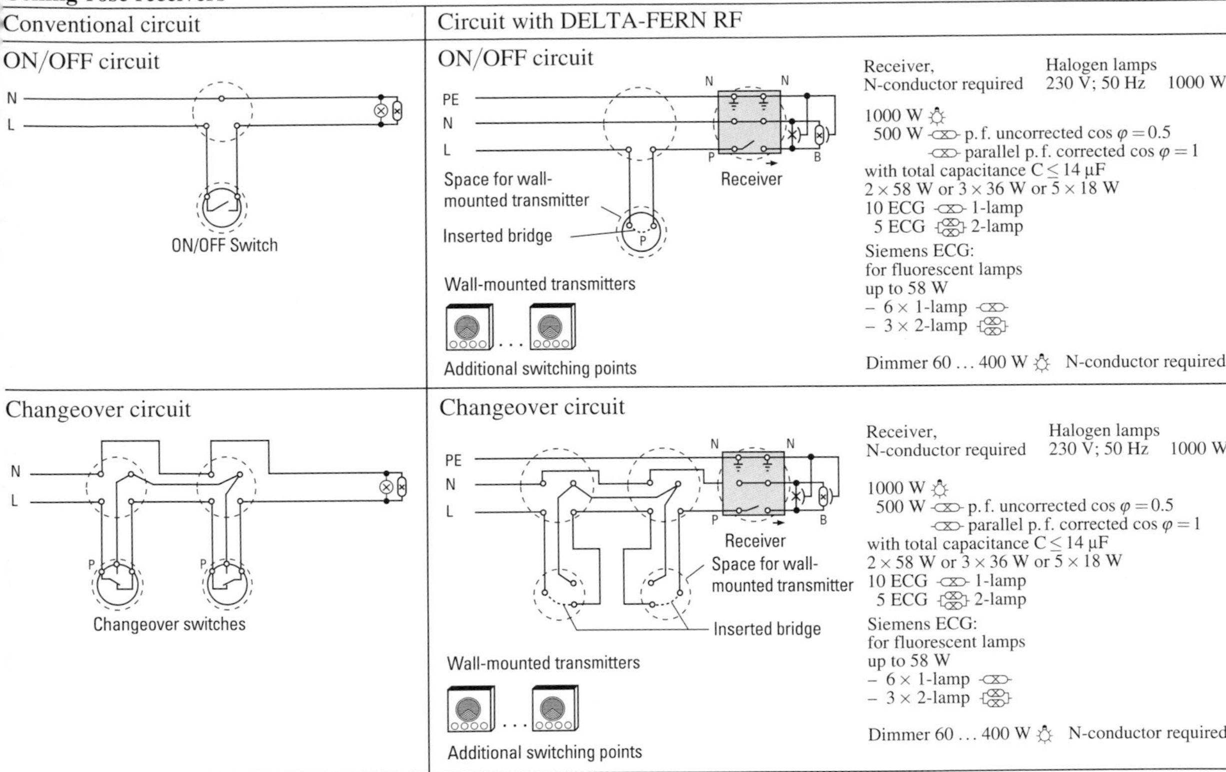

DELTA-FERN RF in new buildings

The DELTA-FERN RF radio remote-control system offers particular advantages when installed in a new building, since time-consuming pre-planning of switching points is no longer necessary.

Only the luminaires and the connection points for the other current-using devices are fitted in the cable run of the electrical installation. When building work is complete and the apartment has been fitted out and furnished, the wall-mounted transmitters as well as the plug-in socket receivers can be positioned according to the tenants' or owners' individual requirements. This type of electrical installation has proven particularly expedient in buildings with fair-faced masonry, exposed concrete, or large glass walls (Fig. 4.3/79).

Use in functional buildings

DELTA-FERN RF for rewiring and new installations

The DELTA-FERN RF radio remote-control system is particularly suitable for use in buildings with complex lighting and louver blind systems.

Using DELTA-FERN RF in this way facilitates subsequent implementation of an energy management system, and offers flexibility with regard to later changes in the use of a room or its structure.

A distributed lighting power supply system also enables fire loads to be considerably reduced.

The built-in receivers for lighting systems with electronic control gear (ECG) as well as lighting systems with conventional con-

Ceiling-rose receivers

Circuit with DELTA-FERN RF

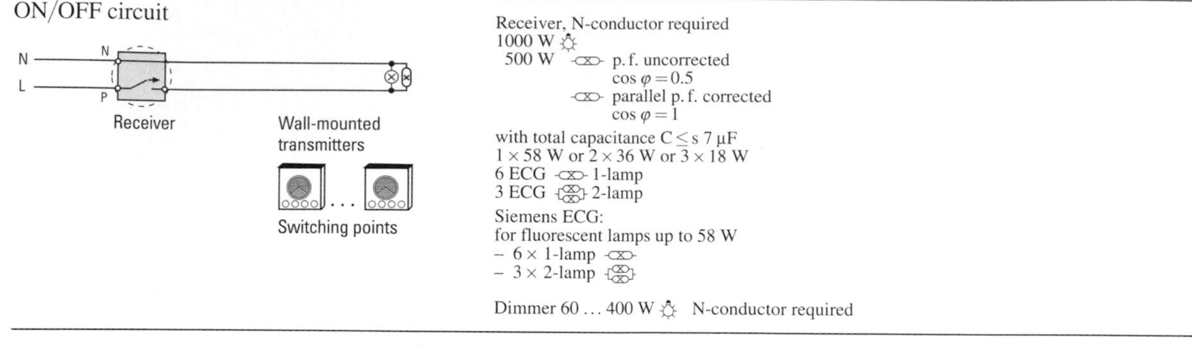

Built-in receivers

Circuit with DELTA-FERN RF

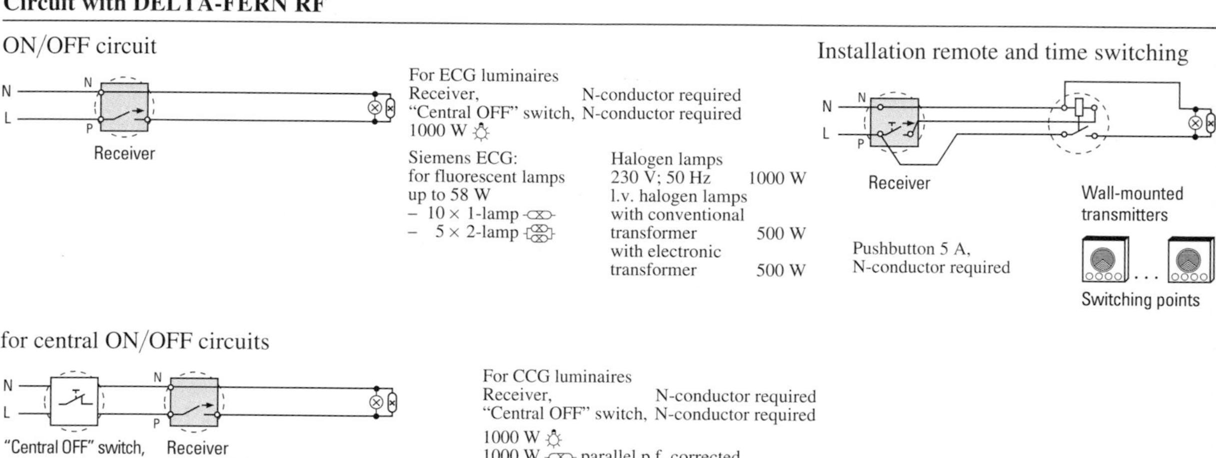

Fig. 4.3/79 Typical circuit diagrams for new installations using the DELTA-FERN RF radio remote-control system

trol gear (CCG) and low-loss control gear (LLCG) are particularly suitable for this type of installation.

The built-in receiver is available as a switch/dim module (Fig. 4.3/80) for use in lighting systems with dimmable electronic control gear (ECG Dynamic). All the built-in receivers can be incorporated in fluorescent luminaires.

DELTA-FERN RF for controlling roller shutters/louver blinds and projection screens

Louver blinds or roller shutters for windows, and projection screens can also be controlled using DELTA-FERN RF (Fig. 4.3/81).

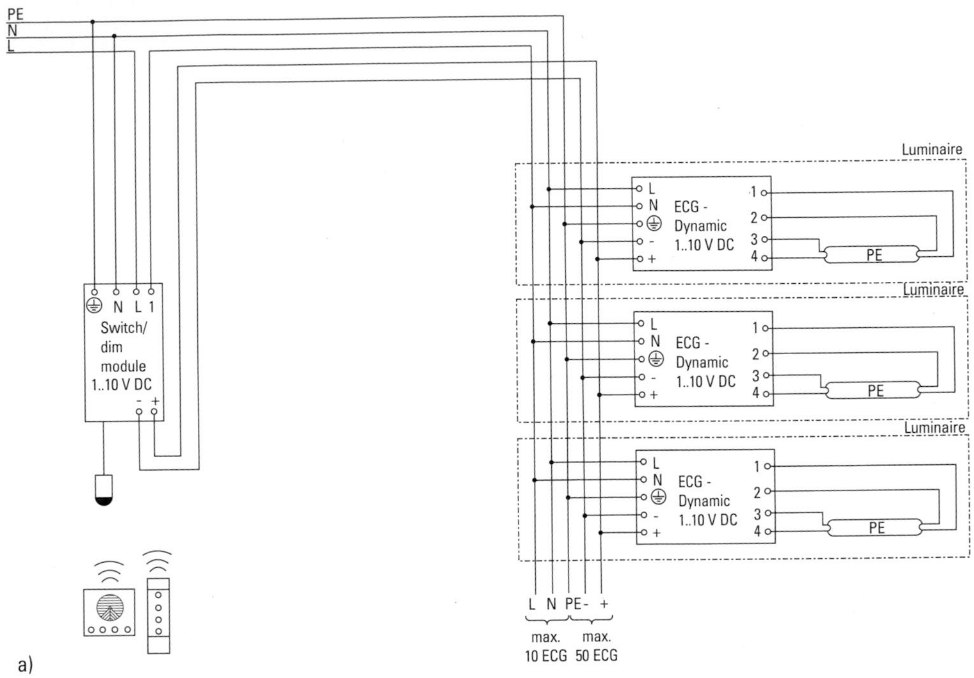

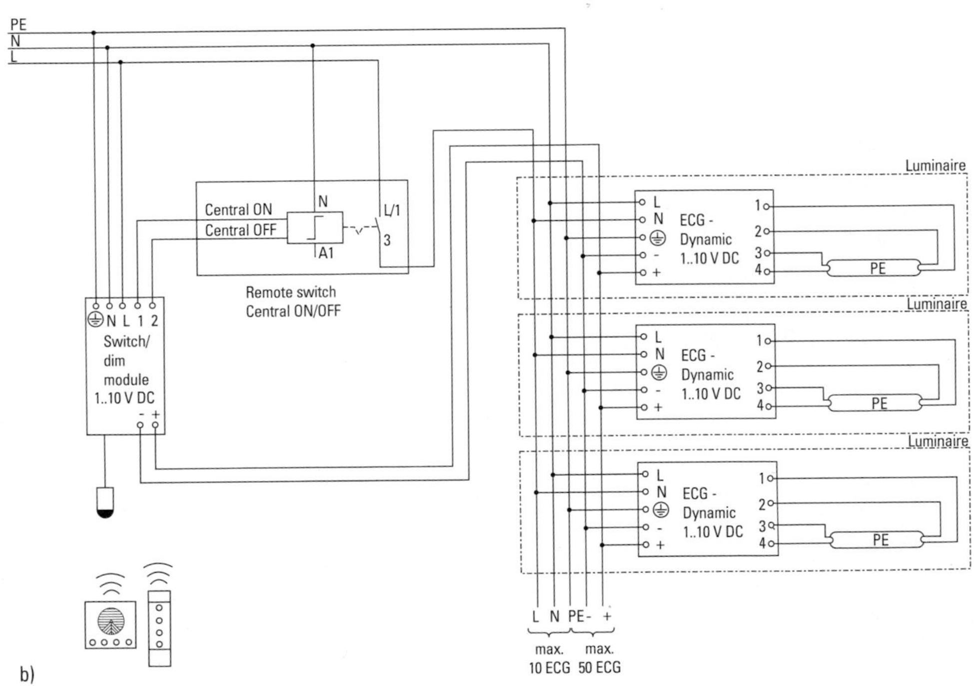

Fig. 4.3/80
Typical circuit diagrams for new installation of the DELTA-FERN RF radio remote-control system in functional buildings
a) Wiring diagram for remote "central OFF" switch b) Wiring diagram for remote "central ON/OFF" switch

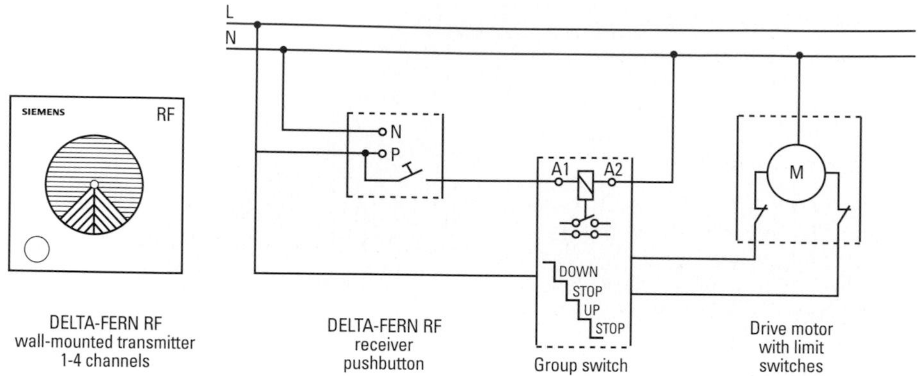

Fig. 4.3/81
Circuit diagram for controlling roller shutters, louver blinds, or projection screens, etc. using the DELTA-FERN RF radio remote-control system

DELTA-FERN RF link to *instabus EIB*

A media coupler is available for linking DELTA-FERN RF to the *instabus EIB* control bus system (see Section 4.4). The coupler performs the following functions:

▷ Media coupler for transmitting switching commands from the DELTA-FERN RF radio remote-control system to *instabus EIB* (Fig. 4.3/82),

▷ Media coupler for transmitting switching commands from the *instabus EIB* to the DELTA-FERN RF radio remote-control system.

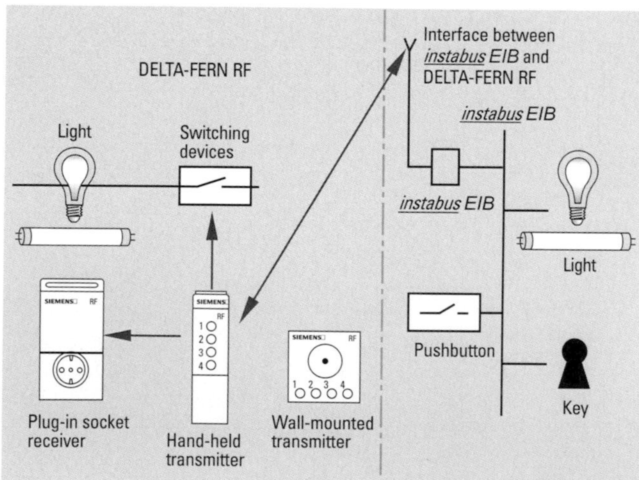

Fig. 4.3/82 DELTA-FERN RF radio remote-control system linked to the *instabus EIB* control bus system

4.3.2.3 Luminous call systems

se

Senior managers of hospitals, clinics, nursing homes or homes for the elderly expect luminous call systems to reduce both the workload of the staff, as well as the operating costs of the institution.

Luminous call systems have been widespread in the above environments for decades, and are continually being used for new applications requiring fast and reliable contact between persons or groups of persons.

Luminous call systems are still used mainly in institutions that care for the elderly and the sick. The prevailing conditions and circumstances within these environments that determine the technical and functional design of modern systems include:

- ▷ Varying organizational and technical concepts and requirements of different institutions/buildings, which demand a high level of flexibility from the systems,
- ▷ Fluctuations in staffing and the generally limited technical knowledge of users (patients, residents, and staff), who require simple, clearly-structured systems, with easy-to-understand functionality and operation,
- ▷ Cost considerations, which demand attractively-priced systems that are easy to install and inexpensive to maintain.

Simple logic circuits in luminous call systems have now been replaced by microprocessor-based control units. The main advantages of the newer systems are that they are easy to program and can be configured in accordance with individual requirements.

Standards

According to DIN 41 050 Part 1 "Luminous call systems – Definitions", luminous call systems are systems with "visual" call indications, to which attention can be drawn by means of an acoustic signal, and which are cancelled at the point of origin of the call, or after the call has been answered.

DIN 41 050 Part 2 "Luminous call systems – Installations, devices, call indication" stipulates the requirements relating to luminous call systems and their design.

In addition to this, the standard specifies colors for call (red) and cancel (green) pushbuttons, and provides a detailed description of visual and acoustic call indications.

Definitions and types of system

The term "luminous call system" is defined in DIN VDE 0834. As a regulation specifically governing the erection and operation of luminous call systems, this standard sets out the safety requirements for such systems with reference to other relevant standards (e.g. DIN VDE 0800, DIN VDE 0804).

In accordance with the current state of the art and good engineering practice, a distinction is drawn between two basic types of system:

1. Luminous call systems without speech facility

According to DIN 41 050 Part 1, 1.2, a luminous call (call for short) is a visual indication denoting, for example, the location at which a call has been initiated or the whereabouts of a person; according to DIN 41 050 Part 1, 1.3, a call with emergency call status is a call with a special signal requesting additional staff after a presence has been set.

DELTA plus "clino opt 99"

It is nowadays usual for these two types of call to be incorporated in a single system, such as DELTA plus "clino opt 99".

2. Luminous call systems with speech facility

According to DIN 41050 Part 1, 1.4 and 1.5, in addition to call functions, these systems feature a speech link (either press-to-talk or duplex) between the main, additional or central call answering units and the room/ward or bed.

DELTA plus "clino phon 95 HS/WIN"

Luminous call installations of this type can be implemented in the form of a system, such as DELTA plus "clino phon 95 HS/WIN", which allows speech to be transmitted freely from two locations by means of modern duplex technology, and without manual assistance.

Buildings in which luminous call systems are used are generally subdivided into wards, departments, etc. In terms of the standards relating to luminous call systems, this structure corresponds to the definition of a zone as "a series of interconnected luminous call devices tailored to the local or organizational conditions" (DIN 41 050 Part 1, 3.18).

Distributed and central luminous call systems

The distinction between distributed and central luminous call systems depends on whether calls are transferred and handled exclusively in independent zones, or whether non-independent zones are controlled from a central station. Both types of system can also be installed as a custom-designed "combined system".

Supplementary systems

Luminous call systems can generally be expanded by adding supplementary systems, such as pagers, door and telephone call systems, as well as electro-acoustic, detection and monitoring systems, etc.

Functional principles and tasks

Users and staff can easily familiarize themselves with the functional principles and tasks of luminous call systems by learning a few basic rules:

▷ The patient (home resident) initiates a call by means of a red call pushbutton; this is the only control operation that he or she has to carry out.
The call is signaled by means of the visual indicators at its point of origin.

▷ When members of staff enter or leave a room, they must press a green cancel pushbutton. This activates four functions:
- Call is cancelled,
- Presence lamp lights up,
- Call is transferred,
- Emergency call prepared.

When a nurse is present, this is indicated by the green presence lamp in the corridor lamp.

▷ Emergency calls or enhanced calls (see page 507) are signaled by means of short, rapid visual and acoustic signals.

Luminous call systems without speech facility

Layout and circuitry

Fully-electronic luminous call systems, such as DELTA plus "clino opt 99", combine low power demand and robust components (see Section 4.3.1). The system can be used either for call indication only (application range I) or for indication of calls and faults (application range II) to DIN VDE 0834.

Figs. 4.3/83 to 4.3/85 illustrate the layout and main components of this type of system.

Zone controller

The central component in luminous call systems is the zone controller (Fig. 4.3/86), which controls the flashing phases of the corridor, zone, and signal lamps, as well as acoustic call indication.

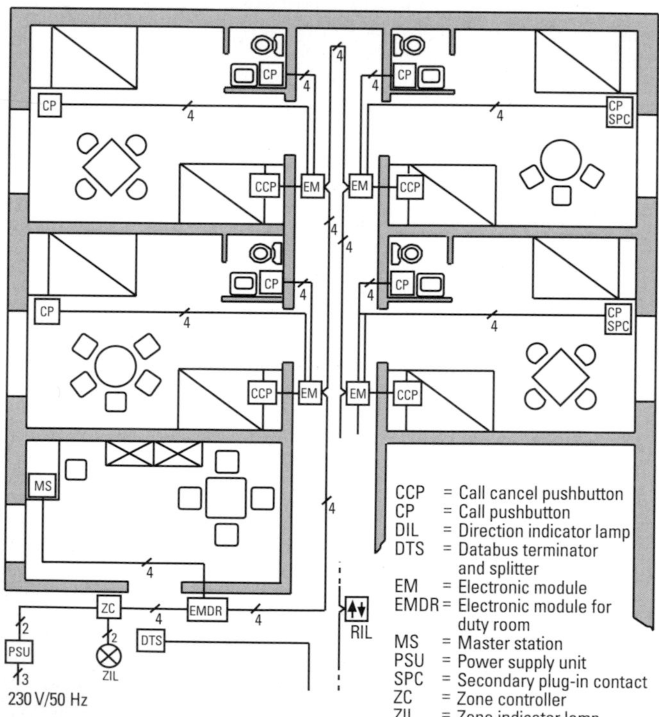

Fig. 4.3/83
Circuit diagram of a DELTA plus "clino opt 99" luminous call system without speech facility in a hospital ward unit

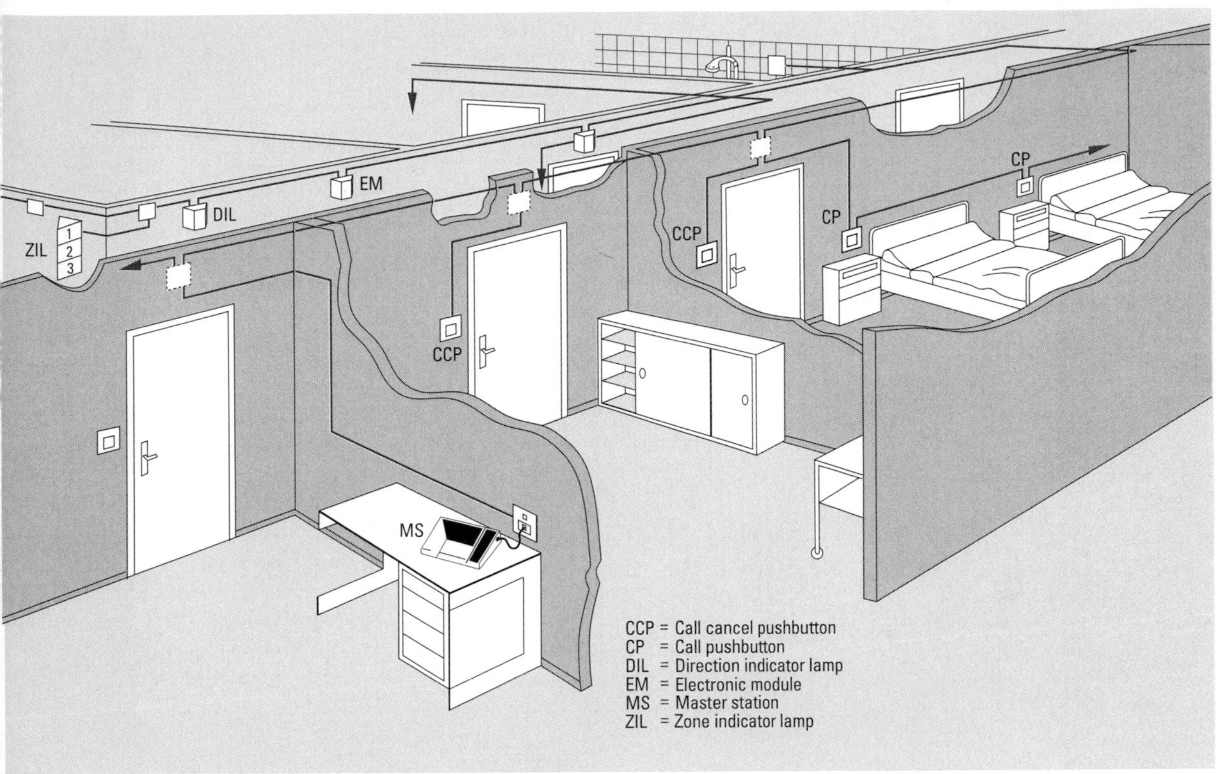

Fig. 4.3/84 Luminous call system without speech facility in a hospital ward unit

Databus

A maximum of 127 rooms (electronic modules) can be controlled via the databus of a zone controller.

Up to 16 zone controllers can be used throughout the system.

A matching power unit supplies the system with 24 V DC via the zone controller.

Data line

The bus line required by the "clino opt 99" system should be implemented by means of a simple communication cable J-Y(St)Y. A twisted core pair must be used for the data line.

Electronic module

The adapter box with the terminals is attached to a device box, which is mounted in the corridor outside the rooms. All the cables for the electronic module are connected here (Fig. 4.3/87). All the electronics required to identify and indicate calls are housed in this unit, which is also the central connecting point for the wiring in the different rooms.

The electronic modules with integrated corridor lamps can be used throughout the entire system. The configuration data is stored in the memory of the electronic module, and thus remains stored even if the electronic module is replaced. Special electronic modules with an additional memory for the configured zone linking data are used in the duty room.

Call and cancel pushbuttons

The call and cancel pushbuttons (Figs. 4.3/88 and 4.3/89) in the rooms are housed in standard switch boxes with a diameter of 55 mm, and secured by means of claws or by tightening the mounting ring.

Call pushbuttons can be permanently installed in the wall, in ducting, or in a bedside

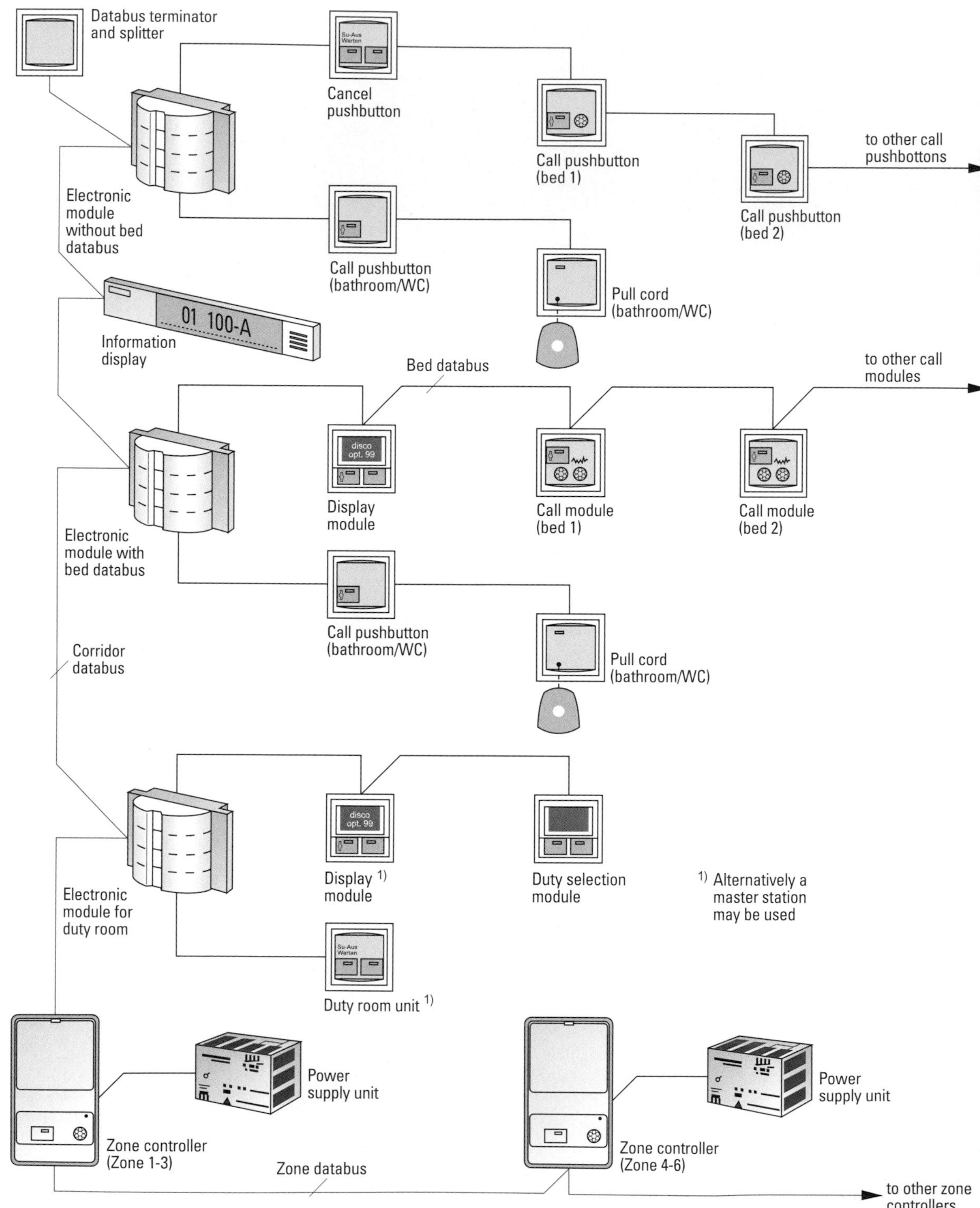

Fig. 4.3/85 Layout of a DELTA plus "clino opt 99" luminous call system without speech facility

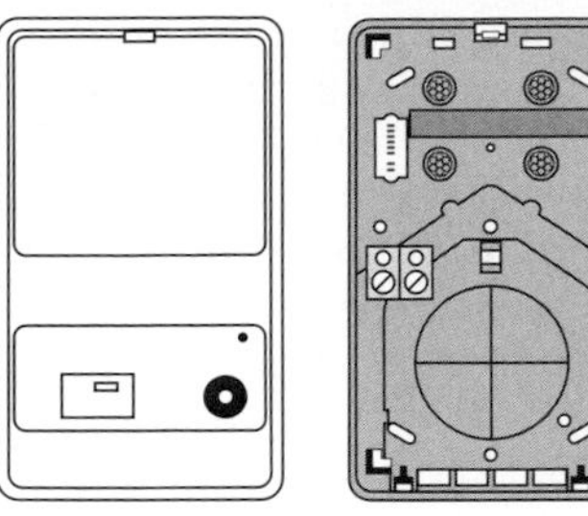

Fig. 4.3/86
Zone controller

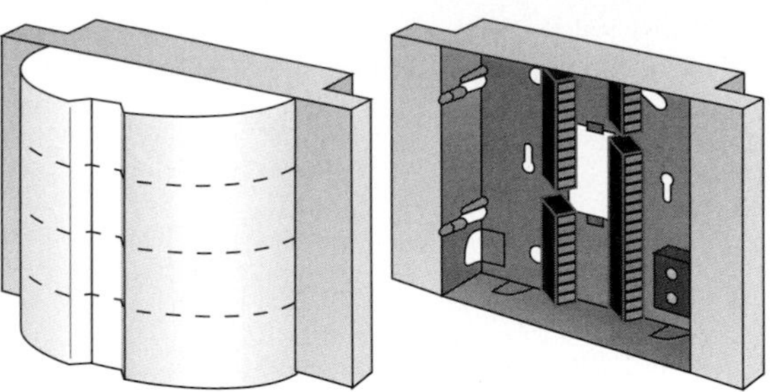

Fig. 4.3/87
Electronic module and adapter box housing

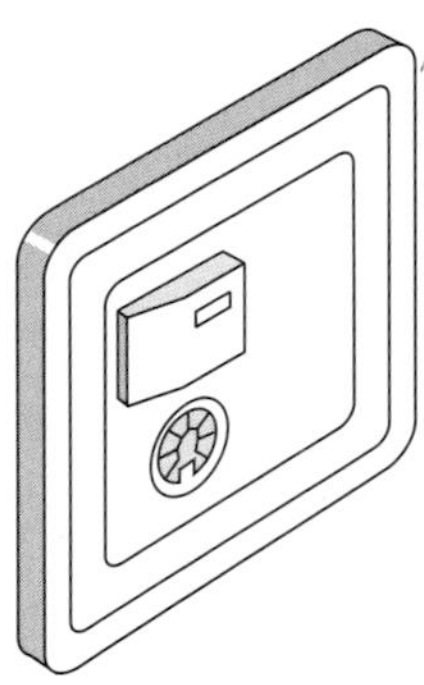

Call module with additional plug-in contact (7 pole)

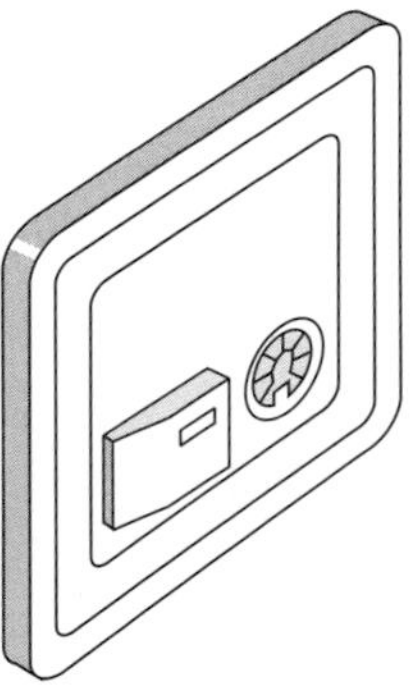

Call unit with call pushbutton and additional plug-in contact (7 pole)

Fig. 4.3/88
Call pushbuttons from the DELTA plus "clino opt 99" system

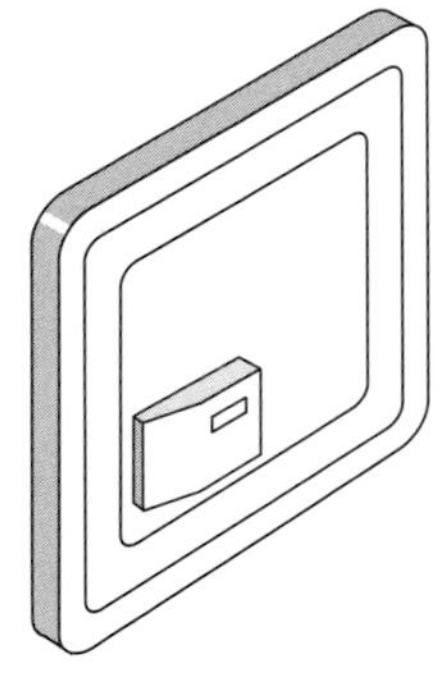

With cancel or presence pushbutton and reminder lamp (LED)

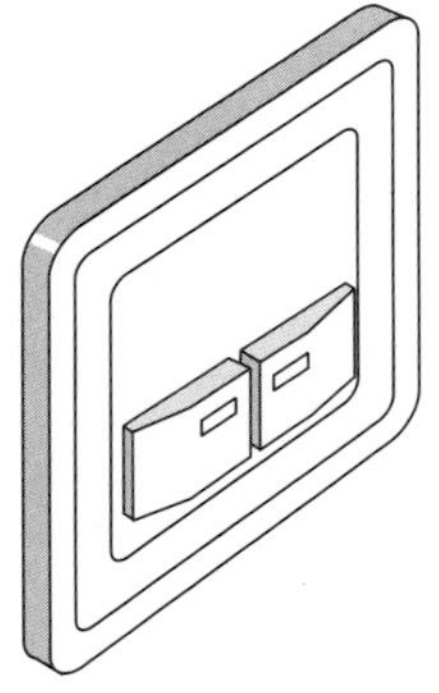

With cancel or presence pushbutton and reminder lamp (LED), call pushbutton and reassurance lamp

Fig. 4.3/89
Cancel pushbuttons from the DELTA plus "clino opt 99" system

table. They are also available as moveable units, such as pear pushbuttons, multiple pushbuttons, or patient handsets (in conjunction with the wall interface electronics for "clino opt 99") (Fig. 4.3/90).

In rooms with en-suite WCs, the cancel or presence pushbuttons are connected in parallel.

Zone indicator lamp

Fig. 4.3/91 shows a zone indicator lamp with 2 lamp sections for indication by ward.

Light bulbs

Light-emitting diodes

Light bulbs to DIN 41050 Part 2, 4.2 are used in corridor, zone, and direction indicator lamps for visual signaling. All other visual signaling devices (e.g. reassurance and reminder lamps, status indicators, etc.) are equipped with light-emitting diodes (LEDs) as standard. These have a long service life, and thus ensure a high level of signaling reliability.

Direction indicator lamps

Direction indicator lamps (Fig. 4.3/92) are used to indicate the location at which a call has been initiated. They can also be retrofitted in the corridor databus, without additional wiring.

Zone and direction indicator lamps can also be replaced with information displays. These are equipped with red LEDs for the luminous alphanumeric display, and an integrated tone generator. Messages appear in plain text in accordance with the configured room designation.

Operational features

Normal calls

Normal calls are initiated when the patient presses the call pushbutton in a pushbutton unit with integrated reassurance lamp.

The reassurance and corridor lamps (red), as well as the zone and direction indicator lamps (white) remain continuously lit to indicate this type of call.

Nurse presence

Calls that are transferred to rooms where the presence of a nurse has been set ("nurse pre-

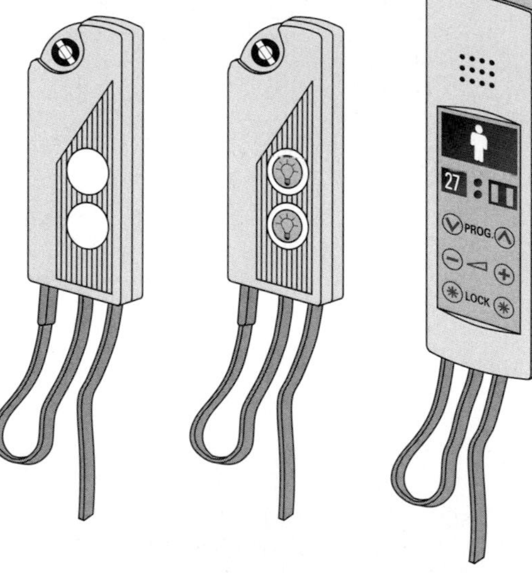

Pear pushbutton with call pushbutton and reassurance lamp

Multiple pushbutton with call pushbutton, reassurance lamp and light pushbutton

Patient handset with call pushbutton, reassurance lamp, light pushbutton, function keys for TV/radio channel and volume control

Fig. 4.3/90
Pear pushbutton, multiple pushbutton, patient handset

Fig. 4.3/91
Zone indicator lamp unit with two lamp sections

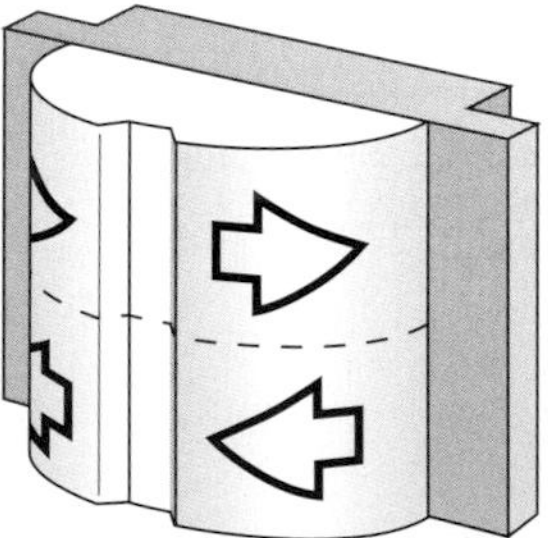

Fig. 4.3/92
Direction indicator lamp unit (right/left), with two direction lamps

sence") are indicated by long, repeated acoustic signals. Display modules also allow calls to be transferred visually to rooms where a presence has been set. The nursing staff cancel the call by pressing the green cancel pushbutton at the point of origin.

Setting a presence

A presence is set by pressing the green pushbutton in the cancel or presence pushbutton unit (e.g. in the duty room unit, Fig. 4.3/93).

Presence lamp

Reminder lamp

This switches on the green presence lamp in the corridor lamp outside the room, activates the buzzer to indicate that the call has been transferred to the room, and enables the emergency call channel. The green reminder lamp in the cancel or presence pushbutton is also lit.

On leaving the room, the nurse resets all functions by pressing the pushbutton again.

Emergency calls

Emergency calls are initiated by nursing staff in the patient's room after a presence has been set.

In the corridor lamp, these calls are indicated visually by the *flickering* red call lamp and the *continuously lit* green presence lamp. The white lamp in the zone and direction indicator lamps also flickers. In rooms in which a presence is set, the call is also transferred acoustically (visually only with a display module) by means of short, rapid signals. The emergency call is cancelled by pressing the green cancel pushbutton at the point of origin.

Enhanced calls

Enhanced calls are special types of call, which can be made available for patients requiring special care. This kind of call is also referred to as a priority call, and is indicated visually by flickering call lamps and acoustically in the same way as an emergency call. Enhanced calls can be allocated to one room on a permanent basis or for a given period of time. Priority calls can be set without additional wiring via the master station (Fig. 4.3/94).

Bathroom/WC calls

Depending on how the zone controller is programmed, calls made from a bathroom or WC are indicated as bathroom/WC calls (continuous light), as bathroom/WC emergency calls (flickering light only when a presence is set) or as enhanced bathroom/WC calls (flickering light only).

In the corridor lamp, this type of call is indicated visually by means of a white lamp. In the case of an en-suite WC, the call functions for the WC are integrated in the electronic module and cancel pushbutton for the room. If there are several entrances to the room and bathroom/WC, the cancel pushbuttons are connected in parallel to all functions.

Call transfer

Calls are transferred by relaying acoustic call signals to rooms in which a presence is set, either within or outside a zone. (Calls can also be transferred visually if display modules are used.) The electronic buzzer is integrated in the cancel or presence pushbutton unit, as well as in the display module.

Diagnostic calls

Diagnostic calls are specially designated calls initiated, for example, by a patient-monitoring device. They are indicated by

Fig. 4.3/93
Duty room unit with presence pushbutton (green), reminder lamp, and acknowledge pushbutton (white)

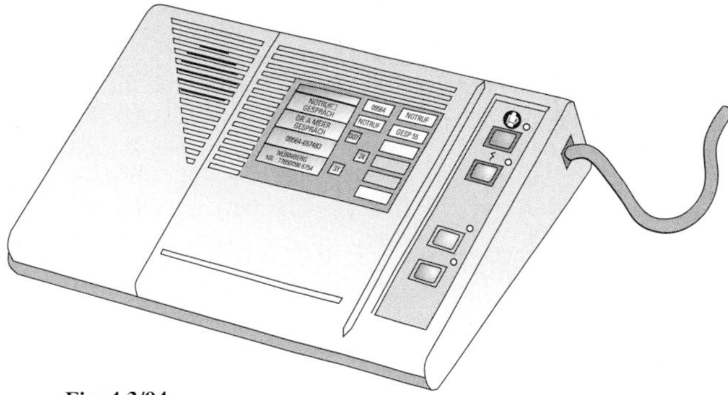

Fig. 4.3/94
Master station for the duty room. Operated via a touch screen by touching the appropriate areas of the display

means of a flickering red light in the corridor lamp.

Doctor calls

Doctor calls are calls requesting medical assistance. They are indicated by rapidly flickering white, red, green (and yellow) lights in the corridor lamp.

Meal calls

Meal calls are acoustic calls initiated by the nursing staff.

Staff calls

Staff calls are acoustic calls in all rooms in which a presence is set.

Zone linking

Zones are linked during quiet periods, e.g. during the night. Depending on the way in which the system is programmed, zones can be linked in any combination via the memory module of the duty room electronic module. The zones to be linked are selected via the duty selection module. This is used for direct or delayed call transfer to other zones. Zones that have been linked can be changed at any time, without installing additional cabling.

Zone indicator lamps

Zone indicator lamps indicate calls from neighboring zones and are generally installed in corridors (see Fig. 4.3/91).

Selective call answering

Selective call answering is implemented by means of call modules in the patient's room. Each of these call modules can be assigned an individual alphanumeric text, which can then be displayed on the display modules in the duty room or when calls are transferred.

Telephone calls

Telephone calls are indicated by means of a special lamp in the zone indicator lamp unit. Telephone calls are redirected with a distinctive acoustic signal.

Pagers

Pagers that distinguish normal, emergency, and doctor calls can be linked to the system.

Priority for emergency and enhanced calls

The visual and acoustic signals of normal calls are overridden by emergency and enhanced calls issued after these. The normal calls are not signaled until the higher-priority calls have been dealt with and cancelled.

Call storage

In accordance with DIN VDE 0108 Section 6.1.1.2, calls are stored when a power failure occurs in the luminous call system. The call is stored for at least the prescribed 15 s startup time for the stand-by power supply. The "clino opt 99" luminous call system enables calls to be stored for approximately 24 hours.

"clino opt 90 WIN" computerized luminous call system

The "clino opt 90 WIN" computerized luminous call system is extremely flexible and user friendly. It integrates all the standard functions described above, and can be easily adapted to accommodate organizational changes, without requiring new software.

All calls are listed according to urgency and chronological order on a monitor of the computerized central unit, and remain displayed in plain text until they have been dealt with (Fig. 4.3/95). Calls are indicated in each room via the LCD on the room terminal.

This helps to eliminate unnecessary journeys between the patient's room and the duty room. The computer continuously monitors data transmissions to ensure that they are error free.

"clino call" paging system

Connecting the "clino call" paging system to the luminous call system enables medical staff to be contacted efficiently wherever they are. The call receivers can also be freely allocated to the central control point of the luminous call system.

Luminous call systems with speech facility

Functional principles and tasks

Luminous call systems with speech facility support the same high level of functionality as the "clino opt 99" and "clino opt 90 WIN" systems discussed earlier. In addition to this, they feature a speech link between the patient and nursing staff, as well as between members of the nursing staff, thus

Fig. 4.3/95
Operating computer used as a main/central call answering unit in a hospital

helping to reduce their workload considerably. The only additional operator component required is the appropriate answer pushbutton on the master station, on the room terminal or the central control point.

Central call answering unit

The operating computer in the duty room is used to selectively display and answer calls from the patients' rooms or beds. It can also be used to link several areas so that the calls can also be answered from these.

Additional call answering unit

The room terminals in the patients' rooms are used as additional call answering units. When calls are transferred, the terminals indicate the calls from other rooms to which a speech link can be established via the answer pushbutton.

The efficiency of luminous call systems, as well as their user-friendliness, has been optimized by the use of state-of-the-art computer technology, allowing even very uncommon organizational structures to be taken into account.

"clino phon 95 WIN"

In the following, the effectiveness and performance of computerized luminous call systems with unrestricted duplex communication will be discussed using the "clino phon 95 WIN" nurse call system as an example.

Nurse call control unit

"clino phon 95 WIN" is a microprocessor-controlled nurse call system with address-coded transmission which operates in the time-division multiplexing mode. Independent computer systems are used as the functional units of the nurse call control unit and operating computers. The required software runs on a multi-user multi-tasking real-time operating system, and allows several application programs to be run simultaneously. The nurse call control unit is installed in a wall-mounted housing that also contains the interface to building systems. "clino phon 95 WIN" allows the computers in the luminous call system to be networked with those of the hospital administration, ultimately enabling a complete communications network to be set up (Figs. 4.3/96 and 4.3/103).

Wiring

Only 6 cores are required for the power supply, speech, and signal lines (Fig. 4.3/97).

Self-monitoring

The system is self-monitoring. A check bit checks each data packet for error free transmission in both directions. Any errors are reported to the persons responsible.

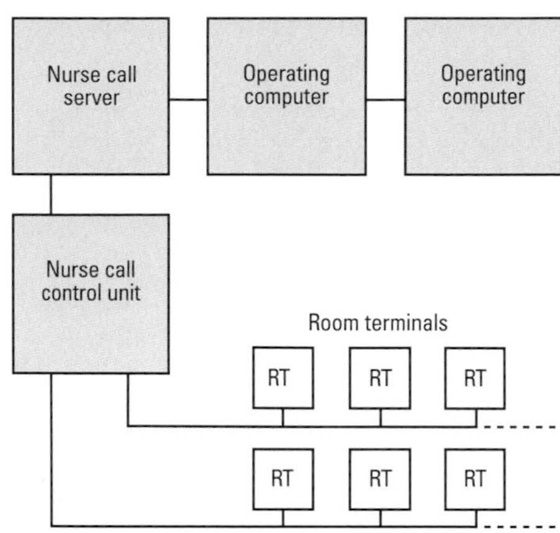

Fig. 4.3/96
Schematic representation of a computerized luminous call system

Speech links

Speech links are established on the basis of unrestricted duplex communication. This is the most convenient form of speech transmission, since it does not require hand-held devices, which means that staff and patients alike have both hands free during conversations (Fig. 4.3/98). Conversation is possible not only between the central and additional call answering units/operating computers and the patients' rooms, but also from room to room via the auxiliary call answering function of the room terminals by means of unrestricted, private duplex communication.

Normal calls that have been answered can be cancelled from any of the call answering units. Emergency and enhanced calls, however, can only be cancelled at the point of origin.

Talking to patients

Staff can talk to patients in their rooms or beds from the central call answering units and operating computers, thereby establishing a speech link with them. The staff cannot hear patients, however, until the patients have cancelled the "private communication" function by pressing the answer pushbutton.

General address call

Using the general address function, calls can be made from the central call answering unit or operating computers in a number of rooms simultaneously via the speech channels of the luminous call system.

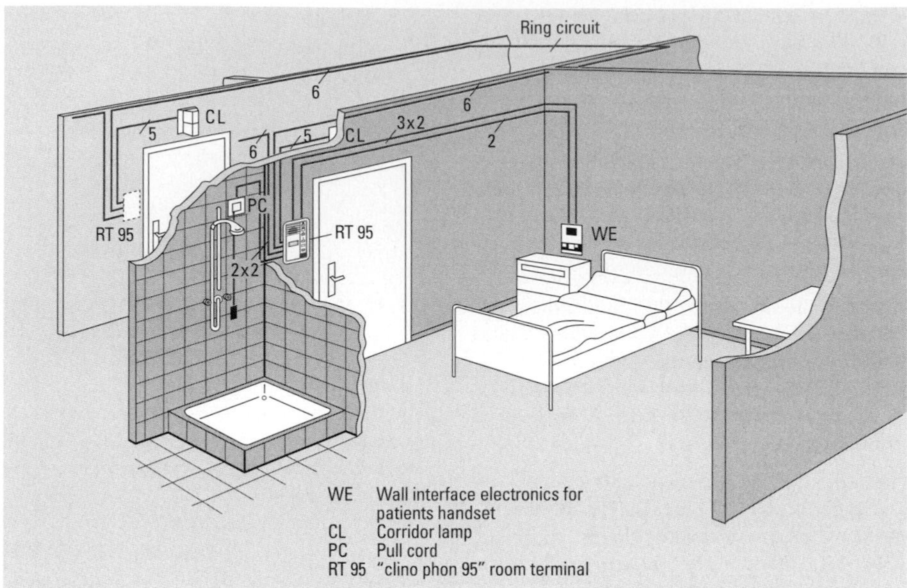

Fig. 4.3/97
Luminous call system with speech facility in a hospital ward. 6-core ring circuit, signaling circuits (2 × 2 core, 3 × 2 core and 5 core), speech circuits 2 core (shielded)

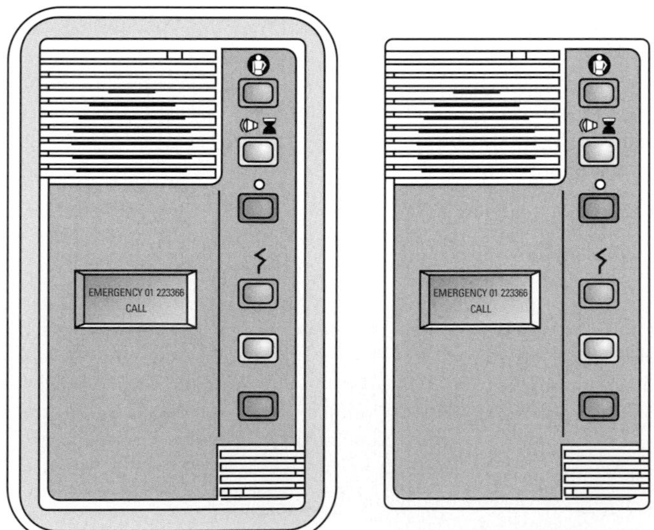

Fig. 4.3/98
Room terminal for surface and flush mounting as an additional call answering unit for unrestricted duplex communication

Private communication The "private communication" function prevents patients from being listened to without their consent.

Nurse presence In this system, nurse presence (see also page 506) is divided into two presence categories: A I for qualified nurses, and A II for trainees. Other criteria can be used to allocate these categories to members of staff, e.g. doctors or nurses.

Nurse presence is identified according to room number and presence category, and displayed on the screen of the central call answering unit/operating computer and at the room terminals. At the same time, a set presence permits speech links to be established between rooms, as well as between the room and the duty room.

Presence category II is indicated by yellow presence lamps.

Call transfer Calls are transferred as described on page 507.

Remind mode The equipment configuration of the system also enables calls identified by origin and type to be recognized and indicated, and normal calls to be parked in remind mode or cancelled remotely. In the remind mode, an answered call is indicated and stored until it can be dealt with by the staff. Emergency and enhanced calls can be cancelled only at their point of origin.

Functions of the operating computer All information (including information accessible via the database) appears in plain text in the GUI on the monitors of the operating computers and the central call answering units (Fig. 4.3/99). Calls are indicated according to priority and time of receipt, and are listed as follows:

- ▷ Type of call,
- ▷ Zone number,
- ▷ Room number,
- ▷ Room characteristic.

The following entries or answering functions are carried out using the computer keyboard or mouse:

- ▷ Call answering / cancellation / reminder function, and call transfer to nursing staff (addressing set presences),
- ▷ Direct call to rooms or beds (via GUI), allocation of room numbers, assignment of priorities, general address calls,
- ▷ Access control via user identification, service functions,
- ▷ Call transfer control, selection of information displays, operation by mouse or keyboard, parallel use of other Windows applications.

Enhanced functionality The standard functionality described above can be easily expanded to include the following:

- ▷ Short-term remind function for each call, e.g. medication, drinks,
- ▷ Long-term remind function with different types of reminder,
- ▷ Output of all system data via a printer, clipboard function for relaying messages between shifts,
- ▷ Patient data entry and management including bed occupancy management, link to existing computer system for transferring patient data,
- ▷ Connection to the "clino call" paging system, information calls to pagers (text and recipient numbers can be entered and sent to the call receivers),
- ▷ Modem for remote diagnosis and importing software via telecommunications systems.

Functions of the room terminals The room terminals house the electronics for data transmission between the room and the central control unit. They also contain the following operator and control components required for the additional call answering

Fig. 4.3/99
Operating computer in a "clino phon 95 WIN" luminous call system

unit, including the short-circuit-proof drivers for the corridor lamps:

Functional scope

Inputs

Outputs

- ▷ 8 call circuits (4 × bed, 1 × diagnostic, 1 × room, 1 × bathroom/WC, and 1 × doctor call),
- ▷ 2 presence circuits,
- ▷ 1 input for cancel pushbutton (bathroom/WC),
- ▷ 4 inputs for bed microphones,
- ▷ 1 *corridor lamp*, 4 reassurance lamps on the beds,
- ▷ 1 bathroom/WC lamp,
- ▷ 1 presence lamp A I,
- ▷ 1 presence lamp A II,
- ▷ 1 buzzer, 1 lamp for conversation, 1 service lamp,
- ▷ 4 outputs for connecting bed loudspeakers.

Control units for patients in the form of patient terminals, patient handsets or withdrawable units for bedside cabinets, are available for all luminous call and speech functions (Fig. 4.3/100). Depending on their component configurations, the control or withdrawable units, which can be connected by means of a plug, can include the following:

- ▷ Call pushbutton with reassurance lamp,
- ▷ Loudspeaker for speech link (electro-acoustic system, if required),
- ▷ Microphone, volume control and channel selector switch,
- ▷ Light pushbutton(s),
- ▷ Additional socket-contact for diagnostic call,
- ▷ Additional socket-contact for pear or multiple pushbuttons,
- ▷ Additional socket-contact for headphones.

Connecting luminous call systems to building systems using *instabus* EIB

Connection to *instabus EIB*

instabus EIB couplers as well as operating, indicating, and signaling components with the appropriate *instabus* interfaces are available for connecting "clino opt" and "clino phon" luminous call systems to building systems via *instabus EIB* (see Section 4.4).

This connection allows:

- ▷ Flexible system control,
- ▷ Data transfer of various system messages or signals,
- ▷ Wiring to be simplified,
- ▷ Fast and easy modifications to switching functions,
- ▷ Incorporation of lighting and louver blind control, as well as electro-acoustic systems.

Fig. 4.3/101 shows an example of how the luminous call system "clino opt 99" can be connected to *instabus EIB*.

Systems and components for enhancing or adapting standard functionality

The following additional components are available to tailor the luminous call system to the particular organizational structure and nursing practices of an institution, and for special applications:

- ▷ Paging systems for integration in the luminous call systems described above (Fig. 4.3/102), **"clino call"**
- ▷ Accounting system for TV and telephone. **"clino tax"**

By integrating individual components in an existing infrastructure, it is possible to gradually build up a modern, comprehensive communications system. Fig. 4.3/103 shows the configuration of a complex open communications network.

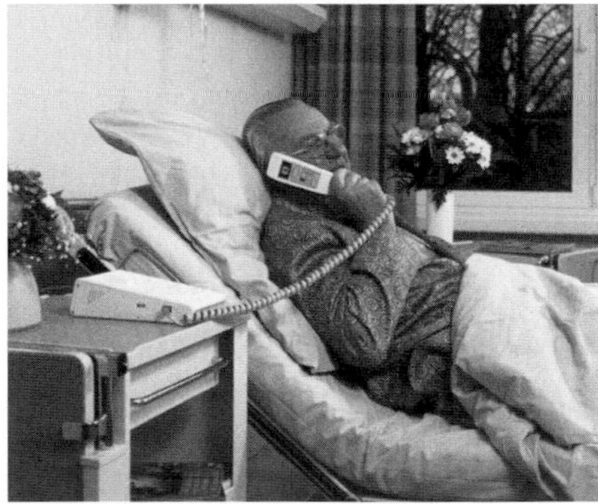

Fig. 4.3/100
Patient control unit for unrestricted or direct duplex communication

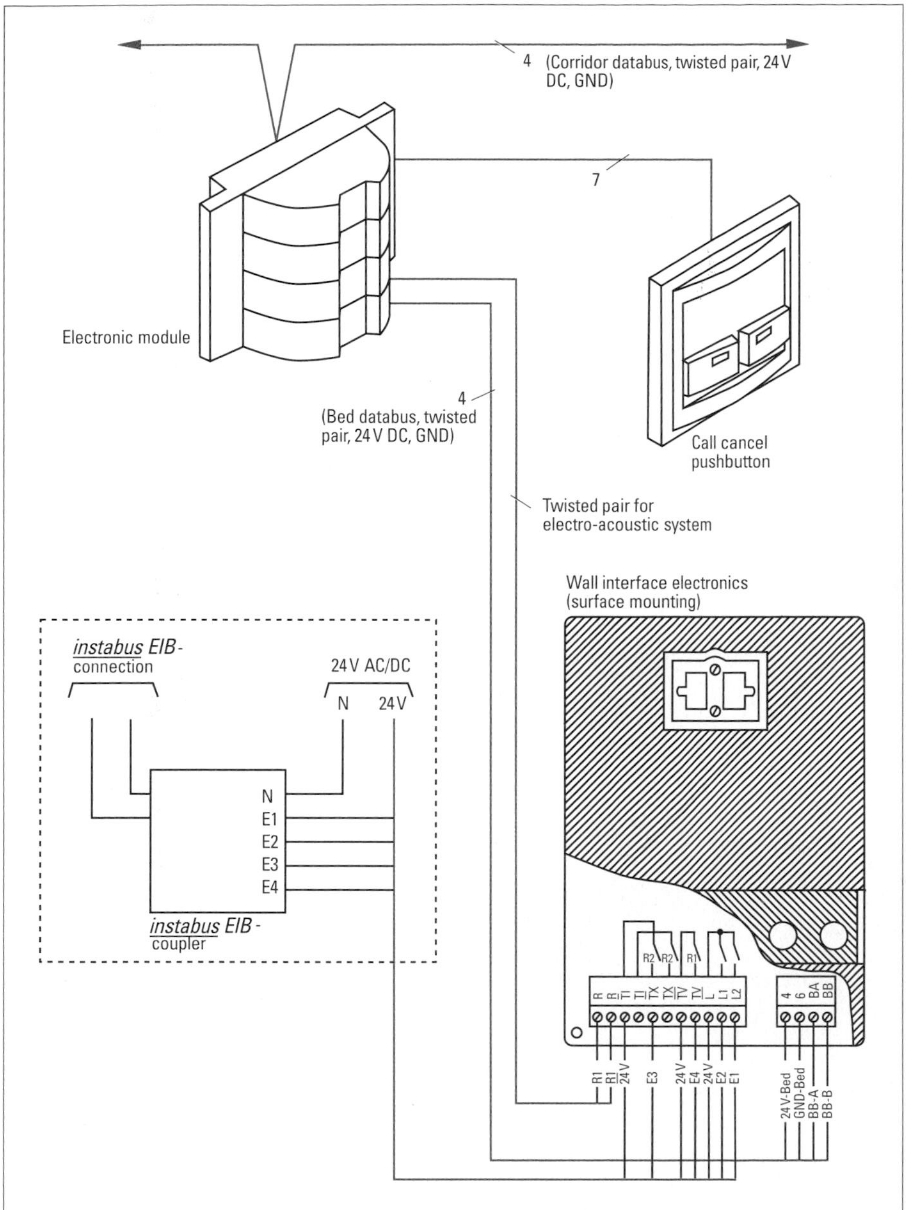

Fig. 4.3/101 Example of a luminous call system "clino opt 99" connected to *instabus EIB*

Fig. 4.3/102 "clino call" radio receiver, tone/display pager with and without speech facility

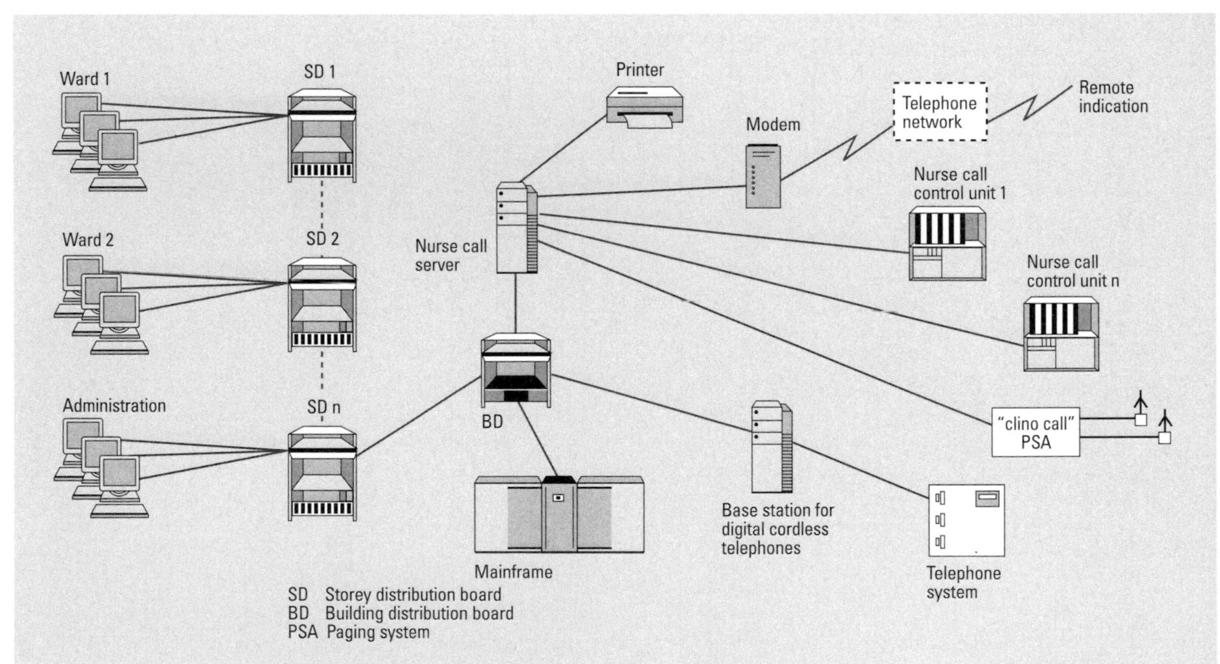

Fig. 4.3/103 Structure of a luminous call system, integrated in an open communications network

4.3.2.4 Signaling systems, door and building call systems, video house call systems

General

Conventional and two-wire signaling systems, door and building call systems, as well as video house call systems are used in residential and office buildings, schools, and industrial enterprises. Although these systems are found predominantly in residential buildings, their configurations differ only very slightly when they are used for other applications.

Systems of this type are generally required to allow visitors to announce their presence by means of acoustic signaling devices, or to request entry to the building via speech systems. By contrast, systems installed outdoors are primarily used to signal breaks, sound alarms, etc.

Standards

The requirements for signaling systems, as well as door and building call systems (which are determined, to a certain extent, by the particular installation) are specified in the following standards:

▷ "Selection and erection of equipment; Low-voltage generation sets": HD 60384/IEC 60364-1...7, DIN VDE 0100-551,
▷ "Isolating transformers and safety isolating transformers": EN 60742/IEC 60742/VDE 0551,
▷ "Particular safety requirements for equipment to be connected to telecommunication networks": EN 41003/VDE 0804 Part 100,
▷ "Limits and methods of measurement of radio disturbance characteristics of industrial, scientific and medical (ISM) radio-frequency equipment": EN 55011/IEC:CISPR 11/VDE 0875 Part 11.

In Germany, the standards of the former "Fernmeldetechnisches Zentralamt" (Central Telecommunications Authority) must also be observed (e.g. "Interfaces for door call systems" FTZ 123 D12).

Acoustic, visual and combined signaling devices

The most widely-used acoustic signaling devices include:

▷ Indoors: electric bells, buzzers, door chimes, and small sirens
▷ Outdoors: alarm bells, horns, and sirens.

In addition to these, visual signaling devices will also be encountered, such as rotating mirror lamps and strobe lights, as well as combined signaling devices that emit acoustic and visual signals.

Power supply

Signaling systems, and door and building call systems are supplied with 8 V AC or 12 V AC by means of bell transformers or power supply units. While many signaling devices can be operated with 8 V DC, this voltage is seldom used. The same applies to the operating voltage 24 V AC/DC. Due to the higher volume and very long cable lengths required, signaling systems for outdoor use are generally operated at 230 V AC. In these cases, too, operating voltages such as 220 V DC, 110 V AC/DC, and 60 V AC/DC are uncommon.

4.3.2.4.1 Signaling systems

Use

The simplest and most cost-effective means of generating acoustic call signals is the bell system, which has been widely used for some time. Bell systems are used to announce visitors in a wide variety of environments, such as single-family houses, villas, and medical practices, as well as in offices, shops, two-family houses, and apartment buildings. This type of system can comprise any number of acoustic signaling devices.

Bell system

A bell signaling system consists of a bell transformer to generate the extra-low voltage power supply of 8 V AC, the bell pushbutton at the house entrance, and the bell, buzzer, or door chimes in the house or apartment.

Door chimes

In modern residential and functional buildings, simple bells are very often replaced by electromechanical or electronic door chimes. With electromechanical door chimes, the acoustic signal is generated mechanically. In electronic door chimes, by contrast, the signal is generated by means of an electronic circuit, and transmitted via a loudspeaker.

While electromechanical door chimes are supplied via bell transformers, electronic door chimes can be powered either via bell transformers or by an internal 9 V monobloc battery in the door chime unit itself.

With versions supplied via transformers, an initial acoustic signal is output when the voltage is applied (when the bell pushbutton is pressed) and a second signal when the voltage is disconnected (when the pushbutton is released). A continuous signal is generated if the voltage remains switched on for a prolonged period of time. These versions are particularly suitable for use in monitoring systems, since the signaling components are not subject to mechanical wear and tear, which makes them extremely reliable when operated continuously.

With battery-powered door chimes, the series of signals continues to sound after the bell pushbutton is released. The bell pushbutton merely provides a start pulse, following which the series of signals is sounded automatically.

The start pulse can be generated either by means of an extra-low voltage supply (bell transformer) or by means of a battery.

Function

The signaling device is activated by pressing the bell pushbutton. The occupier of the building/apartment has to open the door of the building/apartment himself to let the visitor in.

Other types of system

In addition to simple bells and door chimes, systems enhanced to include electric door openers are becoming increasingly widespread. When a door opener pushbutton installed in the apartment is operated, voltage is applied to the magnet system of the door opener, releasing the latch and allowing the visitor to open the gate or door himself.

In apartment buildings, it is particularly usual to install a second bell pushbutton outside the door of the apartment so that visitors from within the building can also announce their presence directly. In cases where additional pushbuttons are installed, it is useful to provide two distinct signals so that the occupier can tell from the sound of the signal whether the visitor is at the door of the building or of the apartment.

4.3.2.4.2 Door and building call systems

Use

Door and building call systems are more expensive than the systems described above, but do offer considerable advantages for the user. They enable the occupier of a house/apartment to find out the identity of visitors and the purpose of their visit before granting them access via the door opener. This greater emphasis on security has led to the widespread use of door and call systems, and a corresponding decline in the use of traditional signaling systems. Door and building call systems are used in single-family houses, villas, and terraced houses, as well as in apartment buildings, residential complexes, and high-rise buildings. They are also suitable for use in medical practices, legal offices, business premises, offices, and industrial plants. Table 4.3/103 provides an overview of the door and call systems described in this section.

Structure

Door and building call systems can be divided into two basic categories: non-simultaneous press-to-talk systems, and duplex systems. The difference between these systems is that press-to-talk allows speech in only one direction at a given time, whereas the duplex system, like the public telephone system, is open for speech transmission in both directions simultaneously. Both call systems comprise one or more door stations, a power supply unit, and the house stations. With press-to-talk systems, a bell transformer is also required for the power supply unit.

Press-to-talk systems

Press-to-talk speech systems permit only private speech communication between the door station and one particular house station (Fig. 4.3/104 a). The house stations are called as required from the door station. The call signal is generated by the integrated buzzer in the house station, by external door chimes or by a bell. The occupier establishes the speech link by pressing the speech pushbutton. When the speech pushbutton is released, the speech direction is automatically reversed, and the visitor can respond. A door-opener pushbutton is incorporated in each house station.

Duplex systems

Duplex systems permit both private as well as non-private speech communication between the door station and a particular house station (Fig. 4.3/104 b and c). In addition to this, communication between a given number of house stations is possible. House stations are called from the door station. The call signal is generated by the internal buzzer of the house station, by external door chimes or by a bell. With internal communication between house stations, the buzzer is used as an internal call signal which means that external door chimes or a bell must be provided for calls from the door station. The speech link to the door station or the other house stations is established by lifting the receiver of the house station. During internal conversations between house stations, the speech link to the door station is automatically disabled. A door-opener pushbutton is

Table 4.3/3 Overview of door and building call systems

Type of system	**Press-to-talk**	**Duplex**		**Hands-free**
Type of circuit	Parallel	Parallel	Meshed system	Radial system
Speech function	Only in one direction at a time	In both directions simultaneously	In both directions simultaneously	In both directions simultaneously
Acoustic signaling device	Integrated	Integrated	Integrated for internal communication	Integrated
Call differentiation (e.g. gate, storey door)	Possible via external door chimes or bell	Possible via integrated bell or external door chimes/ bell	Possible via external door chimes with call differen-iation	Integrated
Power supply	Power supply unit and bell transformer	Power supply unit	Power supply unit	Power supply unit and central amplifier
House stations possible number type	 Any number Surface wall-mounted unit, can be mounted on device boxes	 Any number Surface wall-mounted unit, can be mounted on device boxes	 1 to 8 Surface wall-mounted unit, can be mounted on device boxes	 1 to 8 Surface wall-mounted unit, tabletop device
Door-opener button	Integrated	Integrated	Integrated	Integrated
Private speech function	Integrated	Possible	Possible	Integrated
Wiring material	Indoors In the ground	J-Y(St)Y A-2YF(L)2Y		
Range in meters with conductor diameter 0.6 mm 0.8 mm 1.0 mm Bracketed values for use of additional external door chimes	 90 (40) 150 (65) 250 (105)	 160 (40) 280 (70) 450 (110)	 160 (40) 280 (70) 450 (110)	 50 (50) 70 (70) 100 (100)

incorporated in each house station. Duplex speech systems are being equipped to an increasing extent with video screens, allowing the visitor to be seen as well as heard.

Hands-free systems

In comparison to press-to-talk and duplex systems, hands-free systems have the advantage of being more convenient and considerably more flexible (Fig. 4.3/104 d). With these systems, the user can talk freely in a room, without having to constantly press a pushbutton during the conversation or hold a receiver in his hand.

In addition to this, hands-free systems allow the user to freely select group calls, to monitor rooms, disconnect stations, and restrict call signals.

Types of system

With press-to-talk systems (see Fig. 4.3/105 for circuit diagram), the door station is connected in parallel to any number of house stations. The house stations can only establish direct contact with the door station. An existing speech link can be neither interrupted nor interfered with by selecting a different house station. The house stations are designed for surface mounting on device boxes.

With duplex systems, a basic distinction is drawn between two configurations:

▷ All house stations are connected in parallel, as in a press-to-talk system,
▷ Meshed system.

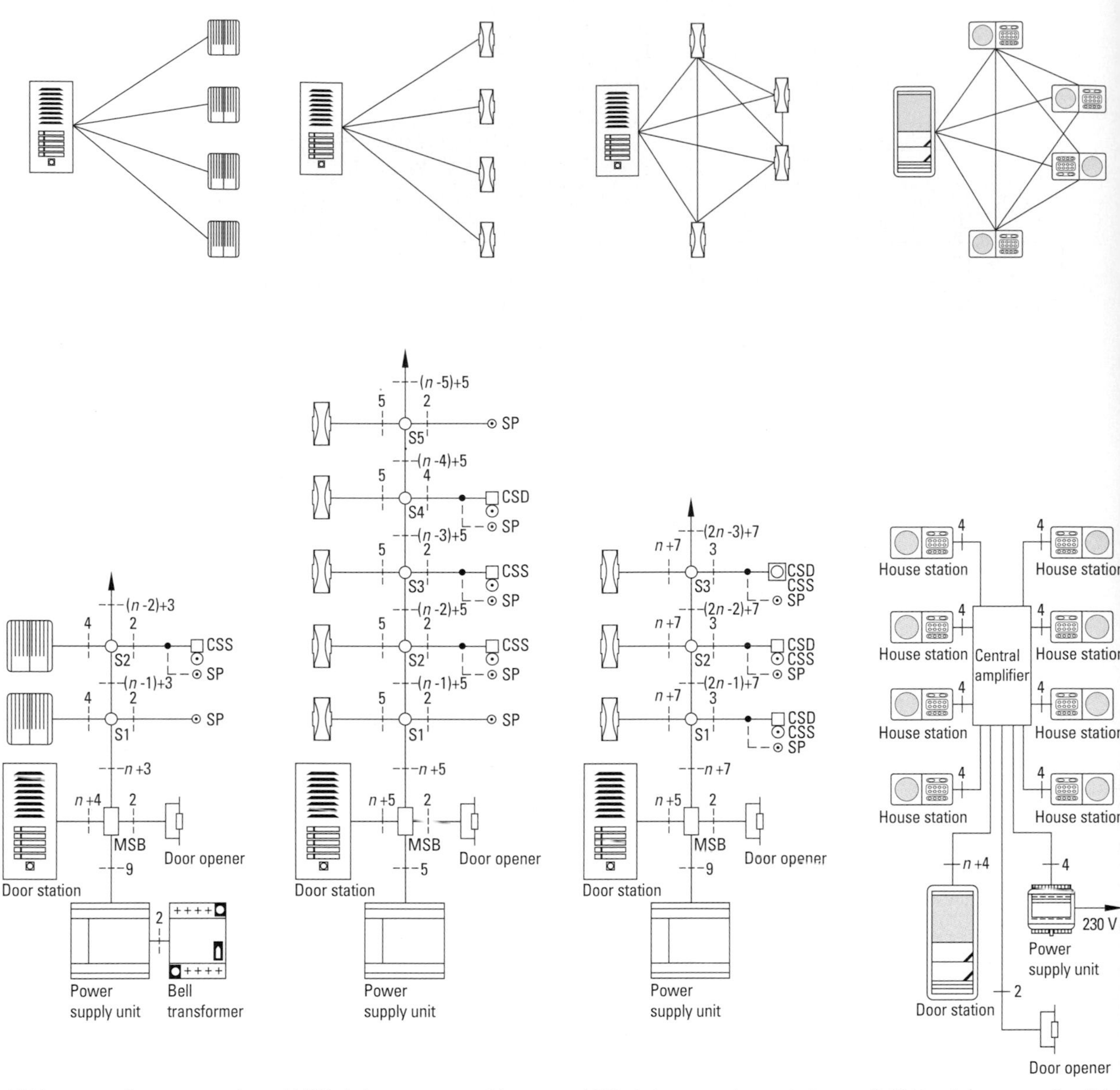

a) With press-to-talk systems, speech is possible in only one direction at a given time. Speech communication is private.

b) With duplex systems, speech is possible in both directions simutaneously. When systems are connected in parallel, private and non-private speech communication is possible.

c) With duplex systems in a meshed system cnfiguration, the number of house stations is restricted.

d) With hands-free systems in radial system configurations, up to eight house stations and one or more door stations can be connected. The system can be expanded at any time.

CSD Call signal from door station
CSS Call signal from storey
n Number of house stations
MSB Main splitting box
S Splitting box in storey
SP Storey pushbutton

Fig. 4.3/104 Layout diagrams to determine the number of conductors per wire

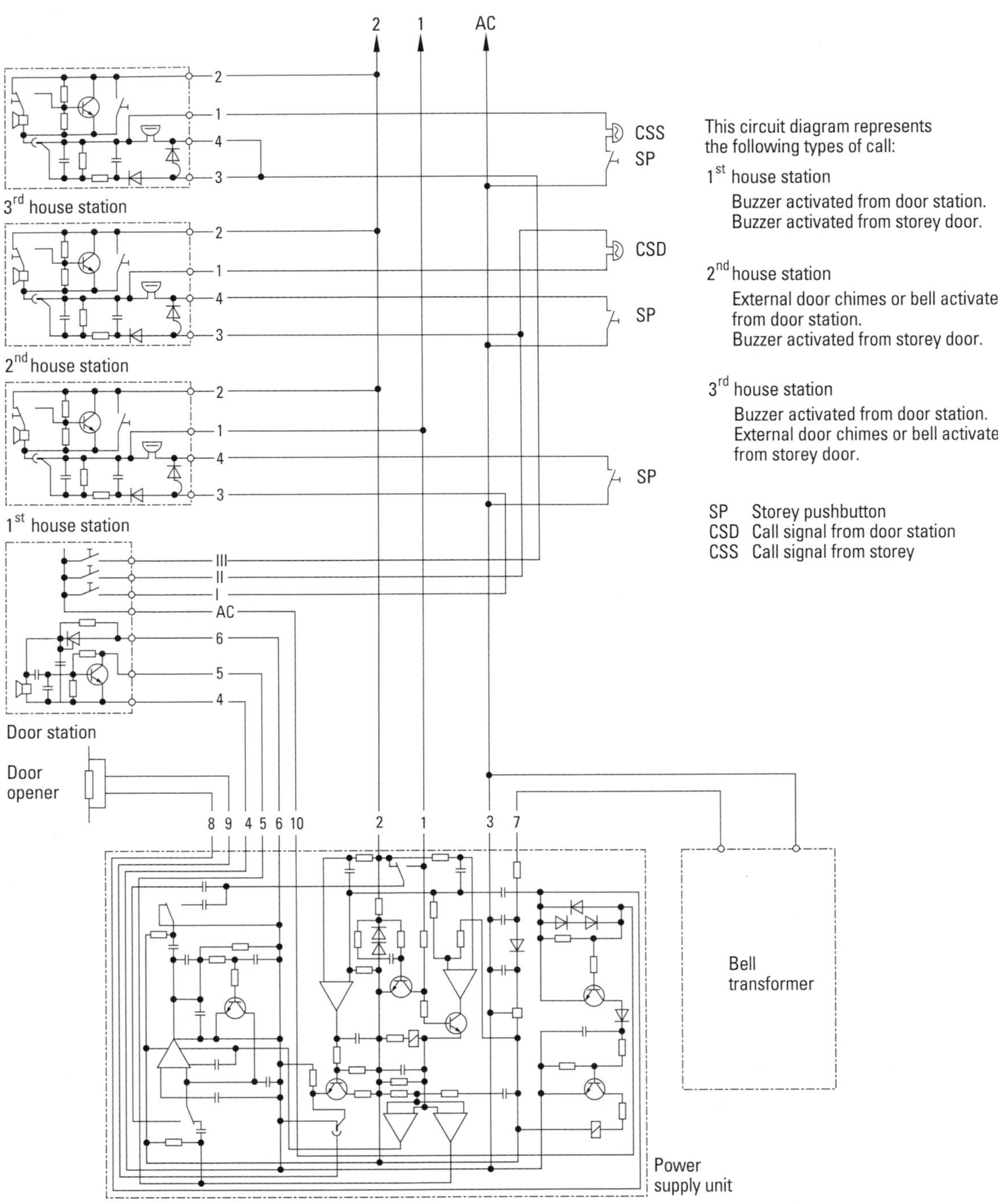

This circuit diagram represents the following types of call:

1st house station

Buzzer activated from door station.
Buzzer activated from storey door.

2nd house station

External door chimes or bell activated from door station.
Buzzer activated from storey door.

3rd house station

Buzzer activated from door station.
External door chimes or bell activated from storey door.

SP Storey pushbutton
CSD Call signal from door station
CSS Call signal from storey

Fig. 4.3/105 Circuit diagram of a press-to-talk system

When all the door stations are connected in parallel (e.g. DOORSET simplex), each house station can only establish direct contact with the door station (Fig. 4.3/106). The speech link is basically non-private, but the system can be retrofitted for private speech communication by adapting each of the house stations. There is no restriction on the number of house stations.

Duplex systems based on a meshed system configuration (Fig. 4.3/107) not only permit direct contact between the house station and door station, but can also be used for communication within the building. The number of house stations is restricted with these house call systems, since the number of call pushbuttons depends on the size of the house station. DOORSET duplex systems are designed for eight house stations. They can be expanded to comprise nine house stations, if the wiring for the door-opener pushbutton is installed separately.

Depending on the type and size of the installation, duplex systems in parallel and meshed system configurations can be combined to form a single call system. If two or more door stations are required in a large residential complex, installing an additional changeover device ensures that the speech link can only be established with the door station at which the visitor activated the call signaling device. The door opener is released in a similar way.

Hands-free systems are designed as radial systems (see Fig. 4.3/104 d), and are equipped with a central amplifier, to which up to eight house stations and one or more door stations can be connected. Each house station is connected to the central amplifier via a 4-core cable (see Table 4.3/3), thereby permitting all types of external and internal speech communication.

Hands-free systems can be expanded at any time.

Planning notes

No special type of wiring (stranded, screened, etc.) is required for extra-low voltage signaling systems. This means that the following standard cables can be used for installations

in conduits Y,
indoors YR,
in the ground A-2Y (L) 2 Y.

With door and building call systems, however, it is advisable to use only shielded cables, e.g.

indoors J-Y(St) Y,
in the ground A-2YF (L) 2Y.

To ensure a high level of speech quality, a conductor diameter of 0.8 mm or even 1.0 mm is recommended. It is also advisable to include several spare conductors in order to accommodate subsequent system modifications, or, if necessary, to extend the range by connecting the conductors in parallel.

4.3.2.4.3 Two-wire door and building call systems (bus technology)

House call system

Two-wire technology (bus technology)

Although the conventional technology described above continues to be used in the majority of door and house call systems for residential and functional buildings, two-wire technology (also known as bus technology) is now becoming increasingly important.

Advanced bus technology enables complete house call systems to be designed using only two wires (conductors). This means that all the information required for communicating between the door station and house station is exchanged via only two wires, thereby permitting call-signaling, speech-link, door-opening, or stairwell lighting functions to be implemented without additional wiring (Fig. 4.3/108).

Structure

This technology not only enables costs to be reduced thanks to simplified wiring and installation, it also improves the quality of communication through the use of electronic devices – from the power supply unit right through to the speech stations.

Based on the modular system, the communication devices are designed in such a way that the various functions required by the user such as private communication, speech control, call recognition (door station/storey call), or parallel connection of two house stations can be implemented via basic devices and pluggable, enclosed supplementary components. It is also possible to connect two house stations in parallel. With expansion modules, the standard system can be extended to include up to 99 bell pushbuttons or residential units (Fig. 4.3/109).

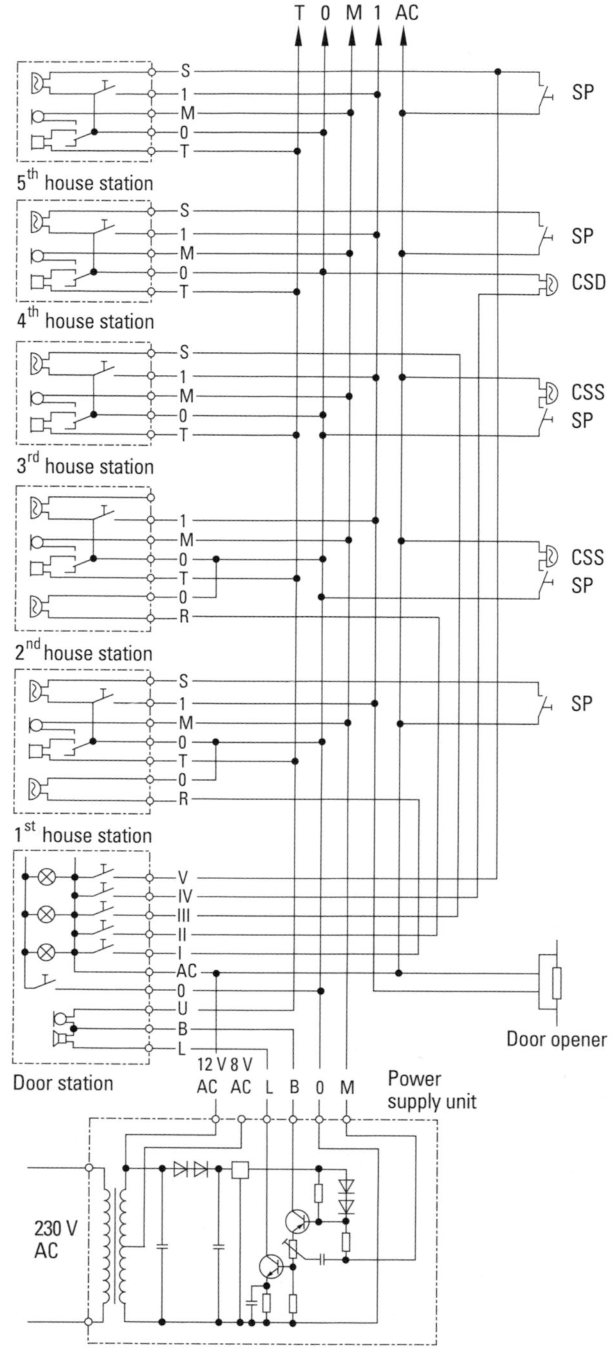

Fig. 4.3/106 Circuit diagram of a duplex system (parallel connection)

This circuit diagram represents the following types of call:

1st house station
Additional bell integrated in house station activitated from door station.
Buzzer activated from storey door.

2nd house station
Additional bell integrated in house station activitated from door station.
External door chimes activated from storey door.

3rd house station
Buzzer activated from door station.
External door chimes or bell activated from storey door.

4th house station
External door chimes or bell activated from door station.
Buzzer activated from storey door.

5th house station
Buzzer activated from door station.
Buzzer activated from storey door.

SP Storey pushbutton
CSD Call signal from door station
CSS Call signal from storey

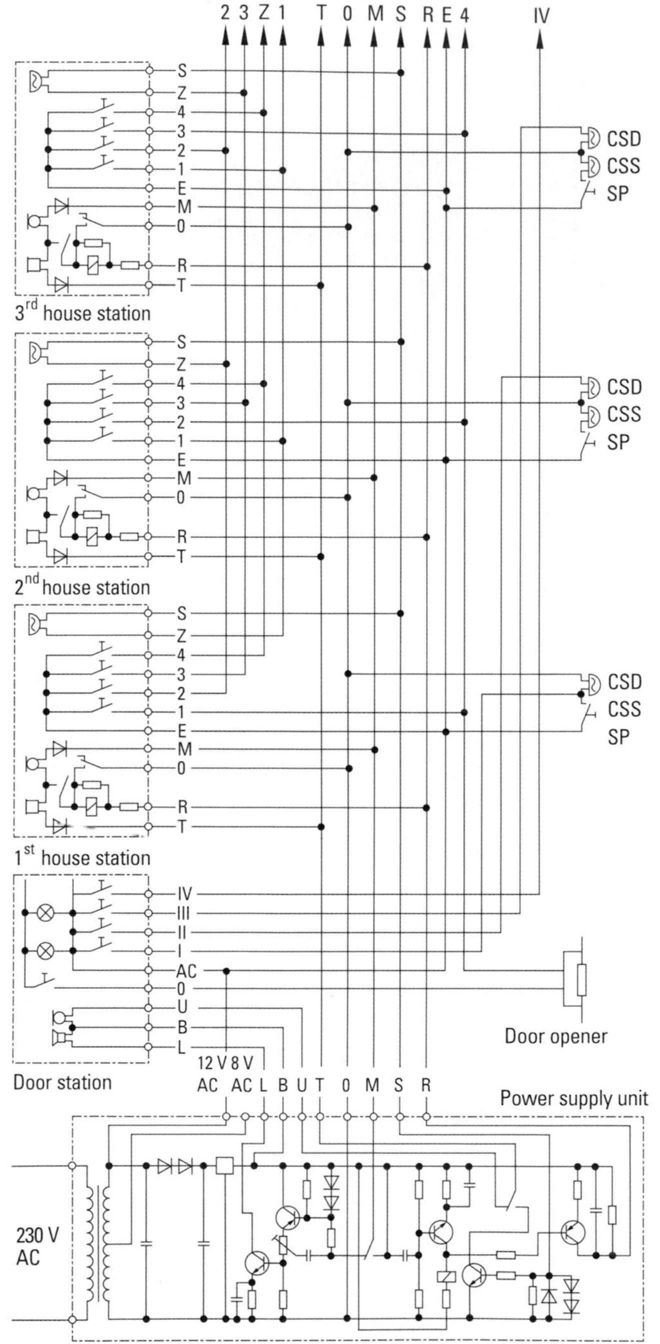

This circuit diagram represents the following types of call:

1st house station

External door chimes or bell activated from door station.
Same external door chimes or bell as before activated from storey door.
House station buzzer for calls within the building.

2nd and 3rd house station

External door chimes with call differentiation activated from door station and storey door.
House station buzzer for calls within the building.

SP Storey pushbutton
CSD Call signal from door station
CSS Call signal from storey

Fig. 4.3/107 Circuit diagram of a duplex system (meshed system configuration)

Features:

- Private communication
- Speech controlled
- Door chime call system with call differentiation for storey call
- Parallel connection of 2 house telephones
- Wear-proof microphone
- Several door stations can be connected
- Short-circuit-proof power supply unit standard door opener
- Existing wiring can be used
- Relay group for external signaling devices

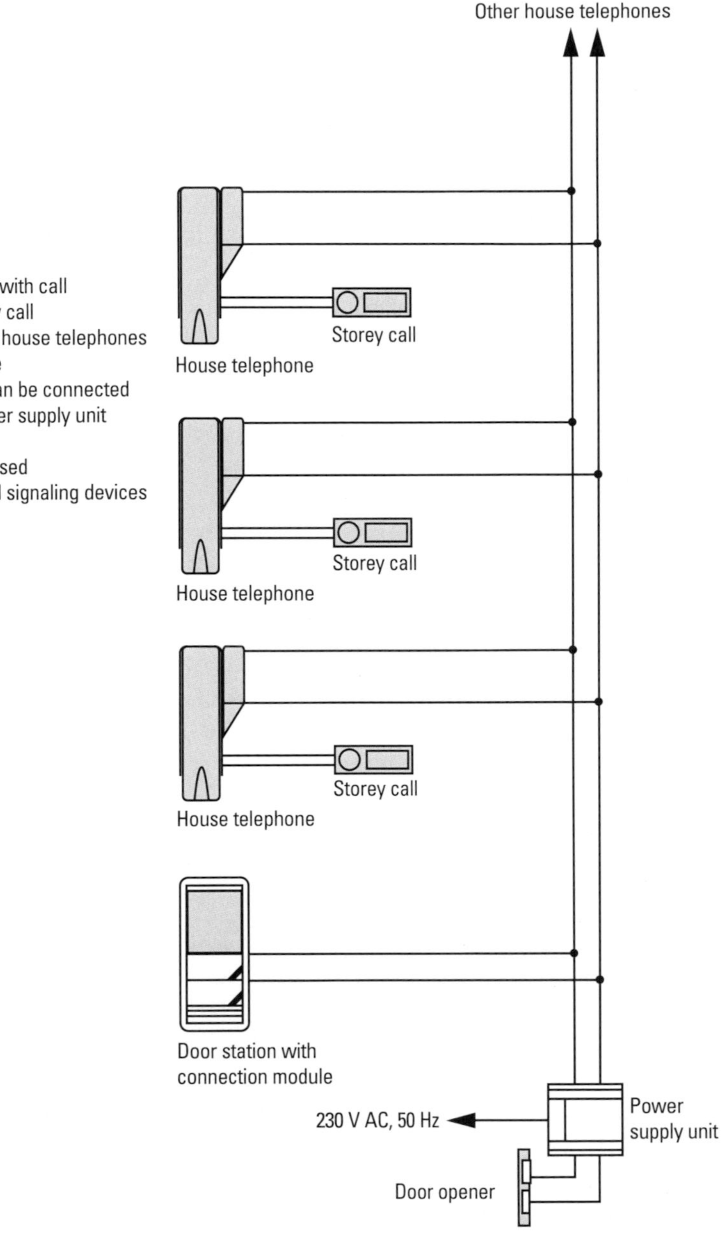

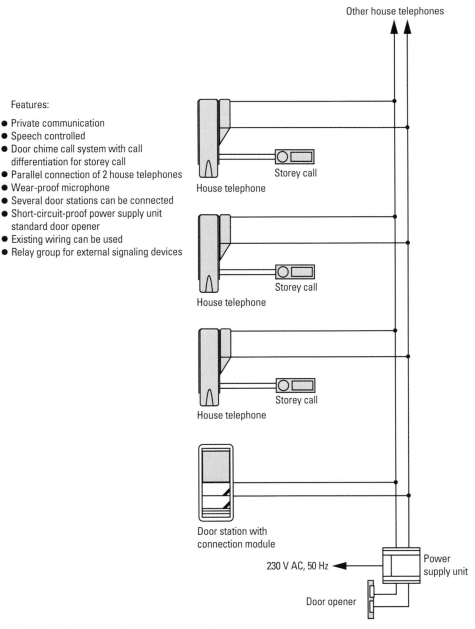

Fig. 4.3/108 Structure and features of a house call system based on two-wire technology

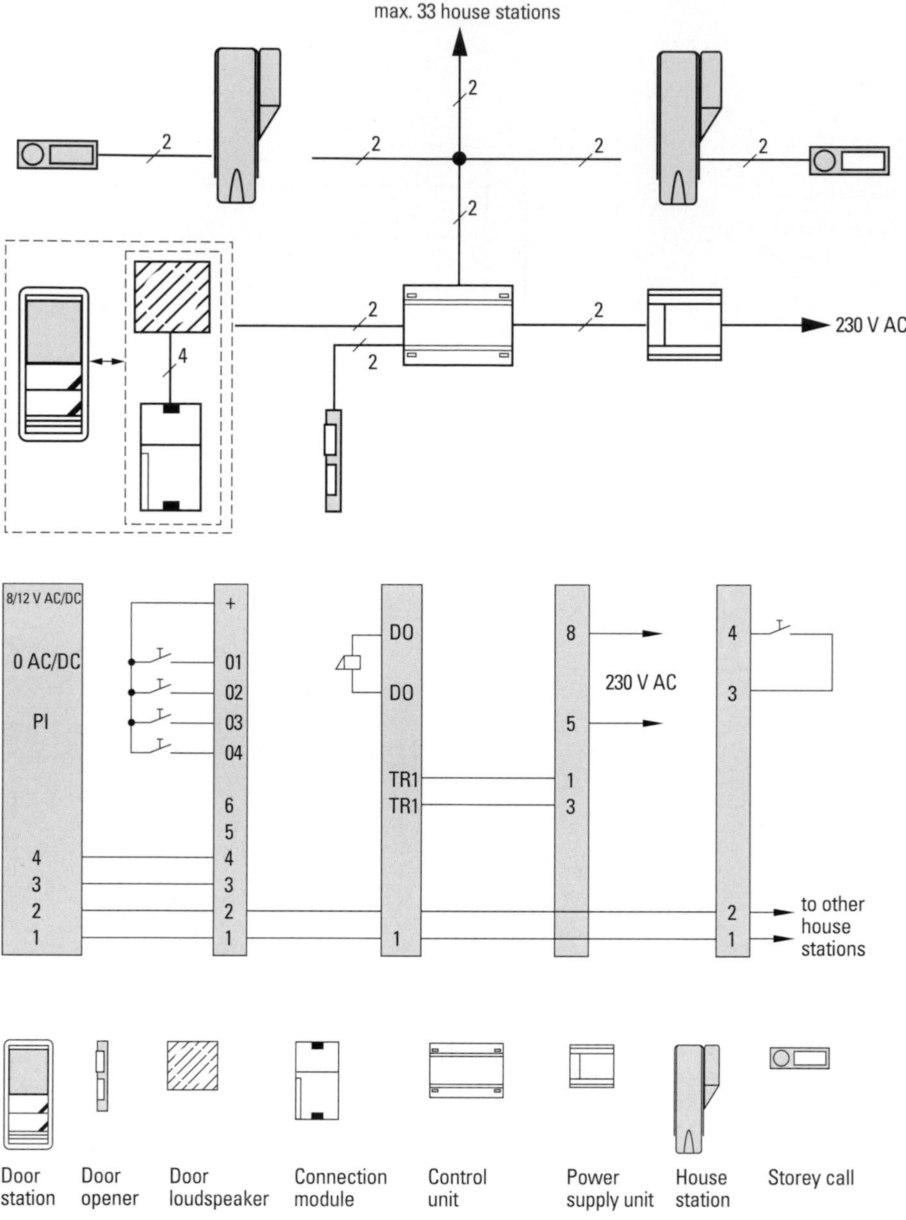

DO Door opener contact
PI Pushbutton illumination
TR1 Transformer 1

Fig. 4.3/109 Circuit diagram of house call system based on two-wire technology

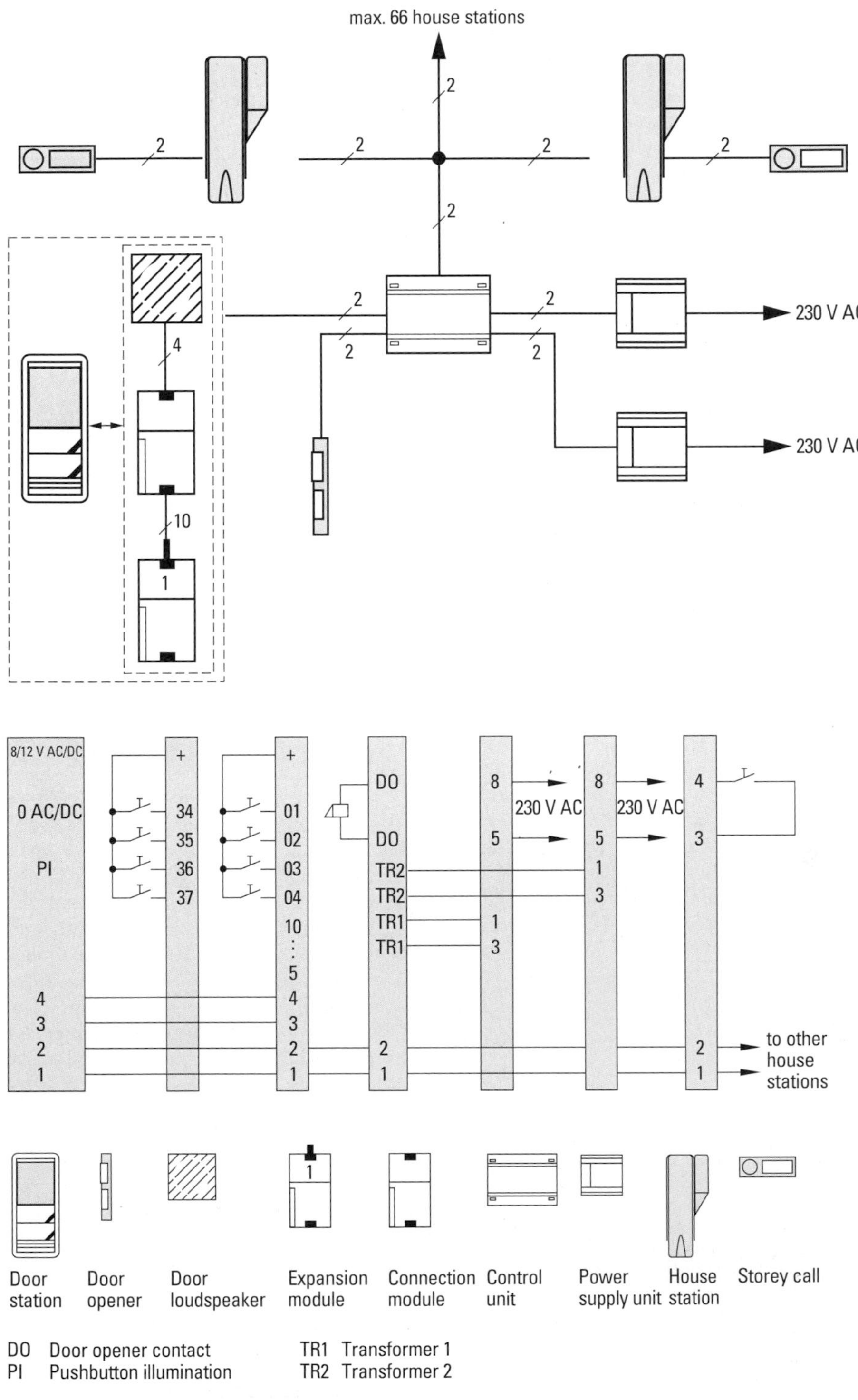

Fig. 4.3/109 (Continued)

Special systems Special systems with an even greater number of door stations, door openers, and house stations can be implemented using the appropriate supplementary components.

A supplementary relay module permits the use of external sensor switches with their own power supply.

All the devices are equipped with flexible connection cables with non-interchangeable line connections in the form of Western plug-and-socket connectors (see Section 4.3.1.2), allowing them to be installed in a system simply and quickly and by just one person.

Since there are many possible applications for door and building systems based on two-wire modular technology, complete assembly kits are available for both single-family houses, as well as for residential complexes of various sizes. For special applications, these can be expanded by means of supplementary components.

Function When the system is installed, each house station has to be allocated an address, similar to a house number. Addresses can be allocated quickly and simply via two encoding switches numbered 0 to 9.

If a bell pushbutton is pressed, the door station generates signals, which are coded differently depending on the type of bell pushbutton. Once a signal has been transmitted throughout the system, every connected house station compares it with its own address to determine whether the two match.

If this is the case, a call signal is sounded at the appropriate house station, and the speech link is established simultaneously to the visitor at the door station.

System components

Door station The door station is designed for hands-free speech communication. The modules of the functional units are housed in frames for flush and surface mounting, which are available in various versions with different dimensions. This allows a series of combinations of door loudspeakers, microphones, bell pushbuttons with nameplates, as well as door code locks, motion detectors, and house numbers, according to the number of residential units. The different versions range from lightweight to robust and vandal-proof variants. It is also possible to integrate this modular system in mailbox systems.

Connection module The connection module is used to connect between 1 and 33 bell pushbuttons. It converts the location numbers of the bell pushbuttons to an address for the bus system, thereby ensuring that the selected house station is activated. The location strip is marked with the numbers 01 to 33. The desired address is set by choosing the appropriate number on the house station.

Expansion module The connection module also allows modern bus technology to be combined with bell distributor panels tailored to the particular type of building. This enables existing distributor panels to be integrated in the two-wire house call system (e.g. during renovation). It is also possible to extend the house call system by adding expansion modules for 34 to 66, (Fig. 4.3/109) and 67 to 99 bell pushbuttons. These are connected by means of a ribbon cable with plugs.

Storey call The "storey call" is available as a bell pushbutton in various designs, from the simple plastic variant to elegant, polished or hammered gunmetal with replaceable or engraved nameplates. It is installed on the appropriate storey outside the door of the apartment.

House station The house station contains an electronic module for the bus connection, with rotary encoding switches for addressing and private speech communication, as well as the call generator for door calls (electronic call signal 1) and storey call (electronic call signal 2). Other replaceable electronic modules are available for various applications. The volume level of the acoustic call can be set during installation; the door-opener pushbutton can also be used as a stairwell lighting switch, depending on the particular version. External door chimes or additional loudspeakers can also be connected. The pluggable terminals reduce the time involved in pre-installing the bus connection, the storey call, and the additional relay or loudspeaker.

Door opener If the house call system is equipped with a door opener, the house door can be opened by pressing the door-opener pushbutton on the house station. If the system is equipped with an additional stairwell lighting function, the door-opener pushbutton has a dual function. Pressing the door-opener pushbutton when the receiver is replaced switches on the stairwell lighting; when the receiver is lifted, the door-opener pushbutton opens the door of the house.

The door openers have built-in mechanical latches that can be unlocked to open the door once, for a specified period of time, or constantly, depending on the version. Standard types of door opener can be used in a house call system with two-wire technology.

Control unit

With up to 33 house stations connected, the control unit is supplied by one bell transformer, and with 34 or more house stations by two bell transformers. The control unit comprises a rectifier circuit, switching controller, door-opener relay, and plug pins to accommodate the additional stairwell-lighting printed circuit board. A green LED indicates that the short-circuit-proof bus voltage is present. The control unit is designed as a modular device to DIN 43880 for snap-on mounting on DIN rails to DIN EN 50022, and has a pitch of 6 modular units (1 MU = 18 mm).

A suitable separate transformer should normally be used for illuminated bell pushbuttons.

Changeover device

The changeover device allows a second door station or a second door opener to be connected. The power for this is supplied by means of a safety-separated transformer.

Additional loudspeaker

An additional loudspeaker can be connected so that the call signal can be heard more easily in other rooms.

Technical data

Rated output voltage of bus:	19.5 V DC
Zero-signal current per house station:	approx. 7 mA
Operating current per house station:	approx. 30 mA
Call signal at house door:	electronic call signal 1
Call signal at apartment door (storey):	electronic call signal 2
Volume of call signal:	max. 86 dB (A)

Standards

The requirements specified in the standards HD 60384/IEC 60364-1...7/DIN VDE 0100-551 must be observed for connections to 230 V AC supply systems. A separate circuit-breaker with a rated current of 16 A (tripping characteristic C) must be used for the incoming supply.

4.3.2.4.4 Video house call systems (two-wire technology)

Speech and video links

In addition to the usual speech link, video house call systems also support a video link between the door station and the house station. Systems of this type enable the entrance area of a house to be monitored unobtrusively. This is made possible by a video camera, which is either integrated in the door station or mounted externally, in conjunction with a monitor incorporated in the house video station. Video transmission can be set to begin either as soon as the bell pushbutton is pressed, or when the receiver is lifted. The transmission ends a few seconds (approx. 5 s) after the speech link has been canceled by replacing the receiver.

If several house video stations are installed in a residential complex (e.g. as secondary video stations or "follow-on" monitors), pressing the bell pushbutton activates all the video monitors, and an additional speech link to the door video station can be established from the house video station and all the secondary stations. During communication from the door video station and one of the various house video stations, it is generally not possible for the other house video stations to "see" or "hear" anything; in other words, the only link is between the door video station and the selected residential unit.

Design

Modern two-wire technology enables a video house call system to be erected very simply, without costly and time-consuming coaxial cables having to be laid (except for the external video camera). Even using only two standard bell wires (type Y or YR, with a diameter of 0.8 mm or 1.0 mm) ensures high-quality, interference-free transmissions of speech and video signals for distances of up to 100 m.

The video house call systems are available as modular assembly kits for villas, single-family houses (Fig. 4.3/110) and two-family houses. In addition to this, system solutions with individual and supplementary components are also available for larger buildings, such as apartment blocks and functional buildings.

Door video station and house station

Supplementary components

The simplest basic system for a single-family house comprises a door video station (camera, loudspeaker, microphone, bell pushbutton) and a house video station (main monitor in conjunction with the house telephone) (Fig. 4.3/111). This basic system can be expanded to accommodate individual user requirements by means of additional components. The additional components can be used to connect either another door video station, or another monitoring camera (external video camera), which can be activated alternately by pressing a button on the house video station. Up to 7 secondary video stations (follow-on monitors) can also be connected to each house video station in a residential unit. Only two cables are required to connect each device to the system (Fig. 4.3/112). It is also possible to expand a house video station to comprise a total of 12 house video stations (main monitors, equivalent to 12 residential units) (Fig. 4.3/113).

System components

Door video station

The door video station comprises a wall box, a robust, vandal-proof aluminum front plate, on which the modular functional units are mounted: for example, loudspeaker with high-quality capacitor microphone, electronics board, and a high-resolution video camera that can be rotated through 45°. Depending on the size of the system, it is also possible to accommodate one or more bell pushbuttons with engravable metal nameplates, and a pushbutton for switching on the house or external lights. The two-hole screws used to secure the front plate can only be unscrewed using a special screwdriver thus preventing unauthorized removal.

House video station

The house video station (main monitor) comprises a high-resolution flat screen for extremely clear images, and a telephone receiver. In addition to the connection for the door video station, the house video station also contains a central control unit with the corresponding outputs for connecting sec-

Fig. 4.3/110
Video house call system: kit for a single-family house

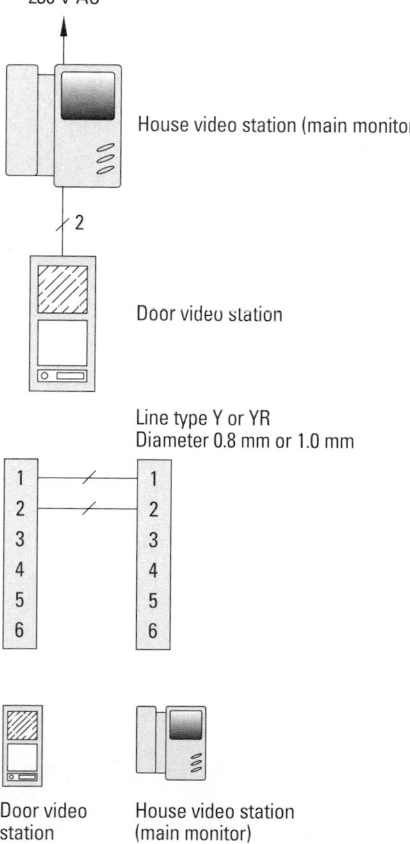

Fig. 4.3/111
Video house call system with one door video station and one house video station; installation drawing for a single-family house

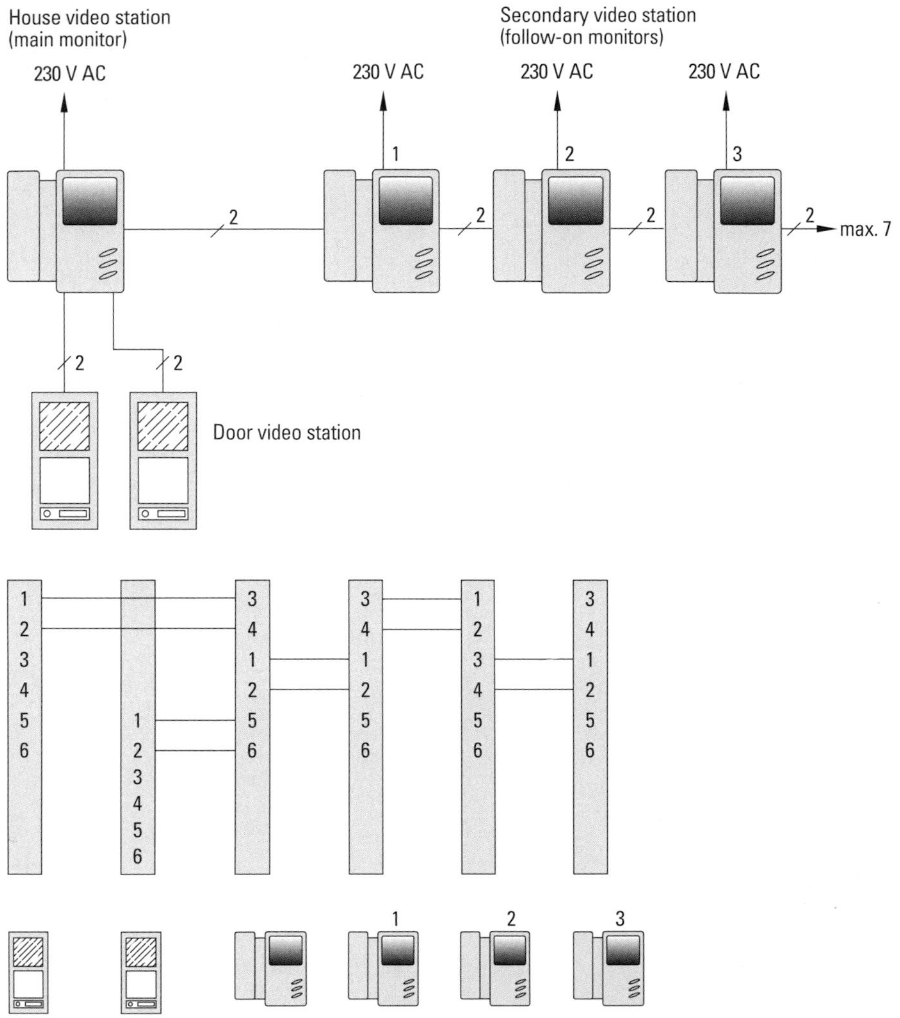

Fig. 4.3/112
Video house call system with two door video stations and one house video station with three secondary video stations; installation drawing for a single-family house

ondary video stations (follow-on monitors), as well as an additional video camera or door video station. Together with the plug-in power supply unit, it also ensures that the door video station is fed with the power it requires.

The bottom of the monitor is fitted with two rotary switches. The rotary switch on the left regulates the volume of the door chimes, while the one on the right adjusts the image contrast. The front of the house station is equipped with a vertical row of three function buttons:

▷ Monitor button for switching on the second door video station or the external video camera,
▷ Call button for communication within the house, with visual signal (small red light),
▷ Door-opener pushbutton which permits visitors to enter the building, with visual signal (small green light).

Secondary video station

In terms of design and functionality, the secondary video station (follow-on monitor) is identical to the house video station (main monitor). Video and speech information arriving at the main monitor can be transmitted

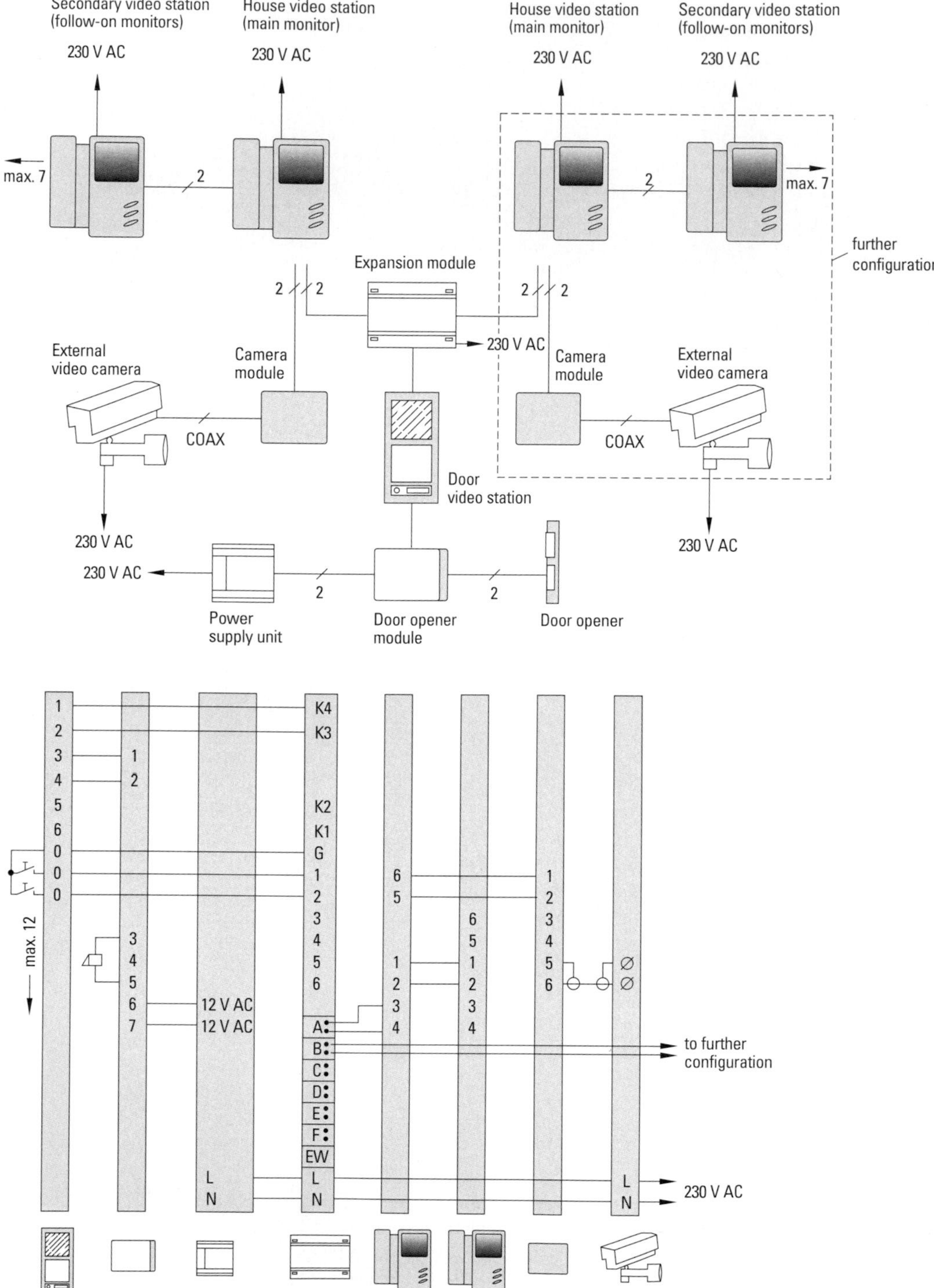

Fig. 4.3/113
Video house call system with one door video station, two additional external video cameras, a door opener, two house video stations, each with one secondary video station; installation drawing for an apartment building

to other rooms by means of a simple two-wire connection. The follow-on monitors also permit communication within the apartment itself. Overall control, however, is assumed by the main monitor.

Expansion module

The expansion module is required wherever several residential units (and thus several main monitors) are connected.

Systems with two to six main monitors require one expansion module. For seven to twelve main monitors, a further module is necessary.

Door-opener module

The door-opener module, together with the bell transformer, is used to control the electric door opener, which must, at the very least, be designed for an operating voltage of 12 V DC. The opening time of the door opener can be set to 1 second or 15 seconds. For special applications, door openers can be connected that are locked when a voltage is applied (e.g. escape doors).

External video cameras

External video cameras can be installed using a connection module which is linked to the house video station, from which it is also supplied. The cameras must be weatherproof and robust if they are to be installed outdoors. To avoid interference, a non-aging coaxial cable to DIN 47 252 Part 2 is required between the connection module and the external video camera.

Location of the built-in camera

When the door video station is fitted and installed, particular attention must be paid to the location of the built-in camera, since this is essential for good picture quality. The video cameras (including externally mounted cameras) should not be directed towards strong backlighting (sunlight), a highly reflective background, or lamp beams. The installation dimensions specified in Fig. 4.3/114 are essential to ensure that the visitor can be recognized properly from the door video station.

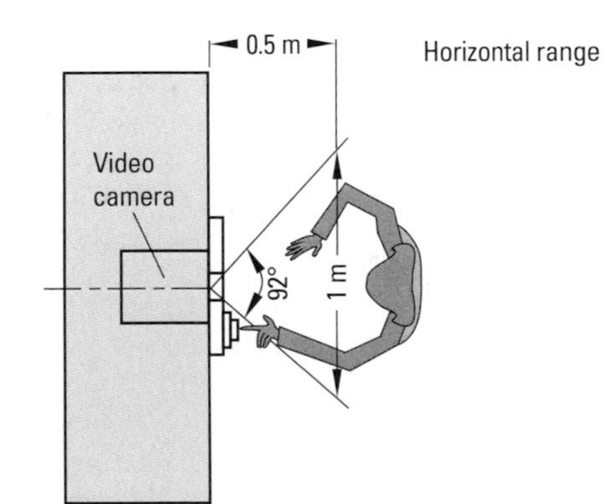

Fig. 4.3/114 Location of the video camera

Technical data

House video station/secondary station:

Supply voltage via power supply unit:	18 V DC
Picture tube:	4″ B/W flat screen
Resolution:	420 lines
Call signal:	melodic door chimes
Duration of monitor display:	approx. 180 s
Power requirements: stand-by/operational:	0.2 W / 8.0 W
Two-wire connection:	any polarity
wire diameter (distance):	max. 1.0 mm (length 100 m)

Door video station/camera:

Resolution:	360 lines
Pixel:	1/3″ CCD 512 × 582
Angular field:	92°, F 1.8
Minimum illumination level:	0.2 lux

With the exception of the door video station and the external video camera, all the system components must be used in dry rooms.

4.4 Building control system with *instabus EIB*

A redefinition of the functionality of buildings, and the modern holistic concept of efficiency, encompassing both the installation and operational phases, require a comprehensive approach to integrating different trades.

Event-controlled, distributed control bus system

Based on the international standard "European Installation Bus (*EIB*)", the event-controlled distributed bus system *instabus EIB* is designed to meet precisely these demands. The system is tailored to handle all the applications encountered in a building, thereby eliminating the disadvantages of conventional installation technology, such as the increasing number of control lines, time-consuming and costly planning and design, high fire loads, and the unmanageable complexities of maintenance and troubleshooting.

The *instabus EIB* is easy to plan and design, and is extremely versatile. This means that the system can be used efficiently to carry out a wide range of tasks in both functional and residential buildings. The applications that can be handled by the system include:

▷ Lighting control,
▷ Roller shutter/louver blind control,
▷ Room control for heating, ventilation and air-conditioning,
▷ Load management,
▷ Indication, signaling and monitoring, right up to the interfaces to
  - Household devices,
  - Telecommunications equipment,
  - Safety equipment.

Separating the power and data transmission paths offers considerable potential for

▷ reducing wiring by up to 60%,
▷ reducing energy costs by up to 30% using distributed, individual room control.

HomeAssistant

The cross-trade technology of the system supports a unified control concept tailored to the particular application – from pushbuttons for activating individual functions, to multi-function pushbuttons, infrared remote control, radio remote control, or ambient control by means of speech commands for controlling all room functions, through to the HomeAssistant and visualization software for controlling and monitoring all the functions performed in a building.

Transmission medium

In addition to the bus lines (**t**wisted **p**air, TP), the system can also support other transmission media such as **p**ower **l**ines (PL), **i**nfra**r**ed (IR), and **r**adio **f**requency (RF).

The development of new devices and functions for operational and energy management (in building systems management) is constantly expanding the range of applications of the modern *instabus EIB* installation system.

4.4.1 Basics of *EIB* technology

The *EIB* is essentially a line-conducted bus installation system (TP) based on a safety extra-low voltage SELV (24 V DC), which can be enhanced by integrating further *EIB*-compatible transmission media such as power lines (PL), radio (RF) or (in future) infrared (IR). Since no wiring has to be added to the power supply system for PL, RF and IR, these *instabus EIB* media can be used in any situation where separate wiring cannot be installed, as is the case when the system is upgraded or any renovation work is carried out.

4.4.1.1 *instabus EIB* and twisted pairs (TP)

Power supply, control functions

In contrast to conventional electrical installations, the 230 V power supply of the current-using devices (e.g. luminaires) and the control functions (e.g. switching) is separate (Fig. 4.4/1) when a bus line is used. The power supply can be routed directly from one current-using device to the next. All the switching, control, monitoring, and signaling information is transmitted via the specified bus line in accordance with stipulated standardized rules. A $2 \times 2 \times 0.8$ mm measuring and control wire is used for this purpose. Serial commands, signals, and data are transmitted directly from bus device to bus device in the form of telegrams via a twisted pair. Transmissions of this type are triggered by actions or events (e.g. when a light switch is actuated, or a preset temperature is exceeded, etc.).

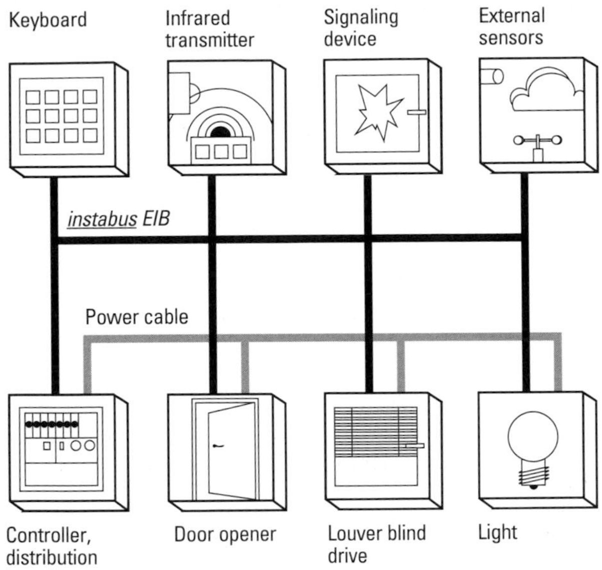

Fig. 4.4/1
Separating power and control cables (*instabus EIB*) cuts down the amount of cabling, and simplifies cabling arrangement

Telegram structure

Elements of the telegram

The telegram comprises user information, such as brightness and temperature levels or switching commands, and the bus-specific information required by *instabus EIB* to send the telegram. The bus-specific information is similar to the information required when letters or telegrams are sent via the postal service (address, sender, express delivery etc.). A telegram contains the following basic elements (illustrated in Fig. 4.4/2):

▷ Control field with priority identification code for graded transmission of information according to urgency,
▷ Address field with the address of the sender and the receiver or group of receivers,
▷ Data field with the actual user information,
▷ Security field for checking error-free transmission,
▷ Acknowledgement field, e.g. to confirm error-free reception.

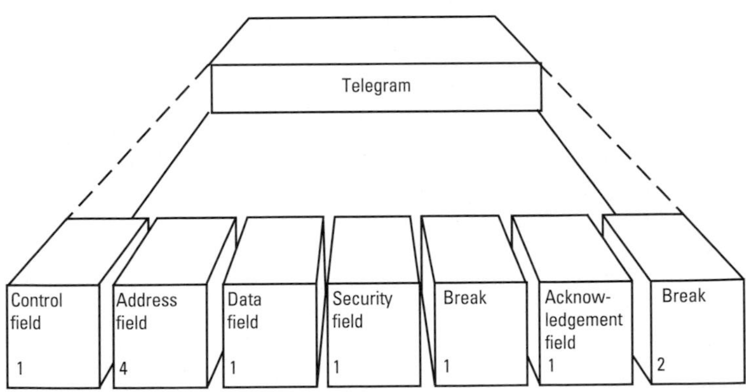

Fig. 4.4/2 Telegram structure of the *instabus EIB* control bus system

Addressing

Bus devices
Sensors

Bus devices that can communicate with each other via telegrams are bus-capable sensors, such as:

▷ Pushbuttons,
▷ Motion detectors,
▷ Light barriers,
▷ Temperature sensors,
▷ Brightness sensors,

Actuators

and bus-capable actuators, such as:

▷ Installation switches or dimmers for luminaires,
▷ Mechanical switches for louver blind drives or heating devices.

Device address/ physical address

Each bus device is assigned a unique device address (also known as a physical address). This address is used to identify a transmitting bus device (sender), thereby ensuring that the origin of a telegram can always be determined with absolute certainty. By contrast, the destination address in the telegram usually references a function of a receiver or group of receivers. This communication or group address is assigned to all the sensor and actuator functions that operate together and therefore have to communicate with each other. When a bus device receives a telegram, it checks whether it is assigned a function with a group address that matches the destination address in the telegram. The content of the telegram is evaluated only by bus devices that are assigned this type of function. Several group addresses can be assigned to one function. Since many functions can belong to the same group address, one sensor is capable of addressing many actuators simultaneously. This enables room-specific functions, such as "all ON" or "continuous row of luminaires at window OFF", to be carried out by sending only one telegram. It is, of course, also possible for several sensors to address the same function of one actuator.

Group address

A comparison with the conventional method of wiring functions individually shows that the *instabus EIB* control bus system implements function-specific wiring (e.g. system control center) by means of one shared wire and "logical connections" via the group addresses.

Transmission procedure

Twisted pair

The transmission procedure used by the *instabus EIB* twisted pair is asynchronous and character oriented. The information is transmitted in the baseband. For this purpose, the extra-low-voltage pulse sequences are superimposed symmetrically.

Advantages

This has the following advantages:

▷ High common-mode rejection factor,
▷ Low noise radiation,
▷ Good transient protection,
▷ Low-impedance line connections unsusceptible to interference.

Transmission System

As far as the transmission rate, and pulse generation/reception are concerned, the entire transmission system is designed to allow cables to be laid freely in a tree structure without line termination. The bus devices can be installed in distribution cabinets together with standard circuit-breaking devices, and in the current-using devices themselves.

Installation of bus devices

Serial transmission

Since only one wire pair is used for communication between the bus devices, telegrams can only be transmitted serially. If transmission requests occur simultaneously, they are regulated by a CSMA/CA[1)] bus access procedure, which ensures that neither time nor telegrams are lost – even when the bus has a particularly large volume of information to transmit. Since transmission requests can only occur in connection with events or actions, the transmission rate of 9.6 kbit/s (equivalent to approximately 40 to 50 telegrams per second) makes it possible to configure control bus systems comprising a large number of bus devices.

Transmission rate

Structure

Structure of the control bus system

The transmission procedure described above applies to the smallest independent unit of the *instabus EIB*, the bus line. The maximum length of a bus line is 1000 m. The required wiring can be installed in a line, ring, star, or tree configuration, as required (Fig. 4.4/3).

Terminating resistors are not necessary.

In accordance with modern installation practice, however, a tree or radial configuration is preferable.

1) Carrier Sense Multiple Access with Collision Avoidance

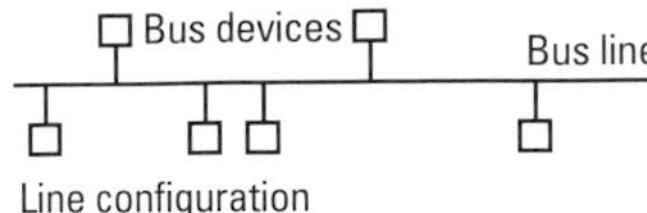

Line configuration

Ring configuration

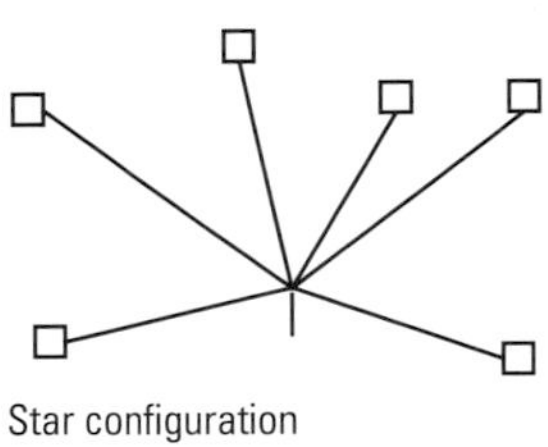

Star configuration

Tree configuration

Fig. 4.4/3
Different configurations for the *instabus EIB* bus line

Power supply

Up to 64 bus devices can be operated on one line without additional amplifiers. A power supply must be provided for each line. If power demand is high, or if greater redundancy is required, a second power supply can be connected in parallel. In this case, the power supplies can also be connected to different external conductors.

Line couplers

Line couplers (LCs) connect up to 12 lines via a main line (Fig. 4.4/4). When lines are connected in this way, the line coupler isolates them electrically from each other to prevent chains of errors.

It is not permissible to connect several lines via line couplers in such a way that ring structures are created.

Program-controlled distribution board

The node created by connecting several lines or line segments, is also known as a program-controlled distribution board (Fig. 4.4/4).

Backbone couplers

Backbone couplers (BC) can be used to combine up to 15 zones to form an *instabus EIB* system comprising over 12,000 bus devices. Backbone and line couplers are identical devices, which are assigned different addresses only when the *instabus EIB* system is commissioned.

Gateways, media couplers

EIB NET

The system can be expanded or connected to other systems by means of gateways[1] or media couplers (PL, RF). Gateways can be connected at any point in the *EIB* network. It is advisable to connect gateways at the line or backbone coupler level (as is mandatory with media couplers), since this usually allows the additional bus load in the system as a whole to be minimized. This principle is illustrated in Fig. 4.4/4, where the link to PROFIBUS or ISDN and *EIB* NET is implemented via the *EIB* backbone line. *EIB* NET is an integrated expansion option that enables several *EIB* systems to be connected quickly. The transmission technology is based on ISO/IEC 8802 (e.g. Ethernet).

By contrast, the gateway (infrared **dec**oder [DEC]) is a simple interface for non-*EIB*-compatible infrared remote-control systems for wireless switching and control, and is connected to a line at the lowest hierarchy level.

In addition to *instabus*, it is also possible to use *EIB* NET and, in principle, any bus system for telecommunication, data processing and data communication as a backbone line. In contrast to an *instabus EIB*/*instabus EIB* link, protocol conversion must be carried out in the gateways if the above systems are used. With *EIB* NET this has been reduced to a minimum.

4.4.1.2 *instabus pl EIB*, Powerline (PL)

With Powerline (PL), the 230 V network of a building is also used to transmit *instabus EIB* telegrams.

Addressing

With PL, addressing is essentially the same as for the TP system.

[1] Communication unit for establishing connections

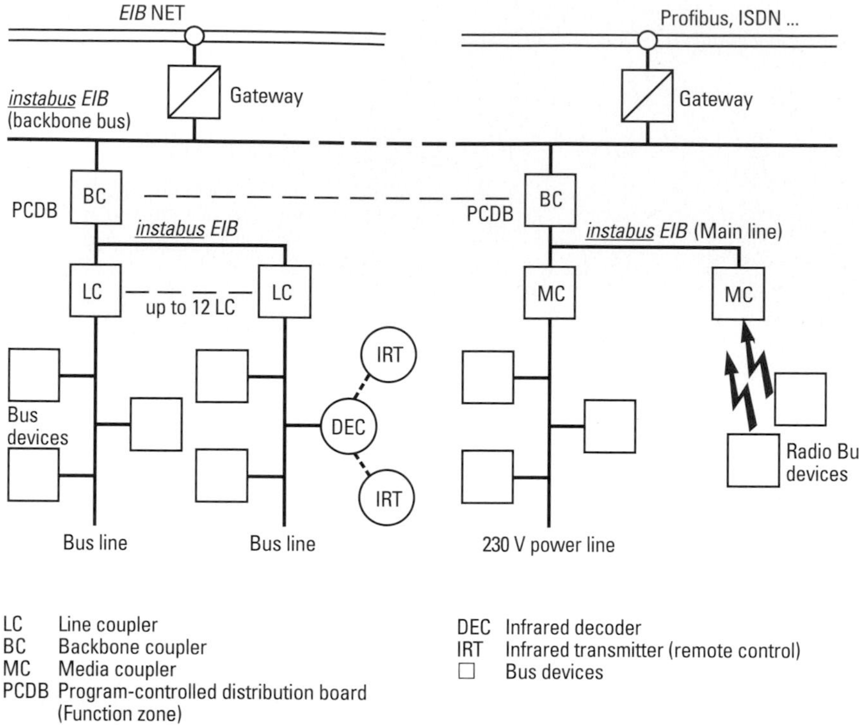

Fig. 4.4/4 Topology of the *instabus EIB* control bus system

An additional system designation is required, however, to distinguish telegrams from the various apartments in an apartment block.

Transmission procedure

Spread Frequency Shift Keying

The transmission procedure is based on spread **f**requency **s**hift **k**eying (SFSK). This involves transmitting the digital signals as differential signals between the L conductor and PEN conductor in the form of two frequencies, which have a greater interval in the permissible frequency band than with conventional FSK.

Special procedures are used to compare signal patterns and correct errors, ensuring that transmission is, as far as possible, interference free. The frequency band used is 95 kHz to 125 kHz. The transfer rate is 1200 bit/s, allowing approximately 6 telegrams to be transmitted per second.

Topology

The PL topology, unlike the TP topology, is dictated essentially by the characteristics of the 230 V power supply installed. The 230 V wires usually converge in distribution boards at point-to-point connections, and are expanded for specific rooms or storeys by means of junctions to form a tree configuration. Additional functions, such as repeaters, phase couplers, and band-stop filters, are important to avoid interference caused by current-using devices, signal damping and potential interference between different PL systems.

Band-stop filters

Since the PL is an open medium, every device connected to the mains power supply is always linked to all the others. Band-stop filters isolate the signals of PL systems from each other, as well as the signals of PL systems and the public utility. The filters also eliminate conducted interference from other areas, and prevent overreaching of the signals of the PL system in which they are installed.

Phase couplers

PL devices are generally connected only to one L-conductor of the power supply. If the wires of the three L-conductors are installed in such a way that capacitive coupling is very low, additional coupling is necessary to enable cross-phase data transmission to take place. Phase couplers passively increase the capacitive coupling of the L-conductors by means of a delta connection of capacitors between the L-conductors. In smaller and medium-sized systems, this is generally sufficient if interference levels are relatively low.

Repeaters

In larger systems, an active phase coupler must be used with a repeater. As the name suggests, a repeater repeats invalid telegrams that have been received at an L-conductor on all three L-conductors. Only one repeater may be used within a system. The device should, if possible, be connected at the neutral point or at the center of the system.

4.4.1.3 *instabus rf EIB*, the radio frequency (RF) solution

Power supply

Radio devices can be positioned freely, since they do not require any wiring of their own. In such cases, however, batteries must be used if a 230 V connection is not available. Power-saving design, deliberately restricted functionality, and the latest battery technology permit battery lifetimes of up to five years.

Addressing

Addressing is the same as for the TP system.

Transmission procedure

Frequency range

Radio transmission takes place in the frequency range 868 to 870 MHz. This non-chargeable usable range is permissible only for devices with limited transmitter power and a correspondingly short range. The maximum transmission time per hour is also regulated. This means that devices operated in this frequency range are, on the whole, more resistant to interference than at 433 MHz, for example.

Radio channels

Within the frequency range, the bands 868.0 to 868.6 MHz (band 1) and 868.7 to 869.2 MHz (band 2) are used for radio transmission. Each band can be assigned up to three radio channels. The permissible upper limits for averaged radio channel loading are 1%/hour for band 1 (**L**ow **D**uty **C**ycle [LDC]), and 0.1%/hour for band 2 (**V**ery **L**ow **D**uty **C**ycle [VLDC]). These extremely short radio channel loading times require a data transfer rate of 19.2 kbit/s, which is relatively high in comparison with those required by the other *EIB* media.

Data transmission devices

The modulation procedure used with radio is **F**requency **S**hift **K**eying (FSK).

Range

The range in open areas is approximately 100 m.

The range indoors can be reduced to 30 m, depending on ceilings and walls.

Topology

Radio channels

The topology of the radio system is essentially the same as that of the twisted-pair system. With radio systems, areas or lines are generally formed by means of a separate radio channel, instead of contiguous wiring networks. A total of six radio channels with various frequencies (3 × LDC and 3 × VLDC) are available. If the radio channels are not sufficient for large systems, an additional area/line structure can be implemented by means of address assignment. As in the TP system, radio lines with different radio channels are linked to each other via radio line couplers (routers) or, if they are integrated in a TP system, via media couplers. Up to 64 devices can be operated per radio line. To increase transmission reliability, up to 3 of these devices can function as retransmitters.

Radio line couplers, media couplers

Retransmitters

In the *instabus rf EIB* system, retransmitters have the same function as repeaters in the *instabus pl EIB* system.

4.4.1.4 Modular design of *instabus EIB* devices

Modules

All *instabus EIB* devices consist of three modules:

- ▷ Bus coupling unit,
- ▷ Terminal, such as sensors or actuators (hardware [HW]),
- ▷ Application program (software [SW]).

Fig. 4.4/5 illustrates two modules for a flush-mounting installation pushbutton.

The two modules are interconnected by means of built-in plug-and-socket devices.

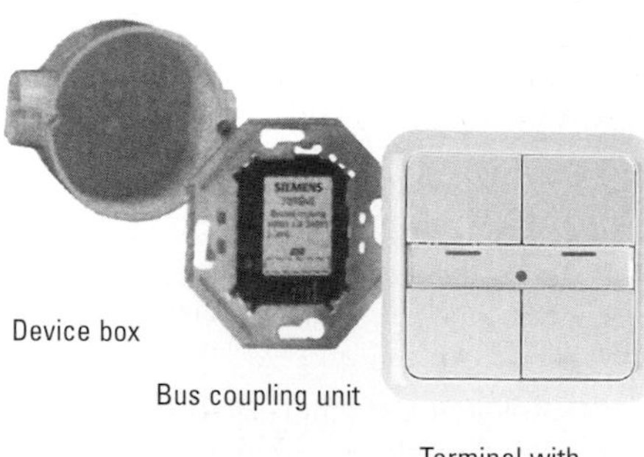

Fig. 4.4/5
Modular design of instabus *EIB* devices: light push-button for flush mounting

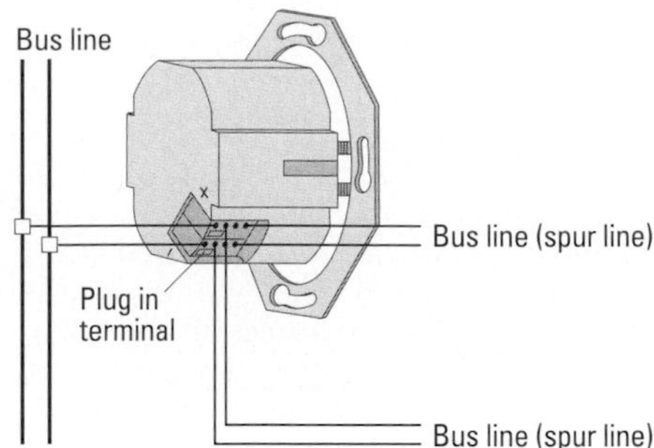

Fig. 4.4/6
Connecting the bus coupling unit to the instabus *EIB* control bus system

Depending on the application, the bus coupling unit can be combined with a diverse range of terminals (e.g. dimmer, louver blind switch, thermostat, and display unit).

Bus coupling unit

In addition to the transmitter and receiver modules for communication via the common core pair, the bus coupling unit also includes the voltage regulator for the electronics, and a microcontroller for processing the communication protocols. It is also possible to store application-specific programs in the bus coupling unit (e.g. the time-delay switching function for circuit-breakers, including the logic for energizing relays).

Since the instabus *EIB* can be operated with other transmission media in addition to the twisted pair cable, it is particularly important to retain the modular design between the bus unit and terminal device. This enables, for example, the same control interface to be used with the flush-mounting bus coupling unit for the media instabus *pl EIB*[1)] and instabus *rf EIB*[2)].

Bus coupling unit for flush mounting

The bus coupling unit for flush mounting is designed for installation in device boxes with a diameter of 60 mm. It is connected to the bus line by means of a 2-pole screwless plug-in terminal, which can accommodate up to four solid wires with a diameter from 0.6 mm to 0.8 mm (Fig. 4.4/6). The plug-in terminal permits the bus line to be looped in or tapped off at this point. A 10-pole interface allows application-specific terminals and their control interfaces (e.g. pushbuttons, motion detectors, brightness sensors) to be connected to the bus coupling unit.

The bus coupling units for twisted-pair connection and flush mounting are available in the following three versions:

- ▷ Installation depth of 32 mm for screw fixing and with a BCU1[3)],
- ▷ Same mechanical design as above, but with BCU2 for enhanced functionality of the bus terminal,
- ▷ Installation depth of 32 mm with claw fixing. This variant can also be secured in device boxes that are not equipped with a screw fixing facility.

Terminal

Application programs

The function of the device is determined not only by the electronics of the application module in the terminal, but also by the application software in the bus coupling unit (Fig. 4.4/7). The application programs for bus terminals of one type are identical for all instabus *EIB* media. Since these application programs (SW) can be modified by setting parameters (i.e. are parameterizable) or downloaded via the bus, the equipment configurations can be defined or altered even at the commissioning stage. This also considerably facilitates "re-installation" to accommodate subsequent changes in use.

This modular design is a feature of all instabus *EIB* devices, irrespective of whether twisted pair, Powerline, or radio is used as the transmission medium.

The differences between these media will not, therefore, be discussed in the following.

---
1) instabus *pl EIB* = **P**ower**l**ine (PL)
2) instabus *rf EIB* = **R**adio **F**requency (RF)
3) **B**us **C**oupling **U**nit

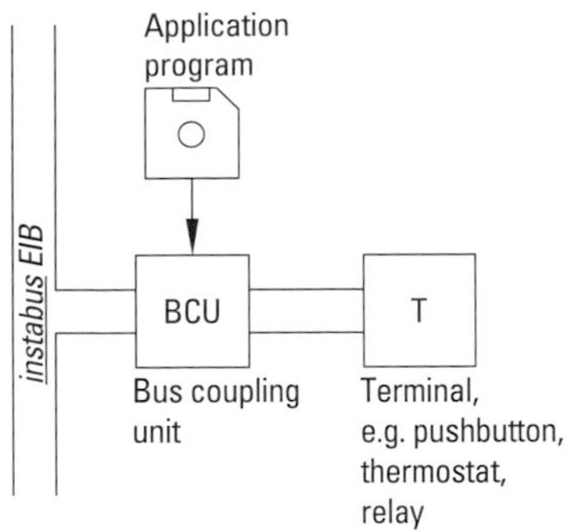

Fig. 4.4/7
Functional principle of *instabus EIB* devices

Instead, the components and potential applications of the system will be illustrated with reference solely to the twisted pair. The information presented below applies equally to the Powerline and radio systems.

4.4.2 *instabus EIB* components

The *instabus EIB* control bus system comprises basic devices, system devices, and various application-specific devices. The shape, type of fixing, and connection levels of the individual bus components have been tailored to match those of tried-and-tested electrical installation equipment.

Types

According to use and type, the devices can be divided into the following categories:

▷ Modular devices for distribution cabinets or dado ducts for snap-on mounting on DIN rails to DIN EN 50022,
▷ Devices for flush or surface mounting enhancing existing operator communication, switching, control, and signaling systems for installation engineering,
▷ Built-in devices for direct installation in current-using devices, such as luminaires and louver blind drives.

4.4.2.1 Basic devices

Basic equipment

The basic devices (Fig. 4.4/8) are mandatory in order to be able to operate the *instabus EIB* control bus system.

The basic devices are:

▷ Power supply unit,
▷ Reactor,
▷ Data rail,
▷ Connector.

Power supply unit

The power supply unit generates the 24 V DC safety extra-low voltage required by the *instabus EIB* control bus system. The power supply unit incorporates a reactor. Each bus line requires at least one power supply unit, which can be mounted on distribution boards (see Section 1.4.5) or in dado ducts (see Section 5.1.4.7). The technical specifications of the unit are as follows:

Rated voltage (primary):	230 V AC
Rated voltage (secondary):	Safety extra-low voltage (SELV) 24 V DC ± 1 V
Rated current (secondary):	640 mA
Current limiting (secondary):	1.5 A
Connection, primary:	screwless plug-in terminals for 1 mm^2 to 2.5 mm^2,
secondary:	voltage with reactor: pressure contacts on data rail, output voltage without reactor: safety extra-low voltage terminal for conductors with diameter 0.6 mm to 0.8 mm.

Reactor

The reactor is used to supply a further bus line if the output voltage without reactor is tapped from a power supply unit. It also prevents the electrical signals on the bus line from short-circuiting.

The technical specifications of the reactor are:

Rated input voltage:	24 V DC ± 1 V,
Rated current:	0.5 A,
Fault current:	1 A,
Connection to power supply unit without reactor and connection to bus line:	Pressure contacts on data rail.

Data rail

The data rail is bonded to the DIN rail (to DIN EN 50022, 35 mm × 7.5 mm). The data rail connects the bus devices via one of its printed conductors by means of the pressure contacts of the bus devices on the distribution board or in the dado duct.

Connector

The connector is snapped onto the DIN rail with the data rail, and connects the data rails

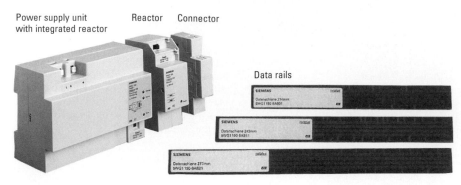

Fig. 4.4/8 Basic devices of the *instabus EIB* control bus system

on a distribution board, or the data rails and the bus lines installed in the building.

4.4.2.2 System devices

Examples of system devices include:

- ▷ Line coupler/backbone coupler/line amplifier,
- ▷ Interface.

These devices are used to expand a bus line to include up to 12 bus lines for one function zone. They can also be used to expand a control bus system to a maximum of 15 function zones, or as an interface to the system.

Line coupler/backbone coupler/line amplifier

The line/backbone coupler[1)] functions as a type of data flow filter and allows only telegrams addressed to bus devices in other bus lines/zones to pass through. This helps to reduce the data traffic in the individual sections, and electrically isolates the bus lines/zones from each other, thereby ensuring that any interference remains within the bus line/zone in which it occurred.

The line/backbone coupler can also be used as a line amplifier downstream of junctions in line segments.

One line can comprise up to four line segments (Fig. 4.4/9.).

[1)] If the line coupler is assigned a zone address instead of a line address when it is parameterized, it can be used as a backbone coupler

Interface

The interface enables external auxiliary equipment (e.g. personal computers) to be linked via a 9-pole D subminiature plug-and-socket connector for addressing and parameterizing, operating and monitoring, as well as for troubleshooting purposes (see Section 4.3.1.2).

Interfaces are available as modular devices and as flush-type devices that match switch/socket units in the DELTA studio and DELTA profil ranges (see Section 4.3.1).

4.4.2.3 Application-specific devices

A large number of different application-specific devices are available for the *instabus EIB* control bus system.

Manual: "Building Management Systems with *instabus EIB*"

The functions and applications of these devices are described in detail in the Siemens technical manual "Building Management Systems with *instabus EIB*".

Order No. E 20001-P 311-A857-X-7600.

4.4.3 Functions

The *instabus EIB* control bus system is extremely easy to install and allows a range of diverse functions to be implemented using just a small number of devices.

In the following sections, a lighting system will be used to illustrate the functions "switching", "dimming", and "control".

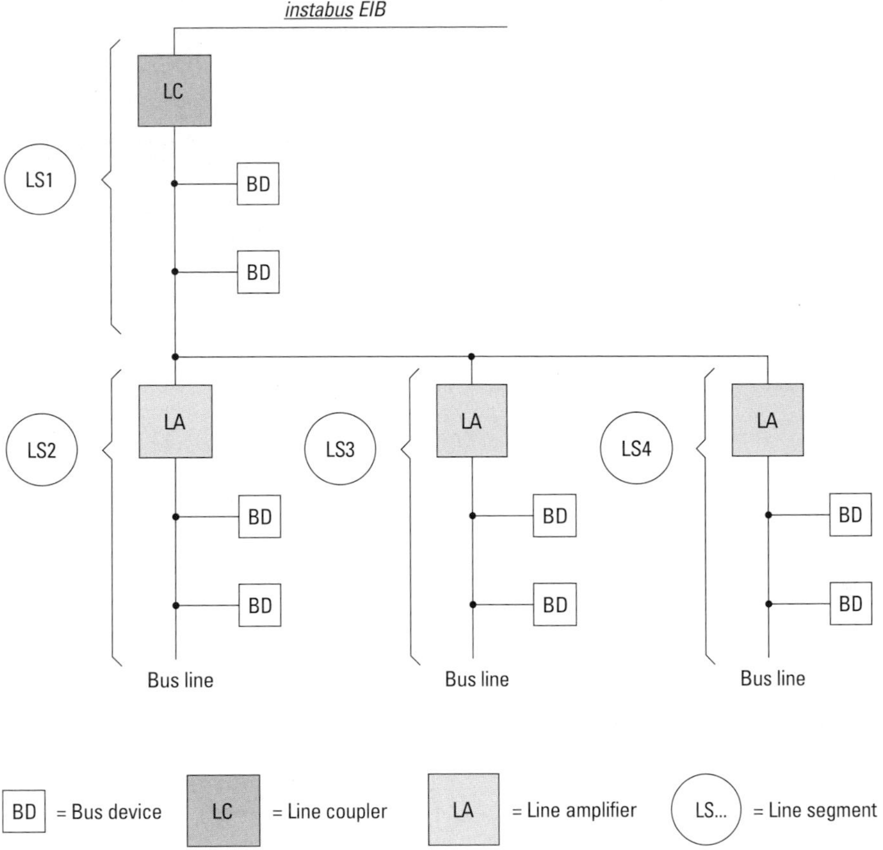

Fig. 4.4/9 System structure, line segments (up to four)

4.4.3.1 Switching

Switching

Fig. 4.4/10 illustrates three commonly-used circuits, and shows that the number of conductors in the power current circuit can vary greatly when conventional installation methods are used. The *instabus EIB*, however, considerably reduces the amount of work required to design and plan, install, and modify systems based on conventional methods. With the *instabus EIB* system, only three conductors are required for the 230 V load current circuit. Control functions are generally carried out via the two wires of the *instabus EIB* system.

4.4.3.2 Dimming

Dimming

A distinction is made between dimming fluorescent lamps with dimmable **e**lectronic **c**ontrol **g**ear (ECG), and dimming incandescent and low-voltage halogen lamps with conventional or electronic control gear. Fluorescent lamps are dimmed by means of switch/dim actuators.

When a VDU workstation is being set up, for example, and the switching function is to be replaced by a dimming function, it is not necessary to modify the installation in any way if *instabus EIB* is used. The only change

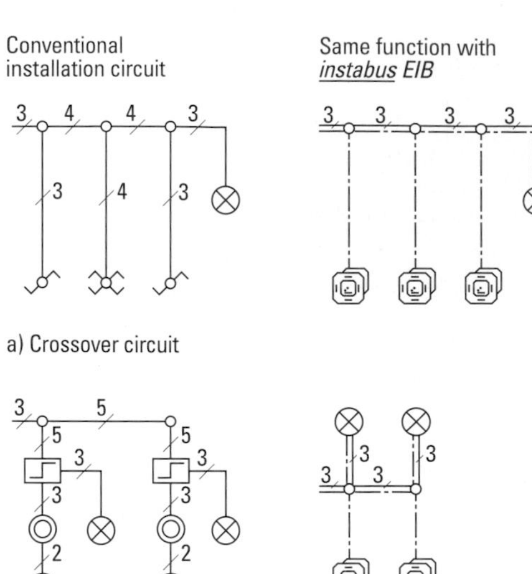

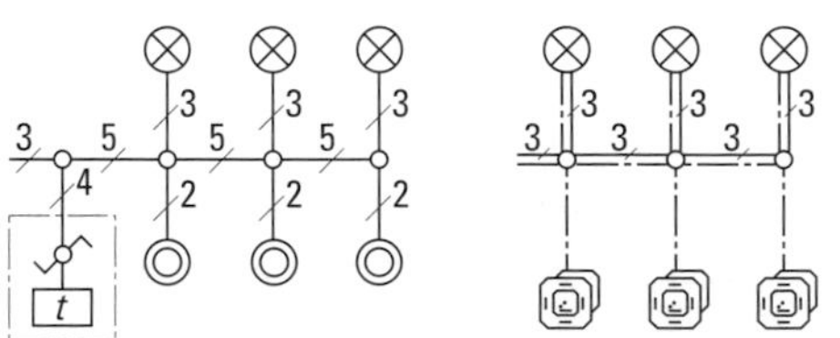

Fig. 4.4/10
Three commonly-used circuits in conventional installation engineering and in the *instabus EIB* control bus system

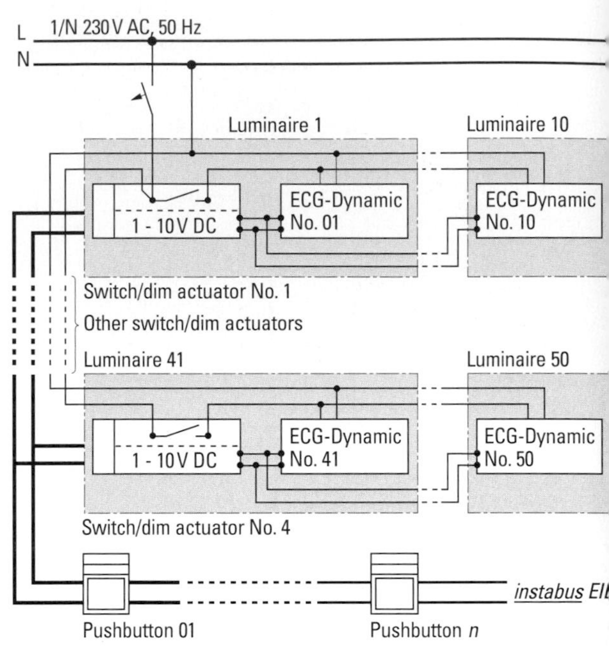

Fig. 4.4/11
Circuit diagram for controlling dimmable electronic control gear (ECG Dynamic) in a lighting system

required is at the actuator level, where a switch/dim device has to be retrofitted. The existing pushbutton can be retained. The new dimming function is parameterized in the bus coupling unit of the pushbutton. Fig. 4.4/11 shows a circuit diagram for controlling dimmable electronic control gear ECG (ECG Dynamic) in a lighting system.

Incandescent and low-voltage halogen lamps can be dimmed directly by means of a universal dimmer (Fig. 4.4/12).

4.4.3.3 Control

Control

Using the basic installation as a foundation (i.e. 230 V installation with three conductors to the current-using devices, and a two-core bus line to all the sensors and actuators), control systems can be easily implemented for a wide range of applications by simply adding suitable bus-capable sensors and actuators and parameterizing the bus coupling unit accordingly.

Examples of this include systems for:

- ▷ Daylight-dependent lighting control,
- ▷ Scenario lighting,
- ▷ Temperature-dependent heating control,
- ▷ Rate-dependent consumer load control.

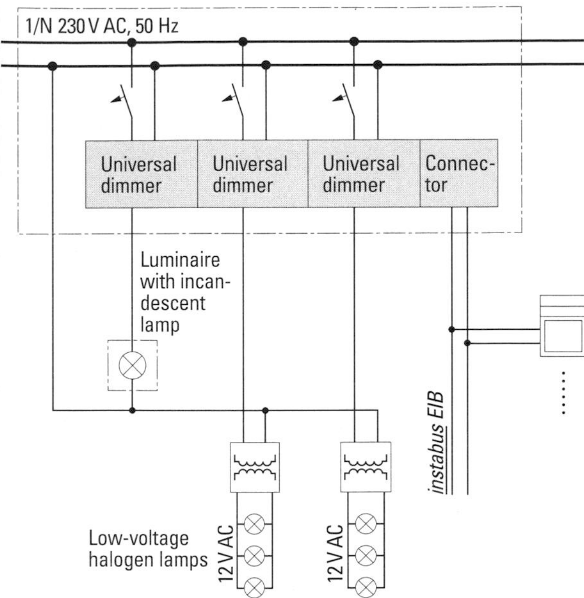

Fig. 4.4/12 Possible application for a universal dimmer

4.4.4 Implementation

Software tools

The software tool ETS (*EIB* Tool Software) is available to assist users in implementing applications (planning, configuration, commissioning and diagnostics) with the *instabus* *EIB* control bus system. This program can be used on personal computers.

4.4.4.1 Planning

Planning and configuration program

The installation drawing required to plan an electrical installation system is created using a planning and configuration program.

The first step involves entering the ground plan of the building in the GUI of a PC and arranging the installation components according to the particular tasks they perform. The resulting device arrangement forms the basis for:

▷ Creating the installation drawing for all the power lines from the individual distribution boards to the individual current-using devices,

▷ Creating the installation drawing for the *instabus* *EIB*, taking into consideration the specified data, such as:
- maximum length of a bus line: 1000 m,
- maximum number of bus devices per bus line: 64,
- maximum distance between two bus devices: 700 m.

Address assignment

The configuration program of the ETS software is then used to assign the unique component identifications (= physical addresses) and to assign the bus devices to functional groups (= assignment of group addresses) in accordance with the specified application requirements (e.g. the point from which a particular current-using device is to be switched, dimmed, or controlled).

The results of the configuration are stored in the form of a project database. These results can be made available as printouts, e.g. as an installation drawing (Fig. 4.4/13) including the designations of the bus components, and as device lists for use in the subsequent phases of the project.

4.4.4.2 Line/cable installation

Power supply cables/installation bus lines

The power supply cables for the current-using devices are installed from the distribution board to one current-using device and then directly to the next. No special tools are required to install the bus lines that run to the points at which the individual sensors, control elements, and actuators are installed or mounted. The bus line is looped through from one installation point to the next. A twisted *EIB* cable (e.g. YCYM $2 \times 2 \times 0.8$ mm) with solid conductors is used for the bus line.

The easy-to-install connection system used with all bus components of the *instabus* *EIB* control bus system allows individual devices to be replaced or expanded subsequently, while the system is online.

4.4.4.3 Mounting

The results of the configuration stage and the component designations in the installation drawing are used as a basis for selecting the bus components, entering their component identifications (= physical addresses), parameterizing them according to their respective applications, and mounting and connecting them at the appropriate lo-

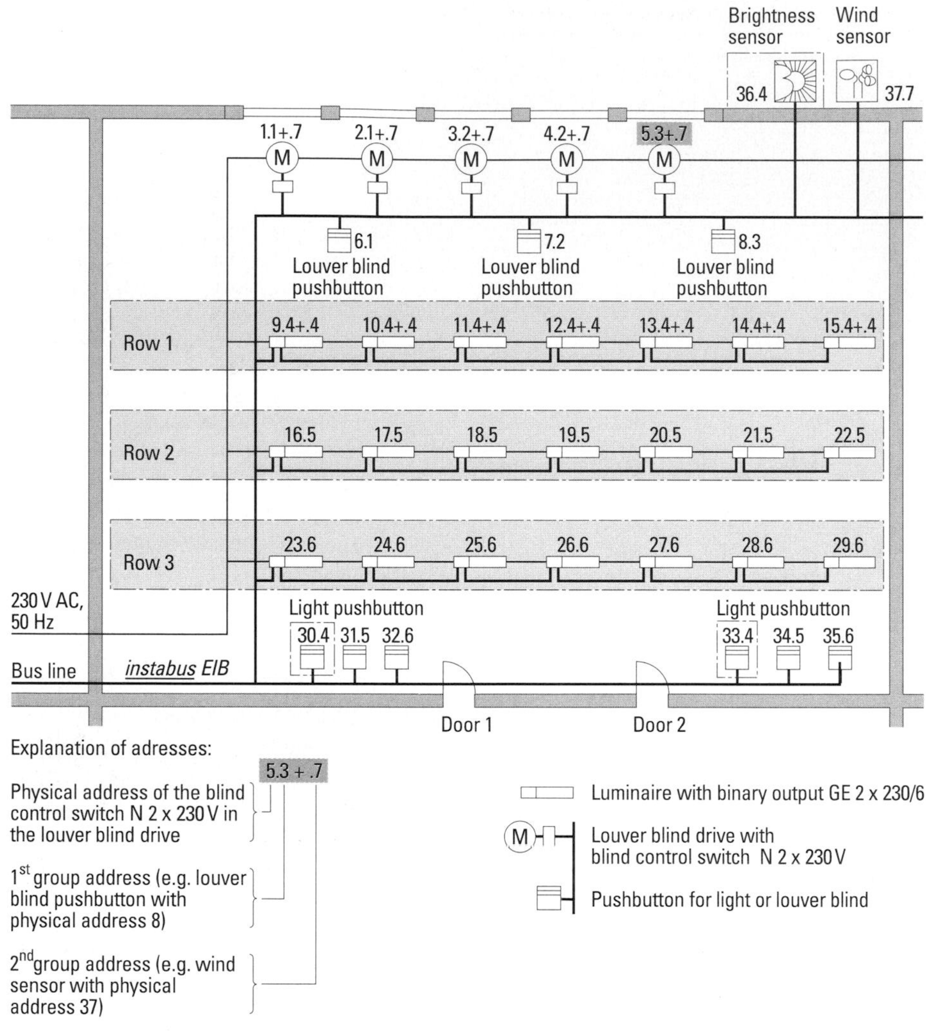

Fig. 4.4/13 Installation drawing of an open-plan office with the *instabus EIB* control bus system

cation. Modular devices are mounted on the distribution board by snapping them on to the DIN rail with data rail. Devices for surface and flush mounting are connected via the screwless plug-in terminals, and by looping the bus line from device to device.

4.4.4.4 Commissioning

Commissioning software

Once all the system components have been mounted, the current-using devices, sensors, and actuators are organized into groups. This can be carried out at any point of the bus system via the interface provided using the ETS commissioning software and a portable PC (see 4.4.2.2).

This method also facilitates subsequent alterations to the assignments of the bus components, or regrouping.

4.4.4.5 European Installation Bus Association

EIBA

By joining together to form EIBA (**E**uropean **I**nstallation **B**us **A**ssociation), leading European electrical installation companies and companies from other sectors, such as telecommunications and alarm systems, have created a broad foundation for the European installation bus. The EIBA lays down quality requirements, ensures that these are met by all *EIB* products and that users have access to products that are genuinely compatible. Only products that have been certified by EIBA may bear the *EIB* logo.

The position of *EIB* as a standard European system for building installations and building management systems is further reinforced by the national and international standards on which it is based.

4.4.4.6 Possible applications

Combining the *instabus EIB* components discussed above with conventional devices for installation engineering (see Sections 1.4.5 and 4.3) provides an effective means of implementing various applications for using functional and residential buildings flexibly in line with the new form of electrical installation. When their (parameterizable) functionality is exploited to the full, the individual *instabus EIB* products establish an ideal basis for this. Table 4.4/1 provides an overview of products and the applications for which they are used.

This range of products is expanding continuously (e.g. to include temperature sensors, brightness sensors, timers, display units, maximum demand monitors, etc.), making control and operation of installation systems increasingly convenient, and operational and energy management in buildings ever more effective for the user.

4.4.4.6.1 Example of installing a lighting and louver blind control system in an open-plan office

Modern lighting systems are expected to provide an appropriate level of artificial lighting for workplaces strictly according to demand. Louver blind control systems should allow users to operate blinds individually, and automatically raise the blinds in stormy conditions. The applications should be cost effective with regard to installation, operation, and changes in use, and should also ensure a high level of acceptance among users. Fig. 4.4/13 contains the installation drawing for the intended use of an open-plan office.

Description of application illustrated in Fig. 4.4/13:

▷ Lighting control
In order to be able to accommodate changes in the use of the room, each luminaire is equipped with a switching actuator (binary output GE 2 × 230/6). The luminaires can be switched ON and OFF in three continuous rows via pushbuttons at doors 1 and 2. Row 1 is also switched on and off automatically by the brightness sensor, depending on the level of outdoor light.
When an installation is planned, the binary outputs are assigned the physical addresses 9 to 29, the pushbuttons 30 to 35, and the brightness sensor 36. Since provision is made for switching on all the luminaires in a continuous row simultaneously, the binary outputs of the luminaires in row 1 are assigned the same group address (4), and those of the second and third rows are assigned the group addresses 5 and 6 respectively. The brightness sensor, which only affects row 1, is assigned the group address 4, and the binary outputs of the luminaires in this row are assigned the second group address 4. To enable the lighting to be switched from each door, the group addresses 4, 5, and 6 are assigned to each of the pushbuttons.

▷ Louver blind control
A louver blind is provided for each window, with two adjacent louver blinds being actuated via one pushbutton. The switching actuators for the blinds (blind control switch N 2 × 230 V) and the louver blind pushbutton are assigned the physical addresses 1 to 8; each group of two switching actuators and the associated louver blind pushbutton are assigned the same group address (1 to 3). In the event of a storm, the wind sensor, which has the physical address 37, controls all the louver blind switching actuators. For this reason, these actuators are assigned a second group address (7).

Table 4.4/1 Bus components for *instabus EIB* applications

Possible Application Product designation	Version	Lighting	Louver blind	Heating	Ventilation	Air-conditioning	Load-management	Indication and control	Operator control and monitoring
Basic/system devices									
Power supply unit	REG	×	×	×	×	×	×	×	×
Reactor	REG	×	×	×	×	×	×	×	×
Data rails 214/243/277[1]	REG	×	×	×	×	×	×		
Connector	REG	×	×	×	×	×	×		
Line coupler	REG	×	×	×	×	×	×	×	×
Backbone coupler	REG	×	×	×	×	×	×	×	×
Line amplifier	REG	×	×	×	×	×	×	×	×
Interface RS 232	REG	×	×	×	×	×	×	×	×
Interface RS 232	UP	×	×	×	×	×	×	×	×
Application-specific devices									
Binary input N 4 × 230 V	REG	×	×	×	×	×	×	×	×
Binary output N 2 × 230 V	REG	×	×	×	×	×	×	×	
Binary output GE 2 × 230/6	GE	×	×	×	×	×	×	×	
Bus coupling unit	UP	×	×	×	×	×	×	×	×
Pushbutton (1 ×)	UP	×	×	×	×	×			
Pushbutton (2 ×)	UP	×	×	×	×	×			
Pushbutton (4 ×)	UP	×	×	×	×	×			
Blind control switch N 2 × 230 V	REG		×						
Switch/dim actuator GE 230/6	GE	×							
IR wall-mounted transmitter (1 ×)	AP	×	×	×	×	×			
IR wall-mounted transmitter (2 ×)	AP	×	×	×	×	×			
IR wall-mounted transmitter (4 ×)	AP	×	×	×	×	×			
IR receiver	AP	×	×	×	×	×			
IR decoder	REG	×	×	×	×	×			
Room temperature controller	UP			×	×	×			
Motion detector	UP/AP	×		×	×	×		×	
Timer (DCF 77)	AP	×	×	×	×	×	×	×	×
Brightness sensor	GE	×	×					×	
Fire sensor	AP							×	
Valve control drive	V			×		×			
Water sensor	UP							×	
Telephone interface	Telecontrol/remote indication								
Communication socket	Connection for household appliances								
Function module/Controller	Control functions for all applications								
Weather station	Acquisition of all weather data, e.g. wind / rain / solar radiation								

REG modular device, UP flush, AP surface, GE built-in, V valve-mounting

[1] Length in mm

If the open-plan office is divided up to create three separate rooms (A, B and C – as shown in Fig. 4.4/14), only the assignments (group addresses) of individual bus devices have to be changed either at the point of installation, or centrally.

Description of the modified application shown in Fig. 4.4/14:

▷ Lighting control
In each room (A, B, and C), the continuous rows of luminaires 1, 2, and 3 are to be switched ON an OFF by means of three pushbuttons.
The brightness sensor controls row 1, which is situated close to the window, across all three rooms (A, B, and C) – but it does not affect luminaires 14 and 15.

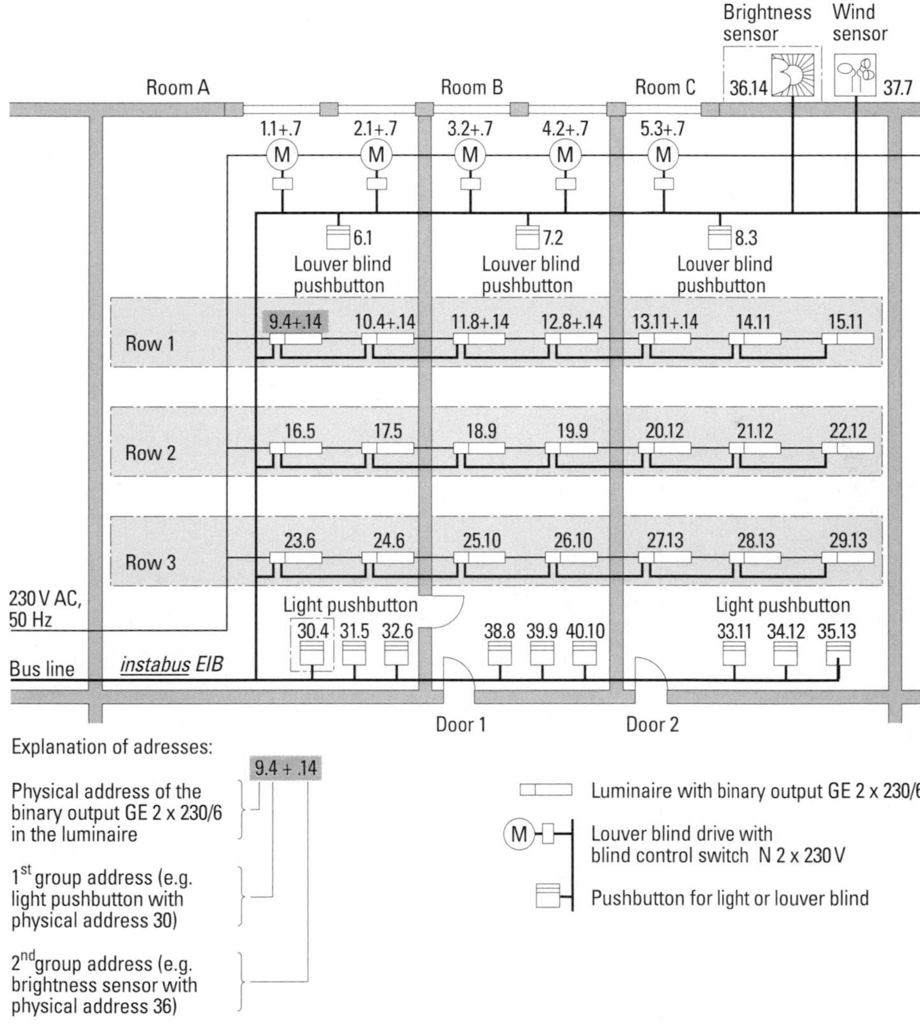

Fig. 4.4/14 Installation drawing of the open-plan office (application 2) as three independent rooms

The following changes are required for the new application (2):

Room A: For luminaires 9 and 10, the second group address is changed from 4 to 14.

Room B: Three additional pushbuttons are installed and connected to the bus line (physical addresses 38 to 40). The luminaires and pushbuttons are assigned group addresses (8, 9, and 10), and the second group address for luminaires 11 and 12 is changed from 4 to 14.

Room C: Luminaires and pushbuttons are assigned new group addresses (11, 12, and 13), and the second group address for luminaires 14 and 15 is no longer required.

▷ Louver blind control
No changes required.

Fig. 4.4/15 uses the open-plan office to illustrate (in tabular form) the bus device assignments and the modified group address assignments for the new application.

Sensors \ Actuators		*instabus EIB* switching devices: Louver blind drives			FL luminaires Row 1				FL luminaires Row 2			FL luminaires Row 3		
	Physical address	1 2	3 4	5	9 10	11 12	13	14 15	16 17	18 19	20 21 22	23 24	25 26	27 28 29
		Group address												
Louver blind push-buttons	6	1 / 1												
	7		2 / 2											
	8			3 / 3										
B. S. W. S	36				4 / 14	4 / 14	4 / 14	4 / –						
	37	7 / 7	7 / 7	7 / 7										
Light Push-buttons	30				4 / Room A: 4	4 / –	4 / –	4 / –						
	31								5 / Room A: 5	5 / –	5 / –			
	32											6 / Room A: 6	6 / –	6 / –
	33				4 / –	4 / –	4 / Room C: 11	4 / Room C: 11						
	34								5 / –	5 / –	5 / Room C: 12			
	35											6 / –	6 / –	6 / Room C: 13
Light Push-buttons new	38					– / Room B:8								
	39									– / Room B: 9				
	40												– / Room B: 10	
Group addresses (total)		1+7	2+7	3+7	4	4	4	4	5	5	5	6	6	6
		1+7	2+7	3+7	4+14	8+14	11+14	11	5	9	12	6	10	13

(In each cell: value before "/" = Group addresses for application 1; value after "/" (shaded) = Group addresses for application 2. In the totals, first row = application 1, second row (shaded) = application 2.)

Group addresses for application 1

Group addresses for application 2

B.S. Brightness sensor
W.S. Wind sensor

Fig. 4.4/15 Allocation of group addresses to the bus devices

4.4.5 Software user interfaces

4.4.5.1 Residential buildings: HomeAssistant

Indication and control in apartments and houses

Home Electronic System (HES)

In an apartment or single-family house equipped with an *EIB* installation or **H**ome **E**lectronic **S**ystem (HES), indication and control can be carried by means of various display and control elements.

Display unit

A display unit (LCD) matching the design of the switches and plug-in sockets is often used. This unit can display up to eight messages with two lines of text, each containing up to 20 characters. The display unit also features a built-in acoustic alarm generator, as well as a key that is used to cancel acoustic alarms and select the next message to be displayed. Installed beside a bed, for example, the unit can be used to indicate whether the door of the house is locked and the roller shutters on the first floor are lowered. It can also indicate whether a fire alarm has been activated, or whether intruders are in the house.

Control unit

In addition to the switch-sized display units, indication and control units are also available in A6 format and larger. These are equipped either with several control keys, the functions of which can be fixed or variable (soft keys), or a touch-sensitive display surface (touch screen).

IR remote-control

The *instabus EIB* can also be linked to a television set by means of an auxiliary device, allowing signals, statuses, and measured values to be displayed on the TV screen. An IR remote-control device can be used to display this data from other rooms, or to switch devices and functions via the *instabus EIB*.

HomeAssistant (HA)

The most efficient way of controlling and monitoring an HES is via the HomeAssistant (HA), a software package for use with conventional multi-media PCs. The user interfaces and control processes are designed to be operated using a mouse and touch screen. No previous training is required to operate the system, since the desired functions are selected by clicking them or simply touching the "control button" icons and additional explanatory text.

The HomeAssistant initial screen (Fig. 4.4/16) comprises nine entry points (Table 4.4/2) via which the user can adjust, monitor, and parameterize all the *EIB* devices/appliances provided for visualization purposes.

Overview/Back navigation keys

The control elements *Overview* and *Back* are navigation keys. *Overview* always returns the user to the HomeAssistant initial screen. If the initial screen is already displayed, this key calls up an overview function that can be used to access the house status. *Back* calls up the screen previously displayed.

Every device/appliance can be accessed and set via two different paths:

Appliance view

▷ Via the *Appliance* view: when this path is used, the search for the particular device/appliance the user wishes to set is initially restricted to the device/appliance type. In the next step, the device/appliance is selected from a room list containing devices/appliances of the same type (e.g.

Table 4.4/2
Entry points for the HomeAssistant and the functions or devices that can be accessed via these

Entry points	Functions/devices
Safety	• Monitoring functions, • Alarm signaling systems
Calendar	• Set time/date, • Task management (calendar function, day/event programs)
Communication	• Telephone, telecontrol
Devices	• All appliances/devices, sorted according to appliance/device classes
Home	• All appliances/devices, sorted according to rooms
Light/temperature	• All appliances/devices relating to "light/climate", e.g.: – lighting, – louver blind and roller shutter control, – room temperature control, – air-conditioning
Health	• "Family doctor" (on CD), • Important telephone numbers
Entertainment/CD	• Integrated television
System	• Configuration functions for the HomeAssistant (configure interface, set password)

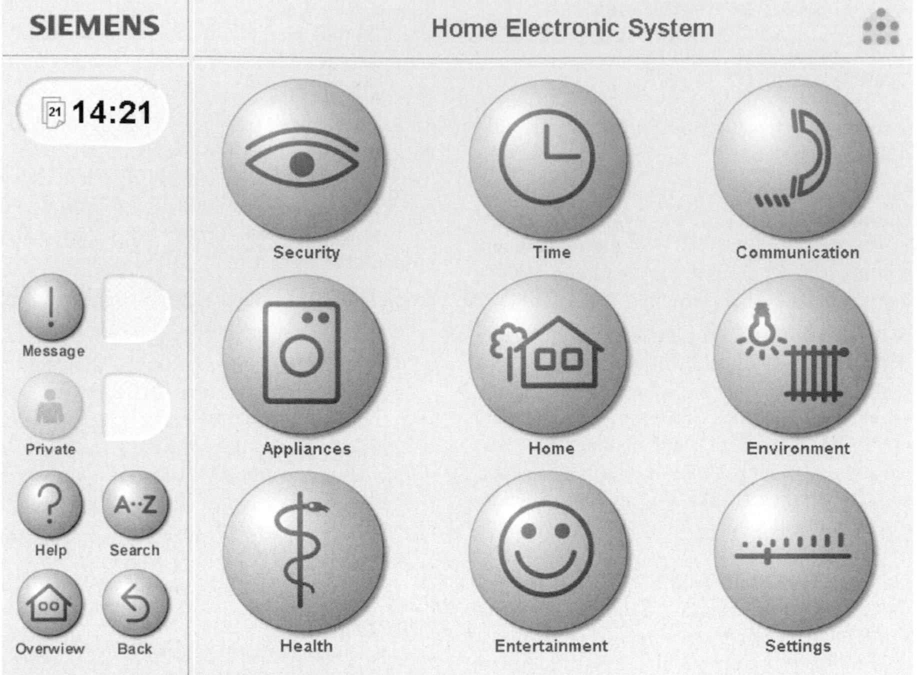

Fig. 4.4/16 Initial screen of the HomeAssistant

"Household appliances" and "Dishwasher"),

Room view

▷ Via the *Room* view: when this path is used, the search is initially restricted to a particular room. The user then selects the desired device/appliance from the devices/appliances in this room (e.g. "Home", "Kitchen" and "Dishwasher").

The user can easily tailor the interface of the HomeAssistant to his own house or apartment. For example, if a room is used for a different purpose (the "bedroom" becomes the "nursery"), or if devices/appliances are replaced ("chandelier" by "halogen spotlight set"), these changes can be implemented while the HomeAssistant is online. The user first selects the room to be changed, replaces the appropriate icon, and then enters the new room name.

Applications in the home

Devices and functions in the home

The following are some of the devices/appliances and functions supported by the HomeAssistant:

- Household appliances (continuous-flow water heater, cooker, dishwasher, washing machine, storage center),
- Television via PC,
- Lighting control,
- Shutter control,
- Temperature control,
- Task management.

Day and event programs

Task management

If an *EIB* time/event module is installed, the task management function can be used to set time as well as event-controlled tasks and store them in the time/event module. They continue to be processed here, even when the HomeAssistant is switched off.

Day program

The day programs are used to automate actions that are repeated at regular intervals (such as raising the roller shutters every morning). Many other appliances (e.g. luminaires, switchable plug-in sockets, room temperature controllers, alarm systems, etc.) can also be integrated in day programs (Fig. 4.4/17).

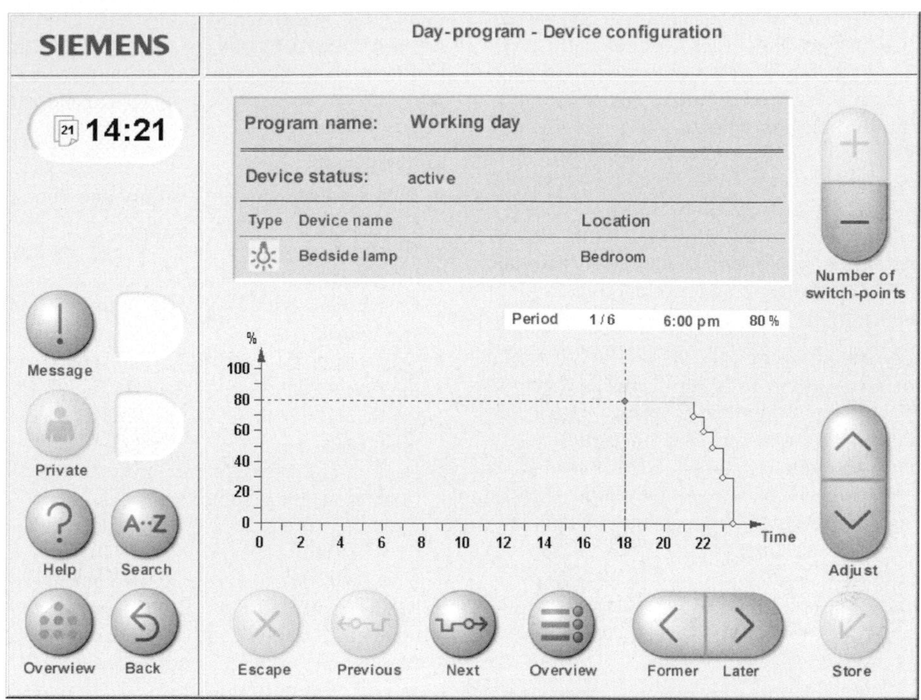

Fig. 4.4/17 Lighting control via a day program

The user can create up to 10 day programs, which are then entered in a calendar enabling them to be activated on the exact day required. A chronological list of the tasks set for the current day can be called up at any time.

Event program

The event programs are an extension of the scenario function. In event programs, the waiting time between consecutive tasks can be set, together with the desired device/appliance statuses. For example, after the event program has been started, a luminaire can be switched on initially at 100 % by means of a sensor or pushbutton, then dimmed automatically to 50 % after five minutes, and switched off completely after 10 minutes.

The user can redefine the link between an event program and the actuating sensor or pushbutton at any time. This means that any *EIB* sensor installed in a house (e.g. door and window contacts, motion detectors, brightness sensors, and operating rockers defined as scenario pushbuttons) can be used to initiate event programs. The sensors are also assigned via the HomeAssistant.

4.4.5.2 Functional buildings: Visualization

Purpose of visualization

Indicating faults

In functional buildings (such as office buildings, shopping centers, hospitals, indoor swimming pools, sports halls, theaters, manufacturing plants, power stations, schools, universities, etc.) it is absolutely essential that faults in the building services systems for power supply and distribution, as well as faults in other systems (e.g. lighting, heating, ventilation, air-conditioning, sanitation, transport, and communication) are indicated immediately at a central location. This ensures that any faults detected can be rectified quickly, and extensive consequential damage prevented.

Indicating operational statuses and values

As a result of the increasing use of computers for handling process data, the measuring instruments and signal lights formerly used to monitor the status of systems and indicate faults have now been replaced by visual display units. Visualization involves representing a building services system schematically on a computer screen, together with the current operational statuses and values for the system. The occurrence, location and type of faults can be indicated in a number of ways – via certain areas of the display or operating data that flash or change color; via an acoustic signal, such as a horn (where appropriate); and via additional explanatory texts displayed in the event of a fault or alarm. As a result, the operating staff are able to monitor the entire system at a glance, and to decide which measures to take to contain damage or rectify faults.

Visualization not only allows each building services system to be displayed in the form of a system overview or several partial displays; it is also possible to represent the building with its storeys and rooms, or an area comprising several buildings, thereby enabling faults to be located as soon as they occur. If the operating staff have selected a particular display and a fault occurs elsewhere in the system, the display for the faulty part of the system can be called up on the screen automatically and the current operating data displayed.

In the past, visualization was carried out via screens connected to a process computer. Personal computers (PCs) are now used for this purpose.

Visualization with *instabus EIB*

Visualization software

In buildings in which the electrical installations have been implemented on the basis of the *instabus EIB* building management system, all the sensors and actuators for lighting control and protection against sunlight, as well as for heating, ventilation, and air-conditioning can be connected via the *instabus EIB* to form a network. A PC can be connected to the bus line by means of the *EIB*/RS 232 interface, allowing buildings, rooms, and systems to be represented graphically and current operational statuses to be displayed via the visualization software.

Additional functions

Software packages for *EIB* visualization are now offered by a number of manufacturers. The operating system required for this software is Windows 3.11, Windows 95, or Windows NT. Since PCs are capable of more than just representing systems schematically, most manufacturers offer additional functions for visualization, such as a scheduler program or a program for archiving measured values acquired via the *instabus EIB*.

Features

In the following section, the visualization software from Siemens for the *instabus EIB* will be used to illustrate and explain the capabilities and features of a modern visualization system.

These include:

- ▷ Display and operation via only one PC, or via several networked PCs,
- ▷ Display, operation, and logging in any language,
- ▷ User access with assignable authorization levels/priorities,
- ▷ Transfer of *EIB* addresses and data from the project database of the *EIB* Tool Software (ETS) to generate the process points,
- ▷ Background displays can be imported as pixel or vector graphics,
- ▷ Integrated image editor for vector graphics with image library,
- ▷ Numerous dynamic display elements (image variables) for displaying current statuses, values, texts, and graphics,
- ▷ Manual and priority-controlled automatic call up of displays,
- ▷ Printing of display copies, event, overview, and parameter logs.

Additional functions

The additional functions supported include:

- ▷ Dropout monitoring of all the *EIB* devices installed,
- ▷ Event programs,
- ▷ Scheduler programs,
- ▷ Historical value monitoring,
- ▷ Event monitoring,
- ▷ Calculations/logic operations,
- ▷ Counter/consumption statistics,
- ▷ Video window with record and play functions,
- ▷ Remote display and control.

Several operator terminals

Different displays on different PCs

If users wish to call up different displays on several PCs at the same time, and operate the systems from these computers, the desired number of PCs can be connected to the bus line via the *EIB*/RS 232 interface (one per PC). The drawback of this solution, however, is that changes to a display, or a new time-controlled task can only be accessed via the PC used to create them. Export and import functions would be required to copy these changes from one PC to all the others by means of an external data carrier (e.g. a floppy disk).

Network version

To avoid this time-consuming process, a special (network) version of the visualization software is available. This allows one PC to be connected to the *instabus EIB* as a server.

This computer and all the other visualization PCs also have to be networked via Ethernet or ISDN. Different displays can be called up or different actions carried out on each PC. If a display or a time-controlled task is changed, for example, the changes are tracked automatically on all the other visualization PCs.

Language versions

Setting languages

It is also important that display, operation, and logging are carried out in the appropriate language. Siemens visualization software for the *instabus EIB* comprises several language versions, and can be easily expanded to accommodate other languages. When the program is started, the user is asked which language he wishes to use. This function also allows the language to be changed quickly if the operating staff change.

User access and authorization

Logging on with a personal code word

Before a user can work with the visualization software, he has to log on by entering a personal code word. Depending on the authorization he has been allocated, each operator can access one, several, or all of the individual functions of the visualization program. It is, therefore, possible, for example, to regulate the number of operators who can generate and change displays, enter and change time-controlled tasks, or switch systems on and off.

If an operator has logged on and been granted access, this is locked again automatically if the operator has not carried out any actions within a parameterizable time.

EIB data points

Entering data

To display *EIB* data points in a system display, the visualization software has to be told the corresponding group addresses and the type of data point belonging to a group address (EIS = ***EIB*** **I**nterworking **S**tandard). Entering this information manually is time-consuming and a potential source of errors. For this reason, a function is available that allows this information to be transferred from the project database of the ETS. If the database cannot be accessed, an additional function of the visualization software can be used to read this data from the online devices via the *instabus EIB*.

The large number of group addresses makes it very difficult for an operator to remember which information belongs to a particular group address. For this reason, Siemens visualization software allows each *EIB* data point to be assigned a meaningful "name" consisting of up to 32 alphanumeric characters.

Since several group addresses can affect one actuator output, it is also possible to link several group addresses logically so that the result of the logic operation behaves in the same way as the actuator output. The result of the logic operation is processed in the same way as an *EIB* data point (i.e. it can also be assigned a name).

Static displays

Static displays (or background displays) are displays or display elements that cannot be changed by process events.

A display called up on the screen of a visualization PC can comprise the following display elements on three superimposed levels:

- ▷ A static background display, **Background display**
- ▷ Static display elements. These are generated using the drawing program integrated in the visualization software (*Image Editor*) and superimposed on the background display, concealing the elements of the background display directly beneath them, **Static display elements**
- ▷ Dynamic display elements (*display variables*), which change depending on the process statuses, and conceal all the display elements directly beneath them, unless their current status is "hidden". **Dynamic display elements**

Pixel graphics (e.g. scanned photographs or drawings) or vector graphics can be copied (imported) to the visualization software as background displays.

Static displays can also be created by the user via the drawing program for vector graphics (*Image Editor*) integrated in the visualization software. This means that displays or display elements that are used on a regular basis do not have to be redrawn repeatedly, but can be stored as image macros in a macro library.

Dynamic display elements

Variable dynamic display elements (display variables), which depend on the status or value of an *EIB* data point, can be inserted in the static displays. The appearance (color, form, text, flashing, etc.) of a display variable changes, depending on the status or value of an *EIB* data point.

The following display variables are available:

▷ Output variable,
▷ Bar variable,
▷ Diagram variable,
▷ Picture variable,
▷ Meter variable,
▷ Switch variable,
▷ State variable,
▷ Symbol variable,
▷ Text variable.
▷ Counter variable,
▷ Week variable (only in conjunction with the scheduler program),

Output variables Output variables enable the user to output switching and control commands, to change setpoints and limit values, as well as set or reset count values.

Bar variables Bar variables enable fast visual comparisons to be made between measured and count values. The user can preselect the length, width, color, and alignment of a bar/graph.

Diagram variables Diagram variables, like continuous-line recorders, represent the characteristics of statuses and values as a function of time. Up to eight *EIB* data points can be represented by lines with different colors.

Picture variables Picture variables are used by the operator to call up a particular screen.

Meter variables Meter variables are used to display measured and count values on "pointer instruments" (voltmeter, ammeter, wattmeter, thermometer, etc.).

Switch variables Switch variables can be used in a variety of ways: to call up displays, initiate event programs or an additional function, and to output switching and control commands. They are a combination of the variables "output", "picture" and "symbol". This means that the statuses of a switch variable can also be displayed in the form of bitmap graphics.

State variables State variables enable the user to display texts corresponding to the current state or value of a bit or bit group in a state byte.

Symbol variables With symbol variables, the symbol displayed is replaced by another (e.g. an open switch by a closed switch) depending on the status of an *EIB* data point.

Text variables Text variables enable the user to display texts corresponding to the current status or value of an *EIB* data point (e.g. OFF; ON; OPEN; CLOSED; 21.5 °C).

Counter variables Counter variables enable to be displayed count values (number of operating cycles, operating hours, etc.) or quantities consumed (l, m³, kWh, etc.).

Week variables In conjunction with the additional function "scheduler programs", week variables are used for cyclical, time-controlled switching of *EIB* data points (and the corresponding building services systems), on the individual days of a week, if necessary, several times a day, as well as for displaying operating instants and duration.

Calling up displays

Displays can be called up on the screen of the visualization PC as follows:

▷ **Initial screen** The initial screen (Fig. 4.4/18), which should be displayed after the PC has booted up and a project opened, can be preselected,
▷ **Manual selection** The user can select a display manually from a list,
▷ The operator can call up the display he requires by selecting a picture variable,
▷ **Automatic selection** Displays can be called up automatically when process events occur. To prevent every event from calling up a new display, the operator can specify the level of importance (priority) as of which an event may call up the corresponding display.

A display can usually be called up in a few seconds. The time required depends on the content of the display and the performance of the PC (clock frequency, graphics card, memory expansion, etc.). The static background display is called up first, followed by the updated dynamic display elements.

Printing screen copies and logs

Color printers If a color printer is connected to the PC, a copy of the current display can be printed out (Fig. 4.4/19). The printer can also be used to output overview and parameter logs

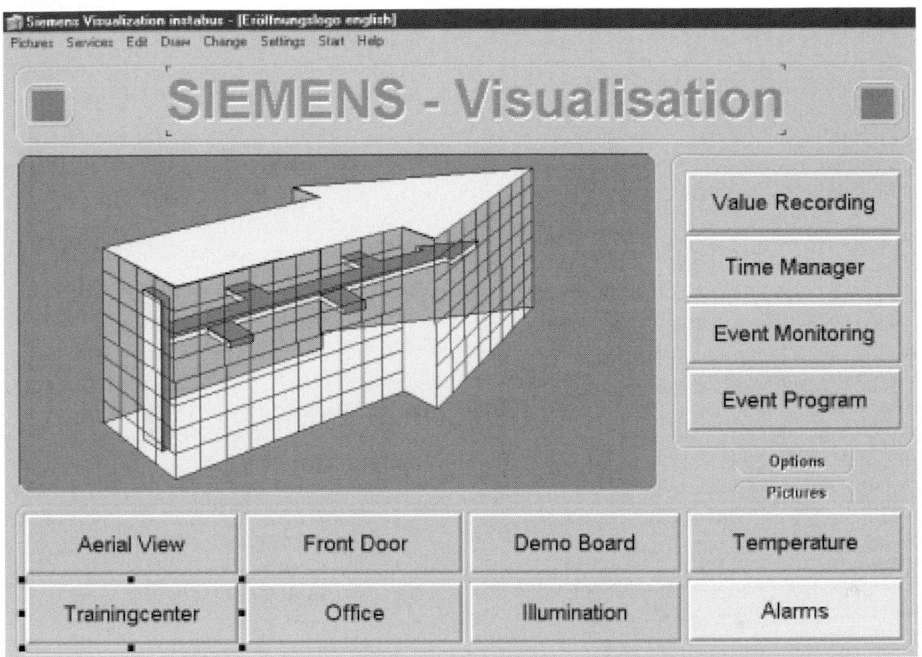

Fig. 4.4/18 Example of an initial screen

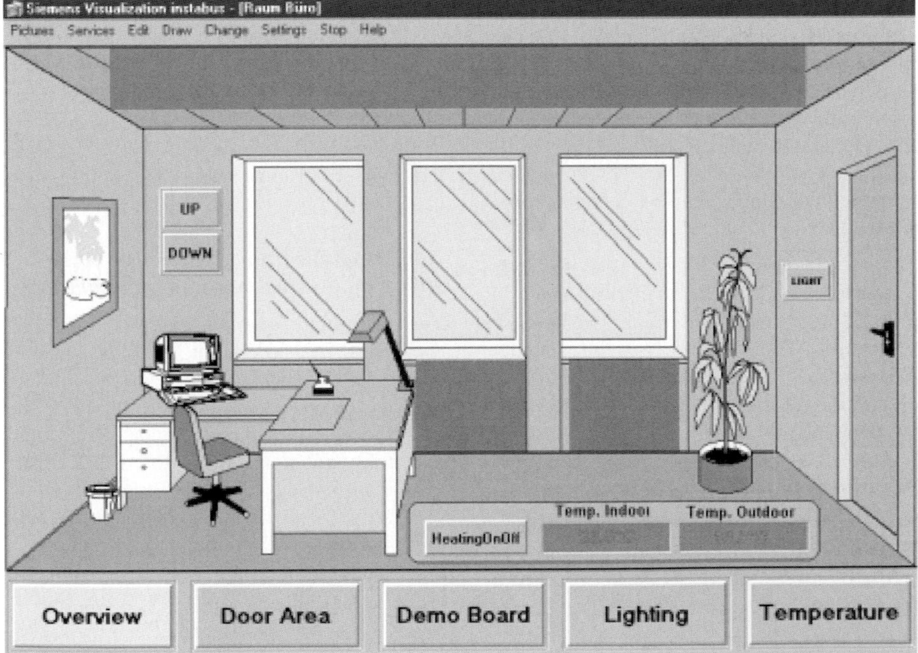

Fig. 4.4/19 Visualization of a room

for the *EIB* data points and additional functions managed by the visualization software.

Event log printers

One or more printers can be connected to the PC as event log printers. These printers log all the status changes of the *EIB* data points for which the event log was released. An event message takes up at least one printed line. It comprises the date and time, the name of the *EIB* data point, and the new status or value of the data point. If required, the event message can also contain a note explaining whether an alarm is an incoming or back-to-normal alarm, or whether a fault or a limit violation is involved. An optional explanatory text of one to four lines can be printed out after the event message. This text tells the operational staff how they are to respond to the event message (i.e. who has to be informed, and what other measures must be taken).

The two most recent event messages are displayed continuously at the bottom of the screen. If required, the user can switch from display mode to full-screen text mode. This screen shows the user all the event messages including the explanatory texts.

Additional function: "Dropout monitoring of all bus devices"

Monitoring all bus devices

This function is used to monitor all the bus devices connected to the *instabus EIB* (i.e. to detect and indicate faulty bus devices). Even with the maximum configuration of an *EIB* system (backbone line, 15 main lines, 12 lines per main line, and 64 devices per line), a faulty device can be detected within six seconds.

To prevent these monitoring activities from overloading the bus, the statuses of one or more bus devices are interrogated only every 15 seconds. The number of bus devices interrogated in 15 seconds, depends on the number of bus devices installed, and is determined automatically by the monitoring program.

Additional function: "Event programs"

Super scenario

An event program is also referred to as a "super scenario". The operator can specify a chain of switching and control commands initiated by a pre-defined event (e.g. a fault indication). The operator can also insert a wait time between two consecutive commands in a chain of commands. He can also initiate one chain of commands from within another automatically.

Additional function: "Scheduler programs"

This additional function comprises the following sub-functions:

Day program, week program, calendar program

- ▷ Day programs,
- ▷ Week programs,
- ▷ Calendar programs (scheduler).

This function must be used in conjunction with the additional function "event programs".

A total of 32 day programs can be created, each of which can be assigned up to 99 instants (switching times), and up to 99 tasks per instant.

Examples of a task include:

- ▷ Call up a particular display,
- ▷ Output a particular explanatory text,
- ▷ Output a switching or control command,
- ▷ Initiate a particular event program,
- ▷ Enable or lock a printer.

A day program can be initiated either on a daily basis, on a certain day of the week (week program), once on a particular day of the year, or only in a particular year (calendar program).

The week program is used to switch a system on the individual days of the week. It provides a particularly clear overview of the switching times for systems that have to be switched cyclically (e.g. air-conditioning system for a lecture theater, stairwell or hall lighting, or sun protection for a building façade).

Additional function: "Historical value monitoring with diagram representation"

Cyclical status recording

The historical database replaces numerous dotted-line recorders. It enables statuses (e.g. system switched off or on, window open or closed) and values (e.g. temperature, brightness, meter count) to be determined cyclically and stored on the hard disk of the visualization PC. Up to 32 groups with up to 64 tracks (= cyclically recorded *EIB* data points) can be determined cyclically with a cycle time of between 15 seconds and 24 hours per group, and stored in a ring buffer with 50 to 10,000 entries per track.

The stored data can be displayed on the screen in graphic form (lines, bars, two-dimensional, three-dimensional) and output on the color printer. Each group can also be exported individually in EXCEL file format, and processed and displayed on a different PC.

Additional function: "Event monitoring database"

Acknowledged and unacknowledged event messages

This additional function enables acknowledged and unacknowledged event messages to be displayed together with the type of acknowledgement (individual or shared). In addition to this, all the event messages for a defined period (e.g. one year) can be stored in a separate event database. Particular events in the database can be selected for statistical evaluations and displayed as a bar chart in accordance with their frequency. An event database can also be printed out in part or entirely in text form. Furthermore, the contents of an event database can be exported as an EXCEL file, and evaluated on a different PC.

Additional function: "Calculations/logic operations"

This additional function must be used in conjunction with the "event programs" function. Together with an editor program, it enables programs to be created in a programming language derived from BASIC. These programs can be initiated by event messages or time-controlled tasks. This makes it possible to determine, for example, whether lights are still on in a building or whether a window is still open. It is also possible to calculate power levels from voltage and current. The result can be assigned to a virtual *EIB* data point and displayed in graphic form.

Additional function: "Counter/consumption statistics"

Daily, weekly, monthly and annual statistics

The counter/consumption statistics enable consumption values (count values) to be recorded cyclically and stored in daily, weekly, monthly, and annual statistics. The data stored in these statistics can be printed out as informative texts or exported. When the data is logged, the user can choose between a detailed, an overview, and a complete log.

Additional function: "Video window with record and play functions"

This function requires a video-capture card, as well as a sound card, if not already installed. Using the video variable monitor, recorder and player, live images from a video camera can be displayed on screen, stored as a Windows film file on the hard disk of the PC and played later. This can be used to document who was in a particular room at a particular time.

Additional function: "Remote display and control"

This additional function requires a modem or a gateway to the public telephone system, and enables fault signals and alarms to be transmitted to a central service point. From the central service point, the service staff can, if necessary, access the *EIB* system in which the fault or alarm was indicated as if their PC were directly linked to the *EIB* system locally.

Indication and control via the Internet

System and room functions can also be executed and displayed via the Internet using an appropriate additional function in the hardware and software.

Visualization versions

To meet the various demands made on visualization software effectively and efficiently, the software is available in a number of different versions:

Mini/training version

▷ A mini or training version manages only a limited number of *EIB* data points, but is particularly attractively priced;

Full version

▷ A full version allows the user to create system displays and to call up updated versions of these displays;

View-only version

▷ A view-only version allows only updated displays to be called up. This version is less expensive than a full version, and is intended for users who do not want to create their own displays;

Configuration version

▷ A configuration version only allows *EIB* data points to be generated and updated displays to be created. This version does not, however, permit the configured displays to be updated. This version is also less expensive than a full version and is intended for service providers who have to create displays on behalf of their customers;

Network version

▷ A network version allows a visualization PC to be connected to an in-house network (e.g. Ethernet or ISDN), permitting access to the systems and data points connected to the <u>*instabus*</u> *EIB* from this PC and other PCs.

5 Application examples of electrical installation engineering

5.1 Functional buildings

5.1.1 General requirements

In addition to the power needed for lighting, functional buildings also require a considerable amount of electrical energy for supplying technical equipment, for example, for air conditioning, elevators, etc.. Although the amount of power needed for lighting is decreasing thanks to energy-saving lighting technology and energy management, the power needed for technical equipment and air conditioning in buildings is increasing.

Standards

The relevant EN standards and harmonization documents published by CENELEC together with the restrictions and footnotes applicable for the country in which the equipment is installed must always be observed. For installations up to 1000 V, this is the **H**armonization **D**ocument 384 or EN 60384. These standards are identical to IEC publication 364 with its individual parts as well as DIN VDE 0100 which is applicable in Germany. Standards for supply systems for safety services (in Germany DIN VDE 0107 for hospitals and DIN VDE 0108 for buildings with communal facilities) must also be observed. Both standards are currently being harmonized and integrated in HD 384.

Connection of buildings

In Germany, the technical supply conditions issued by the respective public utility must be complied with when connecting buildings to the public utility's power supply system. Buildings with a power requirement of approximately 300 kW or higher usually have to be connected to a high-voltage system of the public utility, i.e. a high-voltage substation with transformers is required. This substation is generally installed on one of the lower storeys of the building.

High-rise buildings

High-rise functional buildings often have current-using equipment with high power requirements on their upper storeys, e.g. drive motors or motor-generator systems for elevators, parts of the air-conditioning system, etc.. In the interests of efficiency, the electrical power should be transported to the load centers at the highest possible voltage level (see Section 1.1). The transformers are, therefore, installed on the top storey as well and, if necessary, on an intermediate storey (Fig. 5.1/1).

Nowadays, dry-type transformers with flame-retardant, self-extinguishing cast-resin insulation and oil-free switching devices permit such solutions.

Power supply systems

The type of power supply system depends on the size of the premises, the length of the supply cables, and the types of load. During the project planning stage, a distinction is made between:

▷ main power supply,
▷ load power supply.

The main power supply transports the power from the main low-voltage switchgear to the sub-distribution and storey distribution boards via cables and rising main busbars, while the load power supply transports power on the storeys or in the individual supply sections to the ultimate consumers.

5.1.2 Main power supply

Factors such as the size and shape of the building, room utilization, and the number of storeys or supply sections and their location determine the arrangement of the main power supply cables (Fig. 5.1/2).

Rising main

In long, multi-storey buildings, the main power supply cables are installed starting at the bottom of the building in the form of

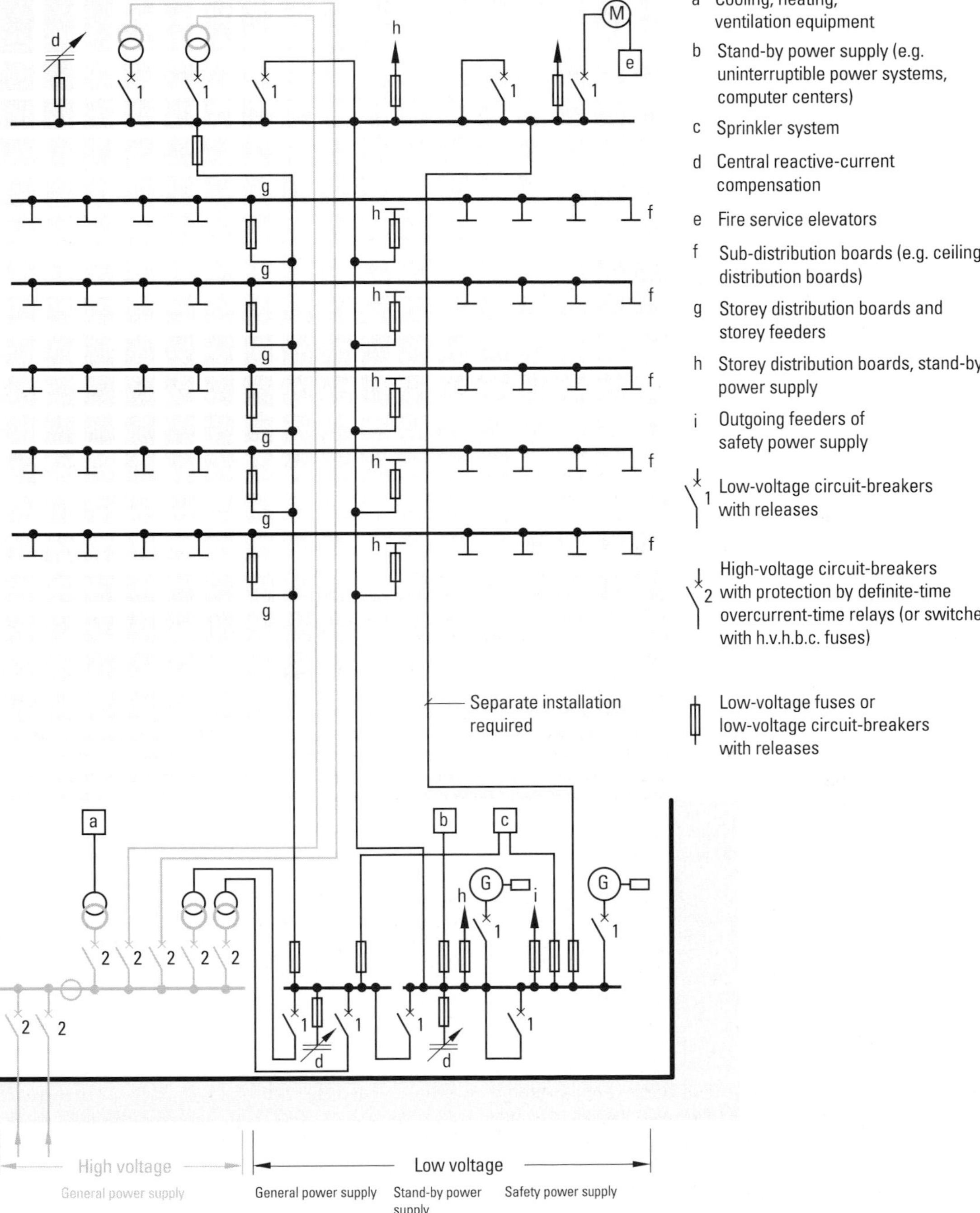
a Cooling, heating, ventilation equipment
b Stand-by power supply (e.g. uninterruptible power systems, computer centers)
c Sprinkler system
d Central reactive-current compensation
e Fire service elevators
f Sub-distribution boards (e.g. ceiling distribution boards)
g Storey distribution boards and storey feeders
h Storey distribution boards, stand-by power supply
i Outgoing feeders of safety power supply
1 Low-voltage circuit-breakers with releases
2 High-voltage circuit-breakers with protection by definite-time overcurrent-time relays (or switches with h.v.h.b.c. fuses)
Low-voltage fuses or low-voltage circuit-breakers with releases
Separate installation required
M
G
High voltage
General power supply
Low voltage
General power supply
Stand-by power supply
Safety power supply

Fig. 5.1/1 Block diagram showing main power supply in high-rise office building

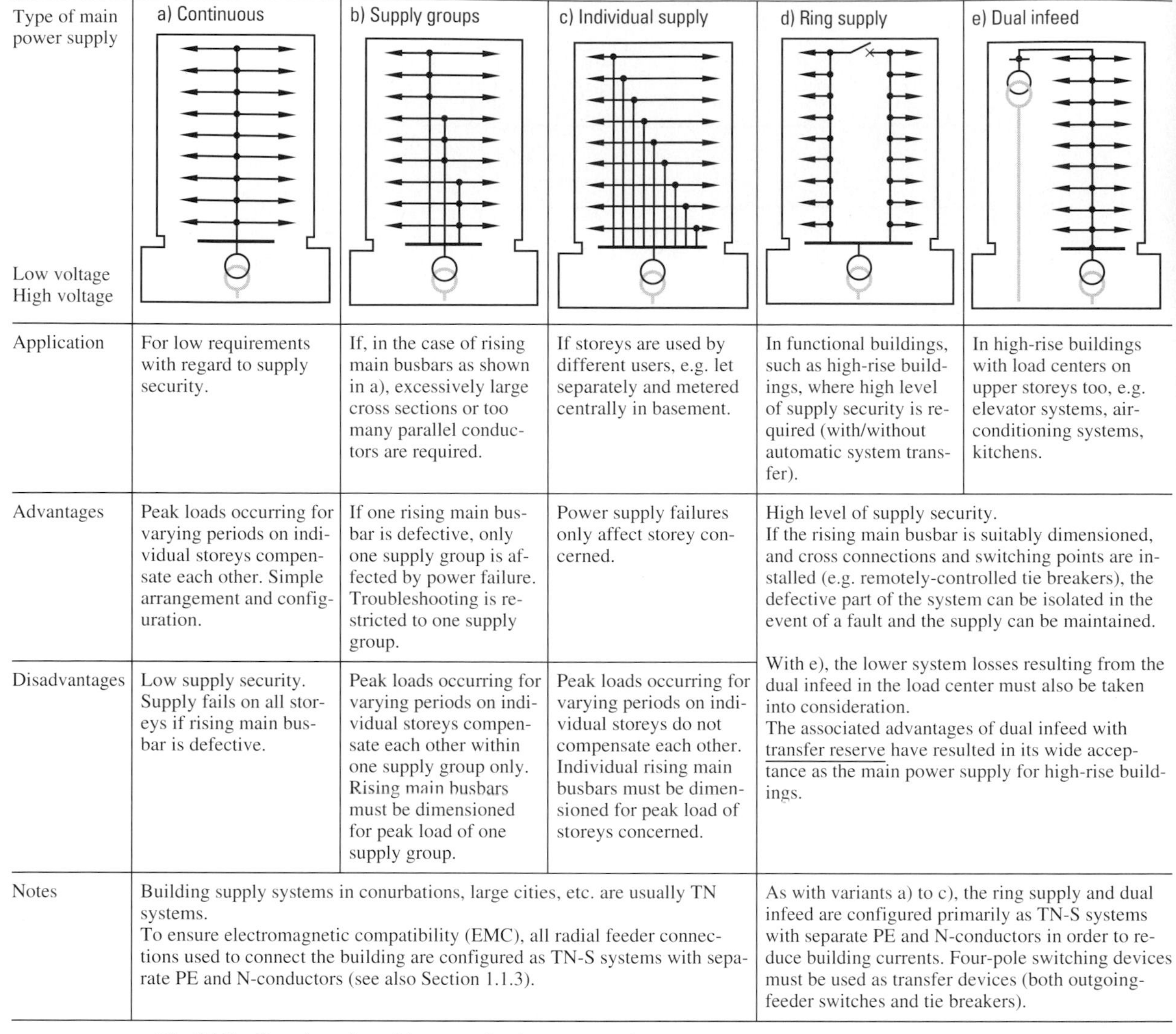

Type of main power supply	a) Continuous	b) Supply groups	c) Individual supply	d) Ring supply	e) Dual infeed
Low voltage High voltage					
Application	For low requirements with regard to supply security.	If, in the case of rising main busbars as shown in a), excessively large cross sections or too many parallel conductors are required.	If storeys are used by different users, e.g. let separately and metered centrally in basement.	In functional buildings, such as high-rise buildings, where high level of supply security is required (with/without automatic system transfer).	In high-rise buildings with load centers on upper storeys too, e.g. elevator systems, air-conditioning systems, kitchens.
Advantages	Peak loads occurring for varying periods on individual storeys compensate each other. Simple arrangement and configuration.	If one rising main busbar is defective, only one supply group is affected by power failure. Troubleshooting is restricted to one supply group.	Power supply failures only affect storey concerned.	High level of supply security. If the rising main busbar is suitably dimensioned, and cross connections and switching points are installed (e.g. remotely-controlled tie breakers), the defective part of the system can be isolated in the event of a fault and the supply can be maintained. With e), the lower system losses resulting from the dual infeed in the load center must also be taken into consideration. The associated advantages of dual infeed with transfer reserve have resulted in its wide acceptance as the main power supply for high-rise buildings.	
Disadvantages	Low supply security. Supply fails on all storeys if rising main busbar is defective.	Peak loads occurring for varying periods on individual storeys compensate each other within one supply group only. Rising main busbars must be dimensioned for peak load of one supply group.	Peak loads occurring for varying periods on individual storeys do not compensate each other. Individual rising main busbars must be dimensioned for peak load of storeys concerned.		
Notes	Building supply systems in conurbations, large cities, etc. are usually TN systems. To ensure electromagnetic compatibility (EMC), all radial feeder connections used to connect the building are configured as TN-S systems with separate PE and N-conductors (see also Section 1.1.3).			As with variants a) to c), the ring supply and dual infeed are configured primarily as TN-S systems with separate PE and N-conductors in order to reduce building currents. Four-pole switching devices must be used as transfer devices (both outgoing-feeder switches and tie breakers).	

Fig. 5.1/2 Overview of possible types of main power supply

Storey feeders and distribution boards

rising main lines at intervals which divide the building into individual supply sections (Fig. 5.1/3). One or more shafts, in which storey feeders or storey distribution boards can be mounted in addition to the rising main lines so that they are physically separate from each other, are provided for the rising main lines.

Busbar trunking systems as rising main lines

Busbar trunking systems are being used to an increasing extent instead of cables or conductors with large cross sections or parallel cables (see Fig. 1.4/45 in Section 1.4.4). Type-tested versions with standard rated currents are preferred.

Stand-by supply and UPS

If, in addition to general loads, particularly important or sensitive loads, e.g. IT systems, etc., also have to be supplied with power if the supplying system has failed or is defective, it is advisable to provide a second rising main busbar system (see Fig. 5.1/1) which is installed separately from the general power supply system and can be supplied with

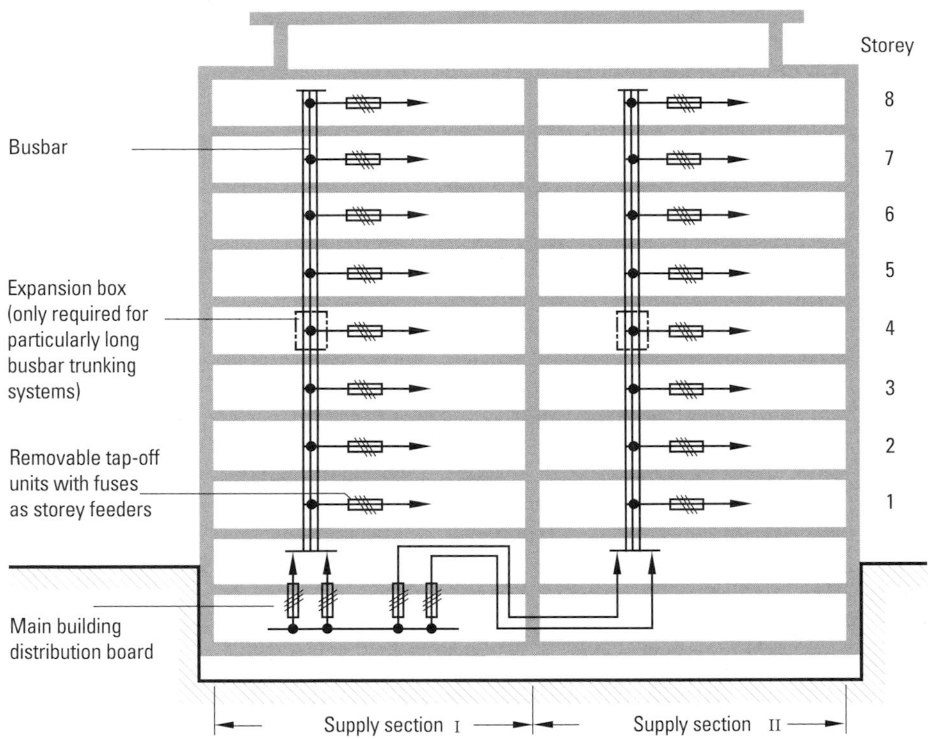

Fig. 5.1/3 Supplying power to long building with individual supply sections via busbar trunking system

power from either a generating set or a central uninterruptible power system (UPS).

Safety power supply

According to DIN VDE 0107 and DIN VDE 0108 (as specified in the harmonization document HD 384 Part 7), an independent safety power supply is required to supply important safety equipment, e.g. safety lighting, sprinkler systems, fire service elevators, etc.. An independent power supply system, which must be installed separately from the general power supply, must be provided for this purpose. Equipment with functional endurance must be used as transmission devices (e.g. cables, conductors, busbars). See Sections 1.1.2 and 1.8 for further information.

Dimensioning transmission equipment

The following criteria must always be taken into consideration when dimensioning the transmission equipment between the main switchgear and the sub-distribution and storey distribution boards:

- ▷ protection against overloading,
- ▷ protection in the event of short circuit,
- ▷ protection against electric shock in the event of indirect contact,
- ▷ compliance with the maximum permissible voltage drop,
- ▷ selective behavior with respect to upstream and downstream protective devices.

The maximum prospective load current (I_B) and the rated current, for which the main power supply lines and rising main busbars must be dimensioned as a minimum requirement, are calculated by multiplying the connected loads of the individual loads by the coincidence factors of the individual system components (see Section 1.1). It is important to ensure that certain load groups are not switched on simultaneously. It is often advisable to record the individual load values over 24 hours in the form of a graph.

Overload/ overcurrent protective device

An overload protective device in compliance with HD 384.4.43 / DIN VDE 0100-430 is, therefore, always required at the beginning of the circuit. An overload protective device may only be omitted if the circuit is configured for the sum of the connected loads to be supplied. In practice, this is recommended only in special cases since low-voltage systems are particularly prone to short-circuit-like faults with very high impedance values at the fault location.

If, as is the case with ring supply or dual infeed with transfer reserve, a number of sub-distribution boards have to be supplied in the event of a fault, the rated current that must be dimensioned for the system is the sum of the load currents of both circuits.

Reserve capacity in event of fault/failure

In addition to ensuring that adequate reserve capacity is provided in the event of a fault or failure, future conversions and expansions must be taken into consideration when the supply system is planned.

The rated current (I_n) of the required overcurrent protective device for protection against overload (fuse or circuit-breaker with releases) must be greater than or equal to the maximum incidental load current I_B

$I_n \geq I_B$.

Overload protection of parallel cables

If, on account of the load current, parallel cables are required for the main power supply lines and rising main busbars, these must be protected against overloads individually or, under certain conditions, by means of a common protective device.

Fuses should only be used at the beginning *and* end of the supply lines if three or more parallel cables are installed.

The permissible current-carrying capacity *(I_z)* of the conductors must be greater than the rated current I_n of the overcurrent protective device used; this, in turn, must be greater than the maximum incidental load current

$I_Z \geq I_n \geq I_B$.

Current-carrying capacities

Current-carrying capacities are specified in the literature supplied by the manufacturer or the relevant national and EN specifications (Germany: e.g. DIN VDE 0298 Part 2 for cables, DIN VDE 0298 Part 4 for conductors, or DIN VDE 0100-430 Supplement 1). These standards contain reduction factors which must be taken into consideration when determining the permissible current-carrying capacity of cables and conductors under non-typical operating and installation conditions. Current-carrying capacities and reduction factors can also be found in the Siemens publication *"Power Cables and their Application"*.

Short-circuit protection

If protective devices are used, the tripping current I_2 (large test current) of which differs from the rated current by a factor of 1.45 or higher, the conductor must be suitable for the following current-carrying capacity, bearing in mind the reduction factors that result from non-typical installation and operating conditions:

$I_Z \geq I_2/1.45 \geq I_n \cdot X/1.45$

$I_n \cdot X$ Tripping current of protective device,

$X \geq 1.45$.

The conductor cross section to be selected must satisfy the following equation:

$K^2 \cdot S^2 \geq I_k^2 \cdot t_k$

K Material coefficient for insulation and conductor material

S Cross section of conductor in mm^2

$I_k^2 \cdot t_k$ Let-through power of protective device during short-circuit breaking time t_k.

Since the above equation is only defined for the time interval 0 to 5 s, the lowest possible short-circuit current at the end of the conductor must be high enough to ensure that the protective device operates reliably and then disconnects within 5 s. If the N and PE-conductors have a reduced cross section, this must be taken into account when the system is dimensioned.

Protection against electric shock with indirect contact

Protection against electric shock in the event of indirect contact in accordance with HD 384.4.41 / IEC 60364-4-41 / DIN VDE 0100-410 by means of automatic disconnection must be ensured for the lowest fault current which can occur at the end of the circuit (see Section 2.4).

If the lowest incidental fault currents are too low, the impedance of the conductor loop along the transmission path must be reduced and the cross section, therefore, increased.

With long cable runs, variants with full PE and PEN-conductors are usually more cost effective compared to variants with reduced PE and PEN-conductors. An additional advantage of this solution is the reduced fire load along the cable routes.

Using circuit-breakers

The use of circuit-breakers with inverse-time or short-time-delay releases instead of fuses is also more cost effective particularly in the case of long cable runs.

Selectivity verification

Partial selectivity

Using selective protective devices across the entire current-time range of possible fault currents considerably restricts the downtimes of the power supply system (fault location located more rapidly) and the affected load sections. Both increase the level of supply security and supply quality. If selectivity is verified for only part of the range, this is referred to as partial selectivity.

If fuses are used in a system with partial selectivity, all affected non-selective sets of fuses should be replaced after a fault has occurred in order to prevent subsequent false tripping caused by previous damage.

Full selectivity with safety power supply

Verification of full selectivity is prescribed for all supply systems for safety services.

Necessary constructional measures

Planning space requirements and installation locations

When the plans for a building are being drawn up, it is essential to ensure that the required electrical system components, their space requirements, and installation locations as well as possible transport routes are clarified with the architect.

The necessary constructional measures (including supply routes, openings for large system components, etc.) must be planned very early on in order to avoid costly alterations at a later stage.

The specifications in the building regulations of the respective country (or, in Germany, the individual federal states) must be complied with, particularly with respect to fire protection, water pollution control, and emissions.

Supplement 1 to DIN VDE 0108 Part 1 (EltBauVo)

Supplement 1 to DIN VDE 0108 Part 1 contains a standard regulation regarding the construction of operating areas for electrical systems (EltBauVo) and the general guidelines regarding fire-protection requirements for cable systems in the Federal Republic of Germany.

Operating areas for transformers and switchgear assemblies

It must be possible to access the electrical operating areas for transformers and switchgear assemblies easily and safely. In the event of an emergency, it must also be possible to leave the operating areas without hindrance. DIN VDE 0100 Part 731 (nominal voltages ≤ 1 kV) and prEN 50179 / DIN VDE 0101 (nominal voltages > 1 kV) must be observed when constructing such operating areas. When the required dimensions of the area are determined, the minimum clearances for walkway widths, walkway lengths, and passage heights must be taken into consideration with respect to the erection of the switchgear assemblies and distribution boards.

Riser ducts and cable routes

The number of riser ducts and their arrangement is based on the concept selected for the power supply system. The shape and size of the riser ducts depend on the type and number of the cables or busbar trunking systems used for the main power supply lines.

Ducts and ceiling openings must be approved by a structural engineer.

Fire resistance rating

In order to prevent fires from spreading and damage resulting from combustion gases, all openings and passages in fire-resistant walls and ceilings must be fully sealed in accordance with the specified fire resistance rating.

Anchor rails

To allow conductors, cables, and busbar trunking systems to be secured easily and quickly, it has proved useful to incorporate horizontal anchor rails in the concrete during construction. The distances between the anchor rails must be calculated and determined, if necessary taking the short-circuit ratios of the main power supply lines into consideration.

Transport routes in buildings

Transport routes and, if necessary, freight elevators with sufficient load-carrying capacity must also be planned to allow transformers, switchgear assemblies, generating sets, etc., to be replaced at a later stage on the upper storeys of the building.

Grounding in building

When the building foundations are laid, the necessary grounding ring conductors can be installed around the building or in the peripheral foundations without additional excavation costs being incurred. A wide-meshed grounding grid in the base foundation is welded to the grounding ring conductors. In addition to this, the grounding ring conductors are connected to the conductive piping that leads into the building; conductive load-bearing supports, concrete-reinforcing iron, etc., are connected to the grounding grid in the base foundation or to the grounding ring conductor. See Section 1.5 for further information on grounding systems and their design.

Electromagnetic compatibility (EMC)

Grounding requirements must be clarified as early as possible. When an electrical installation is being planned, experts must be commissioned to elaborate an electromagnetic compatibility (EMC) concept which includes external and internal lightning protection, shock protection, radio interference suppression, and possible system perturbation, and which takes the required interference immunity for the devices and systems to be used into consideration. A correctly planned power supply system (preferably configured as a TN-S system with a separate N and PE-conductor) is essential in order to ensure electromagnetic compatibility.

Costly improvements (if these are at all possible at a later stage) are thus avoided. Section 1.1 and Chapter 3 contain further information on EMC.

5.1.3 Load power supply

Storey distribution boards

The transition from main power supply to load power supply occurs via storey distribution boards. These distribution boards contain the protective devices, such as fuses and miniature circuit-breakers, required for the load power supply of a certain supply section (e.g. storey) as well as any necessary control devices. The size of the distribution boards primarily depends on the size of the supply section concerned, the type of distribution (central or decentralized), the type of current-using equipment to be supplied and controlled, and the required control equipment. Furthermore, the incorporation of an IT system (e.g. telephone private branch exchanges) is also a key factor influencing the size of the distribution boards.

Standards

These distribution boards must be installed in accordance with the relevant standards for power engineering, e.g. EN 60439-1/-3 / IEC 60439-1/-3 / DIN VDE 0660-500/-504, DIN VDE 0603, DIN VDE 0100, and where appropriate DIN VDE 0107 and DIN VDE 0108, as well as for information technology (where applicable), e.g. DIN VDE 0800 (see Section 1.4.5). In addition, the technical supply conditions of the local public utility, together with the relevant standards (e.g. DIN in Germany), and guidelines in the regional building regulations as well as those issued by the national telecommunications provider, the employers' liability insurance associations, and the property insurers' association must also be taken into consideration.

A distinction is made between two types of load power supply:

▷ central distribution,
▷ decentralized distribution.

5.1.3.1 Central distribution

Central distribution board

In the case of central distribution, all protective devices and a number of control devices for the current-using equipment in a supply section are installed in a central distribution board (e.g. storey distribution board).

In the past, this type of distribution was generally adequate for simple wiring systems with a relatively small number of loads and a few, mainly decentralized control units (e.g. for breaking circuits, two-way circuits, etc.), even if the supply sections were relatively extensive.

The increasingly widespread use of electrical equipment, as well as the rise in the number of electric circuits to provide a sufficiently high level of supply security, and the number of control tasks coupled with the greater demands placed on control convenience in modern buildings means that storey distribution boards have to be extremely versatile. This often results in very large distribution boards as well as complex masses of conductors which increase the fire load and make troubleshooting more difficult (Fig. 5.1/4).

instabus *EIB*

Modern control systems (e.g. *instabus EIB*, see Section 4.4) in which the central distribution boards are largely free from control devices due to the distributed arrangement of sensors and actuators enable the size of the distribution boards to be reduced considerably.

Furthermore, the used bus technology also significantly reduces the number of conductors.

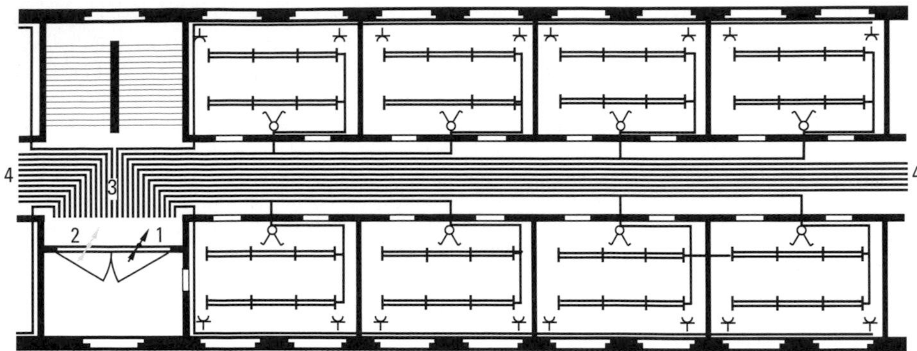

1 Rising main, general
2 Rising main, stand-by power supply
3 Storey distribution board
4 Load power supply circuits

Fig. 5.1/4 Example of central distribution (stand-by power loads not illustrated)

5.1.3.2 Decentralized distribution

Sub-distribution boards

Nowadays, large buildings are often fitted with decentralized distribution systems in which several, mostly identical sub-distribution boards installed at regular intervals along corridors (Fig. 5.1/5) are fed by one storey distribution board. Cable installation is, therefore, considerably simplified.

The advantages of this distribution type with sub-distribution boards are:

- ▷ transparent system structure,
- ▷ short outgoing cables to the loads,
- ▷ lower fire load,
- ▷ straightforward troubleshooting,
- ▷ only a small section of the system is disconnected in the event of a fault.

Flexibility

Conventional decentralized distribution systems allow the horizontal power supply to be adapted relatively easily to changes in the layout of the room or its furnishings since

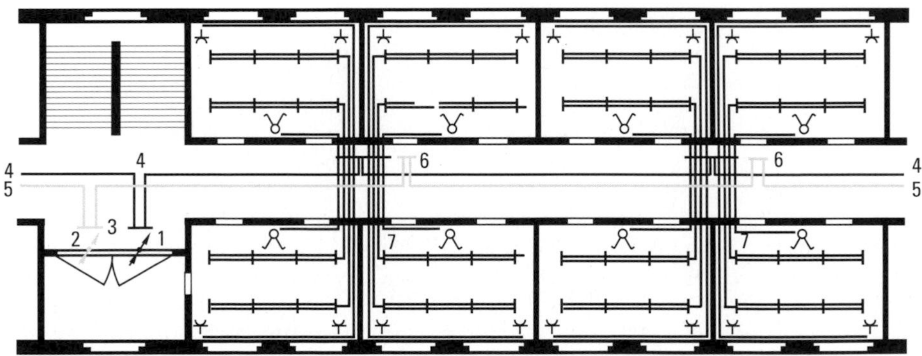

1 Rising main, general
2 Rising main, stand-by power supply
3 Storey distribution board
4 Feeder to sub-distribution boards (e.g. in ceiling), general
5 Feeder to sub-distribution boards, stand-by power supply
6 Sub-distribution board
7 Load circuits

Fig. 5.1/5 Example of decentralized distribution (stand-by power loads not illustrated)

every sub-distribution board has a terminal strip via which all the cables are routed (socket-outlet cables, control cables, luminaire supply cables, and so on). This allows cables to be installed in rooms without tapping boxes. If changes are made, the switches or pushbuttons can be assigned to the luminaires by simply rearranging the terminal connections on the sub-distribution boards.

instabus EIB

Modern, intelligent control systems, such as *instabus EIB*, with a decentralized system structure, support this distribution type because sensors can be freely assigned to actuators via software parameters (usually from a central location) without changes having to be made on the respective sub-distribution board.

Installation of sub-distribution boards

Depending on the structural features of the building, the use of the rooms, and arrangement of the storey distribution boards, the sub-distribution boards are installed in recesses, walls, cavities in suspended ceilings, or sectional floors. It is, however, important to ensure that the distribution boards can be accessed without any difficulty and that the operator stands at a safe position. The most cost-effective solution is to use just one type of distribution board within a building as this will facilitate assembly and maintenance.

5.1.4 Cable installation methods for load power supply

General

Fig. 5.1/6 illustrates the methods normally used for installing cables in functional buildings. These must comply with the applicable standards.

5.1.4.1 Surface installation

Damp locations, cellars, etc.

With surface installation (Fig. 5.1/7), the incoming cables to the loads as well as

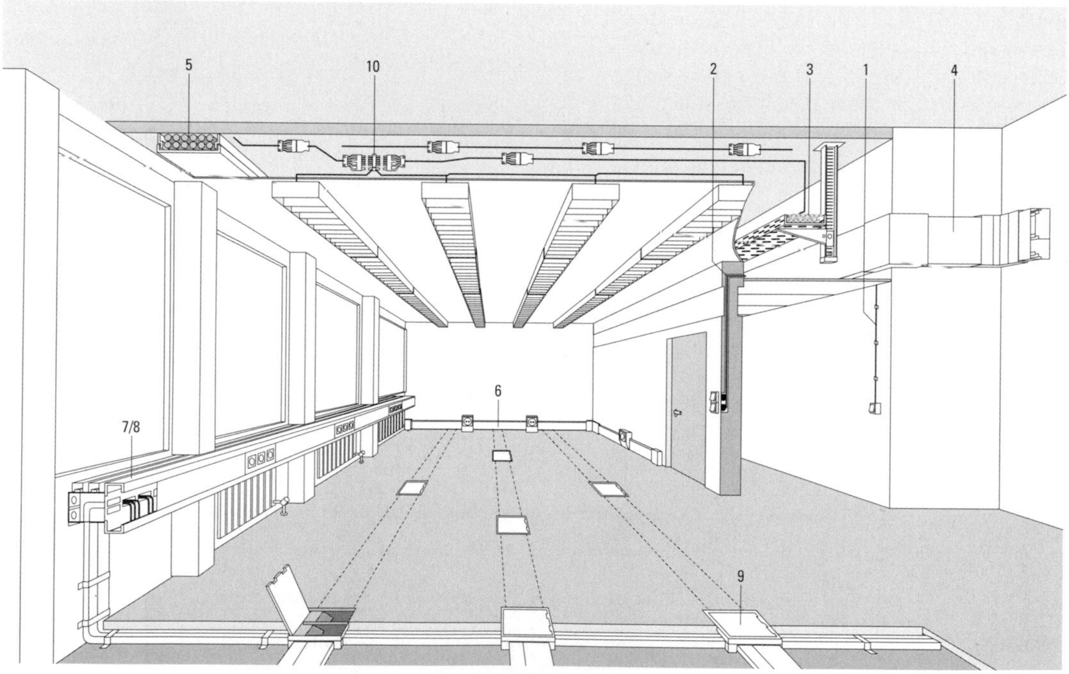

1 Surface installation
2 Flush installation
3 Cable raceways
4 Cable duct for wall
5 Cable duct for ceiling
6 Skirting board duct
7 Sill-type duct
8 Dado duct
9 Underfloor duct system
10 Snap-on system (pre-assembled cable installation system)

Fig. 5.1/6 Overview of cable installation methods for load power supply

switches and socket-outlets of power installations and communications systems are secured directly to the wall of the building using spacing saddles, in plastic/high-strength steel conduits or in duct systems. This cable installation method is used in damp locations, cellars, and corridors.

5.1.4.2 Flush installation

Flat webbed cables

With flush installation (Fig. 5.1/8), the incoming cables are installed directly in the unfinished wall or in conduits and duct systems to protect them against damage and then plastered over. The demand for simplified cable installation led to the development of flat webbed cables which are simply secured using steel nails and with plaster if routed in a different direction. The nail heads are embedded in plastic to protect the cable insulation. Due to the fact that they can be assembled quickly and easily, flat webbed cables have become a very popular and cost-effective installation method in residential buildings, although they are difficult to modify at a later stage.

Communications cables

In contrast, communications cables should only be installed in conduits or duct systems since the systems need to be modified more frequently.

5.1.4.3 Cable raceways

Applications

Cables, which frequently have to be relaid or installed at a later stage in electrical systems of large buildings or in industrial plants, can be installed conveniently in cable raceways, cable ladders, cable troughs, gridways, and cable routes made of hot-dip galvanized steel or plastic. These cable routes are usually laid in cable ducts, under ceilings in corridors, or in suspended ceilings. The cables are, generally, only inserted in the raceways from above. In the case of vertical cable raceways, the cables are secured with integrated strap saddles. The strap saddles are designed so that they can be secured in slots at any position along the cable raceway. The cable raceways can be mounted on wall brackets or are suspended from the ceiling using hangers and clamping brackets.

Design

Cable raceways have a modular design and consist of cable troughs, connectors, and accessories. The direction of the raceway can be changed using right-angle, T, and four-way sections. The individual components are connected to each other using screws or assembled using connectors. Pivoted connectors permit changes in direction from the horizontal to other planes. Partitions allow cables of different voltage levels and current types (AC/DC) and _instabus_ *EIB* cables to be installed in the same cable raceway. The open ends are covered with end plates (Figs. 5.1/9 a and 5.1/9 b).

5.1.4.4 Cable ducts for walls and ceilings

Application

The cable ducts made of impact-resistant plastic or metal, or a combination of the two materials, are used to install cables of power installations and communications systems on walls and ceilings. They permit cables to be rearranged or new cables to be installed if the system is modified.

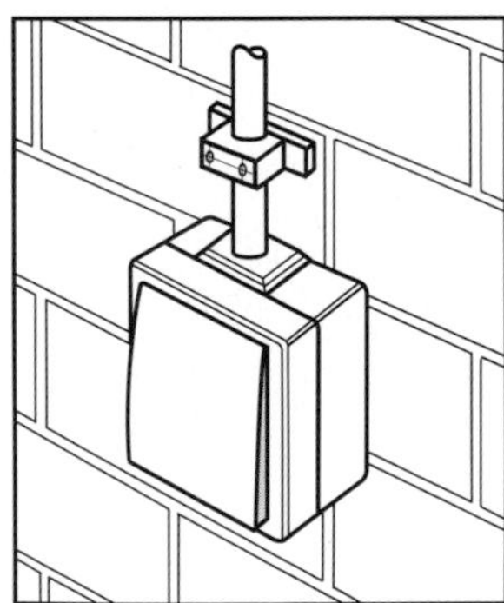

Fig. 5.1/7
Surface installation of on/off switch

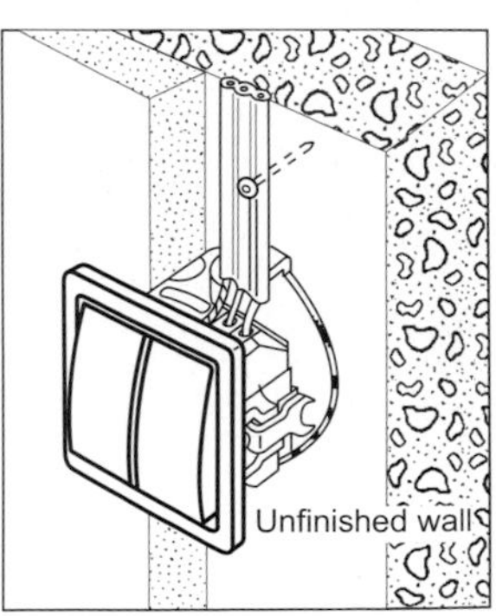

Fig. 5.1/8
Flush installation of two-circuit switch

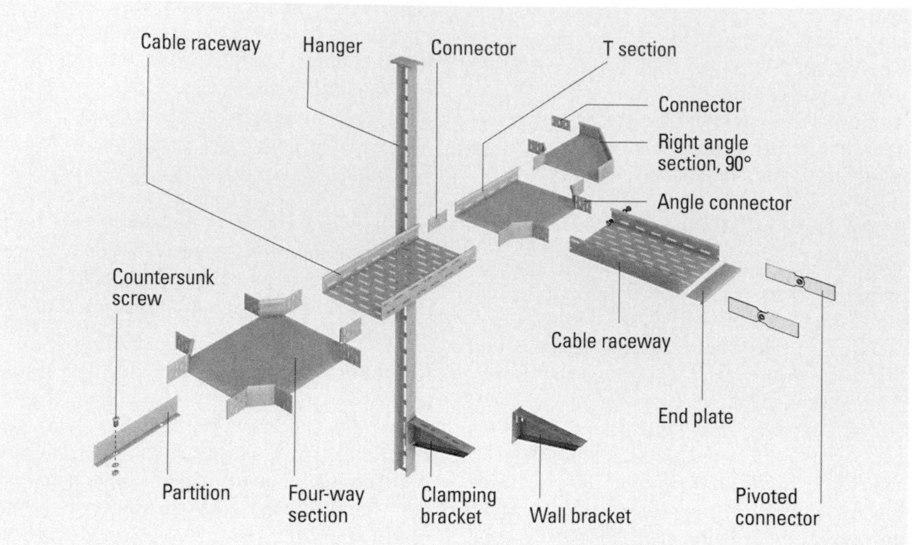

a) Lightweight system

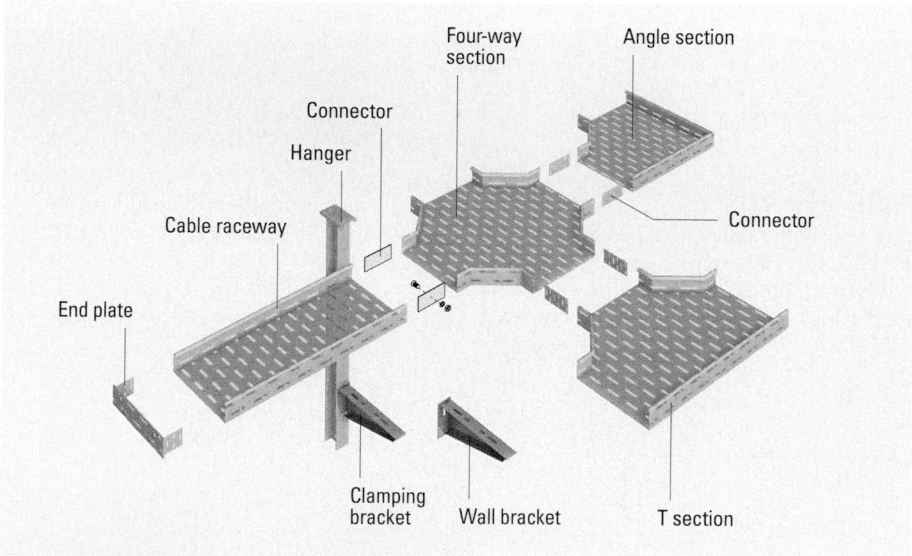

b) Heavy-duty system

Fig. 5.1/9 Cable raceways and accessories

Design

The ducts are available with:

- ▷ welded cross members, standard depth (Fig. 5.1/10),
- ▷ removable retaining clips, low-profile version (Fig. 5.1/11)

Ducts with welded cross members are extremely stable and are, therefore, usually mounted on ceilings in premises used for commercial or industrial purposes.

Standards

Low-profile ducts with retaining clips are preferred for wall mounting in private single-family houses, offices, laboratories, or workshops. Both versions are manufactured according to the standard DIN EN 50085 and protect the cables against mechanical damage and dirt. They are secured to walls and ceilings using dowels.

Design

Ducts with welded cross members consist of a base and a clip-on cover. The welded cross members give the duct the stability of a rectangular pipe. Individual partitions can be inserted in the ducts to permit cables of the power installations and communications system to be installed in compliance with the relevant regulations.

The low-profile cable ducts with retaining clips are supplied with or without partitions, depending on the dimensions.

In both duct systems, changes in direction are achieved using molded sections available as inner corners, outer corners, flat angle sections, T sections, and four-way sections. End plates for sealing the ends of the ducts complete the respective duct programs.

5.1.4.5 Skirting board ducts

Application

Ducts in the form of skirting boards (Fig. 5.1/12) are used to install cables of power installations and communications systems separately from each other. They are mounted in the angle between the wall and floor and are suitable for new, old, and pre-

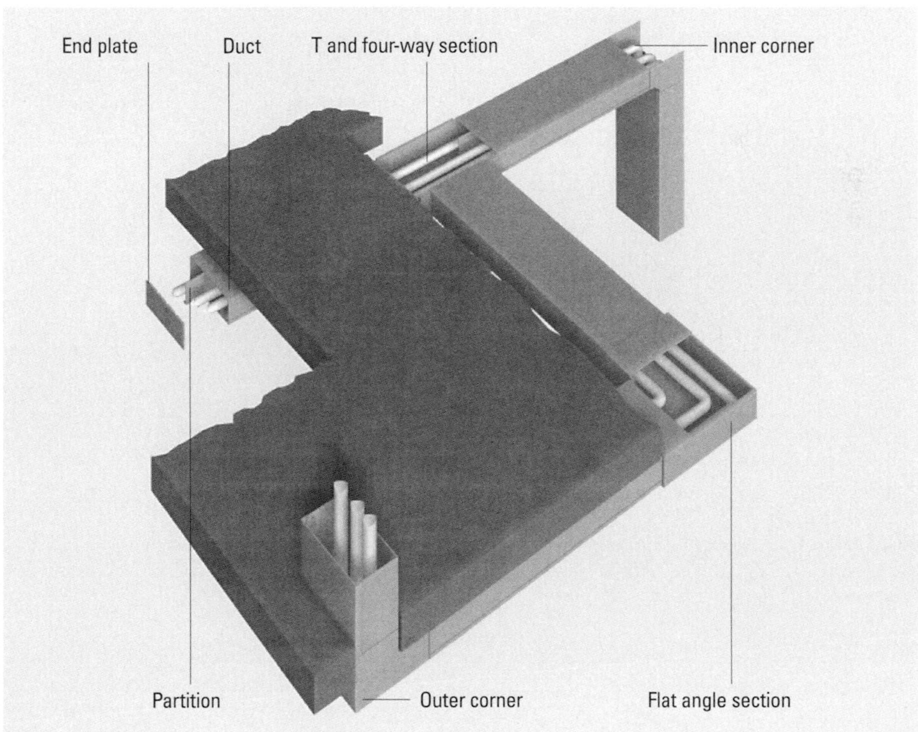

Fig. 5.1/10
Components of cable duct with welded cross members, recommended for use as ceiling duct

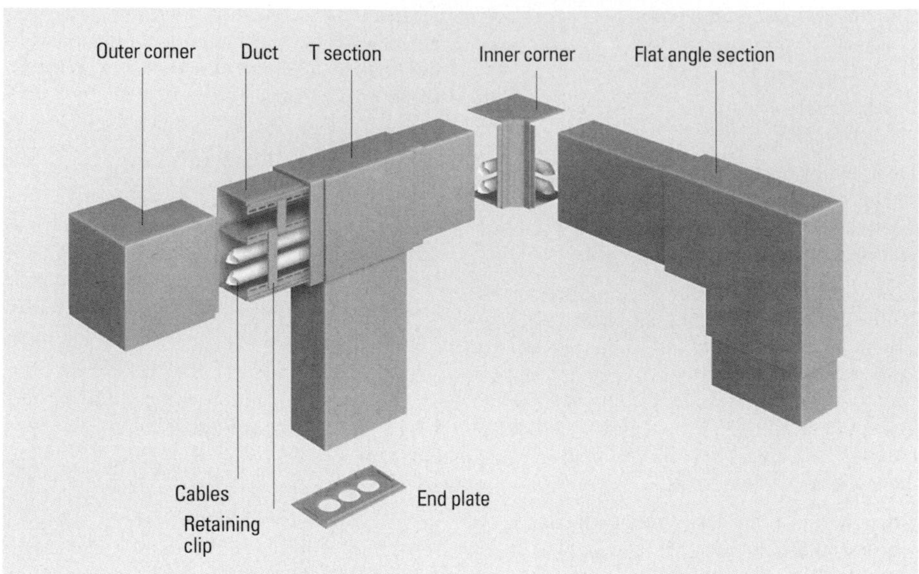

Fig. 5.1/11 Components of cable duct (low-profile version with retaining clips, used as wall duct)

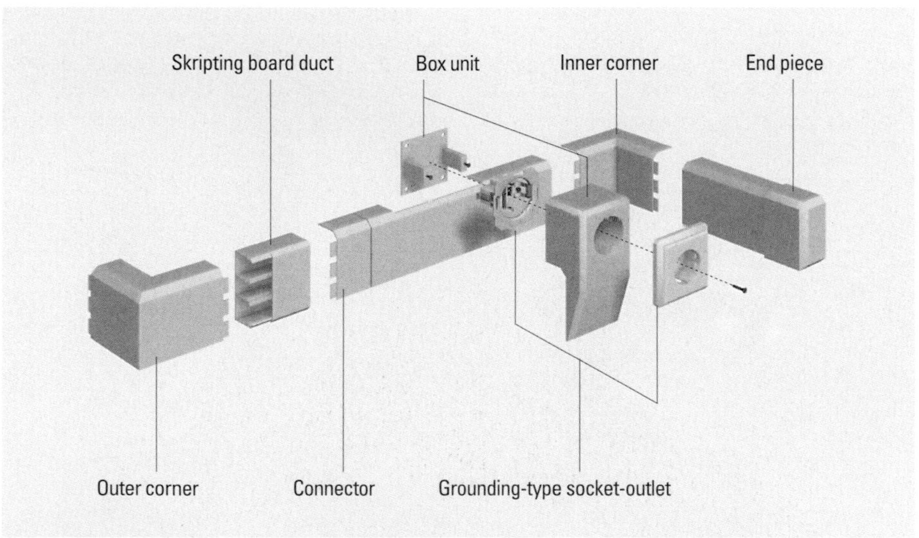

Fig. 5.1/12 Components of skirting board duct with box unit for installation device including accessories

fabricated buildings. A connecting duct is used to bridge doorways.

Design

The skirting board ducts consist of a base and a clip-on cover. Special partitions divide the base into two or three cable runs.

Connectors are used to link individual ducts. One or two installation devices can be mounted at any point along the skirting board duct by means of a box unit. Installation devices with stirrups, fixing dimension 60 mm to DIN 49073, and with center plates measuring 51 mm × 51 mm to DIN 49075 are used (see Section 4.3.1). Changes in direction are achieved using molded sections. The ends of ducts are sealed using end pieces. The skirting board ducts are made of plastic in various colors to match the floor carpeting. Covers are also available which can be bonded to the floor carpeting.

5.1.4.6 Sill-type ducts

Application

Sill-type ducts (Fig. 5.1/13) are used to install cables and installation devices for supplying power to workstations, e.g. in workshops, offices, and laboratories. Cables and outlet boxes for antennas, telephones, and other communications systems can also be installed in these ducts. They are usually mounted underneath window sills.

Design

Sill-type ducts consist of U-section open bases made of metal or plastic with snap-in retaining clips and overlapping clip-on covers. A partition can also be inserted in the central slot of the duct base to permit cables of power installations and communications systems to be installed separately.

The snap-in retaining clips can also be shortened by breaking off the ends. The cables in the individual compartments can then be secured separately via the partition. Equipment is installed using a device box which is snapped into place in the central slot of the duct. A hole saw can be used to make holes in the duct covers to allow equipment to be installed on site. Alternatively, pre-assembled panels are also available for installing one, two, or three devices.

In accordance with DIN VDE 0800, separate panels must be used for installing equipment of communications systems.

5.1.4.7 Dado ducts

Application

As with sill-type ducts, dado ducts (Fig. 5.1/14) are mainly used in administrative buildings, schools, workshops, training centers, laboratories, and similar locations. They are usually mounted underneath the window sill next to the apron wall. This is made possible by the cover which can be snapped on from the front. In addition to cables of power installations and communications systems, *instabus EIB* cables as well as the associated devices can also be installed in dado ducts. All DELTA switch and socket-outlet ranges, *instabus EIB* devices, and standard communications equipment are suitable for installation in these ducts. The installation devices with a center plate measuring 51 mm × 51 mm (to DIN 49075) can be inserted if used

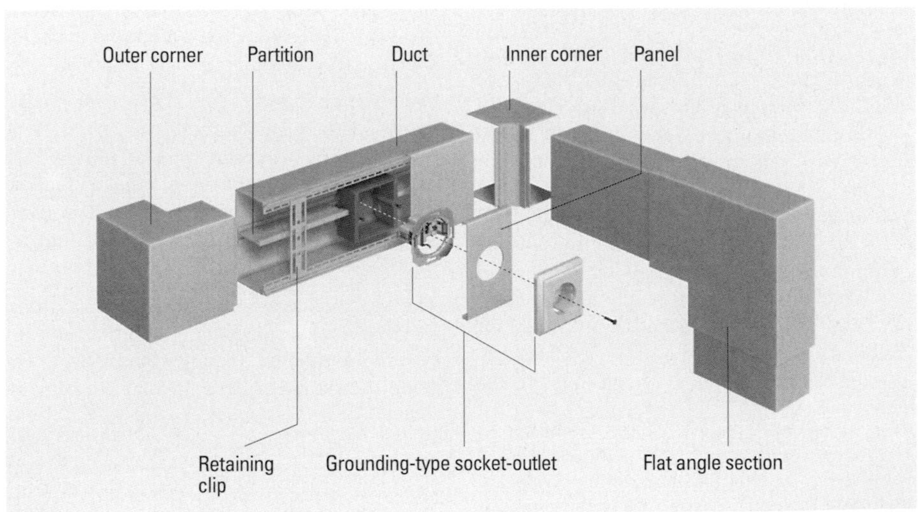

Fig. 5.1/13
Components of sill-type duct for equipment installation

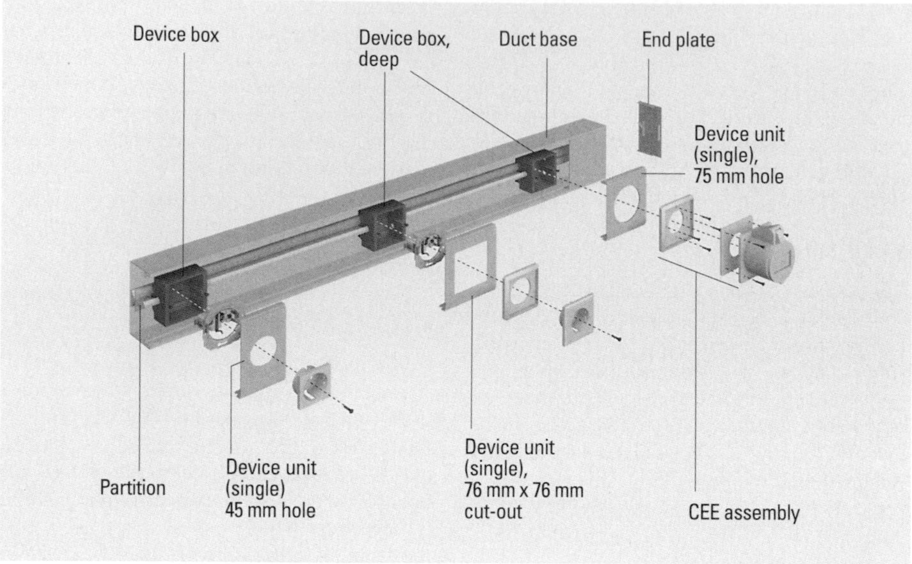

Fig. 5.1/14 Components of dado duct

in conjunction with standard device boxes and device units (single, double, and triple) with 45 mm holes. Installation devices with frames (80 mm side dimension) can be fitted using the deeper device boxes and device units with 75 mm holes or square cut-outs (76 mm side dimension).

Design

Dado ducts consist of three basic components: duct base, duct cover, and the device units. The bottom of the duct base (made of plastic, aluminum, or Sendzimir-galvanized sheet steel) is fitted with continuous rails for mounting device boxes, partitions, terminals, and snap-on modular devices (see Section 4.2). The duct covers, made of either plastic or aluminum, can be snapped into position in the duct base to protect the cable compartment against dirt. Duct covers made of different materials are fully interchangeable, provided that their dimensions are the same.

Duct bases and covers are manufactured in standard 2 m lengths. The device units are available in single, double, or triple versions. Connectors for linking the duct bases, molded sections for changing direction, end plates for sealing the ends of the ducts, assemblies for mounting shrouded plug and socket-outlets, and miniature circuit-breakers as well as accessories for incorporating metal objects in the protective measures complete the product range. Plastic and sheet-steel ducts are available in various colors; aluminum ducts have a natural anodized finish.

Dado trim panels

Dado trim panels (Fig. 5.1/15) are a combination of dado ducts and cover strips. Dado ducts are positioned in front of pillars, over radiators or air-conditioning units and are attached to the wall between the pillars using stand-off brackets made of plate-finished aluminum or sheet steel.

The spaces between the wall and dado duct and between the dado duct and the floor are covered with anodized aluminum strips. The stand-off brackets have special attachments which allow the strips to be snapped into position. The distance between the brackets should not exceed 1 m to ensure that the duct and strips are sufficiently stable.

The standard length of the natural-anodized aluminum strips is 2 m.

Fig. 5.1/15
Dado trim panel between wall and dado duct

5.1.4.8 Underfloor duct systems

Application

In modern functional buildings, both the electrical power for current-using devices and the power supply for communications equipment are distributed via underfloor duct systems.

Installation methods

Seven different installation methods can be used, all of which allow the cables of power installations and communications systems to be installed in the floor. The connecting points should be arranged in the form of a grid so that the plug and socket devices of the various supply systems are within reach of the workstation irrespective of the room layout or furnishings. Only a correctly dimensioned underfloor installation will increase the utility value of a building and, more importantly, allow the communications equipment to be used more efficiently.

5.1.4.8.1 Underfloor installation methods

1. Underfloor installation with under-screed ducts and flush feeder ducts

Use

This installation method (Fig. 5.1/16) is used where power has to be supplied to complex multi-purpose workstations in open-plan, grouped, or individual offices. The unfinished floors are fitted with duct networks (tailored individually to the furnishings or assembled in grids of between 1.2 m × 1.2 m and 1.8 m × 1.8 m), consisting of feeder and under-screed ducts together with height-adjustable floor outlet boxes and accessories (Fig. 5.1/16). Floor outlet boxes with blanking covers can be used as draw-boxes, those with adapter covers for surface outlet boxes as surface-type boxes, and those with flush connections as flush-type boxes. The necessary flush or surface-type outlet boxes must be chosen in accordance with the type of floor construction and the floor thickness.

Thickness

The floor thickness depends on:

- ▷ constructional features
  e.g. sound-proofing, thermal insulation, screed type,
- ▷ technical factors
  e.g. duct depth, screed covering, type of outlet boxes, type and position of the installation device in the accessory frame of a flush outlet box.

Outlet boxes

Surface outlet boxes can be used for floor thicknesses ≥65 mm and flush-type boxes for floor thicknesses of between 75 mm and 100 mm (depending on the dimensions and arrangement of the equipment). If floor heating is installed, the underfloor duct system can also be used for floor thicknesses of between 130 mm and 160 mm.

Modifications

The duct system can be modified at a later stage without costly and complex installation work by simply drilling the under-screed ducts at any location and installing additional outlets.

2. Underfloor installation with flush ducts and removable blanking covers

Floor thicknesses ≤65/75 mm

Flush ducts (Fig. 5.1/17) are recommended if the floor is not thick enough for embedded under-screed ducts (minimum thickness 65/75 mm). The height of the ducts can be adjusted to the level of the screed by leveling the side members (Fig. 5.1/24) thus enabling almost the entire floor thickness to be used. Furthermore, the inside of the ducts can be accessed quickly by opening the duct blanking covers allowing the cables to be installed without any difficulty. Floor thicknesses of approximately 40 mm only permit the use of surface outlet boxes.

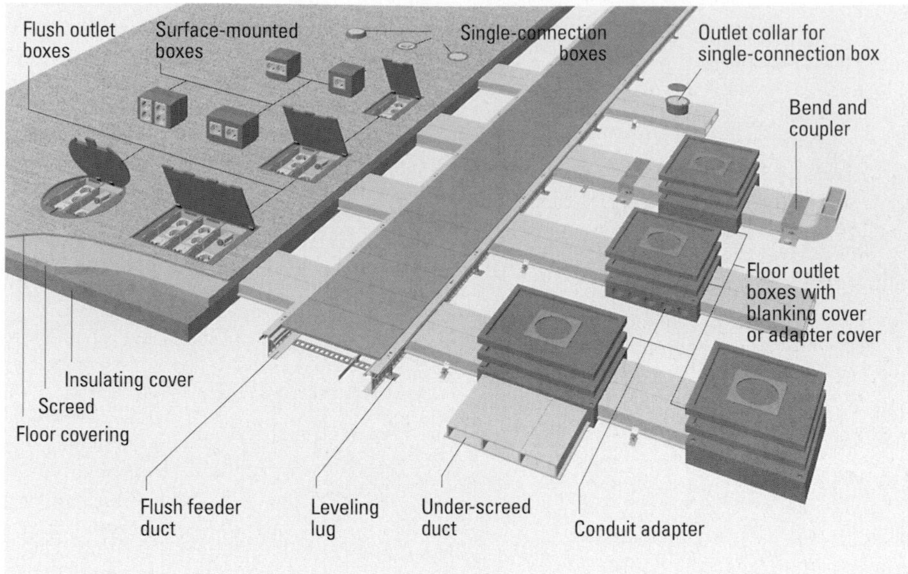

Fig. 5.1/16 Underfloor installation with under-screed ducts and flush feeder ducts

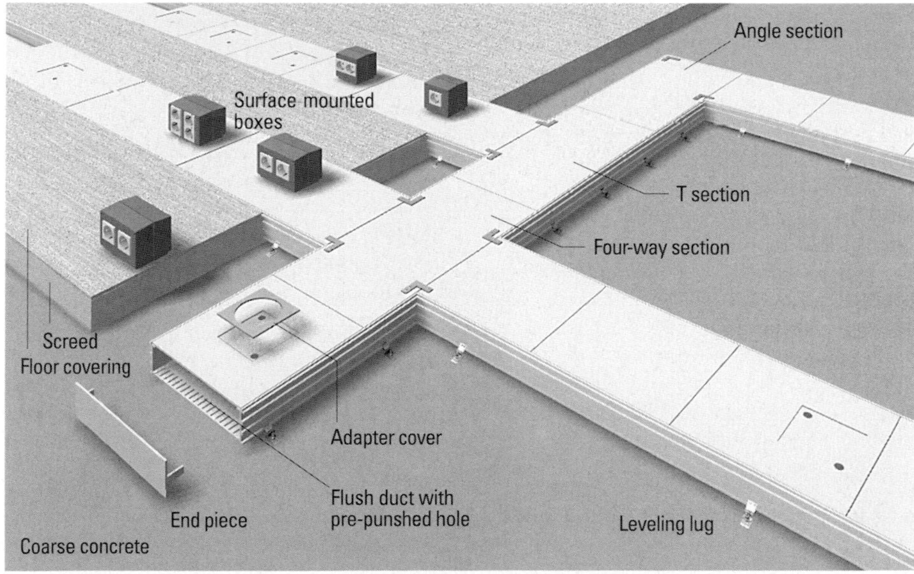

Fig. 5.1/17 Underfloor installation with flush ducts and removable blanking covers

3. Underfloor installation with flush closed ducts

Closed sheet-steel duct

With floor thicknesses ≥40 mm, it is also possible to use the low-cost, flush closed sheet-steel duct system (Fig. 5.1/18). To make full use of the floor thickness available for cable installation, the duct depth is approximately the same as the floor thickness. This means that the surface of the ducts must be aligned with the surface of the screed by means of leveling lugs.

Outlet boxes

Pre-cut duct covers enable the entire range of surface outlet boxes to be installed at relatively low cost. The floor outlet boxes can also be used as draw-boxes, angle boxes, T-boxes, and four-way boxes. Impact noise is reduced to a satisfactorily level using standard floor coverings.

4. Underfloor installation with semi-flush boxes

Casting in concrete layer

If the screed is too shallow to allow the underfloor duct system to be installed with under-screed ducts, an installation method can be selected in which the components are cast directly in the concrete layer (Fig. 5.1/19). Specially designed semi-flush boxes, which are mounted on the shuttering and cast in the concrete layer, are suitable for this purpose.

Infeed
Outlet boxes

The boxes are connected by means of ducts laid on supports. The height of the supports must be such that the ducts lie in the neutral fiber of the unfinished floor. The infeed is routed directly into the duct system. To ensure that no concrete sap can penetrate the semi-flush boxes and ducts during the construction phase, the boxes are wrapped in protective sheeting. The ducts are then inserted in the boxes up as far as the protective sheeting. The semi-flush boxes directly embedded in the unfinished floor permit the use of flush-type boxes with hinged covers despite the relatively shallow screed (35 mm).

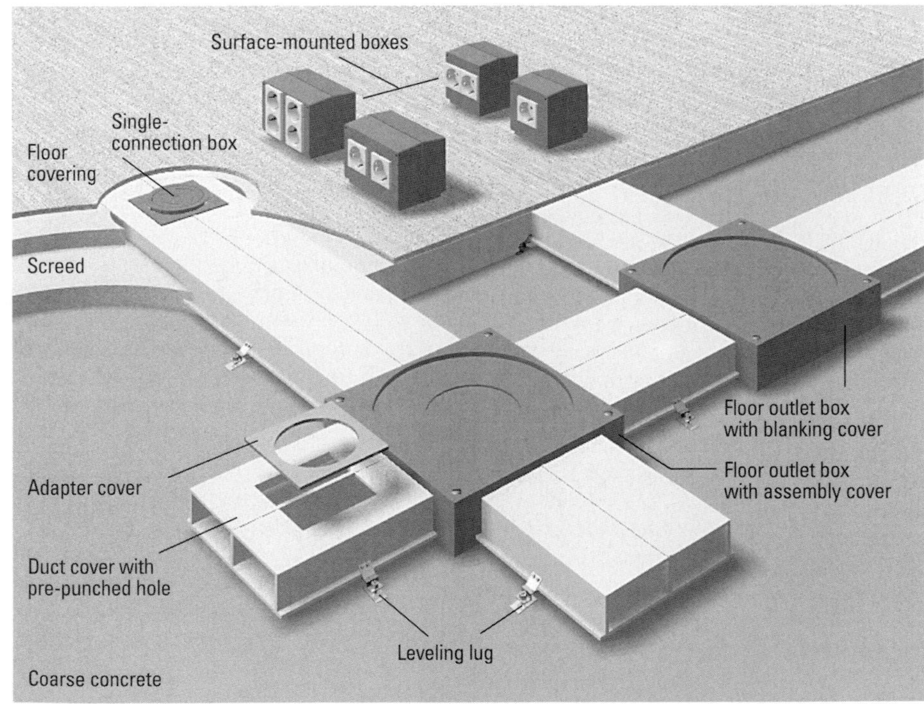

Fig. 5.1/18 Underfloor installation with flush closed ducts

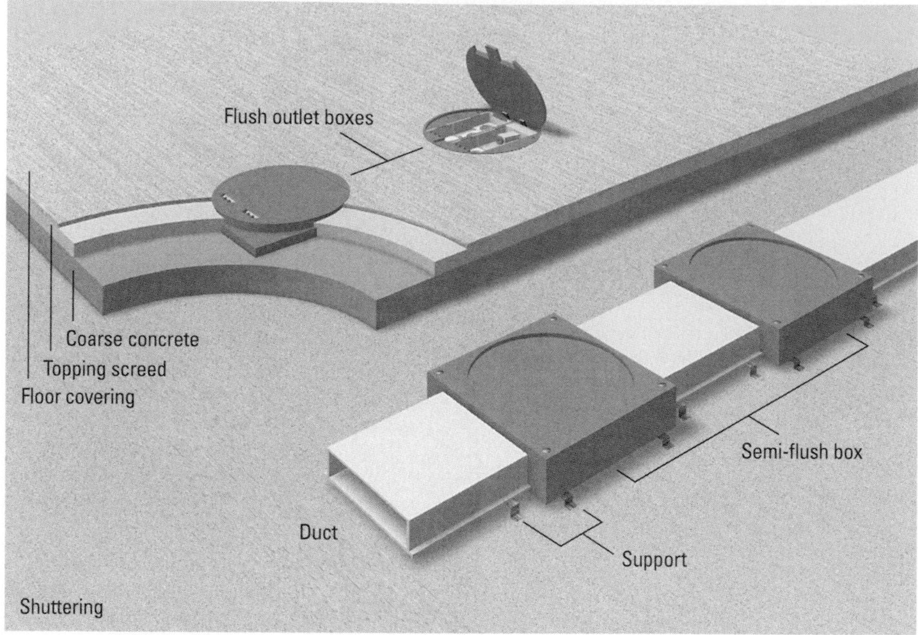

Fig. 5.1/19 Underfloor installation with semi-flush boxes

5. Underfloor installation with ceiling bushings

Casting in coarse concrete layer

This underfloor duct system (Fig. 5.1/20) is either used in unfinished floors with pre-cast holes or is cast directly in the coarse concrete layer. The feeder cables are routed from the intermediate ceiling underneath through the ceiling bushing, into the ceiling-bushing boxes and then on to the installed outlet boxes. The cables from the storey distribution boards are first routed through the intermediate ceiling through the ducting and then into the individual ceiling bushings.

Outlet boxes

Fire partitions

Fire partitions suitable for ceiling-bushing boxes must be used to prevent fire from spreading from one storey to another.

6. Underfloor installation with outlet boxes for false floors

Flexible installation

Installing cables in false floors using underfloor outlet boxes (Fig. 5.1/21) affords a particularly high level of flexibility. Subsequent modification and expansions requiring relatively little time and effort are possible by moving and exchanging the floor panels. The standard side dimensions of the floor panels, 500 mm or 600 mm, result in a relatively closely-spaced connection grid which allows optimum positioning of the outlet boxes in relation to the workstation.

The cavity beneath the false floor is used for cable installation.

Central units

The central units for the low-voltage system and communications equipment are fed from the storey distribution boards. The outlet boxes for flush mounting in the false-floor panels or for surface mounting are supplied by the central units via flexible connecting cables.

Equipment shrouding covers

The undersides of the connecting points must always be covered with the appropriate equipment shrouding covers to provide protection against inadvertent contact with live parts. In view of the harsh environments on construction sites, it is advisable to use equipment shrouding covers made of sheet steel.

Strain-relieving cable entries ensure that cables are connected reliably the first time

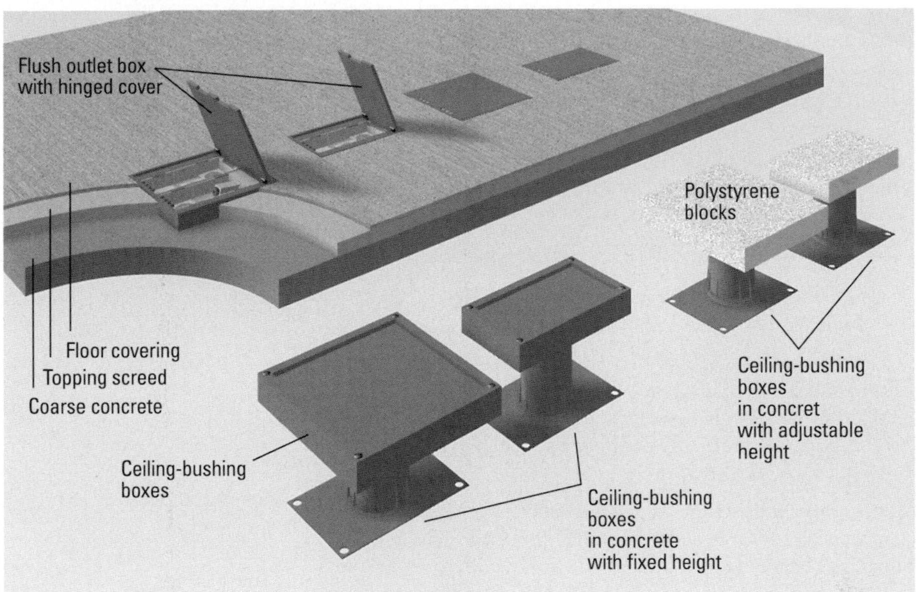

Fig. 5.1/20 Underfloor installation with ceiling bushings

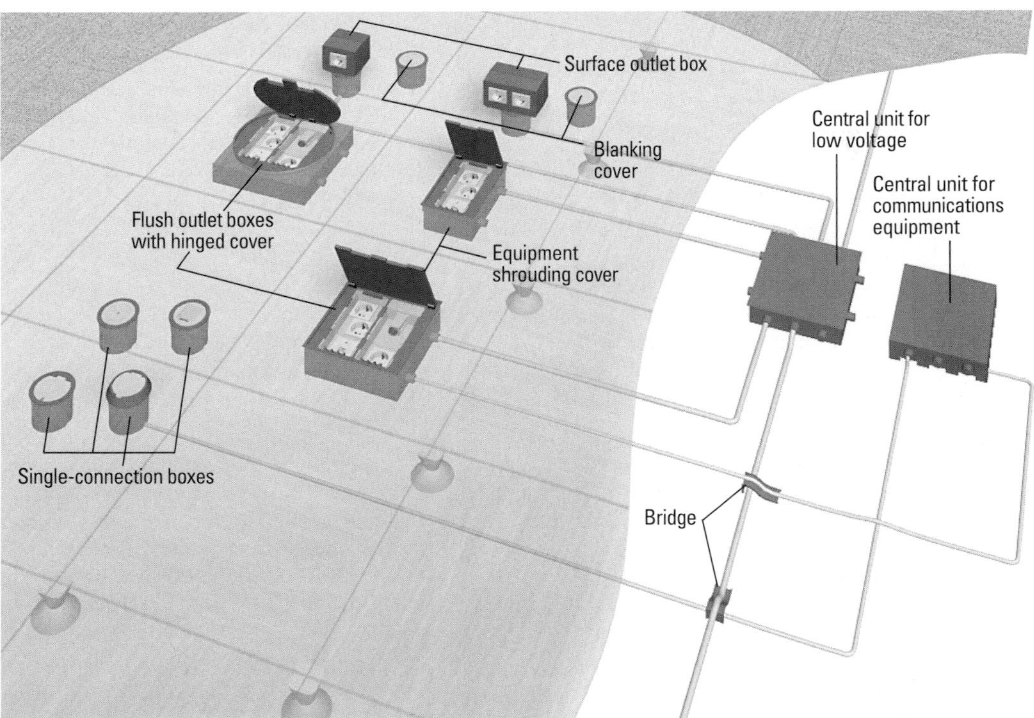

Fig. 5.1/21 Underfloor installation with outlet boxes for false floors

they are installed and whenever modifications are carried out.

Components for snap-on system

Another variant, which can be installed very quickly and conveniently on the construction site, is also available for use with snap-on systems (see Section 5.1.4.9). The accessory frames are supplied via pre-assembled cables which are laid in the false floor and connected to the terminal block.

7. Underfloor installation with outlet boxes for cavity floors

The arched structure of cavity floors is ideal for installing cables of power installations and communications systems. Components of the underfloor duct system are housed in cut-outs in the floors.

The components of the underfloor duct system consist of flush outlet boxes with round hinged covers (Fig. 5.1/22).

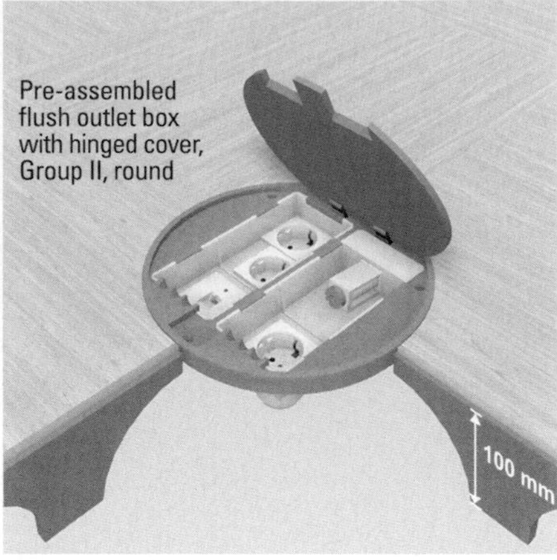

Fig. 5.1/22
Flush outlet box with round hinged cover and accessory frames in cavity floor

Inspection openings

Inspection openings, the position of which must be agreed upon with the floor manufacturer before the screed is laid, must be provided in the cavity floor to enable subsequent modification and expansion of the electrical installation. They are normally positioned at a distance of 3 m from each other. The inspection openings are covered with non-flammable false-floor panels which are set in a square frame.

Modifications

Additional flush outlet boxes can also be installed at a later stage by making holes in the cavity floor.

The holes can be made in two different ways:

- ▷ before the screed is laid, by inserting a suitable polystyrene block in the honeycomb shuttering of the cavity floor,
- ▷ after the screed is laid, by drilling the floor using a suitable diamond drill.

Blanking covers

Unused holes must be sealed with blanking covers.

5.1.4.8.2 Underfloor units

Flush feeder ducts

Application

Flush feeder ducts with removable blanking covers are used to facilitate the installation, modification, and retrofitting of the large number of cables used in power installations as well as communications and IT systems (Fig. 5.1/23). They are routed from the storey distribution board to the center or to the side of the room, in which the cables are to be installed, where they are connected to the under-screed duct system and supply the power installation and communications equipment via terminal units.

Design

The bottomless duct consists of the following parts (Fig. 5.1/24):

- ▷ two height-adjustable aluminum side sections with cross members, perforated knock-out side walls and the necessary connecting lugs,
- ▷ an adjustable snap-in partition,
- ▷ a number of buttable blanking covers,
- ▷ screw-fixed screed anchors,
- ▷ insertable spacers,
- ▷ two plastic angle sections used as edging for the floor covering.

The underside of the duct base is open to allow cables to be routed through the ceiling.

The blanking covers are available in sheet steel and aluminum depending on the width of the duct. The cross members have suitable terminals to allow the ducting to be incorporated in the existing protective system; the blanking covers, on the other hand, must be bolted to the duct so that contact is made with bare metal.

An ideal underfloor ducting system (Fig. 5.1/25) can be constructed using flush feeder ducts and under-screed ducting (see Fig. 5.1/26).

Fig. 5.1/23
Flush feeder duct with under-screed duct and floor outlet box during construction phase

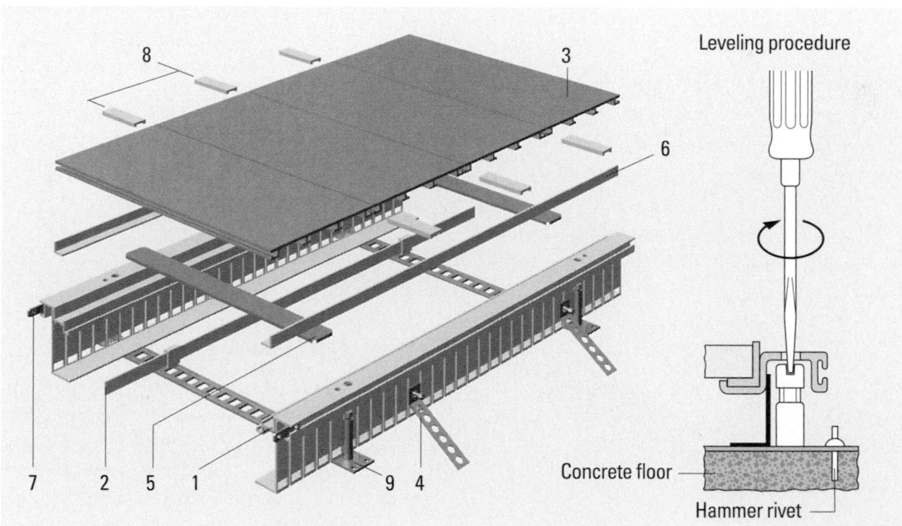

1 Aluminum side section with cross members and perforated side walls
2 Snap-in partition
3 Buttable blanking covers
4 Screw-fixed screed anchors
5 Insertable spacers
6 Plastic angle sections
7 Connecting lugs
8 Blanking cover connectors
9 Leveling lug

Fig. 5.1/24 Components of flush feeder duct

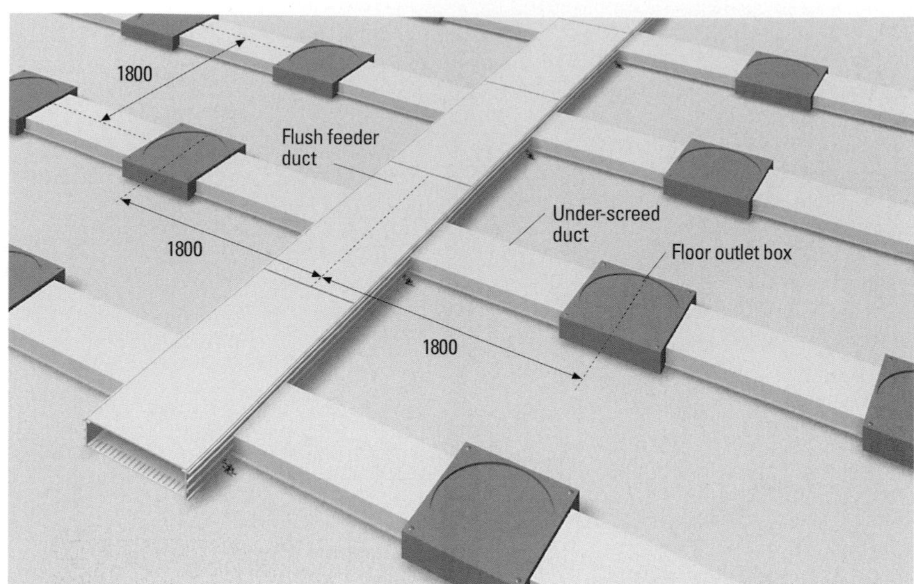

Fig. 5.1/25
Ducting system consisting of flush feeder duct and under-screed ducting with floor outlet boxes (dimensions in mm)

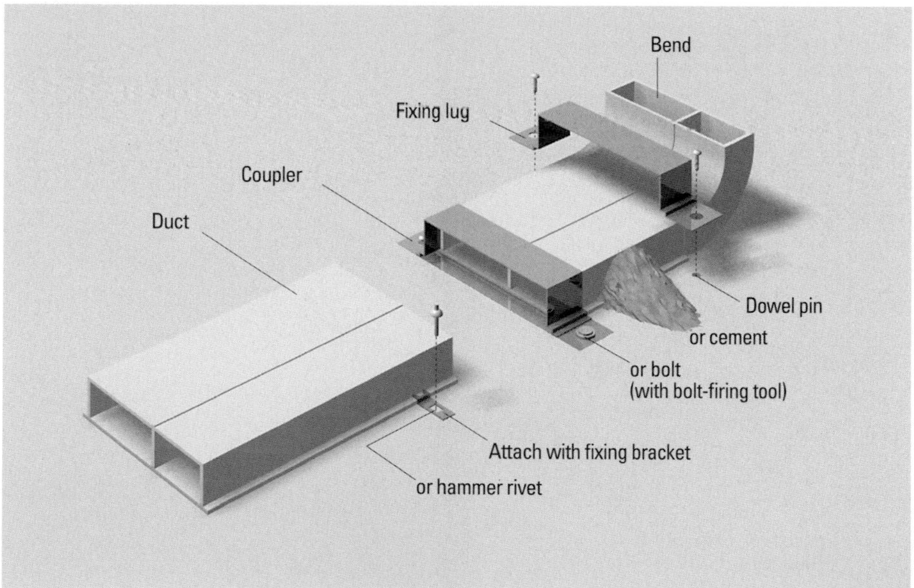

Fig. 5.1/26 Under-screed ducting with accessories allowing attachment to coarse concrete layer

Under-screed ducting

Ducts

Under-screed ducts have a rectangular structure, are closed on all sides, are made of galvanized sheet steel (Fig. 5.1/26), and can have one or more channels. The different channel divisions, i.e. two or three channels, allow cables of power installations and communications systems to be installed separately.

Floor outlet boxes

Application The floor outlet boxes perform many tasks in a ducting system (Fig. 5.1/27).

Fire barriers As draw-boxes, they allow cables to be inserted, retrofitted, and modified in duct runs of between 8 m and 10 m in length. As right-angle, T, and four-way outlet boxes, they are used to change the direction of cables and, as terminal boxes, they securely seal off the ends of ducts. The floor outlet boxes are often fitted with terminal strips to simplify routing, looping, and connection of the cables. The accessibility of the ducting system via the draw-boxes allows adequate fire protection and sound proofing to be provided in the form of fire barriers and sound-proofing elements. As surface-type boxes, they allow surface outlet boxes with various equipment configurations to be mounted and, as flush-type boxes, they permit multiple-component flush-type boxes with hinged covers to be used.

Types The floor outlet boxes are available in a number of sizes and types to suit the various dimensions of the underfloor ducting and required outlet boxes and the different floor thicknesses (Fig. 5.1/28).

The outlet boxes (surface or flush-mounted) together with the accessory frames and the installation devices (Fig. 5.1/27) must be taken into consideration when selecting the floor outlet boxes for the required minimum floor thickness.

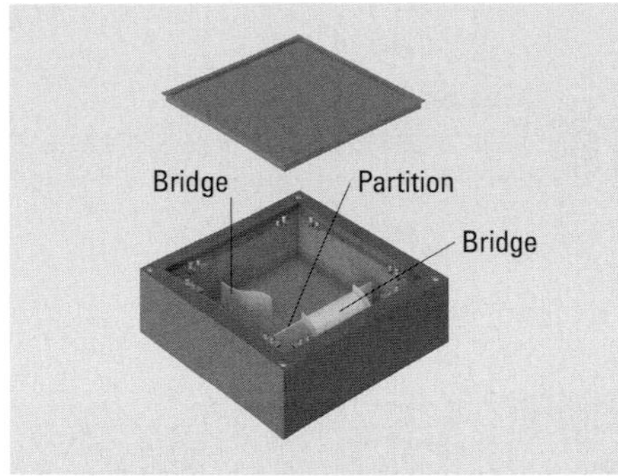

a) Draw-box

b) Surface-type box

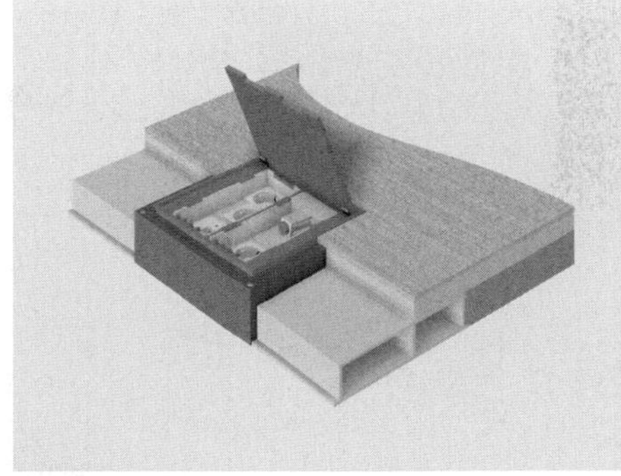
c) Flush-type box

Fig. 5.1/27
Floor outlet boxes for various applications

Terminal units

Flush-type boxes with hinged covers

Application The flush-type boxes with hinged covers, which can house a number of easily installed and combinable accessory frames, have proven successful particularly in open-plan offices. The accessory frames are equipped with the appropriate plug-and-socket systems of power installation and communications equipment (Figs. 5.1/29 and 5.1/30).

Flush-type boxes are available with different dimensions and in various shapes and sizes. They are installed in the appropriate floor outlet boxes and can be used in conjunction

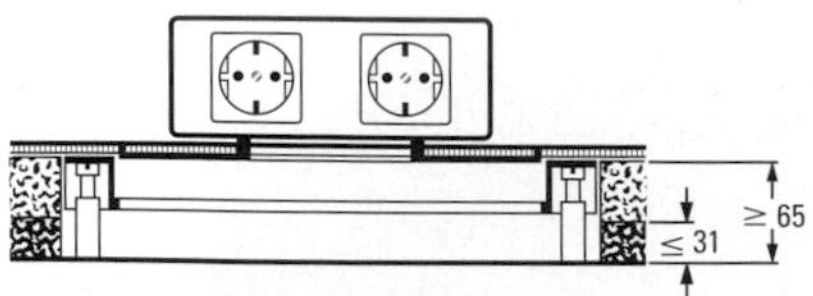

a) Floor outlet box with mounted flat outlet boxes

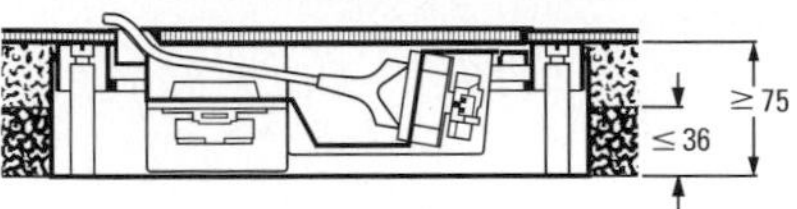

b) Flush outlet box with hinged cover and accessory frame, suitable for all types of grounding-type plug (except right-angle plugs with long molded collars)

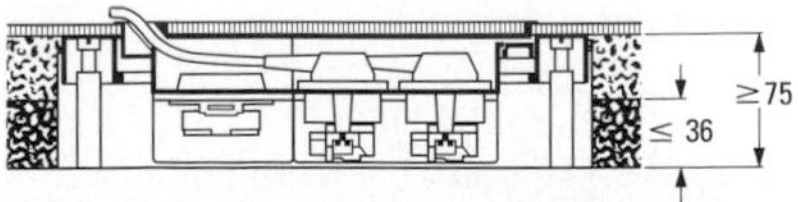

c) Flush outlet box with hinged cover and accessory frame, suitable for right-angle grounding-type plugs

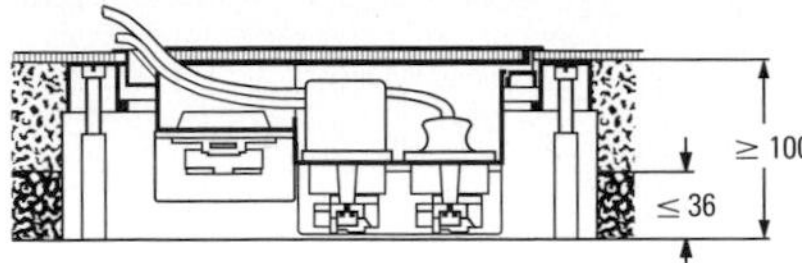

d) Flush outlet box with hinged cover and accessory frame, deep version for right-angle grounding-type plugs, short straight plugs, and small charging units

Fig. 5.1/28
Minimum floor thickness for floor outlet boxes according to surface or flush-mounted outlet box (dimensions in mm)

Fig. 5.1/29 Flush outlet boxes of underfloor ducting system

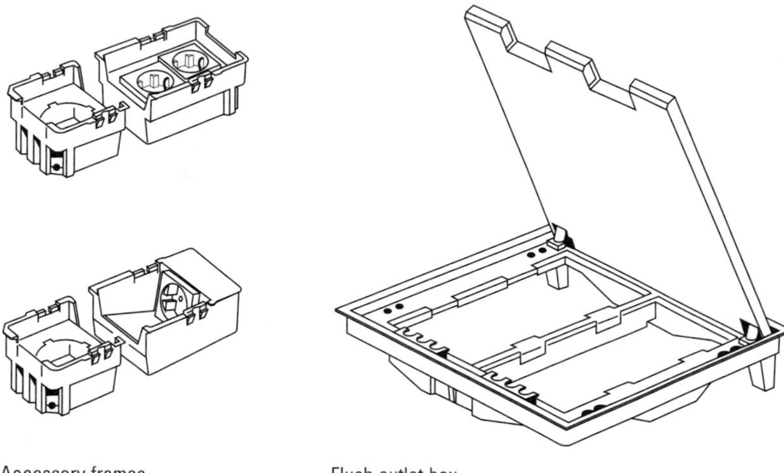

Fig. 5.1/30
Flush outlet box with hinged cover and snap-in accessory frames, equipped with grounding-type socket-outlets and unequipped sections, for example, for installing communications socket-outlets

with the modular snap-in accessory frames. This method of equipment installation affords maximum flexibility vis-à-vis assembly and utilization. Since the number of current-using devices and the type of power plug to be used at a workstation are usually not known until shortly before the workstation is set up, it is essential to fix the floor thickness required for the various possible equipment combinations at the planning stage (Fig. 5.1/28). The most frequently used minimum floor thickness is 75 mm.

Design

The flush outlet box consists of a leveling frame, spring-loaded hinges, hinged cover, and cable outlets. The inside of the frame accommodates the accessory frames, hinges, and sealable cable outlets. Its outer edge serves as protective carpet edging to conceal the roughly cut edges of the floor covering.

Accessory frames

The flush outlet boxes with hinged covers are equipped with devices using snap-in accessory frames made of impact-resistant and flame-retardant plastic, which are supplied by the manufacturer unequipped or pre-wired and ready-equipped with installation devices. An accessory frame consists of two shell halves which permit devices used in power installations and communications systems to be installed separately (Fig. 5.1/30). If an installation location is not used, it can be sealed with a blanking cover. If various frame accessories are combined, it is possible to connect the standard plug types (e.g. right-angle and central plugs with long molded collars) even with a floor thickness of just 75 mm. If small charging devices, e.g. for desktop calculators, have to be installed in flush outlet boxes with hinged covers, deeper accessory frames must be used. In such cases, a minimum floor thickness of 100 mm is necessary.

Installation devices with square stirrups, plug-and-socket systems for radio and communications equipment, and 16 A grounding-type socket-outlets can be installed without any difficulty in the flush outlet boxes with hinged covers and the accessory frames.

Surface outlet boxes

Application

Appropriate adapter covers are used to mount surface outlet boxes protruding above floor level, on the floor outlet boxes, or directly on the outlet collars of the ducts (see Fig. 5.1/31 a). They are used wherever the floor is not thick enough for flush-type boxes to be installed. Surface outlet boxes are suitable for housing devices such as grounding-type or communications socket-outlets. The devices must be fitted with a stirrup (fixing dimension 60 mm to DIN 49073) and a center-plate cover in compliance with DIN 49075. Surface outlet boxes pre-equipped with grounding-type socket-outlets are frequently used as these represent a more cost-effective solution.

Design

The surface outlet boxes consist of two plastic shell halves, a mounting panel with brackets to carry the devices, and a mounting ring (Figs. 5.1/31a and b). The mounting panel is used to separate the devices for power installations and IT systems.

Fig. 5.1/31c shows three surface outlet boxes equipped with different devices.

Single outlet boxes and outlet collars

Assembly

Single outlet boxes, e.g. flush covered outlet boxes, hood-type outlet boxes, and cover-type cable outlet boxes, can be installed at any position along the under-screed duct. Outlet collars serving as adapters must be embedded in the screed for this purpose (Fig. 5.1/32).

Before the screed is laid, they are secured to the holes in the ducting sections. The holes are drilled at the factory or on site.

If the screed has already been laid, it must be drilled using a suitable boring tool; the hole in the duct must be made with a hole saw. The drilling work that must be carried out here using special tools is extremely expensive and is nowadays only used when modifying existing installations.

5.1.4.8.3 Planning underfloor ducting systems

Building ground plans

The architect draws up the most economical ground plan on the basis of the purpose and utilization of a building and its rooms.

Grouped, combined, and individual offices

In the case of functional buildings, he first organizes the ground plans into open-plan, grouped, combined, and individual offices according to the size of the rooms and the number of people that will work in them.

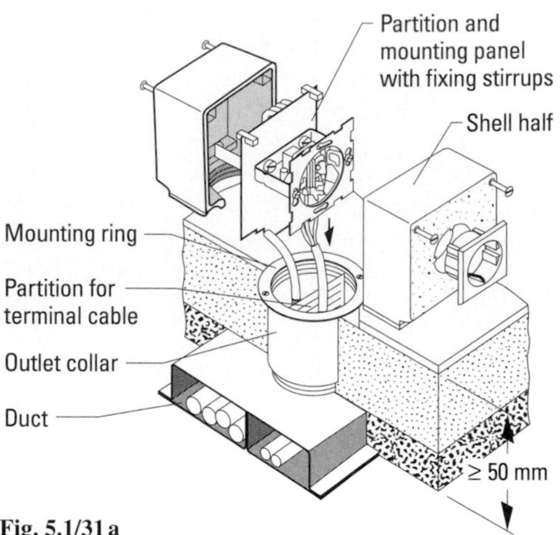

Fig. 5.1/31 a
Surface outlet box, mounted on outlet collar with mounting ring

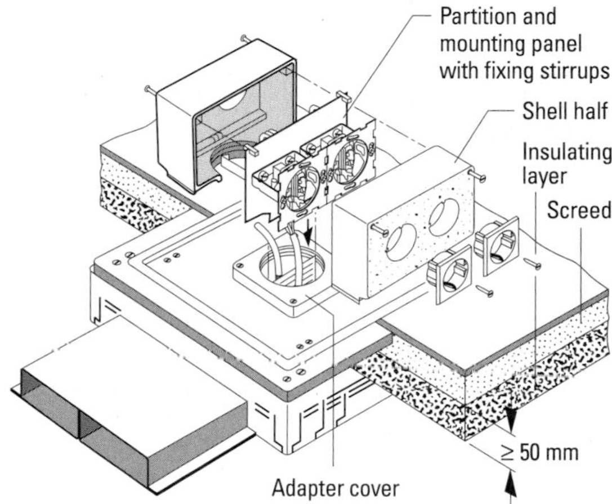

Fig. 5.1/31 b
Surface outlet box, mounted on floor outlet box with adapter cover

Fig. 5.1/31 c
Surface outlet boxes equipped with different devices

Multi-purpose rooms

Depending on how they are used, the rooms are then grouped again into office, lecture, exhibition, training, sales, working, demonstration, conference rooms and so on. Some of the rooms may be multipurpose which, for economic reasons, have to fulfill a number of functions or have to be modified to perform a series of various short-term functions. Fig. 5.1/33 shows just a few typical plans from the infinite number of possible room layouts.

Distribution grid

The ducting may, for example, start at the storey distribution board and then run into the interiors of the offices via lobbies, corridors, and aisles. The duct network connected to it, the selected connection grid, and the number of electrical outlets depend on the particular functions and equipment of the workstations and the possible need for future alterations. The ducting systems can also be arranged individually according to a specific furnishing plan. This means, however, that the range of possible furnishing arrangements is restricted which must be taken into consideration if the ducting system has to be modified at a later stage.

A grid layout should be used wherever possible since this permits the electrical wiring system to be adapted more easily to changing requirements and workstation arrangements. As illustrated in Figs. 5.1/34 and 5.1/35, the individual workstations can then always be positioned immediately next to a connecting point.

Planning aids

Equipment symbols

All items of equipment must be converted into individual symbols before planning of the underfloor ducting system can begin. Equipment symbols (Fig. 5.1/36), which simplify the planning procedure, help to ensure accuracy, and can be easily recognized by the technician, have been developed in line with the relevant standards and the longstanding planning practices of engineering companies.

CAD programs for configuration planning

Planning can be carried out easily on PCs using CAD programs for drawing building ground plans and ducting networks, as well as special stored component libraries.

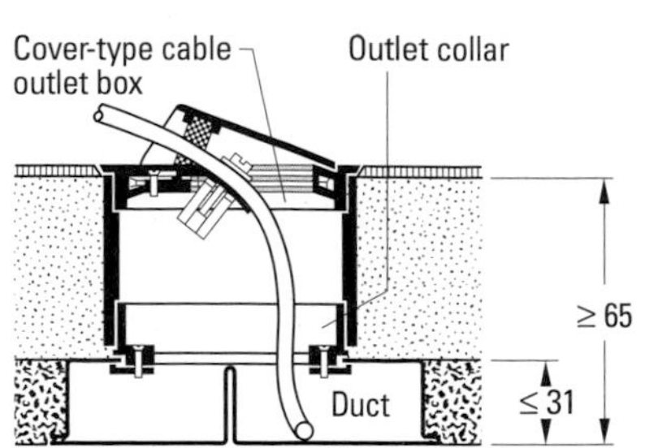

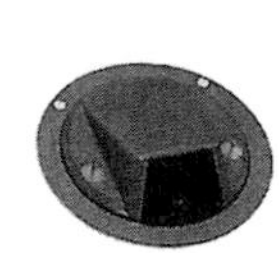

a) Cover-type cable outlet box for maximum of two individual cables

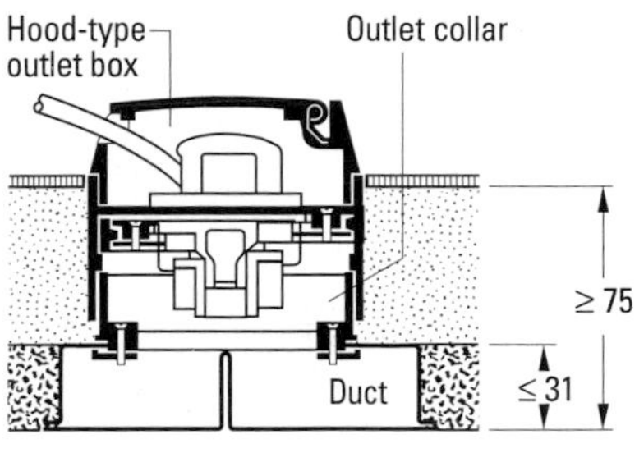

b) Hood-type outlet box for grounding-type socket-outlet with grounding-type right-angle plug

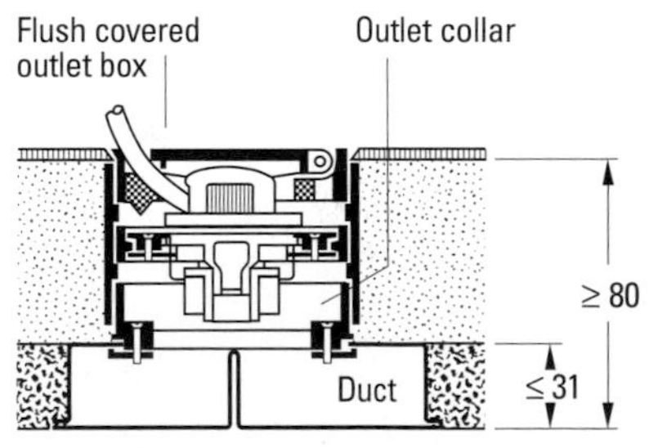

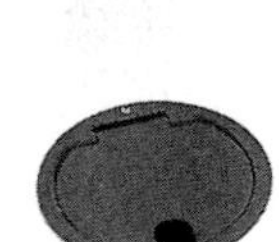

c) Flush covered outlet box for grounding-type socket-outlet with grounding-type right-angle plug

Fig. 5.1/32
Single outlet boxes mounted in outlet collar on under-screed duct together with different outlets (dimensions in mm)

Choice of system

The underfloor ducting system is selected according to the type of floor construction (thickness) and the plans drawn up by the project engineer with respect to the use of the room and the communications equipment used at the workstations. The underfloor ducting systems that can be used for the different floor thicknesses are listed in Table 5.1/1.

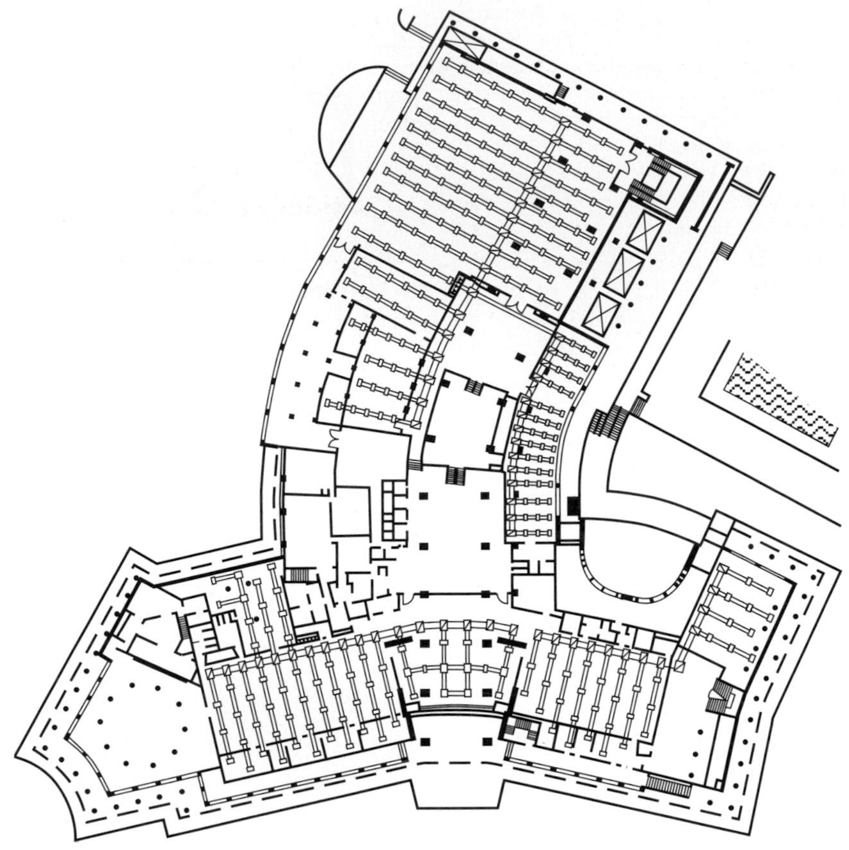

Fig. 5.1/33
Administrative building with open-plan offices; underfloor ducting system arranged in grid pattern

Table 5.1/1
Floor thicknesses: underfloor duct systems required for various underfloor ducting systems

Floor thickness	Underfloor ducting system with	See Fig.
≥ 35 mm	Semi-flush boxes	5.1/37
40 mm to 65 mm	Flush, closed ducts	5.1/38
40 mm to 60 mm	Flush ducts with removable blanking covers	5.1/39
65 mm to 95 mm and greater	Under-screed ducts and flush feeder ducts	5.1/40
65 mm to 95 mm and greater	Ceiling bushings	5.1/41
≥ 80 mm	Outlet boxes for sectional floor systems	5.1/42
≥100 mm	Outlet boxes for cavity floor systems	5.1/43

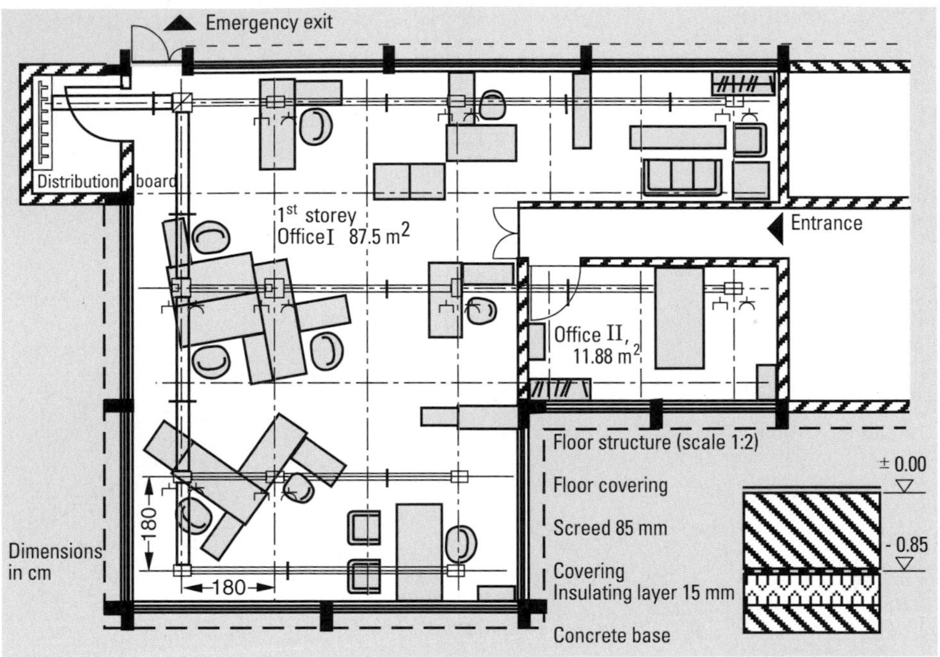

Fig. 5.1/34
Open-plan and individual office with underfloor installation according to furnishing and use (furnished)

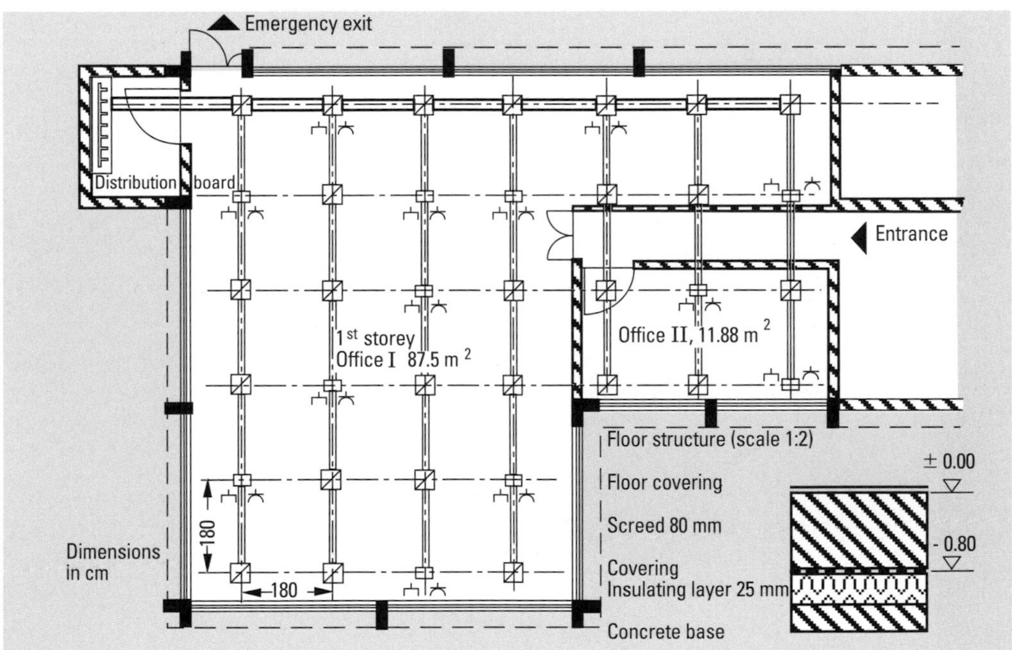

Fig. 5.1/35
Open-plan and individual office with underfloor installation irrespective of furnishing and use (unfurnished)

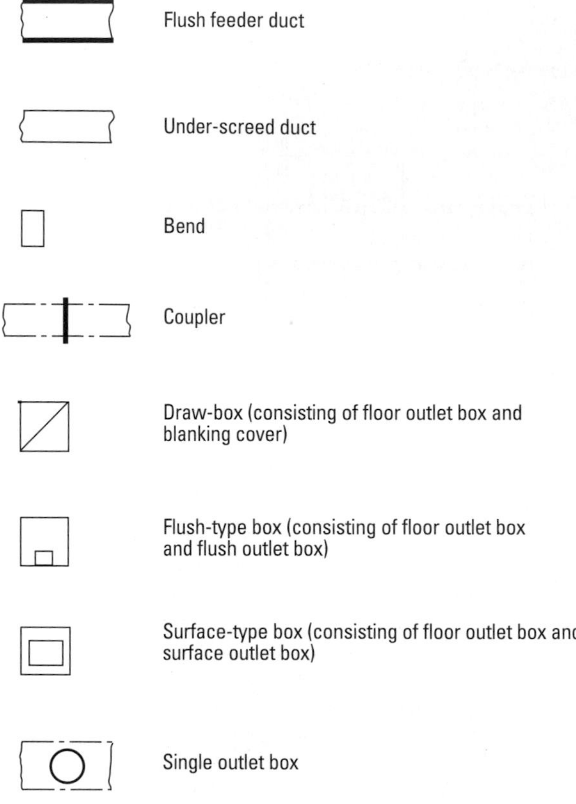

Fig. 5.1/36
Symbols for components and devices for underfloor ducting systems

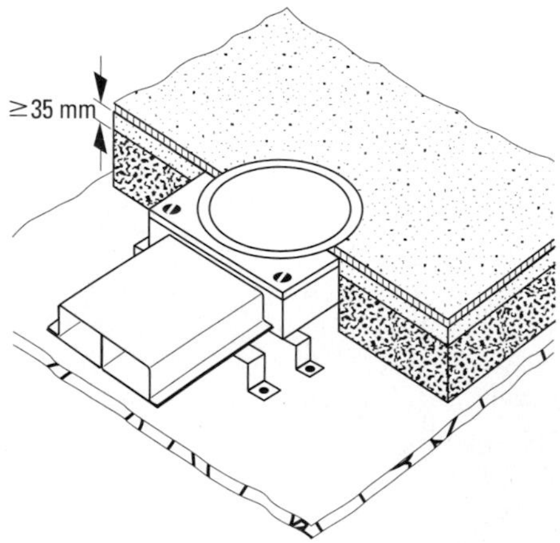

Fig. 5.1/37 Semi-flush box

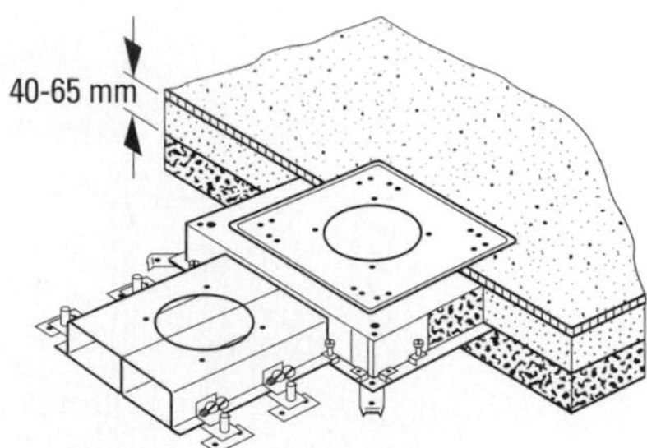

Fig. 5.1/38 Flush, closed duct

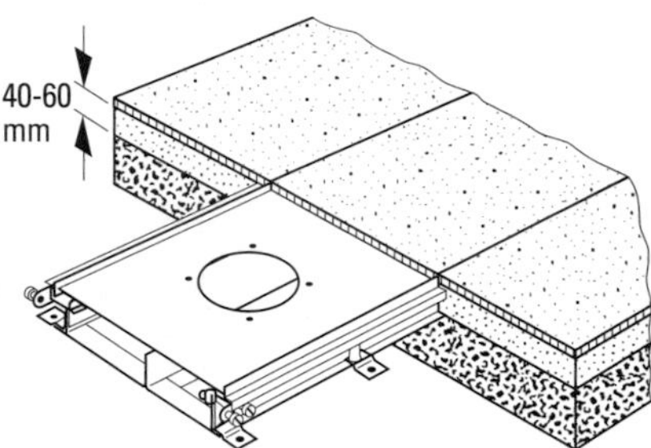

Fig. 5.1/39
Flush duct with removable blanking cover

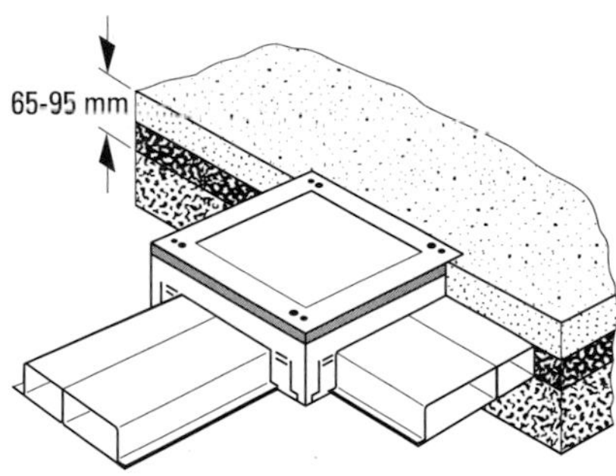

Fig. 5.1/40 Under-screed duct

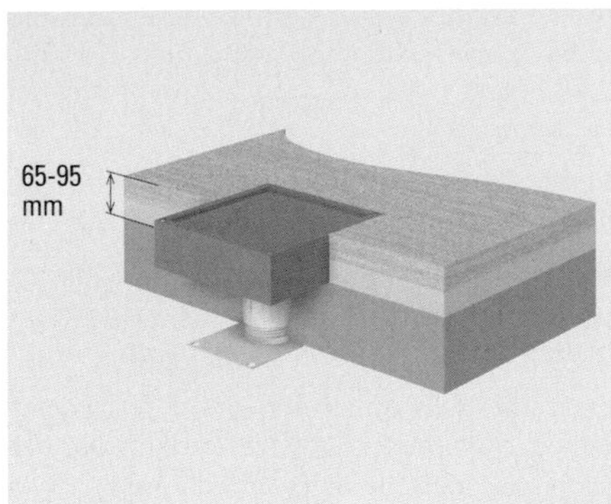

Fig. 5.1/41 Ceiling bushing

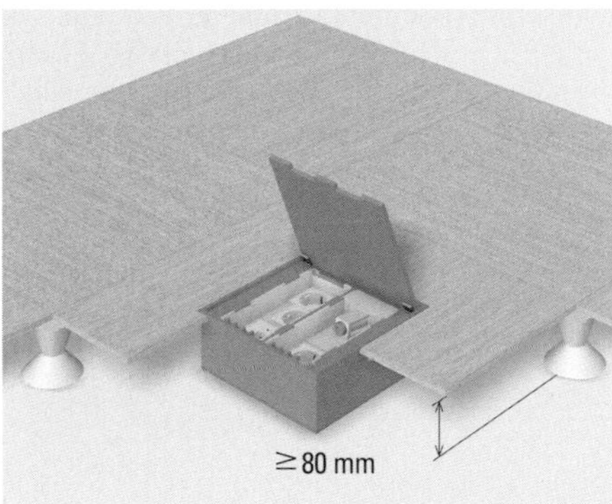

Fig. 5.1/42 Outlet box for sectional floor systems

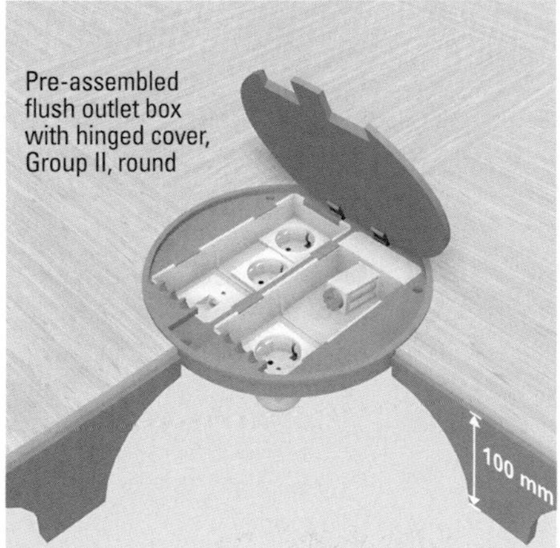

Fig. 5.1/43 Outlet box for cavity floor systems

Choice of equipment according to floor thickness

Floor thickness

The underfloor ducting system is selected in accordance with the construction and thickness, specified by the architect, of the floor in which the ducting network is to be installed. The method of duct installation and the choice of terminal units, e.g. flush-mounted or raised, must be taken into consideration.

When the ducting system, installation method, and terminal units are selected, it is important to ensure that the constructional height of the equipment is compatible with the specified floor thickness.

Duct heights

The required floor thickness depends on the duct height and the thickness of the covering screed. The standard duct heights (external dimensions) are 31 mm and 36 mm.

Surface outlet boxes can, for example, be used with the shallowest ducts for a minimum floor thickness ≥ 65 mm and flush-type boxes with hinged covers for a minimum floor thickness ≥ 75 mm (see Fig. 5.1/28). Owing to the way in which they are mounted on the duct, single outlet boxes must be considered separately since the duct heights, which can be used for different purposes, are included in the floor thickness (see Fig. 5.1/32).

Selection of connection grid

When the underfloor ducting system is being planned, a ground plan of the building or rooms should be provided which shows the layout, architectural style, and the number and position of the shafts carrying the rising main busbars and distribution systems.

Grid module

To determine the most appropriate grid, transparent overlays with different modules, which should be between 1.20 m and 1.80 m apart, are placed over the ground plan. The most suitable grid pattern for the feeder duct and the duct network to which it is connected is determined starting at the storey distribution board – taking into consideration structural features such as escape routes, walkways, room divisions, stairwells, supporting pillars, utility rooms, ventilation and power ducts as well as other fixed installations.

One reference point for the grid is the distance between the center of the windows.

The number of outlets may be subject to official safety regulations. If this is the case, the layout should be based on the smallest possible grid, especially if surface outlet boxes are used.

Determining duct dimensions and sizes of floor outlet boxes

Floor thickness

The desired underfloor ducting system must be installed in the floor thickness – defined as the distance between the surface of the concrete base and the surface of the screed – specified by the architect. Flush ducts and floor outlet boxes are adjusted to the floor thickness by means of their leveling devices; the height and width of under-screed ducts, however, must be determined according to the number of cables to be carried by the ducts while ensuring that an appropriate covering screed is provided.

Since the floor thickness is prescribed and the minimum screed thickness is determined by the compressibility of the insulating material underneath, the maximum possible duct height is calculated using the following equation:

Determining duct height

Floor thickness – Minimum screed thickness = Max. possible duct height.

Cable capacity

If a duct run includes a bend, the cable capacity of the duct is reduced by approximately 30%.

Once the duct dimensions have been determined, the appropriate sizes of the floor outlet boxes must be selected according to the terminal units to be fitted.

5.1.4.9 "SMS-Universal" snap-on system

Application

Thanks to their cables, which are pre-assembled by the manufacturer, snap-on systems (pre-assembled cable installation systems) are ideal for fast and low-cost cable installation in the cavities of intermediate ceilings (Fig. 5.1/44), sandwich walls, and

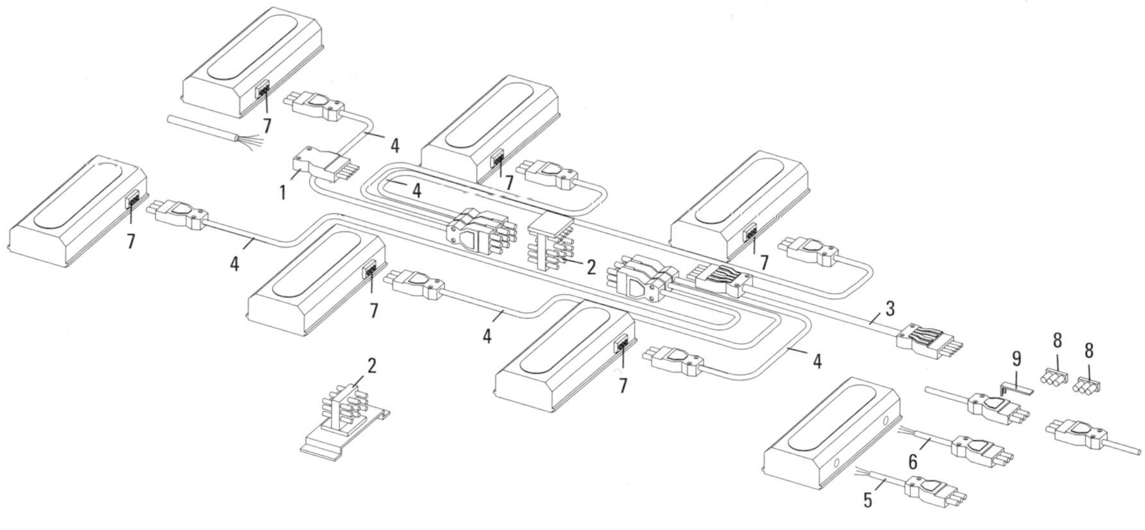

1 Plug/socket, 5 pole, screw-type terminal
2 Terminal block, 2 x 5 pole/6 x 3 pole, 2 x L1, L2, L3
3 Extension cable, 5 x 2.5 mm², plug and socket, or with 1, 2 an 5-pole cable e.g. NYM 5 x 1.5 mm², assembled by customer
4 Extension cable, 3 x 1.5 mm², plug an socket
5 Connecting cable, 3 x 1.5 mm², socket
6 Connecting cable, 3 x 1.5 mm², plug
7 Snap-in connector, 3 pole, screwless
8 Fastener for 3-pole sockets, IP 40
9 Interlock for 3/5-pole cables on plug and socket

Fig. 5.1/44 16 A three-phase snap-on system for use in intermediate ceiling.

false floors. They simplify and speed up installation on site.

Connectors

Advantages

Since all of the cables, with the exception of the feeder cables, are fitted with connectors at the connection and distribution points, a cable network can be created, modified, and expanded as required by simply connecting the components together. In contrast to conventional electrical installation on construction sites, the cables do not need to be cut to length, stripped of insulation, or connected to terminals. Furthermore, they do not require strain relief devices, and housings do not have to be opened and closed.

Versions

The snap-on system is available in a three-phase current version for rated currents ≥ 16 A and rated voltages of 230/400 V AC/DC, together with terminal blocks and a range of associated accessories. Alternating-current loads, e.g. recessed luminaires or grounding-type socket-outlets, can be connected evenly distributed across the three external conductors L1, L2, and L3 via the terminal blocks.

Cable lengths

The cables are available in different standard lengths (up to 8 m).

5.1.4.9.1 Snap-on system components

Terminal blocks

Terminal blocks are used to connect and extend the pre-assembled, plug-in load cables of the snap-on system. They are available with various plug-and-socket devices, e.g. 2 × 5 pole, 4 × 3 pole, 6 × 3 pole, 2 × L1, L2, L3.

Load cables

Terminal blocks and current-using equipment, e.g. socket-outlets or luminaires, can be connected with pre-assembled, plug-in cables of the snap-on system, or with NYM PVC-sheathed cables. The pre-assembled, plug-in cables have conductor cross sections of 3×1.5 mm^2, 3×2.5 mm^2, or 5×1.5 mm^2. Load cables can be connected to pre-assembled luminaires with plug connectors and couplings, or to grounding-type socket-outlets in flush-mounting units for false floors using plug connectors or stripped conductor ends.

Plug connectors

The plug connectors of the snap-on system consist of the plug and coupling, which are available as 3 or 5-pole versions depending on their use. Coding splines on the plugs and slots along the couplings prevent incorrect connection. The 5-pole plug cannot, for example, be inserted in a 3-pole coupling or a 3-pole feeder on the terminal block.

Plug-in luminaires

Snap-in luminaire connection

Sectional ceiling systems are used mainly in functional rooms and workrooms. The predefined grid patterns of these ceilings enable different recessed luminaires with compatible dimensions to be installed. Pre-assembled recessed luminaires complete with lamps and specular louver unit are ready for installation with the snap-in luminaire connection. A plastic film protects the specular louver units against dirt during the construction phase.

5.1.4.9.2 Project planning

Planning

Before the project planning phase begins, a plan to a scale of 1:50 must be available on which the storey distribution boards, the cable ducts, and the current-using equipment are indicated. Information regarding the ceiling, wall, and floor construction as well as the dimensions of the associated grid structure is also useful.

Positions of terminal blocks

Starting from the storey distribution board, the feeder cables together with the extension and main cables are routed via cable raceways (see Fig. 5.1/9) or cable ducts (see Fig. 5.1/10) into the interior of the room. The most suitable position for the terminal blocks is determined taking the locations of the loads into consideration. It is important to ensure that the terminal block positions can be accessed by opening up the ceilings, walls, or floors at a later stage to permit modifications.

Extension cables

The connections from the terminal blocks to the current-using equipment, e.g. luminaires or grounding-type socket-outlets, are then drawn and the lengths measured. For greater clarity, the load groups are indicated accordingly and all of the materials required are entered in a parts list. The individual cable lengths are determined by "mating" the extension cables. Cables are available in lengths of 2 m, 4 m, 6 m, and 8 m.

Advantages

Figs. 5.1/45 and 5.1/46 show two examples of the planning procedure for the SMS-Uni-

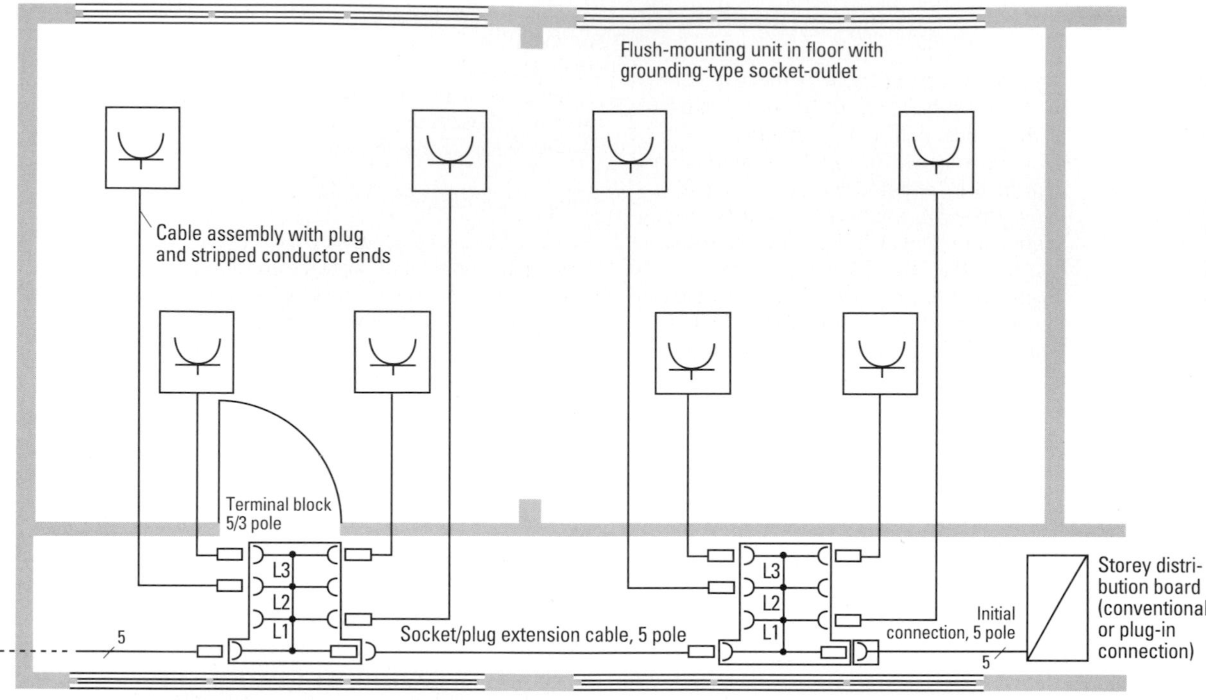

Fig. 5.1/45
Planning example: Snap-on system (16 A three-phase current version) for installing grounding-type socket-outlets in flush-mounting units of underfloor ducting systems for false and cavity floors

versal snap-on system. The ready-to-use, fully-equipped devices and the snap-on system components significantly reduce the amount of time required for installation work on the construction site and make difficult overhead work much easier. Overall costs and the costs that result from subsequent modifications are also considerably reduced.

5.1.4.10 *instabus EIB* snap-on assembly

Application

The basic electrical installation of a building must be flexible enough so that it can be expanded, modified, and adapted at a later stage within a relatively short space of time and without having to modify the building's walls and fixtures. It is, therefore, essential to provide an adequate number of electrical connections and a reserve network with ample capacity for additional connecting cables.

Since large buildings are nowadays for the most part constructed using a steel and reinforced concrete superstructure (foundation, service riser duct, pillars, ceiling construction, and external façades made of large glass panels or lightweight panels made of plastic, aluminum, and so on) and have variable rather than fixed interior walls, only the floors and ceilings can be used for electrical installation purposes.

Building control system with *instabus EIB*

Furthermore, today's modern electrical loads are not only supplied with power but also have to be controlled in order to reduce energy costs, and to regulate their functional processes. Control bus systems such as the *instabus EIB* are, therefore, also used in parallel to the load power supply.

Many loads, e.g. in load management applications (indication, signaling, operation, and monitoring), and lighting, heating, air-conditioning, and roller shutter/louver blind systems are controlled either separately or interdependently and are influenced by highly diverse environmental criteria. Consequently,

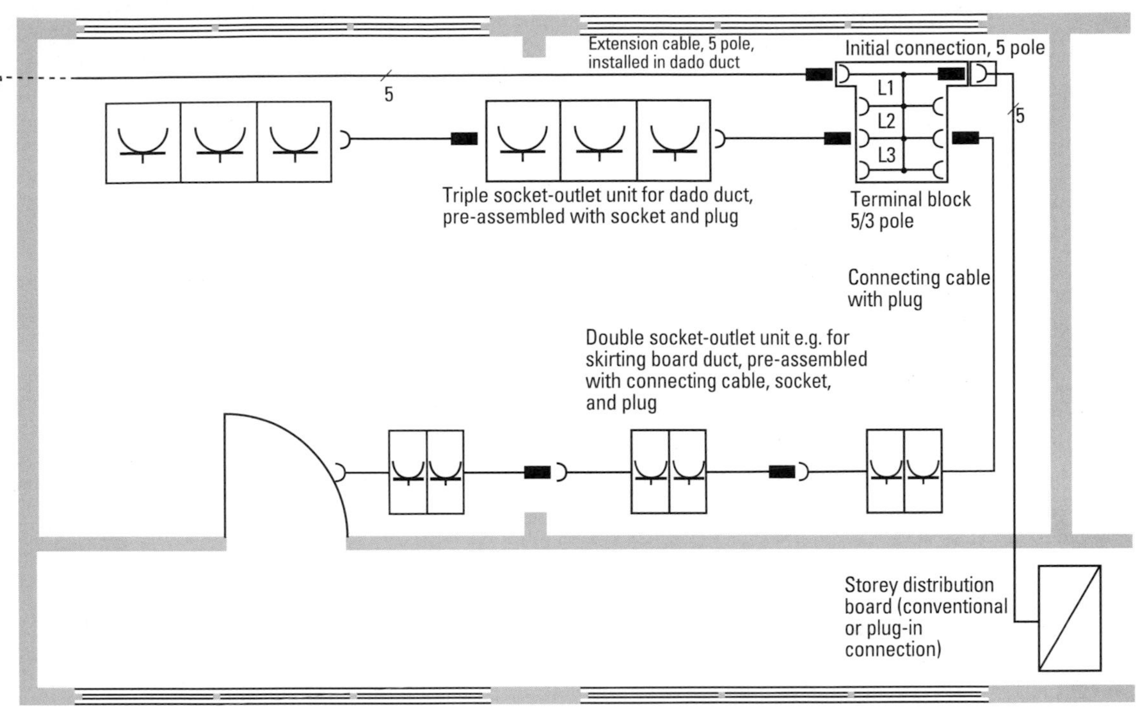

Fig. 5.1/46
Planning example: Snap-on system (16 A three-phase current version) in dado duct for connecting grounding-type socket-outlets

lighting systems in offices, for example, can be controlled according to time, light-intensity, motion, and temperature criteria, either centrally or locally, or by means of a primary control system. The *instabus EIB* control bus system takes care of these very varied and changing switching tasks. The control bus system consists of the cables of the load power supply, the bus cables, the bus components connected to these cables (see Section 4.4), and the plug-in *instabus EIB* snap-on devices.

5.1.4.10.1 *instabus EIB* snap-on devices

Bus coupling unit (BCU), switch relays, plug connectors

The circuit board, on which the bus coupling unit (BCU), all-or-nothing relays, and plug connectors are mounted, is attached to a retaining device inside the insulating case (168 mm × 156 mm × 56 mm) (Fig. 5.1/47).

Coded plug connectors

The power and bus cables are wired in such a way that, if necessary, *instabus EIB* snap-on devices can be replaced without disconnecting the installation bus. The switching and control components in the devices are connected in the incoming circuit and equipped with the mechanically coded plug-and-socket connection elements so that they only have to be snapped on during assembly, thus saving time. The coded plug-and-socket connectors prevent incorrect connection.

Various device versions are available, for example:

Device versions

- ▷ snap-on device with one control input and three switching outputs to operate luminaires or luminaire groups (Fig. 5.1/47)
- ▷ snap-on device with two switching outputs to operate luminaires or luminaire groups and one output with changeover contact, e.g. to actuate a roller shutter motor (Fig. 5.1/48).

Thanks to the low-profile insulating casing, the *instabus EIB* snap-on devices are also suitable for installation in the small cavities in intermediate ceilings.

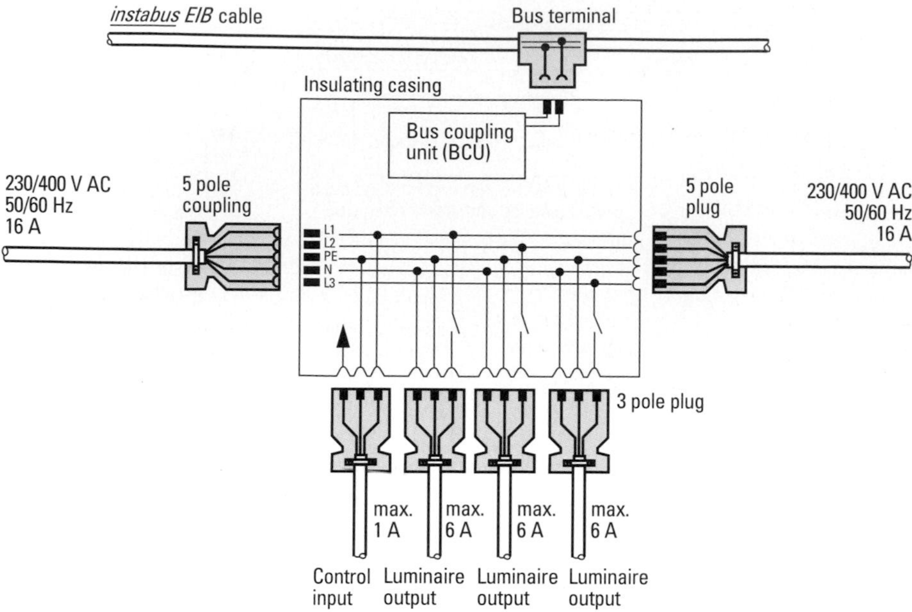

Fig. 5.1/47
instabus EIB snap-on device for switching loads, e.g. three luminaires/luminaire groups, including control input

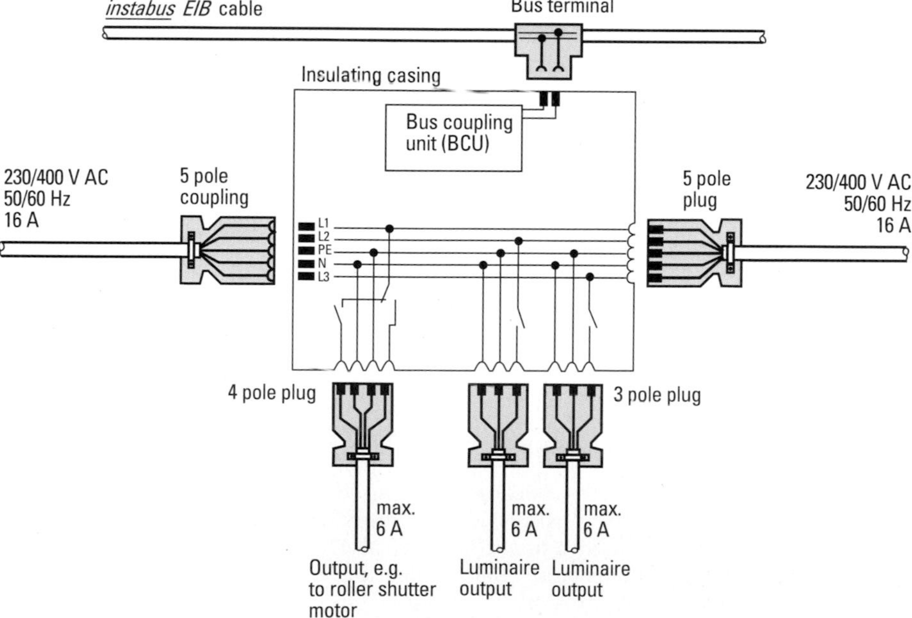

Fig. 5.1/48
instabus EIB snap-on device, e.g. two outputs for connecting luminaire groups and one output for roller shutter motor

5.1.4.10.2 Planning and configuration

Planning aids

Planning and configuring a conventional installation engineering system used to be a very time-consuming and labor-intensive process because of the complicated wiring. The _instabus_ *EIB* control bus system and the associated planning aids (the planning and configuration software ETS [EIB Tool Software]) together with prepared tender texts considerably reduce the time and effort involved in planning and configuring a system and help eliminate errors. The design engineer merely has to enter the desired installation components, as well as their assignments and functionalities.

Computer-assisted planning and configuration

The components are selected together with their functions and assignments at a VDU-based workstation (PC) or on a laptop via on-screen menus. The components and entries can be changed easily at a later stage. The ETS is an indispensable tool when planning and configuring the _instabus_ *EIB* control bus system since it permits the optimum cable lengths to be determined and enables the bus components to be located when modifications or maintenance are performed in the intermediate ceiling (Fig. 5.1/49).

Bus components

The tool software enables the bus components

▷ to be determined precisely,
▷ to be costed accurately,
▷ to be arranged storey by storey,
▷ to be ordered simply,
▷ to be supplied as and when required on the construction site,
▷ to be checked easily after installation.

The installation locations of the devices and the simple connection procedures, which can also be carried out by untrained auxiliary personnel without fear of incorrect connection, are also indicated on the configuration plan. The range of possible applications is

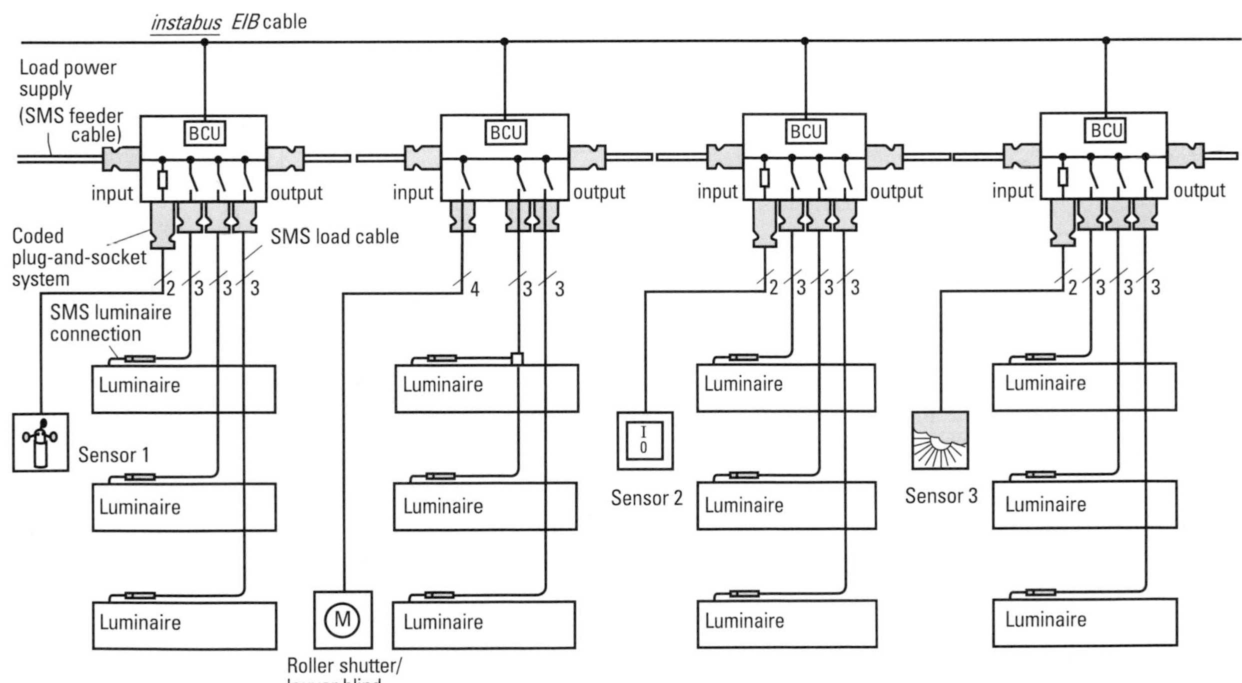

Fig. 5.1/49 _Instabus_ *EIB* snap-on system, circuit diagram with different loads

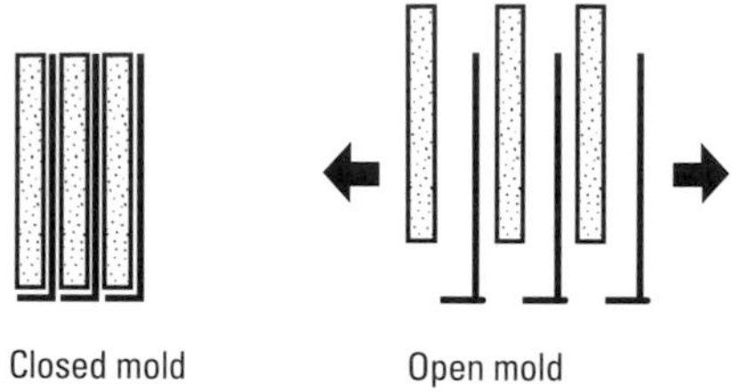

Fig. 5.2/2 Vertical (battery) production

To help the concrete set more quickly, the panels are often precured, e.g. under steam hoods, in ovens, or on heated molding tables. The temperatures involved in this process must not damage the plastic materials used for the electrical installation.

Vertical (battery) production method used very seldom

With the vertical production method, also referred to as battery production, a number of molds are placed together and the wall panels are cast vertically. Once the concrete has set (this usually takes place without heating), the panels are lifted out of the molds, which are in most cases fanned out or pulled apart (Fig. 5.2/2). The concrete is poured into the mold and compacted by means of an internal vibrator. In this case, both sides of the panels are smooth because the mold is double-sided.

In-situ casting

With the in-situ method, the walls and ceilings are shuttered, reinforced and cast in position. In contrast to the large-panel construction method, it is the molds that are removed here and not the prefabricated wall panels. The type and form of mold vary, depending on the equipment used by the building contractor. The surface finish of the ceilings and walls is often of such a high quality that plastering is unnecessary. Heating is not normally used to accelerate setting in this case either.

5.2.3 Planning

Large-panel construction

Cable links

Fitting cable or insulating conduit links on site from panel to panel (wall to wall, or wall to ceiling) is time-consuming, and should be kept to a minimum. It is helpful, therefore, to arrange all the unequipped boxes (e.g. for socket outlets) and lighting points in a panel in one circuit. If the unequipped boxes and insulating conduit links are incorporated in the mold, it can take a relatively long time to secure the junction boxes and switch boxes.

Combined wall and joint boxes

Combined wall and joint boxes can accommodate switches or socket outlets as well as the connection terminals, which means that normal junction boxes are not necessary. In order to facilitate production and storage, it is more economical to work with only one type of box.

Panel layout drawings

All details must be entered on the panel layout drawing (Fig. 5.2/3) (e.g. recesses for cable links).

Installation in panels with horizontal production

From a production point of view, wires and cables can be routed as desired in horizontal molds. It is advisable, however, to attach the insulating conduits to the reinforcing bars. The wall and ceiling links between panels can be established by means of recesses at the edges of the panels (via cut-out elements) or with suitable transfer boxes (Fig. 5.2/4).

Installation in panels with vertical (battery) production

In vertical molds, insulating conduits or cables laid horizontally may be damaged when the concrete is poured in unless they can be attached to reinforcing bars. Insulating conduits must not, however, be attached too rigidly, as there is a danger of them being broken if no movement is possible.

If no reinforcing bars are available, insulating conduits and wires or cables should only be laid vertically. The horizontal connections then have to be established on site in the floor, the ceiling, or in skirting boards.

Connections from wall to wall

Connections from wall to wall can be established wherever it is most convenient from the point of view of the cable run. The recess and the joint between the panels have to be filled in and finished off after the connections have been made. Subsequent finishing is not required if the recess is formed at the lower edge (foot) of the panel. The connection to the adjacent panel is then in the screed or in the groove between the wall panel and the ceiling (floor) panel which is provided for grouting during assembly. Unprotected wires and cables should not be laid across the floor, since experience has shown that they are often damaged during the course of building. To protect the wires and cables, they should be laid across the floor in conduits. End and intermediate

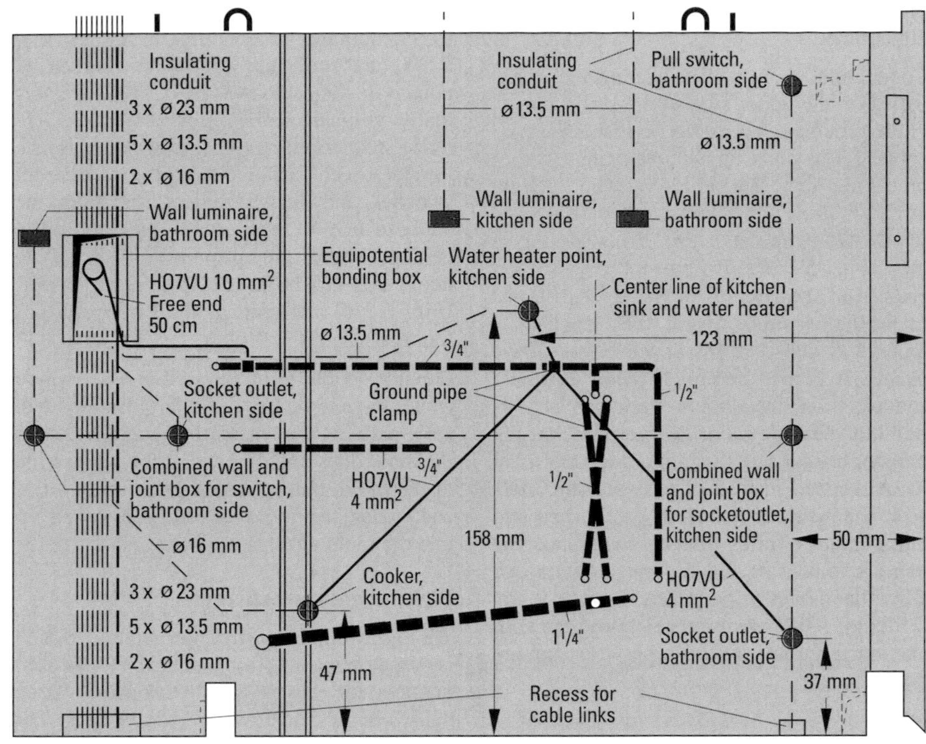

Fig. 5.2/3 Example of a panel layout drawing for a kitchen/bathroom wall section (minimum configuration)

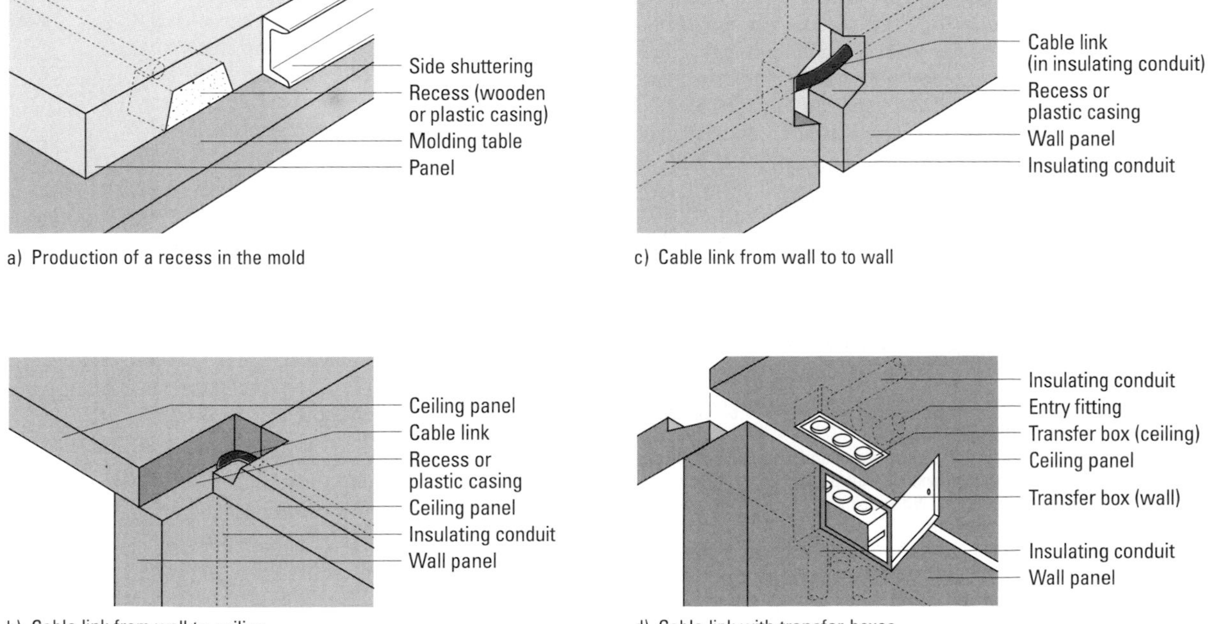

Fig. 5.2/4 Wall and ceiling links for cables/wires with the horizontal panel production method

Fixing boxes: horizontal production

With horizontal construction, if wooden or plastic molds are used and the type of wall to be produced changes frequently, the boxes can be nailed or attached to the mold (Fig. 5.2/8a). If a considerable number of walls of the same type are to be produced, it is preferable to fit the box using "threaded dowels" which are driven into the mold (Fig. 5.2/8b). Magnetically attached locators or adhesive film can be used for steel molds which are rearranged frequently (Fig. 5.2/8c). Additional measures, however, have to be implemented in this case when the concrete is vibrated.

Fixing boxes: vertical production

Since the molds in vertical production are also used for a wide range of different panels, it is not possible to damage the mold (e.g. by drilling).

Either a magnetic or adhesive fixture is used for the boxes in these situations. The boxes are fixed between the mold walls by means of spacer tubes and outer supports (Fig. 5.2/9). The distance between the fixed components can be adapted in accordance with the panel thickness by changing the dimensions of the tubes.

The system is fixed when the mold is "closed". The spring action of the rear of the box compensates for any inaccuracies.

Since extreme mechanical stress can occur when the concrete is poured in, the boxes can also be secured to the reinforcing bar with wire via holes at the front.

In-situ casting

Conduits and cables

With the in-situ casting method, either flexible corrugated insulating conduits or heavy-duty flexible plastic conduits are cast into the panels, depending on the mechanical stress to be expected. Refer to the large-panel construction method for details on fixing.

Wooden and plastic molds

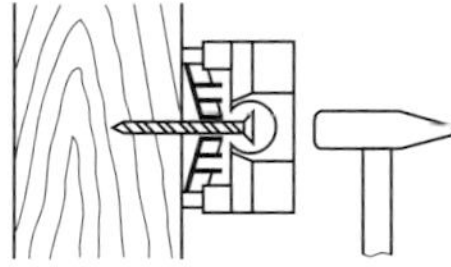

Box fixed with nail or wooden screw

a) If wall types to be produced change frequently (in-situ casting)

Steel molds

Magnet on mold

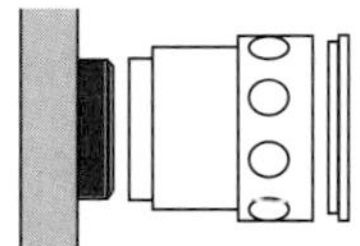

Press box onto magnet

Box fixed with magnet/adhesive film

c) In large-panel construction with steel molds that are rearranged frequently

Wooden and plastic molds

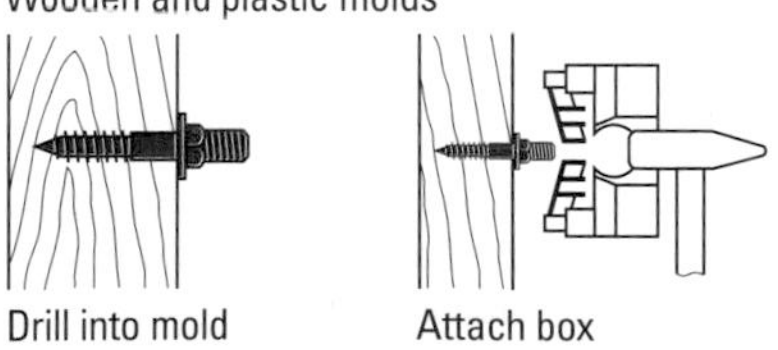

Fixing with twin screw/threaded dowel

b) In large-panel construction (series production) as permanent fixing

Steel molds

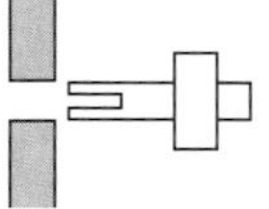

Drill 5.5 mm hole
Insert dowel

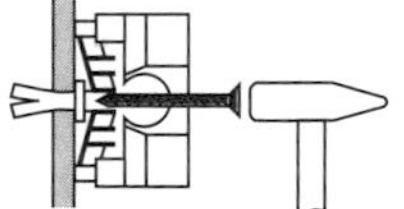

Place box on dowel and drive in steel nail

Fixing with expansion nails/plugs

b) In in-situ construction with closed molds (e.g. tunnel molds). For series production only

Fig. 5.2/8
Fixing methods for switch and combined wall and joint boxes with different concrete construction methods

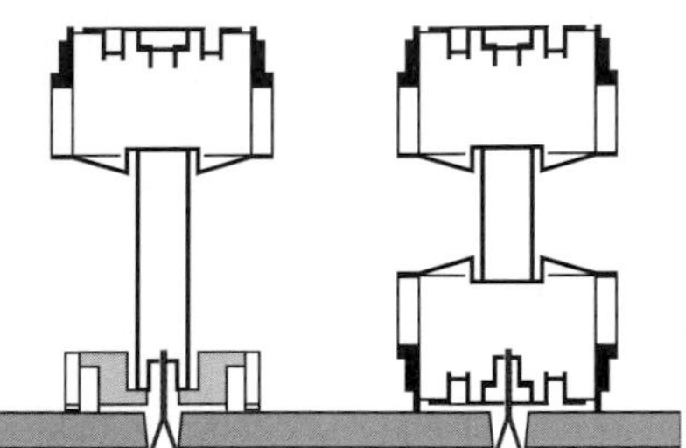

Box at top, secured with outer support and help in place with spacer tube
Two opposite boxes, with spacer tubes
a) Horizontal mold

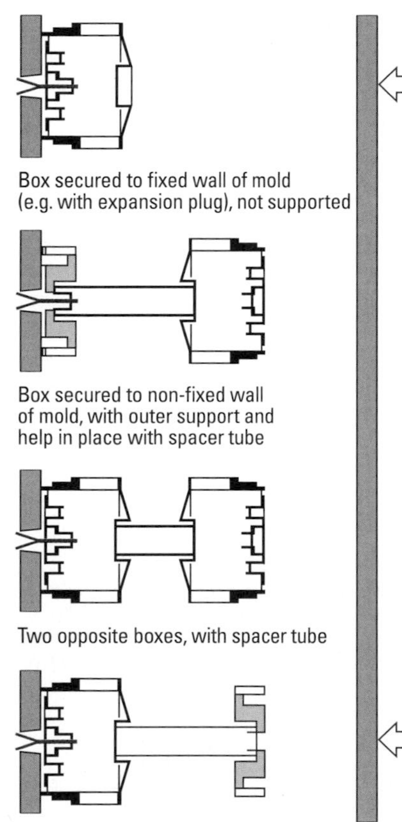

Box secured to fixed wall of mold (e.g. with expansion plug), not supported
Box secured to non-fixed wall of mold, with outer support and help in place with spacer tube
Two opposite boxes, with spacer tube
Box secured to fixed wall of mold, supported for extreme stress with outer supports and spacer tube
b) Vertical mold

Fig. 5.2/9
Fixing methods for combined wall and joint boxes in large-panel construction and in-situ casting system

Tunnel molds

If the wiring is to be accommodated in closed molds (e.g. tunnel molds), cables or conduits are suspended from above. The combined wall and joint boxes are fitted from the outside through openings in the mold and fixed with expansion plugs (Fig. 5.2/8d). For this purpose, suitable cut-outs must be provided in the mold. These also serve as hand-holes for pulling the cables or conduits through the mold. The cut-outs are sealed by the combined wall and joint boxes before the concrete is poured in (Fig. 5.2/10).

Lightweight construction

Conduits and cables

In wall and ceiling sections produced using the lightweight construction method, it is usual to incorporate insulating conduits. The sheathed cables are then drawn into these on site.

Combined wall and joint boxes

Special combined wall and junction boxes designed for the lightweight system have special lugs on the sides which ensure that the boxes are not torn out of the wall.

5.2.5 Meter cabinets and distribution boards

Meter cabinets and distribution boards are available in standard designs (see Section 1.4.5). In the case of the large-panel and in-situ construction methods, the openings required for these can be formed by means of re-usable cores (metal or wooden).

Cast-in boxes for distribution boards

Cast-in boxes are more appropriate for distribution boards. Openings for the conduits can be provided as required. Self-locking conduits stay firmly in place and cannot be pulled out of the openings of the boxes.

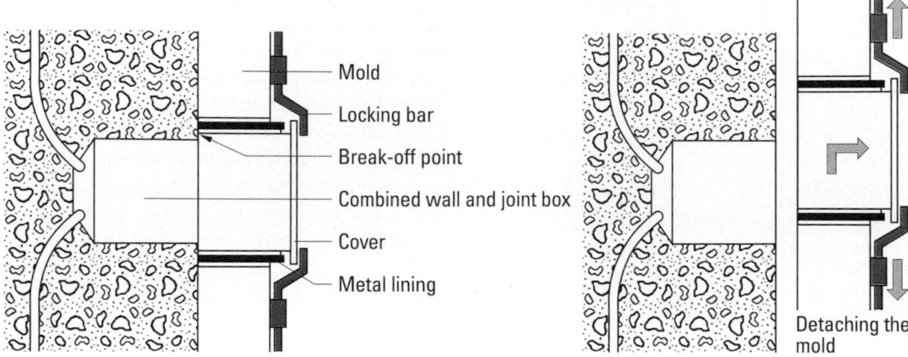

Fig. 5.2/10 Fixing methods for combined wall and joint boxes in in-situ casting system

5.3 Office buildings

Load profile

Office buildings normally include a large number of functional areas with very different and specific requirements with respect to the supply and distribution of electrical power. Apart from the actual offices themselves (open-plan, grouped, and individual offices), office buildings nowadays also contain functional areas such as underground garages, large kitchens and canteens, computer centers, printshops, lecture theaters, etc. The load profile that results from the procedures and processes carried out in these functional areas and from ensuring that the required ambient conditions for these rooms are fulfilled, e.g. temperature (warm/cool), ventilation, transport, etc., is often extremely complex and varies according to the time of day.

When the electrical supply system for such a building is planned, it is, therefore, important to elaborate a load profile and to determine the anticipated simultaneity factors for the individual load groups. This load profile can then be used as a basis for optimizing system configurations and in-house energy consumption.

Load centers

Depending on the structure of the building (high-rise, low-profile, or a combination of both), load centers can be formed with main distribution boards (possibly with voltages > 1kV), from which the final loads are supplied at low power loss. This is particularly desirable in the case of high-power load groups such as elevators, central air-conditioning systems (warm/cool, ventilation), and kitchens.

Distribution zones

For the "surface loads", i.e. lighting and office equipment connected to the socket-outlet system, it is advisable, for example, to form a distribution zone on each storey. If, however, a storey consists of several fire compartments, a distribution zone should be provided for each fire compartment.

Supply security

In accordance with the importance of the individual final loads (to protect personnel in emergencies and to maintain the power supply to important production equipment, e.g. computer systems, in the event of a power failure), a distinction must be made between the following loads:

- ▷ loads of the safety power supply system (see Section 1.8),
- ▷ loads of the stand-by power supply system (uninterruptible or with permissible interruptions),
- ▷ loads of the general power supply system.

Safety power supply

Loads, which require a safety power supply, include:

- ▷ safety lighting,
- ▷ systems for transmitting danger and alarm signals,
- ▷ pumps for fire extinguishing systems,
- ▷ smoke and heat extraction systems,
- ▷ fire service elevators and elevators with evacuation function,
- ▷ other equipment which must continue to function to ensure the safety of personnel in emergencies, particularly in the event of a fire, and must remain operational if the general power supply fails.

Stand-by power supply

A stand-by power supply to HD 384.5.55 / IEC 60364-5-55 / DIN VDE 0100-551 is recommended for elements of load installations without which it would be impossible to continue working. These include, in particular, elements of lighting, telecommunications, and data processing systems. Depending on the installed equipment, the power required by current-using devices connected to the stand-by power supply is generally between 25% and 35% of the total load of the building.

Operating areas

Areas which are primarily used for housing the independent power sources for the safety power supply are dealt with in Section 1.8.2. National regulations may, however, also apply to the installation locations of central equipment associated with the stand-by power supply and general power supply. These regulations usually specify requirements regarding the arrangement and the fire-protection characteristics of these operating areas.

Separate PE conductor

If TN systems are used to supply power to electrical equipment in office buildings, a TN-S system should usually be installed starting at the main building distribution board (see Section 2.4, Fig. 2.4/3). The reason for this is that, owing to the marked increase in the use of networked IT systems, the sensitive computer and peripheral equipment tends to malfunction if circulating currents flow through the PEN conductor and thus through the exposed conductive parts of these devices.

Electrical supply routes

Wherever possible, the main electrical supply routes to the sub-distribution boards on the individual storeys should not pass though the commonly-used corridors and stairwells of the building. If the routes run vertically through the building, it is advisable to install accessible riser ducts; if the routes run horizontally through the individual storeys, the cables should be installed in non-combustible floors (in conduits in the floor base or screed) or along the outside of the corridors in the adjacent rooms.

If combustible cables and wiring are installed in the corridors or stairwells of the building, additional fire-protection measures, which are specified in national regulations or guidelines, may also have to be taken into consideration.

Functional endurance

HD 384.5.56 / IEC 60364-5-56 / DIN VDE 0100-560 specifies that functional endurance over a certain period of time must be ensured for the supply of electrical power to the necessary safety equipment from the safety power supply. These stipulations apply to the infeed of power to the safety equipment and refer to both the cable connections and the required distribution boards. Refer to Section 1.8.2 for information on how these requirements can be satisfied.

Cable installation

Modern construction methods do not permit cables used for power distribution to be installed in external walls or in adjustable partitions. Cable ducts underneath windows or underfloor ducting systems (see Section 5.1.4.8) are used for installing the cables and the associated installation equipment. Cable systems for lighting and other supply systems are routed along the inside of intermediate ceiling cavities. If holes have to be made in fire-protection walls to allow cables to be installed, they must be sealed again afterwards in compliance with the fire-protection category of the fire-protection walls.

Illumination, specific load

prEN 12464 / DIN 5035-2 specify average values for the illumination intensity required for different visual tasks. Nowadays, when the power supply system is planned, an average load of 15 W/m^2 at an illumination intensity of 750 lux can be assumed if specular louver luminaires with fluorescent lamps and electronic control gear are used.

prEN 12464 specifies an average illumination intensity of 500 lux for office workstations. This value is relatively high as are the associated operating costs if artificial light sources are to be used. A more acceptable solution would be to control the artificial lighting according to the daylight intensity in addition to utilization. This optimizes operating costs with respect to power consumption and the service life of the lamps.

Lighting control system

Remote switches, which are also available as a multi-pole version, are suitable for use as simple lighting control systems.

In addition to local switching, operation is also possible by means of higher-level switching commands – doorman switch, time switch, etc. – which can be implemented, for example, using remote switches with "Central ON" and "Central OFF" functions (see Section 4.2). Programmable controllers are available for linking several interdependent functions. Linking functions in this way may be necessary in open-plan offices of administrative centers, for example, particularly if daylight-dependent lighting controllers have to be combined with sunshade systems.

Building control systems with *instabus EIB*

instabus EIB building control systems permit an even broader range of switching and control applications (see Section 4.4).

Socket-outlet circuits

Under normal circumstances, one socket-outlet circuit should be used to supply power to six permanently-installed grounding-type socket-outlets. A single socket-outlet circuit should not supply more than eight permanently installed grounding-type socket-outlets.

This restriction allows for expansions that are often made to the permanently-installed system by means of extensions and flexible multiple couplers. In this case, a power failure caused by a fault in the branch circuit is restricted to a relatively small area.

Protective device

Experience has shown that, under normal conditions, the daily average load current per socket-outlet is 1 A. 10-A miniature circuit-breakers with C tripping characteristic are recommended for socket-outlet circuits in view of the high making currents necessary for a large number of office devices as well as the requirements regarding fast disconnection in the event of a fault and restriction of the voltage drop along the cable.

5.4 Hotels

Electrical systems make an important contribution to the amenities that hotel guests have come to expect. In certain cases, the effects of these amenities are perceived directly, e.g. lighting and air conditioning. The hotel guest does, however, benefit from all the other types of equipment in the hotel, whether this is used in the kitchen or laundry room, or for telecommunications purposes. A very extensive and reliable power supply system is, therefore, essential.

Building control systems with *instabus EIB*

The high level of comfort and convenience that the guests expect and the highest possible degree of operational efficiency that the hotel owner strives for can both be achieved with *instabus EIB* (see Section 4.4).

Load structure

As far as the load structure is concerned, a hotel usually consists of the following areas:

▷ reception area,
▷ guest rooms,
▷ conference rooms/function halls,
▷ swimming pools,
▷ fitness studios,
▷ hospitality areas, including the associated kitchens, refrigeration rooms, and storage rooms,
▷ technical facilities (elevators, fire extinguishing equipment, ventilation),
▷ hotel car parks.

Hotels often attach a great deal of importance to their "corporate image" which also shapes the requirements vis-à-vis the electrical installation of the building.

Specific load values

Owing to the enormous range of equipment and facilities that can be used in hotels, it is impossible to provide any detailed information regarding the specific load values of the individual areas. Experience has shown, however, that an average specific load (including all current-using equipment) of 3500 W/room for hotels without air conditioning and 4500 W/room for hotels with air conditioning can be assumed when planning the electrical system.

Official regulations and technical standards

Since hotels are used almost exclusively by people who are unfamiliar with the buildings, they are usually subject to a number of special safety and official regulations and standards.

Apart from constructional specifications for walkways (corridors and stairwells), certain safety devices (see Section 1.8) must also be provided as of a specific building size. In addition to power failures and internal faults in the electrical power supply system, these specifications also consider fire as a particular hazard for hotel users.

Limit values, which determine whether or not the necessary safety devices must be provided, are often specified, for example:

▷ the number of guests in the hospitality area,
▷ the number of beds in the accommodation area,
▷ the number of seats in the conference and function rooms,
▷ the size and type of underground garage (if applicable).

Safety devices

The unfamiliarity of the hotel guests with the building makes it particularly important to ensure that the safety devices in the hotel have a reliable power supply.

This applies to:

▷ the safety lighting along the escape routes in the building and in large guest and function rooms,
▷ systems used to raise the alarm and to issue instructions in emergencies,
▷ booster pumps for fire extinguishing systems,
▷ elevator systems,
▷ parts of the ventilation and smoke extraction systems.

HD 384.5.56 / IEC 60364-5-56 / DIN VDE 0100-560 specifies the required type of safety power supply. Section 1.8 contains information on the requirements regarding electrical safety and functional endurance which must be ensured even in fire scenarios.

Stand-by power supply unit

In large hotels, it is advisable to install a stand-by power supply unit so that, in addition to supplying the above-mentioned load groups, parts of the kitchen, the refrigeration rooms, part of the room lighting, and power supply units for telecommunications, data transmission, and signaling systems can also be supplied in the event of a power failure.

Guest rooms

To ensure that malfunctions caused by electrical faults in a guest room do not affect other guest rooms, it is advisable to provide each guest room with a small distribution board containing the overcurrent protective devices for the branch circuits.

It should be possible to operate the general lighting in guest rooms (Fig. 5.4/1) from the door and from the bed. Fluorescent lamps, e.g. mounted behind the curtain pelmet, are suitable. Warm-tone lamps, which can be used together with filament lamps, are recommended on account of their excellent color-rendering properties. The lighting can also be complemented by decorative wall, standard, or table lamps and downlighting flush-mounted ceiling luminaires. The various lighting elements allow the guests to adapt the lighting to their particular requirements.

Apart from the connections for radio, TV, video, and cleaning equipment, additional socket-outlets as well as telecommunications outlets must be provided next to the desk and bed for electrical devices that the guests may wish to use in the room.

The lighting and socket-outlets in the guest room should be supplied from two different circuits.

The shaver socket-outlet with integrated isolating transformer installed in the bathroom increases the level of safety by providing protection against indirect contact and allows shavers with different plug systems to be connected.

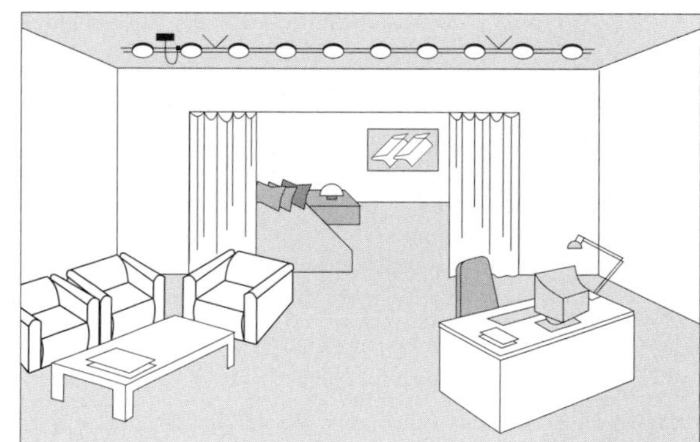

Fig. 5.4/1 Guest-room lighting

Reception hall The reception hall is the calling card of the hotel. The lighting in the reception hall, which is the most important point of contact between the hotel staff and the guests, must be bright but not dazzling. The general lighting of the hall can be relatively subdued so that the supplementary lighting for groups of chairs, e.g. standard lamps or downlighting ceiling luminaires, imparts a feeling of well-being and security. Accentuated lighting can emphasize the architectural design of the hall.

Restaurant In the restaurant, downlighting ceiling luminaires should be used to light the individual tables. Filament lamps and low-voltage halogen spotlights, e.g. with a rated voltage of 12 V, are preferred for this part of the lighting system, while fluorescent lamps are suitable for the complementary general lighting. When lamps for restaurants are selected, the color-rendering properties of the light must be chosen carefully to make sure that the food looks natural and appetizing.

Cafés and bars In cafés and bars, guests expect a particularly inviting and comfortable atmosphere which should be enhanced by the lighting of the room. The subdued general lighting can be complemented by a variety of effects achieved using low-voltage halogen spotlights and miniature projector lamps with color filters, ultraviolet projectors (directed towards fluorescent pictures and decorations), dance floors with underfloor lighting, etc. A lighting control system is recommended for these applications.

Conference rooms Conference and assembly rooms should have controllable general lighting to cater for different visual tasks.

Film, slide, or overhead projection as well as video presentations require adjustable levels of illuminance. Fluorescent lighting is recommended for these applications. It is important to ensure that the luminaires around the projection area can be switched off separately.

It should also be possible to control both the lighting and the window blinds centrally (possibly from the speaker's desk).

An adequate number of socket-outlets must be provided for the projectors.

Banqueting halls Banqueting halls are used for many purposes: balls, fashion shows, conferences, exhibitions, or concerts. All make different demands on the lighting and other electrical equipment. A different lighting arrangement is used for each event. Lighting control systems should also be provided, for example, for concerts or stage presentations. If stage floodlights or effect spotlights are not part of the permanent equipment, a sufficient number of adequately-rated controllable socket-outlet circuits must be provided to allow these types of lighting to be connected temporarily.

Kitchens Kitchen, storage, and refrigeration installations are usually located next to each other; they are connected to a common distribution board which is supplied directly from the low-voltage main distribution board because of the high power demand. In the event of a power failure, some of the electrical kitchen appliances, the refrigeration system, and the lighting should be supplied with power from the stand-by power supply so that the guests can continue to be catered for without any significant restrictions.

In addition to the circuits for permanently-installed current-using equipment, socket-outlet circuits for alternating and three-phase current must be provided for connecting hand-held appliances.

According to HD 384.4.46 / IEC 364-4-46 / DIN VDE 0100-460, it must be possible to disconnect large kitchens from the power supply by means of an emergency-off switch. This specifically applies to electrical devices and not to the kitchen lighting.

Corridors and stairs The lighting in corridors and stairwells should be both functional and decorative. The luminaires must be arranged and connected in such a way that part of the lighting can remain switched on during the night. The supply for this reduced lighting could be connected to the safety lighting so that adequate illumination is provided even in the event of a power failure.

The switches or pushbuttons in the corridors and stairwells must have illuminated symbols to enable the guests to locate them easily. Corridors and stairwells should be provided with an adequate number of socket-outlets to allow cleaning appliances to be connected.

The lighting and socket-outlet circuits in the corridors and stairwells must be independent of those serving the guest rooms.

5.5 Hospitals and medical practices

Hospitals and rooms used for medical purposes outside hospitals are special structural installations. They are ranked among buildings which contain some of the most extensive and complex technical and electrical equipment available. Two important features that distinguish hospitals from many other types of building are all-year-round operation without down-times and the direct use of electrical apparatus on people. This places enormous demands on the electrical equipment with regard to its installation and operation.

Building control systems with *instabus EIB*

These demands can be fully satisfied using building control systems with *instabus EIB* (see Section 4.4).

The breakneck pace of developments in the field of medical engineering has also called for a new holistic approach when considering the electrical safety requirements for hospitals and locations used for medical purposes.

Official regulations and technical standards

Although international technical safety standards have existed for a relatively long time for electrical medical apparatus, e.g. the EN 60601-1 / IEC 60601-1 and EN 60601-2 / IEC 60601-2 series, the structural engineering specifications vis-à-vis the rooms and buildings, including the electrical supply and safety requirements themselves, are still laid down at a national level in official regulations and standards. An international safety standard for the electrical installations of these buildings and building areas is, however, currently being prepared. Many of the differences between the current national safety standards can be attributed to tradition and, when considered more closely, are often relatively minor.

The following points are of particular importance in all the applicable standards:

▷ protection of the patient, and of the doctor and personnel when electromedical apparatus is being used,
▷ security of supply for vital, electrically-powered apparatus,
▷ general protection of patients with varying degrees of mobility, e.g. in an emergency.

The detailed specifications regarding the configuration of the power supply for these types of service installation are explained using the current German standard "DIN VDE 0107 – Electrical installations in hospitals and locations for medical use outside hospitals" as an example.

Utilization category

According to DIN VDE 0107 (applicable since 1989), locations or buildings used for medical purposes are categorized according to the way in which the individual rooms/locations are utilized:

▷ Rooms used for non-medical purposes
These include corridors, stairwells, ward duty rooms, bathrooms and toilets, kitchenettes, restrooms, administrative rooms and operating areas.
▷ Rooms used for medical purposes, utilization category 0
Rooms where patients are accommodated and nursed. Mains-dependent electromedical apparatus is not used here.
▷ Rooms used for medical purposes, utilization category 1
Rooms where patients are examined and treated (including minor surgery) using mains-dependent electromedical apparatus. The apparatus is not, however, inserted into the body of the patient or used for operations on internal organs. Failure or disconnection of this apparatus as a result of a power failure or malfunction does not represent a danger to the patients.
▷ Rooms used for medical purposes, utilization category 2
Rooms where patients are examined, operated on, monitored, or undergo intensive medical treatment. Failure of the necessary vital electromedical apparatus is unacceptable since this would put the patients at risk.

Table 5.5/1 contains examples illustrating the assignments of rooms to the utilization categories, as defined in DIN VDE 0107, according to their intended use. The purpose of this is to facilitate dialog between the doctor and electrician. In cases of doubt, a higher utilization category should always be selected. The final decision as to which utilization category the room in question belongs must, however, be taken by the operator of the hospital or of the rooms used for medical purposes.

Table 5.5/1
Examples illustrating the assignments of room types to utilization categories in accordance with DIN VDE 0107, Table 1

Utilization category	Room type according to specified use	Type of medical utilization
0	– Wards, – Operating theater sterilization rooms, – Operating theater washrooms, – Dental and human medicine surgeries	No electromedical apparatus used
1	– Wards, – Physiotherapy rooms, – Hydrotherapy rooms, – Massage rooms, – Dental and human medicine surgeries, – Rooms for radiological diagnosis and therapy, – Endoscopy rooms, – Dialysis rooms, – Intensive examination rooms, – Delivery rooms, – Surgical outpatient departments, – Heart-catheter rooms for diagnosis	Electromedical apparatus used on or inside the patient's body inserted via body orifices or for minor operations (not surgical) Examinations using flow-directed catheters
2	– Pre-operation rooms, – Operating theaters, – Recovery rooms, – Casting rooms, – Intensive examination rooms, – Intensive monitoring rooms, – Endoscopy rooms, – Rooms for radiological diagnosis and therapy, – Heart-catheter rooms for diagnosis and therapy (except rooms in which flow-directed catheters only are used), – Clinical delivery rooms, – Rooms for emergency and acute dialysis	All types of operation on organs (major surgery), insertion of heart catheters, surgical insertion of operating apparatus, all types of operation, life-support with electromedical apparatus, open-heart surgery

Room types are assigned to utilization categories according to their intended medical use and the medical equipment used in them. For this reason, certain room types can be assigned to several utilization categories. When power installations in hospitals are planned, it is not usually possible to predict whether electromedical apparatus will be used at certain locations, e.g. in wards. Utilization category 0 should, therefore, not be used in cases of doubt.

Power supply

The topology of the functional units, the requirements vis-à-vis the medical equipment, and the amount of power supplied to the building are important factors that determine the configuration of the power supply and distribution system in a hospital.

Criteria for power supply

The following criteria must be taken into consideration:

▷ the size and layout of the individual buildings,
▷ expansive or compact construction,
▷ the amount of equipment installed in the functional units,
▷ district heating or internal source of heating, e.g. electrically-generated heating,
▷ internal kitchen and laundry or use of external facilities.

Specific power values

On account of the very different requirements regarding power supply, no generally applicable specific power values can be provided for hospitals.

Safety power supply

DIN VDE 0107 and the German building regulations for hospitals, however, contain very specific requirements regarding the reliable supply of power to important electrical equipment necessary to run a hospital. According to these specifications, a safety power supply, which is independent of the mains supply and which is activated automatically within 15 s if the mains supply fails, must be provided for the following equipment:

▷ safety lighting along escape routes, in rooms used for medical purposes, as well as operating and working areas crucial for ensuring that the hospital continues to function normally,

- ▷ fire service elevators and required bed elevators,
- ▷ ventilation systems for smoke extraction,
- ▷ public-address and paging systems,
- ▷ alarm and warning systems,
- ▷ fire extinguishing systems,
- ▷ medical equipment,

particularly:

- operating theater lamps and similar lighting elements,
- electrical devices in rooms of utilization category 2,
- electrical equipment for medical gas supply, including compressed air, vacuum supply, and anesthetic gas extraction apparatus.

Stand-by power supply

Furthermore, the power supply for the equipment and installations listed below, which are essential in ensuring that the hospital continues to function normally, must be safeguarded by a stand-by power source which is independent of the general power supply:

- ▷ sterilization equipment,
- ▷ internal systems, particularly the heating, ventilation, supply, and disposal systems,
- ▷ refrigerating systems,
- ▷ catering equipment,
- ▷ other important elevators and equipment.

The transfer time must not exceed 15 seconds for the equipment listed above. It is advisable to provide a second independent stand-by power supply for these loads.

It should, however, also be possible to switch this power supply to the safety power supply system (so that a stand-by power supply is available if maintenance work has to be carried out or if faults occur in the safety power source).

Safety power sources

In practice, reciprocating internal combustion motors (diesel motors) with synchronous generators conforming to ISO 8528-12 / DIN 6280 Part 13 have proven successful as power sources for safety power supply systems with a transfer time of 15 seconds. Gasoline motors are not permitted as drive units because of the potential dangers associated with gasoline storage and misgivings concerning starting reliability.

Engine-based cogenerating stations conforming to DIN 6280 Part 14 are equally suitable as safety power sources if they are intended for use throughout the entire year, if they consist of several modules, and if they provide the same degree of electrical supply security as the diesel drive unit. As far as supply security is concerned, this means reliable fuel supply, reliable heat dissipation, and suitably qualified personnel who ensure that the stations are operating correctly.

When the power output of the generating sets is determined, it is important to remember that the necessary safety devices and medical apparatus must be supplied with power within the transfer time of 15 seconds. Since this short power transfer – which must take place within the specified voltage and frequency tolerances – may, in certain situations, cause problems, coordination with the manufacturer of the generating set is essential in all cases. The current-using load may have to be divided into a number of load blocks which are connected to the supply at different times.

If only one safety power source is provided for the safety power supply, suitable measures should be implemented to ensure that the power supply is maintained if, for example, maintenance work has to be carried out. One possibility would be to temporarily install a mobile power generator which is connected via an external connection (fire service connection).

"Supplementary safety power supply"

In addition to the safety power supply, a supplementary safety power supply with an integrated safety power source conforming to DIN VDE 0107 is required for:

- ▷ operating theater lighting,
- ▷ electromedical apparatus for operations as well as vital operating and intensive therapy measures (rooms of utilization category 2). This may be specified in regional regulations or apply in individual cases.

Power source of supplementary safety power supply

Since the power source of the supplementary safety power supply constitutes a security measure for the most important electromedical equipment, it should be located as close as possible to the devices to be supplied with power. If this is not possible, it must at least be installed in the building in which the current-using equipment is located.

For this reason and in view of the specified short transfer time (max. 0.5 s), battery systems with static inverters or three-phase inverter outputs conforming to DIN VDE 0107 have gained general acceptance.

Capacity of supplementary safety power supply

The supplementary safety power supply should be designed to supply the connected operating theater lighting and the planned electromedical equipment with power for at least one hour.

The power rating must be determined together with the operator of the medical installation. Adequate power reserves should be provided.

General power supply

The general power supply from the public mains network should be fed via an incoming ring or double-feeder unit in the case of hospitals, which means that a switching reserve is provided if a fault occurs in the public utility network.

If power is fed from the high-voltage network, several transformers should be provided, even in the case of small hospitals. This ensures that electrical equipment continues to operate if a transformer fault occurs.

Distribution system

The distribution system can be configured as described in Section 5.1.

The locations at which electrical equipment for power generation (generating sets and batteries) and distribution (switchgear >1 kV and transformers) is installed must comply with the requirements of HD 384.5.56 / IEC 60364-5-56 / DIN VDE 0100-560. The information in Section 1.8.2 regarding fire protection for this central equipment must also be observed. Furthermore, the fire-protection requirements specified in DIN VDE 0107 must be taken into consideration for rooms in which the main distribution boards of the electrical system are housed.

Permissible network systems and protective measures

Essential requirements to be taken into consideration when installing electrical systems in hospitals are:

▷ the power supply to all current-using equipment must be safe and reliable,
▷ essential apparatus and equipment must continue to be supplied even in the event of faults,
▷ hazards resulting from shock currents must be avoided.

To ensure that these requirements are fulfilled, DIN VDE 0107 permits only certain systems, protective measures, and supply types in the individual rooms/areas depending on the respective utilization category (Table 5.5/2).

Separate and selective system installation

The distribution and load systems of the safety power supply must be installed separately from the systems of the general power supply (see also HD 384.5.56 / IEC 60364-5-56 / DIN VDE 0100-560). This prevents harmful interference, for example, caused by faults in less important parts of the system, renders the system more transparent, and improves maintainability. The two systems are separated as of the main distribution board of the safety power supply. The main distribution board is supplied with power redundantly from the main distribution board of the general power supply and from the safety power source. DIN VDE 0107 specifies that the safety power supply must have a selective system configuration. It is, therefore, necessary to carefully match all of the system's characteristic values, beginning with the system infeed and safety power source, the cable and conductor connections including the intended overcurrent protective devices, and finally the branch circuits of the loads. According to DIN VDE 0107, this requires the prospective short-circuit currents to be calculated (see Section 1.2), the correct disconnection point to be determined by means of the disconnection/tripping characteristics of the intended protective devices, and selective tripping of the series-connected protective devices to be determined (see Section 1.3).

With the safety power sources, the internal resistance or short-circuit characteristic must also be known. According to IEC 60909 / DIN VDE 0102, a conductor temperature of 80 °C is assumed for cable connections (this corresponds to a resistance 1.24 times higher than for a conductor temperature of 20 °C).

IT system for vital and medical equipment

DIN VDE 0107 specifies that an IT system is mandatory for the load circuits used to supply power to vital electromedical equipment in rooms of utilization category 2 (operating theaters and rooms used for intensive monitoring and examination).

The reason for this is that the faulty circuit must not be disconnected if a short circuit to an exposed conductive part occurs, i.e. the

Table 5.5/2 Table of requirements for permissible systems and protective measures

<table>
<tr><th rowspan="3">Permissible systems and protective measures for:</th><th colspan="3">Room types</th></tr>
<tr><th>Not used for medical purposes</th><th colspan="2">Used for medical purposes: utilization category</th></tr>
<tr><th>0</th><th>1</th><th>2</th></tr>
<tr><td>Loads of general power supply</td><td>Systems:
– TN-S system
– TT system
– IT system

Protective measures:
all to HD 384.4.41/ IEC 60364-4-41/ DIN VDE 0100-410</td><td rowspan="3">Systems:
– TN-S system
– TT system
– IT system

Protective measures:
– Protection by means of Class II equipment
– Extra-low voltage SELV and PELV, but restricted to max. 25 V
– TT system
– TN system with RCDs and restrictions
– IT system with alarm
– Safety separation and restriction to one current-using device only

Requirement:
– Supplementary equipotential bonding in patient area</td><td rowspan="3">Requirements for utilization category 1 are permissible for all loads *except* vital equipment;

Requirements for vital equipment:

System:
– TT system

Protective measures:
– IT system with special alarm

Double infeed (redundancy of general power supply and safety power supply up to sub-distribution board of room in utilization category 2)

Requirement:
– Supplementary equipotential bonding in patient area</td></tr>
<tr><td rowspan="2">Loads of safety power supply</td><td>Infeed from *general power supply*
Systems:
– TN-S system
– TT system
– IT system

Protective measures:
– all to HD 384.4.41/ IEC 60364-4-41/ DIN VDE 0100-410</td></tr>
<tr><td>Infeed from *safety power source*
Systems:
– TN-S system
– TT system
– IT system

Protective measures:
– Protection by means of Class II equipment
– Extra-low voltage SELV, PELV
– Safety separation
– IT system with alarm
– TT system/TN-S system with restrictions</td></tr>
</table>

electrical equipment can continue to function normally.

According to EN 60742 / IEC 60742 / DIN VDE 0551, isolating transformers must be used in IT systems. Since operation with low-capacity safety power sources is also possible, the electrical characteristics of these transformers must be as follows:

▷ Rated short-circuit voltage $u_{kr} \leq 3\ \%$,
▷ No-load current $I_0 \leq 3\ \%$,
▷ Maximum rush current $8 \cdot I_n$.

Since overload protection of the isolating transformers by disconnection is not permissible, an appropriate signal must be output in the event of an overload. A current relay (see Section 4.2.9) serving as a quick-response monitoring device, possibly combined with a thermal release, would be suitable for this purpose.

According to DIN VDE 0107, IT systems are formed in the distribution boards for rooms of utilization category 2. The purpose of this is to ensure that they remain transparent and can be inspected easily and that any faults that occur affect a limited area only. For the same reason, the maximum rated output of the isolating transformers was fixed at 8 kVA and the rated current at 36 A.

IT systems must have a redundant power supply which is fed directly from the building's main distribution boards and the supplementary safety power source via two independent infeeds.

The voltage monitoring device and transfer device must be installed in the distribution board of utilization category 2, or directly in the operating theater lighting in the case of lights with direct infeed.

The transfer device must be designed in such a way that possible faults will not cause both infeeds to fail.

Each IT system must have an integrated high-sensitivity ground-leakage monitor and an indicator unit for monitoring the system status. Normal operating states are indicated visually. Hazardous situations are indicated both visually and acoustically. The indicator unit also has a test button which is used to check whether the system is functioning correctly.

Functional endurance of cables and conductors of safety power supply

The cables and conductors of the safety power supply should usually be routed along their own carrier systems and separated from the cables and conductors of the general power supply. The functional endurance of the cables and conductors must comply with HD 384.5.56 / IEC 60364-5-56 / DIN VDE 0100-560 (see Section 1.8.2) or the relevant regional building laws.

With incoming-feeder cables of IT systems which lead to distribution boards of utilization category 2, at least one of the infeeds, preferably the one with the highest level of electrical reliability (supplementary safety power supply), must have a functional endurance of over 90 minutes (see Section 1.8.2).

The area supplied by a sub-distribution board should be restricted to a maximum of one fire compartment.

PE (equipment grounding) conductor

In hospitals, it is now forbidden to use a common PEN conductor downstream of a building's main distribution board. This applies to all supply systems.

The PE (equipment grounding) conductor should be routed together with the other conductors in a common covering. It can, however, also be installed separately. A separate equipment grounding conductor must always be provided for each circuit. Requirements regarding the cross-sectional area and labeling of the conductor are specified in HD 384.5.54 / IEC 60364-5-54 / DIN VDE 0100-540.

Supplementary equipotential bonding

According to DIN VDE 0107, supplementary equipotential bonding is required in the area surrounding the patient (patient environment to EN 60601-1-1 / IEC 60601-1-1 / DIN VDE 0750-1-1) to prevent differences in potential – which could endanger patients – from occurring in the event of a fault.

This means that all extraneous conductive parts, which are located within a radius of 1.5 m around a patient being examined or undergoing treatment with mains-operated electromedical apparatus, must be incorporated in the supplementary equipotential bonding.

In rooms of utilization category 2 in which open-heart surgery is performed, terminal studs must also be provided to permit the potential of mobile electromedical apparatus to be equalized.

If barriers against electrical interference fields and arrester systems of electrostatically conductive floors are fitted, they must also be connected to the supplementary equipotential bonding.

The equipotential bonding conductors must be routed along an equipotential bonding strip in each room. The equipotential bonding strip must be connected to the protective conductor bar of the distribution board (see also Fig. 5.8/2 in Section 5.8.1).

Requirements regarding the cross-sectional area and labeling of the supplementary equipotential bonding conductors are specified in HD 384.5.54 / IEC 60364-5-54 / DIN VDE 0100-540. A minimum cross-sectional area of 4 mm² must be used.

Interference suppression

Sensitive patient-monitoring systems which contain a large number of measuring and recording devices are used in operating theaters, intensive-monitoring and therapy wards, intensive-examination rooms, as well as ECG, EEG, and heart-catheter rooms.

Electric fields, caused by the power cable network, or magnetic fields of transformers, reactors, and motors can have a negative effect on the measuring accuracy of these systems. Interference suppression measures are, therefore, recommended.

Electric fields

Interference caused by electric fields can usually be prevented by using steel conduits or metal-covered cables for the power cables in the above-mentioned rooms. The steel conduits or metal coverings must be conductively connected to each other and to the equipotential bonding conductor. Plastic socket-outlets along the cables and conduits do not have any significant effect on the interference suppression measures implemented.

Cables for carrying high currents – rising mains cables or supply cables of elevators – must be routed past these rooms with a clearance of at least 6 m.

More extensive measures may be required in rooms which, for example, are intended for EEG. A screening mesh must be laid under the plaster of all the surfaces of the room, except doors and windows, and connected to the equipotential bonding conductor.

A phosphor-bronze wire fabric with a mesh size of 50 mm to 100 mm and a wire diameter of 0.8 mm has proven successful in practice. It is bonded to the surfaces using adhesives and then plastered over. The individual sections of fabric must be soldered together and connected to the equipotential bonding conductor at a number of suitable points.

Magnetic fields

Interference caused by magnetic fields of electrical power installations can be prevented by arranging the interference-generating devices in a suitable manner and ensuring that there is an adequate clearance between them and the measuring equipment to be protected. This must be taken into account by the architect in the room-layout plan. The distances between the source of interference and the measuring equipment specified in Table 5.5/3 are generally sufficient.

The type and extent of the interference suppression measures should be discussed and agreed upon with the manufacturer of the electromedical measuring equipment.

Table 5.5/3
Distances between source of interference and measuring equipment

Equipment	Minimum distance m
Ballasts for fluorescent lamps	0.75
Transformers, motors	6
Multi-strand conductors or cables Cross-sectional area of external conductors:	
$10\ mm^2$ to $70\ mm^2$	3
$95\ mm^2$ to $185\ mm^2$	6
$> 185\ mm^2$	9

5.6 Industrial buildings and exhibition halls

5.6.1 Industrial buildings

Requirements for electrical power supply systems

The structure of electrical power supply systems in industrial plants is largely determined by the type of manufacturing involved, as is the selection of the electrical equipment. Maximum possible security of supply, together with a high degree of redundancy of the equipment used are, in addition to the general conditions that have to be fulfilled when designing the electrical distribution system (see Section 1.1), important basic requirements from the point of view of avoiding interruptions to production.

Minimum security of supply

The so-called "(n–1)-criterion" is often specified as the minimum requirement for preventing production stoppages caused by interruptions to the power supply. It specifies that the total supply capacity must be ensured for the production facilities, even if one system component associated with power supply fails.

This means that two independent infeeds (e.g. via ring-cable feeders) are necessary for connection to the public power supply system in order to avoid interruptions to production in the event of a fault.

Standby power supply

In the event of a sudden failure of the entire public power supply, a standby power supply system ensures that the supply is maintained to important loads, such as IT systems for production control, and that critical production facilities or production processes are shut down in an orderly manner.

Industrial power stations and engine-based cogenerating stations

In cases where steam is required for industrial processes, it may prove economical to use an industrial power station for generating induced current (cogeneration of power and heat). Engine-based cogenerating stations are being used to an increasing extent to meet the demand for electrical energy and heat in industrial plants. They are usually operated in parallel to the public supply, with the agreement of the public utility responsible. *(Important: short-circuit current load increases).*

Uninterruptible power systems

To avoid interference caused by short-time voltage dips > 15% (e.g. with automatic reclosing or when short circuits are interrupted in the high-voltage system), it may be necessary to use uninterruptible power systems (see Section 1.1.4) to supply voltage-sensitive loads, e.g. process computers for controlling production, programmable controllers, etc.

Load per unit area

Industrial loads can be roughly divided into groups, which differ according to the type, the scale and the degree of automation of the manufacturing processes, as well as the effect on the power supply system and the power quirements. Values for the expected average load per unit area P_m in W/m² for the entire power demand, including approximately 17–22 W/m² for lighting, can be found in Table 5.6/1 for outline planning purposes.

Structure of the high-voltage system

Prerequisites

Medium-sized industrial plants are usually connected to the public 10-kV or 20-kV network of the public utility responsible. Smaller-sized plants must also be connected to these voltage levels if the system perturbations caused by the industrial plant (when connected to the low-voltage system) have an adverse effect on the public loads.

Given the considerable size of the production areas and the high load density, it is im-

Table 5.6/1 Average load per unit area

Type of production facility	Examples	Average load per unit area P_m W/m²	at cos φ
– Production facilities with low to medium power requirements. – Loads that are distributed more or less uniformly over the production area. – Power requirements that do not vary very much with time. – Relatively low level of automation.	Repair workshops, automatic lathe shops, spinning and weaving mills, precision-engineering production	50 to 100	0.6
– Loads that are distributed more or less uniformly over the production area. – Considerable differences in the connected loads of the individual consumers. – Non-coincident power requirements to momentary rhythmic impulse loads. – Medium to high level of automation.	Toolmakers' shops, mechan. workshops, welding plants	70 to 100 170 to 500 150 to 300	0.6 0.6 0.7
– Very high power loads (e.g. concentrated loads, such as ovens, presses or other large machines), – Small loads that are insignificant in terms of the total supply.	Press shops, Hardening shops, metallurgical plants, rolling mills	200 to 450 200 to 500	0.5 0.9

portant, in the interests of efficiency, that power is transmitted to the load centers using a high-voltage and not a low-voltage current.

Transformer load-center substation in the load center

For decades now, transformer load-center substations with GEAFOL encapsulated-winding dry-type transformers, that can be set up in the manufacturing area of a production facility, have proved their value as cost-effective components for supplying low-voltages to industrial systems.

Connecting the transformer load-center substation

The high-voltage connection of the transformer load-center substations is established via:

▷ radial cables (Fig. 5.6/1 a),
▷ ring cables (Fig. 5.6/1 b).

In both cases, the high-voltage switching station is installed at a suitable location in a closed electrical operating area.

In the case of ring cables, the ring cable load transfer switch panels are set up locally as part of the transformer load-center substation, directly in the manufacturing area.

A high-voltage radial cable connection via load transfer switches or switch-disconnectors with high-voltage high-breaking-capacity fuses is the preferred solution here. This is due to the following reasons:

▷ smaller cable cross-sectional areas for radial cables on the supply side of the h.v. h.b.c. fuses,
▷ lower costs for the provision of switchgear,
▷ more straightforward system management through the central switching station,
▷ smaller number of switching devices required (lower outlay for maintenance),
▷ considerably less space required for the transformer load-center substation in the manufacturing area, since the high-voltage fields are not required,
▷ considerably less short-circuit energy released in the event of faults on the high-voltage side of the transformer load-center substations and in their feeder cables thanks to the current-limiting effect of the h.v. h.b.c. fuses.

The last point in particular is very important with regard to the safety of personnel in the immediate vicinity of the transformer load-center substations, (which are free standing in the manufacturing area) and their feeder cables.

When circuit-breakers with time-overcurrent protection relays are used for the radial cables instead of load transfer switches or switch-disconnectors with h.v.h.b.c. fuses,

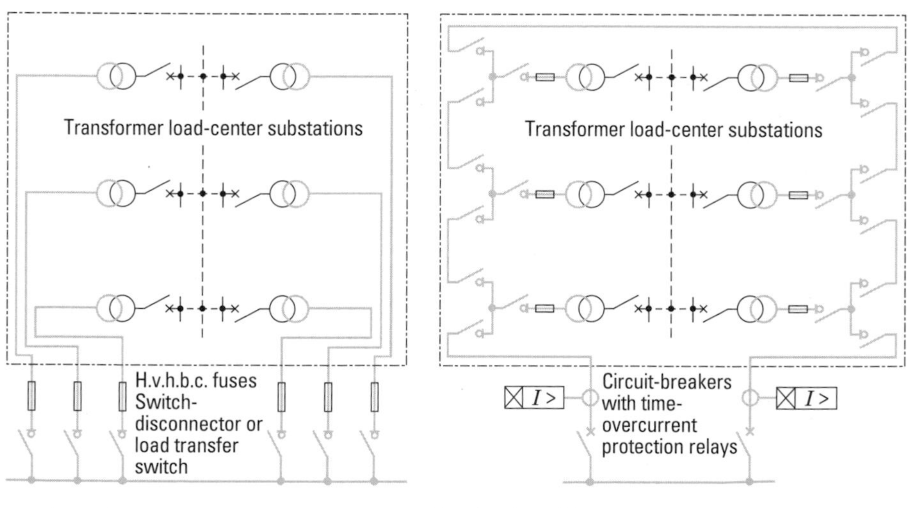

Fig. 5.6/1 Connection of the transformer load-center substation

all the advantages associated with the current limiting effect of the h.v. h.b.c. fuses are lost as a result of the operating time (at least 70 ms ≙ 3 ½ periods).

Requirements for low-voltage systems

Power requirements

Coincidence factor

The power requirements, in accordance with which the supply system for a production area is to be designed, only rarely correspond to the sum of the individual ratings of all the connected production facilities. In practice, a simultaneity factor is, therefore, taken into account for the variable capacity utilization of the connected production facilities or production lines and for the load cycle during a production cycle (see Section 1.1).

Maximum power requirements

The maximum power requirements of the industrial network can be calculated with:

$$P_{\max} = \sum_{i=1}^{n} (P_i \cdot g_i)$$

P_i Installed active power in kW
P_{max} Maximum power requirements in system in kW
g_i Coincidence factor

Capacity utilization factor

If predominantly motorized drives are supplied, e.g. in the case of precision engineering production (Table 5.6/1), a capacity utilization factor a_i also has to be taken into account for the open-circuit time and the mechanical load of the motors which is usually below the rated output. The efficiency of the motors also has to be included in the calculation.

The maximum power requirements in the system are then calculated approximately as follows:

$$P_{\max} = \sum \frac{P_{\mathrm{Mot}} \cdot g_i \cdot a_i}{\eta_i}$$

P_{max} Maximum power requirements in system in kW
P_{Mot} Rated output of motor in kW
η_i Efficiency of motor
g_i Coincidence factor
a_i Capacity utilization factor

P_{max}, P_{Mot}, g_i, a_i and η_i have to be taken into account as values for a machine or as average values for a production facility or production line.

The rated output (apparent power S in kVA) of the required infeed power must be determined from the maximum power requirement P_{max} while taking account of the mean power factor $cos\ \varphi$, a reactive power compensation device that may be installed, and the required redundancy [e.g. (n–1)-principle].

Impulse loading due to large loads

The magnitude of the infeed power can also be influenced by sudden loading caused by large individual ratings (e.g. press drives). In this case, the equivalent thermal load (P_{th}) must be determined from the load cycle as a root-mean-square value using the following equation:

Thermally equivalent load P_{th}

$$P_{\mathrm{th}} = \sqrt{\frac{P_1^2 \cdot t_1 + P_2^2 \cdot t_2 + \cdots + P_n^2 \cdot t_n}{t_1 + t_2 + \cdots + t_n}}$$

P_{th} Thermally equivalent load in kW
$P_1, P_2, \ldots P_n$ Individual ratings of large loads in kW
$t_1, t_2, \ldots t_n$ Time

If current impulses such as these overlap, this also has to be taken into account when the thermally equivalent load is calculated.

Voltage stability

Impulse loads influence the voltage stability of the system (see Section 1.1.2). The short-circuit power of the system at the point of common coupling is a measure of the voltage stability. The amplitude of the voltage variations as a function of the frequency per unit of time can be seen in Fig.1.1/6 of Section 1.1.2.

Special requirements for welding equipment

When electrical welding equipment is in operation, rhythmic voltage variations and harmonic loading are likely to have a considerable effect on the system.

In the case of spot-welding machines – the most commonly used welding equipment in the automotive industry (for welding thin sheet metal) – welding times of only 5 to 15 supply cycles (≙ 100 ms to 300 ms) are favored in order to minimize heat loss. In practice, this results in a relatively low duty ratio (DR, c.d.f.) of approximately 1% to 7% for single-spot-welding equipment, and up to 20 % for multiple-spot-welding equipment. The cycle time is the sum of the ON and OFF periods (Fig 5.6/2).

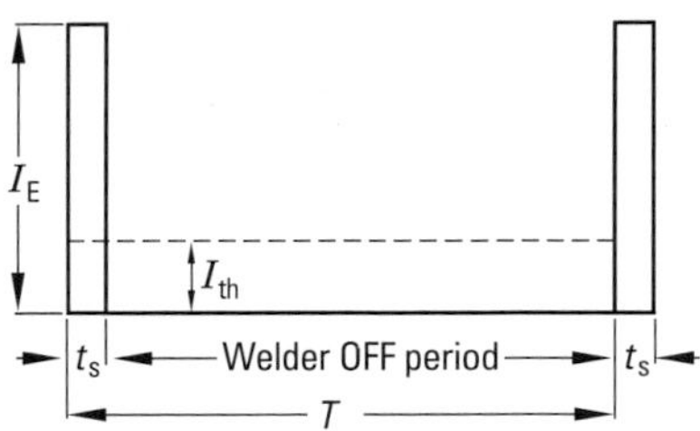

I_E R.M.S. welding current of the individual machine
I_{th} Thermally equivalent current
t_s ON period
T Cycle time

Fig. 5.6/2
Power system loading by a welding machine

Example A
No overlapping of welding machines

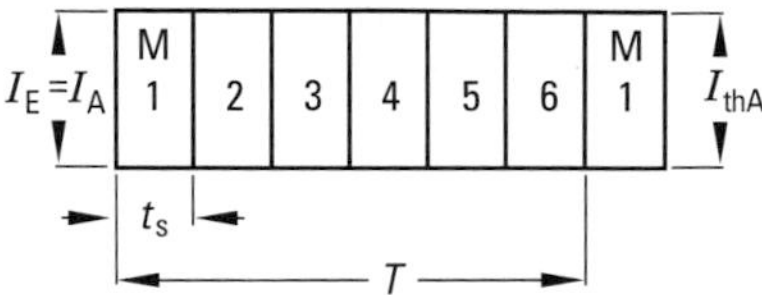

Example B
All machines operating simultaneously

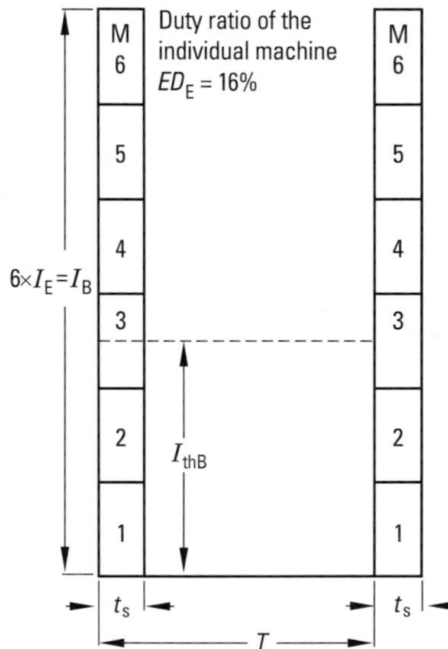

Fig. 5.6/3
Power system loading by six welding machines ($n = 6$)

Thermally equivalent current

When designing system components with relatively long thermal time constants, such as transformers, cables etc., the thermally equivalent current can be determined as a root-mean-square value from the impulse currents of the welding machines by using the following equation:

$$I_{th} = I_E \cdot \sqrt{\frac{t_s}{T}} = I_E \cdot \sqrt{\frac{ED}{100\%}}$$

If several welding machines are connected to the system, the magnitude of the impulse current and thus the thermally equivalent current, varies according to the length of time for which they are used.

If, for example, in a group of six individual machines the welding operations occur in sequence (Fig. 5.6/3, example A), the r.m.s. value of the thermally equivalent current is:

$$I_{thA} = I_E \cdot \sqrt{n \cdot \frac{ED_E}{100\%}} = I_A;$$

$$\text{since } \frac{n \cdot t_s}{T} = 1 \text{ and } I_E = I_A.$$

If, on the other hand, the welding periods of all the machines coincide (Fig. 5.6/3, example B), the r.m.s. value is given by:

$$I_{thB} = I_B \cdot \sqrt{\frac{ED_E}{100\%}} = \sqrt{n} \cdot I_E;$$

$$\text{since } I_B = n \cdot I_E \text{ and } ED_E = \frac{100\%}{n}$$

I_{th} Thermally equivalent current (A)
I_{thA} Thermally equivalent current with non-overlapping operation (A) (example A)
I_{thB} Thermally equivalent current with simultaneous operation of all machines (A) (example B)
I_E R.M.S. welding current of an individual machine (A)
I_A R.M.S. welding current of all machines (A) (example A)
I_B R.M.S. welding current of all machines (A) (example B)
n Number of machines
ED Duty ratio (%)
ED_E Duty ratio of an individual machine (%)
t_w Welding time (s)
T Cycle time (s)

Comparing these two extreme situations reveals that, in example A, the thermally equivalent current is equal to the r.m.s. welding current drawn by the individual machines, whereas in example B it assumes the value

$$\sqrt{n} \cdot I_E \; (= 2.45 \cdot I_E \text{ for } n = 6)$$

Thermal loading in practice

In practice, thermal loading will always occur between the extreme values for non-overlapping and fully overlapping welding. The fact that welding machines usually have very different rated outputs, ON/OFF periods and cycle times and are operated with different phase control (conduction interval per half-cycle <0.01 s) should also be taken into account.

Connection to the 400-V supply system

In the past, single-point-welding machines were always connected to two outer conductors of a 400-V supply system. The objective when planning a welding system is to distribute the welding machines as evenly as possible among the three outer conductors so that loading is as uniform as possible.

Connection to the 690-V supply system

In order to reduce costs, the automotive industry in a number of countries also uses higher voltages for supplying welding systems (e.g. 500 V in Germany). However, in order to comply with IEC 38, these are being replaced to an increasing extent with a voltage level of 690 V. In this case, the welding machines are usually connected between the outer conductor and the neutral conductor (N-conductor). The supply systems are designed as TN-C-S systems according to HD 384.3/IEC 60364-3/DIN VDE 0100-300.

Cross section of the N-conductor

While the 50-Hz fundamental components of the welding current are, to a large extent, balanced by several welding machines that are distributed evenly among the outer conductors (N-conductor current becomes very low), the 150-Hz harmonic current, which occurs with single-phase welding machines, is the aggregate of the individual values in the N-conductor. Under extremely unfavorable conditions, the neutral conductor can be subjected to a much higher thermal load than the outer conductors. It should, therefore, be designed with at least the same cross section as the outer conductors.

Voltage drop and transformer rating

In low-voltage systems loaded predominantly by welding machines, the thermally equivalent current is normally of little significance from the point of view of determining the rating of the infeed transformer. The transformer rating must, above all, be chosen on the basis of an acceptable voltage drop. With regard to this, a further important criterion is whether a permissible impulse load is single-phase (only in 690-V supply systems), two-phase or three-phase. With the same welding current, a two-phase impulse load produces approximately twice the voltage drop of a balanced three-phase impulse load of the same magnitude.

Voltage drop due to two-phase loads

Fig. 5.6/4 shows the voltage drops that result from two-phase impulse loads in a 1000-kVA transformer with $u_{kr} = 6\%$ and $u_{kr} = 4\%$. The characteristics are calculated for $\cos\,\varphi = 0.7$. For transformer ratings other than 1000 kVA, the voltage drop can be converted proportionally as an approximate value if the ratio R_T/X_T, which depends on the size of the transformer, is disregarded.

Voltage drop due to three-phase loads

The voltage drops for three-phase impulse loads Δu can be estimated as follows if the impedances of the high and low-voltage systems are disregarded:

$$\Delta u \approx n \cdot u' + \frac{1}{2} \cdot \frac{(n \cdot u'')^2}{100} \text{ in } \%$$

$u' = u_{Rr} \cdot \cos\varphi + u_{Xr} \cdot \sin\varphi$

$u'' = u_{Rr} \cdot \sin\varphi - u_{Xr} \cdot \cos\varphi$

u_{Rr} Rated value of the ohmic voltage drop (%)

u_{Xr} Rated value of the inductive voltage drop (%)

$$n = \frac{\textit{Apparent power} \quad S_M}{\textit{Rated output of transformer} \quad S_T}$$

If the welding machines are not interlocked (i.e. they can all be put into operation simultaneously), asymmetric three-phase impulse loads will occur as a result of the randomness of the overlap, even if the machines are distributed evenly among the three outer conductors.

If no values are available for the overlapping impulse current in a welding system, this can be estimated by using the probability formula in order to calculate the voltage drop for welding machines that are not interlocked.

To ensure satisfactory welding quality in spot-welding systems, a maximum voltage dip of $< 10\%$ is generally permissible. Only standard transformers which have a rated impedance voltage (u_{kr}) of 4% are used for supplying welding systems. The thermal uti-

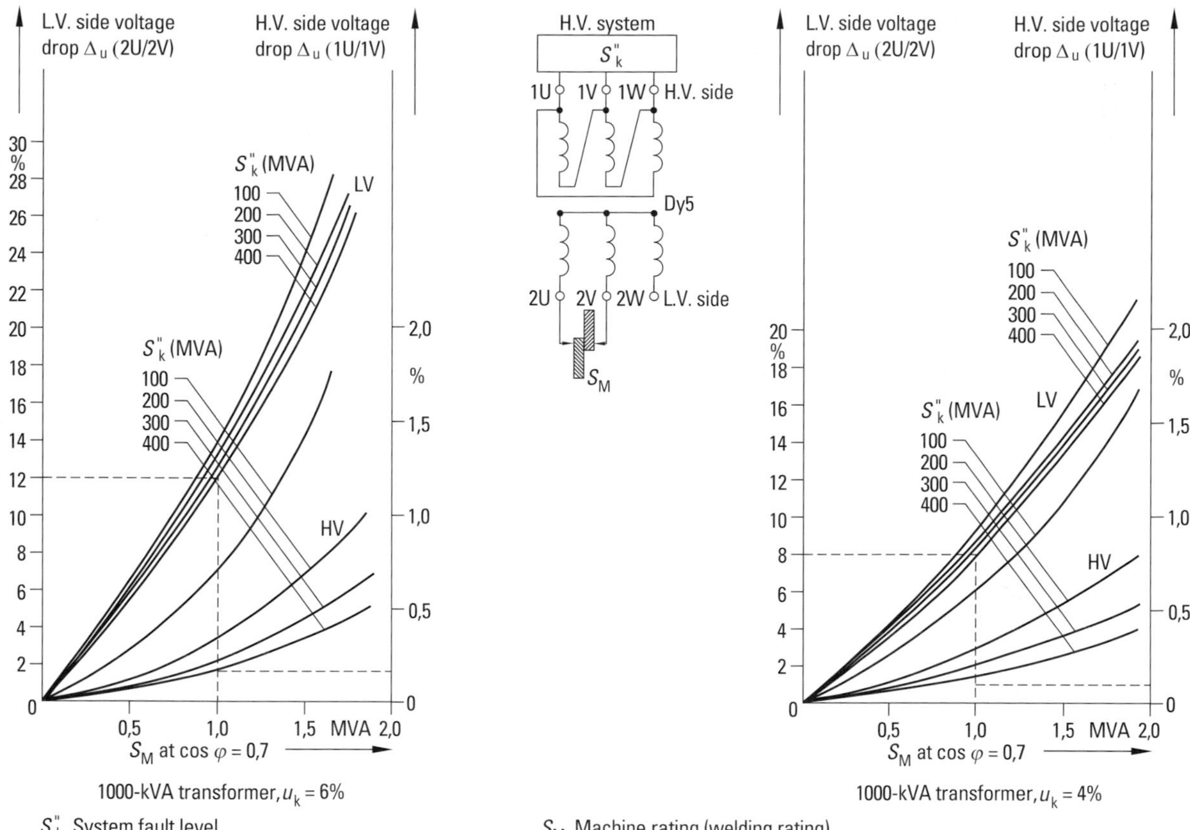

Fig. 5.6.4
Voltage drops at the high and low-voltage terminals of a supply transformer due to two-phase impulse loading by resistance welding machines in relation to the high-voltage fault level.

lization of the transformers is, therefore, usually below 40% of the rated output. In order to utilize the thermal capacity of the transformers more effectively, three-phase loads that are not voltage sensitive (e.g. asynchronous motors) are often connected, in addition to the welding machines.

Three-phase welding systems with voltage-dependant control are being used to an increasing extent in the automotive industry.

The system voltage fluctuations caused by overlapping current impulses are corrected within 2 – 4 ms.

Provision for high-power loads (> 200 kVA)

High-power loads are often encountered in the primary and chemical industries (e.g. for compressors, electric ovens etc.). They are not as common in the metal-working industry. These loads are preferably supplied from the high-voltage system either directly or via transformers.

Supply from the high-voltage system

Supply voltage 400/690V

A supply voltage of 400/690 V is being used to an increasing extent for industrial supply systems. When high-power loads are supplied, the higher system voltage enables more efficient solutions to be implemented as far as the supply-system components and transmission losses are concerned. High-voltage connections are, therefore, usually unnecessary.

Separate 230/400-V systems are, however, required for 230-V loads (e.g. lighting).

Fault level in the low-voltage system

Fault level and voltage stability

As already mentioned in Section 1.1, the voltage stability of the system increases in proportion to the system fault level. A very stable voltage is extremely important in industrial systems with irregular system operation (starting and switching off motors, impulse loading due to welding machines and presses, arc furnaces etc).

Fault level of the equipment

In the case of the transformers normally used in industrial systems with a rated u_{kr} of 6%, or 4% in welding systems, the short-circuit current is approximately 16 to 25 times the rated transformer current.

The rated short-circuit breaking capacity I_{cn} of the low-voltage circuit breakers manufactured nowadays amounts to more than 30 or 40 times the rated current. This breaking capacity is, therefore, never fully utilized for the transformer incoming-feeder disconnectors. It is only important for dimensioning the maximum permissible short-circuit current in the system when several transformers are operated in parallel (see Section 1.1) or for outgoing circuit-breakers with low rated current intensities. In the case of l.v. h.b.c. fuses, the rated short-circuit breaking capacity is, at present, over 100 kA.

Whereas the high-voltage system has to be dimensioned in accordance with the system fault level S_k'' specified by the public utility, in low-voltage systems, full advantage should be taken of the fault level as governed by the equipment used. Automatic disconnection for providing protection against electric shock caused by indirect contact, as specified in HD384.4.41/IEC 60364-4-41/DIN VDE 0100-410 (see Section 2.4), can, therefore, be implemented more easily even in extensive fused outgoing circuits.

Low-voltage system configuration

System configuration – features

A distinction is made between the following types of system (see Section 1.1.3):

Radial system

- ▷ Simple system design, transparent and easy to manage.
- ▷ No special requirements with regard to security of supply, voltage stability or flexibility.

Radial system with part-load or full-load reserve

- ▷ High security of supply through additional switched reserve capacity, usually corresponding to a fairly large proportion of the load, depending on the design of the transformers and connecting cables (see Section 1.1.3, Fig 1.1/11, System with transfer reserve).

Interconnected radial systems

- ▷ More effective utilization of investment costs through continuous parallel operation of transformers in an interconnected system.
- ▷ Redundant supply [(n–1)-principle] is maintained if a transformer or a trunk cable connection fails between the transformer stations (see Section 1.1.3, Fig. 1.1/11, System with instantaneous reserve).

Advantages of interconnected radial systems

As a result of the stringent requirements with regard to security of supply and voltage stability in large industrial installations, the only system that represents a viable option, at present, is the interconnected radial system. The costs for radial systems in closed interconnected systems and radial systems with transfer reserve are approximately the same. Interconnected radial systems have the following distinct advantages:

- ▷ better voltage stability due to the higher fault level,
- ▷ the regulations for preventing electric shock by means of automatic disconnection, as specified in HD384.4.41/IEC 60364-4-41/DIN VDE 0100-410, and for the selective behavior of power system protection equipment, particularly in the case of fuses with high rated currents and long cables, can be adhered to more easily,
- ▷ more balanced utilization of the transformers, and therefore lower transmission losses,
- ▷ lower system simultaneity factor as a result of a larger integrated system, and hence lower transformer ratings.

Closed meshed systems

If still greater security of supply to the loads is required, a closed meshed system, as shown in Fig. 5.6/5, must be used.

Advantages of meshed systems

Additional outlay (which depends on the mesh size) is required for this system type. In comparison to the interconnected radial system, it offers further advantages:

▷ greater flexibility if loads in the system are redistributed,
▷ lower voltage drops as far as the ultimate consumers,
▷ instantaneous reserve capacity as far as the sub-distribution boards or important loads.

The meshed system in its conventional form, as shown in Fig 5.6/5, with node spaces corresponding to the loads per unit area, and adapted to the dimensions of the production area and its unit spacing, is seldom used. This is due to the complex structure of the system which becomes apparent when fuses in the system fail or if production methods change (e.g. production lines instead of connecting individual loads).

Connected radial systems

Interconnected radial systems are nowadays the preferred solution.

Only radial systems with and without transfer reserve may be designed as TN-S systems (separate PE and N-conductors). Transfer devices then have to have 4 poles. Closed systems can only be constructed as TN-C systems with combined equipment grounding and neutral conductors (PEN).

Determining efficient transformer ratings

The chosen transfer rating has a considerable influence on the overall cost effectiveness of the supply concept for an industrial plant. The following information is based on the most commonly used interconnected radial systems. The example given here is a large production area with relatively homogeneous load distribution in a production hall.

Interconnected systems: total permissible transformer rating

When several transformers are operated in parallel via a low-voltage system, the permissible short-circuit current limit is generally between 3 and 4.5 MVA. If the supply is to be maintained with n–1 transformers, the various options available are shown in Table 1.1/5, Section 1.1.3, if the influence of asynchronous motors and resistance in the system is not taken into account. The percentage utilization values of the transformers under normal operating conditions listed in this table show that the ratio of installed transformer capacity in the system S_r to the secured system capacity S_{n-1} becomes less favorable as the size of the transformer increases.

Effect of transformer size on cable outlay

The transformer rating also has a considerable effect on the cable outlay for the low-voltage system connected on the load side. With the exception of the meshed system, the transformer load-center substations supply the sub-distribution boards via a radial

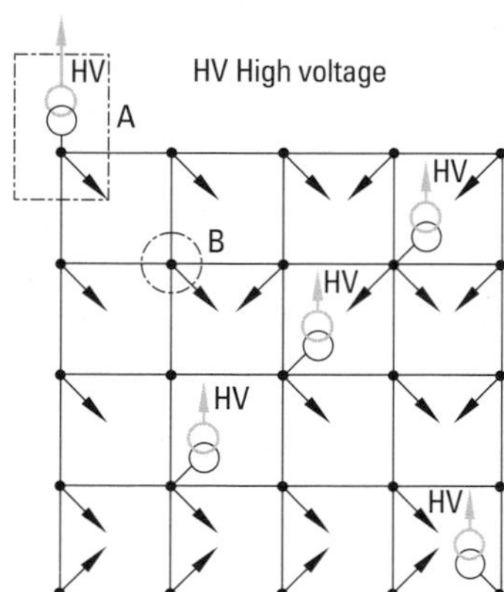

Closed system:
Instantaneous reserve up to sub-distribution boards
Low voltage drops

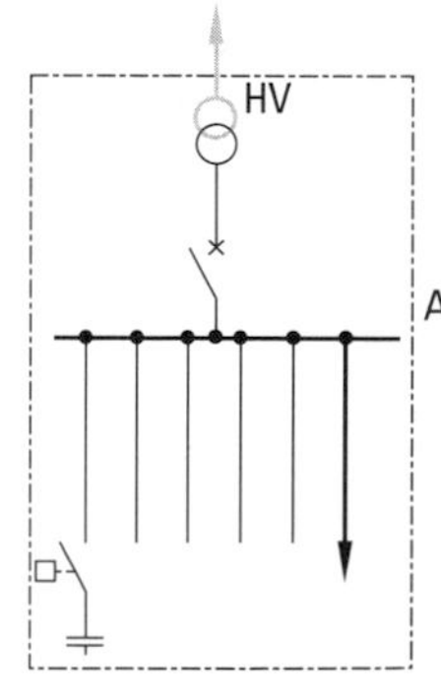

Transformer load-center substation

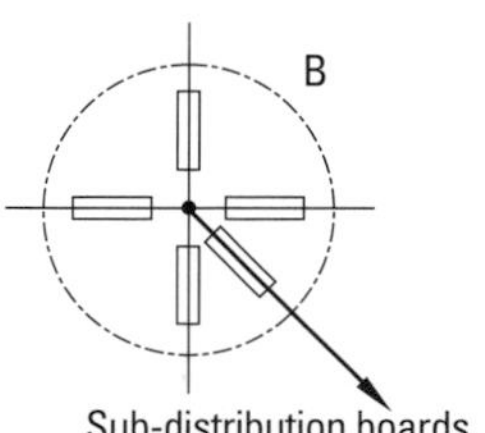

Nodal point of a closed system

Fig. 5.6/5 Closed system

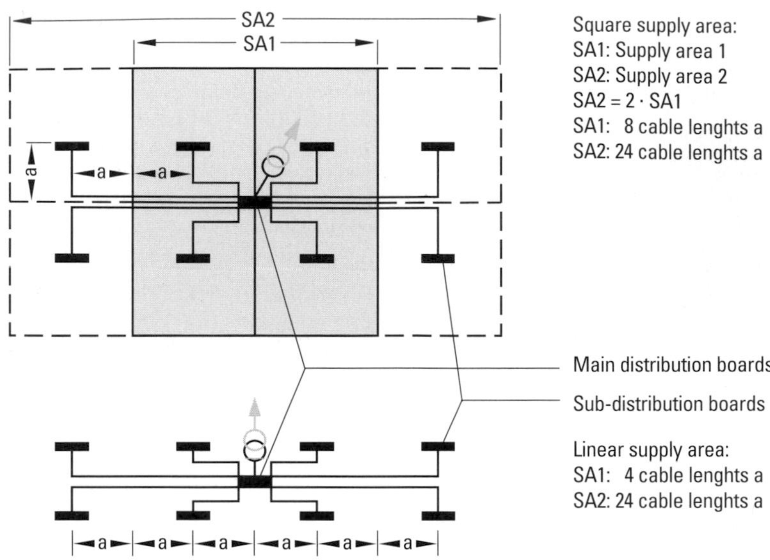

Fig. 5.6/6 Cable outlay with different transformer supply areas

system. In addition, cable connections are necessary between the stations for backup purposes in the event of a transformer failure.

If a given load per unit area is assumed, the supply area of a station increases as the rated power of the transformer increases. Fig. 5.6/6 shows the cable outlay involved in connecting 4 or 8 sub-distribution boards with the same connected load in square and linear supply areas if the transformer rating (and therefore the supply area) is doubled.

Supply area, square and linear

From the geometrical relationships assumed in Fig. 5.6/6, it can be seen that when the transformer rating and supply area double, and the load per unit area remains the same, the cable outlay, expressed by the factor SA2/(2 · SA1), is

$$\frac{24}{2 \cdot 8} = 1.5$$

for the square supply area, and

$$\frac{16}{2 \cdot 4} = 2$$

for the linear supply area.

In practice, the supply area will usually be somewhere between these two types (and values). It was assumed here that the transformer was set up in the load center of the respective supply area. If the transformer is set up at the edge of the supply area, the factor continues to increase.

The same factors apply if the transformer rating is doubled for the cable connections between the individual transformer substations in the interconnected system (Fig. 5.6/7).

Variation of total system cost

If the total system costs are considered, the variations in the total costs are determined in relation to the permissible number of transformers in a supply area with an average load per unit area of 300 to 350 VA/m^2 and 400 V rated voltage (Fig. 5.6/8).

The costs components (DM/kVA) for the high-voltage terminal (panel and cable) and the transformer load-center substation tend to fall as the substation rating increases, while the cost component for the low-voltage system increases by the factor 1.5 to 2 when the size of the substation doubles (see Fig. 5.6/6 and 5.6/7).

Variation of cost for the individual system components

The variation of total cost is represented by curves a and b in Fig. 5.6/8. Curve a shows the variation of total cost referred to the total installed substation capacity for interconnected radial systems. If the total costs are referred to the security of supply if a transformer feeder failure occurs [(n –1)-principle], the cost variation is shown by curve b.

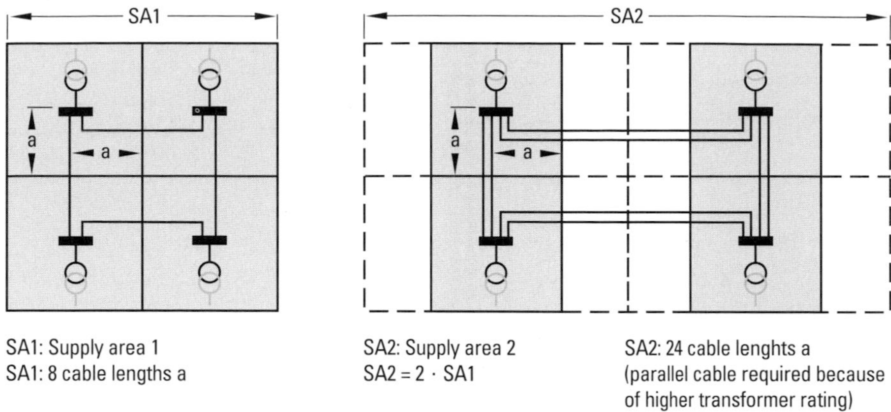

Fig. 5.6/7 Outlay for connecting cables between substations

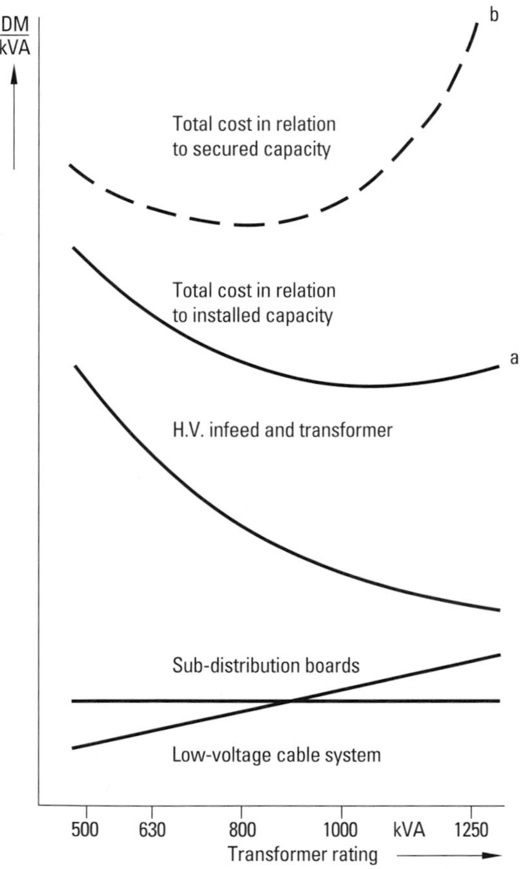

Fig. 5.6/8
Variations in investment costs of system structure with transformer size for an average load per unit area of 300 to 350 VA/m^2

In the example under consideration, the cost is at a minimum with a transformer rating of 800 kVA. Even if depreciation and the cost of losses are included in the equation, this value is approximately the same.

Variation for other loads per unit area and system voltages

With lower and medium loads per unit area at higher rated system voltages, the cost minimum shifts (as a result of the lower cost component for the low-voltage cable system) to lower transformer ratings, and with lower rated system voltages or higher loads per unit area, to higher transformer ratings.

System protection

Purpose

The purpose of system protection in low-voltage industrial systems is to protect the equipment against impermissible loads caused by overload and short-circuit currents and to disconnect defective equipment selectively.

L.v.h.b.c. fuses

Selectivity in radial systems

In low-voltage cable systems, l.v.h.b.c. fuses are usually used for this task. Information on choosing the rated current of the fuses and on selectivity in radial systems (series-connected fuses in series circuits) can be found in Section 1.3.3.

Selectivity in meshed systems

When l.v.h.b.c. fuses are connected in parallel, e.g. in low-voltage meshed systems, selectivity is generally ensured if, with the same fuse type and rated current, the highest short-circuit current of a branch is not more than 60% – 90% (Fig. 5.6/9) of the sum of all the short-circuit currents ($\Sigma I_{\text{k part}}$).

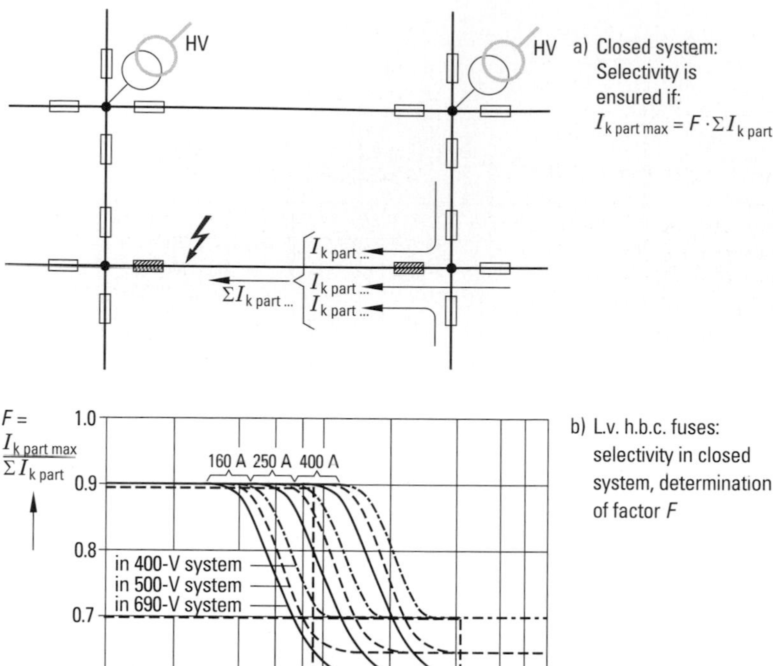

Fig. 5.6/9 Selectivity in closed systems

For 160, 250 and 400-A l.v. h.b.c. fuses, the diagram in Fig. 5.6/9 can be referred to for precise information on selective behavior in meshed systems.

Example

$$\frac{I_{\text{k part max}}}{\Sigma I_{\text{k part}}} = 0.9 \rightarrow \begin{array}{ll} I_{\text{n fuse}} & \leq 400\ \text{A} \\ U_{\text{n system}} & \leq 500\ \text{V} \\ \Sigma\, I_{\text{k part}} & < 10\ \text{kA} \end{array}$$

$$\frac{I_{\text{k part max}}}{\Sigma I_{\text{k part}}} = 0.7 \rightarrow \begin{array}{ll} I_{\text{n fuse}} & \leq 400\ \text{A} \\ U_{\text{n system}} & \leq 400\ \text{V} \\ \Sigma\, I_{\text{k part}} & > 40\ \text{kA} \end{array}$$

Circuit-breakers in the ring cable connections

In interconnected radial systems (see Section 1.1.3, Fig. 1.1/11), circuit-breakers are being used to an increasing extent in the cables that link transformer substations. Selective disconnection of faults or the ring cables is, however, possible if the circuit-breakers are equipped with short-time-delay overcurrent releases that can be switched from the usual definite-time delay to an inverse-time delay (e.g. I^2t characteristic).

If circuit-breakers with normal short-time-delay (not delayed) overcurrent releases are used in ring cable connections, the system is divided up among the individual radial systems in the event of ring cable faults. This affects troubleshooting and the quality of supply (lower system fault level) and must be taken into account when the required transformer incoming supply per subsystem (forecast for the peak load of the subsystem) is determined (see Sections 1.1.2 and 1.1.3).

Transformer protection

Transformers are usually protected or monitored for overcurrents and internal short circuits.

The rated values specified in the relevant standards and guidelines for constructing transformers usually apply for continuous operation at defined limit values for the temperature of the coolant and for a normal service life. When industrial buildings are supplied, however, the time-related loading of the transformers changes in relation to the production type and cycle. In addition, the coolant temperature often deviates from the values specified in the regulations.

Permitted loading cycle for oil-immersed transformers (DIN VDE 0536)

The constructional requirements (DIN VDE 0536 in Germany) specify the permissible loads for oil-immersed transformers for normal daily use of life for load cycles (previous load/overload for limited period), for continuous operation at different coolant temperatures and normal use of life, and for overloading with increased use of life for emergency loading.

Detection of thermal overload

Overcurrents occur in the range of 1 to 1.5 times the rated transformer current, and can be detected by means of the thermal release of the circuit-breaker on the low-voltage side of the transformer. Thermal protectors and thermally-delayed maximum-indicating ammeters, incorporating bimetallic measuring elements and non-return pointers, are also used.

Protection against internal faults and short circuits

Short-circuit fault currents are reliably detected by means of h.v. h.b.c. fuses in conjunction with load transfer switches or switch-disconnectors, or by means of circuit-breakers with definite-time overcurrent-time or inverse-time overcurrent protective relays, and are interrupted on the high-voltage side. Internal transformer faults with a low current intensity are not detected by either of the protective devices.

If the internal transformer fault develops into an internal short circuit, this is interrupted by the h.v. h.b.c. fuses.

Buchholz relays

In the event of a two or three-pole short circuit evolving from an internal transformer fault, circuit-breakers cannot, on account of their relatively long operating times (minimum operating time > 70 ms), usually prevent the transformer tank from bursting. If circuit-breakers are used on the high-voltage side of the transformer sub-stations, Buchholz relays are recommended for liquid-cooled transformers.

Monitoring resin-cast transformers

GEAFOL resin-cast transformers, on the other hand, are generally equipped with temperature-monitoring systems and the appropriate tripping unit.

These perform the following functions:

- ▷ output temperature warnings,
- ▷ perform temperature tripping,
- ▷ activate the ventilation system for separately-ventilated transformers with forced-air cooling (AF).

Effect of reserve capacity on critical faults

In a meshed system for industrial buildings, a suitable basis for project planning is the provision of reserve capacity for critical faults, e.g. (n–1)-principle in the event of a transformer failure.

The way in which the transformers are cooled results in different operating conditions:

Permissible overload values

- ▷ if liquid-cooled transformers or dry-type transformers with natural air-to-air cooling (AN) are used, they should not be loaded (under fault conditions) to values that exceed their respective ratings or the overload values that are permitted for a limited period of time (see DIN VDE 0536 in Germany).
  Under normal operating conditions, average permissible loading is, therefore:
  $$\frac{n-1}{n} \cdot S_r,$$
- ▷ if separately-cooled resin-cast transformers are used, the system should be planned in such a way that loading under fault-free conditions does not exceed the respective transformer rating and that the capacity of the transformer is utilized by employing forced-air cooling (AF) in the event of a substation failure. In the case of 8FA transformer load-center substations with GEAFOL transformers, this capacity can be as much as 140% of the rated capacity.

These operating conditions have to be taken into account when the thermal releases of the transformer incoming-feeder disconnectors are set on the low-voltage side in order to avoid false tripping and system collapses if a fault that could normally be rectified occurs.

Power-factor correction

Low-voltage-side power-factor correction

Depending on the nature of the loads involved, for example in the metal-working industry (with a large number of motors), allowances have to be made for a relatively high lagging reactive-power demand. Power-factor correction (see Section 1.6) is, with a few exceptions, applied almost invariably on the low-voltage side, since it is preferable – from a technical and cost-related point of view – to compensate for reactive power as closely as possible to the loads themselves.

This means that:

▷ transmission losses in cables and electric lines are reduced, as are losses in the transformers as of the correction point ($P_v \approx I^2 R$),

▷ investment costs for this equipment are lower because the rated capacity of the equipment can usually be reduced,

▷ energy costs are reduced.

Individual correction

Power-factor correction of individual loads is hardly ever used nowadays, because of the large number of loads with different ratings and ON periods. It is also very expensive, since each load has to be compensated for its maximum reactive-power demand.

It is not possible to adjust the system or reference power factor sufficiently to provide a technically viable and cost-effective solution.

Individual correction is used in industrial buildings, for lighting systems and, more rarely, for compensating the no-load reactive power in motors.

Group correction with automatic controllers

The system or reference power factor (cos φ approx. 0.95 to 1) can be adjusted satisfactorily by means of group correction in the transformer load-center substations with automatic power-factor correction units (see Section 1.6). The reactive power flow in low-voltage systems is, as a result, restricted to short cable routes. This applies in particular to interconnected radial systems with transformer load-center substations that have been set up locally in the load center of a supply area.

Harmonic interference

The increase in the use of regulated drives in the production industry has lead to a rise in the levels of harmonic interference in industrial systems.

When power capacitors are used to compensate for this, parallel resonance phenomena can occur in the system (anti-resonant circuit capacitors/power transformers). If the natural frequency of this oscillating circuit coincides with the frequency of a current harmonic, the harmonic interference is amplified and the equipment is subjected to an increase in thermal stress. In individual cases, this can lead to overloading or cause overcurrent protective devices to be triggered.

Reactor-connected capacitor control units

By using reactor-connected capacitor control units (series resonant circuit with $f_r < f_n$ of the smallest line current harmonic), amplified points of resonance are avoided in the system, and, depending on the level of imbalance, some of the harmonic current is filtered (see Section 1.6).

Filter circuits

Filter circuits must be used if a high level of harmonics is generated (see Section 1.6).

Blocking circuits

If ripple control systems are used in the public utility system, interference can occur in the system if filter circuits with center frequencies close to the ripple control frequency are used. Blocking circuits must be installed in these cases.

Cables

Preferred cables

PROTODUR cables are normally used for low-voltage systems in industrial buildings.

The use of these cables enables the locations at which the switchboards, distribution boards and loads are installed to be easily adapted to the structural factors of a manufacturing area and facilitates rapid alterations in the event of production-related changes or extensions.

Cable installation

Grouped cables, or cables with conductors of large cross-sectional areas, are best laid in cable ducts, in duct blocks or on cable racks (see Section 5.1.4.3).

Busbar trunking systems and conductor bar systems

Instead of cable systems with sub-distribution boards, busbar trunking systems can also be used to connect loads and carry the balancing currents between the transformers (see Section 1.4.4). Depending on the loading capacity and the application, a distinction is made between busbar trunking systems (≤800 A, approx.) and busway systems (≥1000 A, approx.) (see Section 1.4.4).

Busbar trunking systems ≤ 800 A

Busbar trunking systems are used instead of sub-distribution boards for connecting loads in manufacturing areas. This method of connection is particularly advantageous when, for example, machines are installed in rows. With a busbar trunking system running along the length of the row, the electrical loads can be connected to the supply by means of plug connectors.

Busway systems ≥1000 A

Busway systems are used in industrial buildings to convey power between the transformer load-center stations, in some cases to connect large concentrated loads, and to supply the busbar trunking systems or sub-distribution boards. Connecting loads that draw less than 200 A directly to the conductor bar system is usually too complex and too expensive.

Complete systems

Using both systems, it is possible to install supply systems that are spatially tailored to the manufacturing premises. Basic systems of this type, which are constructed on a grid corresponding to the layout of the building (Fig. 5.6/10) and are generously dimensioned, are largely independent of the load distribution within the supply areas at any particular time.

Supply systems with busbar trunking systems can, therefore, be erected at an early stage, often before the exact installation plans for the machines and terminals have been drawn up.

If the production facilities are modified or replaced, this usually has no or little effect on the supply system. These advantages, however, require a higher level of investment, at least when the system is expanded for the first time.

Transformer load-center substation

As in the case of a cable system, transformer load-center substations should be set up in the load center of their supply area in order to reduce transmission losses. If the load centers are moved within a production area, the transformer load-center substations should, if possible, be moved together with the production facilities.

Integrating production automation and building control system with *instabus EIB*

Automation units and communications systems

Nowadays, it is hard to imagine industrial buildings without production or process automation systems. In order to control and monitor the sequence of processes that are becoming increasingly complex in modern production facilities, powerful automation units and communications systems are required. Siemens responded to this challenge by developing its SIMATIC programmable controllers and the PROFIBUS (**PRO**cess **Fi**eld **BUS**) communications system (conforming to the international standard EN 50 170 and DIN 19245), which are now used in many countries throughout the world.

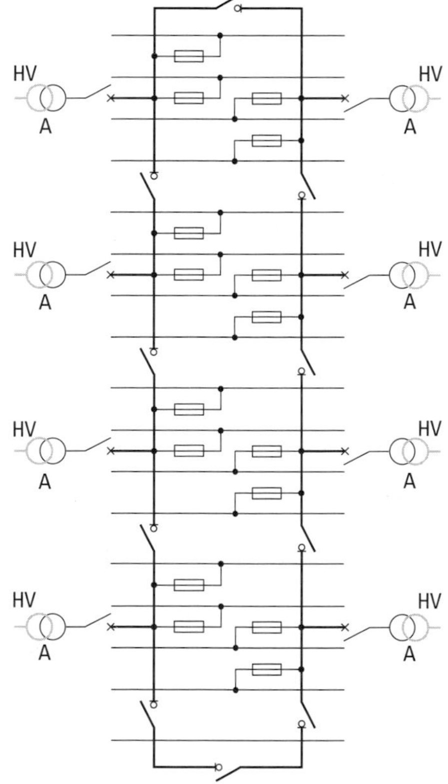

— Busway system
— Busbar trunking system

A Transformer load-center substation with transformer 1000 kVA, 400 V $u_k = 6\%$

HV High voltage

Fig. 5.6/10
Example of a low-voltage distribution system with busbar trunking systems. A maximum of four transformers can be connected together.

Building system applications

Integrating building system applications and functions in process automation on the basis of the *instabus EIB* (see Section 4.4) allows production environments to be adapted to production processes.

It is, therefore, possible to integrate demand-related, process-specific control systems for lighting, heating, air conditioning and ventilation in one single control concept.

The link between PROFIBUS and *instabus EIB* is established via gateways. The result of this is integrated project planning, a common database for all the systems involved, and numerous other advantages during on-line operation. Further information on building system engineering can be found in Section 4.4.

Lighting

Lighting systems

In the past, rows of luminaries incorporating fluorescent lamps were used in large production shops with a luminaire mounting height of up to 8 m (but usually ≥6 m). High-bay reflector luminaries fitted with metal-vapor lamps were used for greater mounting heights.

The increasing level of automation in production means that employees in industrial buildings require a level of lighting that is similar to that in offices.

High-bay area luminaires

Modern high-bay area luminaries with fluorescent lamps provide the quality of light that is required as well as:

- ▷ high level of color vision and visual contrast,
- ▷ high level of illumination with optimum glare reduction,
- ▷ ideal distribution of brightness within the room,
- ▷ good modeling characteristics.

Up to a mounting height of approximately 14 m, the specific power requirements $[W/(m^2 \cdot lx)]$ of modern high-bay area luminaires with fluorescent lamps is nowadays lower than that of narrow-angle high-bay reflector luminaries with HQL lamps of the same illuminance.

Regulating the level of illuminance

Using electronic control gear for dimming fluorescent lamps (ECG Dynamic), together with light sensors (that detect daylight and regulate the required illuminance) and the *instabus EIB*, it is possible to further reduce lighting costs in specific production areas, depending on the level of production, utilization and daylight.

DIN 5035 Part 2

DIN 5035 Part 2 specifies the minimum guide values for rated illuminance, luminous color, level of color rendering properties, and quality class of direct glare reduction, on which the planning and evaluation of lighting systems for indoor production and office areas are based.

Emergency lighting to DIN 5035 Part 5

In the case of emergency lighting for production and office areas, DIN 5035 Part 5 distinguishes between stand-by lighting and safety lighting.

Stand-by lighting

The stand-by-lighting system performs the tasks of the normal artificial lighting system for a limited period of time. Installation depends on the requirements of the individual company and is, therefore, the responsibility of the owner.

Safety lighting

The safety lighting system is necessary for general safety and accident prevention. According to DIN 5035 Part 5, it is divided into:

- ▷ Safety lighting for escape routes,
- ▷ Safety lighting for workplaces that are particularly hazardous.

Illuminance for safety lighting

The local minimum illuminance value at the end of the service life of the lamp is as follows:

- ▷ for escape routes:
  1 lx (lux) measured in a plane 0.2 m above floor level along the axis of the escape route,
- ▷ for workplaces with special hazards:
  10% of the rated illuminance required for the workplace, as specified in DIN 5035 Part 2, but not less than 15 lx.

Requirements for equipment

The electrotechnical requirements for safety lighting equipment are specified in DIN VDE 0108: They apply for:

- ▷ stand-by power sources,
- ▷ system type and protective measures,
- ▷ distribution boards,
- ▷ cables and line systems,
- ▷ current-using equipment.

Further guidelines and regulations

If further official guidelines and regulations apply for additional safety devices, the requirements of the highest authority must, in each case, be adhered to. In Germany, these guidelines and regulations include:

▷ ordinances in workplaces,
▷ guidelines pertaining to workplaces,
▷ regulations for the prevention of accidents,
▷ relevant building regulations,
▷ decrees pertaining to individual commercial and industrial sectors (e.g. Trade Supervisory Authority).

5.6.2 Exhibition halls

Exhibition halls are used to display a variety of objects and are frequented by a large number of visitors from outside the locality. The very diverse requirements encountered here make particular demands on the design of the electrical supply system and on security of supply. Exhibition halls are usually built without provision for natural lighting.

The load centers are:

Specific power demand

▷ The general lighting
with a load per unit area of roughly 10 W/m^2 for an illuminance of approximately 300 lx,
▷ The ventilation and extraction system
with a connected load per unit volume of 5 kW/1000 m^3 of room volume,
▷ The stand supplies
with a load per unit area of 50 to 100 W/m^2 with two-dimensional loading (individual objects may, in some cases, require many times this power).

General requirements

Fluorescent lamps, metal-vapor lamps or a mixture of the two are used for general lighting in the halls, depending on the ceiling height. High-bay area luminaires with fluorescent lamps can be used up to a mounting height of approximately 14 m. In comparison to high-bay reflector luminaires with HQL lamps, they provide a better quality of light and have a lower connected load per unit volume for the same illuminance per m^2. Extremely cost-effective solutions can be implemented using electronic control gear. One additional advantage offered by high-bay area luminaires with fluorescent lamps is that they switch on again immediately if momentary voltage dips occur. The different levels of illuminance required for erecting and dismantling stands, as well as for cleaning and exhibition purposes etc. can be catered for by selective switching (e.g. quarter-half-full switching).

Flexibility and adaptability with building control system

Building control system with *instabus EIB* (see Section 4.4) is particularly suitable for use in exhibition halls, since it provides the degree of flexibility required of the electrical wiring system with regard to adaptability (when erecting and dismantling stands, for cleaning and exhibition purposes etc.), and also with regard to allocating space to stand "tenants" (who change from exhibition to exhibition).

Using building control system with *instabus EIB*, it is possible, for example, to fulfil the different lighting, heating and air conditioning requirements for the various operating phases at the touch of a button. No reinstallation is necessary when stand space is reassigned after an exhibition. Reallocation is carried out by changing the group addresses of the sensors and actuators via the *EIB* Tool Software (ETS).

Visualization

In comparison with Fig. 4.4/14, Fig. 4.4/13 shows the different utilization scenarios, using offices as an example. Visualization software (see Section 4.4.5.2) also makes it possible to display and evaluate all the operating states and values, including fault reports and consumption statistics, at a central location.

Safety lighting

As a result of the large number of people visiting exhibition halls and the fact that many of these visitors are not familiar with the locality, electrical installations in exhibition halls, e.g. for hosting trade fares and other events, are subject to the regulations governing required safety devices according to DIN VDE 0108 (this is currently being harmonized and integrated in HD 384). These regulations specify that exhibition halls, whose rooms (either singly or collectively) have a total useable floor space of more than 2000 m^2, must be equipped with an emergency lighting system. The escape routes in and between the halls must be lit in such a way as to provide an illuminance of at least 1 lx, measured at a height of 20 cm above floor level.

Switch-over time of stand-by power supply

Escape route signs, for example over exits and emergency exits, must always be illuminated (maintained safety lighting). In the event of the normal power supply failing, the switch-over time to the stand-by power supply must be < 1 s. This means that normal stand-by generators alone (with a startup time of 8 to 12 seconds) are not adequate to maintain the power supply.

Safety power supply

An safety power supply is also necessary for the following safety devices:

- ▷ systems providing water for fire fighting,
- ▷ elevators for fire fighters,
- ▷ alarm systems,
- ▷ systems for extracting smoke and heat.

Information on planning and designing the required installations and equipment can be found in Sections 1.8 and 5.9.9 or in DIN VDE 0108.

Supplies to stands

The stands can be supplied by either radial or meshed networks (see Section 1.1.2 and 1.1.3), normally at low voltage (230/400 V AC), with low-voltage main distribution boards that have been set up either centrally or locally with the transformers.

Due to the lower costs involved, transformer substations that are set up locally are almost always used in large exhibition halls.

The reasons for this include:

- ▷ Reduction of the costs associated with transmission losses in the low-voltage transmission systems.
- ▷ Low investment costs for low-voltage transmission systems such as cables or busbars.

The stand space can be utilized effectively in two different ways:

- ▷ small exhibition halls: cables with service distribution boards laid in the false floor, or busbar trunking systems.
- ▷ larger exhibition halls: accessible service ducts that contain the busbar system for the low-voltage-side connection of individual stands, cables, electric lines and distribution boards for supplying the media used, as well as measuring devices and meters.

The current-using equipment in the stands is, in both cases, connected via suitable floor outlets (similar to Fig. 5.6/11 for false floors).

Both connection types facilitate subsequent installation of supply lines. They also make it easier to adapt the cabling and wiring to the supply requirements, e.g. for current-using equipment with very high connected loads.

System according to type of terminal connection

TN-S system

As already mentioned in Section 1.1.2, it is particularly important to make sure that electrical installations in exhibition halls are electromagnetically compatible. The electrical supply system should always be designed as a TN-S system with separate PE conductors (equipment grounding conductors) and N-conductors (neutral conductors). Switching devices between different supply sources (e.g. normal supply system and stand-by power supply) should always be 4 pole.

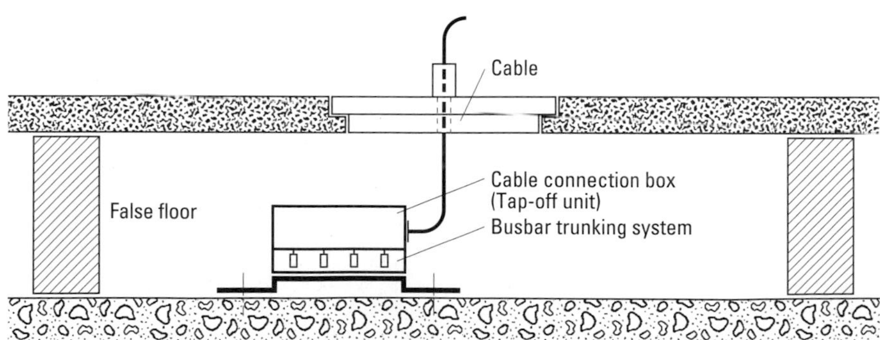

Fig. 5.6/11 Busbar trunking system accommodated in a false floor

5.7 Garages

Electrical systems are indispensable for garages, particularly in those over a certain size. In addition to the equipment that ensures that the garage operates correctly, these electrical systems also include lighting and ventilation installations. Medium-sized and large garages, in particular, also require safety equipment such as safety lighting, CO warning systems, and fire alarms to ensure safe and reliable operation.

Regulations, standards

In Germany, regulations defined by the responsible authorities contain regional specifications which lay down the minimum requirements for basic and safety equipment used in garages. The requirements for electrical systems in garages are explained below on the basis of the German standard garage regulation (1993) and the standards prEN 12464 and DIN VDE 0108-6.

Useful area of garage

In line with the German standard garage regulation (GarVO), garages are categorized according to their useful area as follows:

- ▷ small garages up to 100 m^2,
- ▷ medium-sized garages over 100 m^2 to 1000 m^2,
- ▷ large garages over 1000 m^2.

The useful area is the sum of all adjoining areas formed by the garage parking lots and driveways (excluding parking lots on roofs).

Small garages

The standard garage regulation does not specify any requirements for small garages. The Bavarian garage regulation, however, contains the following general requirements:

- ▷ garages must have electrically-powered lighting only,
- ▷ the electrical systems must be installed, modified, maintained, and operated in accordance with the recognized electrotechnical guidelines.

Medium-sized and large garages

Lighting

According to the standard garage regulation (GarVO), medium-sized and large garages must be equipped with a general lighting system, which must have two switching levels.

Minimum illuminance

With the first lighting level, all useful areas and driveways must be illuminated to a minimum illuminance of 1 lux, and with the second level, to a minimum illuminance of 20 lux.

The required signs indicating the way to the stairs and garage exits must be illuminated in large garages.

Average illuminance

The European draft standard prEN 12 464 for applied lighting technology recommends the following average illuminance values for medium-sized and large garages:

- ▷ for entrance and exit ways
  - – day-time 300 lx
  - – night-time 75 lx,
- ▷ for driveways inside building 75 lx,
- ▷ for parking lots 75 lx.

Apart from illuminance, great importance is also attached to the uniformity of lighting in order to give garage users a suitable feeling of security; dark areas must be avoided at all costs.

Building control systems with *instabus EIB*

A constant-lighting regulator with *instabus EIB*, for example, fully satisfies these requirements and also reduces the amount of power needed for the lighting system.

Furthermore, all of the installations described below (including visualization) can be integrated with *instabus EIB* to allow all of the building services systems to be monitored and controlled from a central location.

Safety lighting

According to the standard garage regulation, large closed garages must also be equipped with safety lighting (conforming to HD 384.5.56 / IEC 60364-5-560 / DIN VDE 0108-6) to illuminate the escape routes. The minimum illuminance along the center line of the escape routes must be 1 lux.

Ventilation

According to the standard garage regulation, medium-sized and large closed garages, except those with a low volume of entry and exit traffic (e.g. garages belonging to residential buildings), must be equipped with a mechanical ventilation system which consists of at least one mechanical exhaust air extraction unit. A mechanical air intake unit must be provided if the garage does not have adequate air intake openings.

The ventilation systems must be dimensioned so that the average half-hourly CO value does not exceed 100 ppm ($= 100\ cm^3/m^3$).

In the case of the exhaust air extraction units, at least two fans of equal size, which generate the required total volumetric flow when operated simultaneously, must be installed in each ventilation system. If the fans are not to be operated simultaneously, it is important to ensure that the second fan is activated automatically if the first fan fails.

Separate circuits, starting at the distribution board for the ventilation system, must be provided for the power supply of the air intake and air extraction fans.

CO warning system

CO warning systems, which issue visual and acoustic signals to inform garage users of hazardous CO levels, are required in large closed garages, except in those with a low volume of entry and exit traffic.

Safety power supply

Apart from the general power supply, an independent stand-by power supply (safety power supply conforming to HD 384.5.56 / IEC 60364-5-560 / DIN VDE 0108-6) must be provided for the CO warning system and the indicator and warning devices.

Fire alarm system

According to the standard garage regulation, fire alarm systems must be provided in medium-sized and large garages if these adjoin other structural installations or rooms for which fire alarm systems are required.

Socket-outlets

In medium-sized and large closed garages, socket-outlets in the vicinity of garage parking lots and driveways must have dedicated electric circuits. They must not be connected to the lighting circuits.

Power sources for safety purposes

Since a switchover time of 15 seconds is adequate for the safety lighting in garages, central batteries (see Section 1.8.1) and, in particular, generating sets which also allow the ventilation systems to continue operating in the event of a power failure can be used as power sources for safety purposes.

Installation

According to DIN VDE 0108-6, garage parking lots and the associated driveways are regarded as damp and wet locations. According to DIN VDE 0100-737, devices and materials in such locations must have at least degree of protection IP 21 (drip proof to EN 60529 / IEC 60529 / DIN VDE 0470-1).

If there is a risk of mechanical damage, lamps should be protected by a protective device (troffer, conduit, etc.). Installation devices, such as switches and socket-outlets, should usually be protected mechanically by being mounted in recesses, for example.

The risk of explosive atmospheres forming where exhaust air exits ventilation systems is particularly high. The requirements specified in EN 60079-14 / IEC 60079-14 / DIN VDE 0165 must be observed for electrical systems in these locations.

The operating status of important technical equipment should be displayed at a permanently monitored, central location. Faults should be indicated visually and acoustically to ensure fast response in potentially hazardous situations.

5.8 Residential buildings

General

Before the foundation stone is laid, the owner of the building, the architect, the electrical system planner, and the electrician must have agreed upon the design and scope of the electrical system, the electrical equipment to be provided, and the associated connection points. Apart from ensuring that there are sufficient socket-outlets and wall outlets for the devices, it is also important that the cross-sectional areas of the conductors are correctly dimensioned and that an adequate number of circuits are provided. Three-phase circuits may be necessary for large items of equipment, e.g. cookers and high-temperature water heaters. An installation, which has been designed to take future requirements into consideration, is more cost-effective than laying cables in an occupied dwelling at a later stage – an undertaking that can be very expensive and unavoidably produces large amounts of dust and dirt.

Building control systems with *instabus EIB*

Problems encountered with changes in room utilization can be avoided using the *instabus EIB* system (see Section 4.4). This technology allows the electrical installation to be adapted at any time to the changing lifestyle and requirements of the inhabitants by simply reprogramming and/or replacing application modules.

5.8.1 Service entries

Service connection for overhead lines

In overhead systems, wall connections or service entrance masts can be used as service entries. The cable between the entrance and the service fuse should always be as short as possible. DIN VDE 0100–732 must be complied with in all cases. The technical design details are defined by the public utility.

Location

The location of the service entrance mast or wall connection and the service fuse should be specified when the building is planned and indicated on the structural drawing. In the interests of safety, the selected locations must not be in rooms which are damp or in which readily flammable objects are stored. Locations for accommodating an antenna and an overhead line connection for IT systems must also be provided.

Furthermore, an entrance for underground cables must be installed in cellars and a reserve conduit with an inside width of at least 36 mm provided for the main line (rising main) so that the service connection can be easily converted from overhead lines to underground cables at a later stage (Fig. 5.8/1). Appropriate provisions should also be made for communications cables.

Service connection via underground cables

Cable distribution systems are reliable, do not require maintenance, and are not unsightly in residential areas.

According to DIN 18015 Part 1, the cable must be routed into the service entrance equipment room or, if such a room is not provided, into another suitable and generally accessible room.

Dimensions of service entrance equipment room

DIN 18012 specifies that the service entrance equipment room (Fig. 5.8/2) should be in the cellar against one of the exterior walls of the building adjacent to or opposite the stairwell or at the bottom of the stairwell itself. The recommended minimum dimensions for the service entrance equipment room for a building with up to 30 apartments are:

	For the room	For the door
Length	2.00 m	–
Width	1.80 m	0.65 m
Height	2.00 m	1.95 m

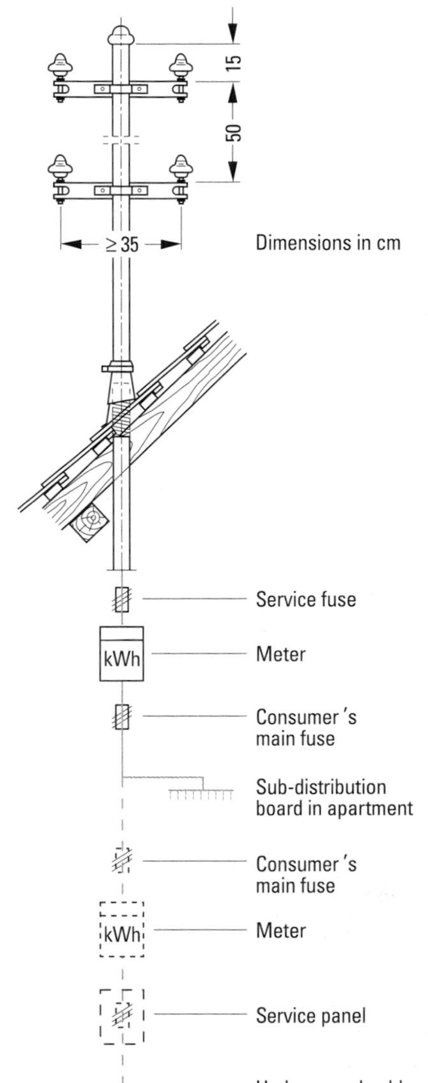

Fig. 5.8/1
Example of service connection with service entrance mast, with provision for conversion to cable entrance

Cable entries

Power cables and IT cables are routed separately; power cables are installed between 60 and 80 cm and IT cables between 35 and 60 cm below ground level. A conduit with an inside width of at least 8 cm must be provided in the exterior wall in accordance with the diameter of the cable. The type and size of the conduit are determined by the public

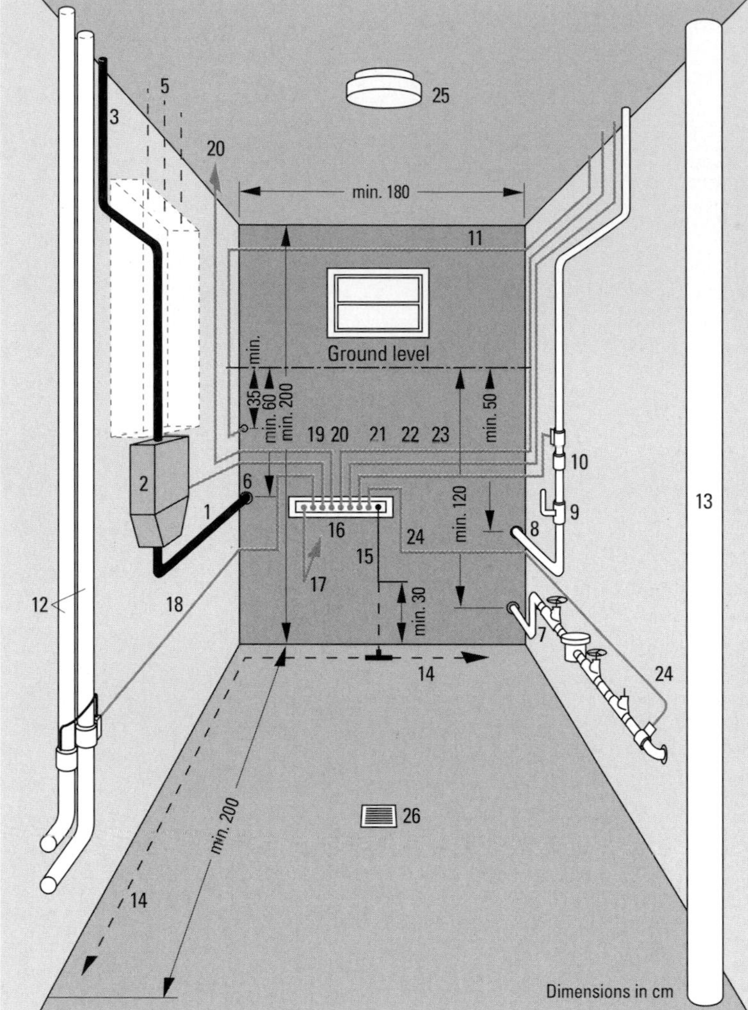

1 Service entrance cable for power current
2 Service panel
3 Main power cable
4 Meter cabinet (if required)
5 Outgoing power cable from meter to sub-circuit distribution boards
6 Cable conduit
7 Water pipe with meter
8 Gas pipe
9 Main gas shut-off valve
10 Insulating section
11 Communications cables
12 Heating pipes
13 Sewage pipe
14 Foundation grounding electrode
15 Terminal lug of foundation grounding electrode
16 Equipotential bonding strip
17 Equipotential bonding conductor to lightning protection system
18 Equipotential bonding conductor to heating pipe
19 Connecting cable for protective measure in TN system
20 Equipment grounding conductor for protective measure in TT system
21 Equipotential bonding conductor to communications system
22 Equipotential bonding conductor to antenna system
23 Equipotential bonding conductor to gas pipe
24 Equipotential bonding conductor to water pipe
25 Ceiling luminaire
26 Floor drain (if required)

Fig. 5.8/2
Example of service entrance equipment room to DIN 18 012 with main equipotential bonding conductor

Table 5.8/1 Standards for service panel

DIN standard	Type	Degree of protection to EN 60 529 / IEC 60 529 / DIN VDE 0470–1
43 627	Cable service panel	IP 40/IP 54
43 636	Overhead line service panel	IP 40
43 637		IP 54

utility. If underground cables, which are installed in conduits, are used for the telecommunications connection, it is important to observe any special requirements of the national telecommunications provider.

In Germany, for example, Deutsche Telekom stipulates that an 8 cm × 8 cm cable block must be inserted into an entry opening measuring 12.5 cm × 12.5 cm in this case. Further information can be found in the FTZ standard 731 MA 1 "Rohrnetze aus Isolierrohr für Fernmeldeleitungen in Gebäuden; Montageanweisung Telekom" and the "Empfehlungen für die Fernmeldeinstallation in Gebäuden", both published by the "Bundesamt für Zulassung in der Telekommunikation (BZT)" (Federal Bureau for Approval in Telecommunications).

Service panel

The service panel must have at least degree of protection IP 40 or IP 54 to EN 60 529 / IEC 60 529 / DIN VDE 0470-1, corresponding to the ambient conditions of the room or the location where it is installed. Service panels installed outdoors or in damp rooms must have degree of protection IP 54. The standards specified in Table 5.8/1 are obligatory for panels.

Stairwell as service entrance equipment room

If a separate service entrance equipment room is not available, the bottom of the stairwell can be used instead.

5.8.2 Main lines (rising main)

Dimension of chase

The main line (rising main) is routed from the service panel to the meters (Fig. 5.8/3). The chase provided for the main line should have a cross-sectional area of 60 mm × 60 mm. The width dimension must be increased accordingly if several main lines are to be installed (see Section 5.1.2). The technical supply conditions of the local public utility contain specifications regarding the arrangement and installation of meters and must be taken into consideration during planning. It is always advisable to contact the responsible public utility very early on, particularly if an electric heating system, e.g. electric storage heaters, is to be installed. The following information regarding the number and rating of main lines is based on residential buildings without electric heating.

Conductor cross-sectional area

The following conductor cross-sectional areas are usual for main lines:

▷ Single-family home, 63 A
 • Thermoplastic single-core non-sheathed cables (H07V-U), 16 mm^2 Cu in conduit,
 • Light plastic-sheathed cables (NYM or PROTODUR NYY), 10 mm^2 Cu.

▷ Two-family home, 80 A
 • Thermoplastic single-core non-sheathed cables (H07V-U), 25 mm^2 Cu in conduit,
 • Light plastic-sheathed cables (NYM or PROTODUR NYY), 16 mm^2 Cu.

H07V-U thermoplastic single-core non-sheathed cables in conduit, or NYM or PROTODUR NYY light plastic-sheathed cables are available. One or more control cores with a cross-sectional area of 1.5 mm^2 are often

a Meter cabinet in stairwell with consumer's main fuse

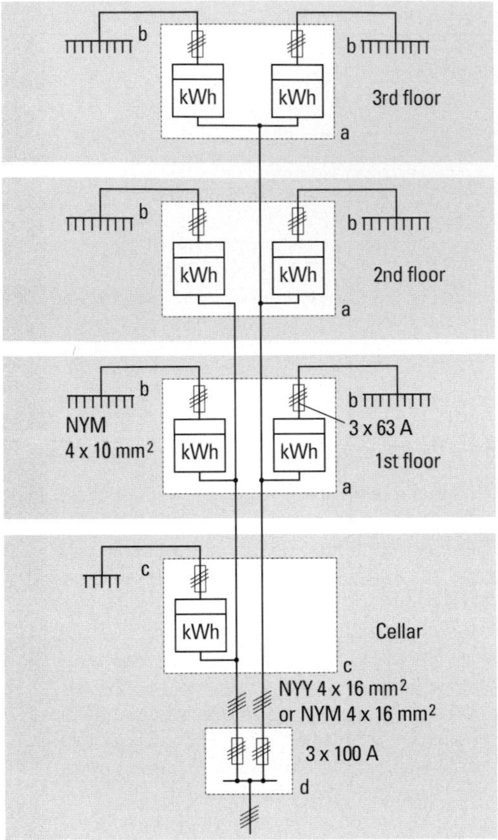

b Sub-distribution board in apartment

c Meter cabinet and sub-distribution board for communal facilities (stairwell lighting, cellar and loft lighting, cellar and loft lighting, antenna booster, utility room, etc., power supply units for signaling systems as well as door and building call systems)

d Service panel

Fig. 5.8/3
Apartment house with two main lines:
Meter cabinet in stairwell, sub-distribution board in apartment
(Cross-sectional areas based on DIN 18015 Part 1; Fig. p. 5 Curve A)

required for control purposes, e.g. to control two-rate meters. If the number of control cores is not known at the time of installation, a reserve conduit should be provided.

Main line for three apartments

One main line with a service fuse of 100 A is sufficient for a house with three apartments. Main lines with service fuses of up to 400 A are possible if service entrance boxes with l.v. h.b.c. fuse-bases are used.

Main lines for four to six apartments

Two main lines should be provided for four to six apartments (Fig. 5.8/3). If one main line only is installed initially, a reserve conduit must also be provided so that conversion to two rising main busbars is relatively straightforward.

Main line for eight or more apartments

Ring main

Larger conductor cross-sectional areas with suitably rated service fuses could be used for eight or more apartments (high-rise building). Main lines configured as ring mains with conductor cross-sectional areas of 25 mm^2 would be a more cost-effective solution. In Fig. 5.8/4, each group of eight apartments is connected to one ring main.

The two infeeds to the NYM 4 $\times$ 25 mm^2 ring main can be protected with a 100 A fuse thus permitting a total current drain of 200 A per external conductor. This load would require a conductor cross-sectional area of 50 mm^2 in the case of a spur line. Installing two NYM 4 x 25 mm^2 light plastic-sheathed cables as a ring main is, in practice, more straightforward than installing one NYM 4 $\times$ 50 mm^2 light plastic-sheathed cable. Another advantage of the ring main is the higher level of supply security (see Section 5.1.2).

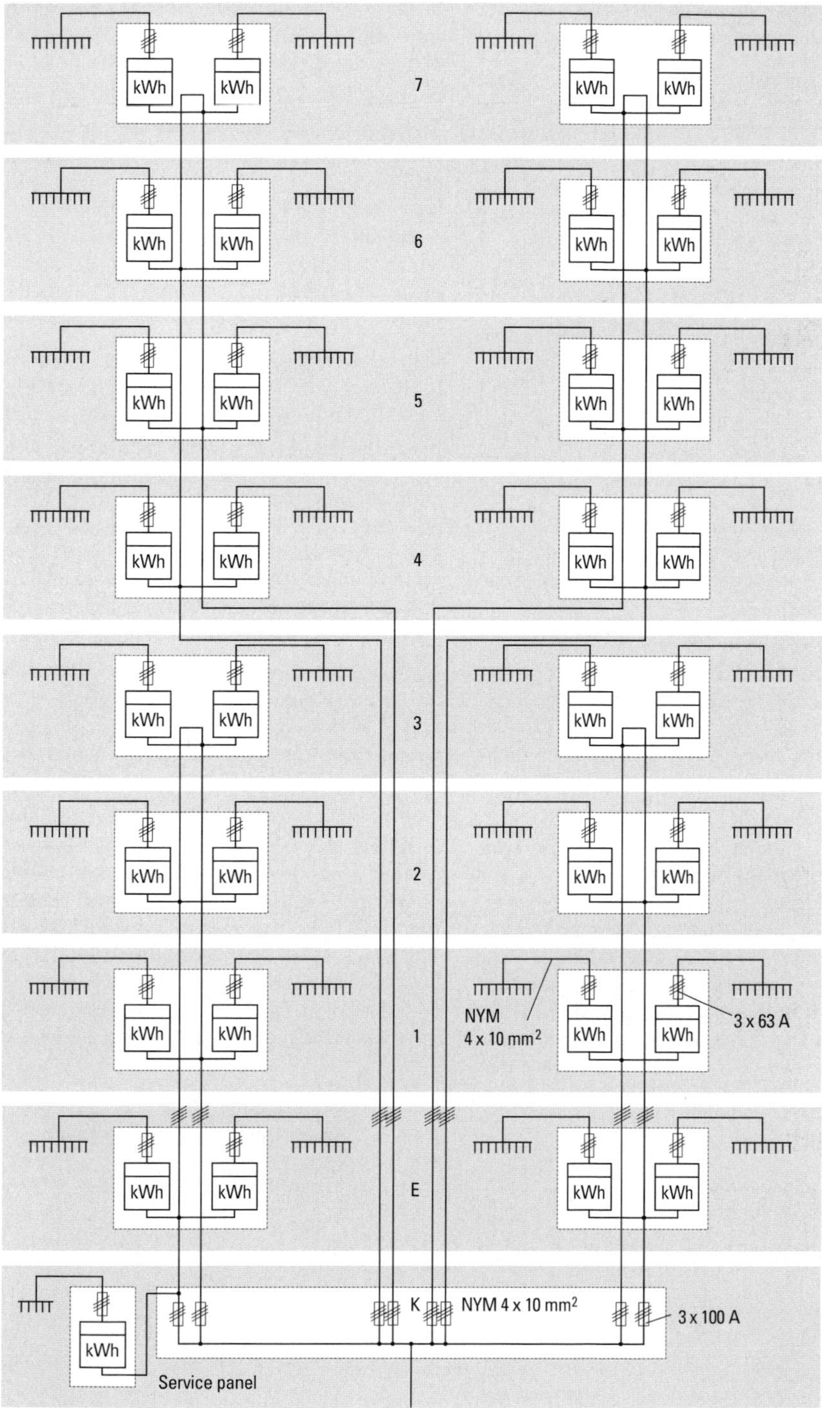

Fig. 5.8/4
Main lines in high-rise residential building:
Meter cabinet in stairwell, sub-distribution board in apartment;
Meter cabinet and sub-distribution board for communal facilities in cellar

5.8.3 Arrangement of meters and sub-distribution boards

Meters

Butt-mounted meters

Meters in double-tier

In single-family homes, the meters are usually installed in the cellar or in the building entrance (e.g. porch, hall). In apartment houses, meters are also installed in the cellar or in a meter cabinet in the corridors and stairwells on each floor (see Section 1.4.5 and 5.2.5) (Fig. 5.8/5).

Meter cabinets should not be installed along escape routes unless appropriate permission has been granted by the building authorities (in Germany, however, this is forbidden by the building product guideline).

Adequate space must also be provided for rate changing devices (this is a requirement specified by public utilities in Germany).

When the installation location is selected, the local public utility should always be consulted and its technical supply conditions complied with.

Meters in cellar, sub-distribution boards in apartment

A number of public utilities prefer meters to be installed at a central location, e.g. in cellars next to the service fuses, since this facilitates meter reading (Fig. 5.8/6). A cable is routed from each of the meters to the sub-distribution boards in the individual apartments.

Position of meter cabinet

The meter cabinet must be positioned with the meter displays at a reading height of between 1.10 m and 1.85 m above floor level. Furthermore, it is essential to ensure that the meter cabinets are easily accessible. A separate meter, which is installed in the meter cabinet in the cellar or on the first floor, is required for the communal facilities in apartment houses, e.g. the general lighting in stairwells, lofts, and cellars as well as doorbell and intercom systems, etc.

Sub-distribution boards

Residual-current and miniature circuit-breakers as well as fuses for the individual circuits of the apartments are installed in a sub-distribution board (STAB wall-mounting distribution boards or SIMBOX 63 small distribution boards [see Section 1.4.5]) which should be provided at the load center of the apartment (close to the kitchen/bathroom).

In single-family homes, service entrance frames in wall recesses are a particularly cost-effective solution. All connection and metering devices for electricity, water, gas, and communications are installed in the service entrance frame (Fig. 5.8/7). The function panels for the individual metering devices are arranged in a predefined layout. The meter cabinet must not exceed a height of 950 mm and a width of 550 mm. The wall recess can be closed off using a standard door.

Electrical connection between meter cabinet and sub-distribution boards

The cables leading from the meter cabinet to the sub-distribution boards must always be rated for three-phase current, even if alternating-current meters are required initially. The technical supply conditions of the local public authority must be observed.

Installing the meters in the cellar and the protective devices in sub-distribution boards of the apartments is no more expensive than installing them together in a meter cabinet. The main advantage of separate installation is that, in the event of a fault, the protective devices can be operated or replaced in or in the vicinity of the apartment (in the corridor

Dimensions of wall recesses

No. of meters	1	2	3	4	5	6	7 / 8	9 / 10
Width of butt-mounted meters mm	330	580	830	1080	1330	–	–	–
Meters in double-tier arrangement mm	–	330	–	580	–	830	1080	1330
Height mm	980, 1130, 1280, 1430							
Depth mm	210 for flush mounting							

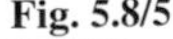

Fig. 5.8/5
SIPRO meter cabinets to DIN VDE 0603 and DIN 43870, and dimensions required for wall recesses

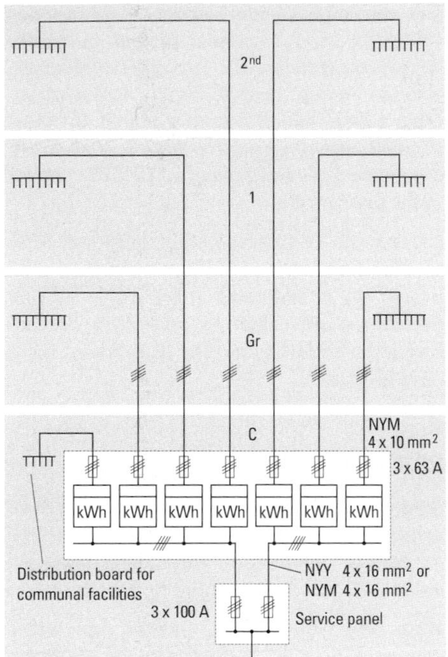

Fig. 5.8/6
Central location of meters, e.g. in cellar; sub-distribution board in apartments

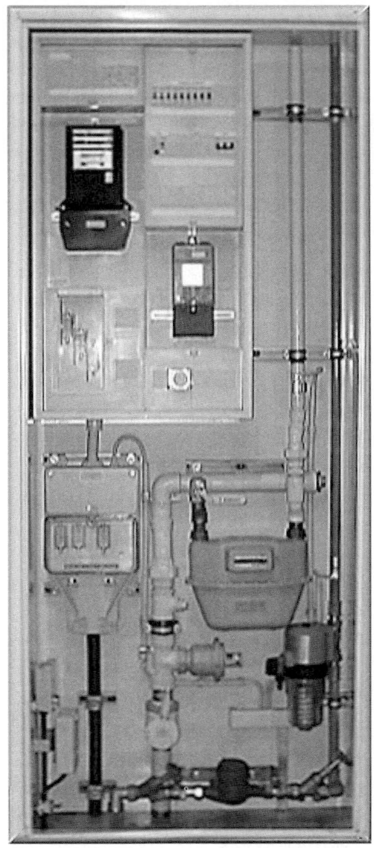

Fig. 5.8/7
Service entrance recess fitted with service entrance frame and connection and metering devices for electricity, water, gas, and communications

or stairwell) and the meters can be read at an easily accessible location near the building entrance.

Meters and protective devices in a single cabinet

If the meter is located in the vicinity of the load center (kitchen or bathroom), meters and protective devices can be installed together in a single cabinet (meter distribution cabinet). This solution is often convenient for single-family homes, small dwellings, and apartments. A second meter location can be provided if, for example, a room (or adjoining apartment) is let at a later stage (Fig. 5.8/8).

The type and location of the meter distribution cabinet is again determined by the public utility.

Concealed cabinets are preferred in cases where meters are installed in stairwells. In cellars, meter cabinets can be mounted on the wall. SIPRO meter cabinets are suitable for both installation methods.

Protective measure for meter cabinets

DIN VDE 0606 specifies total insulation for meter cabinets as a supplementary protective measure in the event of indirect contact (see also Section 2.4.8).

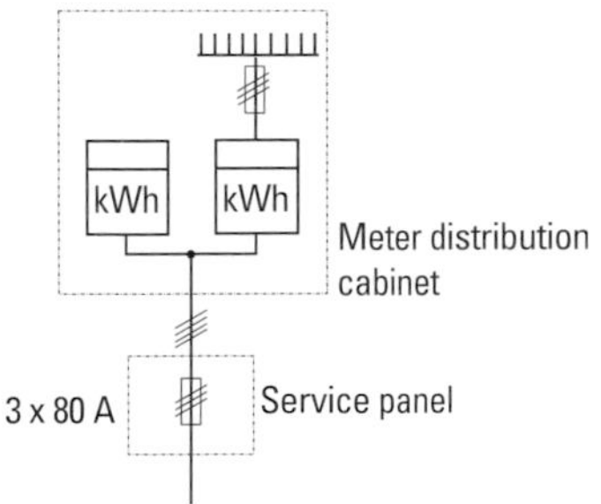

Fig. 5.8/8
Meter distribution cabinet

STAB wall-mounting distribution boards

STAB 160/400 wall-mounting distribution boards are ideal as sub-distribution boards for apartments. The modular assemblies of the boards permit a wide range of different configurations while, at the same time, reducing the number of component parts that need to be stocked. Section 1.4.5 contains information on SIPRO meter cabinets and STAB wall-mounting distribution boards.

Miniature circuit-breakers or fuses

It is always advisable to reserve adequate space for additional overcurrent protective devices or *instabus EIB* modular devices to facilitate expansions or conversions to three-phase current and *instabus EIB* installations. Miniature circuit-breakers, instead of fuses, enable immediate recovery after tripping (see Section 4.1.2 and 4.4).

Domestic electrical installation

Domestic electrical installations include electric low-voltage systems for apartments with a nominal voltage ≤ 250 V AC with respect to ground as well as other low-voltage systems with a nominal voltage ≤ 250 V AC with respect to ground, which are the same size and have the same configuration as low-voltage systems for apartments.

Cellar and loft lighting

The lighting circuits for stairwells, corridors, and communal lofts and cellars are connected to the meter for the general power supply, while the circuits for the cellars and lofts belonging to individual apartments are usually connected to the relevant apartment meters. Special circuits for switching over to the appropriate apartment meter are available for communal, laundry, and ironing facilities; coin-operated meters can also be installed. Further information can be obtained from the relevant public utility.

Circuit-phase distribution in apartment

The power supply for the apartment is distributed among individual circuits. A separate circuit should be provided for the lighting and the socket-outlets in each room. This also applies to adjoining rooms, such as the hall, bathroom, WC, and storage rooms.

Lighting and socket-outlet circuits should be installed separately. The advantage of this is that, if one circuit fails, another circuit is available in the same room.

To provide protection against electric shock (protection in the event of indirect and against direct contact), one or more residual-current protective devices with a rated residual current of $I_{\Delta n} \leq 30$ mA should be connected on the line side of both the lighting and the socket-outlet circuits (see Section 4.1.3). If several residual-current protective devices are installed, the circuits should be distributed among these in such a way that another circuit, which is protected by a different residual-current protective device, is available if a residual-current protective device operates in the same room.

Large items of current-using equipment with a connected load >2 kW are supplied with power by a dedicated three-phase circuit. Suitable residual-current protective devices should be connected on the line side of these circuits as well.

5.8.4 Installation of wiring system

Semi-flush installation

Semi-flush installation of SIFLA flat webbed cables is nowadays the most common method of laying conductors with cross-sectional areas ≤ 4 mm^2 in residential buildings.

With this installation method, no cable chases have to be caulked into the masonry: it is, therefore, particularly suitable for thin partition walls.

Concealed installation

With cross-sectional areas > 4 mm^2, concealed NYM light plastic-sheathed cables are used or insulating conduits, into which thermoplastic single-core non-sheathed cables are pulled. Chases in the masonry should be made when the basic shell of the building is being constructed (chase plan!).

The devices (switching devices, socket-outlets) used for semi-flush installation are identical to those for concealed installation.

Device boxes made of insulating material are used to secure the devices by means of claws or screws; screws are preferred for socket-outlets.

Semi-flush or concealed cables must only be installed horizontally or vertically in accordance with DIN VDE 0100-520 and DIN 18015 Part 3. If cables have to be routed around windows (this should be avoided wherever possible), the intended fixing points for curtain hooks and planned roller shutter boxes must be taken into consideration.

Surface-type installation

Surface-type installation is nowadays only used if subsequent work is carried out in lofts, cellars or wooden buildings, barracks, sheds and rooms where visible cables can be tolerated.

Surface-mounted cables should preferably be installed in conduits or ducts. Alternatively, inconspicuous light plastic-sheathed cables can be used. If there is a risk of mechanical damage, switching devices and socket-outlets are flush-mounted or recessed.

Joint boxes and combined device and joint boxes

Figs. 5.8/9 and 5.8/10 illustrate cable installation using joint boxes or combined device and joint boxes. The latter provide space for the terminals of the cable connections and outgoing device feeders. They are also recommended for the socket-outlet ring main (with reserve outlet boxes).

The combined device and joint boxes offer the following advantages compared with standard joint boxes:

- ▷ convenient installation from floor,
- ▷ no visible, unsightly outlet box covers,
- ▷ easy access to connecting points without damage to paint or wallpaper,
- ▷ short cable runs to socket-outlets and wall luminaires mounted at a later stage.

Building control systems with *instabus EIB*

As a preparatory measure for implementing building control system with *instabus EIB*, it is advisable to install the *instabus EIB* cable parallel to the power cables in all tapping and device boxes. Furthermore, all distribution boards should be interconnected via an *instabus EIB* cable. This ensures that all applications can be incorporated in the building management system at a later stage (see Section 4.4).

Reserve outlet boxes

A "socket-outlet ring main" or a reserve box cable with additional device boxes for subsequent installation of socket-outlets is recommended for all living areas. The reserve boxes, which contain terminals for cable connections, are sealed with a cover. Socket-outlets and reserve boxes should be installed at intervals of approximately 1.25 m.

If an adequate number of reserve boxes are provided, socket-outlets can be fitted where they are needed if, for example, the furniture is rearranged.

Antenna socket

If television sets and radio equipment are placed next to each other, more than two socket-outlets are required in addition to the antenna socket (e.g. for an additional CD player, video recorder, TV lamp).

Multiple socket-outlet

Instead of single socket-outlets, it is also possible to fit double socket-outlets at a later stage (Fig. 5.8/11).

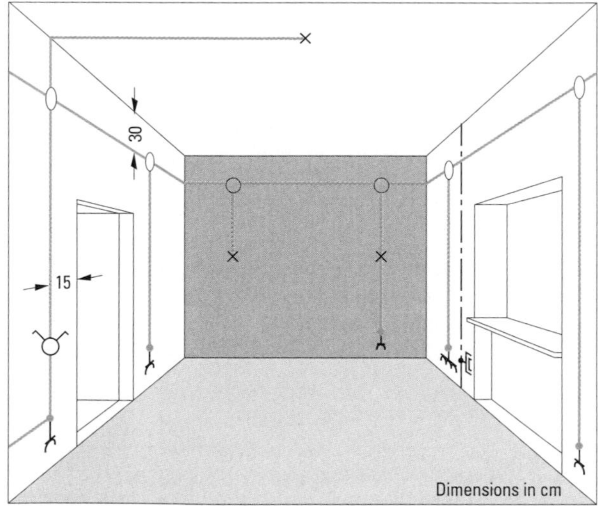

Fig. 5.8/9
Example of cable installation with joint boxes

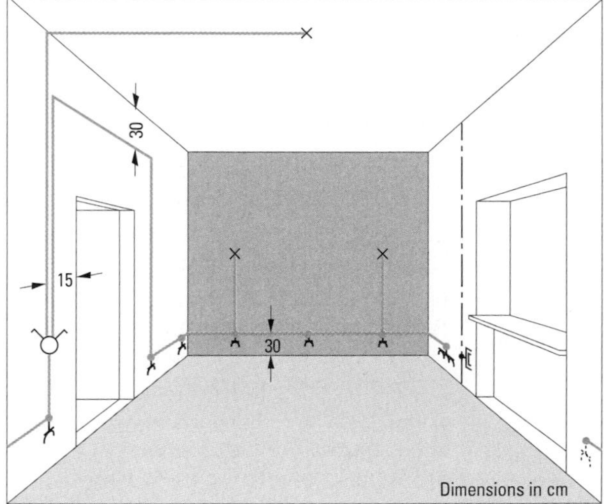

Fig. 5.8/10
Example of cable installation with combined device and joint boxes

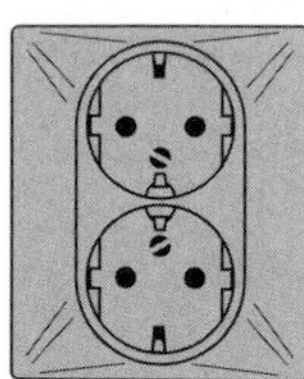

Fig. 5.8/11
Double socket-outlets for device box (60 mm diameter)

Wall luminaires

The wall boxes for device luminaires must be installed according to the instructions of the occupant in line with the fixed furnishings fitted in the apartment. If a socket-outlet ring main is provided and the occupant only needs to be able to switch the lighting on and off locally, the cable can be run from every socket-outlet or reserve box to the wall luminaires along the shortest routes and, therefore, at minimum cost.

Remote switches

It is, however, often necessary to switch the lighting on and off from several locations. If more than two switching points are involved or the cable runs are relatively long, remote switches which can be operated from a number of pushbuttons are recommended (see Section 4.2).

Mirror lighting

Two luminaires, one on either side of the mirror, should be installed to provide optimum mirror lighting (in the bedroom, bathroom, hall, and WC).

Locating lamp

Locating lamps in the actuating devices make them easier to find in the dark. They can be fitted easily in any switch or pushbutton.

Pilot lamp

If switches, e.g. for lighting and heating devices, are not fitted in the room itself but rather outside the room, switches with integrated pilot lamps to indicate the status of these devices are recommended. The pilot lamp lights up if a device is switched on. Provision for switches with pilot lamps must be made when the cables are installed since an N-conductor is required.

Reserve conduits

If the cables for connecting a washing machine, dishwasher, infrared thermal radiator in the bathroom, living room heaters, etc., are not to be installed immediately, it is advisable to provide reserve conduits to allow these to be connected at a later stage.

Outlets in living room

In large living rooms, provision should be made not only for lighting in the middle of the room, but also for wall luminaires for seating areas and luminaires for flower windows.

There should be a sufficient number of socket-outlets so that additional standard and table lamps as well as electrical appliances (TV and radio, CD player, video recorder, toaster, vacuum cleaner, etc.) can be connected. It is also particularly convenient if the socket-outlet circuits provided for lighting purposes can be switched on and off at the door.

An adequate number of antenna sockets to allow antennas, and broadband or SAT equipment to be connected should be fitted so that the position of TV and radio appliances can be varied.

Outlets on balcony and terrace

The balcony and terrace should have at least one socket-outlet, which is protected by a residual-current circuit-breaker, for a table lamp and an outlet for an infrared thermal radiator.

Outlets in bedroom and nursery

A socket-outlet ring main with additional reserve boxes is also useful in the bedroom and nursery to enable the furnishings to be rearranged if desired.

Two or three socket-outlets next to the beds are recommended so that reading lamps, radio receivers, heated pillows, electric blankets, etc. can be connected. Antenna sockets should also be fitted in the bedroom.

Outlets in bathroom and WC

A three-phase supply lead sufficient for the maximum possible load (storage heaters or continuous-flow water heaters) is recommended for water heating in the bathroom.

If there is no separate utility room, outlets must be provided for connecting a washing machine and, where appropriate, a tumble dryer. Three-phase current is not usually required since modern household washing machines are designed so that they can be operated from an alternating-current circuit protected with a 16 A fuse.

In addition to the lighting, a socket-outlet for electric razors should also be installed in the bathroom and WC. Razor outlets with integrated isolating transformers also provide protection against direct contact with live parts. Electrical installation in rooms with a bath tub or shower is subject to the requirements specified in HD 384.7.701 / IEC 60364-7-701 / DIN VDE 0100-701 (see Section 5.9.1). According to this standard, socket-outlets may only be installed in Zone 3 which means that only isolating transformers, safety extra-low voltage, and residual-current protective devices with a rated residual current of $I_{\Delta n} \leq 30$ mA must be provided for protection against shock currents.

Outlets in kitchen

Since the work surfaces in kitchens often run along the sides of the room, it is not possible to provide optimum illumination using central lighting. Luminaires above the working areas are, therefore, always required.

The luminaires and an adequate number of socket-outlets above the working areas for small household appliances (beater, food mixer, coffee grinder, etc.), refrigerators, extractor hoods above the cooker, or fans mounted in walls or windows can be connected to one circuit. A further socket-outlet circuit is required for household appliances with a higher connected load, e.g. automatic coffee machines, grills, and high-speed kettles. A water boiler above the kitchen sink, for which a socket-outlet must always be provided, is more practical than a high-speed kettle. Fully-automatic washing machines and dryers are installed in kitchens, particularly in small apartments in which there is no utility room and the bathroom is too small. In this case, a separate circuit is necessary. The dishwasher also requires its own electric circuit. Water heaters with high connected loads and the cooker each require a separate three-phase circuit.

It is important to ensure that switches and socket-outlets are not concealed by hinged covers that may be fitted to the cooker.

instabus EIB

HomeAssistant

If *instabus-EIB*-compatible appliances are not installed straightaway, the appropriate connections for these appliances should still be provided. An outlet for the HomeAssistant (see Section 4.4) should also be installed so that this can be operated in the kitchen.

Communications systems (telecommunications)

Wherever possible, the main line of the communications system should be installed in the stairwell, either semi-flush or concealed under plaster. If the cable is installed in a conduit, for which DIN 18015 recommends an internal diameter of at least 29 mm, chases at least 50 mm deep must be provided. The size of the chases must be increased accordingly if several main lines are to be installed.

Recesses for junction boxes positioned approximately 2.25 m above the finished floor must be provided on each floor. A conduit with an internal diameter of 16 mm, which terminates in a special telecommunications junction box, is laid from the junction box to carry the cables of the communications system to the installation location of the terminal unit. The chase for the conduit must be at least 30 mm deep.

Even if a communications connection is not required until some point in the future, subsequent costly and dirty installation work can be avoided by laying the conduits as reserve conduits.

If power cables and cables of the communications system are laid in parallel or if they cross, they must be at least 1 cm apart, unless a separating web is fitted. (Further information can be found in DIN VDE 0800-1, DIN 18015, and the regulations issued by the national telecommunications provider.)

Teleindication and telecontrol functions

In order to support teleindication and telecontrol functions via the communications network with *instabus EIB*, an *instabus EIB* outlet must be fitted at the installation location of the communications gateway (see Section 4.4).

Signaling systems

The cables for the call-signal generator, bell, and door opener pushbuttons can be installed in the power cable chases both in the stairwell and in the apartments. The bell transformer is installed in the associated distribution board.

The bell pushbuttons with nameplates and the illuminated pushbutton to switch on the stairwell or garden lighting are grouped together on a central metal panel at the entrance to the building or at the garden gate. If illuminated nameplates are used, the required tubular filament lamps should be connected to the stairwell and garden lighting.

An additional pushbutton with a nameplate is installed for the call-signal generator at the apartment door. This pushbutton has a bell symbol, whereas the pushbutton for the door opener inside the apartment has a key symbol.

Further information on door and building call systems and acoustic signaling devices (especially regarding their use, circuitry, and the required number of cables) can be found in Section 4.3.2.4.

Door and building call systems

Door and building call systems establish a speech connection between the apartment and building entrance or garden gate and can also be used as internal communication systems inside the apartment (see Section 4.3.2.4). The necessary cables must be installed in both the stairwell and apartments in chases which are physically separate from power cables. The power supply unit required for each system is installed in the appropriate distribution board. The door station, installed at the building entrance, with bell pushbuttons and the pushbuttons for the stairwell or garden lighting is connected to the appropriate number of house stations in the apartments.

Another pushbutton with a nameplate is installed for the call-signal generator at the door to each floor. Different call signals should be used so that the occupants of the apartment can tell whether the visitor is at the building entrance or the apartment door.

With press-to-talk systems, the speech connection between the door station and a house station is always private so that the conversation can be neither interrupted nor overheard by another house station. The house stations are suitable for surface mounting.

With duplex systems, the speech connection between the door station and a house station may or may not be private, depending on the type of system. In contrast to press-to-talk systems, they are also suitable for internal communication between the individual house stations. The house stations resemble the telephones used for the public telephone network. They are suitable for surface mounting and can also be converted to table-top units. All house stations have a door-opener pushbutton.

5.9 General information on special areas, locations, and installations

5.9.1 Rooms with bath tub or shower basin

Areas with special requirements

In bathrooms and shower rooms, a person with bare, wet feet has a very low contact resistance with respect to the ground potential. In these rooms, it is, therefore, particularly important to provide protection against electric shock. For this reason, rooms in which bath tubs or shower basins are located have been divided into zones which require special attention. The installation of electrical equipment such as current-using devices and cables in these zones is subject to certain restrictions so that no live parts can be drilled and no hazardous vagabond voltages can form when soap dishes, towel rails and so on are mounted at a later stage.

Zoning, permissible equipment, required degrees of protection

Fig. 5.9/1 shows the zones into which a room with a bath tub is divided, and Fig. 5.9/2 the areas for a room with a shower basin.

According to HD 384.7.701 / IEC 60364-7-701 / DIN VDE 0100-701, the following electrical equipment is permitted in the respective zones:

Zone 0	Only equipment which is approved for use in bath tubs or shower basins;
Zone 1	Only permanently installed high-temperature water heaters and air extraction devices;
Zone 2	Only permanently installed high-temperature water heaters, air extraction devices, and luminaires;
Zone 0, 1, 2	No switching devices or socket-outlets, except switches integrated in current-using equipment in Zone 1 and 2;
Zone 3	Socket-outlets, only permissible in conjunction with residual-current protective devices (RCD) with $I_{\Delta n} \leq 30$ mA or transformers with safety separation or with safety extra-low voltage (SELV);

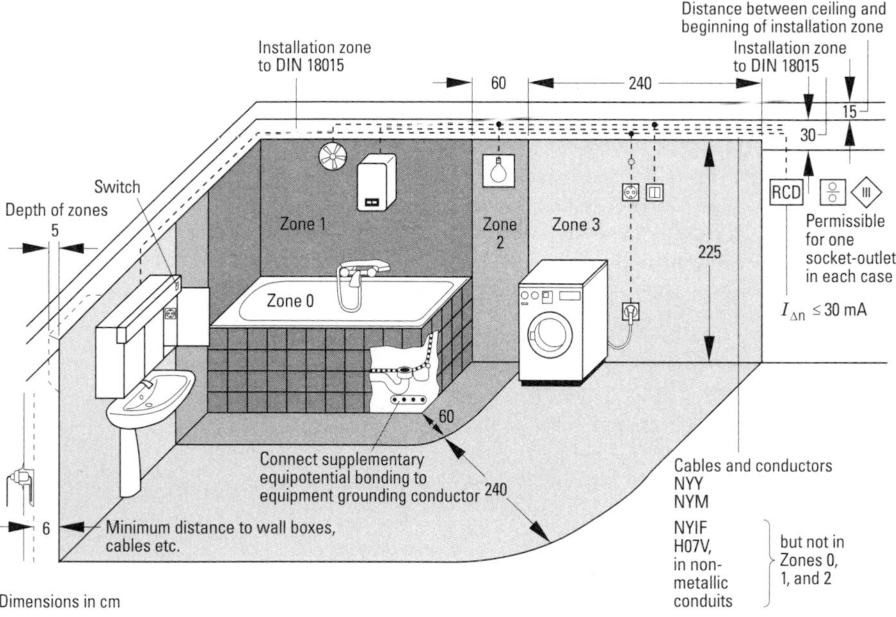

Recommended:
The permanently installed water heater and air extraction devices approved for Zone 1 should be installed outside this area
- in the interests of safety
- to facilitate maintenance
- to provide unrestricted freedom of movement for the users

Joint boxes
- • permissible in installation zone
- ° in Zone 3, only permissible if made of plastic; not permissible in Zones 0, 1, and 2

Fig. 5.9/1
Zoning with special requirements in rooms with bath tubs or shower basins to HD 384.7.701 / IEC 60364-7-701 / DIN VDE 0100-701

Zone 1, 2	Only call and signaling systems with safety extra-low voltage [SELV] (≤25 V AC, ≤ 60 V DC);
Zone 0, 1, 2	No cables in/under plaster, behind covers, except for permanently installed current-using equipment in Zone 1 and 2; only applies to cables installed in walls at depths ≤ 5 cm;
Zone 1, 2	Between rear side of wall and wall surface in bath or shower: minimum distance of 6 cm to cables and recess-mounted wall boxes;

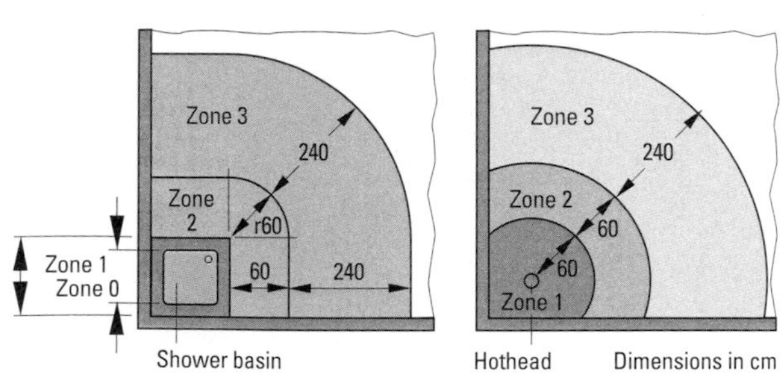

Fig. 5.9/2
Example of zoning in rooms with shower basin or shower

Zone 0, 1, 2, 3	No cables or conductors for supplying other rooms or locations;
Zone 0, 1, 2	No joint boxes, outlet boxes made of insulating material permissible in Zone 3.

Degrees of protection

The required degrees of protection to EN 60529 / IEC 60529 / DIN VDE 0470-1 for the above-mentioned electrical equipment are specified in Table 5.9/1.

Shock protection and protection against ingress of solid foreign bodies: at least IP 2X.

Switching devices/ socket-outlets

As illustrated in Fig. 5.9/1, switching devices and socket-outlets may only be mounted in Zone 3. All of the socket-outlets in Zone 3 must be protected by residual-current protective devices with $I_{\Delta n} \leq 30$ mA to ensure that any mobile equipment used in the bathroom and supplied with power via socket-outlets cannot cause injury.

Furthermore, safety separation and safety extra-low voltage are theoretically also permissible as protective measures. These two measures are, however, of little relevance in practice. The requirement that plug-and-socket devices in Zone 3 must be protected by residual-current protective devices with $I_{\Delta n} \leq 30$ mA only applies to new installations.

Socket-outlets with **r**esidual-**c**urrent protective **d**evices (SRCDs) can be fitted in existing installations. They can be used to replace the existing socket-outlets which are not protected by residual-current protective devices.

Equipotential bonding

In order to prevent touch voltages, all extraneous conductive parts in the bathroom which can conduct voltage must be interconnected by means of an equipotential bonding conductor. The equipotential bonding conductor is also required even if there is no electrical equipment in rooms with a bath tub or shower. Equipotential bonding is not necessary for plastic bath tubs and shower basins or plastic drain pipes. Mobile bath tubs and shower basins must also be connected to the equipment grounding conductor via an equipotential bonding conductor.

Cables and conductors

Cables and conductors which supply power to other rooms or locations must not be routed through Zones 0 to 3.

Since, however, protection is the prime objective, this should apply to the entire room containing a bath tub or shower.

Rooms with shower basin/ shower

Fig. 5.9/2 shows an example with zoning for rooms with a shower basin or shower. Only a shower head is fitted in the case of the shower. In this case, the zones are measured from the center of the shower head.

Table 5.9/1
Degrees of protection for electrical equipment in bathrooms

	Zone 0	1	2	3
Bathrooms in which moisture condensation frequently forms, e.g. in public bathrooms and bathrooms in sports centers	IP X7	IP X5	IP X5	IP X5
Bathrooms in which moisture condensation seldom forms, e.g. domestic bathrooms	IP X7	IP X4, IP X5[1)]	IP X4	IP X1[2)]

[1)] The degree of protection IP X5 must be selected if jet-water is to be expected, e.g. from massage showers.
[2)] Degree of protection IP X0 according to HD 384.7.701 / IEC 60364-7-701 / DIN VDE 0100-701

5.9.2 Indoor and outdoor swimming pools

Swimming pools and swimming complexes represent a major hazard for people if they are exposed to the effects of electricity. Particularly stringent protective measures are, therefore, required in these cases.

Zones are defined for swimming pools and foot-baths in swimming pools (Fig. 5.9/3 and Table 5.9/2).

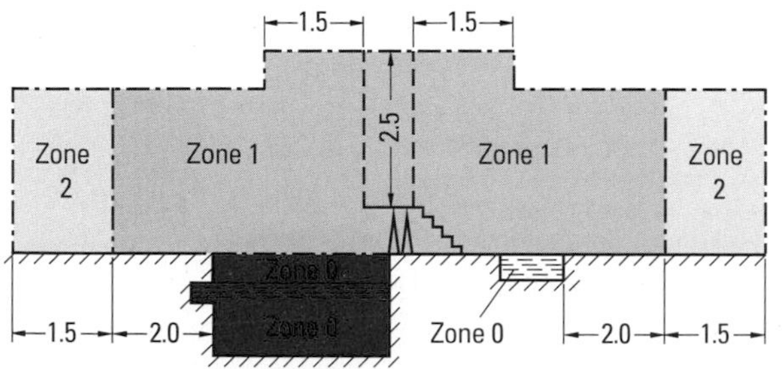

a) Protection zones for swimming pools and foot-baths

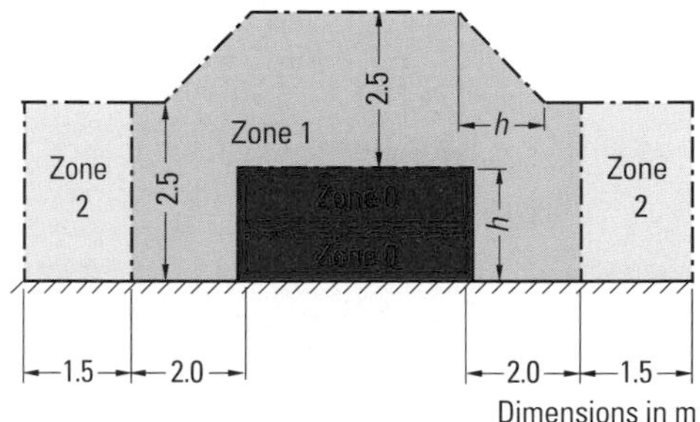

b) Protection zones for swimming pools which can be erected on a surface (e.g. ground)

Fig. 5.9/3 Zoning for swimming pools and foot-baths to HD 384.7.702 / IEC 60364-7-702 / DIN VDE 0100-702

Table 5.9/2
Electrical installation requirements in swimming pools to HD 384.7.702 / IEC 60634-7-702 / DIN VDE 0100 Part 702

Zone	Protective measure to HD 384.4.41/ IEC 60364-4-41/ DIN VDE 0100-410	Installation equipment	
		Arrangement	to degree of protection EN 60529 / IEC 60529 / DIN VDE 0470-1
0	Safety extra-low voltage (SELV) with max. 12 V AC, 30 V DC. The protective measures "Protection by means of obstacles, clearances, non-conductive locations, and ungrounded local equipotential bonding" must not be applied.	None. Except small swimming pools in which it is not possible to arrange the devices outside Zone 1. Socket-outlets in Zone 1 only if they are located at least 1.25 m (normal arm's reach) from Zone 0 and 0.3 m above the floor. Protection by means of a residual-current protective device with $I_{\Delta n} \leq 30$ mA or by dedicated, safety-separated transformers must be provided.	–
1			Protection against ingress of water: IP X5; for small swimming pools in buildings which are not usually cleaned with jet-water IP X4.
2	Safety separation with dedicated transformers with electrical separation, safety extra-low voltage (SELV), residual-current protective device with $I_{\Delta n} \leq 30$ mA	Switching devices, socket-outlets, and other installation equipment are permitted if they are protected by one of the specified protective measures.	Protection against ingress of water: IP X2 for indoor swimming pools (swimming complexes), IP X4 for outdoor swimming pools, IP X5 if jet-water is used for cleaning purposes.

5.9.3 Construction sites

Feed points

Electrical equipment, which is used on construction sites, must be supplied with power from special feed points. The following can be used as feed points:

- ▷ construction site distribution boards to EN 60439-4 / IEC 60439-4 / DIN VDE 0660-501,
- ▷ plug-in distribution units with minimum degree of protection IP 43 to EN 60529 / IEC 60529 / DIN VDE 0470-1 with 2-pole (max.) grounding outlets to DIN 49440 or DIN 49442, which are protected by residual-current protective devices with $I_n \leq 30$ mA and are equipped with their own grounding electrode, e.g. miniature construction site distribution boards,
- ▷ outgoing feeders of existing permanently installed distribution boards which are specifically assigned to the construction site,
- ▷ stand-by power supply systems to DIN VDE 0100-728 (in future HD 384.5.551 / IEC 60364-5-551 / DIN VDE 0100-5-551),
- ▷ transformers with electrically isolated windings.

Permissible systems

Only the TT, TN-S, or IT systems with insulation monitoring may be used for the distribution system downstream of the feed points. TN-C systems may only be used downstream of construction site distribution boards if the cables and conductors have cross sections of at least 10 mm^2 Cu or 16 mm^2 Al, are not moved under normal

Table 5.9/2
(continued)

Current-using equipment		Arrangement of power source/ safety-separated transformer	Arrangement of cables and conductors	Arrangement of joint boxes
Arrangement	to degree of protection EN 60529 / IEC 60529 / DIN VDE 0470-1			
E.g. luminaires, high-temperature water heaters, only if they are permanently installed, and designed and approved specifically for use in swimming pools.	Protection against ingress of water: IP X8	Outside Zones 0, 1, and 2	Surface-mounted or embedded in plaster up to 5 cm. Only used for supplying the devices in the zone, without conductive covering and not in conductive conduits.	Not permitted
	Protection against ingress of water: IP X5; small swimming pools which are not usually cleaned with jet-water: IP X4.		Heating cables for electric room heaters, embedded in the floor, only permissible if they are covered with a grounded metal grid or have a grounded metallic covering. Supplementary local equipotential bonding must also be provided.	
Class 2 luminaires, other Class 1 devices if they are protected by residual-current protective devices with $I_{\Delta n} \leq 30$ mA or supplied with power from a safety-separated transformer.	Protection against ingress of water: As for installation equipment		Surface-mounted or embedded in plaster up to 5 cm. Not in accessible conductive conduits. Heating cables for electric room heaters, embedded in the floor, only permissible if they are covered with a grounded metal grid or have a grounded metallic covering. Supplementary local equipotential bonding must also be provided.	As for installation equipment

operating conditions, and are mechanically protected so that they can be regarded as being fixed. In TT and TN-S systems with socket-outlets up to 16 A, all circuits with socket-outlets must be protected by means of residual-current protective devices with $I_{\Delta n} \leq 30$ mA for single-phase operation.

Other socket-outlets can be protected by means of residual-current protective devices with $I_{\Delta n} \leq 500$ mA.

Requirements for equipment

To ensure that electrical systems on construction sites can be isolated quickly in emergencies, the systems must be fitted with switching devices which switch all ungrounded conductors simultaneously. Switchboards and distribution boards must have at least degree of protection IP 43, in accordance with EN 60529 / IEC 60529 / DIN VDE 0470-1.

The degrees of protection specified in EN 60439-4 / IEC 60439-4 / DIN VDE 0660-501 apply to construction site distribution boards. H07RN-F and A07RN-F rubber-insulated flexible cables which conform to HD 22.1.53 / IEC 60245-1 / DIN VDE 0282-1 or equivalent cable types must be used as flexible cables.

Cables and conductors must be installed with mechanical protection or with mechanically fixed covers at locations where they are exposed to a particularly high degree of mechanical stress. When overhead, suspended cables are connected, suitable strain relief devices must be provided at the connecting points. Installation switches, plug-and-socket devices, tapping boxes, and so on must have at least degree of protection IP X4 in compliance with EN 60529 / IEC 60529 / DIN VDE 0470-1.

Supply connection

Only the following devices may be used for supply connections on construction sites:

▷ plug-and-socket devices (2-pole) with ground contact to DIN VDE 0620, types conforming to DIN 49440 or DIN 49442,

▷ plug-and-socket devices (2-pole) with ground contact to DIN VDE 0620 for severe operating conditions, types conforming to DIN 49440, DIN 49441, DIN 49442, and DIN 49443,

▷ plug-and-socket devices (CEE plug-and-socket devices) to EN 60309 / IEC 60309 / DIN VDE 0623 for severe operating conditions, types conforming to DIN 49462 or 49465.

The casing of the plug-and-socket devices must be made of insulating material. Switching and control devices, starting and regulating resistors, as well as electrical machines outside switchgear assemblies and distribution boards must have at least degree of protection IP 44 in compliance with EN 60529 / IEC 60529 / DIN VDE 0470-1.

Residual-current protective devices must be suitable for operation at low temperatures (down to –25 °C).

5.9.4 Agricultural and horticultural estates

Terms

Agricultural estates are buildings and installations which are primarily used for keeping livestock.

Horticultural estates include greenhouses, hothouses, and the associated equipment.

The term "livestock" refers to large animals such as horses, cattle, sheep, and pigs, and small animals such as poultry, rabbits, and so on.

Ambient conditions

People and livestock are subject to an exceptionally high risk of injury on agricultural and horticultural estates (both outdoors and indoors) as a result of particular ambient conditions, e.g. the effects of humidity, dust, highly-aggressive chemical vapors, acids, or salts on the electrical equipment.

Increased risk of fire

Readily flammable materials, which are used at these locations, also increase the risk of fire.

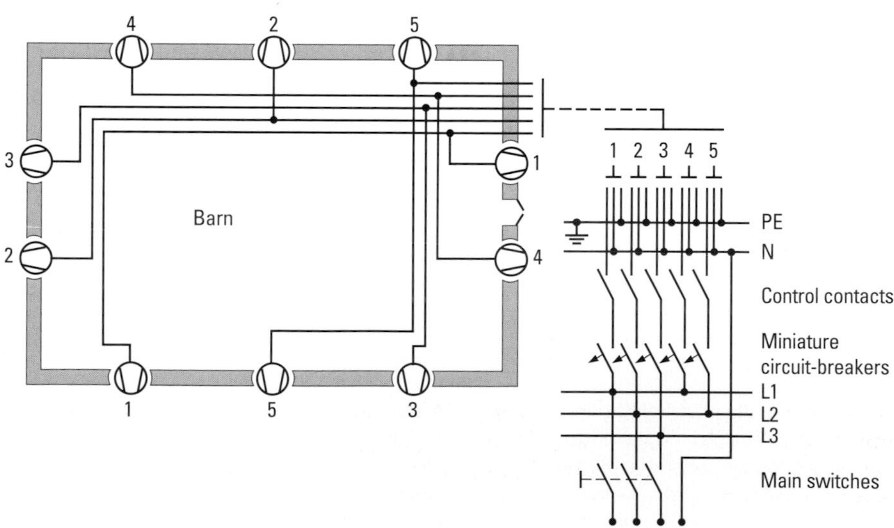

Fig. 5.9/4 Safe circuit-phase distribution for barn ventilation in factory-farming installations

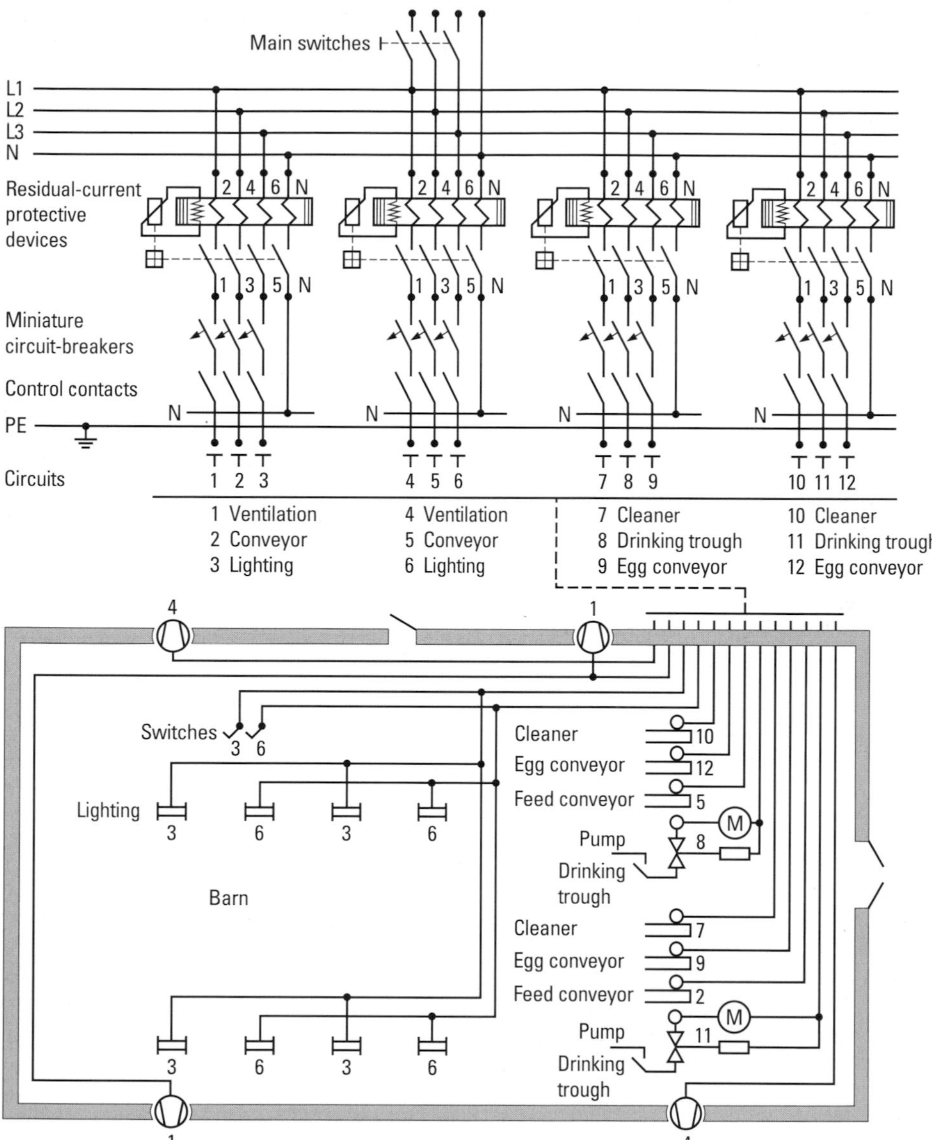

Fig. 5.9/5
Safe decentralization by assigning residual-current protective devices to circuits of vital supply and disposal systems

With factory-farming installations, hazards resulting from the failure of vital systems, e.g. air supply, must be avoided (see Figs. 5.9/4 and 5.9/5).

Residual-current protective system

Depending on the degree of danger present on agricultural or horticultural estates, fire protection measures incorporating residual-current protective devices must also be provided, particularly for barns (and poultry sheds), areas and rooms adjacent to barns, rooms used for factory-farming, storage and supply rooms for hay, chaff, concentrated feed and fertilizer, as well as rooms and areas in which grain, fodder, and potatoes, for example, are prepared (drying, steaming, and so on).

Permissible touch voltage

Since livestock is particularly sensitive to the effects of voltage, a maximum touch voltage of 25 V AC and 60 V DC only is permitted.

Protection against electric shock

To provide protection against electric shock, circuits with socket-outlets in TN, TT, and IT systems must be protected by means of residual-current protective devices with $I_{\Delta n} \leq 30$ mA. The protective measure for indirect contact must ensure that touch voltages higher than 25 V AC and 60 V DC cannot persist.

It is also advisable to protect all branch circuits, i.e. load circuits, using a residual-current protective device with a maximum rating of $I_{\Delta n} \leq 30$ mA.

All accessible exposed conductive parts of electrical equipment and all extraneous conductive parts located in areas where livestock is kept must be connected to each other by means of a supplementary equipotential bonding conductor and then to the equipment grounding conductor of the installation. It is advisable to embed a metal grid, which is connected to the equipment grounding conductor, in the ground.

Fire protection

In order to provide adequate fire protection, it is essential to ensure that no dust deposits settle on the electrical equipment. Such deposits prevent heat dissipation and, thus, represent a fire hazard.

Locations exposed to fire hazards

Switchgear and switching devices for protective measures, as well as control and isolation purposes must be installed outside of locations exposed to fire hazards.

This measure is not necessary if the switchgear/switching devices are installed in an enclosure which has the following minimum degree of protection in compliance with EN 60529 / IEC 60529 / DIN VDE 0470-1:

▷ IP 4X in dust-free environments,

▷ IP 5X in dust-laden environments.

Overload and short-circuit protection

When the switchgear/switching devices are installed, it is important to ensure that fire cannot be propagated by cables and conductors. This requirement can be satisfied by means of PVC-sheathed cables and conductors, for example. Overload and short-circuit protection is a general requirement. The necessary protective devices must be installed outside the rooms and locations exposed to fire hazards.

It is advisable to install cables and conductors, which are not actually used to supply power to locations exposed to fire hazards, outside of these areas.

PEN conductors must not be used in rooms or locations exposed to fire hazards; circuits which have to be routed through these rooms and locations are the exception.

Motors

Motors, which are remotely controlled or which start automatically (i.e. do not need to be monitored continuously), must be protected against excessive increases in temperature by means of a manually-resettable protective device, e.g. motor starter.

Luminaires

Luminaires in damp rooms must have degree of protection IP 44 and those in rooms exposed to dust IP 5X. Lamps and components of luminaires must be protected against mechanical stress.

Limiting fault currents

In order to limit the flow of fault currents with a view to reducing the risk of fire, all circuits must be protected by residual-current protective devices with $I_{\Delta n} \leq 0.3$ mA. If it is not possible to use a residual-current protective device, an insulation monitoring device (with continuous monitoring) must issue a visual and acoustic signal if an insulation fault occurs. The cables or conductors of the circuits to be protected can be monitored by means of an insulated pilot wire, e.g. an equipment grounding conductor.

Selecting and using electrical equipment

When items of electrical equipment are selected and used, it is important to ensure that they have at least degree of protection IP 44. Higher degrees of protection must be provided according to the external influences. Cables and conductors in barns must be installed in such a way that they cannot be reached or damaged by livestock.

NYM conductors or NYY cables with plastic sheaths must be used for permanent installation.

If disconnectors and switching devices for emergency switching including emergency stopping are installed, it is important to ensure that they cannot come into contact with livestock and that they are not installed at locations where access is hindered by livestock.

Furthermore, activities which could cause the livestock to panic must also be avoided.

5.9.5 Fire protection for exceptional risks and hazards

Operating areas exposed to fire hazards

Operating areas exposed to fire hazards are areas where a fire risk is present due to:

▷ the type of materials processed or stored there,

▷ the processing and storage of combustible materials,

▷ the accumulation of dust, e.g. in wood working shops, agricultural installations, textile factories, paper mills, and so on.

Operating areas exposed to fire hazards also include rooms or locations for which predominantly combustible building materials such as timber are used, e.g. prefabricated houses, since they represent an increased fire risk.

In future, rooms or areas where irreplaceable goods are stored or exhibited will also be classified as exceptional fire risks.

The electrical equipment in the rooms and locations mentioned above must be selected and installed taking these external influences into consideration so that the heat they generate under normal operating conditions and the anticipated increase in temperature in the event of a fault cannot cause a fire.

This can be achieved by selecting suitably designed equipment or by implementing additional protective measures when the equipment is installed. These measures are not necessary if there is no danger of the surface temperature of the equipment causing adjacent combustible materials to ignite.

Electrical equipment

When the electrical system is planned and installed, it is important to ensure that as few items of electrical equipment as possible are installed in rooms and locations of this type. Only those items of electrical equipment which are absolutely essential should be installed in these rooms and locations.

If electrical equipment is used in these rooms and locations, it must satisfy and comply with the following requirements.

If there is a danger of large amounts of dust accumulating on the enclosures of electrical equipment, measures must be taken to prevent the enclosures from reaching excessively high temperatures.

Degree of protection

If dust is likely to accumulate on electrical equipment installed at locations exposed to fire hazards, the equipment must have at least degree of protection IP 5X.

According to the relevant international standards, a higher degree of protection is not necessary if there is no risk of dust accumulating. In the Federal Republic of Germany, however, a preface to a national standard recommends that degree of protection IP 4X at least be satisfied (except in the case of electric heating appliances).

Failing to provide this degree of protection will result in a fire hazard due to heat accumulation caused by the ingress of foreign bodies (diameter of up to 12 mm).

Cables and conductors

If cable and conductor systems are not fully installed under the surface of the plaster or are not protected against fire by non-combustible materials such as concrete, the cables and conductors must be flame retardant in compliance with the relevant building and inspection regulations. Cables and conductors with improved characteristics in the event of a fire are recommended, especially where there is a high risk of flame propagation, e.g. in long vertical ducts or cable bundles.

If cables and conductors pass through operating areas exposed to fire hazards but are not used to supply power to equipment within these rooms, connections or terminals must not be installed in the operating areas, unless they are housed in suitable enclosures which satisfy the fire resistance tests for installation socket-outlets to DIN VDE 0606-1 (test temperature 850 °C).

Overload and short-circuit protection

It is always important to ensure that all cable and conductor systems, which supply power to and pass through operating areas exposed to fire hazards, are protected against overloads and short circuits. The appropriate protective devices must be installed outside these operating areas. If the feed point is located within the operating area exposed to fire hazards, the overload and short-circuit protective devices must be installed at the feed point of these circuits.

Protection against insulation faults

In TN and TT systems, the cable/conductor systems and installations, with the exception of mineral-insulated cables and busbars, must always be protected against insulation faults by means of an RCD (residual-current protective device) with $I_{\Delta n} \leq 300$ mA.

Ceiling heaters with panel heating elements must be protected by means of RCDs with $I_{\Delta n} \leq 30$ mA because these elements can ignite at fault currents as low as 100 mA. In both cases, only RCDs (see Section 4.1.3) without an auxiliary power source must be used, i.e. RCDs which comply with EN 61 008-1, EN 61 009-1 / IEC 61 008-2-1, IEC 61 009-2-1 / DIN VDE 0664.

If IT systems are used in operating areas exposed to fire hazards, the insulation monitoring device must issue both visual and acoustic alarm signals. If a second fault occurs, the disconnecting time of the overcurrent protective device must not exceed 5 seconds. It is important to ensure that manual disconnection can be performed as quickly as possible after the first fault.

Connection to equipment grounding conductor

It is always advisable to use only cables with concentric conductors within operating areas or locations exposed to fire hazards. The concentric conductors must be connected to the equipment grounding conductor so that, in the event of a fault, it can be assumed that the fault current flows via the concentric conductor to the grounding system and trips the line-side residual-current protective device (RCD) before the cable/conductor can overheat and cause a fire.

When the cable and conductor systems are installed, it is important to remember that PEN conductors are not permitted.

Cable and conductor systems, which are routed through operating areas exposed to fire hazards, should also be protected against insulation faults by means of a residual-current protective device (RCD).

Each neutral conductor must be connected to a terminal with disconnecting device to allow the insulation of the neutral conductor to be measured with respect to ground at a later stage when the system is in service. The terminal is usually in the distribution board.

Bare electrical conductors must never be installed. Suitable precautions must be taken to prevent electric arcs, sparks, or hot components from igniting adjacent combustible materials.

If flexible cables are used, they should preferably be suitable for onerous operating conditions and severe loads, e.g. H07RN-F or other protected cables.

Wherever possible, switching devices should be installed outside the operating areas exposed to fire hazards. If this is not possible, they must have the appropriate degree of protection.

Motors, which are controlled automatically or remotely, or are not monitored continuously, must be protected against excessively high temperatures by means of an overload protective device with manual reset or an equivalent overload protective device. Motors with delta-wye starting must also be protected against excessively high temperatures at the wye stage if they are operated without continuous monitoring.

The surface temperature of luminaires used in these operating areas must not exceed a specific limit.

The surface temperature is limited to 90 °C under normal operating conditions, and to 115 °C under fault conditions.

Luminaires

The luminaires must bear the ▽F ▽F identification in the triangle.

Luminaires, particularly those in operating areas in which a potential fire hazard exists as a result of dust or fibers, must be constructed in such a way that their surface cannot exceed a specific temperature limit in the event of a fault and that potentially dangerous quantities of dust and fibers cannot accumulate on them.

Spotlights, projectors

If no manufacturer specifications are available regarding the use of small spotlights and projectors, they must be positioned at the following minimum distances from combustible materials:

- ▷ <100 W 0.5 m,
- ▷ ≥100 W to 300 W 0.8 m,
- ▷ ≥300 W to 500 W 1.0 m.

Lamps and other components of luminaires must be protected against the anticipated mechanical loads. The protective elements must not be attached to the lamp holders, unless they form an essential part of the luminaire.

Measures must be taken to ensure that components such as lamps or hot parts cannot fall out of the luminaire.

Electric heating and ventilation systems

If electric heating and ventilation systems are used, the dust concentration and air temperature must not give rise to a fire risk. Thermal cut-outs integrated in these devices must have a manual reset only. The heaters must be attached to non-combustible bases. Heaters, which are erected or mounted close to combustible materials, must be fitted with suitable covers in order to prevent the materials from igniting. If storage heaters are used, it is important to ensure that the hot core cannot ignite combustible dust or fibers. The storage heaters normally used in residential and office buildings are not suitable for operating areas exposed to fire hazards.

The enclosures of electric heating devices, such as heaters, resistors, etc., must not reach temperatures higher than those specified above for luminaires. These devices must be constructed and installed in such a way that heat dissipation cannot be prevented as a result of dust etc. accumulating.

Rooms and locations with combustible building materials, e.g. prefabricated houses

The installation technician must take suitable precautions and ensure that electrical equipment cannot ignite combustible walls, floors, and ceilings. As with operating areas exposed to fire hazards, protection against insulation faults must be provided here as well. This means that residual-current protective devices are an essential requirement.

Selecting electrical equipment

When electrical equipment for cavity walls is selected and installed, it is important to ensure that suitable strain and compression relief devices are fitted at the cable entry openings. The electrical equipment must bear the ▽ symbol indicating that it is suitable for use in cavity walls. With distribution boards, it is important to note that only cavity wall distribution boards conforming to DIN VDE 0603 are used. The tapping boxes must comply with DIN VDE 0606.

If electrical equipment which does not satisfy the conditions of the above-mentioned standards are used, the equipment must be covered with a 12 mm coating of silica fibers, with an equally non-flammable material, or with a 100 mm coating of glass or rock wool. If such materials are used, the heat dissipation capacity of the covered electrical equipment, especially cables and conductors, will be reduced which means that a lower current load must be selected.

Electrical equipment, such as socket-outlets and switches in device boxes, must be secured with screws only, and not with claws. It is always important to ensure that only cables, conductors, electric wiring conduits and ducts, which have been verified as fire proof, are used in cavity walls and prefabricated houses. They must satisfy the specifications for fire-risk testing in the associated standards. Cables installed externally, which are connected to joint boxes in cavity walls, must also be fitted with strain relief devices.

5.9.6 Electrical systems on camping sites

The requirements described below apply to electrical systems on camping sites which supply habitable recreational vehicles (including caravans) or tents with a maximum rated system voltage of 440 V AC.

Terms

A *mobile leisure vehicle* is a dwelling unit which is used for leisure purposes temporarily or seasonally and which may also be suitable for use in highway traffic.

A *caravan* is a towed mobile leisure vehicle which is used for vacations and which is suitable for use in highway traffic.

A *motor caravan* is a mobile leisure vehicle with integrated drive unit, which is used for vacations and which is suitable for use in highway traffic.

A *mobile home* is a dwelling unit which is equipped with portable equipment, but does not comply with either the construction or application-related regulations for highway traffic vehicles.

A *caravan lot* is a lot on a camping site which is intended for a mobile leisure vehicle.

A *camping site* is an area of land which has two or more caravan lots.

An *electrical feed point* for a caravan lot is an electrical device used to connect the supply cables of mobile leisure vehicles to the mains system.

Protection by means of obstacles or clearance is not suitable as protection against electric shock (see Section 2.3). These two measures must not be applied on camping sites. Protection by means of non-conductive locations is also impermissible.

Feed point

The feed point for supplying power to the caravan lot must be immediately next to the lot. The distance to the connecting point of the mobile leisure vehicle or tent must not be greater then 20 m when the vehicle or tent has been parked or pitched on the allocated lot.

Socket-outlets

The design of socket-outlets, which are used for supplying power to mobile leisure vehicles, must comply with EN 60309-2 / IEC 60309-2 / VDE 0623 Part 20. The enclosure of the socket-outlet must be made of flame-retardant material which must be tested to EN 60695-2-1 / IEC 60695-2-1 / VDE 0471 Part 2-1. A test temperature of 850 °C must be used for constructional elements which carry conducting parts, and 650 °C for other parts. The socket-outlet must be mounted at a height of 0.8 m to 1.5 m above ground level (measured from the lower part of the socket-outlet). The rated current of the socket-outlet must satisfy the power requirements and be no less than 16 A. At least one socket-outlet, which is protected by means of a single, dedicated overcurrent protective device, must be available for each mobile leisure vehicle.

Residual-current protective device

The socket-outlets must be protected by means of residual-current protective devices in compliance with EN 61008-1, EN 61009-1 / IEC 61008-2-1, IEC 61009-2-1 / DIN VDE 0664. No more than three socket-outlets should be assigned to one residual-current protective device.

Electrical connection

The electrical connection between the socket-outlet of the caravan lot and the mobile leisure vehicle must comprise:

▷ a grounding-type plug conforming to EN 60309-2 / IEC 60309-2 / VDE 0623 Part 20,

▷ a flexible cable (H07RN-F type or equivalent) with an equipment grounding conductor and the following characteristics:
length 25 m; minimum cross section 2.5 mm^2; color coding: equipment grounding conductor green/yellow, neutral conductor light blue,

▷ a coupling grounding-type socket-outlet conforming to EN 60309-2 / IEC 60309-2 / VDE 0623 Part 20.

As far as power supply systems for caravan lots are concerned, the cables required for the distribution circuits should be laid underground wherever possible. It is important to note that the cables should only be laid underground without suitable additional mechanical protection at locations where there are no caravan lots; i.e. at locations where tent pegs and spikes are not driven into the ground.

If power is supplied via overhead cables, these must always be insulated. Masts and other supports for overhead cables must be erected and protected in such a way that they cannot be damaged by moving vehicles. The overhead cables must be mounted at a height of at least 6 m above ground level in areas where vehicles are driven, and at a height of 3.5 m in all other areas.

5.9.7 Power supply for boats and yachts at berths

A ground-type socket-outlet conforming to DIN 49 462 Part 1 with the following ratings must be provided for supplying power to boats and yachts at berths:

Rated voltage	220 V – 240 V AC,
Rated current	16 A,
Number of poles	2 + PE,
Type	Splash-proof (IP X4).

The socket-outlets must be installed so that they are within a 20 m radius of each berth. Up to six socket-outlets can be combined to form a socket-outlet group. A residual-current protective device with $I_{\Delta n} < 30$ mA must be connected on the line side of each socket-outlet group. It is, however, advisable to connect a residual-current protective device on the line side of each individual socket-outlet.

Every socket-outlet must be protected by means of an overcurrent protective device (see Section 4.1) with a maximum rated current of 16 A. Two-pole protection may also be required depending on the power supply system. The boat units must be connected using a grounding-type connector socket conforming to DIN 49 462 Part 2 with the above-mentioned ratings.

An equipment grounding conductor (PE) must be incorporated in the electric circuits in order to provide protection against electric shock (see Chapter 2). This means that Class II devices (totally insulated devices) can also be used. Only grounding-type socket-outlets must be used. Cables must be laid in such a way that mechanical damage caused by the motion of the vessel and pedestrians/vehicles around the berth is prevented.

Mechanically protected boxes must be used for all cable connections.

5.9.8 Classrooms with experiment/demonstration benches

Safety extra-low voltage/functional extra-low voltage with safety separation

An experiment/demonstration bench is an area used for performing experiments with electrical equipment or devices in classrooms. Only safety extra-low voltage (SELV) or functional extra-low voltage (FELV) with safety separation should be used, provided that the intended purpose of the experiment/demonstration bench permits this. Since this is not, however, always the case in practice, a residual-current protective device with $I_{\Delta n} \leq 30$ mA must be provided in TN and TT systems or with functional extra-low voltage without safety separation supplied by TN or TT systems if protection against both direct and indirect contact is required (see also Section 2.4).

Extraneous conductive parts

Extraneous conductive parts, which are located within normal arm's reach of the experiment/demonstration bench, must be insulated, fitted with an insulating cover or enclosure or, alternatively, connected together via an equipotential bonding conductor and then to the equipment grounding conductor.

5.9.9 Low-voltage generating plants

Standards

The following information applies to low-voltage and extra-low voltage systems with generators for supplying power either continuously or temporarily to the entire installation or just part of it in compliance with DIN VDE 0100-728 (in future: HD 384.5.551 / IEC 60364-5-551 / DIN VDE 0100-5-551).

This section discusses the requirements for the following variants:

▷ power supplied by a system which is not connected to the public mains supply (in Germany, a distinction is not made between public and non-public mains supply systems, i.e. the requirements specified below apply equally to both public and non-public mains supplies in Germany),

▷ power supplied by a system as an alternative to the public mains supply,

▷ power supplied by a system parallel to the public mains supply,

▷ suitable combination of the above-mentioned power supply alternatives.

A generating plant must be erected in accordance with HD 384.5.551 / IEC 60364-5-551 / DIN VDE 0100-5-551. In addition, it is also advisable to take account of the requirements of the public utility and to adapt the plant accordingly.

The following requirements do not apply to independent parts of electrical extra-low voltage equipment which contains both the power source and the power-consuming load and which is not subject to any special equipment standards with regard to electrical safety.

Power sources

Generating plants with the following power sources are considered:

▷ internal-combustion engines,

▷ turbines,

▷ electric motors,

▷ photo-voltaic cells,

▷ electrochemical batteries,

▷ other suitable power sources.

Electrical characteristics

Generating plants with the following electrical characteristics are considered:

▷ mains-excited and independently-excited synchronous generators,

▷ mains-excited and self-excited asynchronous generators,

▷ mains-commutated and self-commutated static inverters with or without bypass devices.

Generating plants for the following applications are considered:

▷ power supply for permanently installed systems,

▷ power supply for temporarily installed systems,

▷ power supply for portable equipment which is not connected to permanently installed equipment.

General requirements

Dimensioning

The excitation and commutation equipment must be dimensioned for the intended use of the generating plants. The generating plant must not prevent other power sources from operating safely and correctly.

The prospective short-circuit current and ground-fault current must be determined for each power source or for each combination of power sources which can be used independently of the other power sources or power source combinations. The specified value of the rated short-circuit breaking capacity of the protective device in the system, which – if appropriate – is connected to the public mains supply, must not be exceeded for any of the intended operating modes of the power sources.

If the generating plant is intended to supply power to a system which is not connected to the public mains supply or a power supply other than the public mains supply, the capacity and operating characteristics of the generating plant must be such that the equipment cannot be damaged if an intended load is connected or disconnected as a result of the voltage or frequency deviating from the specified tolerance range. Suitable devices must be provided which, if necessary, disconnect the system components automatically if the capacity of the generating plant is exceeded.

Special attention must be paid both to the rating of the individual loads in relation to the capacity of the generating plant and to the motor starting currents, as well as to the specified breaking capacity for the protective device in the system.

Protection against electric shock

"*Protection against electric shock under fault conditions*" (see Section 2.4), taking into consideration each power source or combination of power sources which can be used independently of other power sources or combinations of power sources, must be provided for the system.

Protection by automatic disconnection

Protection by automatic disconnection of the power supply must be provided. This must not depend on the grounding system of the public mains supply if the generating plant is used as a switchable alternative supply to a TN system. A suitable grounding electrode (see Section 1.5) must be provided. The residual-current protective device (RCD without auxiliary power source) is ideal for this protective measure (see Section 4.1.3).

Supplementary equipotential bonding

If protection against electric shock under fault conditions for parts of a system, which is supplied by a static inverter, is based on the automatic closing of the bypass circuit-breaker, and if the protective devices on the feed side of the bypass circuit-breaker do not operate within the specified time, a supplementary equipotential bonding conductor must be provided between exposed conductive parts, which can be accessed simultaneously, and extraneous conductive parts on the load side of the static inverter. The resistance of the required supplementary equipotential bonding conductor between the simultaneously accessible conductive parts must satisfy the following condition:

$$R \leq \frac{50\ V}{I_a}\ \Omega.$$

I_a is the maximum current which flows with respect to ground in the event of a fault and which can be supplied by the static inverter for a period of up to 5 seconds.

Portable generating plants

The following requirements apply to portable generating plants and generating plants which are designed to be transported to any location and used temporarily or for short-time duty:

▷ equipment grounding conductors, which are part of a suitable connecting conductor or cable, must be provided between separately installed items of equipment,

▷ in TN, TT, and IT systems, an RCD with $I_{\Delta n} \leq 30$ mA must be installed to allow automatic disconnection.

> *Note*
> In IT systems, an RCD does not have to disconnect, unless one of the ground faults

occurs in part of the system on the feed side of the residual-current protective device.

Overcurrent protection

If devices for detecting overcurrents in the generating plant are provided, they must be installed as closely as possible to the generator terminals.

Parallel operation

Stray harmonic currents

If a generating plant is intended for parallel operation with a public mains supply, or if two or more generating plants are operated in parallel, the stray harmonic currents must be limited so that the thermal rating of the cables and conductors is not exceeded.

The effects of the stray harmonic currents can be limited by:

▷ selecting generating plants with compensating-field windings,

▷ providing a suitable impedance along the connection to the neutral point of the generating plant,

▷ providing switches which interrupt the circuit carrying stray currents, but which are interlocked in such a way that protection against electric shock under fault conditions is not impaired at any time,

▷ providing filters,

▷ providing other suitable devices.

Additional requirements for systems for which the generating plant constitutes a switchable alternative supply to the public mains supply (stand-by power supply system)

Disconnection requirements

Precautions in compliance with the relevant disconnection requirements must be taken to ensure that the generating plant cannot operate in parallel to the public mains supply.

Suitable precautions include:

▷ an electrical, mechanical, or electromechanical interlock between the operating mechanisms or the control circuits of the transfer devices,

▷ a system of interlocks with only one key,

▷ a three-position transfer switch which first disconnects and then connects,

▷ an automatic transfer switch with suitable interlock,

▷ other devices which provide an appropriate level of operational safety.

With TN-S systems, in which the neutral conductor is not disconnected, RCDs must be installed to prevent malfunctions caused as a result of a parallel connection between the neutral conductor and ground.

In TN systems, it may be advisable to disconnect the neutral conductor of the system from the neutral conductor of the public mains supply in order to prevent malfunctions, e.g. in the case of impulse voltages caused by lightning.

Additional requirements for systems in which the generating plants can be operated in parallel to a public mains supply

Selection and use

When a generating plant for parallel operation with a public mains supply is selected and used, it is important to avoid any negative effects on the mains supply and other systems with respect to the power factor, changes in voltage, non-linear distortions, load and symmetry as well as starting, synchronization, and flicker effects. The public utility must be consulted regarding special requirements.

Synchronization

If synchronization is required, automatic synchronization systems which take account of the frequency, phase angle, and voltage are recommended.

Voltage and frequency deviation

Protective devices must be provided which disconnect the generating plant from the public mains supply if the power supply is interrupted or if the voltage or frequency at the terminals of the mains supply deviates from the values which have been defined for normal supply.

The type, sensitivity, and operating times of the protective devices depend on the protective measure implemented for the public mains supply and must be coordinated with the public utility. Devices must be provided which prevent a generating plant from being connected to the public mains supply if the voltage and frequency of the public mains supply are outside the range of operating values which comply with the above-mentioned specifications.

Disconnection

Devices must be provided which enable the generating plant to be disconnected from the public mains supply. The disconnecting devices must be accessible to the public utility at all times. If a generating plant can also be operated as a switchable alternative supply to the public mains supply, the system must also satisfy the supplementary requirements for systems for which the generating plant constitutes a switchable alternative supply to the public mains supply.

If, in the case of the stand-by power supply, there is any uncertainty as to whether the protective measure implemented in TN systems for the general power supply will remain effective, the outgoing circuits must be configured as TN-S systems and RCDs must be used to fulfil the requirements for protection by automatic disconnection of the power supply.

If the normal power supply is fed via a TT system, the neutral conductor of the load system must also be switched over if the stand-by power supply is activated. If, when the stand-by power supply is activated, there is any uncertainty as to whether the protective measure implemented for the general power supply will remain effective, RCDs must be used to fulfil the requirements for protection by automatic disconnection of the power supply.

5.9.10 Erecting and connecting low-voltage switchboards and distributions boards

Standards

Low-voltage switchboards and distribution boards must be manufactured in accordance with the relevant standards (see Section 1.4). EN 60439-1 / IEC 60439-1 / DIN VDE 0660-500, together with additional standards for the respective types of switchboard and distribution board, applies to low-voltage switchgear assemblies. DIN VDE 0603 applies to small distribution boards and meter mounting boards up to 250 V with respect to ground. EN 60439-4 / IEC 60439-4 / DIN VDE 0660-501 applies to construction site distribution boards for rated voltages of up to 400 V AC and for rated currents of up to 630 A.

Degree of protection

Following installation, switchboards and distribution boards must have the degree of protection specified for solid bodies and the ingress of water (IP degree of protection to EN 60529 / IEC 60529 / DIN VDE 0470-1) and protection against direct contact (see Section 2.3). The protective measures must also be effective against indirect contact (see Section 2.4).

According to DIN VDE 0100-731, degrees of protection lower than IP 2X are only permissible in electrical operating areas or closed electrical operating areas.

Installation

Switchboards and distribution boards must be installed, assembled, and secured so that they are resistant to torsional strain, e.g. by using base rails, supporting frames, or racks. The characteristics of the switchboards and distribution boards should not be impaired, e.g. by the effect of dust and/or dampness at the installation location during intermediate storage and assembly. It is essential to follow the instructions supplied by the manufacturer. Switchboards and distribution boards must be arranged and installed in such a way that the width and height of gangways and aisles comply with the dimensions specified in DIN VDE 0100-729.

Cable entries

Identification

Once all the connection work has been completed, the cable entries must be sealed (if suitable fittings have not been provided by the manufacturer) so that the intended degree of protection for switchboards and distribution boards is satisfied. The assignments of incoming conductors to the circuits must be identified clearly and permanently, e.g. by arranging or labeling the conductors according to the circuit documentation. This is best achieved by attaching the circuit number to the respective equipment grounding conductor [PE] (green/yellow) or neutral conductor [N] (light blue) using strips of numbered adhesive tape. This ensures that conductors can be re-assigned to the correct circuits even if maintenance work is carried out at a later stage or the conductors have to be disconnected.

5.9.11 Electrical operating areas and closed electrical operating areas

Layout of system components

Socket-outlets, luminaires, and other devices must be mounted in such a way that, if work has to be carried out on them, live parts in the system cannot be touched. In electrical operating areas and closed electrical operating areas, it is important to ensure that the systems and system components are clearly allocated. This can be achieved, for example, by means of clearly organized circuit diagrams or by labeling the system components.

Lockable doors or covers

Access to closed electrical operating areas should only be possible via lockable doors or covers. Access doors of closed electrical operating areas must open outwards and the door locks must be designed to prevent access by unauthorized personnel. Personnel located in the system must, however, be able to exit it without hindrance. Doors between different sections of closed electrical operating areas do not require a lock. Exits must be positioned accordingly so that the escape route inside the room is no more than 40 m in length. Windows must be secured against unauthorized entry.

Protective measures

With rated system voltages above 50 V AC or 120 V DC, measures conforming to HD 384.4.41 / IEC 60364-4-41 / DIN VDE 0100-410 must be implemented to provide protection against electric shocks (see Section 2.4). If, for technical reasons, these protective measures cannot be implemented for certain items of equipment, the equipment in question must be identified accordingly.

5.9.12 Damp/wet areas and rooms

In damp/wet areas and rooms, electrical equipment must be at least drip-proof (degree of protection IP X1 to EN 60529 / IEC 60529 / DIN VDE 0470-1). In areas and rooms where jet-water is used and electrical equipment does not normally come into contact with the jet-water, e.g. for cleaning purposes, the equipment must be at least splash-proof (degree of protection IP X4 to EN 60529 / IEC 60529 / DIN VDE 0470-1). In areas and rooms where jet-water is used, equipment which is directly exposed to the water jet must have an adequate degree of protection to prevent the ingress of jet-water or, alternatively, suitable supplementary protection which does not prevent the equipment protected in this way from functioning correctly. Metal parts which are exposed to corrosive vapors or fumes must be protected against corrosion, e.g. by means of protective paint or by using corrosion-resistant materials. In protected outdoor installations, equipment must be at least drip-proof (degree of protection IP X1 to EN 60529 / IEC 60529 / DIN VDE 0470-1). In unprotected outdoor installations, the equipment must be spray-water-protected (degree of protection IP X3 to EN 60529 / IEC 60529 / DIN VDE 0470-1).

5.9.13 Applying protective measures for socket-outlets

Installed outdoors or used for supplying outdoor equipment

According to IEC 364-4-471.2.3: 1981, outdoor socket-outlets with a rated current of up to 20 A which have been installed in TN and TT systems must be protected by means of residual-current protective devices with $I_{\Delta n} \leq 30$ mA. Socket-outlets inside rooms from which power is occasionally supplied to portable electrical equipment intended for outdoor use, must also be protected by means of residual-current protective devices. This type of protection does not have to be provided for socket-outlets in areas which are only accessible to electricians or suitably instructed personnel, or for electrical installations and electrical equipment which are inspected on a regular basis.

If a system which allows portable electrical equipment to be used outdoors is planned, one or more socket-outlets should be installed outdoors in accordance with the above-mentioned requirements.

Installation in residential buildings

In TN and TT systems in apartments, residual-current protective devices with $I_{\Delta n} \leq 30$ mA must be provided in the cases specified below as supplementary protection against electric shock under fault conditions:

▷ in circuits with socket-outlets (≤ 32 A) to which hand-held electrical equipment is usually connected (in rooms such as kitchens, hobby rooms, utility rooms, and workshops),

▷ in old systems in which, according to previous standards, protection against indirect contact by means of disconnection was not required,

▷ in circuits with socket-outlets in existing systems for which, according to Group 700 of the German standard DIN VDE 0100, residual-current protective devices with $I_{\Delta n} \leq 30$ mA are obligatory if these socket-outlets are re-installed at a later stage. This requirement also applies if the systems are used by electricians only or undergo regular inspection.

Fig. 5.9/6 shows various possible arrangements of the residual-current protective devices.

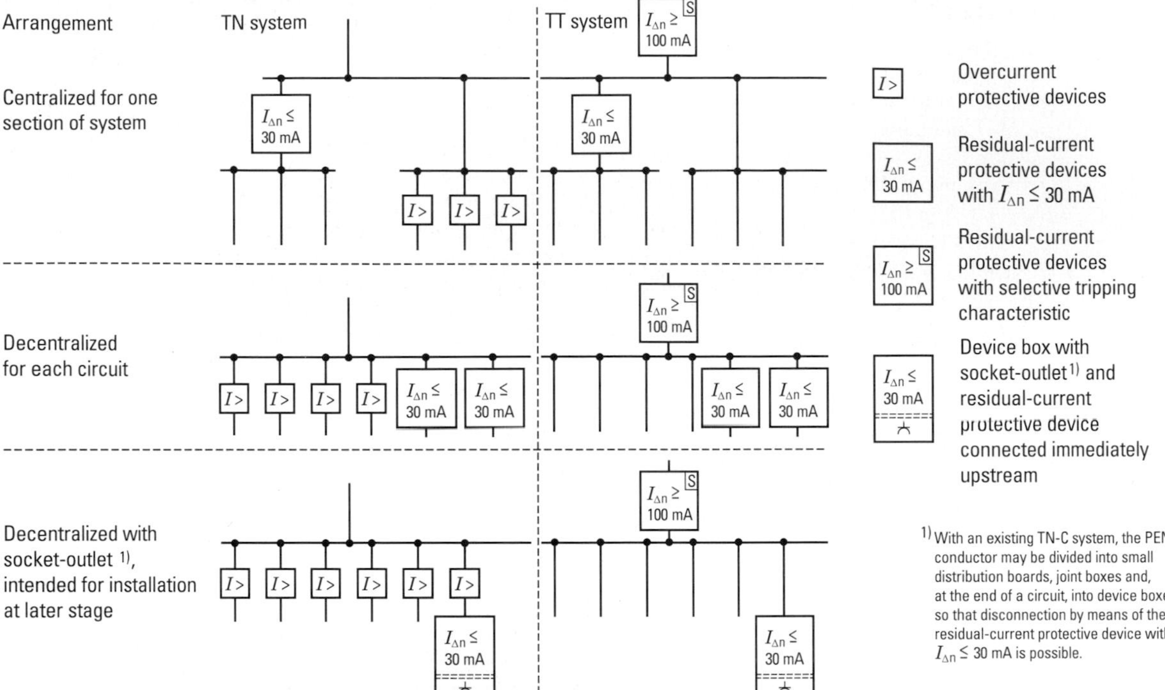

The illustration only shows the protective devices for protection against indirect contact and supplementary protection against direct contact by means of residual-current protective devices conforming EN 61008-1/IEC 61008 with $I_{\Delta n} \leq$ 30 mA

Fig. 5.9/6 Possible arrangements of residual-current protective devices in TN and TT systems

6 Building management system

6.1 Terms and hierarchical structure

Changing terminology

Development of technology and terms

Developments in building control systems have lead to changes in the requirements, technical solutions, and terminology associated with this area. This section, therefore, begins with a brief introduction to these developments from an historical perspective.

Large buildings and building complexes require extensive building services systems (Fig 6.1/1). These include systems for heating, ventilation and air conditioning, electrical engineering, lighting, blinds and shutters, sanitary engineering, transport (elevators, escalators, document transportation systems), and information and communications (telephone systems, paging systems, office and data systems technology). Security systems are also required, e. g. for controlling access, and indicating intrusions and fires. In addition to these, administrative systems are used for the day-to-day running of buildings, e. g. for acquiring system operating data, maintenance, time management, room planning, as well as cost management and control.

Building services system

In the past, central fault indication systems (which indicated and signaled faults centrally) were used to quickly detect faults in building services systems. To enable the central location to respond quickly to the signaled faults, this system was developed further to create the centralized instrumentation and control system for buildings and the building services management system. Increasingly powerful control room computers made it possible to develop and implement energy management functions for reducing energy requirements, thereby bringing about a reduction in building operating costs. Systems of this type were known as building and energy management systems.

Building management

Facility management

All these terms have now been replaced by the term "building management system". This designation was introduced officially in Germany in 1993 in the standards relating to building costs (DIN 276). Nowadays, building management entails integrating not only the building services systems, but also the security and administrative systems in one comprehensive management system. The term "facility management" is used for systems that operate and manage entire buildings or building complexes by means of computer-assisted control. Building control systems are, for the most part, developed and marketed by specialist companies that manufacture measuring and control systems for heating, ventilation and air conditioning systems.

Building control systems

Due to the costs involved, building control systems were formerly used only in large buildings and building complexes. Nowadays, however, these systems are being used to an increasing extent in smaller buildings. This is due to efficient, state-of-the-art microprocessors and personal computers, as well as building system engineering based on the European Installation Bus (*EIB*). In contrast to system automation in residential buildings (i.e. automation of supply and waste disposal systems in apartments and apartment blocks (see Section 4.4), building system automation is used in functional buildings, e. g. administrative and industrial buildings, department stores and commercial buildings, hospitals and retirement homes, leisure complexes and swimming pools, airports and railroad stations, as well as schools and universities.

Tasks and hierarchical structure

Tasks

The main tasks of building system automation include monitoring, controlling and

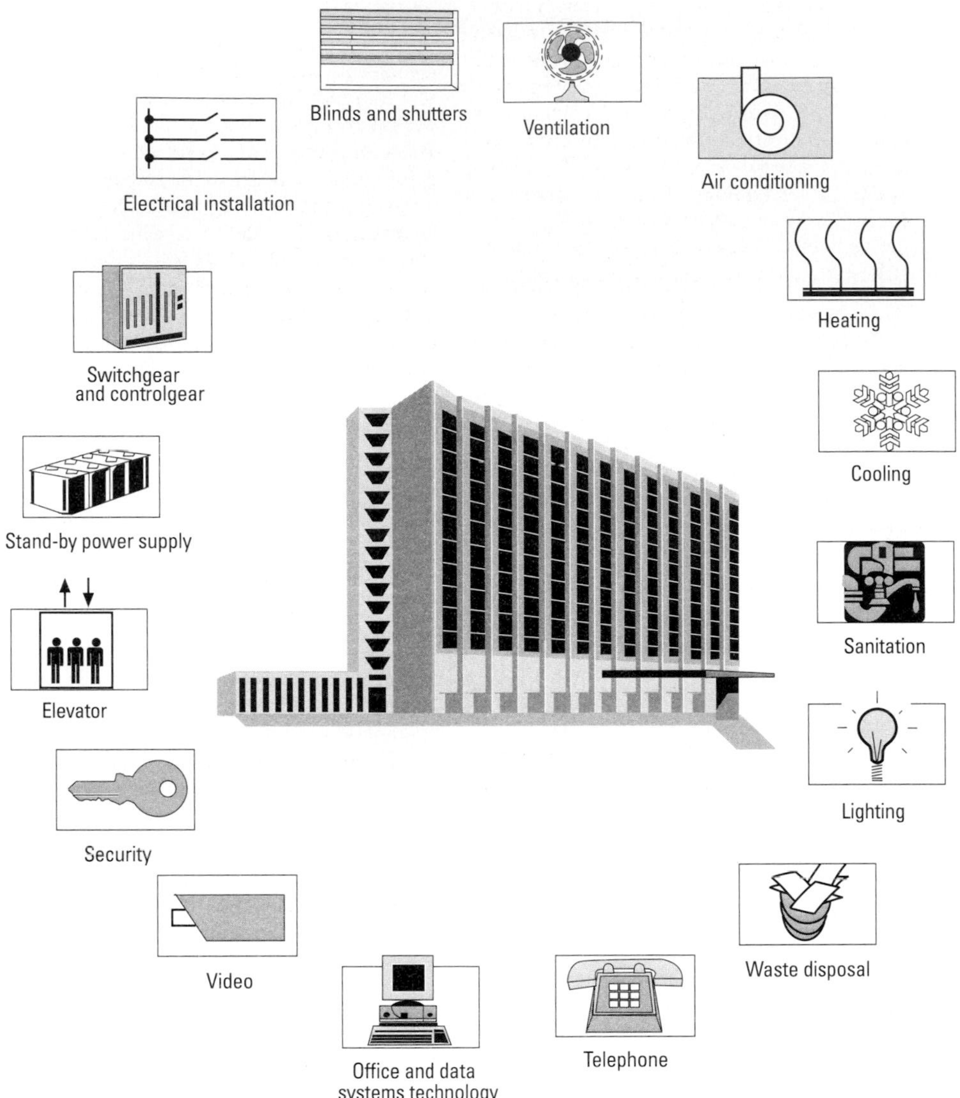

Fig. 6.1/1 Building services systems

optimizing (i.e. running the building and its systems using the lowest possible amount of energy) the building services systems, recording and representing energy requirements and all other consumption values in the form of statistics, displaying operating states and technical faults at a central location, as well as providing centralized control and intervention during normal operation or if a fault occurs.

Due to the considerable size of the building systems that have to be monitored, building control systems are structured in the form of a hierarchy. These systems were formerly divided into the following levels:

▷ control rooms,
▷ sub-control rooms,
▷ substations.

Substations

The substations formed the lowest processing and hierarchy level. Their task was to acquire signals and measured values, and issue commands to the building services systems. These were controlled by means of separate devices that belonged to these systems, but not to the substations. As microprocessor technology became more advanced, however, digital control of the building services systems was carried out by DDC substations (**D**irect **D**igital **C**ontrol).

Sub-control rooms, control rooms

The sub-control rooms reduced the workload of the control room, at a time when the computers used there were not as powerful as their modern equivalents. The control room signaled and displayed faults. It was equipped with a computer, printers, and display devices and processed all the monitoring, display, operating, statistical, and energy management functions. As a result of further technical developments, the control room functions were transferred to the subordinate hierarchy levels in order to increase availability.

Standardization

Standards

To date, guidelines and standards pertaining to building system automation applied only at a national level. For this reason, a group of international experts at CEN TC247, WG3 has been working for several years on a European standard with standardized terminology and functions for building system automation. In addition to this, another expert group at ISO TC205 has started work on an international standard for building system automation. Fig. 6.1/2 shows the current hierarchical structure and standardized terms in accordance with CEN TC247.

In building system automation, a distinction is made between the:
▷ management level,
▷ control and automation level,
▷ field level.

The standards for building system automation elaborated by CEN TC247 will be described in more detail below, insofar as they are known or have been published.

6.2 Field level

Tasks and topology

Field level

The field level is used to enter data from (or output data to) the building services system for signal conditioning and, if necessary, for transferring data to the next highest processing level (Fig. 6.2/1). The tasks of the field level also include room management, i.e. monitoring and controlling room functions, such as lighting, blinds and shutters, heating, ventilation, etc.

Field devices

The field level takes its name from the field devices for inputting and outputting data, the sensors (e.g. measuring sensors) and actuators (e.g. all-or-nothing relays, valve actuators, and flap servo drives) installed in the building services system.

Data input

When data is input from the building services systems, a distinction is made between:
▷ signals (binary values),
▷ measured values (analog values acquired via passive or active sensors),
▷ count values (total consumption pulses).

Data output

When data is output to the building services systems, a distinction is made between:
▷ switching commands,
▷ control commands.

Switching commands

Switching commands are divided into single-stage or multi-stage (two-stage, three-stage) switching commands, as well as pulse commands or persistent commands.

Control commands

With control commands, a distinction is made between pulse-width-modulated binary signals, three-step control commands (higher, off, lower) and analog control commands (0/2... 10 V DC or 0/4... 20 mA).

Direct/service operation units

To enable the system to be operated manually in the event of a control system failure (e.g. to open or close a valve, switch a system or system component on or off), direct/service operation units (SOUs) are often installed at the field level.

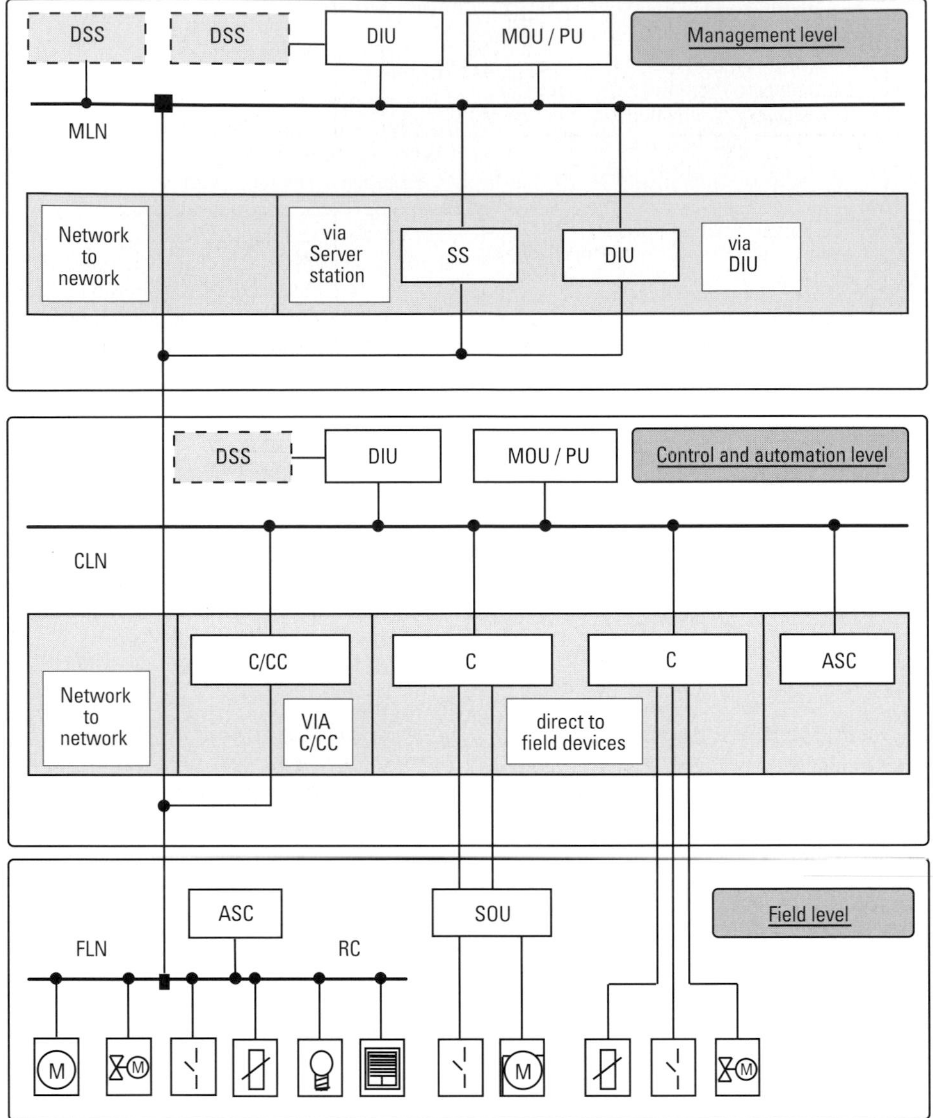

Fig. 6.1/2 Building system automation – levels (source: CEN TC247, WG3)

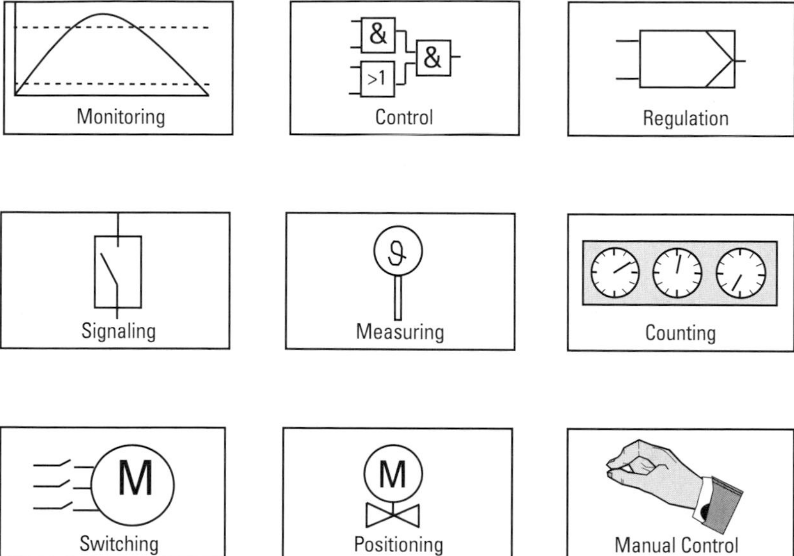

Fig. 6.2/1 Functions at the field level

Communication at the field level

Field level network

The sensors and detectors for acquiring data and the interface modules for issuing commands are connected either directly to the inputs and outputs of the controllers, or linked via a communications network – the **f**ield **l**evel **n**etwork (FLN). Intelligent devices for distributed room function control (**r**oom **c**ontrol **s**ystem RCS) are also connected to the field level network. This network must be competitively priced, immune to interference, and easy to install. Since a very large number of sensors and actuators can be connected, the data transmission rate must be low and the telegrams short for event-driven data transmission to be used effectively. This ensures that data transmission is cost effective, and interference free, and that the response times are short.

6.3 Control and automation level

Tasks and topology

Control and automation level, controllers

The control and automation level carries out all the monitoring, controlling, and optimizing functions in the building services system (Fig. 6.3/1). These tasks are performed either by distributed devices (**a**pplication **s**pecific **c**ontrollers – ASC) or by modular controllers (C) installed in the engineering control rooms. These devices detect and signal technical malfunctions, monitor measured and consumption values to ensure that they do not fall below or exceed limit values, and determine the operating hours of systems and system components in order to derive the maintenance signals on the basis of this information. Their primary task, however, is to digitally control (**d**irect **d**igital **c**ontrol-DDC) the connected building services system. The controllers are programmable devices. Programs and data can be loaded locally to the ASCs and Cs by means of a **p**rogramming **u**nit (PU), and can also be downloaded from (or uploaded to) the management level.

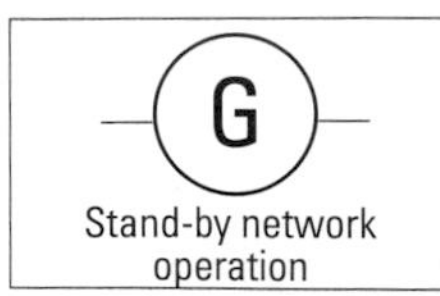

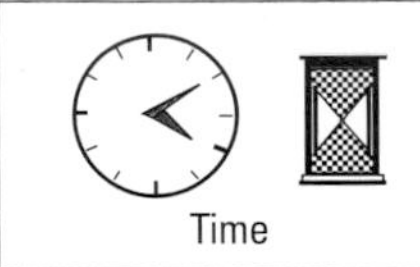

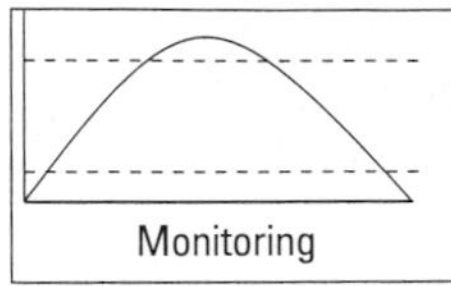

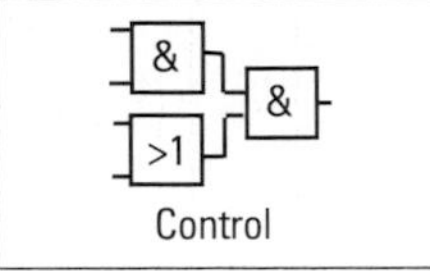

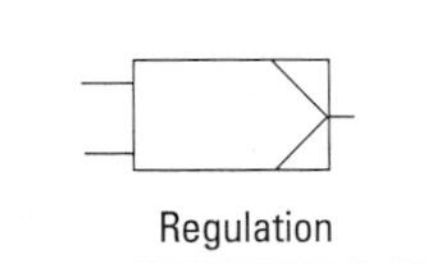

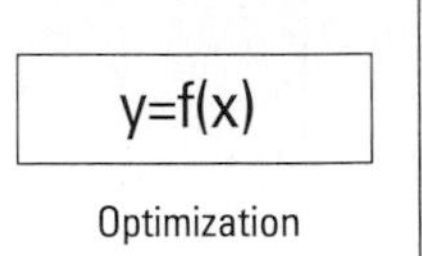

Fig. 6.3/1 Functions at the control and automation level

Monitoring and operator units/stations

The control and automation level can be displayed and operated locally by connecting a monitoring and operator unit/station (MOU).

Energy management

Optimization programs

Optimization programs are used to reduce energy requirements and thus cut operating costs. Modern programs of this type are executed by distributed controllers, thus making energy management more efficient. The optimization functions include:

▷ event programs,
▷ scheduler programs,
▷ variable switching,
▷ cyclic switching,
▷ night cooling,
▷ peak load limiting,
▷ enthalpy regulation.

Event programs

Event programs are used to switch systems and parts of systems on and off according to demand. For example, the event message “temperature too high” initiates a program that activates fans. The event message “temperature normal” triggers another program that deactivates the fans.

Scheduler programs

Scheduler programs ensure that heating, ventilation, and lighting systems are only switched on when rooms or buildings are actually being used. Programs of this type comprise day programs, week programs, and calendar programs. Day programs contain all the related switching times and commands for a specific day. Week programs consist of the day programs for the individual days of each week or of a specific week. Calendar programs enable a specific day program to be started on a specific day of the year, or a specific command to be carried out at a specified time of day.

Variable switching

Variable switching (Fig. 6.3/2) is used to reduce the energy required for heating or cooling purposes. When the heating system is in use, the current outdoor and room temperatures are used to calculate the latest possible time at which the heating has to be switched on (before the room is used) so that the room temperature is at a comfortable level as soon as the room is required. The system also determines the earliest possible time at which the heating can be switched off, so that a comfortable temperature is maintained until the room is vacated. In contrast to the fixed switching times used in scheduler programs (which would have to be based on the times when the room is used and the lowest external temperature), variable switching reduces the period for which the room heating is switched on, thereby reducing heating costs.

Cyclic switching

Cyclic switching is mainly used to reduce the operating times of systems that have a storage effect. For example, the air extraction systems used in stairwells and restrooms are not switched on continuously when the building is in use, but are operated in cycles throughout the day. They might, for example, be switched on initially for 10 minutes, then switched off for 5 minutes. In this scenario, the ventilation system would, therefore, be in operation for only 6 hours every day, as opposed to 9 hours if it were operated continuously.

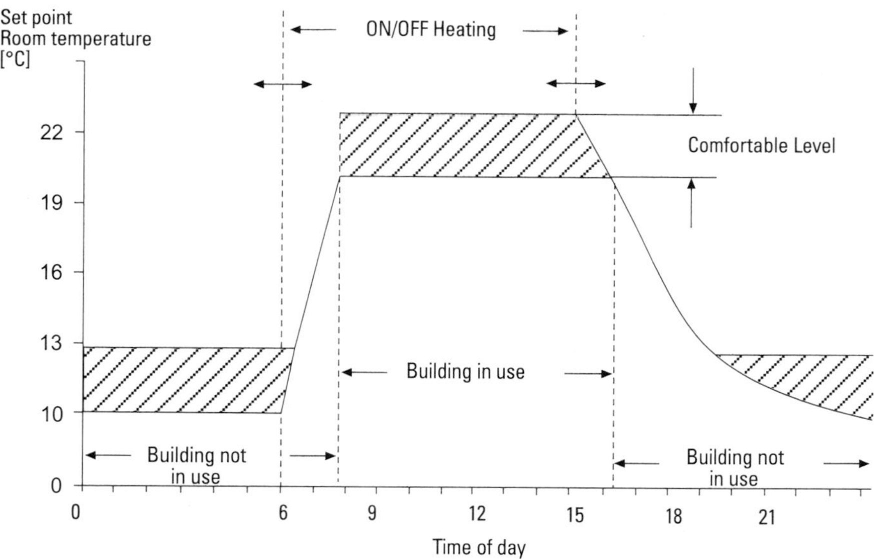

Fig. 6.3/2 Variable switching

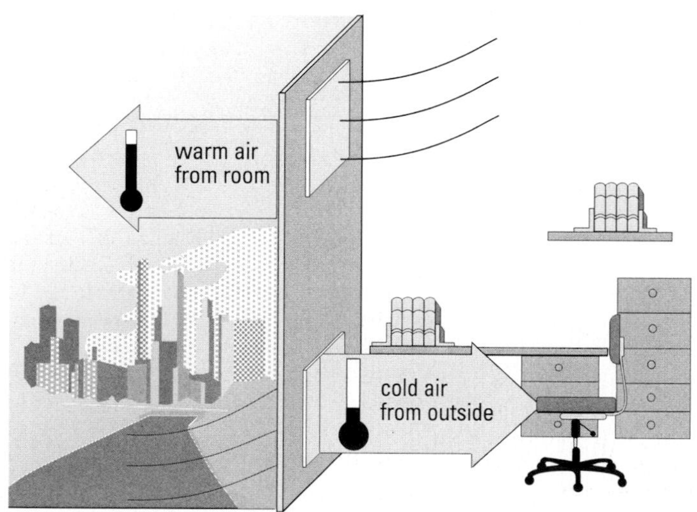

Fig. 6.3/3 Night cooling

Night cooling

Night cooling (Fig 6.3/3) is used to reduce the amount of costly energy required for cooling purposes. During the day, rooms and buildings are warmed by the sun and store this heat. If the external air temperature falls below the temperature in the building during the night, the ventilation system fans are switched on to blow cool air from outside through the building, thereby cooling it without the use of air conditioning systems.

Peak load limiting

Companies or organizations that use large amounts of electrical power have to pay two different kinds of cost to the public utility for the energy that they have actually used and for the maximum power consumption during a specified measuring interval (e. g. 15 minutes). The task of the program for peak load limiting is, therefore, to switch electrical loads off and on automatically during the day, thus avoiding high peak loads and the associated expense of maximum power consumption (Fig. 6.3/4).

Enthalpy regulation

With enthalpy regulation, the interaction between air temperature and humidity represented in the hx diagram is used to calculate the minimum power required to adjust the temperature and humidity of the intake air to the setpoints for the supply air. This calculation takes into account the fact that it is more cost effective to warm or humidify air than it is to cool or dehumidify it.

Communication at the control and automation level

Control and automation level network (CLN)

The controllers are connected via the control and automation level network (CLN), thus enabling them to communicate with each other and with the management level at the top of the hierarchy. Since it is necessary to upload and download data to and from the management level, the control and automation level network must be capable of trans-

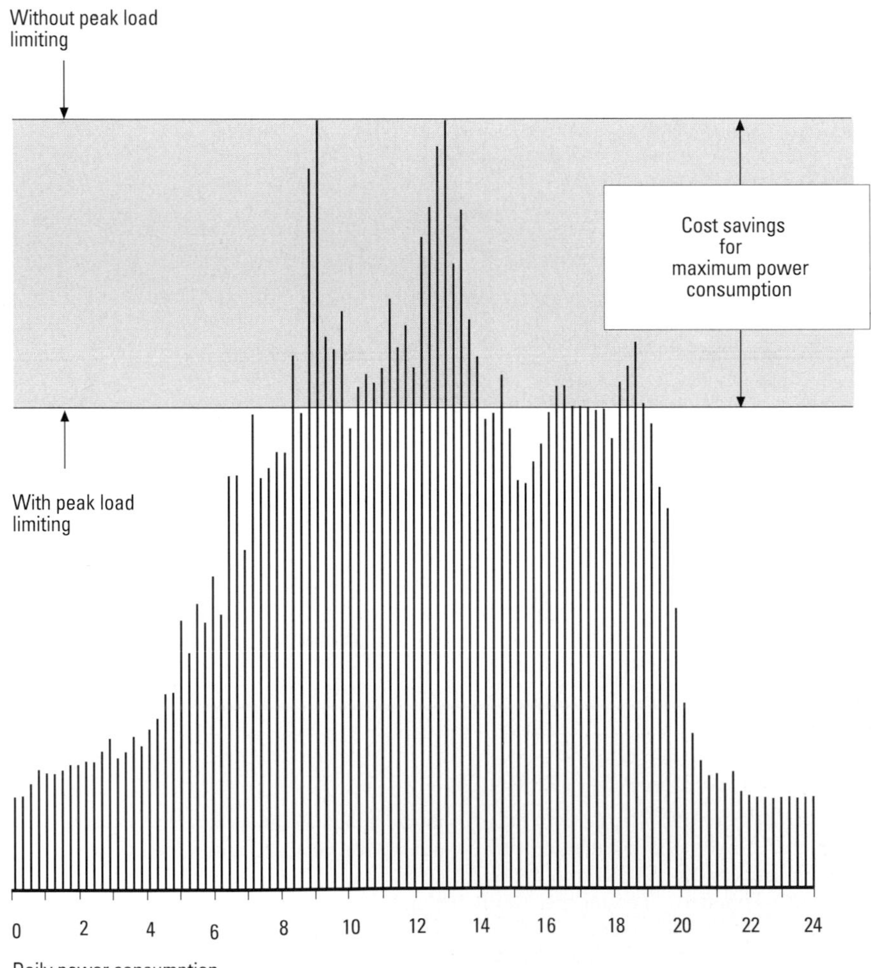

Fig. 6.3/4 Peak load limiting

ferring larger volumes of data at higher speeds than the field level network.

Controllers can be equipped with two **c**ommunication **c**ontrollers (CC) for communicating via both the field level and control and automation level networks.

Dedicated **s**pecial **s**ystems (DSS), such as alarm systems, can be connected at the control and automation level via a **d**ata **i**nterface **u**nit (DIU), i.e. a gateway to the control and automation level network, and, therefore, to the building control system.

6.4 Management level

Tasks and topology

Management level

The management level (Fig. 6.4/1) is the highest hierarchy level of a building control system. The statistics programs and all the functions for man-machine communication (MMC) are processed here. These functions include:

Man-machine communication

- ▷ displaying and printing event messages,
- ▷ adding plaintext notes to event messages (that explain how to respond to these),
- ▷ controlling annunciators,
- ▷ graphical representation of buildings, rooms and systems, which also display operating states, measured and consumption values,
- ▷ controlling paging systems if technical faults occur.

Operations logs, statistics

Operations logs and statistics enable operating staff to detect weak points in the building services system and to verify that energy management is being used effectively to reduce costs. The statistics at the management level include, for example:

- ▷ fault statistics,
- ▷ daily heating log,
- ▷ consumption statistics,
- ▷ power statistics (peak load graph).

Control room

In the past, powerful process computers (to which several printers, and alphanumeric and graphics display units were connected) were used in control rooms. Nowadays, however, control rooms comprise several networked **p**ersonal **c**omputers (PCs) and **w**orkstations (WS) connected to a central server (**s**erver **s**tation SS). Each of the PCs connected to the server usually processes a specific task. They are used, for example, as **m**onitoring and **o**perator **u**nits/stations (MOUs) or as **p**rogramming **u**nits (PU).

Communication at the management level

Management level network (MLN)

The control room devices are interconnected via the **m**anagement **l**evel **n**etwork (MLN). Since most of these devices are personal computers and workstations, which maintain large quantities of statistics and process large volumes of data for man-machine communication, the network must be capable of transferring large volumes of data at very high speeds.

Connecting external systems

External systems (dedicated special systems), such as systems for controlling access, detecting fires and intruders, and systems for maintenance management or facility management, are usually connected via a gateway (**d**ata **i**nterface **u**nit (DIU)) at the management level. The main purpose of this type of connection is to ensure that data from the different external systems and sub-systems is displayed using a standard format in the building system automation control room (where all the information converges).

The control and automation or field level networks can also be connected directly to the management level network via a gateway (DIU).

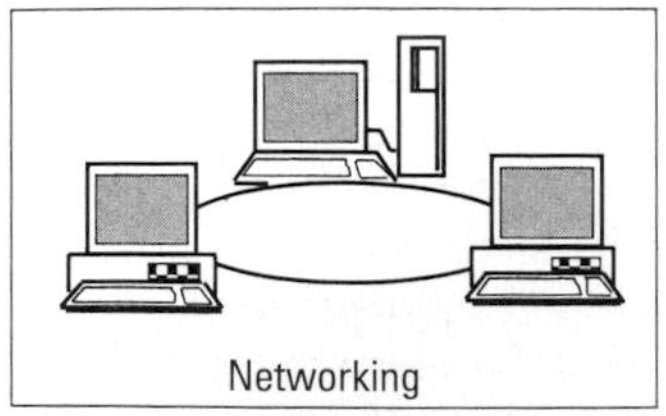

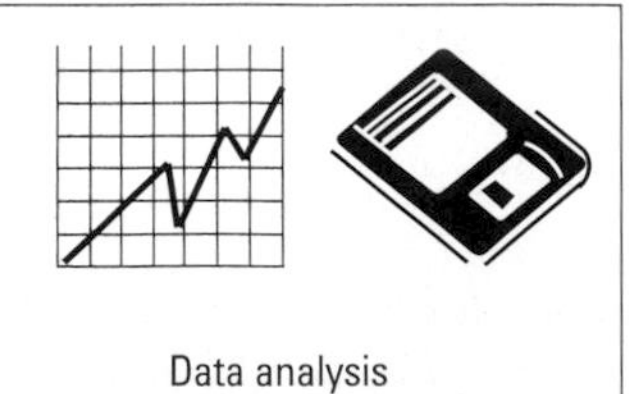

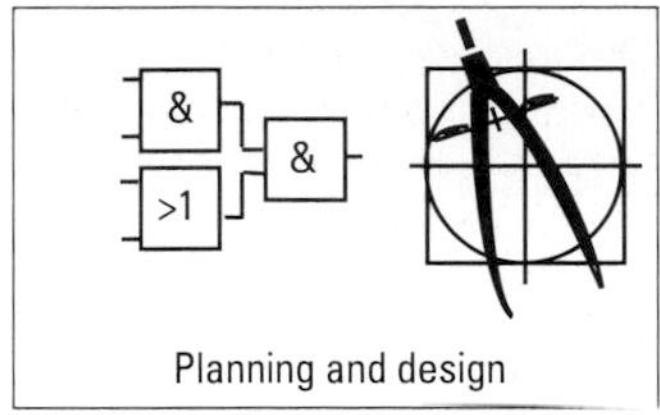

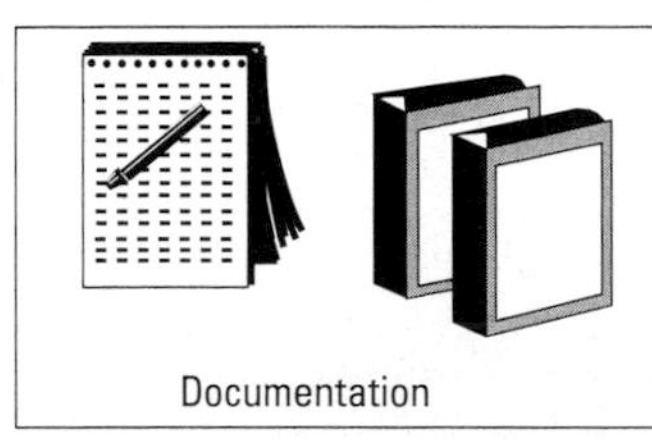

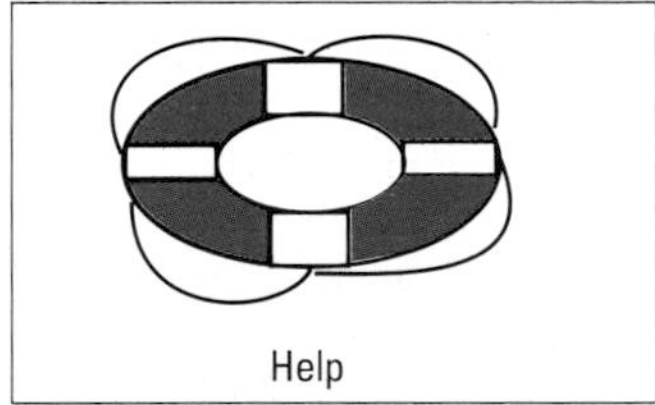

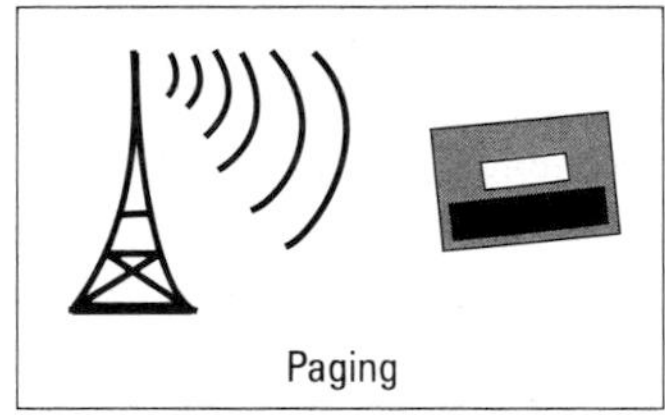

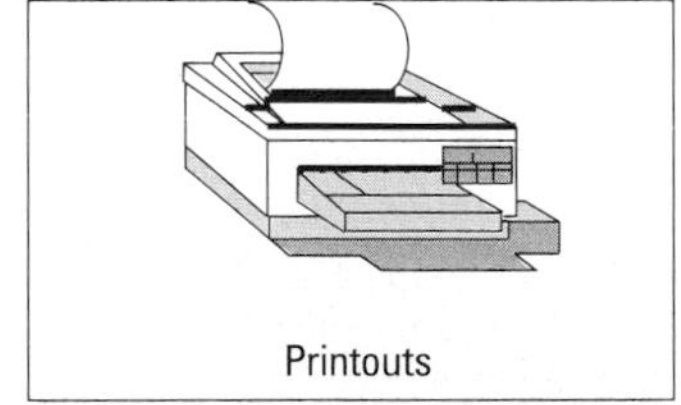

Fig. 6.4/1 Functions at the management level

6.5 Communications standards

Manufacturer-specific or open communication

Information exchange

Most monitoring and control devices in a building control system are nowadays equipped with microprocessors and communications interfaces for exchanging information with other devices. Information can either be exchanged between devices at the same hierarchy level ("peer-to-peer" communication), with another hierarchy level, or with an external system. In the past, the hardware and protocols used for data transmission in building control systems differed from manufacturer to manufacturer.

Standardized protocols and communications interfaces

This meant that users were forced to purchase all the components of a building control system from the same manufacturer, since components and systems produced by different manufacturers could communicate only via expensive gateways. Public bodies in particular wanted the option of connecting controllers made by different building control system manufacturers to a common control room. This is, however, only feasible, if all manufacturers use standardized protocols and communications interfaces.

Communications standards

A group of experts at CEN TC247, WG4 has, therefore, recommended the following communications standards for building system automation:

▷ management level:
BACnet (Ethernet), FND 1.0
▷ control and automation level:
BACnet (LON), *EIBnet*, PROFIBUS, WorldFIP
▷ field level:
BatiBUS, EHS, *EIB*, LONTalk.

BACnet

BACnet (**B**uilding **A**utomation and **C**ontrol **Net**work) was developed in the USA by a working committee at ASHRAE (**A**merican **S**ociety of **H**eating, **R**efrigerating and **A**ir-Conditioning **E**ngineers). The data transmission principle (ISO/OSI layers 1 and 2) corresponds to ISO 8802/3 (Ethernet) and the LON technology (**L**ocal **O**perating **N**etwork) developed by Echelon. BACnet (Ethernet) has been adopted by CEN as European Draft Standard ENV 1805-1, and is regarded by experts as the future international standard for communicating in building system automation. BACnet (LON) has been adopted as European Draft Standard ENV 13321-1.

FND 1.0

FND 1.0 (non-company-specific data transmission) was developed in Germany (DIN V 32735) and is used by public bodies in a number of installations. It has not, however, become established, either within Germany or at an international level. FND 1.0 has been adopted by CEN as European Draft Standard ENV 1805-2.

EIBnet

EIBnet is an extension of the field level *EIB* protocol for use at the control and automation level. The data transmission principle is also based on existing standards, such as ISO 8802/3 (Ethernet). *EIBnet* has been adopted as European Draft Standard ENV 13321-2.

PROFIBUS

PROFIBUS (**Pro**cess **F**ield **Bus**) was developed in Germany (DIN 19245, EN 50170 Vol. 2). An additional application profile has been elaborated specifically for building system automation. Functional verification was provided for various building control systems. PROFIBUS has been adopted as European Draft Standard ENV 13321-1.

WorldFIP

WorldFIP (EN 50170 Vol. 3) was developed in France for process automation and has not, as yet, been used in building system automation. WorldFIP has been adopted as European Draft Standard ENV 13321-1.

BatiBUS

BatiBUS is a protocol developed in France (NFC 46.620 to 46.629) for building system automation and system automation in residential buildings. It is used in France and Southern Europe.

EHS

EHS (**E**uropean **H**ome **S**ystem) is a protocol developed as part of the EU-funded ESPRIT program and is intended for system automation in residential buildings. The main feature of EHS is its use of plug-and-play technology. It has not, as yet, been used in building system automation.

EIB

EIB (**E**uropean **I**nstallation **B**us) is an open protocol developed by European manufacturers (DIN VDE 0829, UTE C 46620-628) for building system engineering (see Section 4.4), i.e. for cross-trade monitoring and control of building services systems at the field level. It is widely used for system automation in residential and functional buildings in Europe, and is being used to an increasing extent throughout the rest of the world. *EIB* is the only communications standard that uses a certification procedure to ensure cross-trade and multi-vendor interworking of commercially available products, and allows systems to be planned, designed and commissioned by means of a single standardized tool – ETS (***EIB*** **T**ool **S**oftware).

LON

LON (**L**ocal **O**perating **N**etwork) is a patented transmission technology for building and industrial system automation based on the LONtalk protocol. It was developed in the USA by Echelon. The availability of the required chips and transceivers has led to LON being widely used in the USA with several transmission media, as well as throughout the rest of the world. In Europe, it is used for building control systems.

BatiBUS, EHS, *EIB*, and LON have been adopted as European Draft Standard ENV 13154-1.

6.6 Planning, invitation to tender, installation

Planning

Planning a building control system is extremely challenging, since it requires experienced planners with expertise in electrical engineering, heating, ventilation, air conditioning and communications technology. The planners must determine which data points in the building services system are to be used by building system automation, and how to control, monitor, and operate the supply and waste disposal systems as efficiently as possible.

Invitation to tender (functional)

Invitation to tender (product based)

Cooperation between several specialist companies

The planners compile a list of specifications for the required equipment and functions and invite tenders for these. Companies submit tenders for the products they provide on the basis of this list. The invitation to tender can be either functional or product based. Public bodies prefer functional invitations to tender in order to avoid having to specify a specific manufacturer or product. Evaluating quotations received in reply to a functional invitation to tender (and comparing costs) is expensive and time consuming, and the solutions offered are seldom of the same technical standard. With product-based invitations to tender, planners specify the solution required by their clients, usually on the basis of tried-and-tested systems and products. Although this can limit the effects of competition, planning is more effective and tenders from the various bidders can be compared more easily. A building control system is usually installed by several specialist companies working together. Switchgear cubicles are constructed by switchgear and controlgear specialists, all the required sensors, actuators and power supply lines are connected by electricians, while engineers employed by the vendor (of the building control system) are responsible for creating and testing the control and optimization programs for building control systems. The operating staff are then trained, and the system undergoes an acceptance test (carried out by the engineer who planned it) before being handed over to the operator.

Certified devices and programs

If certified devices and programs for cross-trade building system engineering are used with standardized *EIB* communication at the field level, electricians can perform most of the tasks for which engineers were formerly required. *EIB* Tool Software (ETS), which is not manufacturer specific, plays an important part here. ETS allows certified *EIB* products made by various manufacturers to be parameterized and commissioned by several electricians working together in the same building. This is made possible by coordinating and exchanging the standardized *EIB* project database created using ETS.

It is also important that operators continue to work on their building control system after it has been handed over to them. The purpose of building control systems is to enable buildings to be operated more efficiently and cost effectively. It is, therefore, essential that the system be maintained and adapted to take account of changes in the building services systems and technical developments in building system automation and engineering.

7 Appendix

7.1 International, regional, and national standards

7.1.1 Creation of international, regional, and national standards,

The need for standardization

The need for standardization arises as a result of technological progress. Individual standards usually originate in suggestions put forward by specific specialist groups (primarily at a national level); anyone can, however, apply individually to the responsible international institution to have work initiated on a standard.

Conceptualization

Standards are created in a similar way to laws and ordinances in federal systems, where an idea, a requirement, or a need serves initially as a set objective. In the case of laws and ordinances, these originate with the legislator or the authorities responsible for legislation. Before a law or ordinance can be voted on, it has to go through a long process of conceptualization, during which it passes through all the various legal instances.

A similar democratically regulated process of cooperation takes place when technical standards are created. In the following, the creation of German standards for electrotechnical installations and products will be presented to illustrate how standards are elaborated at an international level.

Applications for standardization

Technical standards committee

In principle, anyone can apply to the **D**eutsches **I**nstitut für **N**ormung (DIN, the German Institute for Standardization) to initiate work on a standard. In the case of electrotechnical standards, the responsible DIN technical standards committee (the German Electrotechnical Commission (DKE) at DIN and VDE[1)]) begins by establishing whether a need for a standard exists or is anticipated; whether the interested groups are willing to participate; and whether other European or international standards organizations are already working on plans for comparable standards.

The following organizations are responsible for elaborating standards relating to electrical installations of buildings:

▷ ***international:***
IEC: **I**nternational **E**lectrotechnical **C**ommission based in Geneva; IEC

▷ ***regional:***
CENELEC: Comité **E**uropéen de **N**ormalisation **Elec**trotechnique European Committee for Electrotechnical Standardization based in Brussels with representatives for the European Economic Area, consisting of the countries: Belgium, Denmark, Germany, Finland, France, Greece, Ireland, Iceland, Italy, Luxembourg, Netherlands, Norway, Austria, Portugal, Sweden, Switzerland, Spain, Czech Republic, and the United Kingdom. CENELEC

▷ ***national***
DKE: Deutsche **E**lektrotechnische **K**ommission im DIN und VDE
(German Electrotechnical Commission at DIN and VDE) based in Frankfurt/Main; DKE

The following organizations are responsible for elaborating standards in non-electrotechnical areas:

▷ ***international:***
ISO: **I**nternational **S**tandard **O**rganization ISO

▷ ***regional:***
CEN: Comité **E**uropéen de **N**ormalisation European Committee for Standardization CEN

[1)] VDE **V**erband **D**er **E**lektrotechnik Elektronik Informationstechnik e.V. (Association of German Electrical Engineers).

DIN

▷ ***national:***
DIN: **D**eutsches **I**nstitut für **N**ormung
German Institute for Standardization

EITS

There is also the **EITS** (**E**uropean **I**nstitute for **T**elecommunications **S**tandards), which is responsible for standards in the field of telecommunications.

Standardization proposals

European directive

If this preliminary examination reveals that there is a need for a national standard, the other members of the European standards organization must be notified of the standardization proposal. Notification must be carried out in accordance with the procedure specified in European Council Directive 83/189/EEC.

CENELEC members

When they are notified, CENELEC members are also asked if other national committees would be interested in participating in the elaboration of the standard. This is intended to avoid standards organizations unnecessarily duplicating work, and to avoid problems regarding trade restrictions.

This can lead to the following scenarios:

▷ *A large number* of member countries are interested in the standardization proposal.

If this is the case, CENELEC sets up a working group (task force) for this project. This group works at European level (in other words, under the direction of CENELEC), and elaborates a draft European standard. This task can also be assigned to an existing CENELEC committee, as is often the case where electrical installations of buildings are concerned. This area is handled by the CENELEC/TC 64 committee ("Electrical installations of buildings") and its subcommittees, CENELEC/SC 64 A ("Electrical installations of buildings: protection against electric shock"), and CENELEC/SC 64 B ("Electrical installations of buildings: protection against thermal effects").

▷ Only a *small number* of members are interested in the standardization proposal.

If this is the case, the standards organization that made the application assumes overall responsibility for coordinating a draft European standard with the participation of the interested member countries.

▷ *No* other member countries express interest in the standardization proposal.

If this is the case, the national standards organization that made the application may continue to work on the proposed standard alone, and elaborate a national standard.

A number of standardization proposals arise directly from the work of the European or international standards organizations.

Draft standards and processing

In Germany (as in other countries), the planned version of a standard must be made available to the public for comment before the standard itself is published. This usually takes the form of a draft.

Publication of a draft standard

In Germany, the publication of draft standards is normally announced in the "Normenanzeiger" in the information bulletins from DIN. Drafts with a VDE classification are also announced in the periodical "etz". Draft standards are available to everyone. Draft (as well as valid) standards can be obtained from:

▷ VDE-Verlag GmbH, D 10625 Berlin and

▷ Beuth Verlag, D 10772 Berlin.

Anyone can raise objections to statements in a draft standard. These objections have to be submitted to the standards committee specified in the cautionary note on use contained in the draft. If an objection is rejected, the objector is entitled to request a grievance procedure and, if necessary, to refer the dispute to arbitration.

Cautionary note on use

Draft standards are a declaration of intent for the future standard. They therefore include a cautionary note on use, which can be formulated as follows:

This draft standard is presented to the public for examination and comment.
Since the intended standard may deviate from the present version, this draft may be used only by special agreement.
Comments should be addressed to (address of the body responsible for the standard).

Other regional and international standards organizations have similar procedures for developing standards, although the time frame involved may differ from the German requirements. The translations necessary for many standardization projects, as well as the time-consuming public opposition proceedings, can often lead to considerable scheduling problems when the results are being transferred between the national, regional, and international standards bodies.

Opposition proceedings

In the case of participation in elaborating regional and international standards, the outcome of national opposition proceedings, which are handled by the national working committee, must be taken into account in the German national committee's comments on the regional and international draft standards.

The deliberations of the international and regional bodies, which bring together a broad range of experience, as well as different practices and customs, ultimately lead to consensus, although proposals made at a national (German) level are not always accepted. The voting procedures that conclude the draft phase vary between IEC and CENELEC.

Voting procedure

In the case of IEC, each technical committee has countries that are registered as active participants ("P members", P = participate actively), and countries whose status is more that of observers, ("O Members", O = observer). An international draft (DIS = Draft International Standard) is accepted by the IEC, if

- ▷ a two-thirds majority of the technical committee has voted in favor of it

and

- ▷ no more than a quarter of the total number of votes are against it.

Abstentions and rejections without technical justification do not count. Every country has a vote, irrespective of its size and population.

P members

The following countries are P members (active participants) of IEC/TC 64 (February 1996):
Austria, Australia, Belgium, Canada, China, Czech Republic, Denmark, Finland, France, Germany, Hungary, Indonesia, Ireland, Italy, Japan, Netherlands, New Zealand, Norway, Poland, Russia, South Africa, South Korea, Spain, Sweden, Switzerland, United Kingdom, U.S.A., Yugoslavia.

O members

The following countries are O members (observers who can make comments) of IEC/TC 64 (June 1995):
Bulgaria, Croatia, Greece, India, Israel, Malaysia, Portugal, Rumania, Singapore, Slovakia, Ukraine.

Weighted votes

In the formal votes for CENELEC, the member countries have the following weighted votes (weighting of vote in brackets) (April 1996):

- ▷ France (10), Germany (10), Italy (10), United Kingdom (10),
- ▷ Spain (8),
- ▷ Belgium (5), Greece (5), Netherlands (5), Portugal (5), Switzerland (5),
- ▷ Austria (4), Sweden (4),
- ▷ Czech Republic (3), Denmark (3), Finland (3), Ireland (3), Norway (3),
- ▷ Luxembourg (2),
- ▷ Iceland (1).

A **E**uropean **N**orm (EN) or a European **H**armonization **D**ocument (HD) is passed, if 71% or more of the weighted votes cast (not counting abstentions) are in favor of the standard.

When a **E**uropean **N**orm (EN) or a **H**armonization **D**ocument (HD) is passed, all members are obligated to accept it. Exceptions are made for countries that do not belong to the **E**uropean **E**conomic **A**rea (EEA), such as Switzerland.

Publication of the standard

If the responsible working committee has dealt with the objections lodged against the draft standard, and reached agreement on the version of the standard for publication, this version is adopted. Before they are accepted, **E**uropean **N**orms (EN) and European **H**armonization **D**ocuments (HD) are ratified by the European standards organization.

European Norm (EN), Harmonization

When all the preparatory work specified in the standards series DIN 820 has been duly completed, the standard is printed and released for sale.

7.1.2 Reviewing existing standards

Technological progress and changing needs mean that standards have to be reviewed. This is carried out at the latest every five years. If a standard is no longer in keeping with the current state of the art or does not conform to the existing basic specifications and the standards cited therein, it must be revised. Revising a standard for a subsequent version follows the same basic procedure (starting with a standardization proposal) as was described for the first version.

7.1.3 Regional standardization

The IEC and CENELEC have reached a basic agreement whereby every application accepted by CENELEC is first offered to the IEC for processing by a fixed deadline. Only if the IEC is not interested in processing the application, or is unable to do so within the specified time frame, does CENELEC process the standardization product.

Regional/ European development

The regional/European development of electrotechnical European Norms (EN) is shown in Fig. 7.1/1.

Around 90% of the European Norms (EN) elaborated by CENELEC are incorporated in the DIN/DKE standards series either without changes, or with mutually agreed changes based on the results of work carried out by IEC (e. g. IEC Standards).

In addition to the European Norm (EN) CEN/CENELEC also publishes Standards as Harmonization Documents (HD). The difference here is that ENs must be incorporated in a structurally identical form in national standards, whereas with HDs it is sufficient to incorporate merely the content.

Incorporation in national standards

This more liberal form of incorporating HDs permits greater editorial freedom in the drafting of national standards.

7.1.4 Structure of standards, "Electrical installations of buildings (with rated voltages below 1000 V AC or 1500 V DC) HD 384"

Structure of standards

HD 384, which is virtually identical to IEC 60364, contains the specifications/standards for erecting power installations below 1000 V alternating current and 1500 V direct current for electrical installations of buildings. These specifications/standards have an extensive area of validity, and are consequently of fundamental importance for all electrical power installations in the low-voltage range. For this reason, reference is made to these specifications/standards throughout this book. The structure of the HD 384 standards is shown in Fig. 7.1/2.

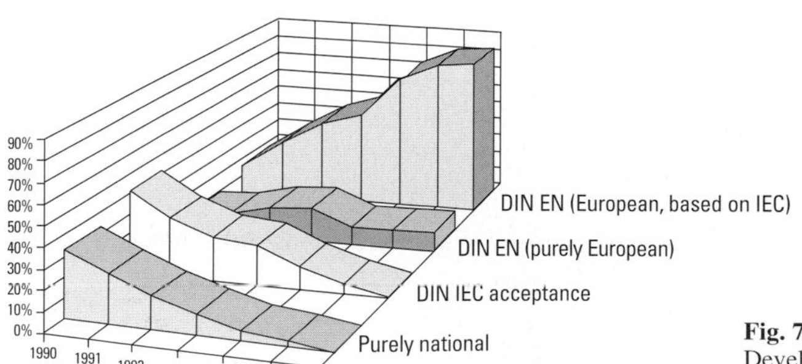

Fig. 7.1/1
Development of electrotechnical standardization (source: DKE)

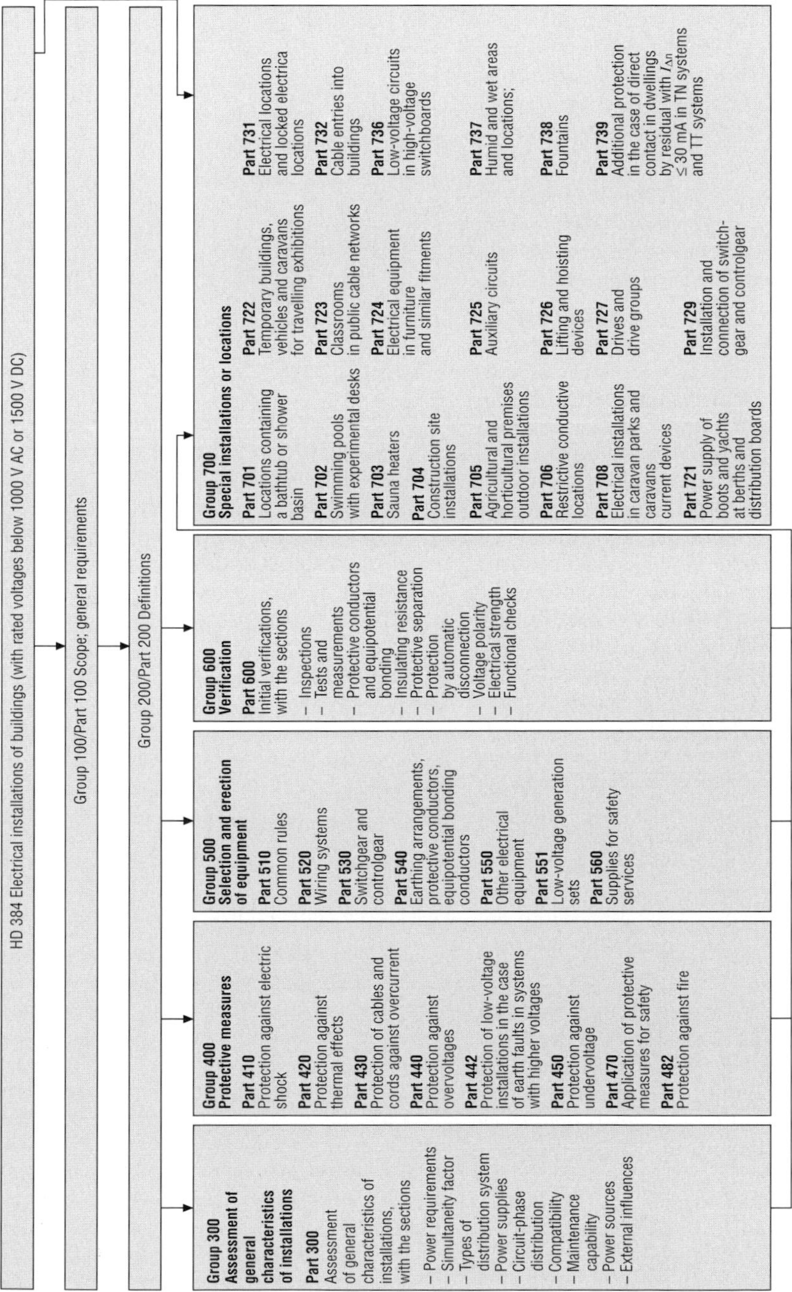

Figure 7.1/2
Structure of HD 384 standards (reproduced by permission of the DIN Deutsches Institut für Normung e.V. and the VDE Verband Der Elektrotechnik Elektronik Informationstechnik e.V.)

7.1.5 Legal validity of standards

Linking legal norms with technical standards

It can generally be assumed that the status of technical standards is not comparable to that of laws in other words, these standards are not in themselves legally binding. They can, however, acquire considerable legal significance, either by being already integrated in the statutory regulations at the law-making level (laws, ordinances, administrative regulations) or as a result of being used by authorities or courts as a source of information for individual decisions. There are various ways in which legal norms can be linked to technical standards.

Incorporation

Incorporation is the verbatim integration of technical rules in legal norms. While this is legally permissible, it is inexpedient, and for this reason is hardly ever put into practice.

Blanket clause method

The *blanket clause method* is the loosest way of linking legal norms and technical rules. The technical rules are only of significance in specific individual cases, where they are used by the authorities and courts to concretize unclear legal terms. They are generally suitable as sources of the specialist technical-scientific knowledge required for a legal decision. Although authorities and courts are not bound to do so, substantiated objections concerning the suitability of a technical standard must always be reviewed. Technical standards acquire greater legal significance when they are referred to in laws or ordinances. Various forms of reference have to be distinguished. A rigid (static) reference refers to technical rules in one specific version, and is unobjectionable in constitutional law. As a method of linking legal norms and technical standards, however, this form of reference is suited only to areas in which technical developments are (provisionally) complete, and no significant innovations are to be expected in the foreseeable future.

Rigid references

Flexible references

A flexible (dynamic) reference refers to the applicable version of one (or more) technical standard(s), thereby avoiding the disadvantage of inflexibility associated with rigid references. Various forms of flexible reference have to be distinguished. "Supplementary" flexible references are direct, binding references to whatever version of a particular technical standard is valid when the law is actually applied. The body of opinion in court rulings and legal literature justifiably regards this type of reference as inadmissible in constitutional law. Since flexible references also cover future, amended versions of the technical rule to which they refer, they contravene the principles of right of law, democracy, and separation of powers embodied in the constitution.

The following example illustrates how and why standards acquire direct legal force in Germany.

Binding legal regulations

The necessity of ensuring the safety of electrical installations and equipment arises from binding legal regulations.

In this context, the following are of importance:

▷ the Energy Resources Policy Act,
▷ the law covering technical equipment and technology at the workplace ("Gerätesicherheitsgesetz" GSG),
▷ the accident prevention regulations of the professional associations (e.g. VBG 4).

The above laws and regulations require that the recognized rules of good safety practice be observed in the following cases:

▷ when erecting electrical installations,
▷ when manufacturing electrical equipment,
▷ when operating installations and equipment.

In principle, the recognized rules of sound engineering practice are not specified by the legislator himself, but rather by reference to the technical rules and regulations contained in:

▷ laws,
▷ ordinances,
▷ administrative regulations.

Technical rules and regulations

In Germany, the series of technical rules and regulations is the DIN standards series, which is elaborated under private law.

The Federal Republic of Germany has concluded a contract with the German Institute for Standardization, DIN. In the terms of this contract, standardization in Germany is a task of the self-government of trade and industry. The central body for standardization is DIN, in its capacity as the responsible standards organization for Germany. The results of work carried out by DIN are the DIN standards ("DIN-Normen"). The procedures used by DIN create generally accepted rules of good engineering practice to which reference can be made in legal provisions.

7.2 Glossary of technical terms

Accidentally dangerous parts

Accidentally dangerous parts are accessible → conductive parts that can easily be touched by the operator and are not live during normal operation, but could be live in the event of a fault.

> *Note*
> Typical parts of this type include housings, and actuators.

Acknowledge switch

Switch with acknowledge device, by means of which the operator can accept an acoustic or visual signal when there is a change in the circuit state.

Active filter

The most important component in an active filter is a power convertor equipped with modern power electronics (e.g. IGBT **I**nsulated **G**ate **B**ipolar **T**ransistor).

Power convertors are used via control software as voltage generators for fundamental-frequency and harmonic voltages with any phase angle and amplitude.

The active filter feeds the appropriate fundamental-frequency or harmonic currents into the node of the system via an LCL coupling filter.

An active filter (such as → SIPCON P) determines the currents and voltages in three-phase systems, and uses these to establish the fundamental and harmonic components, as well as the phase angles and amplitudes.

The active filter, controlled by the control software, then feeds phase-opposed components into the node, and compensates the existing harmonic components either partially or completely.

Active filters have a number of advantages over passive filters comprising capacitors and filter reactors. No other capacitors are connected to the system; the system impedance remains unchanged; ripple-control signals are not influenced; active filters cannot be overloaded; and they are always online with 100 % of their rated output.

Actuating shaft

The part of a → switch by means of which the contact members are moved. The drive engages with the actuating shaft.

Actuator

Device in a system, which can receive/process information and carry out functions.

Adequate clearance as protection against access to hazardous parts

Clearance that prevents a probe (→ test finger) from touching or approaching a dangerous part.

Aging (of fuse-links)

Aging denotes the impairment of quality in a → fuse-link (i.e. a change in its time/current characteristic) due to a change in its material components over a period of time. Resistance to aging is an important criterion of quality in fuses. A fuse must be able to withstand the effects of normal continuous current, cyclical overload, and overvoltage pulses, without impairment of quality leading to unsafe states or faults. Even after a service period of several years without obvious faults, a fuse must be able to disconnect overcurrents or short circuits safely and reliably. The most important factors in determining resistance to aging are material, the design of the → fuse-element, and consistent manufacturing quality. The fuse-links manufactured by Siemens are extremely resistant to aging.

Lower-quality fuse-links can age more quickly as a result of regular overcurrents that almost cause the fuse-element to melt. Parts of the fuse-element can, therefore, be destroyed if the eutectic bonding that develops (for example, between tin deposits on a silver fuse-element) penetrates the fuse-element.

Air humidity, relative

The relationship (in %) between the humidity in the air at a given time and the saturation state at the same temperature.

Altitude

Altitude specifies the installation site in meters above mean sea level (MSL). Most electrical → equipment can generally be used without restrictions up to an altitude of 1000 m above MSL.

Ambient temperature

The temperature (measured under specified conditions) of the air surrounding an item of electrical → equipment (e.g. for enclosed circuit-breakers, the temperature of the air outside the enclosure). The ambient temperature affects heat dissipation, which can make it necessary to reduce the → rated current.

Apparent power

With alternating current, the product of voltage and current irrespective of the power factor → cosφ. Apparent power is a measurable quantity and is specified in VA (→ power).

Applied voltage

Voltage present at the input terminals (usually designated 1/L1, 3/L2, 5/L3) of an item of → equipment before the electrical equipment is switched on.

Arc

Phenomenon that always occurs when an → electric circuit is opened at voltages > 15 V. The factors that determine the intensity of an arc include the type and size of load, the power factor, as well as the phase angle of the voltage waveform at the instant the contact members open. Temperatures of 10,000 °C to 15,000 °C can occur inside an arc, which can cause contact material to vaporize. This means that arcs directly influence the → service life of contact members, and measures must, therefore, be taken either to keep the arc small, or to extinguish it as quickly as possible.

a-release

→ Inverse-time overload release and → overload relay.

Asymmetrical short-circuit current

→ Short-circuit current that initially runs asymmetrically to the zero line, since it contains a DC component in addition to the symmetrical short-circuit current. The DC current decays to the value zero, and the short-circuit current becomes symmetrical.

Audio frequency

In networks with audio-frequency remote control systems, audio-frequency pulses (which energize the receiver relays connected to the network) are superimposed on the supply system.

Audio-frequency blocking circuit

Virtually all audio-frequency blocking circuits are → parallel blocking circuits comprising capacitors. The parallel resonance frequency of the blocking circuit corresponds to the particular audio frequency, thus representing a very high resistance for this frequency. This ensures that the AF signal voltage does not drop below the required operating value of the AF receiver.

Auxiliary circuit

All the conductive parts of a controlgear and switchgear assembly in an electric circuit (with the exception of the → main circuit) used for controlling, measuring, signaling, regulating, interlocking, data processing, etc.

Auxiliary contact

In an → auxiliary circuit, a switch for signaling, as well as interlocking and actuating electrical → switching devices, etc. It comprises one or more contact elements with a common actuating system.

Back-up protection

Interaction of two carefully matched → overcurrent protective devices connected in series at points where, in the event of a fault, a single device (e.g. a → miniature circuit-breaker) is not capable of switching the → prospective short-circuit current. If a correspondingly high → short-circuit current occurs, the back-up overcurrent protective device relieves the next downstream overcurrent protective device, thus preventing it from being overloaded. Both protective devices have to have an appropriate → switching capacity.

Barrier

A part that provides protection against direct contact from the usual direction of access and, in certain cases, against arcs in switching and similar devices.

Basic insulation

Basic insulation is the insulation used for → live parts that ensures protection against → shock currents.

Binary

Binary means "consisting of two units". The two values are usually designated 0 or L (low), and 1 or H (high). The advantage of binary representation is that the two values can be implemented simply in electronics as "voltage present" and "no voltage present".

Bit

Short for "binary digit". A bit is the basic unit of information in digital transmission systems (→ digital) and the unit for counting binary signals (→ binary). It is synonymous with the decision between two states, usually designated 0 and 1 (or L and H). In digital electronics, bits are represented by means of pulses. One bit corresponds to one binary decision (→ binary), e.g. H (high) or L (low). A group of eight bits is usually called a byte.

Bit rate

Transfer rate of a binary signal whose → bits follow each other at a fixed frequency.

Box-type assembly

Enclosed design of a → controlgear and switchgear assembly, suitable for mounting on a vertical surface.

Breaking capacity

→ Root-mean-square value of the current at a particular → cosφ and voltage, at which a switching device or a fuse can still disconnect reliably under prescribed conditions (→ rated breaking capacity). With alternating current, the root-mean-square value of the symmetrical component applies.

Breaking current

Current in one pole of a switching device or a fuse at the instant of arcing (with alternating current, → root-mean-square value of the AC component).

Break-time

Time from the initiation of a command to open a switching device until the arcs are completely extinguished (→ make-time).

Breaking operation

Interruption of an → electric circuit as a result of the contact members of a → switching device being opened.

Bus

Group signal line between functional units, via which data is exchanged (e.g. for control, signaling, and parameterization purposes).

Bus system

All stations physically linked via a bus cable constitute a bus system.

Busbar

A low-impedance conductor, to which several → electric circuits can be connected separately.

Busbar system

Conductor bar system in which incoming and outgoing lines of the conductor bars are routed together. A busbar system can be part of a → controlgear and switchgear assembly, or of a → distribution board.

Busbar trunking system

Extended enclosed busbars, equipped with outgoing points for supplying machines and other → loads with power via variable tap-off units.

Byte

A sequence of 8 → bits that are regarded as a unit.

Cable entry

Part with openings for routing cables and wiring into → switchgear stations or → housings.

> *Note*
> Some types of cable entry can also be used as cable boxes.

Capacitor switching contactor

Capacitor switching contactors operate according to the "precharging" principle. The capacitor is connected via leading contacts and a resistor (i.e. it is precharged) before the actual main switching contacts close.

c.d.f. in %

The → cyclic duration factor is the relationship between the load period and → cycle duration in loads that are switched on and off frequently.

CEE

International commission that publishes rules for assessing electrical products (**C**ommission de l'**É**quipement **É**lectrique). The CEE has assumed the task of combining national rules and regulations to create versions suitable for international use. These rules are intended to make it easier for manufacturers of electrical products to offer their products throughout the world.

CENELEC

European committee for electrotechnical standardization (**C**omité **E**uropéen de **N**ormalisation **Élec**trotechnique). On the basis of IEC publications (→ IEC), CENELEC reworks the national regulations and specifications of participating countries, and elaborates international standards for use in the European electrical industry.

Central correction

Central correction is implemented mainly by means of → reactive power correction units. These are allocated directly to a → switchgear station, where they are installed centrally.

Central processing unit (CPU)

Main component in a programmable controller, which monitors the entire installation, carries out control functions, and stores the information required for particular tasks until it is required.

Circuit diagram

Representation showing how various parts of a network, installation, group of electrical → equipment, or parts thereof are electrically interconnected and mutually influence each other. Circuit diagrams are used both to plan and to connect external wiring as well as to perform troubleshooting in → switchgear stations, wiring systems, controllers, etc.

Circuit-breaker

An automatic switching device that can make, carry, and break currents in the electric circuit, under normal operating conditions, as well as under certain other specified conditions (→ short circuit).

> *Note*
> Circuit-breakers are generally intended for infrequent switching, although some versions are suitable for more frequent switching operations.

c/k value

The c/k value is the operating value of the → reactive power controller and is determined from the connected capacitor power at the first controller output and the transformation ratio of the infeed → current transformer to which the controller is connected.

Modern controllers determine this value automatically within minutes of the correction unit being started up.

Depending on the c/k value, the control electronics create a neutral zone, which is set at 60% to 80% of the capacitor power of first controller output, in a capacitive as well as inductive direction. The c/k value setting ensures that the controller responds and connects and disconnects the capacitor power only when the reactive power demand is correspondingly high (i.e. when the neutral zone of the controller is exceeded).

When installed in a small correction unit whose power is switched in steps of 10 kVAr, the controller responds far more sensitively than in a large industrial system with 100-kVAr steps.

Clearance (in air)

Shortest distance (measured as thread measure) between two conductive parts. This distance plays an important part in determining the insulation level for an item of electrical → equipment.

Clearance between poles

Clearance between the conductive parts of adjacent poles with different polarities.

Clearance to ground

Clearance between the conductive parts of the poles and parts that are grounded or provided for → grounding purposes.

Climates

In terms of technical standards, the interpretation of the physical and chemical states of the atmosphere including the distinctive influences of local weather, such as the temperature and humidity of the ambient air.

Combined ground resistance

Resistance between the → main ground terminal/bar and → ground.

Compartmentalization

Division or separation implemented by means of partitions (barriers). Compartmentalization is required to prevent damage extending beyond its point of origin, especially if this represents a danger to persons and the environment.

Conductive part

Part that can carry current, even if it is not necessarily used to carry current during operation.

Conductor cross-sectional area

Surface area of circular and flat conductors (→ conductor). This is specified in mm^2 and is crucial for the thermal and short-circuit rating of the electrical → conductor.

(With cables and lines for information technology systems, the diameter of the conductor in mm is generally specified instead of the conductor cross-sectional area.)

Conductors

Structure:
A conductor is the metal part of an electrical line or a cable for transmitting electrical power. Conductors can comprise a single wire or several, stranded individual wires. In controlgear and switchgear assemblies, as well as in distribution boards, → busbars are also used as conductors.

The preferred conductor material is copper, since it has a very high electrical conductivity. For larger → conductor cross-sectional areas, aluminum is normally used because it weighs less and can be worked more easily.

- ▷ Conductor designations:
  External conductors (L1, L2, L3)
  Conductors that connect power sources to → loads, but do not originate from the neutral point.
- ▷ Neutral conductor (N)
  Conductor connected to the neutral point, which also transmits electrical energy.
- ▷ Equipment grounding conductor (PE)
  Conductor used in some → protective measures to connect conductive parts that do not belong to the active circuit to each other or to extraneous conductive parts, as well as to → grounding electrodes.
- ▷ PEN conductor
  Conductor with the combined functions of a neutral and equipment grounding conductor.

Connected load

The sum of the rated inputs of all the electrical → loads in an installation.

Construction site distribution board

An assembly of electrical → equipment (such as → switching devices, → overcurrent protective devices, → residual-current protection devices, and socket-outlets, including all the connecting cables), housed inside a cabinet with lockable doors. Construction site distribution boards are approved for → rated voltages ≤ 400 V and rated currents ≤ 630 A (→ type-tested l.v. controlgear and switchgear assembly). Distribution boards of this type are used as infeeds for supplying power to → equipment on construction sites and in similar operating areas, e.g. where construction equipment, tools, and lights are operated.

Consumer's installation

All of the electrical → equipment downstream of the service entrance box or, if this is not required, downstream of the output terminals of the last → distribution board upstream of the → current-using equipment.

Contact

State that occurs when two parts used to conduct current (e.g. contact members) touch. The term is also used to designate contact elements and points.

Contactor (remote switch)

Switching device with only one position of rest, and normally without a mechanical blocking device. Contactors are not operated manually and can make, carry, and break currents, when the condition of the electric circuit (including operating → overload) is normal. Contactors are designed primarily for high → operating frequencies. A distinction is made between contactors for switching motors, and contactor relays for control functions.

Contactors in wiring systems are also known as remote switches.

Note
Contactors can make and break short-circuit currents if they are designed for this task. They are not generally intended for disconnection purposes. The French term for a contactor with main contact elements that are closed in the position of rest is "rupteur". There is no comparable term in English.

Continuous battery power supply

Battery and → load are constantly connected in parallel, and are supplied via a shared charging device. The charging device must supply the load current and is responsible for → float charging the battery. In the event of a power failure, the battery supplies the load, and the supply source does not have to be switched over.

Control circuit

Part of an → auxiliary circuit comprising all the parts of an electric circuit that do not belong to the → main circuit. Control circuits are electric circuits for:
- Signal generation and input,
- Signal processing including conversion, storage, interlock, and amplification,
- Signal output, and for controlling actuators and signal generators.

Control circuits for safety functions

Electric circuits of this type implement special measures to protect persons and their property.

Controlgear and switchgear assembly

Assembly of low-voltage controlgear and switchgear built and tested to EN 60439-1/ IEC 60439-1/DIN VDE 0660-500. This standard makes a distinction between type-tested (TTA) and partially type-tested (PTTA) l.v. controlgear and switchgear assemblies.

TTA: controlgear and switchgear assembly that does not deviate significantly from the original type or system of l.v. controlgear and switchgear assembly tested in accordance with the standard.

PTTA: l.v. controlgear and switchgear assembly that comprises type-tested and/or non-type-tested subassemblies on which special checks can be performed to verify that the relevant requirements are met.

Control-power transformer

Transformer with electrically separated windings used to feed control and → auxiliary circuits. In accordance with the standards relating to electrical equipment of machines (EN 60204-1/IEC 60204-1/DIN VDE 0113-1), a control-power transformer is recommended in auxiliary circuits with more than five electromagnetic loads.

Conventional touch voltage limit U_L

Highest value of the touch voltage U_B that may be present for an unlimited period of time.

Note
The permissible value depends on the particular external influences, but must not exceed 50 V AC or 120 V DC.

Coordination of overcurrent protective devices

Allocation of two or more overcurrent protective devices in series for the purpose of overcurrent selectivity (→ selectivity) and/ or → back-up protection.

Correction

Measure for increasing the efficiency of electrical installations. Correction involves compensating (correcting) the → reactive power that arises when inductive loads (→ load) are operated. The current in the supply conductor of a corrected installation is lower than in an uncorrected installation. As a result, conductor cross-sectional areas can be reduced, as can the cross-sectional areas of windings in transformers and generators. Depending on the particular application, a distinction is made between: → individual, → group, and → central correction.

cos φ

Power factor. The relationship of effective power to → apparent power in single-phase and three-phase AC systems.

Counter-cell

Counter-cells consist of silicone diodes in series, and are connected between a battery and a → load to limit the load voltage.

Cover

Part of the external → housing/enclosure of → controlgear and switchgear assemblies.

Creepage current

Current that develops on the surface of dry, clean insulating material as a result of conductive contaminants between two live parts.

Creepage distance

Shortest distance between two reference points on a surface made of insulating material, across which a current (→ rated insulation voltage) can flow (taking possible grooves into account).

> *Note*
> A join between two parts made of insulating material is regarded as part of the insulating material surface.

Cubicle-type assembly

Enclosed design of a → controlgear and switchgear assembly, generally for floor installation, which can comprise several → sections, → sub-sections, or compartments.

Current limiter

→ Circuit-breakers and → miniature circuit-breakers, the contact members of which are opened dynamically as soon as a → short circuit occurs, without having to wait for release via the breaker mechanism. The → arc that occurs instantaneously dampens the current considerably.

The current limiting effect is achieved by opening the contact members quickly when the → short-circuit current begins to rise, and by means of an effective arc control device. If a limiting effect is to be achieved, the arc has to be quenched before the natural (prospective) maximum current is reached.

Current limiting

With current limiting, the → peak short-circuit current expected on the basis of the circuit constants (R, L) does not occur, but is limited to a lower value, the → let-through current. This is achieved by means of → fuses or current limiting → circuit-breakers and rapid starters, which disconnect very quickly (in a few milliseconds) in the event of high short-circuit currents.

Current setting (of an → overcurrent tripping element)

Current value to which the → tripping element is set, and to which its operating conditions refer.

Current setting range (of an overcurrent tripping element)

Range between the lowest and the highest current values to which the tripping element can be set.

Current transformer

Current transformers transform currents into easily measured values. They protect measuring instruments against → short-circuit currents and → overvoltages by means of their transformation characteristics.

Current-time characteristic (→ time-current characteristic)

Graphical representation of a tripping characteristic (→ tripping element) as a function of current and time values.

Current-using equipment

→ Equipment designed to convert electrical energy to other forms of energy, e.g. to light, heat, or mechanical energy.

Cycle duration

Sum of the in-service period plus dead intervals (see also → c.d.f. in %).

Cyclic duration factor (c.d.f.)

Relationship of on-load operating time (including starting and braking time) to cycle duration (operating cycle), expressed as a percentage. The c.d.f. influences the thermal characteristics of → equipment, and must be taken into account when the duty types of the motors are defined.

Data transmission

Transmission of information between systems and parts of systems. A distinction is made between:

1. Serial data transmission
   Signals are transmitted in sequence via one cable (bus cable) (e.g. *instabus EIB*, PROFIBUS).
   Advantage: only one cable is required to bridge long distances.
2. Parallel data transmission
   Signals are transmitted simultaneously via several cables.
   Advantage: time-critical signals can be transmitted very quickly.

DC component

Magnitude of the deviation from the zero line of the normal AC sinusoidal oscillation. It occurs briefly in the event of a short circuit, for example. The maximum DC component is 100% of the peak value of the symmetrical short-circuit current (in normal low-voltage systems usually 50%).

DC control

Remotely controlled switching devices can be actuated by means of direct or alternating voltage, irrespective of whether the main contact elements of the device switch direct, single-phase alternating, or → three-phase alternating current.

Dead-front assembly

Open-type assembly with an operating face providing protection against direct contact with live parts of electrical → equipment from the front. → Live parts may be accessed from other directions.

Definite-time-delay overcurrent time release

→ Overcurrent tripping element that functions with a delay that is independent of the level of the → overcurrent, and can be settable.

Degree of protection

The degree of protection of a device specifies the scope of protection it provides. The scope of protection comprises protection of persons against contact with live parts, and protection of electrical → equipment against the ingress of solid foreign matter and water into → housings and → enclosures.

The degree of protection is designated internationally by means of a combination of letters (IP = **I**nternational **P**rotection) and numerals (→ IP code). The first numeral specifies protection against contact and ingress of solid foreign bodies, and the second specifies protection against the ingress of water.

Digital display

Display of a measuring instrument, which shows the result of a measurement as a sequence of digits, generally as a decimal number.

Digital

Representation of a value in the form of characters or numbers. The functional characteristic of an originally variable analog value is emulated in predefined steps, which are assigned specific values.

DIN rail (35 mm)

Standardized mounting rail to DIN EN 50022 for snap-mounting modular devices.

Direct contact

Contact between → live parts and persons or livestock.

Direct current

Electrical current that constantly flows in one direction only.

Direct voltage

Electrical voltage that is constant over time. (Storage) batteries, direct voltage generators, or circuits that operate electronically and rectify and stabilize alternating voltage (→ rectifier) are all sources of direct voltage.

Discharge device

Discharge devices are used to discharge capacitors. This means that devices of this type must be suitable for absorbing the energy stored in the capacitor $W_c = \frac{1}{2}CU^2$ and dissipating it in a matter of seconds.

Disconnected position

Position of a withdrawable unit, in which → isolating gaps are open in the → main circuits and → auxiliary circuits, while the withdrawable unit remains mechanically connected to the → controlgear and switchgear assembly.

Isolating gaps can also be created by operating a suitable → switching device, without the withdrawable unit having to be moved.

Disconnector (isolator)

Mechanical → switching device which, in the open position, disconnects all the poles of an → electric circuit and is equipped with a reliable contact position indicator.

Dissipation resistance

The dissipation resistance of a → grounding electrode is the resistance of the → ground between the grounding electrode and the ground reference plane.

Distribution board

Part of an electrical installation for distributing energy to downstream → loads or groups of loads. The basic components of a distribution board are the → busbar system, as well as → switching devices and → overcurrent protective devices for the outgoing cables and lines.

Distribution network

All the lines and cables from the power generator to the → consumer's installation.

Dynamic reactive power compensation

Dynamic compensation is carried out virtually in real time and steplessly. From a technical point of view, dynamic compensation functions in the same way as an → active filter, the only difference being that the reactive component of the fundamental-frequency current is determined and partially or completely corrected by means of reverse current infeed into the node.

EIB

European → Installation Bus.

EIBA

European Installation Bus Association.

Affiliation of leading electrotechnical companies, whose aim is to provide a highly reliable, standardized installation bus system throughout Europe.

Efficiency

Relationship of power input to power output. Efficiency is always less than 1 or 100%, since the power consumption of a → load means that its power output is always less than its power input.

Electric circuit

An electric circuit comprises all the electrical → equipment protected by the same overcurrent protective device(s) in an installation.

Electric circuit (of an installation)

All the items of electrical → equipment in an installation that are supplied from the same feed point, and protected by the same overcurrent protective device.

Electric shock

Pathophysiological effect caused by an electric current flowing through the human body or that of an animal.

Electrical equipment

All devices used for the purpose of generating, converting, distributing and utilizing electrical energy, e.g. machines, transformers, switching devices, measuring instruments, protective devices, cables and wiring, and → current-using equipment.

Electrical interlock

Electrical interdependence of → switching devices as a result of circuit-engineering measures. This is usual in contactor equipment: e.g. one → contactor may be closed only after another has opened. Electrical interlock is implemented by means of auxiliary contacts or auxiliary switches.

Electrical isolation

Isolation of electrically conductive parts with different potentials by means of insulating material or → clearances.

Electromagnetic compatibility (EMC)

Electromagnetic compatibility (EMC) is the capability of an electrical device to function properly in its electromagnetic environment, i.e. without being influenced by or influencing this environment impermissibly.

Emitted interference

Electromagnetic interference emitted by an item of electrical → equipment or an installation, which can influence other electrical → equipment.

Enclosed assembly

Type of construction for an item of electrical → equipment (e.g. a controlgear and switchgear assembly). Enclosed assemblies are closed on all sides in certain cases, with the exception of the mounting surface thereby providing a minimum → degree of protection IP 2X.

Enclosure

Outer casing for protecting electrical → equipment. An enclosure protects the equipment it contains against detrimental environmental influences (→ degrees of protection), as well as providing protection against direct contact (→ protective measures).

Enclosure

→ Housing

Environmental influences

Effects of the surrounding atmosphere (e.g. humidity, cold, solar radiation) that affect an item of electrical → equipment.

Equipment (electrical)

Technical products or components thereof, insofar as these are functionally and structurally designed to use electrical energy. Equipment includes objects for generating, transmitting, distributing, storing, measuring, monitoring, controlling, and converting electrical energy.

Equipment grounding conductor → PE conductor

Conductor used in a number of → protective measures involving indirect contact. Equipment grounding conductors are used to interconnect conductive parts; connect extraneous conductive parts that do not belong to the electrical wiring system, → grounding electrodes, → grounding electrode conductors and grounded → live parts, → the PEN conductor of the grounded terminal of the current source, or as artificial neutral points.

Equipotential bonding

Electrical connection that brings the → exposed conductive parts of electrical → equipment and extraneous → conductive parts to the same or virtually the same potential.

Explosion protection

Explosion protection is required for electrical → equipment used in potentially explosive atmospheres to EN 50014/DIN VDE 0170/0171. Explosion protection is provided by ensuring that an item of equipment, in which explosive → arcs (plasma) can occur during operation, is explosion proof. The ignitable mixture may be able to penetrate the housing; if there is an explosion inside the housing, however, explosive flames are prevented from escaping.

Exposed conductive parts

Accessible conductive parts of → equipment that are not→ live parts, but could be live in the event of a fault.

External conductors

Conductors that connect power sources to current-using equipment, but which do not originate from a neutral point.

Extraneous conductive part

A → conductive part that does not belong to the electrical installation, but which can introduce an electrical potential, including the potential to ground.

Extraneous conductive parts include conductive floors and walls, if potential to ground can be introduced via these.

Factory-built assembly of l.v. controlgear and switchgear (→ TTA)

The designation FBA *formerly* applied to assemblies of switching devices (including all the connecting cables and, where applicable, busbars and other items of equipment) assembled, wired, and tested by the manufacturer (EN 604391/IEC 60439/DIN VDE 0660-500). In accordance with this standard, all controlgear and switchgear assemblies of this type are now designated → TTA.

Fault current (→ residual current)

Fault current is the current that flows as a result of an insulating fault. It is measured, for example, as the difference between the currents flowing into and out of an installation via a measuring device (EN 61008/IEC 61008/DIN VDE 0664).

Fault voltage

In the event of a fault, the voltage that occurs between accessible conductive parts that do not belong to the active circuit, or between these and the → ground. If a person comes into contact with faulty machines or systems, and if he is standing, for example, on a conductive floor, part of the fault voltage has an effect on him (→ touch voltage).

Fault withstand capability

Resistance of an item of electrical → equipment to the electrodynamic (→ short-circuit strength) and thermal (→ thermal short-circuit rating) loads that occur in the event of a short circuit.

The characteristic value for the dynamic load is the → peak short-circuit current (highest instantaneous value of the short-circuit current).

The characteristic value for the thermal load of the short-circuit current is the root-mean-square value of the short-circuit current throughout its duration (sustained short-circuit current).

Feed point of an electrical installation

The point at which electrical energy is fed into an installation.

Filter circuits

Filter circuits are series resonant circuits consisting of reactors and capacitors. In contrast to reactor-connected capacitors, however, filter circuits are tuned precisely to the frequencies of the harmonic currents, and represent an impedance of virtually zero for these.

Fixed correction

Fixed correction is implemented by means of one or more capacitors permanently connected to a load. Disadvantage: fluctuating reactive power demand can lead to undercorrection and overcorrection.

Fixed part

Subassembly comprising items of → equipment that are assembled and wired on a shared supporting structure for fixed installation.

Float charging

Float charging keeps batteries fully charged by continuously supplying them with a low charging current.

Fouling factor

A conventional classification code. The fouling factor depends on the quantities of conductive or hygroscopic dust, ionized gas or salt, as well as the frequency of relative humidity as a cause of hygroscopic absorption or condensation of humidity leading to a decrease in electric strength and/or surface resistance.

Foundation grounding electrode

Conductor embedded in concrete and with a large area of contact with the → ground.

Functional class

This stipulates the current range that a particular fuse-link can disconnect (→ utilization categories (low-voltage fuses)).

Functional class g:
Full-range fuses that can carry currents up to at least their → rated current continuously, and can break currents from the lowest fusing current up to the rated breaking current.

Functional class a:
Partial-range fuses that can carry currents up to at least their → rated current continuously, and break currents above a certain multiple of their rated current up to the rated breaking current.

Functional extra-low voltage (FELV)

→ Protective measure which enables electric circuits to be operated with a rated voltage ≤50 V AC or 120 V DC, but which does not meet the requirements for → safety extra-low voltage, and is, therefore, subject to additional conditions. A distinction is drawn between functional extra-low voltage with and without → protective separation.

Functional grounding

Grounding of one or more points in a system or item of → equipment, which does not perform an electrical safety function.

Functional group

Group of several functional units that are electrically interconnected in order to perform their respective functions.

Functional test

Test to verify that electrical → equipment functions properly within the limits laid down in the specifications. Manufacturers of electrical equipment are obliged to carry out functional tests.

Functional unit

Part of a → controlgear and switchgear assembly comprising all the electrical and mechanical components that contribute to performing one particular function.

Fuse

Protective device used to open the electric circuit in which it is installed, by means of one or more → fuse-elements being sealed off. If the current in a circuit exceeds a given value for a particular period of time, it is interrupted by the fuse in accordance with the → tripping characteristic.

Note
Fuses comprise all the parts that go together to form a complete protective device.

Fuse-base

The fixed part of a → fuse, on which connections and the parts required to mount the → fuse-link are located.

Fuse-element

Particular form of conductor made of copper (in special cases, of silver) in the fuse body of the → fuse-link. Fuse-elements blow in the event of an → overload or → short circuit caused by Joulean heat. The distinctive features of fuse-elements are the solder area and bottleneck sections. The solder area melts if there is an impermissible overload; the bottleneck sections are defined sealing-off points in the event of a short circuit.

Fuse-link

A fuse-link is the part of a → fuse that contains one or more → fuse-elements, and which has to be replaced before the fuse can be switched on again after it has blown. Current interruption, together with its associated phenomena, takes place inside an enclosed fuse chamber filled with quartz sand.

Grading

The path taken by electrical energy to the → load passes through a number of → overcurrent protective devices (fuses, circuit-breakers, and miniature circuit-breakers) connected in series. The operating currents and operating times of → overcurrent protective devices are selected so that they steadily diminish towards the load. This is known as grading. → Selectivity is achieved by means of grading within an installation.

Ground

Designation for the conductive mass of soil, which is at zero potential. According to HD

384.2 S1 + A1 + A2/IEC 60050 (826)/ DIN VDE 0100 200, the word "ground" designates both the location and the soil itself.

Ground fault

A conductive connection (arising either from a fault or an → arc) between an external conductor or insulated neutral conductor (→ conductor) and the → ground or grounded parts (→ grounding).

Ground-fault current

The current that flows as a result of a → ground fault.

Grounding electrode

→ Conductive part or number of conductive parts that have good contact with the → ground, and form an electrical connection with it.

Grounding electrode conductor

→ Equipment grounding conductor that connects the main ground terminal or bar to the → grounding electrode.

Grounding, ground conductor

Conductive connection between the conductive parts to be grounded and the → grounding electrode via a → grounding system. As conductive connections embedded in the earth, grounding electrodes can be designed as grounding rods, and strip or ring grounding electrodes. Grounding is designated open if overcurrent protective devices (e.g. protective spark gaps) are integrated in the ground conductor (→ equipment grounding conductor).

Grounding system

A defined set of → grounding electrodes that are connected conductively to each other or metal parts with the same effect (e.g. stubs, armoring, metal cable jackets) and → grounding electrode conductors.

Group correction

With group correction, the correction device is assigned to one particular → load group. This can comprise motors or, for example, fluorescent luminaires, which are connected to the power system by means of a shared → contactor or other → switching devices.

Harmonic component

In AC systems, the degree to which the sinusoidal half-wave is distorted by harmonics sources (e.g. saturated oscillating-circuit reactors).

Harmonics

Harmonics are the sinusoidal (harmonic) oscillations in the Fourier analysis of non-sinusoidal, periodic oscillations that oscillate at a frequency which is an integer multiple of the fundamental (= system) frequency. The amplitudes of harmonics are considerably smaller than the fundamental frequency.

Hazardous live part

A → live part that can cause an electric shock as a result of certain external influences.

Hazardous part

A part that is hazardous to approach or touch.

Housing (enclosure)

Part that protects → equipment from specific external influences, provides → protection against direct contact from all directions, and has a minimum → degree of protection of IP 2X.

h.v. h.b.c. fuse

High-**v**oltage **h**igh-**b**reaking-**c**apacity fuse. An item of protective equipment suitable for nonrecurrent breaking in h.v. installations in which the current is interrupted when a → fuse-element embedded in sand blows.

$I^2\,t$ characteristic curve

A curve that represents the values of $I^2\,t$ in relation to the break-times (generally the minimum or maximum values of $I^2\,t$) as a function of the → prospective current under specified operating conditions.
Safety class 2: Total insulation,
Safety class 1: Protective measure for equipment grounding conductor,
Safety class 3: Safety extra-low voltage (SELV).

$I^2\,t$ range

Range between the minimum and maximum $I^2\,t$ characteristic curves.

$I^2\,t$ value

The heat value of a prospective short-circuit current (→ let-through current).

Immunity to interference

Insensitivity of an item of electrical → equipment or an installation to electromagnetic interference.

Impedance

Impedance *(Z)* is the sum of the resistances in an → electric circuit at the rated current. It comprises resistance *(R)* and reactance $(X = \omega L)$. The specified values refer to the → rated frequency.

Impulse voltage withstand level for clearances

Voltage level used to specify the clearances of an item of electrical → equipment with a given → overvoltage category and an assumed → fouling factor.

Impulse withstand current

Peak value of the current that an → electric circuit or → switching device can withstand in the closed position in a specified application.

Indirect contact

Contact between persons or livestock and → exposed conductive parts of electrical → equipment that are live as a result of a fault.

Individual correction

With individual correction, capacitors are connected directly to the terminals of individual items of → current-using equipment, and are switched on together with these via a shared → switching device.

IEC (International Electrotechnical Commission)

Commission that contributes to harmonizing standards by elaborating international specifications. The IEC unifies various national rules, thereby helping to remove trade barriers.

Infeed

Functional unit normally used to supply electrical power to an electrical device.

Inherently ground fault resistant

Items of equipment or circuits are inherently ground fault resistant if suitable measures have been applied, and ground faults (→ short circuits to exposed conductive parts) are, therefore, not anticipated under normal operating conditions (→ inherently short circuit proof).

Inherently short circuit proof

Items of → equipment or circuits are inherently short circuit proof if (where suitable measures have been applied) no → short circuits are anticipated (→ inherently ground fault resistant) under normal operating conditions.

Initial symmetrical short-circuit current I''_k

→ Root-mean-square value of the short-circuit current at the instant a short circuit occurs. With short circuits close to generator terminals, this value decays to that of the sustained short-circuit current I_k. With short circuits remote from generator terminals, it is equal to the sustained short-circuit current, and remains virtually constant throughout the duration of the short circuit.

Installation bus (*EIB*) / *instabus EIB*

European installation bus, designated by the trade mark → *EIB*. A distributed, event-controlled bus system tailored to the electrical installation, for switching, signaling, controlling, monitoring, and indicating; can be used both in functional and domestic buildings.

Instantaneous short-circuit trip

→ Tripping element of a → circuit-breaker that provides short-circuit protection for the downstream → load of the cable or wiring. In the event of a → short circuit, the instantaneous short-circuit trip must disconnect all the poles of the circuit-breaker either instantaneously or after a short-time delay.

Instrument transformer

Instrument transformers are items of electrical → equipment that convert primary electrical quantities (e.g. currents and voltages) to the same type of secondary quantities suitable for the connected devices (measuring instruments, meters, protection relays, etc.). Instrument transformers reach magnetic saturation at low → overcurrents, thereby protecting the connected devices against overload.

Insulation monitoring

Measure that involves using a measuring instrument to monitor the → insulation resistance of an electrical installation in which neither an external conductor nor a neutral point is directly grounded. Any insulation faults that occur are indicated; if necessary, the electrical installation is switched off when a second fault occurs.

Insulation resistance

The lowest resistance value measured between parts insulated from each other, or between these and → ground.

Interlock

Electrical:
Type of circuit in which the auxiliary contacts of various devices are switched in such a way that the circuit states are interdependent. This makes it impossible to switch on one → switching device if another is already switched on.

Mechanical:
Means of achieving mechanical interdependence between several functional units, e.g. to prevent a cover from being removed when a switching device is switched on.

Interrupting current

Current that flows the moment a contact is opened, e.g. also overload and short-circuit current. The interrupting current depends on the voltage as well as the inductance or capacitance of the electric circuit.

Inverse-time overload release (a-release)

Thermal overload release that operates with a time delay that decreases as the current rises.

IP code

A classification system for specifying the degree of protection provided by a → housing against access to → hazardous parts, as well as ingress by solid foreign matter and water. The system also includes additional information related to protection of this type.

ISDN (**I**ntegrated **S**ervices **D**igital **N**etwork)

Standardized digital network in which all types of communication data (speech, text, data, as well as still and moving images) are transmitted via a single connection to the exchange, and a single cable to and from the subscriber.

Isolating gap

Opening travel of the contact members in a → switch, which maintains the prescribed → clearance.

Isolation

In electrical power installations, isolation entails disconnecting or separating a system, part of a system, or an item of → equipment from all the ungrounded conductors.

Isolation

Opening of a circuit, which creates a gap in the circuit adequate to protect persons (→ isolating gap).

IT system (network)

→ Types of network and system

LAN (local area network)

A local network for bit-serial transmission of information between interconnected, independent terminals. The term LAN was introduced to distinguish central processor links (to peripherals) with very high → bit rates over short distances, and public networks covering larger distances.

Leakage current (in systems)

A current that flows to → ground, or to an extraneous → conductive part in a fault-free circuit.

Note
This current can have a capacitive component, which depends particularly on the use of capacitors.

LED

A **l**ight **e**mitting **d**iode used to indicate a signal status. An LED lights up, for example, if the load is switched on.

Let-through current i_D

Highest instantaneous value of the current during the break-time of a switching device or a fuse. Controlled short-circuit currents occur if the switching device reduces the amplitude of the short-circuit current (e.g. by means of resistance, operating delay, or arc voltage). The let-through current of a device (e.g. current-limiting fuse and current-limiting circuit-breaker) is an important factor with regard to the thermal stress (I^2t value) of downstream devices (→ current limiting).

Live part

In electrical equipment, a conductive part that is energized under normal operating conditions. Live parts normally also include the → neutral conductor (N) and → conductive parts connected to it. This does not apply if the neutral conductor is also the → PEN conductor.

Load carrying capacity

Capability of a conductor or a device to carry or switch a maximum permissible current under specific conditions.

The load carrying capacity of a switching device depends on a number of factors, including size, enclosure, and ambient temperature.

Electrical:	→ load current,
Mechanical:	mechanical endurance (→ service life)
Thermal:	→ temperature limit, → rated short-time current,
Dynamic:	→ short-circuit strength, → peak short-circuit current

Load current

Current with which a conducting path (e.g. conductors for electrical → equipment) can be loaded continuously or for a short time, without being thermally (→ temperature limit) or dynamically overloaded.

Load/current-using equipment

Devices or equipment that convert electrical energy to another (non-electrical) form of energy. In AC technology, loads are subdivided into three categories:

Ohmic loads, which do not cause any phase displacement between current and voltage in the system (e.g. heaters, incandescent lamps).

Inductive loads, which cause the current to lag behind the voltage in the system (e.g. motors, coils, and electromagnets).

Capacitive loads, which cause the current to lead the voltage in the system (e.g. capacitors).

Load shedding

Method of using circuit-engineering measures to switch off less important → loads on a leading basis in order to prevent a maximum power or current from being exceeded.

Loop impedance

The impedance of a fault loop is the sum of the impedances in a current loop. These are: the impedance of the current source; the impedance of the external conductor, from one pole of the current source to the measuring point; and the impedance of the return conductor (e.g. → equipment grounding conductor, → grounding electrode, and → ground), from the measuring point to the other pole of the current source.

Low voltage

Designation for the voltage range $\leq$ 1000 V AC, max. 500 Hz or $\leq$ 1200 V DC.

Low-voltage controlgear and switchgear assembly

A controlgear and switchgear assembly is a combination of one or more low-voltage controlgear and switchgear devices with appropriate equipment for controlling, measuring, and signaling purposes, as well as protective and controlling equipment. Low-voltage controlgear and switchgear assemblies (including all the internal electrical and mechanical connections and structural parts) are assembled entirely at the manufacturer's responsibility.

l.v. h.b.c. fuse

Low-**v**oltage **h**igh-**b**reaking-**c**apacity fuses comprise a fuse-base and fuse-link, for industrial and similar installations. Because of their design, the → h.b.c. fuse-links may be replaced by suitably trained personnel only.

l.v. h.b.c. fuse-link

The l.v. h.b.c. fuse-link is a protective device suitable for nonrecurrent breaking. The current is interrupted when a → fuse-element embedded in sand blows as a result of Joulean heat. The design of the fuse-element determines the → breaking capacity in accordance with the → utilization categories.

Main circuit

Electric circuit that comprises equipment for generating, converting, distributing, switching, and consuming electrical energy.

Main ground terminal/bar

A terminal or bar provided to connect the → grounding electrodes to the → equipment´ grounding conductor, the → equipotential bonding conductor, and (if necessary) to the conductors for → functional grounding.

Maintenance gangway within a switchgear station

Area that is accessible only to authorized personnel (skilled persons, instructed persons), and is intended primarily for the maintenance of installed electrical → equipment.

Make-time

Time from the initiation of a closing operation until the instant at which the current begins to flow in the main circuit.

Note
The make-time also includes the time delay of all the items of auxiliary equipment required to make the current, insofar as these are parts of the switching device.

Making capacity

Value of the prospective making current that a switching device can make under prescribed conditions at a specified voltage.

Making current (→ rush current)

Current that occurs immediately after an → electric circuit is closed, e.g. when a transformer or a motor is connected to the system. The magnitude of the making current depends on the instant of closing (phase angle of the voltage). Maximum current flows when a device is switched on at voltage zero, and decays to the rated value, after approximately 20 ms.

Making operation

Establishing an electric circuit by closing the contact element of a → switching device.

Master

Masters (= active bus devices) holding the token may send data to and request data from other devices.

Master system

A master system comprises all the → slaves allocated to read from or write to a → master, as well as the master itself.

MCC

Motor **C**ontrol **C**enter is a low-voltage withdrawable-unit-type switchgear station for motor feeders with a main switch and door interlock.

Mean sea level (MSL)

Point of reference on which altitude specifications are based (average level of the surface of the sea).

Mechanical switch

Device for switching interference-free → equipment and parts of installations on and off. The → switching capacity of a switch is predominantly in the order of magnitude of its → rated current, and is specified by the manufacturer.

Meshed systems

In meshed systems, energy is distributed by means of a cable system connected to form a network. This type of system is generally supplied via several points. In the event of a cable failure, each load is supplied via the remaining cable branches (without switching over) (→ radial system, → ring system).

Miniature circuit-breaker

Mechanical switching device used to manually connect an electric circuit to, or to disconnect it from the system. Miniature circuit-breakers also disconnect the electric circuit from the system automatically when the current exceeds a particular maximum value.

In the event of a connected → load or a feeder being thermally overloaded, a thermally delayed release initiates the trip procedure. If a → short circuit occurs, the → instantaneous short-circuit trip carries out this procedure. Each of the two tripping elements functions independently and protects the other. The trip-free mechanism of a miniature circuit-breaker ensures that the switch trips in the event of overcurrents or short circuits, even if the operating lever is locked in position, or is restrained by the operator. Miniature circuit-breakers are manufactured with different → tripping characteristics (B, C, and D) in order to provide optimum protection for the feeders.

Modular device

Device with specified → mounting dimensions for installation on → DIN rails under a shared cover.

Modular system

A method in which individual and compatible functional modules are used to enhance basic devices, to equip/retrofit them for specific applications, or to assemble different variants of a distribution board using simple auxiliary devices.

Modular unit (MU) a

The width of modular (DIN-rail-mounted) devices is specified as $n \cdot (17.5_{+0.5/0})$ mm, where $n = 0.5; 1.0; 1.5; 2.0; 2.5 \ldots$

A modular unit (MU) a is 18 mm (17.5 + 0.5 mm); modular width available for modular (DIN-rail-mounted) devices (e.g. in distribution boards) is $n \cdot 18$ mm.

Mounting dimensions (DIN 43880, R023-001)

Stipulation of the maximum dimensions of a → modular (DIN-rail-mounted) device to enable various devices to be mounted easily on the same → DIN rail under a shared cover.

Mounting frame

Sectioned structural part designed to accommodate various components for installation in → distribution boards, and → switchgear stations or cabinets. When equipped with devices, a structural part of this kind can constitute an independent → controlgear and switchgear assembly.

Mounting plate

Plate designed to accommodate various components for installation in → distribution boards, and → switchgear stations or cabinets.

Mounting position

The installation position or position of normal use stipulated by the manufacturer of an item of electrical → equipment to ensure that the device functions properly.

Mounting structure

Structure that is not part of the → controlgear and switchgear assembly, and which is designed to carry an enclosed-type controlgear and switchgear assembly.

Network circuit-breaker

→ Circuit-breaker in an assembly with a network master relay and a special open-circuit shunt release (shunt release with capacitor unit), which operates reliably at between 10 % and 110 % of the → rated voltage. The relay monitors the direction of the power flow at the point of installation. If the power flows from the transformer to the load (forward power), the relay does not operate. If a fault in the transformer or the high-voltage system causes the power to flow back to the transformer (reverse power, which can occur in a → meshed system with several infeeds), the relay operates, tripping the network circuit-breaker via the shunt release with capacitor unit.

Neutral conductor (N)

A conductor connected to the neutral point of the system, which is suitable for transmitting electrical energy.

Normal arm's reach

An area extending from any point on a surface where persons stand or move about, to the limits of that area which a person can reach with his or her hand in any direction without assistance.

Normal switching duty

Operation of a switching device, during which limits specified by means of the rating (e.g. → operating frequency, → rated output, → rated current) are complied with.

n-release

Instantaneous electromagnetic overcurrent tripping element (→ short-circuit trip).

Obstacle

Part that provides protection against accidental contact, but does not prevent intentional contact.

ON period

Time during which an → electric circuit remains closed.

Open-type assembly

Switchgear station (TTA or PTTA, → type-tested or → partially type-tested l.v. controlgear and switchgear assembly) in which the electrical → equipment is arranged in a supporting structure in such a way that its live parts are accessible.

Operate

To change from one position to another position when a specific value (→ operating value) has been reached, or to begin an operation that terminates with a transition from one position to another. Values for voltage, current, heat, and time are examples of operating values.

Operating areas

Electrical operating areas are used primarily for operating electrical systems. They do *not* necessarily have to be kept locked. As a rule, only instructed persons may enter these areas.

Closed electrical operating areas are used exclusively for operating electrical systems. They *must* be kept locked *under all circumstances*, and may only be opened by authorized persons. *Only* instructed persons may enter areas of this type.

Operating current, ground fault (g)

Ground-fault current, at or above which a ground-fault tripping element (g), e.g. of a circuit-breaker, operates.

Operating current, instantaneous (n)

Current above which operation occurs instantaneously.

Operating current, overload (a)

Continuous current above which operation occurs within preset times (inverse time-delay operation).

Operating current, short-time-delay (z)

Current above which operation occurs after a pre-determined delay.

Operating frequency

Operating frequency specifies the number of operating cycles that can be carried out by a → switching device within one unit of time (e.g. in 1 hour).

Operating gangway inside switchgear stations

Area that the operator has to enter to operate (e.g. switch, control, set, and monitor) electrical devices under normal conditions.

Operating position

Position of the moveable parts of a → switching device when the drive is activated (opposite: "off" position)

In controlgear and switchgear engineering: position of a withdrawable unit, in which it is fully connected for its intended function (→ disconnected position, → test position).

Operating time

With time relays, the time that elapses between a command being issued and the contacts being actuated.

Operating value

Value of an active quantity (e.g. voltage, temperature, current) at which operation occurs.

Operating voltage

Voltage present between the conductors in an item of equipment or part of a system.

Optical fiber/optical waveguide

Dielectric waveguide with a core consisting of optically transparent, low-damping material, and a cladding of optically transparent material with a lower refractive index than the core. Optical fibers/waveguides are used to transmit signals by means of electromagnetic waves in the optical frequency range, and are generally equipped with protective armoring.

Overcurrent

Any current in an → electric circuit that exceeds the → rated current (→ overload, → short-circuit current). Overcurrent represents a danger to cables and wiring, as well as electrical machinery and devices because of impermissibly high heat generation and mechanical forces that may occur.

Overcurrent protective device

An item of → equipment (e.g. circuit-breaker, fuse, miniature circuit-breaker) that operates when a specified current is exceeded, thereby protecting the appropriate → electric circuit by interrupting the current.

Overcurrent relay

Instantaneous electromagnetic relay that operates in the event of an → overcurrent, e.g. when a short circuit occurs. Remote tripping (e.g. of the allocated → circuit-breaker) is carried out by means of an auxiliary contact in the relay.

Overcurrent selectivity

Overcurrent selectivity entails coordinating the tripping characteristics of two or more → overcurrent protective devices so that, in the event of → overcurrents between specific limit values, only the protective device assigned to this range trips, and the other protective devices do not operate.

Overcurrent tripping element

Tripping element used in conjunction with → switching devices. Overcurrent tripping elements permit a mechanical switching device to be opened after a delay or instantaneously, when the current exceeds a specified value. They are used to protect electrical → equipment against → overload or the detrimental effects of → short-circuit currents.

Overload

Operating conditions in an electrically sound, fault-free → electric circuit that give rise to an → overcurrent. An overload can cause damage if it continues for a prolonged period of time and is not disconnected.

Overload current

→ Overcurrent that occurs in an electrically sound, fault-free electric circuit.

Overload relay

Inverse-time relay which, in the event of an overload, operates according to a → time-current characteristic, thus protecting the → switching device as well as a → load against overloading.

Overload release

Overcurrent tripping element that provides protection against → overload.

Overvoltage category

Division of an electrical wiring system into zones according to immunity to possible overvoltages; e.g. overvoltage category 4 applies from the service entrance to the meter, and overvoltage category 3 from the meter to the socket-outlet.

Overvoltage $U_{ü}$

Peak value of the high-frequency transient recovery voltage; higher than the → rated operating voltage U_e ($U_{ü} > U_e$).

Parallel blocking circuit

→ Audio-frequency blocking circuit

Parallel connection

Type of connection in which several devices, contacts or lines are connected in parallel in the same electric circuit.

Partially type-tested l.v. controlgear and switchgear assembly (PTTA)

A controlgear and switchgear assembly to EN 60439-1/IEC 60439-1/DIN VDE 0660-500, comprising type-tested and/or non-type-tested subassemblies, and for which checks and/or calculations are used to verify that the relevant requirements are met. A subassembly can also consist of a single low-voltage switching device with the associated electrical connections, or of a → housing (enclosure), for example.

Partition

Part of the → housing or → enclosure of a compartment that separates the compartment from others.

PE conductor (→ equipment grounding conductor)

A conductor used in certain protective measures against → indirect contact between → exposed conductive parts and
- other conductive parts,
- extraneous → conductive parts,
- → grounding electrodes, → grounding electrode conductors, and grounded → live parts.

Peak short-circuit current i_p

Highest instantaneous value of the prospective current after a → short circuit has occurred. The peak short-circuit current comprises the short-circuit current and a → DC component, and is specified as a peak value.

PEN conductor

Grounded neutral conductor with protective function in three-phase AC systems.

Plug-type connection

Screwless plug-in connections (→ screwless terminals) designed for plug-type connection. Instead of terminal screws, these screwless connections are fitted with tab connectors for mounting the push-on contacts. The contacts are crimped on to the flexible cables that are to be connected.

Polling (master/slave configuration)

Polling involves one of the stations (the → master) controlling bus access by explicitly allowing other stations (the → slaves) to transmit data for a limited time only.

Power installation

Power installations are electrical installations with → equipment for generating, converting, storing, transmitting, distributing and consuming electrical energy for the purpose of carrying out work e.g. mechanical work, work for generating heat and light, or in electrochemical processes.

Note
It is not always possible to make a clear distinction between power installations and other types of electrical installation. The values of voltage, current, and power are not, in themselves, adequate criteria for such a distinction.

Power

1. Physical
Work performed by a force in a unit of time in watts.

2. Electrical
Electrical energy drawn from the system by a → load per unit of time.

The following, load-dependent types of power occur in AC systems:
- Effective power *(P)*
  The proportion of overall power converted to another form of energy.
  The effective power component of the → apparent power is calculated as follows:
  $P = U \cdot I \cdot \cos\varphi$ (W)
- Reactive power
  The proportion of overall power that cannot be converted to the desired form of energy. Reactive power occurs as a result of induction, giving rise to a shift between the current and voltage curves.
  The reactive power component of the apparent power is calculated as follows:
  $Q = U \cdot I \cdot \sin\varphi$ (var)
- Apparent power (*S*)
  The power drawn from the system, which is calculated from the vector sum of effective and reactive power.
  Apparent power is calculated as follows:
  $S = U \cdot I$ (VA)

These three types of power have a geometrical interdependence, which can be represented in a capability curve.

The relationship of effective power to apparent power is determined from the power factor (→ $\cos\varphi$). The power factor is expressed by means of a trigonometrical function as the cosine of φ.

Prearcing time

Time taken for the fuse-element of a → fuse-link to seal off, from the beginning of the impermissible → overload current or → short-circuit current until the beginning of the interruption procedure.

Priority switch

An electromagnetically operated switch, the break contact elements of which can interrupt the control circuit of a → contactor when a dominant load I (e.g. flow-type heater) is switched on for a length of time. The contactor is then set to the OFF position, and switches off the subordinate load II (e.g. off-peak storage heating) as long as load I is switched on.

PROFIBUS

The PROFIBUS (**Pro**cess **Fi**eld **Bus**) is used to transmit signals from field devices, as well as from process devices and programmable controllers in industrial networks, and meets the requirements of EN 50 170 and DIN 19 245 Part 1 and Part 2.

PROFIBUS-DP

Draft standard PROFIBUS-DP (EN 50 170, DIN 19 245, Part 3). The primary task of PROFIBUS-DP is to use the fastest method of transferring data cyclically between the central DP → master and the peripheral devices.

Prospective peak current

Prospective peak current is the highest value of the current that occurs during the transition period after current has begun to flow.

Note
This definition assumes that the current is established by an ideal switch, i.e. with an instantaneous change in impedance from infinity to virtually zero. For multiphase electric circuits, it is also assumed that the current is established at all poles simultaneously, even if the current in only one pole is considered.

Prospective short-circuit current I_p

Anticipated solid → short-circuit current that would flow in an electric circuit in the event of a fault if there was no protective device in the cable run to damp this current.

Prospective touch voltage

The highest touch voltage that can occur in an electrical installation in the event of a fault with negligible → impedance.

Protection against direct contact (protection against electric shock)

Measures to protect persons (and livestock) from hazards that can arise from contact with → live parts of electrical → equipment.

Protection of this type can be either complete or partial. The latter only provides protection against accidental contact.

Protection in the event of indirect contact

Measures to protect persons (and livestock) from hazards that can arise from contact with → exposed conductive parts or extraneous → conductive parts in the event of a fault.

Protective grounding

Grounding of one or more points in a network, installation, or item of → equipment for the purposes of electrical safety.

Protective measures

Measures to protect persons and livestock against → touch voltages that can occur as a result of insulation faults in electrical → equipment.

Protective separation

Protective separation is achieved if individual faults do not cause the voltage in one electric circuit to spill over into another. This is necessary above all if → safety extra-low voltage and other voltages $\leq$ 1000 V AC are switched in one device.

PTC thermistor detector

Part of the thermistor protective device built into the winding of the motor or encapsulated-winding dry-type transformer it protects. As the temperature rises within a particular (narrow) temperature range, the PTC thermistor detector increases its resistance by several powers of ten.

PTTA

→ **P**artially **t**ype-**t**ested l.v. controlgear and switchgear **a**ssembly (see also → TTA).

Public utility

Suppliers of electrical power (power stations) are known as public utilities.

Pushbutton

Switch with a spring return device and actuator operated by applying pressure (generally with the finger or palm), e.g. light pushbuttons, momentary-contact pushbuttons, mushroom buttons, twist switches.

Radial system

The wiring configuration for this kind of system resembles the branches of a tree. Power is fed in from the "trunk", and only one specific path ("branch") is available for each → load ("leaf") (→ meshed system, → ring system).

Radio interference

Radio interference can occur when inductance-laden electric circuits are switched off e.g. as clicks. This can be remedied by means of a suitable RC element for the → isolating gap.

Radio interference level

A frequency-dependent limit for → radio interference (continuous disturbance and clicks). This level applies to the radio interference voltage, radio interference power, and interference-field strength.

Radio interference suppression

Radio interference suppression is the measure used to prevent or reduce the high-frequency electromagnetic oscillations from electrical → equipment and installations that can cause → radio interference (DIN VDE 0875 Part 11, EN 55011, DIN VDE 0871, 06.78 remained in force until 12/31/1995.)

Rated breaking capacity

Highest current that a → switching device can disconnect under specific conditions.

Rated breaking capacity I_{cn}

The rated breaking capacity of a → circuit-breaker is the value of the → short-circuit current it can disconnect at the → rated operating voltage, → rated frequency and specified power factor → $\cos\varphi$ (or specified time constant). The value of the prospective current applies (with alternating current: → root-mean-square value of the AC component); these values are specified by the manufacturer.

Rated current I_n

Current for which the item of electrical equipment is rated, and to which specific characteristic data, e.g. rated output S_n refers.

Rated diversity factor

Relation between the greatest sum of all the prospective currents in a particular → main circuit at any instant and the actual installed value.

When several electric circuits are grouped in a shared enclosure, the rated diversity factor is the factor by which the current in each circuit must be reduced in order to prevent an impermissibly high temperature rise.

Rated frequency

Rated frequency is the system frequency for which a particular item of electrical → equipment is intended, and to which its other characteristic data refers.

Rated impulse withstand voltage (U_{imp})

Peak value of an impulse voltage of a specified waveform and polarity with which an item of electrical → equipment can be loaded without failure under specified test conditions, and to which the clearances refer.

The rated impulse withstand voltage of an item of electrical equipment must be equal to or greater than the transient overvoltages that occur in the system in which it is used.

Rated input voltage U_{1n}

Input voltage (with three-phase current, phase-to-phase voltage) with which a transformer may be excited at the rated frequency.

Rated insulation voltage U_i

Voltage value that specifies the insulation resistance of an item of electrical → equipment or an accessory, and to which the insulation tests, as well as the → creepage distances and clearances refer.

Under no circumstances may the maximum → rated operating voltage exceed the rated insulation voltage.

Note
For electrical equipment without a specified rated insulation voltage, the highest rated operating voltage must be assumed as the rated insulation voltage.

Rated making capacity

Highest current that a switching device can connect under specific conditions.

Rated making capacity I_{cm}

The rated making capacity of a → circuit-breaker is the manufacturer's specification of the short-circuit current it can make at the → rated operating voltage, → rated frequency, and specified power factor → $\cos\varphi$ (or specified time constant). It is expressed in terms of the maximum peak value of the prospective current.

Rated operating current I_e

The rated operating current is the current determined by the conditions under which an item of electrical → equipment is used. It takes into account the → rated voltage and → rated frequency, rated duty, and → degree of protection.

Rated operating voltage U_e

The rated operating voltage of an electric circuit is the voltage value that, together with the rated current, determines the use of the electric circuit.

With multi-phase electric circuits, the voltage between the phases is specified.

Rated output S_{nT} of a transformer

Apparent output power (in VA or kVA) is the product of the → rated output voltage and rated output current; with three-phase transformers, $\sqrt{3}$x the value for which the transformer is dimensioned under the basic reference conditions.

Rated output voltage U_{2n}

Output voltage (with three-phase transformers, the phase-to-phase voltage) at the → rated input voltage, → rated frequency, rated output current, → $\cos\varphi = 1$, and rated temperature.

Rated peak withstand current i_p (→ peak short-circuit current)

Highest permissible instantaneous value (peak value) of the prospective short-circuit current in the most heavily loaded conducting path. This value characterizes the → short-circuit strength of an item of electrical → equipment.

Rated residual current $I_{\Delta n}$

Rated residual current $I_{\Delta n}$ is the → current value for which residual-current circuit-breakers are constructed, and in terms of which they are designated (EN 61008/IEC 61008/DIN VDE 0664).

Rated short-time current I_{cw}

Permissible → root-mean-square value of the AC component of the prospective short-circuit current that an item of electrical equipment (e.g. a switchgear station) can carry for a specific time, e.g. from 0.05 s to 1 s (1-s current) (→ thermal short-circuit rating).

Rated switching capacity

→ Rated making capacity and → rated breaking capacity. The switching capacity for → short-circuit currents is expressed as the sustained prospective short-circuit current of the → circuit-breaker. With alternating current, the root-mean-square value of the symmetrical component applies.

Rated value

Value of a variable that applies for a defined operating condition, and which is generally specified by the manufacturer of the particular component or device. The rated value can refer to any electrical variable, e.g. rated current, rated frequency, peak withstand current, power, etc. The rated value is specified by the index “n”. The proposed index “r” has not become established, and has been used in only a few standards.

Rated voltage for creepage distances

Voltage value used to specify the creepage distances of an item of electrical → equipment with a given insulating material and an assumed → fouling factor.

Rated voltage U_n

Voltage of an item of electrical → equipment to which its other characteristic data refers. With three-phase circuits, the phase-to-phase voltage of the system applies as the rated voltage U_n. For → controlgear and switchgear assemblies, the manufacturer must specify the voltage limits within which correct operation of the → main and → auxiliary circuits is ensured. These limits must always be such that, under normal load conditions, the voltage at the terminals of the → control circuits of installed equipment remains within the limit values stipulated in the applicable standards.

RCD (**R**esidual-**C**urrent Protective **D**evice)

RCD is the generic term for
▷ RCDs with auxiliary source,

and
▷ RCDs without auxiliary source.

Reactive power

→ Power required with single-phase and three-phase alternating current for generating electromagnetic fields, e.g. in electric motors and transformers. Reactive power is essential for operating inductive loads; in contrast to effective power, however, it cannot be converted to any useable form of energy. As a result, reactive power is a "useless" load in cables and systems, particularly the supply systems of → public utilities.

Additional capacitor control systems, reactive-power correction equipment, and capacitors are installed to supply the necessary capacitive reactive power for the loads, thereby compensating the electromagnetic fields, and relieving the power supply system.

Reactive power controller

Reactive power controllers measure the → reactive power present in a system. If deviations occur from the specified value, the controller sends switching commands to the → contactors, which open or close groups of capacitors, as required.

Reactive power correction unit

This comprises capacitors that can be opened and closed to correct variable → reactive power in systems (→ correction).

Reactive power correction unit

Reactive power correction units are used to correct → reactive power centrally.

Reactor-connected capacitors

Capacitors to which reactors are connected on the line side. They are used to prevent a resonant circuit arising from the capacitors and the system inductors (transformers), if → harmonics are present.

Rectifier

Device for converting alternating current to direct current by means of electrical valves, which let current through in one direction while blocking it in the other.

Reduction of cross-sectional area

Reduction in the cross-sectional area of a → conductor in an electrical conductor system.

According to the relevant specifications, a → fuse corresponding to the reduced cross-sectional area must be provided at the beginning of a conductor that has been modified in this way.

Relay (electrical)

→ Switching device which brings about sudden predetermined changes in one or more electric output circuits when specific conditions that control the device arise in the electric input circuit.

Release delay

Defined delay of a breaking operation. In the event of short-time power failures, it is sometimes desirable to prevent closed switching devices from opening immediately, since this would cause the downstream power supply or controller to fail. Remedial measures are provided in the form of → circuit-breakers with delayed undervoltage releases and → contactors with release delay elements.

Release time

Time from the initiation of a command until the instant at which the tripping command can no longer be cancelled.

Remote switch

Switch with electromagnetic remote control. The position of the contact elements in this type of switch is changed via current pulses in the control circuit.

Removed position

Position of a removable part (withdrawable unit), if this is located outside the switchgear station (→ TTA/→ PTTA) and is mechanically and electrically isolated from it.

Residual current

Sum of the instantaneous values of currents that flow through all the active conductors of an electrical system at one point.

With residual-current protective devices to EN 61008/IEC 61008/DIN VDE 0664, this is the vector sum (amount and phase angle of the currents).

Residual-current protective devices

Residual-current protective devices to EN 61008/IEC 1008, DIN VDE 0664 are residual-current circuit-breakers that switch off an electrical installation if a → fault current (residual current) that exceeds the tripping current (→ rated fault current) of the residual-current circuit-breaker flows through grounded conductive parts of the installation that do not belong to the main circuit, or through the human body.

Resistance to creepage

Resistance of insulating material to tracking, and thus to → creepage currents.

Resistance to vibration

Expressed as a multiple of the local acceleration of free fall at which an item of electrical equipment continues to function safely and reliably. The direction of the force, the function (in accordance with which acceleration occurs), and the amplitude value must always be specified together with the g value.

Response time

Time from the initiation of a command until the corresponding movement begins.

Ring system

In a ring system, several transformers feed a self-contained cable ring, to which all the → loads are connected. If the ring is broken at any point, the loads are supplied from one of the two sides, as in a → radial system. A ring system is a form of power supply system (→ meshed system).

Root-mean-square value

Value that corresponds to the root-mean-square of a periodic quantity. In AC theory, current and voltage are specified as root-mean-square values, since the values in AC systems are instantaneous values that depend on the phase angle. Measuring instruments (moving-coil instruments) indicate the root-mean-square value on a continuous basis. The root-mean-square value for the sine curve $= 1/\sqrt{2} \cdot$ amplitude (peak value).

The root-mean-square value of an alternating current corresponds to that of a direct current that generates the same heat (power) as the alternating current.

Rush current

Current peak when transformers are switched on, and when motors are started or switched. Caused by electromagnetic correction procedures (rush effect) when the magnetic field is being set up. The current peak can have the following values for motors at rated voltage:

Starting: $i_{max} = \sqrt{2} \cdot I_{an} \cdot (1.8 \text{ to } 2.0)$,

Star-delta switchover: $i_{max} = \sqrt{2} \cdot I_{an} \cdot (2.1 \text{ to } 3.7)$,

Reversing: $i_{max} = \sqrt{2} \cdot I_{an} \cdot (2.7 \text{ to } 5.0)$,

With transformers, the rush current (→ making current) depends on the particular type of construction, design, winding arrangement, application, power level, etc. As a root-mean-square value, the rush current is approximately 15 to 30 times the primary rated current.

The rush current decays in a few cycles, and after only 20 ms is significantly smaller.

Safety class

Safety classes specify protection against shock currents.

Safety extra-low voltage (SELV)

→ Protective measure with which electric circuits with rated voltages of $\leq$ 50 V AC or 120 V DC can be operated without grounding, and the supply from electric circuits with higher voltages is electrically separated from these.

Safety separation

→ Protective measure in which → equipment is electrically separated from supplying networks and is not grounded.

Scope of protection

→ Degree of protection

Screening

Electrically or magnetically conductive covering used to prevent unwanted signals from being emitted by or interfering with electronic devices.

Screwless terminal

Terminal on electrical installation switches and socket-outlets that allows solid conductors with a cross-sectional area of up to 2.5 mm^2 to be connected without screws. The connection is detachable.

Section

Constructional unit of a switchgear station (controlgear and switchgear assembly) between two adjacent vertical limiting levels.

Selectivity

Combined operation of overcurrent protective devices (circuit-breakers, fuses) connected in series to provide graded disconnection. Selectivity entails the back-up overcurrent protective device closest to the short circuit tripping, while the other overcurrent protective devices in the cable run remain activated. This reduces the spatial and temporal effects of faults to a minimum.

Sensor

Device in a system that can process physical variables and, if necessary, send a → telegram on a → bus.

Series connection

Type of connection in which several devices or contact elements are connected in series in the same electric circuit.

Service life

Period of time over which electrical → equipment functions correctly under normal operating conditions. The service life is specified, for example, in operating hours or operating cycles. A distinction is made between mechanical endurance and electrical service life.

Equipment reaches the end of its *electrical* service life when, for example, the specified number of operating hours is reached or exceeded, or when arc erosion has worn switch contacts to the extent that they can no longer provide reliable contact.

Mechanical endurance is determined from the number of off-load operating cycles specified by the manufacturer that the device can carry out without corrective maintenance or parts being replaced.

Setting range

Range between the highest and lowest → setting value of a scale, in which a → relay (e.g. time-delay relay) or → tripping element can be set to a desired value.

Setting value

Value of a characteristic quantity (e.g. time, current), to which a device is set, and to which its operating conditions refer.

Shielding winding

Copper foil between the primary and secondary winding of a transformer for suppressing interference frequencies (→ screening).

Shock current

A current that flows through the body of a person or an animal, with characteristics that usually cause a pathophysiological (detrimental) effect.

Short circuit

Connection with a negligibly small impedance between conductors that are live during operation. The current in such cases is a multiple of the operating current, which can give rise to thermal (→ rated short-time current) or mechanical (→ rated peak withstand current) overloading of the electrical → equipment and parts of the installation.

Short circuit to exposed conductive part

A conductive connection resulting from an insulation fault between conductive parts that do not belong to the active circuit and operational live parts of electrical → equipment.

Short-circuit current (sustained short-circuit current I_k)

Overcurrent that occurs in the event of a fault in an → electric circuit, e.g. with short-circuited terminals in an item of electrical → equipment, or with faulty jumpering in an electric circuit.

Short-circuit current limiting

One method of limiting short-circuit currents is to disconnect the faulty electric circuit very quickly (within a few ms) so that the prospective short-circuit current does not reach its peak value, but is limited to the considerably lower value of the → let-through current I_D. This can be achieved by means of current-limiting circuit-breakers, for example. Another possible method is to install reactors in the electric circuit.

Short-circuit making or breaking capacity

→ Making or breaking capacity, for which the specified conditions include a → short circuit at the terminals of the → circuit-breaker.

Short-circuit protective device (SCPD)

Device that protects an electric circuit or parts of an electric circuit against short-circuit currents by breaking such currents.

Short-circuit strength

The mechanical resistance to short-circuit stress of switching devices, particularly of busbars in switchgear stations and distribution boards (→ fault withstand capability).

Short-circuit trip

→ Overcurrent tripping element for providing protection against → short circuits.

Short-circuit voltage U_{kn}

Rated value of the voltage as a percentage of the rated input voltage that must be applied at the input terminals of a transformer to ensure that the rated output current (I_{2n}) flows in the event of a short circuit at the output terminals (referred to 20 °C).

Shunt capacitor

Shunt capacitors are usually three-phase capacitors with a high capacitance (= high power), high pulse carrying capacity, at least one internal fuse system, and a high level of capacitive stability.

Shutter

Moveable part which, in one total traveled position, allows the isolating contacts of a removable part (e.g. → withdrawable unit) to connect with the fixed → contact points, and, in the other total traveled position, covers the fixed contact points of the → controlgear and switchgear assembly.

Also used in grounding-type receptacles with increased protection against contact.

SIMOCODE

Siemens **Mo**tor Protection and **Co**ntrol **De**vice, comprising communication-capable devices for use in low-voltage controlgear and switchgear. These intelligent I/O modules communicate with higher-level programmable controllers via the serial → PROFIBUS-DP.

Simultaneity (coincidence) factor

Relationship between the combined → outputs of all → loads in operation at the same time and the sum of the rated outputs of all connected loads. The simultaneity factor is required to determine the line fuses in electrical installations, for example.

Simultaneously accessible parts

Conductors or → conductive parts that can be touched simultaneously by persons and, in certain cases, also by livestock. Examples of simultaneously accessible parts include:

▷ Live parts,
▷ Exposed conductive parts of electrical → equipment,
▷ Extraneous conductive parts,
▷ Grounding electrodes.

SIPCON P

SIPCON = **S**iemens **P**ower **Con**ditioning.
This range comprises components and complete systems that contribute significantly to improving the quality of the network to which they are connected.
SIPCON P is an → active filter, which is used for → dynamic correction, flicker correction, and harmonic filtering.

Site altitude

At site altitudes ≥ 2000 m above MSL (with electronic equipment, ≥ 1000 m above MSL), the lower air density considerably reduces heat dissipation. As a result, it is necessary to reduce the rated current. The electric strength of equipment can also be diminished by the effects of altitude.

SIVACON

Siemens product designation for a range of versatile → type-tested l.v. controlgear and switchgear assemblies with rated busbar currents of ≤ 6300 A. The switchgear cabinets in this range are available for permanently installed and withdrawable units, and comprise only standardized, mass-produced structural parts and modules. The version for withdrawable units allows devices such as → circuit-breakers to be moved from the → connected position to the → test and → disconnected positions, without the cabinet door having to be opened.

Slave

A slave may only exchange data with a → master, if requested to do so by the master.

Snap-on mounting

Method of mounting devices by snapping them onto a standardized mounting rail designed specially for this purpose, e.g. 35 mm DIN rail to DIN EN 50022. Devices of this type can be detached from the rail only by means of a tool.

Sub-section

Constructional unit of a → switchgear station between two adjacent horizontal limiting levels within a → section.

Supply conditions

When electrical power installations with rated voltages ≤ 1000 volts are connected to the low-voltage network of the → public utility, it is essential that the → technical supply conditions published by the public utility are observed.

Supporting structure

Part of a → switchgear station (controlgear and switchgear assembly) and on which the various components and, if necessary, a housing enclosure for the controlgear and switchgear assembly are mounted.

Surge voltage protector

A device that protects electrical equipment against transient overvoltages and limits the duration (often also the level) of the current resulting from such overvoltages.

Switch-disconnector with fuses

Switch-disconnector with motor-control capability, comprising a → switch-disconnector and (connected in series to this) → fuse-bases for inserting → fuse-links.

Switch-disconnector

Switch that meets the requirements for → mechanical switches as well as for → disconnectors.

Switch/switching device

Device for making or breaking a current in one or more electric circuits.

Switchgear station

Electrical installation created by assembling switching devices, overcurrent protective devices, busbars, corresponding connecting cables, and other accessories, e.g. → instrument transformers and measuring instruments.

Switching capacity

Current that a → switching device can make and break under specified conditions (→ breaking and → making capacity).

If a switching device can make higher currents than it can break (or vice versa), the switching capacity can be divided into → breaking capacity and → making capacity (→ short-circuit making or breaking capacity).

Symmetrical short-circuit current

The power-frequency component of the → short-circuit current. With short circuits *close to generator terminals*, the symmetrical short-circuit current decays from the → initial symmetrical short-circuit current I_k'' to the sustained short-circuit current I_k. With short circuits *remote from generator terminals*, the symmetrical short-circuit current is virtually constant throughout the entire duration of the short circuit. This means that the initial symmetrical short-circuit current I_k'' is equal to the sustained short-circuit current I_k.

System grounding

System grounding is the grounding of a point in the active circuit required for equipment or systems to operate correctly. It is designated:

▷ Direct, if it contains no resistors apart from the grounding resistor,
▷ Indirect, if it is established by means of additional ohmic, inductive, or capacitive resistors.

Tap-off unit

Functional unit normally used to supply electrical power to one or more outgoing circuits.

Technical supply conditions

These are published by the → public utility, and must be observed when electrical wiring systems are connected.

Telegram

A bit string used to transmit information from one bus device to another.

Temperature detector

Semiconductor component, the resistance of which changes in accordance with the temperature. It is used, for example, to monitor the winding temperature in three-phase asynchronous motors (→ thermistor protection) and in encapsulated-winding dry-type transformers.

Temperature limit

Highest temperature that the individual parts of an item of electrical → equipment can withstand continuously without being damaged. This temperature is the sum of the → ambient temperature and the permissible self-heating of the electrical equipment. Temperatures in excess of the temperature limit can prevent the equipment from functioning reliably.

Temperature rise limit

The highest permissible temperature rise specified for the individual parts of an item of electrical → equipment.

Terminal marking

Alphabetic and/or numeric name of supply or output terminals (→ terminals) in electrical installations or electrical devices.

Terminals

Connective parts used to link, tap, and connect cables, wiring and solid → conductors.

Test finger

Test apparatus resembling a human finger, which is used to check whether → live parts of an item of → equipment are accessible to touch.

Test position

Position of a withdrawable unit, in which the appropriate → main circuits are open, but the requirements relating to an → isolating gap do not have to be met, and in which the → auxiliary circuits are connected in such a way that a functional test can be performed on the withdrawable unit without it being mechanically disconnected from the → controlgear and switchgear assembly.

The circuit may be opened by operating a suitable → switching device, without the withdrawable unit being moved mechanically.

Test situation

Status of a → switchgear station (→ controlgear and switchgear assembly) or one of its parts, in which the appropriate → main circuits are open, but the requirements relating to an → isolating gap do not have to be met, and in which the corresponding → auxiliary circuits are connected, thus allowing a functional test to be performed on the installed devices.

Thermal overload release

An overload release with an inverse time lag, the function of which (including the time lag) is based on the thermal effect of the current (bimetallic principle) that flows through the tripping element.

Thermal short-circuit rating

→ Fault withstand capability

Thermistor protection

Protection of a motor or an encapsulated-winding dry-type transformer by means of → temperature detectors (PTC or NTC thermistor) integrated in the windings. These detectors monitor the winding temperature directly.

Three-phase current

Three-phase alternating current consisting of three single-phase alternating currents that differ in phase by one third of a cycle (120°). In comparison to single-phase alternating current, three-phase current has the advantage of being able to transport the same amount of energy at a lower cost.

Time-current characteristic

This characteristic specifies the virtual time (e.g.: → prearcing time or → break-time) for certain operational conditions as a function of the prospective → breaking current.

Time-delay relay

→ Switching device with an electronic, electromagnetic time delay, which opens or closes contacts after a set time.

TN, TN-C, TN-C-S, TN-S system

→ Types of network and system

Total clearing time

Time from the initiation of a breaking command until the arcing time has elapsed.

Total insulation

This is a → protective measure implemented by means of

- adding insulation to the basic insulation, or
- reinforcing the basic insulation

in such a way that no shock currents can flow if the basic insulation should fail.

Touch voltage U_B

Voltage (→ fault voltage) that can occur between simultaneously accessible parts in the event of an insulation fault.

Transportable unit

Part of a → controlgear and switchgear assembly or a complete controlgear and switchgear assembly, which is not dismantled for transportation.

Tripping characteristic (graph)

Graphic representation of the relationship between → release time and a particular influencing variable. The time-current graph, for example, shows the time taken for the → tripping element or tripping relay to operate at a particular current.

Tripping characteristic (protection)

→ Miniature circuit-breakers are manufactured with different tripping characteristics.

Tripping characteristic A for protecting instrument transformers in measuring circuits,

Tripping characteristic B for protecting socket-outlet circuits in residential and functional buildings,

Tripping characteristic C for protecting → equipment with higher making currents,

Tripping characteristic D for protecting electrical → equipment with a high level of pulse generation

Tripping current (of an overload release)

Current value at which a tripping element trips within a particular time.

Tripping element (of a mechanical switching device)

Device connected to a mechanical switching device, which trips the latch when specified values (e.g. voltage, current) fall below or exceed a specified level, thereby enabling the switching device to open or close.

TT system (network)

→ Types of network and system

TTA

→ Type-tested l.v. controlgear and switchgear assembly

Type-tested l.v. controlgear and switchgear assembly (TTA)

Assembly of low-voltage controlgear and switchgear that is built and type tested to EN 60439-1/IEC 60439-1/DIN VDE 0660-500, or that does not deviate significantly from the original type or system of l.v. controlgear and switchgear assembly tested in accordance with this standard. → Controlgear and switchgear assemblies can comprise electromagnetic and/or electronic equipment (see also PTTA).

Types of network and system

Electrical networks are categorized according to their voltage, structure, and configuration. With alternating voltage, a distinction is made between:
- Low-voltage networks < 1000 V,
- High-voltage networks ≥ 1 kV to 380 kV,
- Extra-high-voltage networks ≥ 380 kV.

A distinction is made between the following network structures:
- Open networks (e.g. → radial system) and
- Closed networks (e.g. → ring system, → meshed system).

The following low-voltage network configurations exist:

▷ TN system (network):
  In TN systems, a point of the network (neutral point or external conductor) is grounded directly. The exposed conductive parts of the items of electrical → equipment are connected to the grounded point of the network via an → equipment grounding conductor or → PEN conductor. TN systems can have the following variants:
  - TN-S system,
    In this type of system, the → neutral conductor (N) and → equipment grounding conductor (PE) are routed separately throughout the network.
  - TN-C system,
    In this type of system, the functions of the neutral conductor and equipment grounding conductor are combined in a single conductor (the → PEN conductor) throughout the network.
  - TN-C-S system,
    In this type of system, the functions of the neutral conductor and equipment grounding conductor are combined in a single conductor (the → PEN conductor) in only one part of the system; in the remaining part, the → neutral conductor and → equipment grounding conductor are routed separately.

▷ TT system,
  In TT systems, a point of the network is grounded directly; the exposed conductive parts of the → equipment are connected to → grounding electrodes.

▷ IT system,
  In IT systems, there is no direct connection between active conductors and grounded parts; the exposed conductive parts of items of electrical → equipment are grounded.

Utilization categories (low-voltage fuses)

Utilization categories for low-voltage fuses are indicated by means of letters, the first of which designates the → functional class, and the following letter the object to be protected. The following are specified as objects for protection:

L wiring systems, in future, G (general applications)

M switchgear and controlgear,

R semiconductors,

B mining equipment and systems,

Tr transformers.

The following utilization categories are derived from the above:

gG/(gL) full-range cable and wiring protection,

aM partial-range switchgear and controlgear protection,

aR partial-range semiconductor protection,

gR full-range semiconductor protection,

gB full-range mining equipment and systems protection,

gTr full-range transformer protection.

Utilization categories (switching devices)

The utilization category specifies the purpose and loading of → switching devices. It refers to the combinations of requirements stipulated in the applicable device standards (e.g. EN 60947/IEC 60947/DIN VDE 0660), which cover a basic group of practical applications. The utilization category is designated by:

1. Values of the making and breaking currents, expressed as a multiple of the → rated operating current,
2. Values of the voltage, expressed as a multiple of the → rated operating voltage (→ rated voltage),
3. Values of the power factor → cos φ, or time constant t,
4. Behavior regarding → short circuit, → selectivity, and → back-up protection.

Two utilization categories are specified for circuit-breakers:

"A": not designed specifically for → selectivity, e.g. without intentional short-circuit delay for → selectivity under short-circuit conditions, and therefore without specification of the → rated short-time current I_{cw}.

"B": designed specifically for → selectivity under short-circuit conditions with respect to other short-circuit devices connected in series on the load-side. This can be implemented by means of a settable short-time delay, for example. The → rated short-time current I_{cw} must be specified for circuit-breakers of this type.

VBG 4

Accident prevention regulation for "Electrical installations and equipment" published by the **V**erband der **B**erufs**g**enossenschaften (employers' liability insurance association). VBG 4 stipulates protection-related objectives and refers to the relevant standards with regard to implementing appropriate protective measures.

VDE

Verband **D**er **E**lektrotechnik, Elektronik und Informationstechnik e.V. (Association of German Electrical Engineers), formerly **V**erband **D**eutscher **E**lektrotechniker (founded 1893). The VDE is a nonprofit technical/scientific organization, whose aims are: maintaining scientific knowledge and procedures; training its members; elaborating electrotechnical safety rules recognized by the legislative body (DIN VDE specifications); checking/testing electrical products (VDE mark of conformity); and representing the interests of the electrotechnical industry both at home and abroad.

Ventilation

Electrical equipment is enclosed in order to protect it from → environmental influences. The higher the → degree of protection of the enclosure, the less heat exchange occurs. Significant changes in temperature can lead to condensed water forming inside the housing, which can have a detrimental effect on the electrical equipment installed. To prevent condensed water from forming, a ventilator may be required in the housing. In larger switchgear stations, forced ventilation is preferred; the required air circulation is provided by fans.

Voltage drop U_R (V); u_R (%); ΔU (V); Δu (%)

Difference between the no-load voltage U_0 and rated output voltage U_{2n} of a transformer at rated output, referred to an ambient temperature of 20 °C. The voltage drop in cables and lines depends on their length, cross-sectional area, and the material they are made of. Where long distances are involved, the expected voltage drop must be taken into consideration.

Voltage-operated ground-leakage circuit-breaker

Circuit-breaker that operates if the voltage with respect to → ground in an accessible metal part exceeds a specific limit value (→ touch voltage).

Withdrawable circuit-breaker

Designation for a → circuit-breaker that (in contrast to a → withdrawable unit) is not combined with other items of equipment to form a unit.

Withdrawable unit

Replaceable unit that can be installed in a position in which there is an open → isolating gap, while remaining mechanically connected to the switchgear station. This isolating gap may be located in the → main circuits alone, or in both the main and → auxiliary circuits.

Zero-current interrupter

→ Circuit-breaker, the contact member of which opens by means of an → instantaneous short-circuit trip and breaker mechanism in the event of a → short circuit. During the opening time, the → short-circuit current rises to its full value. The → arc that occurs reduces the short-circuit current to the extent that the arc is extinguished only at natural current zero.

z-release

Short-time-delay overcurrent release with lock-out that prevents closing.

7.3 Publications, catalogs, DIN VDE publications, data carriers, etc.

A wide range of informative publications, catalogs, DIN VDE publications, and data carriers are available, providing planners and operators with extensive background material on the topics discussed in this manual. The following table lists these various media together with the numbers of the sections to which they relate.

Section	**Publications on Siemens installation systems**[1]
1.4	Der Universalist – Innovationen bei den Verteilersystemen. Reprint from "produkt profile" 2/98. Order No.: E20001-P311-A868
1.4	Geht alles ganz fix: SIPRO-Universalsystem im Wohn- und Zweckbau. Reprint from "produkt profile" 4/97. Order No.: E20001-P311-A858
1.4	SIKUS 3200: The Distribution Board Program with a System. Order No.: E20001-P311-A830-X-7600
1.4.6 1.4.7 4.4.3	P.I.S.A.A. – Schnelle Angebote, einfache Projektierung von Elektroinstallation mit Siemens *instabus* im Wohnbereich und für Installations- und Isolierstoffverteiler sowie Zählerschränke in allen Einsatzfällen. Order No.: E20001-P311-A843-V1
2.6	Kein Stecker raus bei Gewitter. Koordinierter Einsatz von Blitzstrom- und Überspannungsableitern. Reprint from "drive & control" 3/98. Order No.: E20001-P311-A891
2.6	Thunderstorm – no problem. Reprint from "drive & control" 3/98. Order No.: E20001-P311-A891-X-7600
4.1.1	Doppelt sicher: Die neuen NH-Sicherungseinsätze mit Mittenkennmelder. Order No.: E20001-P311-A853-V1

[1] For details of how to order, see page 721

4.1.3	Alle Fehlerströme im Griff: Allstromsensitive Fehlerstrom-Schutzeinrichtungen für Industrieeinrichtungen. Reprint from "drive & control" 4/93. Order No.: E20001-P311-A722-V1
4.1.3	Fehlerstrom-Schutzeinrichtungen können Schäden verhüten. Reprint from "S+S report" 4/95. Order No.: E20001-P311-A801
4.1.3	Fehlerstrom-Schutzeinrichtungen schützen Menschenleben und Sachwerte. Reprint from "EVU-Betriebspraxis" 6/95. Order No.: E20001-P311-A800
4.1.3	Sicher ist sicher: Die neue FI-Sicherheits-Steckdose. Reprint from "produkt profile" 2 1/98. Order No.: E20001-P311-A869
4.1.3	Weltneuheit: Allstromsensitive Fehlerstrom-Schutzschalter >N< Order No.: E20001-P311-A734
4.1.3	AC/DC Earth-Leakage Circuit-Breaker >N<. Reprint from "drive & control". Order No.: E20001-P311-A889-X-7600
4.2	Da ist Schwung drin – Mehr Marktchancen mit neuen Reiheneinbaugeräten. Reprint from "produkt profile" 1996–1998 Order No.: E20001-P311-A887
4.3.1	Die neuen DELTA Schalter- und Steckdoseneinsätze: Umwerfend einfach in der Handhabung. Order No.: E20001-P311-A861
4.3.1	Schalter und Steckdosen für Plattenbau-Renovierung. Order No.: E20001-P311-A831
4.3.1 4.4.3 4.4.4	Bedienoberflächen und Steckdosen – Bewährt in der Praxis. Reprints from "produkt profile" 1995–1998. Order No.: E20001-P311-A867-V1
4.3.1 4.4.3 4.4.4	Time-Proven User Interfaces and Socket Outlets. Reprints from "product profiles" 1/98–2/98. Order No.: E20001-P311-A867-X-7600
4.3.2.1	DELTA matic für Aufputzmontage. Order No.: E20001-P311-A705-V1
4.3.2.2	DELTA-FERN: Leitungslos schalten mit dem Infrarot-System. Order No.: E20001-P311-A243-V1
4.3.2.2	Infrarot-Fernsteuersystem IR-64K. Order No.: E20001-P311-A777-V1
4.4	Der Siemens *instabus*: Fachberichte. Reprints from "produkt profile" 1991–1996. Order No.: E20001-P311-A787-V1
4.4	The Siemens *instabus*: Reports. Reprints from "product profiles" 1991–1996. Order No.: E20001-P311-A787-V1-7600
4.4 4.4.4.1	Freude am Zuhause: Mit Siemens *instabus*. Order No.: E20001-P311-A765-V2
4.4 4.4.4.1	*instabus EIB* de Siemens: Technique intelligente à domicile. Order No.: E20001-P311-A784-X-7700
4.4 4.4.4.1	*instabus EIB*: Intelligent Buildings for the Future. Order No.: E20001-P311-A893-X-7600
4.4 4.4.4.1	Der HomeAssistant auf Basis des Siemens *instabus*. Reprint from "produkt profile" 4/98. Order No.: E20001-P311-A898
4.4 4.4.4.1	Intelligente Technik fürs Zuhause. Order No.: E20001-P311-A784-V2

4.4 4.4.4.1	Siemens *instabus*: Intelligent Technology for the Home. Order No.: E20001-P311-A784-V1-7600
4.4 4.4.4.1	*instabus EIB* – was er ist, was er kann, was er bringt (Argumentationshilfen). Order No.: E20001-P311-A846
4.4 4.4.4.1	Alten- und behindertengerechtes Wohnen mit dem Siemens *instabus*. Reprint from "produkt profile" 3/96. Order No.: E20001-P311-A837
4.4 4.4.4.1	Building Automation for the Elderly and Disabled. Reprint from "product profiles" 3/96. Order No.: E20001-P311-A837-X-7600
	Catalogs on Siemens installations systems[1]
1.4	SIKUS 3200 Reihenschaltschranksystem. Katalog I 2.34 Order No.: E20001-K2340-P300-V2
1.4	STAB-Wandverteiler, SIKUS-Standverteiler. Katalog I 2.21/2.32 Order No.: E20002-K8231-A101-A1
1.4 1.4.6	Installationsverteiler Katalog I 2.35 Order No.: E20002-K8235-A101-A1
1.4 4.1.1 4.1.2 4.1.3	Products and Systems for Low-Voltage Power Distribution. Catalog NS PS Order No.: E20002-K1801-A101-A-7600
1.4 4.1.1 4.1.2 4.1.3	Produkte und Systeme zur Energieverteilung. Katalog NS PS Order No.: E20002-K1801-A101-A1
1.4 4	Elektrische Installationstechnik. Katalog-CD-ROM ET 01 Order No.: E20002-D8200-A107-A1
1.4 4.1 4.3 4.4.2 4.4 4	Protective Switching and Fuse Systems, Building Management Systems with *instabus EIB*. Catalog I 2.1 Order No.: E20002-K8210-A101-A3-7600
1.4. 4.1 4.3 4.4.2 4.4.4	Schutzschalt- und Schmelzsicherungssysteme, Gebäudesystemtechnik mit *instabus EIB*. Katalog I 2.1 Order No.: E20002-K8210-A101-A4
4.1.1	Kennlinien von Niederspannungssicherungen. Characteristic Curves of LV Fuses. Katalog I 2.21 (German/English/Italian/Swedish) Order No.: E20002-K8221-A101-A2-9D
4.2	Program Overview Modular Devices. Catalog I 2.11 Order No.: E20002-K8211-A101-A1-7600
4.2	Programmübersicht Reiheneinbaugeräte. Katalog I 2.11 Order No. E20002-K8211-A101-A1

[1] For details of how to order, see page 721

4.3.1 4.4.2 4.4.4	DELTA Schalter und *instabus EIB*. Katalog I 2.4 Order No.: E20001-K2400-P300-V1
4.4	Building Management Systems with *instabus EIB* Technical Manual. Order No.: E20001-P311-A857-X-7600
4.4	Gebäudesystemtechnik mit *instabus EIB* Technik Order No.: E20001-P311-A857
	Videos on Siemens installation systems[1]
4.4.1	Take the Bus to Building Management System Technology. Order No.: E20001-P0371-Y130-X-7600
4.4.1 4.4.2	*instabus EIB* (Teil 1–3, Überblick, Technik, Geräte). Order No.: E20001-P0371-Y130
4.4.3	Siemens *instabus*: Der Demo-Shop mit Anwendungen für Wohnung und Haus. Order No.: E20001-P0371-Y136
4.4.3	Siemens *instabus*: The Demo-Shop with Applications for Appartment and House Order No.: E20001-P0371-Y136-X-7600
4.4.3 4.4.4.1	Grenzenlose Möglichkeiten mit dem Siemens *instabus*. Order No.: E20001-P0371-Y137
	Software tools for Siemens installation systems[1]
1.4.6 4.4.3	P.I.S.A.A. – Projektierungshilfe zur Angebotserstellung, Projektierung und Bestellung von Installations- und Isolierstoffverteilern und Komplettzählerschränken und für die Installation mit Siemens *instabus* im Wohnbereich. Tender processing, configuration and ordering of distribution boards and insulated distribution boards and meter cabinets and for electrical installation with Siemens *instabus* for residential buildings.
	Publications from Siemens and Publicis MCD Verlag[1]
1.3 1.4 2 4.1	Schalten, Schützen, Verteilen in Niederspannungsnetzen. (1997) ISBN 3-89578-041-3
1.3 1.4 2 4.1	Switching, Protection and Distribution in Low-Voltage Networks. (1994) ISBN 3-89578-000-6
1.7	Kabel und Leitungen für Starkstrom (1999) ISBN 3-89578-088-X
1.7	Kabel und Leitungen für Starkstrom, Teil 2 (1989) ISBN 3-8009-1524-3
1.7	Power Cables and their Application, Part 1 (1990) ISBN 3-8009-1535-9
1.7	Power Cables and their Application, Part 2 (1993) ISBN 3-8009-1575-8
4.4	Gebäudesystemtechnik mit *EIB* (2000) ISBN 3-89578-076-6

[1] For details of how to order, see page 721

	Selected VDE publications[1]
1.3.1.2 1.3.2	VDE-Schriftenreihe Band 28: Lothar Zentgraf Niederspannungs-Schaltgerätekombinationen – Erläuterungen zu DIN EN 60439-1 (VDE 0660 Teil 500). ISBN 3-8007-2224-0
1.4.5.2 5.9.3	VDE-Schriftenreihe Band 42: Rolf Rüdiger Cichowski Elektrische Anlagen auf Baustellen – Erläuterungen zu DIN VDE 0100-704. ISBN 3-8007-1564-3
1.7	Retzlaff, E.: Dictionary of Code Symbols for Cables and Insulated Cords According to VDE, CENELEC and IEC (German/English). ISBN 3-8007-2288-7
1.7	VDE-Schriftenreihe Band 29: Ewald Retzlaff Lexikon der Kurzzeichen für Kabel und isolierte Leitungen nach VDE, CENELEC und IEC (German/English). ISBN 3-8007-2288-7
1.7	Warner, A.: International Register of Identification Threads and Markings for Cables and Insulated Cords. ISBN 3-8007-2116-3
1.7	VDE-Schriftenreihe Band 32: Heinz Haufe / Heinz Nienhaus / Dieter Vogt Schutz von Kabeln und Leitungen bei Überstrom – DIN VDE 0100-430 mit Beiblatt, DIN VDE 0298, ISBN 3-8007-2072-8
2.3 2.7	VDE-Schriftenreihe Band 140: W. Hörmann / H. Nienhaus / B. Schröder. Schnelleinstieg in die neue DIN VDE 0100-410: Schutz gegen elektrischen Schlag. ISBN 3-8007-2264-X
3	VDE-Schriftenreihe Band 66: Wilhelm Rudolph / Otmar Winter EMV nach VDE 0100 – Erdung, Potentialausgleich, TN-, TT- und IT-System, Vermeiden von Induktionsschleifen ... ISBN 3-8007-2082-5
5.9	VDE-Schriftenreihe Band 67: Herausgeber: Heinz Nienhaus Elektrische Anlagen für Baderäume, Schwimmbäder und alle weiteren feuchten Bereiche und Räume – Anforderungen nach DIN VDE 0100. ISBN 3-8007-2118-X
7.1	VDE-Schriftenreihe Band 1: Wo steht was im VDE-Vorschriftenwerk? Stichwortverzeichnis zu allen DIN VDE-Normen. ISBN 3-8007-2314-X
7.1	VDE-Schriftenreihe Band 13: Ernst Gorenflo u.a. Betrieb von elektrischen Anlagen – Erläuterungen zu DIN VDE 0105-100. ISBN 3-8007-2210-0
7.1	VDE-Schriftenreihe Band 72: Norbert Barz EG-Niederspannungsrichtlinie – Erläuterungen der Richtlinie, ihre Umsetzung in deutsches Recht und Anwendungsfragen. ISBN 3-8007-2192-9

[1] For details of how to order, see page 721

	Other publications
4.4	Handbuch Gebäudesystemtechnik, Grundlagen (German/ English) Published by: – Zentralverband Elektrotechnik- und Elektronikindustrie e.V., Fachverband Installationsgeräte und -systeme, PO Box 701261, 60591 Frankfurt a.M., Germany – Zentralverband der Deutschen Elektrohandwerke, PO Box 900370, 60443 Frankfurt a.M., Germany Address for orders: – Wirtschaftsfördergesellschaft der Elektrohandwerke m.b.H. (WFE) PO Box 900370, 60443 Frankfurt a.M., Germany
4.4	Handbuch Gebäudesystemtechnik, Anwendungen Published by: – Zentralverband Elektrotechnik- und Elektronikindustrie e.V., Fachverband Installationsgeräte und -systeme, PO Box 701261, 60591 Frankfurt a.M., Germany – Zentralverband der Deutschen Elektrohandwerke, PO Box 900370, 60443 Frankfurt a.M., Germany Address for orders: – Wirtschaftsfördergesellschaft der Elektrohandwerke m.b.H. (WFE) PO Box 900370, 60443 Frankfurt a.M., Germany
4.4	Further information on *EIB* is available from EIBA, Brussels. Postal address: EIBA European Installation Bus Association Avenue de la Tauche 5 1160 Brussels Belgium Internet address: www.eiba.be

How to order:

Publications from the Siemens Installation Systems Division (Order No.: E2000. – ...) can be ordered via your regional Siemens office by specifying the appropriate order number.

The software tool P.I.S.A.A. is available from the following Siemens AG service providers:

KUPEK GmbH
Beschaffung und Logistik
Plattenäcker 10d
96450 Coburg
Germany

Publications by Publicis MCD Verlag can be ordered from:

WILEY-VCH
69451 Weinheim
Germany
Telefax + 49-6201-606 184

See also under www.publicis-mcd.de/fachbuecher

VDE publications, see table above.

Note:
The advertising publications, reprints, catalogs, and videos from the Siemens Installation Systems Division are updated on a regular basis.

Index

For those terms explained in the glossary the page numbers are marked with *